# FORMULA INDEX
TO
# NMR LITERATURE DATA

Volume 2

# FORMULA INDEX
TO
# NMR LITERATURE DATA

Volume 2: 1961-1962 References


EDITORS

M. Gertrude Howell
Andrew S. Kende
John S. Webb

CONTRIBUTORS

V. Bauer, R. Brownfield, E. Cohen, H. Corey, T. Fields, W. Fulmor, L. Goldman, J. Hlavka, G. Howell, P. Izzo, A. Kende, H. Kissman, J. Krueger, J. Lancaster, G. Morton, M. Neglia, J. Patrick, M. Sassiver, E. Ullman, J. Webb

*Lederle Laboratories Division of American Cyanamid Company*



IFI/PLENUM • NEW YORK–WASHINGTON • 1966

First Printing – May 1966
Second Printing – November 1968

ISBN 978-1-4684-7160-1 ISBN 978-1-4684-7158-8 (eBook)
DOI: 10.1007/ 978-1-4684-7158-8

Library of Congress Catalog Card Number 64-7756


Softcover reprint of the hardcover 1st edition 1966
A Subsidiary of Plenum Publishing Corporation
227 West 17 Street, New York, N.Y. 10011

# CONTENTS

# ACKNOWLEDGMENTS

THE CONTRIBUTORS wish to express their appreciation to Drs. J. J. Denton and J. H. Clark of the Lederle Laboratories Division for their encouragement and support of this project. The assistance of Dr. Elizabeth M. Hardy and Mrs. Barbara J. Forgione in searching selected portions of the literature, and the help of Mrs. M. B. Jeffries in calculating some of the molecular formulas is gratefully acknowledged. We are particularly pleased to recognize the contribution of Mrs. Susan A. Odell, who has maintained the card files, aided with the proofreading, and attended to numerous secretarial details with exceptional dedication and skill throughout the publication period. Finally, the contributors wish to thank the American Cyanamid Company and its management for permission to publish this compilation and for provision of the facilities necessary for its completion.

# INTRODUCTION TO THE USE OF THIS INDEX

## Origin and Purpose of the Index

THIS VOLUME is the outgrowth of a four-year effort by the abstractors to develop a comprehensive catalog of organic compounds for which proton magnetic resonance data have been reported in the scientific literature. Toward this end, the abstractors have searched page by page through the relevant organic and physical chemistry journals as well as through appropriate review articles and books. References were cataloged if they described data for a single compound (rather than for a mixture) and if they met any one of the following requirements:

Full NMR curve was reproduced in the reference;

Part of an NMR curve was reproduced with proper calibration;

Precise chemical shift or spin-coupling data were reported for one or more specified peaks.

## Journals Covered in Search

In the course of this search the following journals were examined:

*Acta Chemica Scandinavica*
*Analytical Chemistry*
*Angewandte Chemie*
*Annalen der Chemie*
*Arkiv for Kemi*
*Australian Journal of Chemistry*
*Bulletin of the Chemical Society of Japan*
*Bulletin de la Société Chimique de France*
*Canadian Journal of Chemistry*
*Chemische Berichte*
*Chemistry & Industry*
*Experientia*
*Gazzetta Chimica Italiana*
*Helvetica Chemica Acta*
*Journal of the American Chemical Society*
*Journal of Chemical Physics*
*Journal of the Chemical Society*
*Journal of Molecular Spectroscopy*
*Journal of Organic Chemistry*
*Journal of Physical Chemistry*
*Molecular Physics*
*Proceedings of the Chemical Society*
*Recueil des Travaux Chimiques des Pays-Bas*
*Spectrochimica Acta*
*Tetrahedron*
*Tetrahedron Letters*
*Transactions of the Faraday Society*

The following books and review articles were used as secondary sources:

Roberts, John D., *Nuclear Magnetic Resonance. Applications to Organic Chemistry,* McGraw-Hill, 1959

Pople, J. A., Schneider, W. G., and Bernstein, H. J., *High Resolution Nuclear Magnetic Resonance,* McGraw-Hill, 1959

Jackman, L. M., *Applications of Nuclear Magnetic Resonance Spectroscopy in Organic Chemistry,* Pergamon Press, 1959

Brownstein, S., *Chemical Reviews* **59**:463-496 (1959)

Conroy, Harold, "NMR in Organic Structural Elucidation," in: Raphael, R. A., Taylor, B. C., and Wynberg, H., *Advances in Organic Chemistry. Methods and Results,* Vol. II, Interscience, pp. 265-328, 1960

For the convenience of the user we have included compounds listed in other generally available sources. Those in the catalog made available by Humble Oil & Refining Company, which was described by N. F. Chamberlain in *Analytical Chemistry* **31**:56 (1959), and those in the listing made available by G. V. D. Tiers of the Minnesota Mining and Manufacturing Company were included in Volume 1.

The compounds which appear in the Varian NMR Spectra Catalogs, Vols. 1 and 2, are included in Volume 2 of this index, designated by the Varian spectrum number. We are grateful to Dr. J. N. Shoolery and Varian Associates for the permission to do this.

Also included in Volume 2 are the first 121 compounds listed by the Manufacturing Chemists Association Research Project Nuclear Magnetic Resonance Spectral Data. These compounds are designated "M.C.A. No. __."

The editors have attempted to compile and present these references in a comprehensive and consistent manner. However, in any collaborative work involving twenty contributors and a massive volume of material, some omissions and inconsistencies are surely inescapable, and in this connection we solicit the reader's forbearance. Structures have been accepted as given in the references, and no systematic attempt to check or correct them has been made.

## Indexing Order

**1. Compounds.** The location of a compound in this volume is governed by its indexing formula, indicated at the top left of each entry. Except for *salts*, *"complexes,"* and *isotopically-labeled* substances, the indexing formula of a compound is simply its molecular formula with the element sequence written in the following order:

**P, S, N, O, I, Br, Cl, F, C, H, D, then alphabetical**

To illustrate:

| *Compound* | *Indexing Formula* |
|---|---|
| 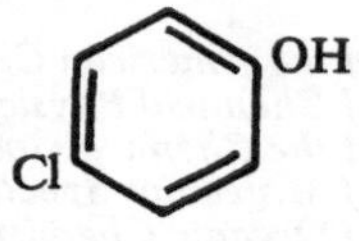<br>*p*-chlorophenol | $OClC_6H_5$ |
| $NH_2$<br>\|<br>$CH_3SCH_2CH_2CHCO_2H$<br>methionine | $SNO_2C_5H_{11}$ |
| $(C_6H_5)_3B$<br>triphenylboron | $C_{18}H_{15}B$ |

The purpose of this element sequence, which differs sharply from the conventional *Beilstein* or *Chemical Abstracts* sequence, is to simplify searches for spectra of particular groups of compounds; e.g., all phosphines or all thiazoles or all guanidine derivatives will tend to be grouped together. Thus, the reader looking for phosphorus-containing compounds will look for them at the beginning of the volume, and the one searching for hydrocarbons will look near the end of the book.

Approximately five percent of the entries in this volume fall into the categories of salts, complexes, and isotopically-labeled substances; these are indexed in the following special way.

**2. Salts.** Those salts in which one ion is organic and the other inorganic are indexed under the parent neutral organic moiety. In the case of metal salts of organic acids, the structural formula is presented as the salt so that the cation can be identified. Salts in which both ions are organic are treated as a single molecule and are indexed under the sum of the component formulas. For the examples below, the indexing formula is underlined:

| *Salt* | *Indexing Formula* |
|---|---|
|  | $\underline{O_4C_4H_4}$ |
| $\overset{\oplus}{(CH_3)_3NH}\ \overset{\ominus}{Cl}$ | $\underline{NC_3H_9}\cdot HCl$ |
| $\overset{\oplus}{(CH_3)_2CHNH_3}\ \overset{\ominus}{OOCCCl_3}$ | $\underline{NO_2Cl_3C_5H_{10}}$ |

A slight exception is made for carbonium ions and carbanions, which are indexed as the charged organic species.

**3. "Complexes."** Within this group are metal-organic compounds in which the organic-to-metal bond is other than a simple covalent linkage. Thus, ferrocene is regarded as a "complex," while tetraethyl lead is not. Coordination compounds such as $(C_2H_5)_2\overset{\oplus}{O}\text{–}\overset{\ominus}{B}F_3$ are indexed in the same way as "complexes." The indexing rule is that the "complex" be indexed under the sum of the organic components in the molecule; carbon monoxide or hydrogen ligands are specifically excluded from the indexing formula, which is underlined in the examples below:

| *Complex* | *Indexing Formula* |
|---|---|
| $(C_5H_5)_2Fe$ <br> ferrocene | $\underline{C_{10}H_{10}}\cdot Fe$ |
| $(C_5H_5)(C_5H_8)Re(CO)_2$ | $\underline{C_{10}H_{13}}\cdot Re(CO)_2$ |

**4. Isotopically-Labeled Compounds.** Deuterated compounds are indexed using the symbol D for deuterium in the element sequence given earlier. All other isotopes, e.g., $C^{13}$ or $O^{17}$, are not differentiated from the normal isotopes in the indexing formula, but a notation is made on or next to the structural formula indicating their presence and location.

| *Isotopically-Labeled Compound* | *Indexing Formula* |
|---|---|
| 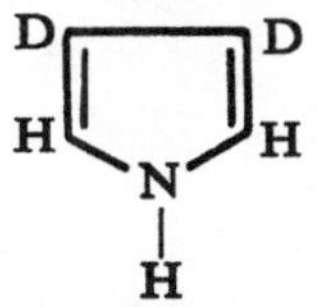 | $\underline{NC_4H_3D_2}$ |
| $C^{13}H_3CH_2OH$ | $\underline{OC_2H_6}$ |

# FORMULA INDEX
TO
# NMR LITERATURE DATA

Volume 2

▶ $PSNO_2C_9H_{20}$

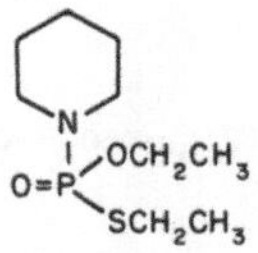

$O{=}P(N(CH_2)_5)(OCH_2CH_3)(SCH_2CH_3)$

Chem. Ber. 95:2896 (1962)

▶ $PSNO_2C_{13}H_{28}$

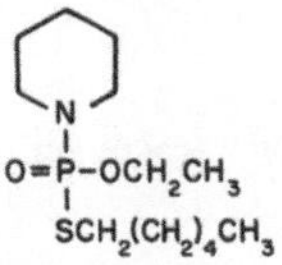

$O{=}P(N(CH_2)_5)(OCH_2CH_3)(SCH_2(CH_2)_4CH_3)$

Chem. Ber. 95:2896 (1962)

▶ $PSNO_2C_{14}H_{22}$

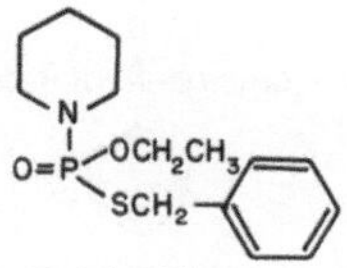

$O{=}P(N(CH_2)_5)(OCH_2CH_3)(SCH_2C_6H_5)$

Chem. Ber. 95:2896 (1962)

▶ $PSNO_2C_{14}H_{30}$

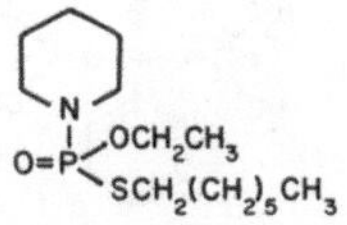

$O{=}P(N(CH_2)_5)(OCH_2CH_3)(SCH_2(CH_2)_5CH_3)$

Chem. Ber. 95:2896 (1962)

▶ $PSNO_4C_5H_{16}$

$(CH_3O)_2P(O)\text{-}SCH_2C(O)\text{-}NHCH_3$

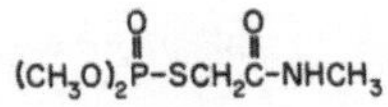

J. Org. Chem. 26:2285 (1961)

▶ $PSOClC_{10}H_{18}$

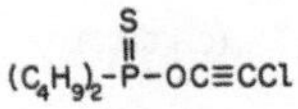

$(C_4H_9)_2\text{-}P(S)\text{-}OC{\equiv}CCl$

J. Org. Chem. 27:2199 (1962)

▶ $PSOCl_2C_2H_5$

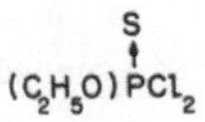

$(C_2H_5O)\,P(S)Cl_2$

Compt. Rend. 254:260 (1962)

▶ $PSOCl_2C_{10}H_{19}$

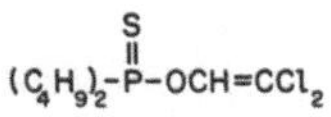

$(C_4H_9)_2\text{-}P(S)\text{-}OCH{=}CCl_2$

J. Org. Chem. 27:2199 (1962)

▶ $PSO_2ClC_4H_{10}$

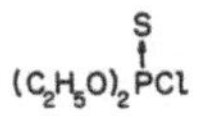

$(C_2H_5O)_2P(S)Cl$

Compt. Rend. 254:260 (1962)

▶ $PSO_2C_{12}H_{19}$

$CH_3CH_2CH_2O\text{-}P(S)(C_6H_5)\text{-}OCH_2CH_2CH_3$

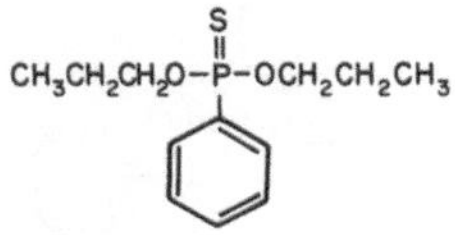

MCA Serial No. 120

▶ $PSO_2C_{12}H_{19}$

$(CH_3)_2CH\text{-}O\text{-}P(S)(C_6H_5)\text{-}O\text{-}CH(CH_3)_2$

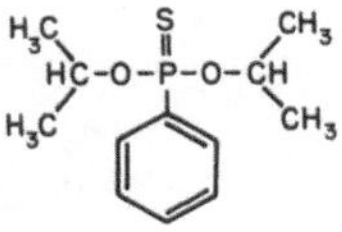

MCA Serial No. 121

▶ $PSO_3C_6H_{15}$

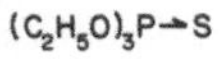

$(C_2H_5O)_3P{\rightarrow}S$

Compt. Rend. 254:260 (1962)

▶ $PSO_3C_6H_{15}$

$(CH_3CH_2O)_3PS$

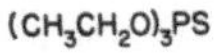

MCA Serial No. 47

▶ $PSO_3C_6H_{15}$

$(C_2H_5O)_2P(O)(SC_2H_5)$

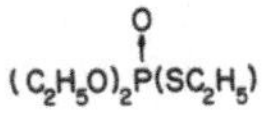

Compt. Rend. 254:260 (1962)

▶ $PSNO_2C_9H_{20}$

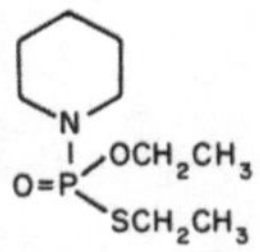

(piperidino)$P(=O)(OCH_2CH_3)(SCH_2CH_3)$

Chem. Ber. 95:2896 (1962)

▶ $PSNO_2C_{13}H_{28}$

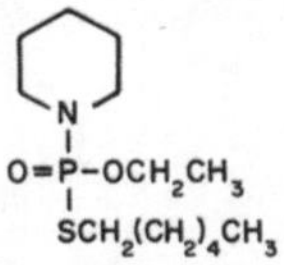

(piperidino)$P(=O)(OCH_2CH_3)(SCH_2(CH_2)_4CH_3)$

Chem. Ber. 95:2896 (1962)

▶ $PSNO_2C_{14}H_{22}$

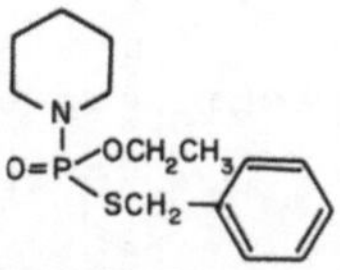

(piperidino)$P(=O)(OCH_2CH_3)(SCH_2C_6H_5)$

Chem. Ber. 95:2896 (1962)

▶ $PSNO_2C_{14}H_{30}$

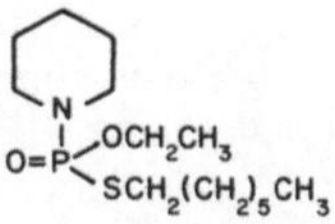

(piperidino)$P(=O)(OCH_2CH_3)(SCH_2(CH_2)_5CH_3)$

Chem. Ber. 95:2896 (1962)

▶ $PSNO_4C_5H_{16}$

$(CH_3O)_2P(=O)-SCH_2C(=O)-NHCH_3$

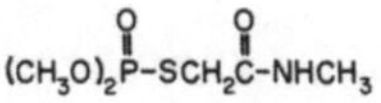

J. Org. Chem. 26:2285 (1961)

▶ $PSOClC_{10}H_{18}$

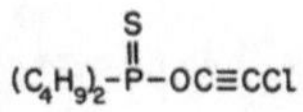

$(C_4H_9)_2-P(=S)-OC\equiv CCl$

J. Org. Chem. 27:2199 (1962)

▶ $PSOCl_2C_2H_5$

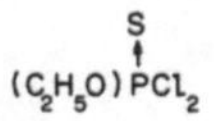

$(C_2H_5O)P(\rightarrow S)Cl_2$

Compt. Rend. 254:260 (1962)

▶ $PSOCl_2C_{10}H_{19}$

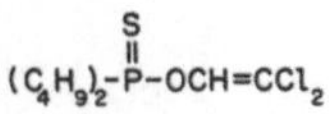

$(C_4H_9)_2-P(=S)-OCH=CCl_2$

J. Org. Chem. 27:2199 (1962)

▶ $PSO_2ClC_4H_{10}$

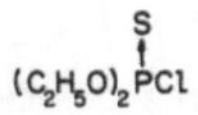

$(C_2H_5O)_2P(\rightarrow S)Cl$

Compt. Rend. 254:260 (1962)

▶ $PSO_2C_{12}H_{19}$

$CH_3CH_2CH_2O-P(=S)(C_6H_5)-OCH_2CH_2CH_3$

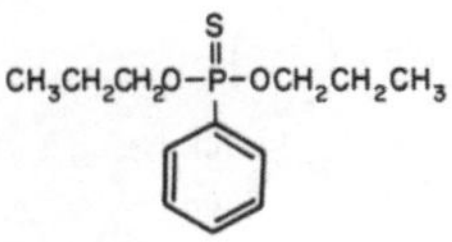

MCA Serial No. 120

▶ $PSO_2C_{12}H_{19}$

$(H_3C)_2HC-O-P(=S)(C_6H_5)-O-CH(CH_3)_2$

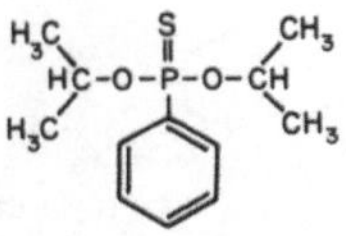

MCA Serial No. 121

▶ $PSO_3C_6H_{15}$

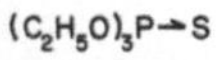

$(C_2H_5O)_3P\rightarrow S$

Compt. Rend. 254:260 (1962)

▶ $PSO_3C_6H_{15}$

$(CH_3CH_2O)_3PS$

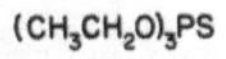

MCA Serial No. 47

▶ $PSO_3C_6H_{15}$

$(C_2H_5O)_2P(\rightarrow O)(SC_2H_5)$

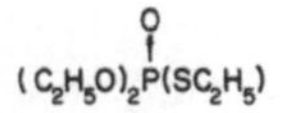

Compt. Rend. 254:260 (1962)

► $PNO_4C_5H_{12}$

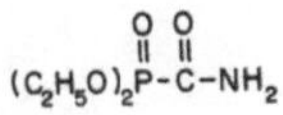

$(C_2H_5O)_2P(=O){-}C(=O){-}NH_2$

J. Org. Chem. 26:2544 (1961)

► $PN_2C_{15}H_{28} \cdot I$

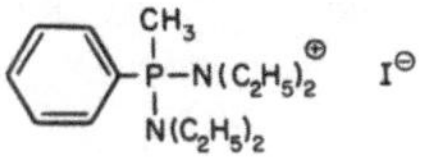


J. Chem. Soc. 3985 (1962)

► $PNC_{18}H_{16}$

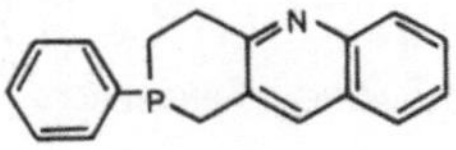


J. Chem. Soc. 5113 (1962)

► $PN_3OC_6H_{18}$

$O{=}P[-N(CH_3)_2]_3$

Varian 145

► $PN_2O_8C_9H_{11}$

URIDINE-2',3'-CYCLIC PHOSPHATE

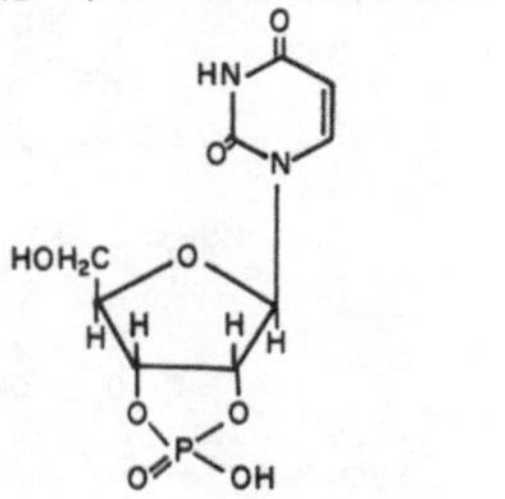


J. Am. Chem. Soc. 84:62 (1962)

► $PN_3O_7C_9H_{12}$

CYTIDINE-2',3'-CYCLIC PHOSPHATE

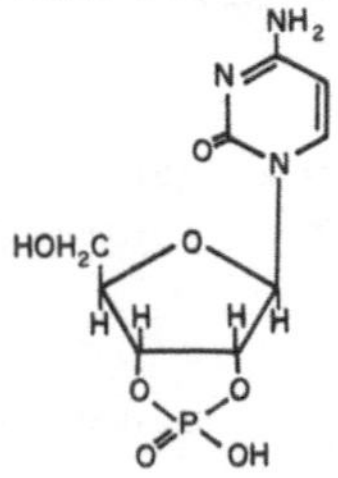


J. Am. Chem. Soc. 84:62 (1962)

► $PN_2O_8C_{10}H_{15}$

5'-THYMIDYLIC ACID

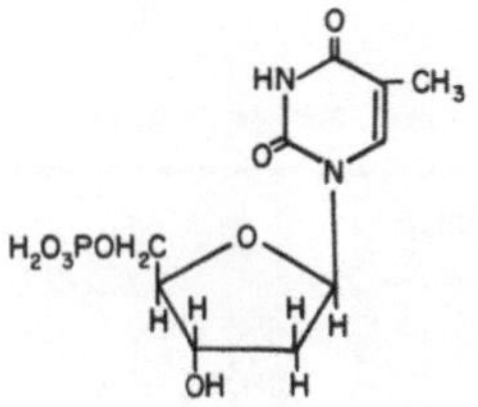


J. Am. Chem. Soc. 83:2919 (1961)

► $PN_3O_7C_9H_{14}$

2'-DEOXYCYTIDINE-5'-PHOSPHATE

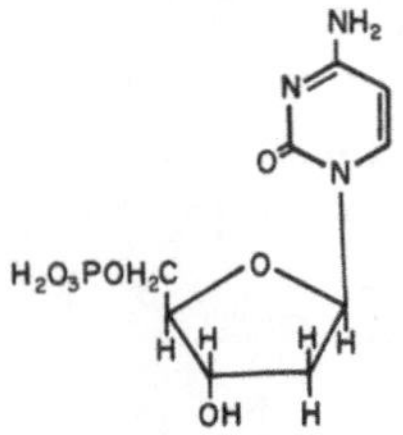


J. Am. Chem. Soc. 83:2919 (1961)

► $PN_2C_{12}H_{13}$

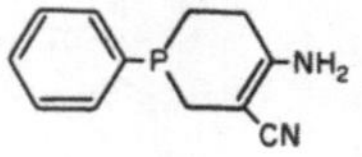


J. Chem. Soc. 5111 (1962)

► $PN_4O_3ClC_6H_{10}$

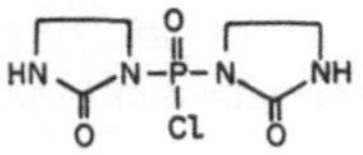


Bull. Soc. Chim. France 2114 (1961)

▶ $PN_5O_3C_{12}H_{16}$

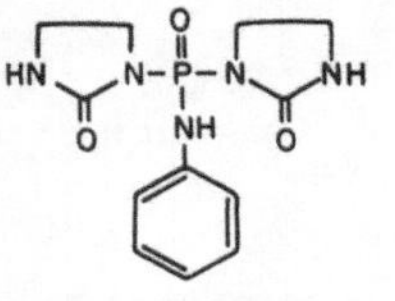

Bull. Soc. Chim. France 2114 (1961)

▶ $PN_5O_6C_{10}H_{12}$

ADENOSINE-2',3'-CYCLIC PHOSPHATE

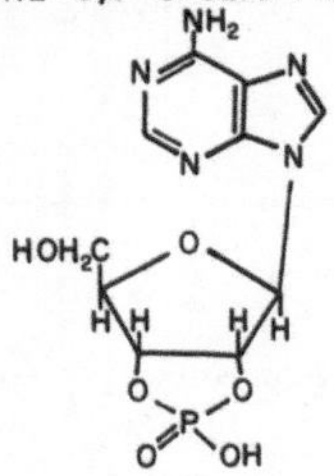

J. Am. Chem. Soc. 84:62 (1962)

▶ $PN_5O_6C_{10}H_{12}$

ADENOSINE-3',5'-CYCLIC PHOSPHATE

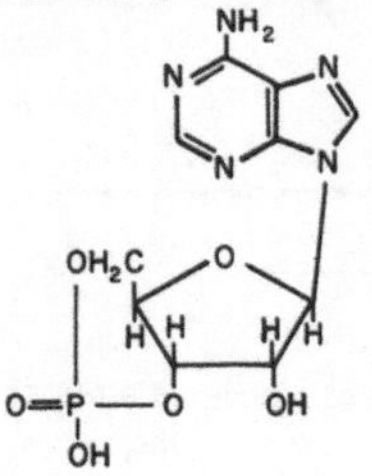

J. Am. Chem. Soc. 84:62 (1962)

▶ $PN_5O_6C_{10}H_{14}$

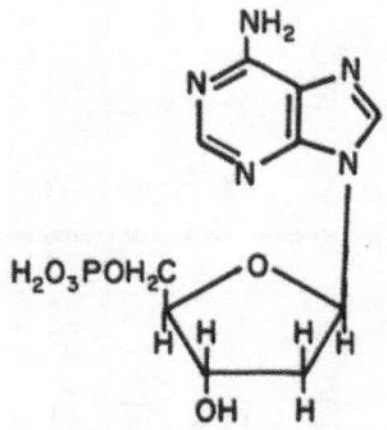

2'-DEOXYADENOSINE-5'-PHOSPHATE

J. Am. Chem. Soc. 83:2919 (1961)

▶ $PN_5O_7C_{10}H_{14}$

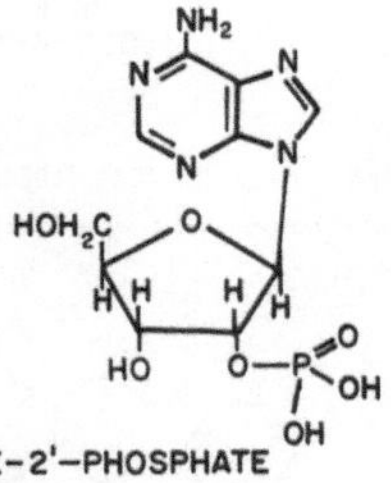

ADENOSINE-2'-PHOSPHATE

J. Am. Chem. Soc. 84:62 (1962)

▶ $PN_5O_7C_{10}H_{14}$

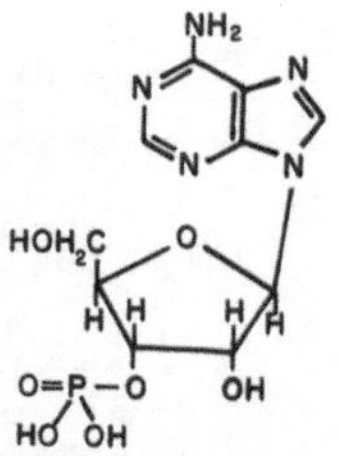

ADENOSINE-3'-PHOSPHATE

J. Am. Chem. Soc. 84:62 (1962)

▶ $POClC_7H_8$

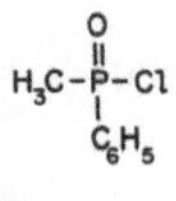

J. Am. Chem. Soc. 84:2503 (1962)

▶ $POClC_8H_{10}$

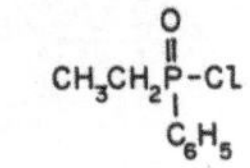

J. Am. Chem. Soc. 84:2503 (1962)

▶ $POClC_9H_{12}$

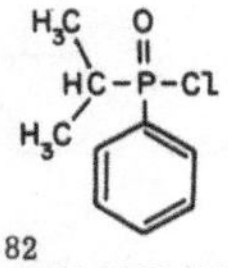

MCA Serial No. 82
J. Am. Chem. Soc. 84:2503 (1962)

▶ $POClC_{10}H_{14}$

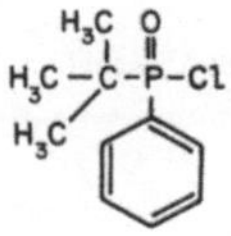

MCA Serial No. 84
J. Am. Chem. Soc. 84:2503 (1962)

▶ $POClC_{10}H_{14}$

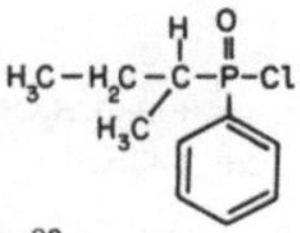

MCA Serial No. 83
J. Am. Chem. Soc. 84:2503 (1962)

▶ $POClC_{11}H_{16}$

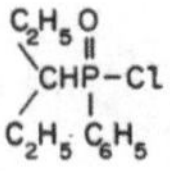

J. Am. Chem. Soc. 84:2503 (1962)

▶ $POCl_2C_2H_5$

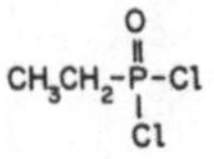

MCA Serial No. 115
J. Am. Chem. Soc. 84:2503 (1962)

▶ $POCl_2C_3H_7$

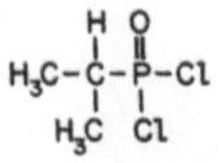

MCA Serial No. 80
J. Am. Chem. Soc. 84:2503 (1962)

▶ $POCl_2C_4H_9$

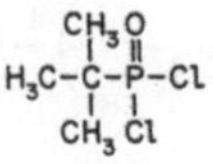

MCA Serial No. 81
J. Am. Chem. Soc. 84:2503 (1962)

▶ $POCl_2C_4H_9$

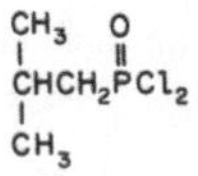

J. Am. Chem. Soc. 84:2503 (1962)

▶ $POCl_2C_4H_9$

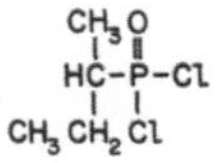

MCA Serial No. 116
J. Am. Chem. Soc. 84:2503 (1962)

▶ $POC_{12}H_{19}$

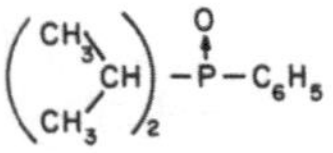

J. Am. Chem. Soc. 84:3467 (1962)

▶ $POC_{12}H_{27}$

$(CH_3CH_2CH_2CH_2)_3PO$

MCA Serial No. 10

▶ $PO_2ClC_4H_{10}$

$(C_2H_5O)_2PCl$

Compt. Rend. 254:260 (1962)

▶ $PO_2ClC_6H_{12}$

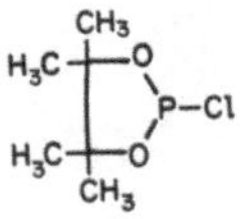

Varian 472

▶ $PO_2Cl_2C_2H_5$

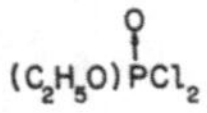

Compt. Rend. 254:260 (1962)

▶ $PO_2Cl_3C_{10}H_{12}$

$(CH_3)_2CH-O-P(=O)(C_6H_5)-CCl_3$

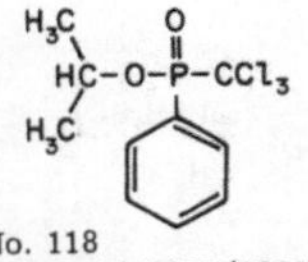

MCA Serial No. 118
J. Am. Chem. Soc. 84:3467 (1962)

▶ $PO_2CH_5$

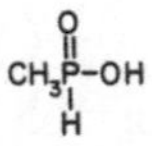

$CH_3-P(=O)(H)-OH$

J. Chem. Soc. 3838 (1962)

▶ $PO_2C_2H_7$

$(CH_3)_2P(=O)-OH$

J. Chem. Soc. 3838 (1962)

▶ $PO_2C_8H_{17}$

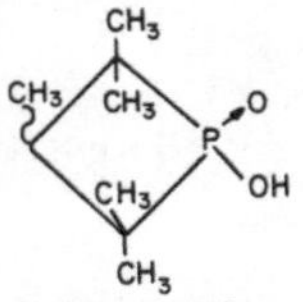

J. Org. Chem. 27:1834 (1962)

▶ $PO_2C_9H_{13}$

$H_3C-CH(CH_3)-P(=O)(C_6H_5)-OH$

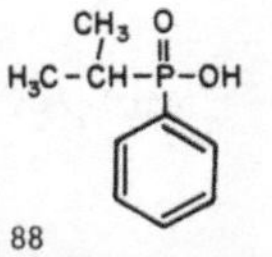

MCA Serial No. 88
J. Am. Chem. Soc. 84:2503 (1962)

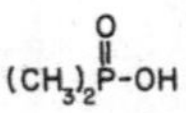

▶ $PO_2C_9H_{19}$

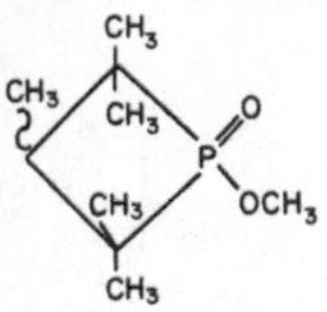

J. Org. Chem. 27:1834 (1962)

▶ $PO_2C_{10}H_{15}$

$(H_3C)_3C-P(=O)(C_6H_5)-OH$

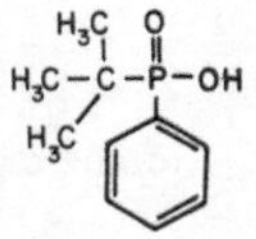

MCA Serial No. 90
J. Am. Chem. Soc. 84:2503 (1962)

▶ $PO_2C_{10}H_{15}$

$(CH_3)_2CH-P(=O)(C_6H_5)-OCH_3$

J. Am. Chem. Soc. 84:2503 (1962)

▶ $PO_2C_{10}H_{15}$

$(C_2H_5O)_2P-C_6H_5$

J. Am. Chem. Soc. 84:3467 (1962)

▶ $PO_2C_{11}H_{17}$

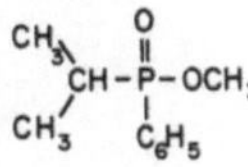

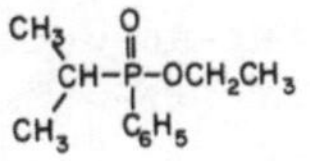

$(CH_3)_2CH-P(=O)(C_6H_5)-OCH_2CH_3$

J. Am. Chem. Soc. 84:2503 (1962)

▶ $PO_2C_{12}H_{19}$

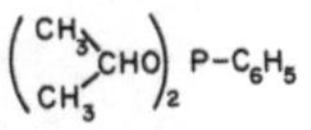

$((CH_3)_2CHO)_2P-C_6H_5$

J. Am. Chem. Soc. 84:3467 (1962)

▶ $PO_2C_{12}H_{19}$

$(H_3C)_2CH-O-P(=O)(C_6H_5)-CH(CH_3)_2$

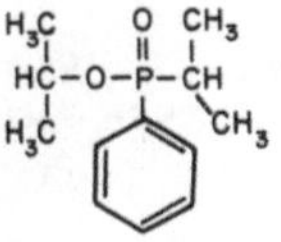

MCA Serial No. 64

▶ $PO_2C_{12}H_{27}$

$(CH_3-CH_2-CH(CH_3)-)_2P(=O)-O-CH(CH_3)-CH_2-CH_3$

MCA Serial No. 12

▶ $PO_2C_{12}H_{27}$

$(CH_3-CH_2-CH(CH_3)-)_2P(=O)-O-(CH_2)_3-CH_3$

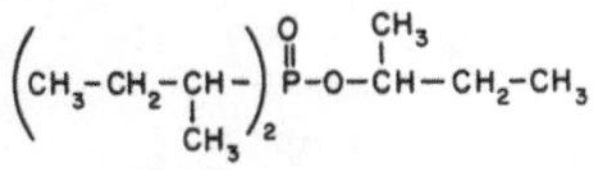

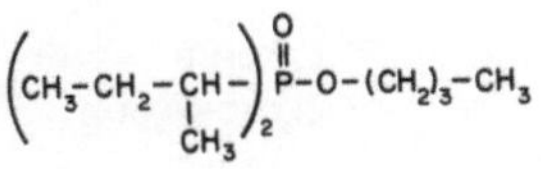

MCA Serial No. 11

▶ $PO_2C_{14}H_{15}$

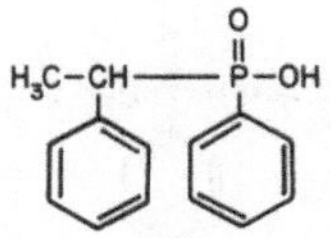

MCA Serial No. 97
J. Am. Chem. Soc. 84:2503 (1962)

▶ $PO_2C_{14}H_{23}$

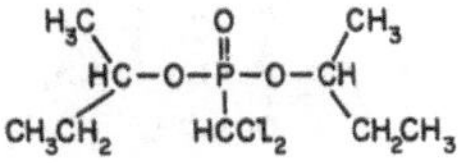

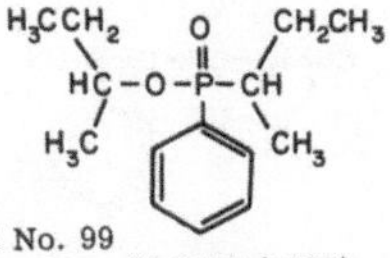

MCA Serial No. 99
J. Am. Chem. Soc. 84:3467 (1962)

▶ $PO_2C_{15}H_{17}$

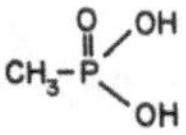

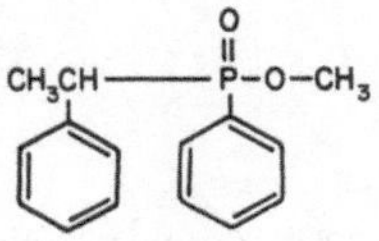

MCA Serial No. 101
J. Am. Chem. Soc. 84:2503 (1962)

▶ $PO_3BrC_7H_{14}$

$(CH_3CH_2O)_2P(=O)-CH_2-CH=CHBr$

Bull. Soc. Chim. France 1616 (1962)

▶ $PO_3BrC_7H_{14}$

$(CH_3CH_2O)_2P(=O)-CH_2-C(Br)=CH_2$

Bull. Soc. Chim. France 1616 (1962)

▶ $PO_3ClC_4H_{10}$

$(C_2H_5O)_2P(O)Cl$

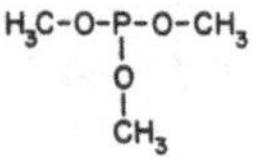

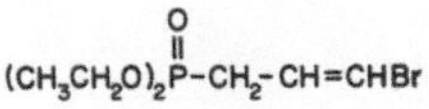

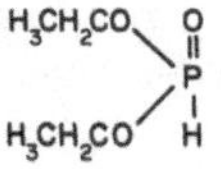

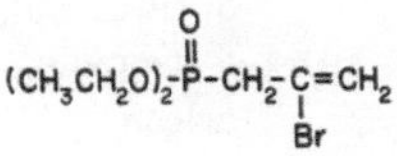

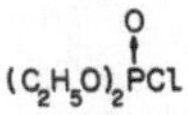

Compt. Rend. 254:260 (1962)

▶ $PO_3ClC_{17}H_{36}$

$[CH_3(CH_2)_3-CH(CH_2CH_3)-CH_2-O]_2(ClCH_2)PO$

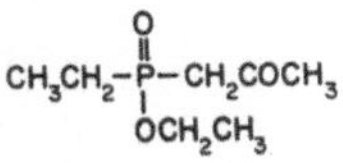

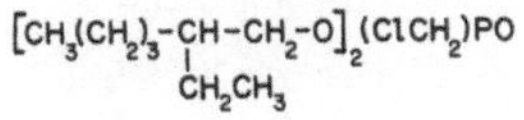

MCA Serial No. 46

▶ $PO_3Cl_2C_9H_{19}$

MCA Serial No. 117
J. Am. Chem. Soc. 84:3467 (1962)

▶ $PO_3CH_5$

$CH_3-P(=O)(OH)_2$

J. Chem. Soc. 3838 (1962)

▶ $PO_3C_3H_9$

$H_3C-O-P(-O-CH_3)-O-CH_3$

MCA Serial No. 86

▶ $PO_3C_4H_{11}$

$(H_3CH_2CO)_2P(=O)H$

J. Chem. Soc. 1365 (1962)

▶ $PO_3C_6H_{15}$

$(CH_3CH_2O)_3P$

MCA Serial No. 5
Compt. Rend. 254:260 (1962)

▶ $PO_3C_7H_{15}$

$CH_3CH_2-P(=O)(OCH_2CH_3)-CH_2COCH_3$

Bull. Soc. Chim. France 1616 (1962)

▶ $PO_3C_8H_{11}$

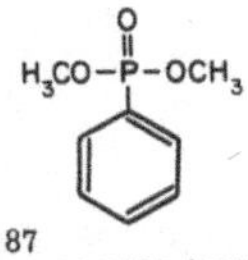

MCA Serial No. 87
J. Am. Chem. Soc. 84:3467 (1962)

▶ $PO_3C_8H_{17}$

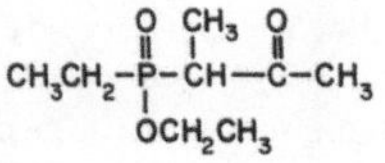

Bull. Soc. Chim. France 1616 (1962)

▶ $PO_3C_8H_{19}$

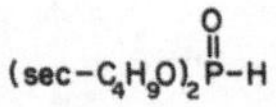

J. Org. Chem. 27:3576 (1962)

▶ $PO_3C_9H_{19}$

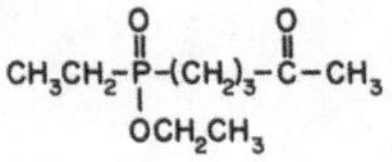

Bull. Soc. Chim. France 1616 (1962)

▶ $PO_3C_9H_{19}$

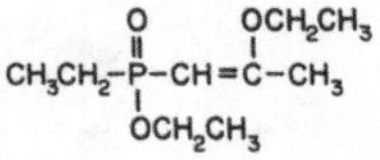

Bull. Soc. Chim. France 1616 (1962)

▶ $PO_3C_9H_{19}$

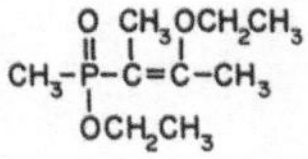

Bull. Soc. Chim. France 1616 (1962)

▶ $PO_3C_9H_{21}$

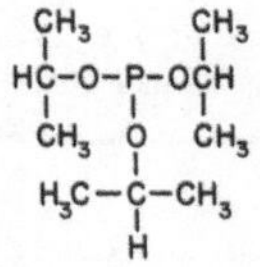

J. Am. Chem. Soc. 84:3467 (1962)
MCA Serial No. 61

▶ $PO_3C_{10}H_{15}$

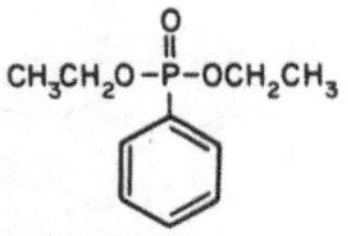

MCA Serial No. 91
J. Am. Chem. Soc. 84:3467 (1962)

▶ $PO_3C_{10}H_{15}$

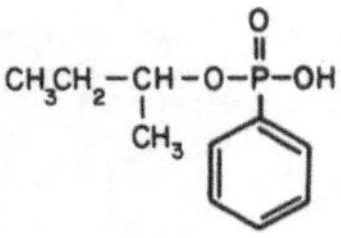

MCA Serial No. 62

▶ $PO_3C_{10}H_{17}$

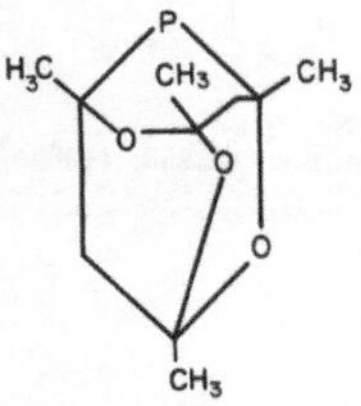

J. Am. Chem. Soc. 83:3279 (1961)

▶ $PO_3C_{10}H_{21}$

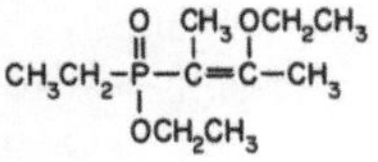

Bull. Soc. Chim. France 1616 (1962)

▶ $PO_3C_{11}H_{17}$

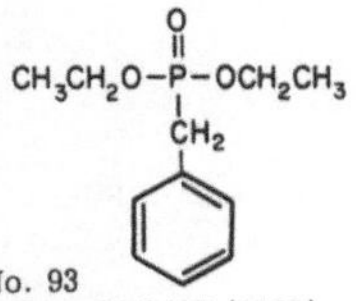

MCA Serial No. 93
J. Am. Chem. Soc. 84:3467 (1962)

▶ $PO_3C_{11}H_{23}$

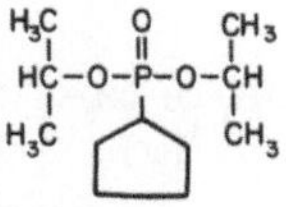

MCA Serial No. 63
J. Am. Chem. Soc. 84:3467 (1962)

▶ $PO_3C_{12}H_{19}$

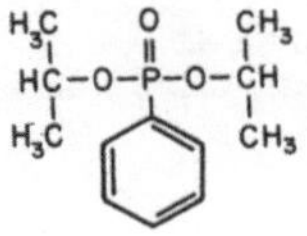

J. Am. Chem. Soc. 84:3467 (1962)
MCA Serial No. 65

▶ $PO_3C_{12}H_{27}$

$[CH_3(CH_2)_3O]_2P(O)(CH_2)_3CH_3$

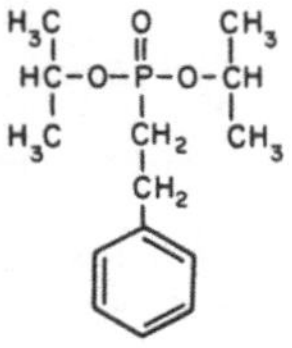

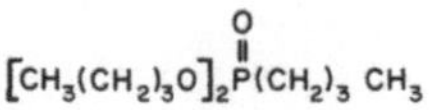

MCA Serial No. 14

▶ $PO_3C_{12}H_{27}$

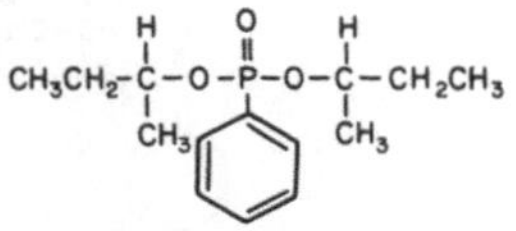

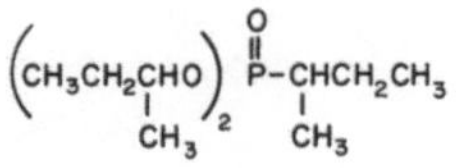

MCA Serial No. 15

▶ $PO_3C_{12}H_{27}$

$[CH_3(CH_2)_3O]_3P$

MCA Serial No. 13

▶ $PO_3C_{13}H_{21}$

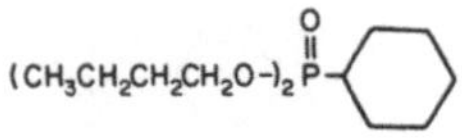

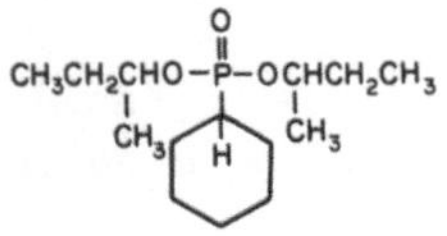

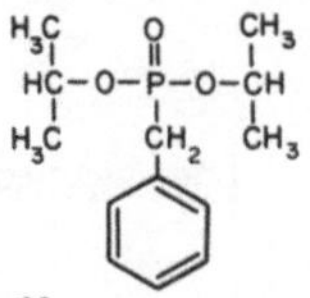

MCA Serial No. 66
J. Am. Chem. Soc. 84:3467 (1962)

▶ $PO_3C_{14}H_{23}$

J. Am. Chem. Soc. 84:3467 (1962)

▶ $PO_3C_{14}H_{23}$

MCA Serial No. 67
J. Am. Chem. Soc. 84:3467 (1962)

▶ $PO_3C_{14}H_{23}$

J. Am. Chem. Soc. 84:3467 (1962)
MCA Serial No. 68

▶ $PO_3C_{14}H_{29}$

$(CH_3CH_2CH_2CH_2O-)_2P(O)-C_6H_{11}$

MCA Serial No. 21

▶ $PO_3C_{14}H_{29}$

MCA Serial No. 69
J. Am. Chem. Soc. 84:3467 (1962)

▶ $PO_3C_{15}H_{25}$

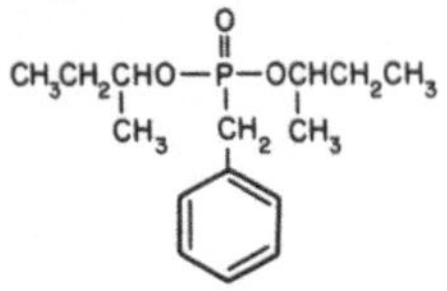

J. Am. Chem. Soc. 84:3467 (1962)
MCA Serial No. 70

▶ $PO_3C_{15}H_{31}$

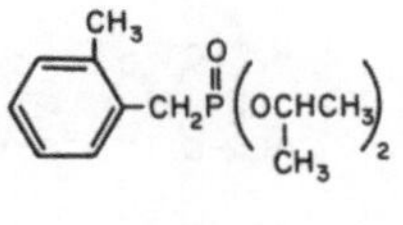

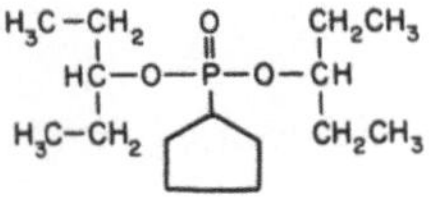

MCA Serial No. 104
J. Am. Chem. Soc. 84:3467 (1962)

▶ $PO_3C_{15}H_{33}$

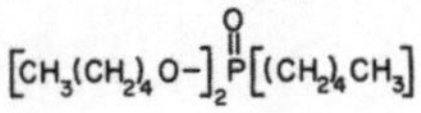

MCA Serial No. 19

▶ $PO_3C_{16}H_{27}$

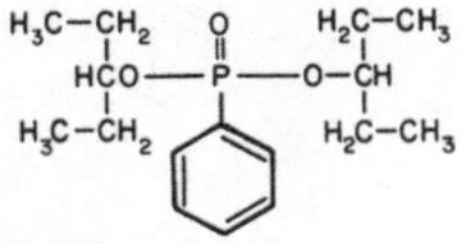

MCA Serial No. 74
J. Am. Chem. Soc. 84:3467 (1962)

▶ $PO_3C_{16}H_{27}$

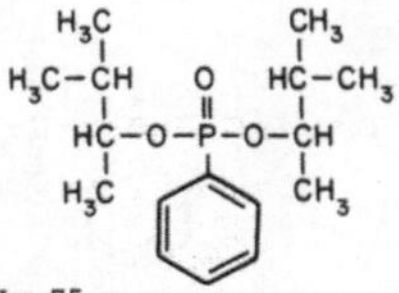

MCA Serial No. 75
J. Am. Chem. Soc. 84:3467 (1962)

▶ $PO_3C_{18}H_{15}$

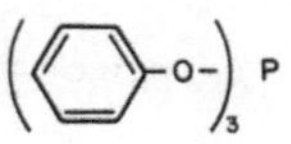

MCA Serial No. 27

▶ $PO_3C_{18}H_{39}$

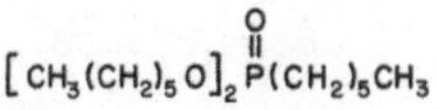

MCA Serial No. 30

▶ $PO_3C_{21}H_{21}$

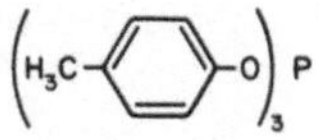

MCA Serial No. 33

▶ $PO_3C_{22}H_{47}$

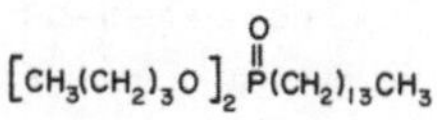

MCA Serial No. 40

▶ $PO_3C_{30}H_{63}$

$[CH_3(CH_2)_8CH_2O]_3P$

MCA Serial No. 43

▶ $PO_4ClC_4H_8$

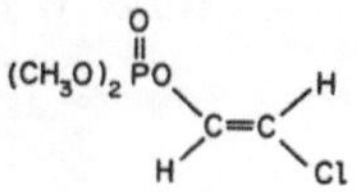

J. Org. Chem. 26:3965 (1961)

▶ $PO_4Cl_2C_9H_{17}$

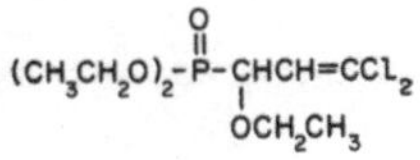

Bull. Soc. Chim. France 1616 (1962)

▶ $PO_4Cl_9C_6H_6$

$(CCl_3CH_2O)_3PO$

MCA Serial No. 45

▶ $PO_4F_{36}C_{21}H_9$

$[CHF_2(CF_2)_5CH_2O]_3PO$

MCA Serial No. 44

▶ $PO_4C_2H_5$

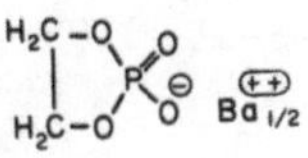

J. Am. Chem. Soc. 83:1102 (1961)

▶ $PO_4C_3H_9$

$(CH_3O)_3PO$

MCA Serial No. 3
J. Chem. Soc. 1372, 3838 (1962)

▶ $PO_4C_6H_{15}$

$(CH_3CH_2O)_3PO$

Varian 482
MCA Serial No. 6
Compt. Rend. 254:260 (1962)

▶ $PO_4C_7H_{15}$

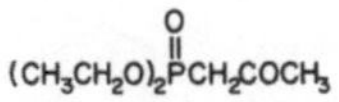

Bull. Soc. Chim. France 1616 (1962)

▶ $PO_4C_8H_{15}$

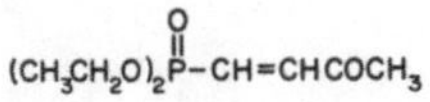

Bull. Soc. Chim. France 1616 (1962)

▶ $PO_4C_8H_{17}$

$(CH_3CH_2O)_2-\overset{O}{\overset{\|}{P}}-CH_2COCH_2CH_3$

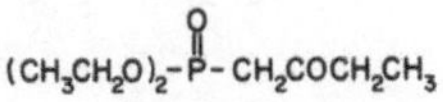

Bull. Soc. Chim. France 1616 (1962)

▶ $PO_4C_8H_{17}$

$(CH_3CH_2O)_2\overset{O}{\overset{\|}{P}}-\underset{\underset{CH_3}{|}}{CH}-COCH_3$

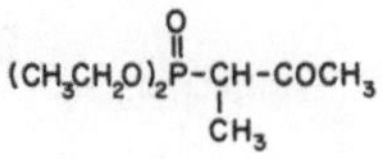

Bull. Soc. Chim. France 1616 (1962)

▶ $PO_4C_9H_{15}$

$(CH_2{=}CH{-}CH_2{-}O)_3PO$

MCA Serial No. 7

▶ $PO_4C_9H_{19}$

$(CH_3CH_2O)_2\overset{O}{\overset{\|}{P}}CH=\underset{\underset{OCH_2CH_3}{|}}{C}-CH_3$

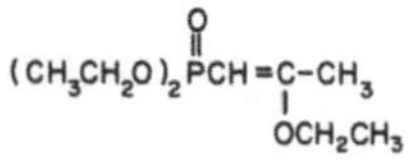

Bull. Soc. Chim. France 1616 (1962)

▶ $PO_4C_9H_{19}$

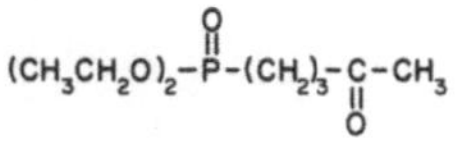

Bull. Soc. Chim. France 1616 (1962)

▶ $PO_4C_9H_{21}$

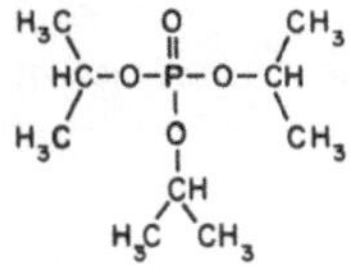

MCA Serial No. 89
J. Am. Chem. Soc. 84:3467 (1962)

▶ $PO_4C_9H_{21}$

$(CH_3CH_2CH_2O)_3PO$

MCA Serial No. 8

▶ $PO_4C_{10}H_{15}$

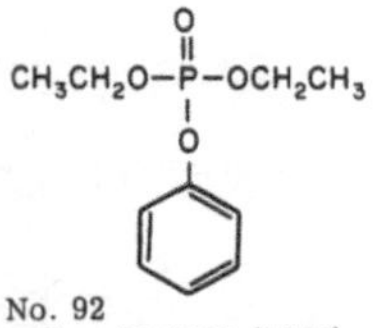

MCA Serial No. 92
J. Am. Chem. Soc. 84:3467 (1962)

▶ $PO_4C_{10}H_{21}$

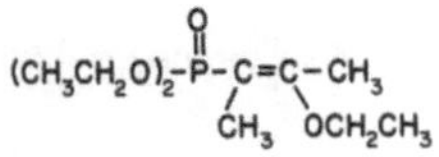

Bull. Soc. Chim. France 1616 (1962)

▶ $PO_4C_{12}H_{19}$

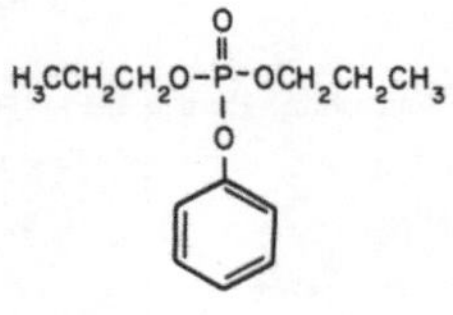

MCA Serial No. 51
J. Am. Chem. Soc. 84:3467 (1962)

▶ $PO_4C_{13}H_{21}$

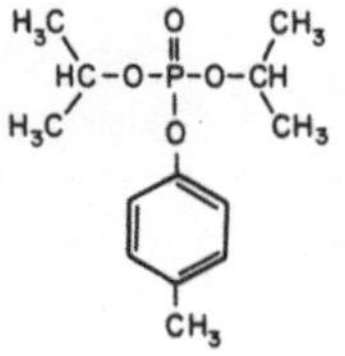

MCA Serial No. 96
J. Am. Chem. Soc. 84:3467 (1962)

▶ $PO_4C_{12}H_{19}$

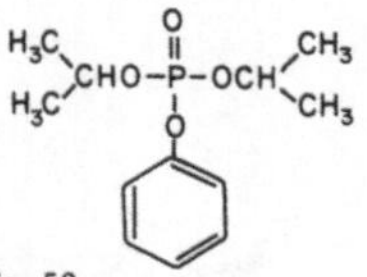

MCA Serial No. 52
J. Am. Chem. Soc. 84:3467 (1962)

▶ $PO_4C_{13}H_{21}$

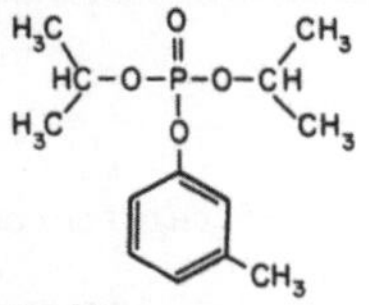

MCA Serial No. 95
J. Am. Chem. Soc. 84:3467 (1962)

▶ $PO_4C_{12}H_{21}$

$(CH_2{=}C(CH_3){-}CH_2{-}O)_3PO$

MCA Serial No. 9

▶ $PO_4C_{13}H_{21}$

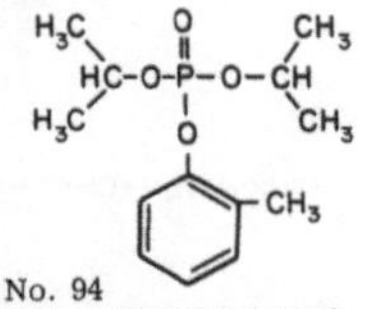

MCA Serial No. 94
J. Am. Chem. Soc. 84:3467 (1962)

▶ $PO_4C_{12}H_{27}$

$(CH_3CH_2CH(CH_3)O)_3PO$

MCA Serial No. 18

▶ $PO_4C_{14}H_{19}$

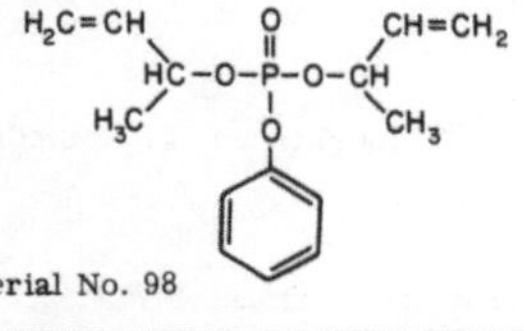

MCA Serial No. 98

▶ $PO_4C_{12}H_{27}$

$[(CH_3)_2CHCH_2O]_3PO$

MCA Serial No. 17

▶ $PO_4C_{14}H_{19}$

$(CH_2{=}CHCH_2CH_2O)_2P(O)OC_6H_5$

J. Am. Chem. Soc. 84:3467 (1962)

▶ $PO_4C_{12}H_{27}$

$[CH_3(CH_2)_3O{-}]_3PO$

MCA Serial No. 16

▶ $PO_4C_{14}H_{23}$

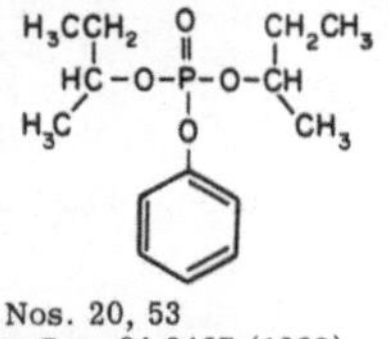

MCA Serial Nos. 20, 53
J. Am. Chem. Soc. 84:3467 (1962)

▶ $PO_4C_{14}H_{23}$

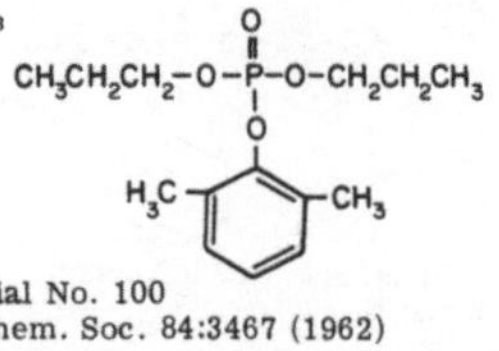

MCA Serial No. 100
J. Am. Chem. Soc. 84:3467 (1962)

▶ $PO_4C_{15}H_{25}$

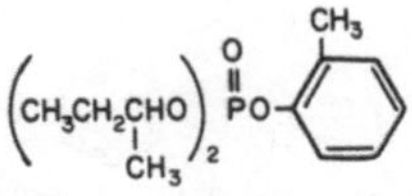

J. Am. Chem. Soc. 84:3467 (1962)

▶ $PO_4C_{15}H_{27}$

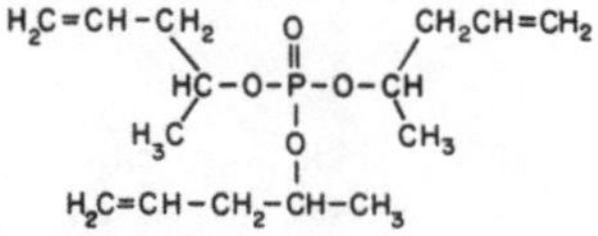

MCA Serial No. 103

▶ $PO_4C_{15}H_{27}$

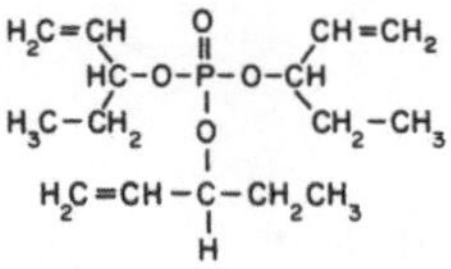

MCA Serial No. 102

▶ $PO_4C_{15}H_{27}$

$[CH_2{=}CH(CH_2)_3]_3PO_4$

J. Am. Chem. Soc. 84:3467 (1962)

▶ $PO_4C_{15}H_{27}$

$(CH_3CH{=}CHCH_2CH_2O)_2\overset{O}{\overset{\|}{P}}OCH_2CH_2CH{=}CHCH_3$

J. Am. Chem. Soc. 84:3467 (1962)

▶ $PO_4C_{15}H_{27}$

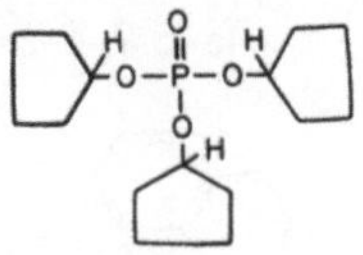

MCA Serial Nos. 22, 71
J. Am. Chem. Soc. 84:3467 (1962)

▶ $PO_4C_{15}H_{23}$

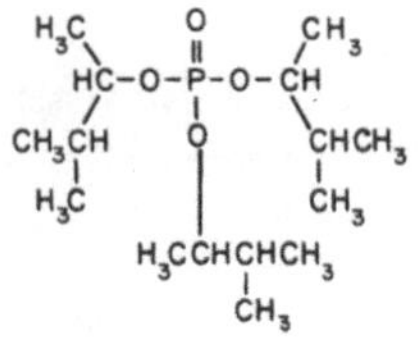

J. Am. Chem. Soc. 84:3467 (1962)
MCA Serial No. 106

▶ $PO_4C_{15}H_{33}$

$[(CH_3CH_2)_2CHO-]_3PO$

MCA Serial Nos. 25, 105
J. Am. Chem. Soc. 84:3467 (1962)

▶ $PO_4C_{15}H_{33}$

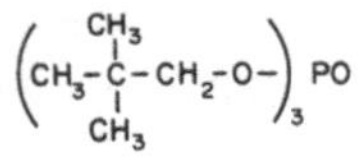

MCA Serial No. 26

▶ $PO_4C_{15}H_{33}$

$(CH_3CH_2CH_2CH_2CH_2O-)_3PO$

MCA Serial No. 23

▶ $PO_4C_{15}H_{33}$

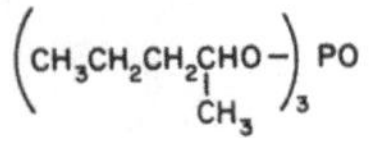

MCA Serial No. 24

▶ $PO_4C_{16}H_{19}$

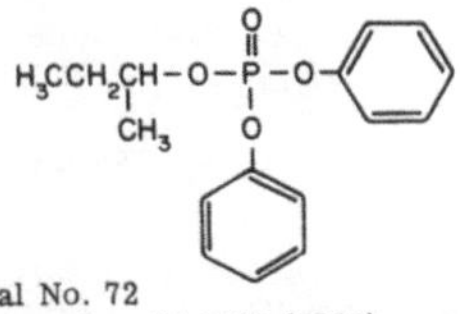

MCA Serial No. 72
J. Am. Chem. Soc. 84:3467 (1962)

▶ $PO_4C_{16}H_{23}$

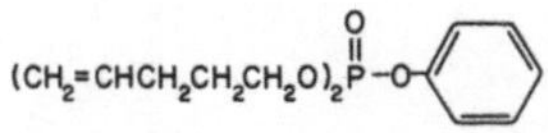

J. Am. Chem. Soc. 84:3467 (1962)

▶ $PO_4C_{16}H_{23}$

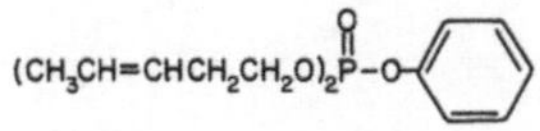

J. Am. Chem. Soc. 84:3467 (1962)

▶ $PO_4C_{16}H_{23}$

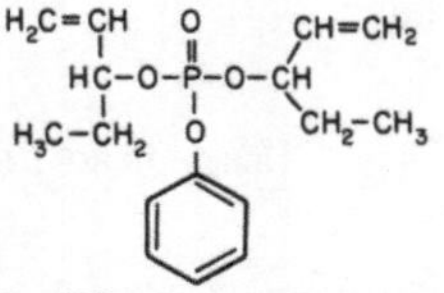

MCA Serial No. 107

▶ $PO_4C_{16}H_{23}$

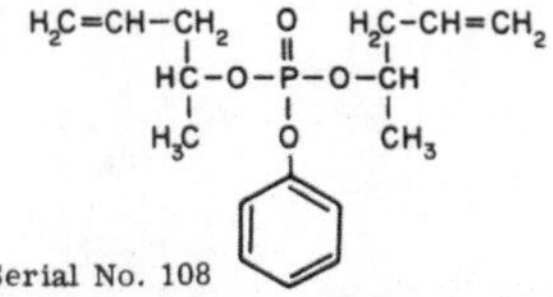

MCA Serial No. 108

▶ $PO_4C_{16}H_{23}$

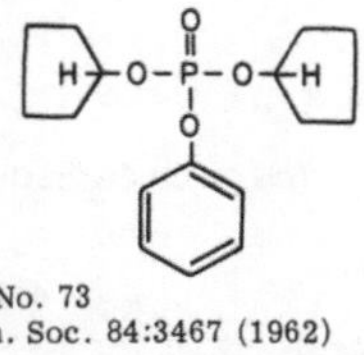

MCA Serial No. 73
J. Am. Chem. Soc. 84:3467 (1962)

▶ $PO_4C_{16}H_{27}$

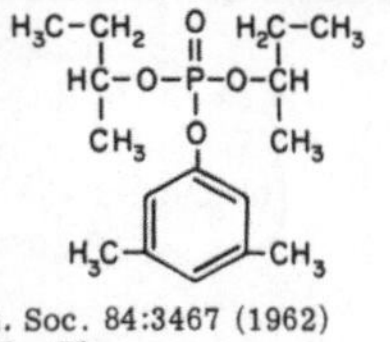

J. Am. Chem. Soc. 84:3467 (1962)
MCA Serial No. 76

▶ $PO_4C_{16}H_{27}$

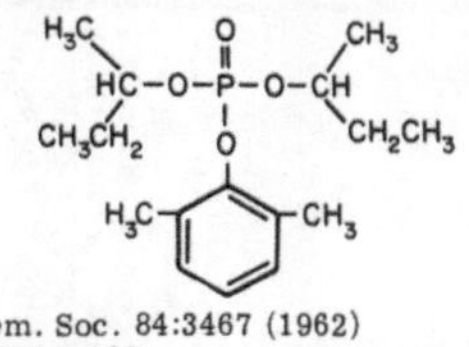

J. Am. Chem. Soc. 84:3467 (1962)
MCA Serial No. 109

▶ $PO_4C_{16}H_{27}$

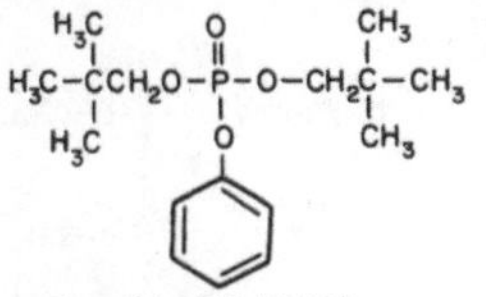

J. Am. Chem. Soc. 84:3467 (1962)
MCA Serial No. 77

▶ $PO_4C_{16}H_{27}$

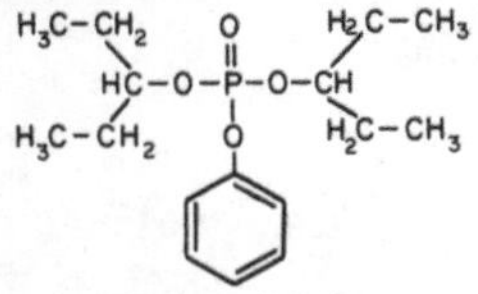

J. Am. Chem. Soc. 84:3467 (1962)
MCA Serial No. 54

▶ $PO_4C_{16}H_{27}$

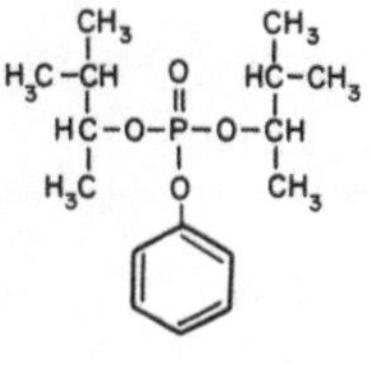

MCA Serial No. 55
J. Am. Chem. Soc. 84:3467 (1962)

▶ $PO_4C_{18}H_{15}$

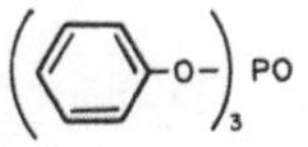

MCA Serial No. 28

▶ $PO_4C_{18}H_{25}$

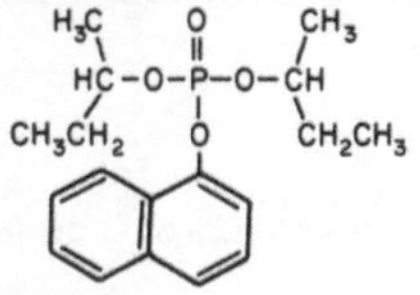

MCA Serial No. 110
J. Am. Chem. Soc. 84:3467 (1962)

▶ $PO_4C_{18}H_{25}$

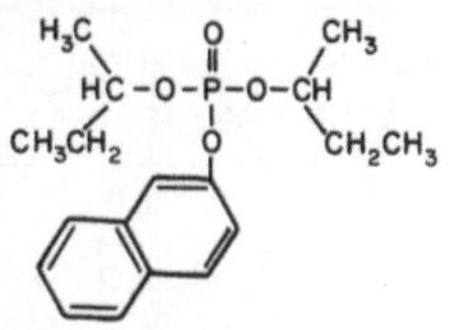

MCA Serial No. 111

▶ $PO_4C_{18}H_{31}$

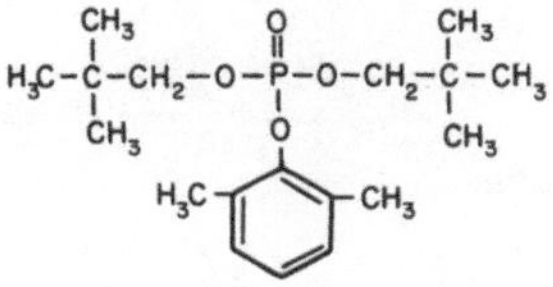

MCA Serial No. 112
J. Am. Chem. Soc. 84:3467 (1962)

▶ $PO_4C_{20}H_{35}$

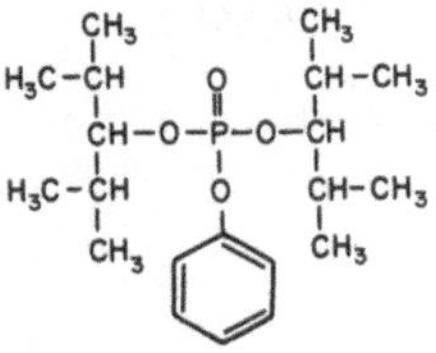

MCA Serial No. 113
J. Am. Chem. Soc. 84:3467 (1962)

▶ $PO_4C_{18}H_{31}$

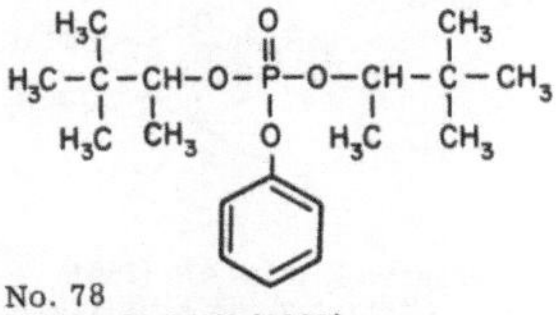

MCA Serial No. 78
J. Am. Chem. Soc. 84:3467 (1962)

▶ $PO_4C_{20}H_{35}$

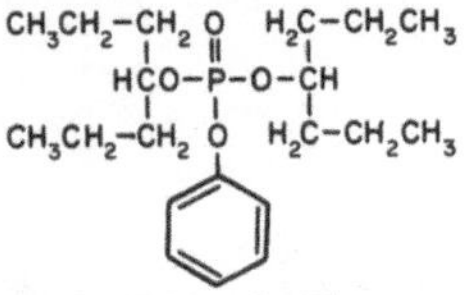

J. Am. Chem. Soc. 84:3467 (1962)
MCA Serial No. 79

▶ $PO_4C_{18}H_{33}$

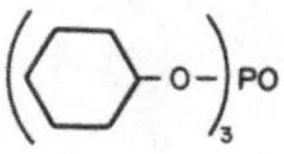

MCA Serial No. 29

▶ $PO_4C_{21}H_{21}$

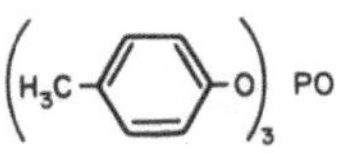

MCA Serial No. 37

▶ $PO_4C_{18}H_{39}$

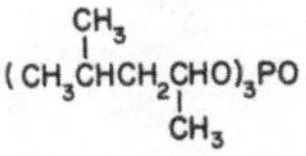

MCA Serial No. 32

▶ $PO_4C_{21}H_{21}$

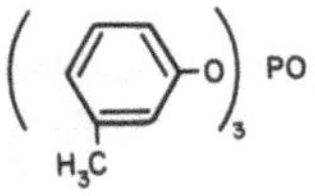

MCA Serial No. 36

▶ $PO_4C_{18}H_{39}$

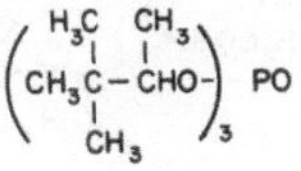

J. Am. Chem. Soc. 84:3467 (1962)

▶ $PO_4C_{21}H_{21}$

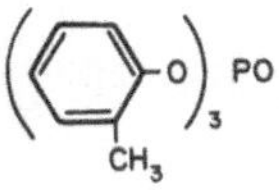

MCA Serial No. 35

▶ $PO_4C_{18}H_{39}$

$[CH_3(CH_2)_5O]_3PO$

MCA Serial No. 31

▶ $PO_4C_{21}H_{21}$

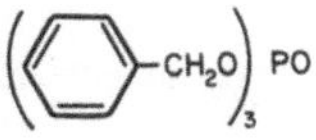

MCA Serial No. 34

▶ $PO_4C_{21}H_{45}$

$$\left(\begin{array}{c}(CH_3)_2CH \\ | \\ CHO \\ | \\ (CH_3)_2CH\end{array}\right)_3 PO$$

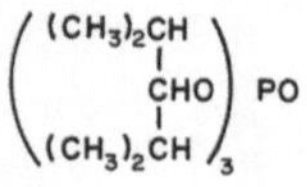

MCA Serial No. 39

▶ $PO_4C_{21}H_{45}$

$$[(CH_3CH_2CH_2)_2CHO]_3PO$$

MCA Serial No. 38

▶ $PO_4C_{22}H_{23}$

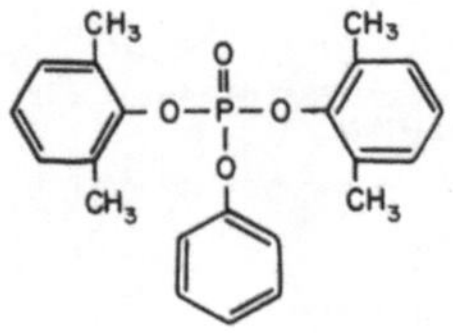

MCA Serial No. 114
J. Am. Chem. Soc. 84:3467 (1962)

▶ $PO_4C_{24}H_{51}$

$$[CH_3(CH_2)_3CH(CH_2CH_3)CH_2O]_3PO$$

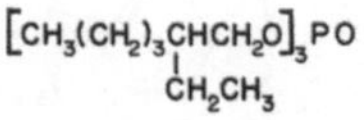

MCA Serial No. 41

▶ $PO_4C_{24}H_{51}$

$$[(CH_3)_2CH(CH_2)_5O]_3PO$$

MCA Serial No. 42

▶ $PO_5C_8H_{17}$

$$(CH_3CH_2O)_2P(O)-CH_2CO_2CH_2CH_3$$

Bull. Soc. Chim. France 1616 (1962)

▶ $PO_5C_{12}H_{15}$

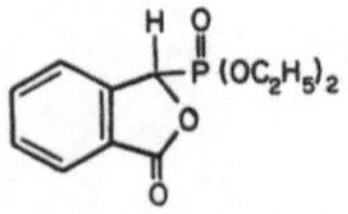

J. Am. Chem. Soc. 84:3467 (1962)

▶ $PO_5C_{14}H_{19}$

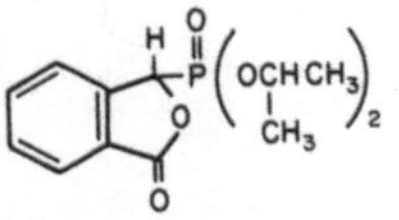

J. Am. Chem. Soc. 84:3467 (1962)

▶ $PO_5C_{32}H_{27}$

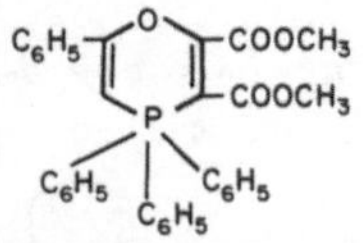

Tetrahedron Letters 477 (1961)
J. Am. Chem. Soc. 83:2018 (1961)

▶ $PO_6C_7H_{13}$

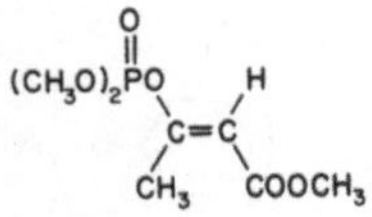

J. Org. Chem. 26:3960 (1961)

▶ $PO_6C_7H_{13}$

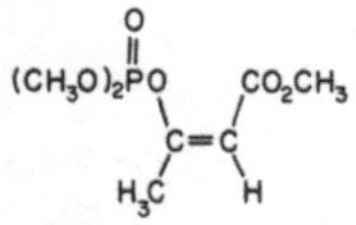

Can J. Chem. 39:1390 (1961)
J. Org. Chem. 26:4620 (1961)

▶ $PO_8C_{30}H_{27}$

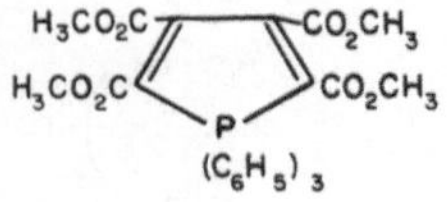

Tetrahedron Letters 477 (1961)

▶ $PO_8C_{30}H_{27}$

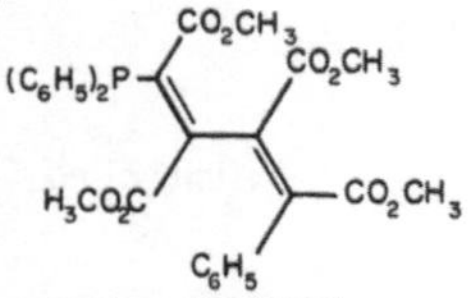

Tetrahedron Letters 480 (1961)

▶ $PBrC_{48}H_{36}$

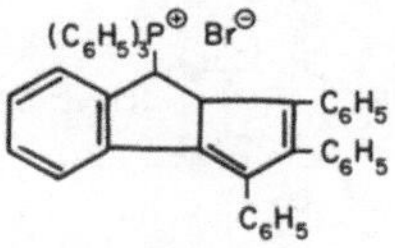


J. Am. Chem. Soc. 84:1505 (1962)

▶ $PC_{18}H_{15}$

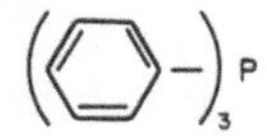


MCA Serial No. 50

▶ $PClC_{21}H_{20}Ni$

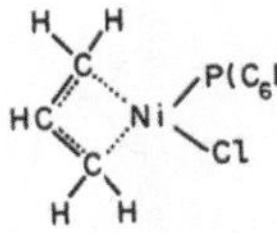


Chem. &Ind. 986 (1961)

▶ $PC_{18}H_{16} \cdot Fe(CO)_4$

$HFe(CO)_4P(C_6H_5)_3^{\oplus}$

J. Chem. Soc. 3653 (1962)

▶ $PClC_{22}H_{22}Ni$

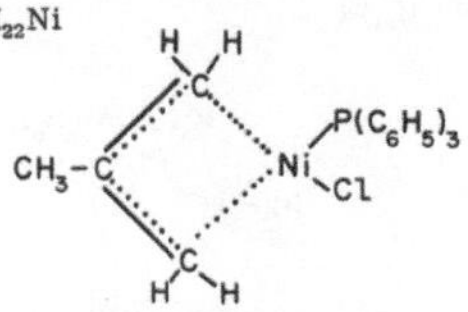


Chem. & Ind. 986 (1961)

▶ $PC_{22}H_{20} \cdot ClO_4$

$(C_6H_5)_3\overset{\oplus}{P}CH{=}CHCH{=}CH_2 \quad ClO_4^{\ominus}$

J. Org. Chem. 26:1434 (1961)

▶ $PC_3H_9 \cdot HCl$

$(CH_3)_3PH^{\oplus} \quad Cl^{\ominus}$

J. Am. Chem. Soc. 83:786 (1961)

▶ $P_2SNC_{13}H_{31}Pt$

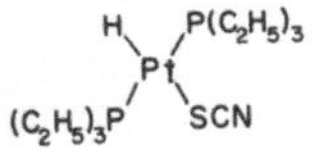


J. Chem. Soc. 5079 (1962)

▶ $PC_6H_{15}$

$(CH_3CH_2)_3P$

J. Chem. Phys. 34:1049 (1961)
J. Am. Chem. Soc. 83:4473 (1961)

▶ $P_2S_2N_2C_{14}H_{16}$

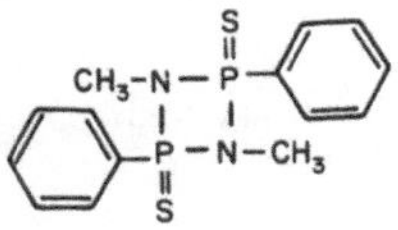


J. Chem. Soc. 4731 (1962)

▶ $PC_{10}H_{19}$

$HC{\equiv}CP(C_4H_9)_2$

Rec. Trav. Chim. 81:635 (1962)

▶ $P_2S_2N_2C_{26}H_{24}$

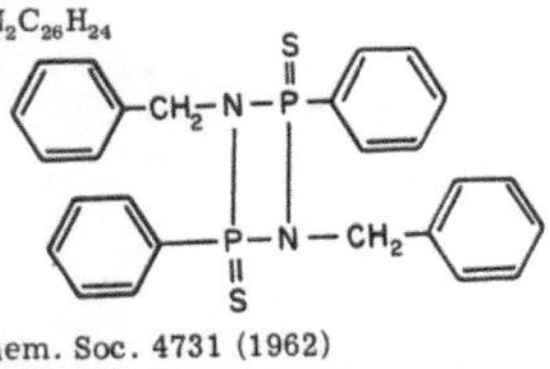


J. Chem. Soc. 4731 (1962)

▶ $PC_{12}H_{27}$

$[CH_3(CH_2)_3]_3P$

MCA Serial No. 49

▶ $P_2S_3O_4C_9H_{22}$

$(C_2H_5O)_2\overset{S}{\overset{\|}{P}}CH_2S\overset{S}{\overset{\|}{P}}(OC_2H_5)_2$

Compt. Rend. 254:260 (1962)

▶ $P_2NO_2C_{12}H_{31}Pt$

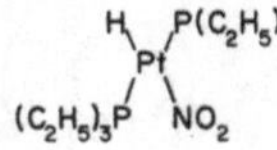

J. Chem. Soc. 5079 (1962)

▶ $P_2NO_3C_{12}H_{31}Pt$

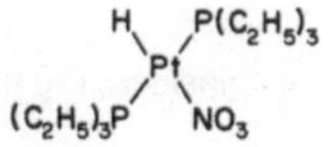

J. Chem. Soc. 5079 (1962)

▶ $P_2NC_{13}H_{31}Pt$

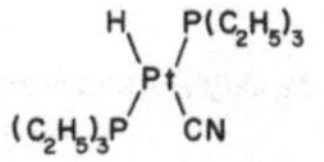

J. Chem. Soc. 5079 (1962)

▶ $P_2N_3C_3H_9$

$P_2N_3(CH_3)_3$

J. Am. Chem. Soc. 83:1334 (1961)

▶ $P_2O_3C_{39}H_{31}Fe$

$HFe(CO)_3[P(C_6H_5)_3]_2^{\oplus}$

J. Chem. Soc. 3661 (1962)

▶ $P_2O_6C_6H_{12}$

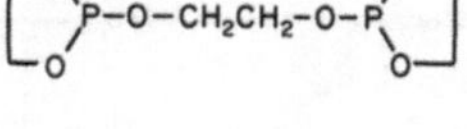

MCA Serial No. 4

▶ $P_2O_6C_9H_{22}$

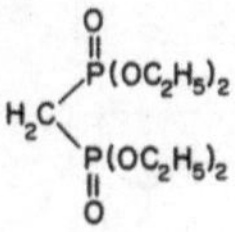

(AND SODIUM SALT)

J. Am. Chem. Soc. 84:4454 (1962)

▶ $P_2IC_{12}H_{31}Pt$

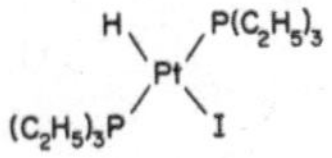

J. Chem. Soc. 5079 (1962)

▶ $P_2BrC_{12}H_{31}Pt$

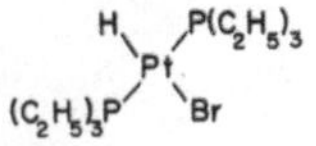

J. Chem. Soc. 5079 (1962)

▶ $P_2ClC_{12}H_{31}Pt$

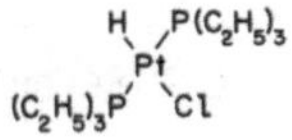

J. Chem. Soc. 5079 (1962)

▶ $P_3S_8C_{12}H_{27}$

$(C_4H_9S)_3P-P_2S_5$

J. Am. Chem. Soc. 84:3054 (1962)

▶ $P_4SNC_{21}H_{49}Ru$

trans-$\{RuH(SCN)[C_2H_4(C_2H_5PC_2H_5)_2]_2\}$

J. Chem. Soc. 2605 (1961)

▶ $P_4IC_{20}H_{49}Os$

trans-$\{OsHI[C_2H_4(C_2H_5PC_2H_5)_2]_2\}$

J. Chem. Soc. 2605 (1961)

▶ $P_4BrC_{20}H_{49}Ru$

trans-$\{RuHBr[C_2H_4(C_2H_5PC_2H_5)_2]_2\}$

J. Chem. Soc. 2605 (1961)

▶ $P_4ClC_{20}H_{49}Os$

trans-$\{OsHCl[C_2H_4(C_2H_5PC_2H_5)_2]_2\}$

J. Chem. Soc. 2605 (1961)

▶ $P_4ClC_{20}H_{49}Ru$

trans-$\{RuHCl[C_2H_4(C_2H_5PC_2H_5)_2]_2\}$

J. Chem. Soc. 2605 (1961)

▶ $SNOC_5H_7$

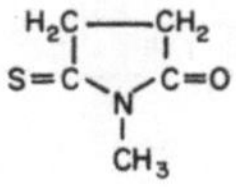

Varian 434

▶ $SNOC_5H_9$

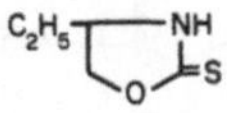

Acta Chem. Scand. 16:591 (1962)

▶ $SNOC_5H_9$

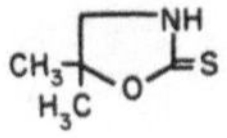

Acta Chem. Scand. 16:591 (1962)

▶ $SNOC_6H_7$

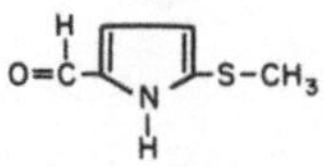

Arkiv Kemi 18:151 (1961)
J. Org. Chem. 26:2616 (1961)

▶ $SNOC_6H_{11}$

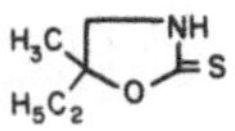

Acta Chem. Scand. 16:591 (1962)

▶ $SNOC_{27}H_{23}$

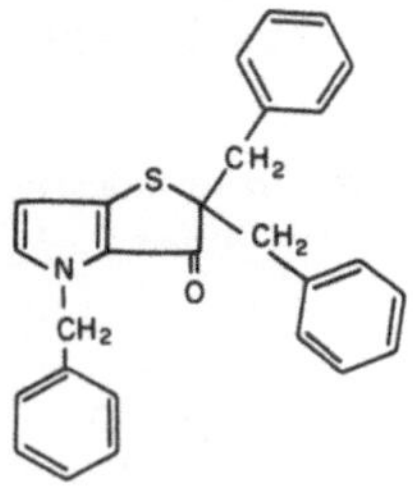

J. Org. Chem. 27:2035 (1962)

▶ $SNOC_{28}H_{45}$

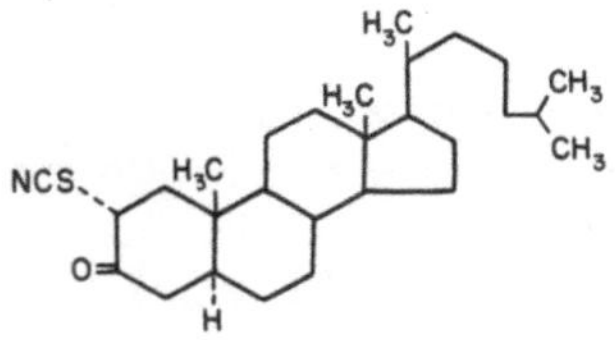

Chem. & Ind. 1793 (1962)

▶ $SNO_2BrC_4H_2$

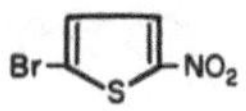

Arkiv Kemi 16:563 (1961)

▶ $SNO_2C_4H_3$

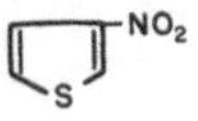

Arkiv Kemi 16:515, 539 (1961)

▶ $SNO_2C_4H_3$

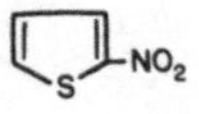


Arkiv Kemi 16:515, 539 (1961)

▶ $SNO_2C_5H_5$

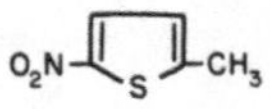


Arkiv Kemi 16:563 (1961); 18:513 (1962)

▶ $SNO_2C_5H_5$

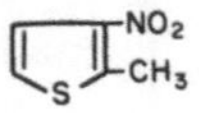


Arkiv Kemi 16:563 (1961)

▶ $SNO_2C_7H_7$

Arkiv Kemi 16:563 (1961)

▶ $SNO_2C_7H_{13}$

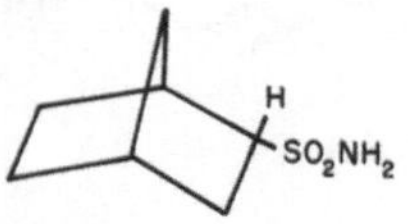


Helv. Chim. Acta 45:1972 (1962)

▶ $SNO_2C_8H_{15}$

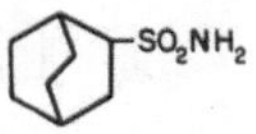


Helv. Chim. Acta 45:1972 (1962)

▶ $SNO_2C_{13}H_{19}$

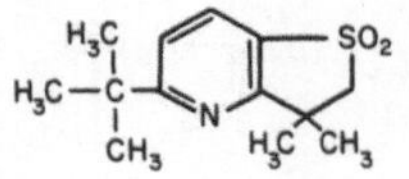


J. Org. Chem. 26:2611 (1961)

▶ $SNO_3C_5H_3$

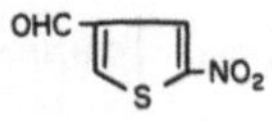


Arkiv Kemi 16:563 (1961)

▶ $SNO_3C_5H_3$

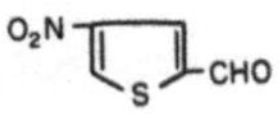


Arkiv Kemi 16:563 (1961)

▶ $SNO_3C_5H_3$

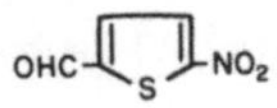


Arkiv Kemi 16:563 (1961)

▶ $SNO_3C_{10}H_9$

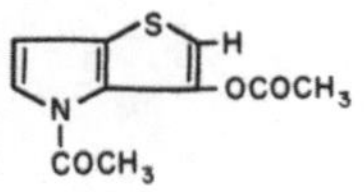


J. Phys. Chem. 65:187 (1961)

▶ $SNO_3C_{11}H_{13}$

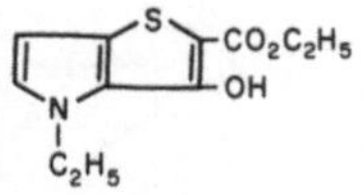


J. Phys. Chem. 65:187 (1961)

▶ $SNO_3C_{11}H_{19}$

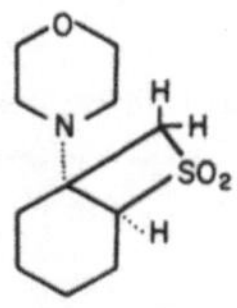


J. Am. Chem. Soc. 84:313 (1962)

▶ $SNO_3C_{12}H_9$

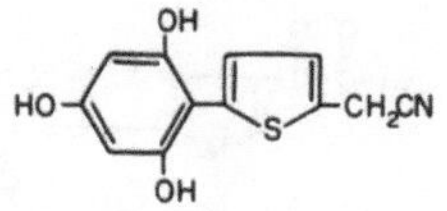


Arkiv Kemi 18:213 (1961)

▶ $SNO_4C_7H_{15}$

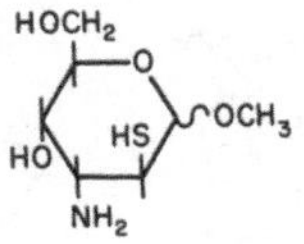


J. Am. Chem. Soc. 83:3827 (1961)

▶ $SNO_3C_{12}H_{15}$

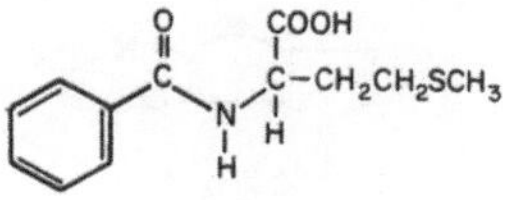


Varian 597

▶ $SNO_4C_{11}H_{11}$

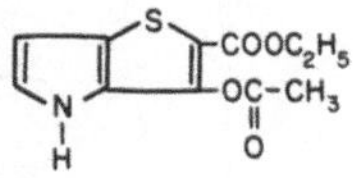


J. Org. Chem. 27:269 (1962)

▶ $SNO_3C_{12}H_{15}$

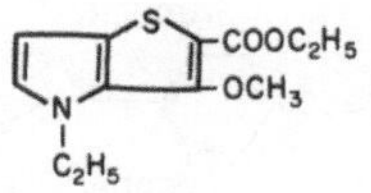


J. Phys. Chem. 65:187 (1961)

▶ $SNO_4C_{11}H_{11}$

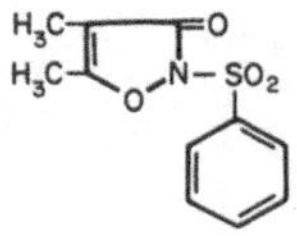


Varian 580

▶ $SNO_3C_{17}H_{15}$

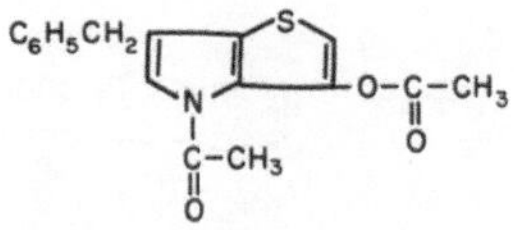


J. Phys. Chem. 65:187 (1961)

▶ $SNO_4C_{12}H_{11}$

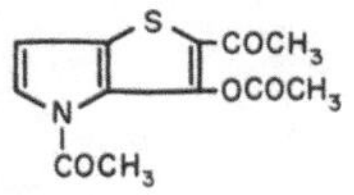


J. Phys. Chem. 65:187 (1961)

▶ $SNO_4C_5H_3$

Arkiv Kemi 16:563 (1961)

▶ $SNO_4C_{13}H_{15}$

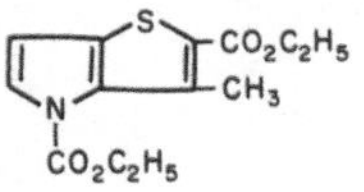


J. Phys. Chem. 65:187 (1961)

▶ $SNO_4C_6H_5$

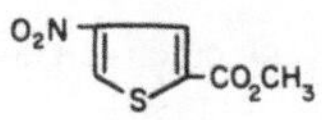


Arkiv Kemi 16:563 (1961)

▶ $SNO_4C_{15}H_{19}D_2$

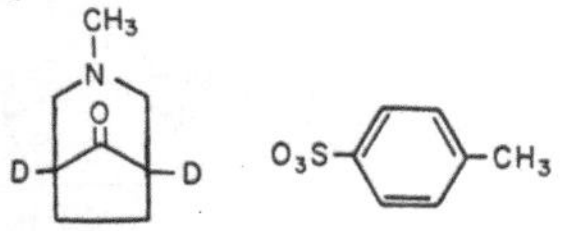


J. Org. Chem. 27:4436 (1962)

▶ $SNO_4C_7H_7$

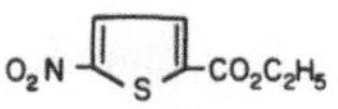


Arkiv Kemi 16:563 (1961)

▶ $SNO_4C_{15}H_{21}$

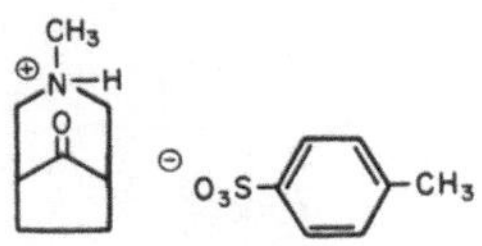


J. Am. Chem. Soc. 84:3139 (1962)

▶ $SNO_4C_{16}H_{23}$

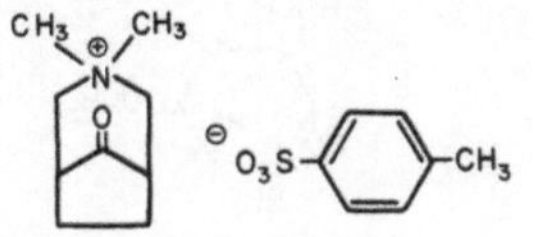

J. Am. Chem. Soc. 84:3139 (1962)

▶ $SNO_4C_{16}H_{23}$

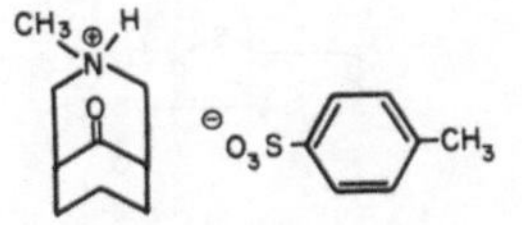

J. Am. Chem. Soc. 84:3139 (1962)

▶ $SNO_4C_{17}H_{25}$

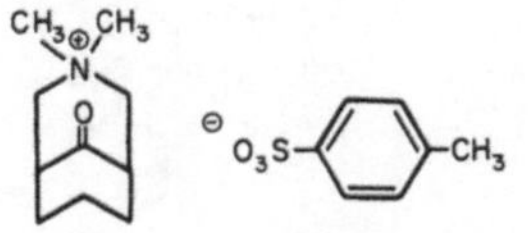

J. Am. Chem. Soc. 84:3139 (1962)

▶ $SNO_4C_{17}H_{25}$

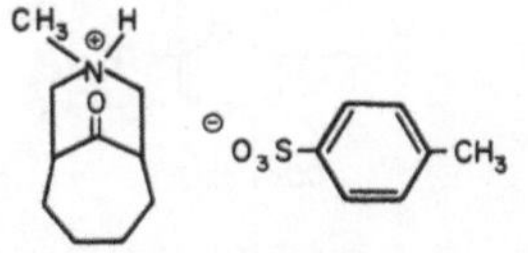

J. Am. Chem. Soc. 84:3139 (1962)

▶ $SNO_4C_{18}H_{27}$

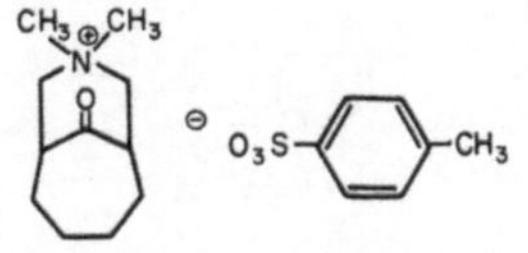

J. Am. Chem. Soc. 84:3139 (1962)

▶ $SNO_4C_{29}H_{33}$

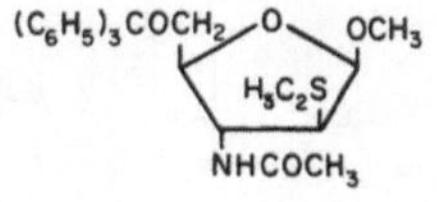

J. Am. Chem. Soc. 83:1900 (1961)

▶ $SNO_5C_{12}H_{13}$

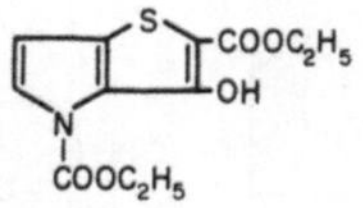

J. Phys. Chem. 65:187 (1961)

▶ $SNO_6ClC_{13}H_{10}$

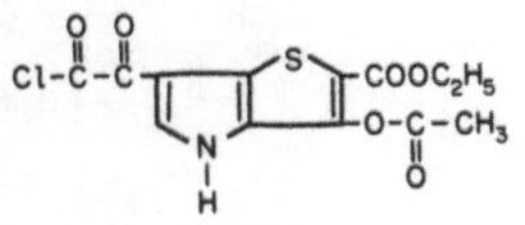

J. Org. Chem. 27:2691 (1962)

▶ $SNO_6C_{14}H_{15}$

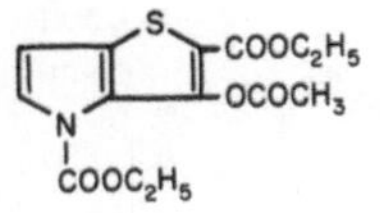

J. Phys. Chem. 65:187 (1961)

▶ $SNO_6C_{18}H_{15}$

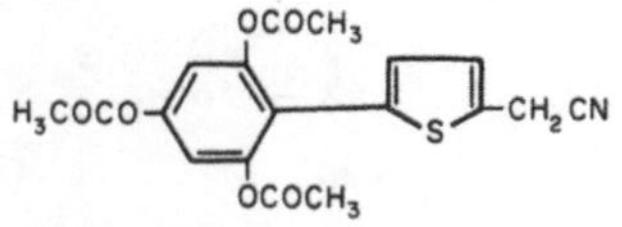

Arkiv Kemi 18:213 (1961)

▶ $SNO_7C_{15}H_{15}$

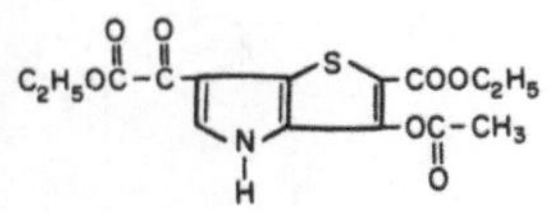

J. Org. Chem. 27:2691 (1962)

▶ $SNO_{11}C_{23}H_{35}$

DESULPHOGLUCOCAPPASALIN
TETRACETATE

Acta Chem. Scand. 16:2066 (1962)

▶ $SNClC_4H_{12}$

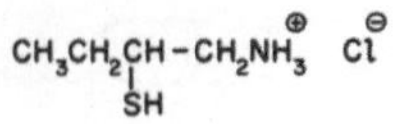

J. Org. Chem. 27:2852 (1962)

▶ $SNCH_3Si$

$SiH_3NCS$

J. Chem. Soc. 4879 (1961)

▶ $SNC_3HD_2$

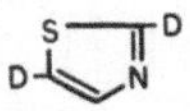

Spectrochim. Acta 18:741 (1962)

▶ $SNC_3H_2D$

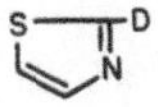

Spectrochim. Acta 18:741 (1962)

▶ $SNC_3H_2D$

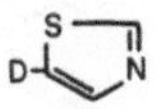

Spectrochim. Acta 18:741 (1962)

▶ $SNC_3H_3$

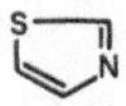

Varian 378
Spectrochim. Acta 18:741 (1962)

▶ $SNC_3H_5$

$CH_3CH_2SCN$

Varian 384

▶ $SNC_4H_5$

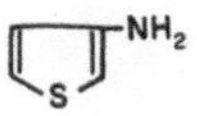

Arkiv Kemi 16:515, 539 (1961)

▶ $SNC_4H_5$

Arkiv Kemi 16:515, 539 (1961)

▶ $SNC_5H_3$

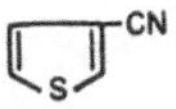

Arkiv Kemi 16:515, 539 (1961)

▶ $SNC_5H_3$

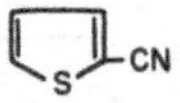

Arkiv Kemi 16:515, 539 (1961)

▶ $SNC_5H_7$

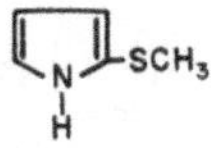

Arkiv Kemi 18:151 (1961)
J. Org. Chem. 26:2615 (1961)

▶ $SNC_5H_7$

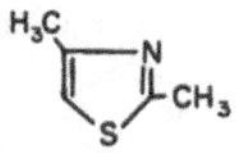

Varian 108

▶ $SNC_6H_5$

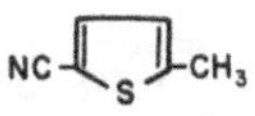

Arkiv Kemi 18:513 (1962)

▶ $SNC_6H_7$

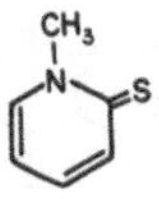

J. Chem. Soc. 2070 (1962)

▶ $SNC_7H_{13}$

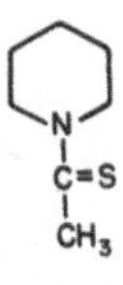

Chem. Ber. 95:2896 (1962)

▶ $SNC_8H_7$

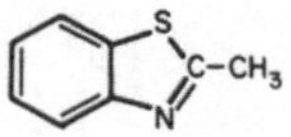

Varian 191

▶ $SNC_8H_7$

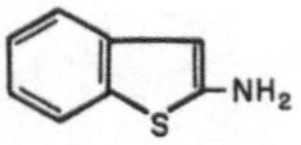

Varian 502

▶ $SNC_8H_9$

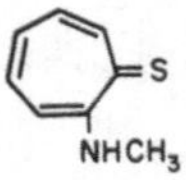

J. Am. Chem. Soc. 83:3135 (1961)

▶ $SNC_8H_{11}$

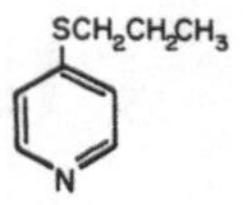

Varian 509

▶ $SNC_9H_{11}$

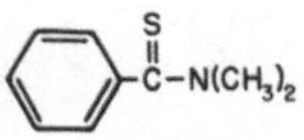

Acta Chem. Scand. 16:1616 (1962)

▶ $SNC_{12}H_9$

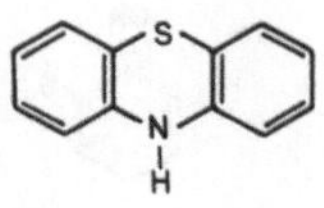

J. Am. Chem. Soc. 84:1020 (1962)

▶ $SNC_{12}H_9$

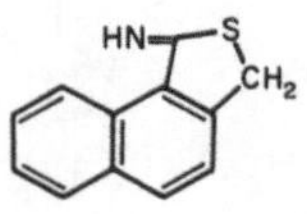

Varian 590

▶ $SNC_{13}H_{11}$

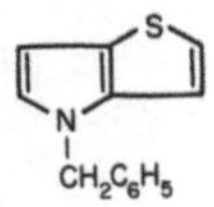

J. Phys. Chem. 65:187 (1961)

▶ $SNC_{13}H_{17}$

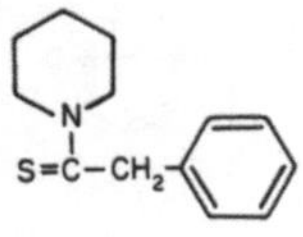

Chem. Ber. 95:2896 (1962)

▶ $SN_2OC_4H_4$

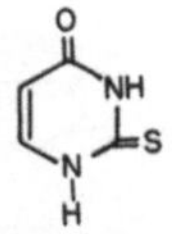

J. Am. Chem. Soc. 84:1042 (1962)
Arkiv Kemi 16:459 (1961)

▶ $SN_2OC_{11}H_{10}$

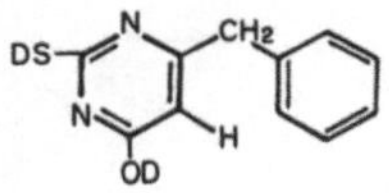

Varian 577

▶ $SN_2OC_{17}H_{14}$

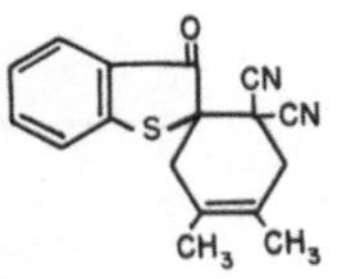

J. Org. Chem. 27:3889 (1962)

▶ $SN_2O_2C_4H_6$

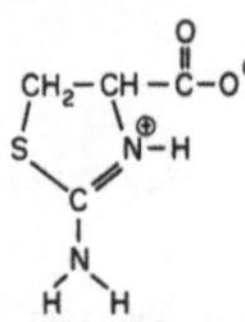

J. Org. Chem. 27:3120 (1962)

▶ $SN_2O_2C_5H_2$

Arkiv Kemi 16:563 (1961)

▶ $SN_2O_2C_5H_2$

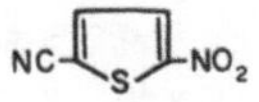


Arkiv Kemi 16:563 (1961)

▶ $SN_2O_2C_{17}H_{20}$

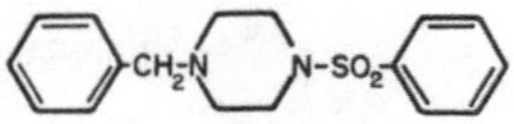


Varian 330

▶ $SN_2O_2C_{30}H_{40}$

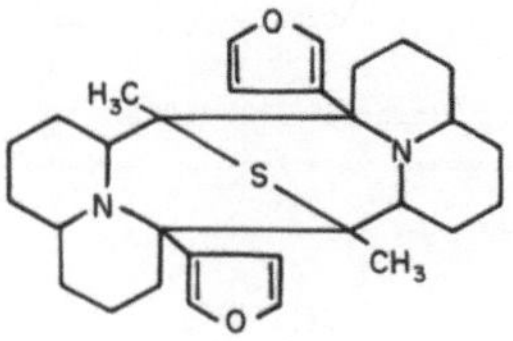


Tetrahedron Letters No. 24:1124 (1962)

▶ $SN_2O_3C_{11}H_{10}$

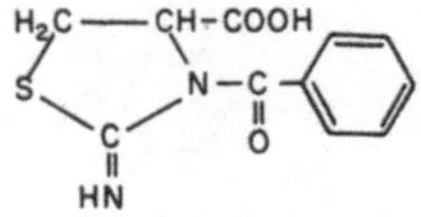


J. Org. Chem. 27:3120 (1962)

▶ $SN_2O_3C_{11}H_{18}$

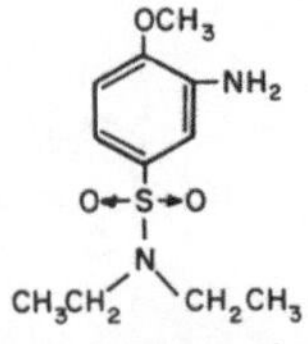


Varian 587

▶ $SN_2O_3C_{12}H_{18}$

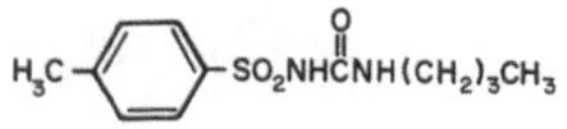


Varian 602

▶ $SN_2O_4C_3H_4$

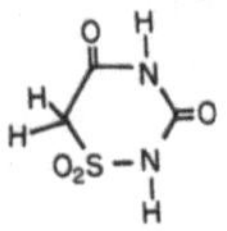


J. Org. Chem. 26:3461 (1961)

▶ $SN_2O_4C_4H_6$

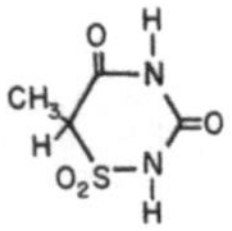


J. Org. Chem. 26:3461 (1961)

▶ $SN_2O_4C_7H_6$

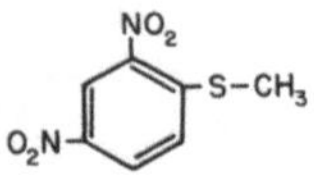


J. Org. Chem. 27:3325 (1962)

▶ $SN_2O_4C_8H_8$

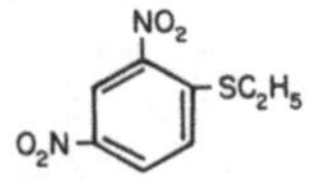


J. Org. Chem. 27:3325 (1962)

▶ $SN_2O_4C_9H_{10}$

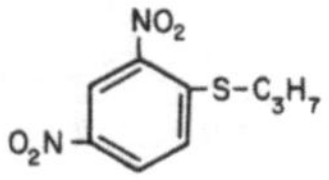


J. Org. Chem. 27:3325 (1962)

▶ $SN_2O_4C_{11}H_{14}$

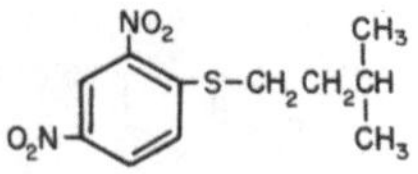


J. Org. Chem. 27:3325 (1962)

▶ $SN_2O_4C_{12}H_8$

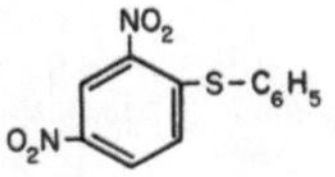

J. Org. Chem. 27:3325 (1962)

▶ $SN_2O_6C_{11}H_{14}$

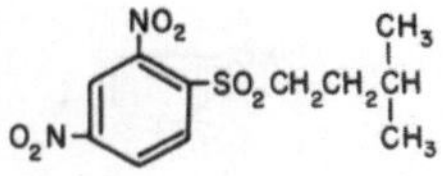

J. Org. Chem. 27:3325 (1962)

▶ $SN_2O_4C_{13}H_{10}$

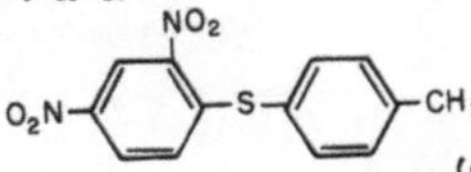

($CH_3$ ORTHO, META, OR PARA)

J. Org. Chem. 27:3325 (1962)

▶ $SN_2O_6C_{12}H_8$

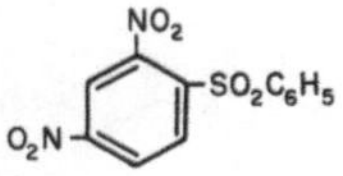

J. Org. Chem. 27:3325 (1962)

▶ $SN_2O_4C_{13}H_{10}$

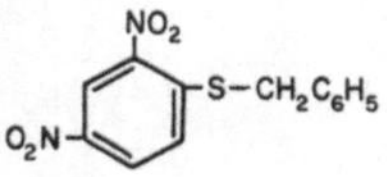

J. Org. Chem. 27:3325 (1962)

▶ $SN_2O_6C_{13}H_{10}$

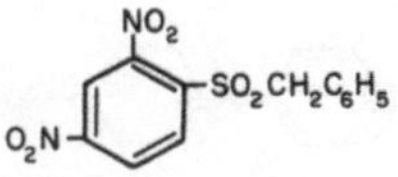

J. Org. Chem. 27:3325 (1962)

▶ $SN_2O_4C_{16}H_{24}$

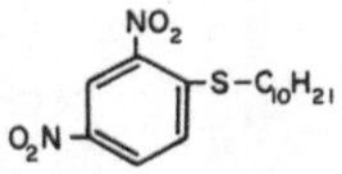

J. Org. Chem. 27:3325 (1962)

▶ $SN_2O_6C_{13}H_{10}$

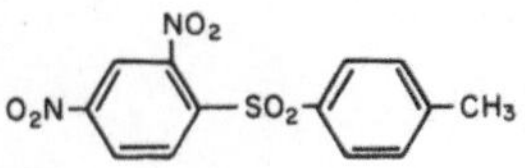

J. Org. Chem. 27:3325 (1962)

▶ $SN_2O_5C_8H_{12}$

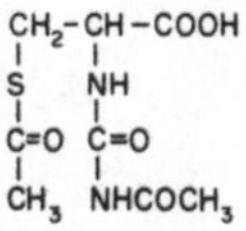

J. Org. Chem. 27:3120 (1962)

▶ $SN_2O_6C_{16}H_{24}$

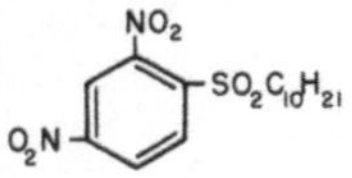

J. Org. Chem. 27:3325 (1962)

▶ $SN_2O_6C_8H_8$

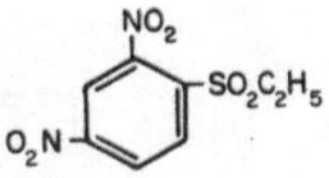

J. Org. Chem. 27:3325 (1962)

▶ $SN_2O_7C_{14}H_{12}$

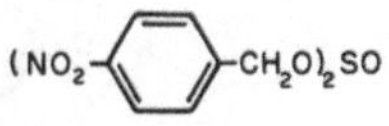

Bull. Chem. Soc. Japan 35:1428 (1962)

▶ $SN_2O_6C_9H_{10}$

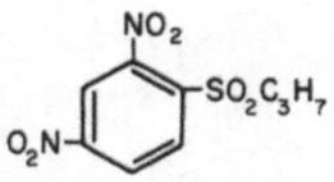

J. Org. Chem. 27:3325 (1962)

▶ $SN_2ClC_6H_7$

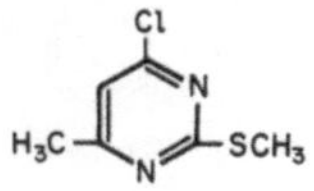

Varian 126

▶ $SN_2C_3H_4$

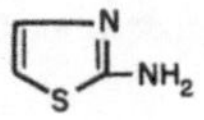

Varian 380

▶ $SN_2C_{12}H_{10}$

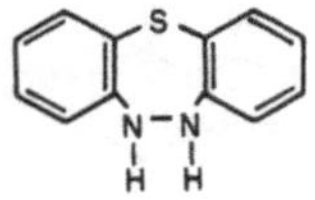

J. Am. Chem. Soc. 84:1020 (1962)

▶ $SN_2C_5H_4$

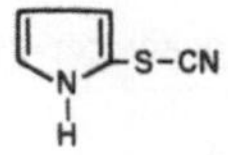

Arkiv Kemi 18:151 (1961)
J. Org. Chem. 26:2615 (1961)

▶ $SN_2C_{17}H_{20}$

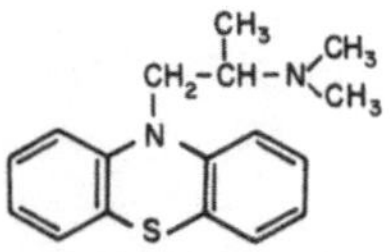

J. Org. Chem. 27:4272 (1962)

▶ $SN_2C_6H_6$

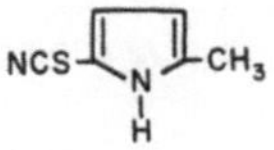

Arkiv Kemi 18:151 (1961)
J. Org. Chem. 26:2616 (1961)

▶ $SN_3OC_9H_3$

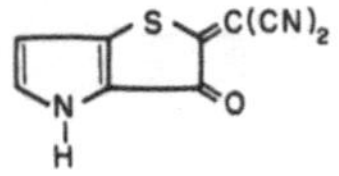

J. Org. Chem. 27:3888 (1962)

▶ $SN_2C_6H_8$

$NCCH_2CH_2SCH_2CH_2CN$

Varian 127

▶ $SN_2C_6H_{14}$

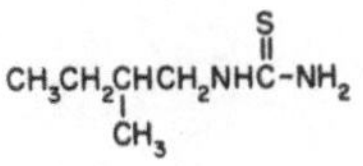

Acta Chem. Scand. 16:936 (1962)

▶ $SN_3OC_{15}H_{13}$

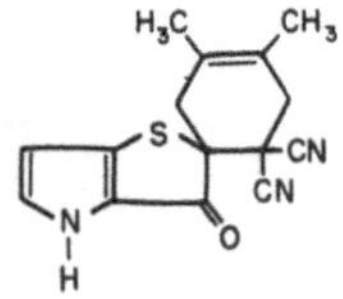

J. Org. Chem. 27:3889 (1962)

▶ $SN_2C_7H_6$

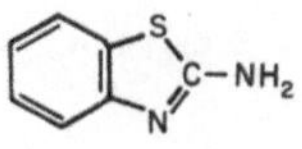

Varian 150

▶ $SN_3O_4C_4H_9$

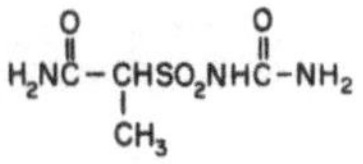

J. Org. Chem. 26:3461 (1961)

▶ $SN_2C_8H_{11}\cdot HCl$

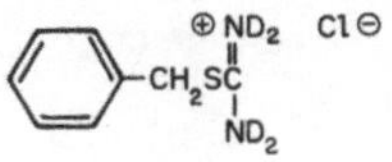

Varian 507

▶ $SN_3O_6C_{11}H_{13}$

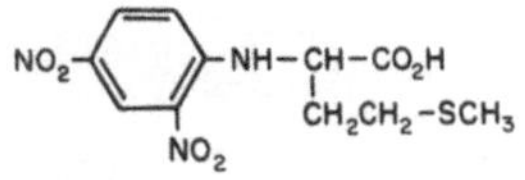

Bull. Chem. Soc. Japan 35:1658 (1962)

▶ $SN_3C_7H_9$

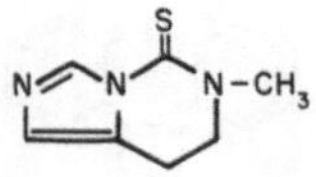

J. Am. Chem. Soc. 83:2023 (1961)

▶ $SN_4OC_6H_6$

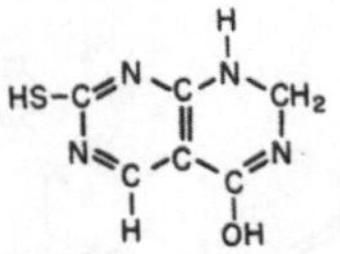

J. Org. Chem. 27:4211 (1962)

▶ $SOBrC_5H_3$

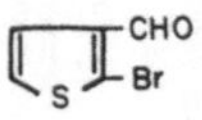

Arkiv Kemi 16:563 (1961)

▶ $SOBrC_5H_3$

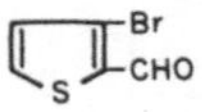

Arkiv Kemi 16:563 (1961)

▶ $SOBrC_5H_3$

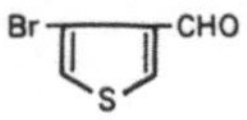

Arkiv Kemi 16:563 (1961)

▶ $SOBrC_5H_3$

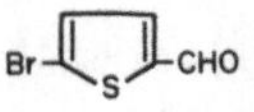

Arkiv Kemi 16:563 (1961)

▶ $SOClC_5H_3$

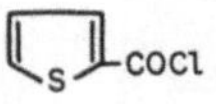

Arkiv Kemi 16:515, 539 (1961); 17:1 (1961)

▶ $SOClC_6H_9'$

$CH_3COSCH_2CH{=}CClCH_3$

J. Am. Chem. Soc. 84:3987 (1962)

▶ $SOClC_6H_9$

$CH_3COSCH_2CCl{=}CHCH_3$

J. Am. Chem. Soc. 84:3987 (1962)

▶ $SOF_6C_4H_4$

$CF_3SCF_2CFHOCH_3$

J. Am. Chem. Soc. 83:843 (1961)

▶ $SOC_2HD_5$

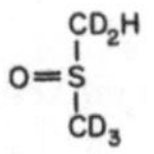

Varian 376

▶ $SOC_2H_4$

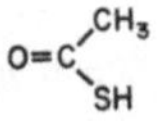

Varian 7

▶ $SOC_2H_6$

$CH_3SOCH_3$

J. Chem. Phys. 37:2198 (1962)

▶ $SOC_3H_9 \cdot ClO_4$

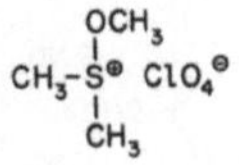

J. Am. Chem. Soc. 84:3701 (1962)

▶ $SOC_3H_9 \cdot ClO_4$

$(CH_3)_3\overset{\oplus}{S}{=}O \quad ClO_4^{\ominus}$

J. Am. Chem. Soc. 84:3701 (1962)

▶ $SOC_5H_6$

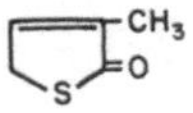

Acta Chem. Scand. 16:789 (1962)

▶ $SOC_5H_4$

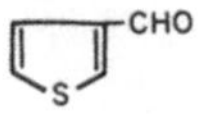

Arkiv Kemi 16:515, 539 (1961)

▶ $SOC_6H_6$

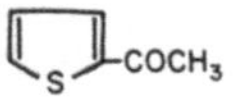

Arkiv Kemi 16:515, 539 (1961); 17:1 (1961); 18:133 (1961)

▶ $SOC_5H_4$

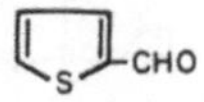

Varian 94
Arkiv Kemi 16:515, 539 (1961); 17:1 (1961); 18:133 (1961)

▶ $SOC_6H_6$

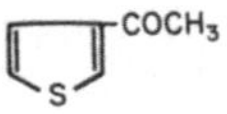

Arkiv Kemi 16:515, 539 (1961)

▶ $SOC_5H_6$

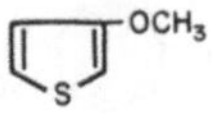

Arkiv Kemi 16:515, 539 (1961)

▶ $SOC_6H_6$

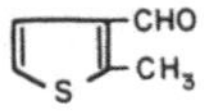

Arkiv Kemi 16:563 (1961)

▶ $SOC_5H_6$

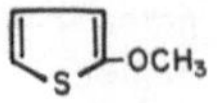

Arkiv Kemi 16:515, 539 (1961); 17:1 (1961)

▶ $SOC_6H_6$

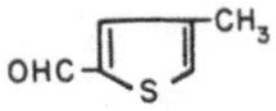

Arkiv Kemi 16:563 (1961); 17:165 (1961)

▶ $SOC_5H_6$

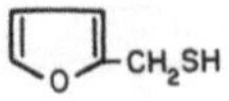

Varian 101
Can. J. Chem. 39:906 (1961)

▶ $SOC_6H_6$

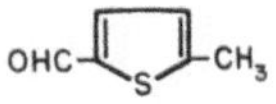

Arkiv Kemi 16:563 (1961); 18:513 (1962)

▶ $SOC_5H_6$

Acta Chem. Scand. 16:789 (1962)

▶ $SOC_6H_6$

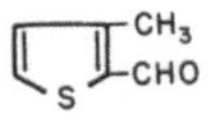

Arkiv Kemi 16:563 (1961); 17:165 (1961); 18:513 (1962)

▶ $SOC_6H_6$

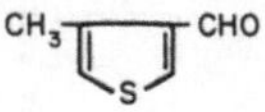


Arkiv Kemi 16:563 (1961); 18:513 (1962)

▶ $SOC_6H_8$

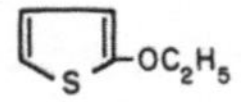


Arkiv Kemi 16:515, 539 (1961)

▶ $SOC_6H_8$

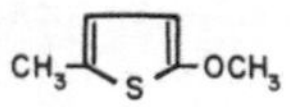


Arkiv Kemi 16:563 (1961); 18:513 (1962)

▶ $SOC_6H_{10}$

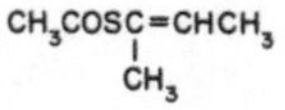


J. Am. Chem. Soc. 84:3987 (1962)

▶ $SOC_7H_6$

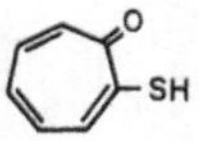


Bull. Chem. Soc. Japan 34:616 (1961)

▶ $SOC_7H_8$

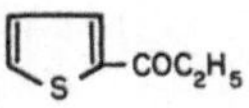


Varian 163
Arkiv Kemi 16:515, 539 (1961)

▶ $SOC_7H_8$

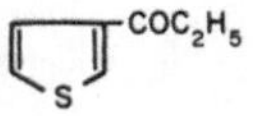


Arkiv Kemi 16:515, 539 (1961)

▶ $SOC_7H_8$

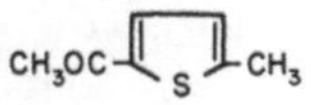


Varian 164
Arkiv Kemi 18:513 (1962)

▶ $SOC_7H_{12}$

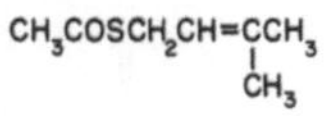


J. Am. Chem. Soc. 84:3987 (1962)

▶ $SOC_7H_{13} \cdot ClO_4$

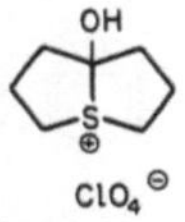


J. Am. Chem. Soc. 84:3701 (1962)

▶ $SOC_8H_8$

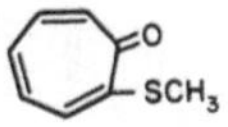


Bull. Chem. Soc. Japan 34:616 (1961)

▶ $SOC_8H_{12}$

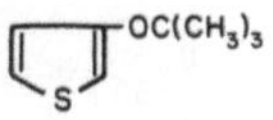


Arkiv Kemi 16:515, 539 (1961)

▶ $SOC_8H_{12}$

Arkiv Kemi 16:515, 539 (1961)

▶ $SOC_8H_{15} \cdot ClO_4$

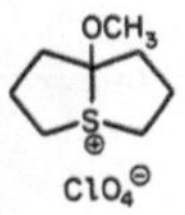


J. Am. Chem. Soc. 84:3701 (1962)

▶ $SOC_9H_{12}$

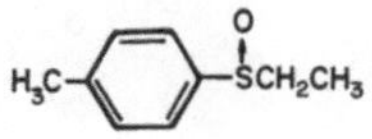

Varian 538

▶ $SOC_{10}H_{12}$

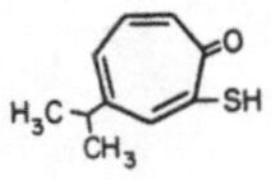

Bull. Chem. Soc. Japan 34:616 (1961)

▶ $SOC_{10}H_{12}$

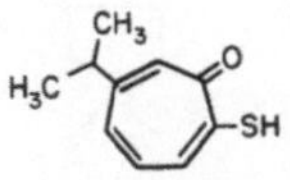

Bull. Chem. Soc. Japan 34:616 (1961)

▶ $SOC_{12}H_{22}$

Tetrahedron Letters No. 3:103 (1962)

▶ $SOC_{12}H_{22}$

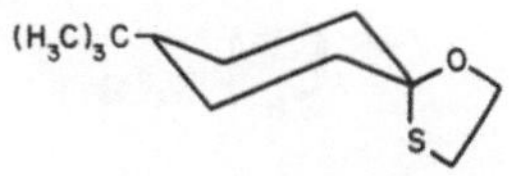

Tetrahedron Letters No. 3:103 (1962)

▶ $SOC_{16}H_{14}$

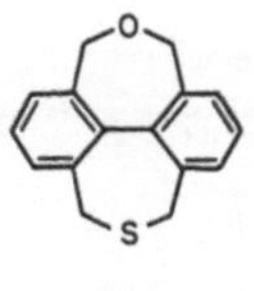

J. Am. Chem. Soc. 83:2780 (1961)

▶ $SO_2BrC_5H_3$

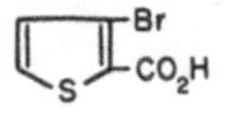

Arkiv Kemi 16:563 (1961)

▶ $SO_2BrC_5H_3$

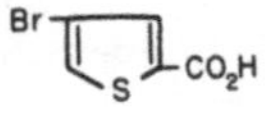

Arkiv Kemi 16:563 (1961)

▶ $SO_2BrC_5H_3$

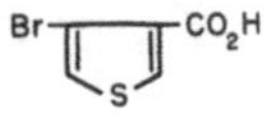

Arkiv Kemi 16:563 (1961)

▶ $SO_2BrC_5H_3$

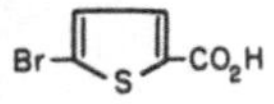

Arkiv Kemi 16:563 (1961)

▶ $SO_2BrC_6H_5$

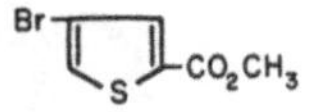

Arkiv Kemi 16:563 (1961)

▶ $SO_2BrC_6H_5$

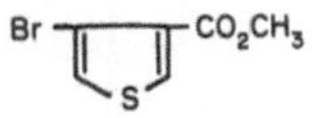

Arkiv Kemi 16:563 (1961)

▶ $SO_2BrC_8H_9$

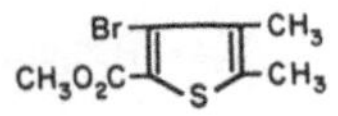

Acta Chem. Scand. 15:1201 (1961)

▶ $SO_2ClC_{11}H_{11}$

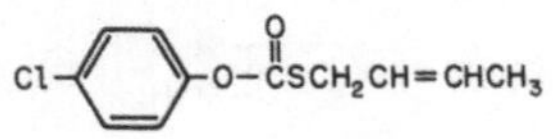

J. Org. Chem. 27:4509 (1962)

▶ $SO_2ClC_{11}H_{11}$

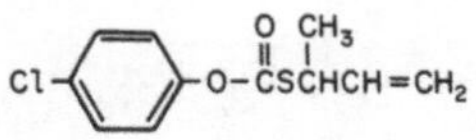

J. Org. Chem. 27:4509 (1962)

▶ $SO_2FC_7H_9$

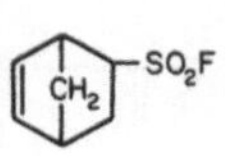

Helv. Chim. Acta 45:1972 (1962)

▶ $SO_2FC_7H_{11}$

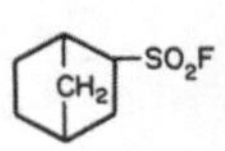

Helv. Chim. Acta 45:1972 (1962)

▶ $SO_2F_3C_8H_5$

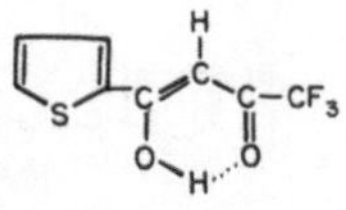

Varian 185

▶ $SO_2C_3H_4$

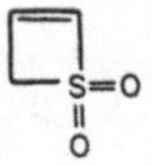

Varian 22
J. Org. Chem. 26:1324 (1961)

▶ $SO_2C_3H_6$

$H_2C=CHSO_2CH_3$

Varian 35

▶ $SO_2C_3H_6$

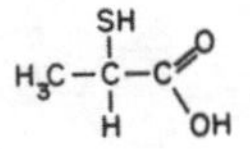

Varian 389

▶ $SO_2C_4H_6$

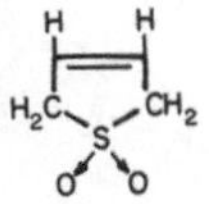

Varian 406

▶ $SO_2C_4H_8$

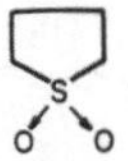

Varian 416

▶ $SO_2C_6H_6$

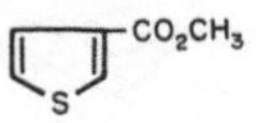

Arkiv Kemi 16:515, 539 (1961); 18:133 (1961)

▶ $SO_2C_6H_6$

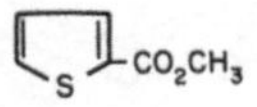

Arkiv Kemi 16:515, 539 (1961); 17:1 (1961)

▶ $SO_2C_6H_6$

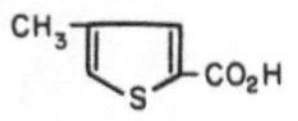

Arkiv Kemi 16:563 (1961)

▶ $SO_2C_6H_6$

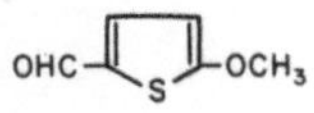

Arkiv Kemi 16:563 (1961)

▶ $SO_2C_7H_8$

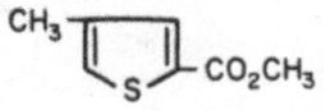

Arkiv Kemi 16:563 (1961)

▶ $SO_2C_9H_{15} \cdot ClO_4$

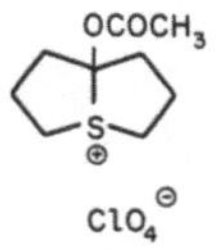

J. Am. Chem. Soc. 84:3701 (1962)

▶ $SO_2C_7H_8$

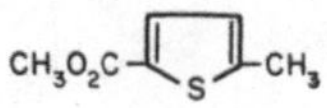

Arkiv Kemi 18:513 (1962)

▶ $SO_2C_{10}H_{10}$

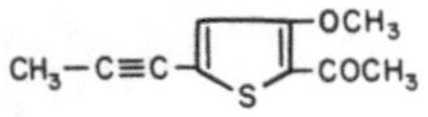

Chem. Ber. 95:2934 (1962)

▶ $SO_2C_7H_{13} \cdot ClO_4$

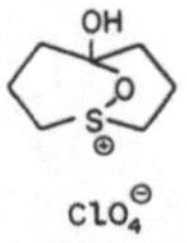

J. Am. Chem. Soc. 84:3701 (1962)

▶ $SO_2C_{13}H_{12}$

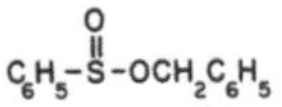

Bull. Chem. Soc. Japan 35:1428 (1962)

▶ $SO_2C_8H_{15} \cdot ClO_4$

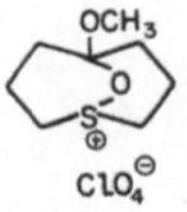

J. Am. Chem. Soc. 84:3701 (1962)

▶ $SO_2C_{13}H_{14}$

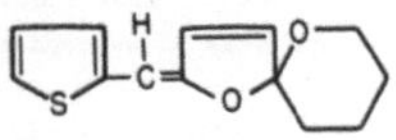

Chem. Ber. 95:1733 (1962)

▶ $SO_2C_9H_8$

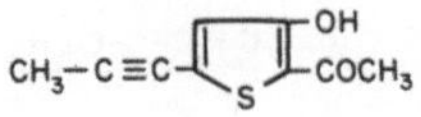

Chem. Ber. 95:2934 (1962)

▶ $SO_2C_9H_{10}$

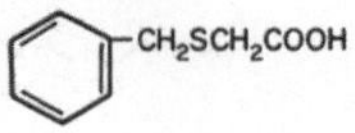

Varian 532

▶ $SO_2C_{14}H_{12}$

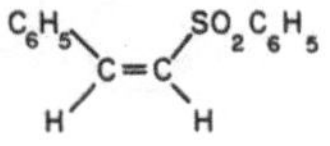

J. Am. Chem. Soc. 83:4636 (1961)

▶ $SO_2C_9H_{12}$

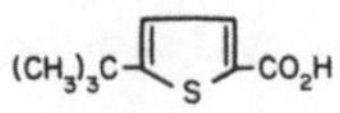

Arkiv Kemi 16:563 (1961)

▶ $SO_2C_{14}H_{12}$

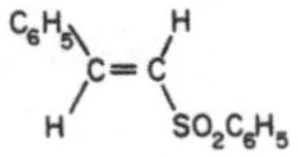

J. Am. Chem. Soc. 83:4636 (1961)

▶ $SO_2C_{15}H_{14}$

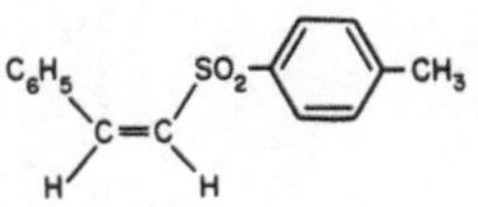

J. Am. Chem. Soc. 83:4636 (1961)

▶ $SO_3C_2H_4$

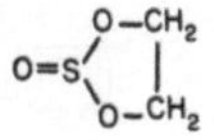

Varian 371
J. Am. Chem. Soc. 83:2105 (1961)

▶ $SO_2C_{15}H_{14}$

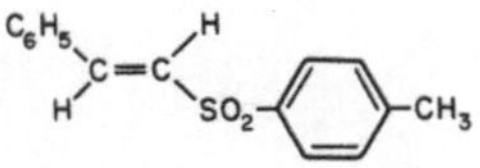

J. Am. Chem. Soc. 83:4636 (1961)

▶ $SO_3C_3H_6$

J. Am. Chem. Soc. 83:2105 (1961)

▶ $SO_2C_{20}H_{24}$

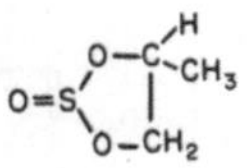

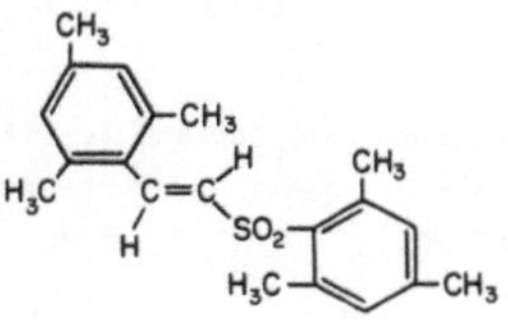

J. Am. Chem. Soc. 83:4636 (1961)

▶ $SO_3C_3H_6$

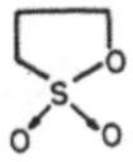

Varian 390

▶ $SO_3C_4H_8$

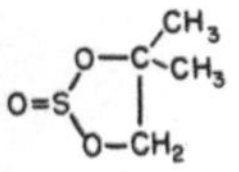

J. Am. Chem. Soc. 83:2105 (1961)

▶ $SO_2C_{20}H_{30}$

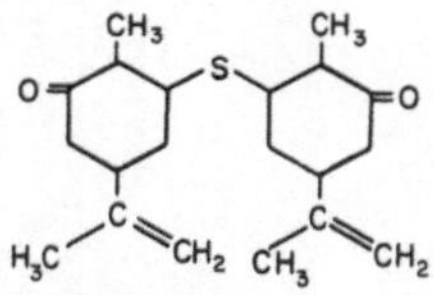

J. Chem. Soc. 3830 (1962)

▶ $SO_3C_4H_{10}$

$C_2H_5O-S(=O)-OC_2H_5$

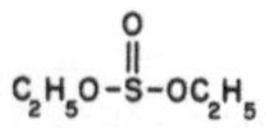

J. Chem. Phys. 37:3012 (1962)
J. Am. Chem. Soc. 83:4666 (1961)

▶ $SO_3Cl_2C_{14}H_{12}$

$(Cl-C_6H_4-CH_2O)_2SO$

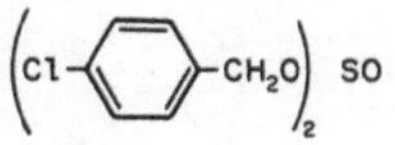

Bull. Chem. Soc. Japan 35:1428 (1962)

▶ $SO_3C_6H_6$

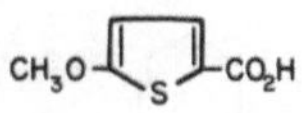

Arkiv Kemi 16:563 (1961)

▶ $SO_3C_2H_4$

$CH_2=CHSO_3H$

J. Chem. Phys. 36:1951 (1962)

▶ $SO_3C_6H_{16}Si$

$(CH_3)_3Si(CH_2)_3SO_3^{\ominus}Na^{\oplus}$

Varian 481

▶ $SO_3C_7H_8$

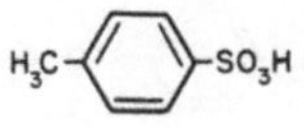

J. Phys. Chem. 66:268 (1962)

▶ $SO_3C_7H_8$

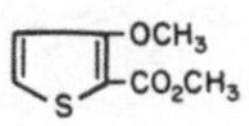

Arkiv Kemi 16:563 (1961)

▶ $SO_3C_9H_{12}$

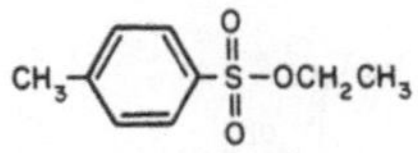

J. Am. Chem. Soc. 83:857 (1961)

▶ $SO_3C_9H_{12}$

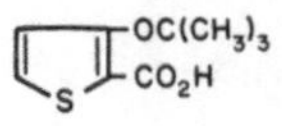

Arkiv Kemi 16:363 (1961)

▶ $SO_3C_9H_{15} \cdot ClO_4$

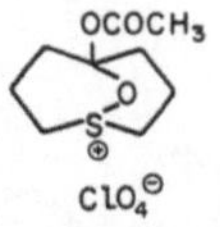

J. Am. Chem. Soc. 84:3701 (1962)

▶ $SO_3C_{10}H_8$

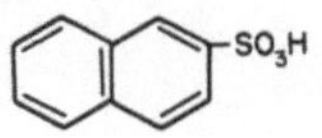

J. Phys. Chem. 66:268 (1962)

▶ $SO_3C_{14}H_{14}$

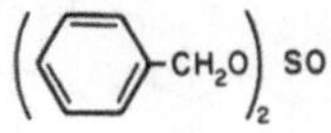

Bull. Chem. Soc. Japan 35:1428 (1962)

▶ $SO_3C_{15}H_{22}$

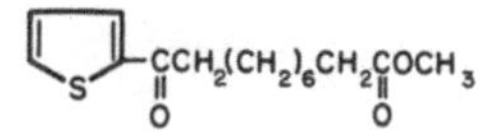

Varian 320

▶ $SO_3C_{16}H_{18}$

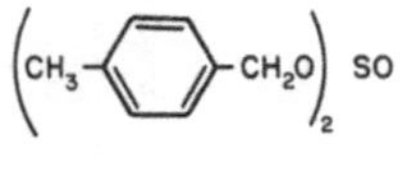

Bull. Chem. Soc. Japan 35:1428 (1962)

▶ $SO_3C_{17}H_{26}$

cis and trans

Tetrahedron Letters No. 3:97 (1962)

▶ $SO_3C_{20}H_{28}$

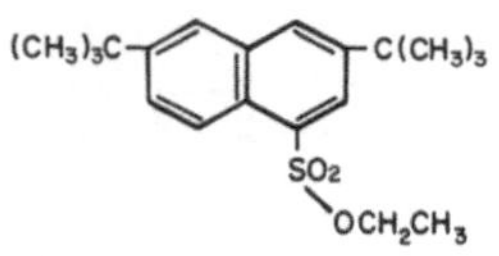

Varian 683

▶ $SO_3C_{23}H_{20}$

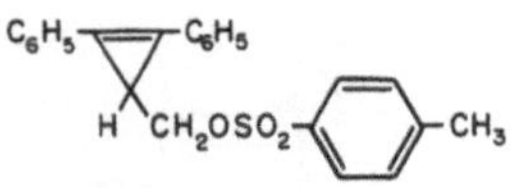

J. Am. Chem. Soc. 84:2793 (1962)

▶ $SO_4F_6C_{10}H_6$

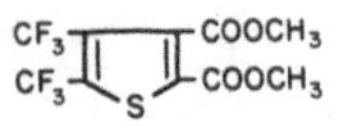

J. Am. Chem. Soc. 83:3438 (1961)

▶ $SO_4C_2H_4$

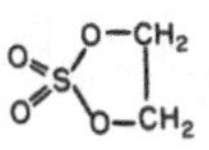

J. Am. Chem. Soc. 83:2105 (1961)

▶ $SO_4C_2H_6$

$(H_3C)_2SO_4$

J. Chem. Soc. 1372 (1962)

▶ $SO_4C_4H_6$

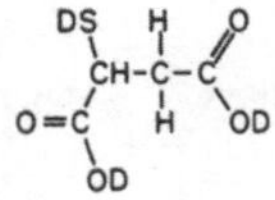

Varian 407

▶ $SO_4C_7H_{14}$

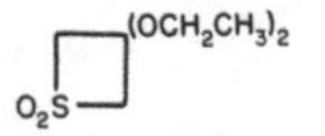

J. Am. Chem. Soc. 84:3030 (1962)

▶ $SO_4C_9H_{10}$

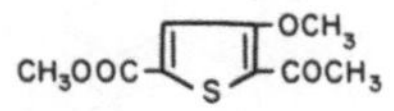

Chem. Ber. 95:2934 (1962)

▶ $SO_4C_9H_{14}$

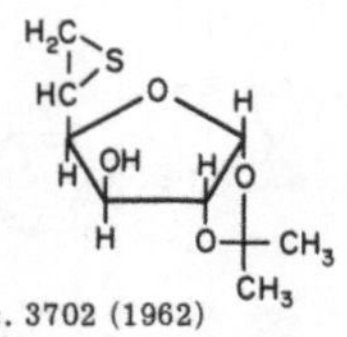

J. Chem. Soc. 3702 (1962)

▶ $SO_4C_9H_{14}$

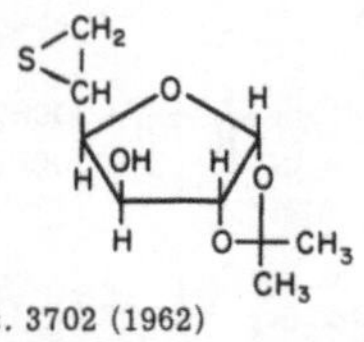

J. Chem. Soc. 3702 (1962)

▶ $SO_4C_{11}H_{12}$

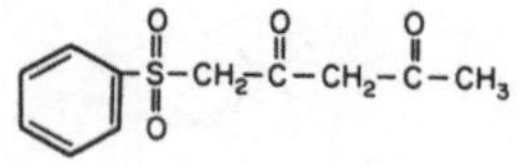

J. Org. Chem. 27:1251 (1962)

▶ $SO_4C_{11}H_{16}$

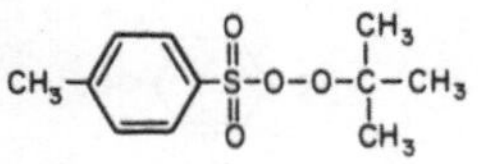

J. Am. Chem. Soc. 83:856 (1961)

▶ $SO_4C_{14}H_{12}$

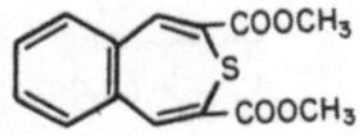

J. Org. Chem. 27:3225 (1962)

▶ $SO_4C_{24}H_{30}$

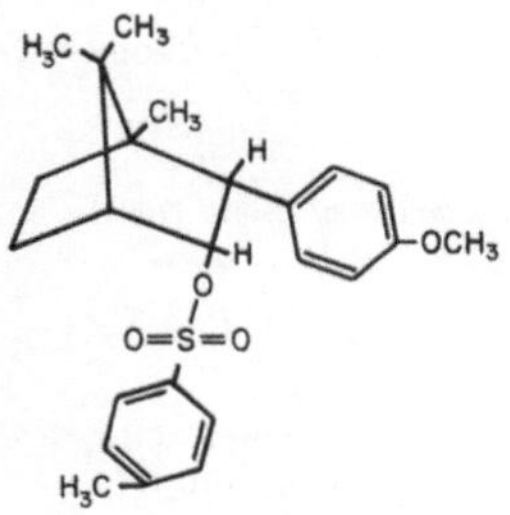

J. Org. Chem. 27:1532 (1962)

▶ $SO_4C_{24}H_{30}$

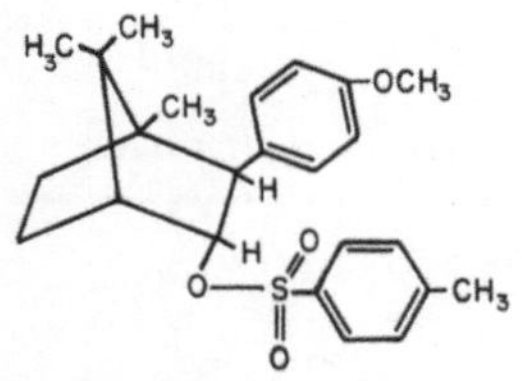

J. Org. Chem. 27:1531 (1962)

▶ $SO_4C_{24}H_{32}$

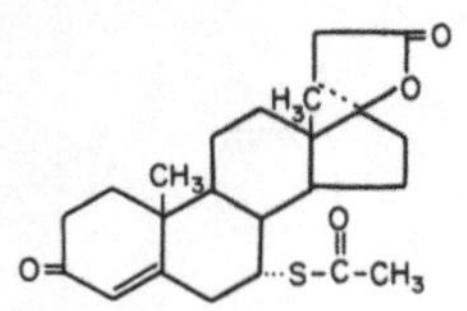

J. Org. Chem. 27:3326 (1962)

▶ $SO_4C_{24}H_{32}$

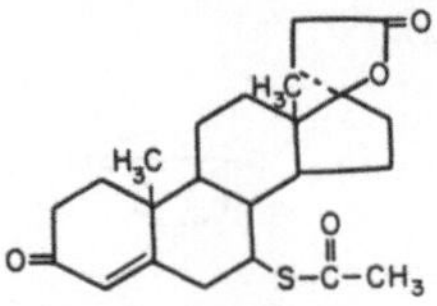

J. Org. Chem. 27:3326 (1962)

▶ $SO_4C_{25}H_{34}$

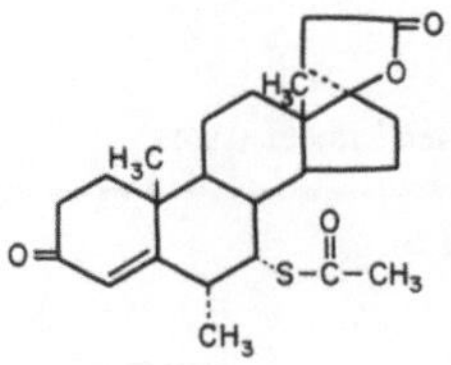

J. Org. Chem. 27:3326 (1962)

▶ $SO_5C_{16}H_{18}$

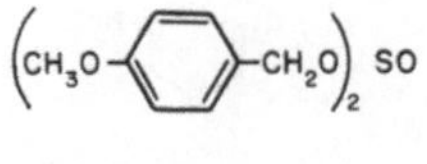

Bull. Chem. Soc. Japan 35:1428 (1962)

▶ $SO_6C_{15}H_{18}$

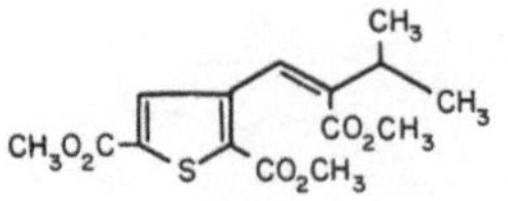

Bull. Chem. Soc. Japan 35:808 (1962)

▶ $SO_6C_{26}H_{30}$

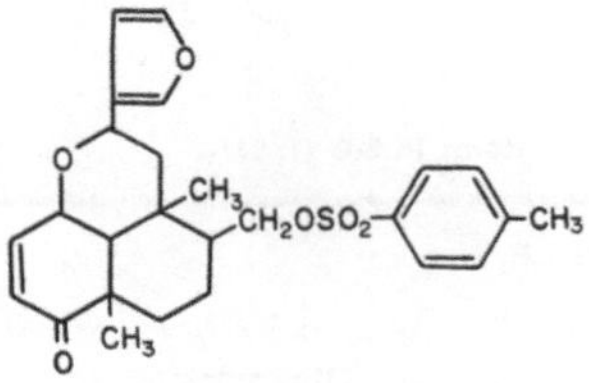

J. Chem. Soc. 4818 (1962)

▶ $SO_7C_{16}H_{20}$

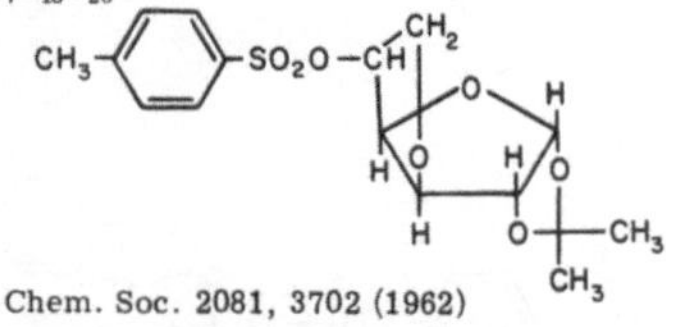

J. Chem. Soc. 2081, 3702 (1962)

▶ $SO_7C_{24}H_{32}$

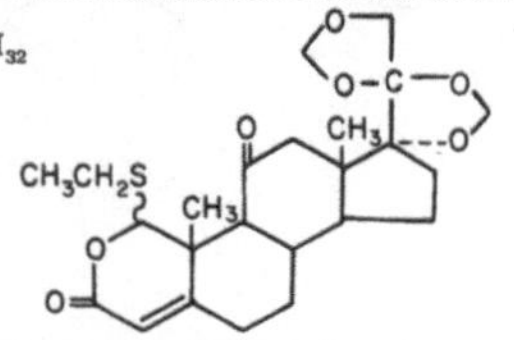

J. Am. Chem. Soc. 84:1270 (1962)

▶ $SO_7C_{26}H_{30}$

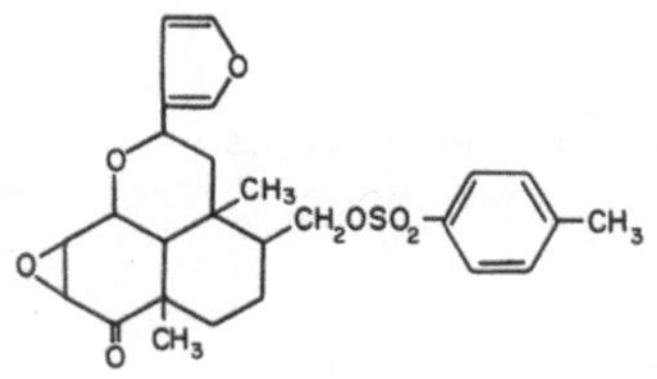

J. Chem. Soc. 4819 (1962)

▶ $SO_8C_{16}H_{18}$

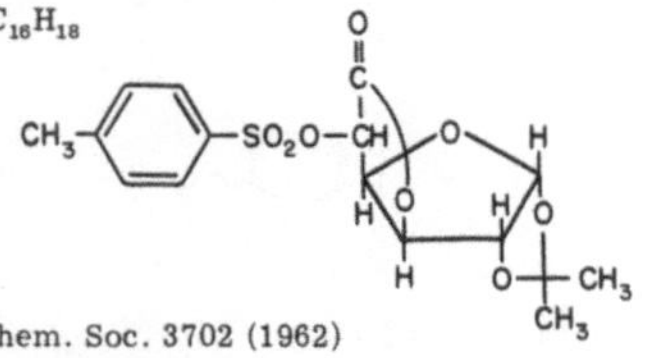

J. Chem. Soc. 3702 (1962)

▶ $SO_8C_{19}H_{26}$

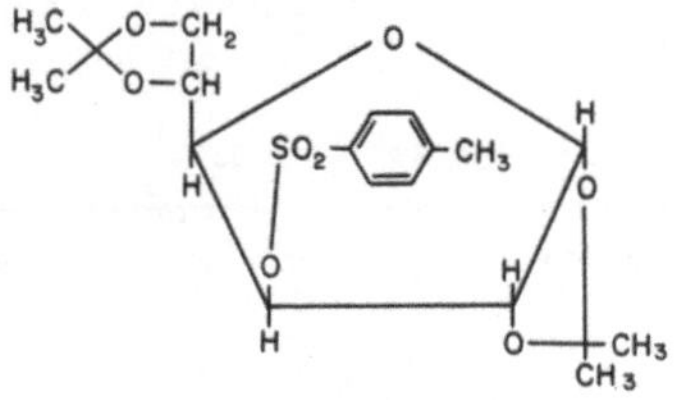

J. Chem. Soc. 3702 (1962)

▶ $SO_8C_{30}H_{34}$

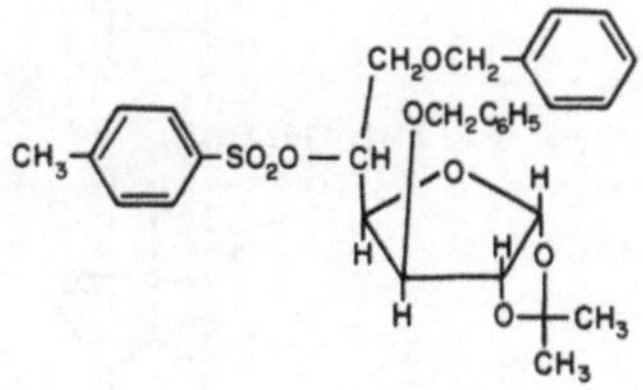


J. Chem. Soc. 3702 (1962)

▶ $SIC_4H_3$

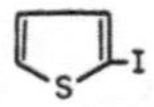


Varian 49
Arkiv Kemi 16:515, 539 (1961)

▶ $SIC_4H_3$

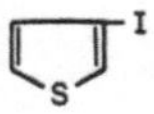


Arkiv Kemi 16:515, 539 (1961)

▶ $SIC_5H_5$

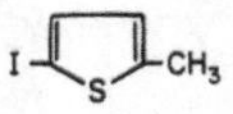


Arkiv Kemi 18:513 (1962)

▶ $SBrClC_4H_2$

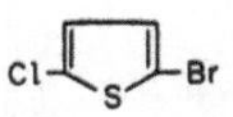


J. Chem. Phys. 37:2053 (1962)

▶ $SBrC_4H_2D$

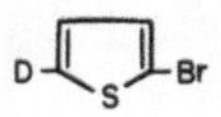


Arkiv Kemi 16:501 (1961)

▶ $SBrC_4H_2D$

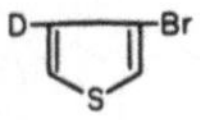


Arkiv Kemi 16:501 (1961)

▶ $SBrC_4H_2D$

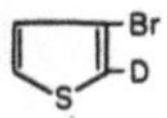


Arkiv Kemi 16:501 (1961)

▶ $SBrC_4H_2D$

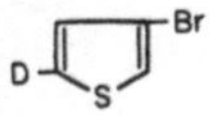


Arkiv Kemi 16:501 (1961)

▶ $SBrC_4H_3$

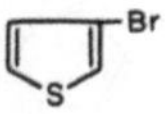


Arkiv Kemi 16:515, 539 (1961)

▶ $SBrC_4H_3$

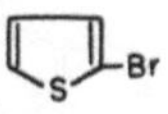


Arkiv Kemi 16:515, 539 (1961); 17:1 (1961)

▶ $SBrC_5H_5$

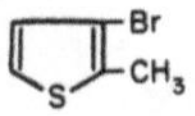


Arkiv Kemi 16:563 (1961)

▶ $SBrC_5H_5$

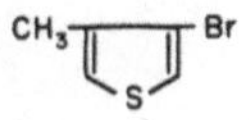


Arkiv Kemi 16:563 (1961); 18:513 (1962)

▶ $SBrC_5H_5$

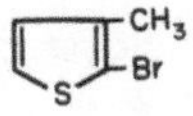

Arkiv Kemi 16:563 (1961); 18:513 (1962)

▶ $SBrC_5H_5$

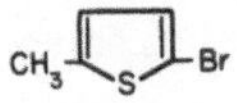

Arkiv Kemi 16:563 (1961); 18:513 (1962)

▶ $SBr_2C_4H_2$

J. Chem. Phys. 36:2644 (1962)

▶ $SBr_2C_4H_2$

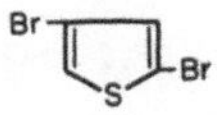

Arkiv Kemi 16:563 (1961)

▶ $SBr_2C_4H_2$

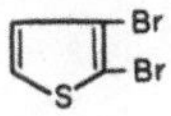

Arkiv Kemi 16:563 (1961)

▶ $SBr_2C_6H_6$

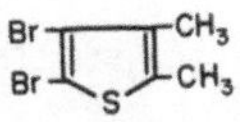

Acta Chem. Scand. 15:1201 (1961)

▶ $SClF_3C_3H_4$

$CH_3SCF_2HCFCl$

J. Am. Chem. Soc. 83:843 (1961)

▶ $SClF_5C_3H_2$

$CF_3SCH_2CF_2Cl$

J. Am. Chem. Soc. 84:3148 (1962)

▶ $SClF_5C_3H_2$

$CF_3SCF_2CH_2Cl$

J. Am. Chem. Soc. 84:3148 (1962)

▶ $SClF_6C_3H$

$CF_3SCFHCF_2Cl$

J. Am. Chem. Soc. 84:3148 (1962)

▶ $SClF_6C_3H$

$CF_3SCF_2CFHCl$

J. Am. Chem. Soc. 83:843 (1961); 84:3148 (1962)

▶ $SClC_5H_5$

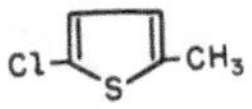

Arkiv Kemi 18:513 (1962)

▶ $SClC_6H_5$

Varian 451

▶ $SCl_2F_3C_3H_3$

$CF_3SCHClCH_2Cl$

J. Am. Chem. Soc. 84:3148 (1962)

▶ $SCl_2F_3C_3H_3$

$CF_3SCH_2CHCl_2$

J. Am. Chem. Soc. 84:3148 (1962)

▶ $SCl_2F_6C_5H_4$

$CH_3SCF_2CFClCF_2CFHCl$

J. Am. Chem. Soc. 83:843 (1961)

▶ $SCl_2F_9C_5H$

$CF_3SCF_2CFClCF_2CFHCl$

J. Am. Chem. Soc. 83:843 (1961)

▶ $SCl_2C_4H_2$

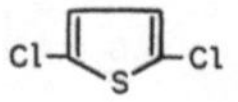


J. Chem. Phys. 36:2644 (1962)

▶ $SCl_2C_{10}H_8$

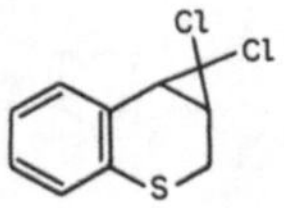


J. Am. Chem. Soc. 83:4034 (1961)

▶ $SCl_2C_{10}H_8$

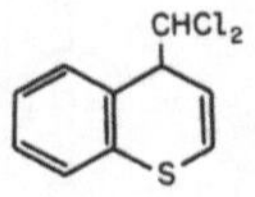


J. Am. Chem. Soc. 83:4034 (1961)

▶ $SCl_2C_{10}H_8$

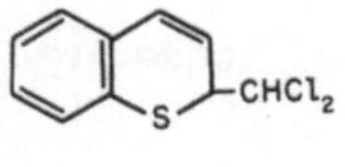


J. Am. Chem. Soc. 83:4034 (1961)

▶ $SF_3C_3H_5$

$CH_3SCF_2CFH_2$

J. Am. Chem. Soc. 83:843 (1961)

▶ $SF_3C_3H_5$

$CH_3SCFHCF_2H$

J. Am. Chem. Soc. 83:843 (1961)

▶ $SF_5C_3H_3$

$CF_3SCH_2CF_2H$

J. Am. Chem. Soc. 83:843 (1961)

▶ $SF_6C_3H_2$

$CF_3SCFHCF_2H$

J. Am. Chem. Soc. 83:843 (1961)

▶ $SF_6C_3H_2$

$CF_3SCF_2CFH_2$

J. Am. Chem. Soc. 83:843 (1961)

▶ $SF_6C_4H_4$

$CH_3SCF_2CFHCF_3$

J. Am. Chem. Soc. 83:843 (1961)

▶ $SF_6C_4H_4$

$CH_3SCF(CF_3)CF_2H$

J. Am. Chem. Soc. 83:843 (1961)

▶ $SF_6C_6H_2$

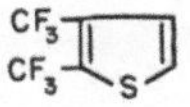

J. Am. Chem. Soc. 83:3438 (1961)

▶ $SF_7C_4H_3$

$C_3F_7CH_2SH$

J. Org. Chem. 26:357 (1961)

▶ $SF_8C_5H_4$

$H(CF_2)_4CH_2SH$

J. Org. Chem. 26:357 (1961)

▶ $SF_9C_4H$

$CF_3SCF(CF_3)CF_2H$

J. Am. Chem. Soc. 83:843 (1961)

▶ $SF_9C_4H$

$CF_3SCF_2CFHCF_3$

J. Am. Chem. Soc. 83:843 (1961)

▶ $SF_9C_5H_3$

$CF_3CH_2SCF(CF_3)CF_2H$

J. Am. Chem. Soc. 83:843 (1961)

▶ $SF_9C_5H_3$

$CF_3CH_2SCF_2CFHCF_3$

J. Am. Chem. Soc. 83:843 (1961)

▶ $SF_{10}C_5H_2$

$(C_2F_5)_2CHSH$

J. Org. Chem. 26:357 (1961)

▶ $SCH_4$

$CH_3SH$

J. Chem. Phys. 37:2198 (1962)

▶ $SC_2H_4$

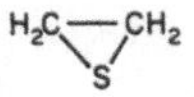


J. Chem. Phys. 36:3097 (1962)

▶ $SC_2H_6$

$(CH_3)_2S$

J. Chem. Phys. 34:1099 (1961); 37:2198 (1962)
J. Am. Chem. Soc. 83:4138 (1961)

▶ $SC_2H_6$

$CH_3CH_2SH$

J. Chem. Phys. 34:1099 (1961)

▶ $SC_2H_6 \cdot BH_3$

$(H_3C)_2S \cdot BH_3$

J. Am. Chem. Soc. 83:4138 (1961)

▶ $SC_3H_6$

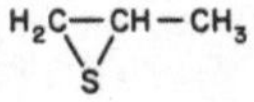

J. Chem. Phys. 36:3097 (1962)

▶ $SC_3H_6$

$H_2C{=}CHSCH_3$

Varian 36

▶ $SC_3H_6 \cdot Fe_2(CO)_6$

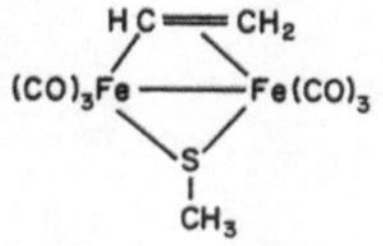

J. Am. Chem. Soc. 83:3600 (1961)

▶ $SC_3H_8$

$CH_3SCH_2CH_3$

Varian 46
Bull. Chem. Soc. Japan 35:1046 (1962)

▶ $SC_3H_8$

$(CH_3)_2CHSH$

Varian 396

▶ $SC_3H_9 \cdot ClO_4$

$(CH_3)_3\,S^{\oplus}\ \ ClO_4^{\ominus}$

J. Am. Chem. Soc. 84:3701 (1962)

▶ $SC_4H_4$

$HC{\equiv}CSCH{=}CH_2$

Rec. Trav. Chim. 81:635 (1962)

▶ $SC_4H_4$

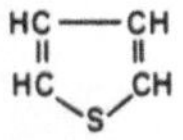

Varian 52
J. Am. Chem. Soc. 83:5020 (1961); 84:4727 (1962)
J. Chem. Phys. 36:2644 (1962)

▶ $SC_4H_6$

$HC{\equiv}CSCH_2CH_3$

Rec. Trav. Chim. 81:313, 635 (1962)

▶ $SC_4H_6 \cdot Fe_2(CO)_6$

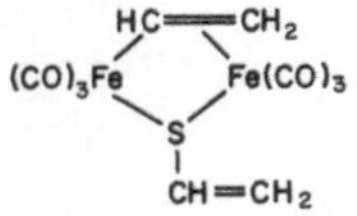

J. Am. Chem. Soc. 83:3600 (1961)

▶ $SC_4H_8$

Varian 80
Tetrahedron 18:791 (1962)

▶ $SC_4H_8 \cdot Fe_2(CO)_6$

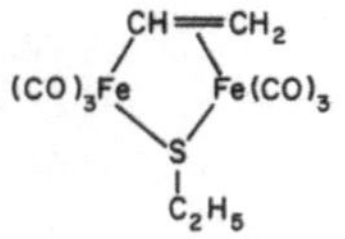

J. Am. Chem. Soc. 83:3600 (1961)

▶ $SC_4H_{10}$

$(C_2H_5)_2S$

Can. J. Chem. 40:1956 (1962)
J. Am. Chem. Soc. 83:4138 (1961)

▶ $SC_4H_{10} \cdot BF_3$

$(C_2H_5)_2S \cdot BF_3$

J. Am. Chem. Soc. 83:4138 (1961)

▶ $SC_4H_{10} \cdot BH_3$

$(C_2H_5)_2S \cdot BH_3$

J. Am. Chem. Soc. 83:4138 (1961)

▶ $SC_5H_5D$

Arkiv Kemi 16:501 (1961)

▶ $SC_5H_5D$

Arkiv Kemi 16:501 (1961)

▶ $SC_5H_5D$

Arkiv Kemi 16:501 (1961)

▶ $SC_5H_5D$

Arkiv Kemi 16:501 (1961)

▶ $SC_5H_6$

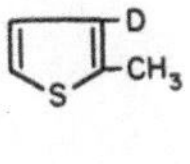

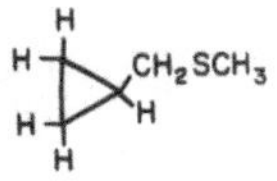

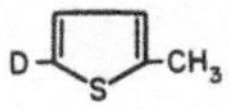

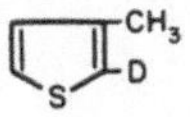

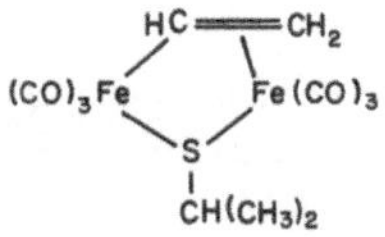

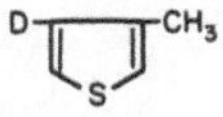

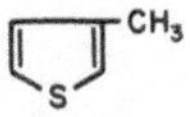

J. Chem. Phys. 36:2644 (1962)
J. Am. Chem. Soc. 83:5020 (1961)
Arkiv Kemi 16:515, 539 (1961); 18:513 (1962)

▶ $SC_5H_6$

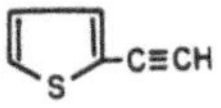

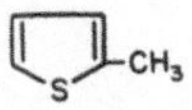

Varian 103
Arkiv Kemi 16:501, 515, 539 (1961); 17:1 (1961); 18:133 (1961); 18:513 (1962)
J. Am. Chem. Soc. 83:5020 (1961)
J. Chem. Phys. 36:2644 (1962)

▶ $SC_5H_8$

$HC{\equiv}CSCH(CH_3)_2$

Rec. Trav. Chim. 81:635 (1962)

▶ $SC_5H_{10}$

J. Org. Chem. 27:2722 (1962)

▶ $SC_5H_{10}$

Varian 118

▶ $SC_5H_{10} \cdot Fe_2(CO)_6$

J. Am. Chem. Soc. 83:3600 (1961)

▶ $SC_5H_{15}B$

$(CH_3)_2S \cdot B(CH_3)_3$

J. Am. Chem. Soc. 83:4138 (1961)

▶ $SC_6H_4$

Arkiv Kemi 16:515, 539 (1961)

▶ $SC_6H_6$

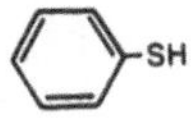

Can. J. Chem. 40:963 (1962)

▶ $SC_6H_8$

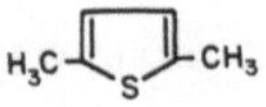


J. Am. Chem. Soc. 83:5020 (1961)
Arkiv Kemi 18:513 (1962)
J. Chem. Phys. 36:2644 (1962)

▶ $SC_6H_8$

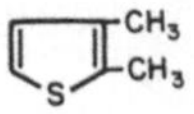


Arkiv Kemi 16:563 (1961)

▶ $SC_6H_{10}$

$H_2C{=}CHCH_2SCH_2CH{=}CH_2$

Varian 136

▶ $SC_6H_{12}$

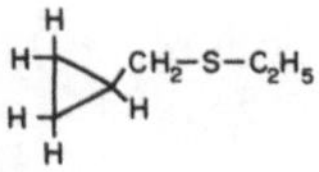


J. Org. Chem. 27:2721 (1962)

▶ $SC_6H_{14}$

$CH_3CH_2CH_2{-}S{-}CH_2CH_2CH_3$

J. Chem. Phys. 34:1094, 1099 (1961)

▶ $SC_6H_{14}$

$CH_3(CH_2)_4SCH_3$

Varian 144

▶ $SC_7H_8$

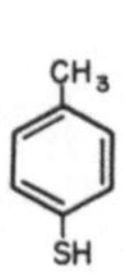


Varian 168

▶ $SC_7H_8$

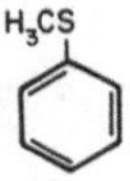


Varian 490

▶ $SC_7H_{19}B$

$(C_2H_5)_2S \cdot B(CH_3)_3$

J. Am. Chem. Soc. 83:4138 (1961)

▶ $SC_8H_6 \cdot Fe_2(CO)_6$

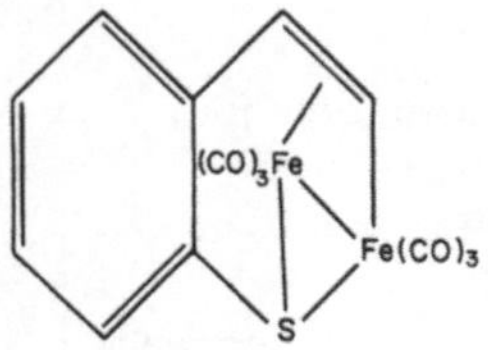


J. Am. Chem. Soc. 83:3600 (1961)

▶ $SC_9H_8$

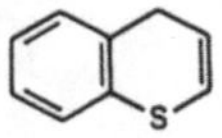


J. Am. Chem. Soc. 83:4034 (1961)

▶ $SC_9H_8$

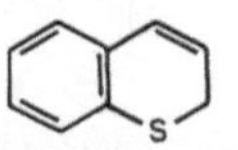


J. Am. Chem. Soc. 83:4034 (1961)

▶ $SC_9H_{10}$

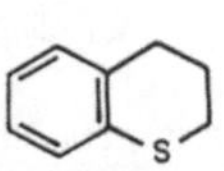


J. Am. Chem. Soc. 83:4034 (1961)

▶ $SC_9H_{10}$

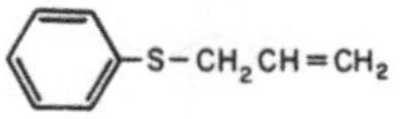

J. Am. Chem. Soc. 84:3987 (1962)

▶ $SC_{11}H_{14}$

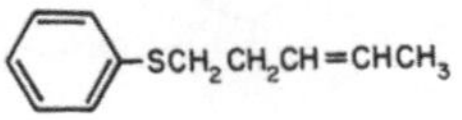

J. Am. Chem. Soc. 84:3987 (1962)

▶ $SC_{11}H_{14}$

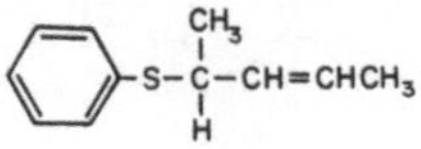

J. Am. Chem. Soc. 84:3987 (1962)

▶ $SC_{11}H_{14}$

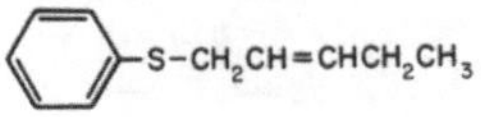

J. Am. Chem. Soc. 84:3987 (1962)

▶ $SC_{11}H_{14}$

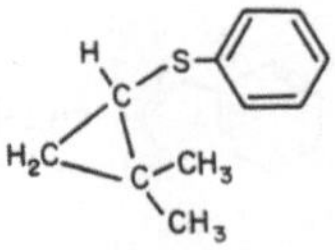

Tetrahedron Letters No. 4:165 (1962)

▶ $SC_{12}H_{16}$

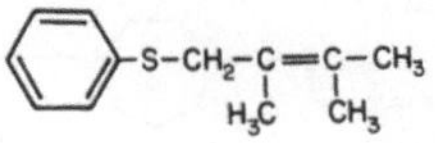

J. Am. Chem. Soc. 84:3987 (1962)

▶ $SC_{14}H_{12}$

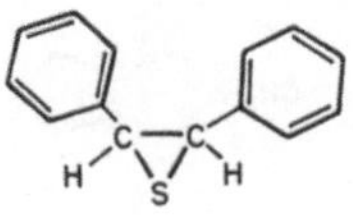

Varian 629

▶ $SC_{14}H_{12}$

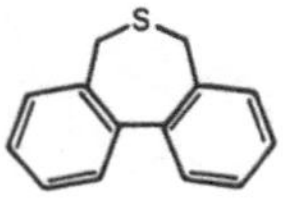

Varian 630

▶ $SC_{16}H_{16}$

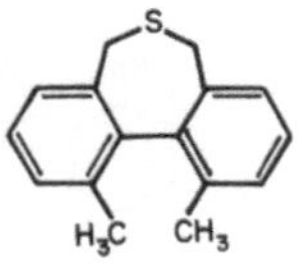

Varian 657

▶ $SC_{16}H_{24}$

cis and trans

Tetrahedron Letters No. 3:97 (1962)

▶ $SC_{20}H_{24}$

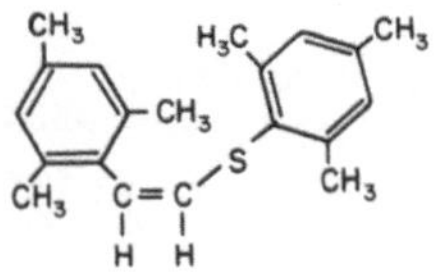

J. Am. Chem. Soc. 83:4636 (1961)

▶ $SC_{21}H_{18}$

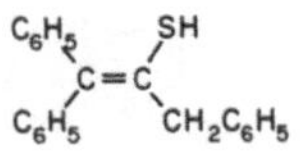

J. Org. Chem. 27:3761 (1962)

▶ $SC_{22}H_{20}$

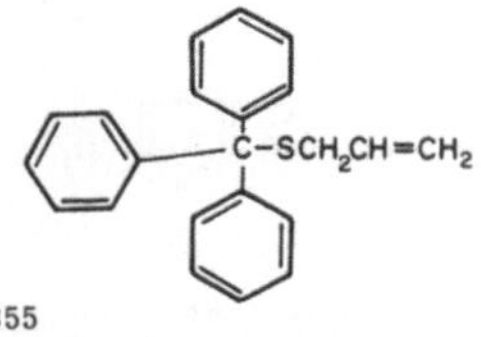

Varian 355

▶ $S_2NOC_{13}H_{21}$

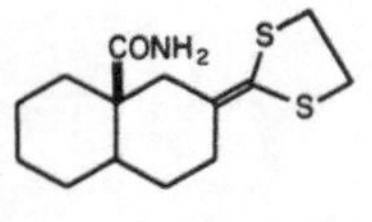

J. Org. Chem. 27:2013 (1962)

▶ $S_2NOC_{13}H_{21}$

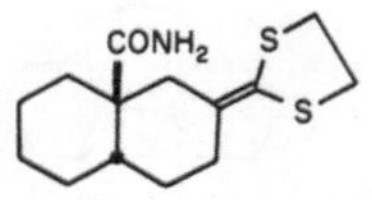

J. Org. Chem. 27:2013 (1962)

▶ $S_2NO_4ClC_4H_2$

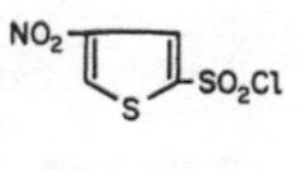

Arkiv Kemi 16:563 (1961)

▶ $S_2NO_4ClC_4H_2$

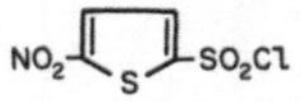

Arkiv Kemi 16:563 (1961)

▶ $S_2NC_5H_3$

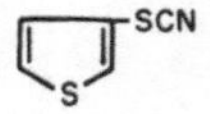

Arkiv Kemi 16:515, 539 (1961)

▶ $S_2NC_5H_3$

Arkiv Kemi 16:515, 539 (1961)

▶ $S_2NC_6H_5$

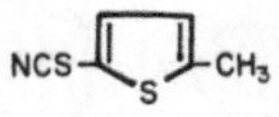

Arkiv Kemi 18:513 (1962)

▶ $S_2NC_6H_9$

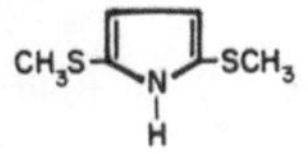

Arkiv Kemi 18:151 (1961)

▶ $S_2NC_9H_7$

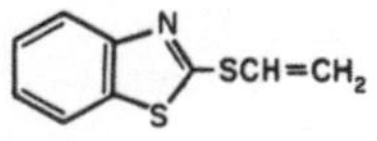

Varian 226

▶ $S_2NC_9H_{15}$

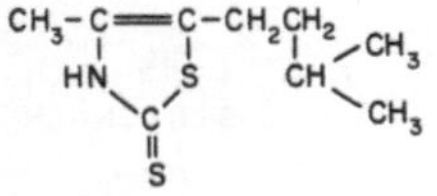

J. Am. Chem. Soc. 83:3096 (1961)

▶ $S_2NC_9H_{17}$

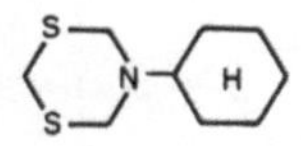

J. Org. Chem. 27:2020 (1962)

▶ $S_2NC_{10}H_9$

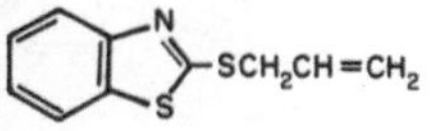

Varian 250

▶ $S_2N_2O_4C_{19}H_{24}$

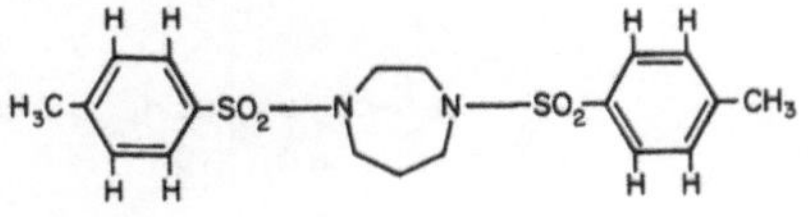

Varian 336

▶ $S_2N_2C_6H_{14}$

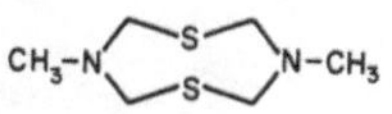

J. Org. Chem. 27:2020 (1962)

▶ $S_2N_2C_{12}H_{16}$

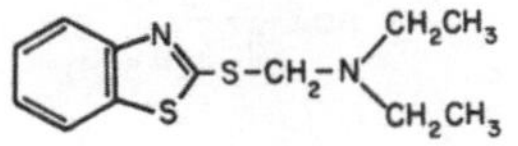

J. Chem. Soc. 2070 (1962)

▶ $S_2OC_6H_6$

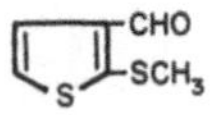

Arkiv Kemi 16:563 (1961)

▶ $S_2N_2C_{12}H_{16}$

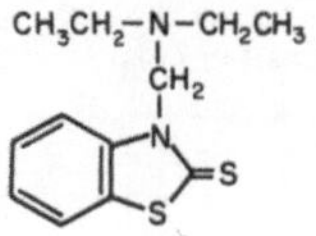

J. Chem. Soc. 2070 (1962)

▶ $S_2OC_6H_6$

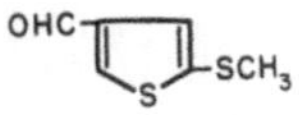

Arkiv Kemi 16:563 (1961)

▶ $S_2OC_6H_6$

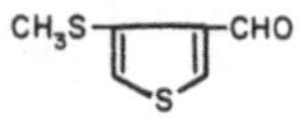

Arkiv Kemi 16:563 (1961)

▶ $S_2N_2C_{16}H_{30}$

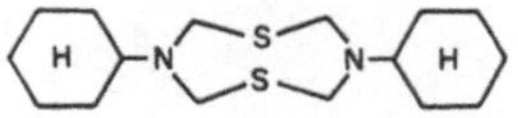

J. Org. Chem. 27:2020 (1962)

▶ $S_2OC_{15}H_{16}$

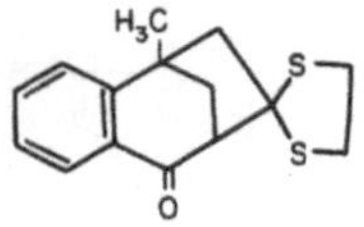

J. Am. Chem. Soc. 84:3517 (1962)

▶ $S_2N_3C_6H_3$

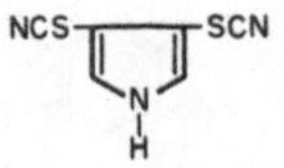

Acta Chem. Scand. 15:227 (1961)

▶ $S_2OC_{16}H_{24}$

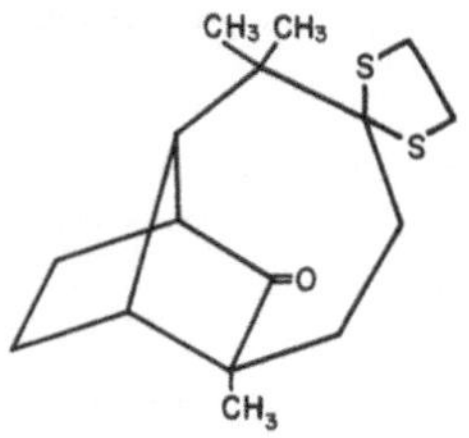

J. Am. Chem. Soc. 83:1251 (1961)

▶ $S_2N_3C_8H_7$

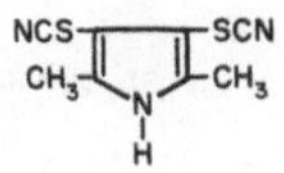

Arkiv Kemi 18:133 (1961)

▶ $S_2N_4C_2H_4$

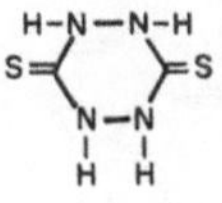

Acta Chem. Scand. 15:1575 (1961)

▶ $S_2O_2BrC_5H_5$

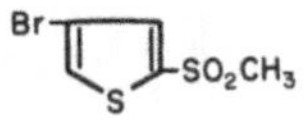

Arkiv Kemi 16:563 (1961)

▶ $S_2O_2ClC_4H_3$

Arkiv Kemi 16:515, 539 (1961)

▶ $S_2O_2ClC_5H_5$

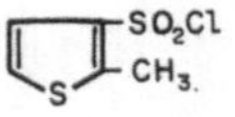

Bull. Chem. Soc. Japan 34:1599 (1961)

▶ $S_2O_2ClC_5H_5$

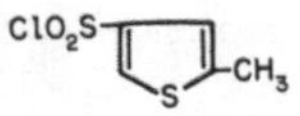

Bull. Chem. Soc. Japan 34:1599 (1961)

▶ $S_2O_2ClC_5H_5$

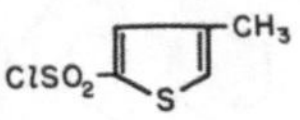

Bull. Chem. Soc. Japan 34:1599 (1961)

▶ $S_2O_2ClC_5H_5$

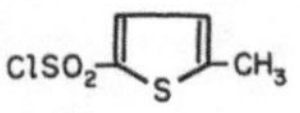

Bull. Chem. Soc. Japan 34:1599 (1961)

▶ $S_2O_2ClC_5H_5$

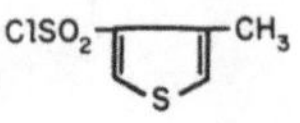

Bull. Chem. Soc. Japan 34:1599 (1961)

▶ $S_2O_2ClC_5H_5$

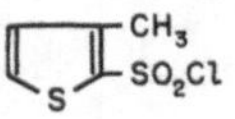

Bull. Chem. Soc. Japan 34:1599 (1961)

▶ $S_2O_2C_6H_6$

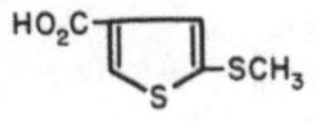

Arkiv Kemi 16:563 (1961)

▶ $S_2O_2C_6H_6$

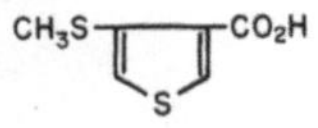

Arkiv Kemi 16:563 (1961)

▶ $S_2O_2C_7H_8$

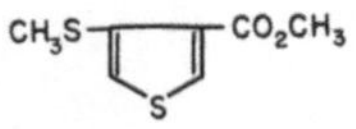

Arkiv Kemi 16:563 (1961)

▶ $S_2O_2C_7H_8$

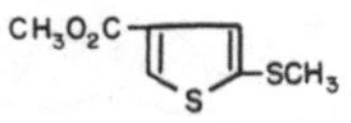

Arkiv Kemi 16:563 (1961)

▶ $S_2O_2C_8H_7D$

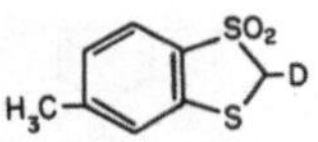

J. Am. Chem. Soc. 84:684 (1962)

▶ $S_2O_2C_{13}H_{11}D$

$C_6H_5SO_2CHDSC_6H_5$

J. Am. Chem. Soc. 84:684 (1962)

▶ $S_2O_2C_{17}H_{20}$

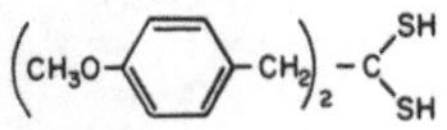

J. Org. Chem. 27:3762 (1962)

▶ $S_2O_2C_{20}H_{24}$

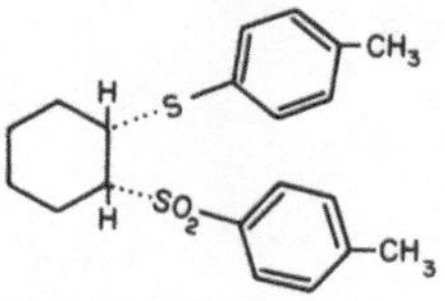

J. Am. Chem. Soc. 83:4641 (1961)

▶ $S_2O_2C_{20}H_{24}$

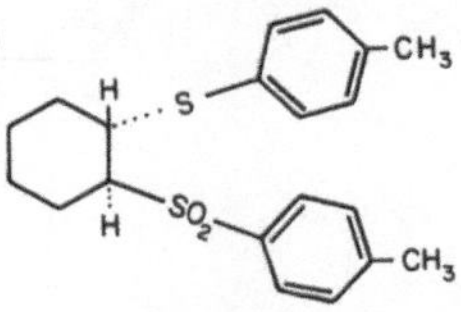

J. Am. Chem. Soc. 83:4641 (1961)

▶ $S_2O_2C_{28}H_{24}$

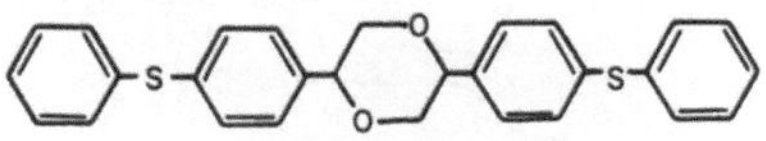

Helv. Chim. Acta 45:982 (1962)

▶ $S_2O_4Cl_4C_8H_4$

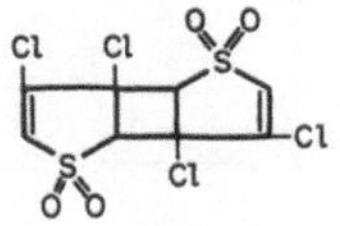

trans–fusion

J. Org. Chem. 26:349 (1961)

▶ $S_2O_4C_8H_{18}$

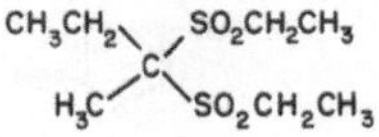

Varian 519

▶ $S_2O_4C_{20}H_{24}$

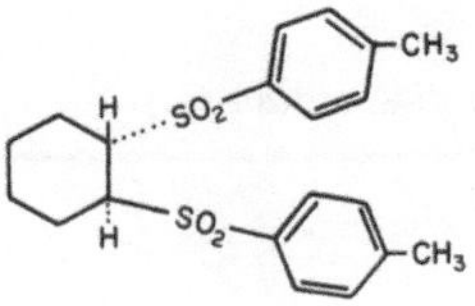

J. Am. Chem. Soc. 83:4641 (1961)

▶ $S_2O_6C_{10}H_8$

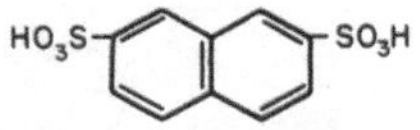

J. Phys. Chem. 66:268 (1962)

▶ $S_2O_6C_{20}H_{28}$

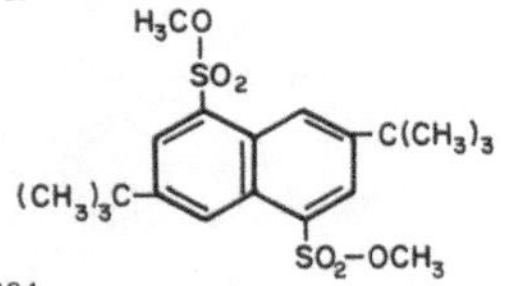

Varian 684

▶ $S_2O_6C_{20}H_{28}$

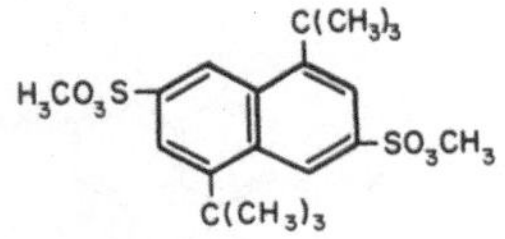

Can. J. Chem. 40:1740 (1962)

▶ $S_2O_7IC_{10}H_6D$

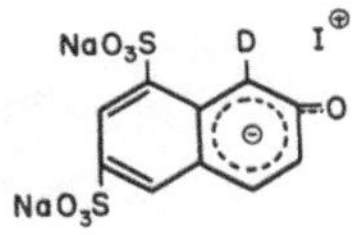

Helv. Chim. Acta 45:2077 (1962)

▶ $S_2O_7IC_{10}H_7$

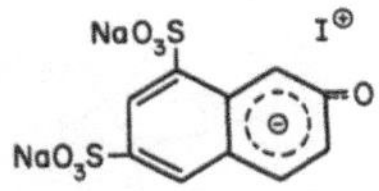

Helv. Chim. Acta 45:2077 (1962)

▶ $S_2O_7BrC_{10}H_6D$

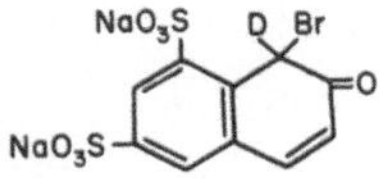

Helv. Chim. Acta 45:2077 (1962)

▶ $S_2O_7BrC_{10}H_7$

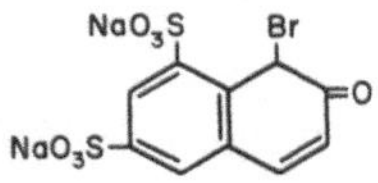

Helv. Chim. Acta 45:2077 (1962)

▶ $S_2O_7BrC_{10}H_7$

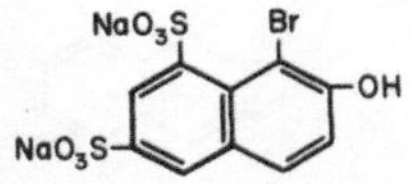


Helv. Chim. Acta 45:2077 (1962)

▶ $S_2O_7C_{10}H_7D$

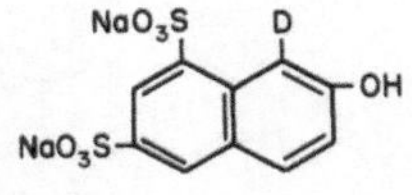


Helv. Chim. Acta 45:2077 (1962)

▶ $S_2O_7C_{10}H_8$

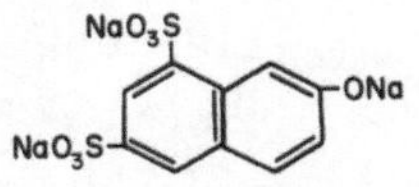


Helv. Chim. Acta 45:2077 (1962)

▶ $S_2O_7C_{10}H_8$

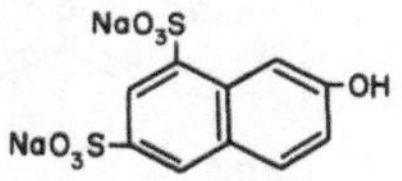


Helv. Chim. Acta 45:2077 (1962)

▶ $S_2O_8C_{13}H_{24}$

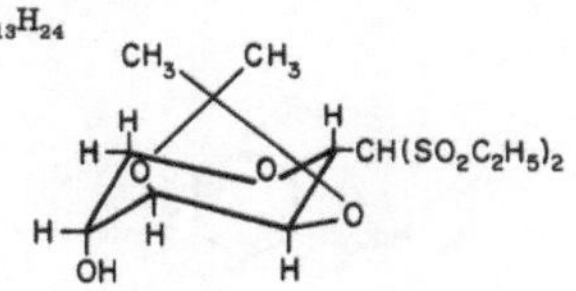


Chem. & Ind. 1465 (1962)

▶ $S_2O_9C_{15}H_{26}$

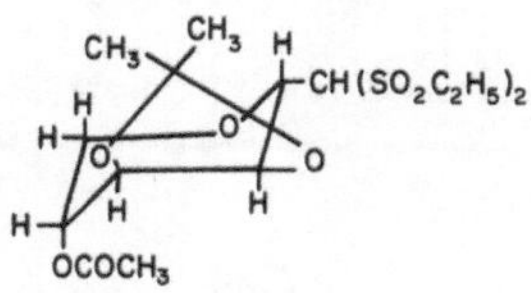


Chem. & Ind. 1465 (1962)

▶ $S_2O_{10}C_{23}H_{28}$

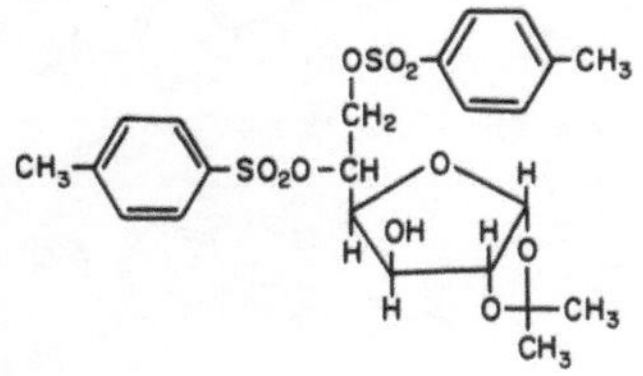


J. Chem. Soc. 3702 (1962)

▶ $S_2O_{11}C_{16}H_{26}$

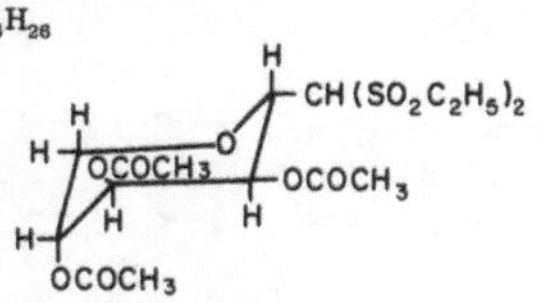


Chem. & Ind. 1465 (1962)

▶ $S_2O_{11}C_{17}H_{28}$

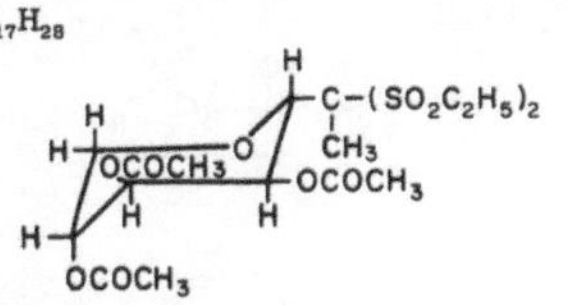


Chem. & Ind. 1465 (1962)

▶ $S_2BrC_4H_3$

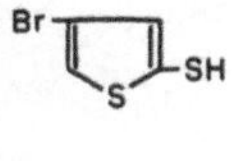


Arkiv Kemi 16:563 (1961)

▶ $S_2BrC_4H_3$

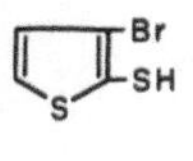


Arkiv Kemi 16:563 (1961)

▶ $S_2BrC_4H_3$

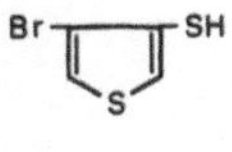


Arkiv Kemi 16:563 (1961)

▶ $S_2BrC_5H_5$

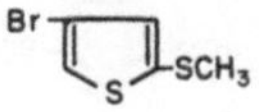

Arkiv Kemi 16:563 (1961)

▶ $S_2BrC_5H_5$

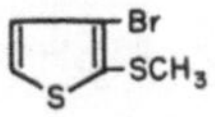

Arkiv Kemi 16:563 (1961)

▶ $S_2BrC_{29}H_{49}$

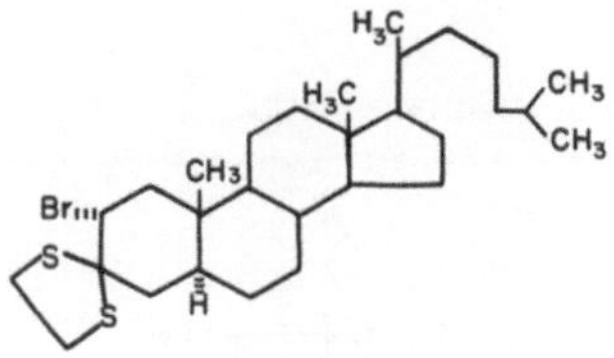

J. Am. Chem. Soc. 83:4623 (1961)

▶ $S_2F_6C_6H_2$

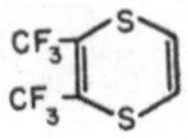

J. Am. Chem. Soc. 83:3438 (1961)

▶ $S_2F_6C_6H_4$

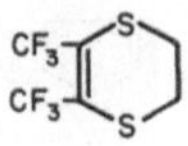

J. Am. Chem. Soc. 83:3438 (1961)

▶ $S_2F_6C_6H_6$

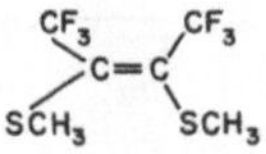

J. Am. Chem. Soc. 83:3438 (1961)

▶ $S_2F_6C_{10}H_{10}$

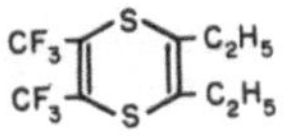

J. Am. Chem. Soc. 83:3438 (1961)

▶ $S_2F_6C_{10}H_{10}$

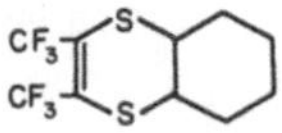

J. Am. Chem. Soc. 83:3438 (1961)

▶ $S_2F_6C_{10}H_{12}$

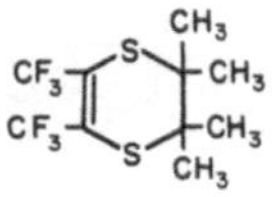

J. Am. Chem. Soc. 83:3438 (1961)

▶ $S_2F_{20}C_{12}H_2$

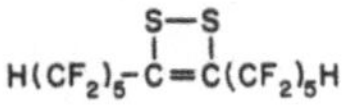

J. Am. Chem. Soc. 83:3434 (1961)

▶ $S_2C_2H_6$

$H_3C-S-S-CH_3$

J. Chem. Phys. 34:1099 (1961)

▶ $S_2C_2H_6 \cdot Fe_2(CO)_6$

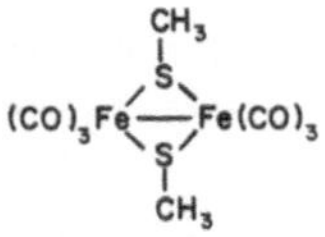

J. Am. Chem. Soc. 83:3600 (1961); 84:2460 (1962)

▶ $S_2C_3H_8$

$HSCH_2CH_2CH_2SH$

Varian 47

▶ $S_2C_4H_4$

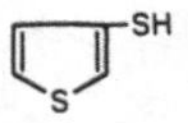

Arkiv Kemi 16:515, 539 (1961)

▶ $S_2C_4H_4$

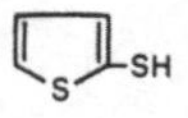

Arkiv Kemi 16:515, 539 (1961)

▶ $S_2C_4H_4D_4$

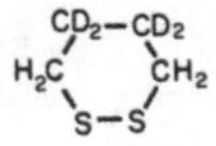

J. Am. Chem. Soc. 83:4357 (1961)

▶ $S_2C_4H_8$

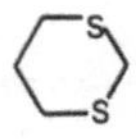

Tetrahedron Letters No. 16:686 (1962)

▶ $S_2C_4H_8$

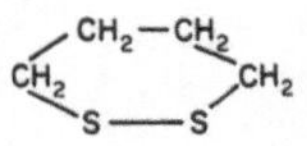

J. Am. Chem. Soc. 83:4357 (1961)

▶ $S_2C_4H_{10}$

$C_2H_5-S-S-C_2H_5$

J. Chem. Phys. 34:1099 (1961)

▶ $S_2C_5H_5D$

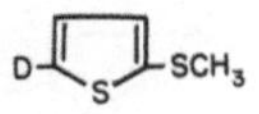

Arkiv Kemi 16:501 (1961)

▶ $S_2C_5H_5D$

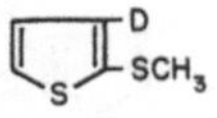

Arkiv Kemi 16:501 (1961)

▶ $S_2C_5H_5D$

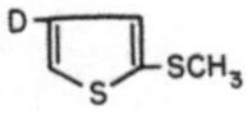

Arkiv Kemi 16:501 (1961)

▶ $S_2C_5H_5D$

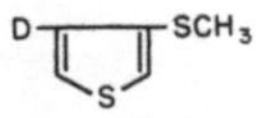

Arkiv Kemi 16:501 (1961)

▶ $S_2C_5H_6$

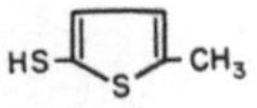

Arkiv Kemi 16:563 (1961); 18:513 (1962)

▶ $S_2C_5H_6$

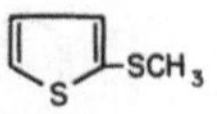

Arkiv Kemi 16:515, 539 (1961)

▶ $S_2C_5H_6$

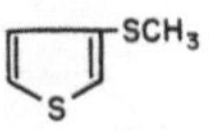

Arkiv Kemi 16:515, 539 (1961)

▶ $S_2C_5H_6$

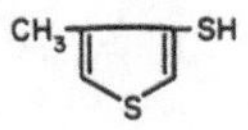


Arkiv Kemi 16:563 (1961)

▶ $S_2C_5H_6$

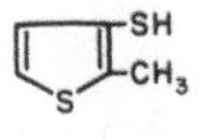


Arkiv Kemi 16:563 (1961)

▶ $S_2C_5H_6$

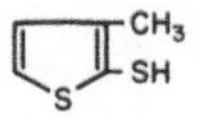


Arkiv Kemi 16:563 (1961)

▶ $S_2C_6H_8$

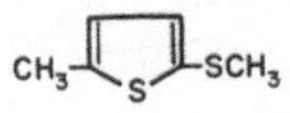


Arkiv Kemi 16:563 (1961); 18:513 (1962)

▶ $S_2C_6H_{12}$

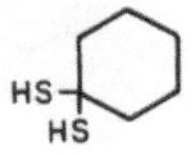


J. Org. Chem. 27:1042 (1962)

▶ $S_2C_6H_{12}$

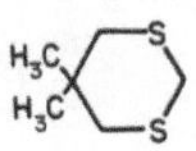


Tetrahedron Letters No. 16:686 (1962)

▶ $S_2C_6H_{12}$

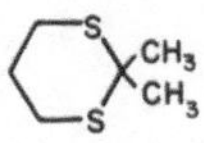


Tetrahedron Letters No. 16:686 (1962)

▶ $S_2C_6H_{14}$

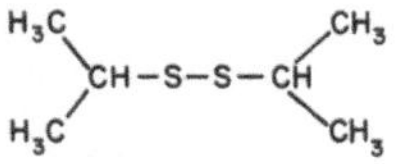


J. Chem. Phys. 34:1099 (1961)

▶ $S_2C_6H_{14}$

$H_3CCH_2CH_2-S-S-CH_2CH_2CH_3$

J. Chem. Phys. 34:1094, 1099 (1961)

▶ $S_2C_8H_6$

Arkiv Kemi 16:515, 539 (1961)

▶ $S_2C_8H_{16}$

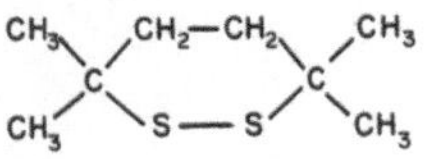


J. Am. Chem. Soc. 83:4357 (1961)

▶ $S_2C_9H_7 \cdot HSO_4$

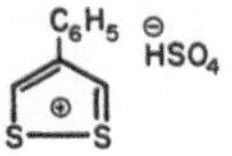


J. Am. Chem. Soc. 83:2934 (1961)

▶ $S_2C_9H_7 \cdot HSO_4$

J. Am. Chem. Soc. 83:2934 (1961)

▶ $S_2C_{12}H_8$

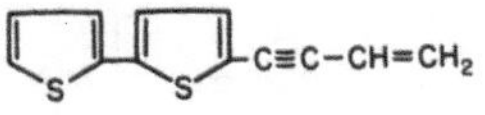


Chem. Ber. 95:2945 (1962)

▶ $S_2C_{12}H_{16} \cdot Fe_2(CO)_2$

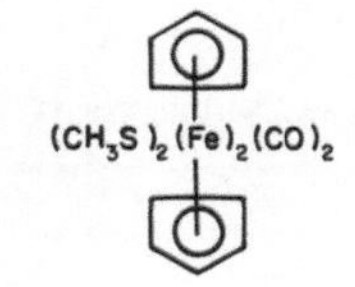

J. Am. Chem. Soc. 83:3600 (1961)

▶ $S_2C_{12}H_{16}Co_2$

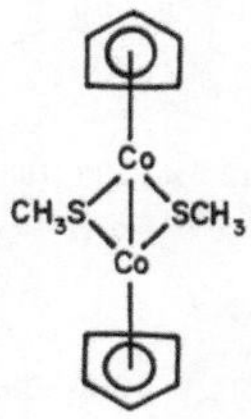

J. Am. Chem. Soc. 83:3600 (1961)

▶ $S_2C_{16}H_{12}$

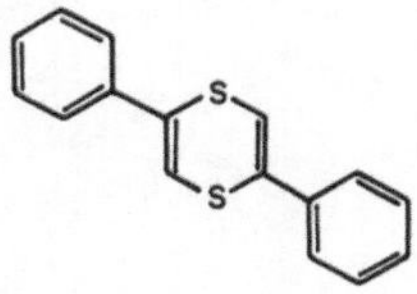

Varian 651

▶ $S_2C_{16}H_{14}$

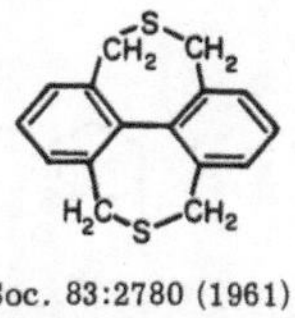

Varian 653
J. Am. Chem. Soc. 83:2780 (1961)

▶ $S_2C_{24}H_{22}$

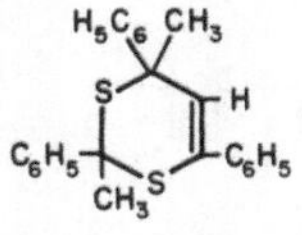

J. Org. Chem. 27:4665 (1962)

▶ $S_2C_{30}H_{26}$

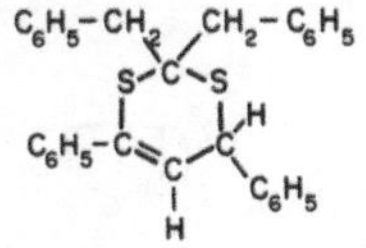

J. Org. Chem. 27:4488 (1962)

▶ $S_3O_4C_{10}H_{14}$

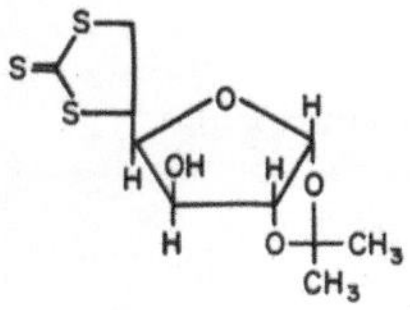

Chem. & Ind. 213 (1962)

▶ $S_3O_{10}C_{14}H_{26}$

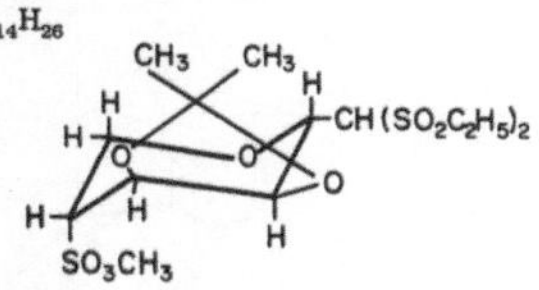

Chem. & Ind. 1465 (1962)

▶ $S_3Cl_3C_6H_9$

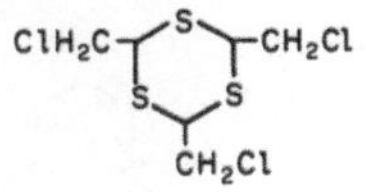

trans

J. Org. Chem. 26:1456 (1961)

▶ $S_3Cl_3C_6H_9$

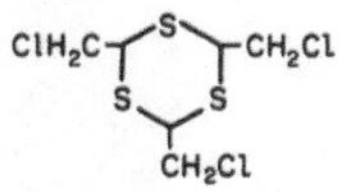

cis

J. Org. Chem. 26:1456 (1961)

▶ $S_3Cl_3C_9H_{15}$

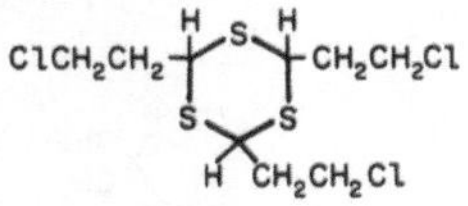

cis

J. Org. Chem. 26:1456 (1961)

▶ $S_3C_4H_4$

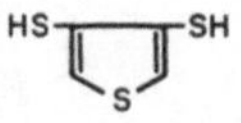


Acta Chem. Scand. 16:105 (1962)

▶ $S_3C_4H_6$

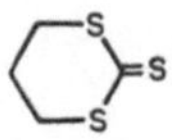


J. Org. Chem. 27:4069 (1962)

▶ $S_3C_5H_6$

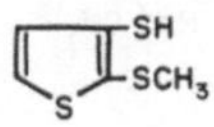


Arkiv Kemi 16:563 (1961)

▶ $S_3C_5H_6$

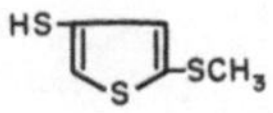


Arkiv Kemi 16:563 (1961)

▶ $S_3C_5H_6$

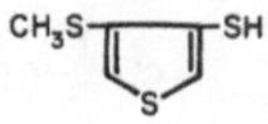


Arkiv Kemi 16:563 (1961)

▶ $S_3C_5H_6$

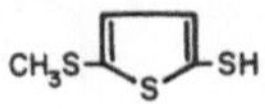


Arkiv Kemi 16:563 (1961)

▶ $S_3C_6H_{12}$

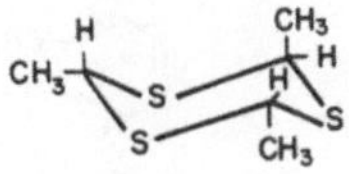


J. Org. Chem. 27:135 (1962)

▶ $S_3C_6H_{12}$

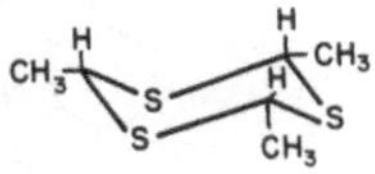


J. Org. Chem. 27:135 (1962)

▶ $S_3C_9H_{12}$

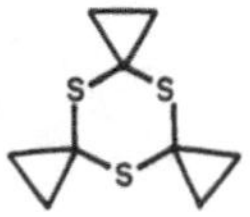


J. Org. Chem. 27:3736 (1962)

▶ $S_3C_{21}H_{18}$

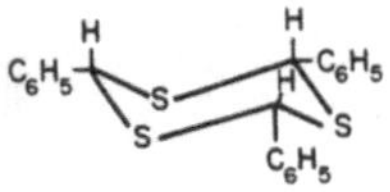


J. Org. Chem. 27:135 (1962)

▶ $S_3C_{21}H_{18}$

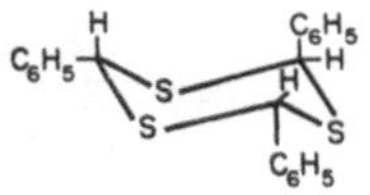


J. Org. Chem. 27:135 (1962)

▶ $S_3C_{22}H_{22}$

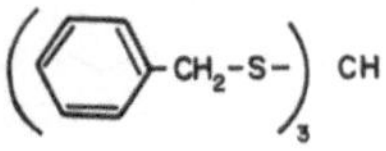


Varian 688

▶ $S_3C_{26}H_{18}$

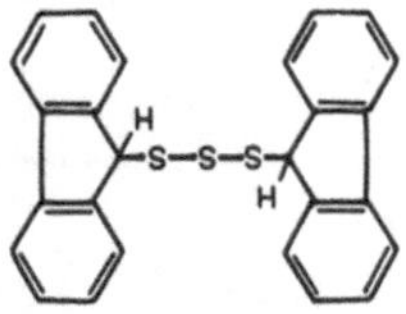


Chem. Ber. 95:2144 (1962)

| | |
|---|---|
| ▶ $S_3C_{27}H_{30}$<br>J. Org. Chem. 27:3762 (1962) | ▶ $NOBrC_5H_4$<br>J. Chem. Soc. 43 (1961) |
| ▶ $S_3C_{29}H_{26}$<br>trans and cis<br>J. Org. Chem. 27:4488 (1962) | ▶ $NOBr_2C_{30}H_{45}$<br>Bull. Soc. Chim. France 1137 (1962) |
| ▶ $S_4F_{12}C_{10}H_2$<br>J. Am. Chem. Soc. 83:3438 (1961) | |
| ▶ $S_4C_7H_{12}$<br>J. Org. Chem. 27:4069 (1962) | ▶ $NOClC_3H_6$<br>$CH_3-C(=NOH)-CH_2Cl$<br>J. Phys. Chem. 65:491 (1961) |
| ▶ $S_6C_{24}H_{24}$<br>trans<br>J. Org. Chem. 26:1456 (1961) | ▶ $NOClC_3H_6$<br>$Cl-C(=O)-N(CH_3)_2$<br>J. Phys. Chem. 66:540 (1962)<br>J. Am. Chem. Soc. 84:13 (1962) |
| ▶ $S_6C_{24}H_{24}$<br>cis<br>J. Org. Chem. 26:1456 (1961) | ▶ $NOClC_5H_4$<br>J. Chem. Soc. 44 (1961) |
| ▶ $S_6C_{27}H_{30}$<br>cis<br>J. Org. Chem. 26:1456 (1961) | ▶ $NOClC_6H_6$<br>J. Chem. Soc. 46 (1961) |

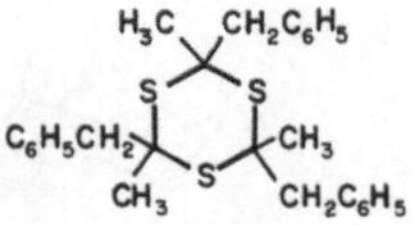

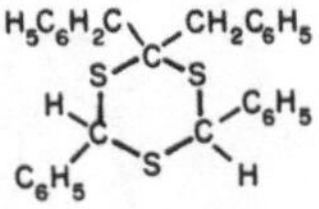

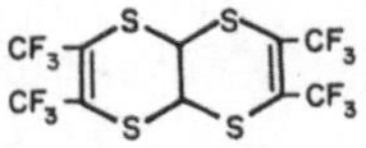

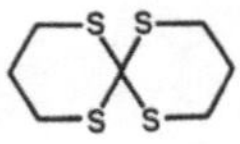

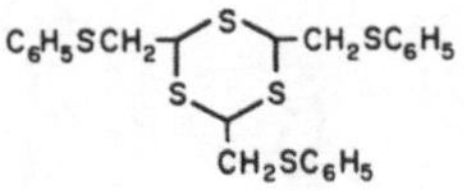

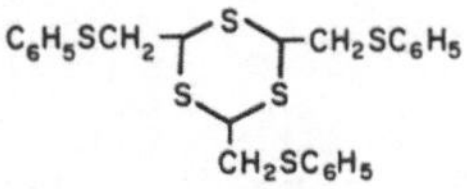

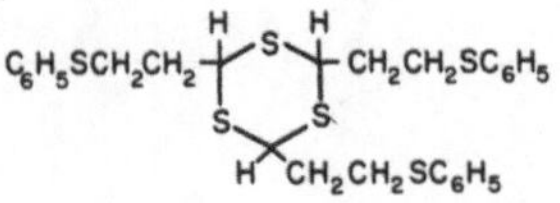

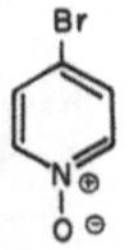

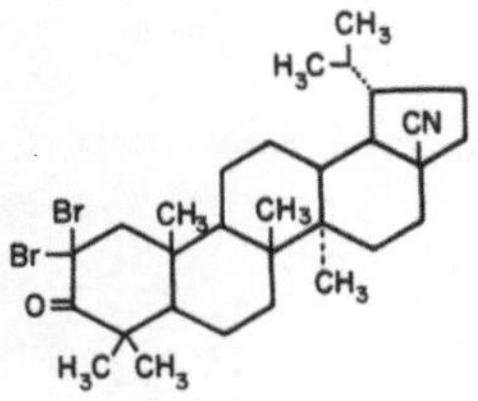

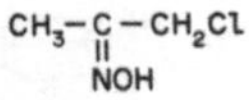

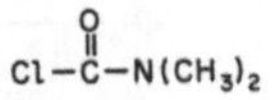

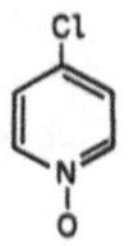

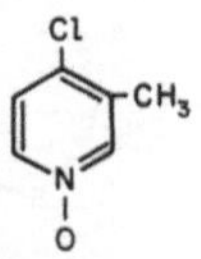

▶ $NOClC_6H_{10}$

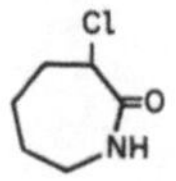

Rec. Trav. Chim. 80:1211 (1961)

▶ $NOClC_{11}H_{17}D$

J. Am. Chem. Soc. 83:2769 (1961)

▶ $NOClC_{11}H_{17}D$

J. Am. Chem. Soc. 83:2769 (1961)

▶ $NOClC_{11}H_{18}$

J. Am. Chem. Soc. 83:2769 (1961)

▶ $NOClC_{11}H_{18}$

J. Am. Chem. Soc. 83:2769 (1961)

▶ $NOCl_2C_6H_9$

Rec. Trav. Chim. 80:1211 (1961)

▶ $NOCl_3C_4H_6$

$CCl_3CON(CH_3)_2$

J. Phys. Chem. 66:540 (1962)

▶ $NOF_2C_2H_3$

$CH_3C(=O)NF_2$

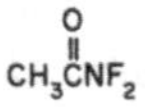

J. Am. Chem. Soc. 83:3912 (1961)

▶ $NOF_2C_{10}H_7$

J. Chem. Soc. 3829 (1962)

▶ $NOF_2C_{10}H_7$

J. Chem. Soc. 3829 (1962)

▶ $NOF_2C_{10}H_9$

J. Chem. Soc. 3829 (1962)

▶ $NOF_3C_4H_6$

$CF_3CON(CH_3)_2$

J. Phys. Chem. 66:540 (1962)

▶ $NOF_3C_7H_{10}$

$CH_3C(=O)CH{=}C(CH_3)NHCH_2CF_3$

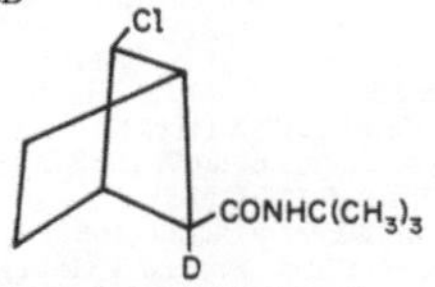

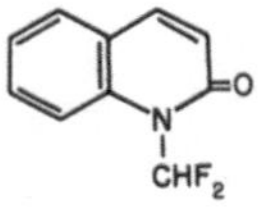

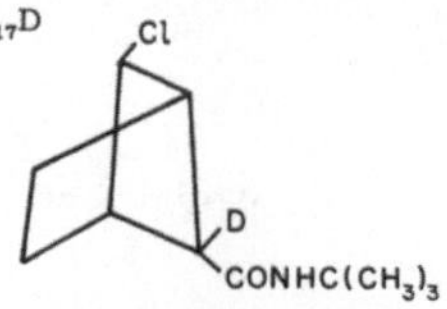

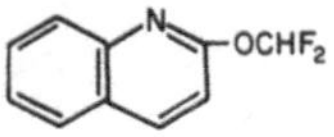

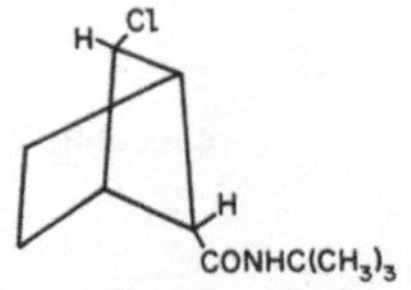

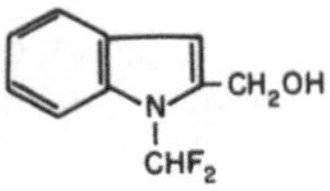

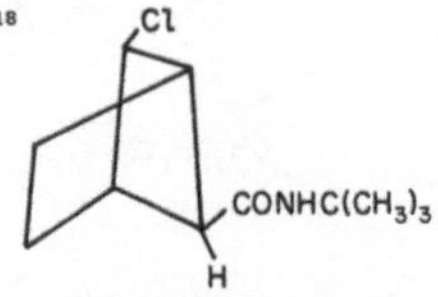

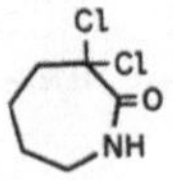

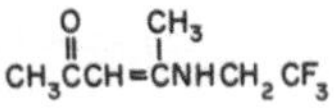

J. Am. Chem. Soc. 84:2691 (1962)

▶ $NOF_3C_8H_6$

$CF_3{-}C({=}NOH){-}C_6H_5$

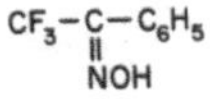

J. Phys. Chem. 65:691 (1961)

▶ $NOF_3C_{10}H_{14}$

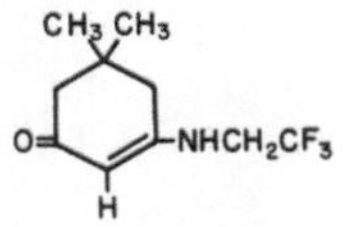


J. Am. Chem. Soc. 84:2691 (1962)

▶ $NOF_3C_{17}H_{16}$

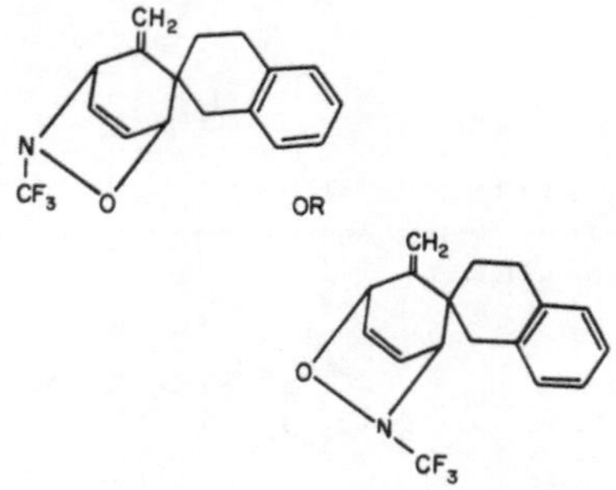


J. Org. Chem. 26:2607 (1961)

▶ $NOF_4C_9H_7$

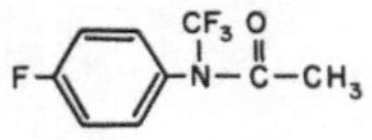


J. Org. Chem. 26:3454 (1961)

▶ $NOC_2H_5$

$HCONHCH_3$

Arkiv Kemi 16:373 (1961)

▶ $NOC_2H_5$

$CH_3{-}CH{=}NOH$

Varian 373
Mol. Phys. 5:153 (1962)

▶ $NOC_3H_5$

$H_3COCH_2CN$

Varian 28

▶ $NOC_3H_7$

$(CH_3)_2N\overset{O}{\overset{\|}{C}}H$

Varian 39
Arkiv Kemi 16:373 (1961)
J. Phys. Chem. 66:540, 2653 (1962)
Mol. Phys. 5:139 (1962)
J. Chem. Phys. 37:2198 (1962)
Bull. Soc. Chim. France 1 (1962)
J. Am. Chem. Soc. 84:571 (1962)

▶ $NOC_3H_7$

$(CH_3)_2N-\overset{O}{\overset{\|}{C^{13}}}-H$

J. Am. Chem. Soc. 84:2649 (1962)

▶ $NOC_3H_7$

$C_2H_5CONH_2$

Varian 40
Can. J. Chem. 40:1522 (1962)

▶ $NOC_3H_7$

$CH_3CH_2CH{=}NOH$

J. Phys. Chem. 65:491 (1961)

▶ $NOC_3H_9$

$H_2N-\underset{CH_3}{\overset{CH_3}{C}}-OH$

Varian 397

▶ $NOC_4H_5$

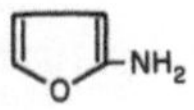


Can. J. Chem. 39:909 (1961)

▶ $NOC_4H_7$

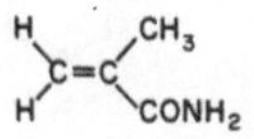

Varian 71
Arkiv Kemi 17:1 (1961)

▶ $NOC_4H_7$

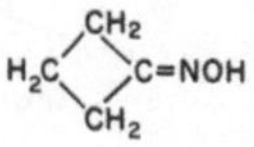

J. Phys. Chem. 65:491 (1961)

▶ $NOC_4H_7$

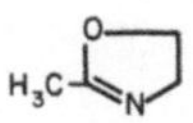

Can. J. Chem. 40:978 (1962)

▶ $NOC_4H_7$

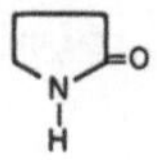

Varian 68

▶ $NOC_4H_7$

$H_3COCH_2CH_2CN$

Varian 69

▶ $NOC_4H_7$

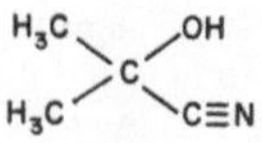

Varian 70

▶ $NOC_4H_9$

$(CH_3)_2C{=}NOCH_3$

J. Phys. Chem. 65:491 (1961)

▶ $NOC_4H_9$

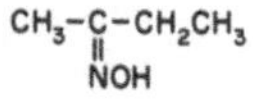

J. Phys. Chem. 65:491 (1961)

▶ $NOC_4H_9$

$CH_3CON(CH_3)_2$

Varian 421
Mol. Phys. 5:139 (1962)
J. Phys. Chem. 66:540 (1962)
Arkiv Kemi 16:373 (1961)

▶ $NOC_4H_9$

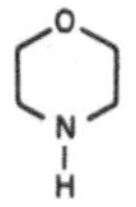

Varian 83

▶ $NOC_4H_9$

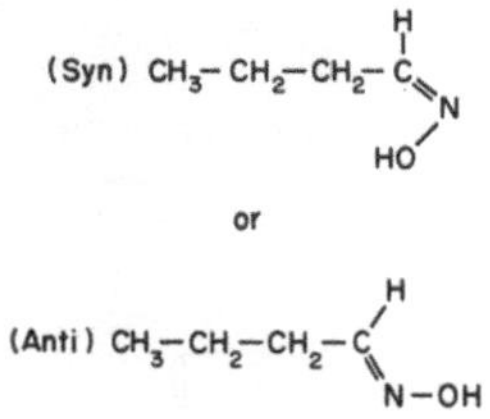


Varian 420

▶ $NOC_4H_{11}$

$(CH_3)_2NCH_2CH_2OH$

Varian 91

▶ $NOC_4H_{11}$

$CH_3CH_2NHCH_2CH_2OH$

Varian 92

▶ $NOC_5H_5$

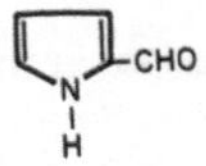

Varian 98
Can. J. Chem. 39:908 (1961)
Arkiv Kemi 18:133 (1961)

▶ $NOC_5H_5$

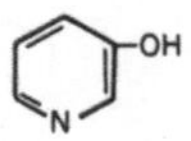

Varian 428

▶ $NOC_5H_5D_9 \cdot Cl$

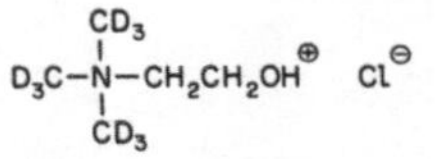

J. Am. Chem. Soc. 83:2484 (1961)

▶ $NOC_5H_7$

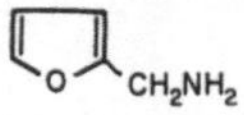

Varian 104

▶ $NOC_5H_9$

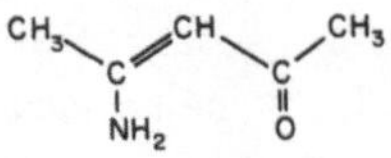

J. Am. Chem. Soc. 83:2099 (1961)

▶ $NOC_5H_9$

$CH_2{=}CHCON(CH_3)_2$

J. Phys. Chem. 66:540 (1962)

▶ $NOC_5H_9$

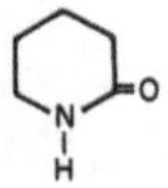

Chem. Ber. 95:2896 (1962)

▶ $NOC_5H_9$

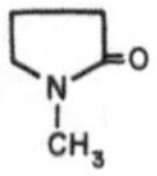

Varian 116

▶ $NOC_5H_{11}$

$(CH_3)_2C{=}NOC_2H_5$

J. Phys. Chem. 65:491 (1961)

▶ $NOC_5H_{11}$

$(CH_3CH_2)_2C{=}NOH$

J. Phys. Chem. 65:491 (1961)

▶ $NOC_5H_{11}$

$CH_3{-}C({=}NOH){-}CH(CH_3)_2$

J. Phys. Chem. 65:491 (1961)

▶ $NOC_5H_{11}$

$HCON(C_2H_5)_2$

Mol. Phys. 5:139 (1962)

▶ $NOC_5H_{11}$

$CH_3CH_2C({=}O){-}N(CH_3)_2$

J. Phys. Chem. 65:1066 (1961); 66:540 (1962)
J. Am. Chem. Soc. 84:13 (1962)

▶ $NOC_5H_{12}D_2 \cdot Cl$

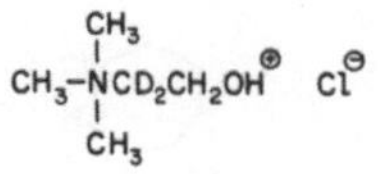

J. Am. Chem. Soc. 83:2484 (1961)

▶ $NOC_6H_7$

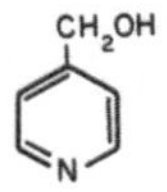

J. Chem. Soc. 44 (1961)

▶ $NOC_5H_{12}D_2 \cdot Cl$

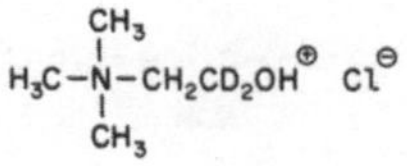

J. Am. Chem. Soc. 83:2484 (1961)

▶ $NOC_6H_7$

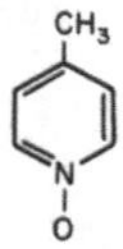

J. Chem. Soc. 44 (1961)

▶ $NOC_6H_5$

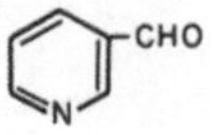

J. Chem. Phys. 36:266 (1962)

▶ $NOC_6H_7$

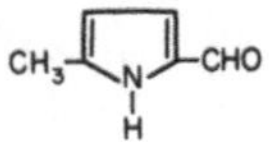

Arkiv Kemi 18:133 (1961)

▶ $NOC_6H_7$

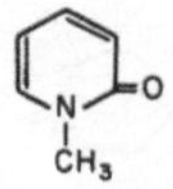

J. Chem. Soc. 860 (1961); 851 (1962)

▶ $NOC_6H_7$

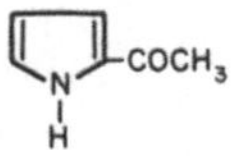

Arkiv Kemi 18:133 (1961)

▶ $NOC_6H_7$

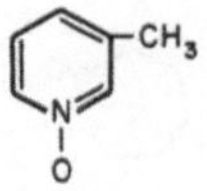

Varian 457
J. Chem. Soc. 46 (1961)

▶ $NOC_6H_7$

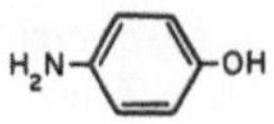

Can. J. Chem. 40:2124 (1962)

▶ $NOC_6H_7 \cdot HCl$

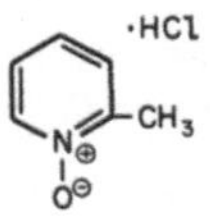

J. Org. Chem. 27:3858 (1962)

▶ $NOC_6H_7$

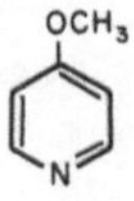

J. Chem. Soc. 44 (1961)

▶ $NOC_6H_9$

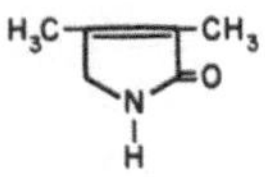

Ann. 654:165 (1962)

▶ $NOC_6H_{11}$

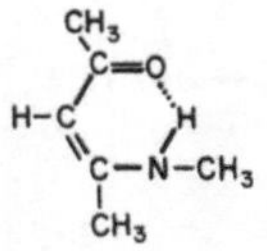

J. Am. Chem. Soc. 83:3914 (1961); 84:2691 (1962)

▶ $NOC_6H_{11}$

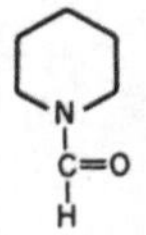

Chem. Ber. 95:2896 (1962)

▶ $NOC_6H_{11}$

Chem. Ber. 95:2896 (1962)

▶ $NOC_6H_{11}$

Chem. Ber. 95:2896 (1962)

▶ $NOC_6H_{11}$

Can. J. Chem. 40:178 (1962)

▶ $NOC_6H_{11}$

Can. J. Chem. 40:178 (1962)

▶ $NOC_6H_{11}$

Rec. Trav. Chim. 80:1211, 1323 (1961)

▶ $NOC_6H_{11}$

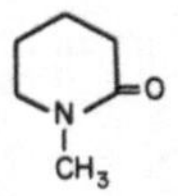

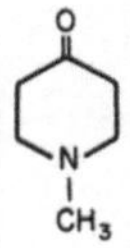

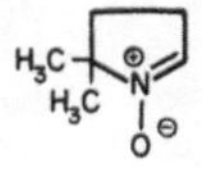

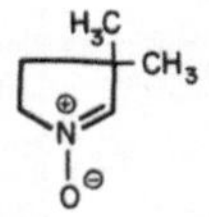

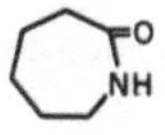

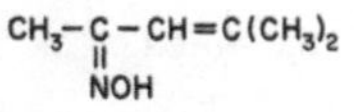

$CH_3-C(=NOH)-CH=C(CH_3)_2$

J. Phys. Chem. 65:491 (1961)

▶ $NOC_6H_{11}$

$(CH_3)_2CHNHC(=O)CH=CH_2$

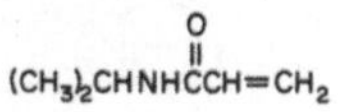

Varian 468

▶ $NOC_6H_{11} \cdot BF_3$

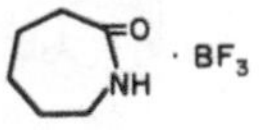

Rec. Trav. Chim. 80:1323 (1961)

▶ $NOC_6H_{11} \cdot HCl$

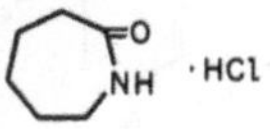

Rec. Trav. Chim. 80:1211 (1961)

▶ $NOC_6H_{13}$

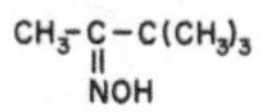

$CH_3-C(=NOH)-C(CH_3)_3$

J. Phys. Chem. 65:491 (1961)

▶ $NOC_6H_{13}$

$CH_3-C(=NOH)-CH(CH_3)CH_2CH_3$

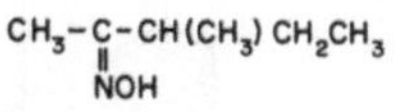

J. Phys. Chem. 65:491 (1961)

| | |
|---|---|
| ▶ $NOC_6H_{13}$<br>$(CH_3)_2CHCH_2CH_2CNH_2$ (C=O)<br>Varian 142 | ▶ $NOC_7H_9$<br>CN, H, H, H, O<br>Varian 173 |
| ▶ $NOC_7H_5$<br>N, CH, O<br>Varian 147 | ▶ $NOC_7H_9$<br>C–CH₃, O, N, CH₃<br>Varian 174 |
| ▶ $NOC_7H_7$<br>COCH₃, N<br>J. Chem. Phys. 36:266 (1962) | ▶ $NOC_7H_9$<br>CH₃, N, O, CH₃<br>J. Chem. Soc. 860 (1961) |
| ▶ $NOC_7H_7$<br>O, C–CH₃, N<br>J. Chem. Soc. 44 (1961) | ▶ $NOC_7H_9$<br>CH₃, N, O, CH₃<br>J. Chem. Soc. 860 (1961) |
| ▶ $NOC_7H_7$<br>N, C–CH₃, O<br>J. Chem. Phys. 37:2603 (1962) | ▶ $NOC_7H_9$<br>CH₃, O, N, CH₃<br>J. Chem. Soc. 860 (1961) |
| ▶ $NOC_7H_9$<br>CH₃, N, CH₃, O<br>J. Chem. Soc. 46 (1961) | ▶ $NOC_7H_9$<br>OCH₃, NH₂<br>Varian 172<br>Can. J. Chem. 40:1866 (1962) |
| ▶ $NOC_7H_9$<br>H₃C, N, CH, O, CH₃<br>Varian 170 | ▶ $NOC_7H_9$<br>NH₂, OCH₃<br>Varian 171<br>J. Chem. Phys. 37:2594 (1962) |

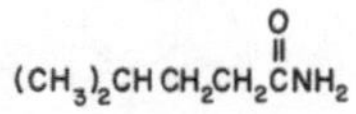

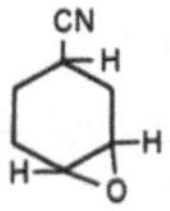

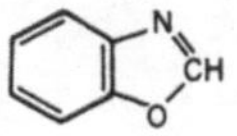

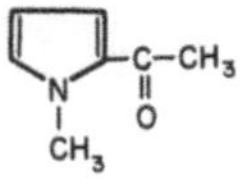

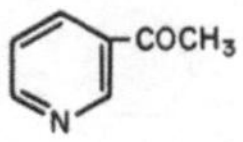

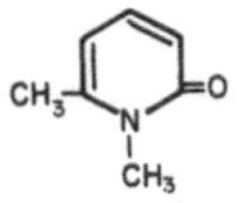

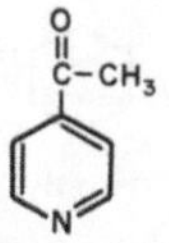

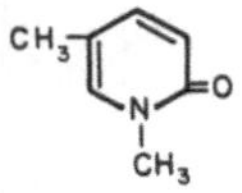

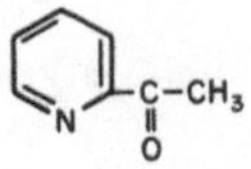

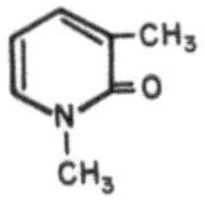

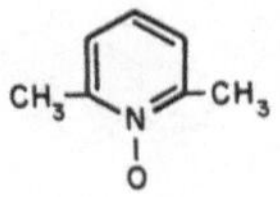

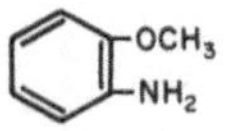

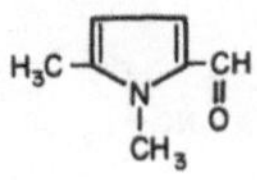

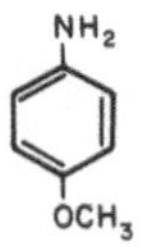

▶ $NOC_7H_{11}$

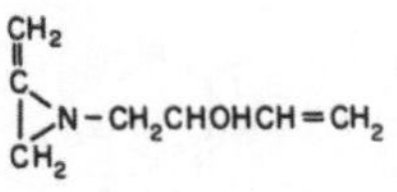

J. Org. Chem. 27:970 (1962)

▶ $NC_7H_{11}$

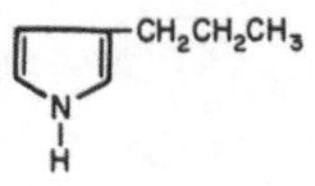

Varian 177

▶ $NOC_7H_{11}$

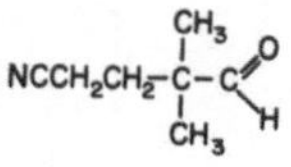

Varian 178

▶ $NOC_7H_{11}$

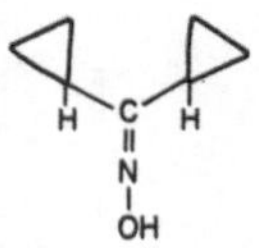

Varian 179

▶ $NOC_7H_{13}$

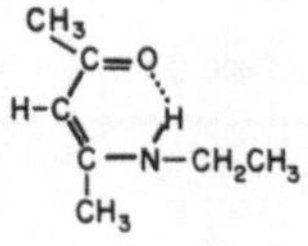

J. Am. Chem. Soc. 83:3914 (1961); 84:2691 (1962)

▶ $NOC_7H_{13}$

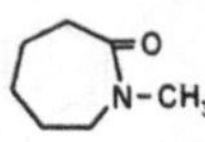

Rec. Trav. Chim. 80:1323 (1961)

▶ $NOC_7H_{13}$

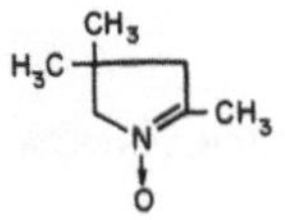

Can. J. Chem. 40:178 (1962)

▶ $NOC_7H_{13}$

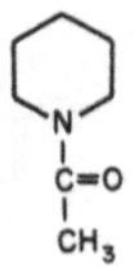

Chem. Ber. 95:2896 (1962)

▶ $NOC_7H_{13} \cdot BF_3$

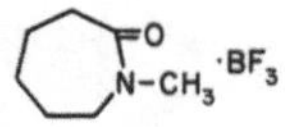

Rec. Trav. Chim. 80:1323 (1961)

▶ $NOC_7H_{15}$

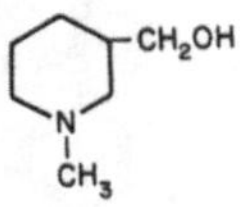

Chem. Ber. 95:2896 (1962)

▶ $NOC_7H_{15}$

$[(CH_3)_2CH]_2C{=}NOH$

J. Phys. Chem. 65:491 (1961)

▶ $NOC_8H_5$

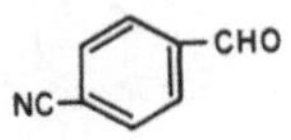

Can. J. Chem. 40:1073, 2331 (1962)

▶ $NOC_8H_7$

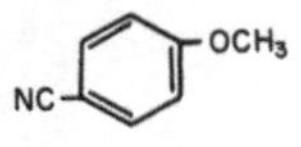

Can. J. Chem. 40:1866 (1962)
J. Chem. Phys. 37:2594 (1962)

▶ $NOC_8H_9$

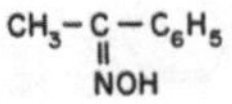


J. Phys. Chem. 65:491 (1961)

▶ $NOC_8H_9$

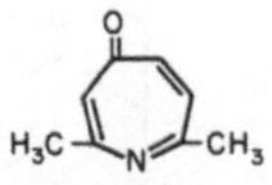


J. Am. Chem. Soc. 84:2260 (1962)

▶ $NOC_8H_9$

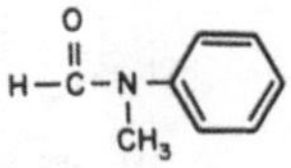


J. Am. Chem. Soc. 83:3729 (1961)

▶ $NOC_8H_{11}$

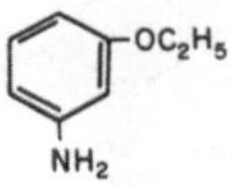


Can. J. Chem. 40:1866 (1962)

▶ $NOC_8H_{11}$

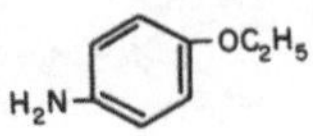


Varian 208
Can. J. Chem. 40:1866 (1962)

▶ $NOC_8H_{11}$

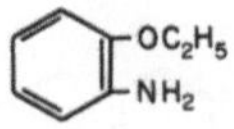


Can. J. Chem. 40:1866 (1962)

▶ $NOC_8H_{11}$

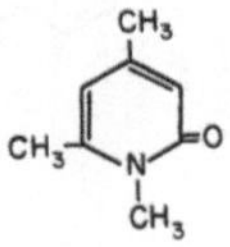


J. Chem. Soc. 860 (1961)

▶ $NOC_8H_{13}$

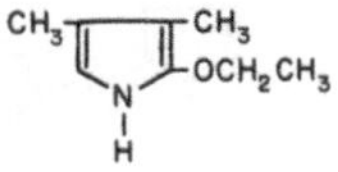


Ann. 654:165 (1962)

▶ $NOC_8H_{13}$

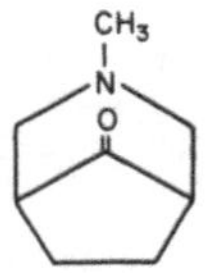


J. Am. Chem. Soc. 84:3139 (1962)

▶ $NOC_8H_{15}$

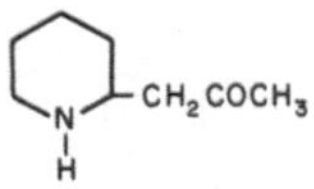


Bull. Soc. Chim. France 1993 (1961)

▶ $NOC_8H_{17}$

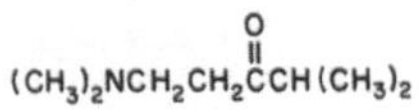


J. Org. Chem. 27:4706 (1962)

▶ $NOC_8H_{17}$

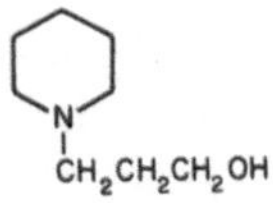


Chem. Ber. 95:2896 (1962)

▶ $NOC_8H_{17}$

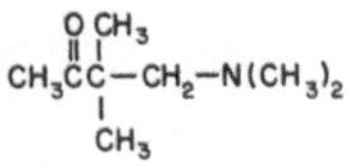


J. Org. Chem. 27:4706 (1962)

▶ $NOC_9H_7$

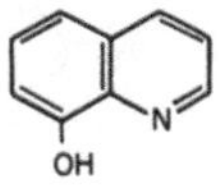


Can. J. Chem. 39:2318 (1961)

▶ $NOC_9H_9$

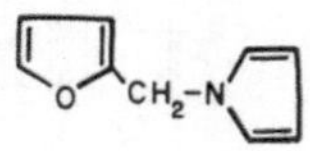

Varian 525

▶ $NOC_9H_{11}$

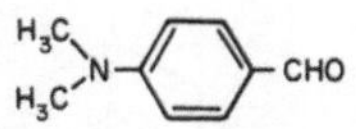

Varian 238
Can. J. Chem. 40:1073, 2331 (1962)
MCA Serial No. 1

▶ $NOC_9H_{11}$

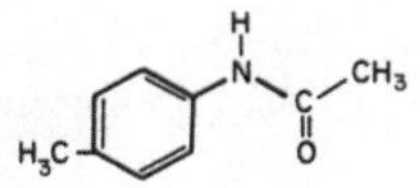

Varian 239

▶ $NOC_9H_{11}$

$C_6H_5CON(CH_3)_2$

J. Phys. Chem. 66:540 (1962)

▶ $NOC_9H_{11}$

$C_2H_5-C(=NOH)-C_6H_5$

Varian 533
J. Phys. Chem. 65:491 (1961)

▶ $NOC_9H_{11}$

$CH_3-C(=NOH)-CH_2-C_6H_5$

J. Phys. Chem. 65:491 (1961)

▶ $NOC_9H_{13}$

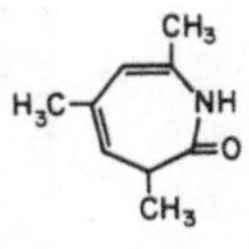

J. Am. Chem. Soc. 84:4988 (1962)

▶ $NOC_9H_{13} \cdot HCl$

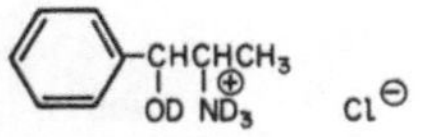

Varian 565

▶ $NOC_9H_{15}$

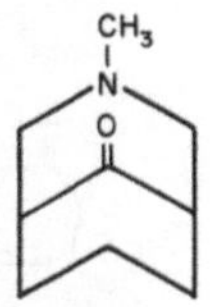

J. Am. Chem. Soc. 84:3139 (1962)

▶ $NOC_9H_{15}$

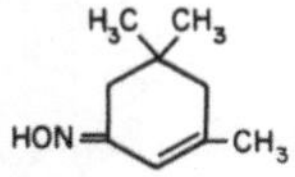

syn and anti

Chem. & Ind. 41 (1962)

▶ $NOC_9H_{17}$

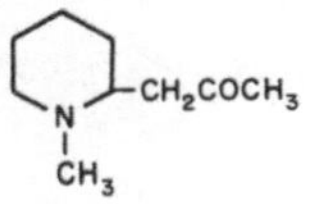

Bull. Soc. Chim. France 1993 (1961)

▶ $NOC_9H_{19}$

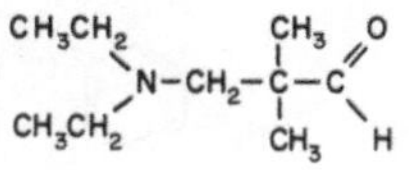

Varian 548

▶ $NOC_{10}H_9$

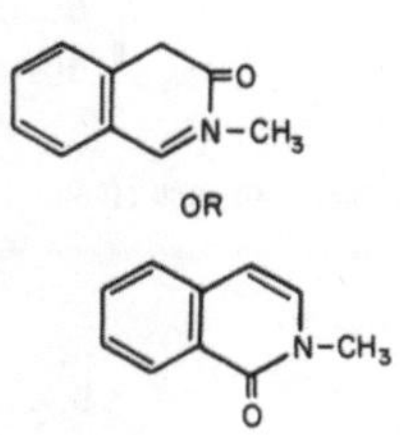

J. Am. Chem. Soc. 83:3729 (1961)

▶ $NOC_{10}H_9$

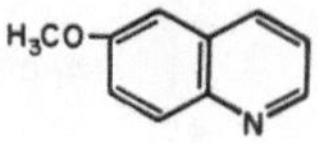


Varian 249

▶ $NOC_{10}H_{11}$

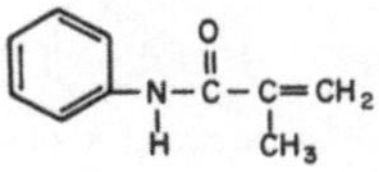


Varian 555

▶ $NOC_{10}H_{13}$

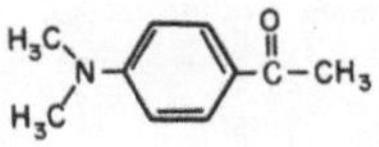


Can. J. Chem. 40:2331 (1962)

▶ $NOC_{10}H_{13}$

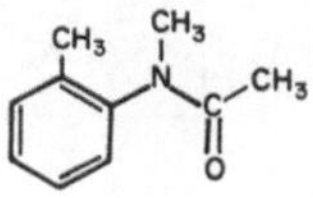


Varian 264

▶ $NOC_{10}H_{13}$

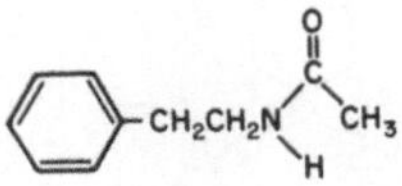


Varian 265

▶ $NOC_{10}H_{13}$

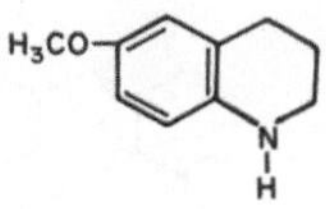


Varian 266

▶ $NOC_{10}H_{15}$

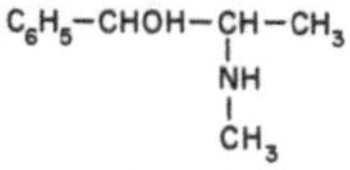


(–) EPHEDRINE
AND
(+)ψ-EPHEDRINE

Can. J. Chem. 39:2536 (1961)

▶ $NOC_{10}H_{15}$

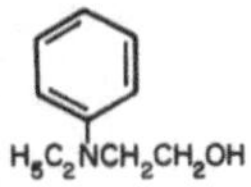


Varian 569

▶ $NOC_{10}H_{17}$

J. Am. Chem. Soc. 84:3139 (1962)

▶ $NOC_{10}H_{17}$

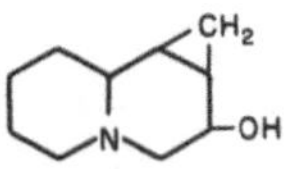


Chem. Ber. 95:2365 (1962)

▶ $NOC_{10}H_{23}$

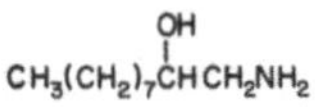


Varian 575

▶ $NOC_{11}H_{11}$

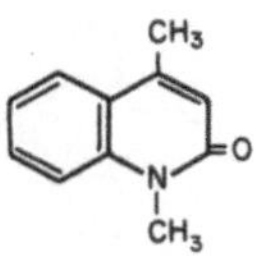


J. Am. Chem. Soc. 83:3729 (1961)

▶ $NOC_{11}H_{13}$

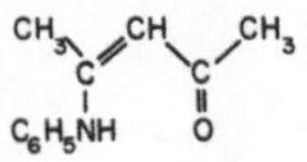

J. Am. Chem. Soc. 83:2099 (1961)

▶ $NOC_{12}H_{15}$

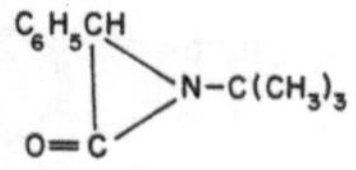

J. Am. Chem. Soc. 84:4975 (1962)

▶ $NOC_{11}H_{15}$

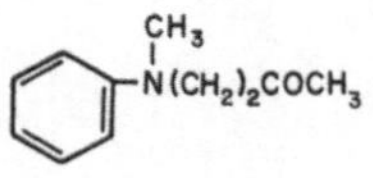

Varian 584

▶ $NOC_{12}H_{17}$

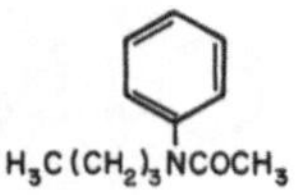

Varian 601

▶ $NOC_{11}H_{15}$

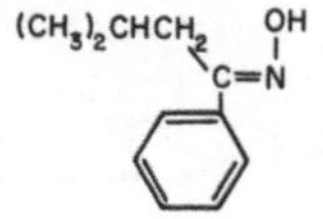

Varian 585

▶ $NOC_{12}H_{19}$

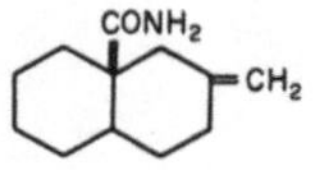

J. Org. Chem. 27:2013 (1962)

▶ $NOC_{11}H_{17}$

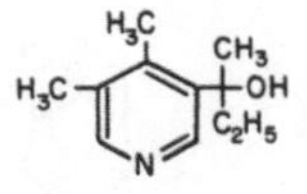

Can. J. Chem. 40:1143 (1962)

▶ $NOC_{12}H_{19}$

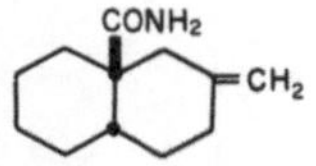

J. Org. Chem. 27:2013 (1962)

▶ $NOC_{11}H_{19}$

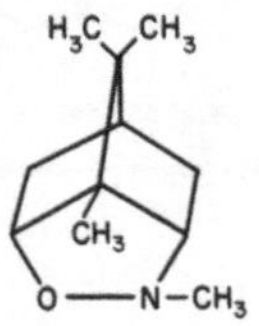

J. Am. Chem. Soc. 84:4360 (1962)

▶ $NOC_{12}H_{21}$

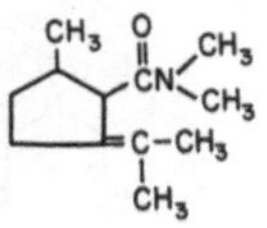

J. Am. Chem. Soc. 84:2775 (1962)

▶ $NOC_{12}H_{23}$

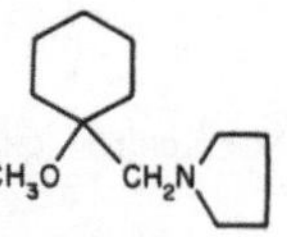

(and PERCHLORATE)

J. Am. Chem. Soc. 84:4806 (1962)

▶ $NOC_{12}H_{15}$

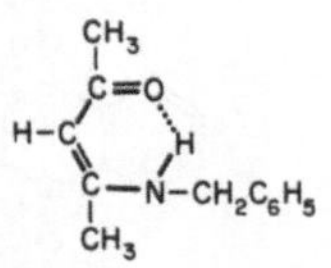

J. Am. Chem. Soc. 83:3914 (1961); 84:2691 (1962)

▶ $NOC_{13}H_{9}$

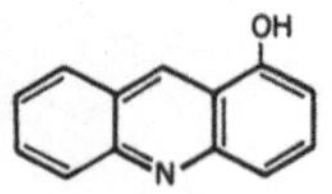

Varian 605

▶ $NOC_{13}H_{13}$

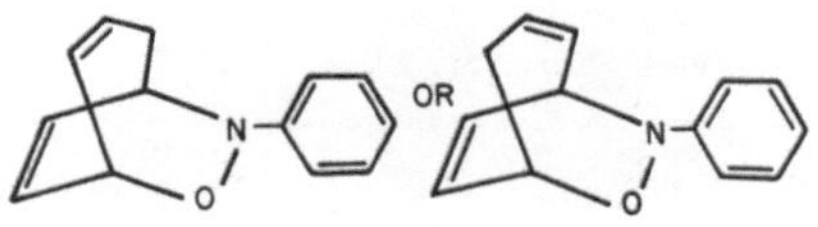


Tetrahedron 12:10 (1961)

▶ $NOC_{13}H_{13}$

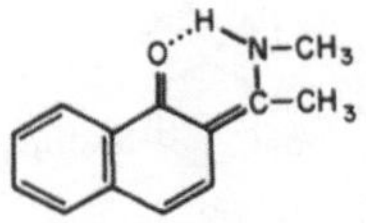


J. Am. Chem. Soc. 83:3914 (1961)

▶ $NOC_{13}H_{13}$

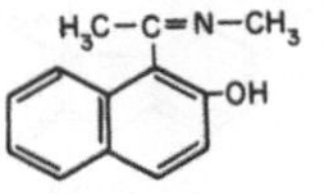


J. Am. Chem. Soc. 83:3914 (1961)

▶ $NOC_{13}H_{13}$

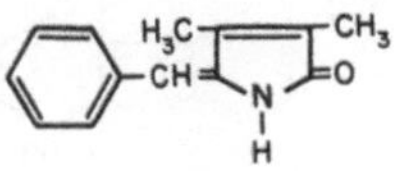


Ann. 654:165 (1962)

▶ $NOC_{13}H_{15}$

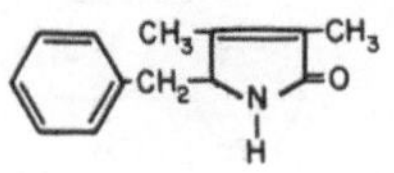


Ann. 654:165 (1962)

▶ $NOC_{13}H_{15}$

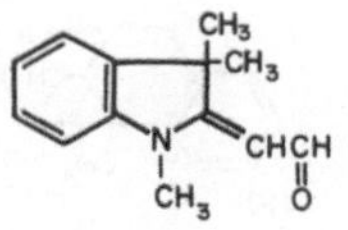


Varian 610

▶ $NOC_{13}H_{17}$

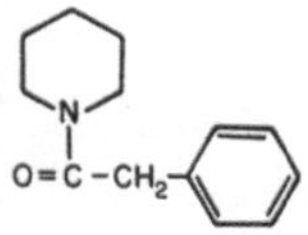


Chem. Ber. 95:2896 (1962)

▶ $NOC_{13}H_{17}$

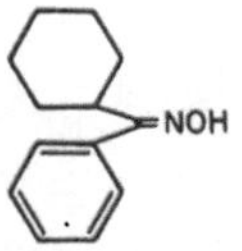


Varian 614
Bull. Chem. Soc. Japan 35:1986 (1962)

▶ $NOC_{13}H_{25} \cdot HClO_4$

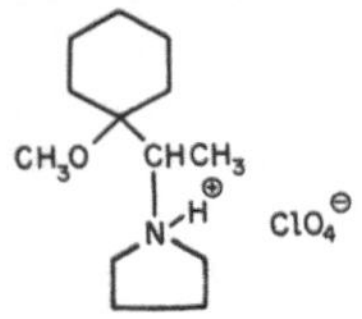


J. Am. Chem. Soc. 84:4806 (1962)

▶ $NOC_{14}H_{11}$

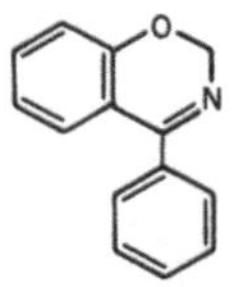


Can. J. Chem. 40:883 (1962)

▶ $NOC_{14}H_{13}$

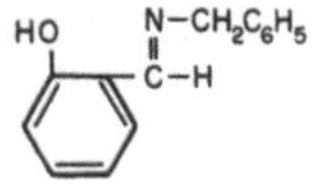


J. Am. Chem. Soc. 83:3914 (1961)

▶ $NOC_{14}H_{13}$

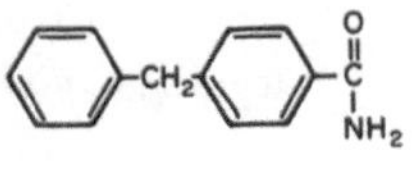


Varian 309

▶ $NOC_{14}H_{15}$

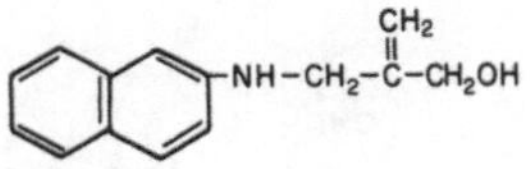

J. Org. Chem. 27:4137 (1962)

▶ $NOC_{14}H_{17}$

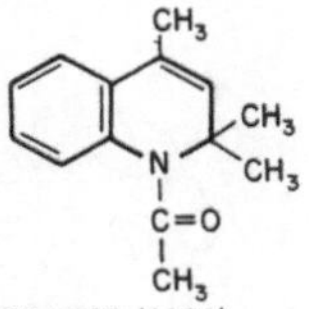

J. Org. Chem. 26:1287 (1961)

▶ $NOC_{14}H_{19}$

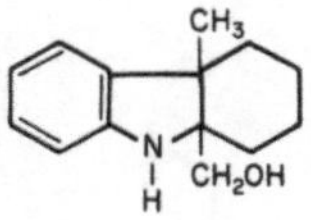

Helv. Chim. Acta 45:1992 (1962)

▶ $NOC_{14}H_{21}$

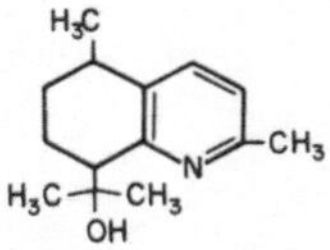

Can. J. Chem. 40:258 (1962)

▶ $NOC_{14}H_{27} \cdot HClO_4$

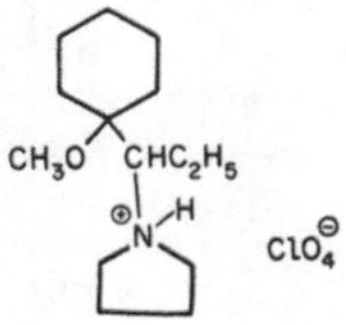

J. Am. Chem. Soc. 84:4806 (1962)

▶ $NOC_{15}H_{15}$

$(C_6H_5CH_2)_2C{=}NOH$

J. Phys. Chem. 65:491 (1961)

▶ $NOC_{15}H_{15}$

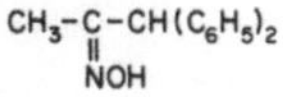

$CH_3{-}C({=}NOH){-}CH(C_6H_5)_2$

J. Phys. Chem. 65:491 (1961)

▶ $NOC_{15}H_{17}$

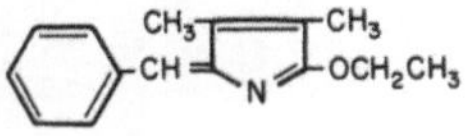

Ann. 654:165 (1962)

▶ $NOC_{15}H_{19}$

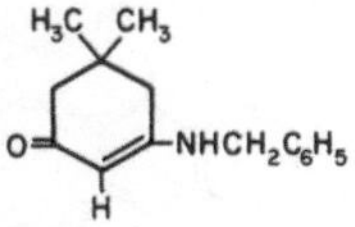

J. Am. Chem. Soc. 84:2691 (1962)

▶ $NOC_{15}H_{19}$

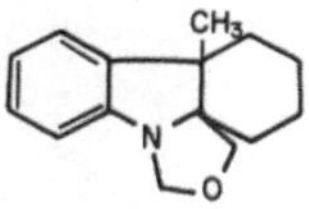

Helv. Chim. Acta 45:1992 (1962)

▶ $NOC_{15}H_{21}$

Helv. Chim. Acta 45:1992 (1962)

▶ $NOC_{15}H_{21}$

J. Org. Chem. 27:2146 (1962)

▶ $NOC_{15}H_{21}$

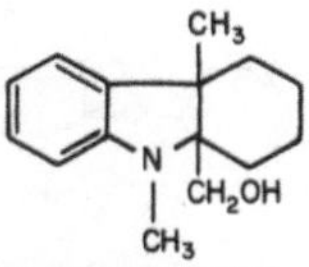

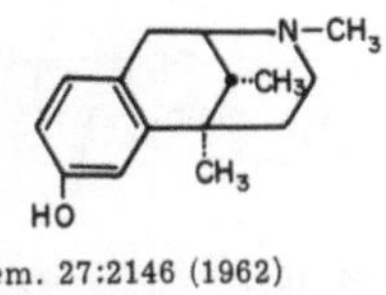

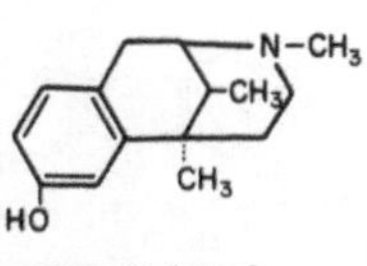

J. Org. Chem. 27:2146 (1962)

▶ $NOC_{15}H_{22} \cdot Cl$

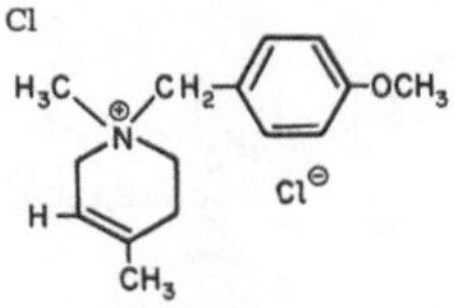

J. Org. Chem. 26:2593 (1961)

▶ $NOC_{16}H_{25}$

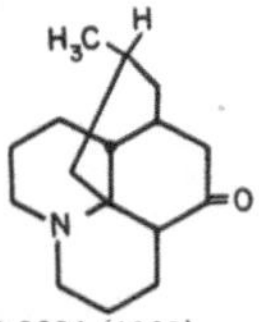

Can. J. Chem. 40:2094 (1962)

▶ $NOC_{15}H_{23}$

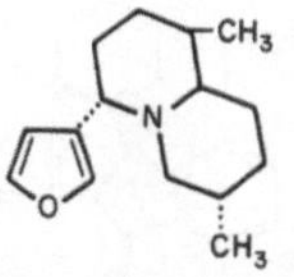

Bull. Chem. Soc. Japan 35:1335, 1494 (1962)

▶ $NOC_{16}H_{25}$

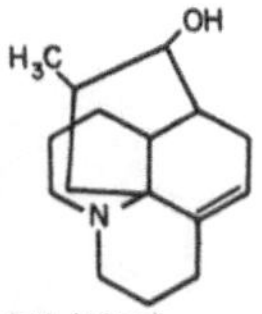

Tetrahedron 15:174 (1961)

▶ $NOC_{15}H_{23}$

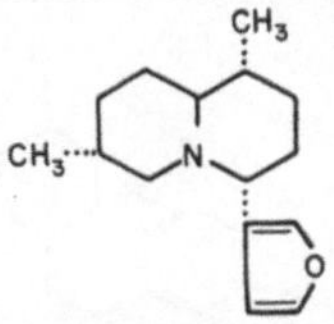

Bull. Chem. Soc. Japan 35:1494 (1962)

▶ $NOC_{17}H_{25}$

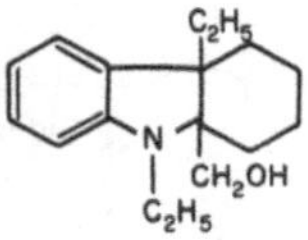

Helv. Chim. Acta 45:1992 (1962)

▶ $NOC_{16}H_{15}$

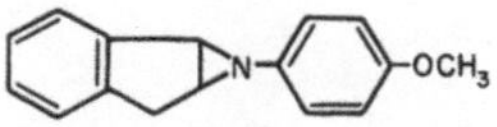

J. Chem. Soc. 1638 (1962)

▶ $NOC_{17}H_{27}$

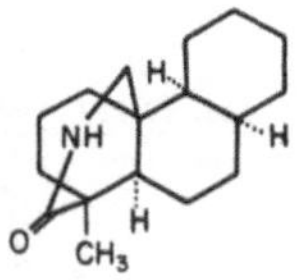

Proc. Chem. Soc. 461 (1961)

▶ $NOC_{16}H_{15}$

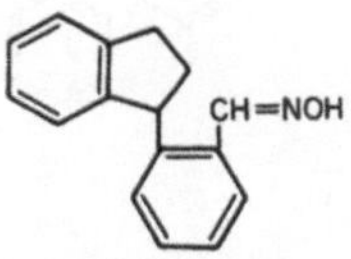

J. Am. Chem. Soc. 84:3531 (1962)

▶ $NOC_{16}H_{15}$

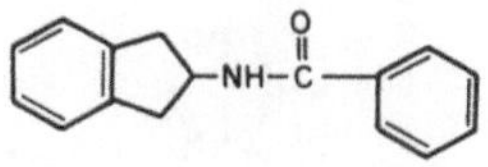

J. Org. Chem. 27:4465 (1962)

▶ $NOC_{18}H_{15}$

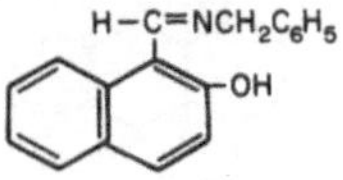

J. Am. Chem. Soc. 83:3914 (1961)

▶ $NOC_{16}H_{17}$

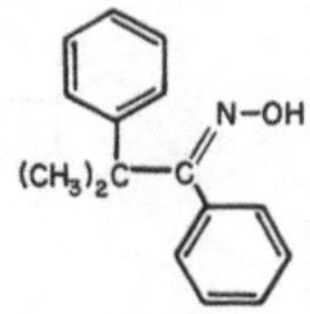

Varian 658

▶ $NOC_{18}H_{19}$

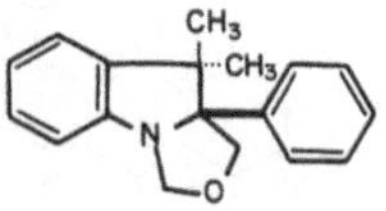

Helv. Chim. Acta 45:1992 (1962)

▶ $NOC_{18}H_{19}$

$CH_3C(=O)CH{=}C(CH_3)NHCH(C_6H_5)_2$

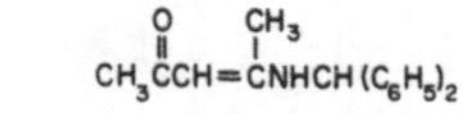

J. Am. Chem. Soc. 84:2691 (1962)

▶ $NOC_{21}H_{19}$

$(CH_3OC_6H_4)(C_6H_5)C{=}NC_6H_4CH_3$

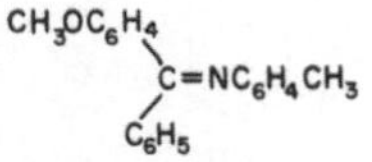

J. Am. Chem. Soc. 83:3474 (1961)

▶ $NOC_{19}H_{17}$

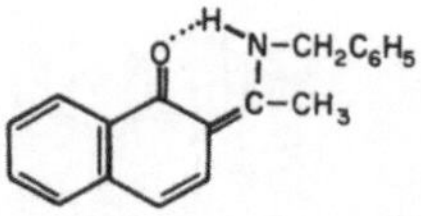

J. Am. Chem. Soc. 83:3914 (1961)

▶ $NOC_{21}H_{21}$

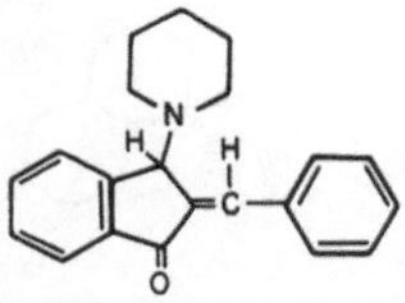

J. Org. Chem. 27:3039 (1962)

▶ $NOC_{20}H_{15}$

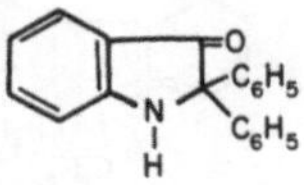

J. Am. Chem. Soc. 83:4792 (1961)

▶ $NOC_{21}H_{27}$

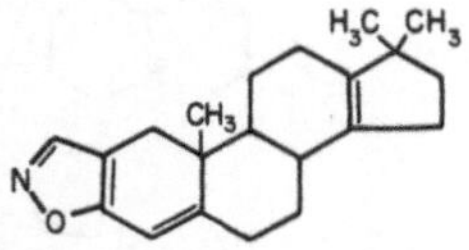

Chem. & Ind. 1984 (1962)

▶ $NOC_{20}H_{15}$

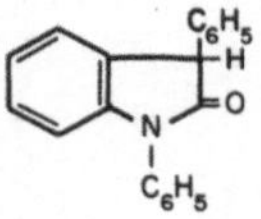

J. Am. Chem. Soc. 83:4792 (1961)

▶ $NOC_{20}H_{15}$

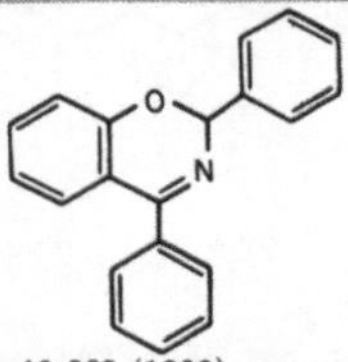

Can. J. Chem. 40:883 (1962)

▶ $NOC_{22}H_{17}$

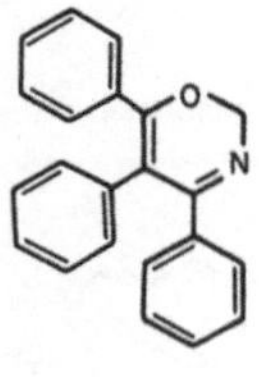

Can. J. Chem. 40:883 (1962)

▶ $NOC_{20}H_{31}$

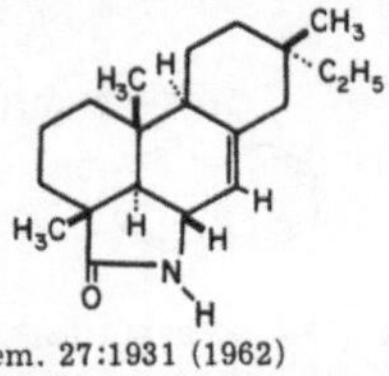

J. Org. Chem. 27:1931 (1962)

▶ $NOC_{22}H_{19}$

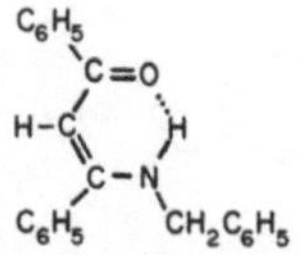

J. Am. Chem. Soc. 83:3914 (1961); 84:2691 (1962)

▶ $NOC_{20}H_{31}$

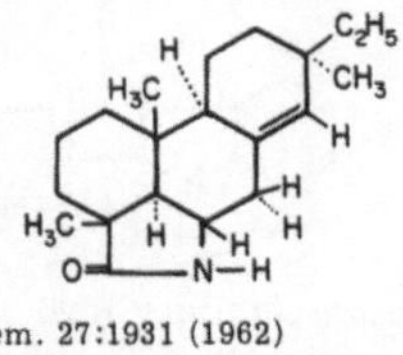

J. Org. Chem. 27:1931 (1962)

▶ $NOC_{22}H_{37}$

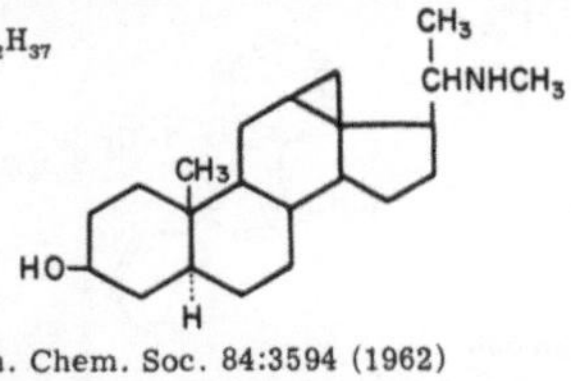

J. Am. Chem. Soc. 84:3594 (1962)

▶ $NOC_{23}H_{39}$

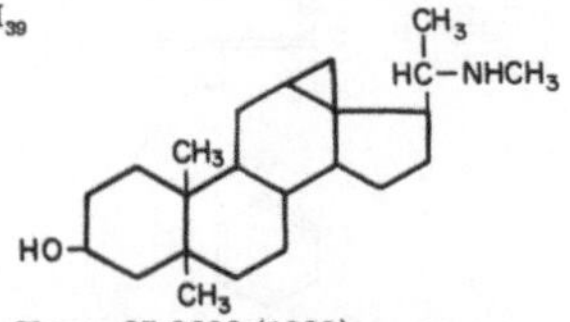


J. Org. Chem. 27:3628 (1962)

▶ $NOC_{24}H_{37}$

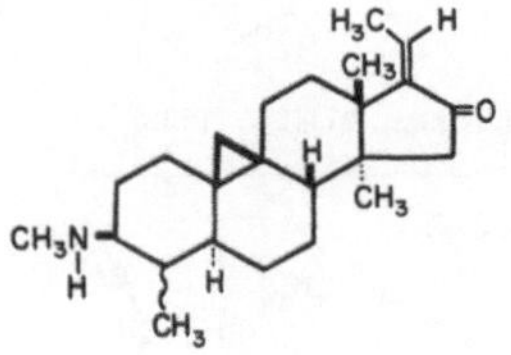


J. Am. Chem. Soc. 84:4590 (1962)

▶ $NOC_{24}H_{37}$

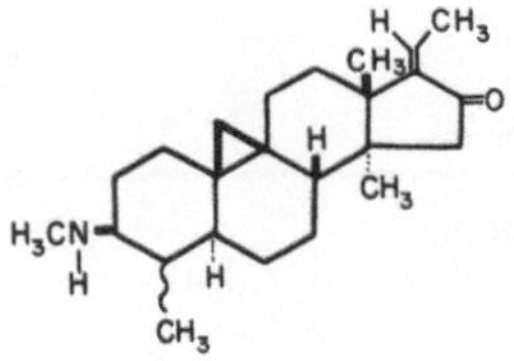


J. Am. Chem. Soc. 84:4590 (1962)

▶ $NOC_{24}H_{39}$

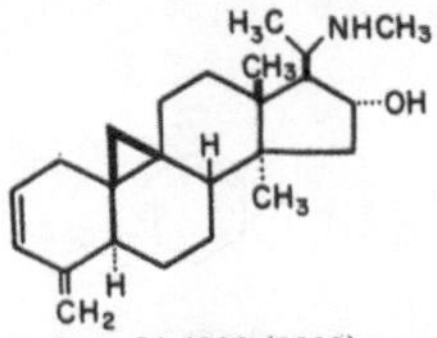


J. Am. Chem. Soc. 84:4590 (1962)

▶ $NOC_{28}H_{45}$

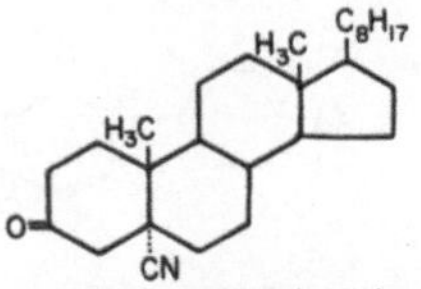


Bull. Soc. Chim. France 2444 (1961)

▶ $NOC_{29}H_{45}$

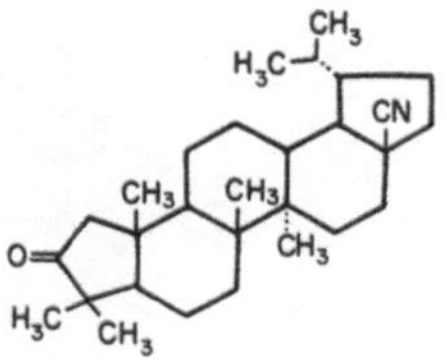


Bull. Soc. Chim. Franc 2444 (1961); 1137 (1962)

▶ $NOC_{30}H_{47}$

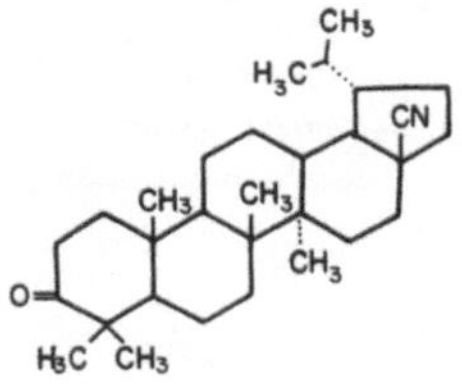


Bull. Soc. Chim. France 2444 (1961); 1137 (1962)

▶ $NOC_{30}H_{51}$

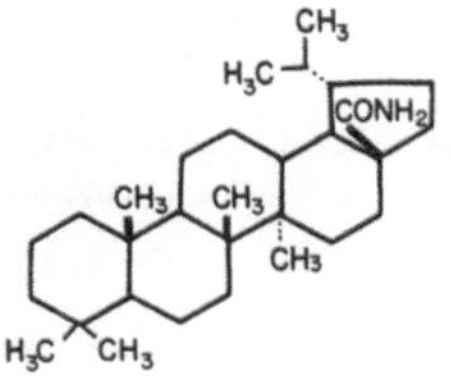


Bull. Soc. Chim. France 1137 (1962)

▶ $NOC_{33}H_{52} \cdot Br$

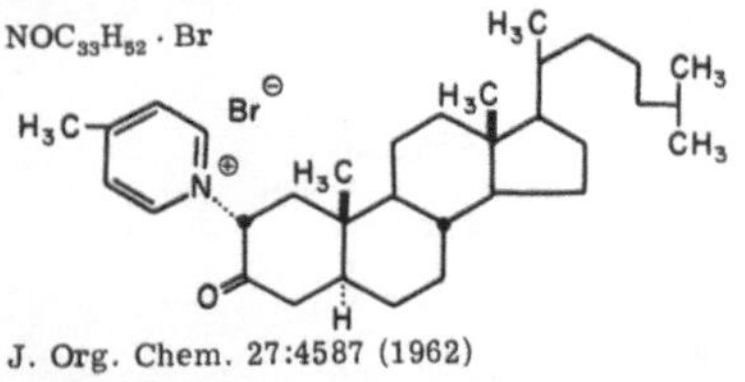


J. Org. Chem. 27:4587 (1962)

▶ $NOC_{33}H_{57}$

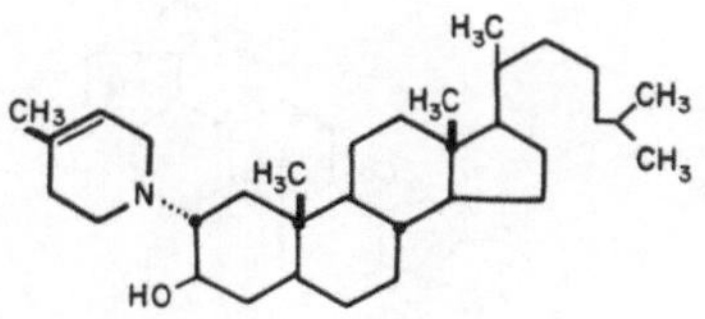

J. Org. Chem. 27:4587 (1962)

▶ $NO_2BrC_{11}H_{10}$

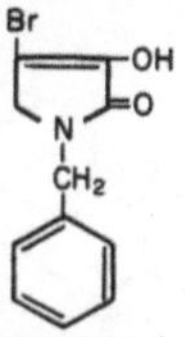

J. Org. Chem. 27:3078 (1962)

▶ $NO_2BrC_{12}H_{12}$

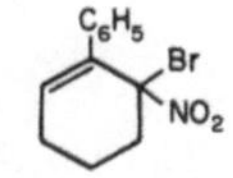

J. Org. Chem. 27:4243 (1962)

▶ $NOC_{33}H_{57}$

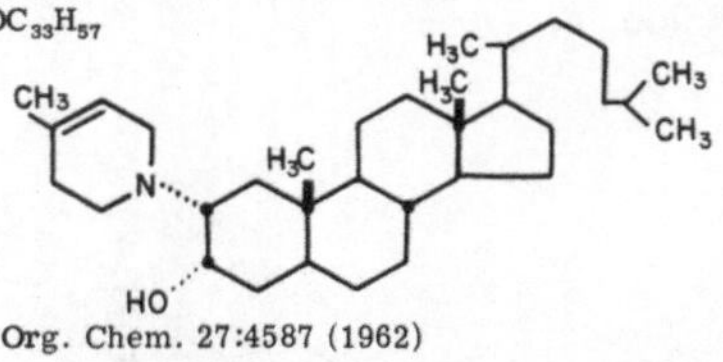

J. Org. Chem. 27:4587 (1962)

▶ $NO_2BrC_{19}H_{20}$

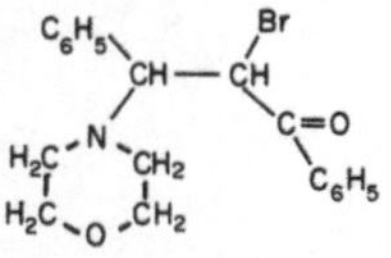

J. Am. Chem. Soc. 84:4299 (1962)

▶ $NO_2BrC_6H_4$

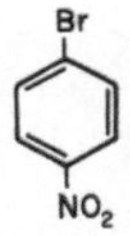

J. Chem. Phys. 37:2594 (1962)

▶ $NO_2ClC_3H_6$

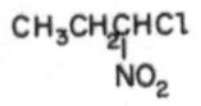

Varian 385

▶ $NO_2BrC_6H_4$

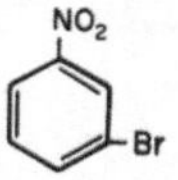

Acta Chem. Scand. 16:1031 (1962)

▶ $NO_2ClC_6H_4$

Varian 122
J. Chem. Phys. 37:2594 (1962)
Helv. Chim. Acta 45:568 (1962)

▶ $NO_2BrC_7H_{10}$

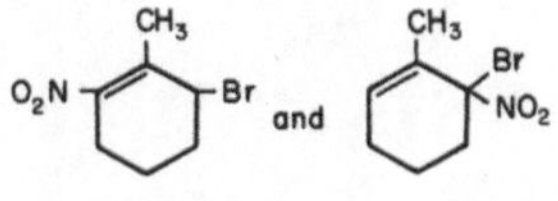

J. Org. Chem. 27:4249 (1962)

▶ $NO_2ClC_6H_4$

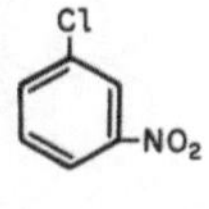

Helv. Chim. Acta 45:568 (1962)

▶ $NO_2BrC_{10}H_8$

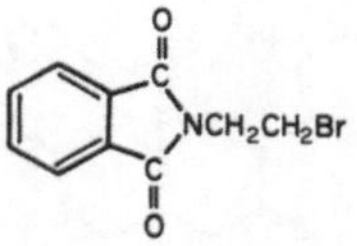

Varian 246

▶ $NO_2ClC_{12}H_{12}$

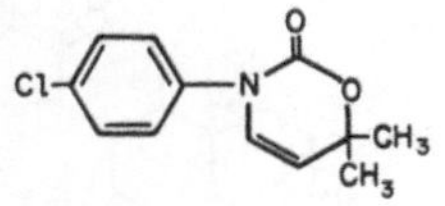

J. Org. Chem. 27:2927 (1962)

▶ $NO_2ClC_{12}H_{14}$

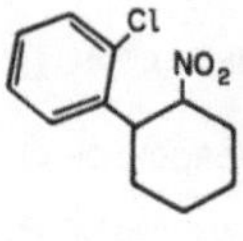


cis and trans

J. Org. Chem. 27:3008 (1962)

▶ $NO_2ClC_{12}H_{14}$

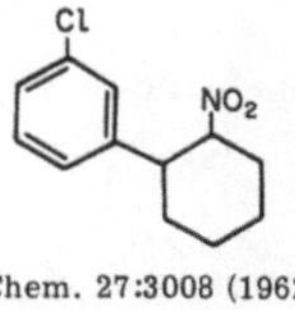


cis and trans

J. Org. Chem. 27:3008 (1962)

▶ $NO_2ClC_{12}H_{14}$

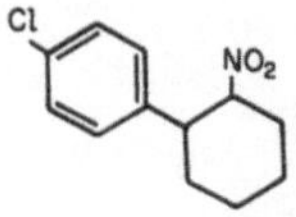


cis and trans

J. Org. Chem. 27:3008 (1962)

▶ $NO_2ClC_{21}H_{26}$

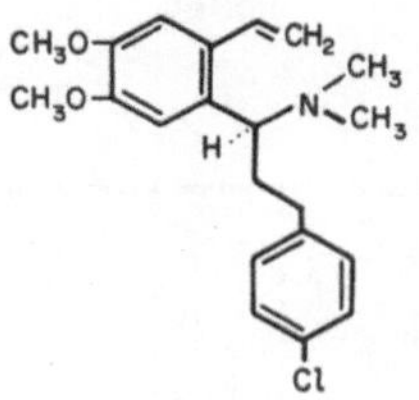


Helv. Chim. Acta 45:2590 (1962)

▶ $NO_2ClC_{21}H_{28}$

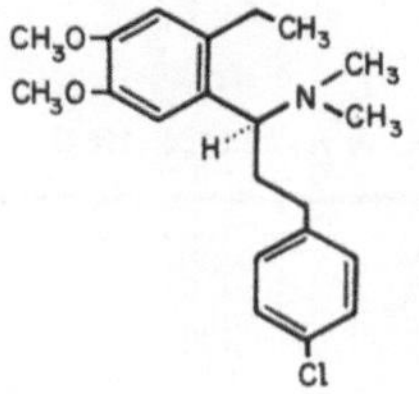


Helv. Chim. Acta 45:2590 (1962)

▶ $NO_2Cl_4C_6H$

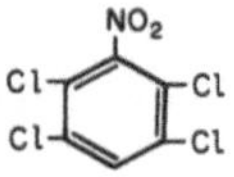


Can. J. Chem. 40:1759 (1962)

▶ $NO_2FC_6H_4$

Helv. Chim. Acta 45:568 (1962)

▶ $NO_2F_2C_{12}H_{11}$

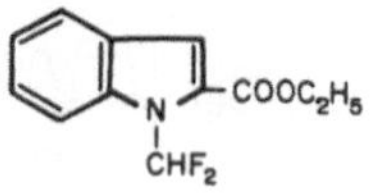


J. Chem. Soc. 3829 (1962)

▶ $NO_2CH_2D$

$CH_2DNO_2$

J. Chem. Phys. 37:3012 (1962)

▶ $NO_2CH_3$

$CH_3{-}NO_2$

J. Chem. Phys. 34:1099 (1961)
J. Am. Chem. Soc. 83:4726 (1961)
Bull. Chem. Soc. Japan 34:143 (1961)
J. Phys. Chem. 66:2653 (1962)

▶ $NO_2C_2H_3$

$CH_2{=}CHNO_2$

Can. J. Chem. 40:2 (1962)
J. Phys. Chem. 66:2653 (1962)

▶ $NO_2C_2H_5$

$H_2NCH_2CO_2H$

J. Am. Chem. Soc. 84:4650 (1962)

▶ $NO_2C_2H_5$

$C_2H_5-NO_2$

Varian 374
J. Chem. Phys. 34:1099 (1961)

▶ $NO_2C_3H_5$

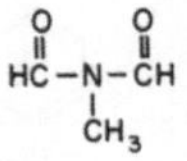

J. Am. Chem. Soc. 84:571 (1962)

▶ $NO_2C_3H_7$

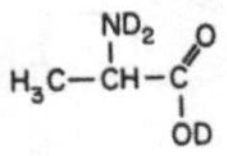

Varian 393

▶ $NO_2C_3H_7$

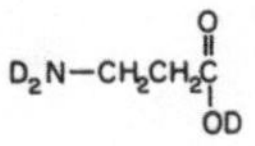

Varian 394

▶ $NO_2C_3H_7$

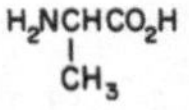

J. Am. Chem. Soc. 84:4650 (1962)

▶ $NO_2C_3H_7$

$CH_3CH_2CH_2-NO_2$

Varian 42
J. Chem. Phys. 34:1094, 1099 (1961)

▶ $NO_2C_3H_7$

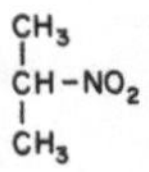

Varian 41
J. Chem. Phys. 34:1099 (1961)

▶ $NO_2C_3H_7$

$CH_3NHCH_2COOH$

AND HYDROCHLORIDE

J. Chem. Phys. 36:3103 (1962)

▶ $NO_2C_4H_5$

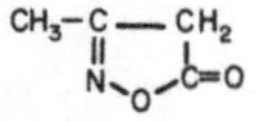

J. Org. Chem. 27:4305 (1962)
Tetrahedron 18:777 (1962)

▶ $NO_2C_4H_5$

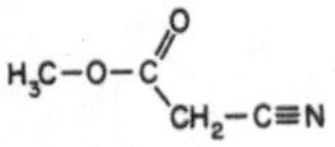

Varian 57

▶ $NO_2C_4H_7$

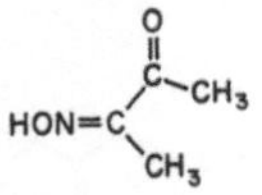

Varian 72

▶ $NO_2C_4H_9$

$(CH_3)_2NCH_2CO_2H$

J. Am. Chem. Soc. 84:4650 (1962)

▶ $NO_2C_4H_9$

$CH_3NH-CH_2COOCH_3$

J. Chem. Phys. 36:3103 (1962)

▶ $NO_2C_4H_9$

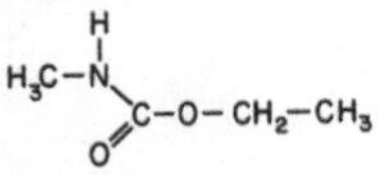

Varian 85

▶ $NO_2C_4H_9$

$H_3C\overset{O}{\overset{\|}{C}}OCH_2CH_2NH_2$

Can. J. Chem. 40:978 (1962)

▶ $NO_2C_4H_9$

$H_3C-\overset{O}{\overset{\|}{C}}-\overset{H}{\overset{|}{N}}-CH_2-CH_2-OH$

Can. J. Chem. 40:978 (1962)

▶ $NO_2C_4H_9$

$CH_3CH_2\underset{NO_2}{\underset{|}{C}}HCH_3$

Varian 84

▶ $NO_2C_5H_5$

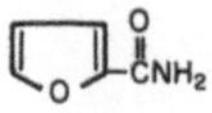


Can. J. Chem. 39:909 (1961)

▶ $NO_2C_5H_6D$

$CN$
$|$
$CHD$
$|$
$COOC_2H_5$

J. Chem. Phys. 37:3012 (1962)

▶ $NO_2C_5H_7$

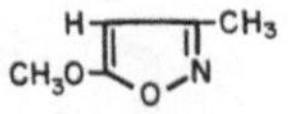


Tetrahedron 18:777 (1962)

▶ $NO_2C_5H_7$

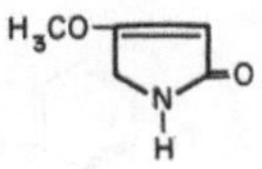


Varian 105

▶ $NO_2C_5H_7$

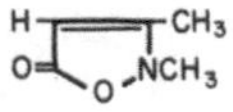


Tetrahedron 18:777 (1962)

▶ $NO_2C_5H_7$

$NCCH_2COOC_2H_5$

J. Am. Chem. Soc. 83:4726 (1961)

▶ $NO_2C_5H_7$

$N\equiv CCH_2CH_2C(=O)OCH_3$

Varian 106

▶ $NO_2C_5H_9$

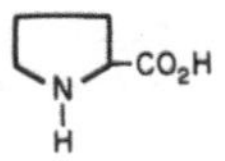


J. Am. Chem. Soc. 84:4650 (1962)

▶ $NO_2C_5H_9$

$CH_3\overset{O}{\overset{\|}{C}}\overset{CH_3}{\overset{|}{N}}—\overset{O}{\overset{\|}{C}}CH_3$

J. Am. Chem. Soc. 84:571 (1962)

▶ $NO_2C_5H_9$

$CH_3ON{=}\underset{CH_3}{\underset{|}{C}}COCH_3$

Varian 117

▶ $NO_2C_5H_9$

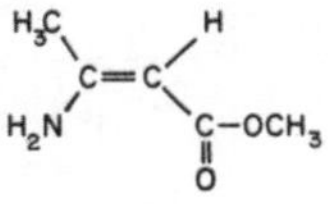

Varian 442

▶ $NO_2C_6H_9$

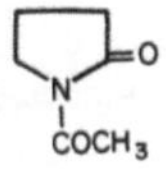

Varian 465

▶ $NO_2C_5H_{11}$

$CH_3CH_2OCON(CH_3)_2$

J. Phys. Chem. 66:540 (1962)

▶ $NO_2C_6H_9$

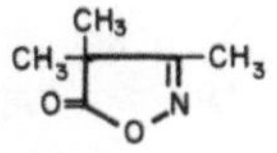

Tetrahedron 18:777 (1962)

▶ $NO_2C_6H_5$

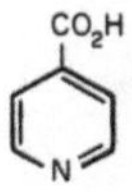

J. Chem. Soc. 44 (1961)

▶ $NO_2C_6H_{13}$

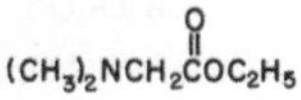

$(CH_3)_2NCH_2\overset{O}{\overset{\|}{C}}OC_2H_5$

Varian 480

▶ $NO_2C_6H_5$

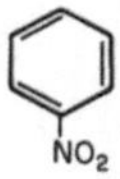

J. Phys. Chem. 66:2653 (1962)

▶ $NO_2C_6H_7$

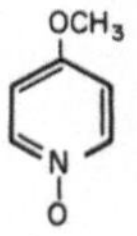

J. Chem. Soc. 44 (1961)

▶ $NO_2C_7H_7$

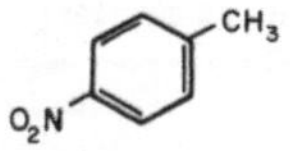

Bull. Chem. Soc. Japan 34:143 (1961)
Can. J. Chem. 40:2331 (1962)
Helv. Chim. Acta 45:568 (1962)
J. Chem. Phys. 37:2594 (1962)

▶ $NO_2C_6H_7$

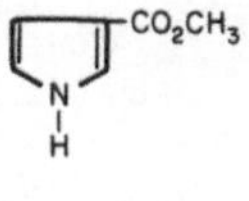

Arkiv Kemi 18:133 (1961)

▶ $NO_2C_7H_7$

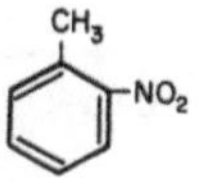

Bull. Chem. Soc. Japan 34:143 (1961)

▶ $NO_2C_6H_9$

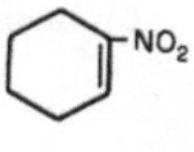

J. Org. Chem. 27:2005 (1962)

▶ $NO_2C_7H_7$

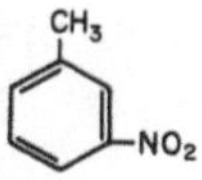

Bull. Chem. Soc. Japan 34:143 (1961)

▶ $NO_2C_7H_7$

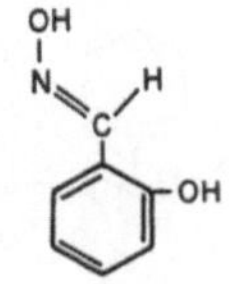

Varian 156

▶ $NO_2C_7H_9$

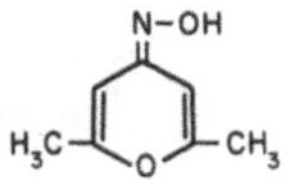

Tetrahedron Letters 670 (1961)

▶ $NO_2C_7H_7$

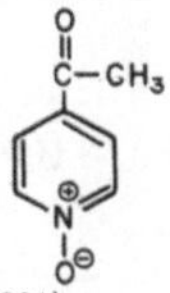

J. Chem. Soc. 44 (1961)

▶ $NO_2C_7H_{11}$

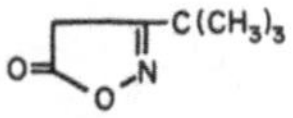

Tetrahedron 18:777 (1962)

▶ $NO_2C_7H_7$

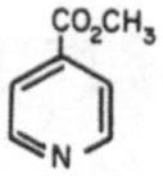

J. Chem. Soc. 44 (1961)

▶ $NO_2C_7H_{11}$

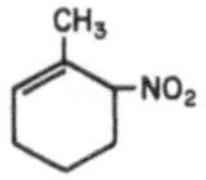

J. Org. Chem. 27:4249 (1962)

▶ $NO_2C_7H_9$

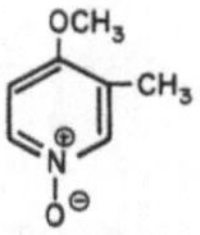

J. Chem. Soc. 46 (1961)

▶ $NO_2C_7H_{11}$

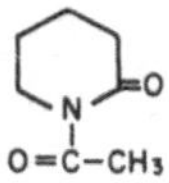

Chem. Ber. 95:2896 (1962)

▶ $NO_2C_7H_9$

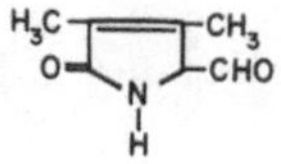

Ann. 654:165 (1962)

▶ $NO_2C_7H_{11}$

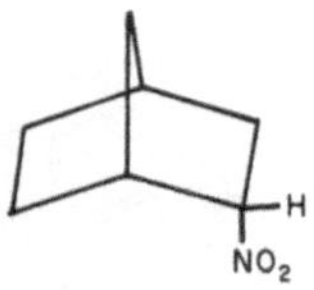

Can. J. Chem. 40:83 (1962)

▶ $NO_2C_7H_9$

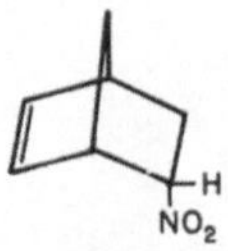

Can. J. Chem. 40:83 (1962)

▶ $NO_2C_7H_{13}$

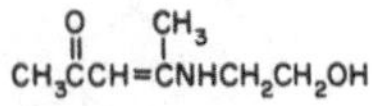

J. Am. Chem. Soc. 84:2691 (1962)

▶ $NO_2C_7H_9$

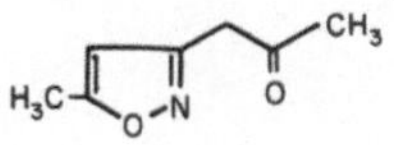

Tetrahedron Letters 670 (1961)

▶ $NO_2C_8H_7$

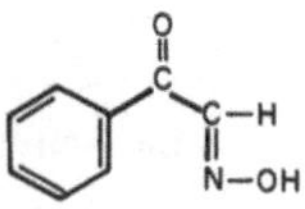

Varian 501

▶ $NO_2C_8H_9$

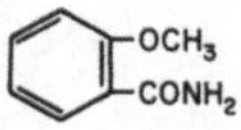

Can. J. Chem. 40:1886 (1962)

▶ $NO_2C_8H_9$

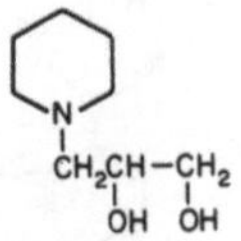

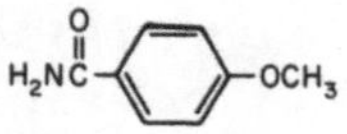

Can. J. Chem. 40:1866 (1962)

▶ $NO_2C_8H_{11}$

Arkiv Kemi 18:133 (1961)

▶ $NO_2C_8H_{11}$

Arkiv Kemi 18:133 (1961)

▶ $NO_2C_8H_{11}$

Varian 508

▶ $NO_2C_8H_{13}$

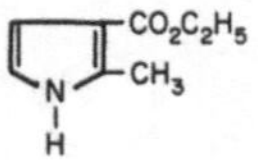

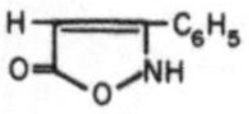

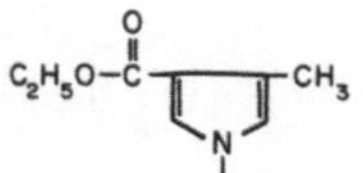

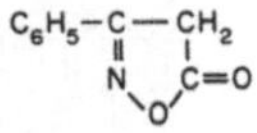

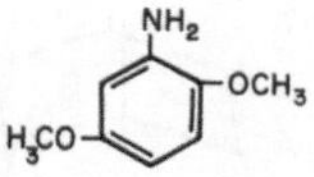

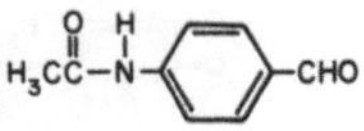

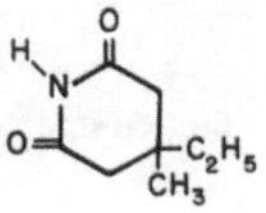

Varian 514

▶ $NO_2C_8H_{15}$

J. Org. Chem. 26:4342 (1961)

▶ $NO_2C_8H_{17}$

Chem. Ber. 95:2896 (1962)

▶ $NO_2C_8H_{19}$

$(CH_3)_3CN\ [(CH_2)_2OH]_2$

Varian 221

▶ $NO_2C_9H_7$

Tetrahedron 18:777 (1962)

▶ $NO_2C_9H_7$

Varian 521
J. Org. Chem. 27:4305 (1962)
Tetrahedron 18:777 (1962)

▶ $NO_2C_9H_9$

Can. J. Chem. 40:1073 (1962)

▶ $NO_2C_9H_{11}$

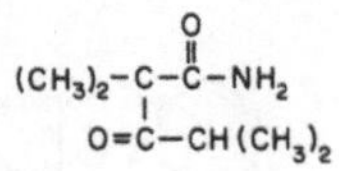

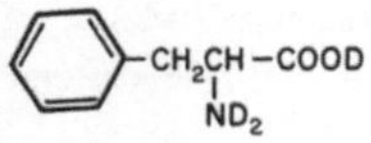

Varian 534

▶ $NO_2C_9H_{13}$

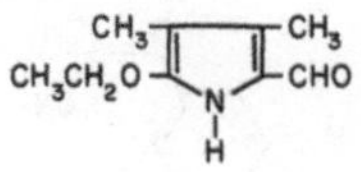

Ann. 654:165 (1962)

▶ $NO_2C_9H_{15}$

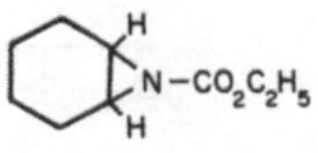

Tetrahedron Letters No. 7:278 (1962)

▶ $NO_2C_9H_{17}$

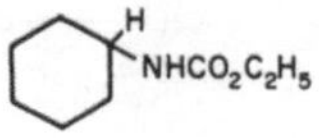

Tetrahedron Letters No. 7:279 (1962)

▶ $NO_2C_{10}H_6D$

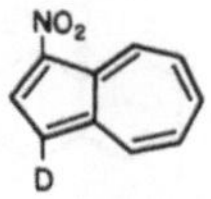

Helv. Chim. Acta 45:1965 (1962)

▶ $NO_2C_{10}H_7$

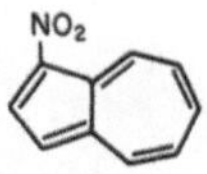

Helv. Chim. Acta 45:1965 (1962)

▶ $NO_2C_{10}H_7$

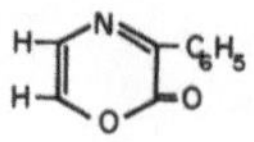

Chem. Ber. 95:1460 (1962)

▶ $NO_2C_{10}H_9$

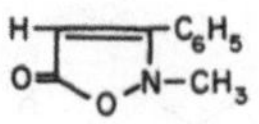

Tetrahedron 18:777 (1962)

▶ $NO_2C_{10}H_9$

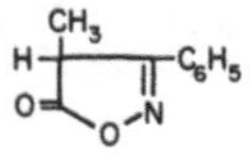

Tetrahedron 18:777 (1962)

▶ $NO_2C_{10}H_9$

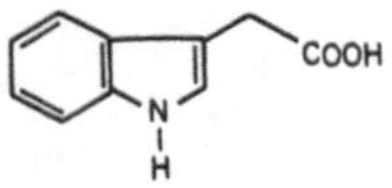

J. Org. Chem. 26:2341 (1961)

▶ $NO_2C_{10}H_9$

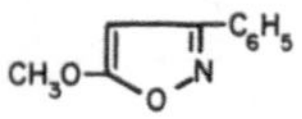

Tetrahedron 18:777 (1962)

▶ $NO_2C_{10}H_9$

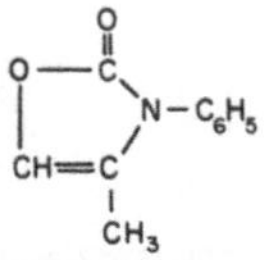

J. Org. Chem. 27:2664 (1962)

▶ $NO_2C_{10}H_{11}$

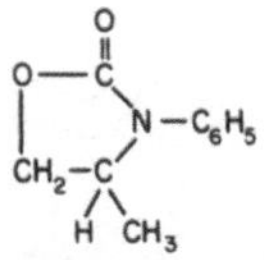

J. Org. Chem. 27:2664 (1962)

▶ $NO_2C_{10}H_{11}$

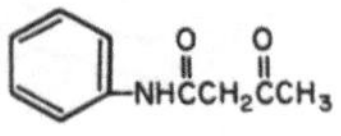

Varian 256

▶ $NO_2C_{10}H_{13}$

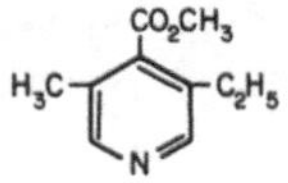

Can. J. Chem. 40:1145 (1962)

▶ $NO_2C_{10}H_{13}$

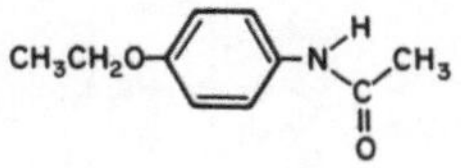

Varian 267

▶ $NO_2C_{10}H_{13}$

Varian 561

▶ $NO_2C_{10}H_{19}$

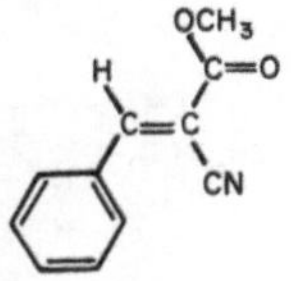

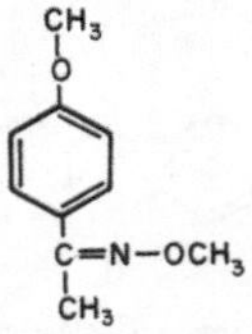

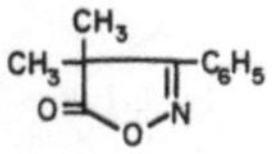

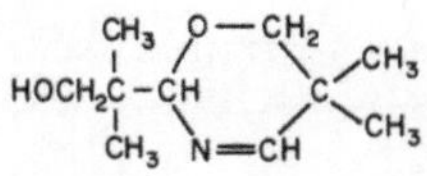

J. Org. Chem. 26:4134 (1961)

▶ $NO_2C_{10}H_{19}$

J. Org. Chem. 27:1926 (1962)

▶ $NO_2C_{10}H_{19}$

J. Org. Chem. 27:1926 (1962)

▶ $NO_2C_{11}H_9$

J. Org. Chem. 27:4105 (1962)

▶ $NO_2C_{11}H_9$

Varian 576

▶ $NO_2C_{11}H_{11}$

Tetrahedron 18:777 (1962)

▶ $NO_2C_{11}H_{11}$

Tetrahedron 18:777 (1962)

▶ $NO_2C_{11}H_{17}$

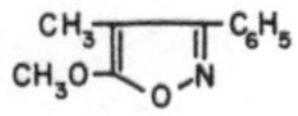

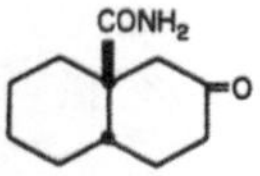

J. Org. Chem. 27:2013 (1962)

▶ $NO_2C_{11}H_{17}$

J. Org. Chem. 27:2013 (1962)

▶ $NO_2C_{12}H_{11}$

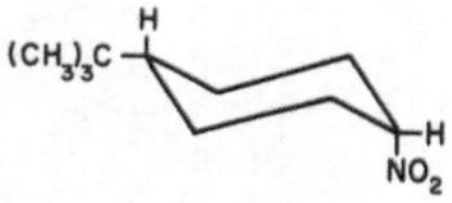

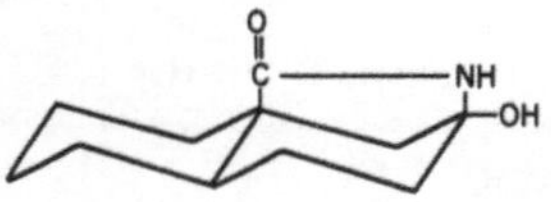

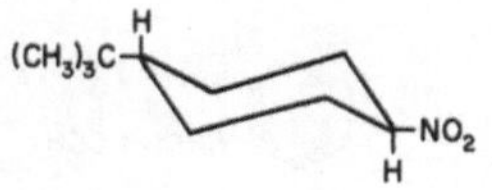

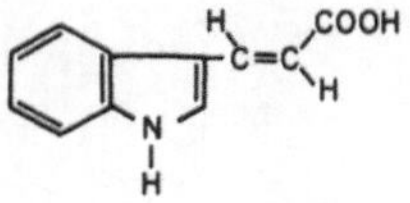

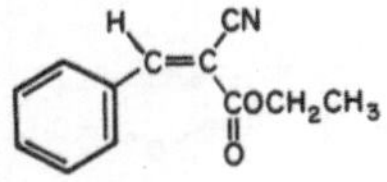

Varian 290

▶ $NO_2C_{12}H_{13}$

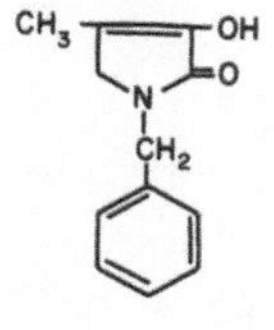

J. Org. Chem. 27:3077 (1962)

▶ $NO_2C_{12}H_{13}$

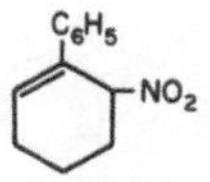

J. Org. Chem. 27:4243 (1962)

▶ $NO_2C_{12}H_{15}$

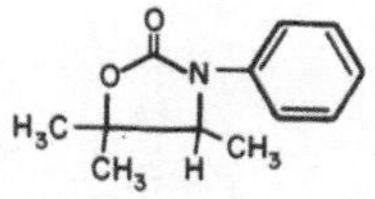

J. Org. Chem. 27:2927 (1962)

▶ $NO_2C_{12}H_{15}$

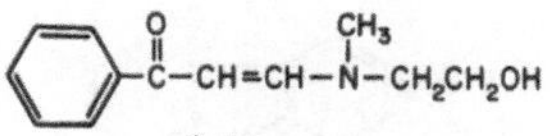

trans

J. Org. Chem. 27:4227 (1962)

▶ $NO_2C_{13}H_{13}$

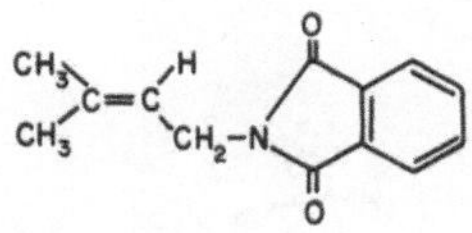

Chem. & Ind. 1020 (1962)

▶ $NO_2C_{13}H_{13}$

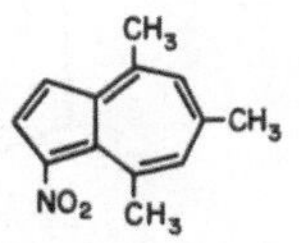

Proc. Chem. Soc. 364 (1962)

▶ $NO_2C_{13}H_{13}$

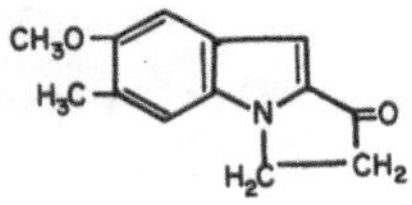

Varian 299

▶ $NO_2C_{13}H_{15}$

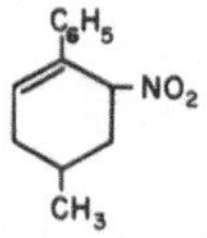

cis and trans

J. Org. Chem. 27:4243 (1962)

▶ $NO_2C_{13}H_{15}$

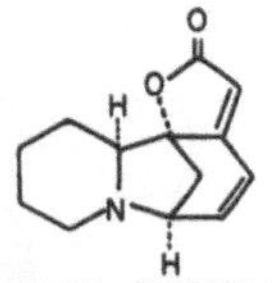

Tetrahedron Letters No. 25:1202 (1962)
Chem. & Ind. 1651, 1652 (1962)

▶ $NO_2C_{13}H_{17}$

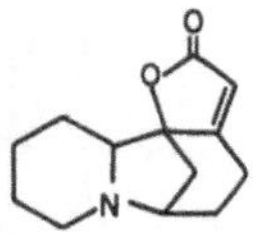

Chem. & Ind. 1651 (1962)

▶ $NO_2C_{13}H_{17}$

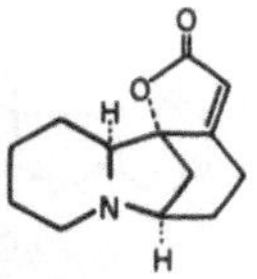

Chem. & Ind. 1652 (1962)

▶ $NO_2C_{13}H_{17}$

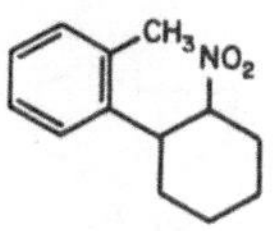

cis and trans

J. Org. Chem. 27:3008 (1962)

▶ $NO_2C_{13}H_{17}$

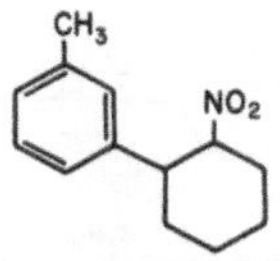

cis and trans

J. Org. Chem. 27:3008 (1962)

▶ $NO_2C_{14}H_{11}$

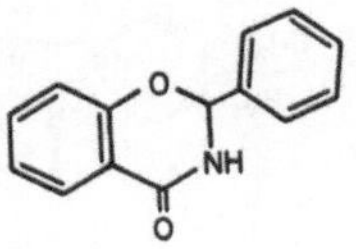

Can. J. Chem. 40:883 (1962)

▶ $NO_2C_{14}H_{17}$

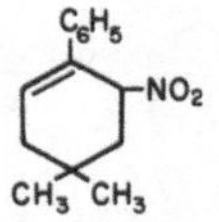

J. Org. Chem. 27:4243 (1962)

▶ $NO_2C_{14}H_{21}$

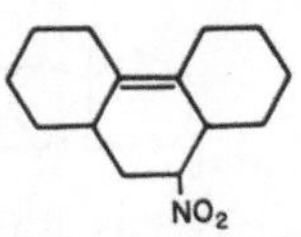

J. Org. Chem. 27:4094 (1962)

▶ $NO_2C_{15}H_{17}$

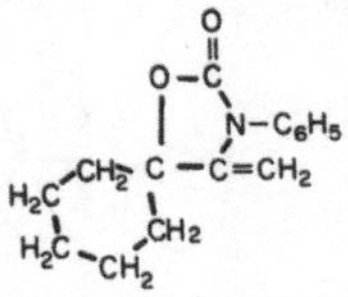

J. Org. Chem. 27:2664 (1962)

▶ $NO_2C_{15}H_{19}$

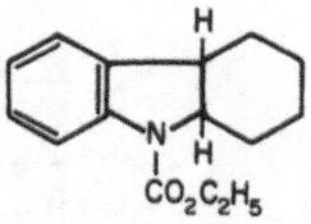

Helv. Chim. Acta 45:854, 1146 (1962)

▶ $NO_2C_{15}H_{19}$

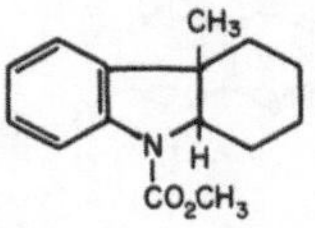

Helv. Chim. Acta 45:854, 1146 (1962)

▶ $NO_2C_{15}H_{21} \cdot HCl$

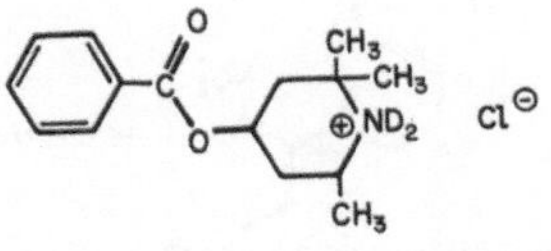

Varian 646

▶ $NO_2C_{15}H_{21} \cdot HCl$

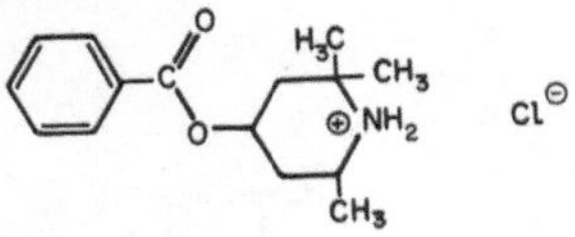

Varian 645

▶ $NO_2C_{15}H_{23}$

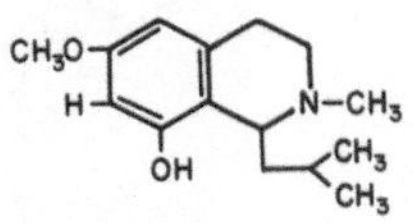

J. Am. Chem. Soc. 84:3210 (1962)

▶ $NO_2C_{16}H_{13}$

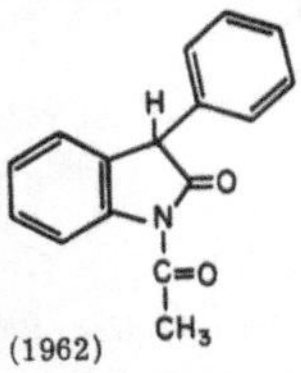

J. Chem. Soc. 5303 (1962)

▶ $NO_2C_{16}H_{13} \cdot Fe$

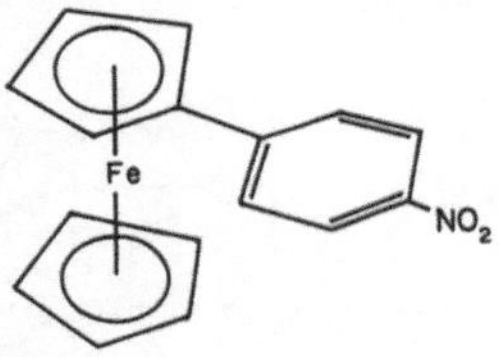

Varian 321

▶ $NO_2C_{16}H_{21}$

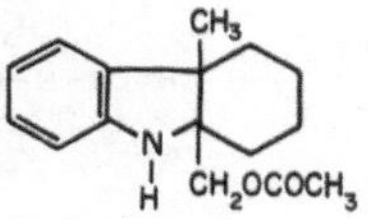

Helv. Chim. Acta 45:1992 (1962)

▶ $NO_2C_{16}H_{25}$

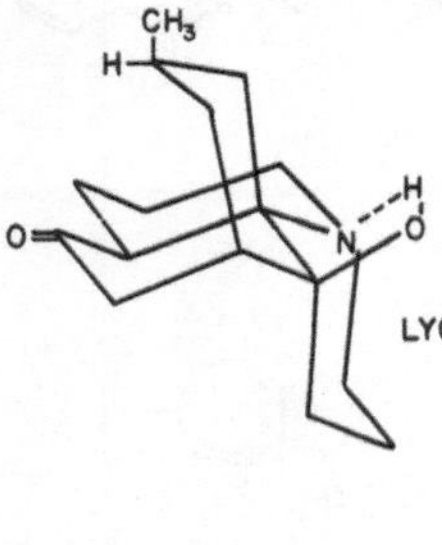

LYCODOLINE

Tetrahedron Letters No. 3:87 (1962)

▶ $NO_2C_{16}H_{25}$

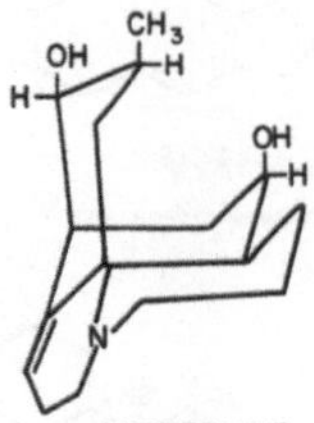

LYCOFOLINE

Can. J. Chem. 40:237 (1962)
Tetrahedron 18:1467 (1962)

▶ $NO_2C_{17}H_{23}$

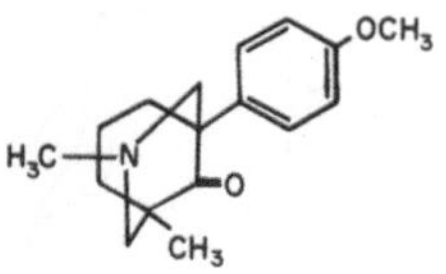

Chem. & Ind. 1288 (1962)

▶ $NO_2C_{17}H_{15}$

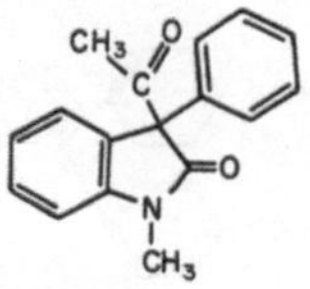

J. Chem. Soc. 5303 (1962)

▶ $NO_2C_{17}H_{23}$

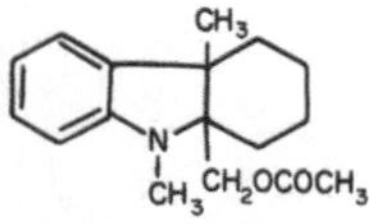

Helv. Chim. Acta 45:1992 (1962)

▶ $NO_2C_{17}H_{17}$

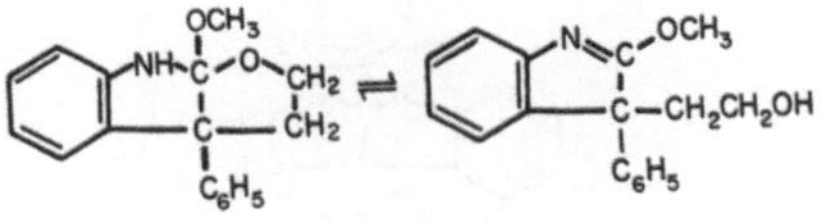

J. Am. Chem. Soc. 84:4574 (1962)

▶ $NO_2C_{18}H_{19}$

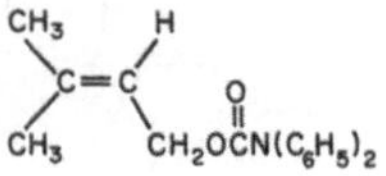

Chem. & Ind. 1020 (1962)

▶ $NO_2C_{18}H_{21}$

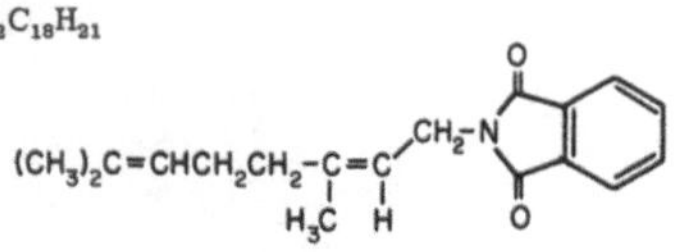

cis and trans

Chem. & Ind. 1020 (1962)

▶ $NO_2C_{17}H_{23}$

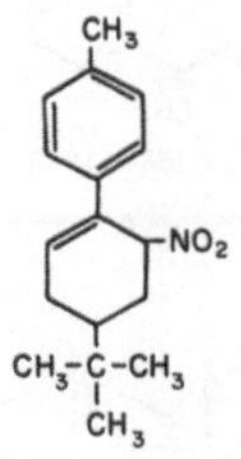

trans

J. Org. Chem. 27:4243 (1962)

▶ $NO_2C_{18}H_{29}$

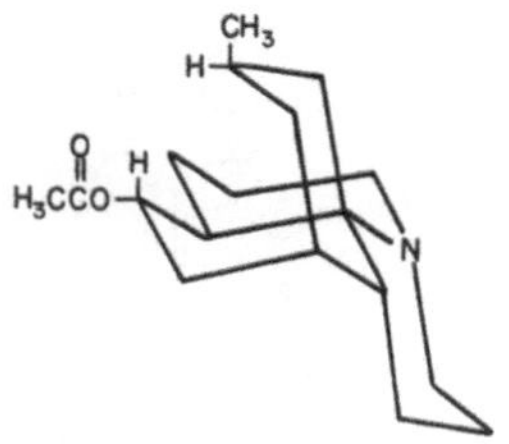

Can. J. Chem. 40:2094 (1962)

▶ $NO_2C_{18}H_{29}$

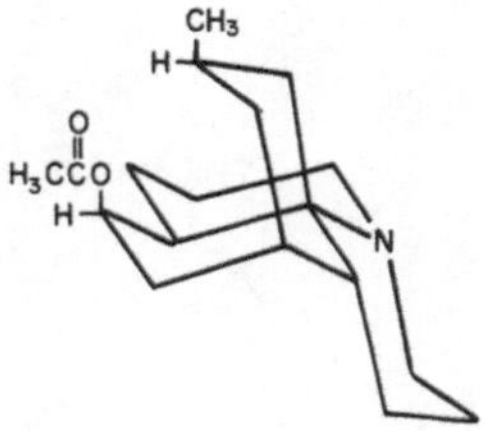

Can. J. Chem. 40:2094 (1962)

▶ $NO_2C_{19}H_{25}$

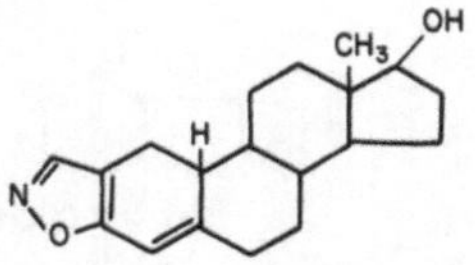

Chem. & Ind. 1984 (1962)

▶ $NO_2C_{19}H_{27}$

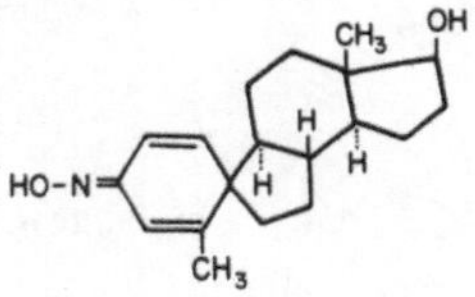

Helv. Chim. Acta 45:2346 (1962)

▶ $NO_2C_{20}H_{17}$

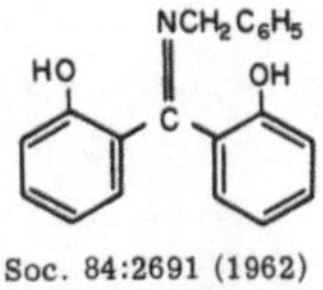

J. Am. Chem. Soc. 84:2691 (1962)

▶ $NO_2C_{20}H_{22} \cdot Cl$

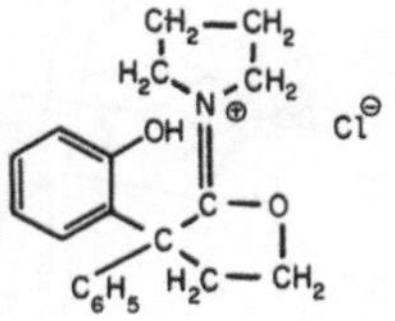

J. Am. Chem. Soc. 84:4567 (1962)

▶ $NO_2C_{20}H_{25}$

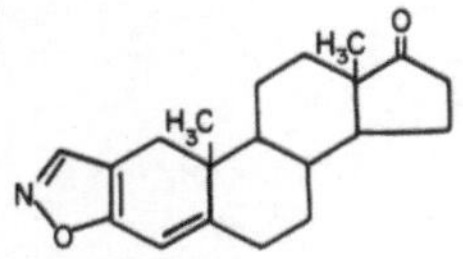

Chem. & Ind. 1984 (1962)

▶ $NO_2C_{20}H_{27}$

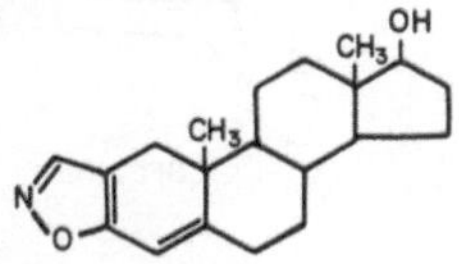

Chem. & Ind. 1984 (1962)

▶ $NO_2C_{21}H_{19}$

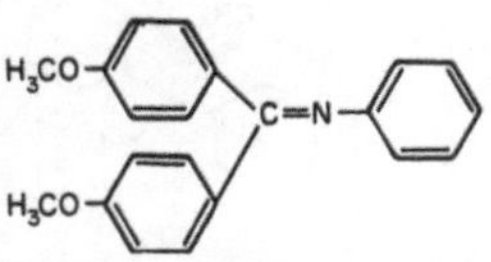

Tetrahedron Letters No. 26:1271 (1962)

▶ $NO_2C_{21}H_{24} \cdot I$

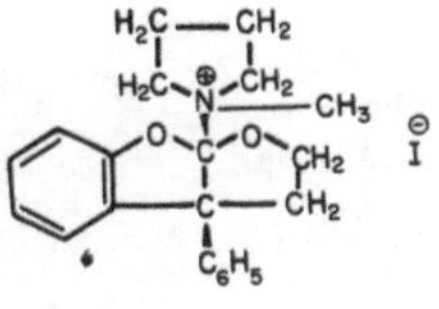

J. Am. Chem. Soc. 84:4567 (1962)

▶ $NO_2C_{21}H_{27}$

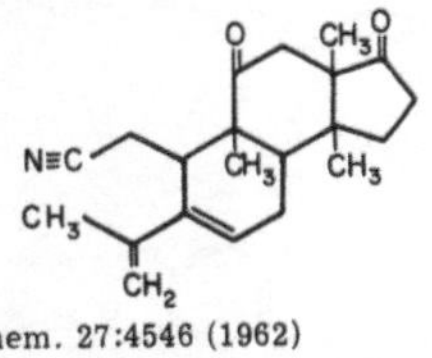

J. Org. Chem. 27:4546 (1962)

▶ $NO_2C_{21}H_{29}$

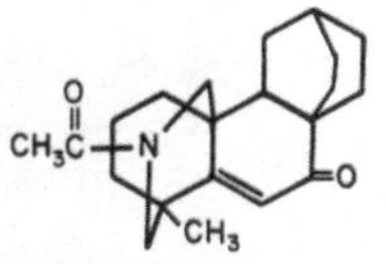

Tetrahedron 14:57 (1961)

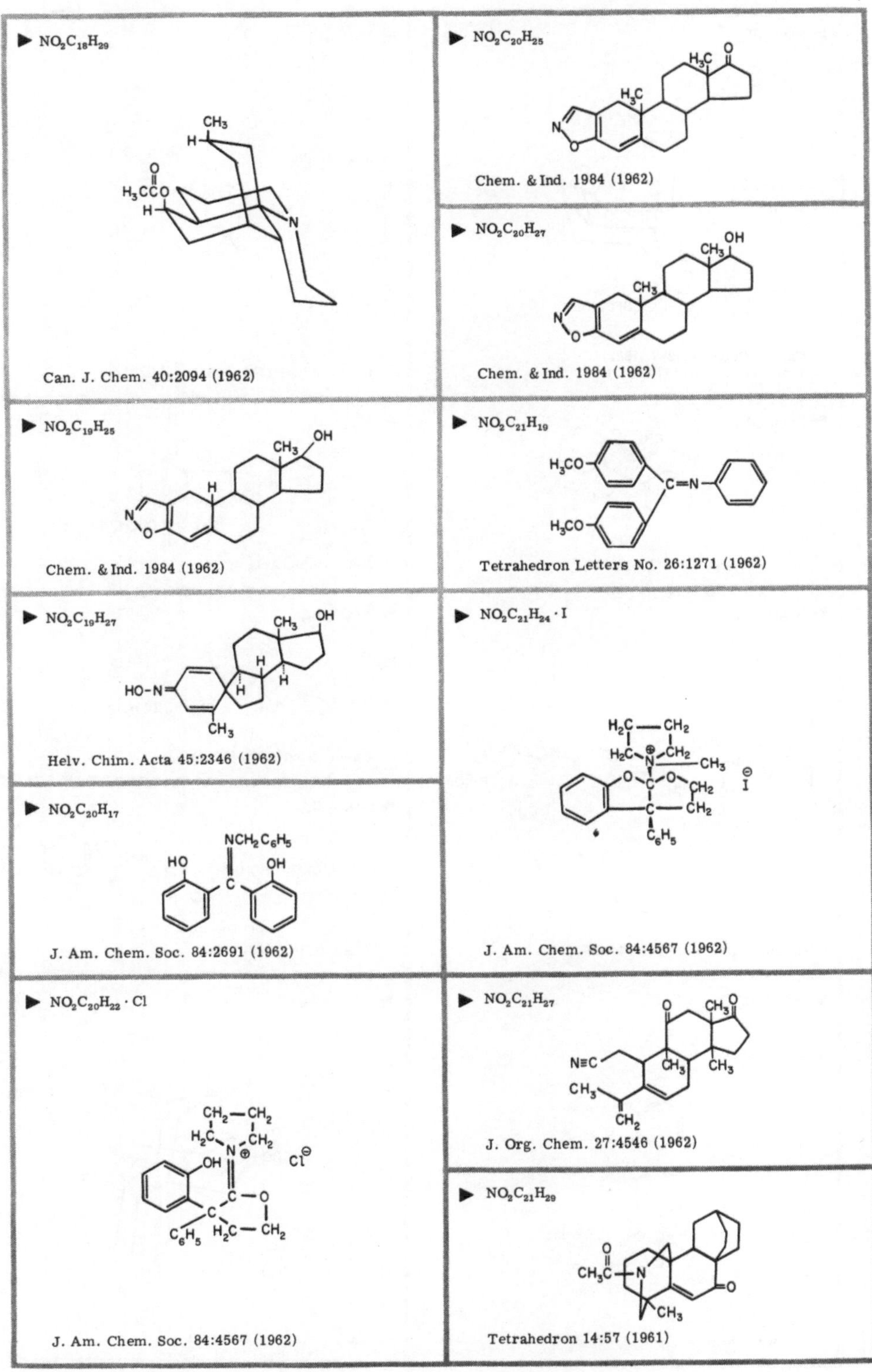

▶ $NO_2C_{21}H_{31}$

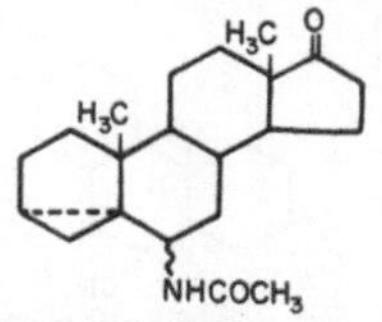

J. Org. Chem. 27:4624 (1962)

▶ $NO_2C_{21}H_{31}$

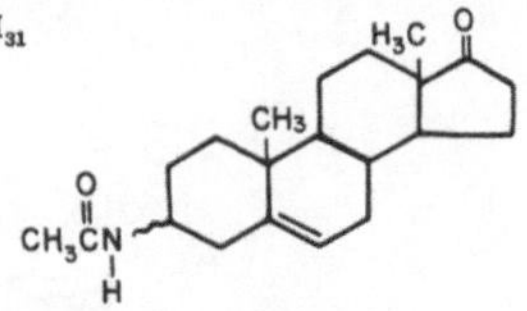

J. Org. Chem. 27:4624 (1962)

▶ $NO_2C_{21}H_{33}$

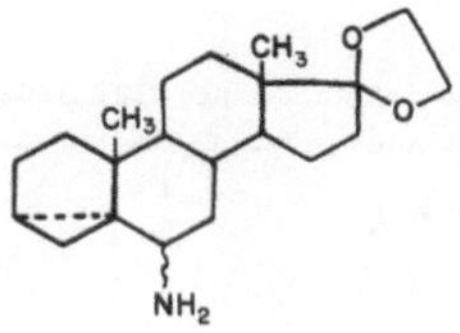

J. Org. Chem. 27:4610 (1962)

▶ $NO_2C_{22}H_{21}$

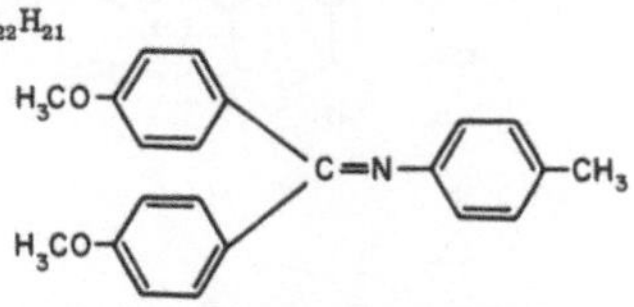

Tetrahedron Letters No. 26:1271 (1962)

▶ $NO_2C_{22}H_{29}$

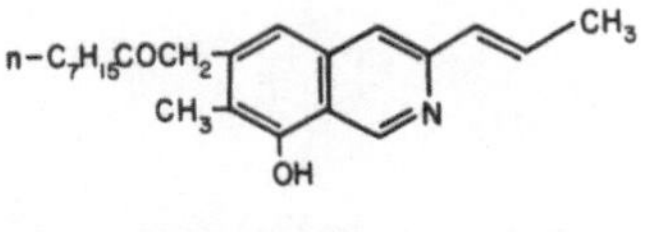

Tetrahedron 18:1173 (1962)

▶ $NO_2C_{22}H_{29}$

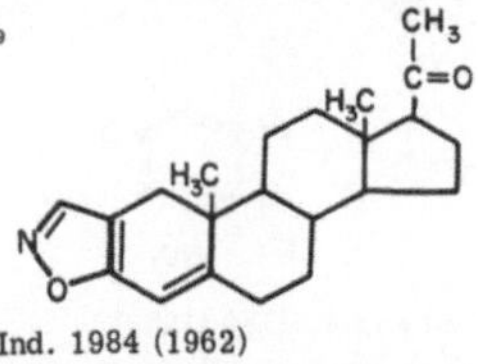

Chem. & Ind. 1984 (1962)

▶ $NO_2C_{22}H_{31}$

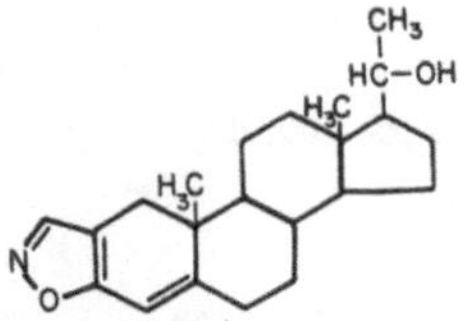

Chem. & Ind. 1984 (1962)

▶ $NO_2C_{22}H_{33}$

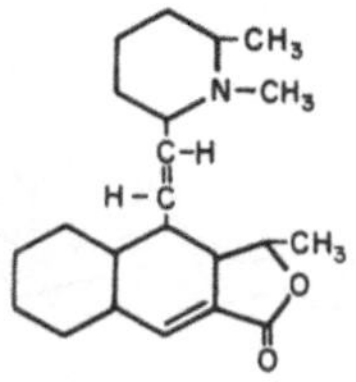

Australian J. Chem. 14:64 (1961)

▶ $NO_2C_{22}H_{35}$

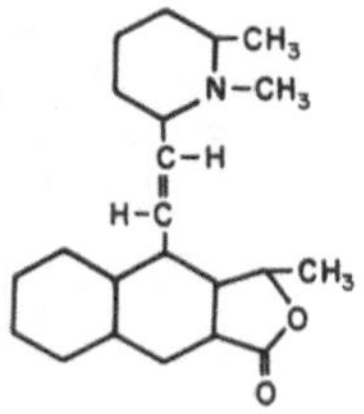

Australian J. Chem. 14:64 (1961)

▶ $NO_2C_{23}H_{21}$

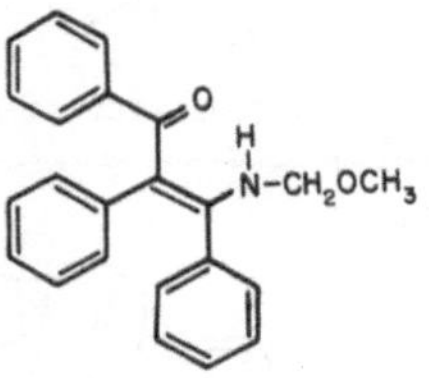

Can. J. Chem. 40:883 (1962)

▶ $NO_2C_{23}H_{27}$

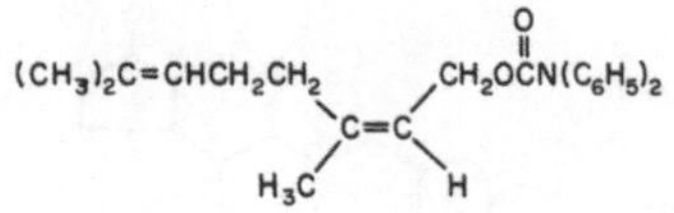

Chem. & Ind. 1020 (1962)

▶ $NO_2C_{28}H_{45}$

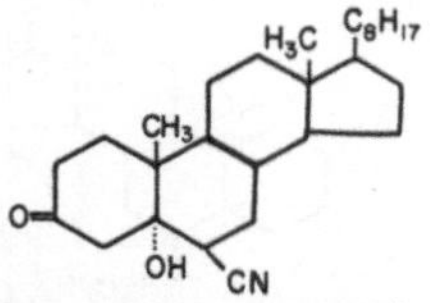

Bull. Soc. Chim. France 2444 (1961)

▶ $NO_2C_{23}H_{33}$

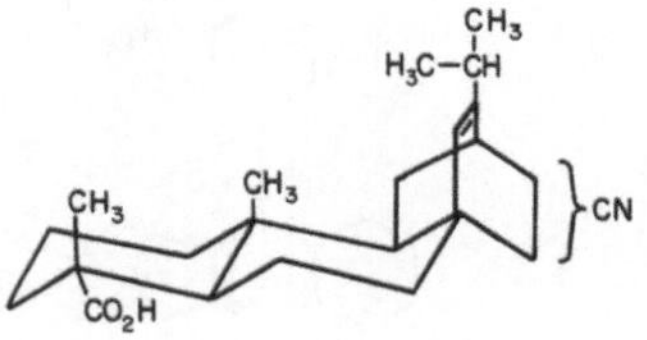

Tetrahedron Letters No. 16:695 (1962)

▶ $NO_2C_{29}H_{47}$

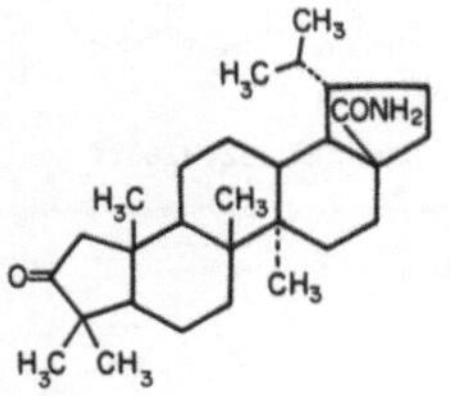

Bull. Soc. Chim. France 1137 (1962)

▶ $NO_2C_{24}H_{35}$

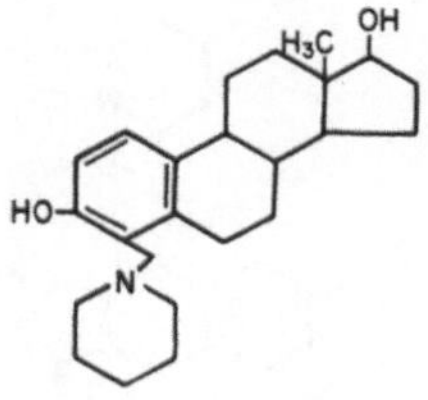

Bull. Soc. Chim. France 1679 (1961)

▶ $NO_2C_{30}H_{49}$

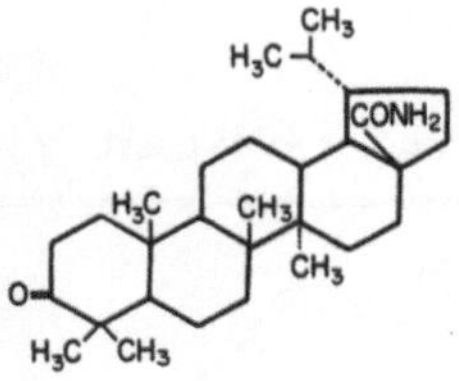

Bull. Soc. Chim. France 1137 (1962)

▶ $NO_2C_{24}H_{35}$

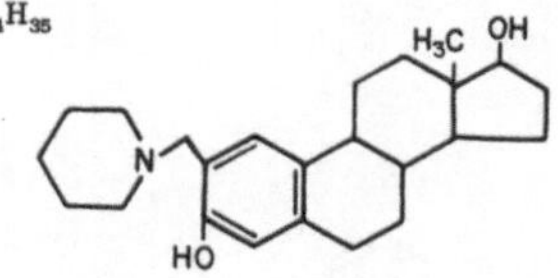

Bull. Soc. Chim. France 1679 (1961)

▶ $NO_3IC_7H_6$

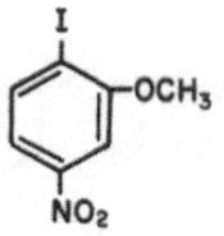

Can. J. Chem. 40:1866 (1962)

▶ $NO_2C_{26}H_{31}$

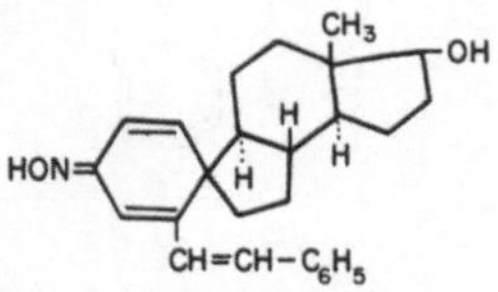

Helv. Chim. Acta 45:2346 (1962)

▶ $NO_3Br_2C_6H_3$

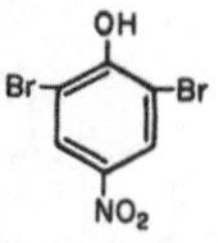

Can. J. Chem. 40:1759 (1962)

▶ $NO_3ClC_9H_{10} \cdot HCl$

HO–C₆H₃(Cl)–CH₂–CH(NH₃⊕)–COOH Cl⊖

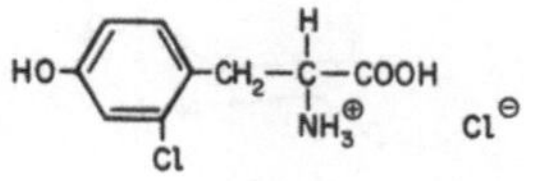

J. Chem. Soc. 3275 (1962)

▶ $NO_3C_4H_7$

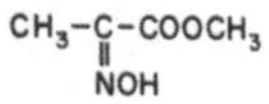

$CH_3–C(=NOH)–COOCH_3$

J. Phys. Chem. 65:491 (1961)

▶ $NO_3Cl_2C_{16}H_{13}$

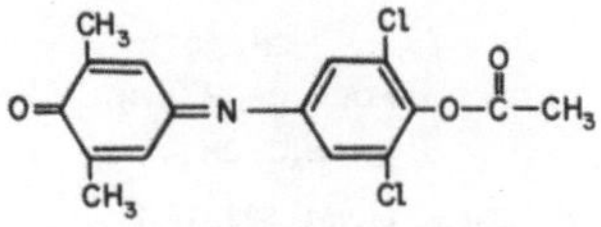

J. Org. Chem. 27:1849 (1962)

▶ $NO_3C_4H_9$

$(CH_3)_2–C(NO_2)CH_2OH$

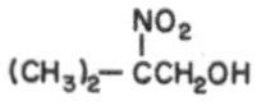

Varian 422

▶ $NO_3CH_3$

$H_3CONO_2$

J. Chem. Soc. 1372 (1962)

▶ $NO_3C_5H_7$

$CH_3CH_2OC(=O)CH_2NCO$

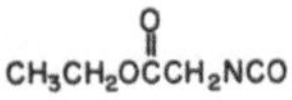

Varian 107

▶ $NO_3C_2H_5$

$CH_3CH_2ONO_2$

J. Chem. Phys. 34:1099 (1961)

▶ $NO_3C_5H_9$

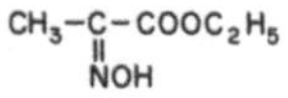

$CH_3–C(=NOH)–COOC_2H_5$

J. Phys. Chem. 65:491 (1961)

▶ $NO_3C_3H_5$

$CH_3–C(=NOH)–COOH$

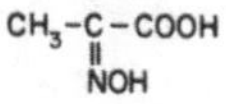

J. Phys. Chem. 65:1491 (1961)

▶ $NO_3C_5H_9$

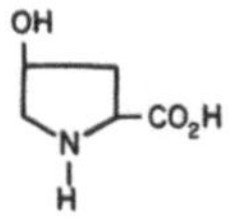

Mol. Phys 5:195 (1962)

▶ $NO_3C_3H_7$

$HOCH_2CH(NH_2)CO_2H$

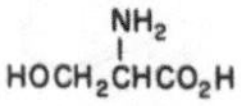

J. Am. Chem. Soc. 84:4650 (1962)

▶ $NO_3C_5H_9$

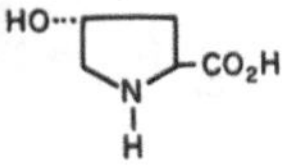

Mol. Phys. 5:195 (1962)

▶ $NO_3C_4H_3$

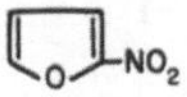

Can. J. Chem. 39:907 (1961)

▶ $NO_3C_6H_5$

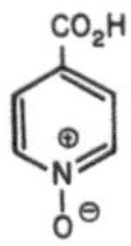

J. Chem. Soc. 44 (1961)

▶ $NO_3C_6H_5$

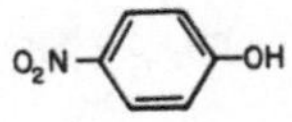

Can. J. Chem. 40:2124 (1962)
Helv. Chim. Acta 45:568 (1962)

▶ $NO_3C_6H_{11}$

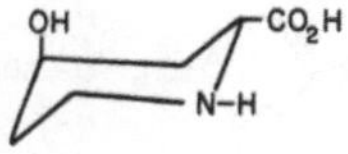

Acta Chem. Scand. 16:2457 (1962)

▶ $NO_3C_6H_{11}$

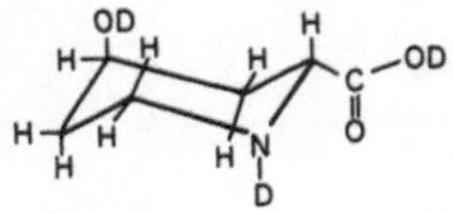

Varian 469

▶ $NO_3C_6H_{13}$

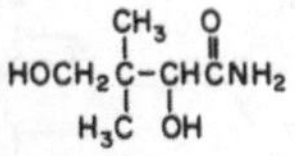

Angew. Chem. 74:751, 998 (1962)
Angew. Chem. Intern. Ed. 1:592 (1962)

▶ $NO_3C_7H_5$

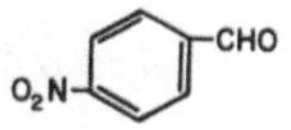

Can. J. Chem. 40:1073, 1897, 2331 (1962)

▶ $NO_3C_6H_{11}$

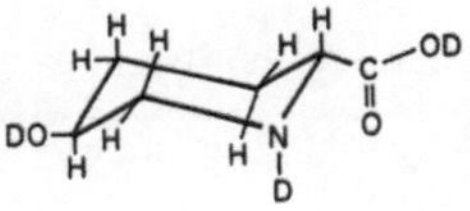

Varian 470

▶ $NO_3C_7H_5$

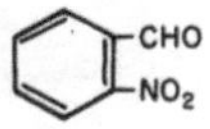

Varian 148
Can. J. Chem. 40:1073, 1897 (1962)

▶ $NO_3C_7H_5$

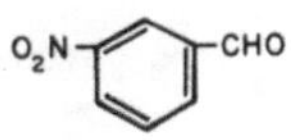

Can. J. Chem. 40:1073, 1897 (1962)

▶ $NO_3C_6H_{11}$

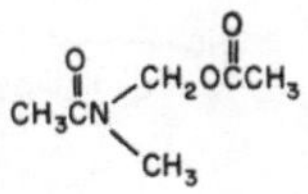

Chem. Ber. 94:2462 (1961)

▶ $NO_3C_7H_7$

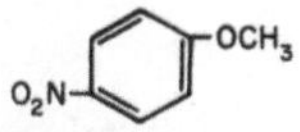

Can. J. Chem. 40:1866 (1962)
Helv. Chim. Acta 45:568 (1962)
J. Chem. Phys. 37:2594 (1962)

▶ $NO_3C_6H_{11}$

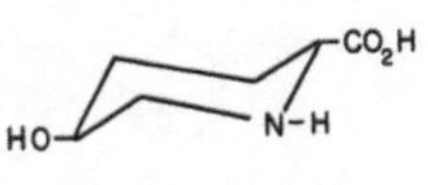

Acta Chem. Scand. 16:2457 (1962)

▶ $NO_3C_7H_7$

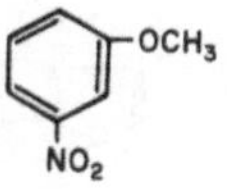

Can. J. Chem. 40:1866 (1962)

▶ $NO_3C_7H_7$

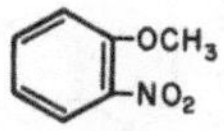

Can. J. Chem. 40:1866 (1962)

▶ $NO_3C_7H_7$

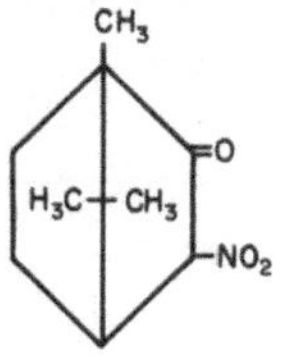

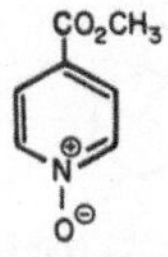

J. Chem. Soc. 44 (1961)

▶ $NO_3C_8H_7$

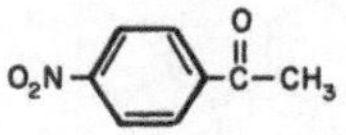

Can. J. Chem. 40:2331 (1962)

▶ $NO_3C_8H_9$

Can. J. Chem. 40:1866 (1962)

▶ $NO_3C_8H_{17}$

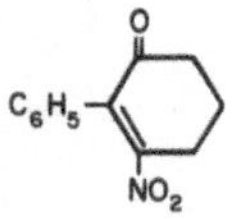

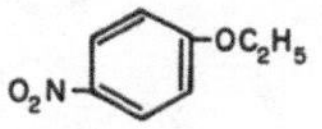

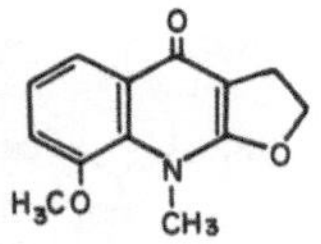

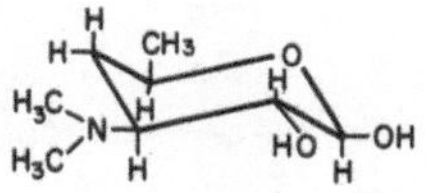

DESOSAMINE

Tetrahedron Letters No. 17:735 (1962)

▶ $NO_3C_9H_9$

Varian 522

▶ $NO_3C_{10}H_{13}$

J. Org. Chem. 27:4057 (1962)

▶ $NO_3C_{10}H_{15}$

Australian J. Chem. 15:431 (1962)

▶ $NO_3C_{12}H_{11}$

Tetrahedron 15:60 (1961)

▶ $NO_3C_{13}H_{13}$

Varian 300

▶ $NO_3C_{13}H_{15}$

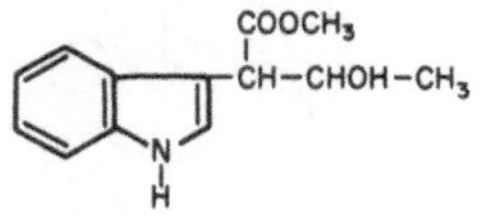

J. Am. Chem. Soc. 83:4678 (1961)

▶ $NO_3C_{13}H_{15}$

Varian 611

▶ $NO_3C_{14}H_{15}$

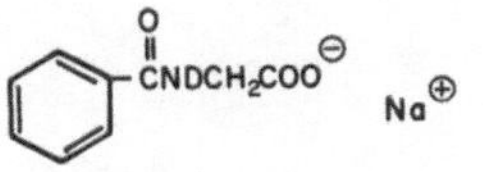

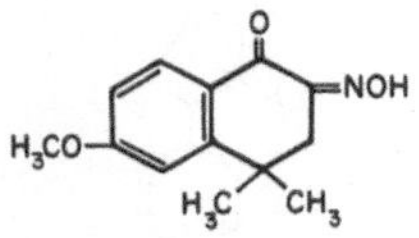

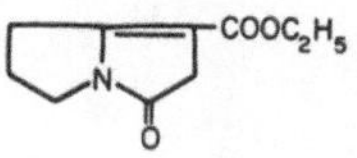

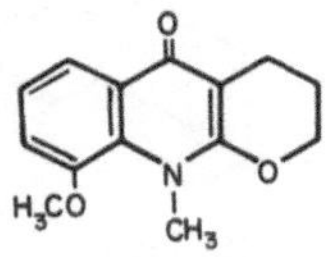

Varian 311

▶ $NO_3C_{14}H_{15}$

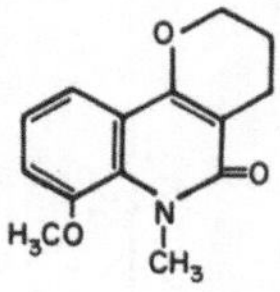

Varian 312

▶ $NO_3C_{14}H_{17}$

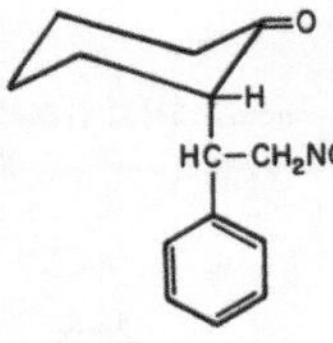

Varian 632

▶ $NO_3C_{14}H_{21}$

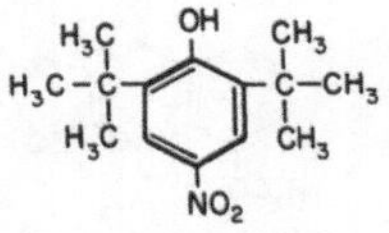

J. Chem. Soc. 954 (1961)

▶ $NO_3C_{15}H_{27}$

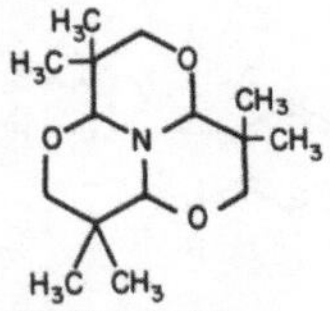

J. Org. Chem. 26:4135 (1961)

▶ $NO_3C_{16}H_{15}$

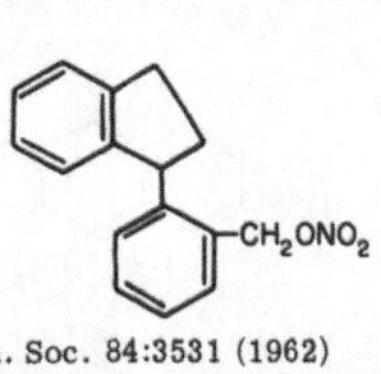

J. Am. Chem. Soc. 84:3531 (1962)

▶ $NO_3C_{16}H_{15}$

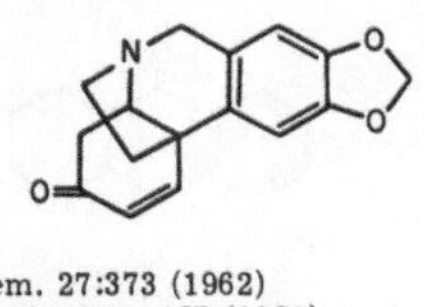

J. Org. Chem. 27:373 (1962)
Tetrahedron Letters 107 (1961)

▶ $NO_3C_{16}H_{15}$

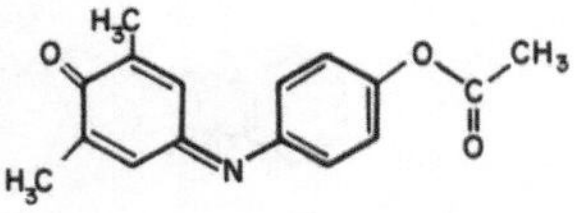

Varian 322

▶ $NO_3C_{16}H_{17}$

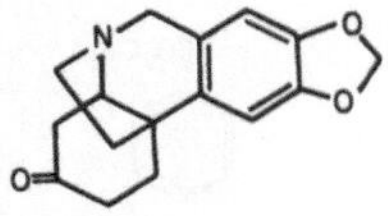

J. Org. Chem. 27:373 (1962)
Tetrahedron Letters 107 (1961)

▶ $NO_3C_{16}H_{19}$

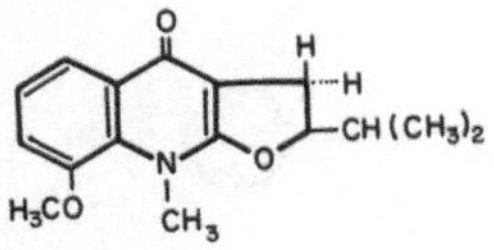

Varian 325

▶ $NO_3C_{17}H_{17}$

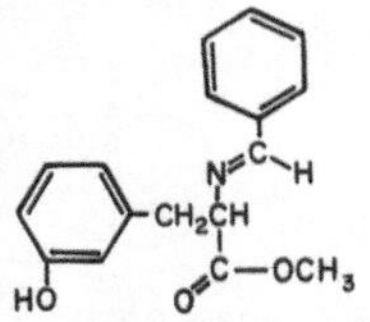

Varian 663

▶ $NO_3C_{17}H_{19}$

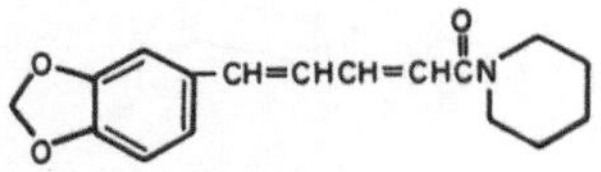

Varian 328

▶ $NO_3C_{18}H_{15}$

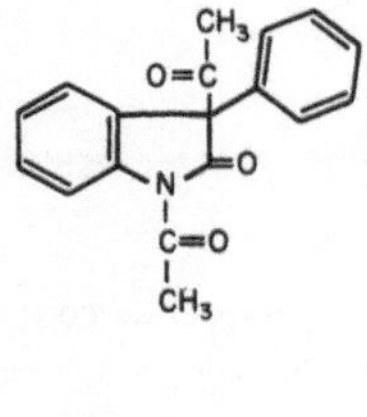

J. Chem. Soc. 5303 (1962)

▶ $NO_3C_{18}H_{19}$

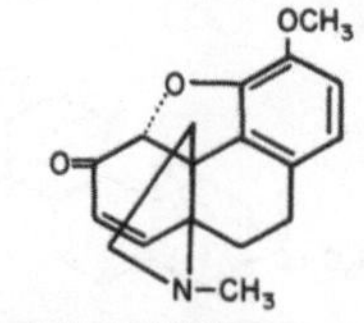

J. Org. Chem. 27:4122 (1962)

▶ $NO_3C_{18}H_{27}$

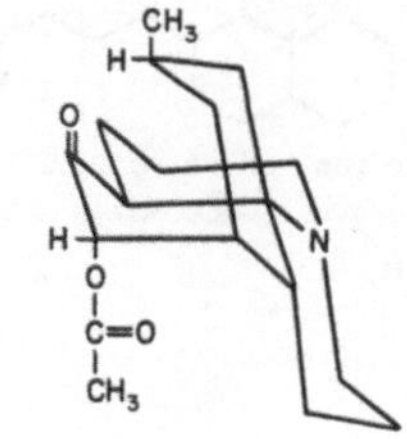

Can. J. Chem. 40:2094 (1962)

▶ $NO_3C_{18}H_{27}$

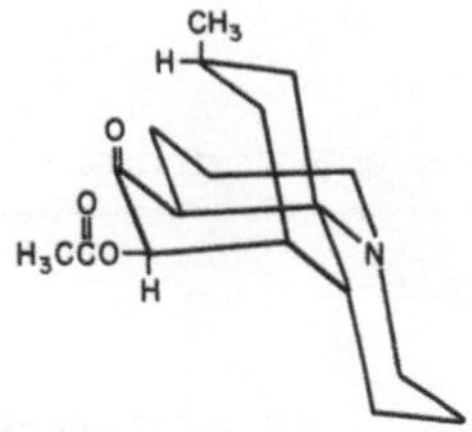

Can. J. Chem. 40:2094 (1962)

▶ $NO_3C_{18}H_{27}$

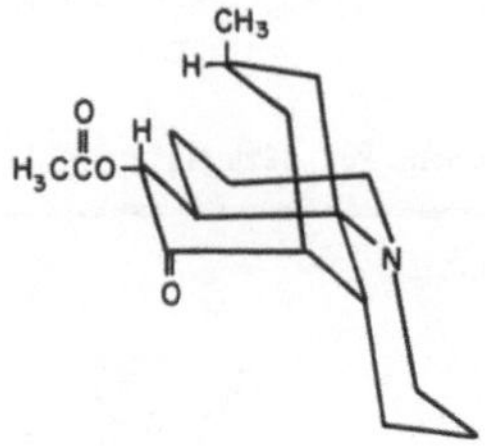

Can. J. Chem. 40:2094 (1962)

▶ $NO_3C_{18}H_{27}$

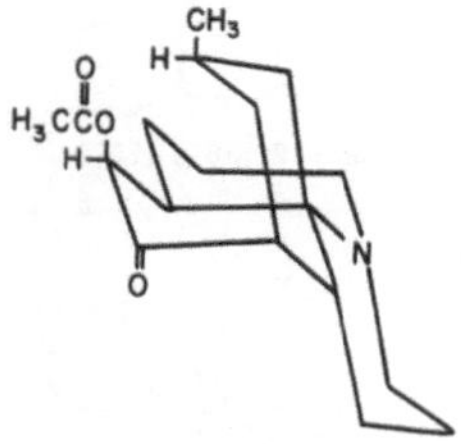

Can. J. Chem. 40:2094 (1962)

▶ $NO_3C_{18}H_{27}$

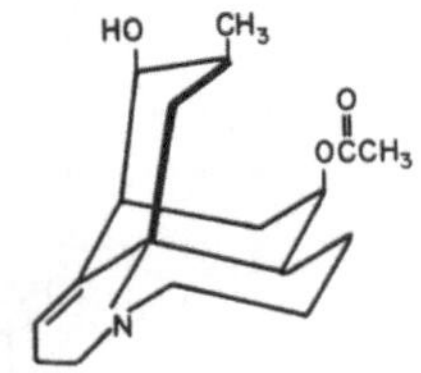

Tetrahedron 18:1467 (1962)

▶ $NO_3C_{18}H_{29}$

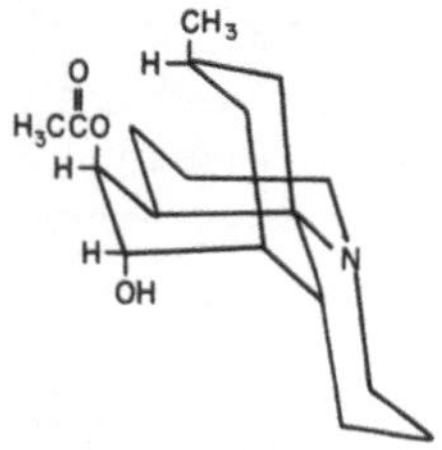

Can J. Chem. 40:2094 (1962)

▶ $NO_3C_{19}H_{27}$

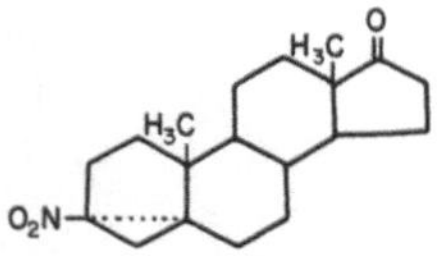

J. Org. Chem. 27:4615 (1962)

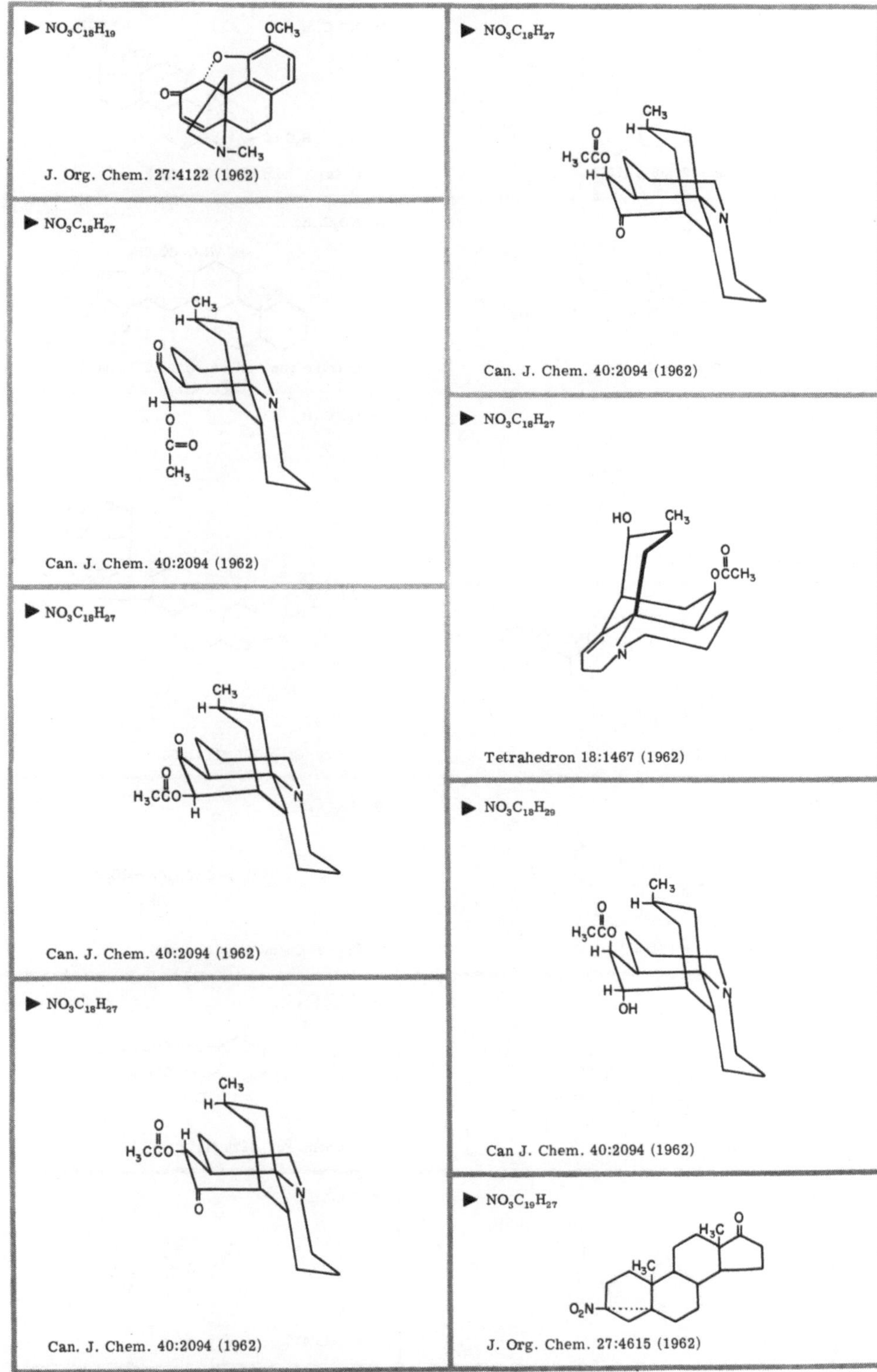

▶ $NO_3C_{19}H_{27}$

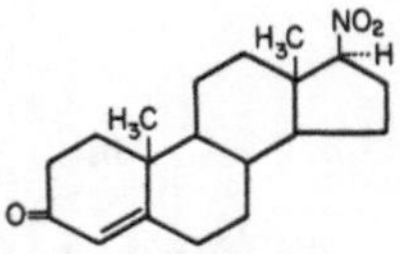

J. Org. Chem. 27:3824 (1962)

▶ $NO_3C_{20}H_{27}$

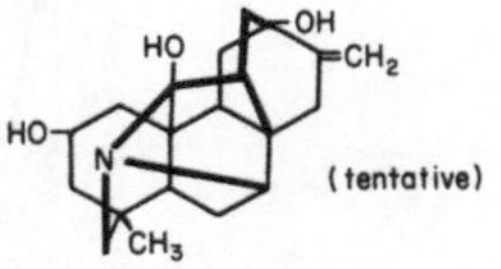

Tetrahedron 14:98 (1961)

▶ $NO_3C_{20}H_{29}$

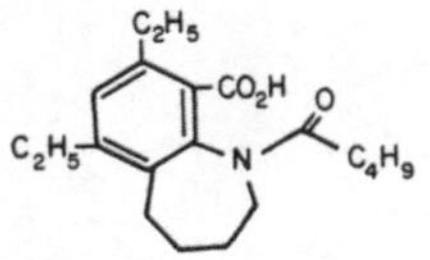

Chem. & Ind. 1191 (1962)

▶ $NO_3C_{21}H_{27}$

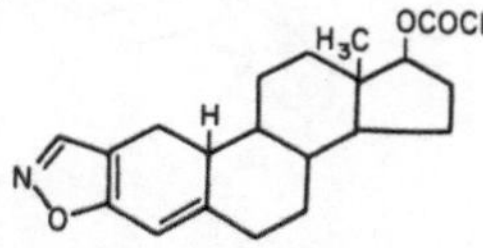

Chem. & Ind. 1984 (1962)

▶ $NO_3C_{22}H_{29}$

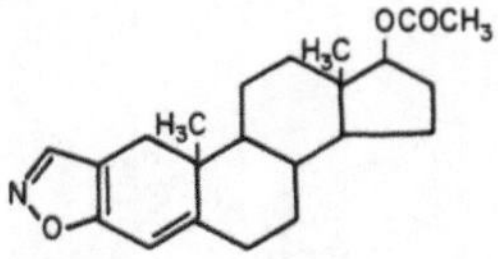

Chem. & Ind. 1984 (1962)

▶ $NO_3C_{23}H_{23}$

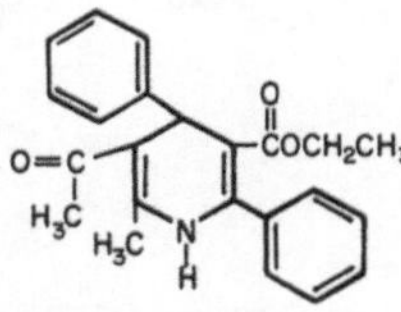

Varian 695

▶ $NO_3C_{24}H_{33}$

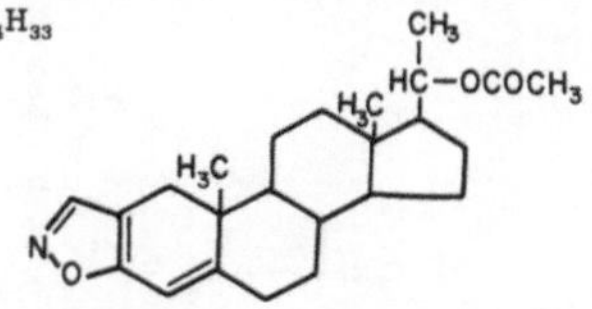

Chem. & Ind. 1984 (1962)

▶ $NO_3C_{26}H_{39}$

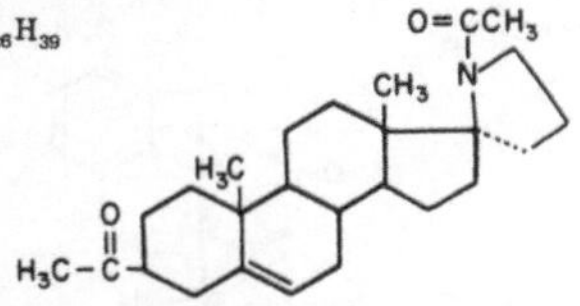

J. Org. Chem. 27:3825 (1962)

▶ $NO_3C_{27}H_{35}$

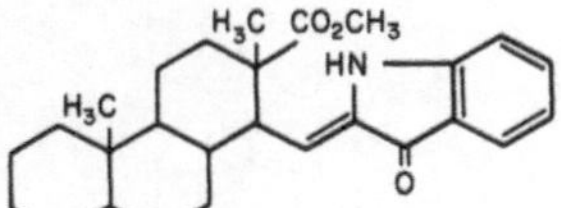

Tetrahedron Letters No. 21:975 (1962)

▶ $NO_3C_{27}H_{39}$

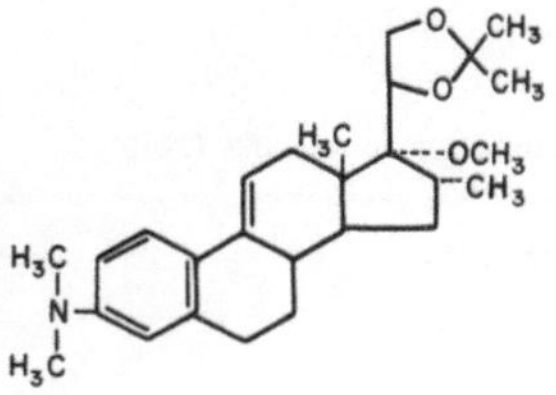

J. Org. Chem. 27:914 (1962)

▶ $NO_4FC_5H_8$

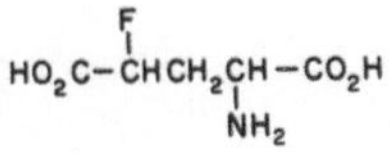

Can. J. Chem. 40:1575 (1962)

▶ $NO_4F_2C_{13}H_{13}$

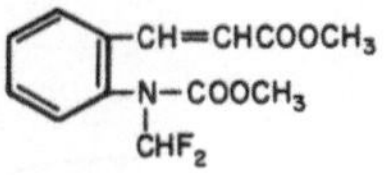

J. Chem. Soc. 3829 (1962)

▶ $NO_4C_4H_7$

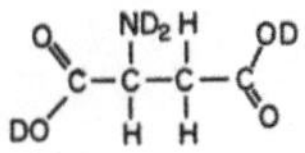

Varian 410

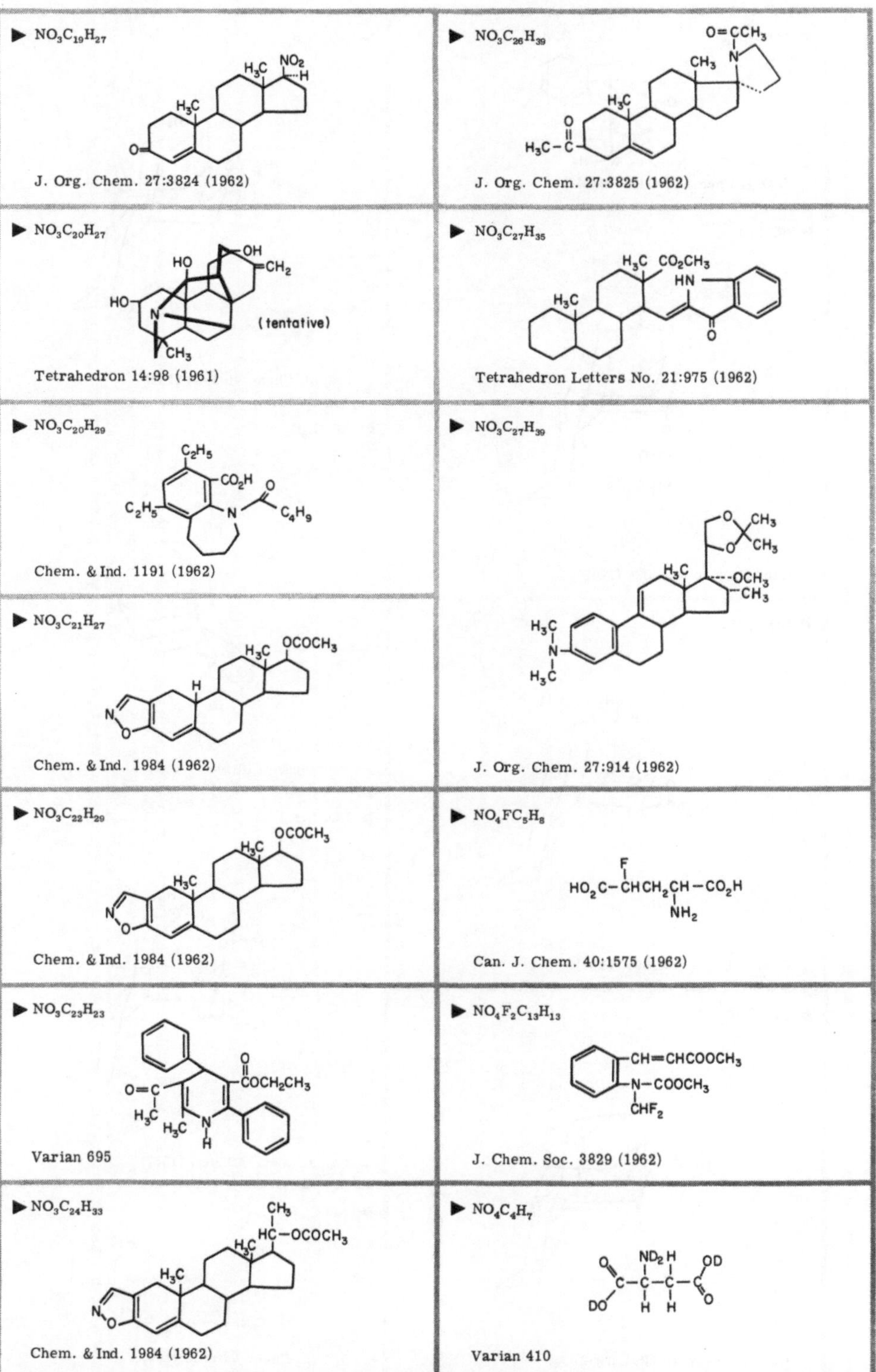

▶ $NO_4C_5H_9$

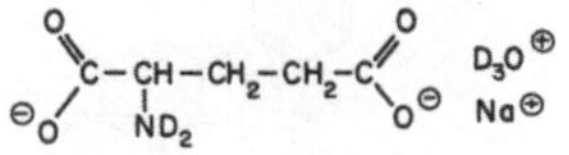

Varian 435

▶ $NO_4C_7H_9$

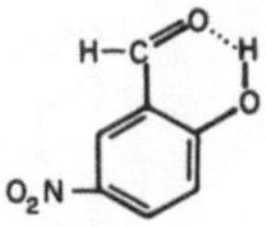

Varian 484

▶ $NO_4C_7H_9$

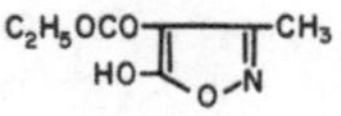

Tetrahedron 18:777 (1962)

▶ $NO_4C_8H_{11}$

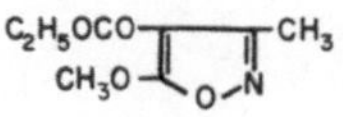

Tetrahedron 18:777 (1962)

▶ $NO_4C_9H_9$

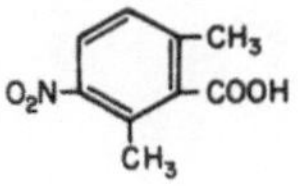

J. Am. Chem. Soc. 83:4457 (1961)

▶ $NO_4C_{10}H_{11}$

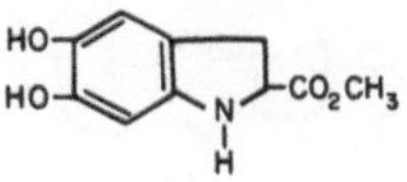

Helv. Chim. Acta 45:638 (1962)

▶ $NO_4C_{10}H_{11}$

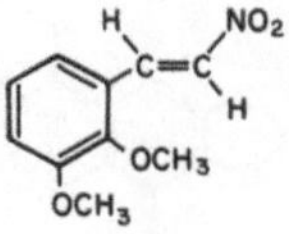

Varian 257

▶ $NO_4C_{11}H_{21}$

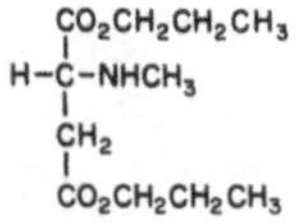

Helv. Chim. Acta 45:2005 (1962)

▶ $NO_4C_{12}H_{11}$

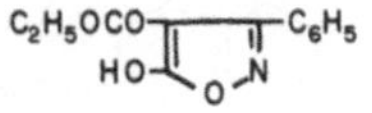

Tetrahedron 18:777 (1962)

▶ $NO_4C_{12}H_{11}$

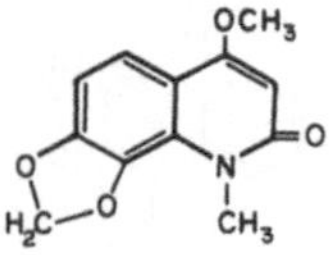

Varian 291

▶ $NO_4C_{12}H_{19}$

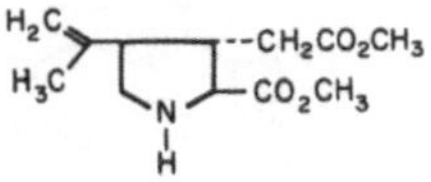

Bull. Chem. Soc. Japan 35:1899 (1962)

▶ $NO_4C_{13}H_{17}$

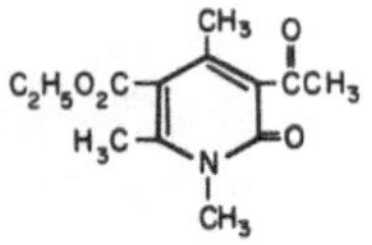

J. Chem. Soc. 860 (1961)

▶ $NO_4C_{13}H_{17}$

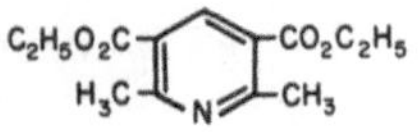

J. Chem. Soc. 860 (1961)

▶ $NO_4C_{13}H_{17}$

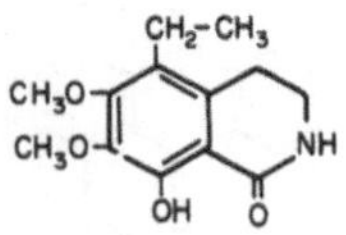

J. Org. Chem. 27:3720 (1962)

▶ $NO_4C_{13}H_{23}$

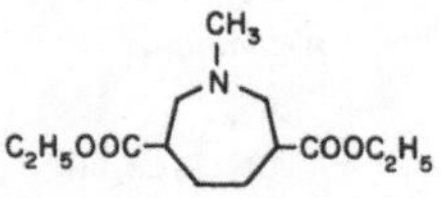

J. Org. Chem. 27:4436 (1962)

▶ $NO_4C_{14}H_{13}$

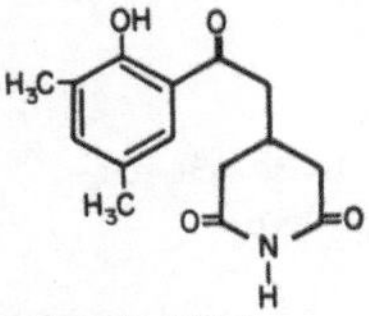

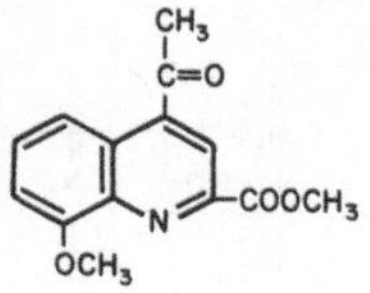

J. Am. Chem. Soc. 83:3906 (1961)

▶ $NO_4C_{14}H_{17}$

Varian 313

▶ $NO_4C_{14}H_{19}$

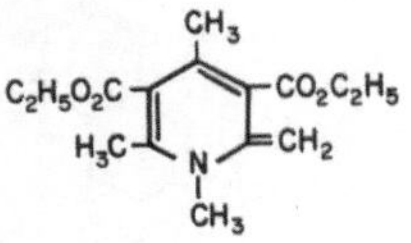

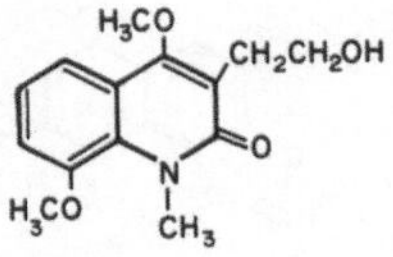

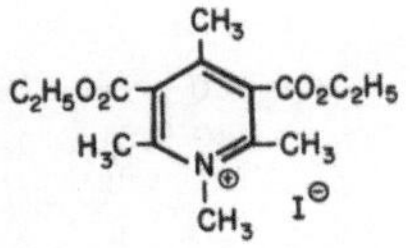

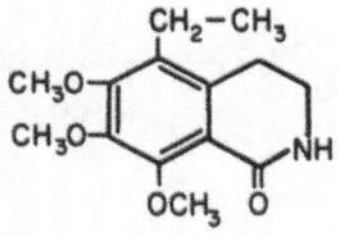

J. Org. Chem. 27:3720 (1962)

▶ $NO_4C_{14}H_{19}$

J. Chem. Soc. 860 (1961)

▶ $NO_4C_{14}H_{21}$

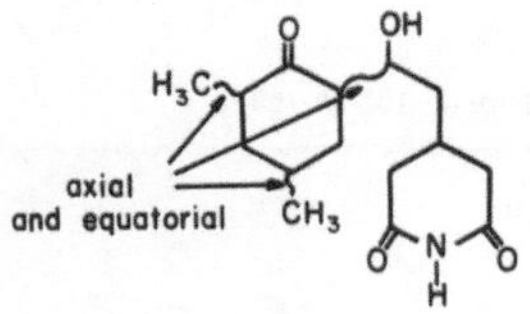

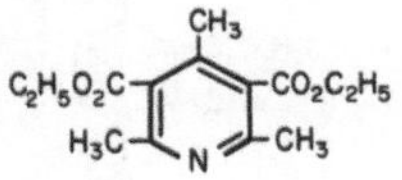

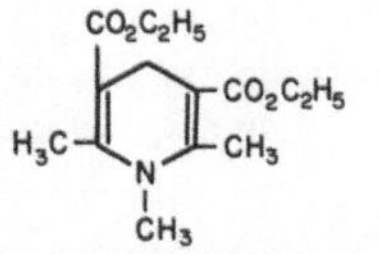

J. Am. Chem. Soc. 84:797 (1962)

▶ $NO_4C_{15}H_{11}$

Varian 635

▶ $NO_4C_{15}H_{17}$

J. Org. Chem. 27:3658 (1962)

▶ $NO_4C_{15}H_{21}$

J. Chem. Soc. 860 (1961)

▶ $NO_4C_{15}H_{22} \cdot I$

J. Chem. Soc. 860 (1961)

▶ $NO_4C_{15}H_{23}$

axial and equatorial

Tetrahedron Letters No. 25:1176 (1962)

▶ $NO_4C_{17}H_{17}$

J. Org. Chem. 27:373 (1962)
Tetrahedron Letters 107 (1961)

▶ $NO_4C_{17}H_{19}$

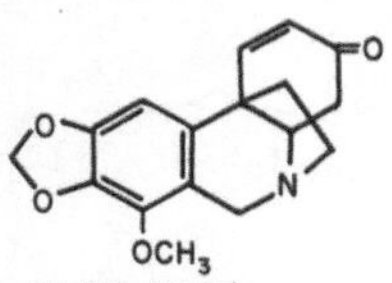

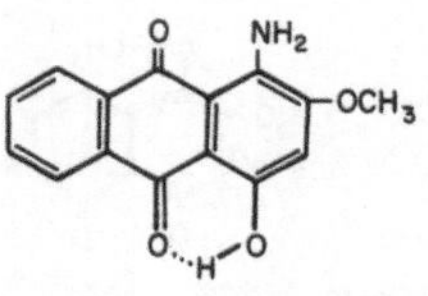

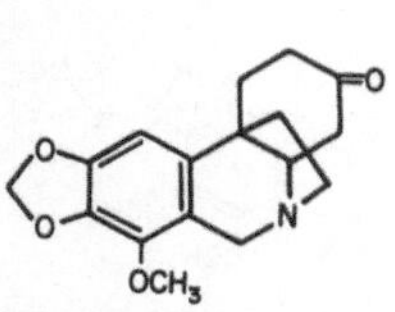

J. Org. Chem. 27:373 (1962)

▶ $NO_4C_{19}H_{19}$

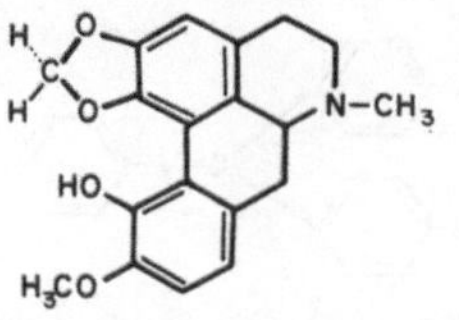

Varian 333

▶ $NO_4C_{19}H_{27}$

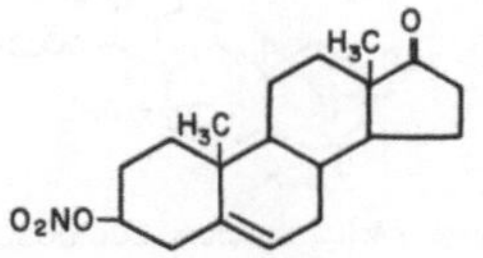

J. Org. Chem. 27:4615 (1962)

▶ $NO_4C_{20}H_{17}$

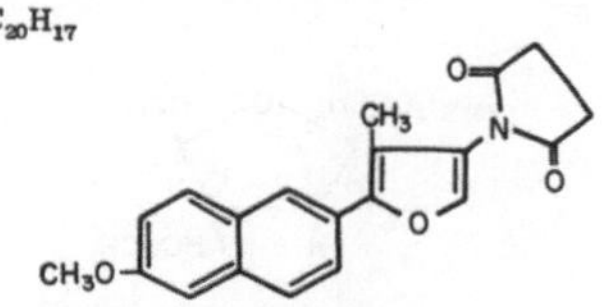

J. Org. Chem. 27:3818 (1962)

▶ $NO_4C_{20}H_{19}$

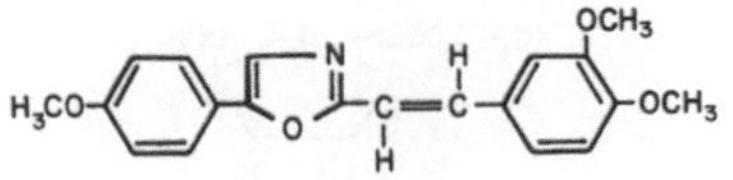

ANNULOLINE

Tetrahedron Letters No. 3:83 (1962)

▶ $NO_4C_{20}H_{21}$

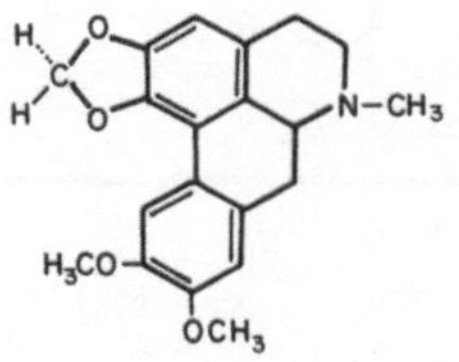

Varian 342

▶ $NO_4C_{20}H_{23}$

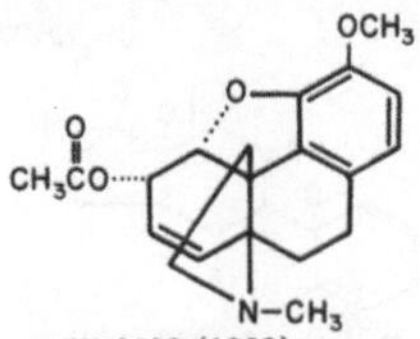

J. Org. Chem. 27:4122 (1962)

▶ $NO_4C_{20}H_{23}$

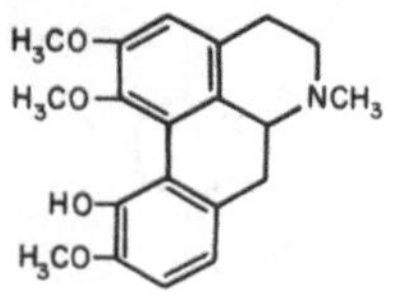

Can. J. Chem. 39:1801 (1961)

▶ $NO_4C_{20}H_{29}$

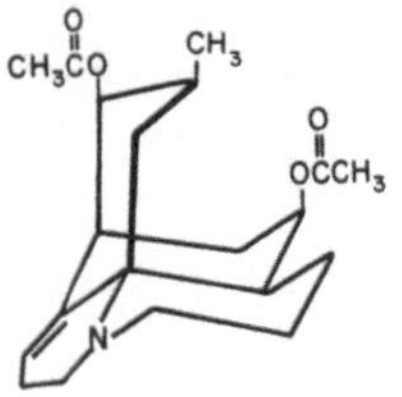

Tetrahedron 18:1467 (1962)

▶ $NO_4C_{20}H_{31}$

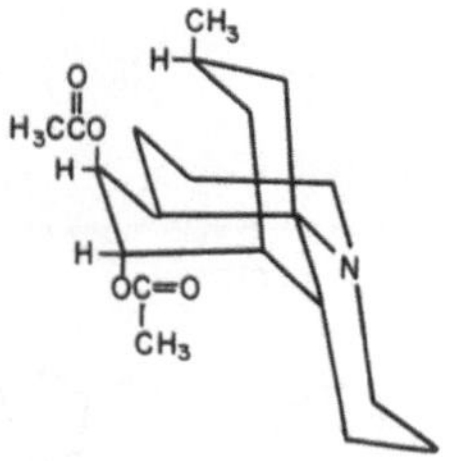

Can. J. Chem. 40:2094 (1962)

▶ $NO_4C_{21}H_{23}$

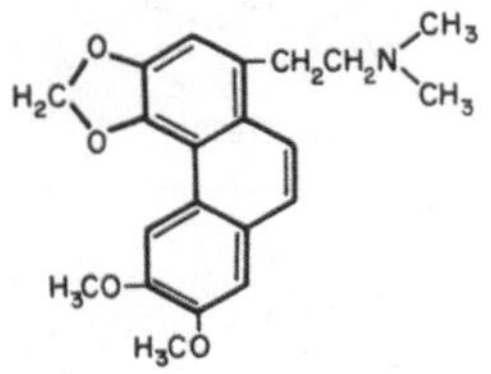

Varian 348

▶ $NO_4C_{21}H_{25}$

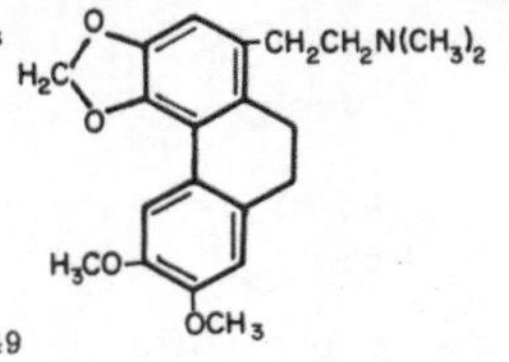

Varian 349

▶ $NO_4C_{21}H_{31}$

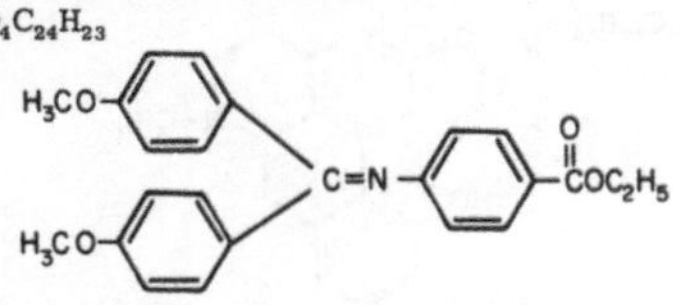

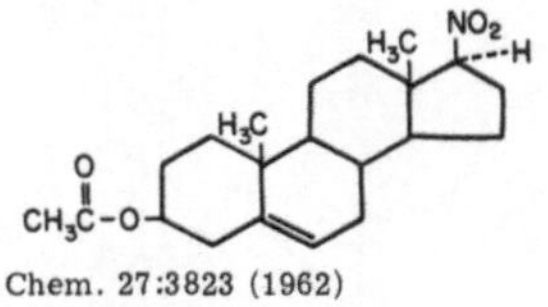

J. Org. Chem. 27:3823 (1962)

▶ $NO_4C_{22}H_{29}$

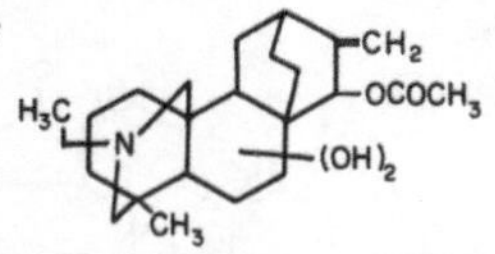

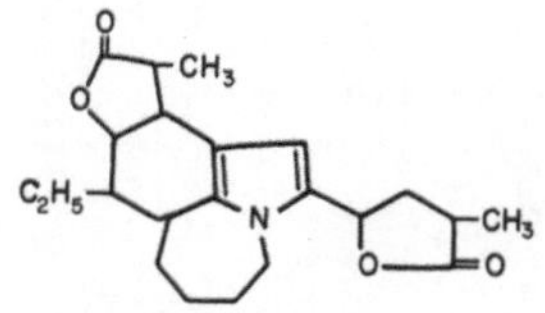

Can. J. Chem. 40:2416 (1962)

▶ $NO_4C_{22}H_{33}$

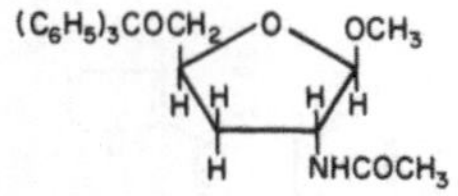

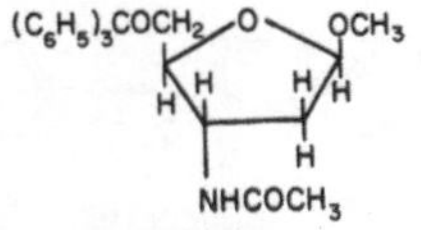

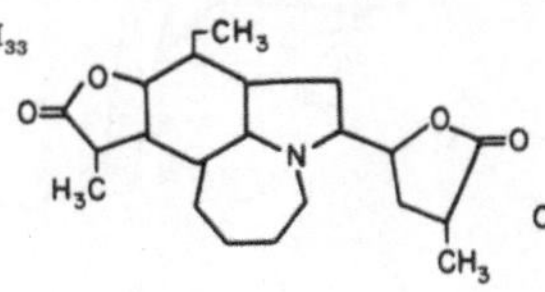

OR ISOMER

Tetrahedron Letters 707 (1961)

▶ $NO_4C_{23}H_{27}$

Tetrahedron 18:1176 (1962)

▶ $NO_4C_{23}H_{29}$

Tetrahedron 18:1173 (1962)

▶ $NO_4C_{24}H_{23}$

Tetrahedron Letters No. 26:1271 (1962)

▶ $NO_4C_{24}H_{37}$

tentative partial structure LUCIDUSCULINE

Bull. Chem. Soc. Japan 34:455 (1961)

▶ $NO_4C_{27}H_{29}$

J. Am. Chem. Soc. 83:1900 (1961)

▶ $NO_4C_{27}H_{29}$

J. Am. Chem. Soc. 83:1900 (1961)

▶ $NO_5ClC_9H_8$

Varian 223

▶ $NO_5Cl_2C_{16}H_{13}$

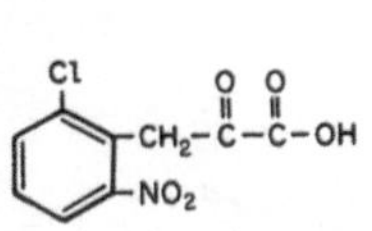

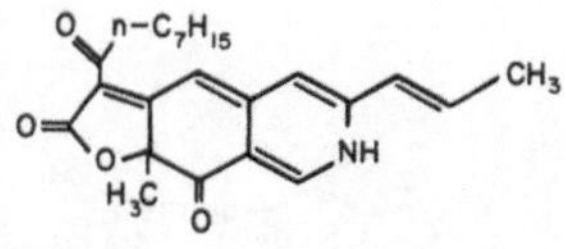

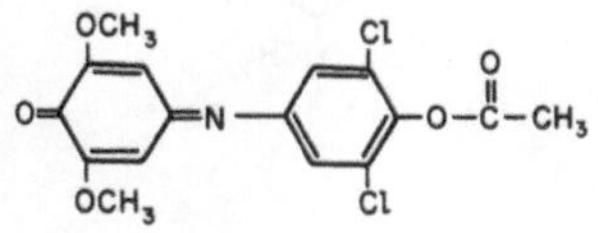

J. Org. Chem. 27:1849 (1962)

▶ $NO_5FC_{23}H_{30}$

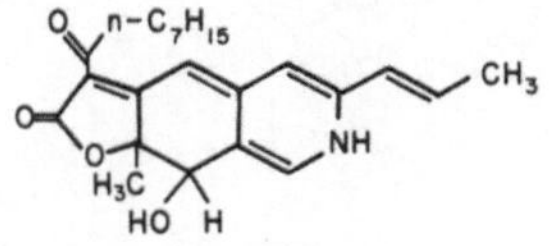

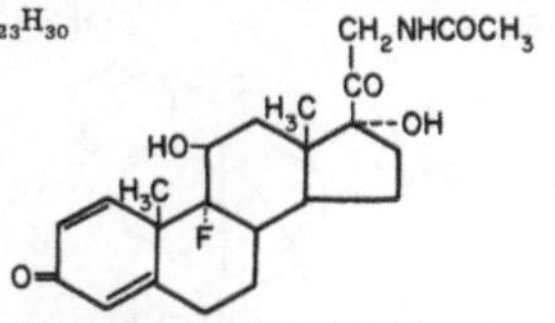

J. Am. Chem. Soc. 84:1265 (1962)

▶ $NO_5FC_{23}H_{32}$

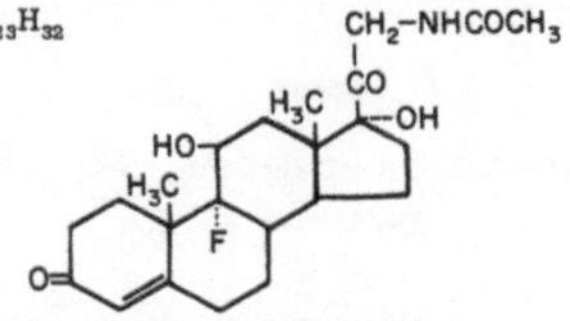

J. Am. Chem. Soc. 84:1265 (1962)

▶ $NO_5C_{12}H_{21}$

DIACETYL DESOSAMINE

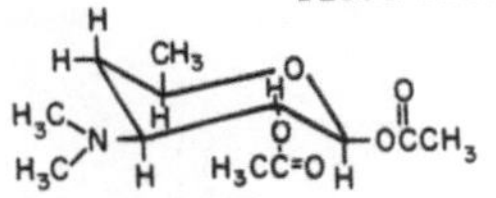

Tetrahedron Letters No. 9:377 (1962); No. 17:735 (1962)

▶ $NO_5C_6H_{13}$

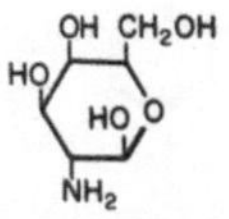

J. Am. Chem. Soc. 84:3216 (1962)

▶ $NO_5C_{14}H_{11}$

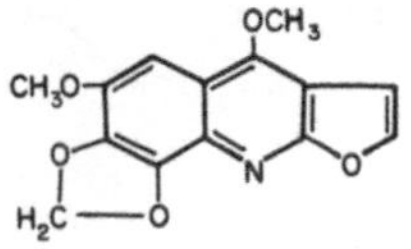

Varian 304

▶ $NO_5C_9H_{15}$

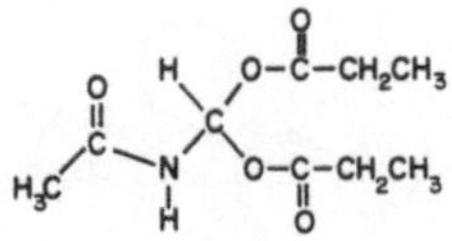

Varian 547

▶ $NO_5C_{14}H_{13}$

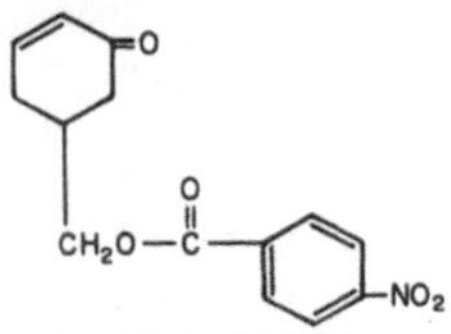

J. Am. Chem. Soc. 84:4125 (1962)

▶ $NO_5C_9H_{17}$

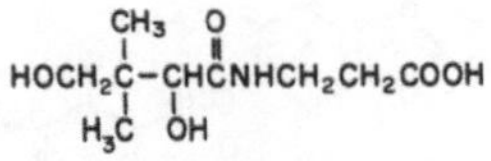

Angew. Chem. 74:751, 998 (1962)
Angew. Chem. Intern. Ed. 1:592 (1962)

▶ $NO_5C_{14}H_{17}$

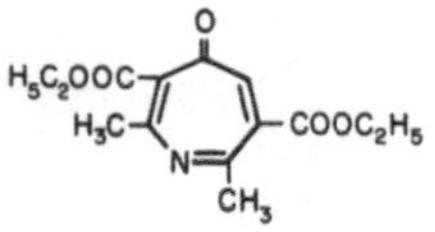

J. Am. Chem. Soc. 84:2260 (1962)

▶ $NO_5C_{10}H_{13}$

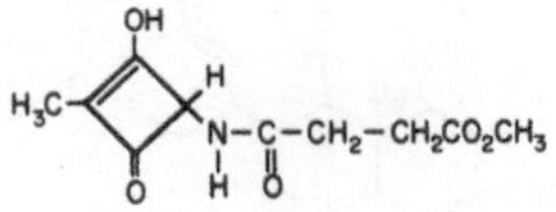

Gazz. Chim. Ital. 91:1315 (1961)

▶ $NO_5C_{14}H_{21}$

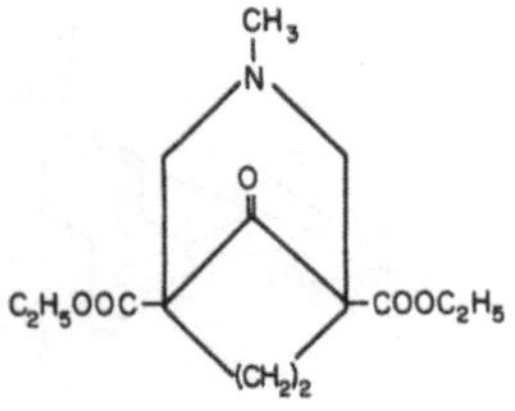

J. Org. Chem. 27:4436 (1962)

▶ $NO_5C_{10}H_{19}$

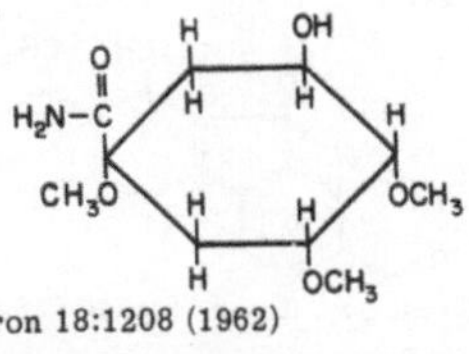

Tetrahedron 18:1208 (1962)

▶ $NO_5C_{11}H_{15}$

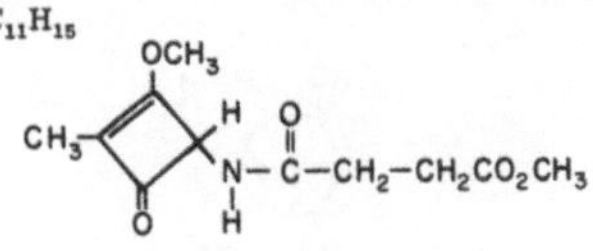

Gazz. Chim. Ital. 91:1315 (1961)
Tetrahedron Letters 543 (1961)

▶ $NO_5C_{15}H_{21}$

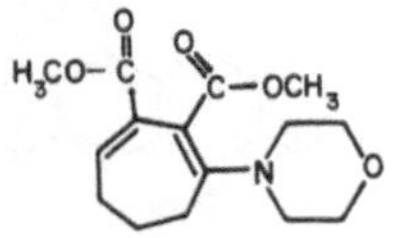

Varian 643

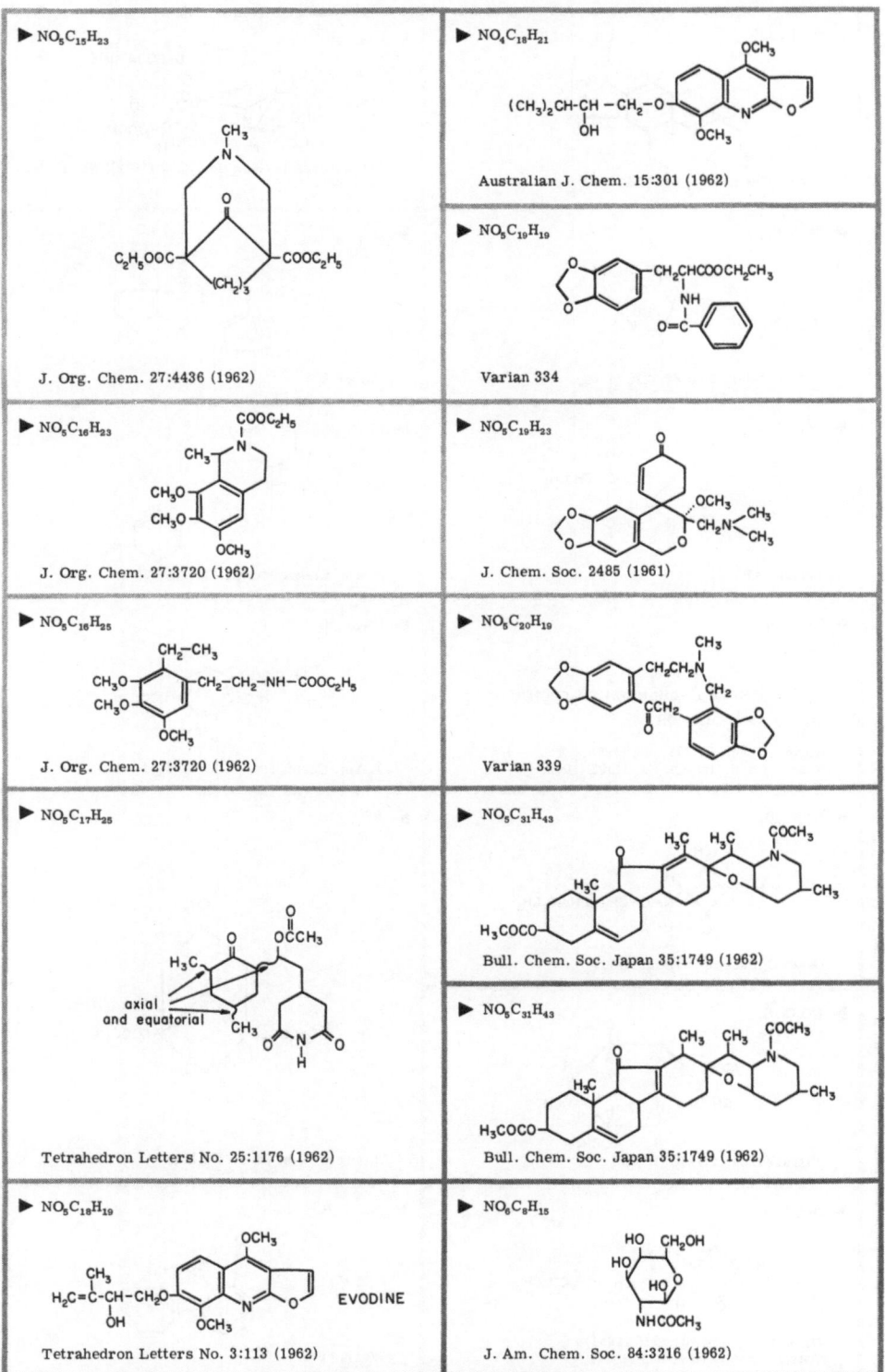

▶ $NO_5C_{15}H_{23}$

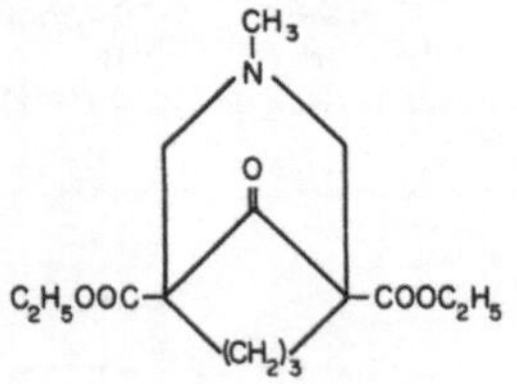

J. Org. Chem. 27:4436 (1962)

▶ $NO_4C_{18}H_{21}$

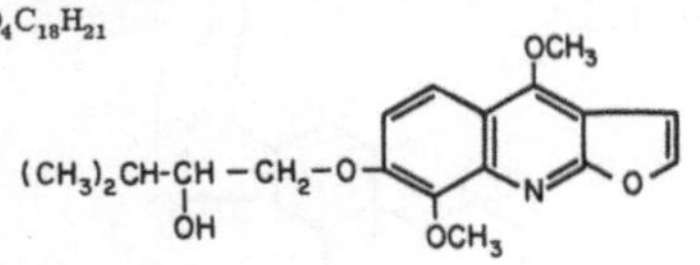

Australian J. Chem. 15:301 (1962)

▶ $NO_5C_{19}H_{19}$

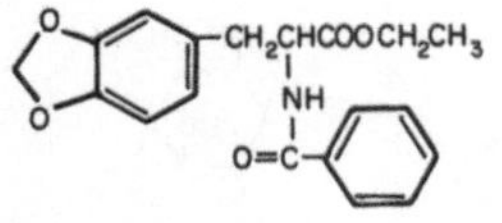

Varian 334

▶ $NO_5C_{16}H_{23}$

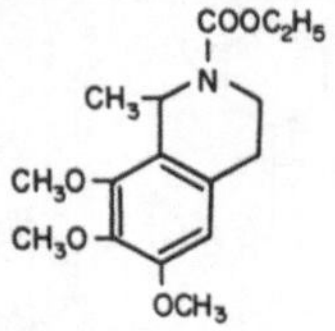

J. Org. Chem. 27:3720 (1962)

▶ $NO_5C_{19}H_{23}$

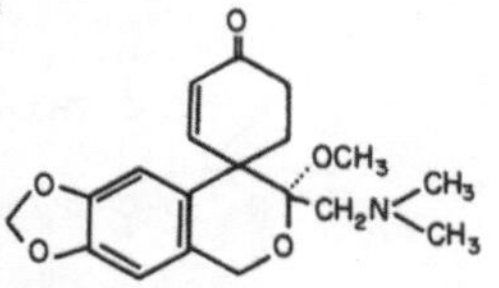

J. Chem. Soc. 2485 (1961)

▶ $NO_5C_{16}H_{25}$

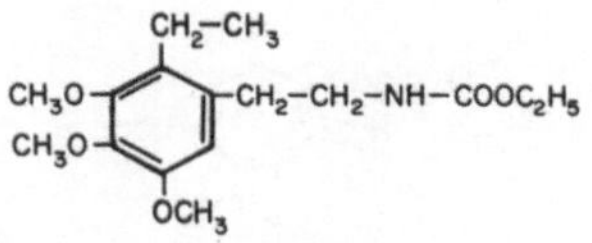

J. Org. Chem. 27:3720 (1962)

▶ $NO_5C_{20}H_{19}$

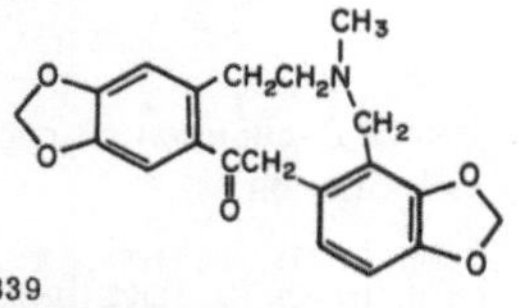

Varian 339

▶ $NO_5C_{17}H_{25}$

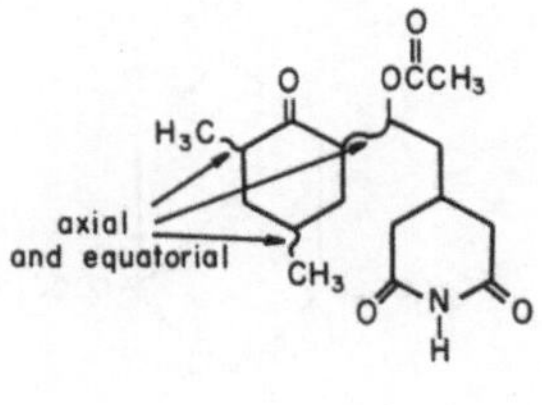


Tetrahedron Letters No. 25:1176 (1962)

▶ $NO_5C_{31}H_{43}$

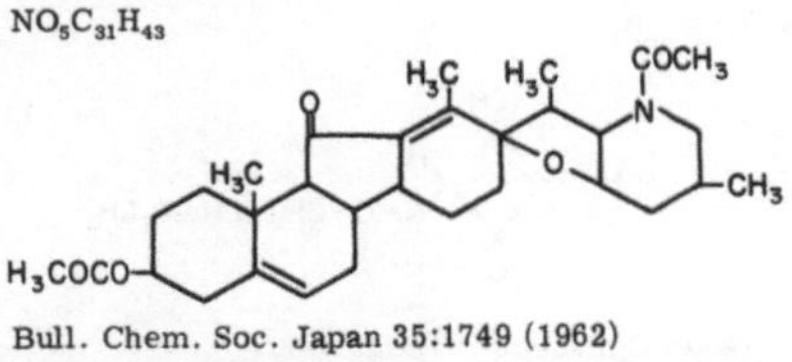

Bull. Chem. Soc. Japan 35:1749 (1962)

▶ $NO_5C_{31}H_{43}$

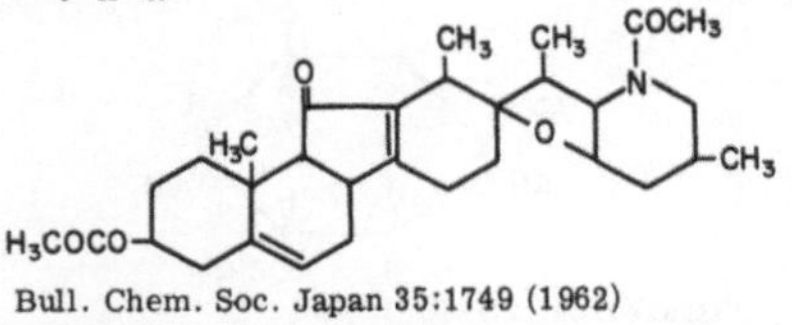

Bull. Chem. Soc. Japan 35:1749 (1962)

▶ $NO_5C_{18}H_{19}$

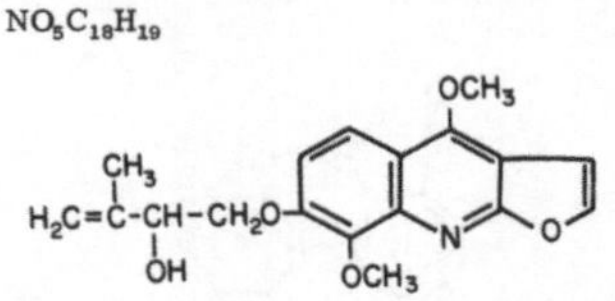

EVODINE

Tetrahedron Letters No. 3:113 (1962)

▶ $NO_6C_8H_{15}$

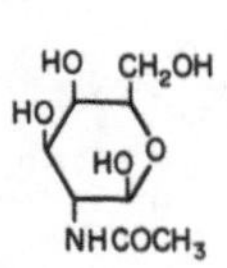

J. Am. Chem. Soc. 84:3216 (1962)

▶ $NO_6C_9H_9$

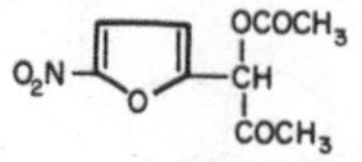


Varian 526

▶ $NO_6C_{10}H_{11}$

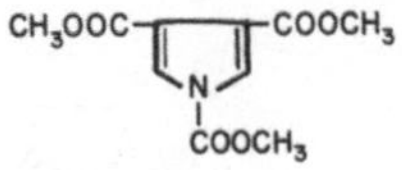


J. Org. Chem. 27:302 (1962)

▶ $NO_6C_{11}H_{13}$

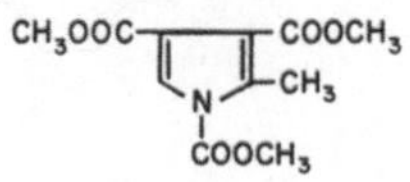


J. Org. Chem. 27:302 (1962)

▶ $NO_6C_{12}H_{15}$

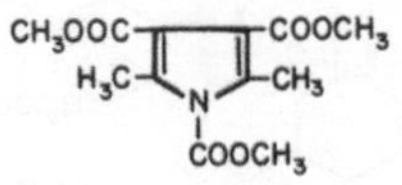


J. Org. Chem. 27:302 (1962)

▶ $NO_6C_{16}H_{23}$

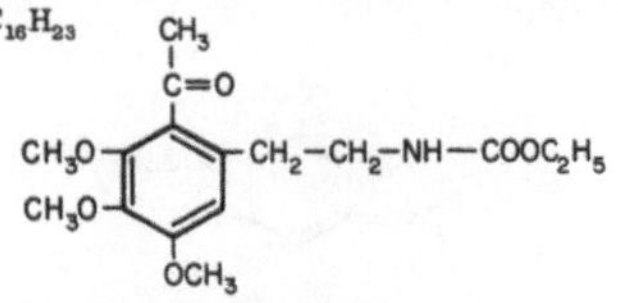


J. Org. Chem. 27:3720 (1962)

▶ $NO_6C_{18}H_{27}$

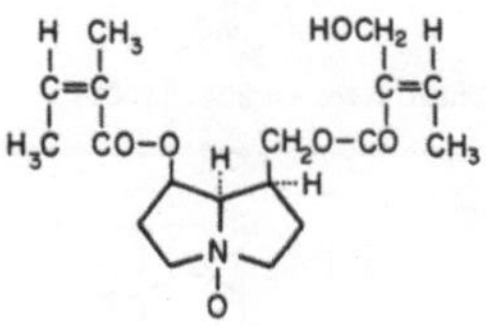


J. Org. Chem. 26:3047 (1961)

▶ $NO_6C_{20}H_{17}$

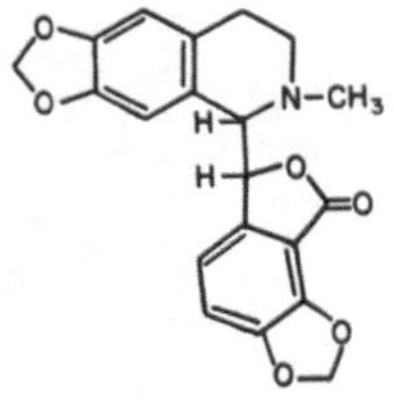


Can. J. Chem. 39:1801 (1961)

▶ $NO_6C_{21}H_{21}$

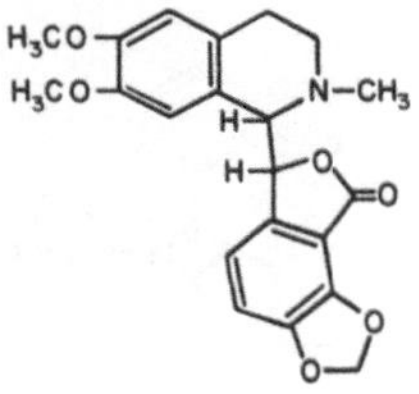


Can. J. Chem. 39:1801 (1961)

▶ $NO_6C_{21}H_{21}$

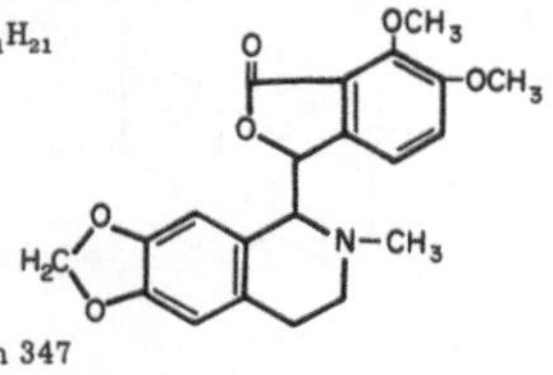


Varian 347

▶ $NO_6C_{22}H_{25}$

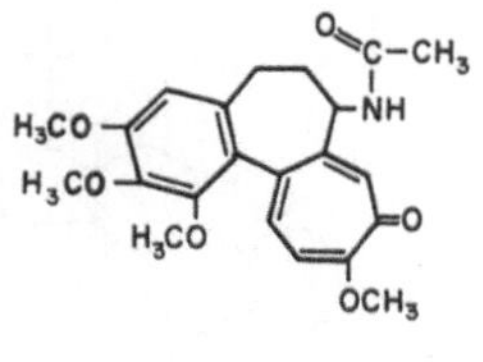


Varian 689

▶ $NO_6C_{22}H_{25}$

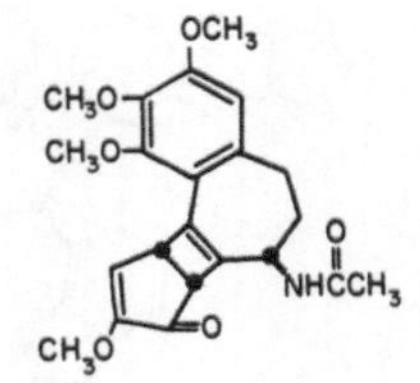

Varian 690
J. Am. Chem. Soc. 83:3914 (1961)

▶ $NO_6C_{22}H_{27}$

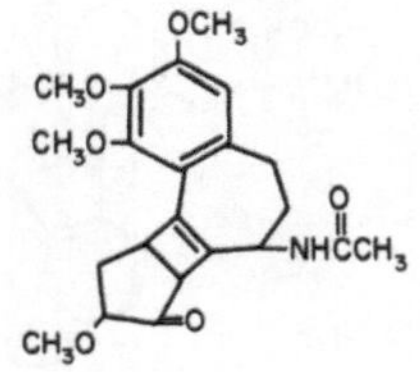

J. Am. Chem. Soc. 83:3914 (1961)

▶ $NO_6C_{23}H_{35}$

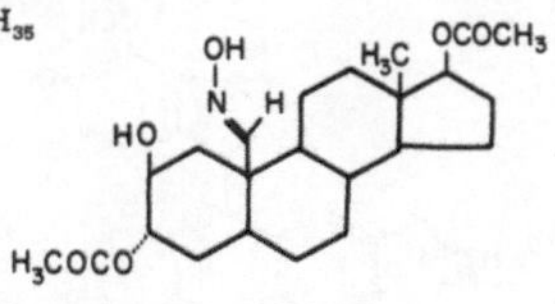

Chem. & Ind. 1192 (1962)

▶ $NO_6C_{24}H_{31}$

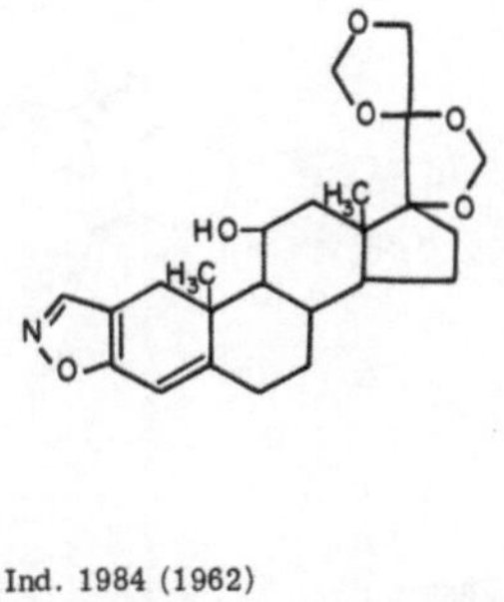

Chem. & Ind. 1984 (1962)

▶ $NO_6C_{24}H_{35}$

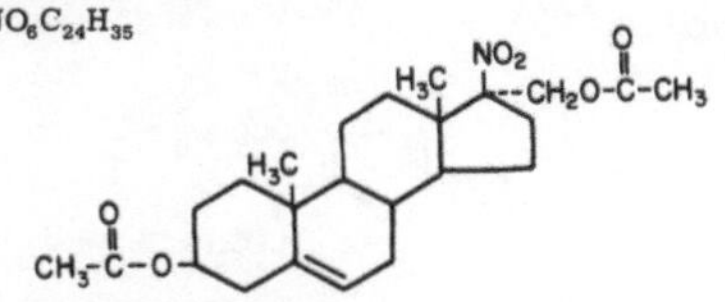

J. Org. Chem. 27:3823 (1962)

▶ $NO_6C_{28}H_{35}$

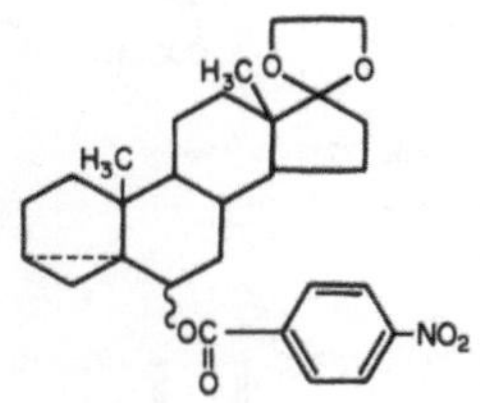

J. Org. Chem. 27:4610 (1962)

▶ $NO_6C_{33}H_{45}$

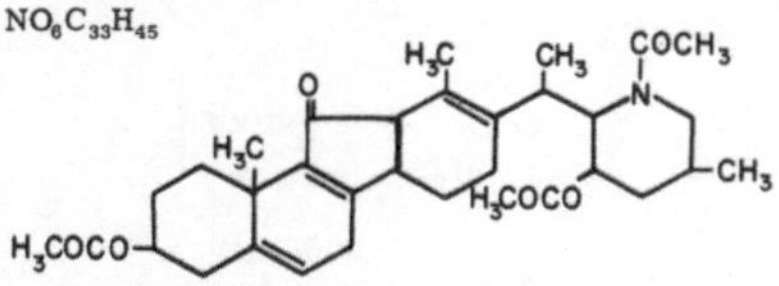

Bull. Chem. Soc. Japan 35:1749 (1962)

▶ $NO_7C_{10}H_{11}$

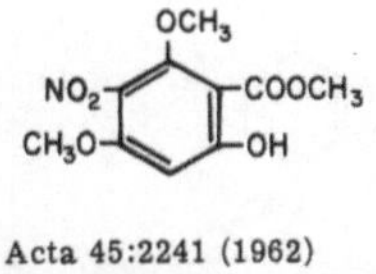

Helv. Chim. Acta 45:2241 (1962)

▶ $NO_7C_{10}H_{11}$

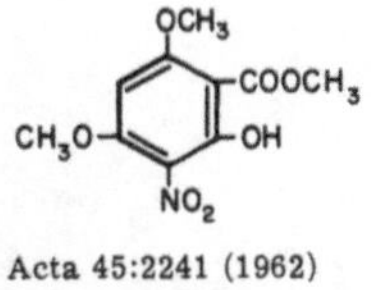

Helv. Chim. Acta 45:2241 (1962)

▶ $NO_7C_{12}H_{17}$

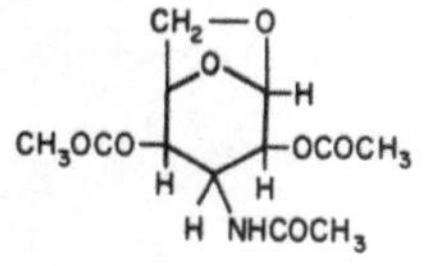

Proc. Chem. Soc. 382 (1962)

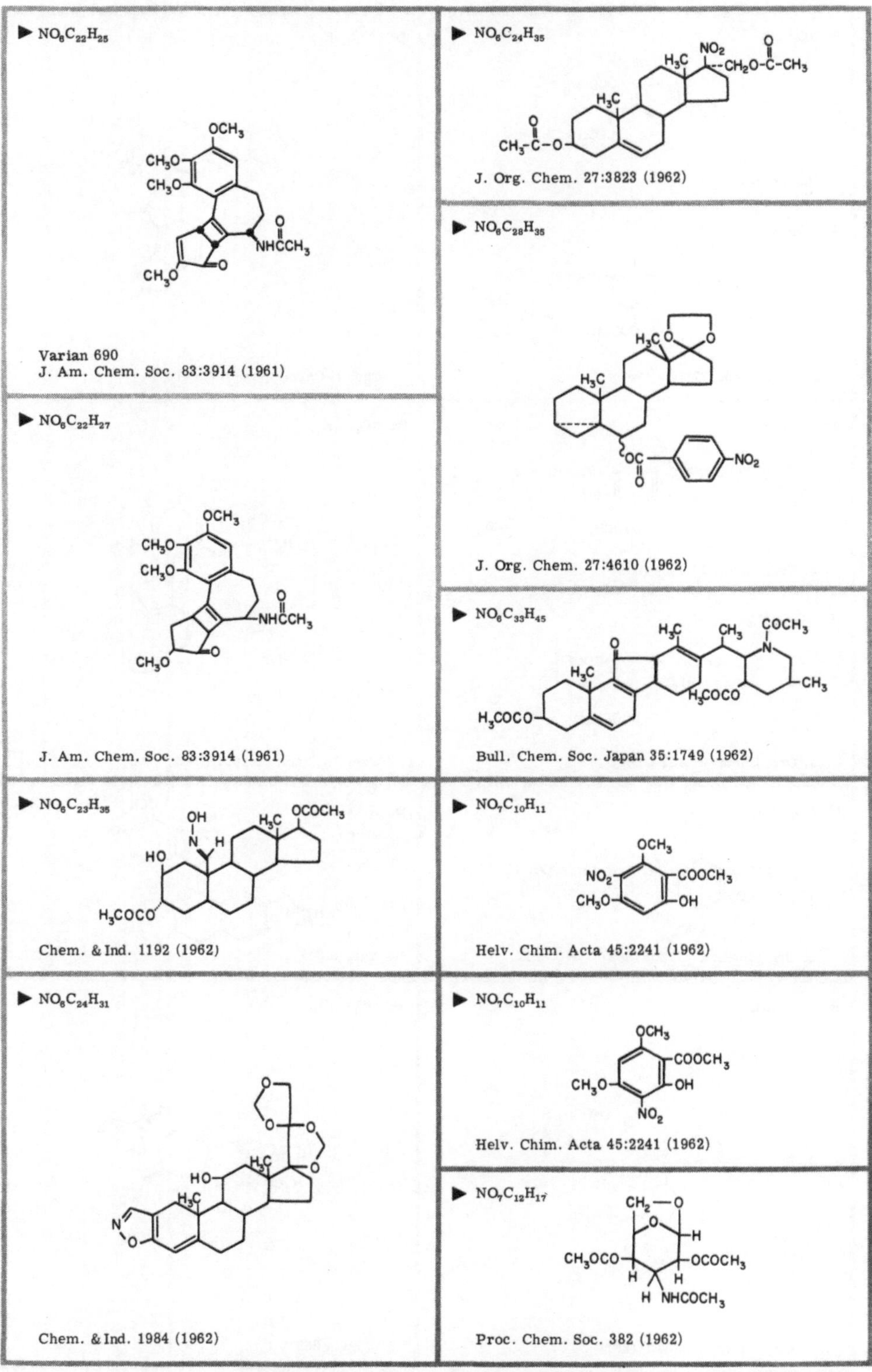

▶ $NO_7C_{13}H_{21}$

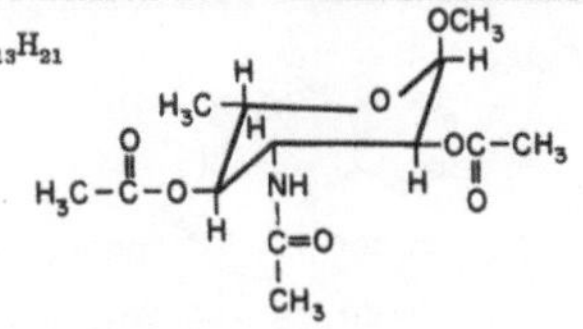

J. Chem. Soc. 2503 (1962)

▶ $NO_7C_{13}H_{21}$

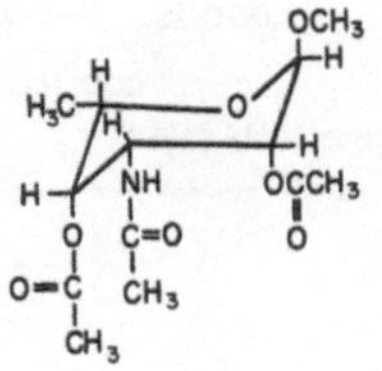

J. Chem. Soc. 2503 (1962)

▶ $NO_7C_{16}H_{17}$

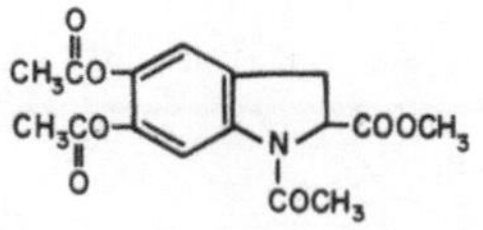

Helv. Chim. Acta 45:638 (1962)

▶ $NO_7C_{21}H_{23}$

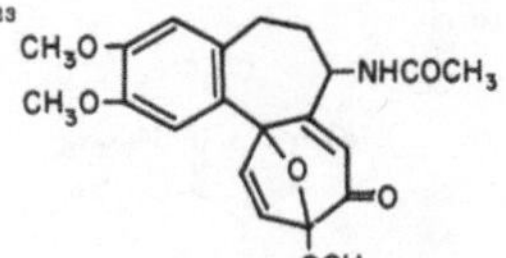

Chem. & Ind. 859 (1962)

▶ $NO_7C_{24}H_{39}$

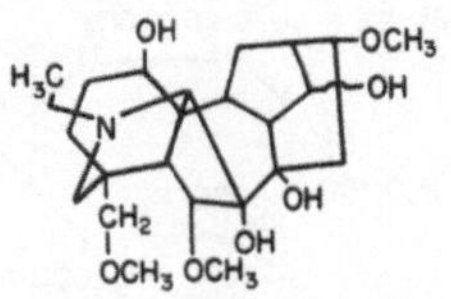

Bull. Chem. Soc. Japan 34:455 (1961)

▶ $NO_7C_{29}H_{31}$

DIACETYLVERTICILLATINE

J. Org. Chem. 27:2987 (1962)

▶ $NO_7C_{29}H_{33}$

DIACETYLDIHYDROVERTICILLATINE

J. Org. Chem. 27:2987 (1962)

▶ $NO_8ClC_{17}H_{16}$

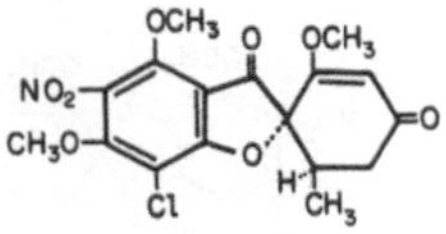

Helv. Chim. Acta 45:2241 (1962)

▶ $NO_8C_{14}H_{21}$

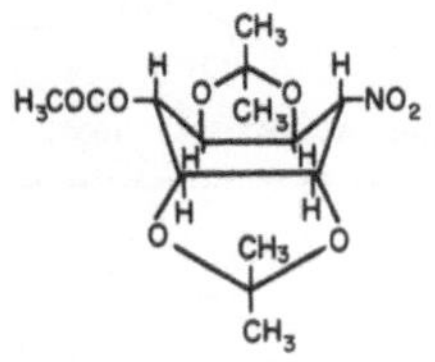

Chem. Ber. 94:3071 (1961)

▶ $NO_8C_{14}H_{21}$

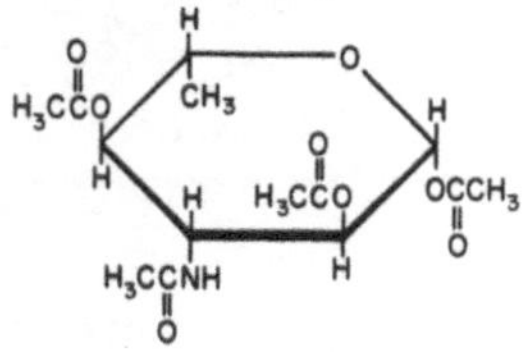

J. Chem. Soc. 2503 (1962)

▶ $NO_8C_{14}H_{21}$

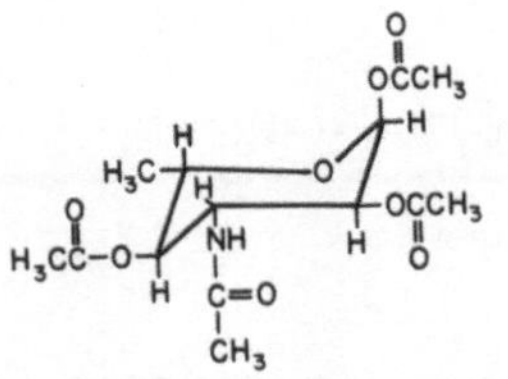

J. Chem. Soc. 2503 (1962)

▶ $NO_8C_{16}H_{15}$

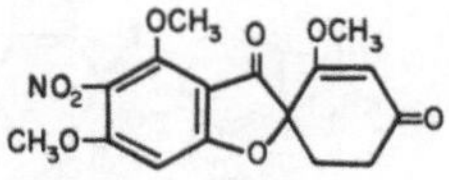

Helv. Chim. Acta 45:2241 (1962)

▶ $NO_8C_{16}H_{15}$

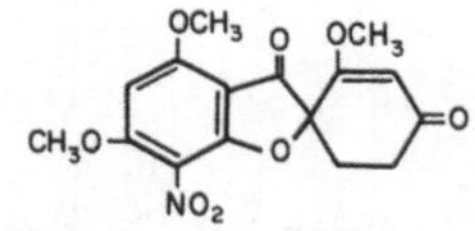

Helv. Chim. Acta 45:2241 (1962)

▶ $NO_8C_{17}H_{17}$

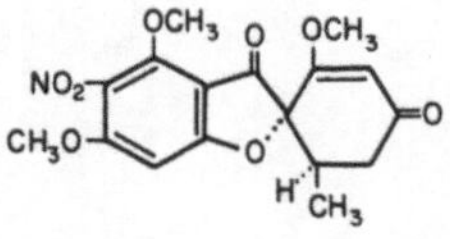

Helv. Chim. Acta 45:2241 (1962)

▶ $NO_8C_{17}H_{17}$

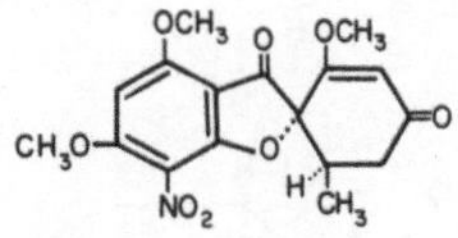

Helv. Chim. Acta 45:2241 (1962)

▶ $NO_8C_{17}H_{19}$

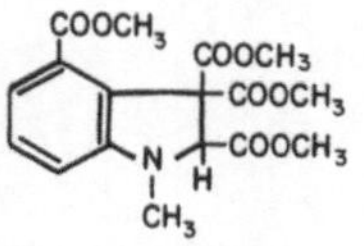

Proc. Chem. Soc. 277 (1962)

▶ $NO_8C_{22}H_{21}$

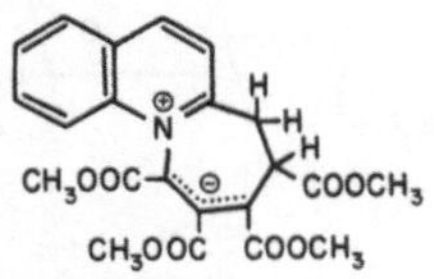

J. Chem. Soc. 4418 (1962)

▶ $NO_8C_{22}H_{21}$

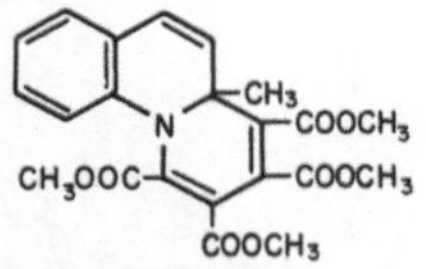

J. Chem. Soc. 4418 (1962)

▶ $NO_8C_{25}H_{25}$

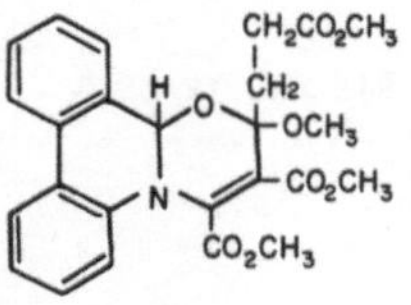

J. Chem. Soc. 3762 (1962)

▶ $NO_9C_{18}H_{19}$

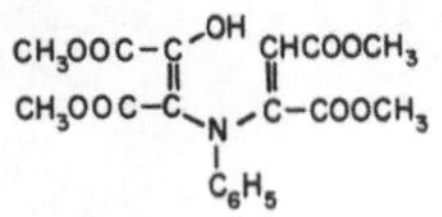

J. Org. Chem. 26:1724 (1961)

▶ $NO_{10}C_{14}H_{13}$

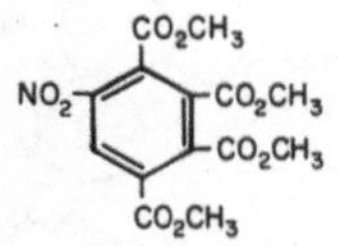

Bull. Soc. Chim. France 1255 (1962)

▶ $NO_{11}C_{15}H_{21}$

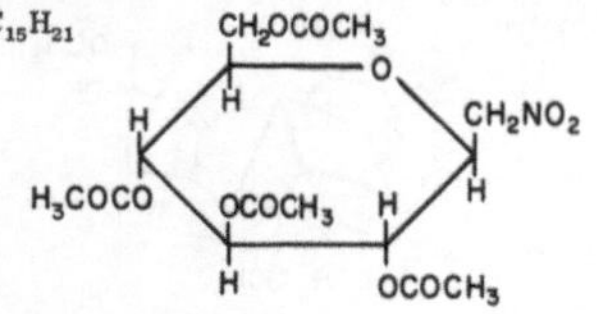

Chem. & Ind. 1827 (1962)

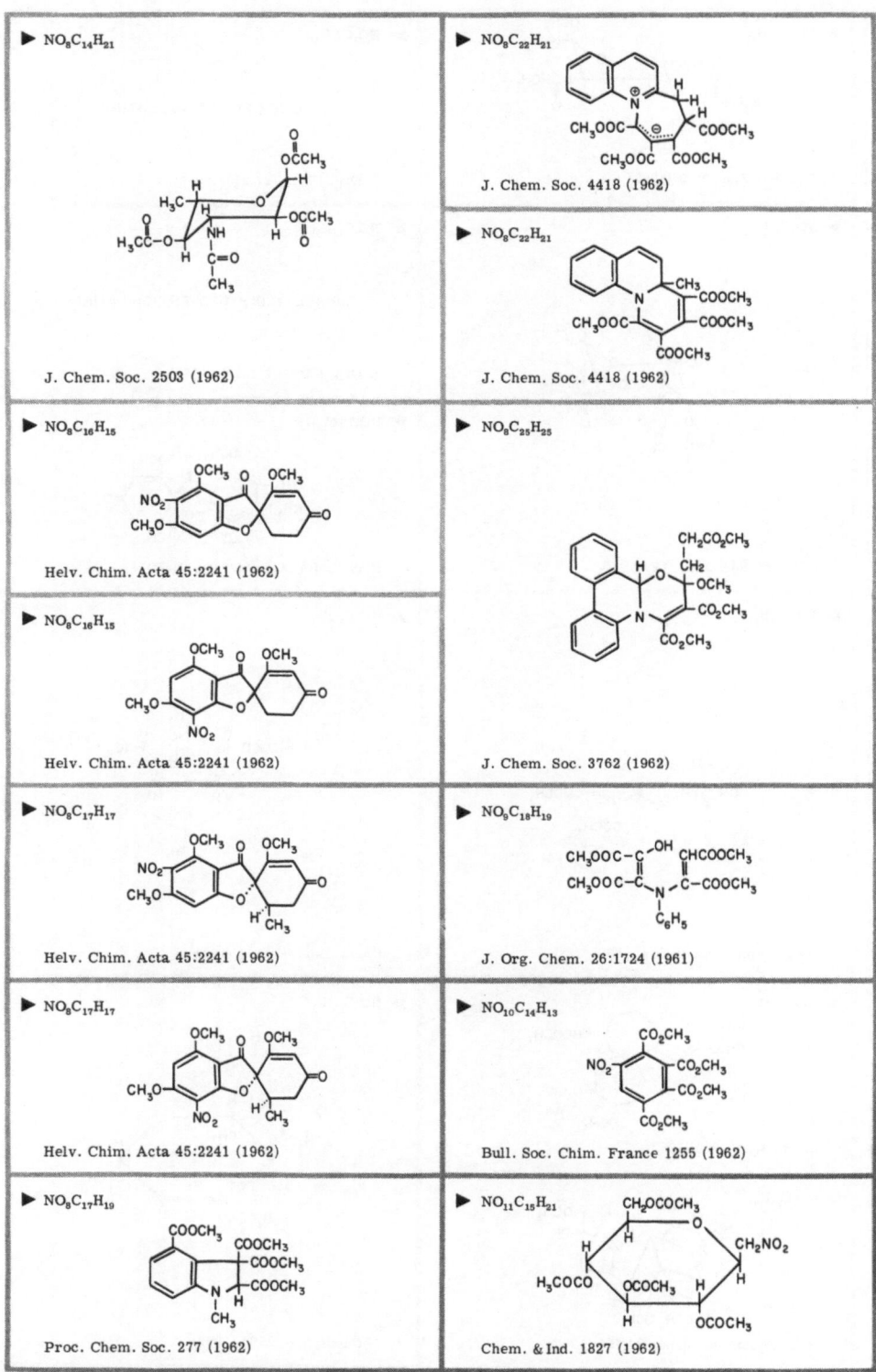

▶ $NO_{11}C_{15}H_{21}$

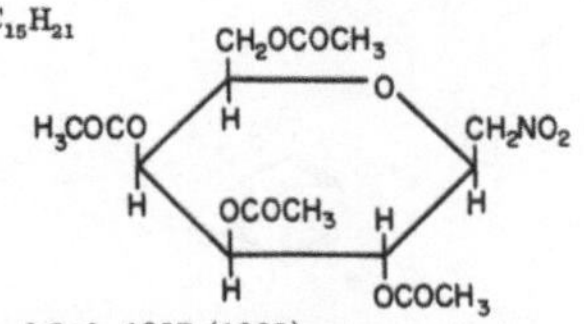

Chem. & Ind. 1827 (1962)

▶ $NO_{11}C_{15}H_{21}$

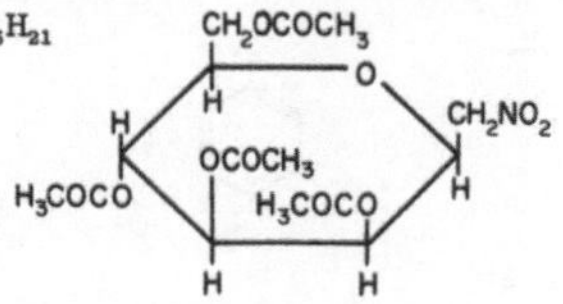

Chem. & Ind. 1827 (1962)

▶ $NO_{11}C_{18}H_{25}$

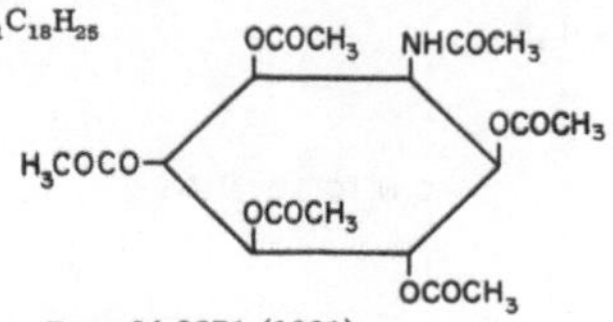

Chem. Ber. 94:3071 (1961)

▶ $NO_{11}C_{18}H_{25}$

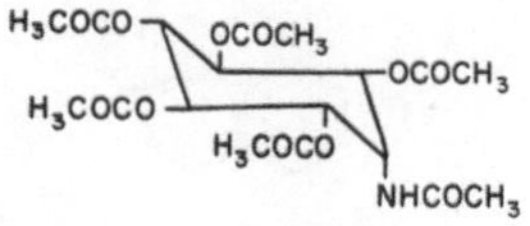

J. Am. Chem. Soc. 83:2005 (1961)

▶ $NO_{11}C_{18}H_{25}$

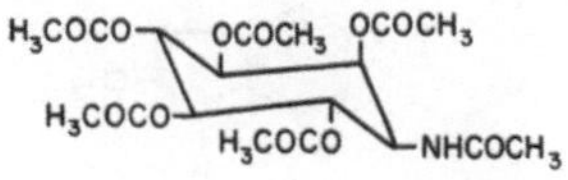

Chem. Ber. 94:3071 (1961)

▶ $NO_{12}C_{16}H_{21}$

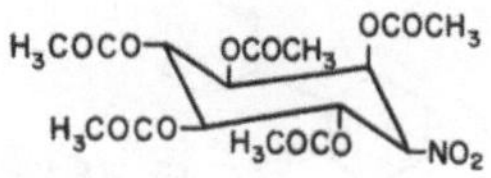

Chem. Ber. 94:3071 (1961)

▶ $NO_{12}C_{16}H_{21}$

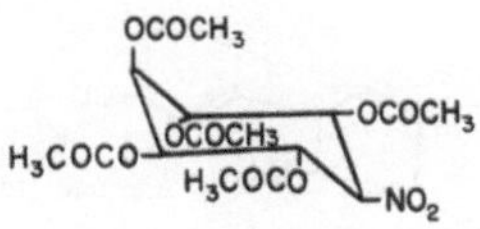

Chem. Ber. 94:3071 (1961)

▶ $NO_{12}C_{16}H_{21}$

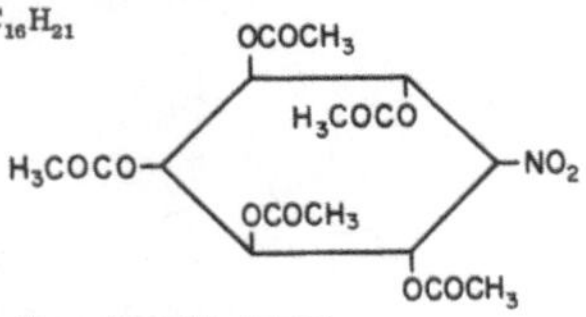

Chem. Ber. 94:3071 (1961)

▶ $NO_{12}C_{38}H_{61}$

ACUMYCIN

Helv. Chim. Acta 45:1396 (1962)

▶ $NIC_6H_6$

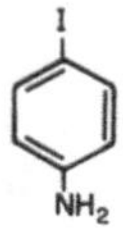

J. Chem. Phys. 37:2594 (1962)
Helv. Chim. Acta 45:568 (1962)

▶ $NBrC_4H_6$

$Br-CH_2-CH_2-CH_2-CN$

Varian 58

▶ $NBrC_5H_4$

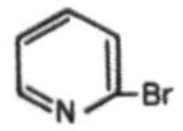

J. Chem. Phys. 37:2603 (1962)

▶ $NBrC_5H_4$

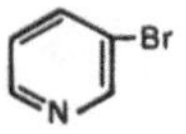

J. Chem. Phys. 36:266 (1962)

▶ $NBrC_6H_6$

J. Chem. Phys. 37:2594 (1962)
Helv. Chim. Acta 45:568 (1962)

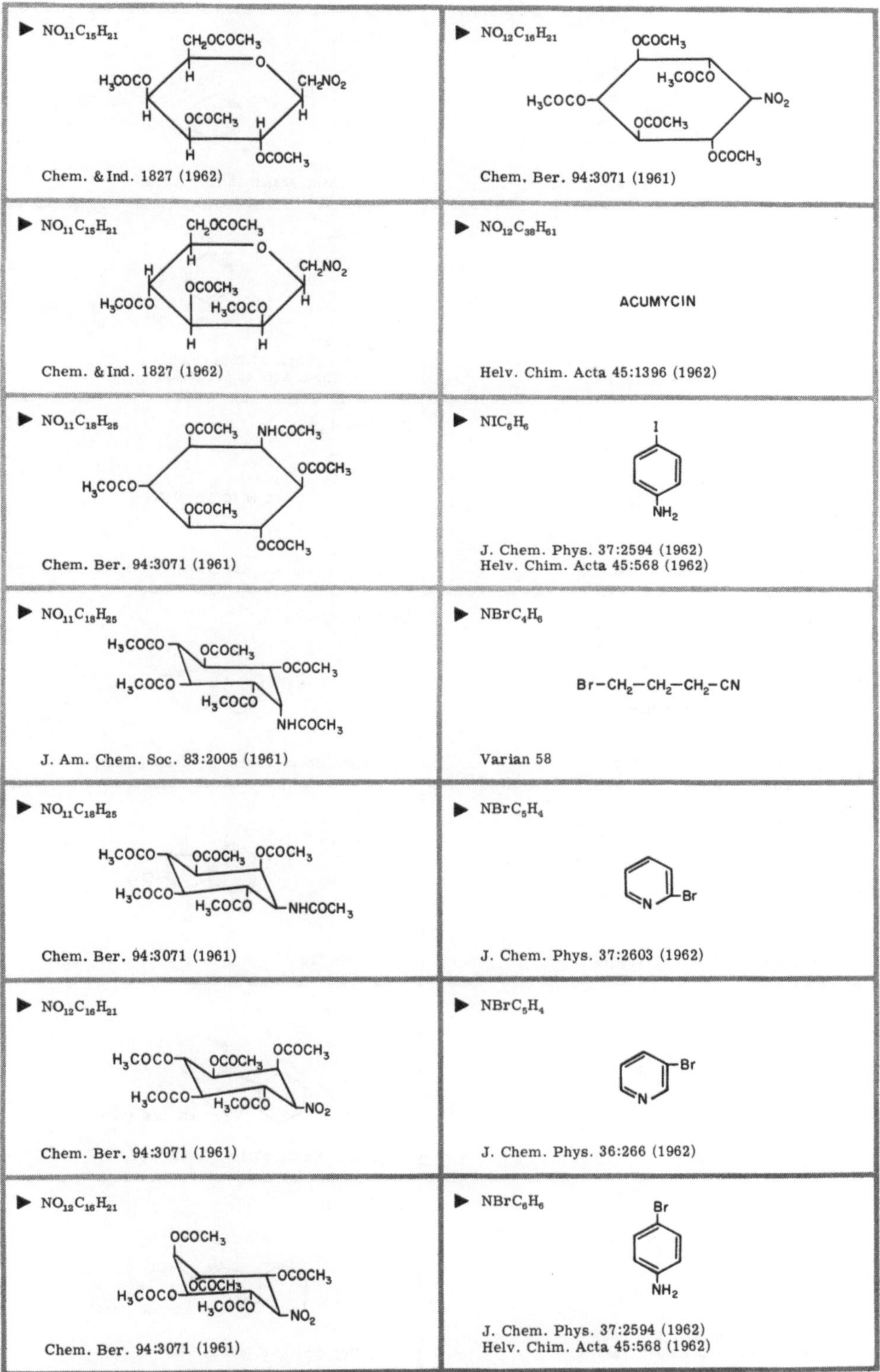

▶ $NBrC_6H_{12}$

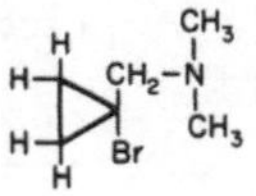

J. Org. Chem. 27:2722 (1962)

▶ $NClC_2H_2$

$C^{13}H_2ClCN$

J. Am. Chem. Soc. 83:4479 (1961)

▶ $NClC_2H_2$

$CH_2ClCN$

J. Chem. Phys. 37:2198 (1962)

▶ $NClC_5H_4$

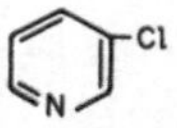

J. Chem. Phys. 36:266 (1962)

▶ $NClC_5H_4$

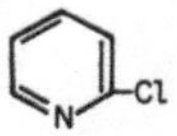

J. Chem. Phys. 37:2603 (1962)

▶ $NClC_5H_{12} \cdot HCl$

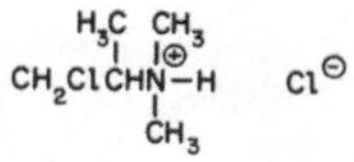

Varian 448

▶ $NClC_6H_6$

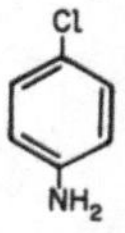

Varian 123
J. Chem. Phys. 37:2594 (1962)
Helv. Chim. Acta 45:568 (1962)

▶ $NClC_6H_6$

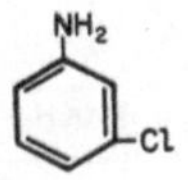

Acta Chem. Scand. 16:1031 (1962)

▶ $NClC_7H_4$

Varian 483
J. Chem. Phys. 37:2594 (1962)
Helv. Chim. Acta 45:568 (1962)

▶ $NClC_8H_{11}B$

$C_6H_5(Cl)B \cdot N(CH_3)_2$

Proc. Chem. Soc. 421 (1961)

▶ $NClC_9H_6$

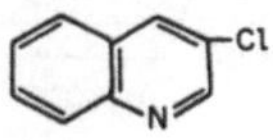

Angew. Chem. Intern. Ed. 1:215 (1962)

▶ $NClC_9H_8$

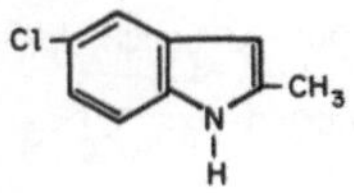

Varian 228

▶ $NClC_{12}H_{16}$

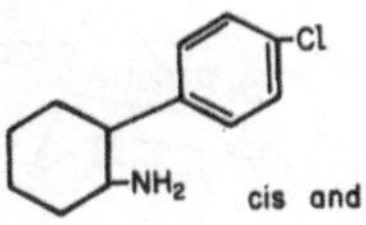

cis and trans

J. Org. Chem. 27:3008 (1962)

▶ $NClC_{12}H_{16}$

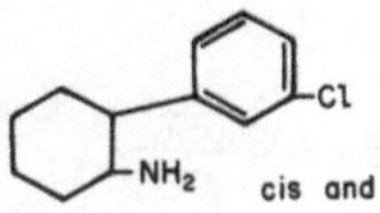

cis and trans

J. Org. Chem. 27:3008 (1962)

▶ $NClC_{12}H_{16}$

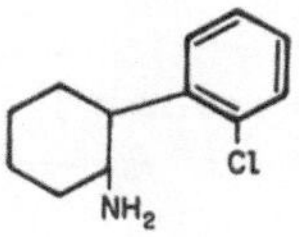


cis and trans

J. Org. Chem. 27:3008 (1962)

▶ $NCl_2C_2H$

$C^{13}HCl_2CN$

J. Am. Chem. Soc. 83:4479 (1961)

▶ $NCl_2C_2H$

$CHCl_2CN$

J. Chem. Phys. 37:2198 (1962)

▶ $NCl_2C_5H_5$

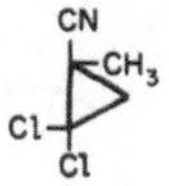


Chem. Ber. 95:2280 (1962)

▶ $NCl_2C_5H_{14}B$

$(CH_3)_3N \cdot BCl_2CH_2CH_3$

J. Am. Chem. Soc. 83:4138 (1961)

▶ $NCl_2C_{12}H_{15}$

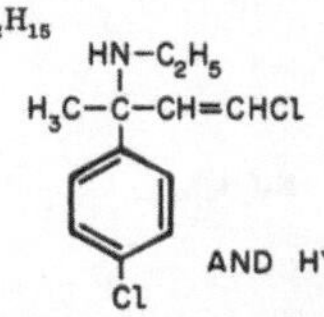


AND HYDROCHLORIDE

J. Org. Chem. 27:2747 (1962)

▶ $NFC_6H_4D_2$

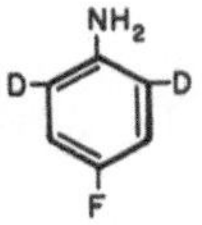


Acta Chem. Scand. 16:1031 (1962)

▶ $NFC_6H_5D$

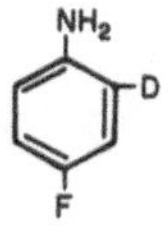


Acta Chem. Scand. 16:1031 (1962)

▶ $NFC_6H_6$

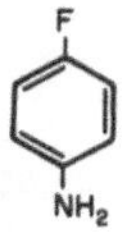


Acta Chem. Scand. 16:1031 (1962)
Helv. Chim. Acta 45:568 (1962)

▶ $NFC_7H_4$

Helv. Chim. Acta 45:568 (1962)

▶ $NF_2C_5H_{14}B$

$(CH_3)_3N \cdot BF_2C_2H_5$

J. Am. Chem. Soc. 83:4138 (1961)

▶ $NCH_3D_2$

$CH_3ND_2$

( AND $CH_3ND_2$ ADSORBED ON ALUMINA )

Bull. Chem. Soc. Japan 35:1545 (1962)

▶ $NCH_5$

$CH_3-NH_2$

J. Chem. Phys. 34:1099 (1961); 37:2198 (1962)
Bull. Chem. Soc. Japan 35:1545 (1962)

▶ $NCH_9Si_2$

$(SiH_3)_2NCH_3$

J. Chem. Soc. 4879 (1961)

▶ $NC_2HD_2$

$CHD_2CN$

J. Am. Chem. Soc. 83:4726 (1961)

▶ $NC_2H_2D$

$CH_2DCN$

J. Chem. Phys. 37:3012 (1962)
J. Am. Chem. Soc. 83:4726 (1961)

▶ $NC_2H_3$

$CH_3CN$

Bull. Chem. Soc. Japan 34:143 (1961)
J. Am. Chem. Soc. 83:4726 (1961)
J. Chem. Phys. 34:1062, 1064, 1099 (1961); 35:1533 (1961); 37:2198 (1962)
J. Phys. Chem. 66:2653 (1962)
J. Chem. Soc. 2576 (1962)

▶ $NC_2H_5$

Varian 372

▶ $NC_2H_7$

$CH_3NHCH_3$

J. Chem. Phys. 37:2198 (1962)

▶ $NC_2H_7$

$C_2H_5NH_2$

J. Chem. Phys. 34:1099 (1961)
Can. J. Chem. 40:1522 (1962)
J. Chem. Soc. 1124 (1962)
Proc. Chem. Soc. 118 (1961)

▶ $NC_2H_9Si$

$SiH_3N(CH_3)_2$

J. Chem. Soc. 4879 (1961)

▶ $NC_3H_2D$

$H_2C{=}CDCN$

J. Am. Chem. Soc. 83:1300 (1961)

▶ $NC_3H_3$

$CH_2{=}CHCN$

Varian 15
Can. J. Chem. 40:2 (1962)
J. Phys. Chem. 66:2653 (1962)
J. Chem. Phys. 34:295 (1961)
J. Am. Chem. Soc. 83:231, 1300, 2045 (1961)

▶ $NC_3H_5$

$C_2H_5CN$

J. Chem. Phys. 34:1099 (1961)
J. Phys. Chem. 66:2653 (1962)

▶ $NC_3H_7$

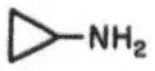


Varian 37

▶ $NC_3H_7$

$H_2C{=}CHCH_2NH_2$

Varian 38

▶ $NC_3H_9$

$N(CH_3)_3$

Bull. Chem. Soc. Japan 35:1317 (1962)

▶ $NC_3H_9$

$CH_3CH_2CH_2NH_2$

J. Chem. Phys. 34:1094, 1099 (1961)

▶ $NC_3H_9$

$CH_3N(CH_3)_2$

J. Chem. Phys. 37:2198 (1962)

▶ $NC_3H_9$

$(CH_3)_2CHNH_2$

J. Org. Chem. 27:4091 (1962)
J. Chem. Phys. 34:1099 (1961)
J. Am. Chem. Soc. 83:1230 (1961)

▶ $NC_3H_9 \cdot BCl_3$

$(CH_3)_3N \cdot BCl_3$

Bull. Chem. Soc. Japan 35:1317 (1962)

▶ $NC_3H_9 \cdot BF_3$

$(CH_3)_3N \cdot BF_3$

J. Am. Chem. Soc. 83:4138 (1961)

▶ $NC_3H_9 \cdot BH_3$

$(CH_3)_3N \cdot BH_3$

J. Am. Chem. Soc. 83:4138 (1961)

▶ $NC_4H_3$

$HC{\equiv}C-CH_2CN$

J. Am. Chem. Soc. 83:4729 (1961)

▶ $NC_4H_4 \cdot Mn(CO)_3$

Proc. Chem. Soc. 326 (1962)

▶ $NC_4H_5$

Chem. Ber. 95:2280 (1962)

▶ $NC_4H_5$

$CH_2{=}CHCH_2CN$

Varian 56
Chem. Ber. 95:2280 (1962)

▶ $NC_4H_5$

$CH_3-CH=CH-CN$

cis and trans

J. Chem. Phys. 36:2644 (1962)
J. Am. Chem. Soc. 83:1300, 2045 (1961)

▶ $NC_4H_5$

$H_2C{=}C(CH_3)-CN$

Varian 97
Arkiv Kemi 17:1 (1961)
J. Chem. Phys. 36:2644 (1962)
J. Am. Chem. Soc. 83:1300, 2045 (1961)

▶ $NC_4H_5$

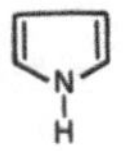


Varian 55
Arkiv Kemi 18:133 (1961)
J. Am. Chem. Soc. 83:5020 (1961)

▶ $NC_4H_7$

$(CH_3)_2CHNC$

Varian 408
J. Chem. Phys. 35:1533 (1961)

▶ $NC_4H_7$

$CH_3CH_2CH_2CN$

J. Chem. Phys. 34:1094, 1099 (1961)

▶ $NC_4H_7$

$CH_3-CH(CN)-CH_3$

J. Chem. Phys. 34:1099 (1961)

▶ $NC_4H_9$

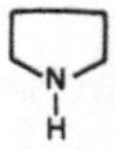

Tetrahedron 18:791 (1962)

▶ $NC_4H_9Si$

$(CH_3)_3SiCN$

J. Chem. Phys. 37:2198 (1962)

▶ $NC_4H_{10}$

$(CH_3)_2CHCH_2NH_2$

Proc. Chem. Soc. 118 (1961)

▶ $NC_4H_{11}$

$(C_2H_5)_2NH$

Proc. Chem. Soc. 118 (1961)
J. Chem. Soc. 1124 (1962)

▶ $NC_4H_{11}$

$CH_3CH_2CH(NH_2)CH_3$

Varian 88

▶ $NC_4H_{11}$

$CH_3CH_2CH_2CH_2NH_2$

Varian 89

▶ $NC_4H_{11}$

$(CH_3)_3CNH_2$

Varian 90

▶ $NC_4H_{11}BF_3$

$C_2H_5(CH_3)_2N \cdot BF_3$

J. Am. Chem. Soc. 83:4138 (1961)

▶ $NC_4H_{11} \cdot BH_3$

$C_2H_5(CH_3)_2N \cdot BH_3$

J. Am. Chem. Soc. 83:4138 (1961)

▶ $NC_5H_5$

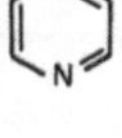

Varian 96
Can. J. Chem. 39:1158 (1961)

▶ $NC_5H_5$

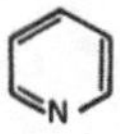

($C^{13}$)

J. Am. Chem. Soc. 84:2649 (1962)

▶ $NC_5H_5D_6$

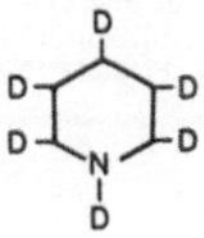

Varian 60
Chem. Ber. 95:2896 (1962)

▶ $NC_5H_7$

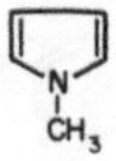

J. Chem. Phys. 36:2644 (1962)

▶ $NC_5H_7$

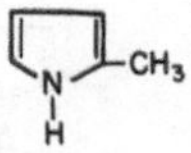

Arkiv Kemi 18:133 (1961)
J. Am. Chem. Soc. 83:5020 (1961); 84:4655 (1962)

▶ $NC_5H_8$

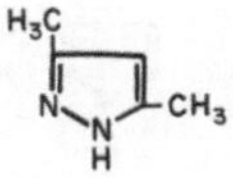

Varian 441

▶ $NC_5H_9$

$(CH_3)_3C-CN$

J. Phys. Chem. 66:2653 (1962)

▶ $NC_5H_9$

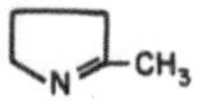

Can. J. Chem. 40:178 (1962)

▶ $NC_5H_9$

$CH_3(CH_2)_3NC$

J. Chem. Phys. 35:1533 (1961)

▶ $NC_5H_9$

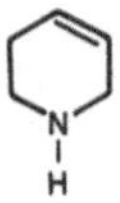

Varian 115

▶ $NC_5H_9$

$(CH_3)_3CNC$

J. Chem. Phys. 35:1533 (1961)

▶ $NC_5H_{11}$

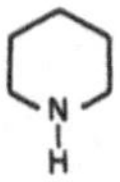

Chem. Ber. 95:2460, 2896 (1962)

▶ $NC_5H_{11}$

$H_3CCH=NC_3H_7$

Can. J. Chem. 40:884 (1962)

▶ $NC_5H_{11}$

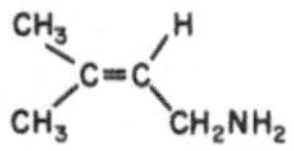

Chem. & Ind. 1020 (1962)

▶ $NC_5H_{11} \cdot BF_3$

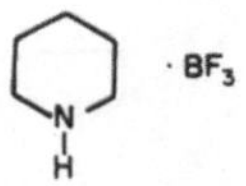

Rec. Trav. Chim. 80:1323 (1961)

▶ $NC_5H_{12}BF_2$

$(CH_3)_3N \cdot BF_2CH=CH_2$

J. Am. Chem. Soc. 83:4138 (1961)

▶ $NC_6H_6D$

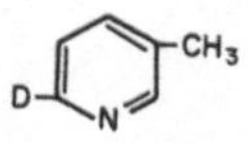

Can. J. Chem. 40:2030 (1962)

▶ $NC_6H_6D$

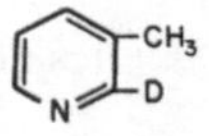

Can. J. Chem. 40:2030 (1962)

▶ $NC_6H_6D \cdot ZnCl_2$

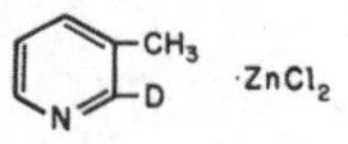

Can. J. Chem. 40:2030 (1962)

▶ $NC_6H_6D \cdot ZnCl_2$

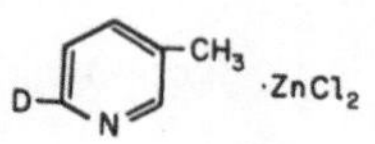

Can. J. Chem. 40:2030 (1962)

▶ $NC_6H_7$

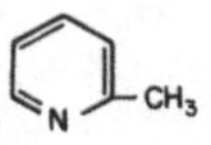

J. Org. Chem. 27:4641 (1962)
Bull. Chem. Soc. Japan 34:143 (1961)

▶ $NC_6H_7$

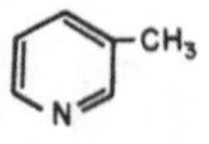

J. Chem. Soc. 46 (1961)
Bull. Chem. Soc. Japan 34:143 (1961)
Can. J. Chem. 40:2030 (1962)

▶ $NC_6H_7$

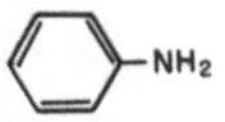

Bull. Chem. Soc. Japan 34:1606 (1961)
Tetrahedron Letters 519 (1961)
Can. J. Chem. 40:963 (1962)
J. Org. Chem. 27:4641 (1962)

▶ $NC_6H_7$

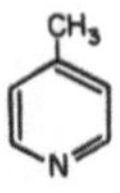

Bull. Chem. Soc. Japan 34:143 (1961)
Mol. Phys. 5:153 (1962)

▶ $NC_6H_7 \cdot ZnCl_2$

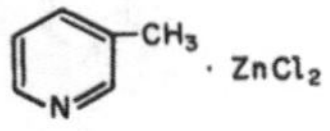

Can. J. Chem. 40:2030 (1962)

▶ $NC_6H_9$

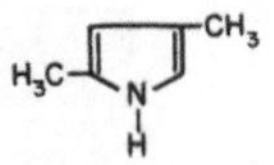

Varian 131
J. Chem. Phys. 36:2644 (1962)

▶ $NC_6H_9$

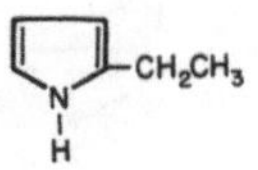

Varian 130

▶ $NC_6H_9$

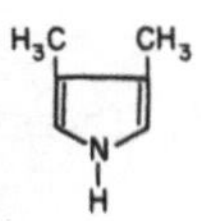

Can. J. Chem. 39:909 (1961)

▶ $NC_6H_{11}$

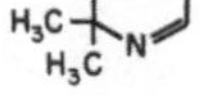

Can. J. Chem. 40:178 (1962)

▶ $NC_6H_{13}$

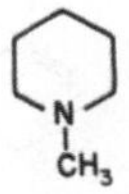

Chem. Ber. 95:2896 (1962)

▶ $NC_6H_{13}$

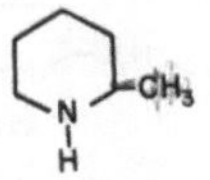

Varian 477
Chem. Ber. 95:2896 (1962)

▶ $NC_6H_{13}$

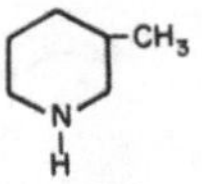

Varian 478
Chem. Ber. 95:2896 (1962)

▶ $NC_6H_{13}$

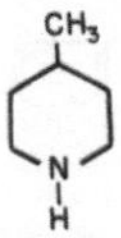

Varian 479
Chem. Ber. 95:2896 (1962)

▶ $NC_6H_{13} \cdot HCl$

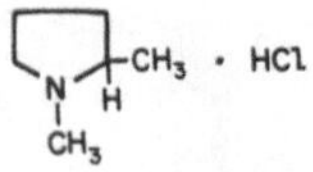

cis and trans

Tetrahedron Letters No. 24:1104 (1962)

▶ $NC_6H_{14} \cdot ClO_4$

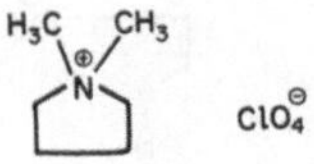

J. Am. Chem. Soc. 84:4806 (1962)

▶ $NC_6H_{15}$

$(CH_3CH_2)_3N$

Bull. Chem. Soc. Japan 35:1317 (1962)
J. Am. Chem. Soc. 83:4138 (1961)

▶ $NC_6H_{15} \cdot BCl_3$

$(C_2H_5)_3N \cdot BCl_3$

Bull. Chem. Soc. Japan 35:1317 (1962)

▶ $NC_6H_{15} \cdot BF_3$

$(CH_3CH_2)_3N \cdot BF_3$

J. Am. Chem. Soc. 83:4138 (1961)

▶ $NC_6H_{15} \cdot BH_3$

$(C_2H_5)_3N \cdot BH_3$

J. Am Chem. Soc. 83:4138 (1961)

▶ $NC_6H_{18}B$

$(CH_3)_3N \cdot B(CH_3)_3$

J. Am. Chem. Soc. 83:4138 (1961)

▶ $NC_7H_7$



Varian 154

▶ $NC_7H_7$

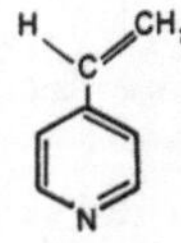

Varian 155

▶ $NC_7H_9$

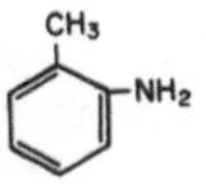

Bull. Chem. Soc. Japan 34:143, 1606 (1961)

▶ $NC_7H_9$

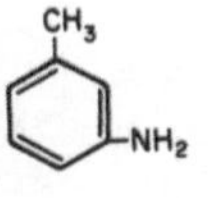

Bull. Chem. Soc. Japan 34:143, 1606 (1961)

▶ $NC_7H_9$

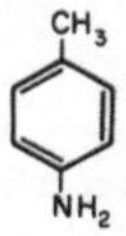

J. Chem. Phys. 37:2594 (1962)
Bull. Chem. Soc. Japan 34:143, 1606 (1961)

▶ $NC_7H_{11}$

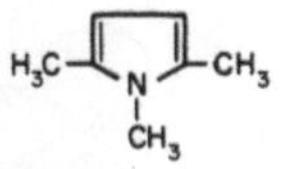

J. Chem. Phys. 36:2644 (1962)
Arkiv Kemi 18:133 (1961)

▶ $NC_7H_9$

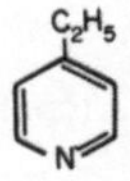

J. Chem. Soc. 44 (1961)

▶ $NC_7H_{11}$

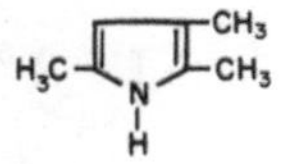

Can. J. Chem. 39:909 (1961)

▶ $NC_7H_9$

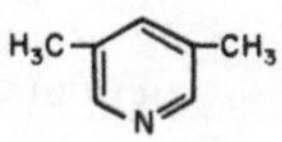

J. Chem. Soc. 46 (1961)
J. Chem. Soc. 860 (1961)

▶ $NC_7H_{11}$

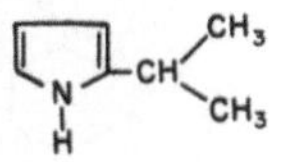

J. Am. Chem. Soc. 84:4655 (1962)

▶ $NC_7H_9$

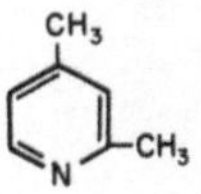

J. Chem. Soc. 860 (1961)

▶ $NC_7H_{11}$

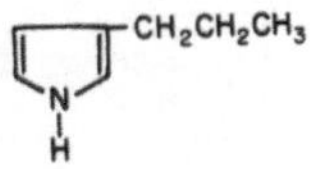

J. Am. Chem. Soc. 84:4655 (1962)

▶ $NC_7H_9$

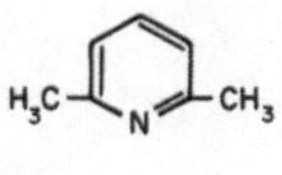

Varian 169
J. Chem. Soc. 860 (1961)

▶ $NC_7H_{11}$

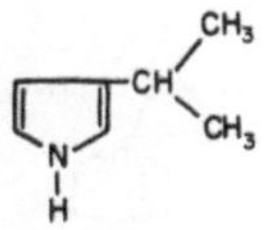

J. Am. Chem. Soc. 84:4655 (1962)

▶ $NC_7H_{11}$

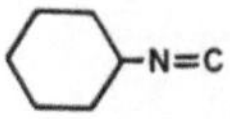

J. Chem. Phys. 35:1533 (1961)

▶ $NC_7H_{13}$

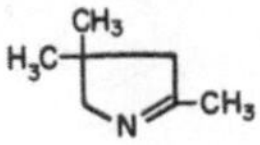

Can. J. Chem. 40:178 (1962)

▶ $NC_7H_{15}$

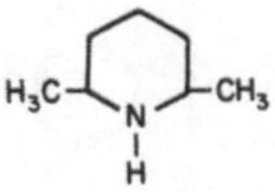

Chem. Ber. 95:2896 (1962)

▶ $NC_7H_{15}$

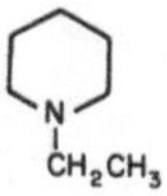

Chem. Ber. 95:2896 (1962)

▶ $NC_7H_{15} \cdot HCl$

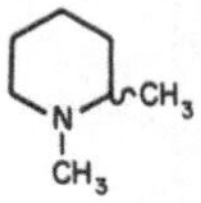

$\cdot HCl$

cis and trans

Tetrahedron Letters No. 24:1104 (1962)

▶ $NC_7H_{20}B$

$(C_2H_5)(CH_3)_2N \cdot B(CH_3)_3$

J. Am. Chem. Soc. 83:4138 (1961)

▶ $NC_8H_7$

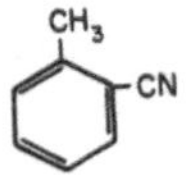

Bull. Chem. Soc. Japan 34:143 (1961)

▶ $NC_8H_7$

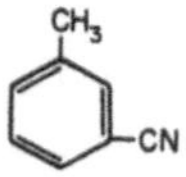

Bull. Chem. Soc. Japan 34:143 (1961)

▶ $NC_8H_7$

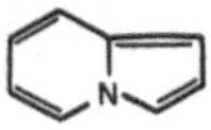

J. Chem. Soc. 3291 (1962)

▶ $NC_8H_7$

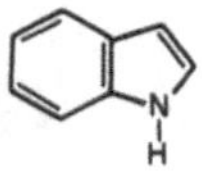

J. Am. Chem. Soc. 84:2534 (1962)

▶ $NC_8H_7$

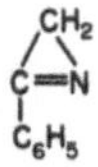

J. Org. Chem. 27:3557 (1962)
J. Am. Chem. Soc. 83:4483 (1961)

▶ $NC_8H_7$

$C_6H_5CH_2N{\equiv}C$

J. Chem. Phys. 35:1533 (1961)
J. Am. Chem. Soc. 84:3553 (1962)

▶ $NC_8H_7$

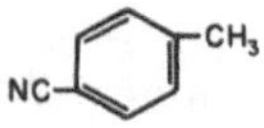

Bull. Chem. Soc. Japan 34:143 (1961)
Can. J. Chem. 40:2331 (1962)
Helv. Chim. Acta 45:568 (1962)

▶ $NC_8H_8 \cdot ClO_4$

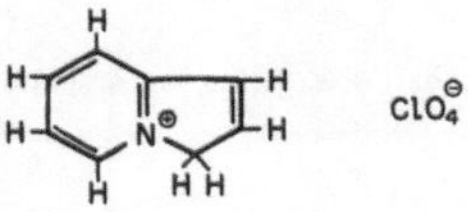

J. Chem. Soc. 3292 (1962)

▶ $NC_8H_{11}$

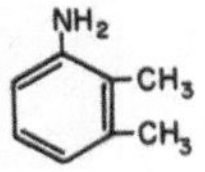

Bull. Chem. Soc. Japan 34:1606 (1961)

▶ $NC_8H_{11}$

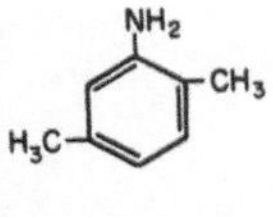

Bull. Chem. Soc. Japan 34:1606 (1961)

▶ $NC_8H_9$

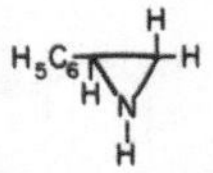

J. Org. Chem. 27:3532 (1962)

▶ $NC_8H_{11}$

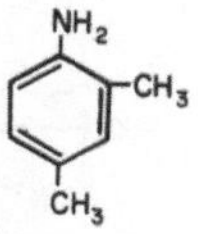

Bull. Chem. Soc. Japan 34:1606 (1961)

▶ $NC_8H_9$

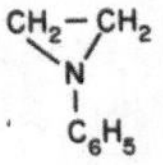

J. Org. Chem. 27:3532 (1962)

▶ $NC_8H_{11}$

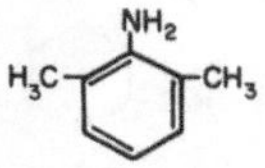

Bull. Chem. Soc. Japan 34:1606 (1961)

▶ $NC_8H_{11}$

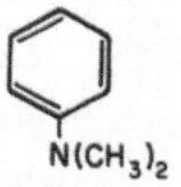

J. Phys. Chem. 66:2653 (1962)

▶ $NC_8H_{11}$

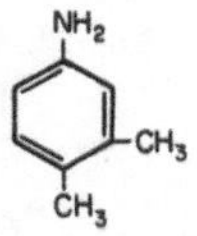

Bull. Chem. Soc. Japan 34:1606 (1961)

▶ $NC_8H_{11}$

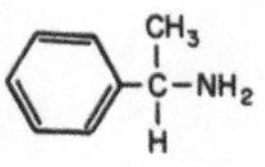

Varian 207
J. Org. Chem. 27:4091 (1962)

▶ $NC_8H_{12} \cdot I$

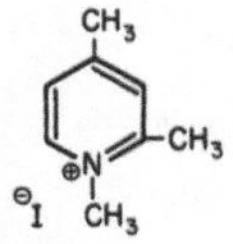

J. Chem. Soc. 860 (1961)

▶ $NC_8H_{11}$

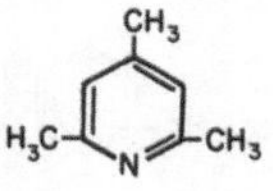

J. Chem. Soc. 46 (1961)
Can. J. Chem. 40:1759 (1962)

▶ $NC_8H_{12} \cdot I$

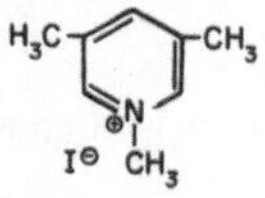

J. Chem. Soc. 860 (1961)

▶ $NC_8H_{13}$

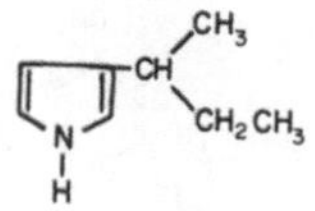

J. Am. Chem. Soc. 84:4655 (1962)

▶ $NC_8H_{13}$

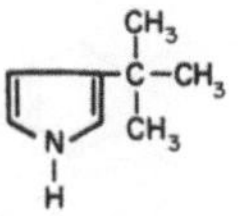

J. Am. Chem. Soc. 84:4655 (1962)

▶ $NC_8H_{13}$

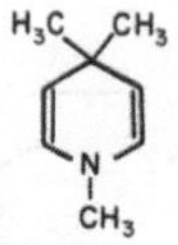

J. Org. Chem. 27:3767 (1962)

▶ $NC_8H_{13}$

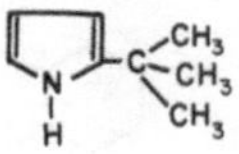

J. Am. Chem. Soc. 84:4655 (1962)

▶ $NC_8H_{13}$

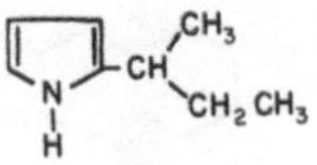

J. Am. Chem. Soc. 84:4655 (1962)

▶ $NC_8H_{15}$

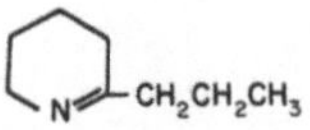

Rec. Trav. Chim. 80:513 (1961)
Chem. Ber. 95:2460, 2896 (1962)

▶ $NC_8H_{15} \cdot HCl$

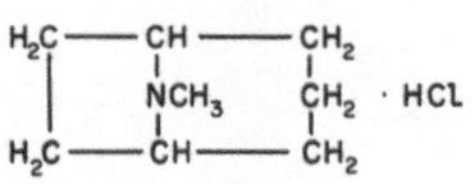

Tetrahedron Letters No. 24:1105 (1962)

▶ $NC_8H_{16} \cdot ClO_4$

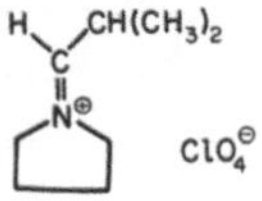

J. Am. Chem. Soc. 84:4806 (1962)

▶ $NC_8H_{17}$

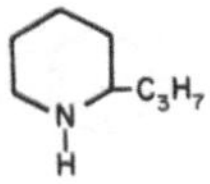

Chem. Ber. 95:2460, 2896 (1962)

▶ $NC_8H_{19}$

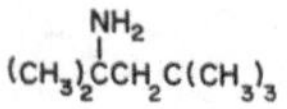

Varian 220

▶ $NC_9H_7$

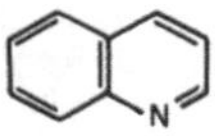

Can. J. Chem. 39:1864 (1961)

▶ $NC_9H_7$

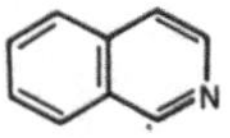

Varian 520

▶ $NC_9H_9$

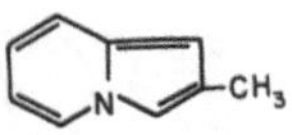

J. Chem. Soc. 3291 (1962)

▶ $NC_9H_9$

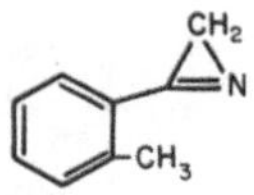

J. Org. Chem. 27:3557 (1962)

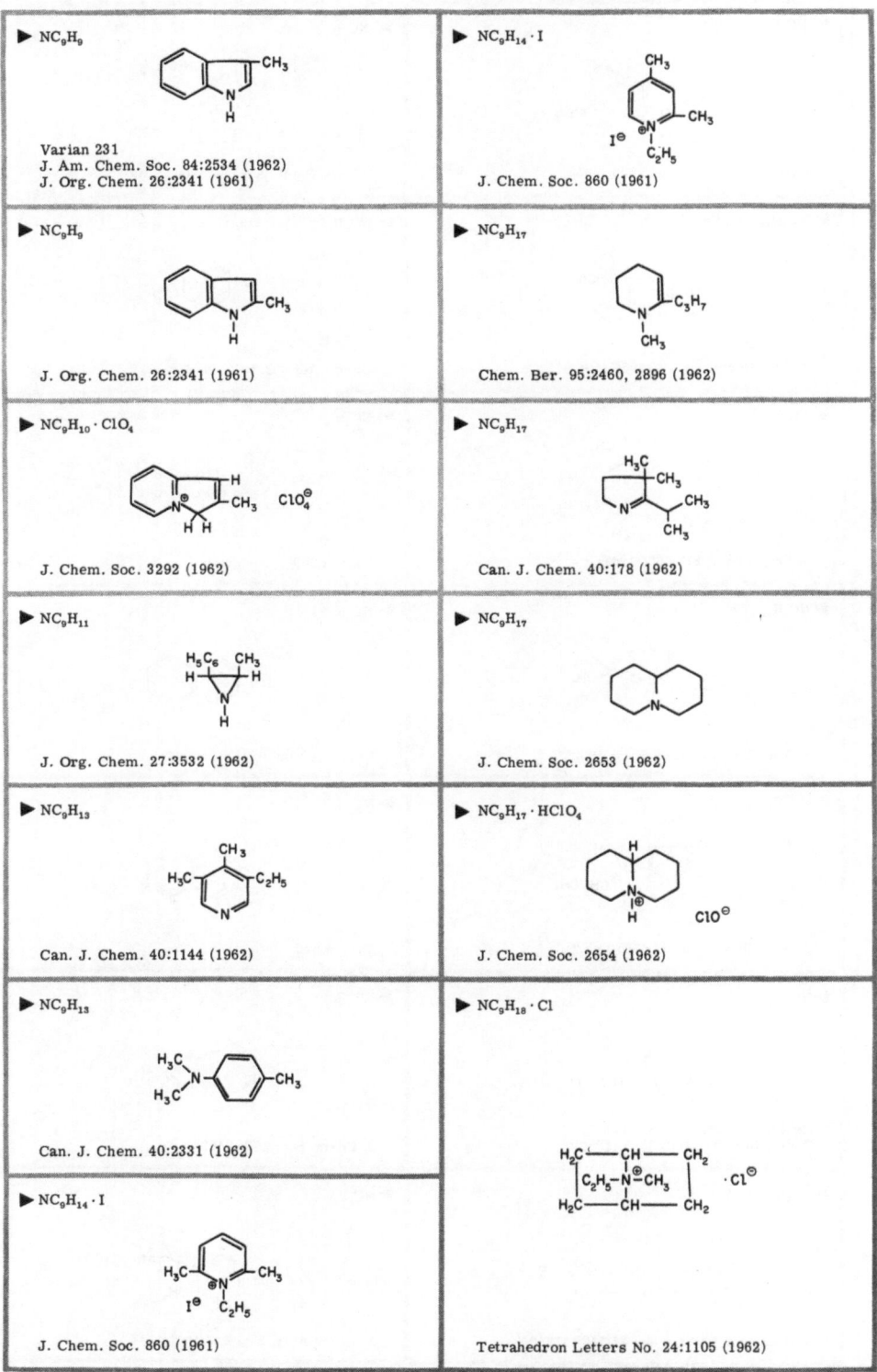

$NC_9H_9$
Varian 231
J. Am. Chem. Soc. 84:2534 (1962)
J. Org. Chem. 26:2341 (1961)
$NC_9H_9$
J. Org. Chem. 26:2341 (1961)
$NC_9H_{10} \cdot ClO_4$
J. Chem. Soc. 3292 (1962)
$NC_9H_{11}$
J. Org. Chem. 27:3532 (1962)
$NC_9H_{13}$
Can. J. Chem. 40:1144 (1962)
$NC_9H_{13}$
Can. J. Chem. 40:2331 (1962)
$NC_9H_{14} \cdot I$
J. Chem. Soc. 860 (1961)
$NC_9H_{14} \cdot I$
J. Chem. Soc. 860 (1961)
$NC_9H_{17}$
Chem. Ber. 95:2460, 2896 (1962)
$NC_9H_{17}$
Can. J. Chem. 40:178 (1962)
$NC_9H_{17}$
J. Chem. Soc. 2653 (1962)
$NC_9H_{17} \cdot HClO_4$
J. Chem. Soc. 2654 (1962)
$NC_9H_{18} \cdot Cl$
Tetrahedron Letters No. 24:1105 (1962)

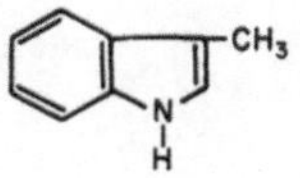

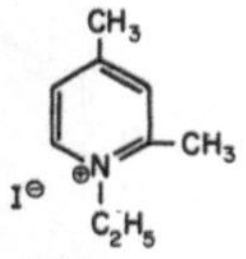

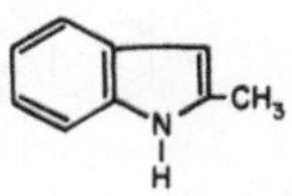

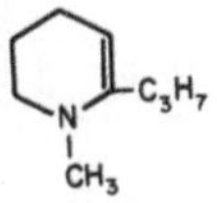

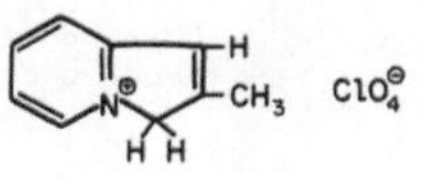

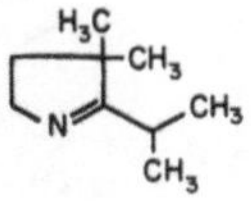

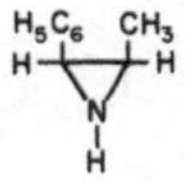

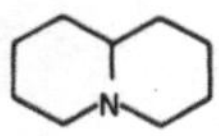

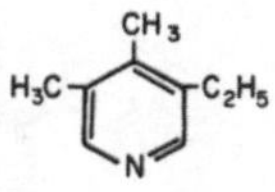

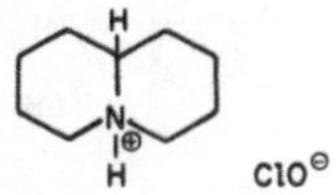

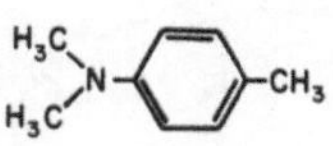

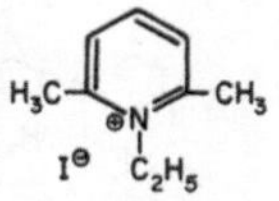

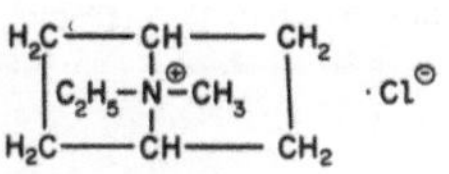

▶ $NC_9H_{19}$

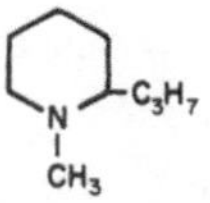

Chem. Ber. 95:2460, 2896 (1962)

▶ $NC_9H_{20} \cdot I$

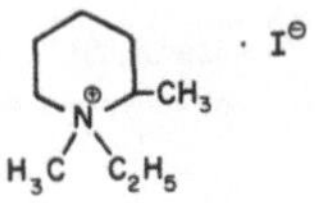

Tetrahedron Letters No. 24:1105 (1962)

▶ $NC_9H_{24}B$

$(CH_3)_3N \cdot B(C_2H_5)_3$

Bull. Chem. Soc. Japan 35:1317 (1962)

▶ $NC_9H_{24}B$

$(C_2H_5)_3N \cdot B(CH_3)_3$

J. Am. Chem. Soc. 83:4138 (1961)

▶ $NC_{10}H_9$

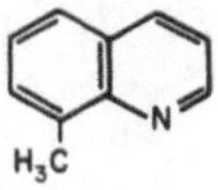

Can. J. Chem. 39:2318 (1961)

▶ $NC_{10}H_9$

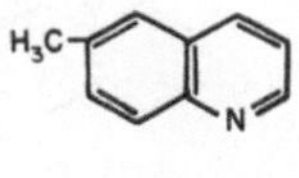

Angew. Chem. Intern. Ed. 1:215 (1962)

▶ $NC_{10}H_9$

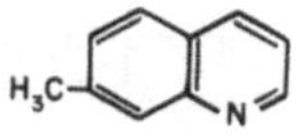

Angew. Chem. Intern. Ed. 1:215 (1962)

▶ $NC_{10}H_{11}$

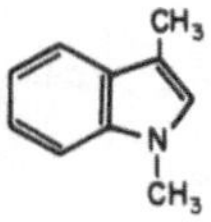

J. Am. Chem. Soc. 84:2534 (1962)

▶ $NC_{10}H_{11}$

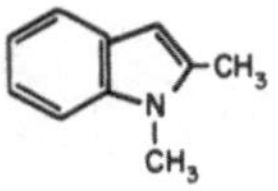

J. Am. Chem. Soc. 84:2534 (1962)

▶ $NC_{10}H_{11}$

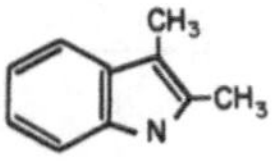

J. Am. Chem. Soc. 84:2534 (1962)

▶ $NC_{10}H_{11}$

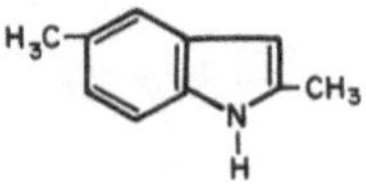

Varian 255

▶ $NC_{10}H_{12} \cdot ClO_4$

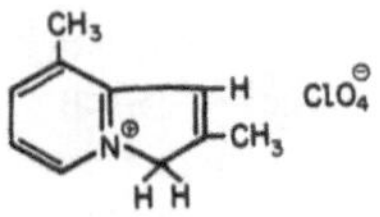

J. Chem. Soc. 3292 (1962)

▶ $NC_{10}H_{12} \cdot ClO_4$

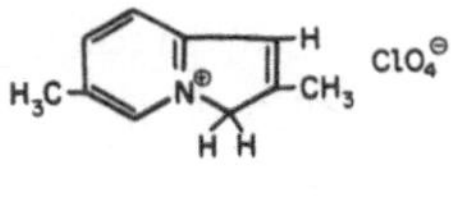

J. Chem. Soc. 3292 (1962)

▶ $NC_{10}H_{12} \cdot ClO_4$

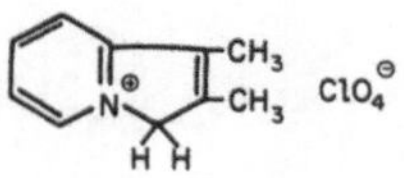

J. Chem. Soc. 3292 (1962)

▶ $NC_{10}H_{17}$

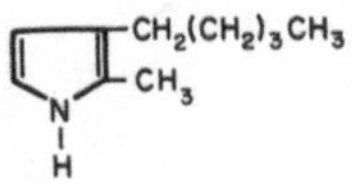

Varian 278

▶ $NC_{10}H_{13}$

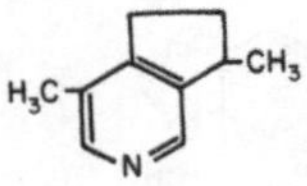

tentative structure

Gazz. Chim. Ital. 91:1037 (1961)

▶ $NC_{10}H_{17}$

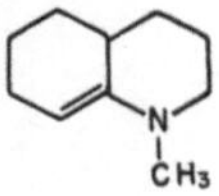

J. Org. Chem. 27:4029 (1962)

▶ $NC_{10}H_{13}$

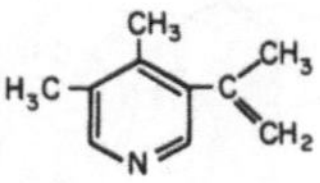

J. Org. Chem. 26:2733 (1961)

▶ $NC_{10}H_{17}$

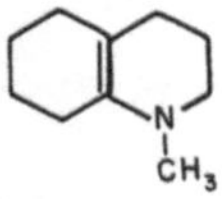

J. Org. Chem. 27:4029 (1962)

▶ $NC_{10}H_{13}$

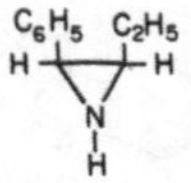

J. Org. Chem. 27:3532 (1962)

▶ $NC_{10}H_{18} \cdot ClO_4$

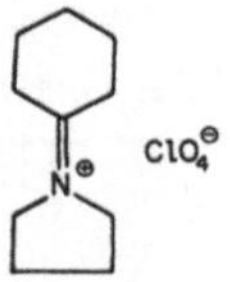

J. Am. Chem. Soc. 84:4806 (1962)

▶ $NC_{10}H_{13}$

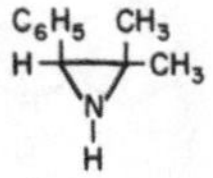

J. Org. Chem. 27:3532 (1962)

▶ $NC_{10}H_{15}$

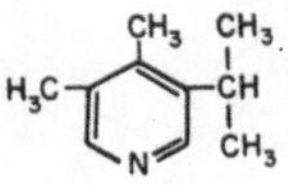

J. Org. Chem. 26:2733 (1961)

▶ $NC_{10}H_{19}$

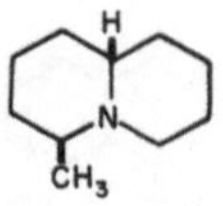

J. Chem. Soc. 2653 (1962)
Bull. Chem. Soc. Japan 35:697, 1335 (1962)

▶ $NC_{10}H_{15}$

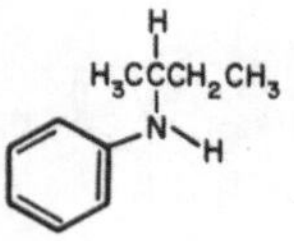

Varian 568

▶ $NC_{10}H_{19}$

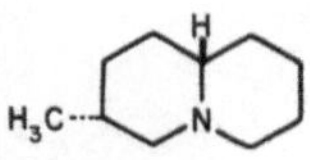

J. Chem. Soc. 2653 (1962)
Bull. Chem. Soc. Japan 35:697, 1335 (1962)

▶ $NC_{10}H_{19}$

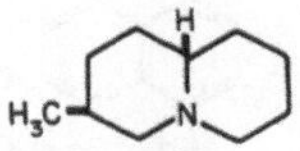


J. Chem. Soc. 2653 (1962)
Bull. Chem. Soc. Japan 35:697, 1335 (1962)

▶ $NC_{10}H_{19} \cdot HClO_4$

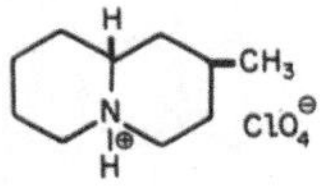


J. Chem. Soc. 2654 (1962)

▶ $NC_{10}H_{19}$

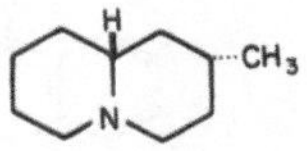


J. Chem. Soc. 2653 (1962)

▶ $NC_{10}H_{19} \cdot HClO_4$

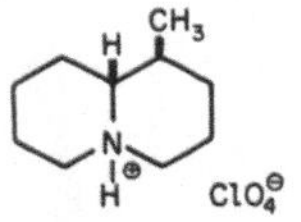


J. Chem. Soc. 2654 (1962)

▶ $NC_{10}H_{19}$

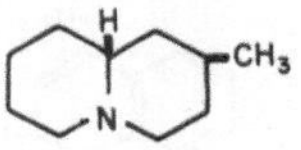


J. Chem. Soc. 2653 (1962)

▶ $NC_{10}H_{19} \cdot HClO_4$

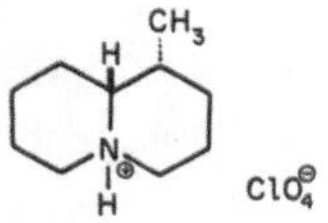


J. Chem. Soc. 2654 (1962)

▶ $NC_{10}H_{19}$

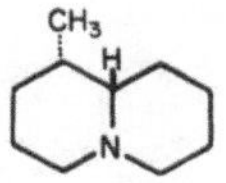


J. Chem. Soc. 2653 (1962)
Bull. Chem. Soc. Japan 35:697, 1335 (1962)

▶ $NC_{10}H_{19} \cdot HClO_4$

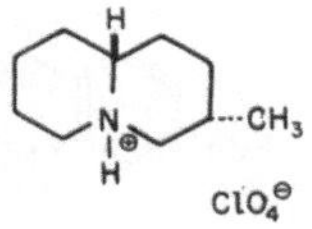


J. Chem. Soc. 2654 (1962)

▶ $NC_{10}H_{19}$

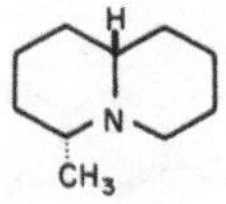


J. Chem. Soc. 2653 (1962)
Bull. Chem. Soc. Japan 35:697, 1335 (1962)

▶ $NC_{10}H_{19} \cdot HClO_4$

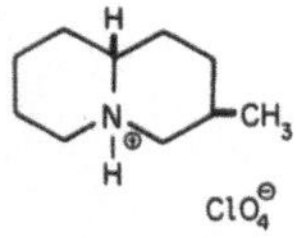


J. Chem. Soc. 2654 (1962)

▶ $NC_{10}H_{19}$

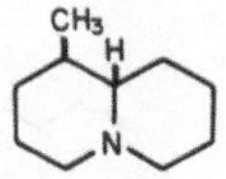


J. Chem. Soc. 2653 (1962)
Bull. Chem. Soc. Japan 35:697, 1335 (1962)

▶ $NC_{10}H_{19} \cdot HClO_4$

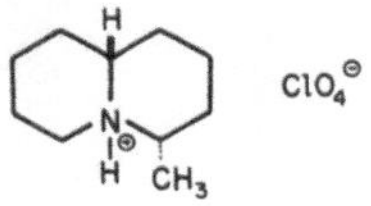


J. Chem. Soc. 2654 (1962)

▶ $NC_{10}H_{19} \cdot HClO_4$

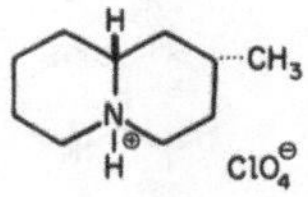


J. Chem. Soc. 2654 (1962)

▶ $NC_{10}H_{19} \cdot HClO_4$

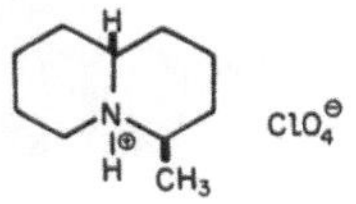


J. Chem. Soc. 2654 (1962)

▶ $NC_{10}H_{20} \cdot I$

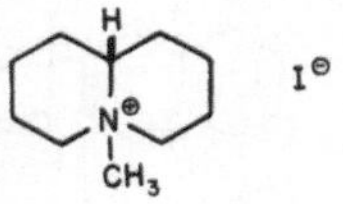

J. Chem. Soc. 2655 (1962)

▶ $NC_{11}H_{11}$

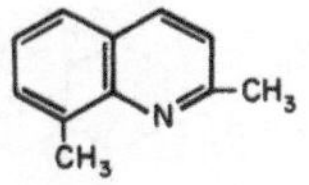

Can. J. Chem. 39:1864 (1961)

▶ $NC_{11}H_{11}$

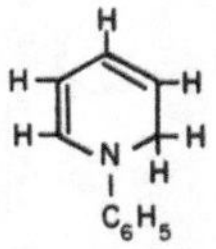

J. Org. Chem. 27:1439 (1962)

▶ $NC_{11}H_{11}$

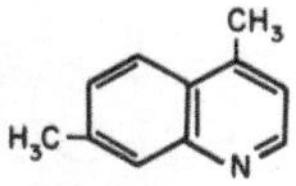

Angew. Chem. Intern. Ed. 1:215 (1962)

▶ $NC_{11}H_{11}$

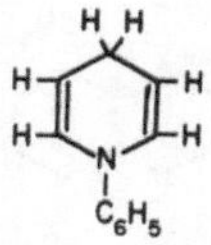

J. Org. Chem. 27:1439 (1962)

▶ $NC_{11}H_{11}$

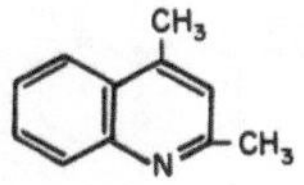

Varian 578
Angew. Chem. Intern. Ed. 1:215 (1962)

▶ $NC_{11}H_{11}$

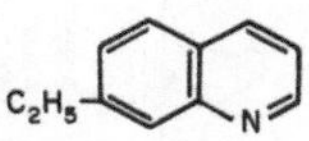

Can. J. Chem. 39:1864 (1961)

▶ $NC_{11}H_{11}$

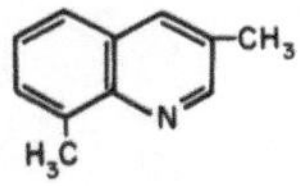

Can. J. Chem. 39:1864 (1961)

▶ $NC_{11}H_{11}$

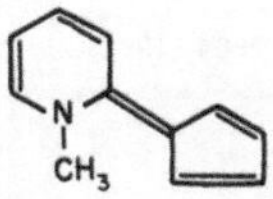

Chem. & Ind. 1362 (1961)

▶ $NC_{11}H_{11}$

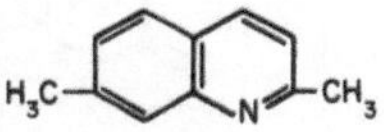

Angew. Chem. Intern. Ed. 1:215 (1962)

▶ $NC_{11}H_{11}$

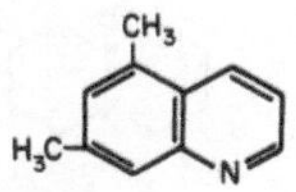

Can. J. Chem. 39:1864 (1961)

▶ $NC_{11}H_{11}$

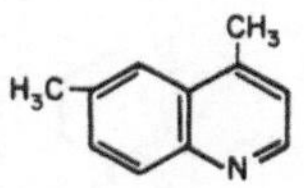

Can. J. Chem. 39:1864 (1961)
Angew. Chem. Intern. Ed. 1:215 (1962)

▶ $NC_{11}H_{11}$

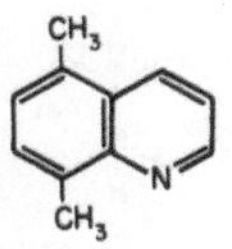

Can. J. Chem. 39:1864 (1961)

▶ $NC_{11}H_{11}$

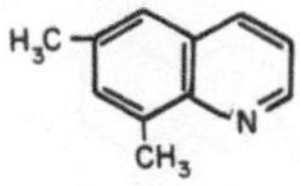

Can. J. Chem. 39:1864 (1961)

▶ $NC_{11}H_{11}$

Varian 579

▶ $NC_{11}H_{12}$

and

68%

32%

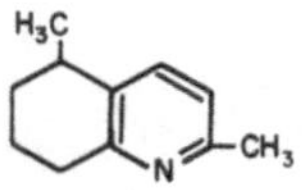

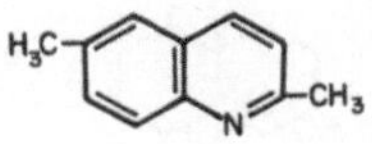

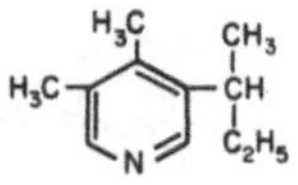

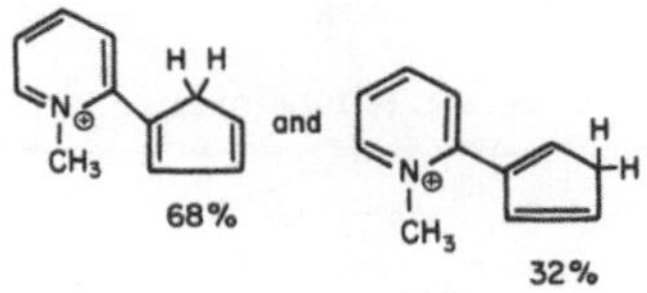

Chem. & Ind. 1362 (1961)

▶ $NC_{11}H_{13}$

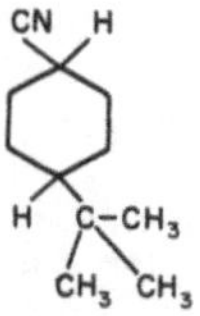

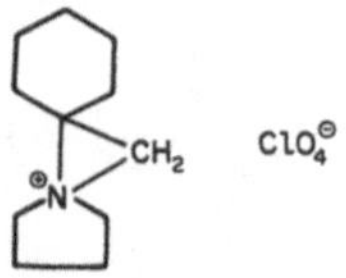

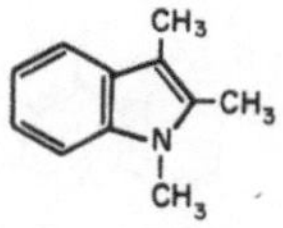

J. Am. Chem. Soc. 84:2534 (1962)

▶ $NC_{11}H_{13}$

Varian 286

▶ $NC_{11}H_{14} \cdot ClO_4$

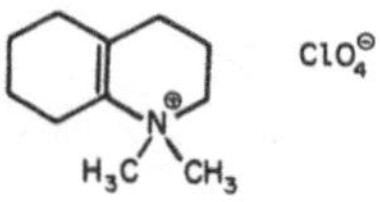

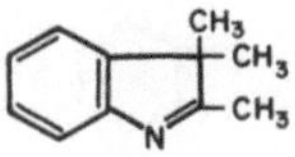

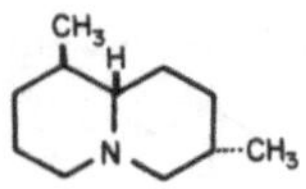

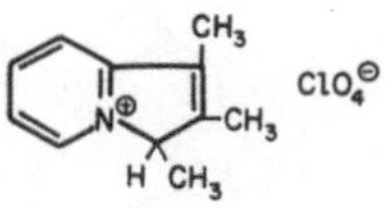

J. Chem. Soc. 3292 (1962)

▶ $NC_{11}H_{15}$

Can. J. Chem. 40:258 (1962)

▶ $NC_{11}H_{17}$

Can. J. Chem. 40:1144 (1962)

▶ $NC_{11}H_{19}$

J. Org. Chem. 27:4601 (1962)

▶ $NC_{11}H_{20} \cdot ClO_4$

J. Am. Chem. Soc. 84:4806 (1962)

▶ $NC_{11}H_{20} \cdot ClO_4$

J. Org. Chem. 27:4029 (1962)

▶ $NC_{11}H_{21}$

Bull. Chem. Soc. Japan 35:697, 1335 (1962)

▶ $NC_{11}H_{21}$

TENTATIVE STRUCTURE

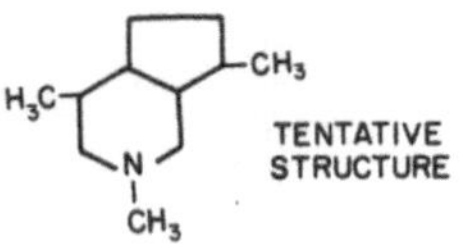

Gazz. Chim. Ital. 91:1037 (1961)

▶ $NC_{11}H_{21} \cdot HClO_4$

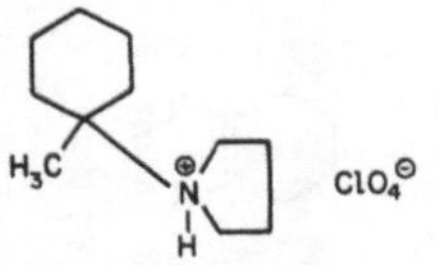


J. Am. Chem. Soc. 84:4806 (1962)

▶ $NC_{11}H_{22} \cdot I$

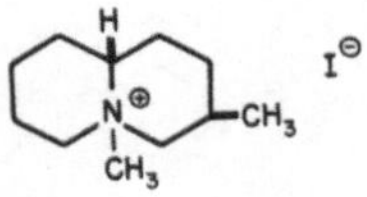


J. Chem. Soc. 2655 (1962)

▶ $NC_{11}H_{22} \cdot I$

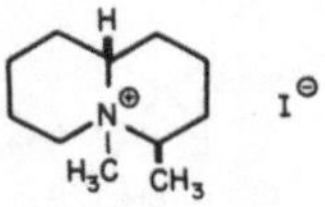


J. Chem. Soc. 2655 (1962)

▶ $NC_{11}H_{22} \cdot I$

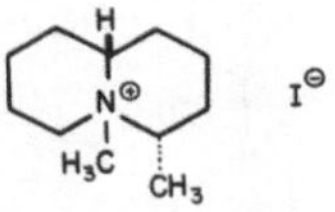


J. Chem. Soc. 2655 (1962)

▶ $NC_{11}H_{22} \cdot I$

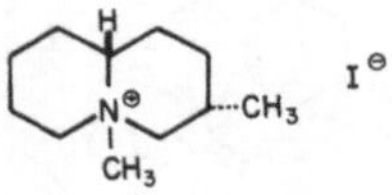


J. Chem. Soc. 2655 (1962)

▶ $NC_{11}H_{22} \cdot I$

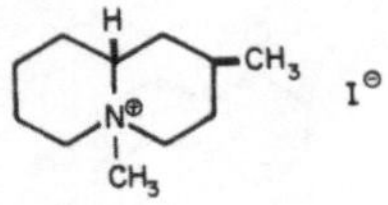


J. Chem. Soc. 2655 (1962)

▶ $NC_{11}H_{22} \cdot I$

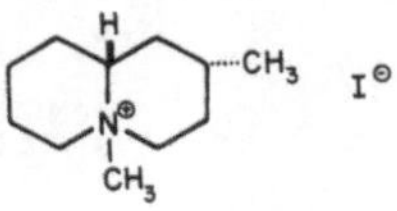


J. Chem. Soc. 2655 (1962)

▶ $NC_{11}H_{22} \cdot I$

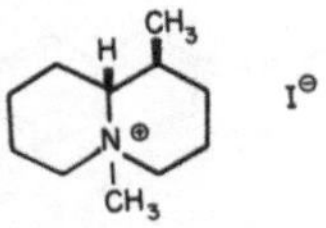


J. Chem. Soc. 2655 (1962)

▶ $NC_{11}H_{22} \cdot I$

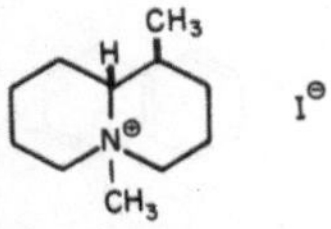


J. Chem. Soc. 2655 (1962)

▶ $NC_{12}H_9$

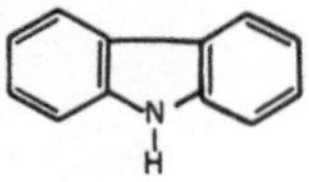


J. Am. Chem. Soc. 84:1020 (1962)

▶ $NC_{12}H_{11}$

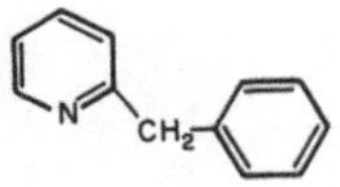


Varian 593

▶ $NC_{12}H_{13}$

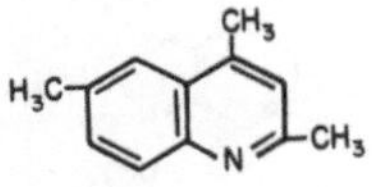


Angew. Chem. Intern. Ed. 1:215 (1962)

▶ $NC_{12}H_{14}D$

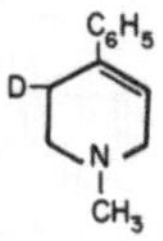


Tetrahedron Letters No. 13:556 (1962)

▶ $NC_{12}H_{15}$

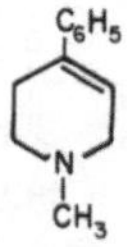


Tetrahedron Letters No. 13:556 (1962)

▶ $NC_{12}H_{15}$

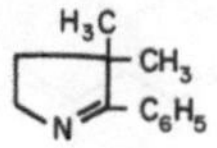

Can. J. Chem. 40:178 (1962)

▶ $NC_{12}H_{15}$

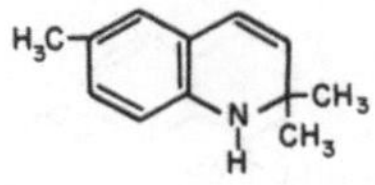

J. Org. Chem. 27:4713 (1962)

▶ $NC_{12}H_{15}$

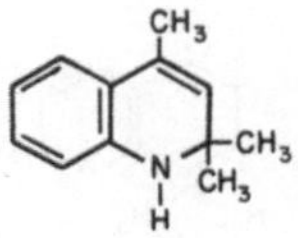

J. Org. Chem. 26:1287 (1961)

▶ $NC_{12}H_{15}$

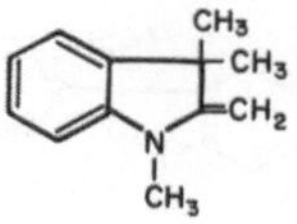

Varian 596

▶ $NC_{12}H_{16} \cdot Cl$

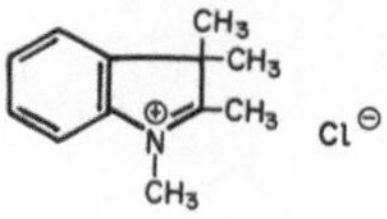

Varian 599

▶ $NC_{12}H_{17}$

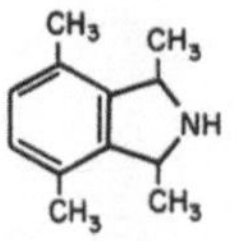

Proc. Chem. Soc. 119 (1961)

▶ $NC_{12}H_{17}$

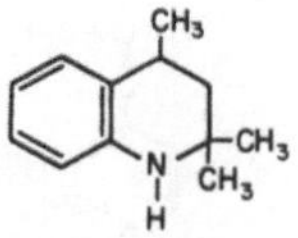

J. Org. Chem. 26:1287 (1961)

▶ $NC_{12}H_{17} \cdot HCl$

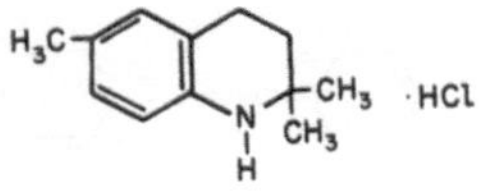

J. Org. Chem. 27:4713 (1962)

▶ $NC_{12}H_{22}$

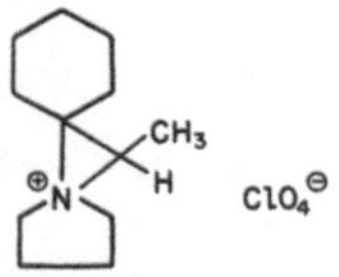

J. Am. Chem. Soc. 84:4806 (1962)

▶ $NC_{12}H_{22} \cdot ClO_4$

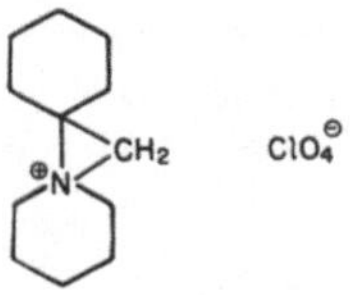

J. Am. Chem. Soc. 84:4806 (1962)

▶ $NC_{12}H_{23}$

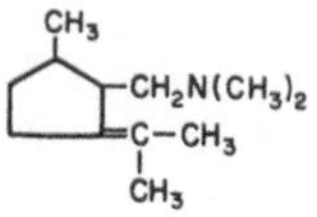

J. Am. Chem. Soc. 84:2775 (1962)

▶ $NC_{12}H_{23}$

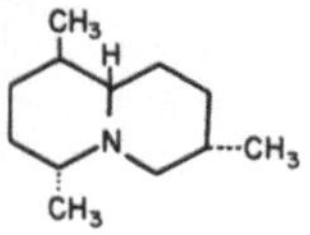

Bull. Chem. Soc. Japan 35:697, 1335 (1962)

▶ $NC_{12}H_{30}B$

$(C_2H_5)_3N \cdot B(C_2H_5)_3$

Bull. Chem. Soc. Japan 35:1317 (1962)

▶ $NC_{13}H_{11}$

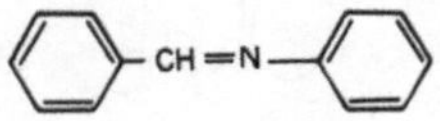

Can. J. Chem. 40:883 (1962)

▶ $NC_{13}H_{11}$

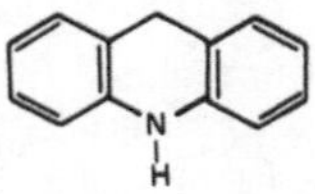

J. Am. Chem. Soc. 84:1020 (1962)

▶ $NC_{13}H_{19}$

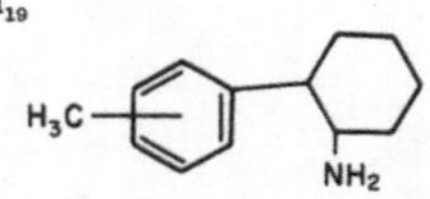

6 ISOMERS—ortho and para cis and trans

J. Org. Chem. 27:3003 (1962)

▶ $NC_{13}H_{22} \cdot I$

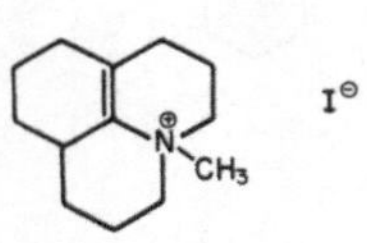

J. Org. Chem. 27:4029 (1962)

▶ $NC_{13}H_{24} \cdot ClO_4$

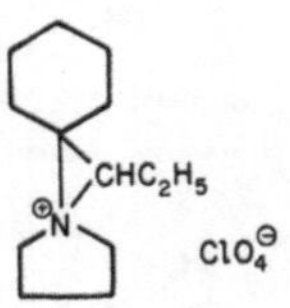

J. Am. Chem. Soc. 84:4806 (1962)

▶ $NC_{14}H_{13}$

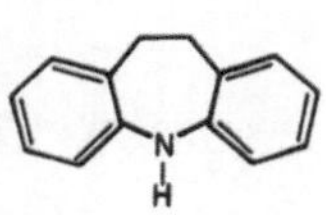

Varian 631

▶ $NC_{14}H_{16}B$

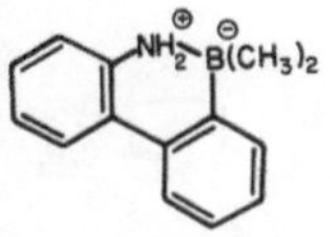

J. Am. Chem. Soc. 83:187 (1961)

▶ $NC_{14}H_{19}$

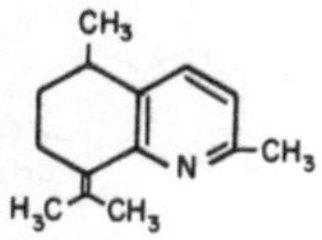

Can. J. Chem. 40:258 (1962)

▶ $NC_{14}H_{19}$

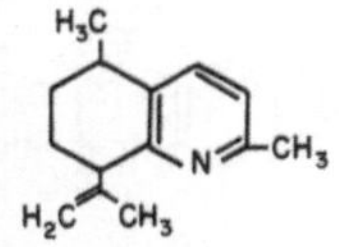

Can. J. Chem. 40:258 (1962)

▶ $NC_{16}H_{15}$

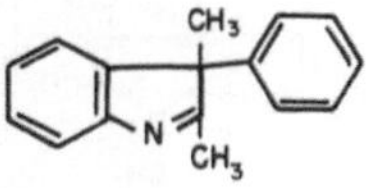

J. Org. Chem. 27:1555 (1962)

▶ $NC_{16}H_{15}$

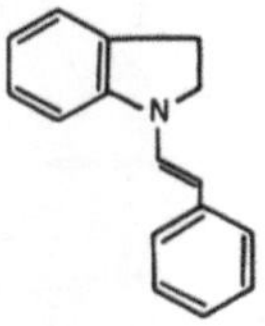

Helv. Chim. Acta 45:640 (1962)

▶ $NC_{16}H_{17}$

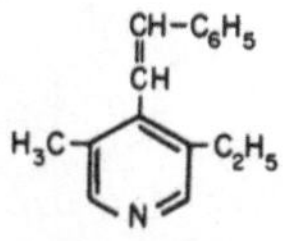

Can. J. Chem. 40:1145 (1962)

▶ $NC_{16}H_{27}$

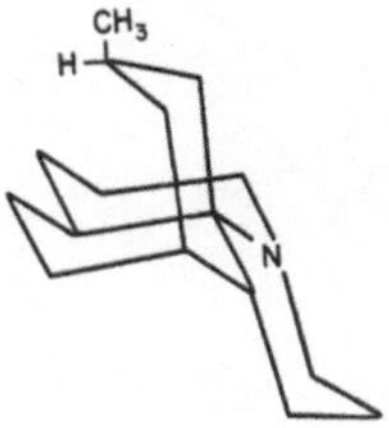


Can. J. Chem. 40:2094 (1962)

▶ $NC_{17}H_{21}$

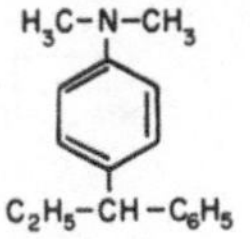


Chem. & Ind. 1570 (1962)

▶ $NC_{17}H_{21}$

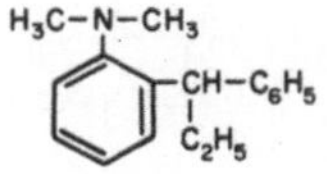


Chem. & Ind. 1570 (1962)

▶ $NC_{19}H_{17}$

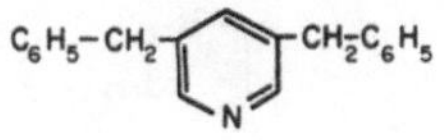


J. Org. Chem. 27:316 (1962)

▶ $NC_{19}H_{29}$

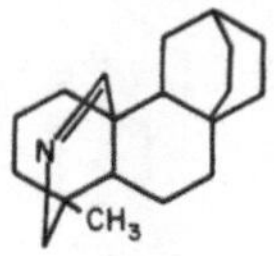


Tetrahedron 14:58 (1961)

▶ $NC_{20}H_{17}$

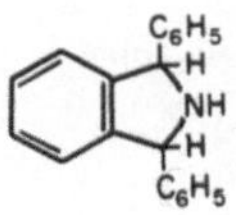


cis and trans

J. Am. Chem. Soc. 84:2196 (1962)

▶ $NC_{20}H_{20} \cdot Cl$

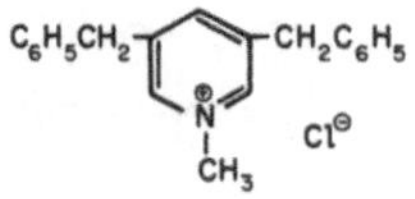


J. Org. Chem. 27:316 (1962)

▶ $NC_{23}H_{35}$

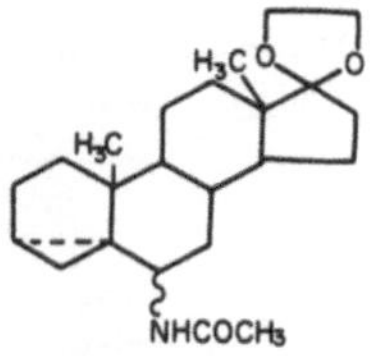


J. Org. Chem. 27:4610 (1962)

▶ $NC_{30}H_{49}$

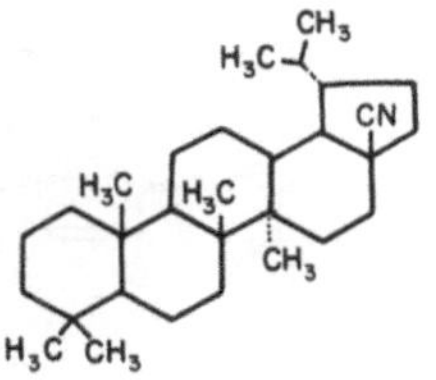


Bull. Soc. Chim. France 1331, 2444 (1961)

▶ $NC_{34}H_{49}$

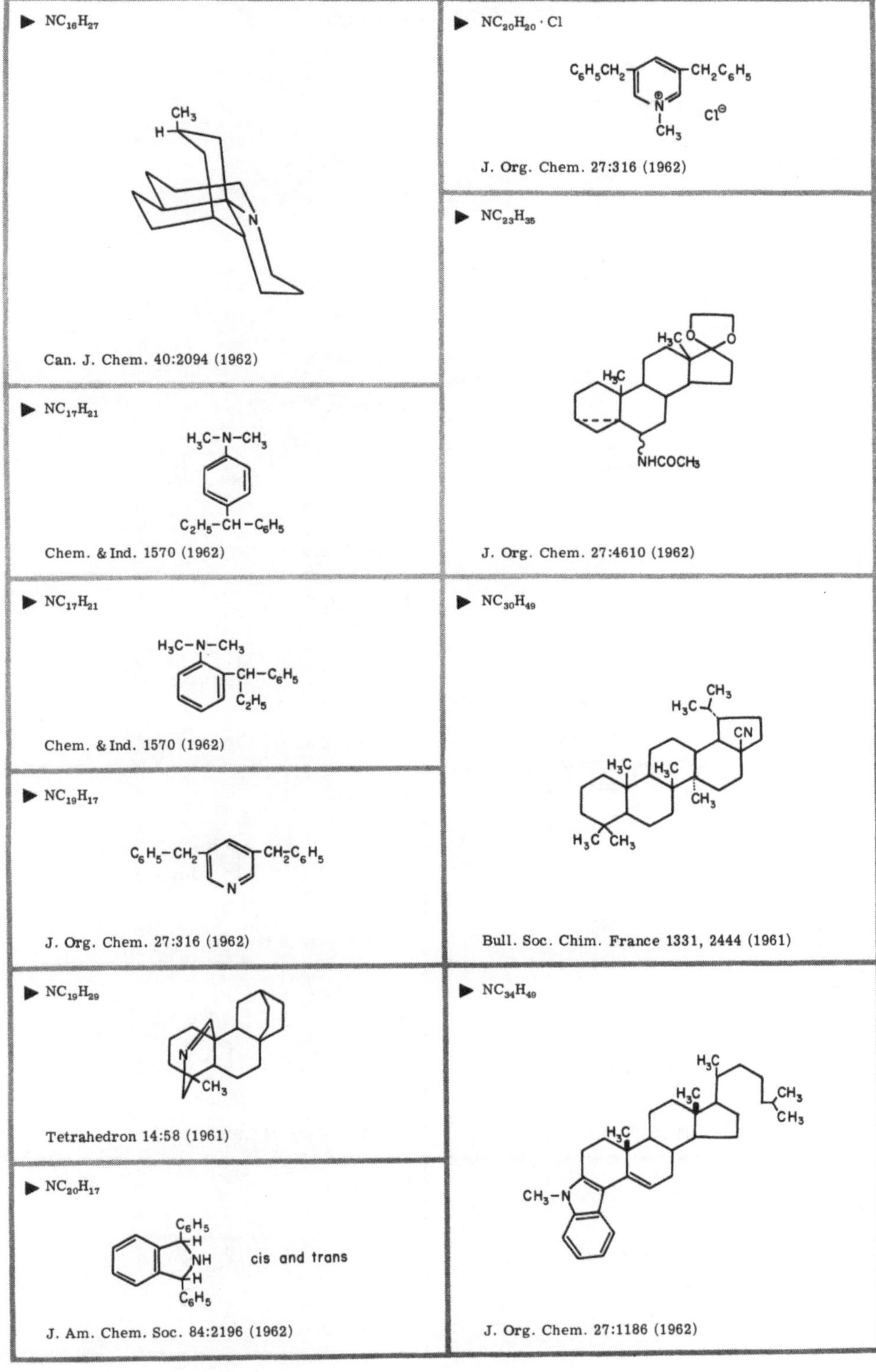

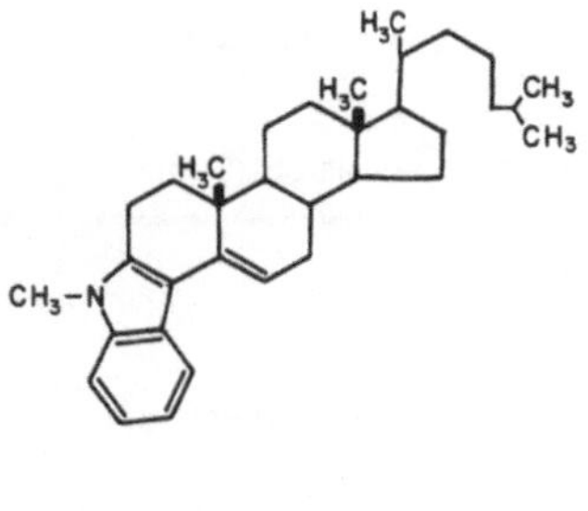


J. Org. Chem. 27:1186 (1962)

▶ $N_2OClC_{10}H_9$

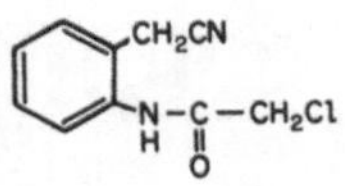

Varian 248

▶ $N_2OClC_{17}H_{19}$

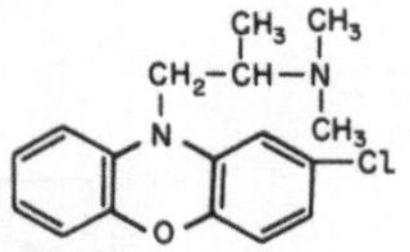

J. Org. Chem. 27:4272 (1962)

▶ $N_2OCl_2C_6H_{10}$

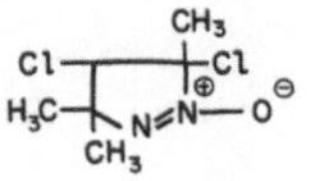

J. Org. Chem. 27:1309 (1962)

▶ $N_2OCl_2C_{13}H_{14}$

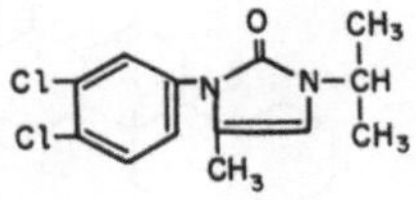

J. Am. Chem. Soc. 84:501 (1962)
J. Org. Chem. 27:3080 (1962)

▶ $N_2OC_2H_2$

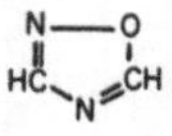

Helv. Chim. Acta 45:446 (1962)

▶ $N_2OC_2H_6$

$NH_2-CH_2-CONH_2$

J. Phys. Chem. 65:1902 (1961)

▶ $N_2OC_2H_6$

$(CH_3)_2N-N=O$

Varian 375

▶ $N_2OC_3H_6$

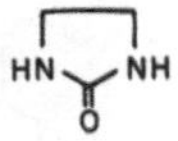

Chem. & Ind. 95 (1962)

▶ $N_2OC_3H_6$

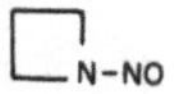

J. Am. Chem. Soc. 83:4417 (1961)

▶ $N_2OC_4H_4$

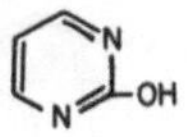

Arkiv Kemi 16:459 (1961)

▶ $N_2OC_5H_6$

J. Chem. Soc. 44 (1961)

▶ $N_2OC_5H_8$

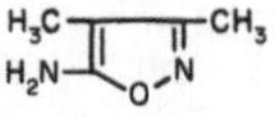

Tetrahedron 12:55 (1961)

► $N_2OC_5H_8$

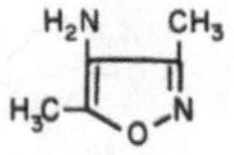

Tetrahedron 12:55 (1961)

► $N_2OC_6H_6$

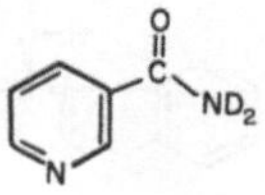

Varian 454

► $N_2OC_6H_6$

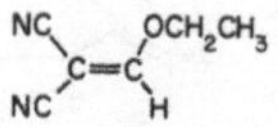

Varian 452

► $N_2OC_6H_6$

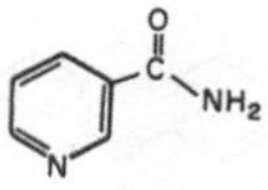

Varian 453

► $N_2OC_6H_{10}$

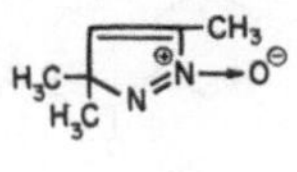

J. Org. Chem. 27:1309 (1962)

► $N_2OC_6H_{10}$

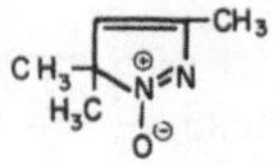

J. Org. Chem. 27:1309 (1962)

► $N_2OC_6H_{12}$

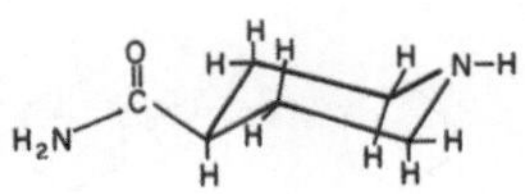

Varian 473

► $N_2OC_7H_8$

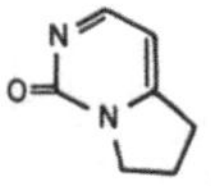

J. Am. Chem. Soc. 84:2266 (1962)

► $N_2OC_7H_8$

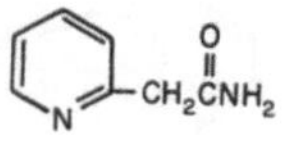

Varian 159

► $N_2OC_7H_8$

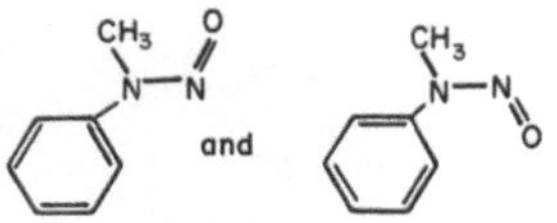


Varian 488

► $N_2OC_7H_{10}$

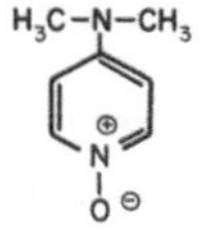

J. Chem. Soc. 44 (1961)

► $N_2OC_8H_{10}$

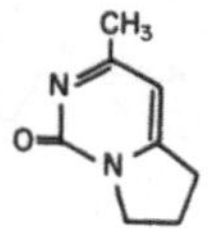

J. Am. Chem. Soc. 84:2266 (1962)

► $N_2OC_8H_{12}$

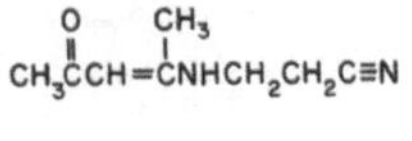

J. Am. Chem. Soc. 84:2691 (1962)

► $N_2OC_9H_{10}$

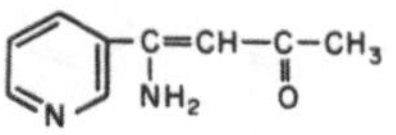

Tetrahedron Letters 672 (1961)
J. Am. Chem. Soc. 83:2099 (1961)

► $N_2OC_9H_{12}$

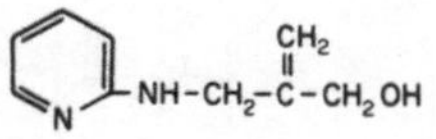

J. Org. Chem. 27:4137 (1962)

► $N_2OC_{11}H_{12}$

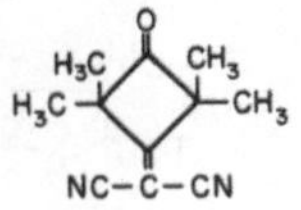

J. Am. Chem. Soc. 83:4867 (1961)

► $N_2OC_9H_{13} \cdot I$

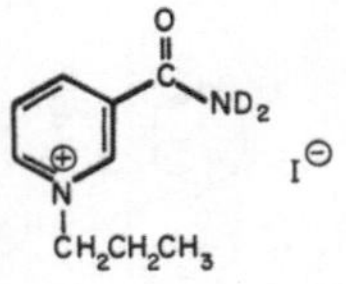

Varian 539

► $N_2OC_{12}H_{12}$

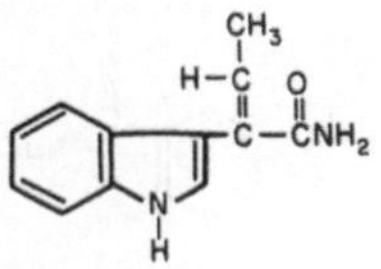

J. Am. Chem. Soc. 83:4678 (1961)

► $N_2OC_9H_{14}$

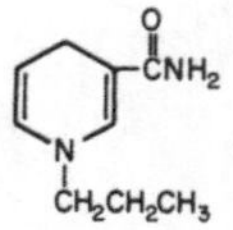

Varian 542

► $N_2OC_{12}H_{14}$

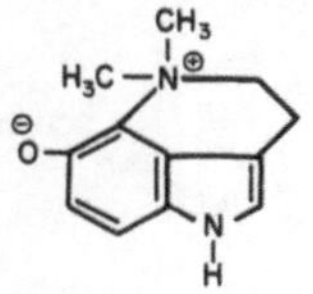

J. Am. Chem. Soc. 83:3341 (1961)

► $N_2OC_9H_{14}$

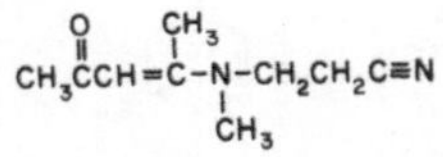

J. Am. Chem. Soc. 84:2691 (1962)

► $N_2OC_{12}H_{16}$

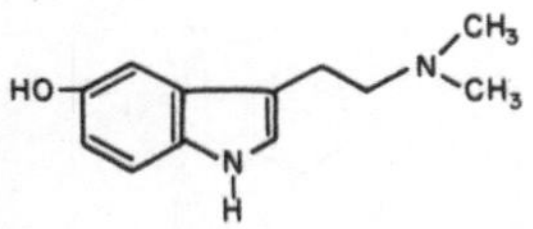

J. Am. Chem. Soc. 83:3341 (1961)

► $N_2OC_{10}H_{20}$

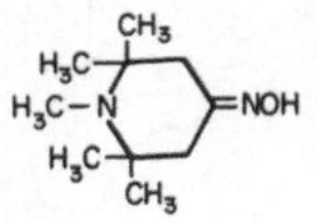

Varian 574

► $N_2OC_{13}H_{10}B$

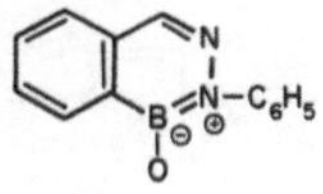

J. Am. Chem. Soc. 84:2648 (1962)

► $N_2OC_{11}H_{12}$

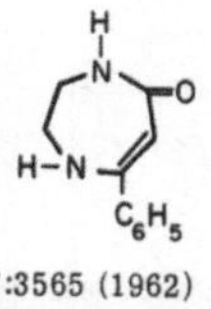

J. Org. Chem. 27:3565 (1962)

► $N_2OC_{13}H_{18}$

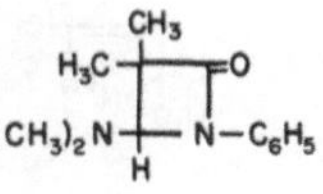

J. Am. Chem. Soc. 84:4988 (1962)

► $N_2OC_{11}H_{12}$

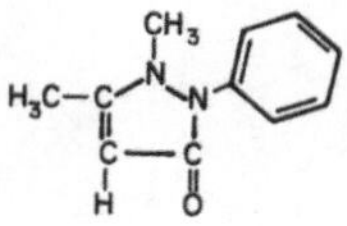

Varian 581

► $N_2OC_{13}H_{24}$

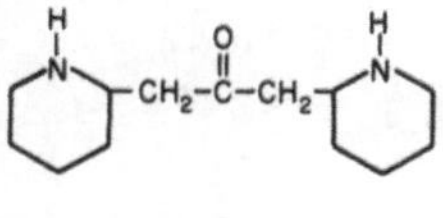

Chem. & Ind. 654 (1962)

▶ $N_2OC_{14}H_{16}$

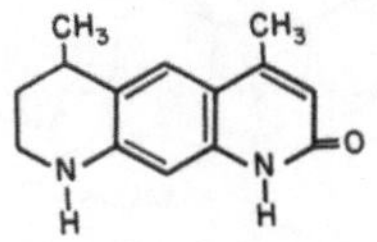

J. Am. Chem. Soc. 83:3729 (1961)

▶ $N_2OC_{14}H_{18}$

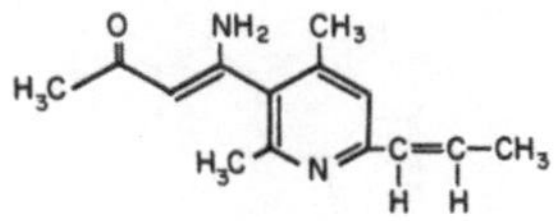

Tetrahedron Letters 671 (1961)

▶ $N_2OC_{14}H_{20}$

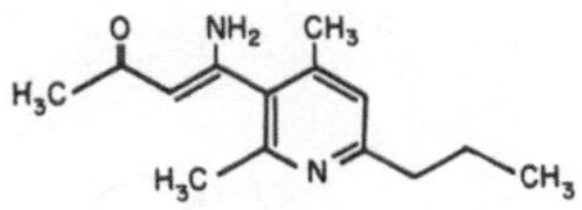

Tetrahedron Letters 672 (1961)

▶ $N_2OC_{14}H_{22}$

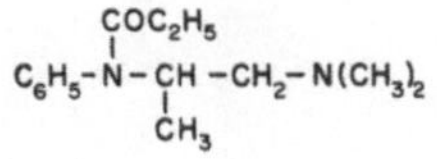

J. Org. Chem. 27:4272 (1962)

▶ $N_2OC_{14}H_{22}$

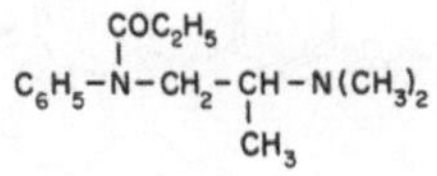

J. Org. Chem. 27:4272 (1962)

▶ $N_2OC_{16}H_{14}$

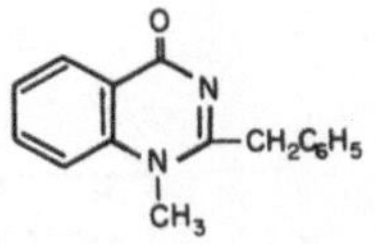

Tetrahedron 16:224 (1961)

▶ $N_2OC_{16}H_{16}$

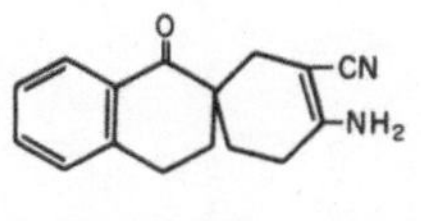

J. Org. Chem. 26:3292 (1961)

▶ $N_2OC_{16}H_{18}$

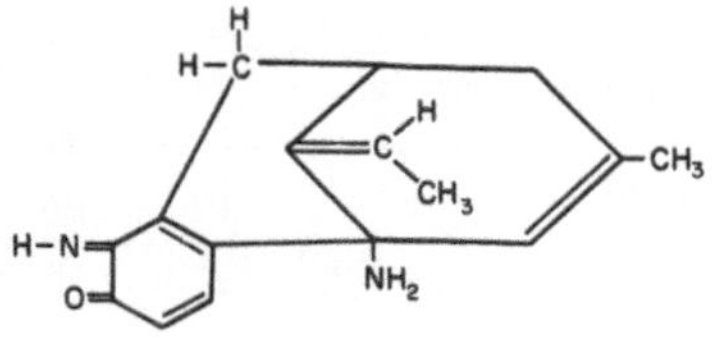

MCA Serial No. 2

▶ $N_2OC_{17}H_{18}$

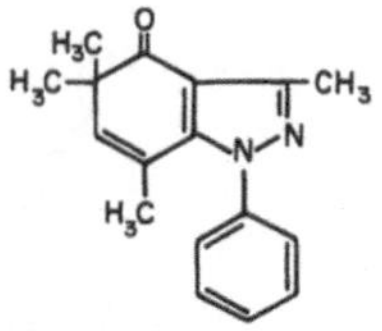

Varian 664

▶ $N_2OC_{17}H_{18}$

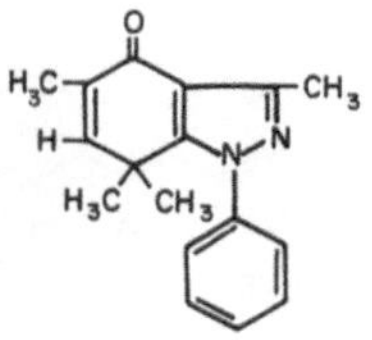

Varian 665

▶ $N_2OC_{17}H_{20}$

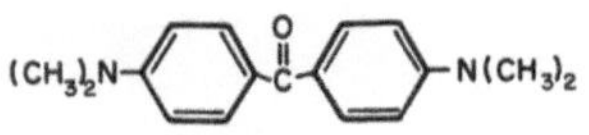

Varian 329

▶ $N_2OC_{17}H_{24}$

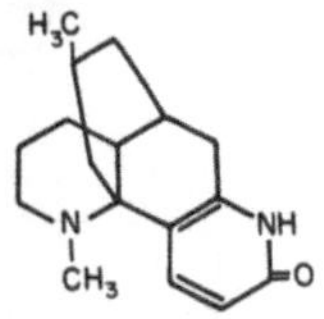

Tetrahedron 18:567 (1962)

▶ $N_2OC_{17}H_{26}$

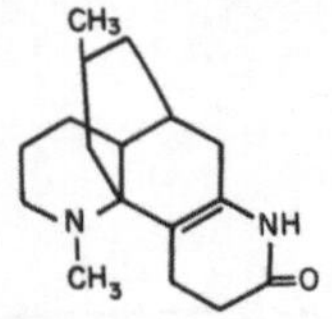

Tetrahedron 18:567 (1962)

▶ $N_2OC_{19}H_{26}$

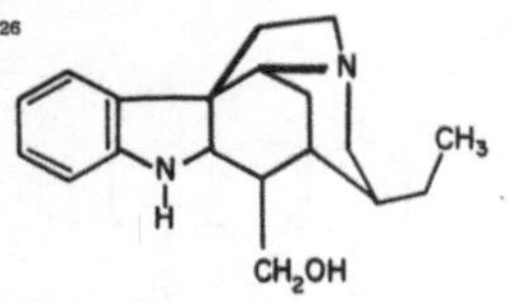

Tetrahedron 14:117 (1961)

▶ $N_2OC_{18}H_{20}$

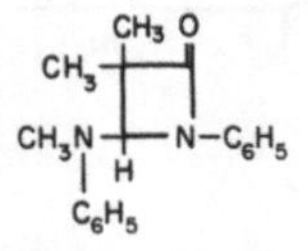

J. Am. Chem. Soc. 84:4988 (1962)

▶ $N_2OC_{19}H_{28}$

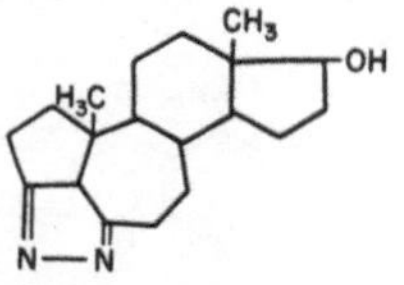

Helv. Chim. Acta 45:1031 (1962)

▶ $N_2OC_{18}H_{22}$

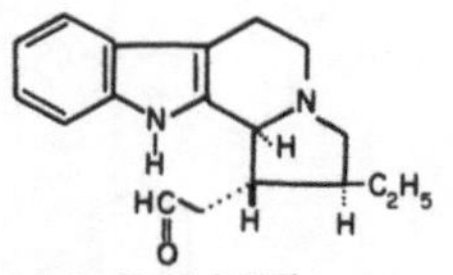

Helv. Chim. Acta 45:62 (1962)

▶ $N_2OC_{20}H_{22}$

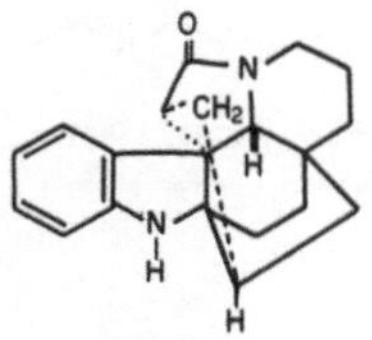

Helv. Chim. Acta 45:1146 (1962)

▶ $N_2OC_{19}H_{20}$

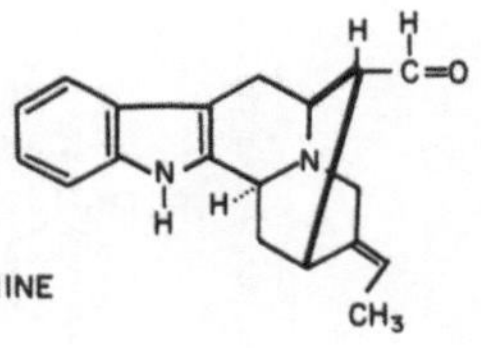

VELLOSIMINE

J. Org. Chem. 27:2982 (1962)

▶ $N_2OC_{20}H_{24}$

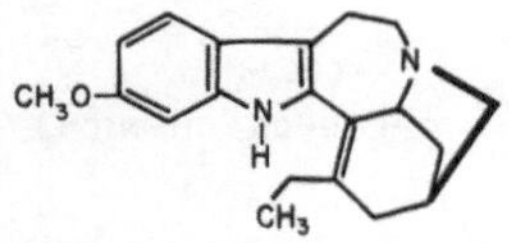

Bull. Soc. Chim. France 1679 (1961)

▶ $N_2OC_{19}H_{26}$

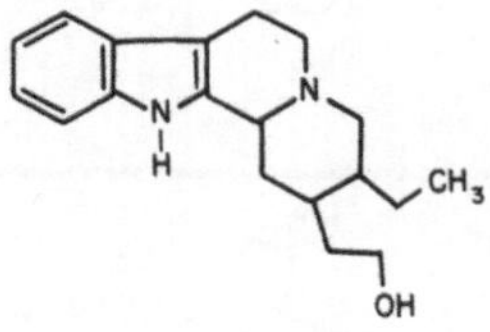

J. Org. Chem. 27:4702 (1962)

▶ $N_2OC_{21}H_{24}$

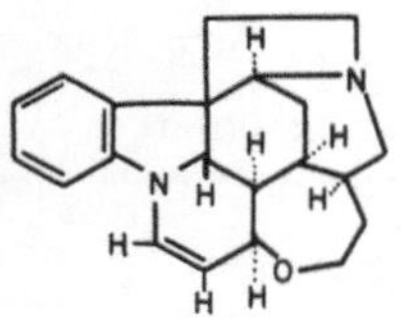

Helv. Chim. Acta XLIV:624 (1961)

▶ $N_2OC_{21}H_{26}$

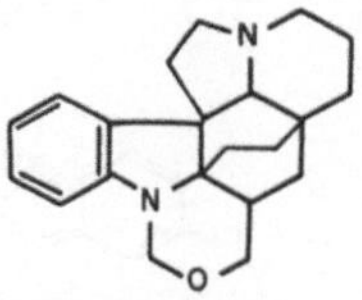

Helv. Chim. Acta 45:854 (1962)

▶ $N_2OC_{22}H_{30}$

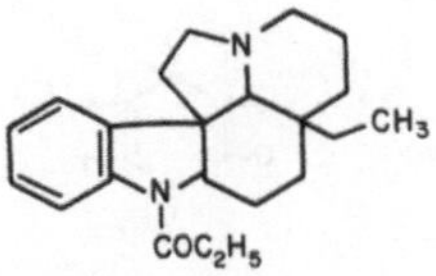

Chem. & Ind. 1949 (1962)

▶ $N_2OC_{24}H_{18}$

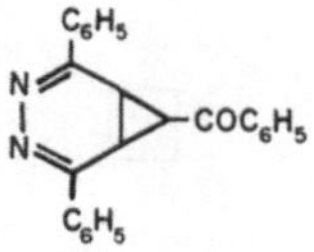

Chem. Ber. 95:611 (1962)

▶ $N_2OC_{24}H_{38}$

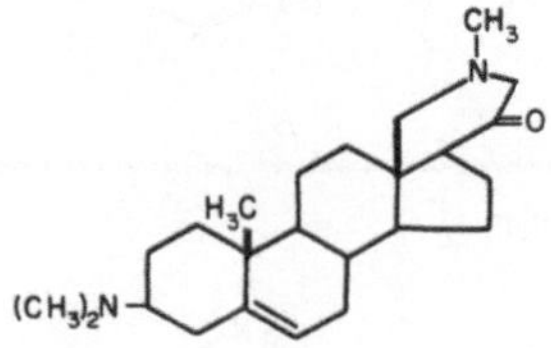

Bull. Soc. Chim. France 2234 (1962)

▶ $N_2OC_{24}H_{40}$

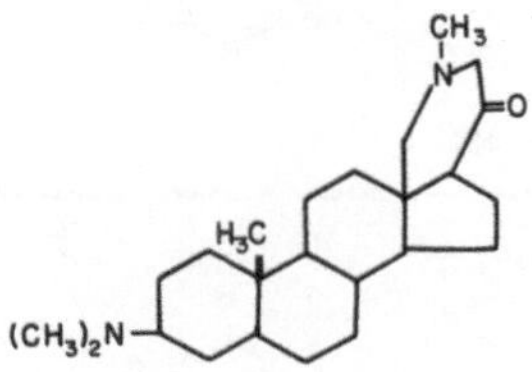

Bull. Soc. Chim. France 2234 (1962)

▶ $N_2OC_{25}H_{42}$

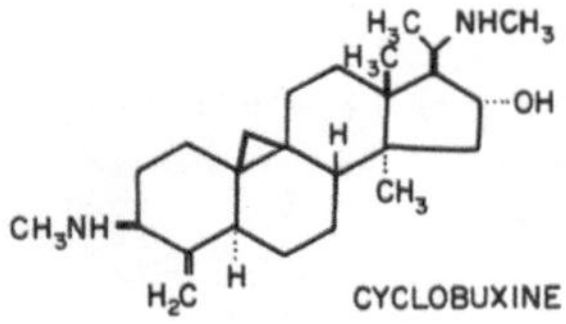

CYCLOBUXINE

J. Am. Chem. Soc. 84:4590 (1962)

▶ $N_2OC_{25}H_{44}$

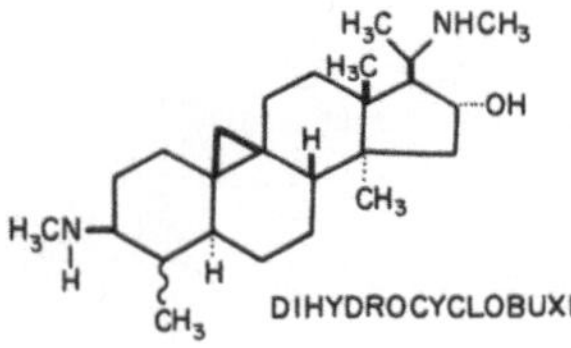

DIHYDROCYCLOBUXINE

J. Am. Chem. Soc. 84:4590 (1962)

▶ $N_2OC_{27}H_{46}$

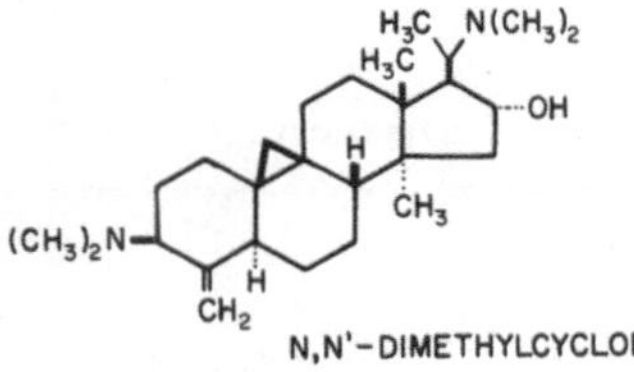

N,N'-DIMETHYLCYCLOBUXINE

J. Am. Chem. Soc. 84:4590 (1962)

▶ $N_2OC_{31}H_{28}$

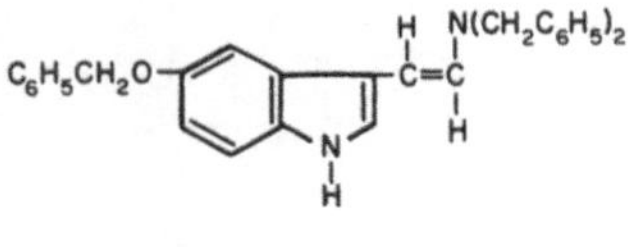

J. Org. Chem. 27:4105 (1962)

► $N_2O_2IC_4H_3$

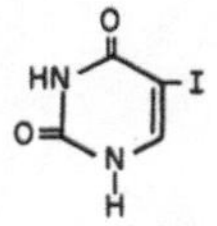

J. Am. Chem. Soc. 84:1042 (1962)

► $N_2O_2BrC_4H_3$

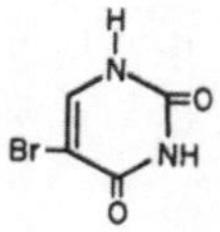

J. Am. Chem. Soc. 83:2909 (1961); 84:1042 (1962)

► $N_2O_2BrC_5H_7$

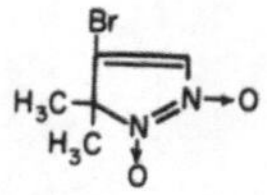

J. Org. Chem. 27:1309 (1962)

► $N_2O_2BrC_6H_3$

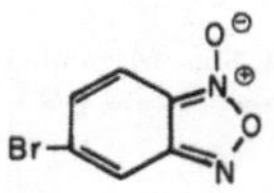

J. Org. Chem. 27:3219 (1962)

► $N_2O_2Br_2C_6H_{10}$

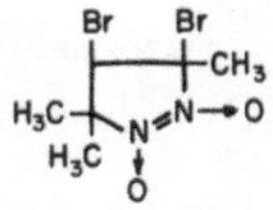

J. Org. Chem. 27:1309 (1962)

► $N_2O_2ClC_5H_5$

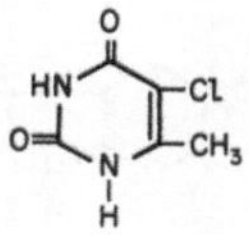

J. Am. Chem. Soc. 84:1042 (1962)

► $N_2O_2ClC_6H_3$

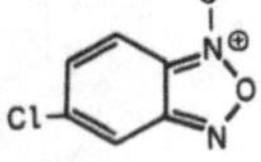

J. Org. Chem. 27:3219 (1962)

► $N_2O_2FC_4H_3$

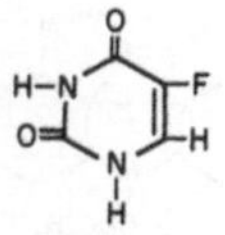

J. Am. Chem. Soc. 84:1042 (1962)

► $N_2O_2C_3H_6$

J. Am. Chem. Soc. 83:4417 (1961)

► $N_2O_2C_4H_4$

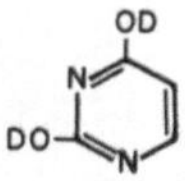

Varian 399

► $N_2O_2C_4H_4$

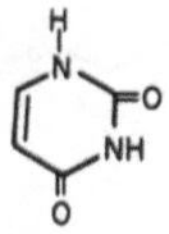

J. Am. Chem. Soc. 83:2909 (1961); 84:1042 (1962)

► $N_2O_2C_5H_6$

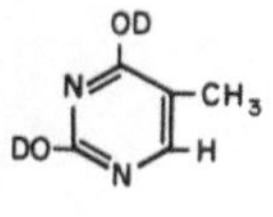

Varian 432

► $N_2O_2C_5H_6$

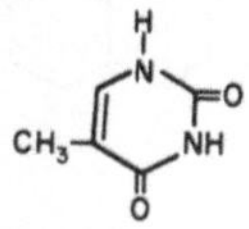

J. Am. Chem. Soc. 84:1042 (1962)
J. Am. Chem. Soc. 83:2909 (1961)

▶ $N_2O_2C_5H_6$

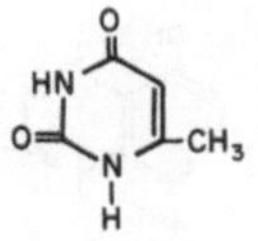

J. Am. Chem. Soc. 84:1042 (1962)

▶ $N_2O_2C_5H_{10}$

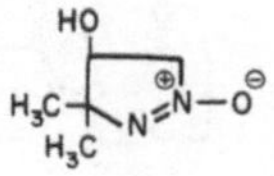

J. Org. Chem. 27:1309 (1962)

▶ $N_2O_2C_6H_4$

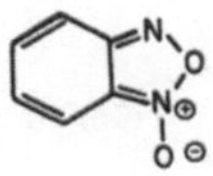

Chem. & Ind. 990 (1961)

▶ $N_2O_2C_6H_6$

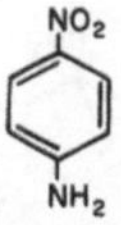

J. Chem. Phys. 37:2594 (1962)
Helv. Chim. Acta 45:568 (1962)

▶ $N_2O_2C_6H_6$

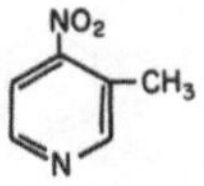

J. Chem. Soc. 46 (1961)

▶ $N_2O_2C_6H_8$

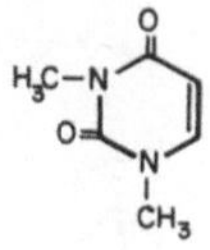

Varian 460

▶ $N_2O_2C_6H_8$

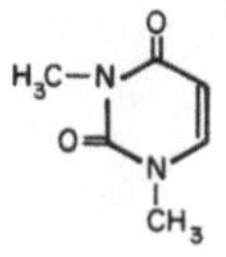

Varian 461

▶ $N_2O_2C_6H_{10}$

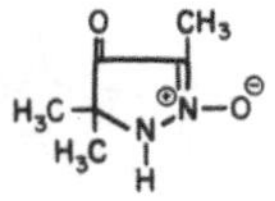

J. Org. Chem. 27:2882 (1962)

▶ $N_2O_2C_6H_{10}$

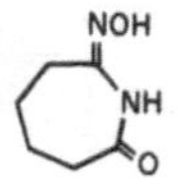

Bull. Chem. Soc. Japan 34:1812 (1961)

▶ $N_2O_2C_6H_{10}$

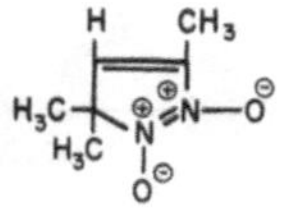

J. Org. Chem. 27:1309 (1962)

▶ $N_2O_2C_6H_{10}$

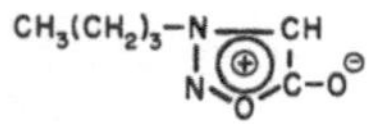

Helv. Chim. Acta 45:2426 (1962)

▶ $N_2O_2C_6H_{12}$

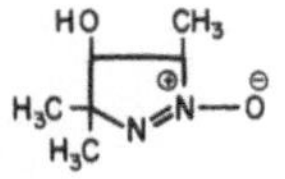

J. Org. Chem. 27:2884 (1962)

▶ $N_2O_2C_6H_{14}$

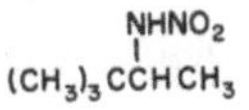

J. Org. Chem. 26:4192 (1961)

▶ $N_2O_2C_7H_6$

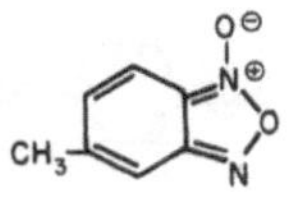

J. Org. Chem. 27:3219 (1962)

▶ $N_2O_2C_7H_8$

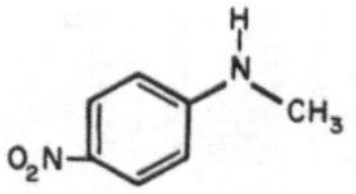

Varian 489

▶ $N_2O_2C_7H_{12}$

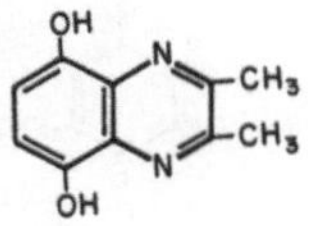

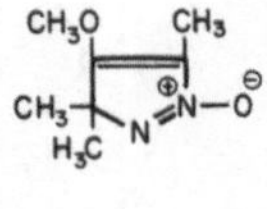

J. Org. Chem. 27:1309 (1962)

▶ $N_2O_2C_7H_{12}$

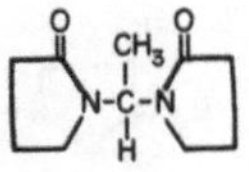

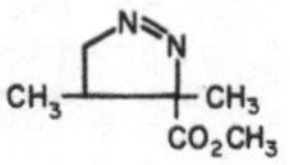

J. Am. Chem. Soc. 84:3736 (1962)

▶ $N_2O_2C_7H_{12}$

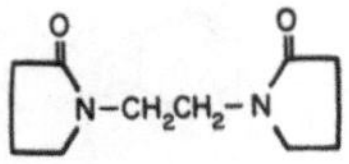

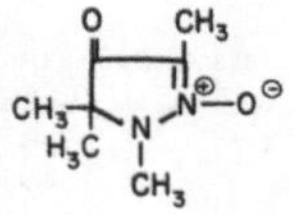

J. Org. Chem. 27:2882 (1962)

▶ $N_2O_2C_7H_{14}$

J. Org. Chem. 26:4192 (1961)

▶ $N_2O_2C_8H_6$

Varian 495

▶ $N_2O_2C_{10}H_{10}$

Varian 553

▶ $N_2O_2C_{10}H_{16}$

Varian 571

▶ $N_2O_2C_{10}H_{16}$

Varian 572

▶ $N_2O_2C_{10}H_{18}$

J. Org. Chem. 26:4192 (1961)

▶ $N_2O_2C_{10}H_{18}$

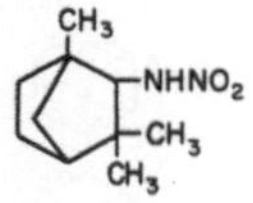

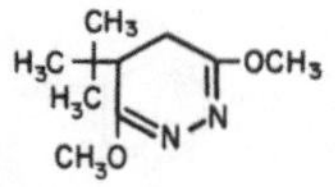

Acta Chem. Scand. 16:1877 (1962)

▶ $N_2O_2C_{10}H_{18}$

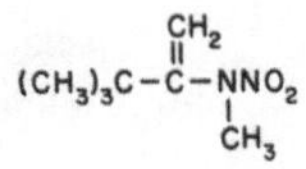

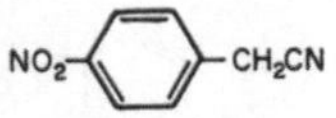

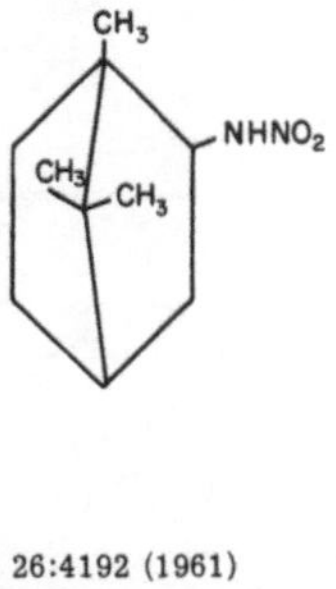

J. Org. Chem. 26:4192 (1961)

▶ $N_2O_2C_{11}H_{12}$

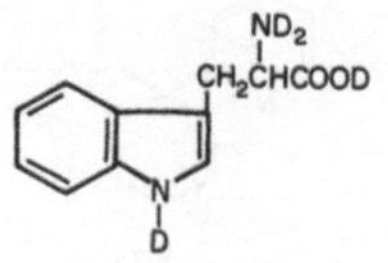


Varian 582

▶ $N_2O_2C_{11}H_{12}$

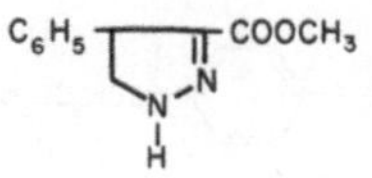


J. Org. Chem. 26:1912 (1961)

▶ $N_2O_2C_{11}H_{12}$

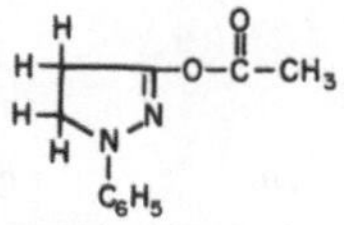


J. Org. Chem. 26:2803 (1961)

▶ $N_2O_2C_{11}H_{17}B$

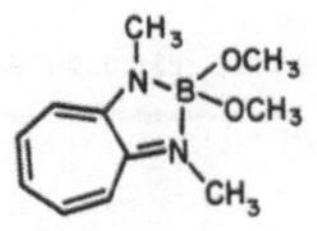


J. Am. Chem. Soc. 84:4720 (1962)

▶ $N_2O_2C_{12}H_{12}$

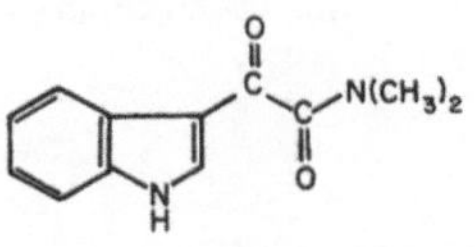


Varian 594

▶ $N_2O_2C_{12}H_{14}$

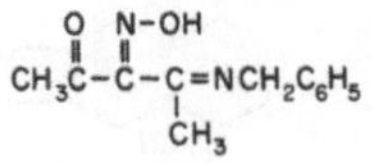


J. Am. Chem. Soc. 84:2691 (1962)

▶ $N_2O_2C_{12}H_{14}$

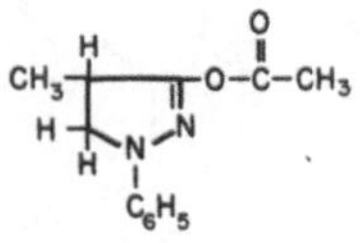


J. Org. Chem. 26:2803 (1961)

▶ $N_2O_2C_{12}H_{14}$

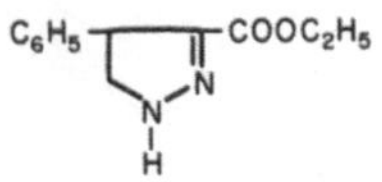


J. Org. Chem. 26:1912 (1961)

▶ $N_2O_2C_{12}H_{14}$

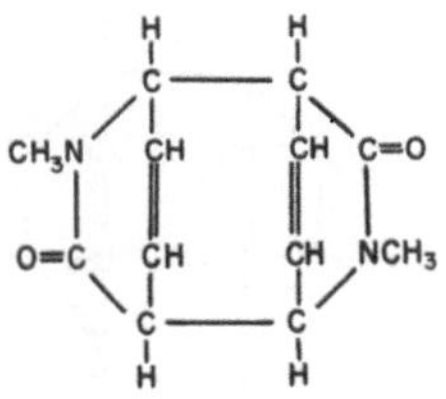


J. Am. Chem. Soc. 83:4472 (1961)

▶ $N_2O_2C_{12}H_{14}$

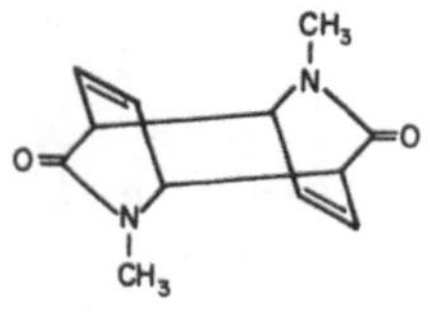


Tetrahedron Letters 650 (1961)

▶ $N_2O_2C_{12}H_{14} \cdot HCl$

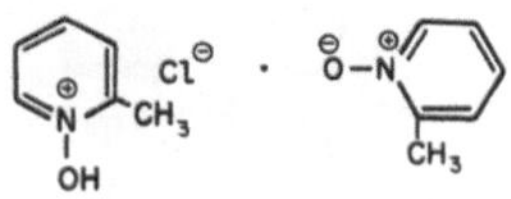


J. Org. Chem. 27:3857 (1962)

▶ $N_2O_2C_{12}H_{16}$

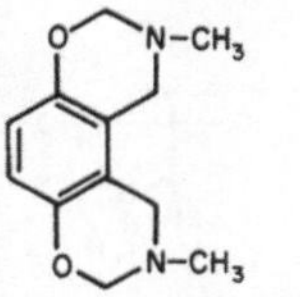

J. Org. Chem. 27:2752 (1962)

▶ $N_2O_2C_{13}H_{18}$

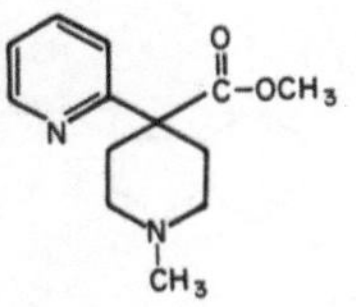

Varian 301

▶ $N_2O_2C_{12}H_{16}$

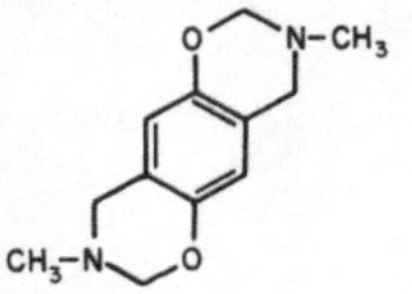

J. Org. Chem. 27:2752 (1962)

▶ $N_2O_2C_{13}H_{20}$

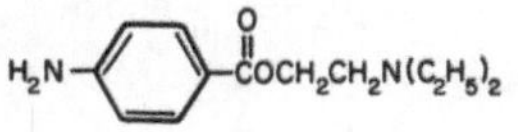

Varian 302

▶ $N_2O_2C_{12}H_{18}$

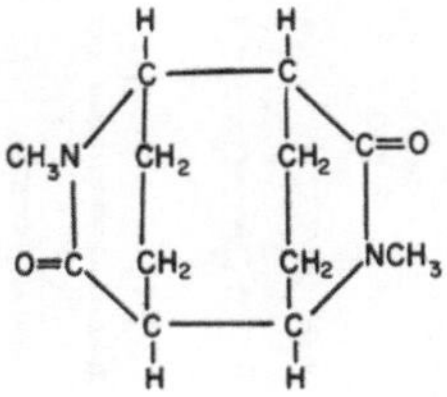

J. Am. Chem. Soc. 83:4472 (1961)

▶ $N_2O_2C_{13}H_{22}$

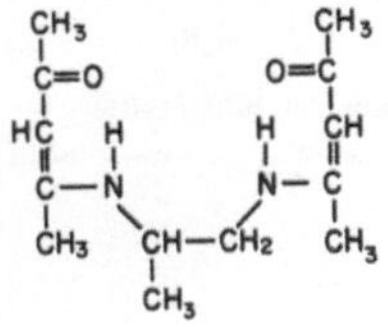

J. Am. Chem. Soc. 83:2099 (1961)

▶ $N_2O_2C_{12}H_{20}$

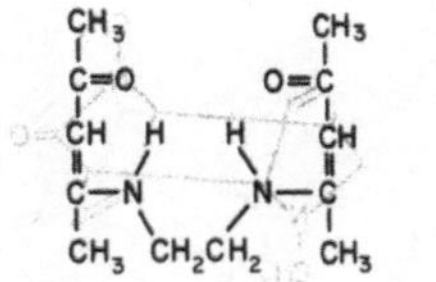

J. Am. Chem. Soc. 83:2099 (1961)

▶ $N_2O_2C_{13}H_{22}$

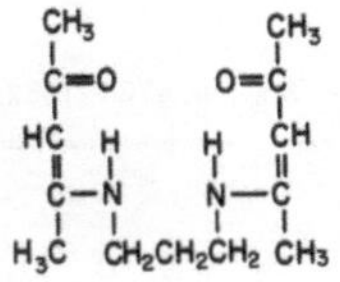

J. Am. Chem. Soc. 83:2099 (1961)

▶ $N_2O_2C_{13}H_{16}$

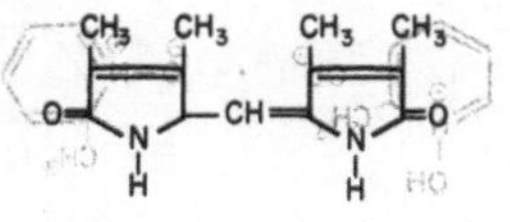

Ann. 654:165 (1962)

▶ $N_2O_2C_{14}H_{10}$

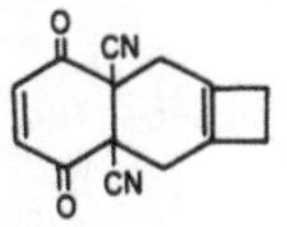

J. Org. Chem. 26:3507 (1961)

▶ $N_2O_2C_{14}H_{12}$

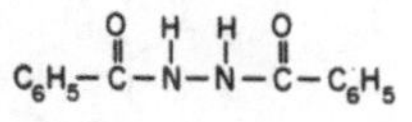

Chem. & Ind. 1961 (1961)

▶ $N_2O_2C_{14}H_{12}$

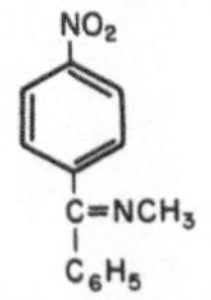

J. Am. Chem. Soc. 83:3474 (1961)

▶ $N_2O_2C_{14}H_{12}$

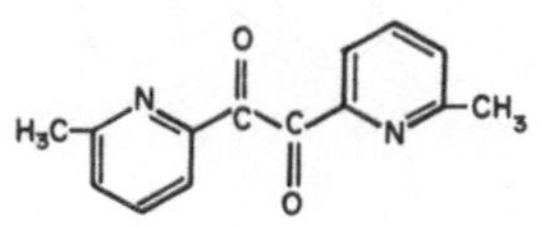

Varian 623

▶ $N_2O_2C_{14}H_{14}$

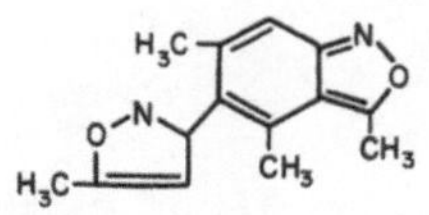

Tetrahedron Letters 672 (1961)

▶ $N_2O_2C_{14}H_{18}$

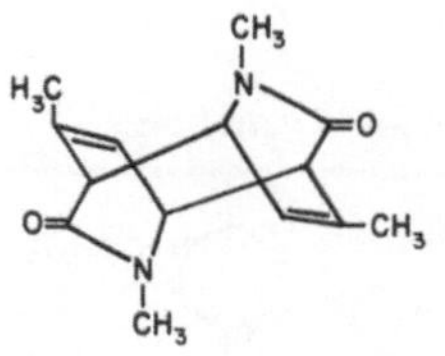

Tetrahedron Letters 652 (1961)

▶ $N_2O_2C_{14}H_{18}$

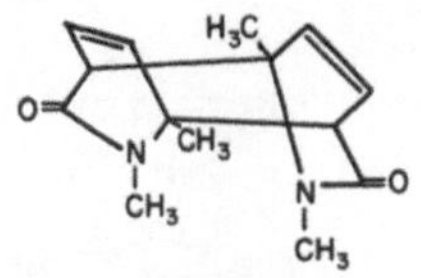

Tetrahedron Letters 650 (1961)

▶ $N_2O_2C_{14}H_{24}$

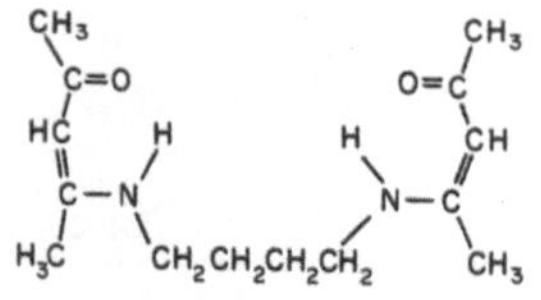

J. Am. Chem. Soc. 83:2099 (1961)

▶ $N_2O_2C_{15}H_{12}$

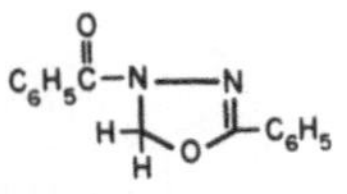

Chem. & Ind. 1961 (1961)

▶ $N_2O_2C_{15}H_{20}$

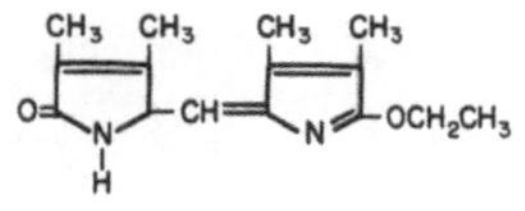

Ann. 654:165 (1962)

▶ $N_2O_2C_{15}H_{26}$

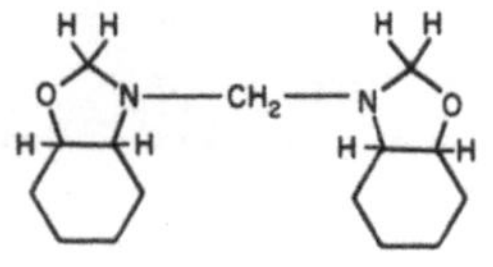

J. Org. Chem. 27:2965 (1962)

▶ $N_2O_2C_{16}H_{14}$

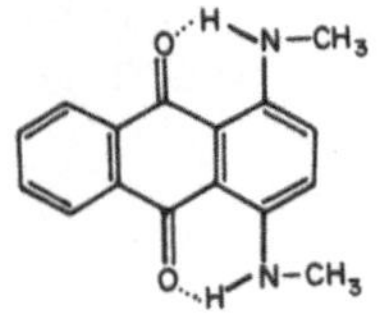

Varian 652

► $N_2O_2C_{16}H_{16}$

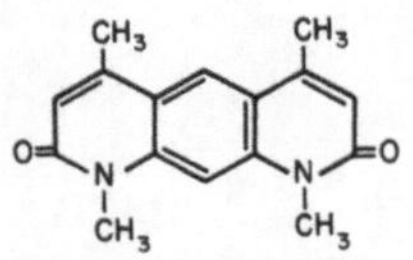

J. Am. Chem. Soc. 83:3729 (1961)

► $N_2O_2C_{16}H_{20}$

J. Org. Chem. 27:2752 (1962)

► $N_2O_2C_{16}H_{20}$

J. Org. Chem. 27:2752 (1962)

► $N_2O_2C_{17}H_{14}$

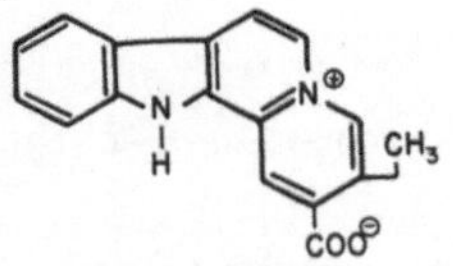

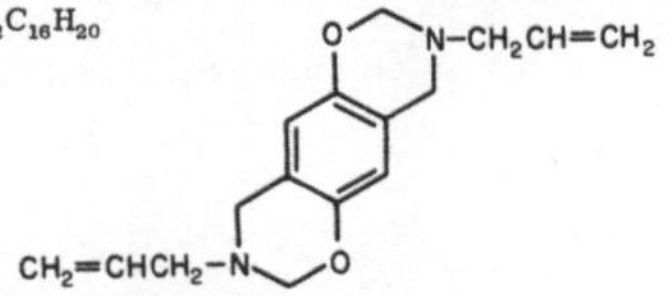

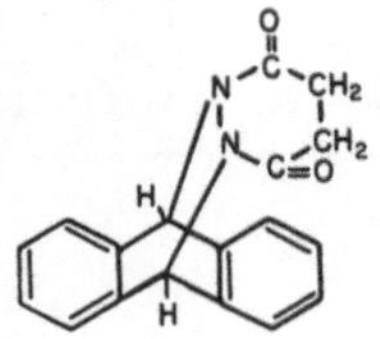

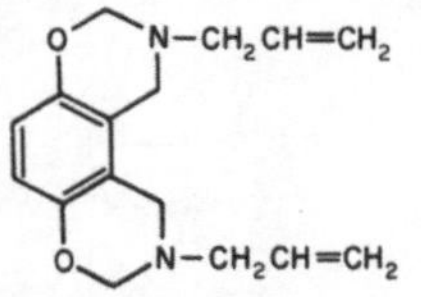

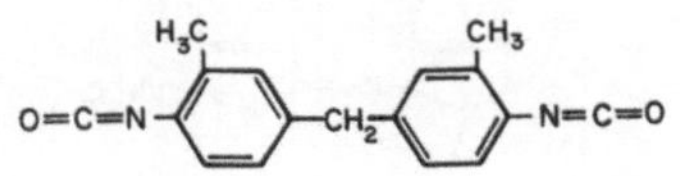

Varian 661

► $N_2O_2C_{17}H_{24}$

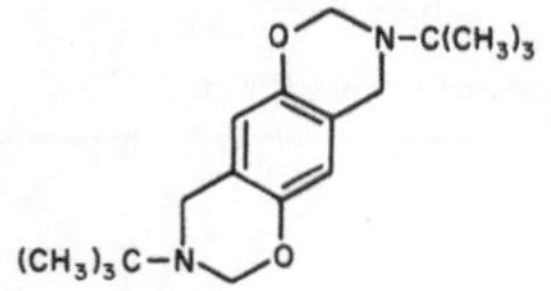

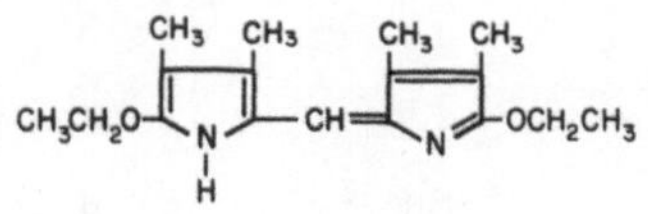

Ann. 654:165 (1962)

► $N_2O_2C_{18}H_{12}$

J. Org. Chem. 27:1115 (1962)

► $N_2O_2C_{18}H_{14}$

J. Am. Chem. Soc. 84:3393 (1962)

► $N_2O_2C_{18}H_{14}$

J. Org. Chem. 27:1115 (1962)

► $N_2O_2C_{18}H_{28}$

J. Org. Chem. 27:2752 (1962)

► $N_2O_2C_{18}H_{28}$

J. Org. Chem. 27:2752 (1962)

► $N_2O_2C_{19}H_{16}$

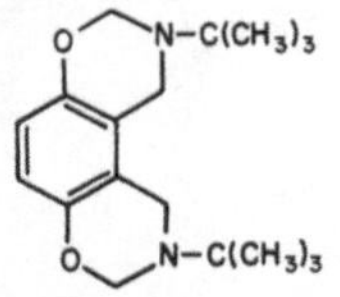

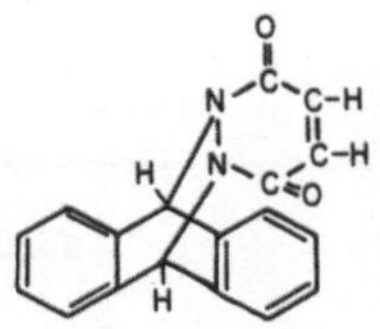

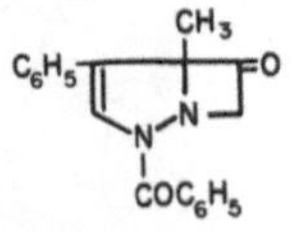

J. Am. Chem. Soc. 84:3022 (1962)

▶ $N_2O_2C_{19}H_{22}$

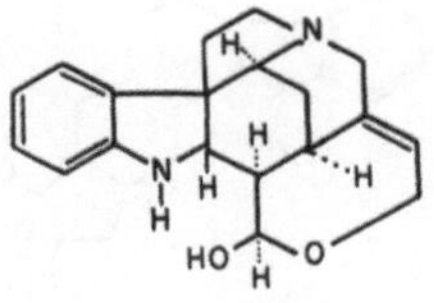

Helv. Chim. Acta 45:2266 (1962)

▶ $N_2O_2C_{20}H_{24}$

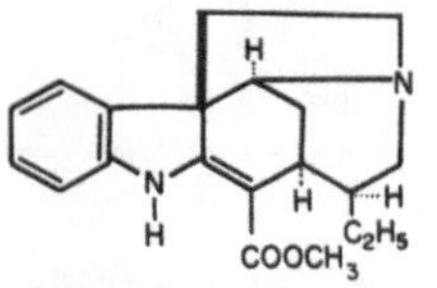

DIHYDROAKUAMMICIN

Helv. Chim. Acta 44:1877 (1961)

▶ $N_2O_2C_{19}H_{22} \cdot HCl$

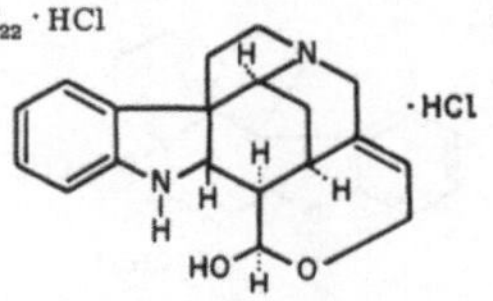

Helv. Chim. Acta 45:2266 (1962)

▶ $N_2O_2C_{20}H_{28}$

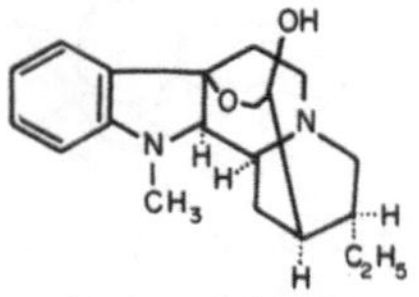

J. Indian Chem. Soc. 38:403 (1961)

▶ $N_2O_2C_{19}H_{24}$

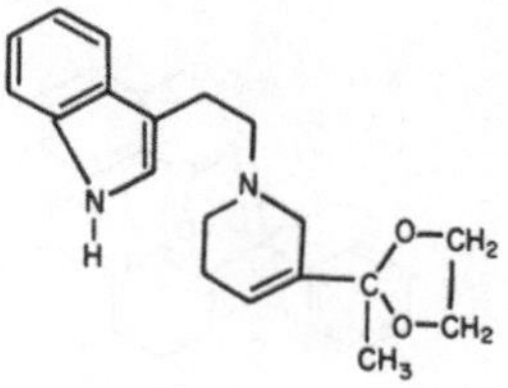

J. Am. Chem. Soc. 84:3732 (1962)

▶ $N_2O_2C_{21}H_{18}$

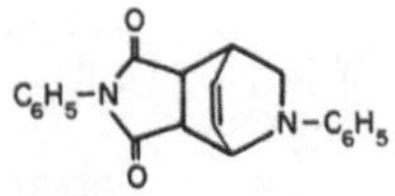

J. Org. Chem. 27:1439 (1962)

▶ $N_2O_2C_{21}H_{24}$

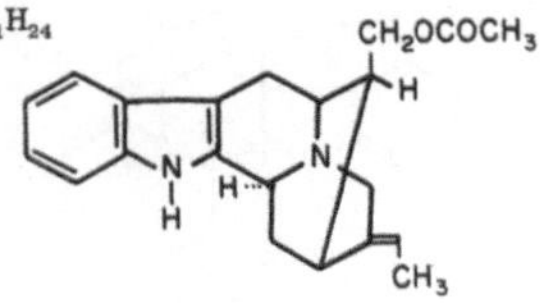

J. Am. Chem. Soc. 84:2161 (1962)

▶ $N_2O_2C_{20}H_{22}$

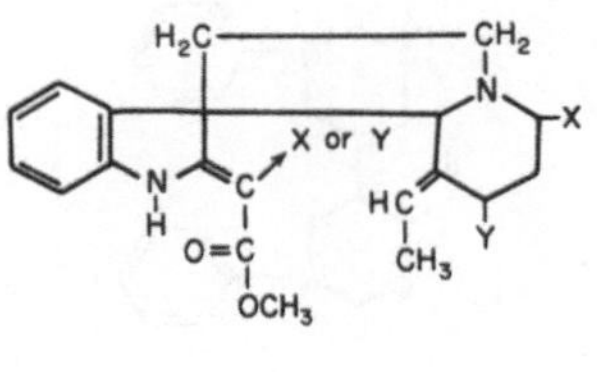

Tetrahedron Letters No. 10:412 (1962)

▶ $N_2O_2C_{21}H_{24}$

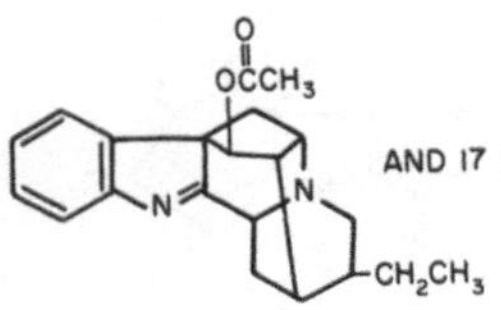

AND 17 EPIMERS

Tetrahedron Letters 367 (1961)

▶ $N_2O_2C_{21}H_{24}$

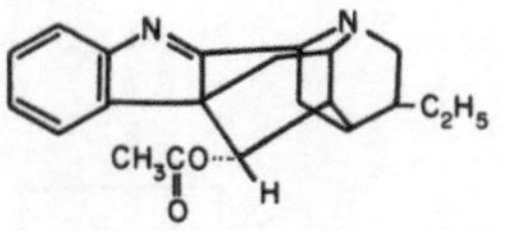

J. Am. Chem. Soc. 84:622 (1962)

▶ $N_2O_2C_{21}H_{24}$

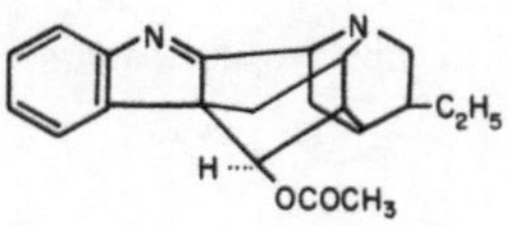

J. Am. Chem. Soc. 84:622 (1962)

▶ $N_2O_2C_{21}H_{28}$

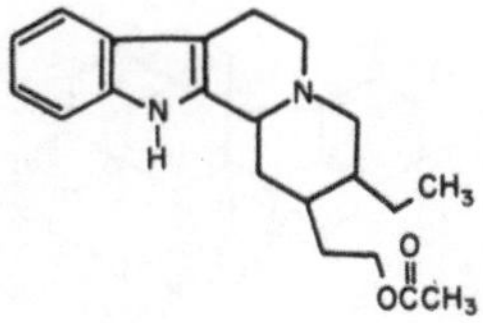

J. Org. Chem. 27:4702 (1962)

▶ $N_2O_2C_{21}H_{24}$

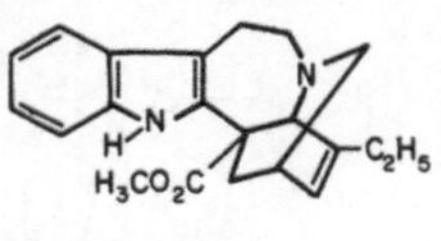

CATHARANTHINE

Tetrahedron Letters 209 (1961)

▶ $N_2O_2C_{21}H_{28}$

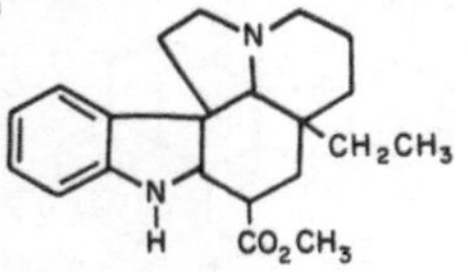

Tetrahedron Letters No. 6:236 (1962)

▶ $N_2O_2C_{21}H_{26}$

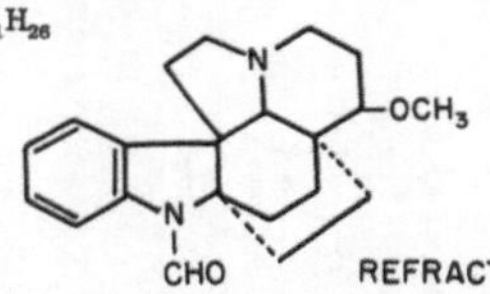

REFRACTIDINE

Tetrahedron Letters No. 1:59 (1962)

▶ $N_2O_2C_{21}H_{26}$

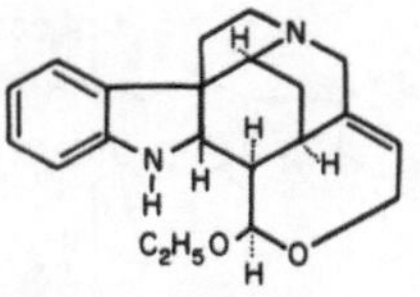

Helv. Chim. Acta 45:2266 (1962)

▶ $N_2O_2C_{22}H_{14}$

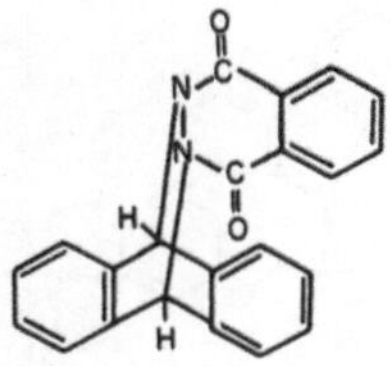

J. Org. Chem. 27:1115 (1962)

▶ $N_2O_2C_{21}H_{26}$

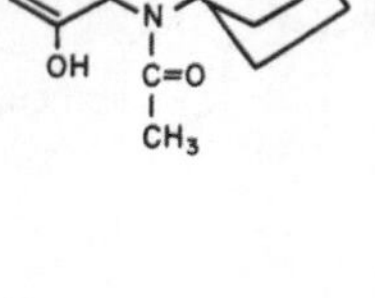

Experientia 18:397 (1962)

▶ $N_2O_2C_{22}H_{20}$

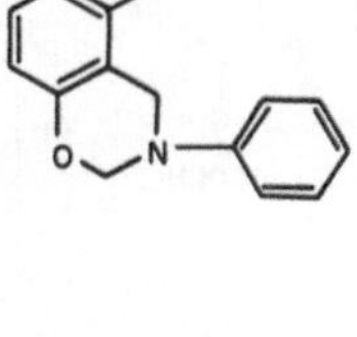

J. Org. Chem. 27:2752 (1962)

▶ $N_2O_2C_{22}H_{20}$

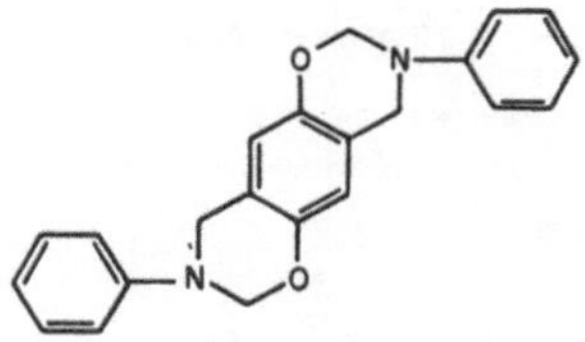

J. Org. Chem. 27:2752 (1962)

▶ $N_2O_2C_{22}H_{28}$

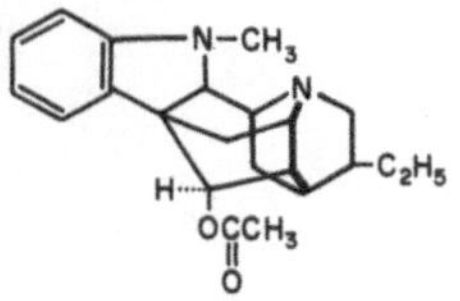

J. Am. Chem. Soc. 84:622 (1962)

▶ $N_2O_2C_{22}H_{28}$

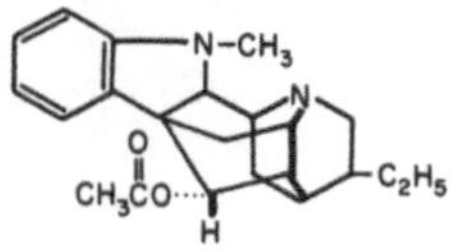

J. Am. Chem. Soc. 84:622 (1962)

▶ $N_2O_2C_{22}H_{24}$

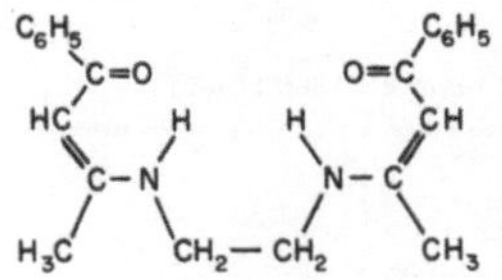

J. Am. Chem. Soc. 83:2099 (1961)

▶ $N_2O_2C_{25}H_{22}$

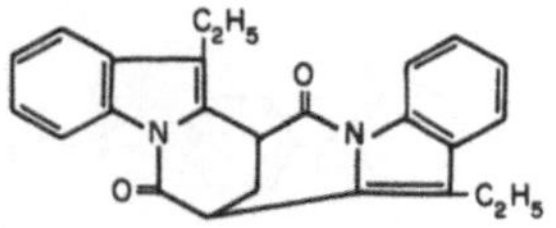

Tetrahedron Letters No. 10:411 (1962)

▶ $N_2O_2C_{28}H_{24} \cdot Ni$

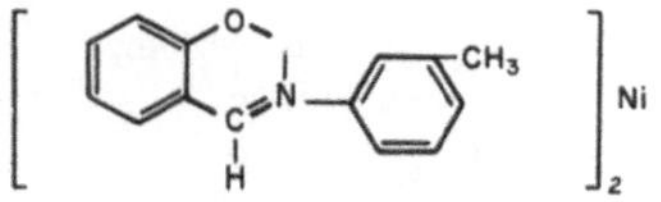

J. Am. Chem. Soc. 84:3968 (1962)

▶ $N_2O_2C_{22}H_{28}$

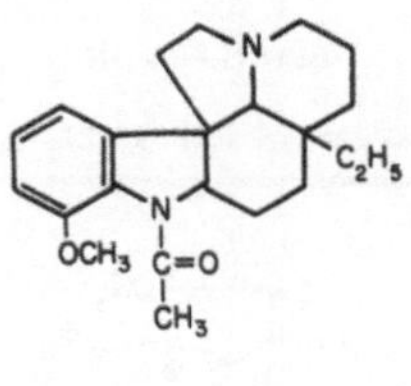

Experientia 17:162 (1961)

▶ $N_2O_2C_{29}H_{48}$

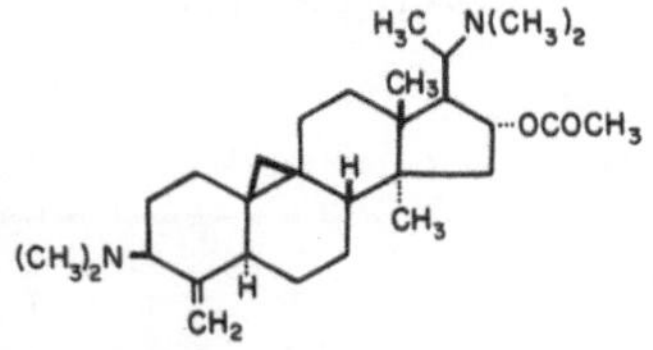

N,N'-DIMETHYLCYCLOBUXINE O-ACETATE

J. Am. Chem. Soc. 84:4590 (1962)

▶ $N_2O_2C_{22}H_{28}$

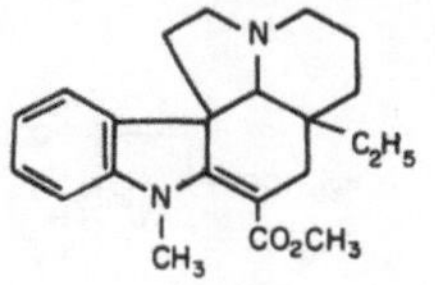

Bull. Soc. Chim. France 2237 (1962)

▶ $N_2O_2C_{30}H_{28} \cdot Ni$

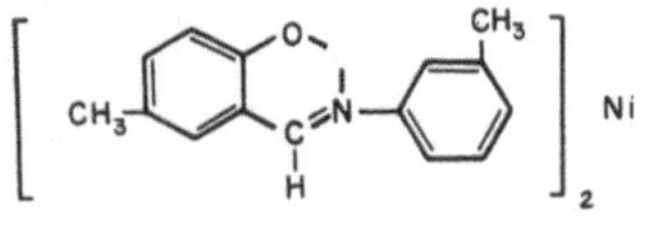

J. Am. Chem. Soc. 84:3968 (1962)

▶ $N_2O_2C_{30}H_{28} \cdot Ni$

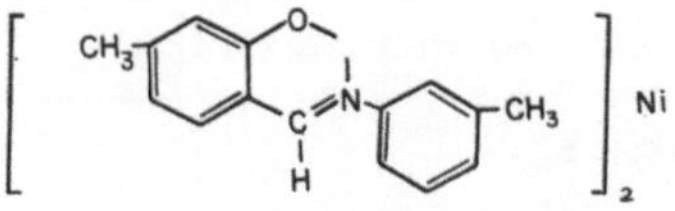

J. Am. Chem. Soc. 84:3968 (1962)

▶ $N_2O_3Cl_2C_{17}H_{10}$

Varian 327

▶ $N_2O_3C_4H_P$

Varian 413

▶ $N_2O_3C_4H_8$

$NH_2CH_2CONHCH_2COONa$

J. Phys. Chem. 65:1902 (1961)

▶ $N_2O_3C_4H_8$

$H_2NCH_2CONHCH_2CO_2H$

GLYCYLGLYCINE

J. Am. Chem. Soc. 84:4650 (1962)

▶ $N_2O_3C_5H_4$

J. Chem. Soc. 44 (1961)

▶ $N_2O_3C_5H_{10}$

$H_2NCH_2CONHCHCO_2H$ ($CH_3$)

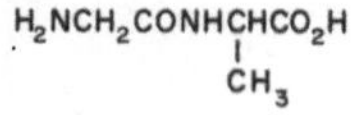

GLYCYL-L-ALANINE

J. Am. Chem. Soc. 84:4650 (1962)

▶ $N_2O_3C_6H_6$

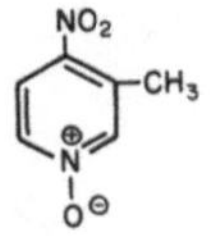

J. Chem. Soc. 46 (1961)

▶ $N_2O_3C_7H_8$

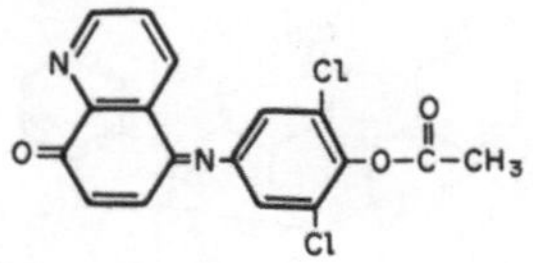

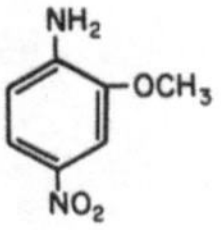

Can. J. Chem. 40:1866 (1962)

▶ $N_2O_3C_7H_{10}$

Varian 491

▶ $N_2O_3C_7H_{12}$

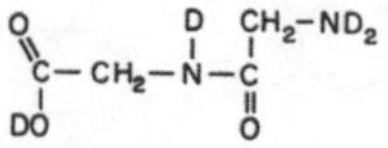

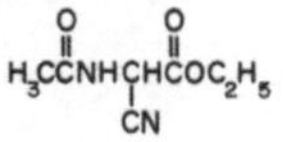

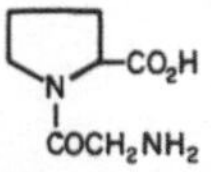

GLYCYLPROLINE

J. Am. Chem. Soc. 84:4650 (1962)

▶ $N_2O_3C_8H_{12}$

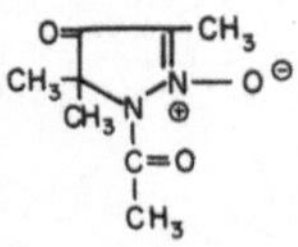

J. Org. Chem. 27:2882 (1962)

▶ $N_2O_3C_9H_{14}$

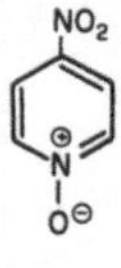

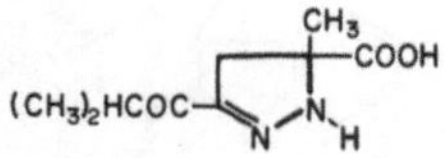

Varian 543

▶ $N_2O_3C_{14}H_{12}$

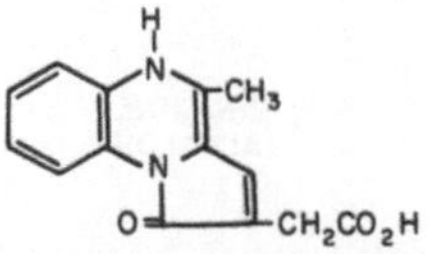

Tetrahedron Letters No. 25:1227 (1962)

▶ $N_2O_3C_{15}H_{20}$

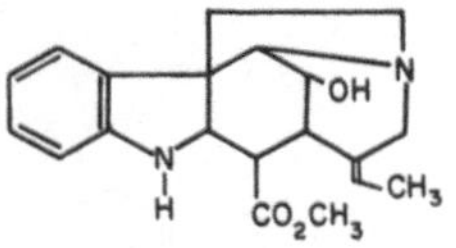

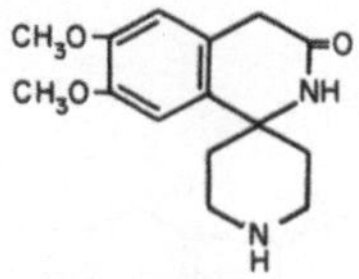

J. Chem. Soc. 3834 (1962)

▶ $N_2O_3C_{16}H_{14}$

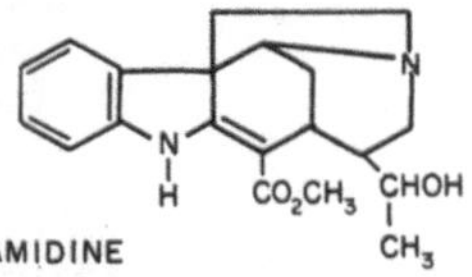

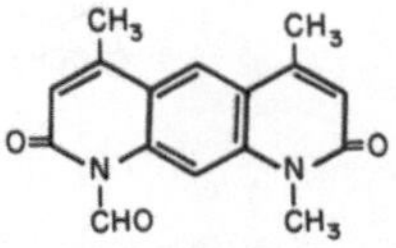

J. Am. Chem. Soc. 83:3729 (1961)

▶ $N_2O_3C_{19}H_{18}$

Varian 677

▶ $N_2O_3C_{19}H_{24}$

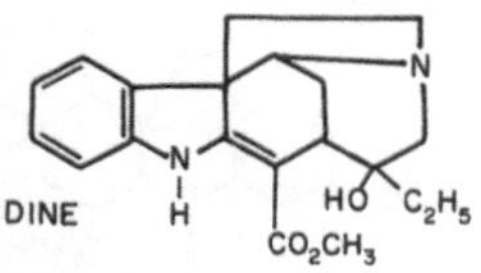

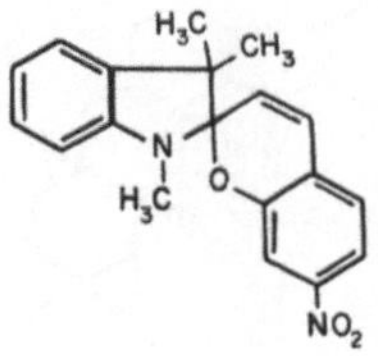

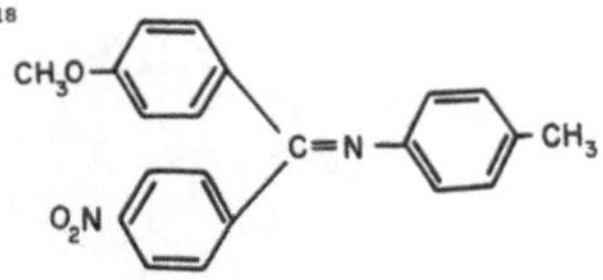

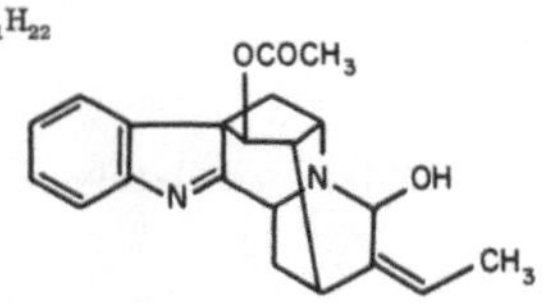

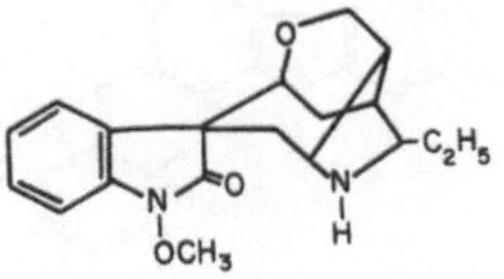

J. Org. Chem. 27:4123 (1962)

▶ $N_2O_3C_{20}H_{24}$

Bull. Soc. Chim. France 1088 (1962)

▶ $N_2O_3C_{20}H_{24}$

ECHITAMIDINE

Tetrahedron Letters No. 15:655 (1962)

▶ $N_2O_3C_{20}H_{24}$

LOCHNERIDINE

Chem. & Ind. 1986 (1962)

▶ $N_2O_3C_{21}H_{18}$

J. Am. Chem. Soc. 83:3474 (1961)

▶ $N_2O_3C_{21}H_{22}$

Helv. Chim. Acta 45:611 (1962)

▶ $N_2O_3C_{21}H_{22}$

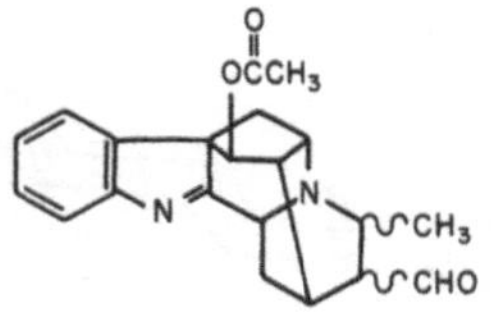

PERAKINE

Tetrahedron Letters 367 (1961)

▶ $N_2O_3C_{21}H_{24}$

MAYUMBINE

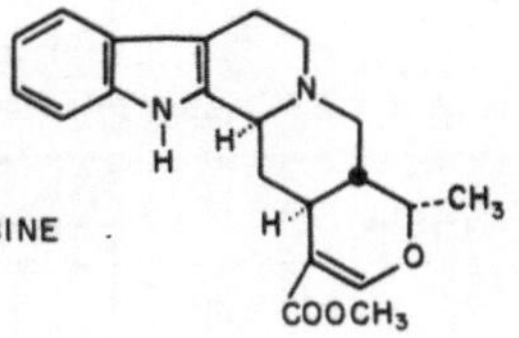


Tetrahedron Letters 823 (1961)
J. Am. Chem. Soc. 83:5037 (1961)

▶ $N_2O_3C_{21}H_{24}$

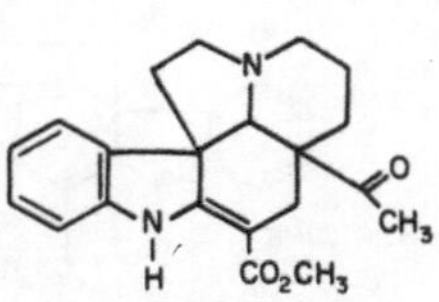


Bull. Soc. Chim. France 2237 (1962)

▶ $N_2O_3C_{21}H_{24}$

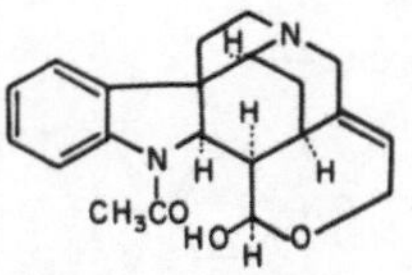


Helv. Chim. Acta 45:2266 (1962)

▶ $N_2O_3C_{21}H_{24}$

APO-ψ-AKUAMMIGINE

J. Chem. Soc. 318 (1962)

▶ $N_2O_3C_{21}H_{24}$

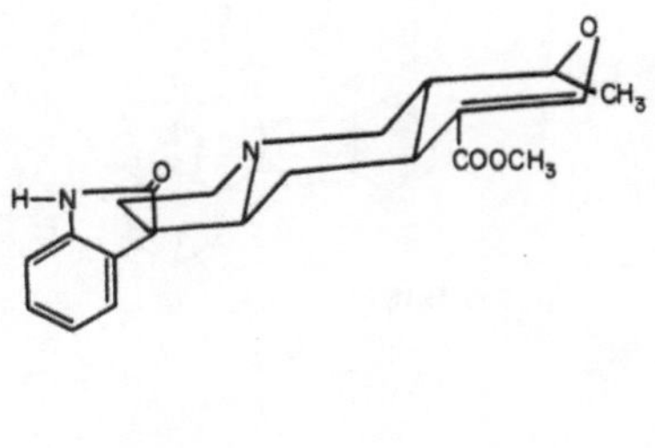


J. Am. Chem. Soc. 84:3871 (1962)

▶ $N_2O_3C_{21}H_{24}$

RHAZINE
(ALKALOID)

Chem. & Ind. 266 (1962)

▶ $N_2O_3C_{21}H_{24}$

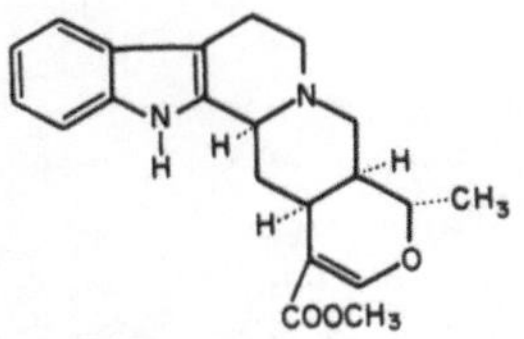


J. Am. Chem. Soc. 83:5037 (1961)

▶ $N_2O_3C_{21}H_{24}$

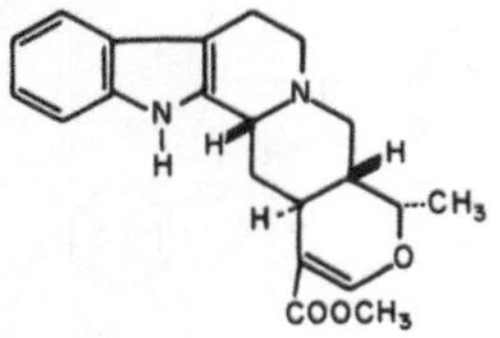


J. Am. Chem. Soc. 83:5037 (1961)

▶ $N_2O_3C_{21}H_{24}$

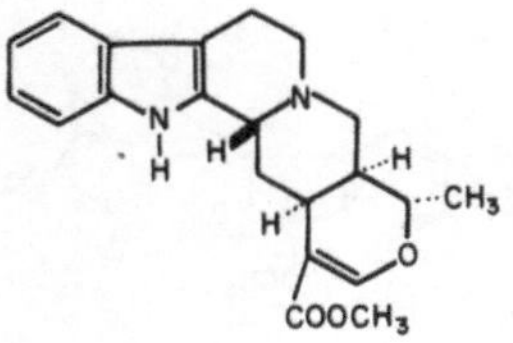


J. Am. Chem. Soc. 83:5037 (1961)

▶ $N_2O_3C_{21}H_{26}$

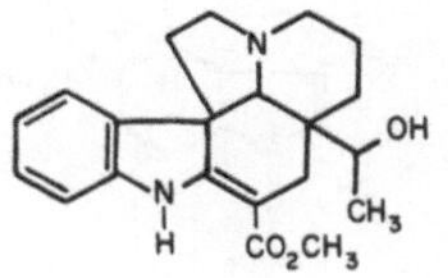

Bull. Soc. Chim. France 2237 (1962)

▶ $N_2O_3C_{22}H_{30}$

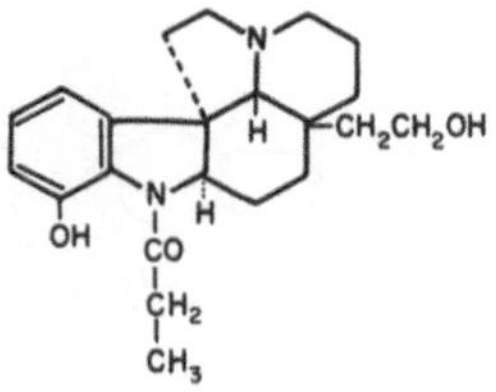

Helv. Chim. Acta 45:2260 (1962)

▶ $N_2O_3C_{21}H_{26}$

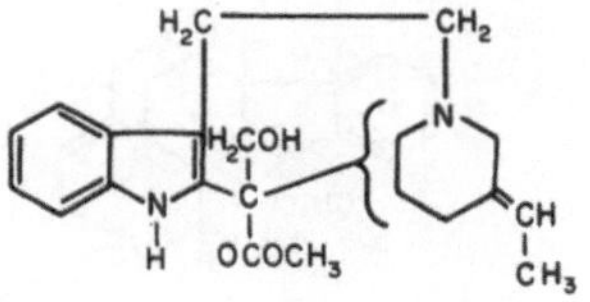

Tetrahedron Letters No. 10:409 (1962)

▶ $N_2O_3C_{23}H_{28}$

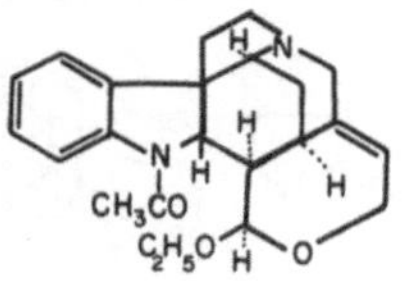

Helv. Chim. Acta 45:2266 (1962)

▶ $N_2O_3C_{21}H_{28}$

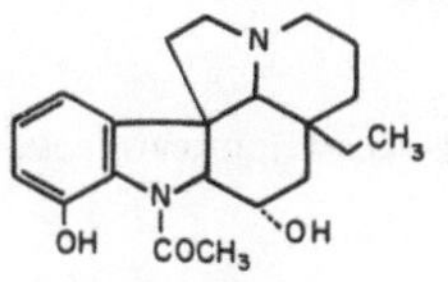

J. Am. Chem. Soc. 84:3480 (1962)

▶ $N_2O_3C_{23}H_{30}$

ψ-AKUAMMIGINEDIHYDROMETHINE

J. Chem. Soc. 318 (1962)

▶ $N_2O_3C_{22}H_{26}$

ψ-AKUAMMIGINE

J. Chem. Soc. 318 (1962)

▶ $N_2O_3C_{22}H_{28}$

APO-ψ-AKUAMMIGINEDIHYDROMETHINE

J. Chem. Soc. 318 (1962)

▶ $N_2O_3C_{23}H_{30}$

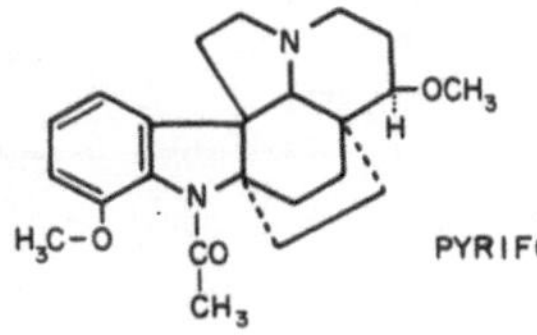

PYRIFOLINE

Tetrahedron Letters No. 1:59 (1962)

▶ $N_2O_3C_{21}H_{26}$

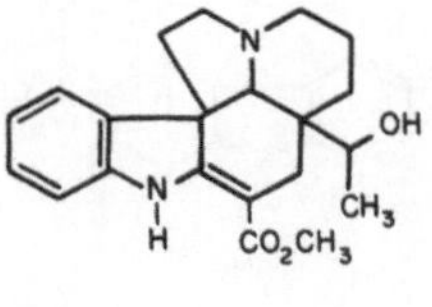


Bull. Soc. Chim. France 2237 (1962)

▶ $N_2O_3C_{22}H_{30}$

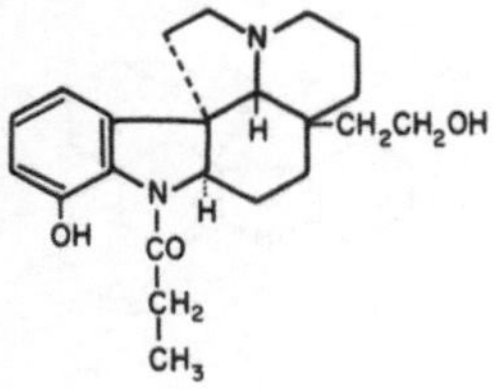


Helv. Chim. Acta 45:2260 (1962)

▶ $N_2O_3C_{21}H_{26}$

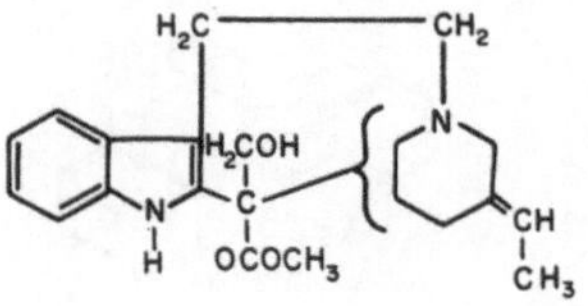


Tetrahedron Letters No. 10:409 (1962)

▶ $N_2O_3C_{23}H_{28}$

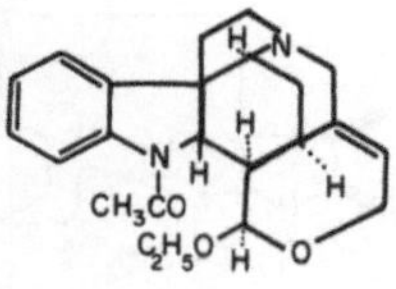


Helv. Chim. Acta 45:2266 (1962)

▶ $N_2O_3C_{21}H_{28}$

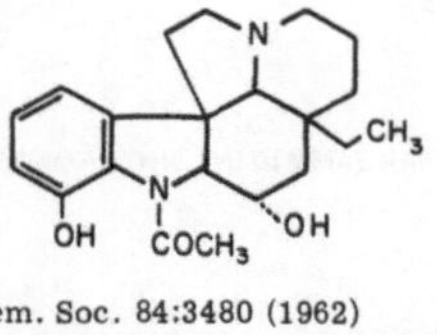


J. Am. Chem. Soc. 84:3480 (1962)

▶ $N_2O_3C_{23}H_{30}$

ψ-AKUAMMIGINEDIHYDROMETHINE

J. Chem. Soc. 318 (1962)

▶ $N_2O_3C_{22}H_{26}$

ψ-AKUAMMIGINE

J. Chem. Soc. 318 (1962)

▶ $N_2O_3C_{22}H_{28}$

APO-ψ-AKUAMMIGINEDIHYDROMETHINE

J. Chem. Soc. 318 (1962)

▶ $N_2O_3C_{23}H_{30}$

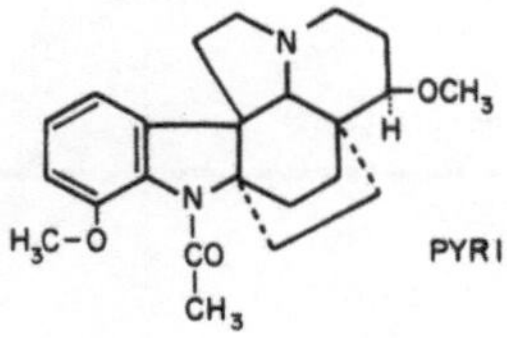


PYRIFOLINE

Tetrahedron Letters No. 1:59 (1962)

▶ $N_2O_4C_6H_4$

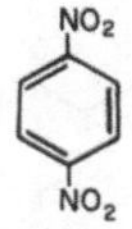

J. Chem. Phys. 37:2594 (1962)
Helv. Chim. Acta 45:568 (1962)

▶ $N_2O_4C_7H_6$

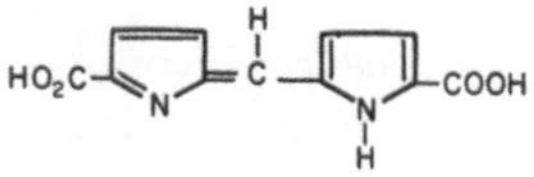

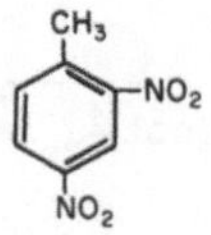

J. Org. Chem. 27:3325 (1962)

▶ $N_2O_4C_8H_{10}$

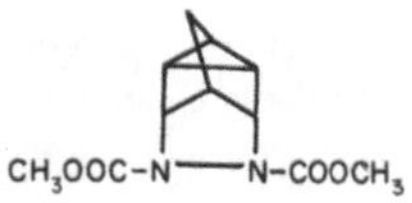

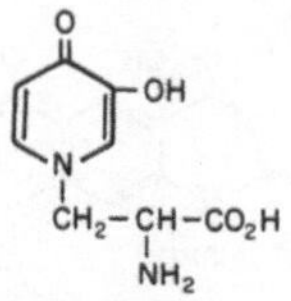

Can. J. Chem. 40:1377 (1962)

▶ $N_2O_4C_8H_{18} \cdot 2HCl$

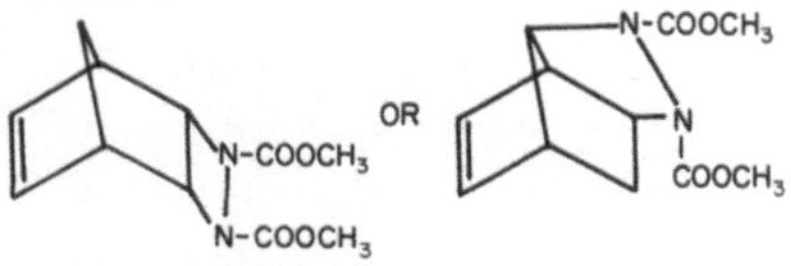

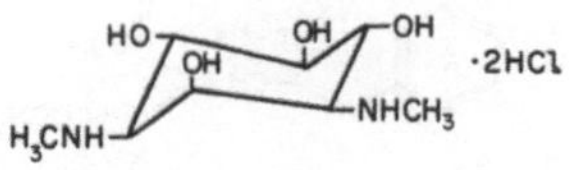

Tetrahedron Letters No. 12:523 (1962)

▶ $N_2O_4C_{10}H_6$

Varian 245

▶ $N_2O_4C_{10}H_{10}$

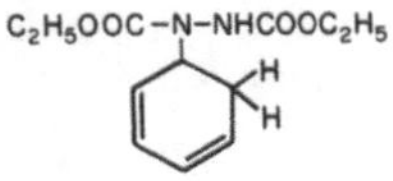

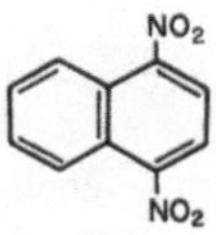

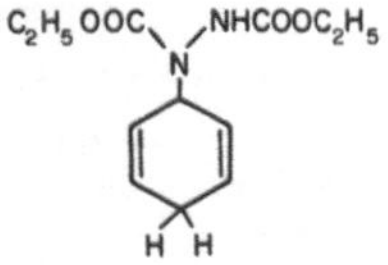

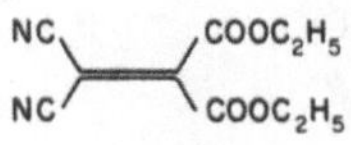

J. Org. Chem. 27:2237 (1962)

▶ $N_2O_4C_{10}H_{22} \cdot 2HCl$

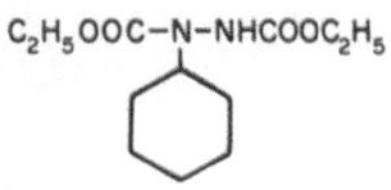

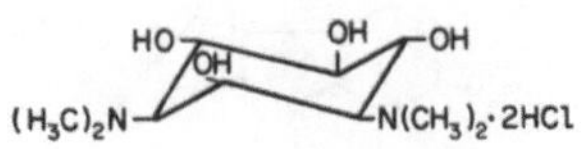

Tetrahedron Letters No. 12:522 (1962)

▶ $N_2O_4C_{11}H_8$

J. Org. Chem. 27:3697 (1962)

▶ $N_2O_4C_{11}H_{14}$

J. Org. Chem. 27:4059 (1962)

▶ $N_2O_4C_{11}H_{14}$

J. Org. Chem. 27:4059 (1962)

▶ $N_2O_4C_{12}H_{18}$

J. Org. Chem. 27:1956 (1962)

▶ $N_2O_4C_{12}H_{18}$

J. Org. Chem. 27:1955 (1962)

▶ $N_2O_4C_{12}H_{22}$

J Org Chem 27:1956 (1962)

▶ $N_2O_4C_{12}H_{22}$

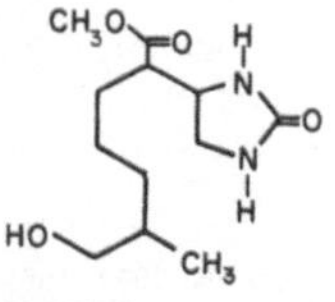

Chem. & Ind. 95 (1962)

▶ $N_2O_4C_{12}H_{26}$

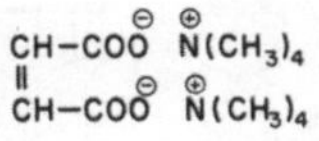

J. Phys. Chem. 66:1702 (1962)

▶ $N_2O_4C_{15}H_{12}$

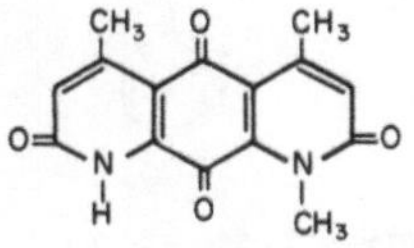

J. Am. Chem. Soc. 83:3729 (1961)

▶ $N_2O_4C_{13}H_{14}$

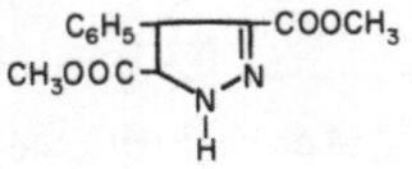

J. Org. Chem. 26:1912 (1961)

▶ $N_2O_4C_{15}H_{18}$

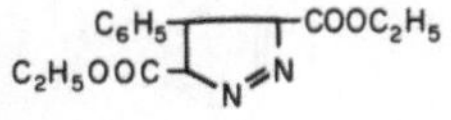

J. Org. Chem. 26:1912 (1961)

▶ $N_2O_4C_{14}H_{16}$

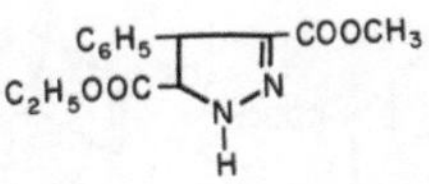

J. Org. Chem. 26:1912 (1961)

▶ $N_2O_4C_{16}H_{14}$

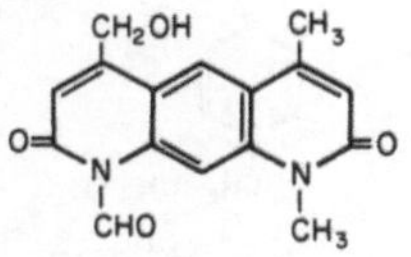

J. Am. Chem. Soc. 83:3729 (1961)

▶ $N_2O_4C_{14}H_{16}$

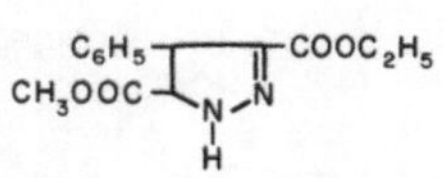

J. Org. Chem. 26:1912 (1961)

▶ $N_2O_4C_{16}H_{16}$

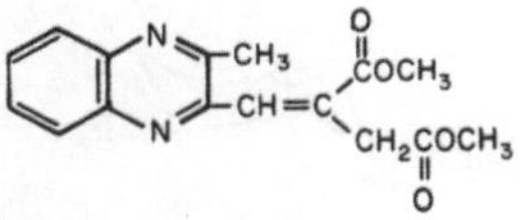

Tetrahedron Letters No. 25:1227 (1962)

▶ $N_2O_4C_{14}H_{18}$

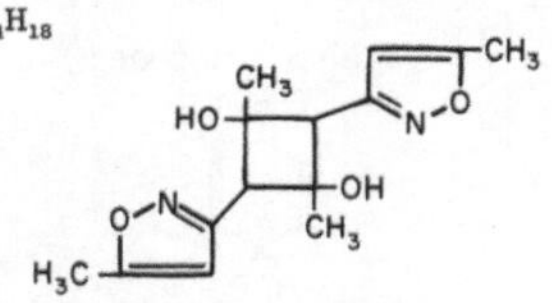

Tetrahedron Letters 671 (1961)

▶ $N_2O_4C_{18}H_{16}$

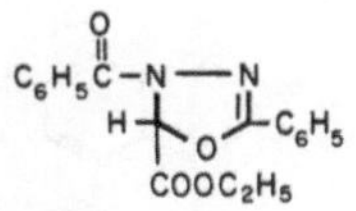

Chem. & Ind. 1961 (1961)

▶ $N_2O_4C_{14}H_{24}$

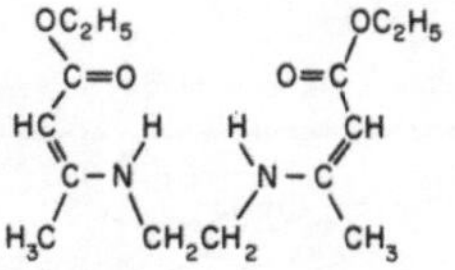

J. Am. Chem. Soc. 83:2099 (1961)
Tetrahedron 18:791 (1962)

▶ $N_2O_4C_{18}H_{16}$

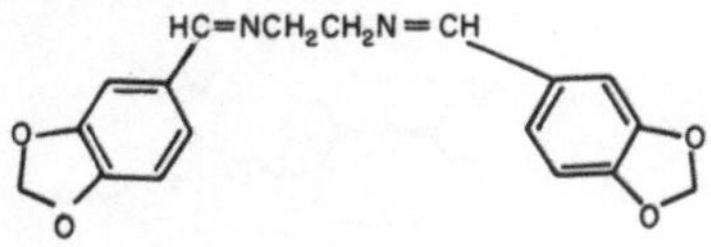

Varian 673

▶ $N_2O_4C_{20}H_{26}$

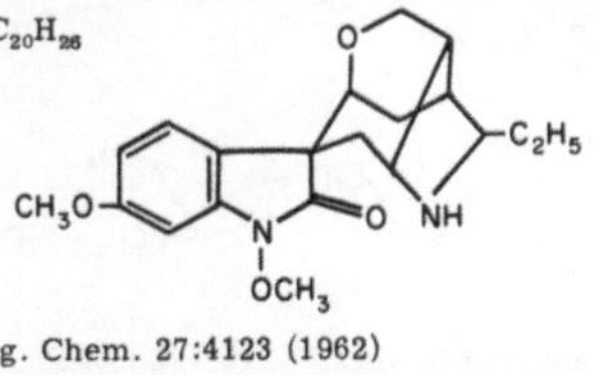

J. Org. Chem. 27:4123 (1962)

▶ $N_2O_4C_{21}H_{24}$

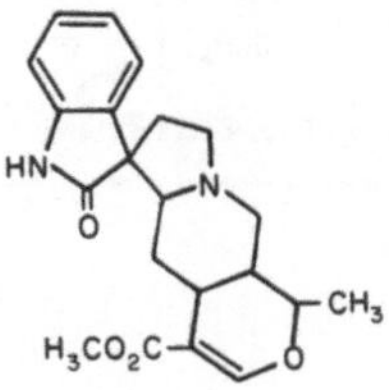


Tetrahedron Letters 823 (1961)

▶ $N_2O_4C_{22}H_{26}$

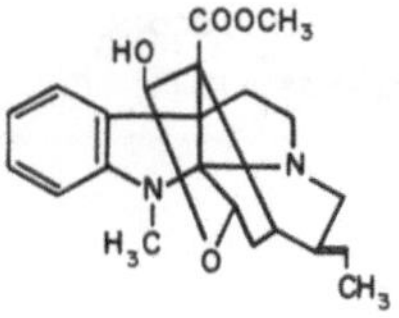


CORYMINE

Proc. Chem. Soc. 298 (1962)

▶ $N_2O_4C_{21}H_{26}$

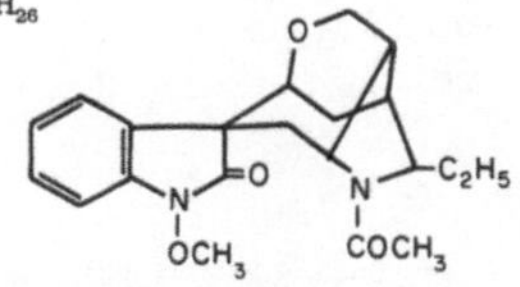


J. Org. Chem. 27:4123 (1962)

▶ $N_2O_4C_{22}H_{26}$

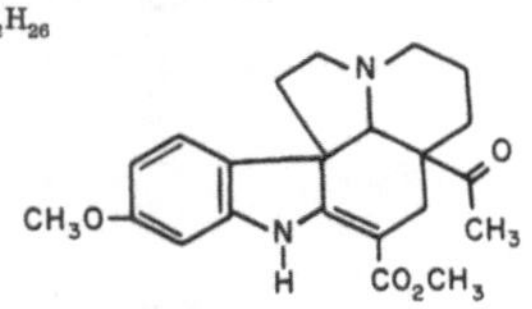


Bull. Soc. Chim. France 2237 (1962)

▶ $N_2O_4C_{21}H_{28}$

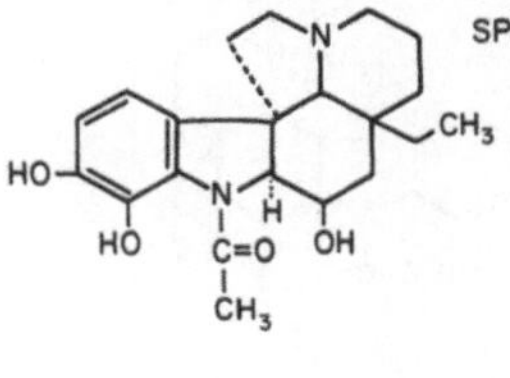


SPEGAZZINIDINE

J. Am. Chem. Soc. 84:3480 (1962)
Experientia 18:113 (1962)

▶ $N_2O_4C_{22}H_{28}$

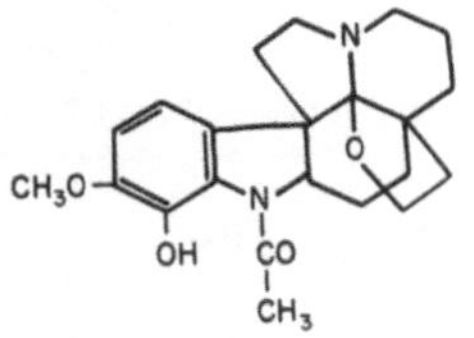


Chem. & Ind. 1949 (1962)

▶ $N_2O_4C_{22}H_{22}$

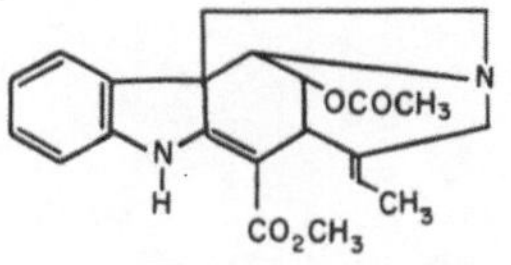


Bull. Soc. Chim. France 1088 (1962)

▶ $N_2O_4C_{22}H_{30}$

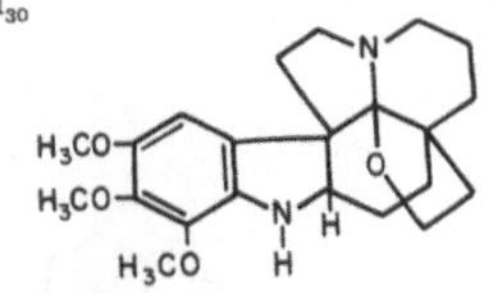


Tetrahedron Letters No. 22:1006 (1962)

▶ $N_2O_4C_{22}H_{24}$

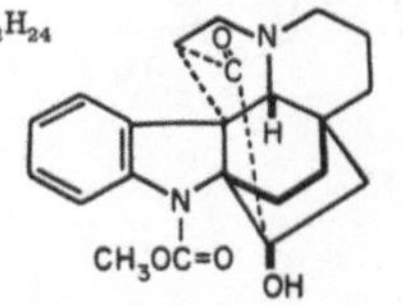


KOPSINE

Helv. Chim. Acta 45:1146 (1962)

▶ $N_2O_4C_{23}H_{26}$

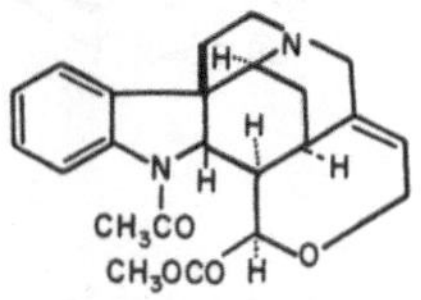


Helv. Chim. Acta 45:2266 (1962)

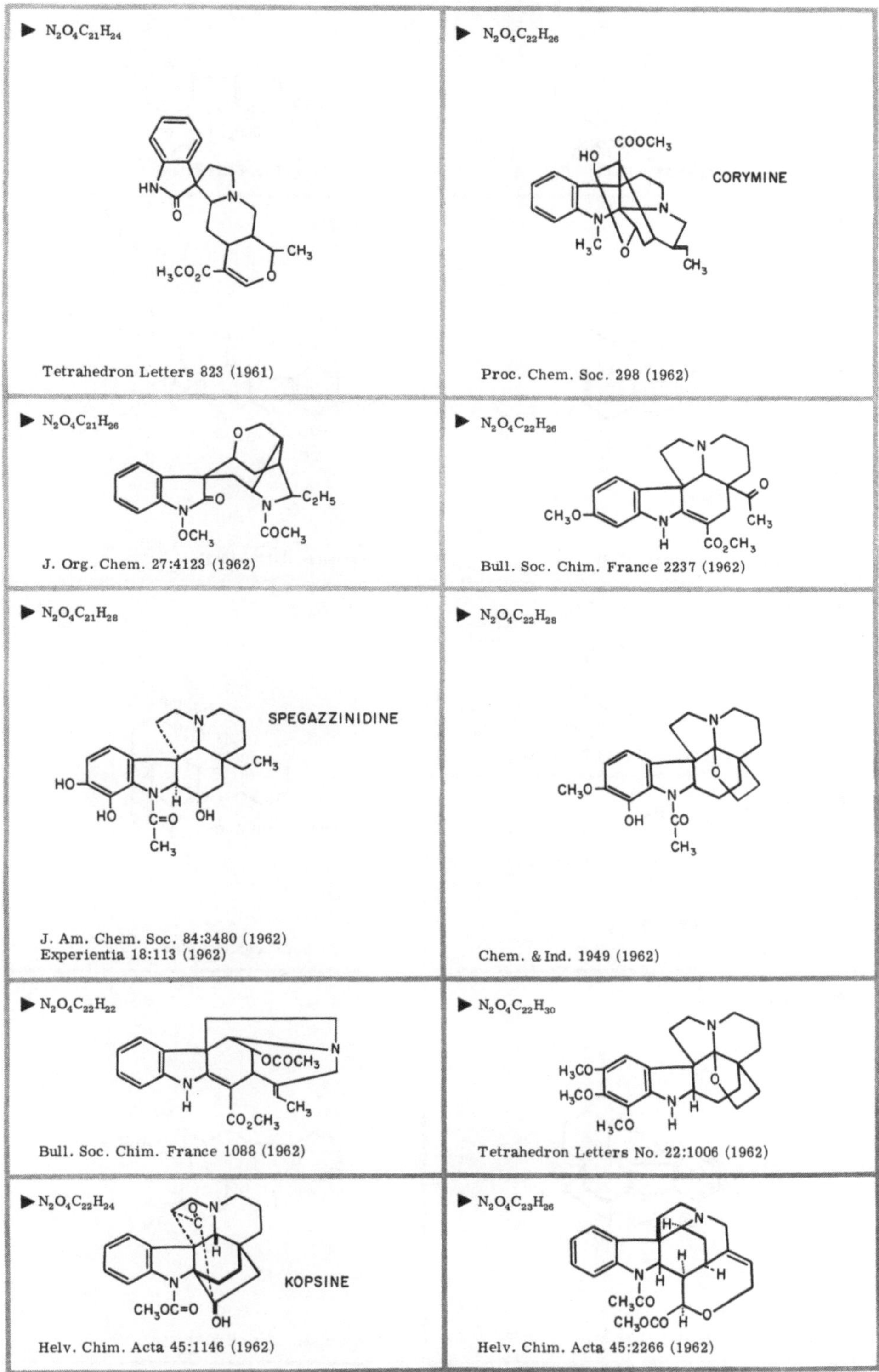

▶ $N_2O_4C_{23}H_{26}$

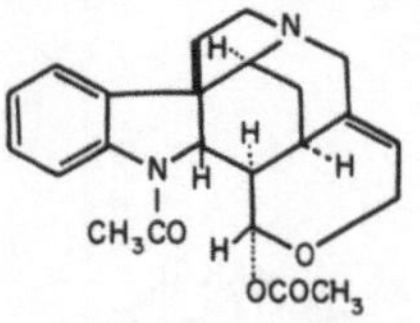


Helv. Chim. Acta 45:2266 (1962)

▶ $N_2O_4C_{23}H_{30}$

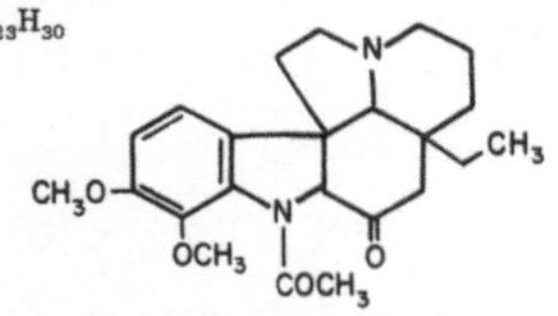


J. Am. Chem. Soc. 84:3480 (1962)

▶ $N_2O_4C_{23}H_{26}$

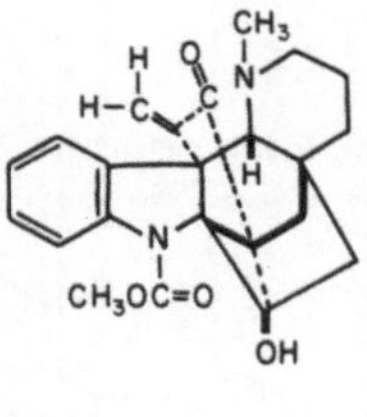


Helv. Chim. Acta 45:1146 (1962)

▶ $N_2O_4C_{23}H_{32}$

SPEGAZZINIDINE DIMETHYLETHER

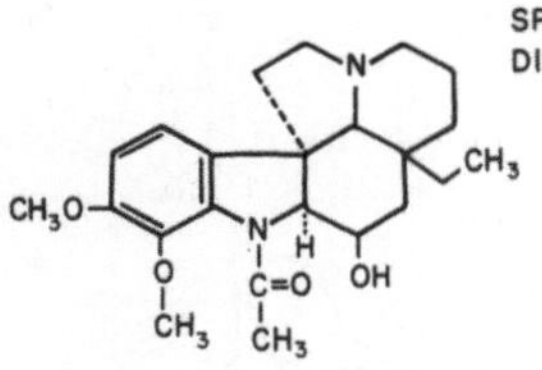


J. Am. Chem. Soc. 84:3480 (1962)
Experientia 18:113 (1962)

▶ $N_2O_4C_{23}H_{28}$

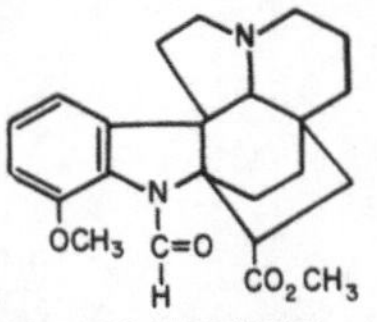


J. Am. Chem. Soc. 84:1499 (1962)

▶ $N_2O_4C_{23}H_{32}$

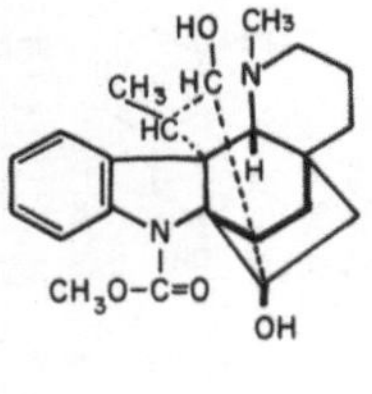


Helv. Chim. Acta 45:1146 (1962)

▶ $N_2O_4C_{23}H_{28}$

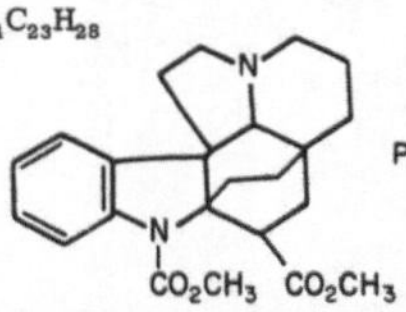


PLEIOCARPINE

Helv. Chim. Acta 45:854 (1962)

▶ $N_2O_4C_{23}H_{28}$

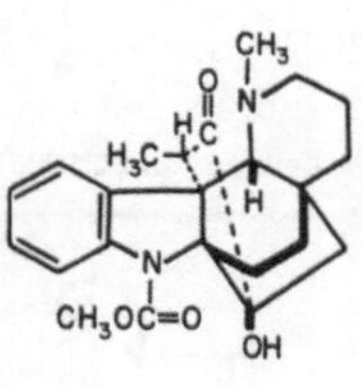


Helv. Chim. Acta 45:1146 (1962)

▶ $N_2O_4C_{24}H_{28}$

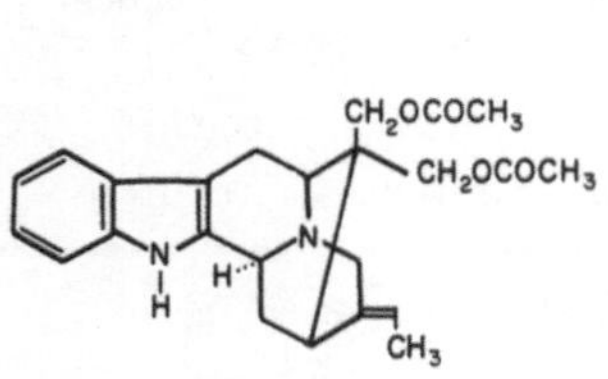


J. Am. Chem. Soc. 84:2161 (1962)

▶ $N_2O_4C_{24}H_{30}$

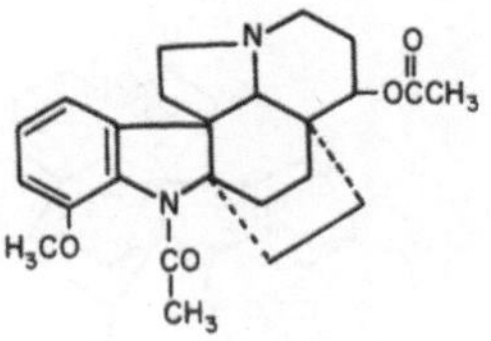

Tetrahedron Letters No. 1:59 (1962)

▶ $N_2O_4C_{24}H_{32}$

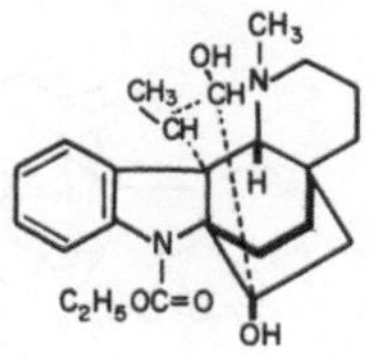

Helv. Chim. Acta 45:1146 (1962)

▶ $N_2O_5ClC_6H_3$

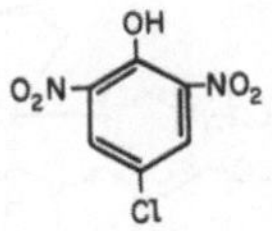

Can. J. Chem. 40:1759 (1962)

▶ $N_2O_5FC_9H_{11}$

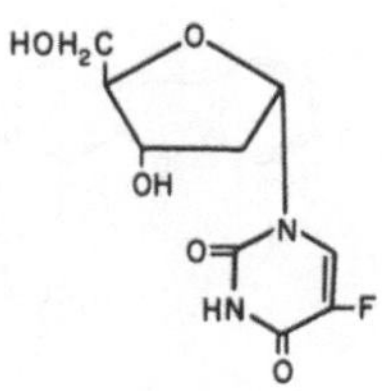

Can. J. Chem. 39:110 (1961)

▶ $N_2O_5FC_9H_{11}$

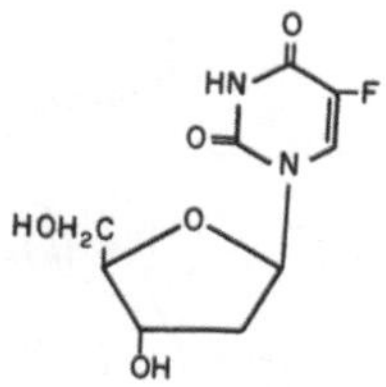

Can. J. Chem. 39:110 (1961)

▶ $N_2O_5F_3C_{14}H_{21}$

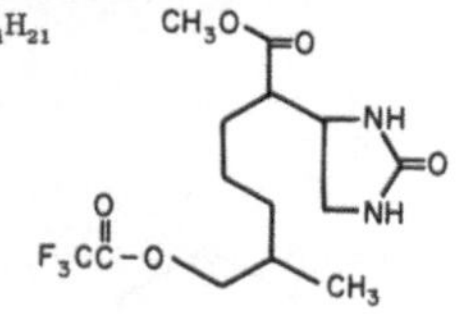

Chem. & Ind. 95 (1962)

▶ $N_2O_5C_7H_6$

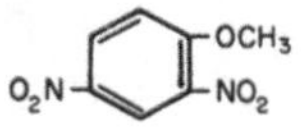

Varian 149
Can. J. Chem. 40:1866 (1962)

▶ $N_2O_5C_9H_{12}$

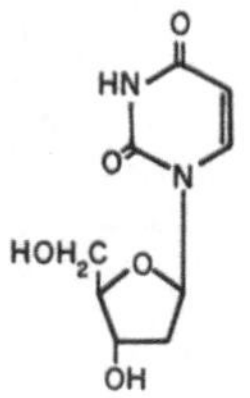

J. Am. Chem. Soc. 83:2919 (1961)

▶ $N_2O_5C_{10}H_{14}$

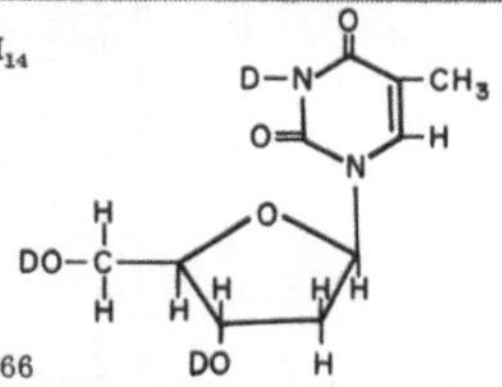

Varian 566

► $N_2O_5C_{10}H_{14}$

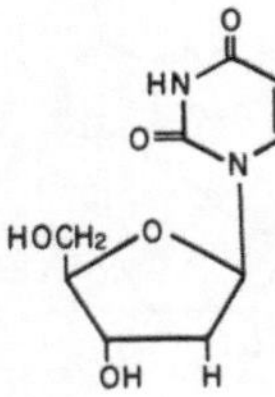


Can. J. Chem. 39:110, 116 (1961)
J. Am. Chem. Soc. 84:4464 (1962)

► $N_2O_5C_{22}H_{22}$

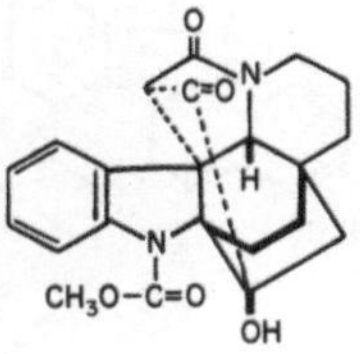

Helv. Chim. Acta 45:1146 (1962)

► $N_2O_5C_{10}H_{14}$

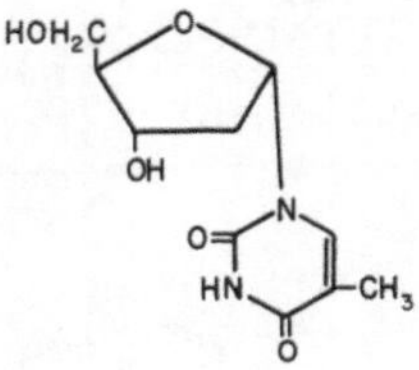

Can. J. Chem. 39:110, 116 (1961)

► $N_2O_5C_{22}H_{24}$

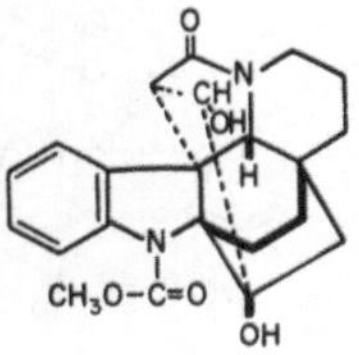

Helv. Chim. Acta 45:1146 (1962)

► $N_2O_5C_{20}H_{22}$

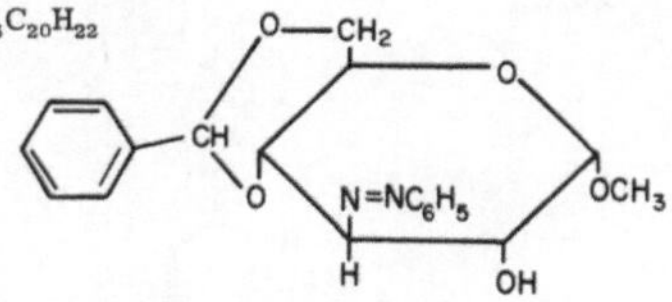

J. Chem. Soc. 851 (1962)

► $N_2O_5C_{23}H_{30}$

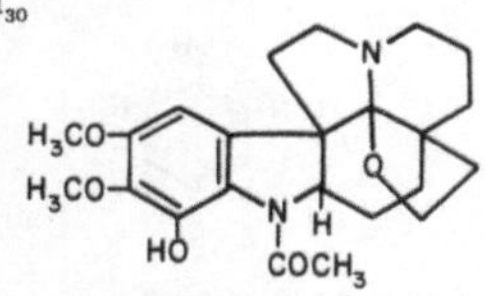

Tetrahedron Letters No. 22:1002 (1962)

► $N_2O_5C_{20}H_{28}$

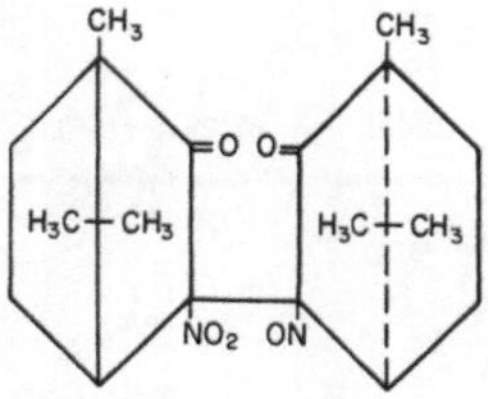

Australian J. Chem. 15:431 (1962)

► $N_2O_5C_{23}H_{30}$

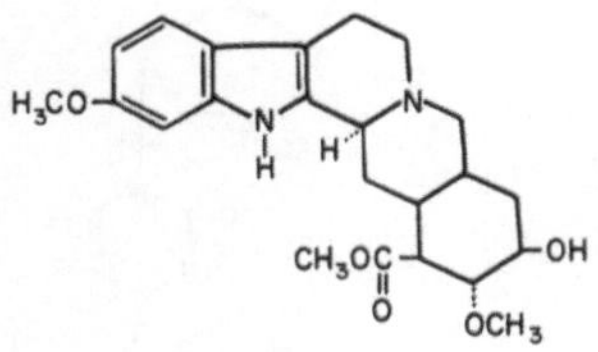

J. Am. Chem. Soc. 83:4816 (1961)

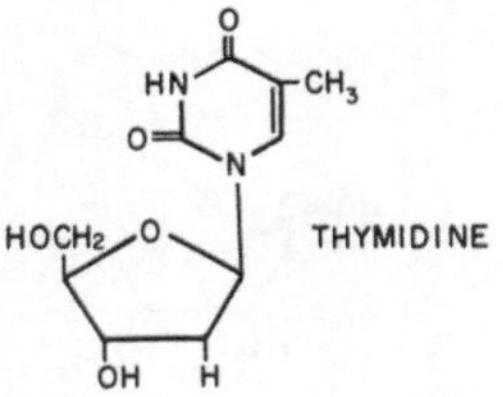

▶ $N_2O_5C_{23}H_{30}$

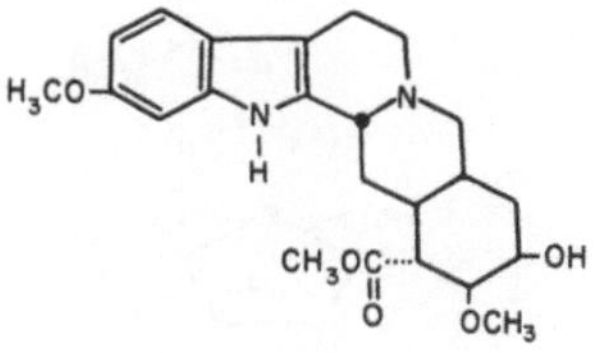


J. Am. Chem. Soc. 83:4816 (1961)

▶ $N_2O_5C_{25}H_{34}$

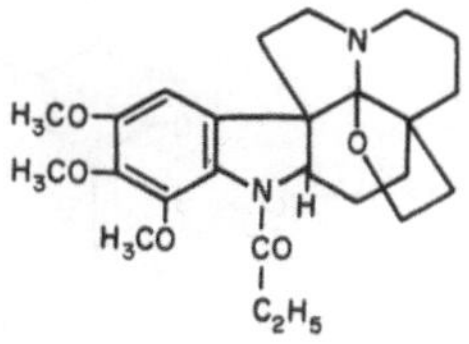


Tetrahedron Letters No. 22:1003 (1962)

▶ $N_2O_5C_{23}H_{30}$

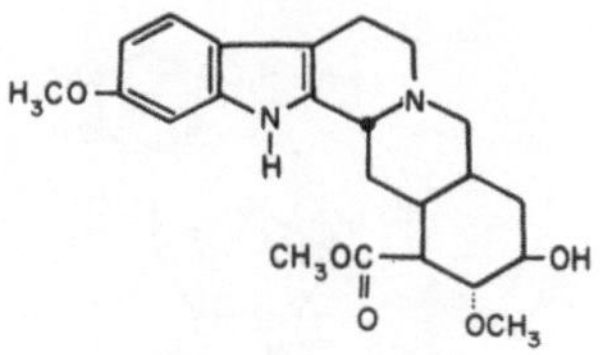


J. Am. Chem. Soc. 83:4816 (1961)

▶ $N_2O_5C_{26}H_{34}$

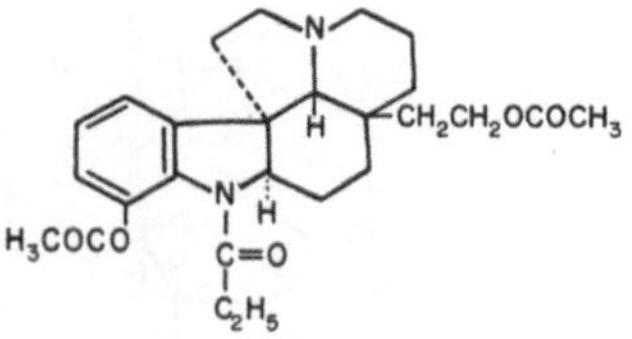


Helv. Chim. Acta 45:2260 (1962)

▶ $N_2O_5C_{24}H_{32}$

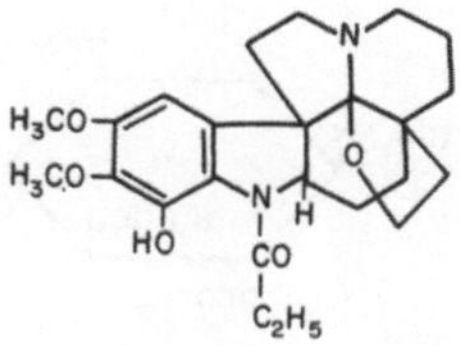


Tetrahedron Letters No. 22:1002 (1962)

▶ $N_2O_5C_{28}H_{36}$

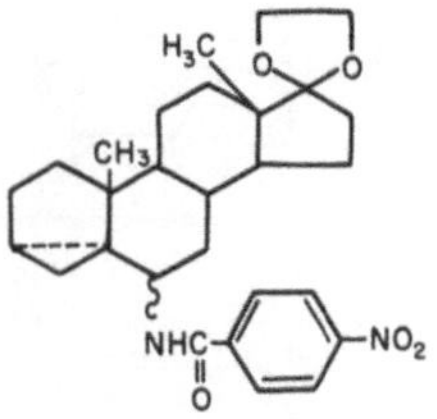


J. Org. Chem. 27:4610 (1962)

▶ $N_2O_5C_{24}H_{32}$

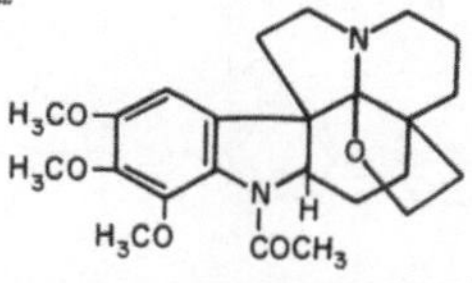


Tetrahedron Letters No. 22:1003 (1962)

▶ $N_2O_6BrC_{11}H_{11}$

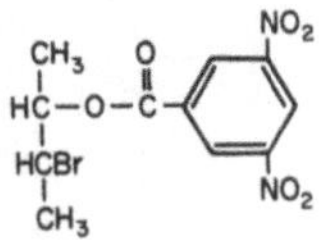


Can. J. Chem. 39:2266 (1961)

▶ $N_2O_6FC_9H_{11}$

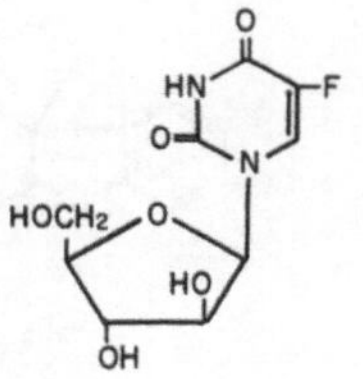

J. Am. Chem. Soc. 83:2208 (1961)

▶ $N_2O_6C_{10}H_{14}$

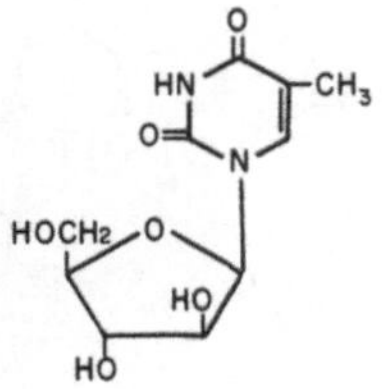

J. Am. Chem. Soc. 83:2208 (1961)

▶ $N_2O_6C_9H_{12}$

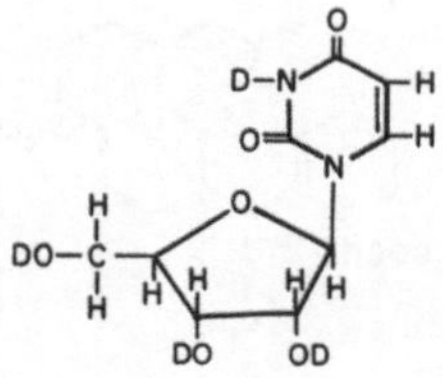

Varian 535

▶ $N_2O_6C_{10}H_{14}$

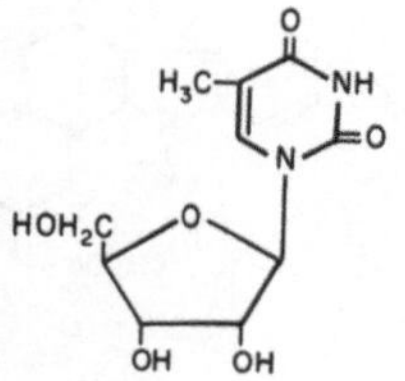

J. Am. Chem. Soc. 83:2909, 2919 (1961)

▶ $N_2O_6C_9H_{12}$

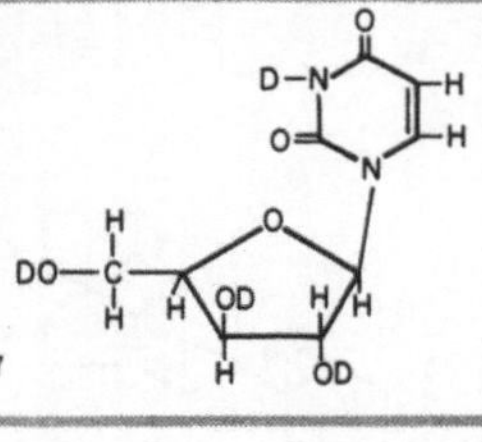

Varian 537

▶ $N_2O_6C_{16}H_{28}$

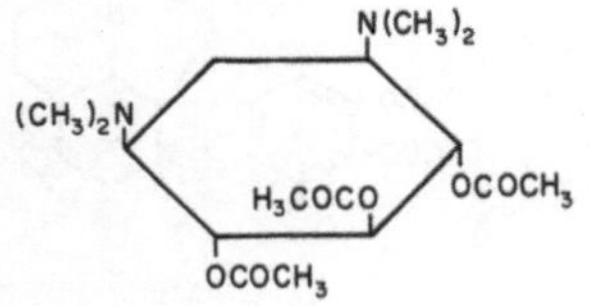

Chem. Ber. 94:3071 (1961)

▶ $N_2O_6C_9H_{12}$

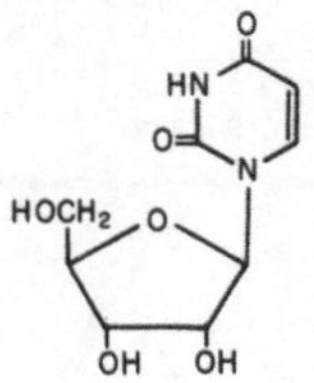

URIDINE

J. Am. Chem. Soc. 83:2909 (1961); 84:4464 (1962)

▶ $N_2O_6C_{17}H_{16}$

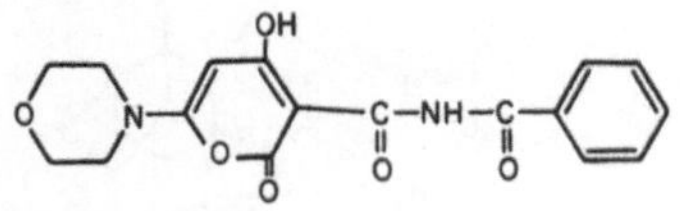

J. Chem. Soc. 3556 (1962)

▶ $N_2O_6C_{22}H_{24}$

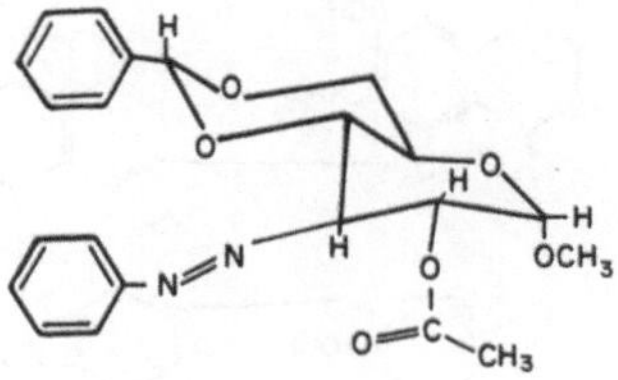


Varian 356

▶ $N_2O_6C_{22}H_{26}$

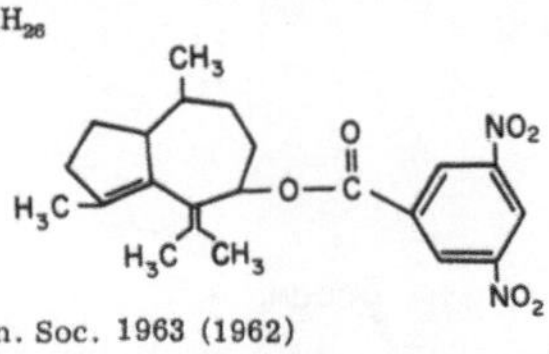


J. Chem. Soc. 1963 (1962)

▶ $N_2O_6C_{22}H_{28}$

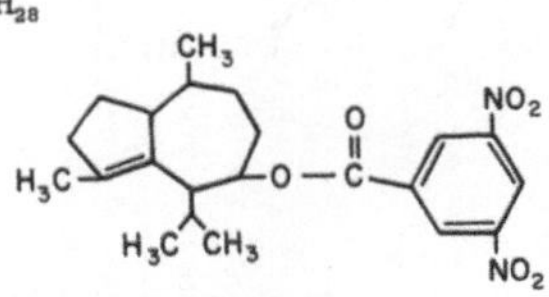


J. Chem. Soc. 1963 (1962)

▶ $N_2O_6C_{25}H_{32}$

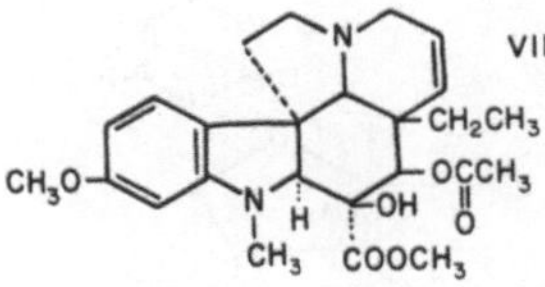


VINDOLINE

J. Am. Chem. Soc. 84:1058 (1962)

▶ $N_2O_6C_{25}H_{34}$

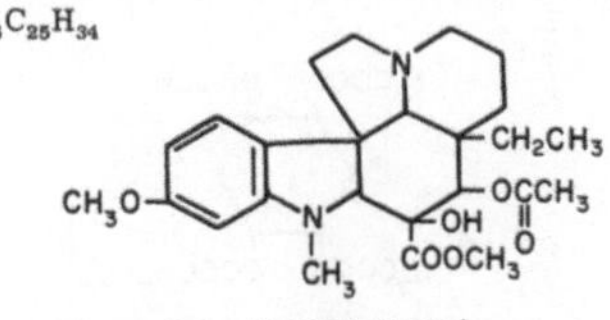


J. Am. Chem. Soc. 84:1058 (1962)

▶ $N_2O_6C_{27}H_{36}$

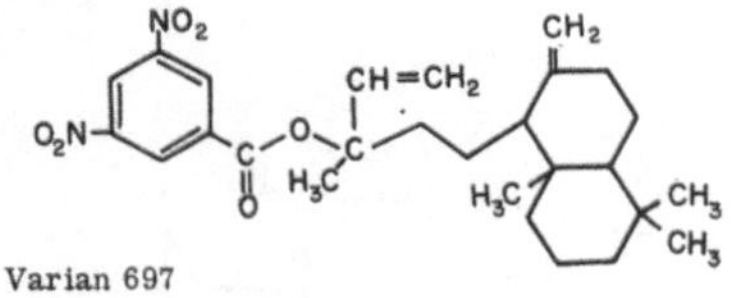


Varian 697

▶ $N_2O_6C_{38}H_{44} \cdot I_2$

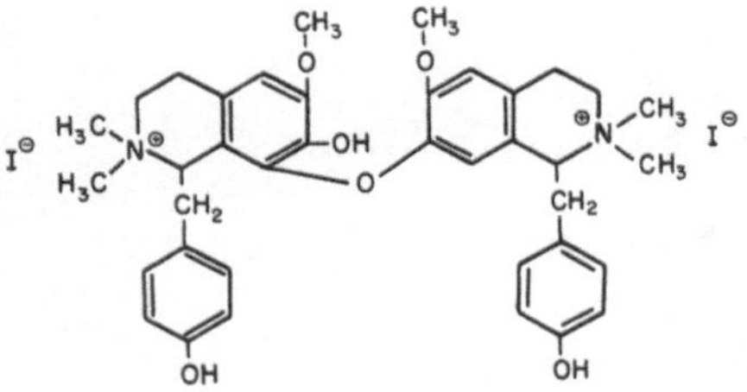


Tetrahedron Letters No. 10:443 (1962)

▶ $N_2O_7FC_{25}H_{23}$

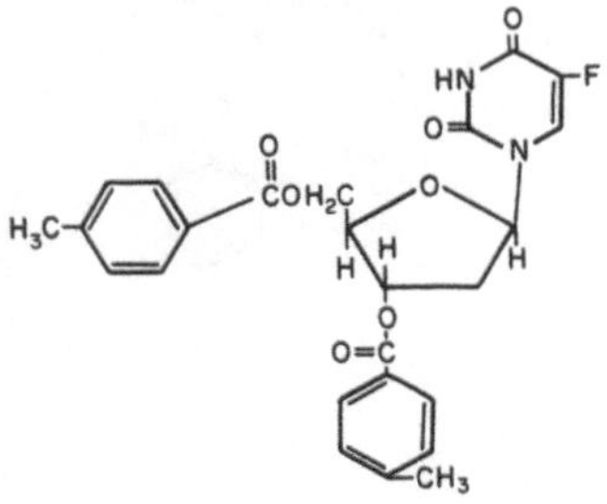


Can. J. Chem. 39:110 (1961)

▶ $N_2O_7FC_{25}H_{23}$

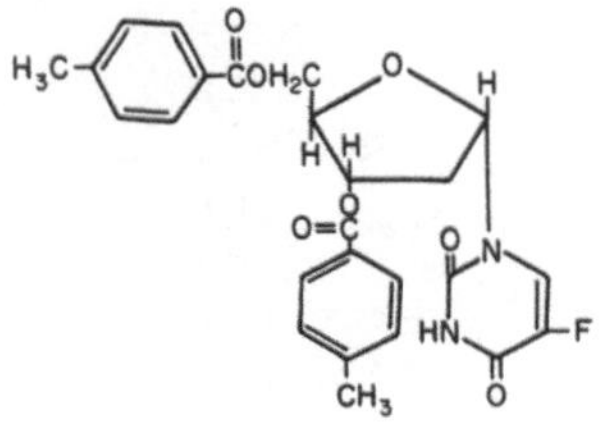


Can. J. Chem. 39:110 (1961)

► $N_2O_7C_{26}H_{26}$

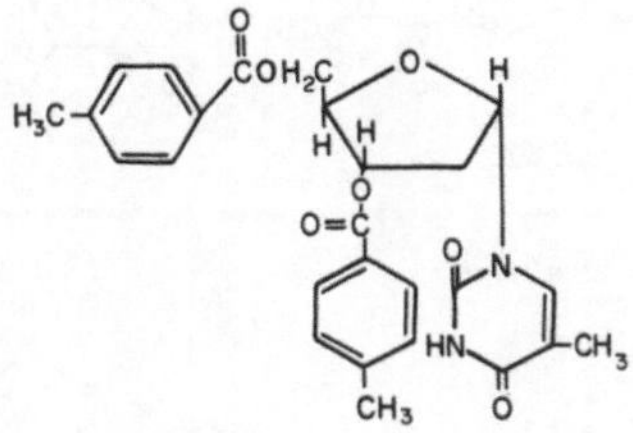

Can. J. Chem. 39:110 (1961)

► $N_2O_7C_{26}H_{26}$

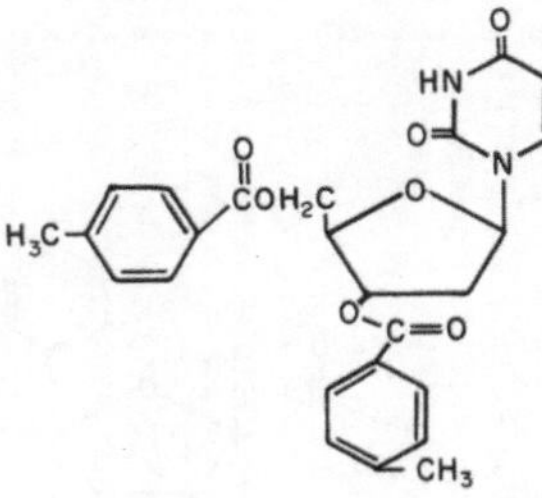

Can. J. Chem. 39:110 (1961)

► $N_2O_7C_{29}H_{30}$

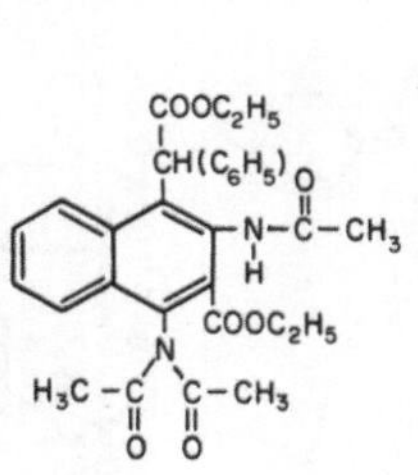

J. Org. Chem. 27:1055 (1962)

► $N_2O_7C_{39}H_{44}$

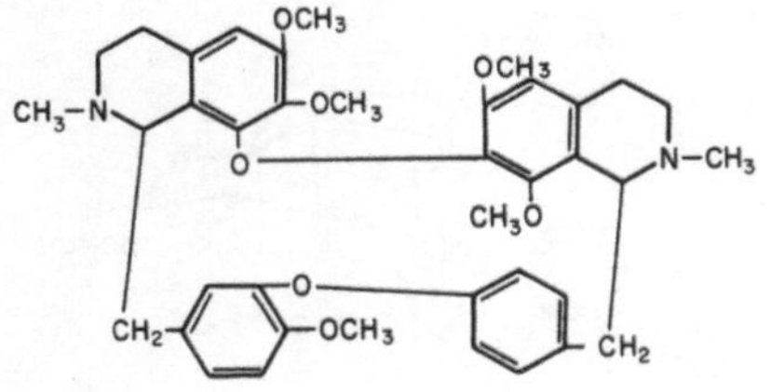

Tetrahedron 18:427 (1962)

► $N_2O_8BrC_{20}H_{23}$

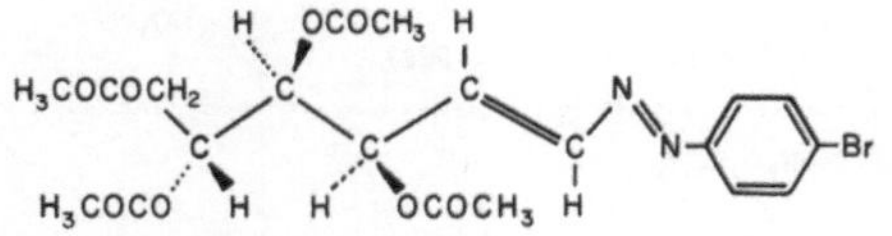

J. Org. Chem. 27:2564 (1962)

► $N_2O_8C_{13}H_{16}$

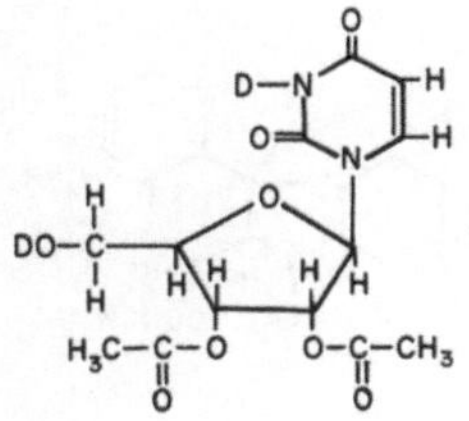

Varian 612

► $N_2O_8C_{14}H_{22}$

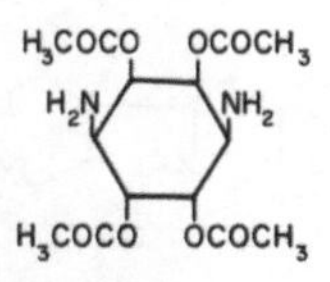

Chem. Ber. 94:3071 (1961)

▶ $N_2O_8C_{18}H_{14-16} \cdot HCl$

BETANIDIN·HCl

Helv. Chim. Acta 45:640 (1962)

▶ $N_2O_8C_{23}H_{23} \cdot O_2F_3C_2$

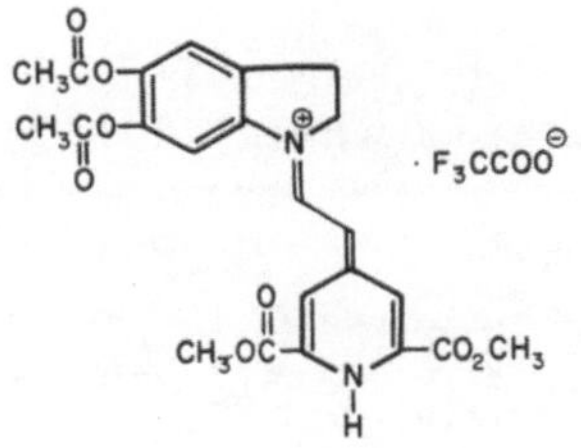

Helv. Chim. Acta 45:640 (1962)

▶ $N_2O_8C_{26}H_{28}$

ADIFOLINE TETRAMETHYLESTER
(STRUCTURE INCOMPLETELY KNOWN)

J. Chem. Soc. 2714 (1961)

▶ $N_2O_9C_{15}H_{18}$

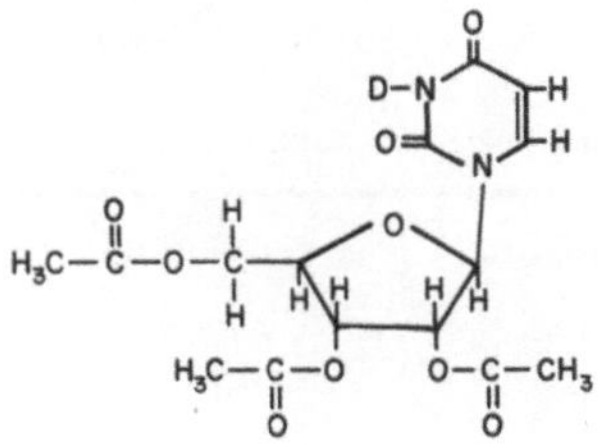

Varian 642

▶ $N_2O_9C_{28}H_{40}$

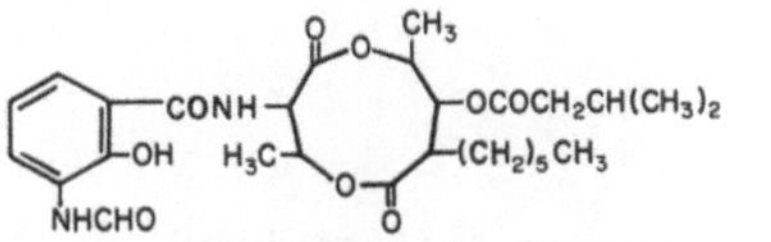

J. Am. Chem. Soc. 83:1639 (1961)

▶ $N_2O_9C_{33}H_{40}$

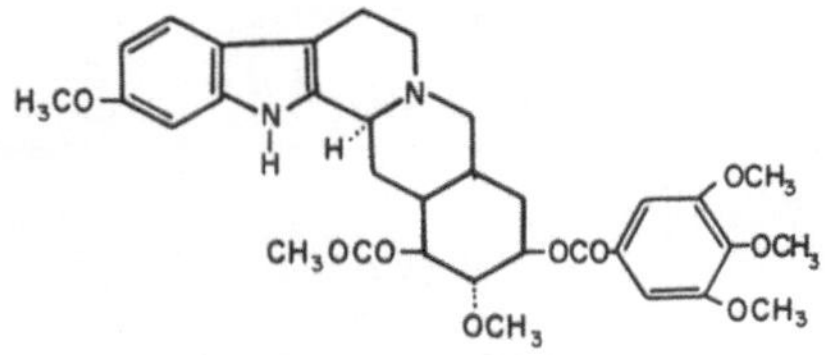

J. Am. Chem. Soc. 83:4816 (1961)

▶ $N_2O_9C_{33}H_{40}$

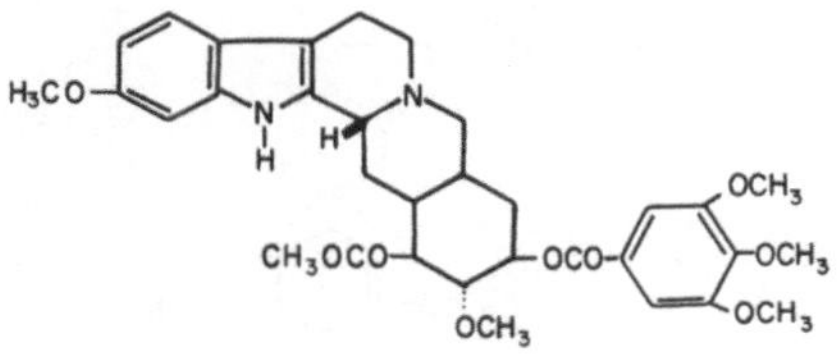

J. Am. Chem. Soc. 83:4816 (1961)

▶ $N_2O_9C_{33}H_{40}$

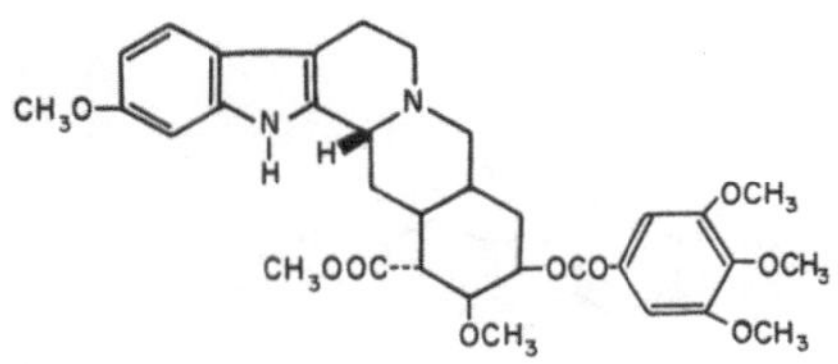

J. Am. Chem. Soc. 83:4816 (1961)

▶ $N_2O_{10}C_{18}H_{26}$

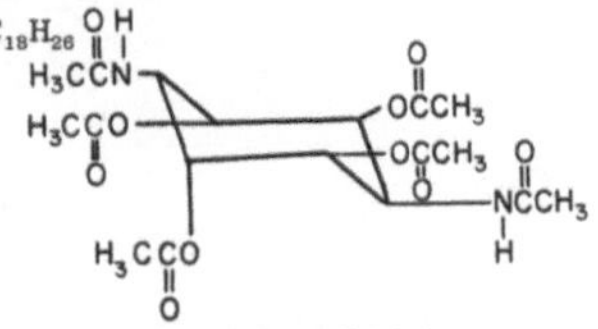

J. Am. Chem. Soc. 83:2005 (1961)

▶ $N_2O_{10}C_{18}H_{26}$

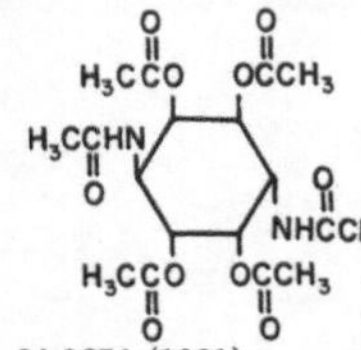

Chem. Ber. 94:3071 (1961)

▶ $N_2O_{10}C_{24}H_{26}$

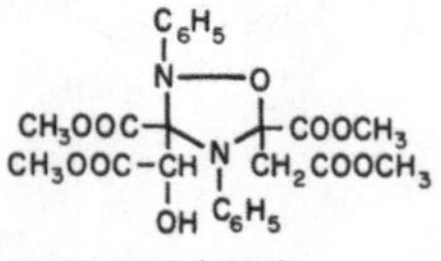

J. Org. Chem. 26:1724 (1961)

▶ $N_2O_{10}C_{25}H_{24}$

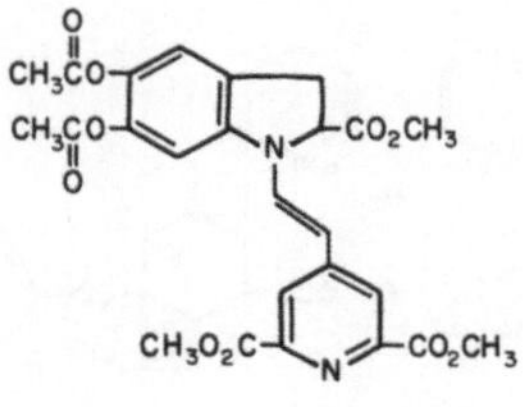

Helv. Chim. Acta 45:640 (1962)

▶ $N_2O_{10}C_{35}H_{42}$

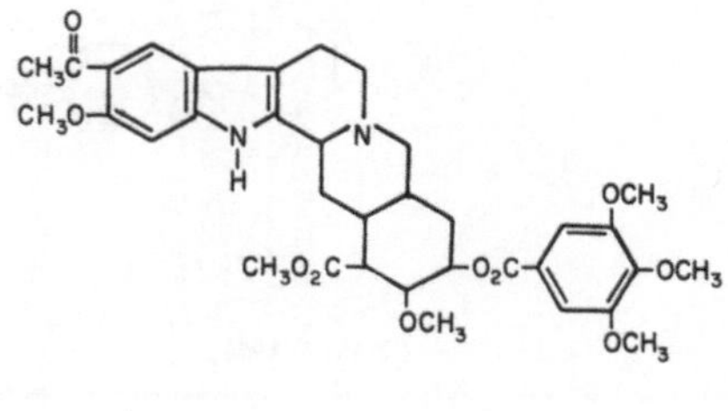

Chem. & Ind. 1863 (1962)

▶ $N_2O_{10}C_{35}H_{42}$

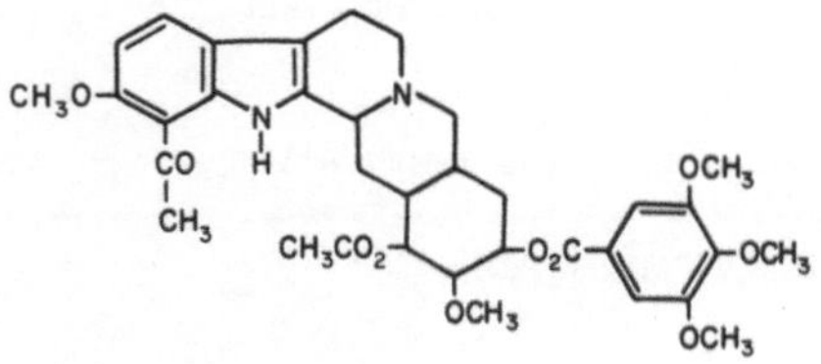

Chem. & Ind. 1863 (1962)

▶ $N_2O_{11}C_{32}H_{44}$

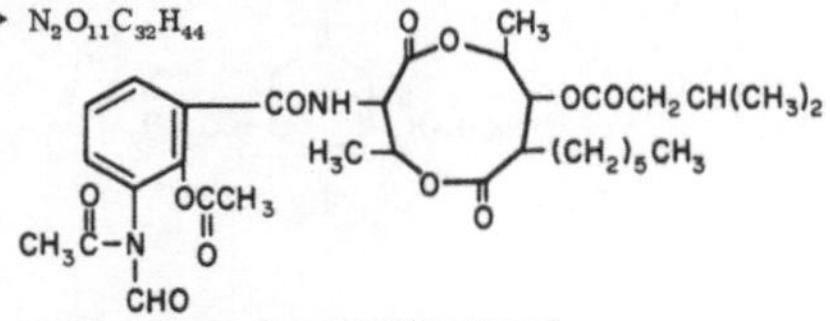

J. Am. Chem. Soc. 83:1639 (1961)

▶ $N_2O_{12}C_{14}H_{18}$

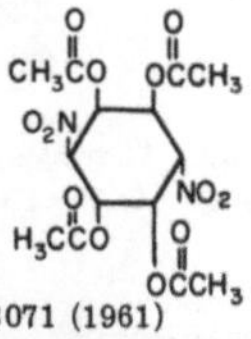

Chem. Ber. 94:3071 (1961)

▶ $N_2O_{12}C_{14}H_{18}$

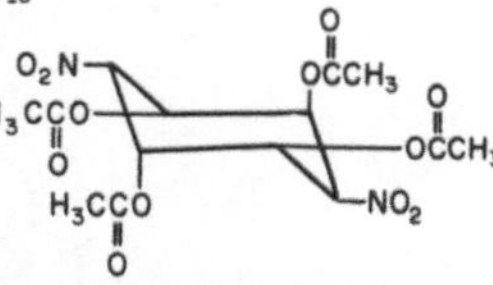

J. Am. Chem. Soc. 83:2005 (1961)

▶ $N_2O_{12}C_{30}H_{34}$

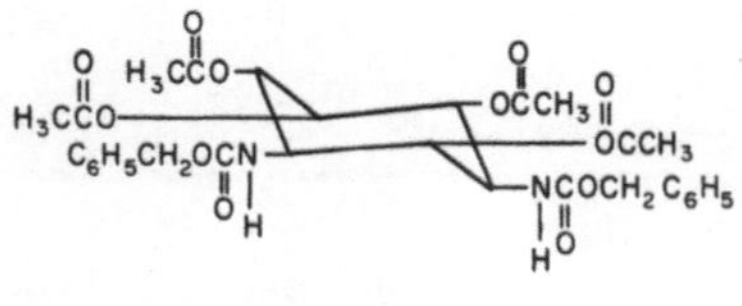

J. Am. Chem. Soc. 83:2005 (1961)

▶ $N_2O_{12}C_{30}H_{34}$

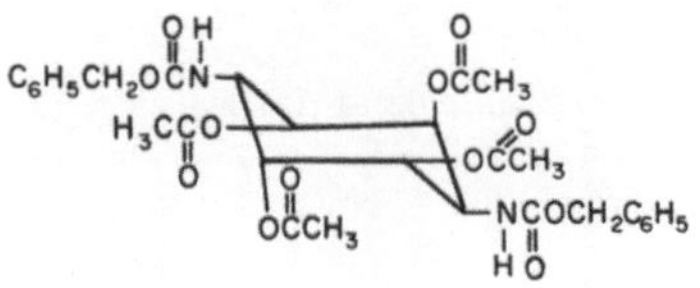

J. Am. Chem. Soc. 83:2005 (1961)

▶ $N_2O_{12}C_{30}H_{34}$

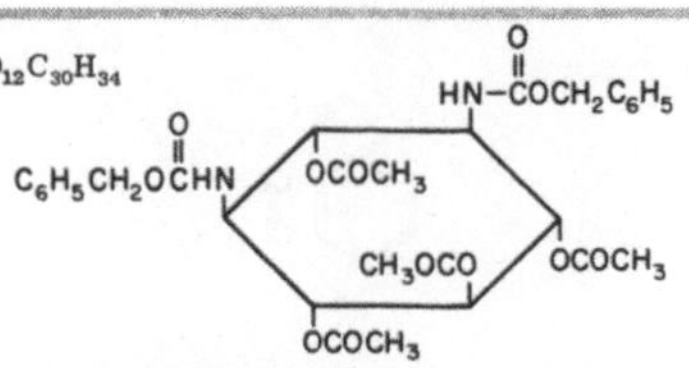

Chem. Ber. 94:3071 (1961)

▶ $N_2O_{12}C_{44}H_{50}$

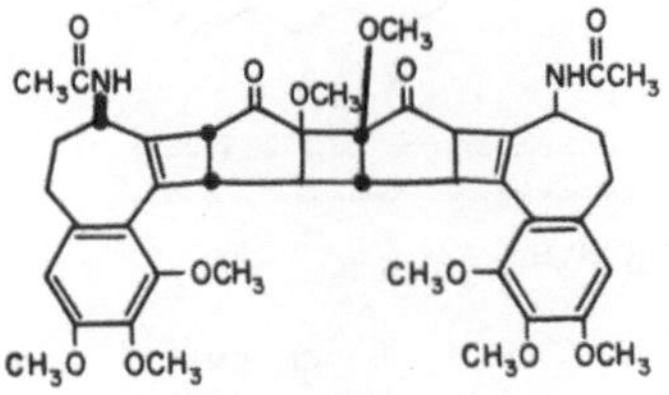

J. Am. Chem. Soc. 83:3914 (1961)

▶ $N_2O_{12}C_{44}H_{52}$

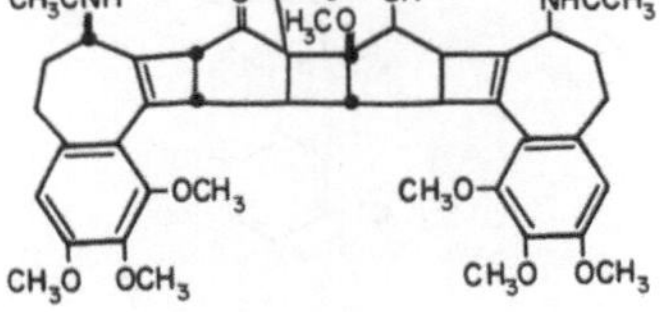

J. Am. Chem. Soc. 83:3914 (1961)

▶ $N_2O_{12}C_{44}H_{54}$

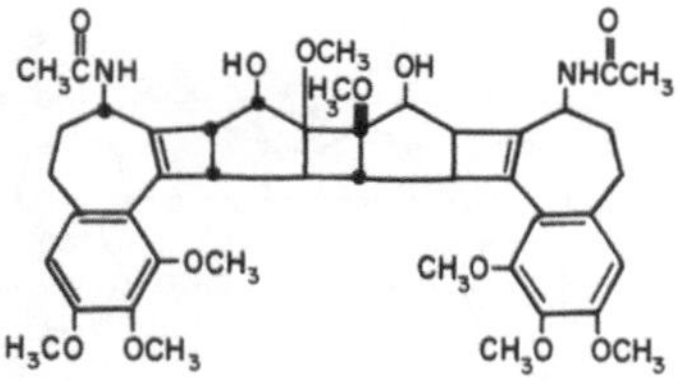

J. Am. Chem. Soc. 83:3914 (1961)

▶ $N_2O_{13}C_{46}H_{54}$

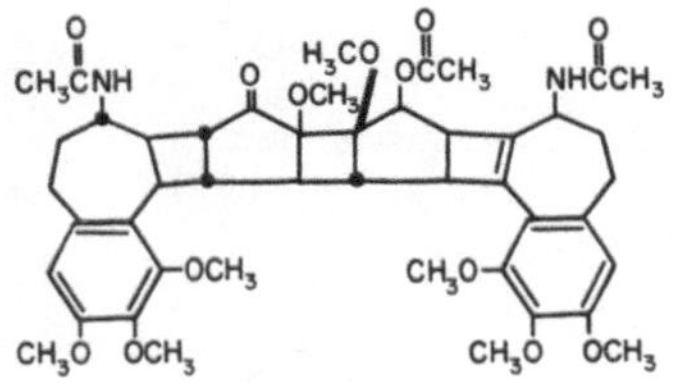

J. Am. Chem. Soc. 83:3914 (1961)

▶ $N_2O_{13}C_{46}H_{56}$

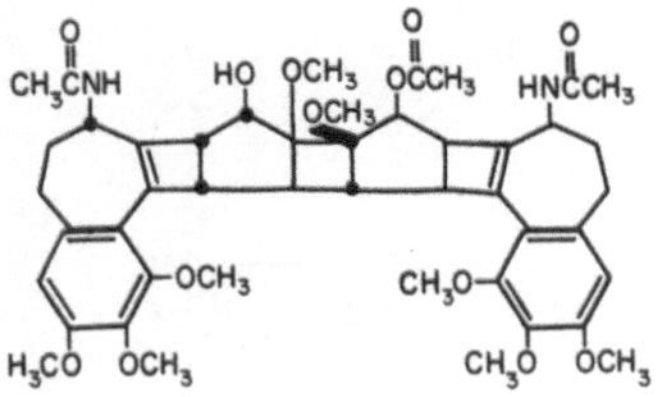

J. Am. Chem. Soc. 83:3914 (1961)

▶ $N_2O_{14}C_{48}H_{58}$

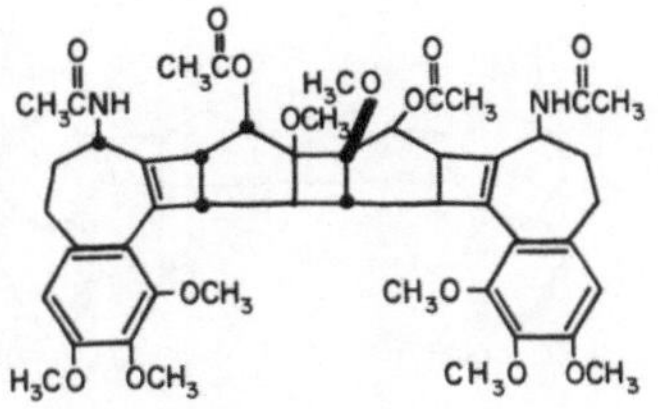

J. Am. Chem. Soc. 83:3914 (1961)

▶ $N_2ClC_5H_5$

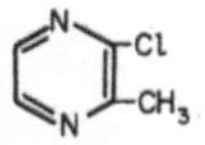

J. Org. Chem. 26:2363 (1961)

▶ $N_2ClC_8H_7$

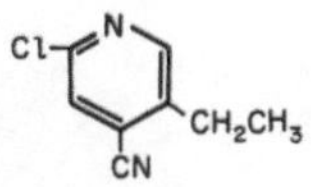

J. Am. Chem. Soc. 84:3393 (1962)

▶ $N_2BrC_{27}H_{21}$

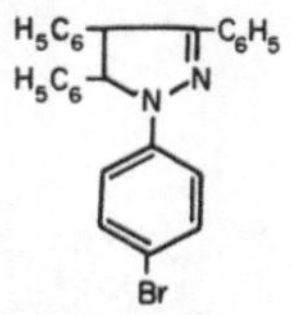

J. Org. Chem. 27:3975 (1962)

▶ $N_2ClC_8H_7$

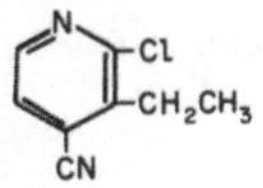

J. Am. Chem. Soc. 84:3393 (1962)

▶ $N_2Br_2C_5H_4$

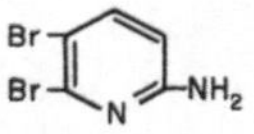

J. Org. Chem. 27:2475 (1962)

▶ $N_2ClC_{14}H_{15}$

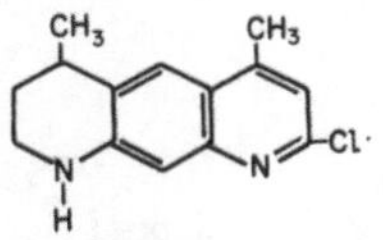

J. Am. Chem. Soc. 83:3729 (1961)

▶ $N_2Br_2C_{13}H_{14}$

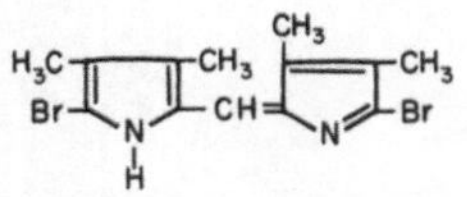

Ann. 654:165 (1962)

▶ $N_2Cl_2C_6H_4$

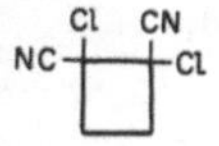

J. Chem. Phys. 37:2725 (1962)

▶ $N_2Br_2C_{27}H_{20}$

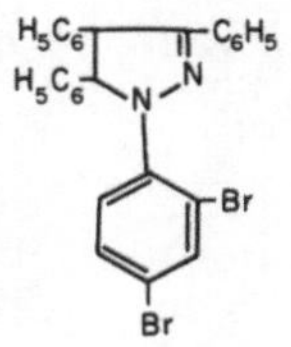

J. Org. Chem. 27:3975 (1962)

▶ $N_2Cl_2C_{21}H_{18}$

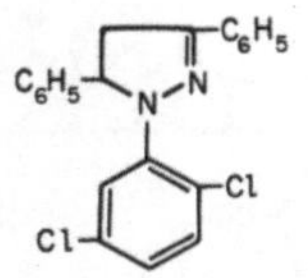

J. Org. Chem. 27:3975 (1962)

▶ $N_2Cl_2C_{27}H_{20}$

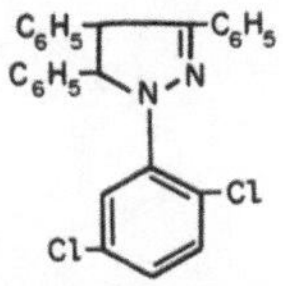


J. Org. Chem. 27:3975 (1962)

▶ $N_2Cl_3C_{27}H_{19}$

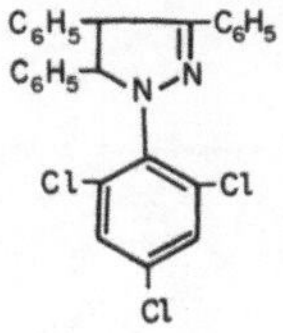


J. Org. Chem. 27:3975 (1962)

▶ $N_2CH_2$

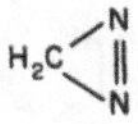


J. Am. Chem. Soc. 84:1063 (1962)

▶ $N_2CH_6Si_2$

$(SiH_3)_2NCN$

J. Chem. Soc. 4879 (1961)

▶ $N_2C_2H_6 \cdot HCl$

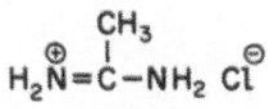


J. Am. Chem. Soc. 84:1506 (1962)

▶ $N_2C_2H_6 \cdot HNO_3$

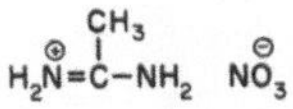


J. Am. Chem. Soc. 84:1506 (1962)

▶ $N_2C_3HD$

$CHD(CN)_2$

J. Chem. Phys. 37:3012 (1962)

▶ $N_2C_3H_2$

$CH_2(CN)_2$

J. Chem. Phys. 37:2198 (1962)

▶ $N_2C_3H_4$

Varian 20
J. Am. Chem. Soc. 84:336 (1962)
Bull. Soc. Chim. France 1710 (1961)
Tetrahedron Letters No. 20:913 (1962)
Bull. Soc. Chim. France 697 (1962)

▶ $N_2C_3H_4$

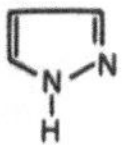


Varian 379
Bull. Soc. Chim. France 1710 (1961)

▶ $N_2C_4H_4$

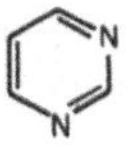


J. Am. Chem. Soc. 84:336 (1962)
Arkiv Kemi 16:459 (1961)

▶ $N_2C_4H_4$

$(C^{13})$

J. Am. Chem. Soc. 84:2649 (1962)

▶ $N_2C_4H_4$

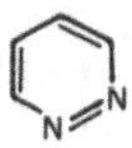


Varian 398

▶ $N_2C_4H_6$

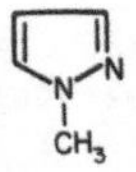

Bull. Soc. Chim. France 1710 (1961)

▶ $N_2C_4H_6$

J. Chem. Phys. 36:2644 (1962)
Bull. Soc. Chim. France 1710 (1961)

▶ $N_2C_4H_6$

Bull. Soc. Chim. France 697 (1962)
J. Chem. Phys. 36:2644 (1962)
J. Am. Chem. Soc. 84:336 (1962)

▶ $N_2C_4H_6$

J. Chem. Phys. 36:2644 (1962)
J. Am. Chem. Soc. 84:336 (1962)

▶ $N_2C_4H_6$

J. Chem. Phys. 36:2644 (1962)
J. Am. Chem. Soc. 84:336 (1962)

▶ $N_2C_4H_{10} \cdot HCl$

$CH_3NH-C(CH_3)=\overset{\oplus}{N}HCH_3 \quad Cl^{\ominus}$

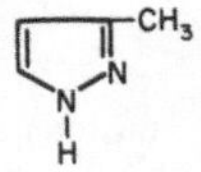

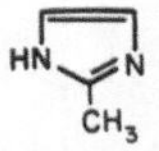

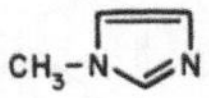

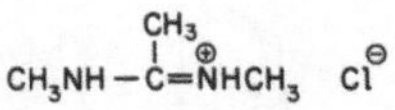

J. Am. Chem. Soc. 84:1506 (1962)

▶ $N_2C_5H_4$

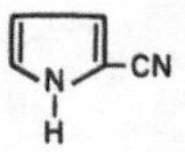

Can. J. Chem. 39:909 (1961)

▶ $N_2C_5H_6$

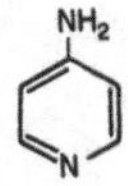

J. Chem. Soc. 44 (1961)

▶ $N_2C_5H_6$

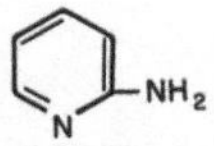

Varian 431

▶ $N_2C_5H_6$

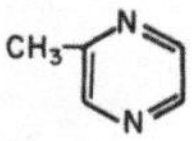

J. Org. Chem. 26:4984 (1961)

▶ $N_2C_5H_6$

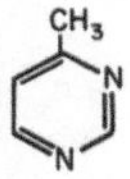

J. Am. Chem. Soc. 84:336 (1962)

▶ $N_2C_5H_6$

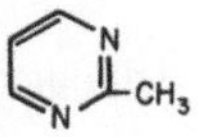

J. Am. Chem. Soc. 84:336 (1962)

▶ $N_2C_5H_8$

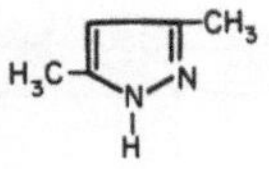

Varian 441
Bull. Soc. Chim. France 1710 (1961)

▶ $N_2C_5H_8$

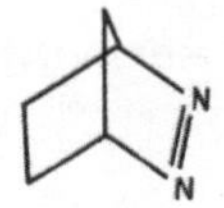

J. Am. Chem. Soc. 84:586 (1962)

▶ $N_2C_5H_8$

Bull. Soc. Chim. France 1710 (1961)

▶ $N_2C_5H_8$

Bull. Soc. Chim. France 1710 (1961)

▶ $N_2C_5H_8$

Bull. Soc. Chim. France 697(1962)

▶ $N_2C_5H_{12}$

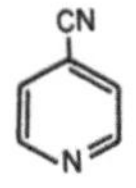

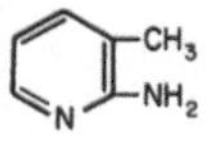

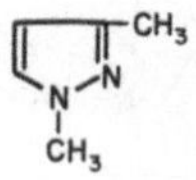

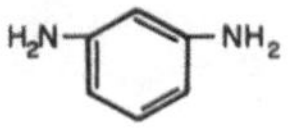

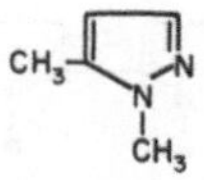

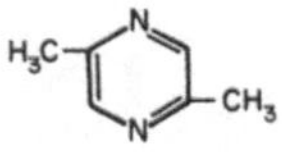

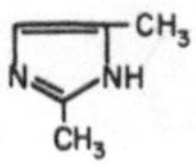

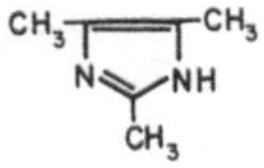

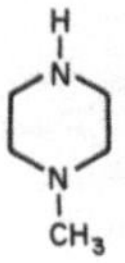

Varian 119

▶ $N_2C_6H_4$

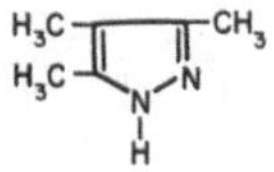

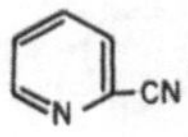

J. Chem. Phys. 37:2603 (1962)

▶ $N_2C_6H_4$

J. Chem. Soc. 44 (1961)

▶ $N_2C_6H_8$

J. Chem. Phys. 37:2473, 2476 (1962)

▶ $N_2C_6H_8$

Varian 458

▶ $N_2C_6H_8$

Varian 459

▶ $N_2C_6H_{10}$

Bull. Soc. Chim. France 697 (1962)

▶ $N_2C_6H_{10}$

Bull. Soc. Chim. France 1710 (1961)

▶ $N_2C_6H_{10}$

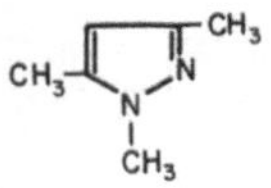

Bull. Soc. Chim. France 1710 (1961)

▶ $N_2C_6H_{10}$

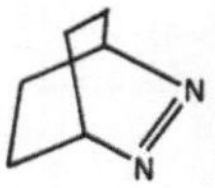

J. Am. Chem. Soc. 84:586 (1962)

▶ $N_2C_6H_{12}$

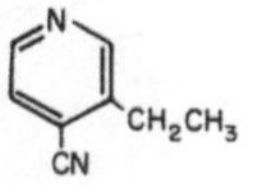

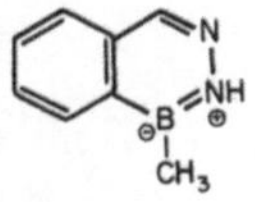

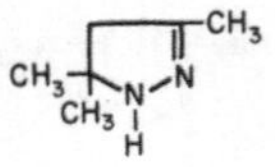

J. Org. Chem. 27:1309 (1962)

▶ $N_2C_7H_6$

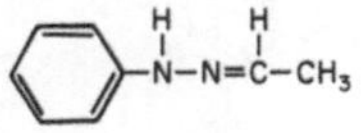

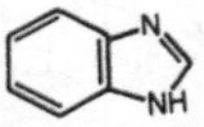

Australian J. Chem. 15:862 (1962)
Tetrahedron Letters No. 20:915 (1962)

▶ $N_2C_7H_{10}$

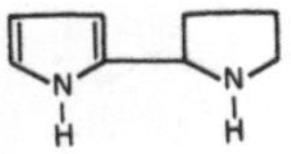

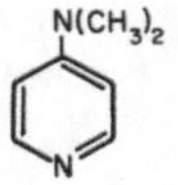

J. Chem. Soc. 44 (1961)

▶ $N_2C_8H_6$

Varian 494

▶ $N_2C_8H_8$

J. Am. Chem. Soc. 84:2178 (1962)

▶ $N_2C_8H_8$ •

J. Am. Chem. Soc. 84:3393 (1962)

▶ $N_2C_8H_9B$

J. Am. Chem. Soc. 84:2648 (1962)

▶ $N_2C_8H_{10}$

J. Org. Chem. 26:4377 (1961)

▶ $N_2C_8H_{12}$

J. Am. Chem. Soc. 84:2178 (1962)

▶ $N_2C_9H_{12}$

J. Org. Chem. 26:4377 (1961)

▶ $N_2C_9H_{12}$

J. Org. Chem. 26:4377 (1961)

▶ $N_2C_9H_{12}$

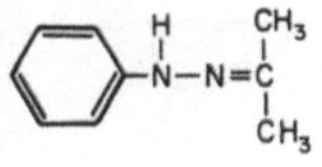

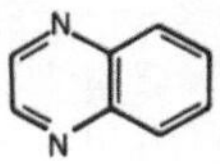

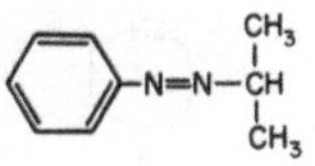

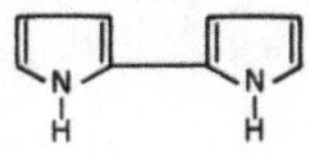

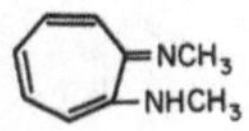

J. Am. Chem. Soc. 83:3125 (1961)

▶ $N_2C_9H_{14}$

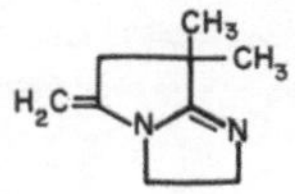

Varian 541

▶ $N_2C_{10}H_{10} \cdot PdI_2$

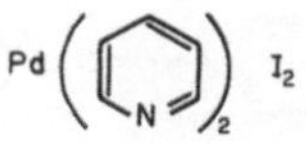

Can. J. Chem. 39:324 (1961)

▶ $N_2C_{10}H_{10} \cdot PdBr_2$

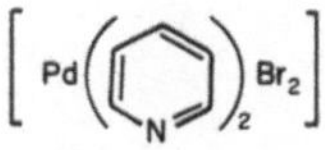

Can. J. Chem. 39:324 (1961)

▶ $N_2C_{10}H_{10} \cdot PdCl_2$

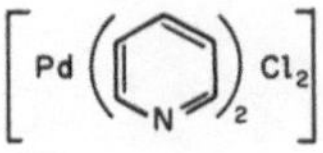

Can. J. Chem. 39:324 (1961)

▶ $N_2C_{10}H_{10} \cdot Pt(NO_2)_2$

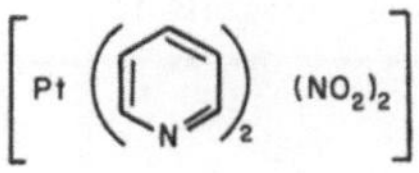

Can. J. Chem. 39:324 (1961)

▶ $N_2C_{10}H_{10} \cdot PtI_2$

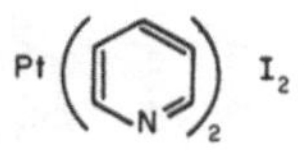

Can. J. Chem. 39:324 (1961)

▶ $N_2C_{10}H_{10} \cdot PtBr_2$

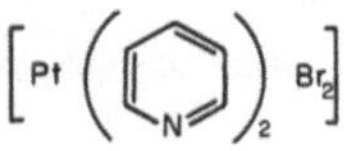

Can. J. Chem. 39:324 (1961)

▶ $N_2C_{10}H_{10} \cdot PtCl_2$

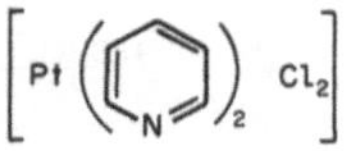

Can. J. Chem. 39:324 (1961)

▶ $N_2C_{10}H_{10} \cdot Pt(NH_3)_2Cl_2$

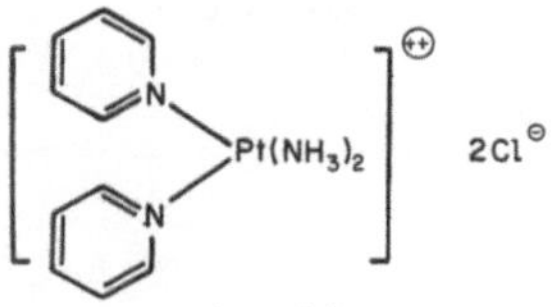

cis and trans

Can. J. Chem. 39:324 (1961)

▶ $N_2C_{10}H_{14}$

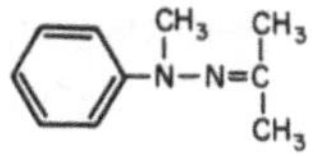

J. Org. Chem. 26:4377 (1961)

▶ $N_2C_{10}H_{14}$

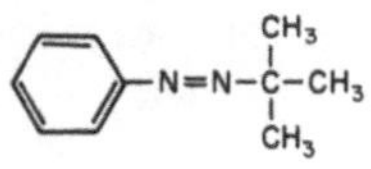

J. Org. Chem. 26:4377 (1961)

▶ $N_2C_{10}H_{14}$

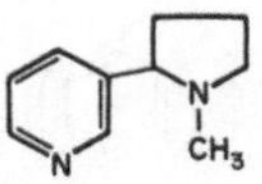

Varian 269

▶ $N_2C_{10}H_{16}$

$N{\equiv}CCH_2(CH_2)_6CH_2C{\equiv}N$

Varian 570

▶ $N_2C_{10}H_{17}B$

$C_6H_5\ B[N(CH_3)_2]_2$

Proc. Chem. Soc. 421 (1961)

▶ $N_2C_{11}H_8$

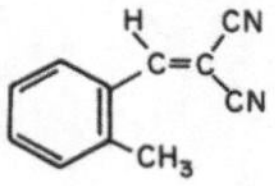

Varian 283

▶ $N_2C_{11}H_{16}$

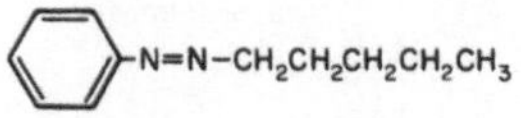

J. Org. Chem. 26:4377 (1961)

▶ $N_2C_{11}H_{18}$

$C_6H_5-NH-CH_2\underset{CH_3}{\underset{|}{C}}H-N(CH_3)_2$

J. Org. Chem. 27:4272 (1962)

▶ $N_2C_{11}H_{18}$

$C_6H_5-NH-\underset{CH_3}{\underset{|}{C}}HCH_2-N(CH_3)_2$

J. Org. Chem. 27:4272 (1962)

▶ $N_2C_{12}H_{10}$

$C_6H_5N{=}NC_6H_5$

J. Am. Chem. Soc. 83:4838 (1961)

▶ $N_2C_{12}H_{16}$

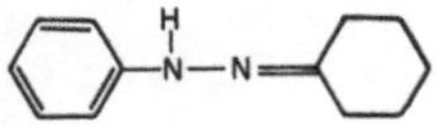

J. Org. Chem. 26:4377 (1961)

▶ $N_2C_{12}H_{16}$

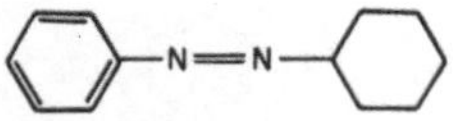

J. Org. Chem. 26:4377 (1961)

▶ $N_2C_{12}H_{18}$

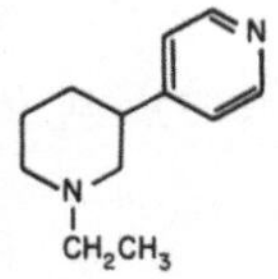

Chem. Ber. 95:2896 (1962)

▶ $N_2C_{12}H_{24}$

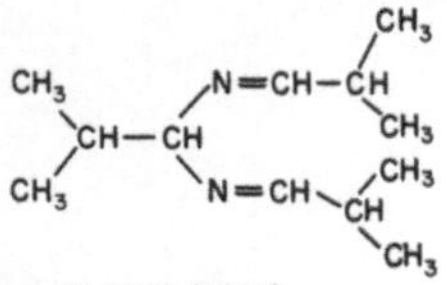

J. Org. Chem. 26:1822 (1961)

▶ $N_2C_{13}H_{12}$

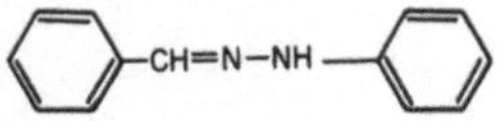

J. Org. Chem. 27:4663 (1962)

▶ $N_2C_{13}H_{12}$

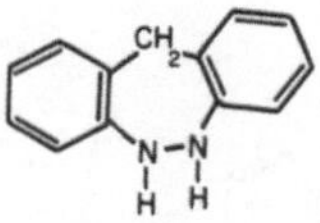

J. Am. Chem. Soc. 84:1020 (1962)

▶ $N_2C_{13}H_{14}$

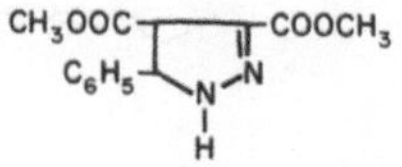

J. Org. Chem. 27:3975 (1962)

▶ $N_2C_{13}H_{16}$

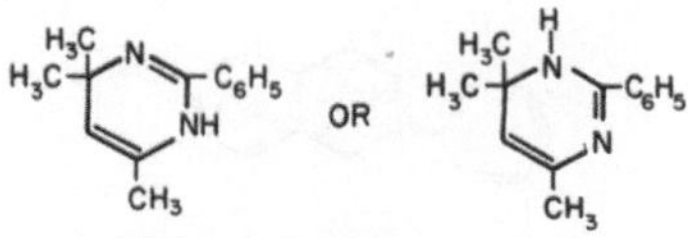

J. Org. Chem. 27:4091 (1962)

▶ $N_2C_{13}H_{16} \cdot PtCl_2$

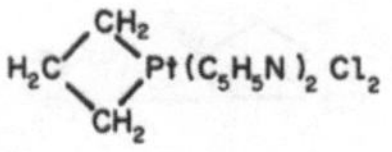

J. Chem. Soc. 4736 (1962)

▶ $N_2C_{13}H_{18}$

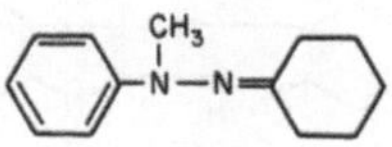

J. Org. Chem. 26:4377 (1961)

▶ $N_2C_{13}H_{20}$

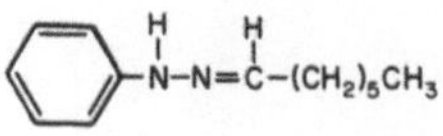

J. Org. Chem. 26:4377 (1961)

▶ $N_2C_{14}H_{14}$

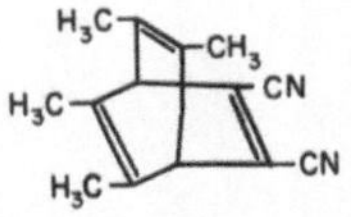

Tetrahedron Letters No. 19:879 (1962)

▶ $N_2C_{14}H_{22}$

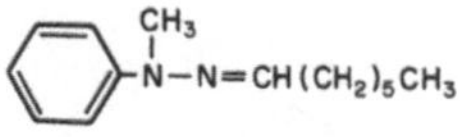

J. Org. Chem. 26:4377 (1961)

▶ $N_2C_{15}H_{18}$

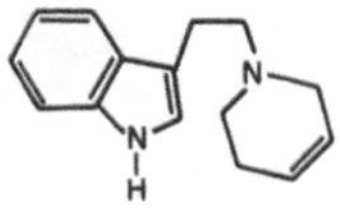

J. Am. Chem. Soc. 84:3732 (1962)

▶ $N_2C_{16}H_{12}$

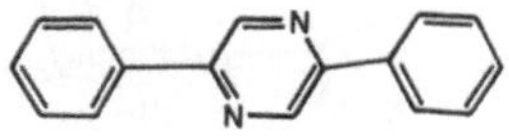

Tetrahedron 18:1139 (1962)

▶ $N_2C_{16}H_{16}$

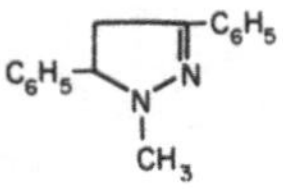

J. Org. Chem. 27:3975 (1962)

▶ $N_2C_{16}H_{16}$

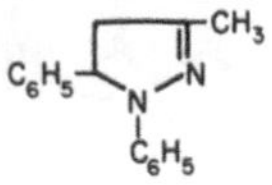

J. Org. Chem. 27:3975 (1962)

▶ $N_2C_{16}H_{20}$

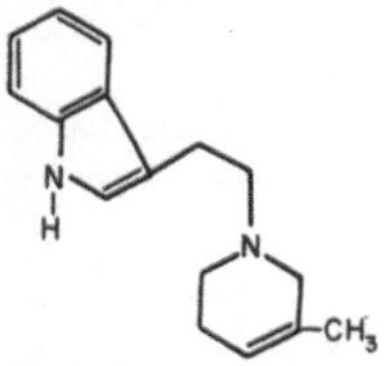

J. Am. Chem. Soc. 84:3732 (1962)

▶ $N_2C_{16}H_{20}$

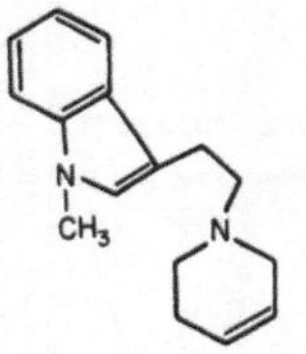

J. Am. Chem. Soc. 84:3732 (1962)

▶ $N_2C_{16}H_{22}$

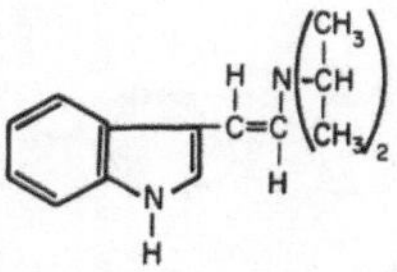

J. Org. Chem. 27:4105 (1962)

▶ $N_2C_{17}H_{22}$

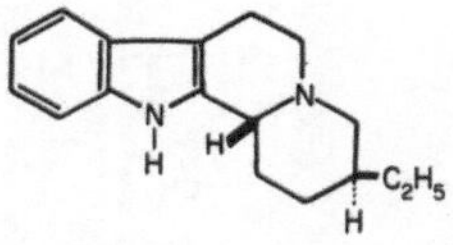

J. Am. Chem. Soc. 84:4914 (1962)

▶ $N_2C_{17}H_{22}$

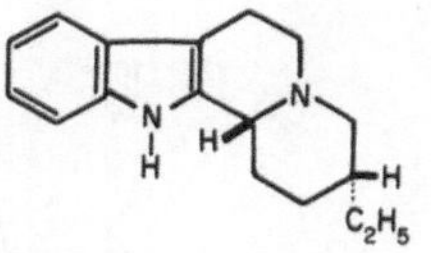

J. Am. Chem. Soc. 84:4914 (1962)

▶ $N_2C_{17}H_{22}$

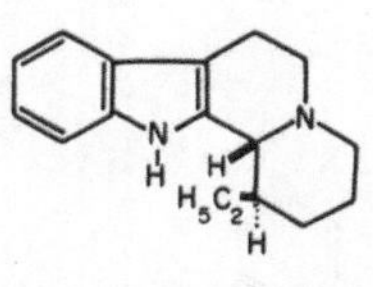

J. Am. Chem. Soc. 84:4914 (1962)

▶ $N_2C_{17}H_{22}$

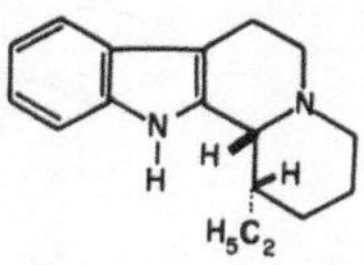

J. Am. Chem. Soc. 84:4914 (1962)

▶ $N_2C_{17}H_{22}$

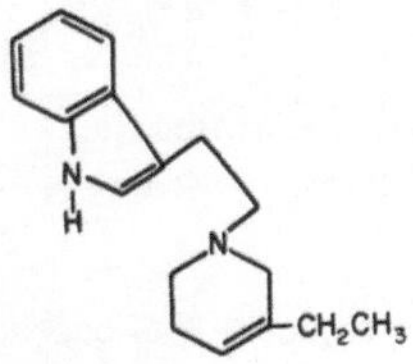

J. Am. Chem. Soc. 84:3732 (1962)

▶ $N_2C_{18}H_{16}$

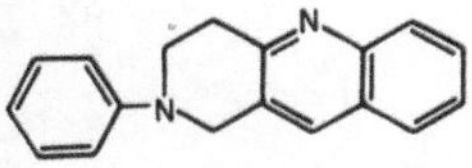

J. Chem. Soc. 5113 (1962)

▶ $N_2C_{18}H_{18}$

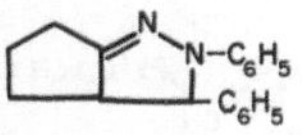

J. Org. Chem. 27:3975 (1962)

▶ $N_2C_{18}H_{18}$

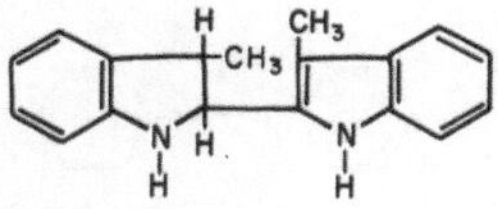

J. Org. Chem. 26:2340 (1961)

▶ $N_2C_{18}H_{18}$

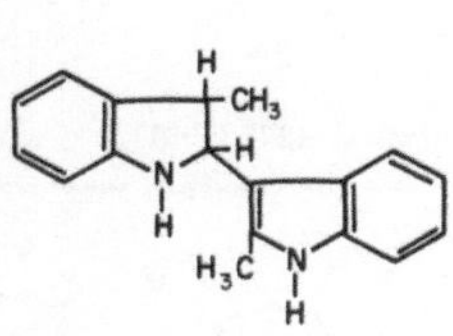

J. Org. Chem. 26:2341 (1961)

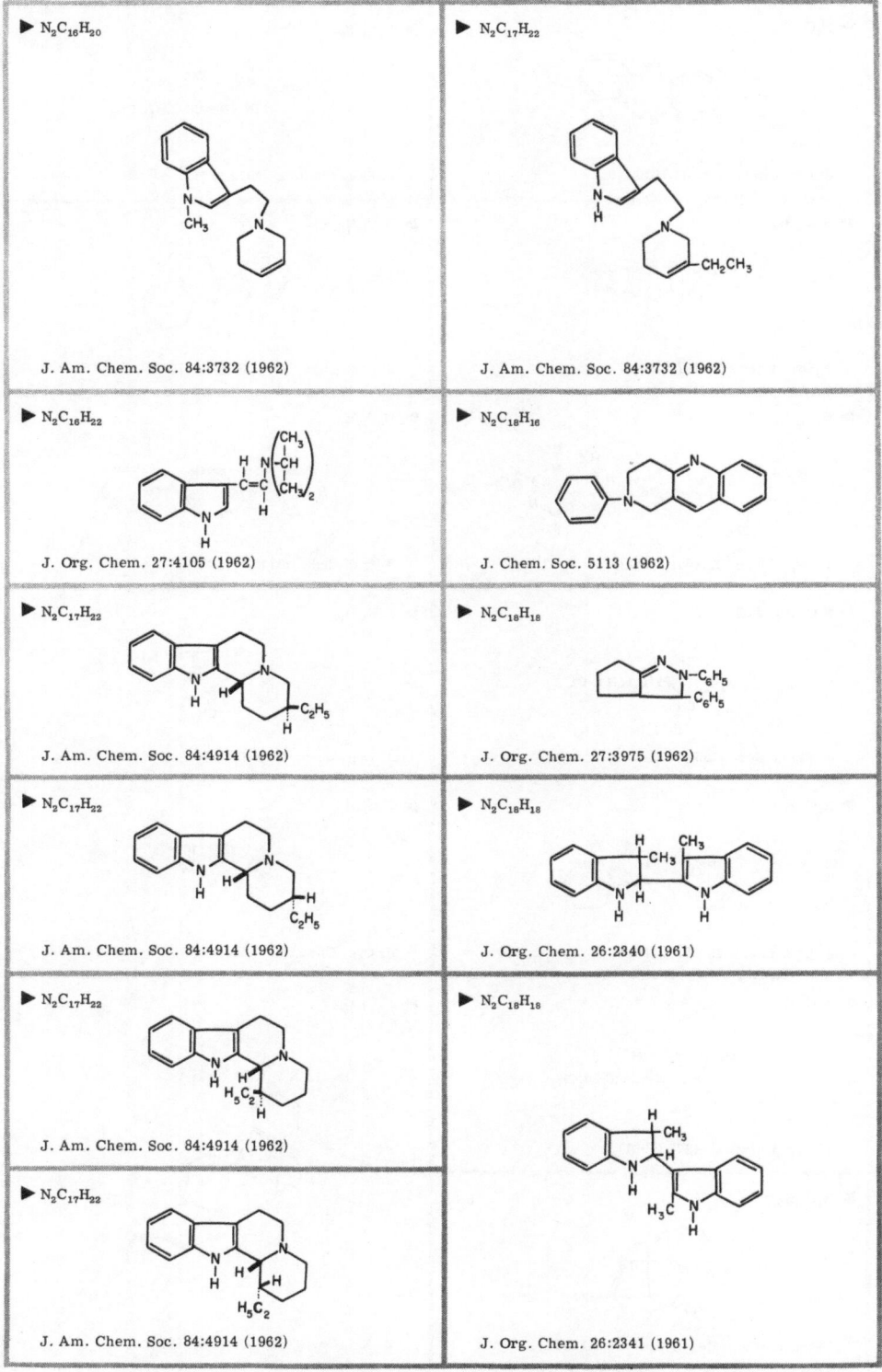

▶ $N_2C_{19}H_{20}$

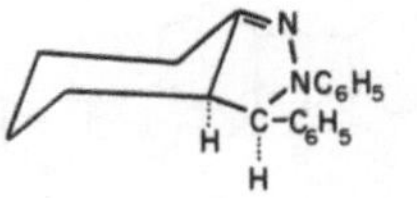

J. Org. Chem. 27:299, 3975 (1962)

▶ $N_2C_{20}H_{24}$

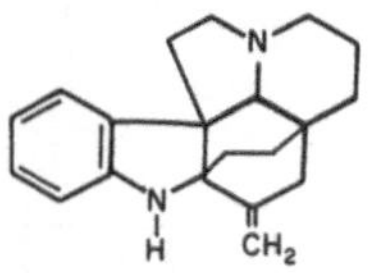

Helv. Chim. Acta 45:854 (1962)

▶ $N_2C_{19}H_{20}$

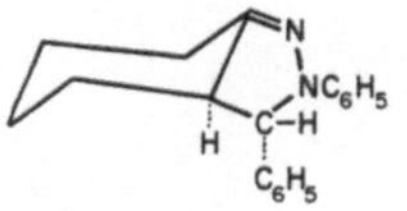

J. Org. Chem. 27:299, 3975 (1962)

▶ $N_2C_{20}H_{26}$

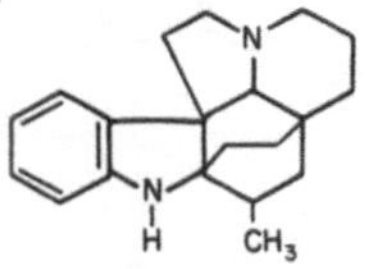

KOPSINAN

Helv. Chim. Acta 45:854 (1962)

▶ $N_2C_{19}H_{24}$

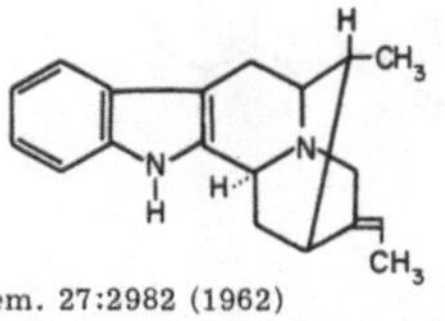

J. Org. Chem. 27:2982 (1962)

▶ $N_2C_{21}H_{18}$

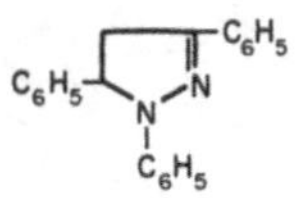

J. Org. Chem. 27:3975 (1962)

▶ $N_2C_{20}H_{18}$

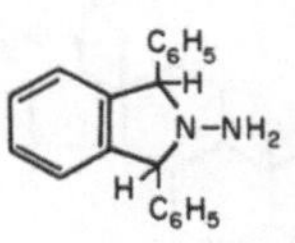

cis and trans

J. Am. Chem. Soc. 84:2196 (1962)

▶ $N_2C_{21}H_{22}$

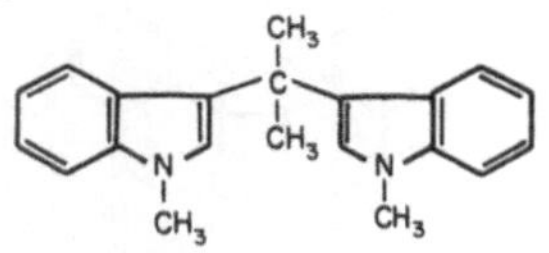

J. Org. Chem. 27:2941 (1962)

▶ $N_2C_{20}H_{22}$

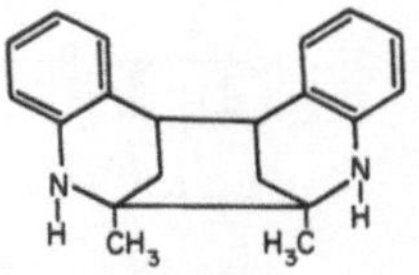

Tetrahedron Letters 783 (1961)

▶ $N_2C_{21}H_{24}$

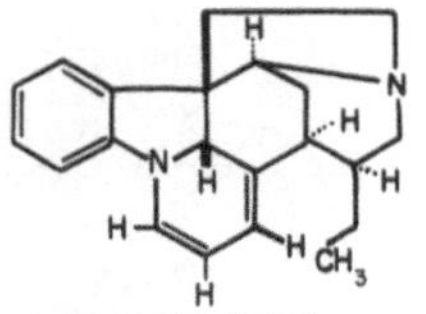

Helv. Chim. Acta 44:624 (1961)

▶ $N_2C_{20}H_{22}$

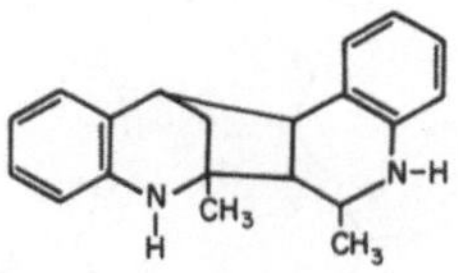

Tetrahedron Letters No. 7:293 (1962)

▶ $N_2C_{21}H_{26}$

$N_{(a)}$-METHYL-ISOKOPSINYLEN

Helv. Chim. Acta 45:854 (1962)

▶ $N_2C_{21}H_{26}$

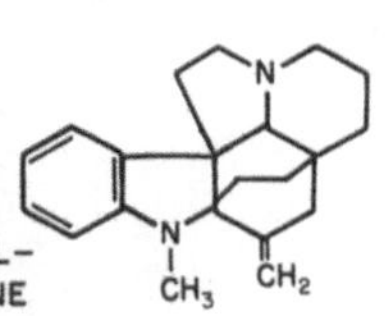

$N_{(a)}$-METHYL-KOPSINYLENE

Helv. Chim. Acta 45:854 (1962)

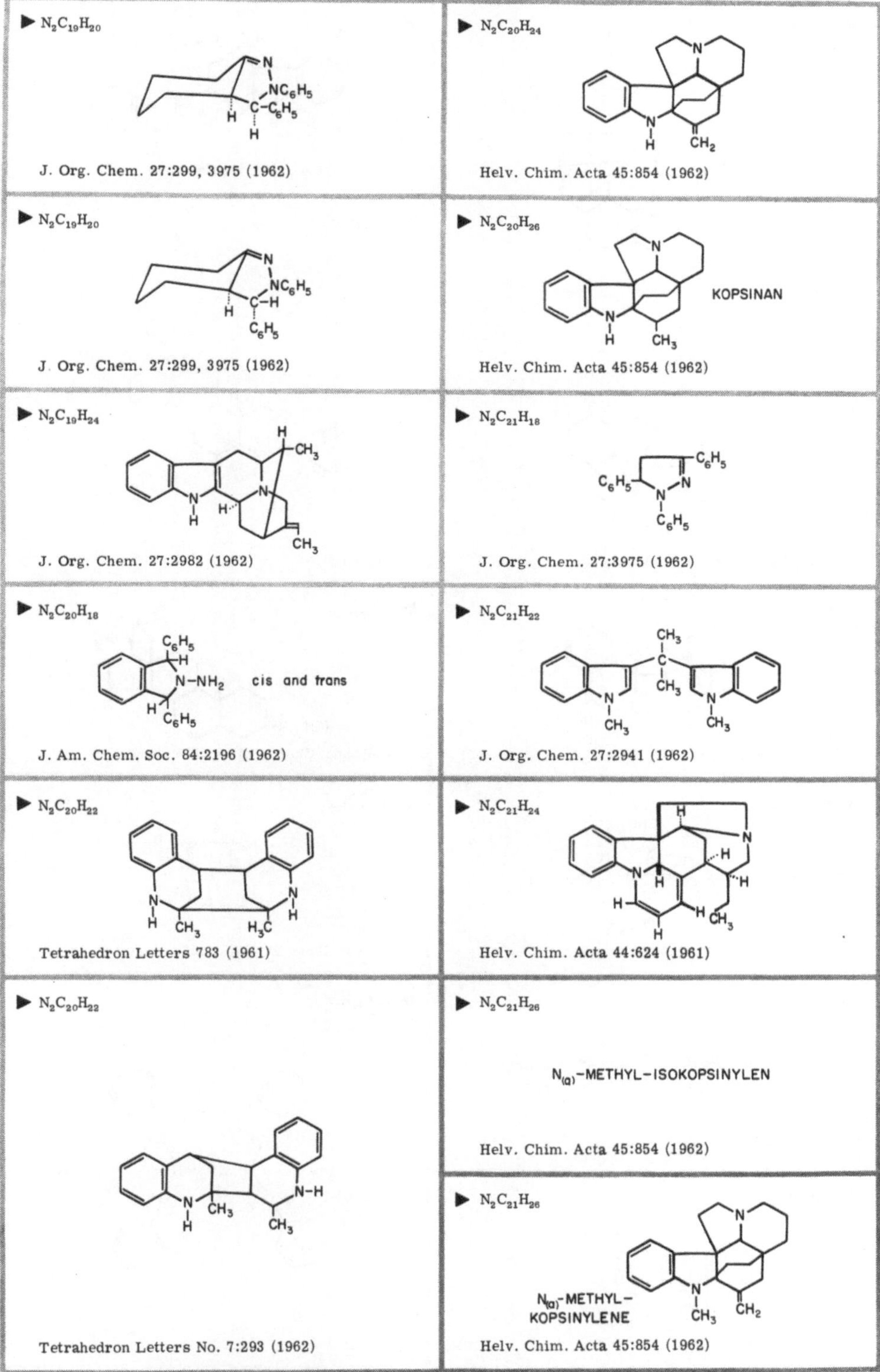

▶ $N_2C_{21}H_{26}$

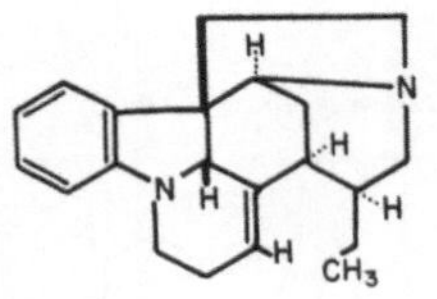

Helv. Chim. Acta 44:624 (1961)

▶ $N_2C_{21}H_{28}$

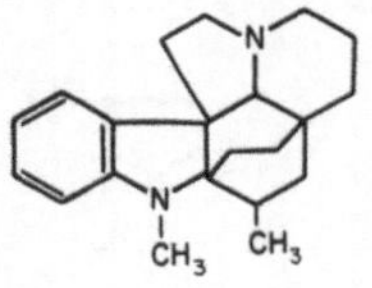

Helv. Chim. Acta 45:854 (1962)

▶ $N_2C_{22}H_{26}$

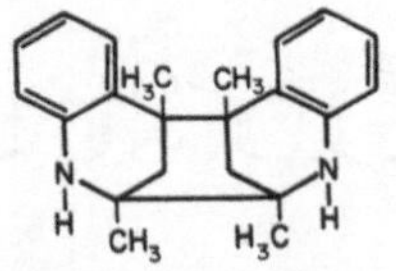

Tetrahedron Letters 783 (1961)

▶ $N_2C_{22}H_{32}B_2$

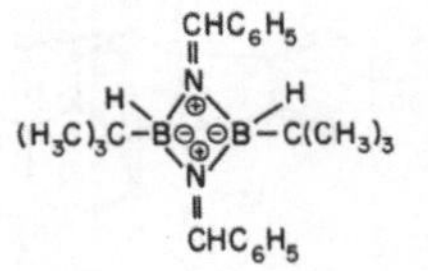

Tetrahedron 17:117 (1962)

▶ $N_2C_{23}H_{20}$

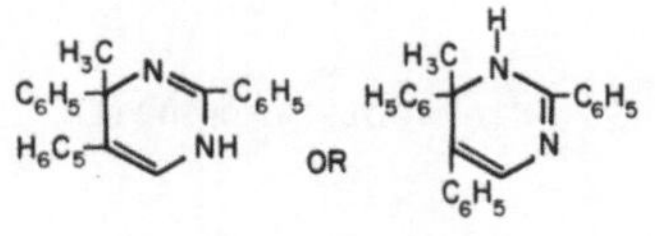

J. Org. Chem. 27:4091 (1962)

▶ $N_2C_{24}H_{22}$

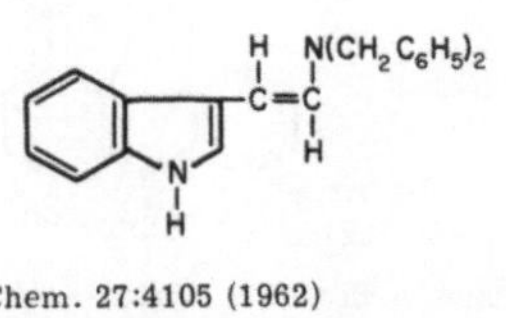

J. Org. Chem. 27:4105 (1962)

▶ $N_2C_{24}H_{26}$

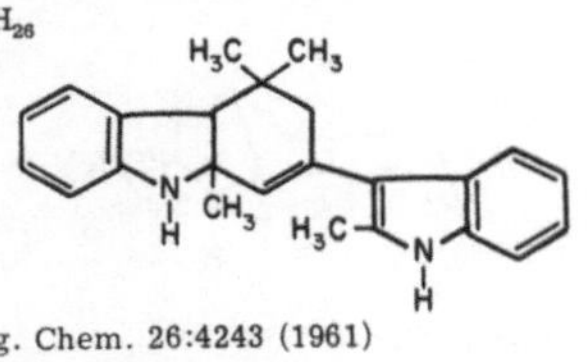

J. Org. Chem. 26:4243 (1961)

▶ $N_2C_{24}H_{40}$

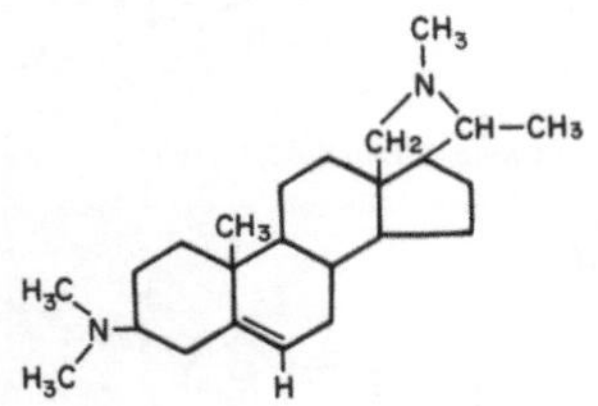

Varian 359

▶ $N_2C_{24}H_{42}$

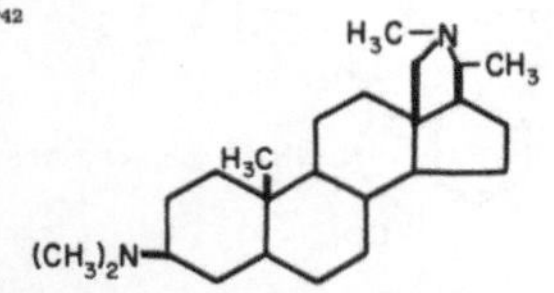

Bull. Soc. Chim. France 2234 (1962)

▶ $N_2C_{27}H_{22}$

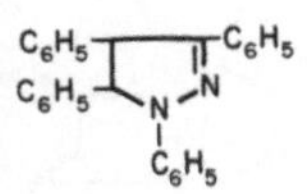

J. Org. Chem. 27:3975 (1962)

▶ $N_2C_{29}H_{30}$

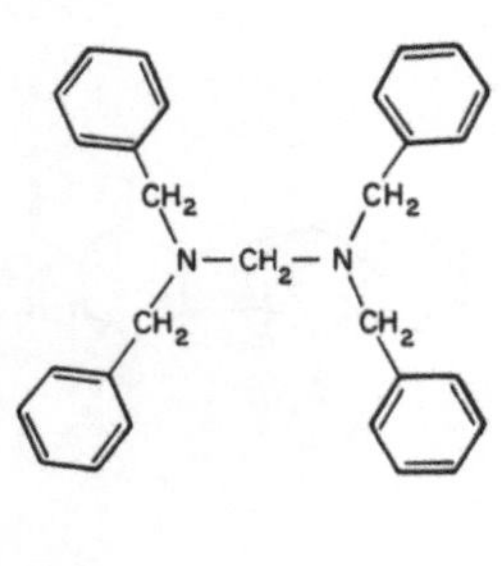

Varian 365

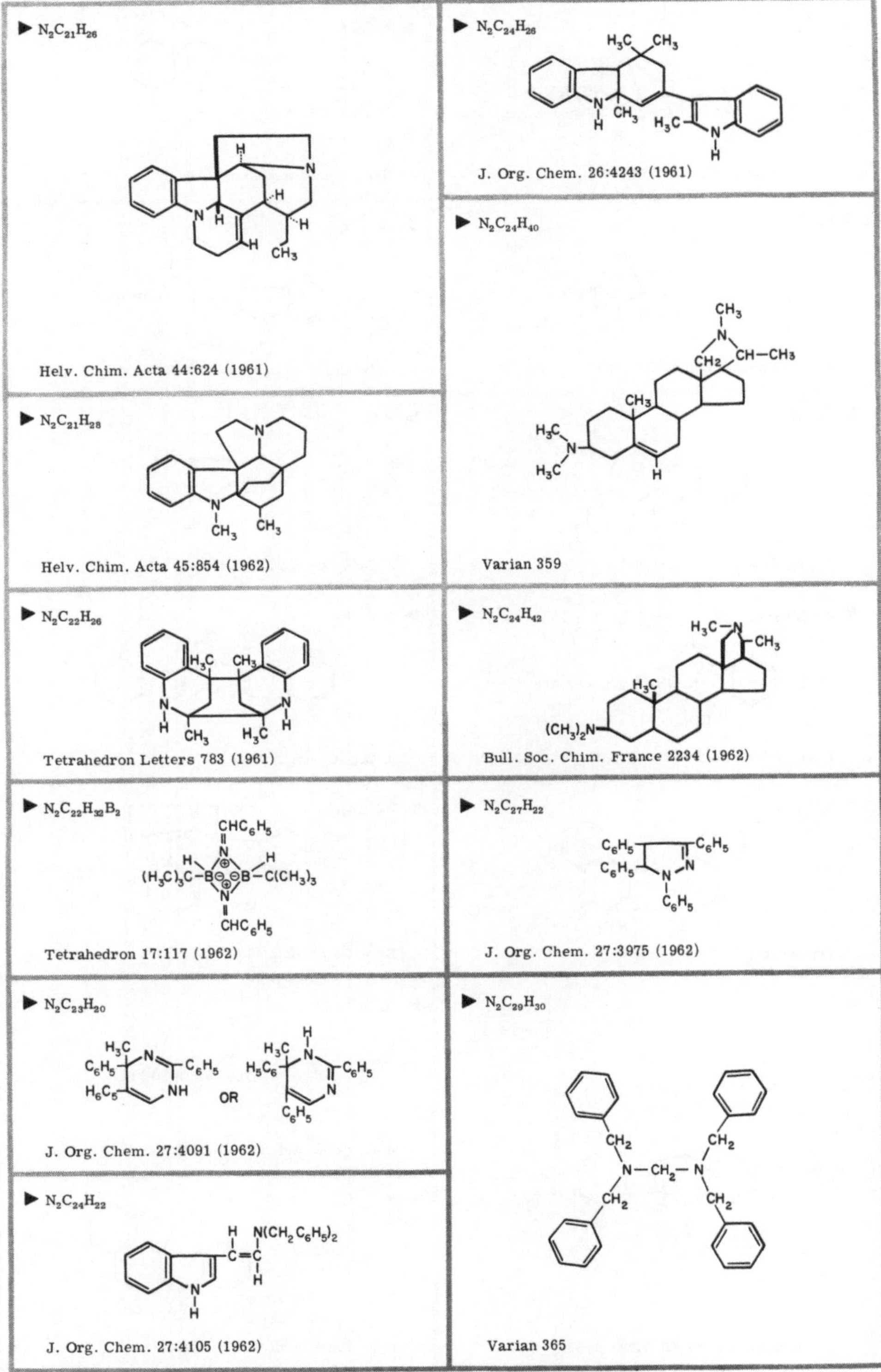

▶ $N_3OC_3H_6 \cdot Cl$

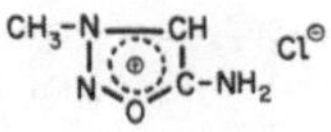

Helv. Chim. Acta 45:2426 (1962)

▶ $N_3OC_4H_5$

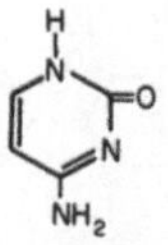

J. Am. Chem. Soc. 83:2909 (1961); 84:1042 (1962)

▶ $N_3OC_4H_5$

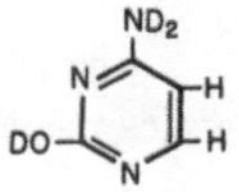

Varian 402

▶ $N_3OC_4H_5$

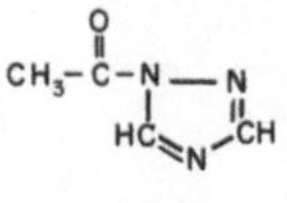

J. Org. Chem. 27:2632 (1962)

▶ $N_3OC_4H_5 \cdot HCl$

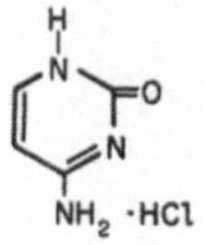

J. Am. Chem. Soc. 83:2909 (1961)

▶ $N_3OC_4H_7$

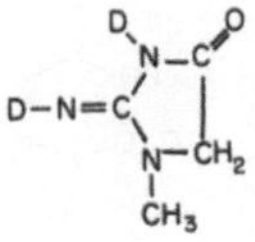

Varian 411

▶ $N_3OC_4H_8 \cdot Cl$

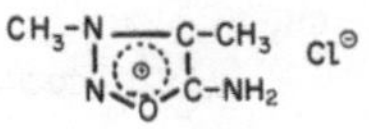

Helv. Chim. Acta 45:2426 (1962)

▶ $N_3OC_4H_9$

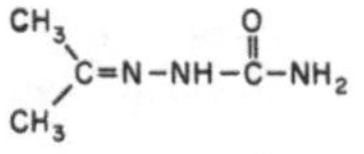

J. Am. Chem. Soc. 84:753 (1962)

▶ $N_3OC_5H_7$

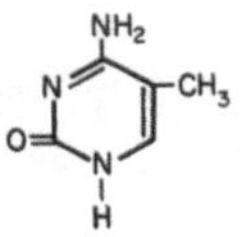

J. Am. Chem. Soc. 84:1042 (1962)

▶ $N_3OC_5H_{11}$

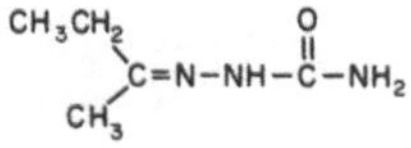

J. Am. Chem. Soc. 84:753 (1962)

▶ $N_3OC_6H_{13}$

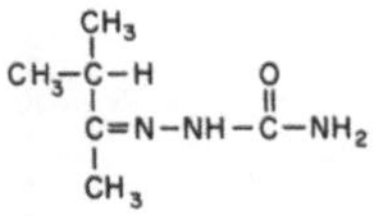

J. Am. Chem. Soc. 84:753 (1962)

▶ $N_3OC_6H_{13}$

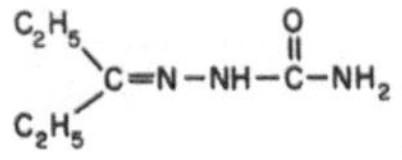

J. Am. Chem. Soc. 84:34, 753 (1962)

▶ $N_3OC_6H_{13}$

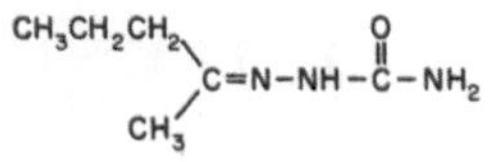

J. Am. Chem. Soc. 84:753 (1962)

▶ $N_3OC_7H_9$

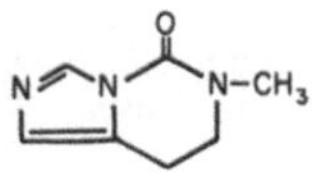

J. Am. Chem. Soc. 83:2023 (1961)

▶ $N_3OC_7H_{15}$

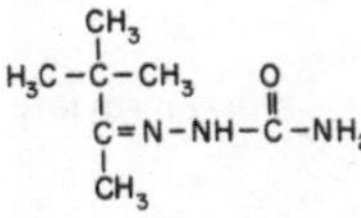

J. Am. Chem. Soc. 84:753 (1962)

▶ $N_3OC_8H_7$

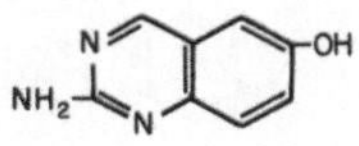

Bull. Chem. Soc. Japan 35:1244 (1962)

▶ $N_3OC_8H_8 \cdot Cl$

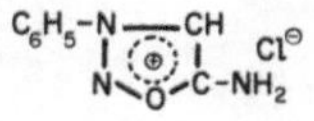

Helv. Chim. Acta 45:2426 (1962)

▶ $N_3OC_8H_{17}$

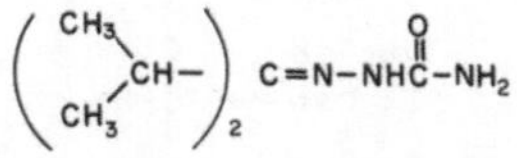

J. Am. Cnem. Soc. 84:753 (1962)

▶ $N_3OC_9H_{10} \cdot Cl$

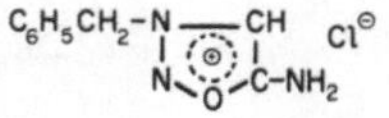

Helv. Chim. Acta 45:2426 (1962)

▶ $N_3OC_{10}H_{11}$

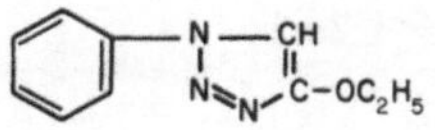

Helv. Chim. Acta 45:2441 (1962)

▶ $N_3OC_{11}H_{14} \cdot Cl$

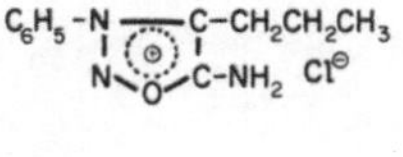

Helv. Chim. Acta 45:2426 (1962)

▶ $N_3OC_{12}H_{16} \cdot Cl$

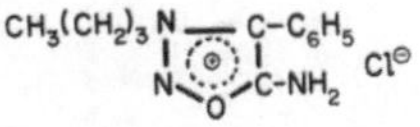

Helv. Chim. Acta 45:2426 (1962)

▶ $N_3OC_{13}H_{17}$

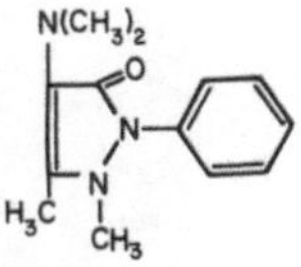

Varian 615

▶ $N_3OC_{14}H_{17}$

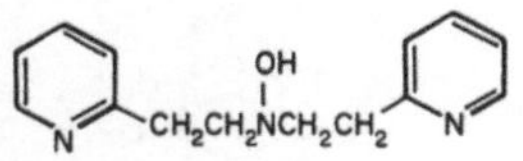

Varian 633

▶ $N_3OC_{16}H_{25}$

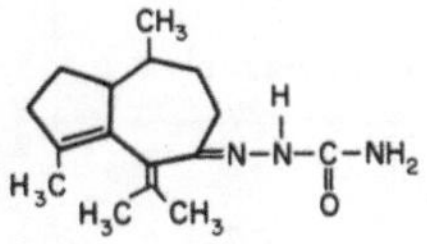

J. Chem. Soc. 1963 (1962)

▶ $N_3O_2C_3H_3$

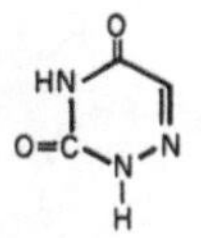

J. Am. Chem. Soc. 84:1042 (1962)

▶ $N_3O_2C_5H_7$

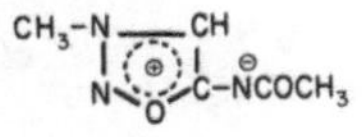

Helv. Chim. Acta 45:2441 (1962)

▶ $N_3O_2C_5H_8 \cdot Cl$

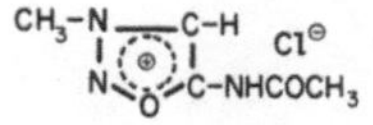

Helv. Chim. Acta 45:2441 (1962)

▶ $N_3O_2C_6H_9$

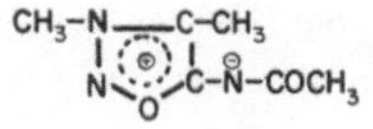

Helv. Chim. Acta 45:2441 (1962)

▶ $N_3O_2C_6H_{11}$

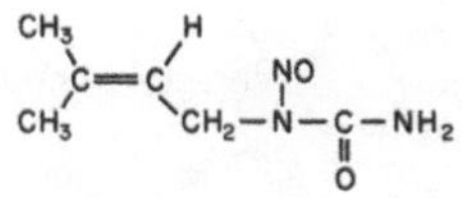

Chem. & Ind. 1020 (1962)

▶ $N_3O_2C_7H_5$

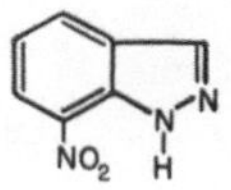

Varian 485

▶ $N_3O_2C_{10}H_9$

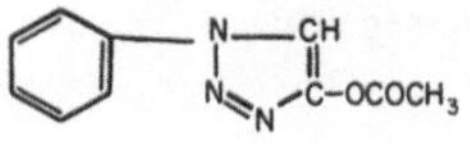

Helv. Chim. Acta 45:2441 (1962)

▶ $N_3O_2C_{14}H_{11}$

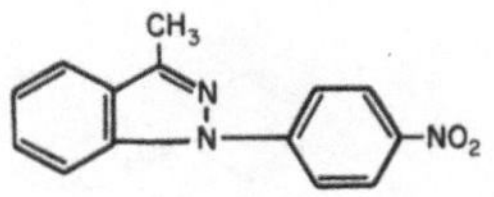

Tetrahedron Letters No. 24:1115 (1962)

▶ $N_3O_2C_{15}H_{21}$

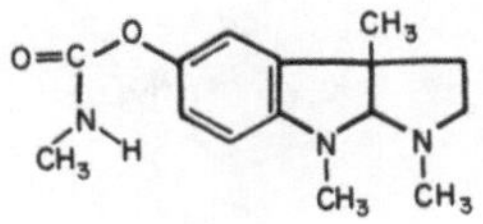

Varian 319

▶ $N_3O_3C_{19}H_{31}$

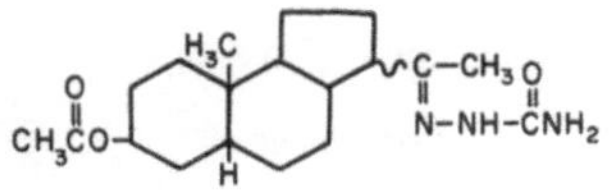

Helv. Chim. Acta 45:943 (1962)

▶ $N_3O_3C_{24}H_{39}$

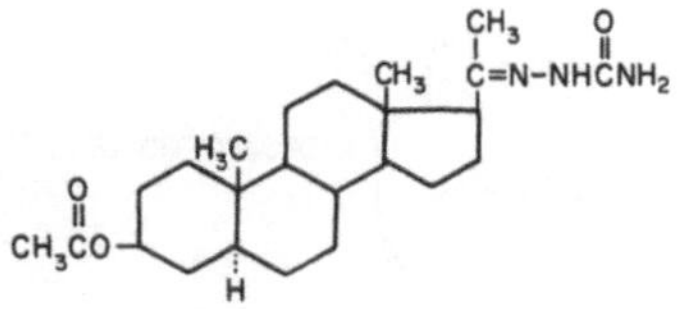

Helv. Chim. Acta 44:1380 (1961); 45:943 (1962)

▶ $N_3O_4C_4H_3$

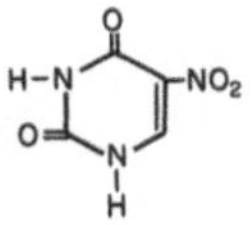

J. Am. Chem. Soc. 84:1042 (1962)

▶ $N_3O_4C_5H_5$

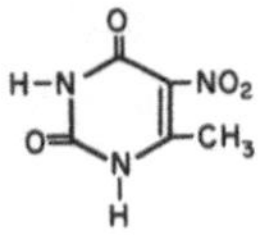

J. Am. Chem. Soc. 84:1042 (1962)

▶ $N_3O_4C_6H_3$

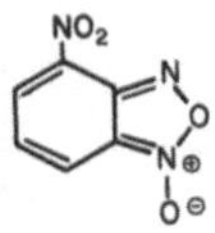

Chem. & Ind. 990 (1961)

▶ $N_3O_4C_6H_5$

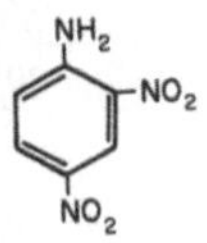

J. Org. Chem. 27:452 (1962)

▶ $N_3O_4C_6H_{11}$

$H_2NCH_2CONHCH_2CONHCH_2CO_2H$

TRIGLYCINE

J. Am. Chem. Soc. 84:4650 (1962)

▶ $N_3O_4C_9H_{13}$

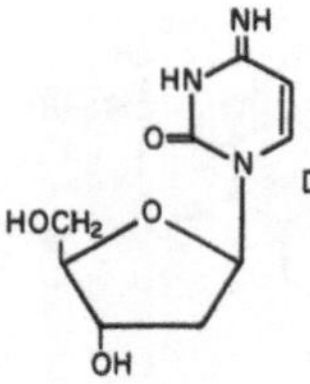

DEOXYCYTIDINE

J. Am. Chem. Soc. 83:2909 (1961); 84:4464 (1962)

▶ $N_3O_4C_{10}H_{13}$

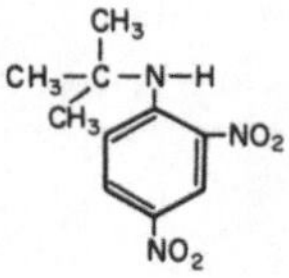

J. Org. Chem. 27:452 (1962)

▶ $N_3O_5C_9H_{13}$

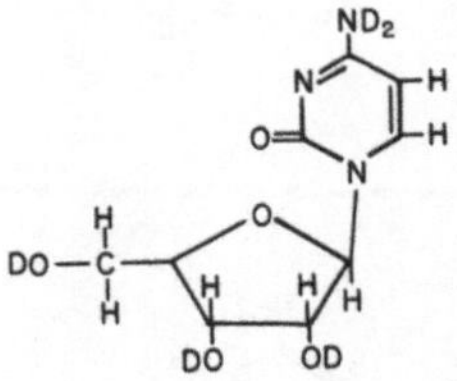

Varian 540

▶ $N_3O_5C_9H_{13}$

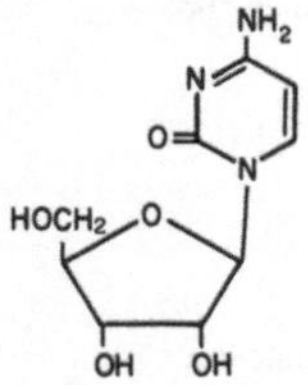

CYTIDINE

J. Am. Chem. Soc. 83:2909 (1961); 84:4464 (1962)

▶ $N_3O_5C_{14}H_{21} \cdot HCl$

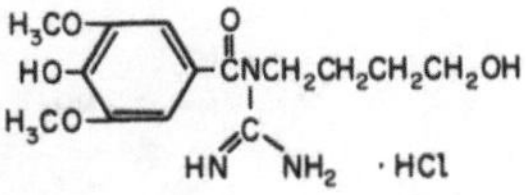

Tetrahedron Letters No. 13:546 (1962)

▶ $N_3O_5C_{32}H_{49}$

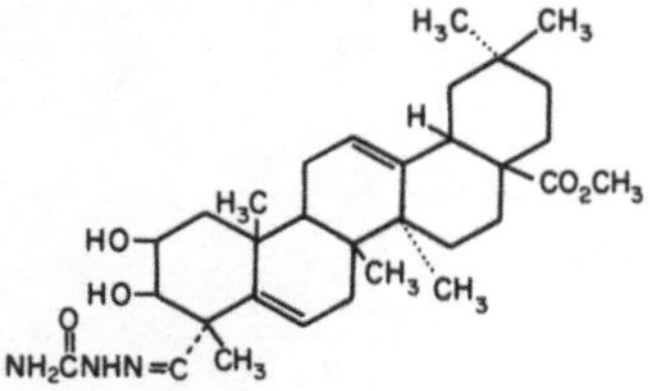

J. Chem. Soc. 4308 (1961)

▶ $N_3O_6C_6H_3$

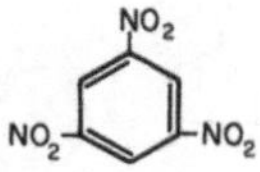

Can. J. Chem. 40:1759 (1962)

▶ $N_3O_6C_7H_5$

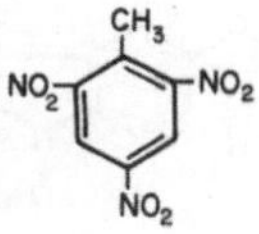

Varian 486

▶ $N_3O_6C_8H_7$

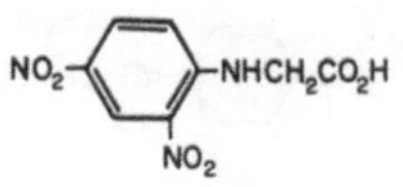

Bull. Chem. Soc. Japan 35:1658 (1962)

▶ $N_3O_6C_8H_{11}$

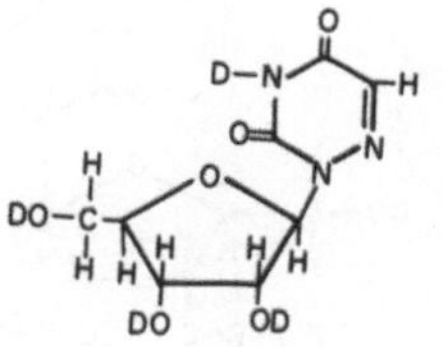

Varian 510

▶ $N_3O_6C_9H_9$

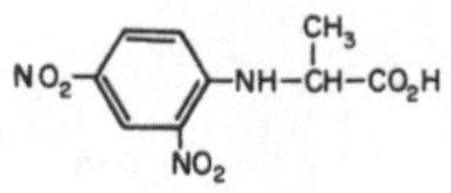

Bull. Chem. Soc. Japan 35:1658 (1962)

▶ $N_3O_6C_{11}H_{11}$

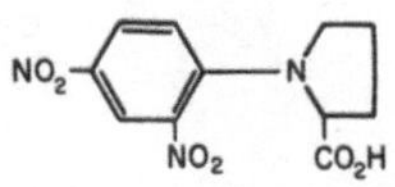

Bull. Chem. Soc. Japan 35:1658 (1962)

▶ $N_3O_6C_{11}H_{13}$

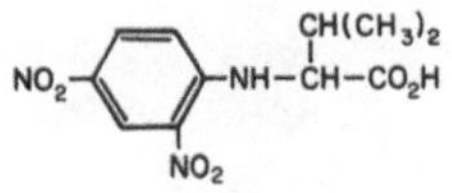

Bull. Chem. Soc. Japan 35:1658 (1962)

▶ $N_3O_6C_{12}H_{15}$

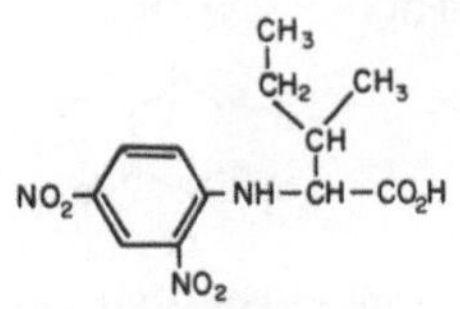

Bull. Chem. Soc. Japan 35:1658 (1962)

▶ $N_3O_6C_{12}H_{15}$

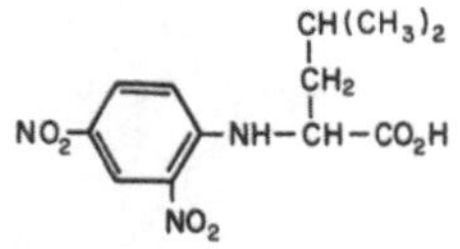

Bull. Chem. Soc. Japan 35:1658 (1962)

▶ $N_3O_6C_{15}H_{13}$

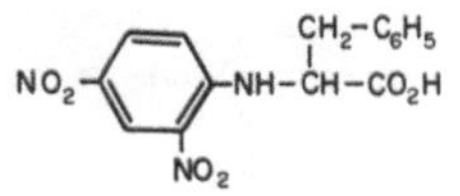

Bull. Chem. Soc. Japan 35:1658 (1962)

▶ $N_3O_6C_{15}H_{17}$

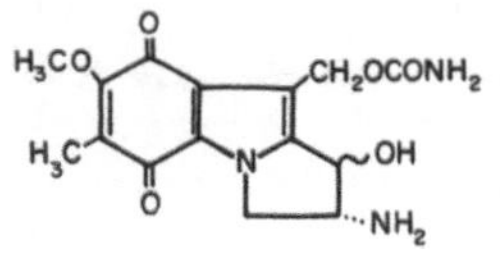

J. Am. Chem. Soc. 84:3185 (1962)

▶ $N_3O_6C_{16}H_{19}$

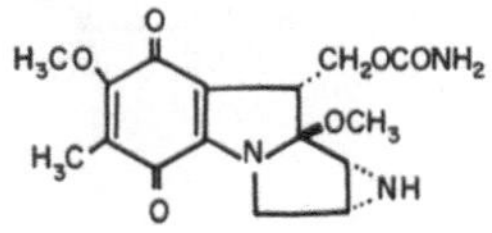

J. Am. Chem. Soc. 84:3185 (1962)

▶ $N_3O_6C_{21}H_{23}$

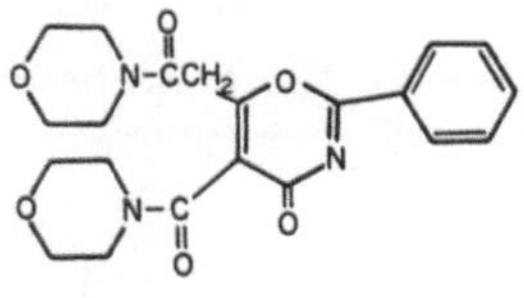

J. Chem. Soc. 3556 (1962)

▶ $N_3O_7C_6H_3$

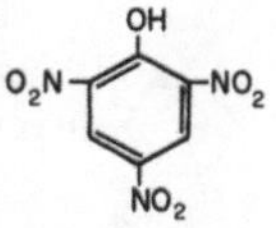


Can. J. Chem. 40:759 (1962)

▶ $N_3O_7C_9H_9$

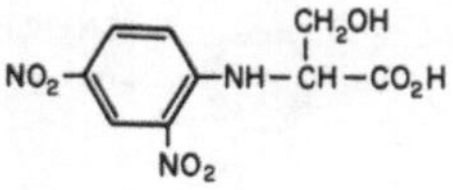


Bull. Chem. Soc. Japan 35:1658 (1962)

▶ $N_3O_7C_{10}H_{11}$

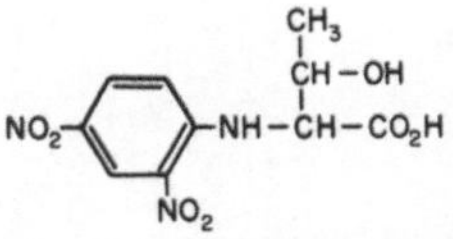


THREONINE
and ALLOTHREONINE

Bull. Chem. Soc. Japan 35:1658 (1962)

▶ $N_3O_7C_{10}H_{13}$

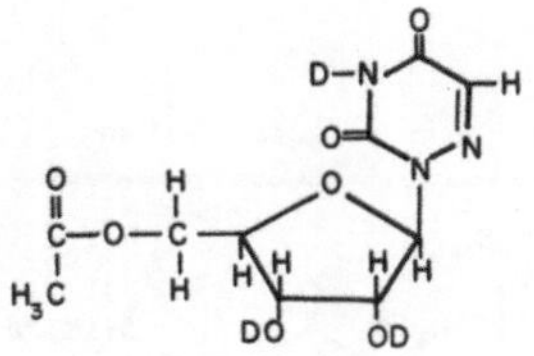

Varian 562

▶ $N_3O_7C_{11}H_{11}$

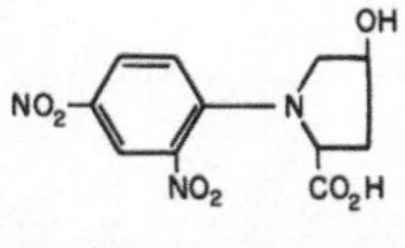


Bull. Chem. Soc. Japan 35:1658 (1962)

▶ $N_3O_7C_{15}H_{13}$

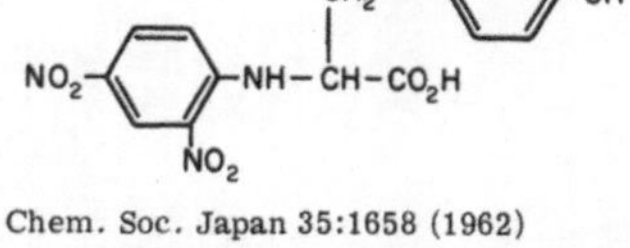


Bull. Chem. Soc. Japan 35:1658 (1962)

▶ $N_3O_8C_{10}H_9$

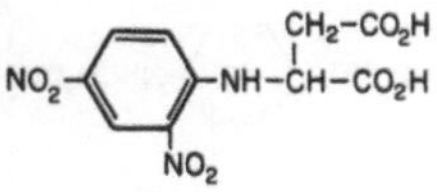


Bull. Chem. Soc. Japan 35:1658 (1962)

▶ $N_3O_8C_{12}H_{15}$

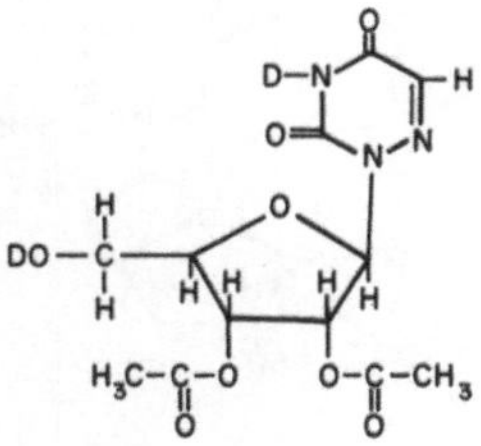

Varian 598

▶ $N_3O_9C_{14}H_{17}$

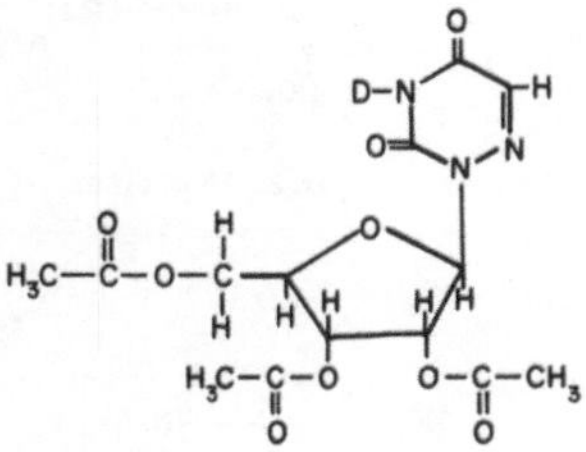

Varian 634

▶ $N_3Cl_3C_3H_9B_3$

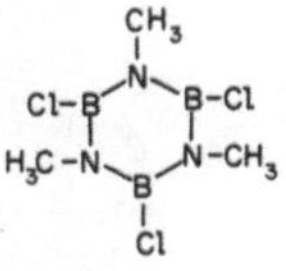

J. Chem. Phys. 34:1043 (1961)

▶ $N_3Cl_3C_6H_{15}B_3$

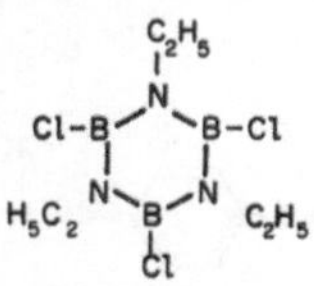

J. Chem. Phys. 34:1043 (1961)
Bull. Chem. Soc. Japan 35:1317 (1962)

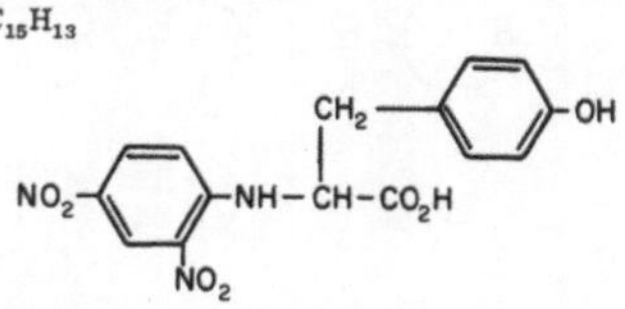

▶ $N_3Cl_3H_3B_3$

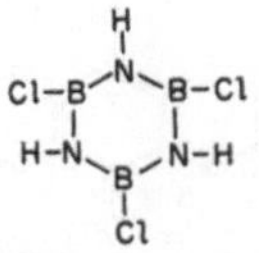

J. Chem. Phys. 34:1043 (1961)

▶ $N_3C_2H_3$

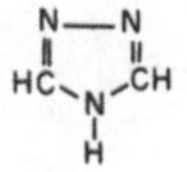

J. Org. Chem. 62:2632 (1962)

▶ $N_3C_3H_{12}B_3$

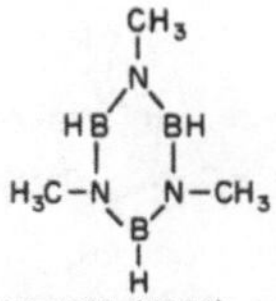

J. Chem. Phys. 34:1043 (1961)

▶ $N_3C_4H_5$

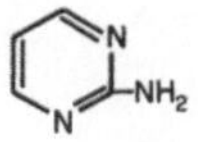

Varian 401
Arkiv Kemi 16:459 (1961)

▶ $N_3C_5H_7$

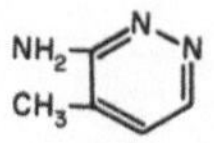

Acta Chem. Scand. 16:2389 (1962)

▶ $N_3C_5H_9$

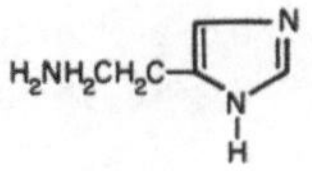

Varian 443

▶ $N_3C_6H_5$

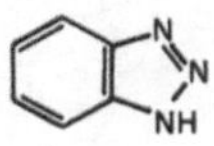

Australian J. Chem. 15:862 (1962)

▶ $N_3C_6H_{18}B_3$

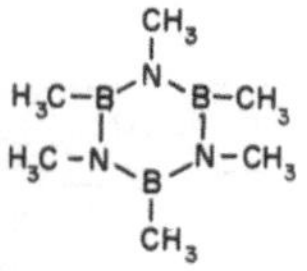

J. Chem. Phys. 34:1043 (1961)

▶ $N_3C_7H_7$

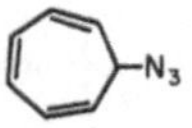

Chem. & Ind. 859 (1962)

▶ $N_3C_8H_7$

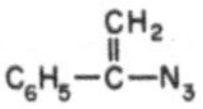

J. Org. Chem. 27:3557 (1962)
J. Am. Chem. Soc. 83:4483 (1961)

▶ $N_3C_9H_9$

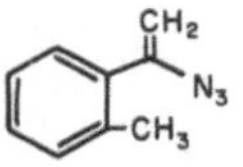

J. Org. Chem. 27:3557 (1962)

▶ $N_3C_9H_{24}B_3$

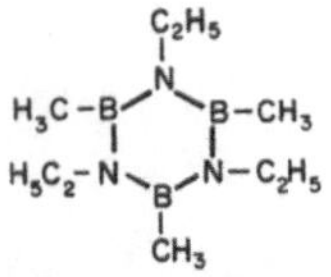

Bull. Chem. Soc. Japan 35:1317 (1962)
J. Chem. Phys. 34:1043 (1961)

▶ $N_3C_9H_{24}B_3$

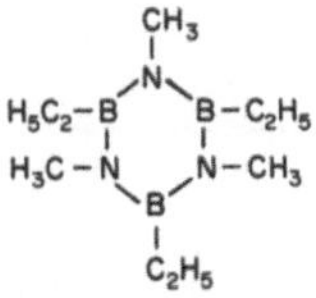

J. Chem. Phys. 34:1043 (1961)

▶ $N_3C_{11}H_{11}$

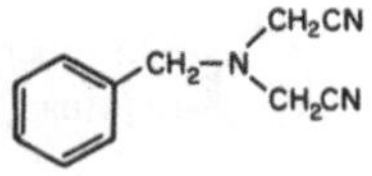

Varian 285

▶ $N_3C_{12}H_{13}$

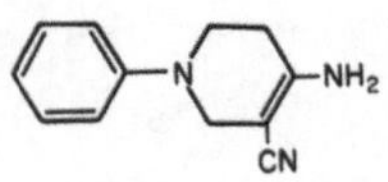

J. Chem. Soc. 5111 (1962)

▶ $N_3C_{12}H_{17}$

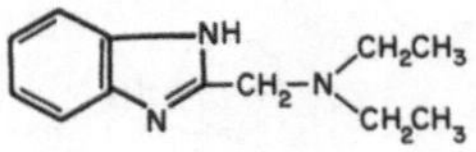

Varian 296

▶ $N_3C_{12}H_{30}B_3$

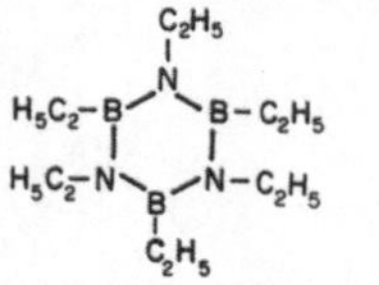

Bull. Chem. Soc. Japan 35:1317 (1962)
J. Chem. Phys. 34:1043 (1961)

▶ $N_3C_{14}H_{11}$

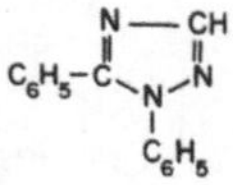

J. Org. Chem. 27:2632 (1962)

▶ $N_3C_{14}H_{11}$

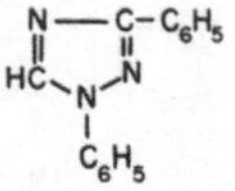

J. Org. Chem. 27:2632 (1962)

▶ $N_3C_{14}H_{23}$

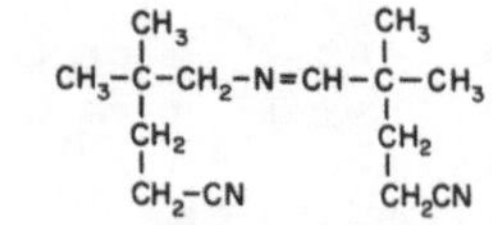

J. Org. Chem. 26:1822 (1961)

▶ $N_3C_{14}H_{25}$

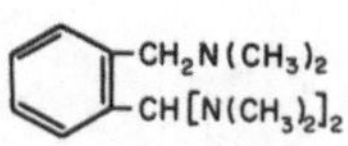

J. Org. Chem. 26:1372 (1961)

▶ $N_3C_{16}H_{14} \cdot ClO_4$

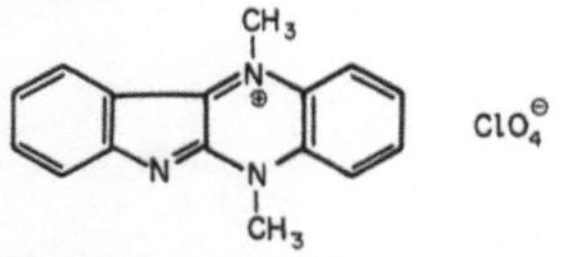

J. Chem. Soc. 3929 (1962)

▶ $N_3C_{18}H_{42}Al$

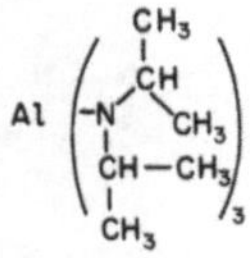

J. Am. Chem. Soc. 83:2835 (1961)

▶ $N_3C_{21}H_{24}B_3$

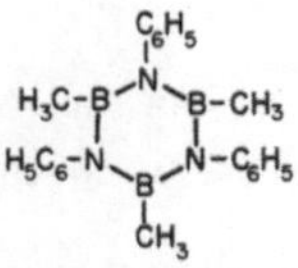

J. Chem. Phys. 34:1043 (1961)

▶ $N_3C_{22}H_{15}$

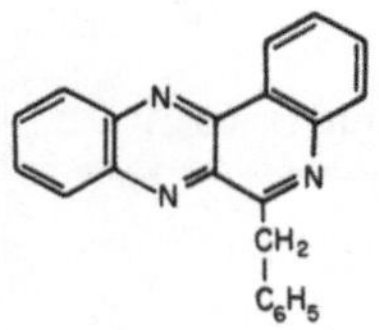

J. Am. Chem. Soc. 84:220 (1962)

▶ $N_3C_{24}H_{30}B_3$

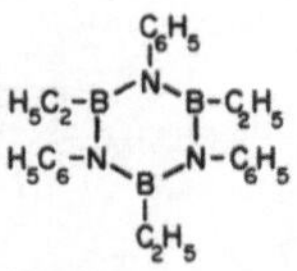

J. Chem. Phys. 34:1043 (1961)

▶ $N_3C_{25}H_{31}$

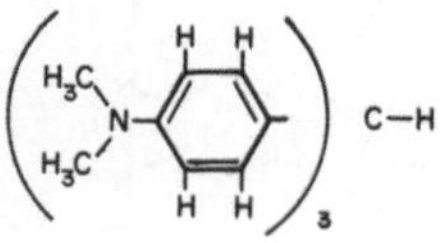

Varian 360

▶ $N_3C_{27}H_{29}$

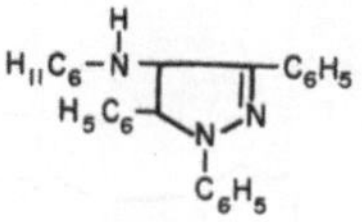

J. Org. Chem. 27:3975 (1962)

▶ $N_4OC_9H_6$

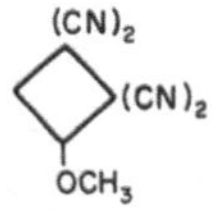

J. Am. Chem. Soc. 84:2210 (1962)

▶ $N_4OC_5H_4$

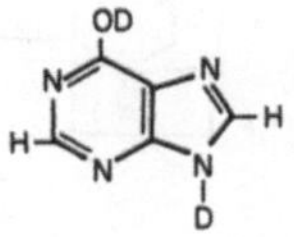

Varian 426

▶ $N_4OC_{10}H_8$

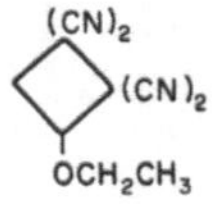

J. Am. Chem. Soc. 84:2210 (1962)

▶ $N_4OC_6H_6$

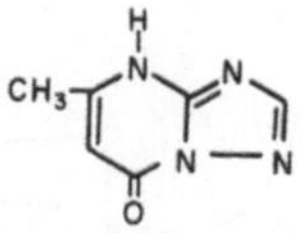

J. Chem. Soc. 3046 (1961)

▶ $N_4OC_{11}H_8$

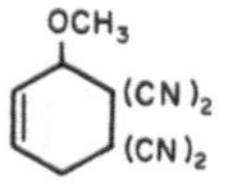

J. Am. Chem. Soc. 84:2210 (1962)

▶ $N_4OC_6H_6$

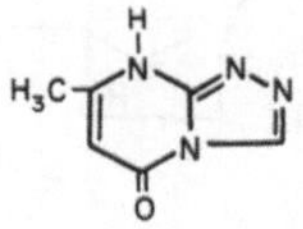

J. Chem. Soc. 3046 (1961)

▶ $N_4OC_6H_6$

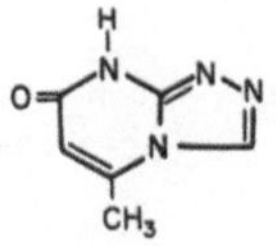

J. Chem. Soc. 3046 (1961)

▶ $N_4OC_{15}H_{10}$

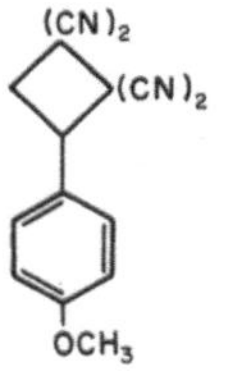

J. Am. Chem. Soc. 84:2210 (1962)

▶ $N_4OC_6H_6$

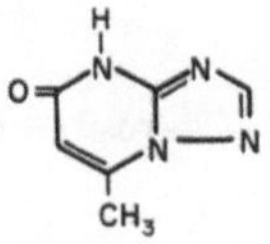

J. Chem. Soc. 3046 (1961)

▶ $N_4OC_6H_6$

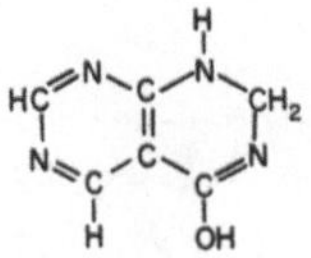

J. Org. Chem. 27:4211 (1962)

▶ $N_4OC_{16}H_{12}$

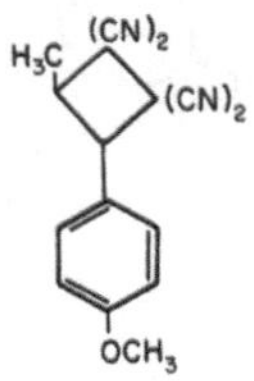

J. Am. Chem. Soc. 84:2210 (1962)

▶ $N_4OC_{38}H_{38}$

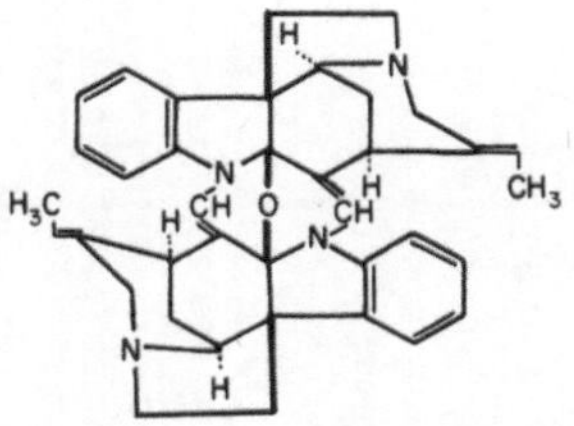

Tetrahedron 14:143 (1961)

▶ $N_4OC_{38}H_{46}$

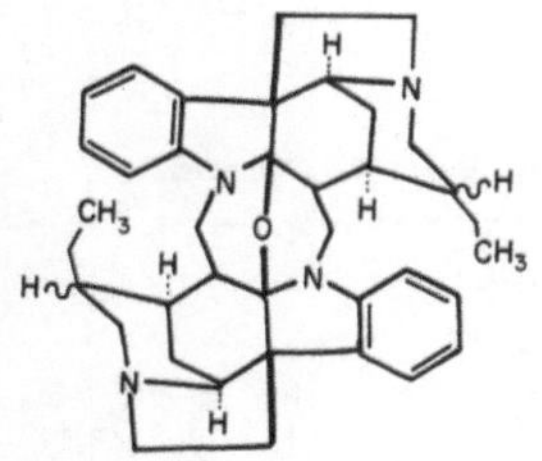

Tetrahedron 14:143 (1961)

▶ $N_4OC_{38}H_{42}$

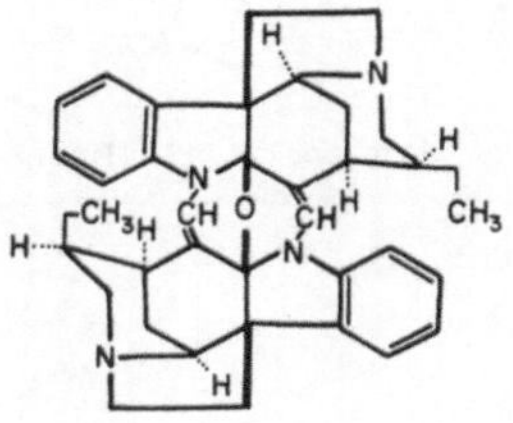

Tetrahedron 14:143 (1961)

▶ $N_4OC_{40}H_{44} \cdot Cl_2$

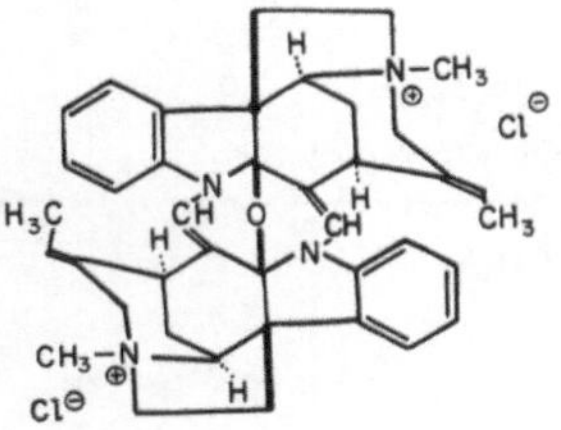

Tetrahedron 14:143 (1961)

▶ $N_4OC_{38}H_{44}$

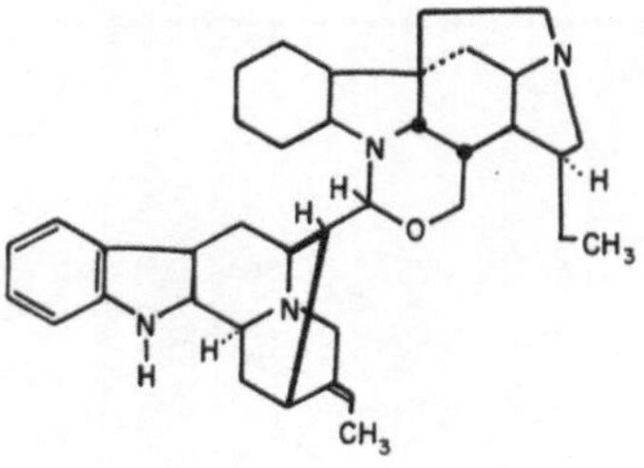

J. Org. Chem. 27:2983 (1962)

▶ $N_4O_2BrC_{13}H_{15}$

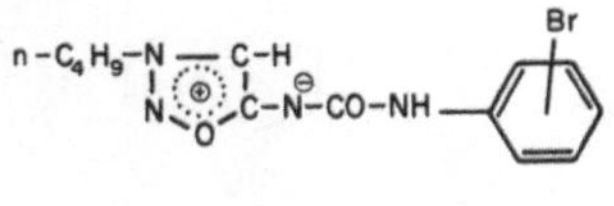

Helv. Chim. Acta 45:2441 (1962)

▶ $N_4O_2Cl_2C_{40}H_{46}$

C-TOXIFERIN I

Helv. Chim. Acta 44:623 (1961)

▶ $N_4O_2C_6H_{10}$

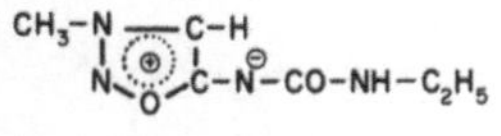

Helv. Chim. Acta 45:2441 (1962)

▶ $N_4O_2C_6H_{11} \cdot Cl$

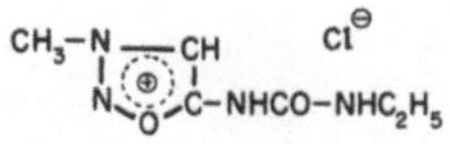

Helv. Chim. Acta 45:2441 (1962)

▶ $N_4O_2C_7H_{13} \cdot Cl$

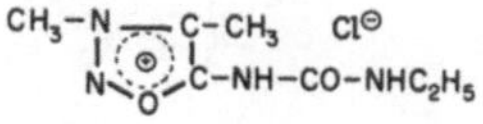

Helv. Chim. Acta 45:2441 (1962)

▶ $N_4O_2C_8H_8$

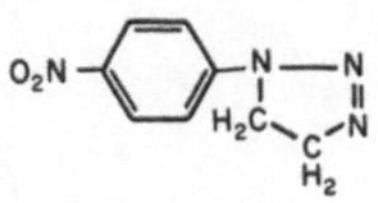

J. Am. Chem. Soc. 84:993 (1962)

▶ $N_4O_2C_8H_{10}$

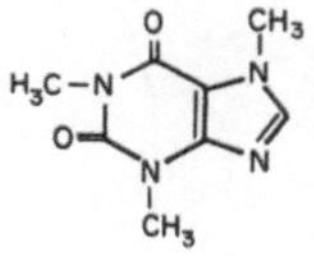

Varian 204

▶ $N_4O_2C_{10}H_{12}$

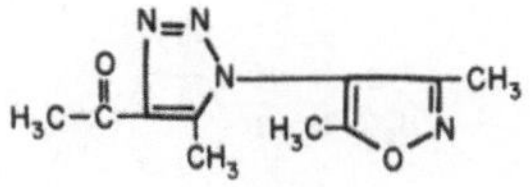

J. Chem. Soc. 2083 (1962)

▶ $N_4O_2C_{16}H_{10}$

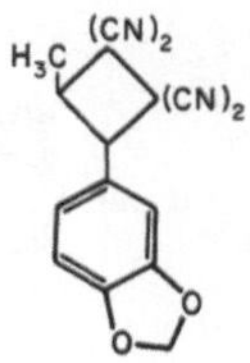

J. Am. Chem. Soc. 84:2210 (1962)

▶ $N_4O_2C_{18}H_{30}$

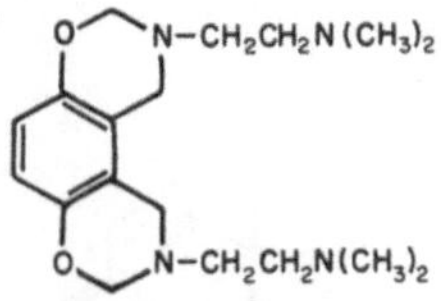

J. Org. Chem. 27:2752 (1962)

▶ $N_4O_2C_{18}H_{30}$

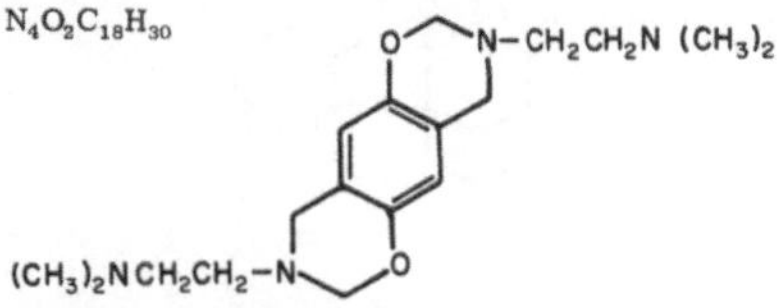

J. Org. Chem. 27:2752 (1962)

▶ $N_4O_2C_{20}H_{18}$

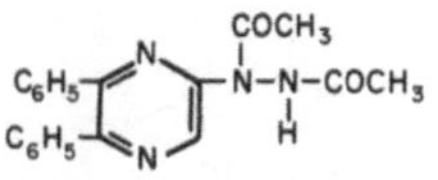

J. Org. Chem. 29:3246 (1962)

▶ $N_4O_2C_{40}H_{46}D_2 \cdot Cl_2$

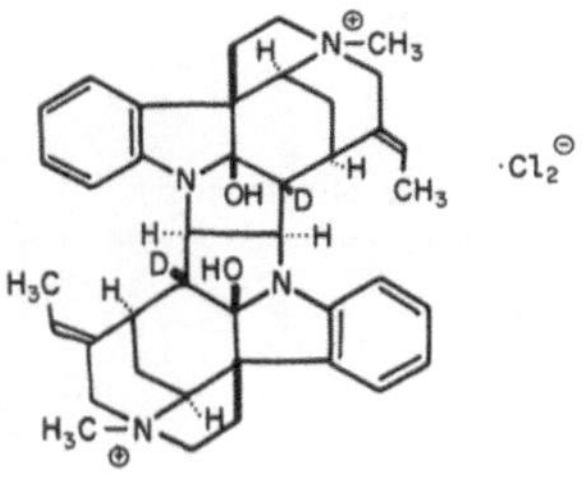

Helv. Chim. Acta 44:2211 (1961)

▶ $N_4O_2C_{40}H_{48} \cdot Cl_2$

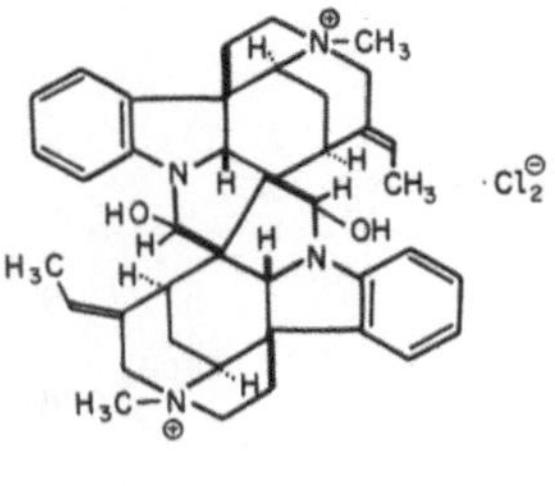

Helv. Chim. Acta 44:2211 (1961)

► $N_4O_2C_{40}H_{48} \cdot Cl_2$

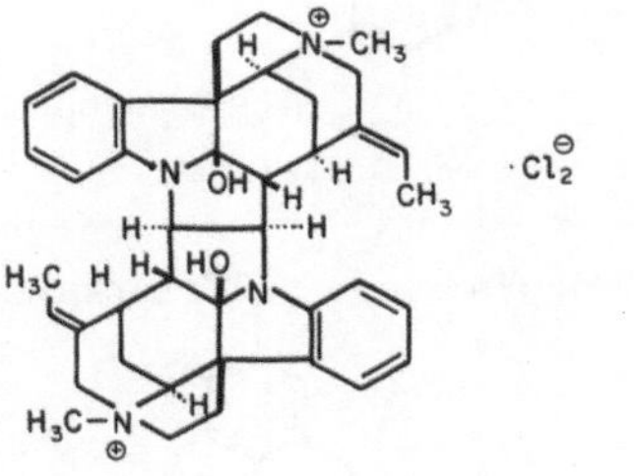

Helv. Chim. Acta 44:2211 (1961)

► $N_4O_3C_{34}H_{36}$

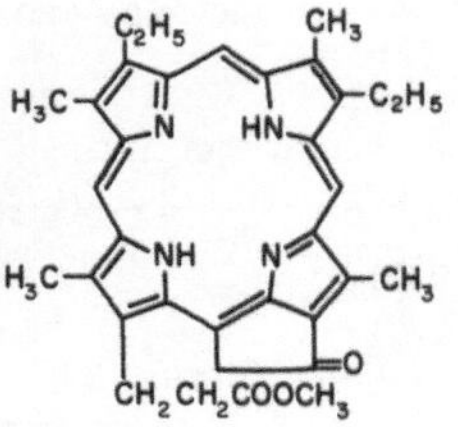

J. Am. Chem. Soc. 83:3743 (1961)

► $N_4O_2C_{42}H_{52} \cdot Cl_2$

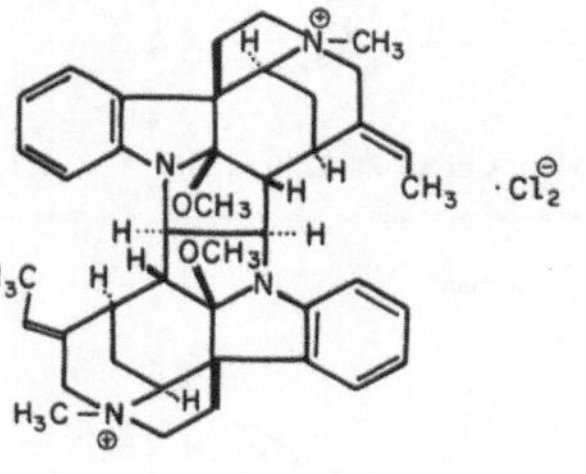

Helv. Chim. Acta 44:2211 (1961)

► $N_4O_4ClC_{13}H_{13}$

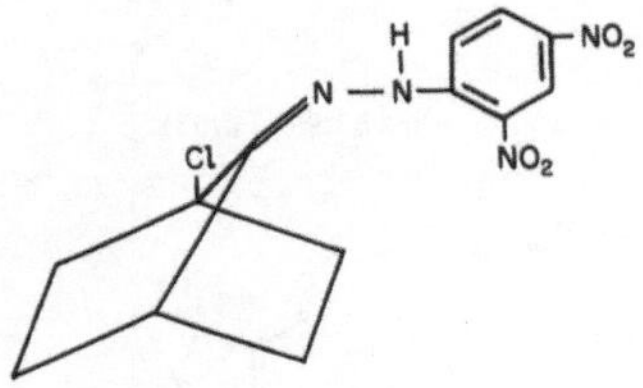

J. Org. Chem. 26:2181 (1961)

► $N_4O_3C_6H_5D_9$

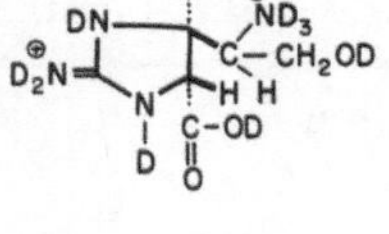

J. Chem. Soc. 1645 (1962)

► $N_4O_4C_7H_6$

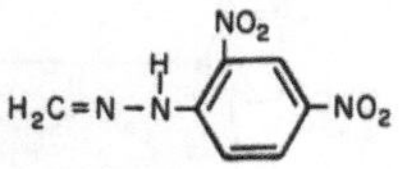

Can. J. Chem. 40:884 (1962)

► $N_4O_3C_{22}H_{20}$

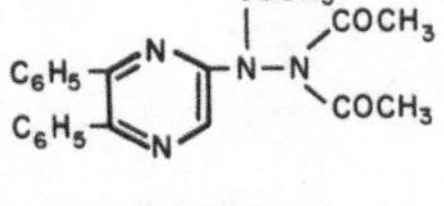

J. Org. Chem. 27:3246 (1962)

► $N_4O_4C_9H_{10}$

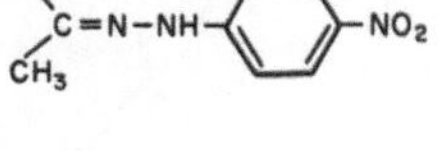

Varian 233
Ann. 653:135 (1962)
J. Am. Chem. Soc. 84:753 (1962)

▶ $N_4O_4C_{10}H_{12}$

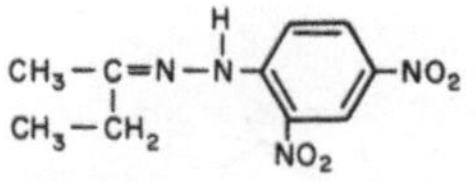

J. Am. Chem. Soc. 84:753 (1962)

▶ $N_4O_4C_{11}H_{14}$

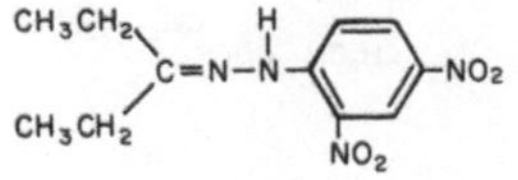

J. Am. Chem. Soc. 84:753 (1962)

▶ $N_4O_4C_{11}H_{14}$

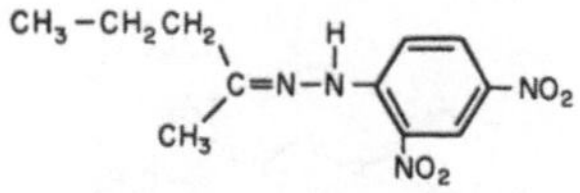

J. Am. Chem. Soc. 84:753 (1962)

▶ $N_4O_4C_{11}H_{14}$

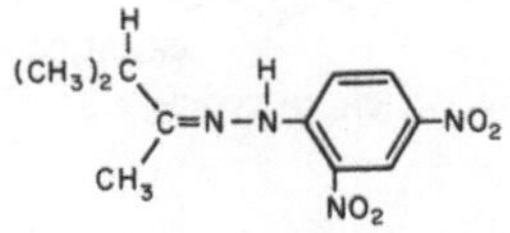

J. Am. Chem. Soc. 84:753 (1962)

▶ $N_4O_4C_{12}H_{14}$

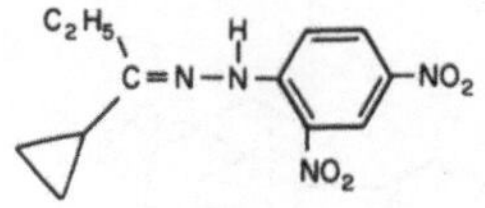

J. Am. Chem. Soc. 84:753 (1962)

▶ $N_4O_4C_{12}H_{16}$

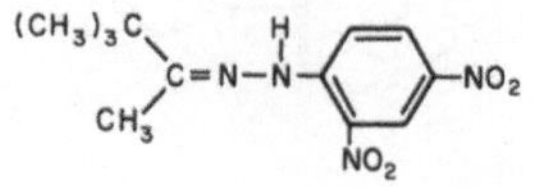

J. Am. Chem. Soc. 84:753 (1962)

▶ $N_4O_4C_{13}H_{10}$

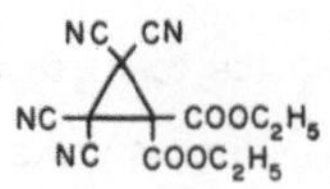

J. Org. Chem. 27:2237 (1962)

▶ $N_4O_4C_{13}H_{16}$

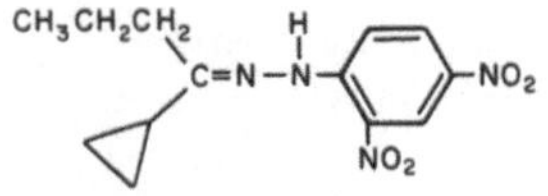

J. Am. Chem. Soc. 84:753 (1962)

▶ $N_4O_4C_{13}H_{16}$

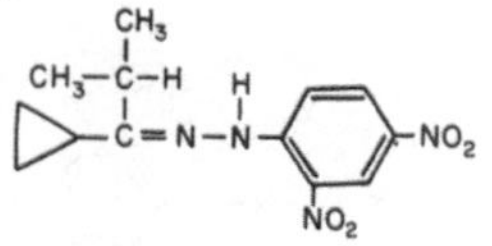

J. Am. Chem. Soc 84:753 (1962)

▶ $N_4O_4C_{13}H_{18}$

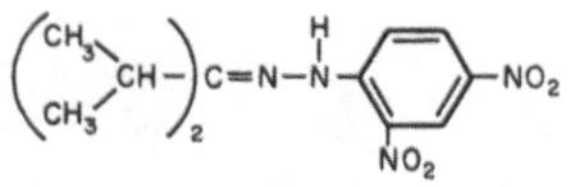

J Am Chem. Soc 84:753 (1962)

▶ $N_4O_4C_{14}H_{12}$

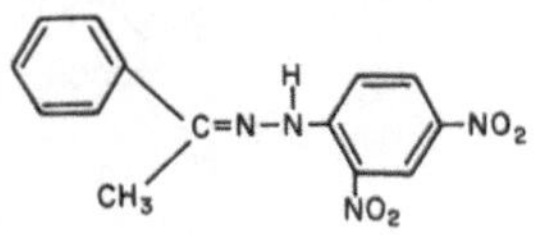

J. Am. Chem. Soc. 84:753 (1962)

▶ $N_4O_4C_{15}H_{14}$

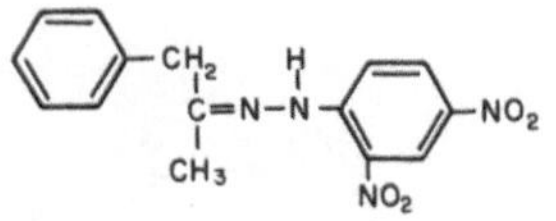

J. Am. Chem. Soc. 84:753 (1962)

▶ $N_4O_4C_{15}H_{22}$

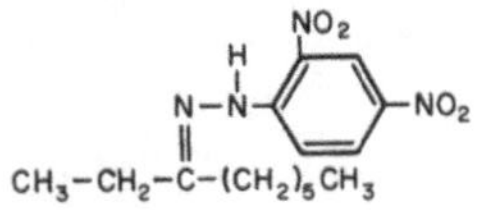

Helv. Chim. Acta 45:129 (1962)

► $N_4O_4C_{16}H_{20}$

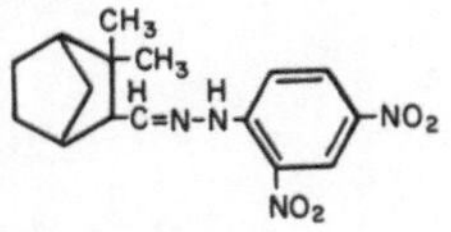

J. Org. Chem. 26:709 (1961)

► $N_4O_4C_{32}H_{28}$

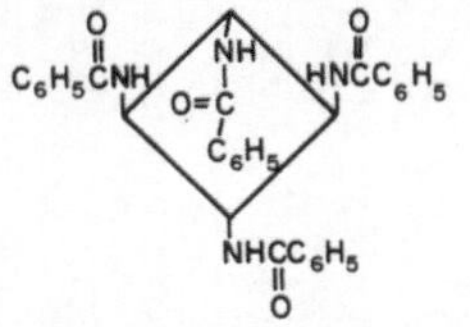

J. Am. Chem. Soc. 84:1008 (1962)

► $N_4O_4C_{33}H_{32}$

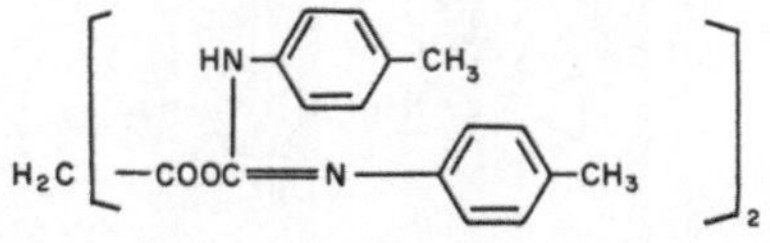

J. Am. Chem. Soc. 84:1310 (1962)

► $N_4O_4C_{34}H_{38}$

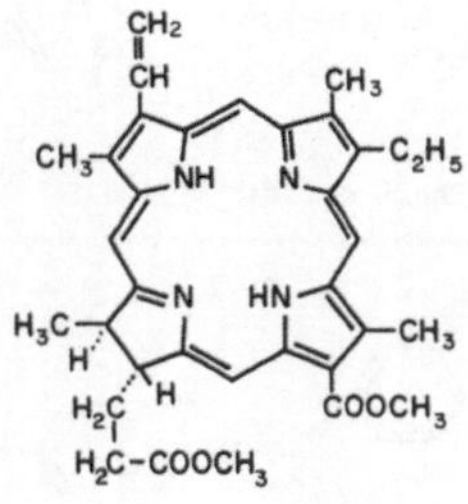

J. Am. Chem. Soc. 83:4676 (1961)

► $N_4O_4C_{34}H_{38}$

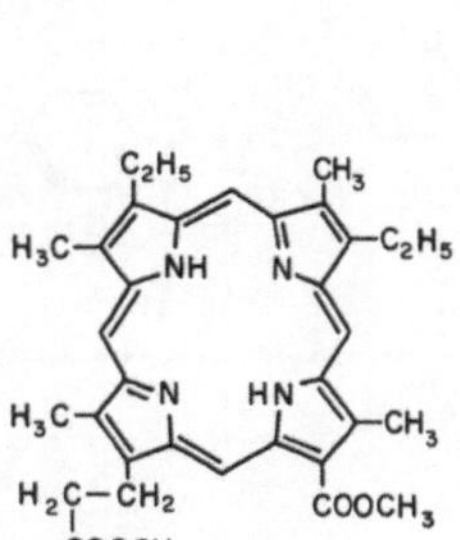

J. Am. Chem. Soc. 83:4676 (1961)

► $N_4O_4C_{36}H_{38}$

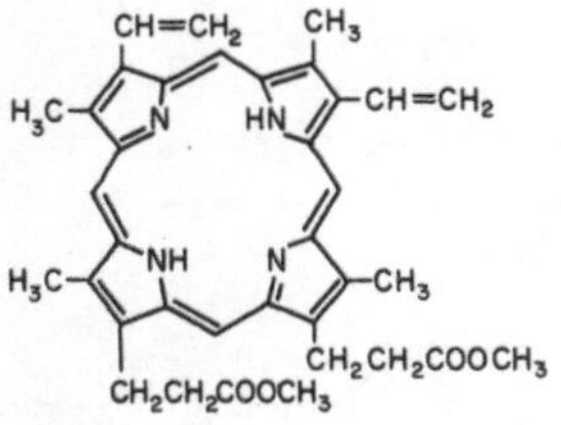

J. Am. Chem. Soc. 83:3743 (1961)

► $N_4O_4C_{36}H_{42}$

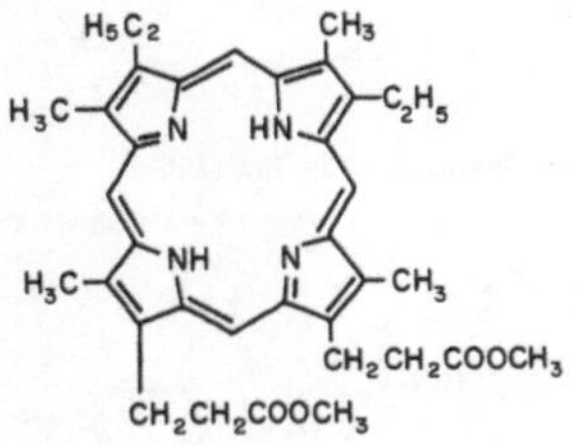

J. Am. Chem. Soc. 83:3743 (1961)

► $N_4O_4C_{40}H_{48} \cdot Cl_2$

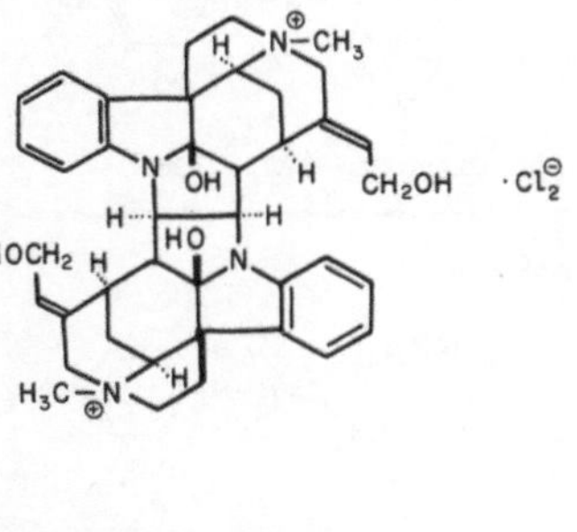

Helv. Chim. Acta 44:2211 (1961)

► $N_4O_5C_8H_{14}$

$H_2N(CH_2CONH)_3CH_2COOH$

TETRAGLYCINE

J. Am. Chem. Soc. 84:4650 (1962)

▶ $N_4O_5C_{55}H_{72}Mg$

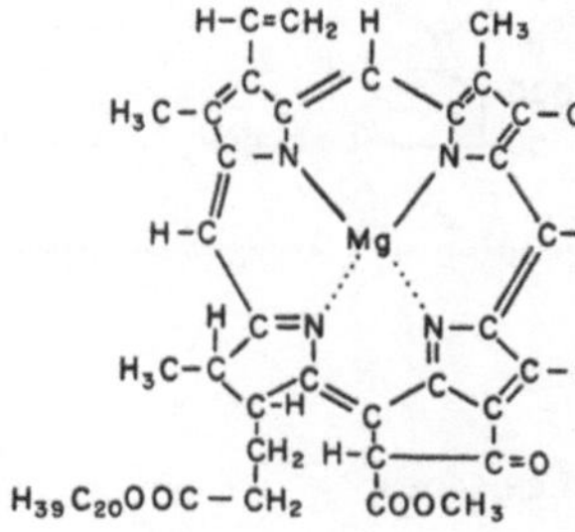

CHLOROPHYLL A

J. Am. Chem. Soc. 84:3587 (1962)

▶ $N_4O_6ClC_{36}H_{39}$

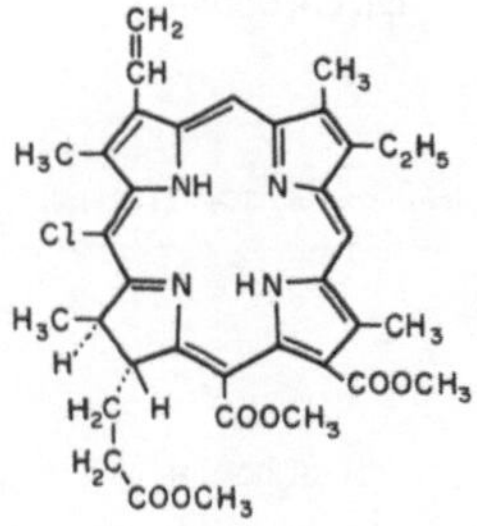

J. Am. Chem. Soc. 83:4676 (1961)

▶ $N_4O_6F_2C_{12}H_{12}$

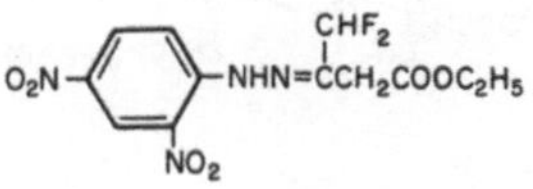

Varian 293

▶ $N_4O_6C_6H_2$

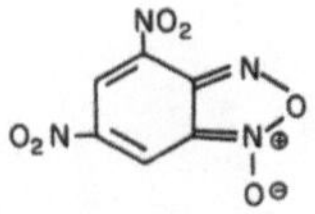

Chem. & Ind. 990 (1961)

▶ $N_4O_6C_{12}H_{16}$

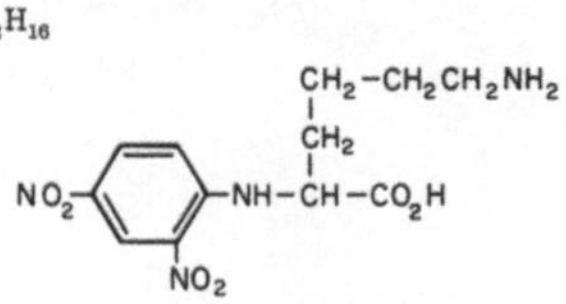

Bull. Chem. Soc. Japan 35:1658 (1962)

▶ $N_4O_6C_{12}H_{26}$

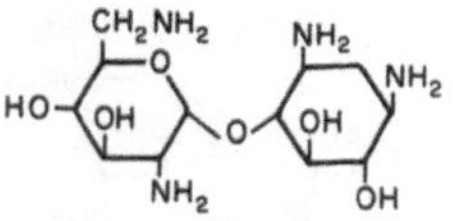

J. Am. Chem. Soc. 84:3216 (1962)

▶ $N_4O_6C_{14}H_{16}$

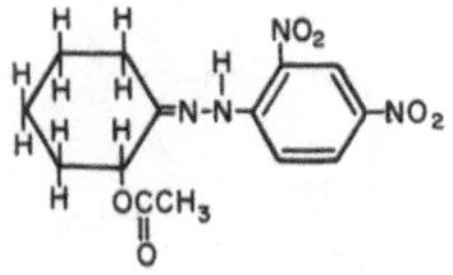

J. Org. Chem. 27:2728.(1962)

▶ $N_4O_6C_{15}H_{18}$

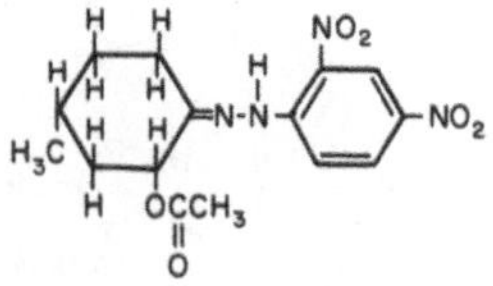

J. Org. Chem. 27:2728 (1962)

▶ $N_4O_6C_{15}H_{18}$

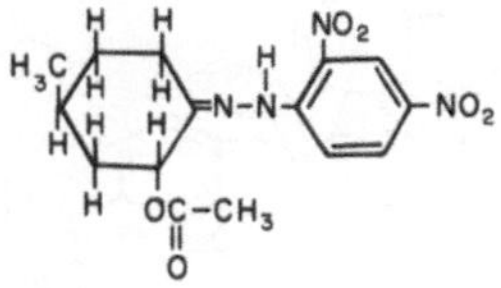

J. Org. Chem. 27:2728 (1962)

▶ $N_4O_6C_{20}H_{42}$

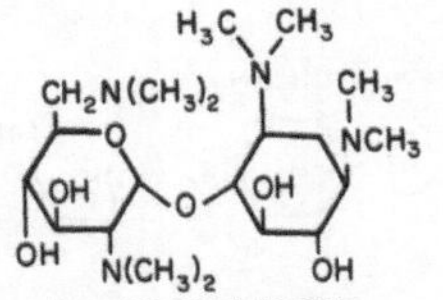

J. Am. Chem. Soc. 84:3216 (1962)

▶ $N_4O_8C_{16}H_{18}$

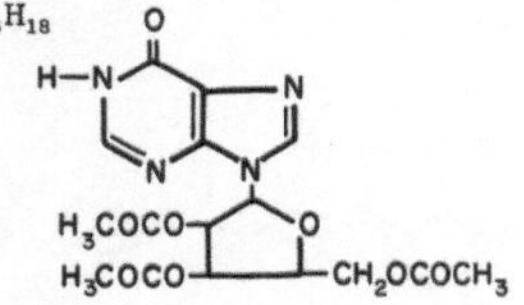

Varian 324

▶ $N_4O_6C_{36}H_{42}$

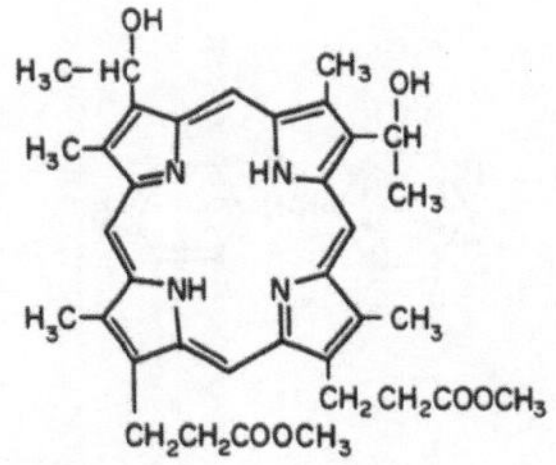

J. Am. Chem. Soc. 83:3743 (1961)

▶ $N_4O_8C_{40}H_{46}$

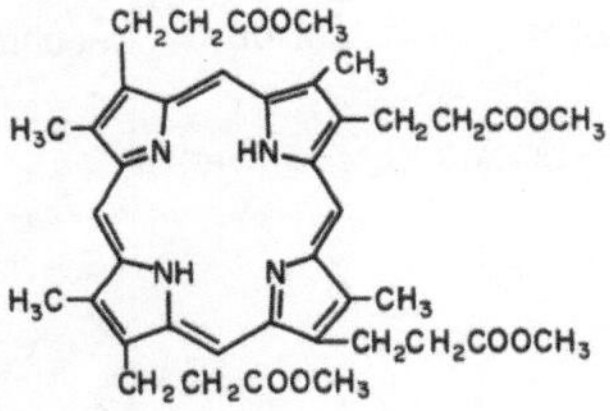

J. Am. Chem. Soc. 83:3743 (1961)

▶ $N_4O_6C_{37}H_{42}$

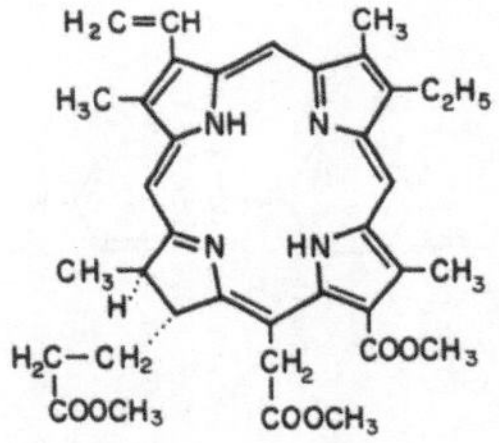

J. Am. Chem. Soc. 83:4676 (1961)

▶ $N_4O_8C_{40}H_{46}$

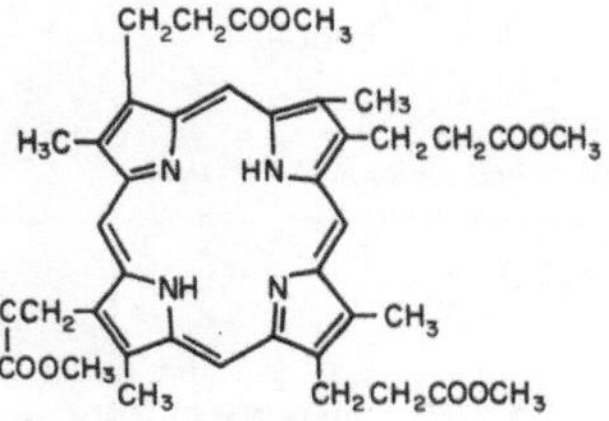

J. Am. Chem. Soc. 83:3743 (1961)

▶ $N_4O_6C_{40}H_{50}$

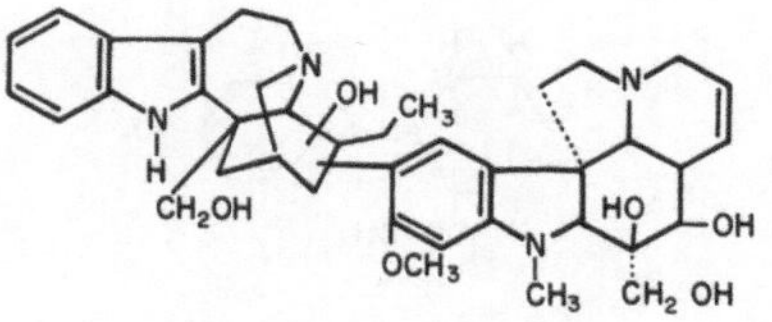

J. Am. Chem. Soc. 84:1509 (1962)

▶ $N_4O_9C_{46}H_{56}$

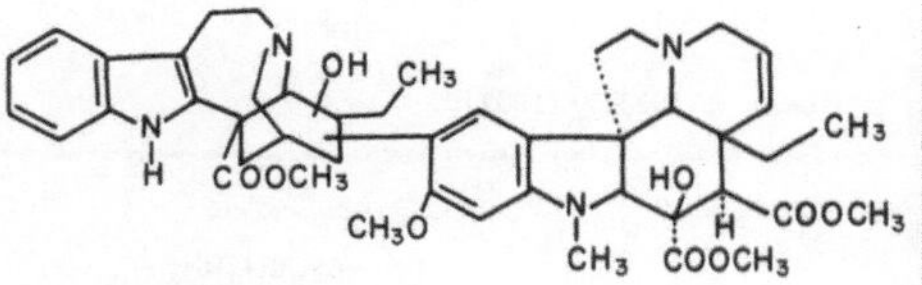

J. Am. Chem. Soc. 84:1509 (1962)

▶ $N_4O_{10}C_{20}H_{34}$

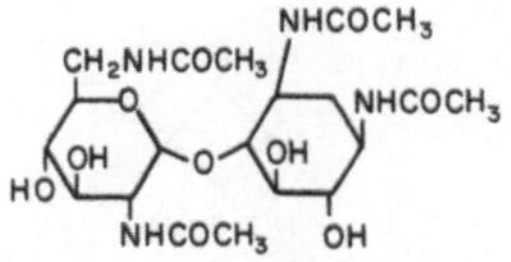

J. Am. Chem. Soc. 84:3216 (1962)

▶ $N_4O_{10}C_{46}H_{54}$

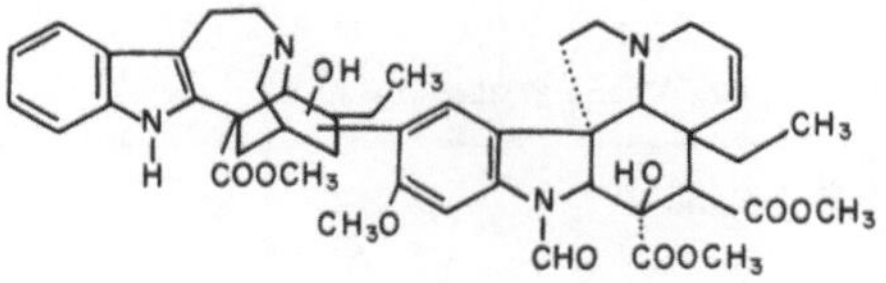

J. Am. Chem. Soc. 84:1509 (1962)

▶ $N_4O_{11}C_{50}H_{60}$

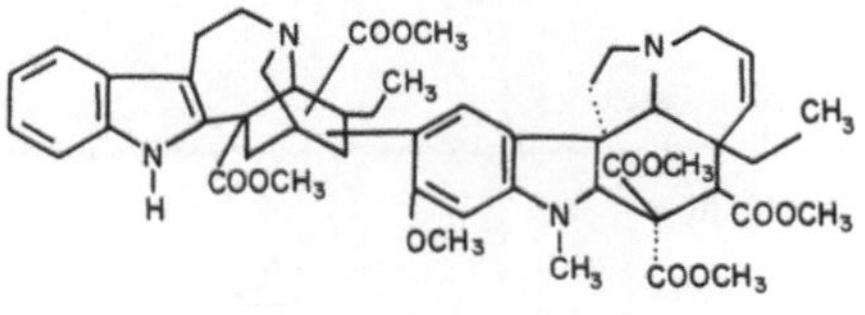

J. Am. Chem. Soc. 84:1509 (1962)

▶ $N_4O_{16}C_{48}H_{54}$

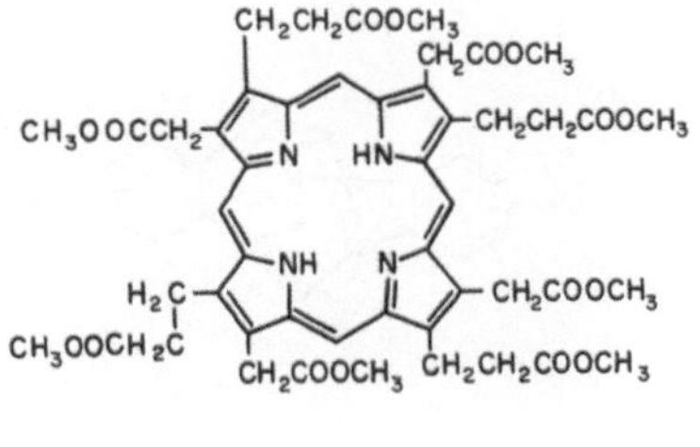

J. Am. Chem. Soc. 83:3743 (1961)

▶ $N_4ClC_{10}H_9$

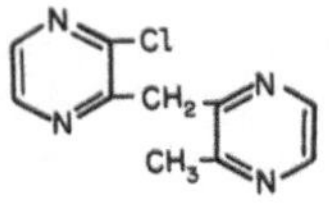

J. Org. Chem. 26:4984 (1961)

▶ $N_4Cl_2C_{40}H_{46}$

C-DIHYDROTOXIFERIN

Helv. Chim. Acta 44:623 (1961)

▶ $N_4F_3C_2H$

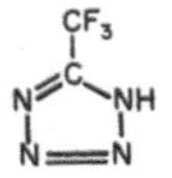

J. Org. Chem. 27:3249 (1962)

▶ $N_4CH_2$

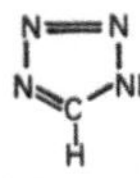

Tetrahedron 14:244 (1961)

▶ $N_4C_2H_4$

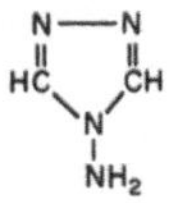

J. Org. Chem. 62:2632 (1962)

▶ $N_4C_3H_6$

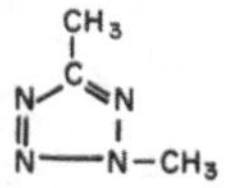

Varian 387

▶ $N_4C_4H_{12}$

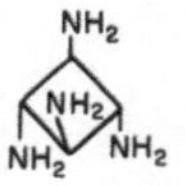

J. Am. Chem. Soc. 84:1008 (1962)

▶ $N_4C_7H_{21}Al$

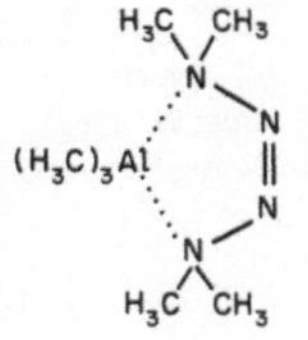

Can. J. Chem. 40:2186 (1962)

▶ $N_4C_8H_{16}$

Chem. Ber. 95:1493 (1962)

▶ $N_4C_{10}H_{12} \cdot 2HCl$

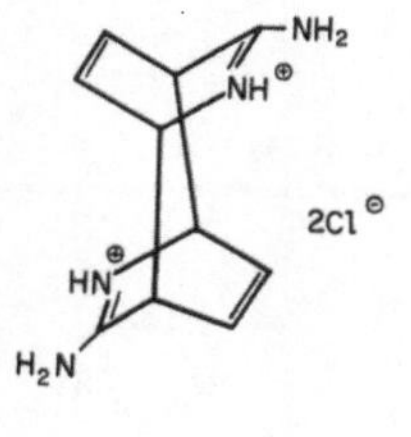

J. Am. Chem. Soc. 83:2967, 4487 (1961)

▶ $N_4C_{14}H_{12}$

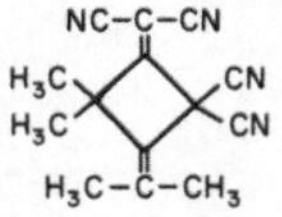

J. Am. Chem. Soc. 83:4867 (1961)

▶ $N_4C_{14}H_{12}$

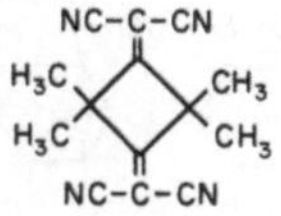

J. Am. Chem. Soc. 83:4867 (1961)

▶ $N_4C_{16}H_{14}$

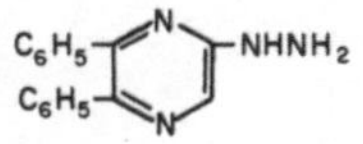

J. Org. Chem. 27:3246 (1962)

▶ $N_4C_{16}H_{16}$

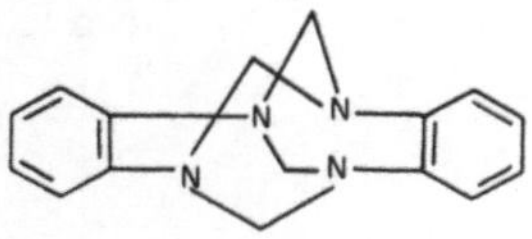

Chem. Ber. 95:1493 (1962)

▶ $N_4C_{16}H_{22}$

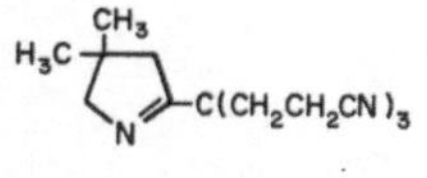

Varian 660

▶ $N_4C_{18}H_{16}$

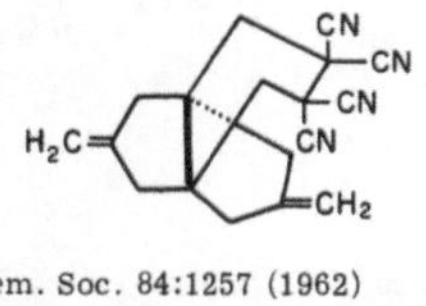

J. Am. Chem. Soc. 84:1257 (1962)

▶ $N_4C_{18}H_{20}$

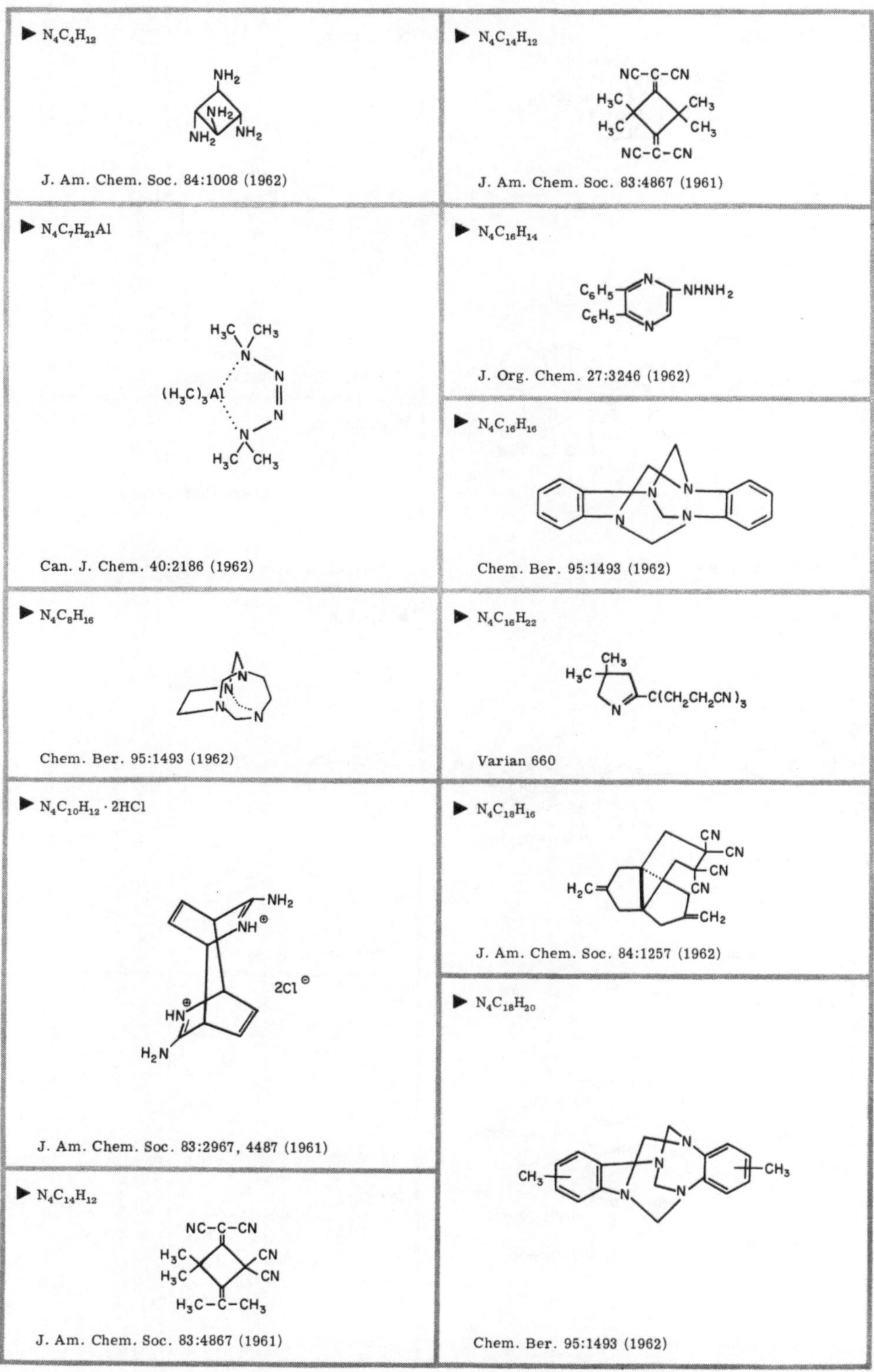

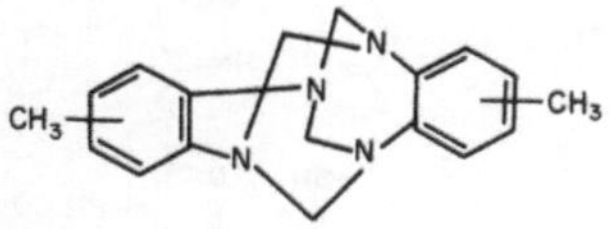

Chem. Ber. 95:1493 (1962)

▶ $N_4C_{18}H_{22} \cdot Ni$

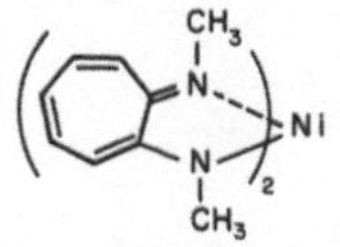

J. Chem. Phys. 37:347 (1962)
J. Am. Chem. Soc. 83:3125 (1961)

▶ $N_4C_{26}H_{22}$

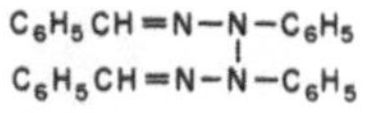

$C_6H_5CH{=}N{-}N{-}C_6H_5$
$C_6H_5CH{=}N{-}N{-}C_6H_5$

J. Org. Chem. 27:4663 (1962)

▶ $N_4C_{20}H_{14}$

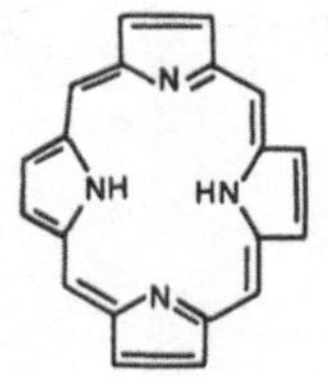

J. Am. Chem. Soc. 84:4307 (1962)

▶ $N_4C_{40}H_{46}D_2 \cdot Cl_2$

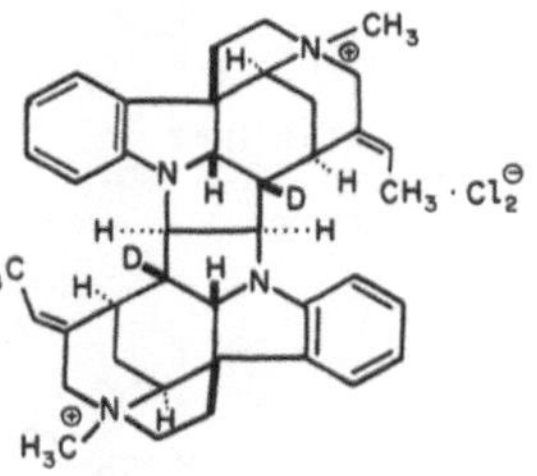

Helv. Chim. Acta 44:2211 (1961)

▶ $N_4C_{22}H_{30} \cdot Ni$

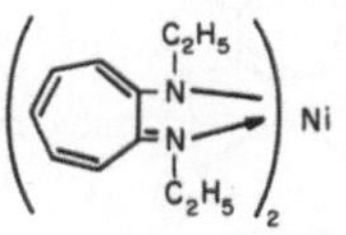

J. Chem. Phys. 37:347 (1962)

▶ $N_4C_{40}H_{48} \cdot Cl_2$

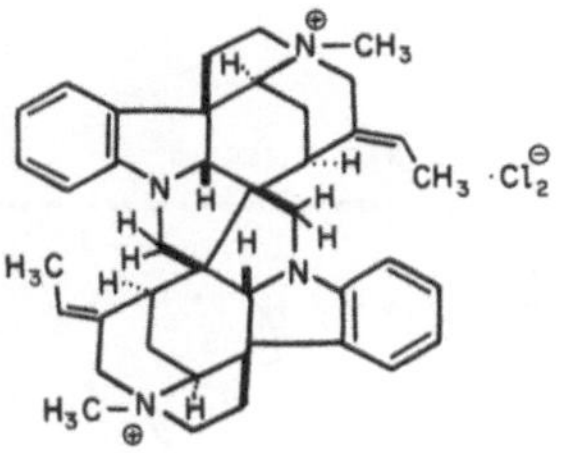

Helv. Chim. Acta 44:2211 (1961)

▶ $N_4C_{26}H_{22}$

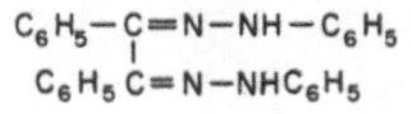

$C_6H_5{-}C{=}N{-}NH{-}C_6H_5$
$C_6H_5C{=}N{-}NHC_6H_5$

J. Org. Chem. 27:4663 (1962)

▶ $N_4C_{26}H_{22}$

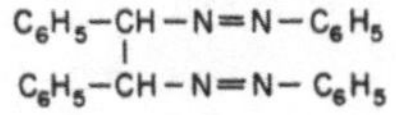

$C_6H_5{-}CH{-}N{=}N{-}C_6H_5$
$C_6H_5{-}CH{-}N{=}N{-}C_6H_5$

J. Org. Chem. 27:4663 (1962)

▶ $N_4C_{40}H_{48} \cdot Cl_2$

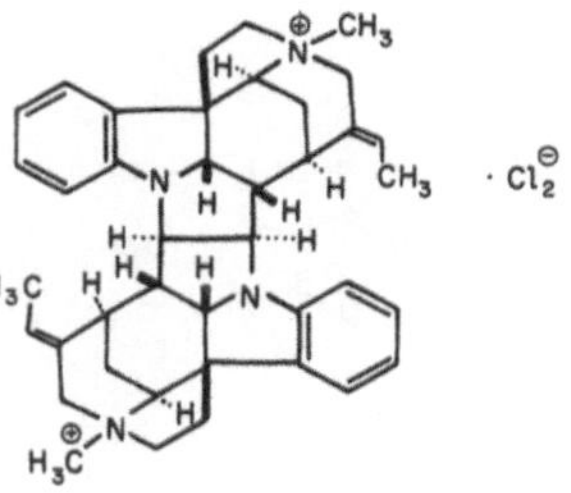

Helv. Chim. Acta 44:2211 (1961)

▶ $N_4C_{54}H_{38} \cdot Ni$

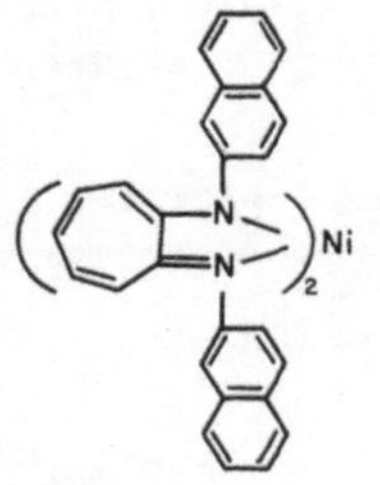

J. Chem. Phys. 37:347 (1962)
J. Am. Chem. Soc. 83:3714 (1961)

▶ $N_4C_{70}H_{54} \cdot Ni$

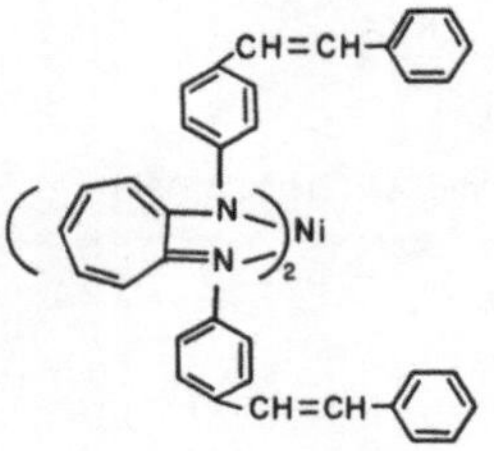

J. Chem. Phys. 37:347 (1962)

▶ $N_5OC_5H_5$

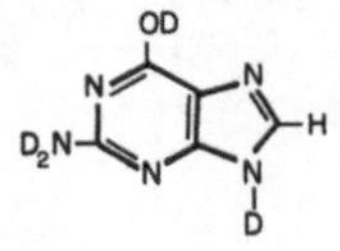

Varian 430

▶ $N_5OC_9H_{15}$

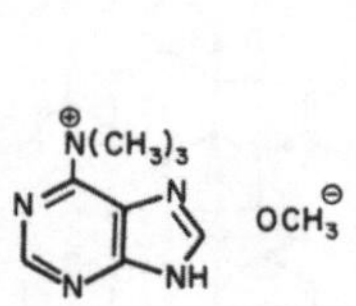

J. Org. Chem. 27:3276 (1962)

▶ $N_5O_3C_{10}H_{13}$

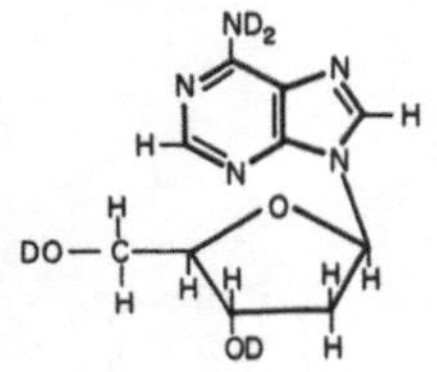

Varian 563

▶ $N_5O_3C_{10}H_{13}$

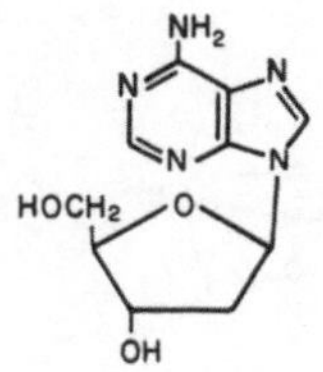

DEOXYADENOSINE

J. Am. Chem. Soc. 83:1906 (1961); 84:4464 (1962)

▶ $N_5O_4C_{10}H_{13}$

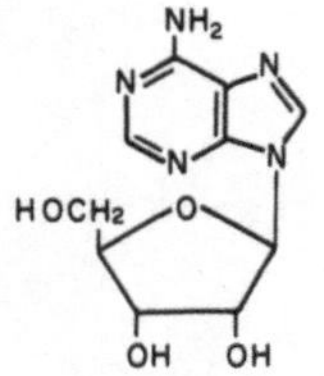

ADENOSINE

J. Am. Chem. Soc. 83:2909 (1961); 84:4464 (1962)

▶ $N_5O_4C_{10}H_{13}$

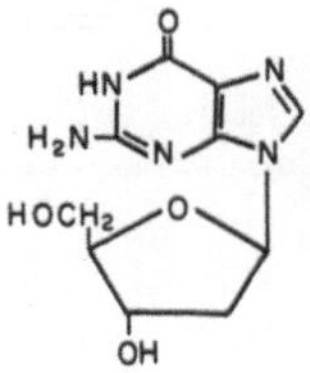


DEOXYGUANOSINE

J. Am. Chem. Soc. 83:2909 (1961); 84:4464 (1962)

▶ $N_5O_5C_{10}H_{13}$

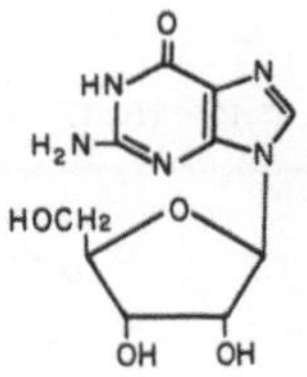


GUANOSINE

J. Am. Chem. Soc. 83:2909 (1961); 84:4464 (1962)

▶ $N_5O_6C_{18}H_{12} \cdot M$

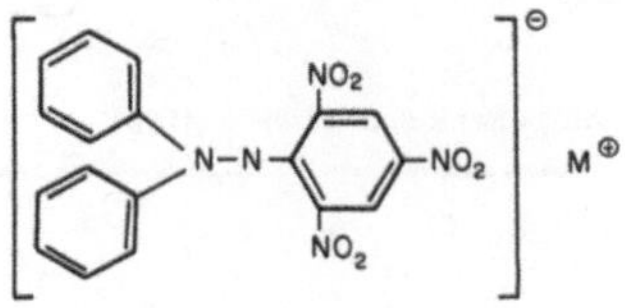


M = Li, Na, K, Rb

J. Org. Chem. 27:1248 (1962)

▶ $N_5O_7C_{16}H_{19}$

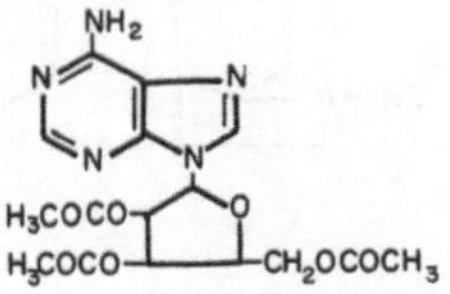


Varian 326

▶ $N_5O_{19}C_{33}H_{55}$

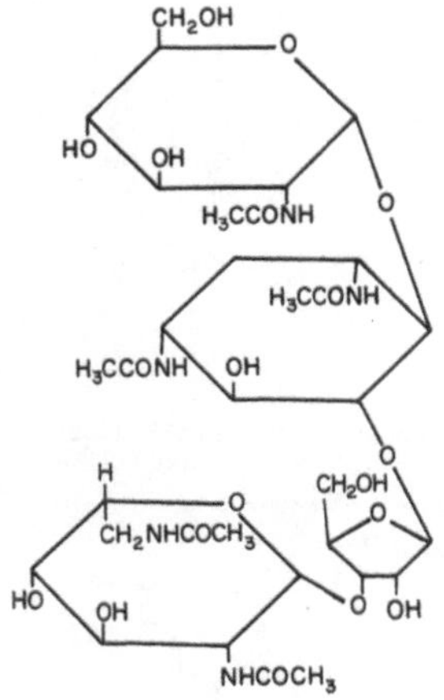


J. Am. Chem. Soc. 84:3216 (1962)

▶ $N_5ClC_9H_{10}$

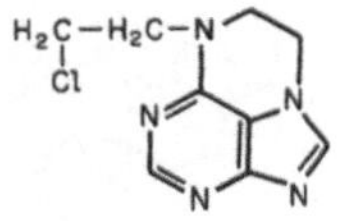


Can. J. Chem. 40:563 (1962)

▶ $N_5C_5H_5$

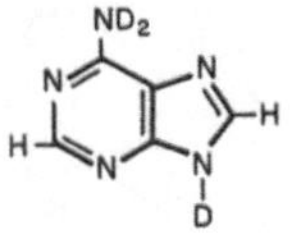


Varian 429

▶ $N_5C_{10}H_{13}$

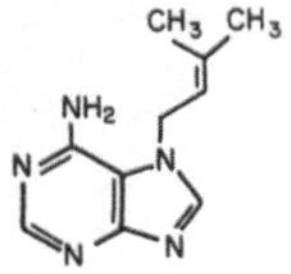


J. Am. Chem. Soc. 84:2148 (1962)

► $N_5C_{10}H_{13}$

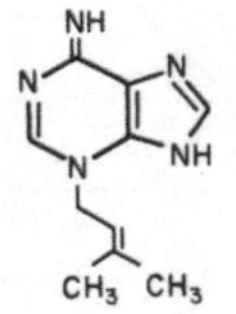

Ann. Pharm. Franc. 20:285 (1962)
J. Am. Chem. Soc. 84:2148 (1962)

► $N_5C_{10}H_{13} \cdot HCl$

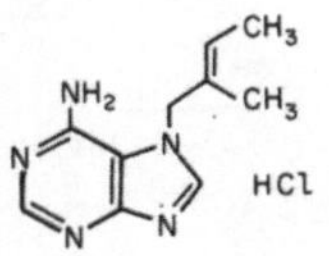

J. Am. Chem. Soc. 84:2148 (1962)

► $N_5C_{10}H_{14} \cdot Cl$

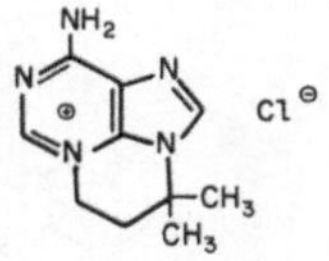

J. Am. Chem. Soc. 84:2148 (1962)

► $N_5C_{10}H_{15}$

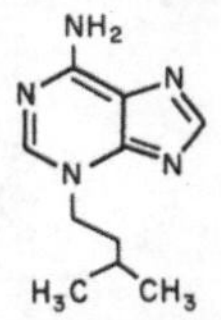

J. Am. Chem. Soc. 84:2148 (1962)

► $N_5C_{17}H_{19}$

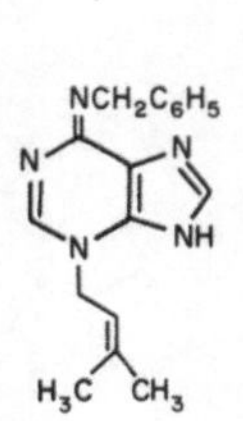

J. Am. Chem. Soc. 84:2148 (1962)

► $N_6OC_5H_4$

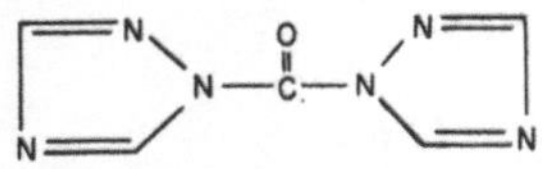

J. Org. Chem. 27:2632 (1962)

► $N_6O_2C_{10}H_{14}$

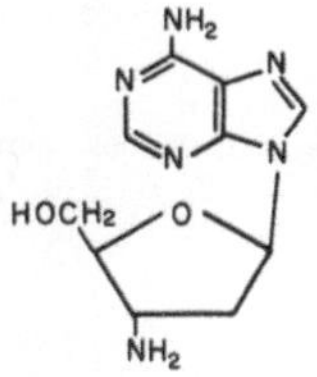

J. Am. Chem. Soc. 83:1906 (1961)

► $N_6O_2C_{10}H_{14}$

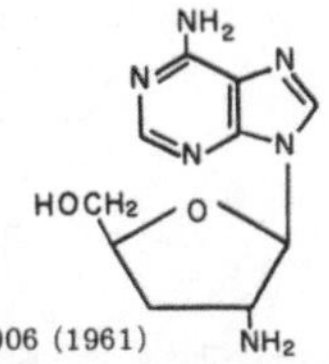

J. Am. Chem. Soc. 83:1906 (1961)

► $N_6O_2C_{40}H_{32} \cdot Ni$

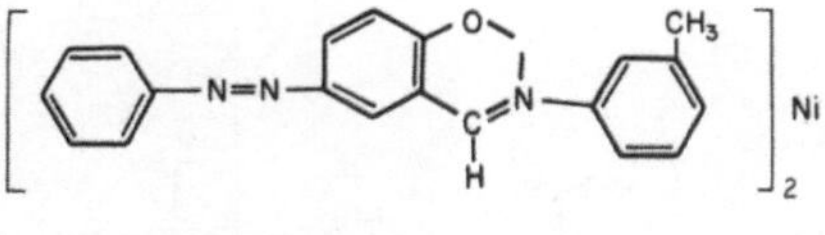

J. Am. Chem. Soc. 84:3968 (1962)

► $N_8O_5C_{40}H_{42} \cdot Cl_2$

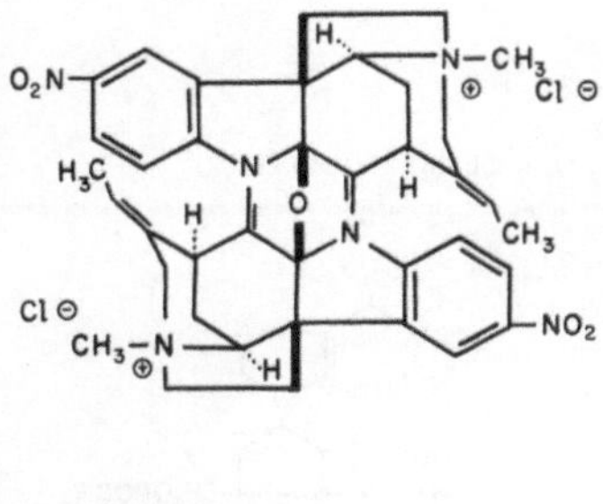

Tetrahedron 14:144 (1961)

▶ $N_6O_{19}C_{35}H_{52}$

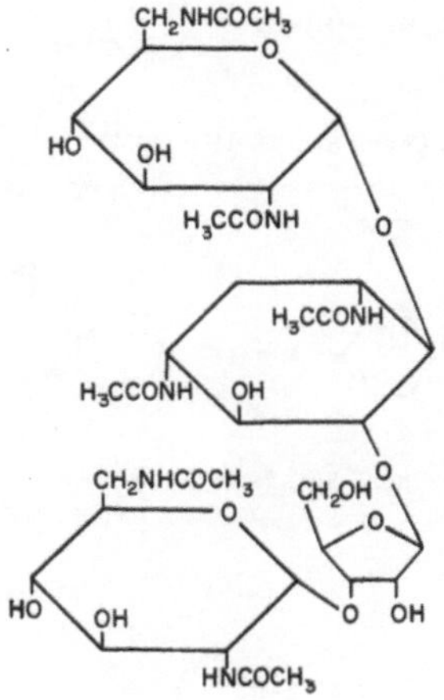


J. Am. Chem. Soc. 84:3216 (1962)

▶ $N_6O_{19}C_{35}H_{58}$

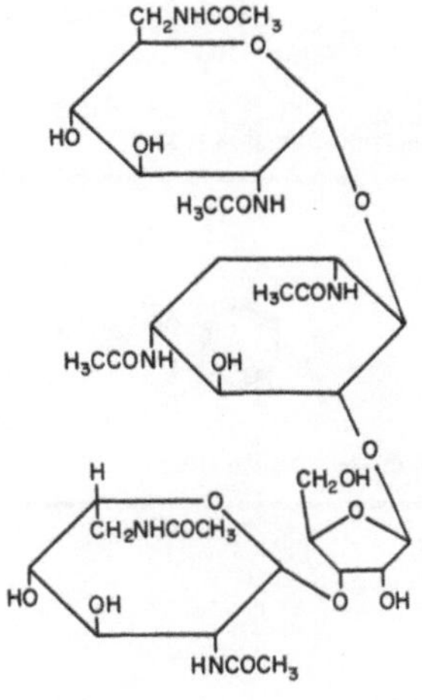


J. Am. Chem. Soc. 84:3216 (1962)

▶ $N_6C_8H_{18} \cdot 2HNO_3$

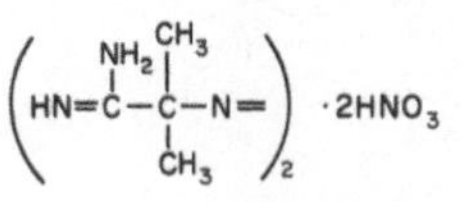


J. Am. Chem. Soc. 84:1506 (1962)

▶ $N_6C_{12}H_{22} \cdot 2HNO_3$

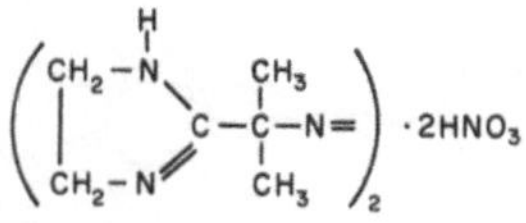


J. Am. Chem. Soc. 84:1506 (1962)

▶ $N_6C_{15}H_{14}$

TRI(METHYLPYRAZINE)

J. Org. Chem. 26:4984 (1961)

▶ $N_7O_5C_{22}H_{29} \cdot 2HCl$

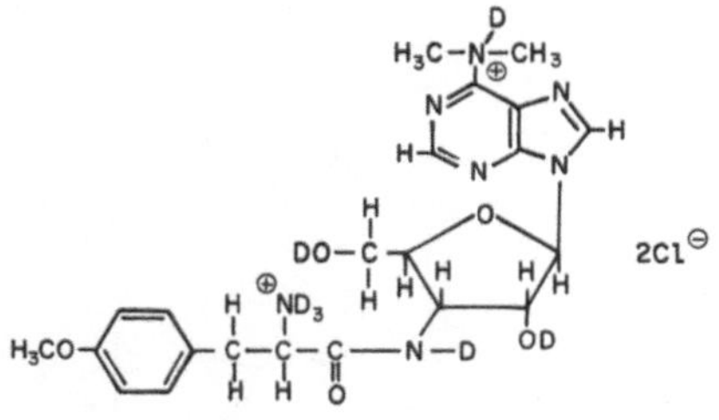


Varian 691

▶ $N_8F_2C_{34}H_{36} \cdot Ni$

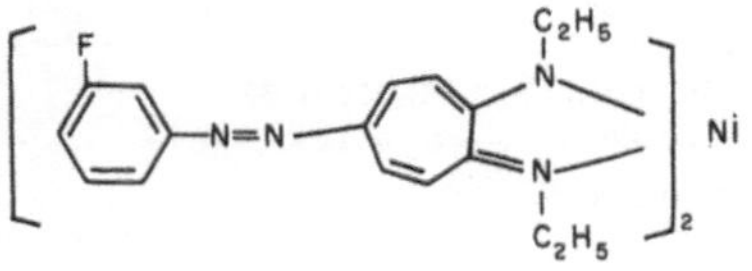


J. Am. Chem. Soc. 84:4100 (1962)

▶ $N_8F_2C_{34}H_{36} \cdot Ni$

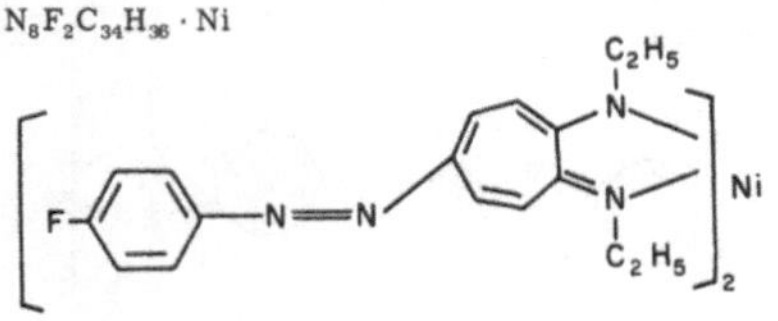


J. Am. Chem. Soc. 84:4100 (1962)

▶ $N_8F_2C_{34}H_{36} \cdot Ni$

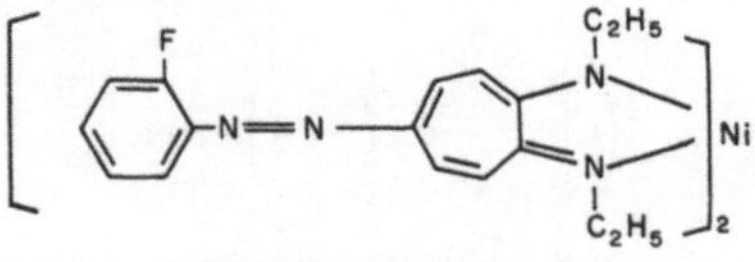

J. Am. Chem. Soc. 84:4100 (1962)

▶ $N_8C_{32}H_{18}$

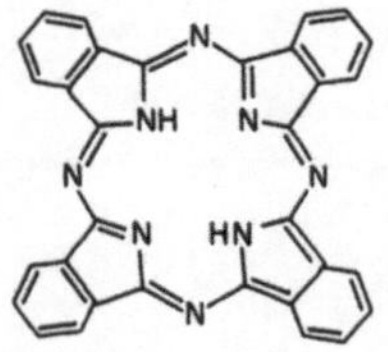

J. Am. Chem. Soc. 83:3743 (1961)

▶ $N_8C_{36}H_{42} \cdot Ni$

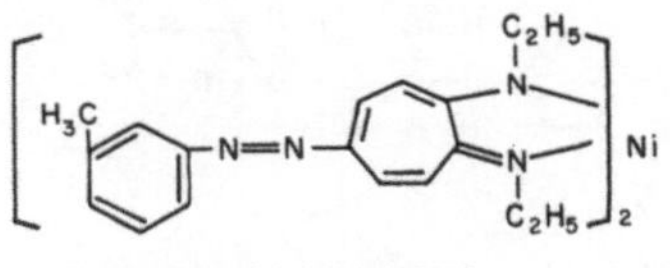

J. Am. Chem. Soc. 84:4100 (1962)

▶ $N_8C_{36}H_{42} \cdot Ni$

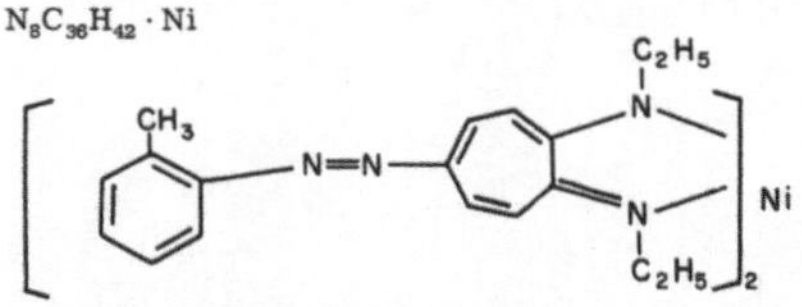

J. Am. Chem. Soc. 84:4100 (1962)

▶ $N_8C_{42}H_{42} \cdot Ni$

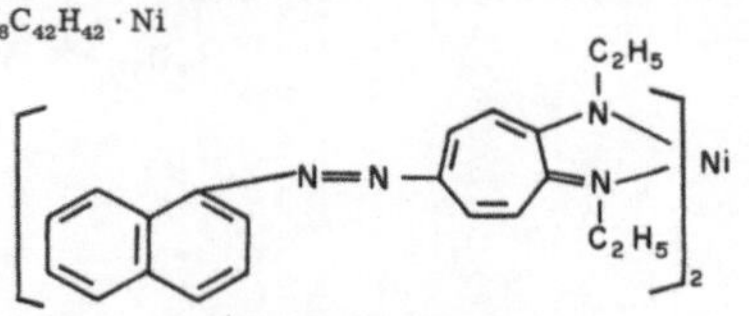

J. Am. Chem. Soc. 84:4100 (1962)

▶ $N_8C_{42}H_{42} \cdot Ni$

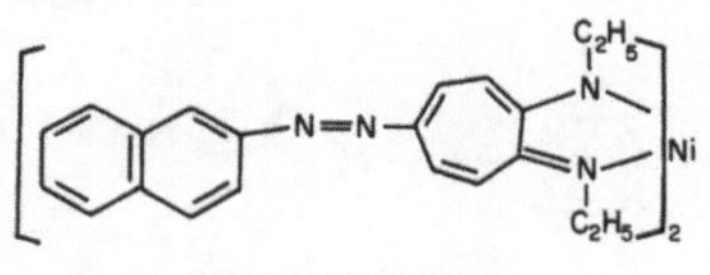

J. Am. Chem. Soc. 84:4100 (1962)

▶ $N_8C_{50}H_{38} \cdot Ni$

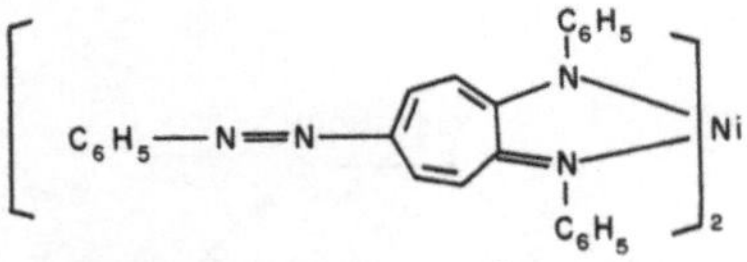

J. Am. Chem. Soc. 84:4100 (1962)

▶ $N_{10}O_4C_{34}H_{36} \cdot Ni$

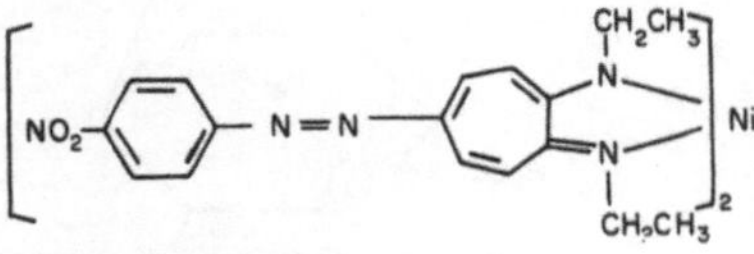

J. Am. Chem. Soc. 84:4100 (1962)

▶ $OIC_6H_5$

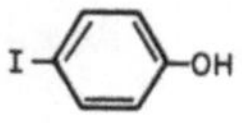

Can. J. Chem. 40:2122 (1962)

▶ $OIC_7H_7$

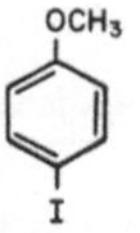

Varian 153
J. Chem. Phys. 37:2594 (1962)

▶ $OIC_7H_7$

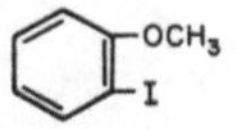

Can. J. Chem. 40:1866 (1962)

▶ $OIC_{21}H_{33}$

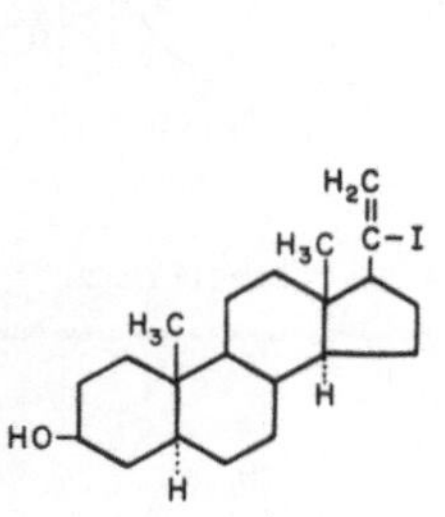

J. Chem. Soc. 474 (1962)

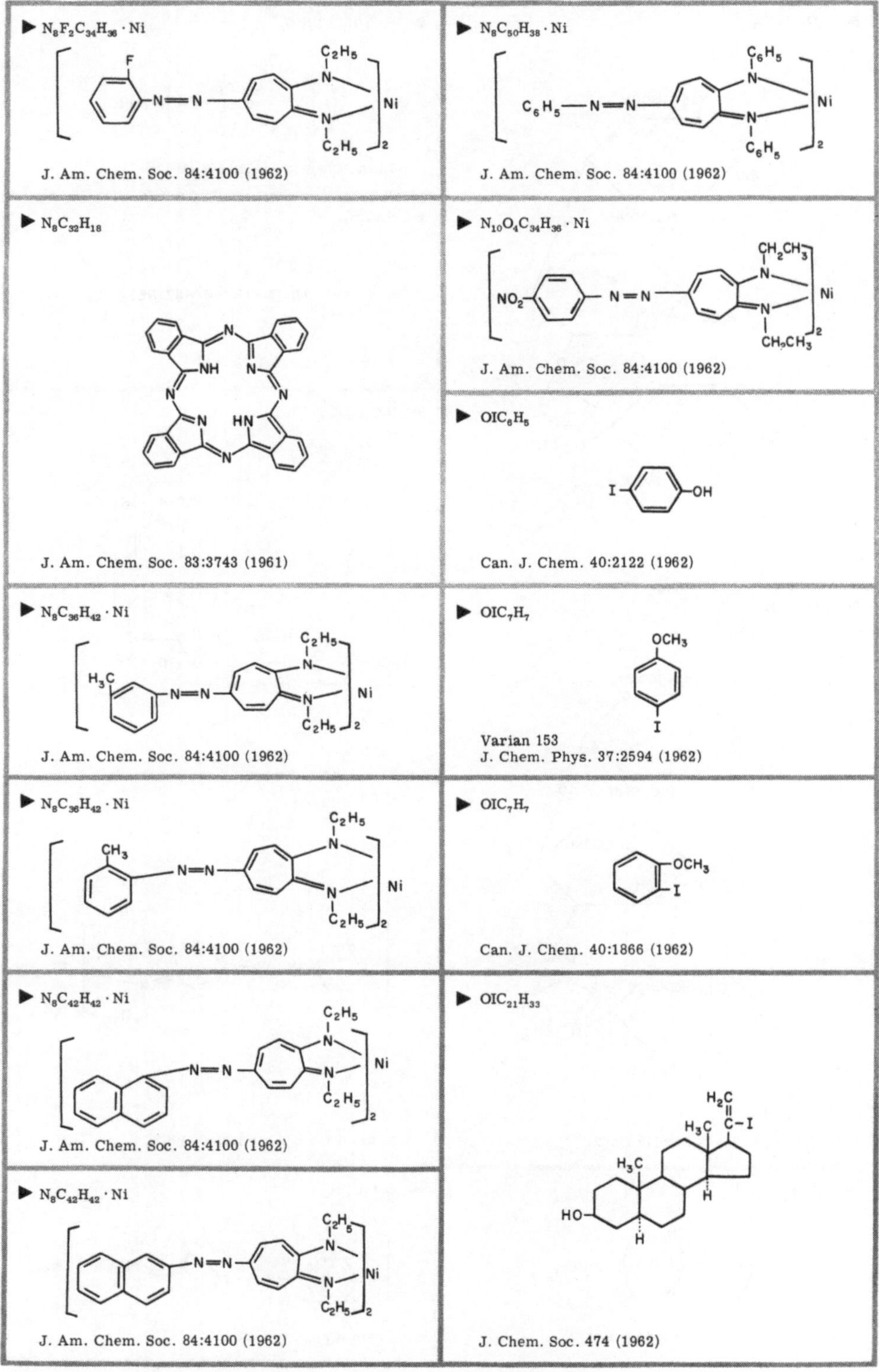

▶ $OBrClC_{27}H_{44}$

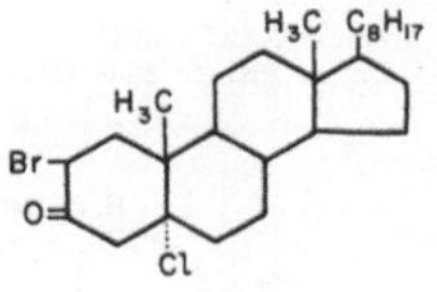


Bull. Soc. Chim. France 2444 (1961)

▶ $OBrC_7H_5$

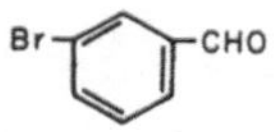


Can. J. Chem. 40:1073 (1962)

▶ $OBrC_7H_7$

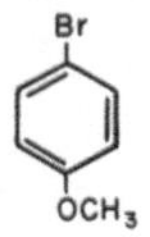


J. Chem. Phys. 37:2594 (1962)
Helv. Chim. Acta 45:560 (1962)

▶ $OBrClC_{27}H_{44}$

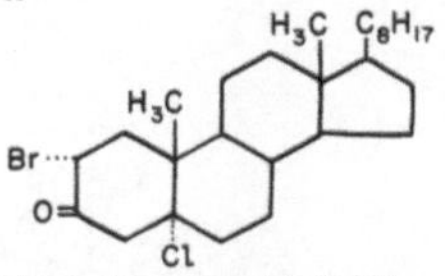


Bull. Soc. Chim. France 2444 (1961)

▶ $OBrC_7H_7$

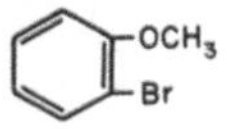


Can. J. Chem. 40:1866 (1962)

▶ $OBrCl_2C_{27}H_{43}$

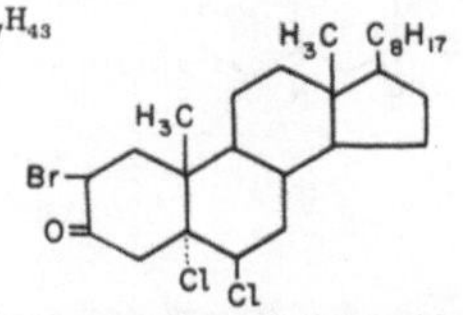


Bull. Soc. Chim. France 2444 (1961); 1866 (1962)

▶ $OBrC_7H_7$

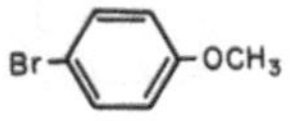


Can. J. Chem. 40:1866 (1962)

▶ $OBrF_6C_4H$

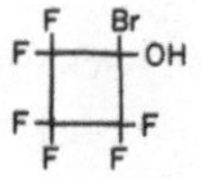


J. Am. Chem. Soc. 83:4670 (1961)

▶ $OBrC_8H_9$

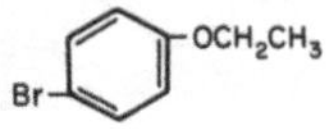


Varian 198

▶ $OBrC_4H_7$

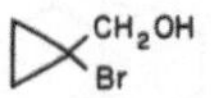


J. Org. Chem. 27:2722 (1962)

▶ $OBrC_8H_{11}$

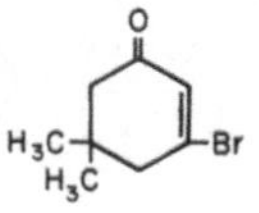


Tetrahedron Letters No. 11:473 (1962)

▶ $OBrC_6H_5$

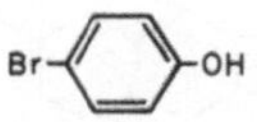


Can. J. Chem. 40:2122 (1962)
Helv. Chim. Acta 45:568 (1962)

▶ $OBrC_8H_{13}$

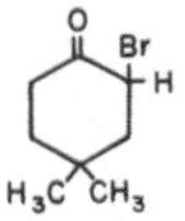


Tetrahedron 18:791 (1962)

▶ $OBrC_{10}H_{15}$

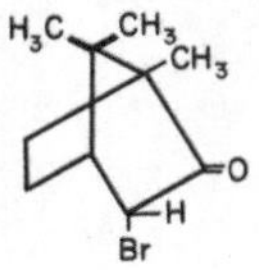

J. Am. Chem. Soc. 83:4095 (1961)

▶ $OBrC_{10}H_{15}$

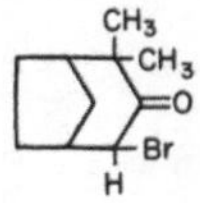

J. Org. Chem. 26:707 (1961)

▶ $OBrC_{10}H_{15}$

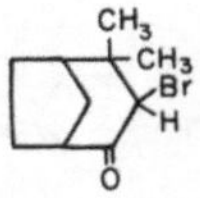

J. Org. Chem. 26:707 (1961)

▶ $OBrC_{14}H_{11}$

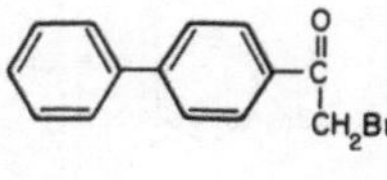

Varian 303

▶ $OBrC_{16}H_{11}$

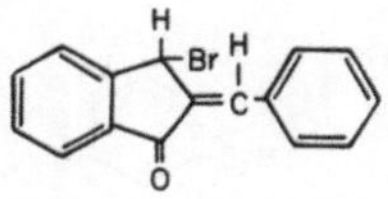

J. Org. Chem. 27:3039 (1962)

▶ $OBrC_{27}H_{44}D$

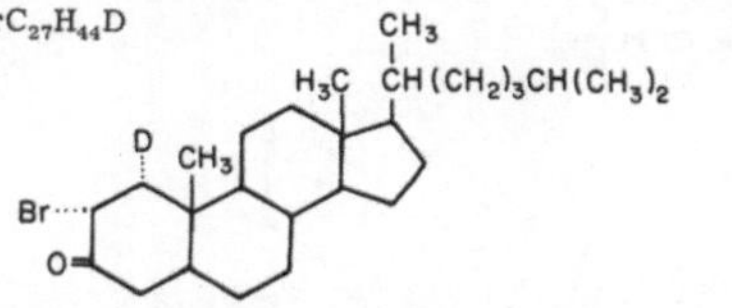

Tetrahedron Letters No. 15:650 (1962)

▶ $OBrC_{27}H_{45}$

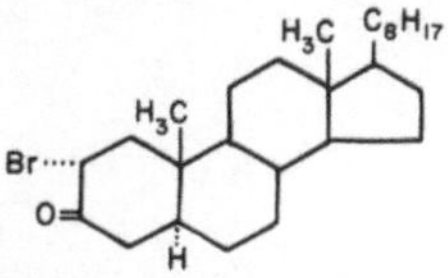

J. Org. Chem. 27:4587 (1962)
Bull. Soc. Chim. France 2444 (1961)

▶ $OBrC_{27}H_{45}$

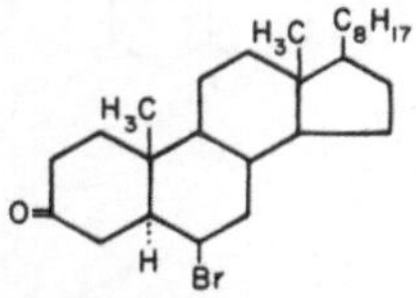

Bull. Soc. Chim. France 2444 (1961)

▶ $OBrC_{27}H_{46}D$

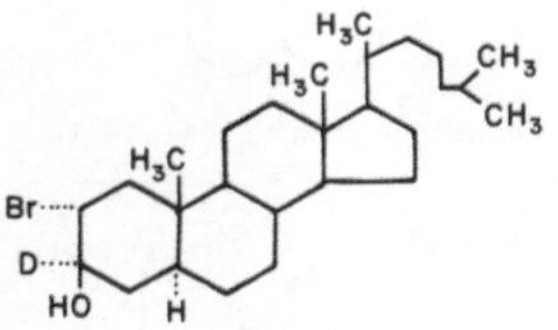

J. Am. Chem. Soc. 83:4623 (1961)

▶ $OBrC_{28}H_{21}$

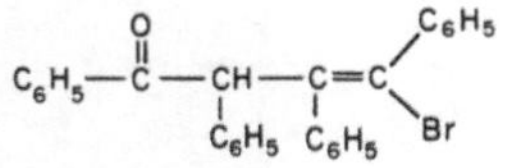

J. Am. Chem. Soc. 84:4166 (1962)

▶ $OBrC_{28}H_{47}$

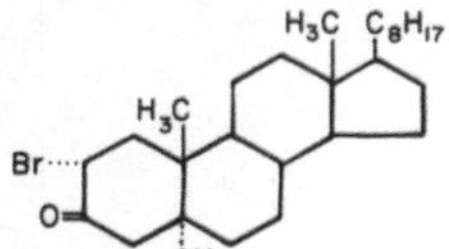

Bull. Soc. Chim. France 2444 (1961)

▶ $OBrC_{28}H_{47}$

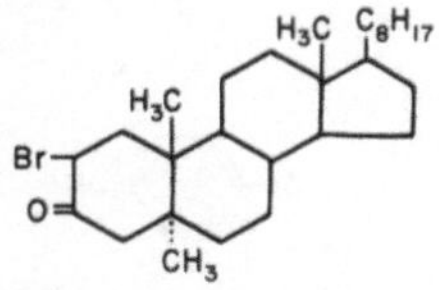

Bull. Soc. Chim. France 2444 (1961)

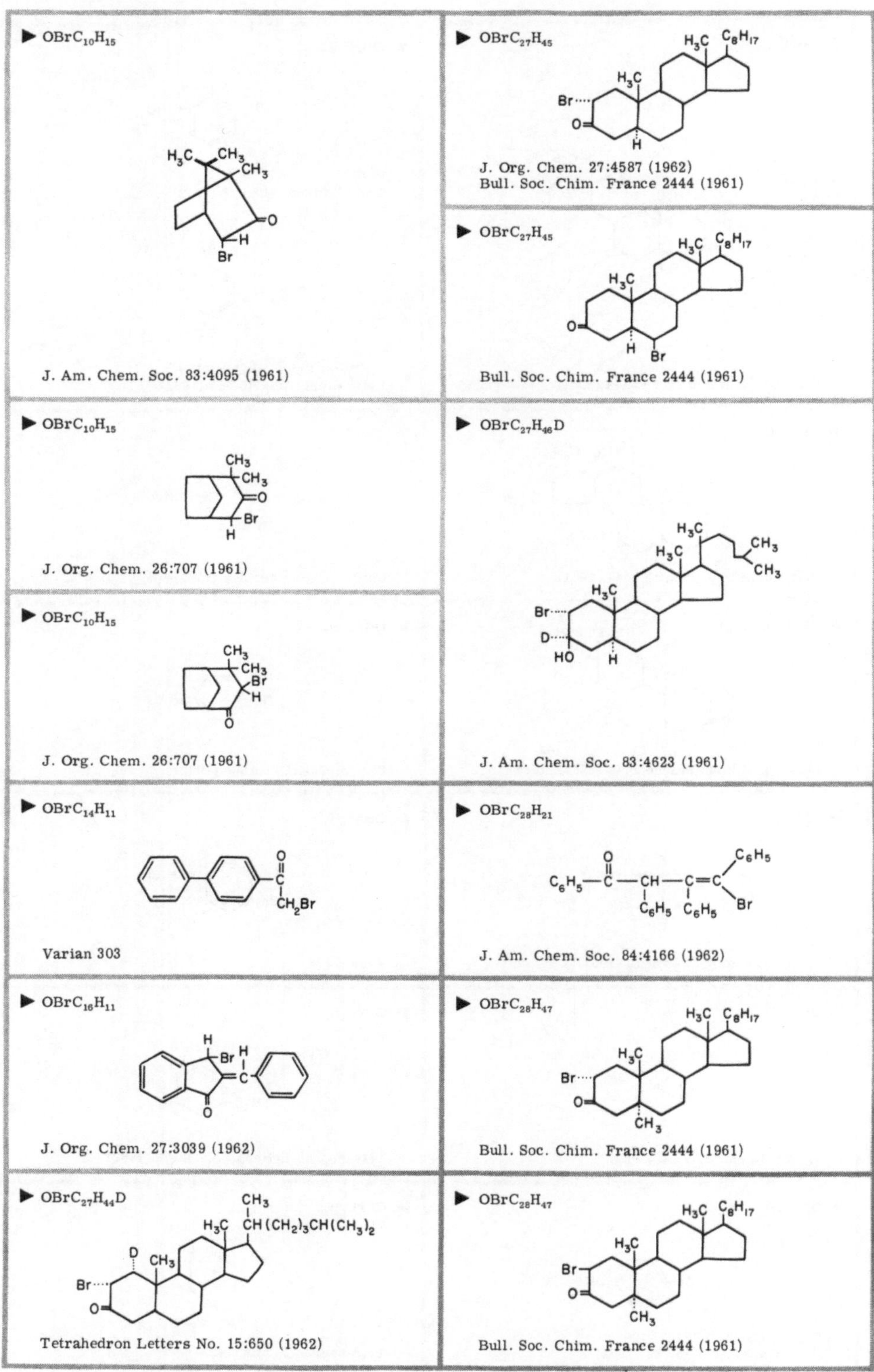

▶ $OBrC_{29}H_{47}$

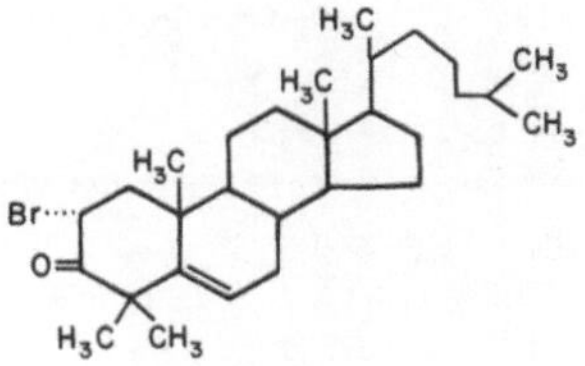


Chem. & Ind. 1667 (1961)

▶ $OBr_2ClC_{27}H_{43}$

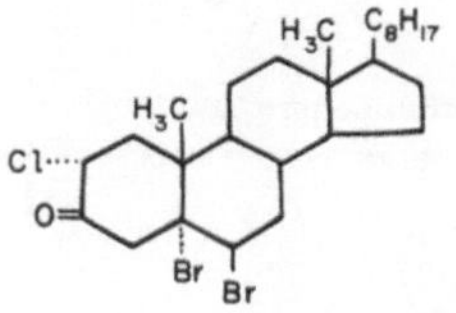


Bull. Soc. Chim. France 2444 (1961); 1866 (1962)

▶ $OBr_2C_5H_8$

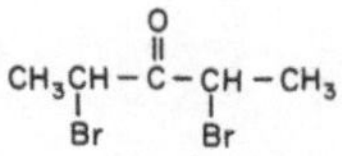


Acta Chem. Scand. 16:2060 (1962)

▶ $OBr_2C_{10}H_{14}$

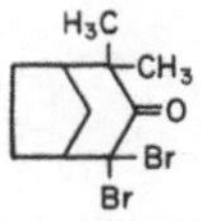


J. Org. Chem. 26:707 (1961)

▶ $OBr_2C_{10}H_{14}$

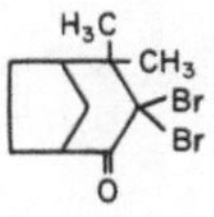


J. Org. Chem. 26:707 (1961)

▶ $OBr_2C_{10}H_{14}$

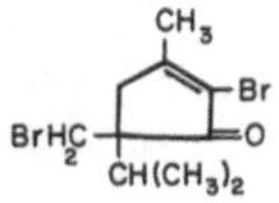


Varian 564

▶ $OBr_2C_{27}H_{44}$

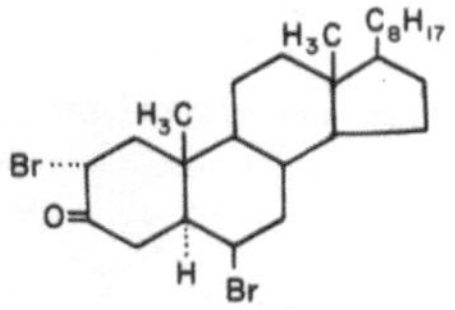


Bull. Soc. Chim. France 2444 (1961)

▶ $OBr_2C_{27}H_{44}$

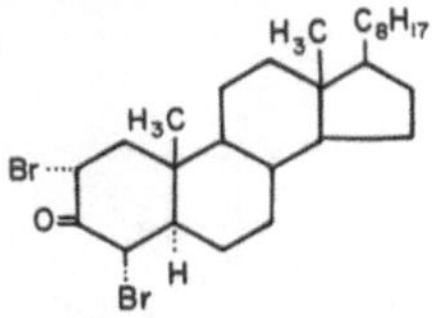


Bull. Soc. Chim. France 2444 (1961)

▶ $OBr_2C_{27}H_{44}$

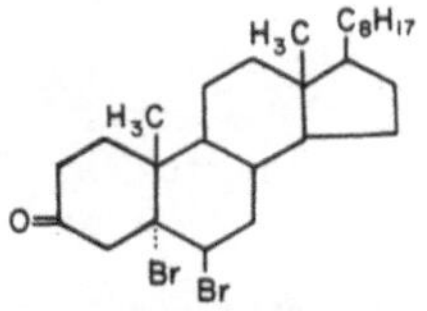


Bull. Soc. Chim. France 2444 (1961); 1866 (1962)

▶ $OBr_2C_{30}H_{48}$

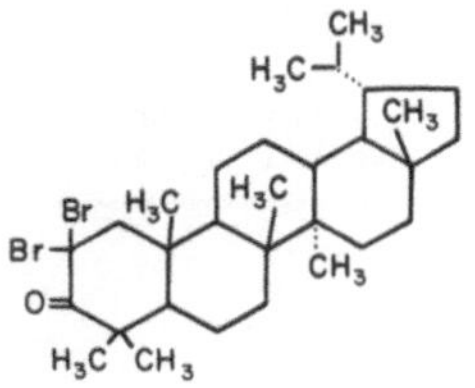


Bull. Soc. Chim. France 1137 (1962)

▶ $OBr_3C_6H_3$

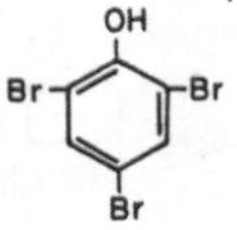

Can. J. Chem. 40:1759 (1962)

▶ $OBr_3C_{27}H_{43}$

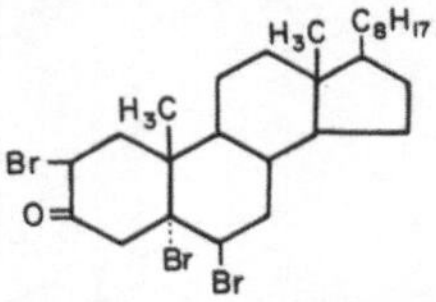

Bull. Soc. Chim. France 2444 (1961); 1866 (1962)

▶ $OBr_3C_{27}H_{43}$

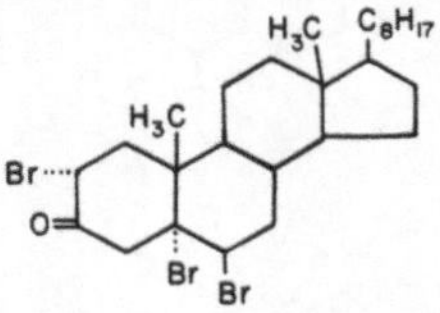

Bull. Soc. Chim. France 2444 (1961); 1866 (1962)

▶ $OClF_6C_4H$

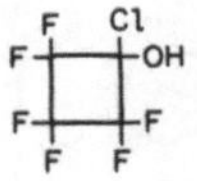

J. Am. Chem. Soc. 83:4670 (1961)

▶ $OClC_2H_5$

$CH_2ClCH_2OH$

Varian 12
Bull. Soc. Chim. France 754 (1962)
J. Mol. Spectr. 7:32 (1961)

▶ $OClC_3H_5$

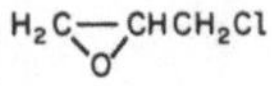

J. Chem. Phys. 35:1521 (1961)
J. Chem. Phys. 37:3012 (1962)

▶ $OClC_4H_5$

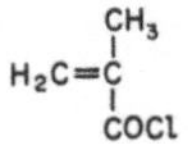

Varian 400
Arkiv Kemi 19:1 (1961)

▶ $OClC_4H_7$

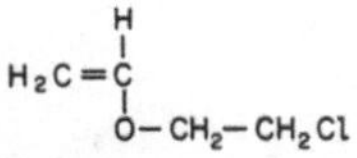

J. Chem. Soc. 2023 (1962)

▶ $OClC_6H_5$

Can. J. Chem. 40:2122 (1962)
Helv. Chim. Acta 45:568 (1962)

▶ $OClC_7H_5$

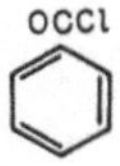

Tetrahedron Letters 519 (1961)

▶ $OClC_7H_5$

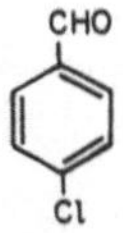

Varian 146
Can. J. Chem. 40:1073, 2331 (1962)
J. Chem. Phys. 37:2594 (1962)
J. Am. Chem. Soc. 84:2649 (1962)

▶ $OClC_7H_5$

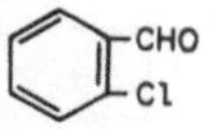

Can. J. Chem. 40:1073 (1962)
J. Am. Chem. Soc. 84:2649 (1962)

▶ $OClC_7H_5$

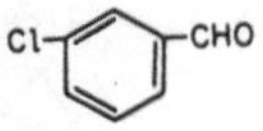

Can. J. Chem. 40:1073 (1962)
J. Am. Chem. Soc. 84:2649 (1962)

▶ $OClC_7H_7$

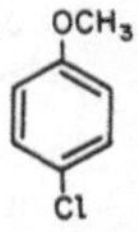


Can. J. Chem. 40:1866 (1962)
J. Chem. Phys. 37:2594 (1962)

▶ $OClC_8H_7$

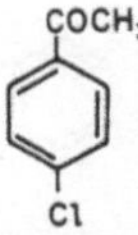


Varian 188
J. Chem. Phys. 37:2594 (1962)

▶ $OClC_8H_9$

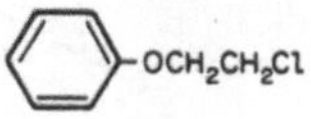


Varian 199

▶ $OClC_9H_{15}$

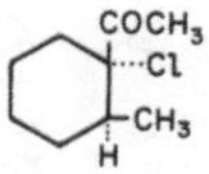


J. Am. Chem. Soc. 83:3980 (1961)

▶ $OClC_9H_{15}$

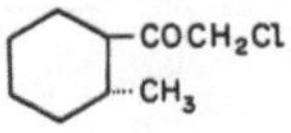


J. Am. Chem. Soc. 83:3980 (1961)

▶ $OClC_{12}H_{15}$

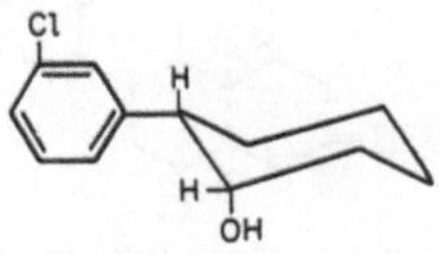


J. Org. Chem. 27:717 (1962)

▶ $OClC_{12}H_{15}$

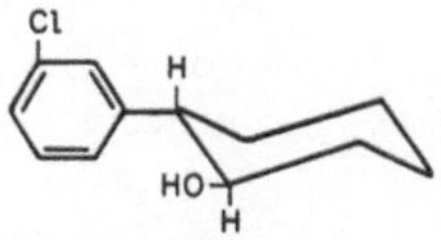


J. Org. Chem. 27:717 (1962)

▶ $OClC_{12}H_{15}$

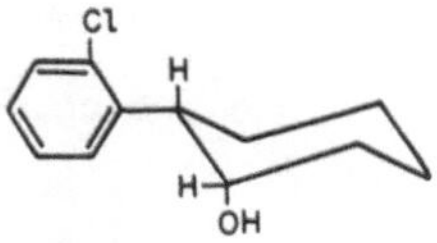


J. Org. Chem. 27:717 (1962)

▶ $OClC_{12}H_{15}$

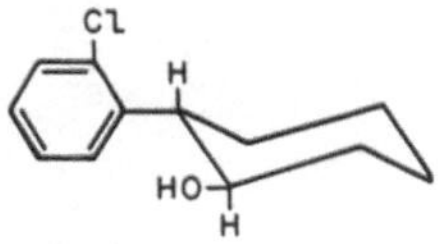


J. Org. Chem. 27:717 (1962)

▶ $OClC_{12}H_{15}$

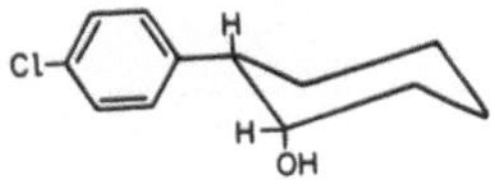


J. Org. Chem. 27:717 (1962)

▶ $OClC_{12}H_{15}$

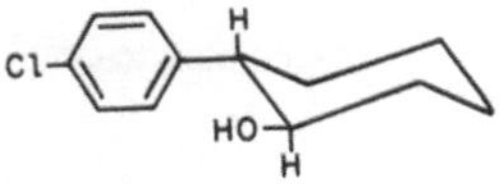


J. Org. Chem. 27:717 (1962)

▶ $OClC_{27}H_{45}$

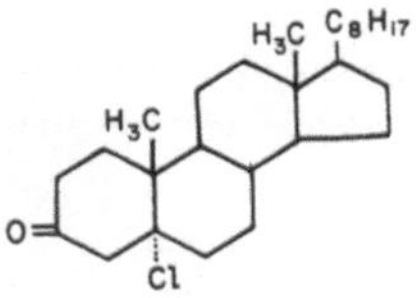


Bull. Soc. Chim. France 2444 (1961)

▶ $OCl_2C_2H_4$

$CHCl_2CH_2OH$

J. Mol. Spectr. 7:32 (1961)
Bull. Soc. Chim. France 754 (1962)

▶ $OCl_2C_3H_6$

$ClCH_2\overset{OH}{\overset{|}{C}}HCH_2Cl$

Varian 386

▶ $OCl_2C_3H_9AlSi$

$(CH_3)_3SiOAlCl_2$

J. Am. Chem. Soc. 84:1069 (1962)

▶ $OCl_2C_4H_6$

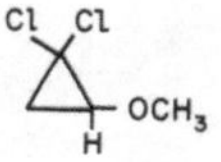

J. Am. Chem. Soc. 84:2249 (1962)

▶ $OCl_2C_4H_8$

$CH_3CH_2OCHClCH_2Cl$

Varian 75

▶ $OCl_2C_5H_8$

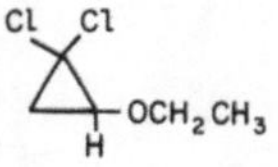

J. Am. Chem. Soc. 84:2249 (1962)

▶ $OCl_2C_7H_4$

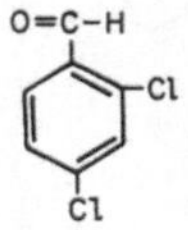

J. Chem. Phys. 37:1009 (1962)

▶ $OCl_2C_7H_4$

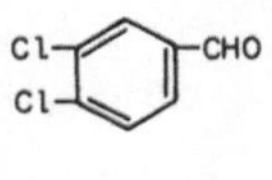

Can. J. Chem. 40:1073 (1962)

▶ $OCl_2C_7H_4$

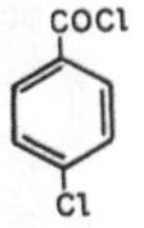

J. Chem. Phys. 37:2594 (1962)

▶ $OCl_2C_7H_6$

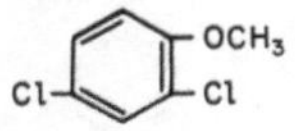

Can. J. Chem. 40:1866 (1962)

▶ $OCl_2C_{10}H_8$

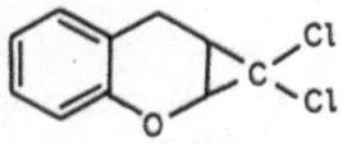

J. Am. Chem. Soc. 84:813 (1962)

▶ $OCl_2C_{10}H_8$

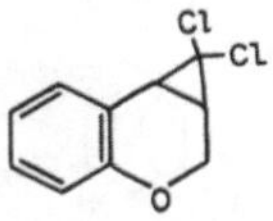

J. Am. Chem. Soc. 84:813 (1962)

▶ $OCl_2C_{10}H_{10}$

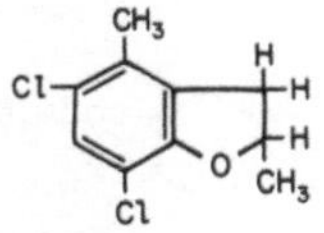

J. Org. Chem. 27:2726 (1962)

▶ $OCl_2C_{27}H_{44}$

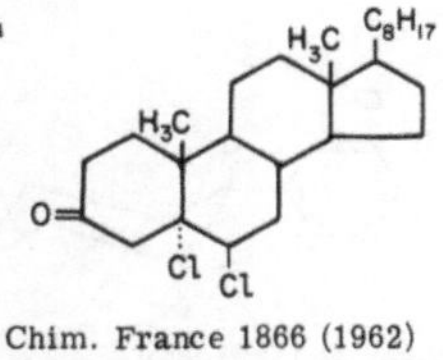

Bull. Soc. Chim. France 1866 (1962)

▶ $OCl_3C_2H$

$\overset{O}{\overset{\|}{HC}}{}^{13}CCl_3$

J. Am. Chem. Soc. 84:2649 (1962)

► $OCl_3C_2H_3$

$CCl_3CH_2OH$

Bull. Soc. Chim. France 754 (1962)

► $OCl_3C_6H_3$

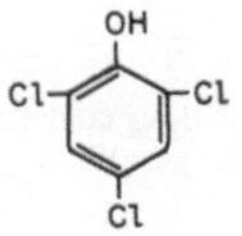

Can. J. Chem. 40:1759 (1962)

► $OCl_3C_{20}H_{33}$

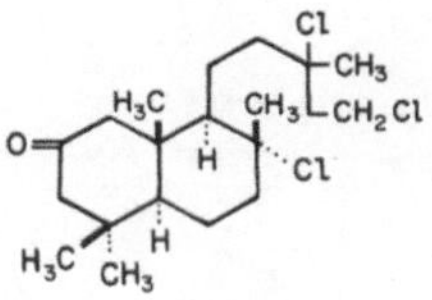

Tetrahedron 18:285 (1962)

► OFCH

$HC^{13}(=O)F$

J. Am. Chem. Soc. 84:2649 (1962)

► $OFC_2H_3$

$CH_3C(=O)F$

J. Am. Chem. Soc. 84:2733 (1962)

► $OFC_2H_3 \cdot AsF_5$

$CH_3COF \cdot AsF_5$

J. Am. Chem. Soc. 84:2733 (1962)

► $OFC_2H_3 \cdot BF_3$

$CH_3C(=O)F \cdot BF_3$

J. Am. Chem. Soc. 84:2733 (1962)

► $OFC_2H_3 \cdot PF_5$

$CH_3C(=O)F \cdot PF_5$

J. Am. Chem. Soc. 84:2733 (1962)

► $OFC_2H_3 \cdot SbF_5$

$CH_3C(=O)F \cdot SbF_5$

J. Am. Chem. Soc. 84:2733 (1962)

► $OFC_3H_3$

$CH_2{=}CH{-}COF$

J. Chem. Phys. 35:1900 (1961)

► $OFC_3H_5$

$CH_3CH_2COF$

J. Am. Chem. Soc. 84:2733 (1962)

► $OFC_3H_5 \cdot AsF_5$

$CH_3CH_2COF \cdot AsF_5$

J. Am. Chem. Soc. 84:2733 (1962)

► $OFC_3H_5 \cdot BF_3$

$CH_3CH_2COF \cdot BF_3$

J. Am. Chem. Soc. 84:2733 (1962)

▶ $OFC_3H_5 \cdot PF_5$

$CH_3CH_2COF \cdot PF_5$

J. Am. Chem. Soc. 84:2733 (1962)

▶ $OFC_3H_5 \cdot SbF_5$

$CH_3CH_2COF \cdot SbF_5$

J. Am. Chem. Soc. 84:2733 (1962)

▶ $OFC_7H_5$

$C_6H_5COF$

J. Am. Chem. Soc. 84:2733 (1962)

▶ $OFC_7H_5$

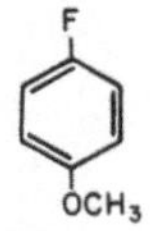

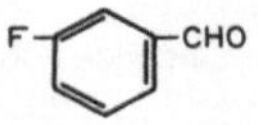

Can. J. Chem. 40:1073 (1962)

▶ $OFC_7H_5$

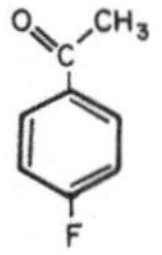

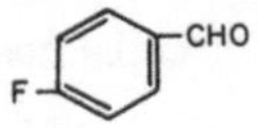

Can. J. Chem. 40:1073, 2331 (1962)

▶ $OFC_7H_5 \cdot AsF_5$

$C_6H_5COF \cdot AsF_5$

J. Am. Chem. Soc. 84:2733 (1962)

▶ $OFC_7H_5 \cdot BF_3$

$C_6H_5COF \cdot BF_3$

J. Am. Chem. Soc. 84:2733 (1962)

▶ $OFC_7H_5 \cdot PF_5$

$C_6H_5COF \cdot PF_5$

J. Am. Chem. Soc. 84:2733 (1962)

▶ $OFC_7H_5 \cdot SbF_5$

$C_6H_5COF \cdot SbF_5$

J. Am. Chem. Soc. 84:2733 (1962)

▶ $OFC_7H_7$

Helv. Chim. Acta 45:568 (1962)

▶ $OFC_8H_7$

Varian 189

▶ $OFC_8H_9$

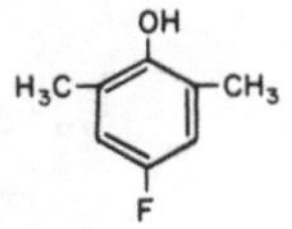

Chem. & Ind. 460 (1962)

▶ $OFC_8H_9$

Varian 200

▶ $OFC_{19}H_{27}$

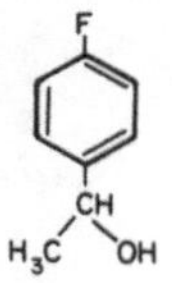

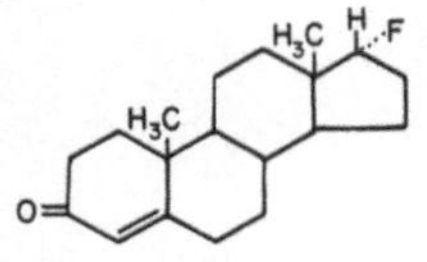

Tetrahedron Letters No. 26:1251 (1962)

▶ $OF_2C_7H_6$

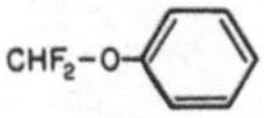


J. Chem. Soc. 3829 (1962)

▶ $OF_2C_8H_8$

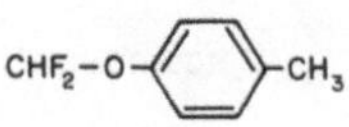


J. Chem. Soc. 3829 (1962)

▶ $OF_2C_{21}H_{32}$

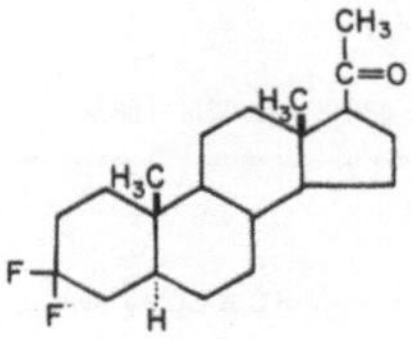


J. Org. Chem. 26:2438 (1961)

▶ $OF_2C_{21}H_{32}$

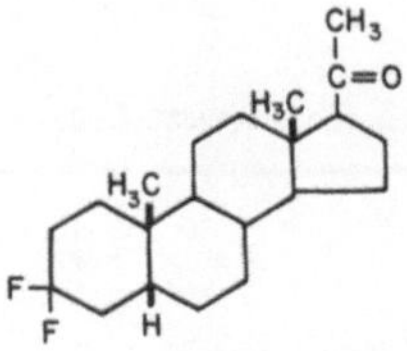


J. Org. Chem. 26:2438 (1961)

▶ $OF_3C_2H_3$

$F_3CCH_2OH$

Varian 4
Can. J. Chem. 40:387 (1962)

▶ $OF_3C_{10}H_9$

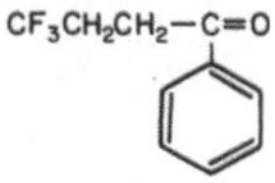


J. Am. Chem. Soc. 84:797 (1962)

▶ $OF_4C_3H_4$

$HCF_2CF_2CH_2OH$

Varian 19

▶ $OF_7C_4H$

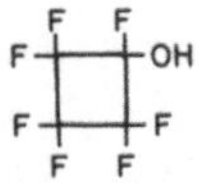


J. Am. Chem. Soc. 83:4670 (1961)

▶ $OF_7C_4H_3$

$C_3F_7CH_2OH$

J. Org. Chem. 26:357 (1961)

▶ $OF_8C_5H_4$

$H(CF_2)_4CH_2OH$

J. Org. Chem. 26:357 (1961)

▶ $OCH_2$

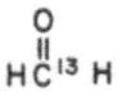


J. Am. Chem. Soc. 84:2649 (1962)

▶ $OCH_2$

$CH_2O$ in $D_2O$ solution

Chem. Ber. 94:3317 (1961)

▶ $OCH_3D$

$CH_2DOH$

J. Chem. Phys. 37:3012 (1962)
J. Am. Chem. Soc. 84:3768 (1962)

▶ $OCH_4$

$CH_3OH$

Varian 1
J. Mol. Spectr. 7:32 (1961)
J. Am. Chem. Soc. 84:4664 (1962)
J. Chem. Phys. 34:1099 (1961); 37:2198 (1962)
Gazz. Chim. Ital. 92:1125 (1962)

▶ $OC_2H_4$

$CH_3CHO$

Varian 6
J. Phys. Chem. 66:2653 (1962)
J. Chem. Phys. 34:1099 (1961); 37:2198 (1962)
J. Am. Chem. Soc. 83:4095 (1961); 84:2649 (1962)

▶ $OC_2H_4$

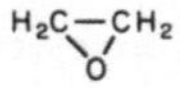


J. Chem. Phys. 36:3097 (1962)

▶ $OC_2H_6$

$CH_3OCH_3$

J. Chem. Phys. 34:1099 (1961)
J. Am. Chem. Soc. 84:3972 (1962)

▶ $OC_2H_6$

$H_3CCH_2OH$

Varian 14
J. Chem. Soc. 849 (1962)
Can. J. Chem. 40:387, 1522, 1956 (1962)
Bull. Soc. Chim. France 1:754 (1962)
Gazz. Chim. Ital. 92:1125 (1962)
J. Chem. Phys. 34:1099 (1961)

▶ $OC_2H_7$

$C_2H_5OH_2^{\oplus}$

J. Chem. Phys. 34:2207 (1961)

▶ $OC_3HD_5$

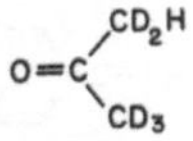


Varian 388

▶ $OC_3H_2$

$HC{\equiv}CCHO$

J. Am. Chem. Soc. 83:1978 (1961)
Rec. Trav. Chim. 81:635 (1962)

▶ $OC_3H_4$

$HC{\equiv}COCH_3$

Rec. Trav. Chim. 81:635 (1962)

▶ $OC_3H_4$

$HC{\equiv}CCH_2OH$

Varian 21
Arkiv Kemi 16:471 (1961)
J. Mol. Spectr. 7:32 (1961)
Rec. Trav. Chim. 81:635 (1962)
J. Am. Chem. Soc. 83:1978 (1961)

▶ $OC_3H_4D_4$

$CHD_2CD_2CH_2OH$

J. Am. Chem. Soc. 84:2838 (1962)

▶ $OC_3H_4D_4$

$CD_3CHDCH_2OH$

J. Am. Chem. Soc. 84:2838 (1962)

▶ $OC_3H_5D$

$CH_2D{-}COCH_3$

J. Chem. Phys. 37:3012 (1962)

▶ $OC_3H_6$

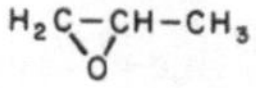

Varian 32
J. Chem. Phys. 36:3097 (1962)

▶ $OC_3H_6$

$H_2C{=}CHOCH_3$

J. Am. Chem. Soc. 83:1300 (1961)
J. Chem. Soc. 2023 (1962)

▶ $OC_3H_6$

$H_2C{=}CHCH_2OH$

Varian 34

▶ $OC_3H_6$

$C_2H_5-CHO$

J. Chem. Phys. 34:1099 (1961)
J. Am. Chem. Soc. 84:2649 (1962)

▶ $OC_3H_6$

$CH_3COCH_3$

J. Am. Chem. Soc. 83:2959, 4726 (1961)
Bull. Chem. Soc. Japan 34:143 (1961); 35:1046 (1962)
J. Chem. Phys. 37:2198 (1962)

▶ $OC_3H_6$

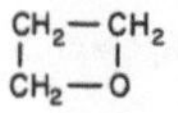

Varian 33

▶ $OC_3H_6D_2$

$CH_3CH_2CD_2OH$

J. Am. Chem. Soc. 84:37 (1962)

▶ $OC_3H_8$

$(CH_3)_2CHOH$

Varian 44
Can. J. Chem. 40:1956 (1962)
J. Chem. Phys. 34:1099 (1961)
J. Am. Chem. Soc. 83:1230 (1961)

▶ $OC_3H_8$

$CH_3CH_2CH_2OH$

Varian 43
J. Chem. Phys. 34:1099 (1961)
J. Mol. Spectr. 7:32 (1961)

▶ $OC_3H_8$

$CH_3CH_2OCH_3$

J. Am. Chem. Soc. 83:4138 (1961)

▶ $OC_3H_8 \cdot BF_3$

$C_2H_5OCH_3 \cdot BF_3$

J. Am. Chem. Soc. 83:4138 (1961)

▶ $OC_4H_4$

$HC{\equiv}COCH{=}CH_2$

Rec. Trav. Chim. 81:635 (1962)

▶ $OC_4H_4$

Varian 50
J. Phys. Chem. 65:1539 (1961)
J. Am. Chem. Soc. 83:5020 (1961); 84:583 (1962)
Can. J. Chem. 39:222 (1961)

▶ $OC_4H_5D$

$CH_2{=}C(CH_3){-}C({=}O)D$

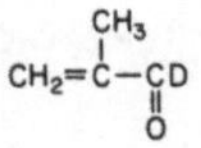

J. Am. Chem. Soc. 83:2198 (1961)

▶ $OC_4H_6$

$H_2C{=}C(CH_3){-}CHO$

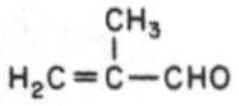

Arkiv Kemi 17:1 (1961)
J. Am. Chem. Soc. 83:2198 (1961)

▶ $OC_4H_6$

$HC{\equiv}CCH_2CH_2OH$

J. Am. Chem. Soc. 83:1978 (1961); 84:347 (1962)
Rec. Trav. Chim. 81:635 (1962)

▶ $OC_4H_6$

$CH_3C{\equiv}CCH_2OH$

J. Mol. Spectr. 7:32 (1961)
Arkiv Kemi 16:471 (1961)

▶ $OC_4H_6$

$CH_3{-}CO{-}CH{=}CH_2$

J. Chem. Phys. 35:1900 (1961); 37:1951 (1962)
J. Phys. Chem. 66:2653 (1962)

▶ $OC_4H_6$

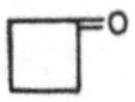

J. Am. Chem. Soc. 83:1226 (1961)

▶ $OC_4H_6$

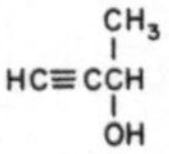

$HC{\equiv}CCH(CH_3)OH$

J. Mol. Spectr. 7:32 (1961)

▶ $OC_4H_8$

$CH_3CH_2CH_2CHO$

Varian 78
J. Chem. Phys. 34:1094, 1099 (1961)

▶ $OC_4H_6$

$HC{\equiv}C{-}CH_2OCH_3$

J. Am. Chem. Soc. 83:1978 (1961)

▶ $OC_4H_8$

Varian 77
Tetrahedron 18:791 (1962)

▶ $OC_4H_6$

$HC{\equiv}COCH_2CH_3$

Rec. Trav. Chim. 81:313, 635 (1962)

▶ $OC_4H_8$

$H_2C{=}CHOCH_2CH_3$

J. Chem. Soc. 2023 (1962)

▶ $OC_4H_6$

$H_3CCH{=}CHCHO$

cis and trans

J. Mol. Phys. 4:385 (1961)
Arkiv Kemi 16:471 (1961)
J. Chem. Phys. 36:2644 (1962)

▶ $OC_4H_8$

$C_2H_5C({=}O)CH_3$

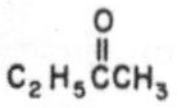

Varian 76
Bull. Chem. Soc. Japan 35:1046 (1962)

▶ $OC_4H_8$

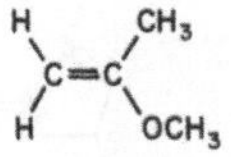

J. Chem. Phys. 36:2644 (1962)
Arkiv Kemi 17:1 (1961)

▶ $OC_4H_8$

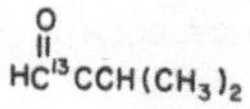

J. Chem. Phys. 34:1099 (1961)
J. Am. Chem. Soc. 84:2649 (1962)

▶ $OC_4H_8$

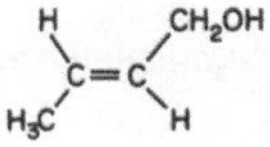

Varian 414

▶ $OC_4H_8 \cdot BF_3$

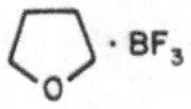

Rec. Trav. Chim. 80:1323 (1961)

▶ $OC_4H_8D_2$

$(CH_3)_2CHC^{13}D_2OH$

J. Am. Chem. Soc. 83:1230 (1961)

▶ $OC_4H_{10}$

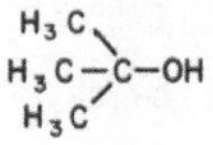

J. Chem. Soc. 2576 (1962)

▶ $OC_4H_{10}$

$(CH_3)_3COH$

Varian 423

▶ $OC_4H_{10}$

$CH_3OCH(CH_3)_2$

Bull. Chem. Soc. Japan 35:1046 (1962)

▶ $OC_4H_{10}$

$(CH_3)_2CHC^{13}H_2OH$

J. Am. Chem. Soc. 83:1230 (1961)

▶ $OC_4H_{10}$

$(CH_3CH_2)_2O$

J. Chem. Phys. 34:1099 (1961)
Can. J. Chem. 40:1522, 1956 (1962)
J. Am. Chem. Soc. 83:4138 (1961)

▶ $OC_4H_{10} \cdot BF_3$

$(C_2H_5)_2O \cdot BF_3$

J. Am. Chem. Soc. 83:4138 (1961)

▶ $OC_4H_{10} \cdot BH_3$

$(C_2H_5)_2O \cdot BH_3$

J. Am. Chem. Soc. 83:4138 (1961)

▶ $OC_5H_6$

J. Am. Chem. Soc. 84:2452 (1962)

▶ $OC_5H_6$

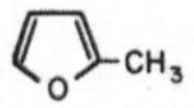

J. Phys. Chem. 65:1539 (1961)
J. Am. Chem. Soc. 83:5020 (1961)
Can. J. Chem. 39:909 (1961)
J. Chem. Phys. 36:2644 (1962) table II

▶ $OC_5H_8$

J. Am. Chem. Soc. 83:1226 (1961)
Can. J. Chem. 39:2316 (1961)

▶ $OC_5H_8$

$HC\equiv COCH_2CH_2CH_3$

Rec. Trav. Chim. 81:635 (1962)

▶ $OC_5H_6$

$HC\equiv CCH=CHOCH_3$

Varian 100

▶ $OC_5H_8$

$H_2C=CHCH_2OCH=CH_2$

Varian 110

▶ $OC_5H_8$

$C_2H_5COCH=CH_2$

J. Chem. Phys. 35:1900 (1961); 37:1951 (1962)

▶ $OC_5H_8$

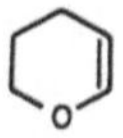

Varian 111

▶ $OC_5H_8$

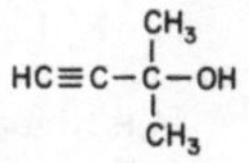

Varian 436
J. Mol. Spectr. 7:32 (1961)
J. Am. Chem. Soc. 83:1978 (1961)

▶ $OC_5H_{10}$

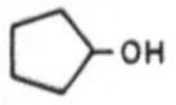

J. Chem. Soc. 1939 (1961)

▶ $OC_5H_8$

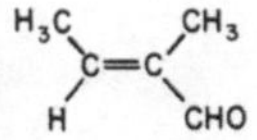

Arkiv Kemi 17:1 (1961)

▶ $OC_5H_{10}$

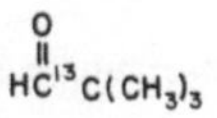

J. Am. Chem. Soc. 84:2649 (1962)

▶ $OC_5H_8$

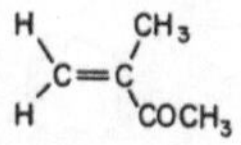

Arkiv Kemi 17:1 (1961)

▶ $OC_5H_{10}$

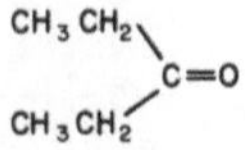

Can. J. Chem. 39:2223 (1961)
J. Am. Chem. Soc. 84:37 (1962)

▶ $OC_5H_{10}$

$CH_3C(=O)CH(CH_3)_2$

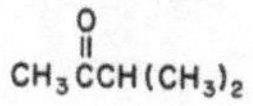

Bull. Chem. Soc. Japan 35:1046 (1962)

▶ $OC_5H_{10}$

$(CH_3)_2C{=}CH(CH_2OH)$

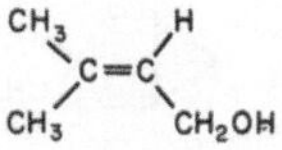

Chem. & Ind. 1020 (1962)

▶ $OC_5H_{10}$

$H_2C{=}CH{-}O{-}CH(CH_3)_2$

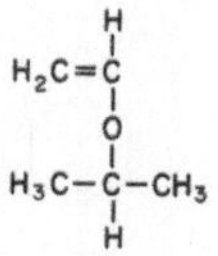

J. Chem. Soc. 2023 (1962)

▶ $OC_5H_{10}$

$(CH_3)_2C(OH)CH{=}CH_2$

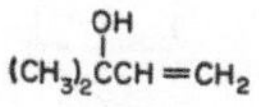

Varian 444

▶ $OC_5H_{10}D_2$

$(CH_3)_3CC^{13}D_2OH$

J. Am. Chem. Soc. 83:1230 (1961)

▶ $OC_5H_{11}D$

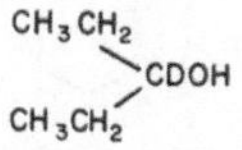

$(CH_3CH_2)_2CDOH$

J. Am. Chem. Soc. 84:37 (1962)

▶ $OC_5H_{12}$

$(CH_3CH_2)_2CHOH$

J. Am. Chem. Soc. 84:37 (1962)

▶ $OC_5H_{12}$

$(CH_3)_3CC^{13}H_2OH$

J. Am. Chem. Soc. 83:1230 (1961)

▶ $OC_5H_{12}$

$(CH_3)_3C^{13}C^{13}H_2OH$

J. Am. Chem. Soc. 83:4479 (1961)

▶ $OC_5H_{12}$

$CH_3CH_2{-}C(CH_3)(OH){-}CH_3$

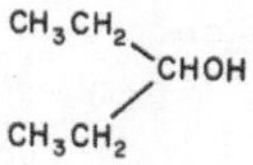

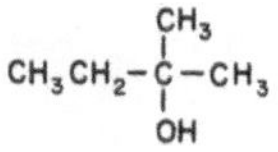

Tetrahedron 16:139 (1961)
J. Am. Chem. Soc. 84:37 (1962)

▶ $OC_5H_{15}AlSi$

$(CH_3)_3SiOAl(CH_3)_2$

J. Am. Chem. Soc. 84:1069 (1962)

▶ $OC_6H_6$

$CH_3C{\equiv}C{-}C{\equiv}CCH_2OH$

Arkiv Kemi 16:471 (1961)

▶ $OC_6H_6$

$C_6H_5OH$

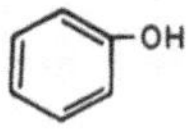

Tetrahedron Letters 519 (1961)
Can. J. Chem. 40:963, 2122 (1962)
J. Chem. Phys. 37:2618 (1962)

▶ $OC_6H_6D_6$

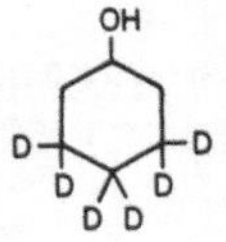

J. Am. Chem. Soc. 84:1053 (1962)

▶ $OC_6H_{10}$

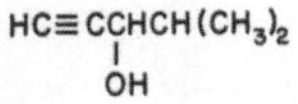

Varian 133

▶ $OC_6H_{10}$

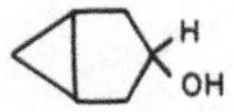

J. Am. Chem. Soc. 83:3251 (1961)

▶ $OC_6H_8$

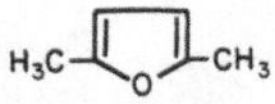

J. Phys. Chem. 65:1539 (1961)
J. Am. Chem. Soc. 83:5020 (1961)

▶ $OC_6H_{10}$

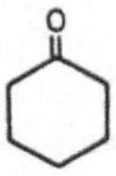

J. Am. Chem. Soc. 83:1226 (1961)

▶ $OC_6H_8$

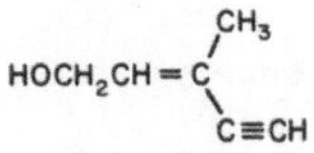

cis and trans

Helv. Chim. Acta. 45:548 (1962)

▶ $OC_6H_{10}$

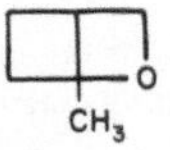

J. Org. Chem. 26:2257 (1961)

▶ $OC_6H_{10}$

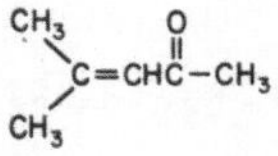

J. Mol. Phys. 5:153 (1962)
J. Org. Chem. 27:4091 (1962)

▶ $OC_6H_{10} \cdot BF_3$

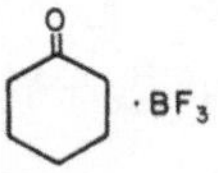

Rec. Trav. Chim. 80:1323 (1961)

▶ $OC_6H_{10}$

$H_2C{=}CHCH_2OCH_2CH{=}CH_2$

Varian 134

▶ $OC_6H_{12}$

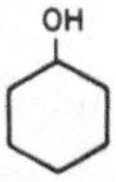

Bull. Chem. Soc. Japan 35:1905 (1962)
J. Org. Chem. 26:2648 (1961)
J. Am. Chem. Soc. 83:1146 (1961); 84:2464 (1962)

▶ $OC_6H_{10}$

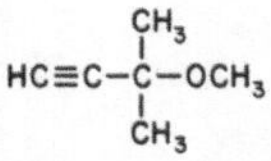

J. Am. Chem. Soc. 83:1978 (1961)

▶ $OC_6H_{12}$

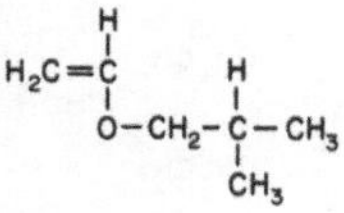

J. Chem. Soc. 2023 (1962)

▶ $OC_6H_{12}$

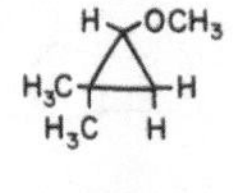

Tetrahedron Letters No. 6:243 (1962)

▶ $OC_6H_{15} \cdot BF_4$

$(C_2H_5)_3O^{\oplus}\ BF_4^{\ominus}$

Can. J. Chem. 40:1956 (1962)

▶ $OC_6H_{12}$

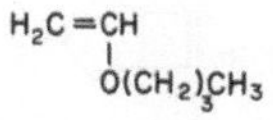

J. Chem. Soc. 2023 (1962)

▶ $OC_6H_{16}Si$

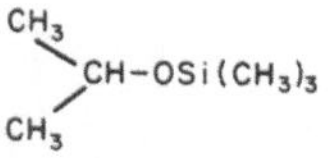

Tetrahedron 18:1147 (1962)

▶ $OC_6H_{12}$

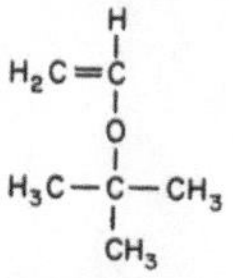

J. Chem. Soc. 2023 (1962)

▶ $OC_6H_{17}B$

$C_2H_5CH_3O \cdot B(CH_3)_3$

J. Am. Chem. Soc. 83:4138 (1961)

▶ $OC_6H_{12}$

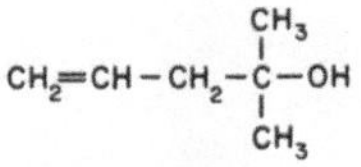

Tetrahedron 16:139 (1961)

▶ $OC_7H_6$

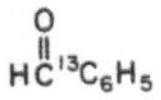

J. Am. Chem. Soc. 84:2649 (1962)

▶ $OC_6H_{12}$

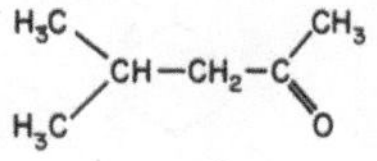

Varian 139

▶ $OC_7H_6$

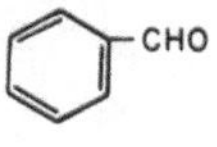

Varian 151
Can. J. Chem. 40:1073, 2333 (1962)

▶ $OC_6H_{14}$

$[(H_3C)_2CH]_2O$

Can. J. Chem. 40:1956 (1962)
J. Chem. Phys. 34:1099 (1961)

▶ $OC_7H_8$

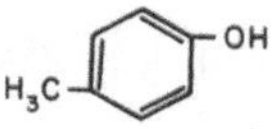

Bull. Chem. Soc. Japan 34:143, 1602 (1961)
Can. J. Chem. 40:2124 (1962)
Helv. Chim. Acta 45:568 (1962)

▶ $OC_6H_{14}$

$(CH_3CH_2CH_2)_2O$

J. Chem. Phys. 34:1091, 1099 (1961)

▶ $OC_7H_8$

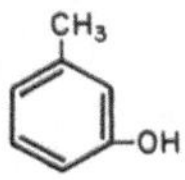

Varian 160
Bull. Chem. Soc. Japan 34:143 (1961)

▶ $OC_7H_8$

$C_6H_5CH_2OH$

Varian 161
J. Mol. Spectr. 7:32 (1961)

▶ $OC_7H_{10}$

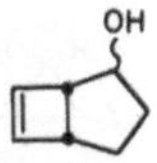


J. Am. Chem. Soc. 84:1220 (1962)

▶ $OC_7H_8$

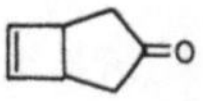


J. Am. Chem. Soc. 84:1220 (1962)

▶ $OC_7H_{10}$

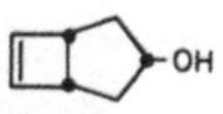


J. Am. Chem. Soc. 84:1220 (1962)

▶ $OC_7H_8$

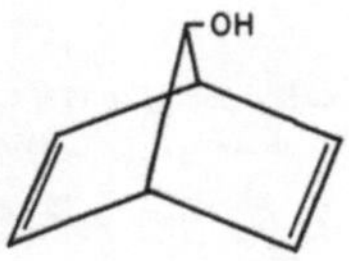


J. Org. Chem. 26:287 (1961)

▶ $OC_7H_{10}$

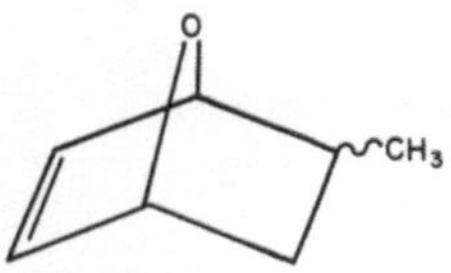


J. Am. Chem. Soc. 84:4120 (1962)

▶ $OC_7H_8$

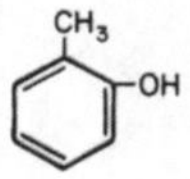


Bull. Chem. Soc. Japan 34:143 (1961)

▶ $OC_7H_{10}$

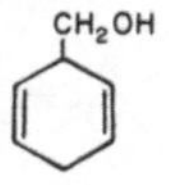


J. Am. Chem. Soc. 83:1005 (1961)

▶ $OC_7H_8$

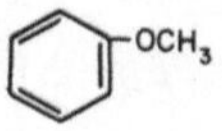


Varian 162
Can. J. Chem. 40:1866 (1962)
J. Chem. Phys. 37:2198 (1962)

▶ $OC_7H_{10}$

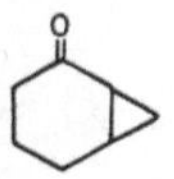


J. Am. Chem. Soc. 83:4678 (1961)

▶ $OC_7H_8D_4$

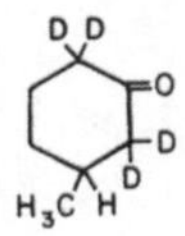


Can. J. Chem. 39:2268 (1961)

▶ $OC_7H_{12}$

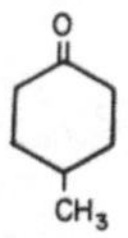


Tetrahedron Letters No. 25:1175 (1962)

▶ $OC_7H_{12}$

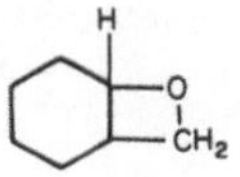

J. Org. Chem. 26:2257 (1961)
J. Am. Chem. Soc. 83:3096 (1961)

▶ $OC_7H_{12}$

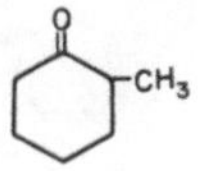

Tetrahedron Letters No. 25:1175 (1962)

▶ $OC_7H_{12}$

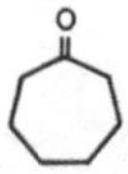

J. Am. Chem. Soc. 83:1226 (1961)

▶ $OC_7H_{12}$

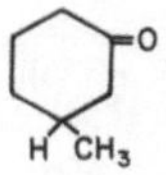

Can. J. Chem. 39:2268 (1961)

▶ $OC_7H_{14}$

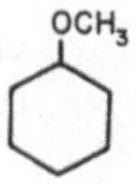

Trans. Faraday Soc. 57:390 (1961)

▶ $OC_7H_{14}$

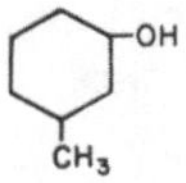

cis and trans

Tetrahedron Letters No. 17:743 (1962)

▶ $OC_7H_{14}$

$CH_2{=}CHCH_2CH_2{-}C(CH_3)_2{-}OH$

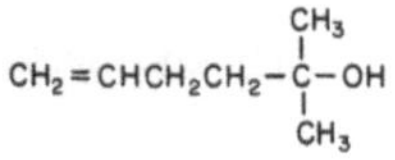

Tetrahedron 16:139 (1961)

▶ $OC_7H_{14}$

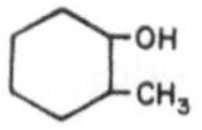

cis and trans

J. Am. Chem. Soc. 83:1146 (1961)
Tetrahedron Letters No. 17:744 (1962)

▶ $OC_7H_{14}$

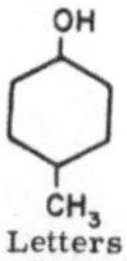

cis and trans

Tetrahedron Letters 17:744 (1962)
J. Am. Chem. Soc. No. 84:2464 (1962)

▶ $OC_7H_{14}$

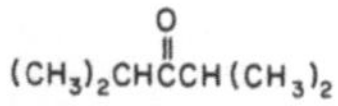

$(CH_3)_2CHC(=O)CH(CH_3)_2$

J. Am. Chem. Soc. 83:1230 (1961)

▶ $OC_7H_{14}$

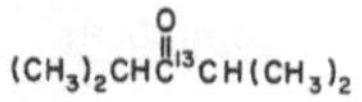

$(CH_3)_2CH{}^{13}C(=O)CH(CH_3)_2$

J. Am. Chem. Soc. 83:1230 (1961)

▶ $OC_7H_{16}$

$(CH_3)_2CH{}^{13}CH(OH)CH(CH_3)_2$

J. Am. Chem. Soc. 83:1230 (1961)

▶ $OC_7H_{16}$

$(CH_3)_2CHCH(OH)CH(CH_3)_2$

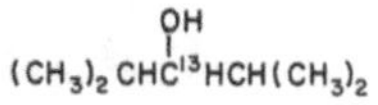

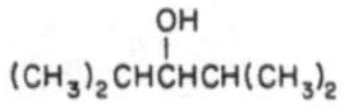

J. Am. Chem. Soc. 83:1230 (1961)

▶ $OC_7H_{16}$

$(CH_3)_3COCH(CH_3)_2$

Varian 183

▶ $OC_7H_{19}B$

$(C_2H_5)_2O \cdot B(CH_3)_3$

J. Am. Chem. Soc. 83:4138 (1961)

▶ $OC_8H_8$

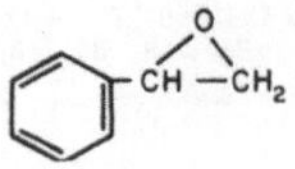


Varian 193
J. Mol. Spectr. 9:477 (1962)

▶ $OC_8H_8$

$C_6H_5COCH_3$

Varian 192
Bull. Chem. Soc. Japan 34:143 (1961)

▶ $OC_8H_8$

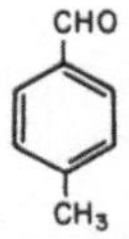


Can. J. Chem. 40:1073, 2331 (1962)

▶ $OC_8H_8$

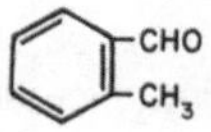


Can. J. Chem. 40:1073 (1962)

▶ $OC_8H_8$

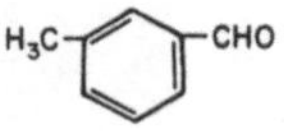


Can. J. Chem. 40:1073 (1962)

▶ $OC_8H_{10}$

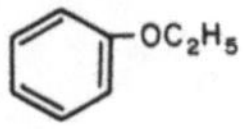


Can. J. Chem. 40:1866 (1962)

▶ $OC_8H_{10}$

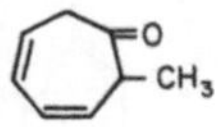


J. Am. Chem. Soc. 83:600 (1961)
Bull. Chem. Soc. Japan 34:619 (1961)

▶ $OC_8H_{10}$

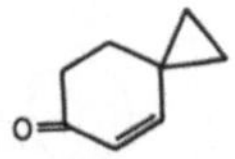


J. Org. Chem. 27:1887 (1962)

▶ $OC_8H_{10}$

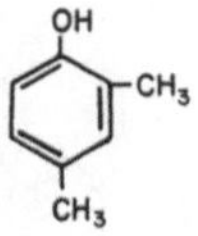


Bull. Chem. Soc. Japan 34:744 (1961)

▶ $OC_8H_{10}$

$C_6H_5CH(OH)CH_3$ (printed as $C_6H_5CHOH$ with $CH_3$ below)

J. Mol. Spectr. 7:32 (1961)

▶ $OC_8H_{10}$

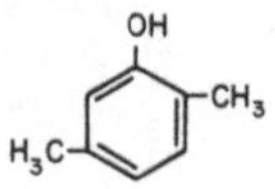


Bull. Chem. Soc. Japan 34:744 (1961)

▶ $OC_8H_{10}$

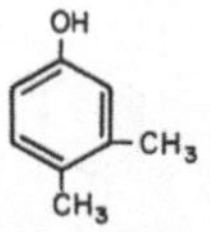


Bull. Chem. Soc. Japan 34:744 (1961)

▶ $OC_8H_{10}$

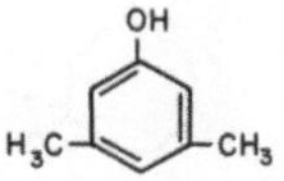


Bull. Chem. Soc. Japan 34:744 (1961)

▶ $OC_8H_{10}$

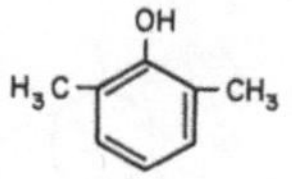


Bull. Chem. Soc. Japan 34:744 (1961)

▶ $OC_8H_{10}$

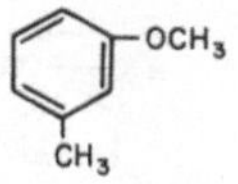


Can. J. Chem. 40:1866 (1962)

▶ $OC_8H_{10}$

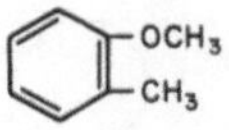


Can. J. Chem. 40:1866 (1962)

▶ $OC_8H_{10}$

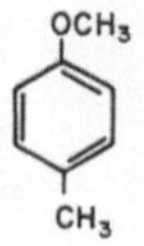


Varian 205
Can. J. Chem. 40:1866, 2331 (1962)
J. Chem. Phys. 37:2594 (1962)

▶ $OC_8H_{10}$

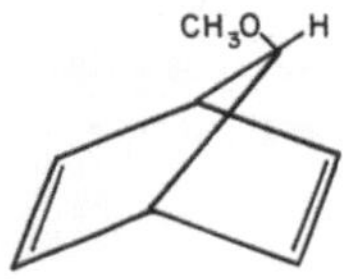


J. Am. Chem. Soc. 84:4876 (1962)

▶ $OC_8H_{12}$

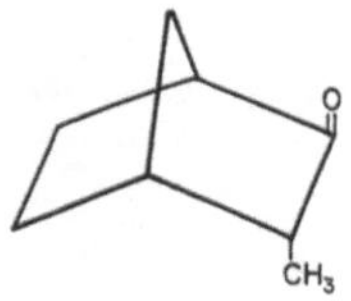


J. Am. Chem. Soc. 84:2611 (1962)

▶ $OC_8H_{12}$

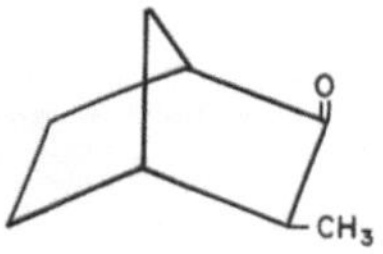


J. Am. Chem. Soc. 84:2611 (1962)

▶ $OC_8H_{12}$

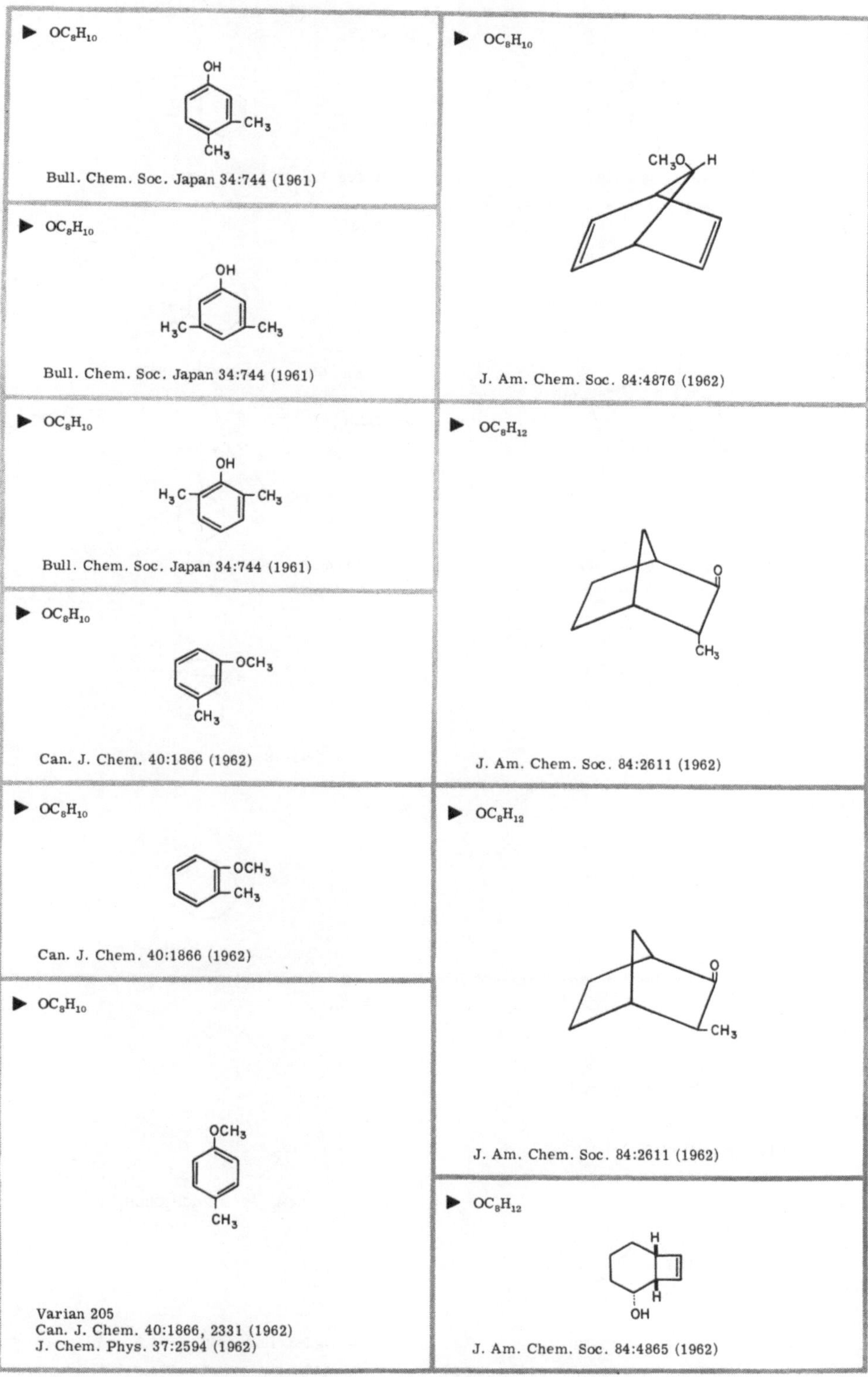

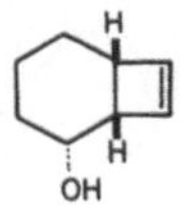


J. Am. Chem. Soc. 84:4865 (1962)

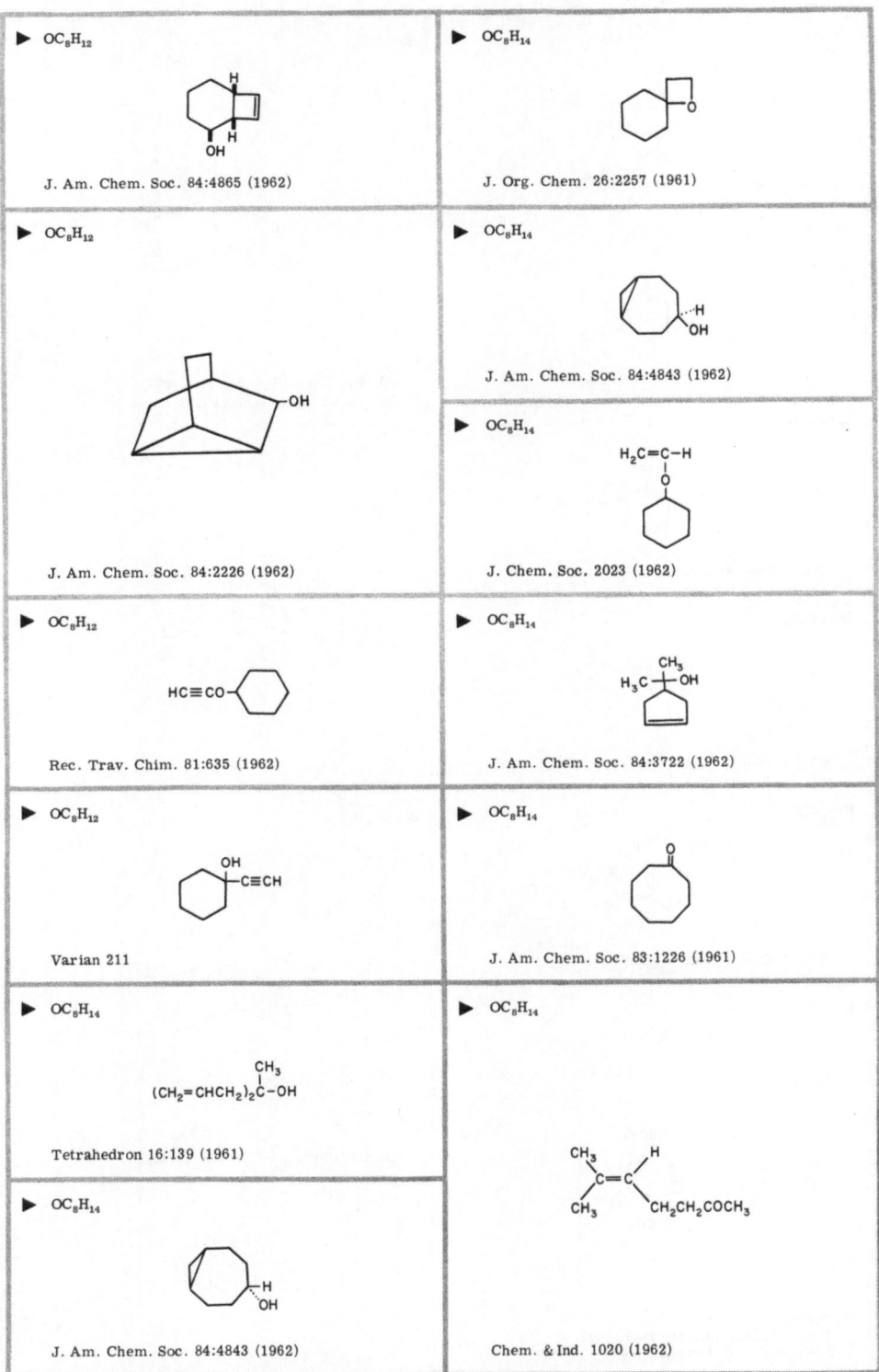

▶ $OC_8H_{12}$

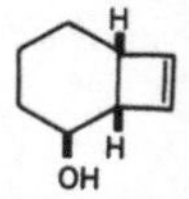

J. Am. Chem. Soc. 84:4865 (1962)

▶ $OC_8H_{14}$

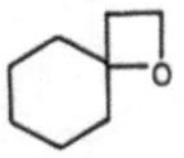

J. Org. Chem. 26:2257 (1961)

▶ $OC_8H_{12}$

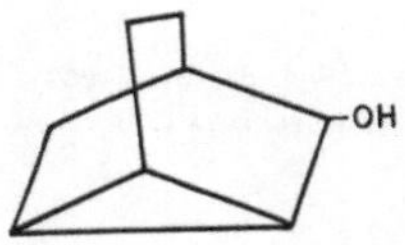

J. Am. Chem. Soc. 84:2226 (1962)

▶ $OC_8H_{14}$

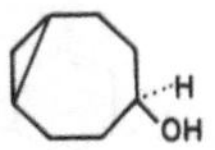

J. Am. Chem. Soc. 84:4843 (1962)

▶ $OC_8H_{14}$

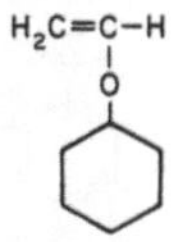

J. Chem. Soc. 2023 (1962)

▶ $OC_8H_{12}$

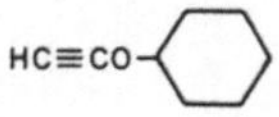

Rec. Trav. Chim. 81:635 (1962)

▶ $OC_8H_{14}$

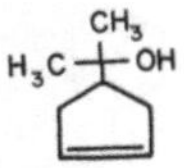

J. Am. Chem. Soc. 84:3722 (1962)

▶ $OC_8H_{12}$

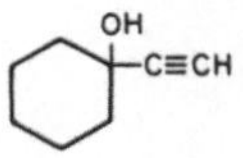

Varian 211

▶ $OC_8H_{14}$

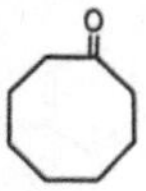

J. Am. Chem. Soc. 83:1226 (1961)

▶ $OC_8H_{14}$

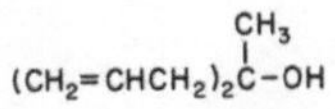

Tetrahedron 16:139 (1961)

▶ $OC_8H_{14}$

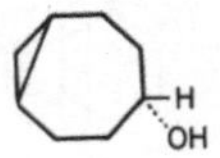

J. Am. Chem. Soc. 84:4843 (1962)

▶ $OC_8H_{14}$

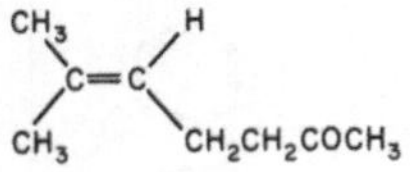

Chem. & Ind. 1020 (1962)

▶ $OC_8H_{14}$

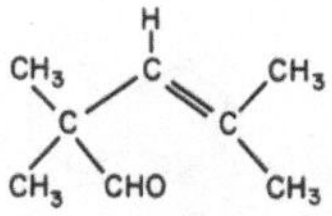


J. Org. Chem. 26:3132 (1961)

▶ $OC_8H_{16}$

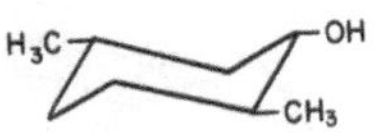


Tetrahedron Letters No. 17:743 (1962)

▶ $OC_8H_{16}$

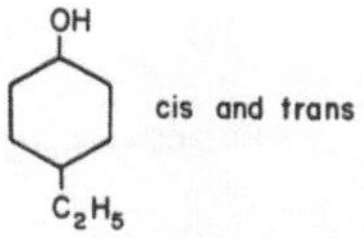


J. Am. Chem. Soc. 84:2464 (1962)

▶ $OC_8H_{16}$

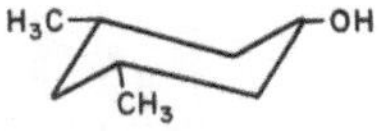


Tetrahedron Letters No. 17:743 (1962)

▶ $OC_8H_{16}$

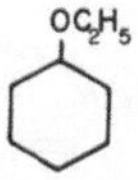


Trans. Faraday Soc. 57:390 (1961)

▶ $OC_8H_{16}$

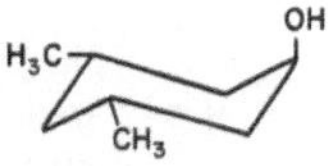


Tetrahedron Letters No. 17:744 (1962)

▶ $OC_8H_{16}$

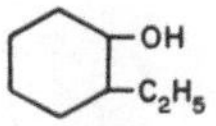


cis and trans

Tetrahedron Letters No. 17:744 (1962)

▶ $OC_8H_{16}$

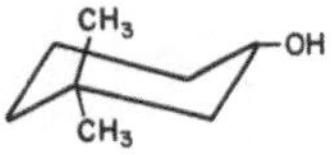


Tetrahedron Letters No. 17:743 (1962)

▶ $OC_8H_{16}$

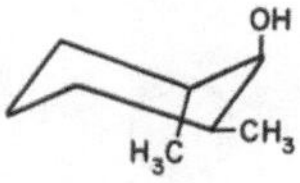


Tetrahedron Letters No. 17:744 (1962)

▶ $OC_8H_{16}$

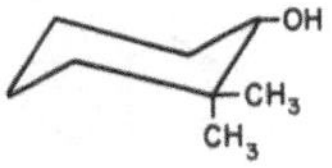


Tetrahedron Letters No. 17:743 (1962)

▶ $OC_8H_{16}$

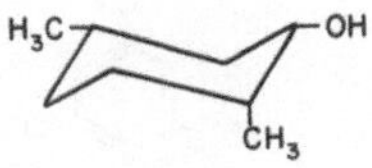


Tetrahedron Letters No. 17:743 (1962)

▶ $OC_8H_{16}$

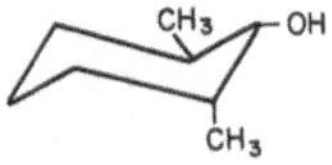


Tetrahedron Letters No. 17:743 (1962)

▶ $OC_8H_{16}$

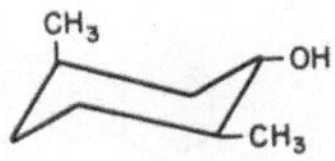


Tetrahedron Letters No. 17:743 (1962)

▶ $OC_8H_{16}$

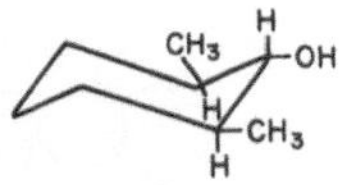


Tetrahedron Letters No. 17:743 (1962)

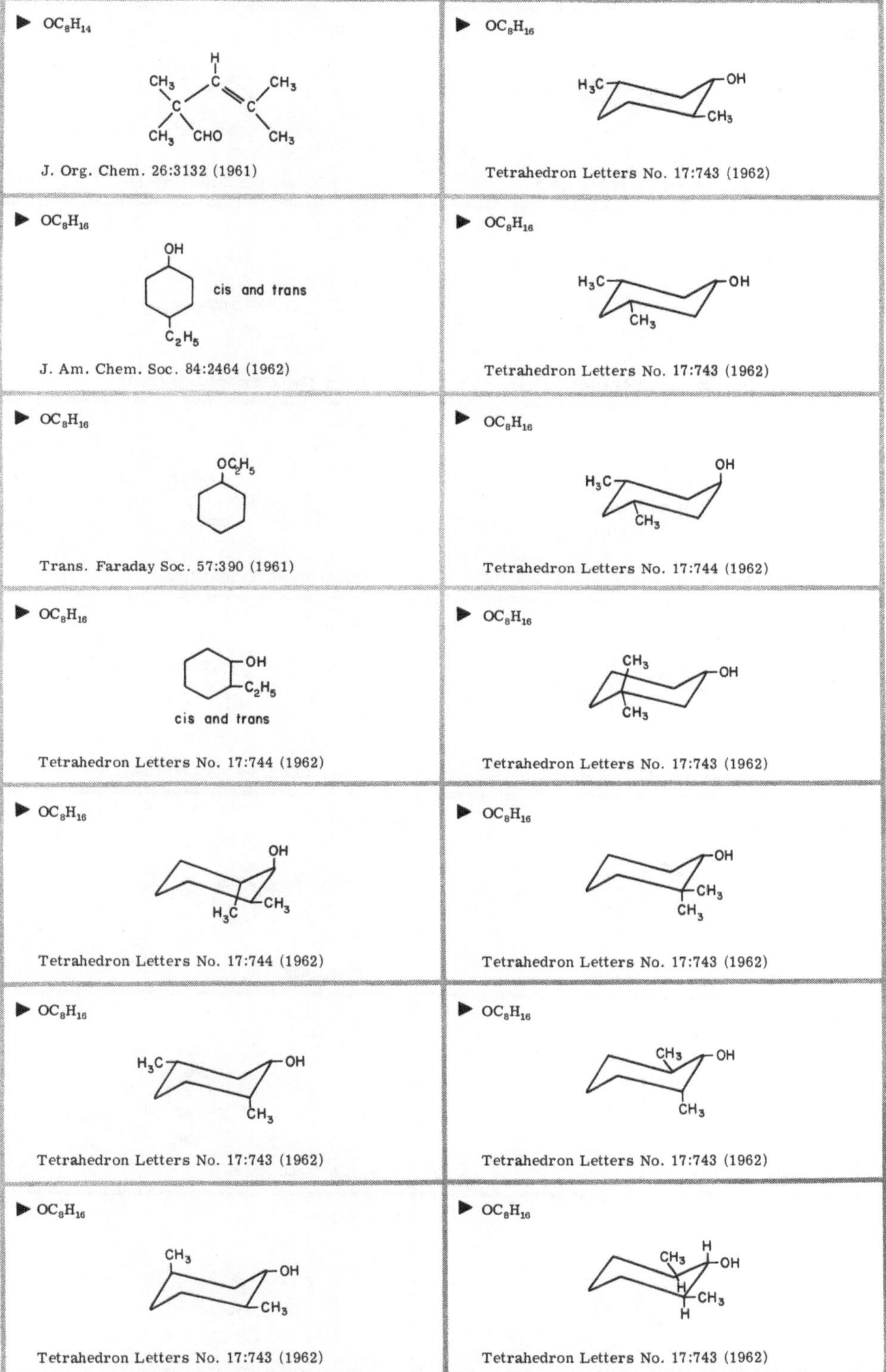

▶ $OC_8H_{16}$

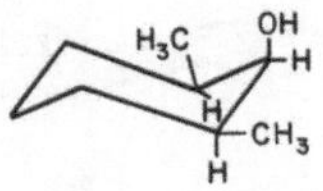

Tetrahedron Letters No. 17:744 (1962)

▶ $OC_8H_{17}B$

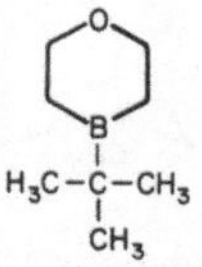

J. Am. Chem. Soc. 83:2541 (1961)

▶ $OC_8H_{18}$

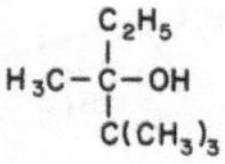

J. Mol. Spectr. 7:32 (1961)

▶ $OC_9H_6$

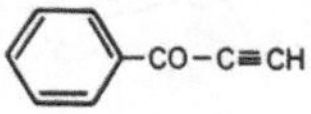

Trans. Faraday Soc. 57:28 (1961)

▶ $OC_9H_8$

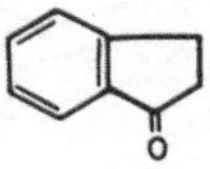

Virian 229

▶ $OC_9H_8$

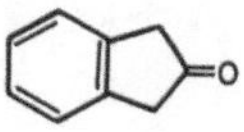

Varian 523

▶ $OC_9H_8$

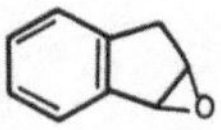

Varian 524

▶ $OC_9H_8$

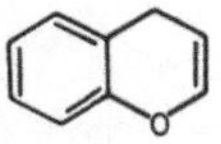

J. Am. Chem. Soc. 84:813 (1962)

▶ $OC_9H_8$

$HC{\equiv}CCH_2OC_6H_5$

J. Am. Chem. Soc. 83:1978 (1961)

▶ $OC_9H_{10}$

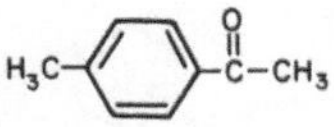

Can. J. Chem. 40:2331 (1962)

▶ $OC_9H_{10}$

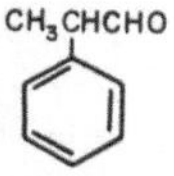

J. Am. Chem. Soc. 84:4824 (1962)

▶ $OC_9H_{10}$

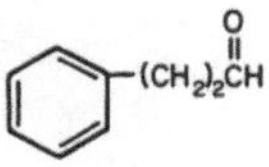

Varian 529

▶ $OC_9H_{12}$

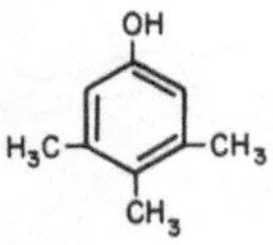

MCA Serial No. 85

▶ $OC_9H_{12}$

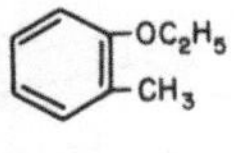

Can. J. Chem. 40:1866 (1962)

▶ $OC_9H_{12}$

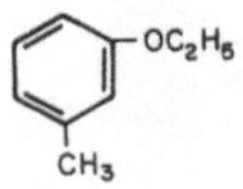


Can. J. Chem. 40:1866 (1962)

▶ $OC_9H_{12}$

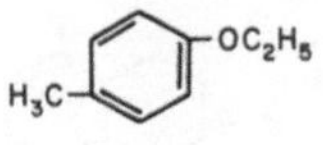


Can. J. Chem. 40:1866 (1962)

▶ $OC_9H_{14}$

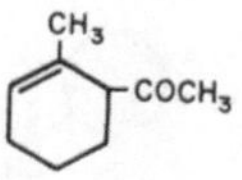


J. Org. Chem. 27:4249 (1962)

▶ $OC_9H_{14}$

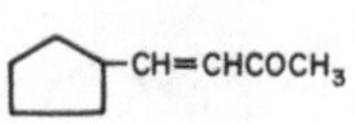


Varian 544

▶ $OC_9H_{14}$

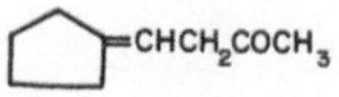


Varian 545

▶ $OC_9H_{16}$

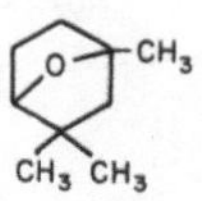


J. Am. Chem. Soc. 84:3913 (1962)

▶ $OC_9H_{18}$

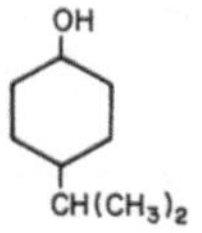


cis and trans

J. Am. Chem. Soc. 84:2464 (1962)
Tetrahedron Letters No. 17:743 (1962)

▶ $OC_9H_{18}$

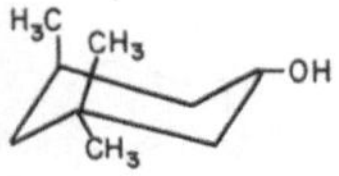


Tetrahedron Letters No. 17:743 (1962)

▶ $OC_9H_{18}$

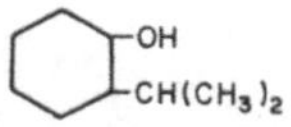


cis and trans

Tetrahedron Letters No. 17:744 (1962)

▶ $OC_9H_{20}Si$

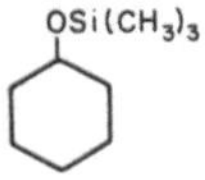


Tetrahedron 18:1147 (1962)

▶ $OC_9H_{21}B$

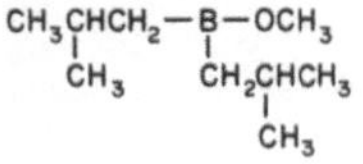


J. Am. Chem. Soc. 84:4362 (1962)

▶ $OC_{10}H_{10}$

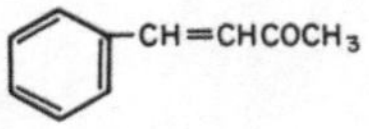

Varian 251

▶ $OC_{10}H_{12}$

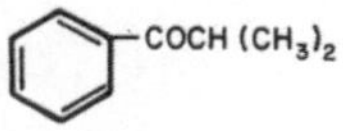

Varian 559

▶ $OC_{10}H_{12}$

$C_6H_5CH_2CH_2COCH_3$

Bull. Soc. Chim. France 842 (1961)

▶ $OC_{10}H_{14}$

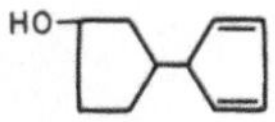

J. Chem. Soc. 1939 (1961)

▶ $OC_{10}H_{12}$

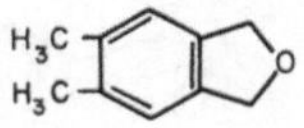

Bull. Chem. Soc. Japan 34:1571 (1961)

▶ $OC_{10}H_{14}$

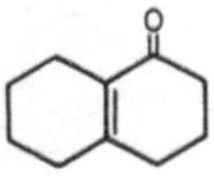

J. Org. Chem. 26:3729 (1961)

▶ $OC_{10}H_{12}$

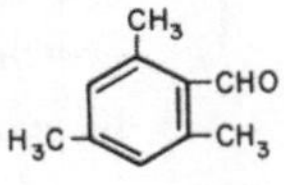

Can. J. Chem. 40:1073 (1962)

▶ $OC_{10}H_{14}$

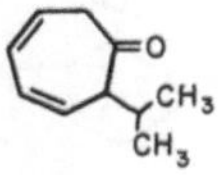

Bull. Chem. Soc. Japan 34:619 (1961)

▶ $OC_{10}H_{12}$

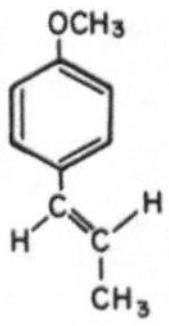

Varian 258

▶ $OC_{10}H_{14}$

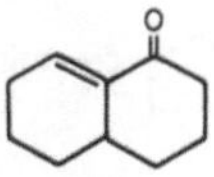

J. Org. Chem. 26:3729 (1961)

▶ $OC_{10}H_{12}$

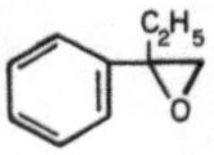

Varian 558

▶ $OC_{10}H_{14}$

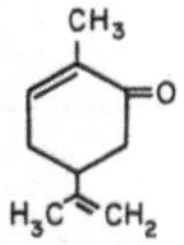

Varian 271
Bull. Chem. Soc. Japan 35:1899 (1962)

▶ $OC_{10}H_{14}$

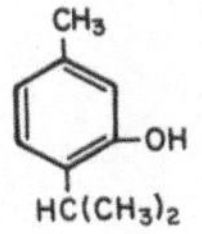


Varian 270

▶ $OC_{10}H_{14}$

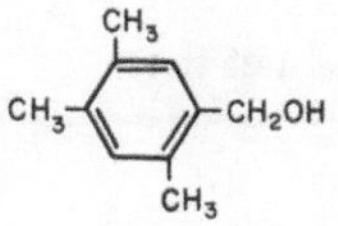


Bull. Chem. Soc. Japan 34:1571 (1961)

▶ $OC_{10}H_{14}$

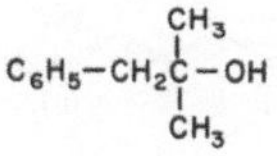


Tetrahedron 16:139 (1961)

▶ $OC_{10}H_{14}$

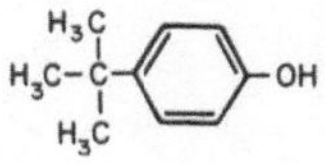


Can. J. Chem. 40:2124 (1962)

▶ $OC_{10}H_{14}$

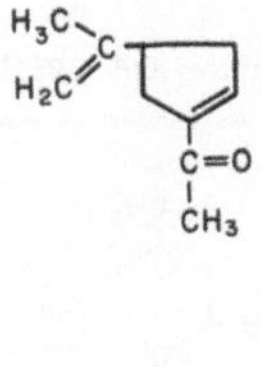


Bull. Chem. Soc. Japan 35:1899 (1962)

▶ $OC_{10}H_{14}$

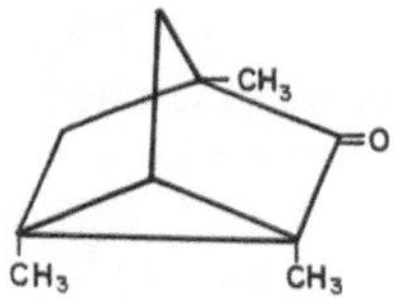


J. Am. Chem. Soc. 84:879 (1962)

▶ $OC_{10}H_{15}D_5$

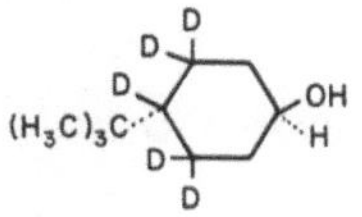


J. Am. Chem. Soc. 84:1053 (1962)

▶ $OC_{10}H_{15}D_5$

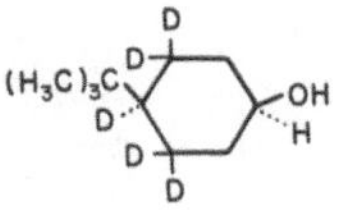


J. Am. Chem. Soc. 84:1053 (1962)

▶ $OC_{10}H_{16}$

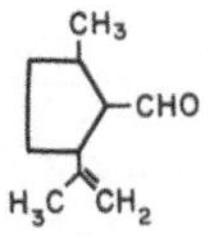


Tetrahedron Letters No. 1:79 (1962)

▶ $OC_{10}H_{16}$

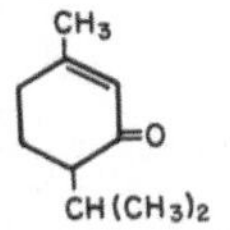


Varian 275

▶ $OC_{10}H_{16}$

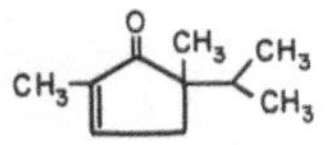


J. Am. Chem. Soc. 84:1224 (1962)

▶ $OC_{10}H_{16}$

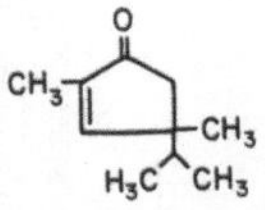

J. Am. Chem. Soc. 84:1224 (1962)

▶ $OC_{10}H_{18}$

$(CH_3)_2C=CH-CH_2CH_2C(OH)(CH_3)CH=CH_2$

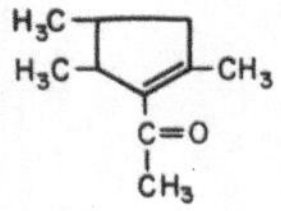

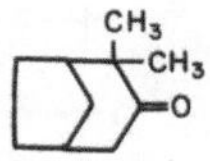

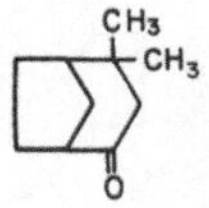

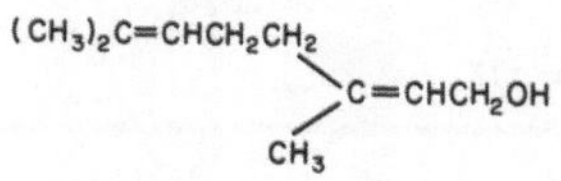

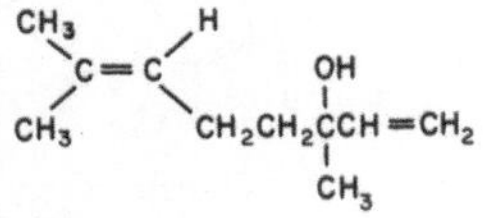

Chem. & Ind. 1020 (1962)

▶ $OC_{10}H_{16}$

J. Chem. Soc. 2255 (1961)

▶ $OC_{10}H_{18}$

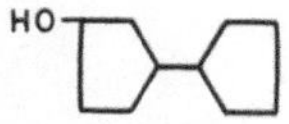

J. Chem. Soc. 1939 (1961)

▶ $OC_{10}H_{16}$

$(CH_2=CH-CH_2)_3C-OH$

Tetrahedron 16:139 (1961)

▶ $OC_{10}H_{18}$

$CH_2=CHCH(CH_3)-C(C_2H_5)_2CHO$

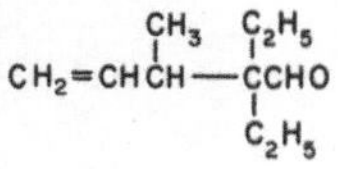

Ann. 649:38 (1961)

▶ $OC_{10}H_{16}$

J. Org. Chem. 26:707 (1961)

▶ $OC_{10}H_{18}$

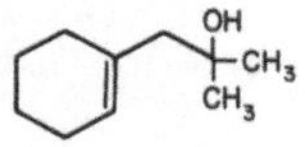

J. Am. Chem. Soc. 84:2775 (1962)

▶ $OC_{10}H_{16}$

J. Org. Chem. 26:707 (1961)

▶ $OC_{10}H_{18}$

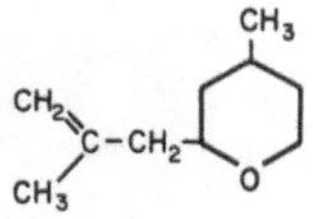

Helv. Chim. Acta 45:397 (1962)

▶ $OC_{10}H_{18}$

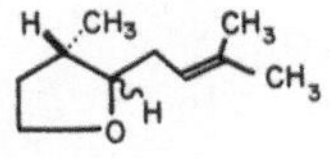

$(CH_3)_2C=CHCH_2CH_2C(CH_3)=CHCH_2OH$

cis and trans

Varian 279
Chem. & Ind. 1020 (1962)

▶ $OC_{10}H_{18}$

Helv. Chim. Acta 44:605 (1961)

▶ $OC_{10}H_{18}$

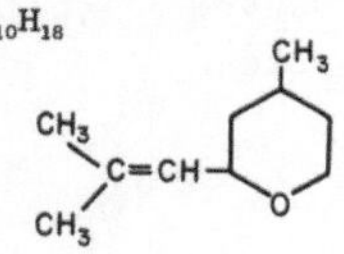

cis and trans

Helv. Chim. Acta 44:604 (1961); 45:397 (1962)

► $OC_{10}H_{18}$

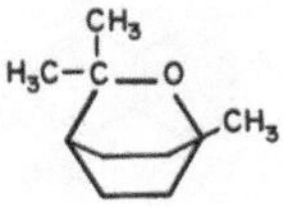


Varian 280

► $OC_{10}H_{19}Al$

$Al(CH{=}CH_2)_3 \cdot (C_2H_5)_2O$

J. Phys. Chem. 65:224 (1961)

► $OC_{10}H_{20}$

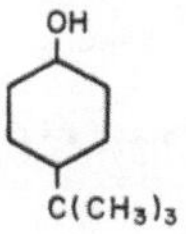


cis and trans

J. Am. Chem. Soc. 83:1146 (1961); 84:2464 (1962)
Tetrahedron Letters No. 3:97 (1962); No. 17:743 (1962)

► $OC_{10}H_{20}$

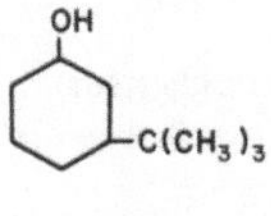


cis and trans

Tetrahedron Letters No. 17:743 (1962)

► $OC_{10}H_{20}$

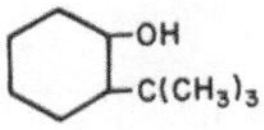


cis and trans

Tetrahedron Letters No. 17:743 (1962)

► $OC_{10}H_{20}$

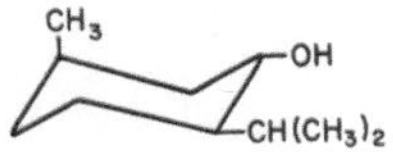


Tetrahedron Letters No. 17:743 (1962)

► $OC_{10}H_{20}$

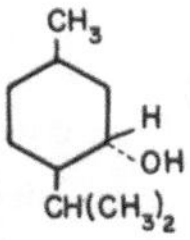


Varian 281

► $OC_{10}H_{20}$

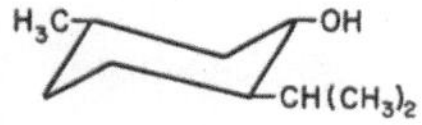


Tetrahedron Letters No. 17:743 (1962)

► $OC_{10}H_{20}$

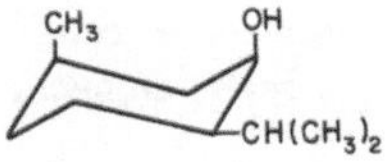


Tetrahedron Letters No. 17:744 (1962)

► $OC_{10}H_{20}$

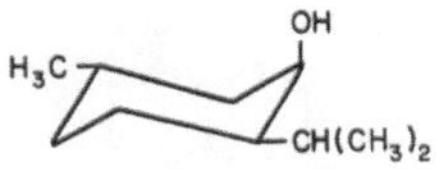


Tetrahedron Letters No. 17:744 (1962)

► $OC_{10}H_{20}$

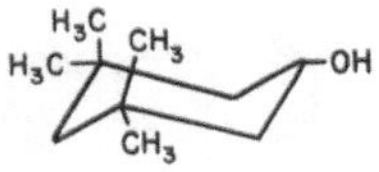


Tetrahedron Letters No. 17:743 (1962)

► $OC_{10}H_{20}$

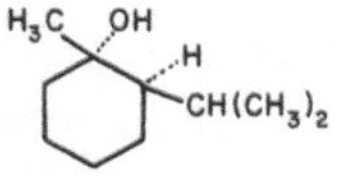


J. Chem. Soc. 4497 (1962)

▶ $OC_{10}H_{20}$

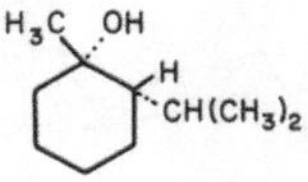

J. Chem. Soc. 4497 (1962)

▶ $OC_{10}H_{20}$

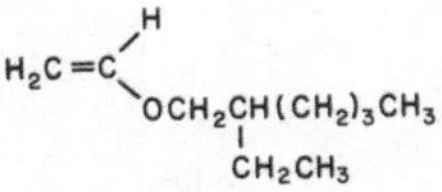

J. Chem. Soc. 2023 (1962)

▶ $OC_{10}H_{22}$

$CH_3(CH_2)_8CH_2OH$

Varian 282

▶ $OC_{10}H_{23}B$

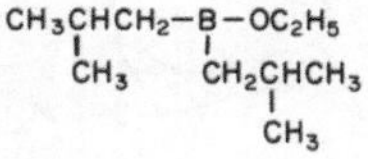

J. Am. Chem. Soc. 84:4362 (1962)

▶ $OC_{11}H_8$

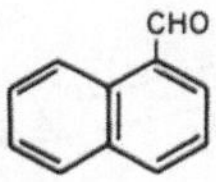

Can. J. Chem. 40:1073 (1962)

▶ $OC_{11}H_8$

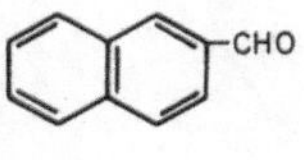

Can. J. Chem. 40:1073 (1962)

▶ $OC_{11}H_{10}$

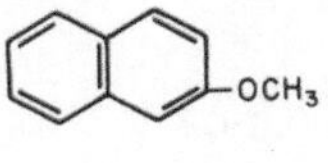

Tetrahedron 18:841 (1962)

▶ $OC_{11}H_{10}$

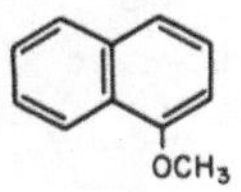

Tetrahedron 18:841 (1962)

▶ $OC_{11}H_{14}$

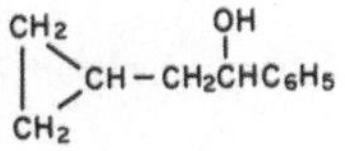

J. Am. Chem. Soc. 84:4295 (1962)

▶ $OC_{11}H_{14}$

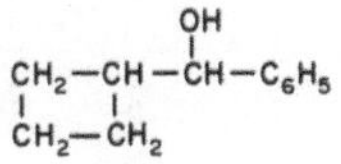

J. Am. Chem. Soc. 84:4295 (1962)

▶ $OC_{11}H_{14}$

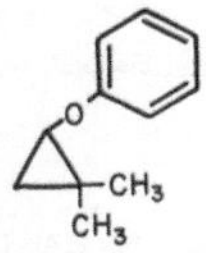

Tetrahedron Letters No. 6:242 (1962)

▶ $OC_{11}H_{16}$

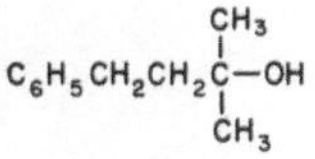

Tetrahedron 16:139 (1961)

▶ $OC_{11}H_{16}$

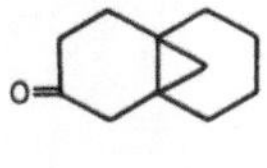

J. Am. Chem. Soc. 83:2784 (1961)

▶ $OC_{11}H_{16}$

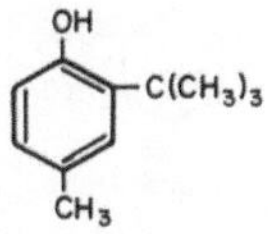

Varian 288

▶ $OC_{11}H_{16}$

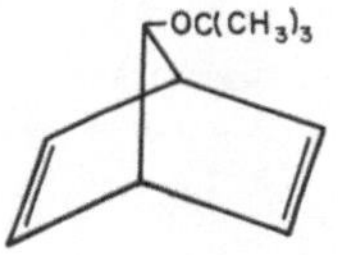


J. Org. Chem. 26:287 (1961)

▶ $OC_{11}H_{20}$

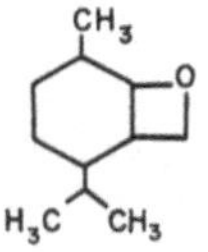


J. Org. Chem. 27:1982 (1962)

▶ $OC_{11}H_{20}$

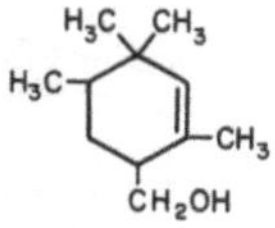


Bull. Chem. Soc. Japan 35:818 (1962)

▶ $OC_{11}H_{18}$

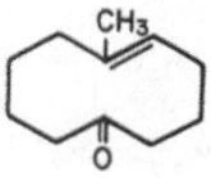


J. Org. Chem. 26:4782 (1961)

▶ $OC_{11}H_{20}$

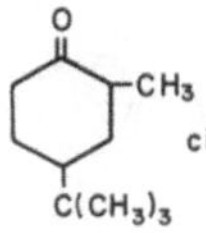


cis and trans

Tetrahedron Letters No. 25:1175 (1962)

▶ $OC_{11}H_{20}$

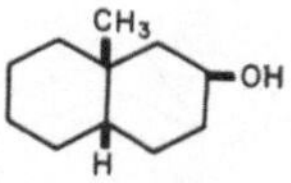


Tetrahedron 18:983 (1962)
J. Am. Chem. Soc. 83:1146 (1961)

▶ $OC_{11}H_{20}$

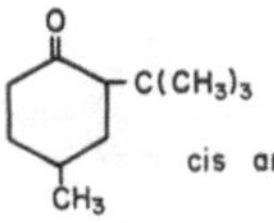


cis and trans

Tetrahedron Letters No. 25:1175 (1962)

▶ $OC_{11}H_{20}$

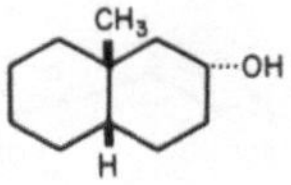


J. Am. Chem. Soc. 83:1146 (1961)

▶ $OC_{11}H_{20}$

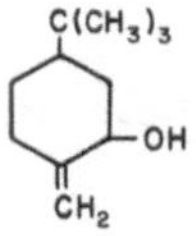


cis and trans

J. Chem. Soc. 1650 (1961)

▶ $OC_{11}H_{20}$

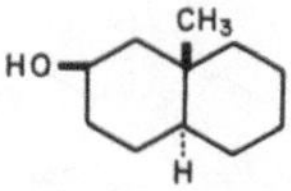


J. Am. Chem. Soc. 83:1146 (1961)
Tetrahedron 18:983 (1962)

▶ $OC_{11}H_{22}$

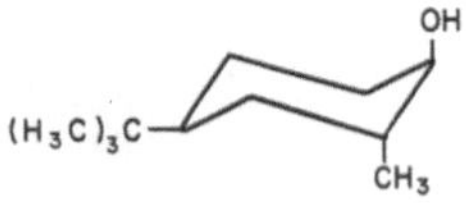


Tetrahedron Letters No. 17:744 (1962)

▶ $OC_{11}H_{20}$

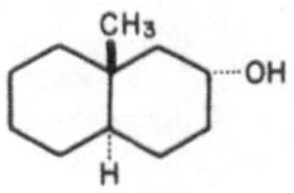


J. Am. Chem. Soc. 83:1146 (1961)

▶ $OC_{11}H_{22}$

Tetrahedron Letters No. 17:743 (1962)

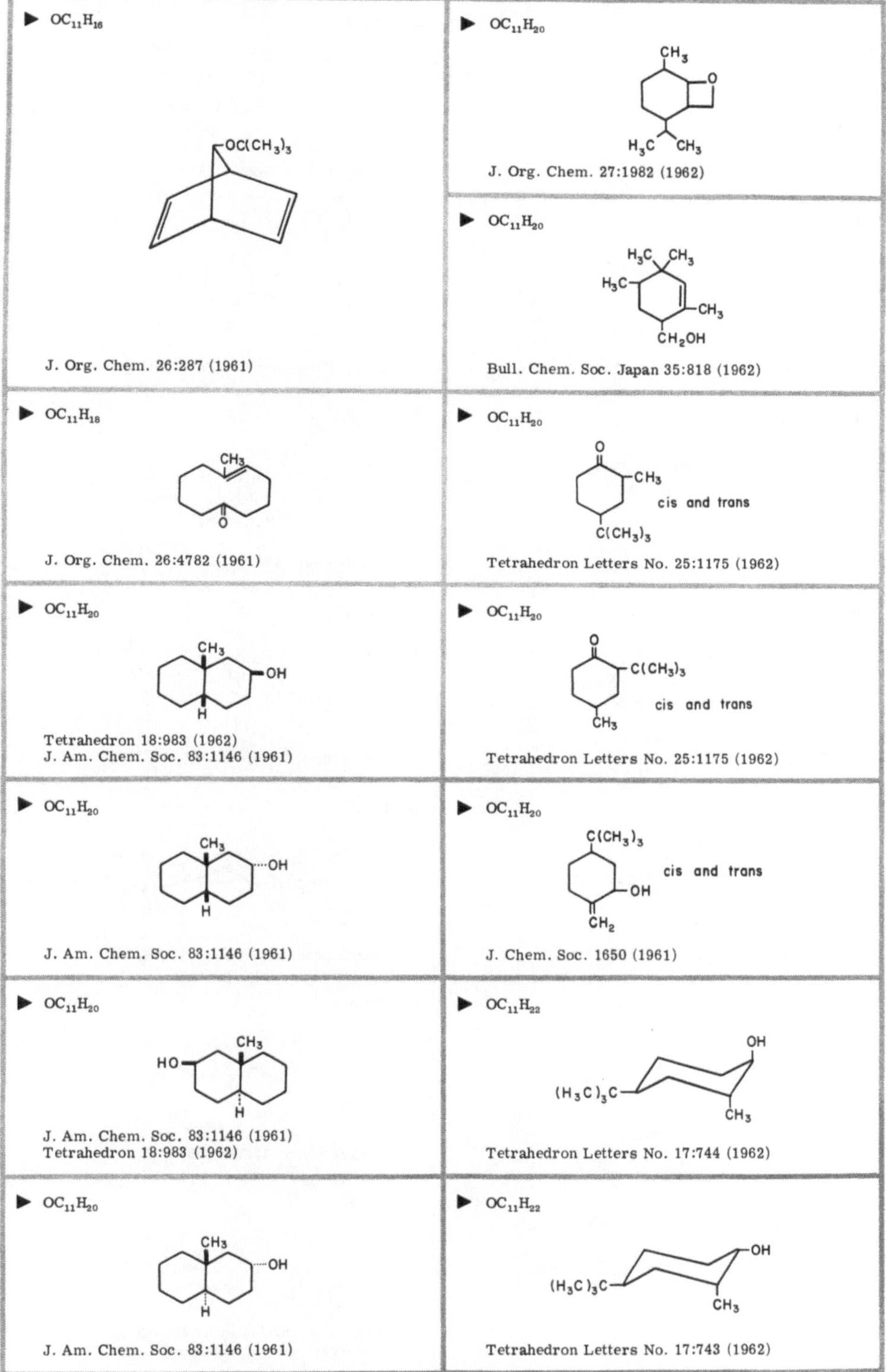

► $OC_{11}H_{22}$

Tetrahedron Letters No. 17:743 (1962)

► $OC_{12}H_{18}$

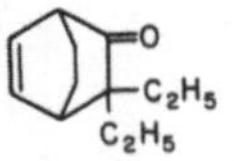

J. Org. Chem. 26:990 (1961)

► $OC_{11}H_{22}$

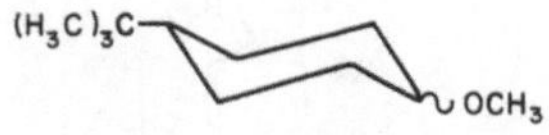

cis and trans

Tetrahedron Letters No. 3:97 (1962)

► $OC_{12}H_{18}$

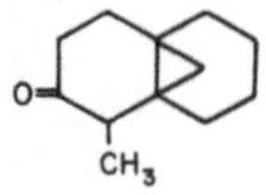

J. Am. Chem. Soc. 83:2784 (1961)

► $OC_{12}H_8$

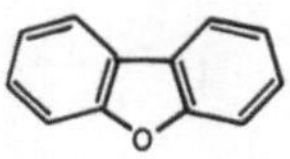

Varian 589

► $OC_{12}H_{20}$

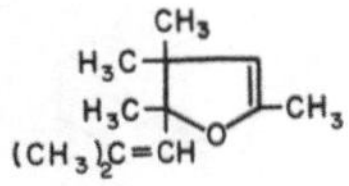

J. Chem. Soc. 2255 (1961)

► $OC_{12}H_{10}$

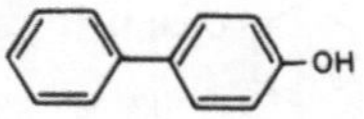

Can. J. Chem. 40:2124 (1962)

► $OC_{12}H_{24}$

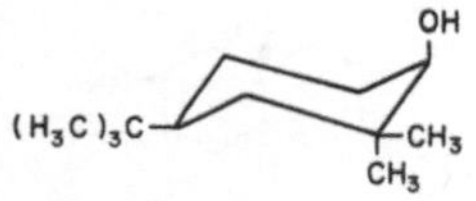

Tetrahedron Letters No. 17:744 (1962)

► $OC_{12}H_{15} \cdot BF_4$

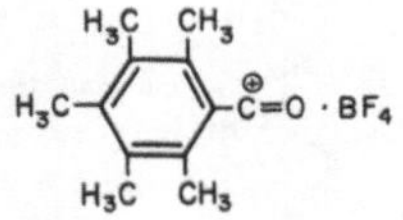

Tetrahedron Letters No. 22:985 (1962)

► $OC_{12}H_{24}$

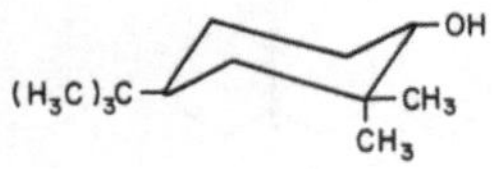

Tetrahedron Letters No. 17:743 (1962)

► $OC_{12}H_{18}$

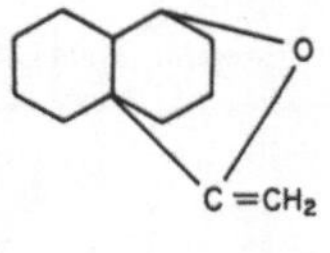

J. Chem. Soc. 2023 (1962)

► $OC_{13}H_{10}$

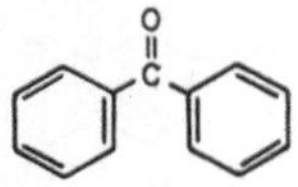

J. Phys. Chem. 65:2023 (1961)

► $OC_{13}H_{12}$

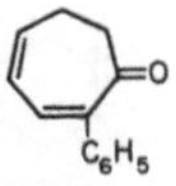

Bull. Chem. Soc. Japan 34:619 (1961)
Rec. Trav. Chim. 81:591 (1962)

▶ $OC_{13}H_{12}$

$(C_6H_5)_2CHOH$

Varian 607
J. Mol. Spectr. 7:32 (1961)

▶ $OC_{13}H_{12}$

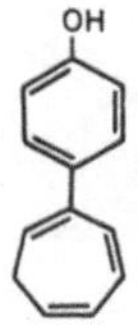

OR TAUTOMER

Chem. & Ind. 1192 (1962)

▶ $OC_{13}H_{12}$

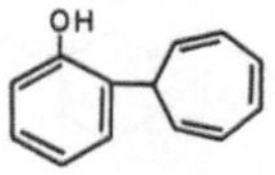

Chem. & Ind. 1192 (1962)

▶ $OC_{13}H_{12}$

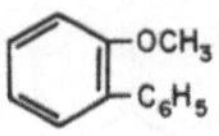

Can. J. Chem. 40:1866 (1962)

▶ $OC_{13}H_{12}$

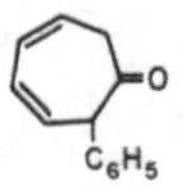

Rec. Trav. Chim. 81:591 (1962)

▶ $OC_{13}H_{14}$

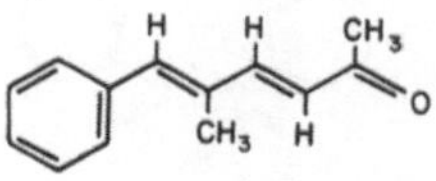

J. Org. Chem. 27:1991 (1962)

▶ $OC_{13}H_{14}$

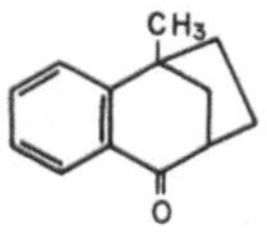

J. Am. Chem. Soc. 84:3517 (1962)

▶ $OC_{13}H_{16}$

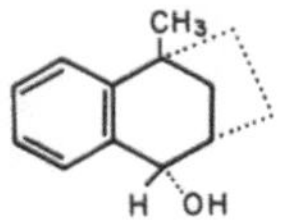

J. Am. Chem. Soc. 84:3517 (1962)

▶ $OC_{13}H_{16}$

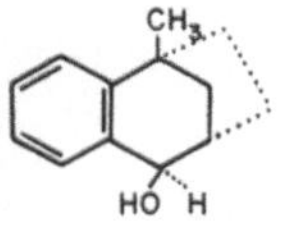

J. Am. Chem. Soc. 84:3517 (1962)

▶ $OC_{13}H_{16}$

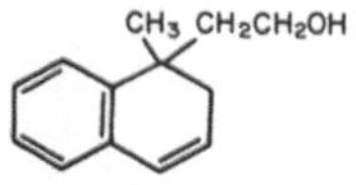

J. Am. Chem. Soc. 84:3517 (1962)

▶ $OC_{13}H_{16}$

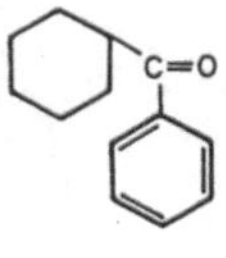

Varian 613

▶ $OC_{13}H_{18}$

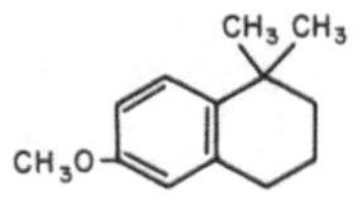

J. Chem. Soc. 4794 (1961)

▶ $OC_{13}H_{18}$

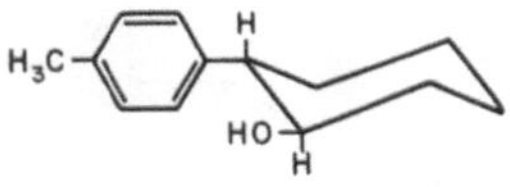

J. Org. Chem. 27:717 (1962)

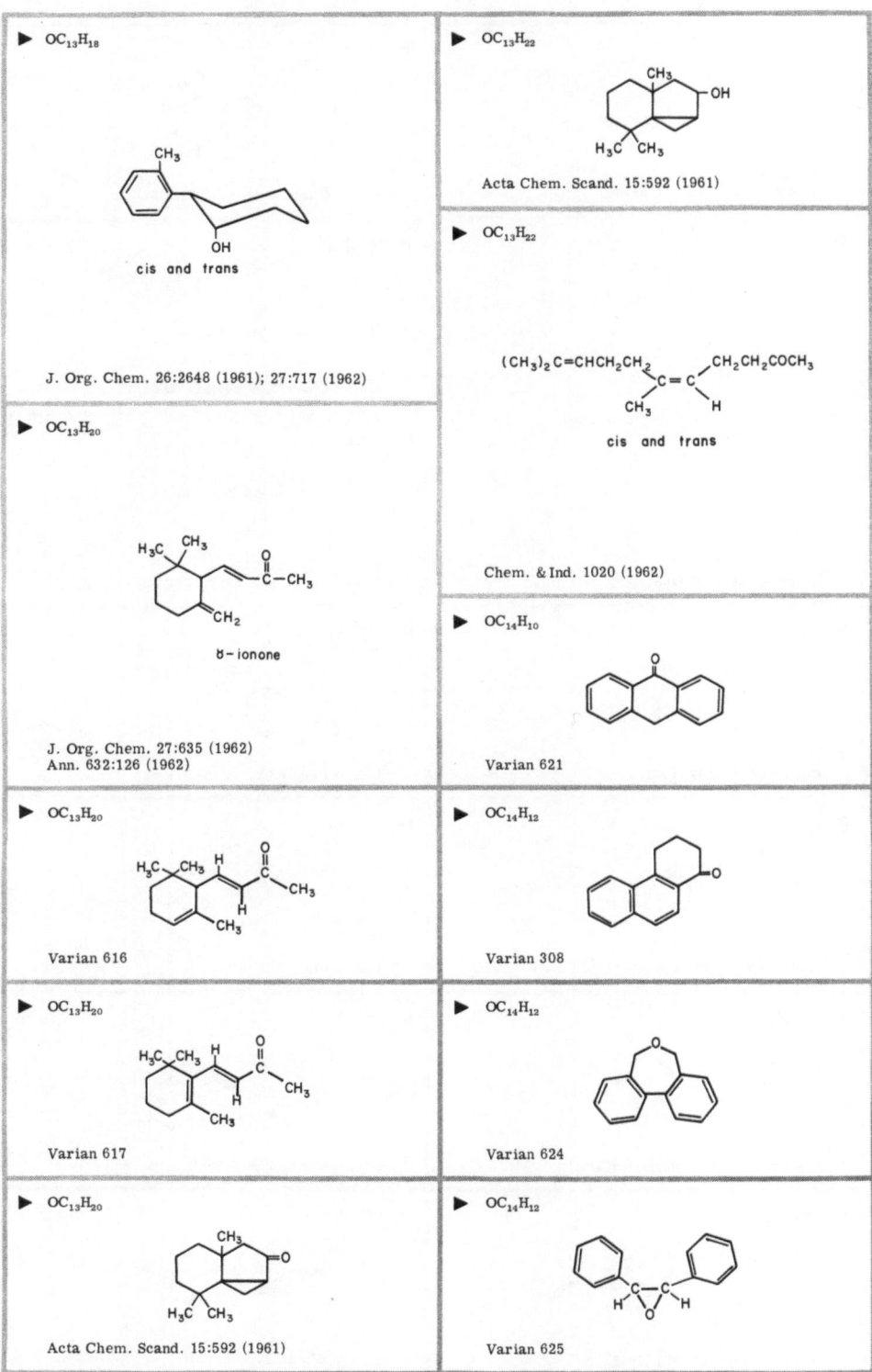

▶ $OC_{13}H_{18}$

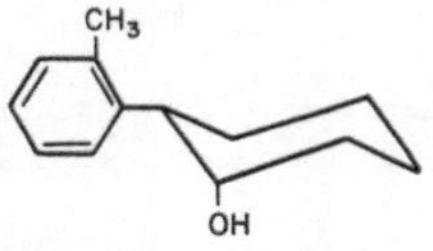


cis and trans

J. Org. Chem. 26:2648 (1961); 27:717 (1962)

▶ $OC_{13}H_{20}$

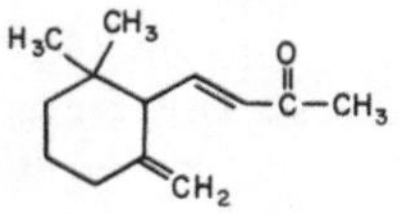


δ-ionone

J. Org. Chem. 27:635 (1962)
Ann. 632:126 (1962)

▶ $OC_{13}H_{20}$

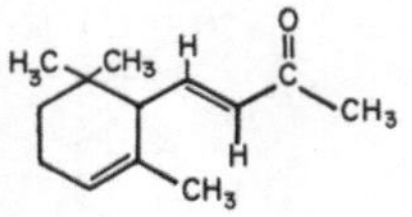


Varian 616

▶ $OC_{13}H_{20}$

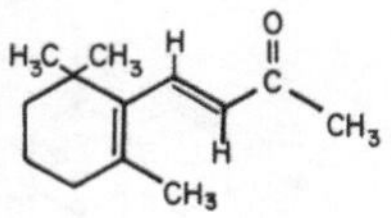


Varian 617

▶ $OC_{13}H_{20}$

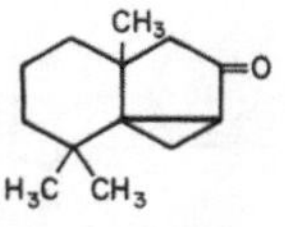


Acta Chem. Scand. 15:592 (1961)

▶ $OC_{13}H_{22}$

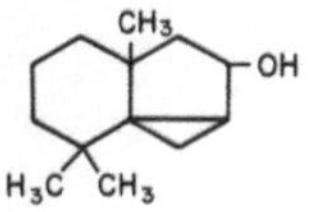


Acta Chem. Scand. 15:592 (1961)

▶ $OC_{13}H_{22}$

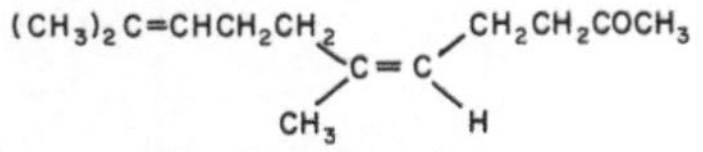


cis and trans

Chem. & Ind. 1020 (1962)

▶ $OC_{14}H_{10}$

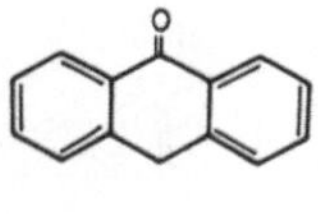


Varian 621

▶ $OC_{14}H_{12}$

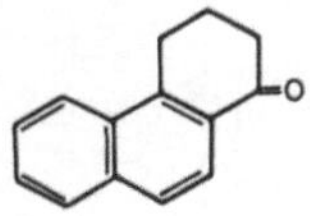


Varian 308

▶ $OC_{14}H_{12}$

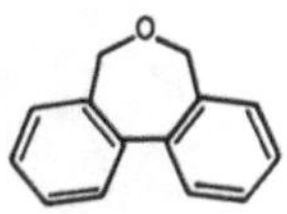


Varian 624

▶ $OC_{14}H_{12}$

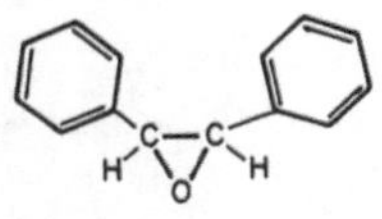


Varian 625

▶ $OC_{14}H_{12}$

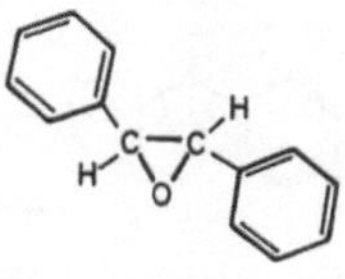


Varian 626

▶ $OC_{14}H_{14}$

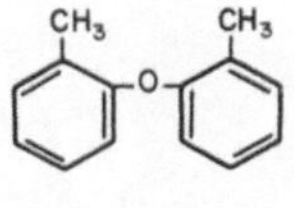


J. Chem. Phys. 37:1564 (1962)

▶ $OC_{14}H_{14}$

$(C_6H_5)_2C(CH_3)OH$

J. Mol. Spectr. 7:32 (1961)
Tetrahedron 16:139 (1961)

▶ $OC_{14}H_{16}$

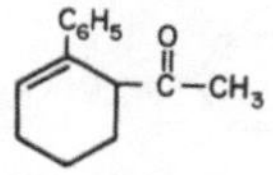


J. Org. Chem. 27:4243 (1962)

▶ $OC_{14}H_{18}$

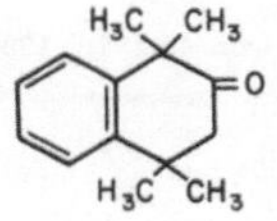


Tetrahedron Letters 244 (1961)
Can. J. Chem. 40:1664 (1962)

▶ $OC_{14}H_{20}$

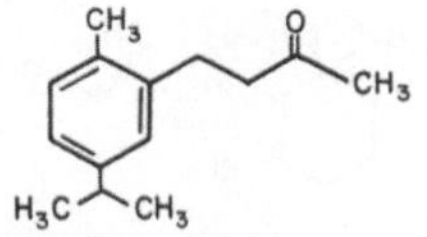


J. Org. Chem. 26:3140 (1961)

▶ $OC_{14}H_{20}$

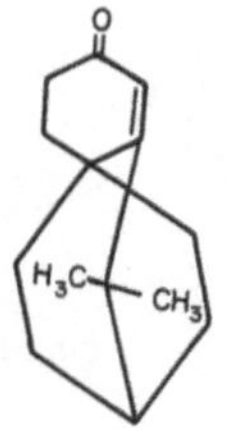


J. Am. Chem. Soc. 84:3205 (1962)

▶ $OC_{14}H_{22}$

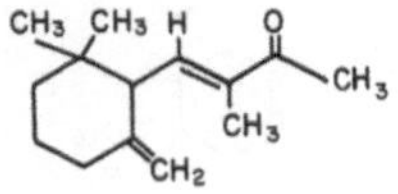


J. Org. Chem. 27:2935 (1962)

▶ $OC_{14}H_{22}$

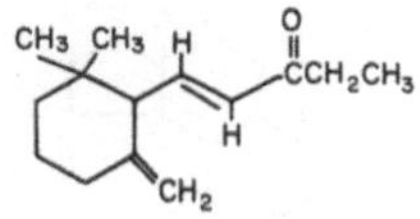


J. Org. Chem. 27:635 (1962)

▶ $OC_{14}H_{22}$

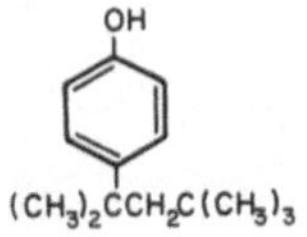


Varian 315

▶ $OC_{14}H_{22}$

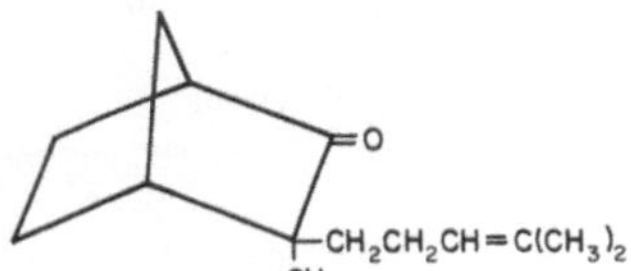


J. Am. Chem. Soc. 84:2611 (1962)

► $OC_{14}H_{22}$

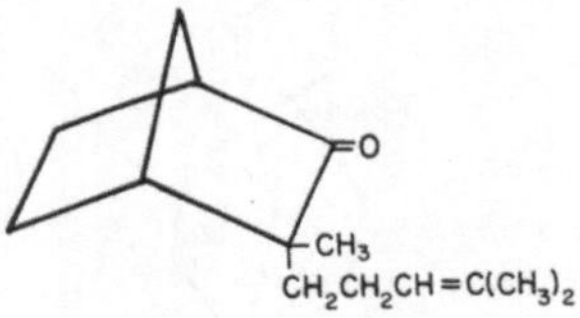


J. Am. Chem. Soc. 84:2611 (1962)

► $OC_{14}H_{22}$

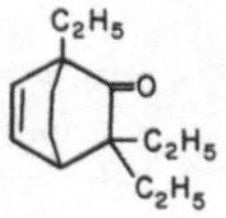


J. Org. Chem. 26:990 (1961)

► $OC_{14}H_{22}$

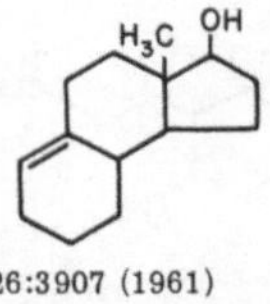


J. Org. Chem. 26:3907 (1961)

► $OC_{15}H_{10}$

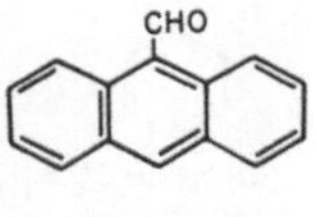


Can. J. Chem. 40:1073 (1962)

► $OC_{15}H_{10}$

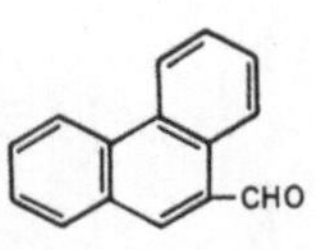


Can. J. Chem. 40:1073 (1962)

► $OC_{15}H_{12}$

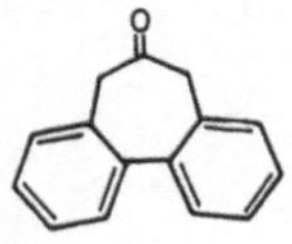


Varian 637

► $OC_{15}H_{14}$

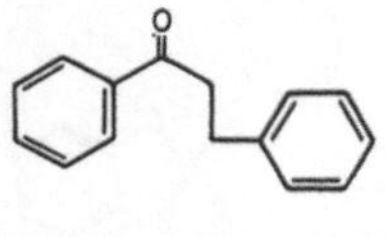


Bull. Soc. Chim. France 1962 (1962)

► $OC_{15}H_{14}$

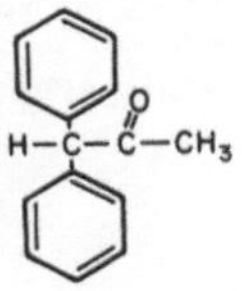


Varian 318

► $OC_{15}H_{14}$

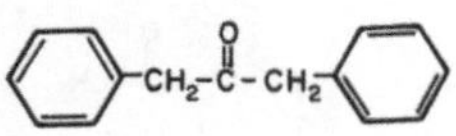


Varian 638

► $OC_{15}H_{16}$

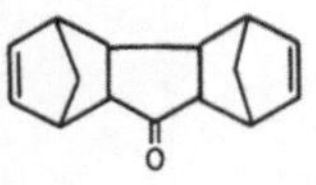


exo-trans-exo

Helv. Chim. Acta 45:1870 (1962)

► $OC_{15}H_{20}$

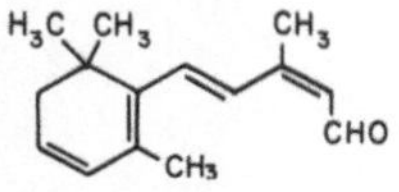


Helv. Chim. Acta 45:528 (1962)

► $OC_{15}H_{20}$

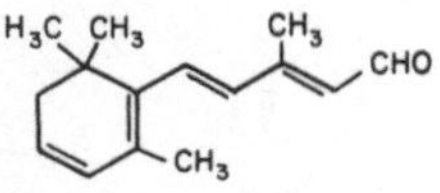


Helv. Chim. Acta 45:528 (1962)

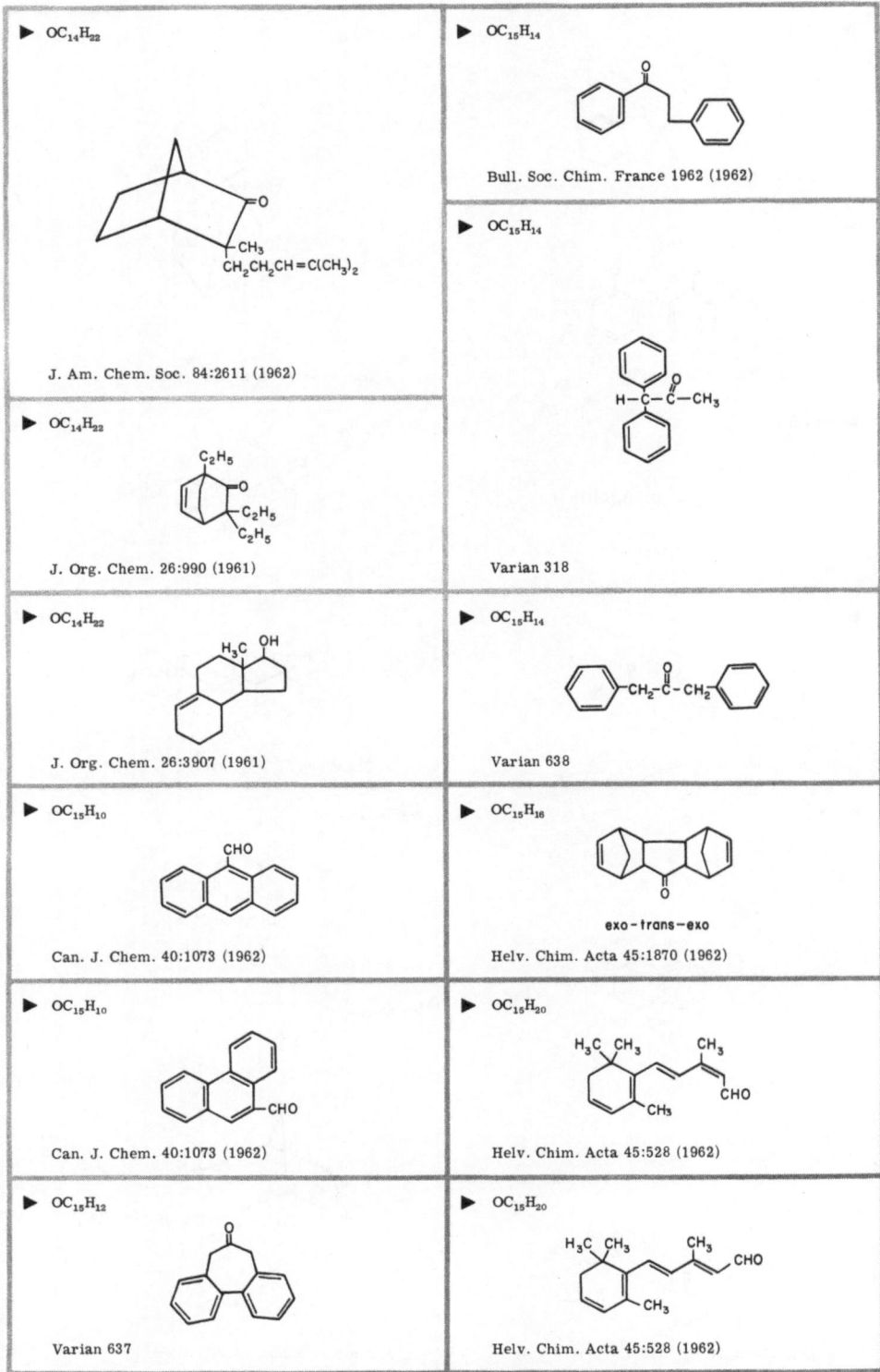

▶ $OC_{15}H_{20}$

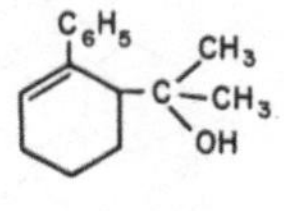

J. Org. Chem. 27:4243 (1962)

▶ $OC_{15}H_{20}$

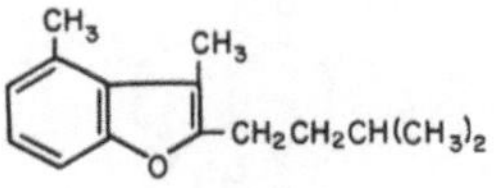

J. Am. Chem. Soc. 83:3096 (1961)

▶ $OC_{15}H_{22}$

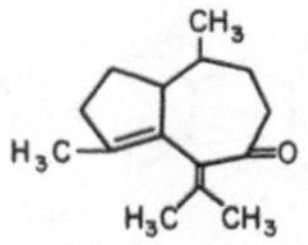

J. Chem.Soc. 795 (1962); 1963 (1962)

▶ $OC_{15}H_{22}$

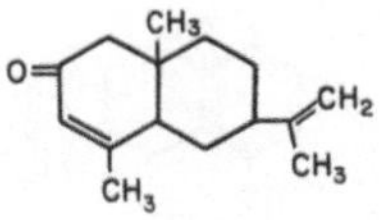

Acta Chem. Scand. 16:1311 (1962)

▶ $OC_{15}H_{22}$

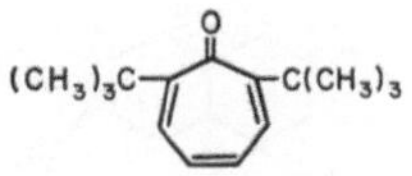

J. Am. Chem. Soc. 83:601 (1961)

▶ $OC_{15}H_{22}$

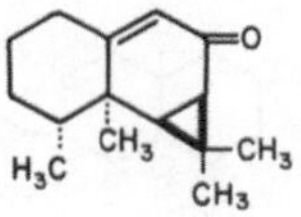

Tetrahedron Letters No. 18:833 (1962)

▶ $OC_{15}H_{22}$

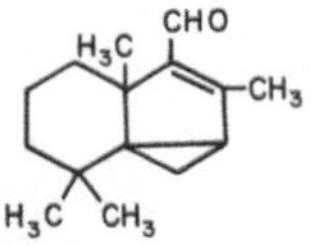

J. Org. Chem. 26:1964 (1961)

▶ $OC_{15}H_{22}$

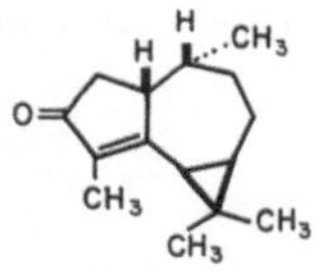

Proc. Chem. Soc. 280 (1962)

▶ $OC_{15}H_{24}$

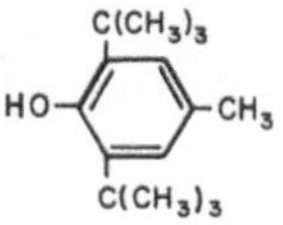

J. Am. Chem. Soc. 84:2258 (1962)

▶ $OC_{15}H_{24}$

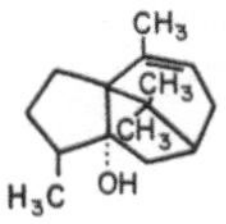

J. Am. Chem. Soc. 84:3205 (1962)

▶ $OC_{15}H_{24}$

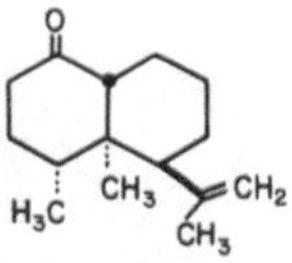

Tetrahedron Letters No. 18:830 (1962)

▶ $OC_{15}H_{26}$

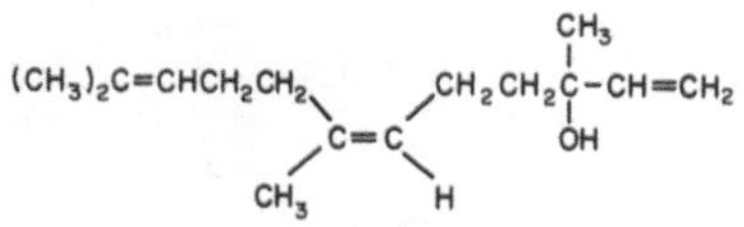

cis and trans

Chem. & Ind. 1020 (1962)

▶ $OC_{15}H_{26}$

$$CH_3\underset{\substack{|\\CH_3}}{C}=CH-CH_2CH_2-\underset{\substack{|\\CH_3}}{C}=CH-CH_2CH_2-\underset{\substack{|\\CH_3}}{C}=CH-CH_2OH$$

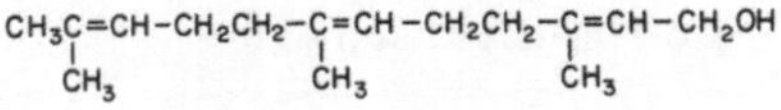

FARNESOLS
trans–trans isomer

J. Am. Chem. Soc. 82:5749 (1960)

▶ $OC_{15}H_{26}$

$$CH_3\underset{\substack{|\\CH_3}}{C}=CH(CH_2)_2\underset{\substack{|\\CH_3}}{C}=CH(CH_2)_2\underset{\substack{|\\CH_3}}{C}=CH-CH_2OH$$

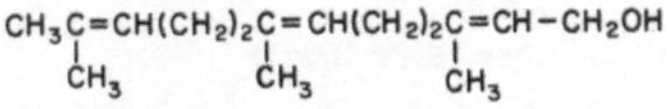

FARNESOLS
trans-cis and cis-cis isomers

Chem.& Ind. 1907 (1961)

▶ $OC_{15}H_{26}$

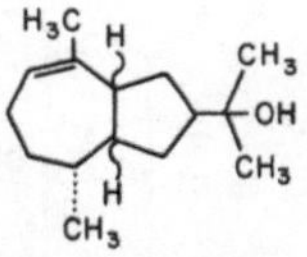

Collection Czech. Chem. Commun. 27:1914 (1962)

▶ $OC_{15}H_{25}$

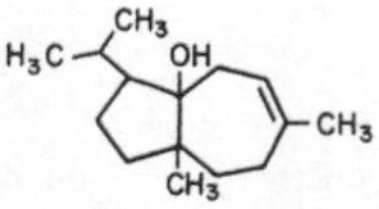

OR

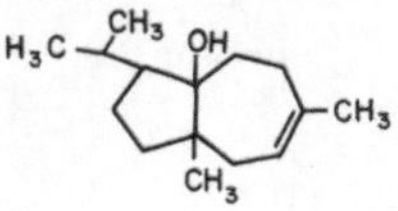

J. Org. Chem. 26:981 (1961)

▶ $OC_{15}H_{26}$

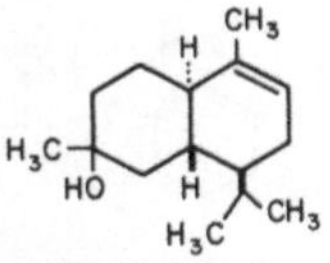

Tetrahedron 15:217 (1961)

▶ $OC_{15}H_{26}$

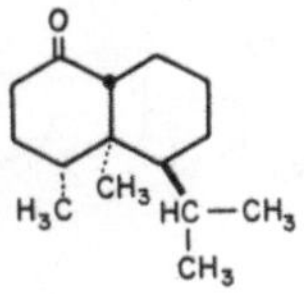

Tetrahedron Letters No. 18:830 (1962)

▶ $OC_{15}H_{26}$

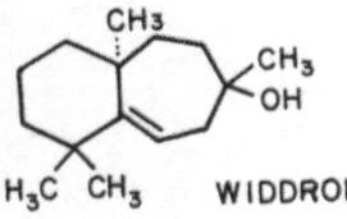

WIDDROL

Chem. & Ind. 1618 (1961)

▶ $OC_{15}H_{26}$

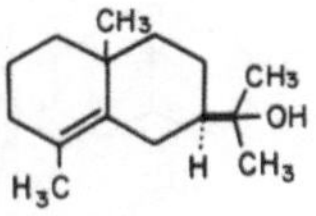

Chem. & Ind. 1759 (1962)

▶ $OC_{15}H_{26}$

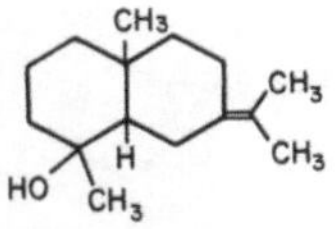

Tetrahedron 18:983 (1962)

▶ $OC_{15}H_{26}$

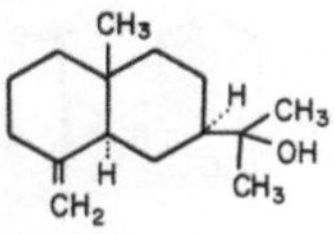

Tetrahedron 18:983 (1962)

▶ $OC_{15}H_{26}$

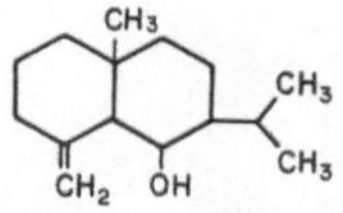


Tetrahedron 18:969 (1962)

▶ $OC_{15}H_{28}$

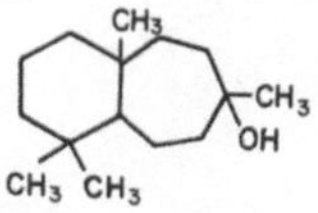


Acta Chem. Scand. 15:1191 (1961)

▶ $OC_{16}H_{12}$

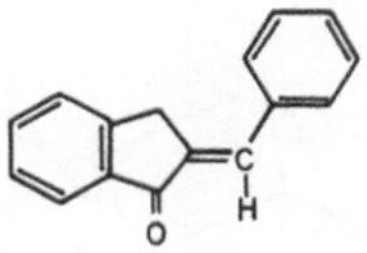


Varian 649

▶ $OC_{16}H_{12}$

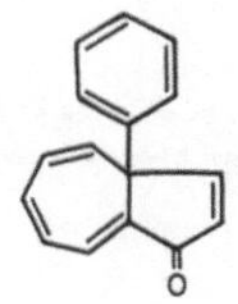


Helv. Chim. Acta 45:600 (1962)

▶ $OC_{16}H_{12}$

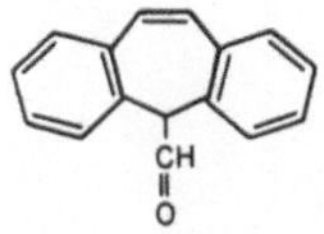


Tetrahedron 18:1007 (1962)

▶ $OC_{16}H_{14}$

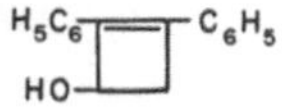


J. Am. Chem. Soc. 84:2793 (1962)

▶ $OC_{16}H_{14}$

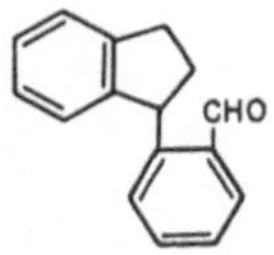


J. Am. Chem. Soc. 84:3531 (1962)

▶ $OC_{16}H_{14}$

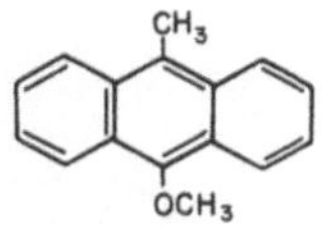


Tetrahedron 18:388 (1962)

▶ $OC_{16}H_{14}$

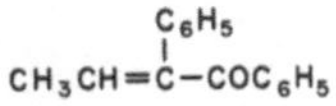


J. Am. Chem. Soc. 84:2793 (1962)

▶ $OC_{16}H_{14}$

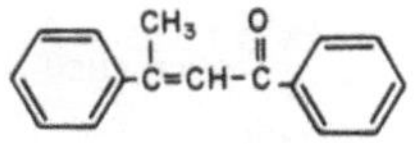


J. Org. Chem. 27:4091 (1962)

▶ $OC_{16}H_{14}$

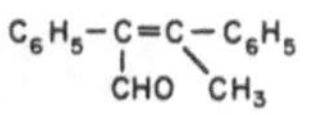


J. Am. Chem. Soc. 84:2793 (1962)

▶ $OC_{16}H_{16}$

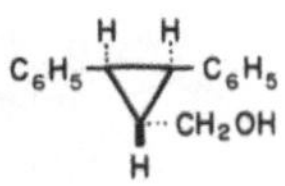


J. Am. Chem. Soc. 84:2793 (1962)

▶ $OC_{16}H_{16}$

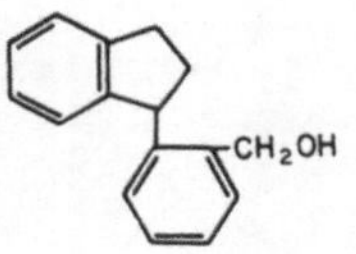


J. Am. Chem. Soc. 84:3531 (1962)

▶ $OC_{16}H_{22}$

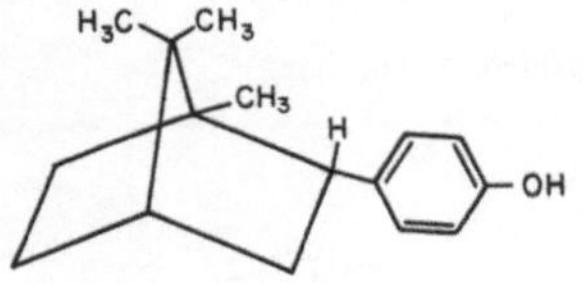


J. Org. Chem. 27:1531 (1962)

▶ $OC_{16}H_{18}$

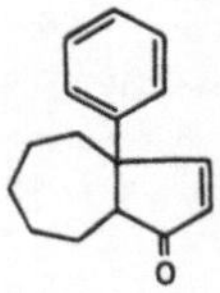


Helv. Chim. Acta 45:600 (1962)

▶ $OC_{16}H_{22}$

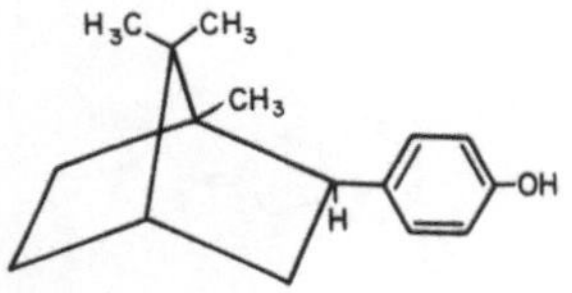


J. Org. Chem. 27:1531 (1962)

▶ $OC_{16}H_{18}$

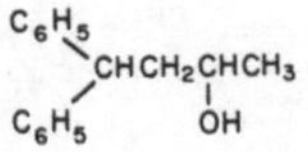


J. Am. Chem. Soc. 84:4831 (1962)

▶ $OC_{17}H_{16}$

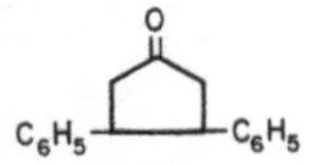


J. Am. Chem. Soc. 83:4838 (1961)

▶ $OC_{16}H_{20}$

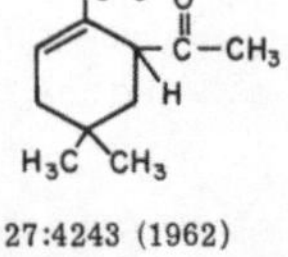


J. Org. Chem. 27:4243 (1962)

▶ $OC_{17}H_{16}$

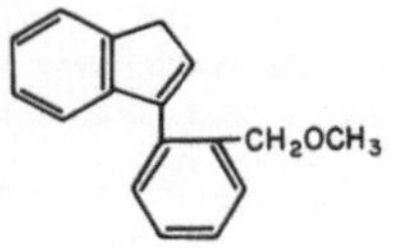


J. Am. Chem. Soc. 84:3531 (1962)

▶ $OC_{16}H_{20}$

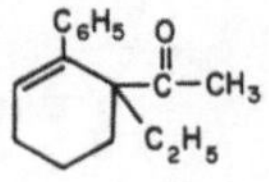


J. Org. Chem. 27:4243 (1962)

▶ $OC_{17}H_{16}$

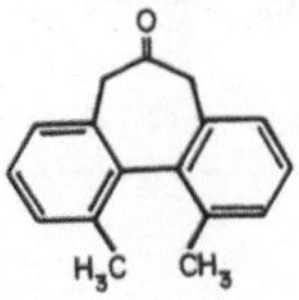


Varian 662

▶ $OC_{17}H_{22}$

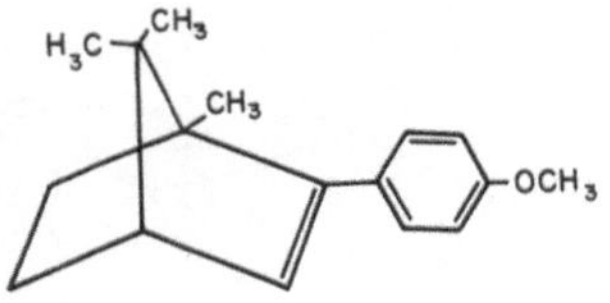


J. Org. Chem. 27:1531 (1962)

▶ $OC_{17}H_{18}$

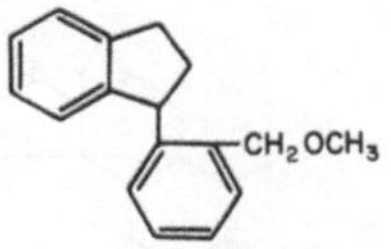


J. Am. Chem. Soc. 84:3531 (1962)

▶ $OC_{17}H_{18}$

$C_6H_5CH_2CH_2COCH_2CH_2C_6H_5$

Bull. Soc. Chim. France 842 (1961)

▶ $OC_{17}H_{22}$

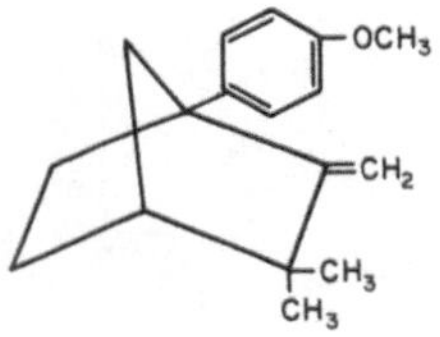


J. Org. Chem. 27:1531 (1962)

▶ $OC_{17}H_{18}$

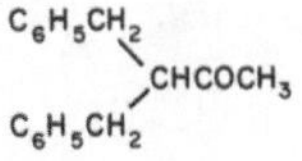


Bull. Soc. Chim. France 842 (1961)

▶ $OC_{17}H_{22}$

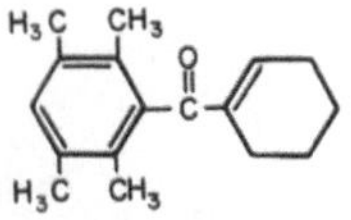


J. Org. Chem. 26:4217 (1961)

▶ $OC_{17}H_{20}$

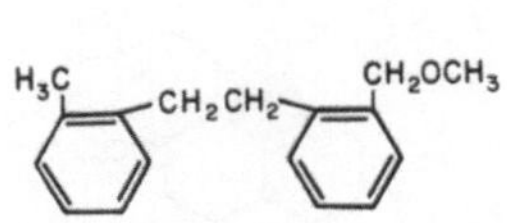


J. Am. Chem. Soc. 83:960 (1961)

▶ $OC_{17}H_{24}$

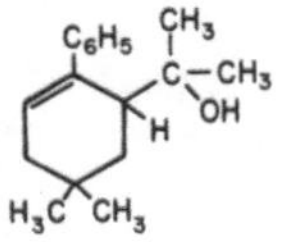


J. Org. Chem. 27:4243 (1962)

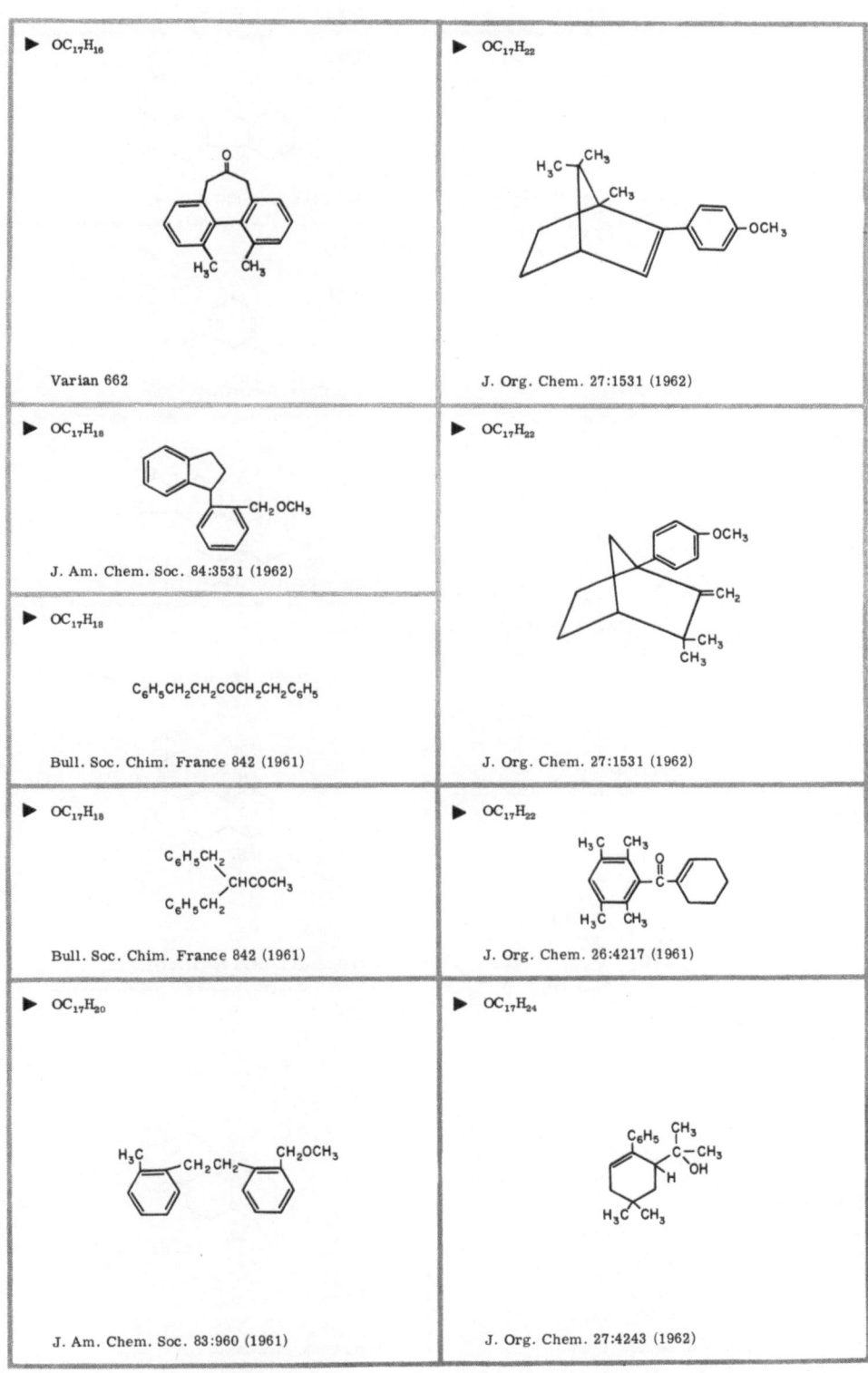

► $OC_{17}H_{24}$

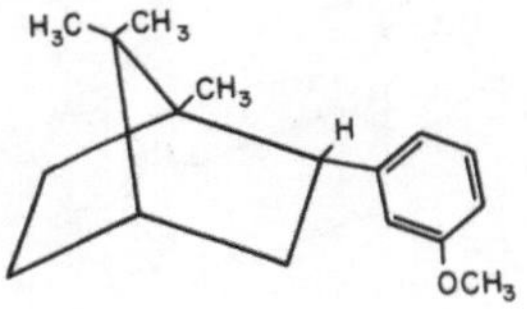


J. Org. Chem. 27:1531 (1962)

► $OC_{17}H_{24}$

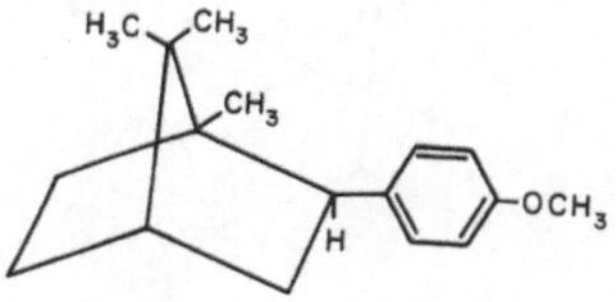


J. Org. Chem. 27:1531 (1962)

► $OC_{17}H_{24}$

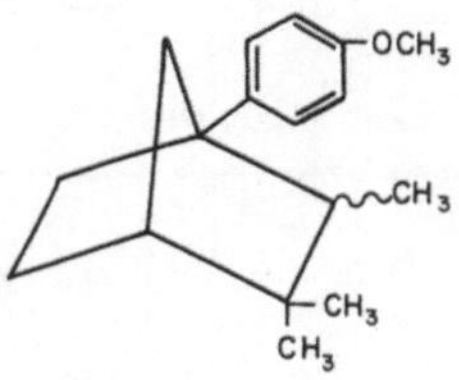


J. Org. Chem. 27:1531 (1962)

► $OC_{17}H_{28}$

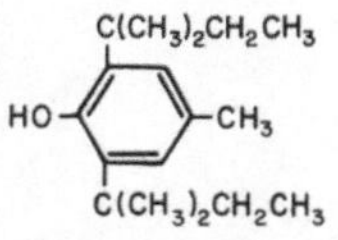


J. Am. Chem. Soc. 84:2258 (1962)

► $OC_{18}H_{12}$

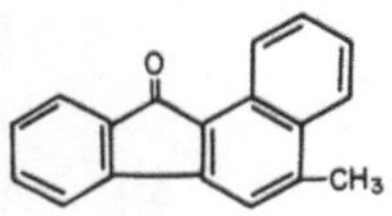


J. Am. Chem. Soc. 83:193 (1961)

► $OC_{18}H_{14}$

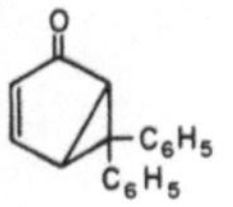


J. Am. Chem. Soc. 84:4527 (1962)

► $OC_{18}H_{22}$

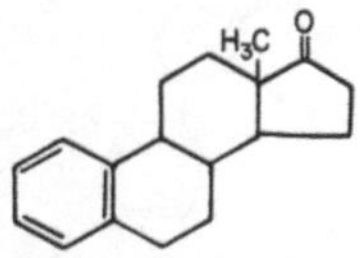


Chem. & Ind. 1716 (1962)

► $OC_{18}H_{26}$

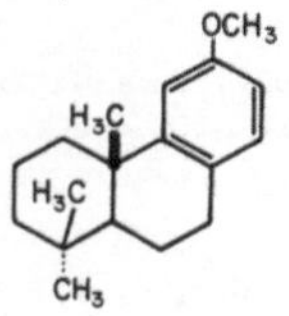


Tetrahedron Letters 360 (1961)

► $OC_{18}H_{32}$

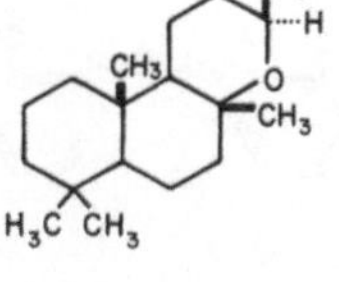


Helv. Chim. Acta 45:400 (1962)

▶ $OC_{18}H_{32}$

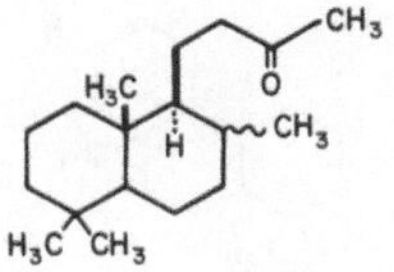


Helv. Chim. Acta 45:400 (1962)

▶ $OC_{19}H_{14}$

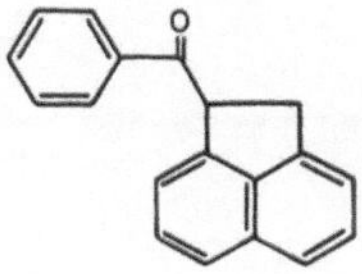


Can. J. Chem. 40:57 (1962)

▶ $OC_{19}H_{16}$

$(C_6H_5)_3COH$

J. Mol. Spectr. 7:32 (1961)
Tetrahedron 16:139 (1961)

▶ $OC_{19}H_{18}$

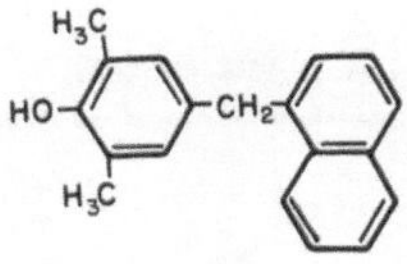


Varian 332

▶ $OC_{19}H_{24}$

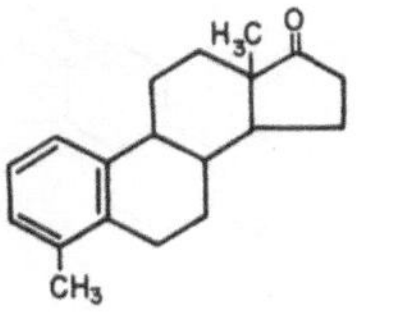


Chem. & Ind. 1716 (1962)

▶ $OC_{19}H_{24}$

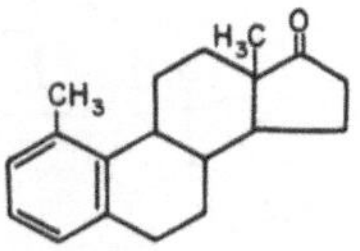


Chem. & Ind. 1716 (1962)

▶ $OC_{19}H_{24}$

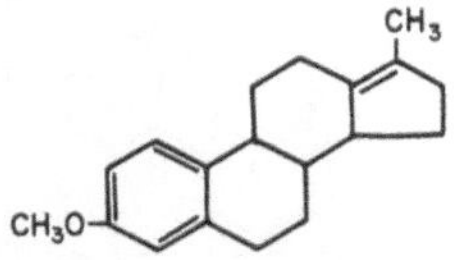


J. Org. Chem. 26:4587 (1961)

▶ $OC_{19}H_{26}$

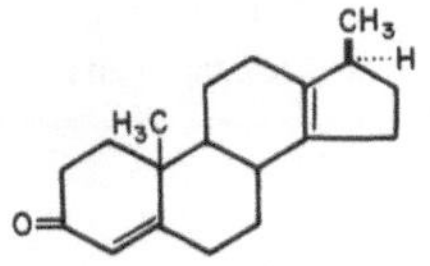


Tetrahedron Letters No. 26:1251 (1962)

▶ $OC_{19}H_{26}$

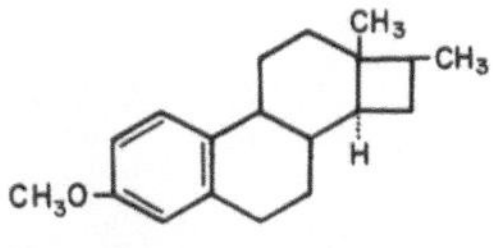


J. Am. Chem. Soc. 84:115 (1962)

▶ $OC_{19}H_{26}$

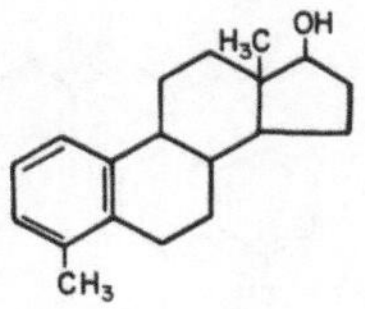


Chem. & Ind. 1716 (1962)

▶ $OC_{19}H_{28}$

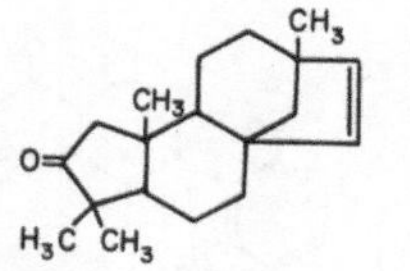


J. Chem. Soc. 4049 (1962)

▶ $OC_{19}H_{30}$

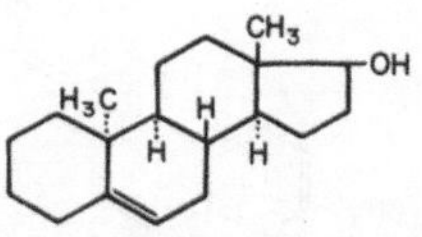


Helv. Chim. Acta 45:2403 (1962)

▶ $OC_{19}H_{30}$

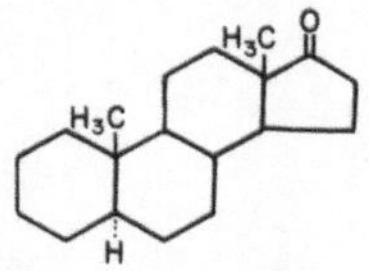


Helv. Chim. Acta 44:1380 (1961)

▶ $OC_{19}H_{30}$

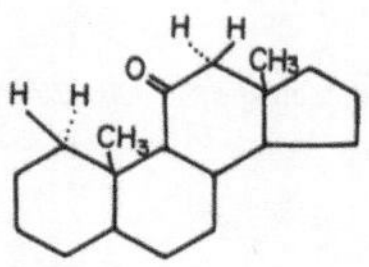


Varian 680

▶ $OC_{19}H_{32}$

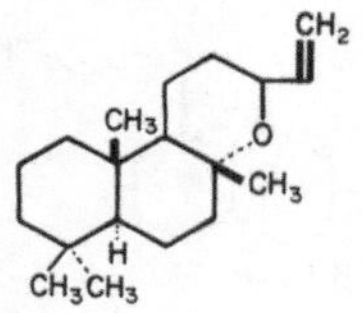


Tetrahedron 18:169 (1962)

▶ $OC_{20}H_{20}$

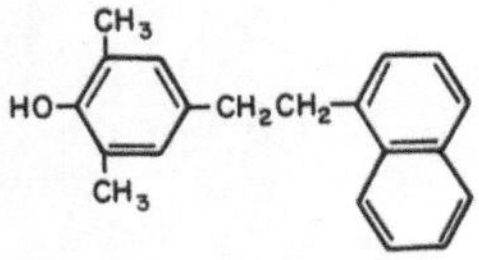


Varian 340

▶ $OC_{20}H_{20}Si$

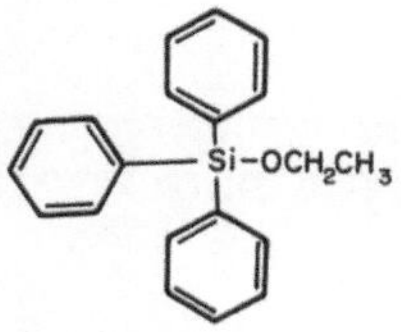


Varian 341

▶ $OC_{20}H_{26}$

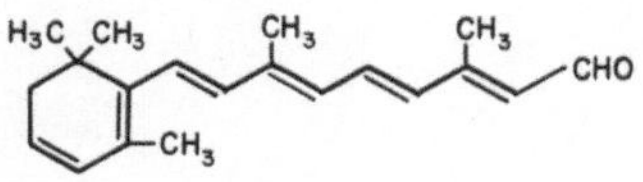


Helv. Chim. Acta 45:548 (1962)

▶ $OC_{20}H_{28}$

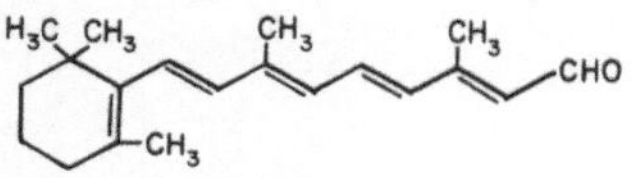


Helv. Chim. Acta 45:548 (1962)

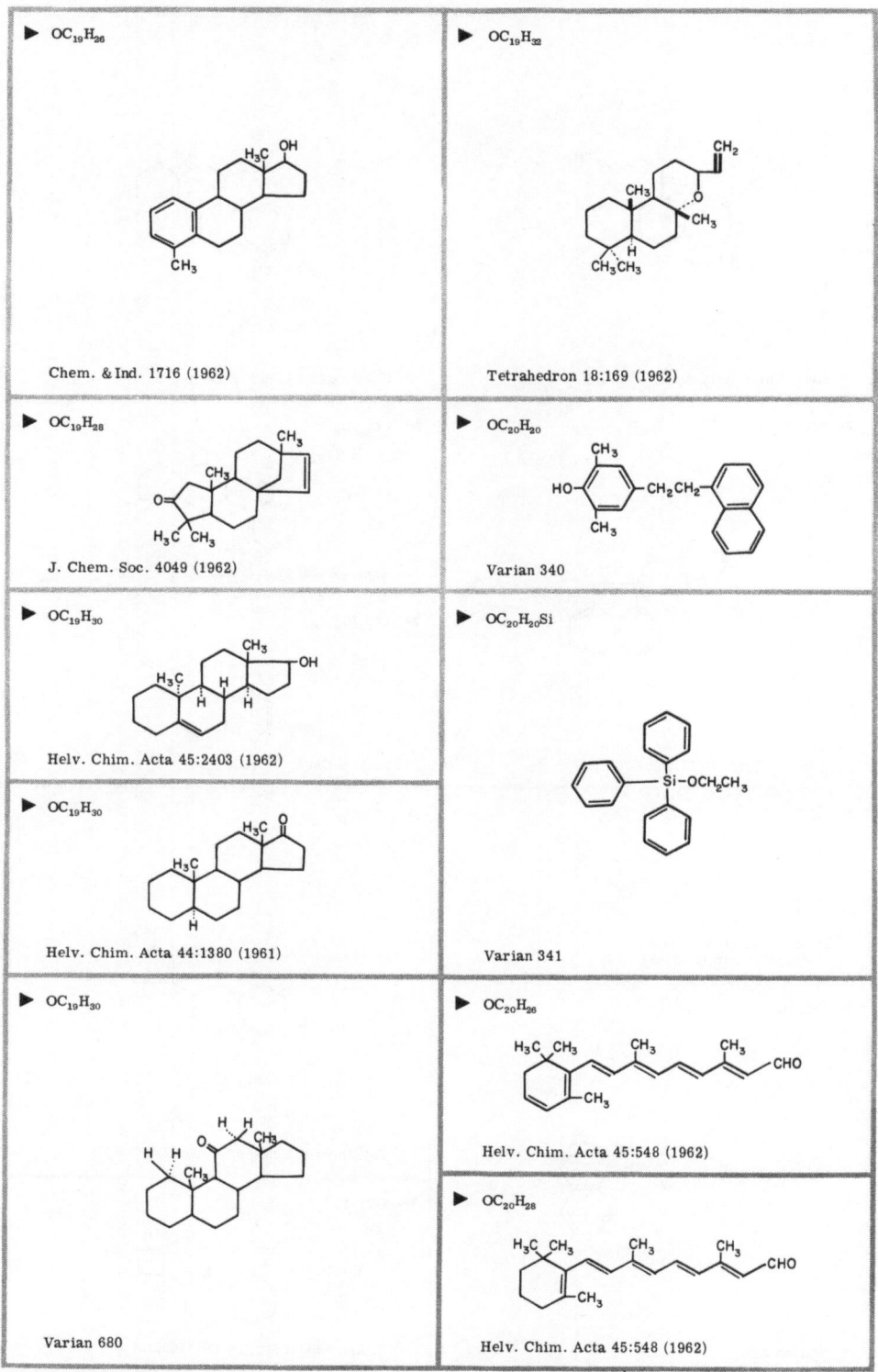

► $OC_{20}H_{28}$

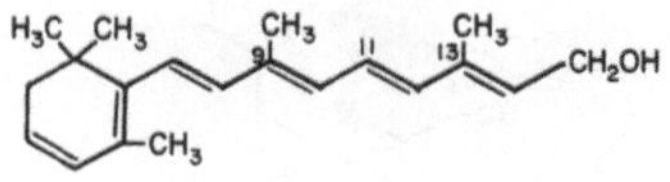


all trans,
13-cis,
11-cis,
9-cis,
9,13-Di-cis and
11,13-Di-cis isomers

Helv. Chim. Acta 45:548 (1962)

► $OC_{20}H_{30}$

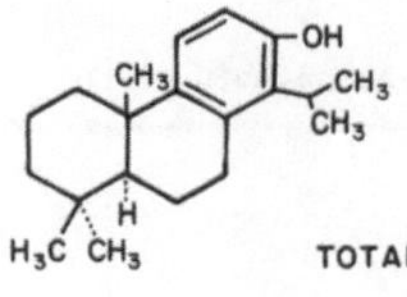


TOTAROL

Tetrahedron Letters 359 (1961)
Tetrahedron 18:465 (1962)

► $OC_{20}H_{30}$

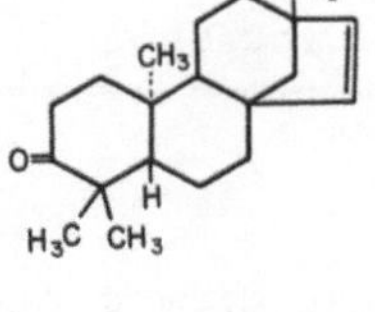


J. Chem. Soc. 4046 (1962)

► $OC_{20}H_{30}$

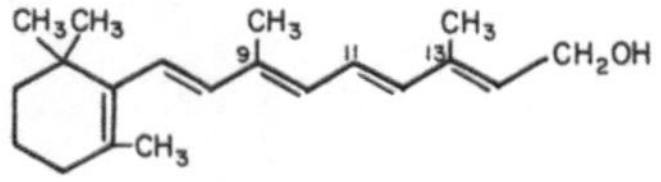


all trans
13-cis,
11-cis,
9-cis
9,13-Di-cis and
11,13-Di-cis isomers

Helv. Chim. Acta 45:548 (1962)

► $OC_{20}H_{30}$

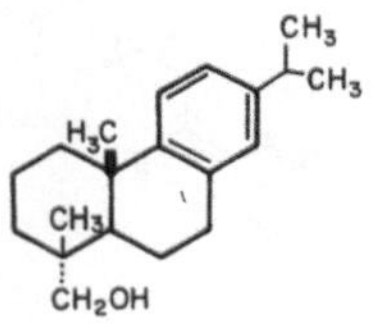


Tetrahedron Letters 360 (1961)

► $OC_{20}H_{34}$

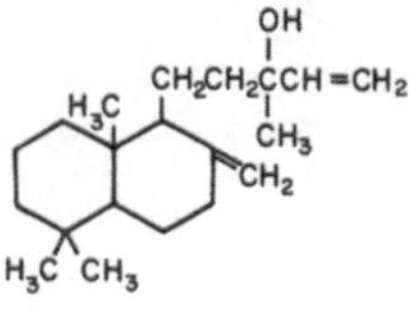


Varian 685

► $OC_{20}H_{40}$

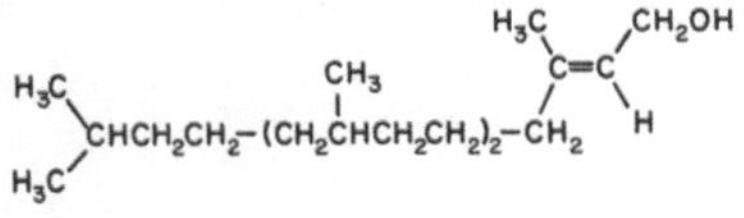


Varian 346

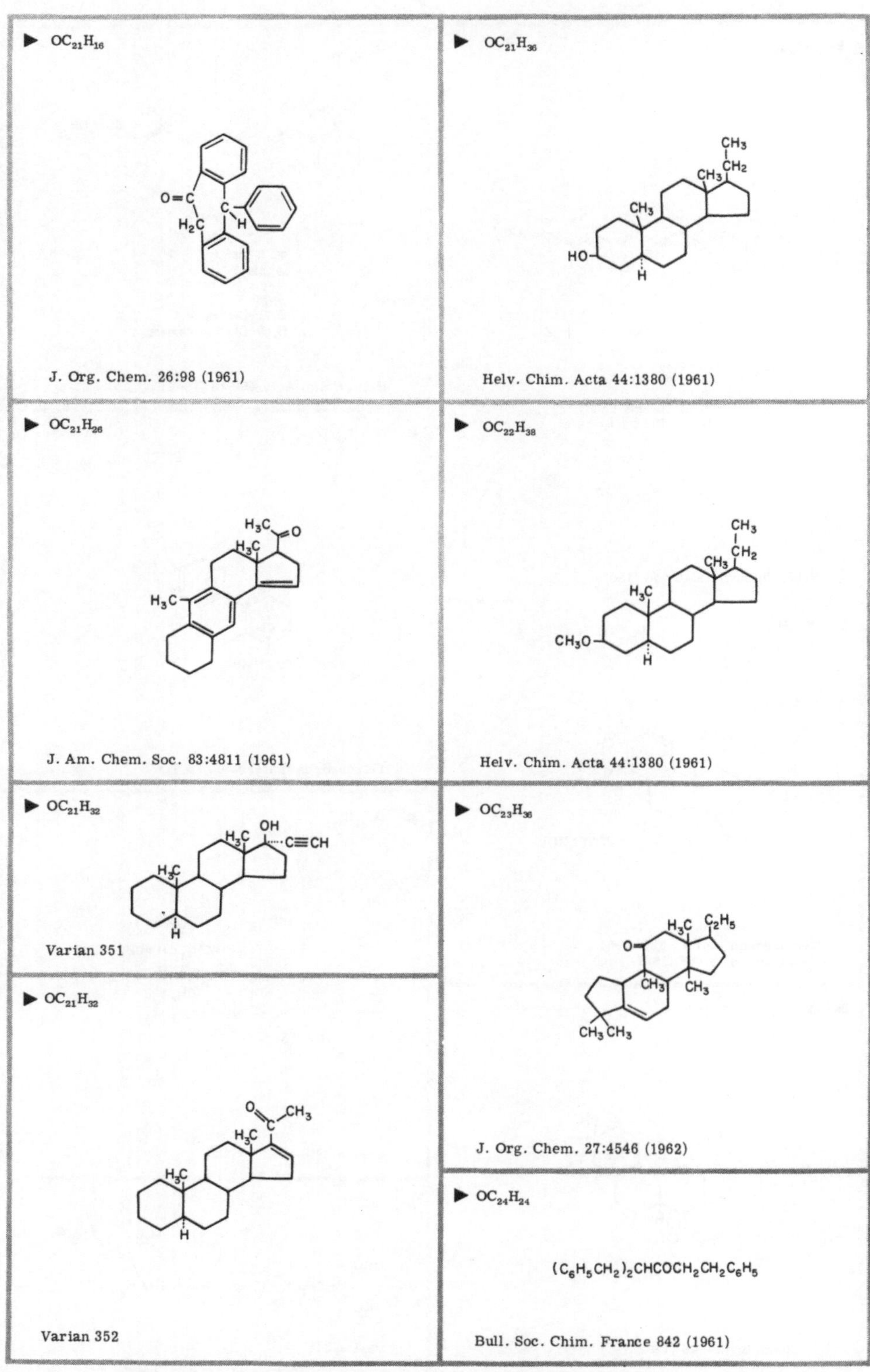

► $OC_{21}H_{16}$

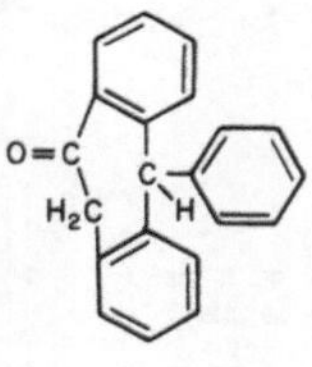

J. Org. Chem. 26:98 (1961)

► $OC_{21}H_{26}$

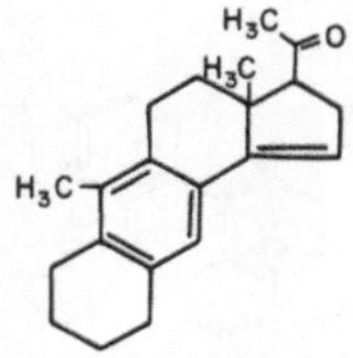

J. Am. Chem. Soc. 83:4811 (1961)

► $OC_{21}H_{32}$

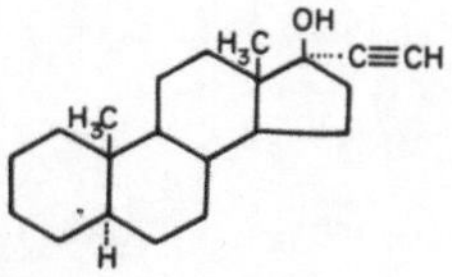

Varian 351

► $OC_{21}H_{32}$

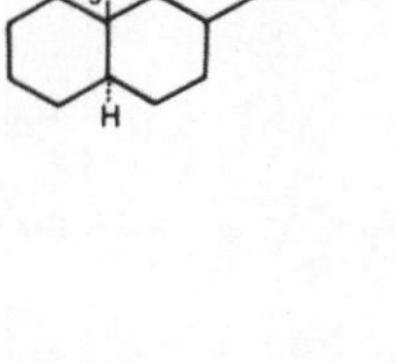

Varian 352

► $OC_{21}H_{36}$

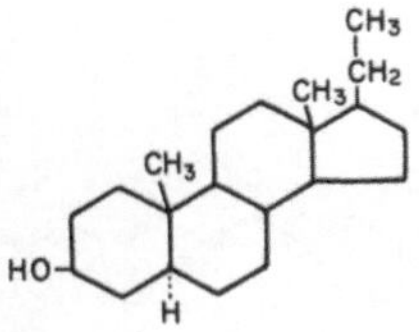

Helv. Chim. Acta 44:1380 (1961)

► $OC_{22}H_{38}$

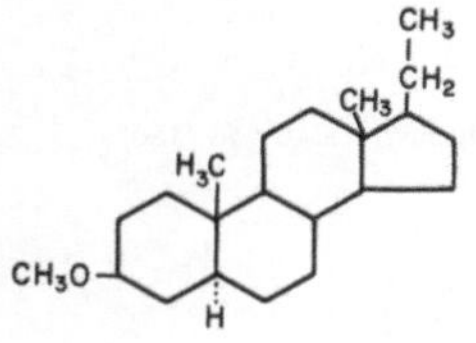

Helv. Chim. Acta 44:1380 (1961)

► $OC_{23}H_{36}$

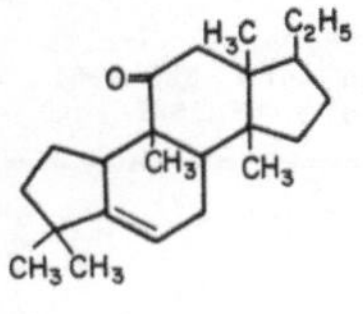

J. Org. Chem. 27:4546 (1962)

► $OC_{24}H_{24}$

$(C_6H_5CH_2)_2CHCOCH_2CH_2C_6H_5$

Bull. Soc. Chim. France 842 (1961)

▶ $OC_{24}H_{24}$

$(C_6H_5CH_2)_3CCOCH_3$

Bull. Soc. Chim. France 842 (1961)

▶ $OC_{24}H_{38}$

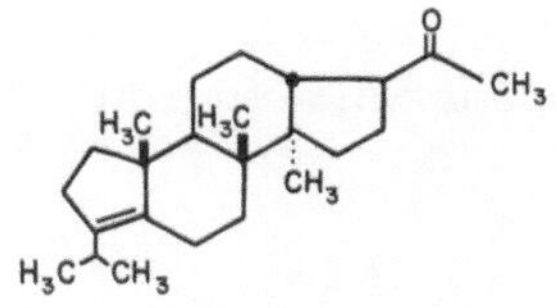


Bull. Soc. Chim. France 1832 (1962)

▶ $OC_{24}H_{38}$

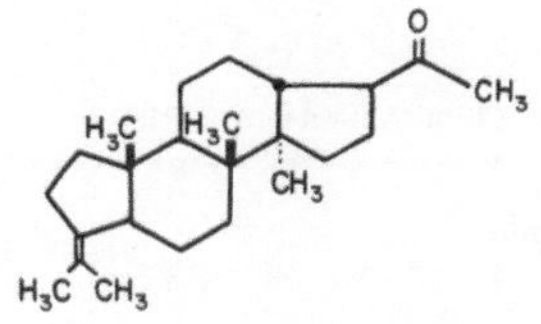


Bull. Soc. Chim. France 1832 (1962)

▶ $OC_{27}H_{44}$

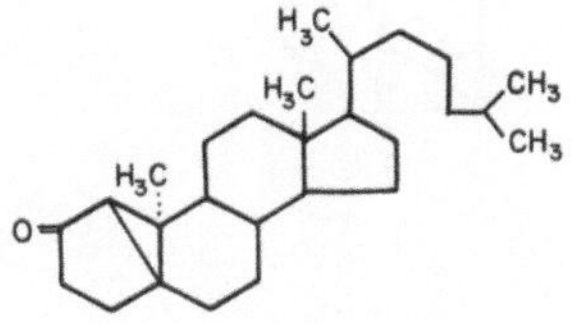


J. Am. Chem. Soc. 84:2268 (1962)

▶ $OC_{27}H_{44}$

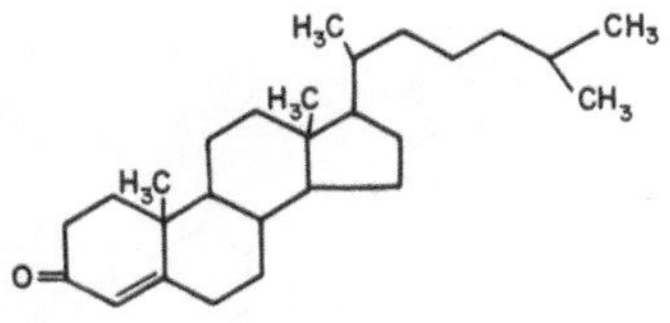


J. Am. Chem. Soc. 84:204 (1962)

▶ $OC_{27}H_{45}D$

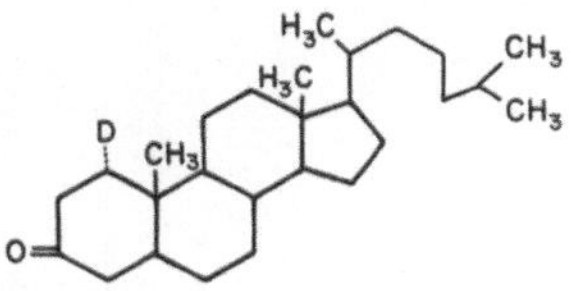


Tetrahedron Letters No. 15:649 (1962)

▶ $OC_{27}H_{46}$

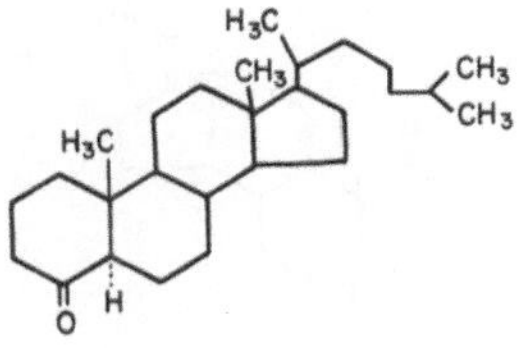


Helv. Chim. Acta 44:1380 (1961)

► $OC_{27}H_{46}$

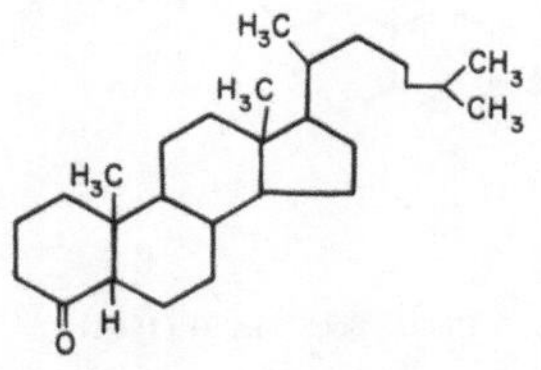


Helv. Chim. Acta 44:1380 (1961)

► $OC_{27}H_{46}$

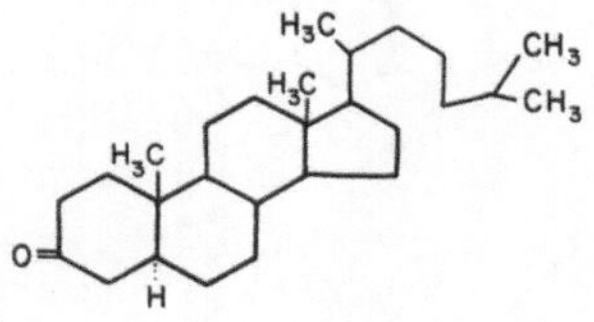


Bull. Soc. Chim. France 2444 (1961)

► $OC_{27}H_{46}$

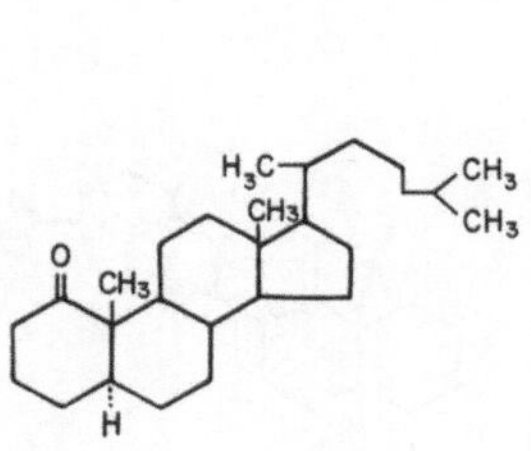


Helv. Chim. Acta 44:1380 (1961)

► $OC_{27}H_{46}$

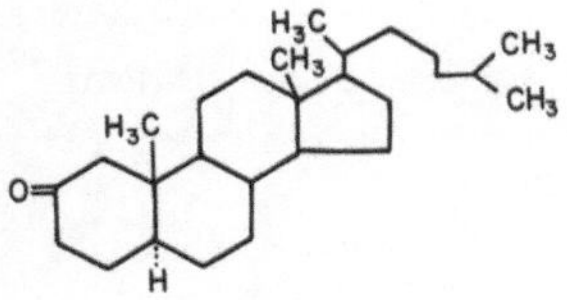


Helv. Chim. Acta 44:1380 (1961)

► $OC_{27}H_{46}$

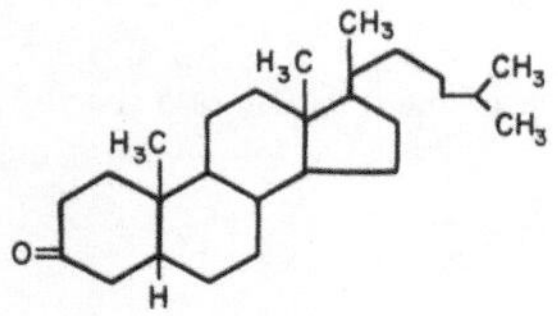


Helv. Chim. Acta 44:1380 (1961)

► $OC_{27}H_{46}$

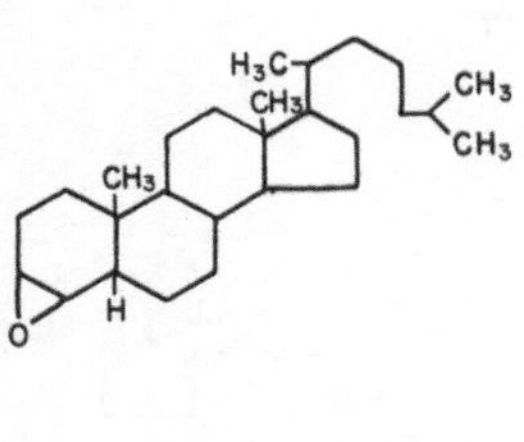


Helv. Chim. Acta 44:1380 (1961)

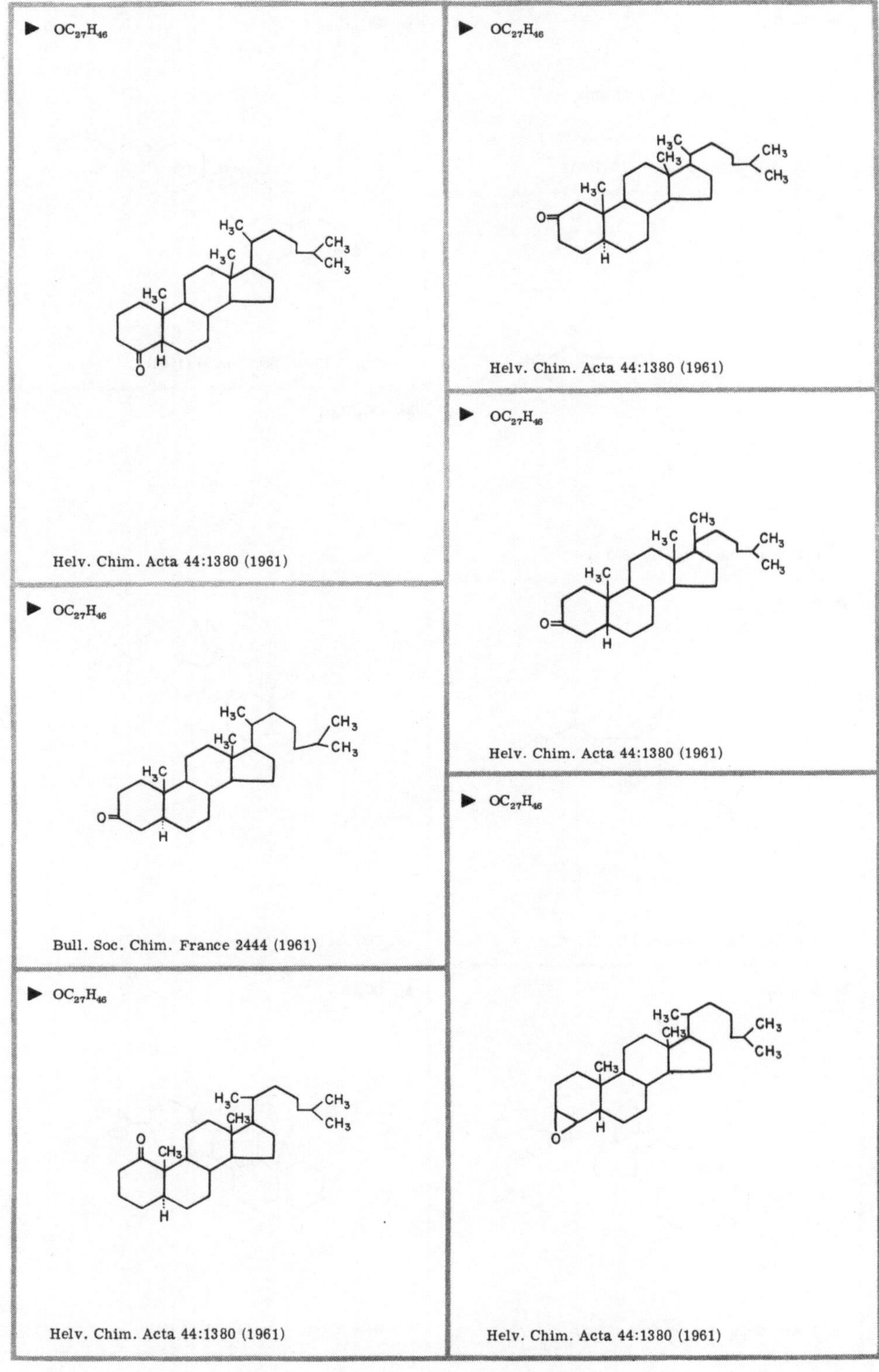

▶ $OC_{27}H_{46}$

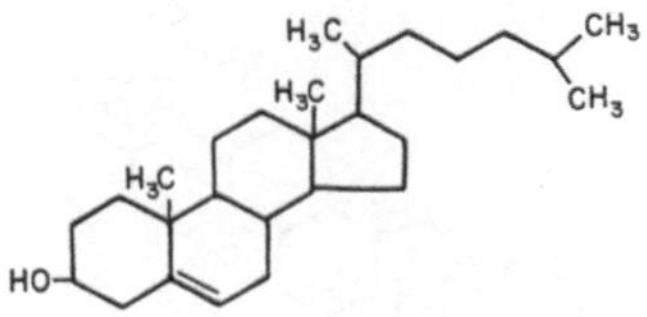

Varian 363
J. Am. Chem. Soc. 84:204 (1962)

▶ $OC_{28}H_{44}$

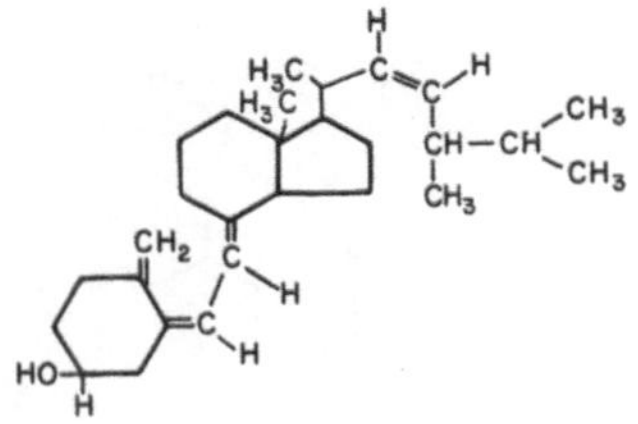

Varian 364

▶ $OC_{27}H_{48}$

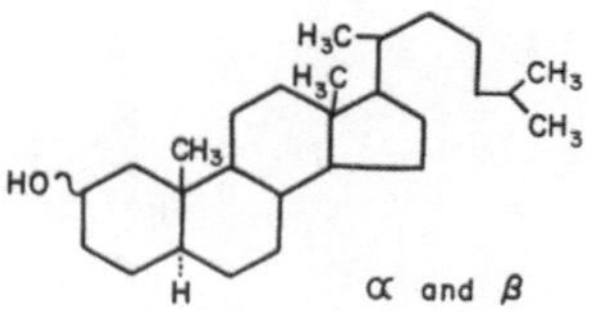


Tetrahedron Letters No. 17:743 (1962)

▶ $OC_{28}H_{46}$

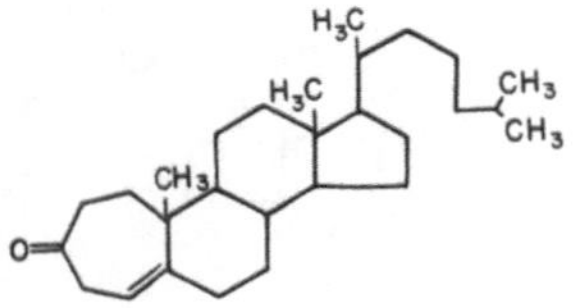

J. Am. Chem. Soc. 84:989 (1962)

▶ $OC_{28}H_{22}$

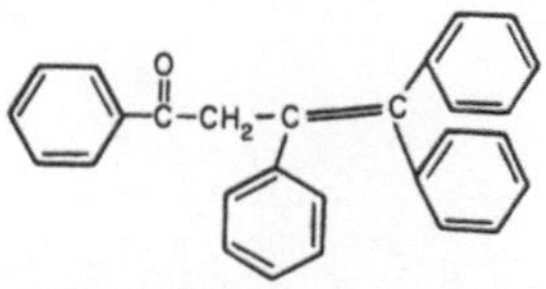

J. Org. Chem. 27:2041 (1962)

▶ $OC_{28}H_{48}$

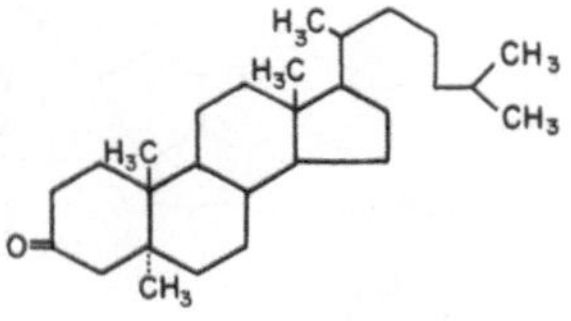

Bull. Soc. Chim. France 2444 (1961)

▶ $OC_{28}H_{42}$

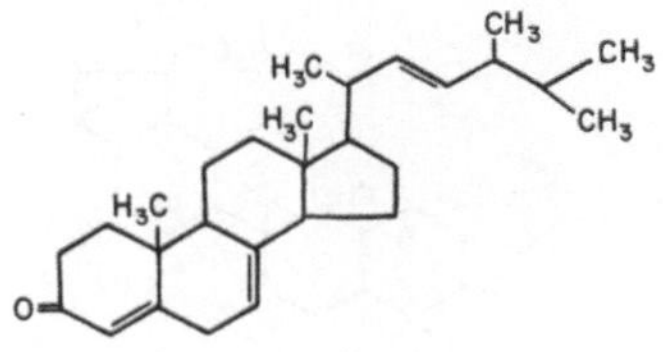

J. Am. Chem. Soc. 84:204 (1962)

▶ $OC_{29}H_{28}$

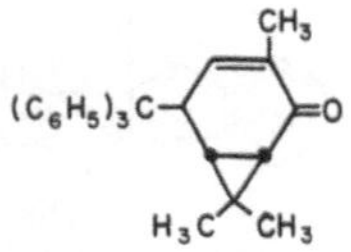

Tetrahedron Letters No. 15:665 (1962)

▶ $OC_{29}H_{48}$

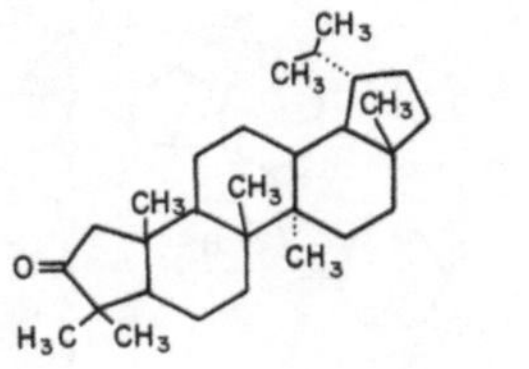

Bull. Soc. Chim. France 2444 (1961); 1137 (1962)

▶ $OC_{29}H_{48}$

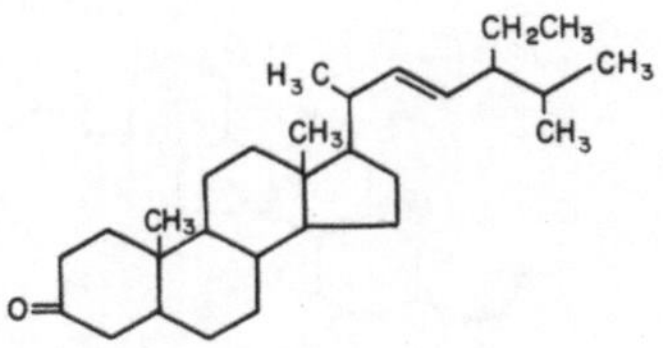

J. Am. Chem. Soc. 84:204 (1962)

▶ $OC_{29}H_{50}$

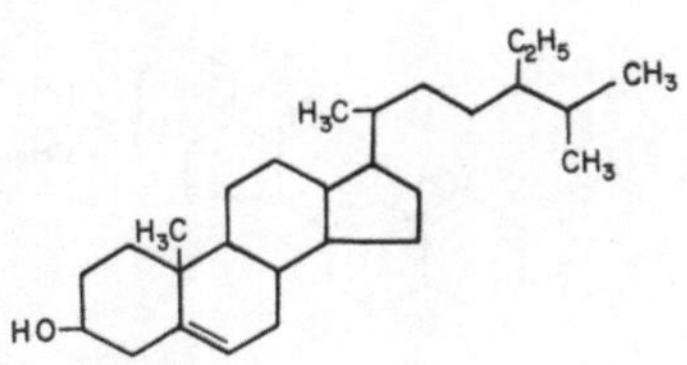

J. Am. Chem. Soc. 84:204 (1962)

▶ $OC_{30}H_{20}$

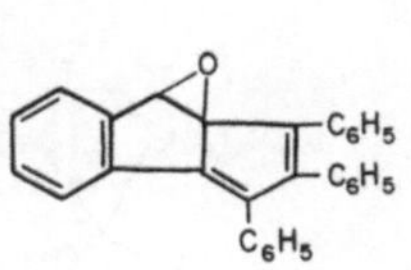

J. Am. Chem. Soc. 84:1505 (1962)

▶ $OC_{30}H_{48}$

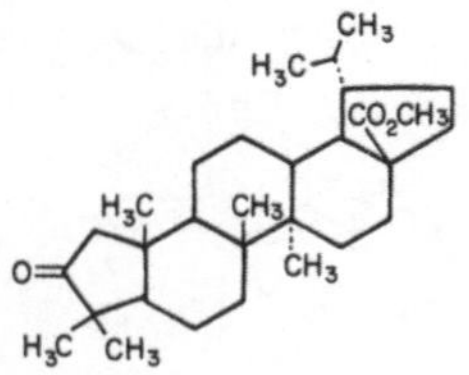

Bull. Soc. Chim. France 1137 (1962)

▶ $OC_{30}H_{48}$

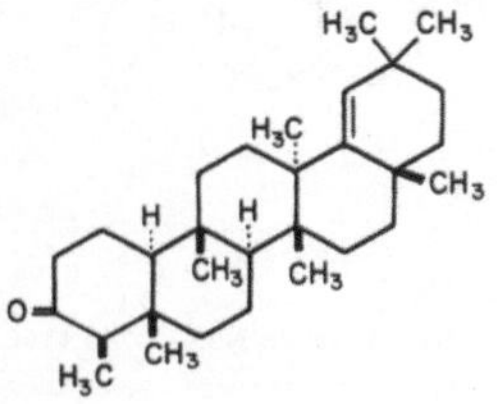

Tetrahedron 15:223 (1961)

▶ $OC_{30}H_{50}$

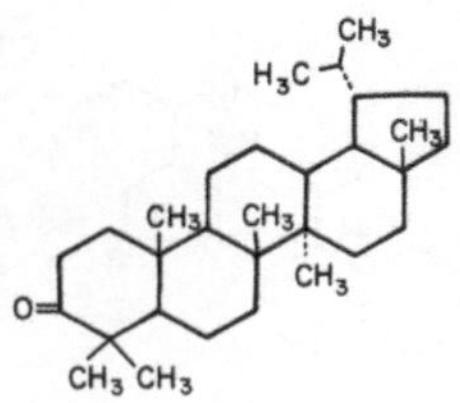

Bull. Soc. Chim. France 2444 (1961); 1137 (1962)

▶ $OC_{30}H_{50}$

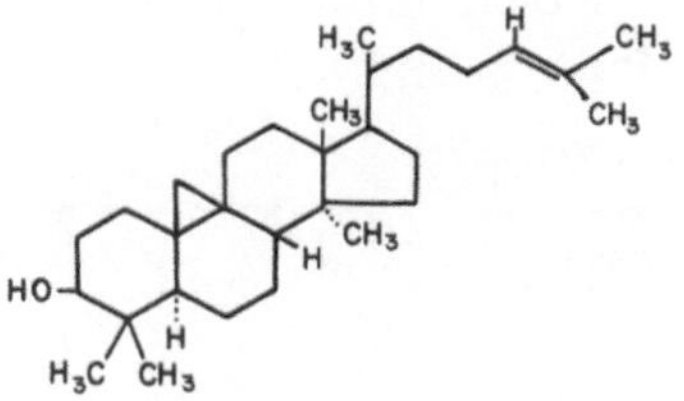


J. Chem. Soc. 4034 (1962)

▶ $OC_{30}H_{52}$

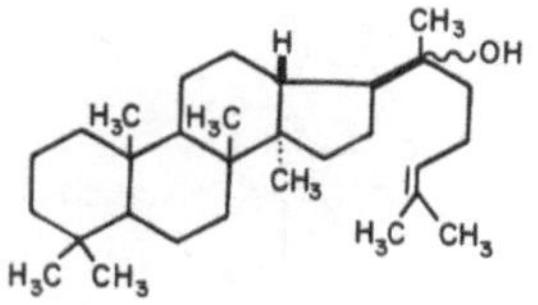


Bull. Soc. Chim. France 1832 (1962)

▶ $OC_{30}H_{50}$

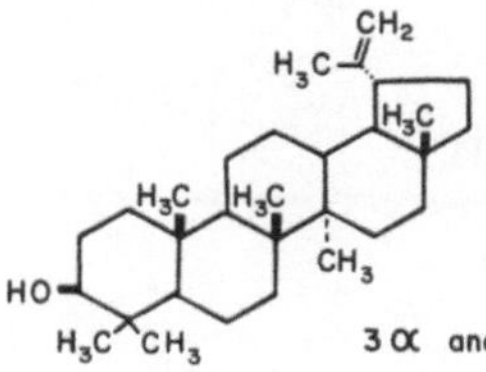


3α and β

Bull. Soc. Chim. France 1137 (1962)

▶ $OC_{30}H_{52}$

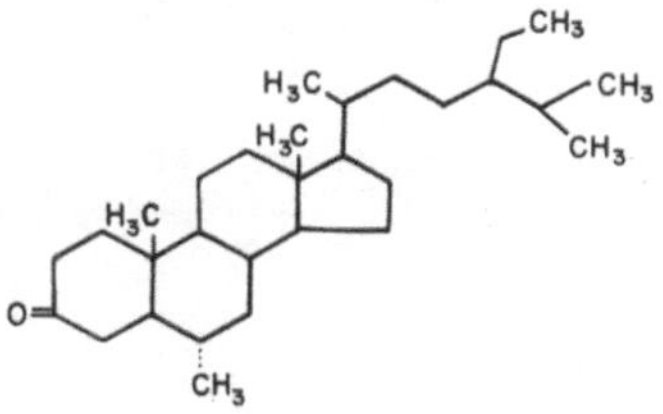


J. Am. Chem. Soc. 84:204 (1962)

▶ $OC_{30}H_{52}$

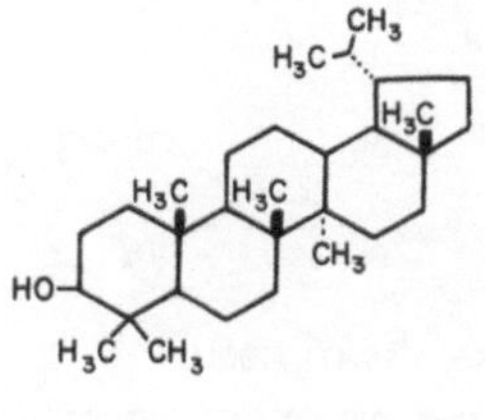


Bull. Soc. Chim. France 1137 (1962)

▶ $OC_{31}H_{22}$

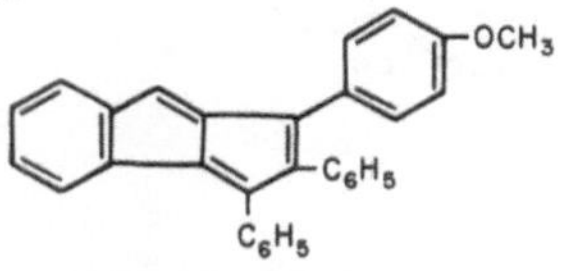


J. Am. Chem. Soc. 84:1505 (1962)

▶ $OC_{31}H_{30}$

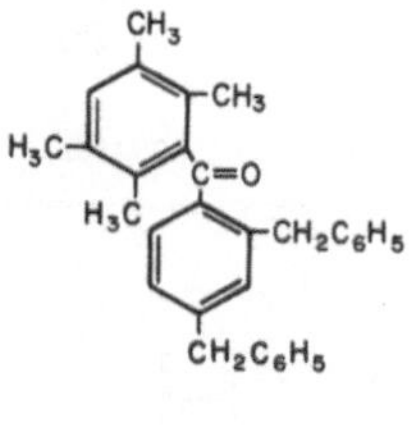


J. Org. Chem. 26:4218 (1961)

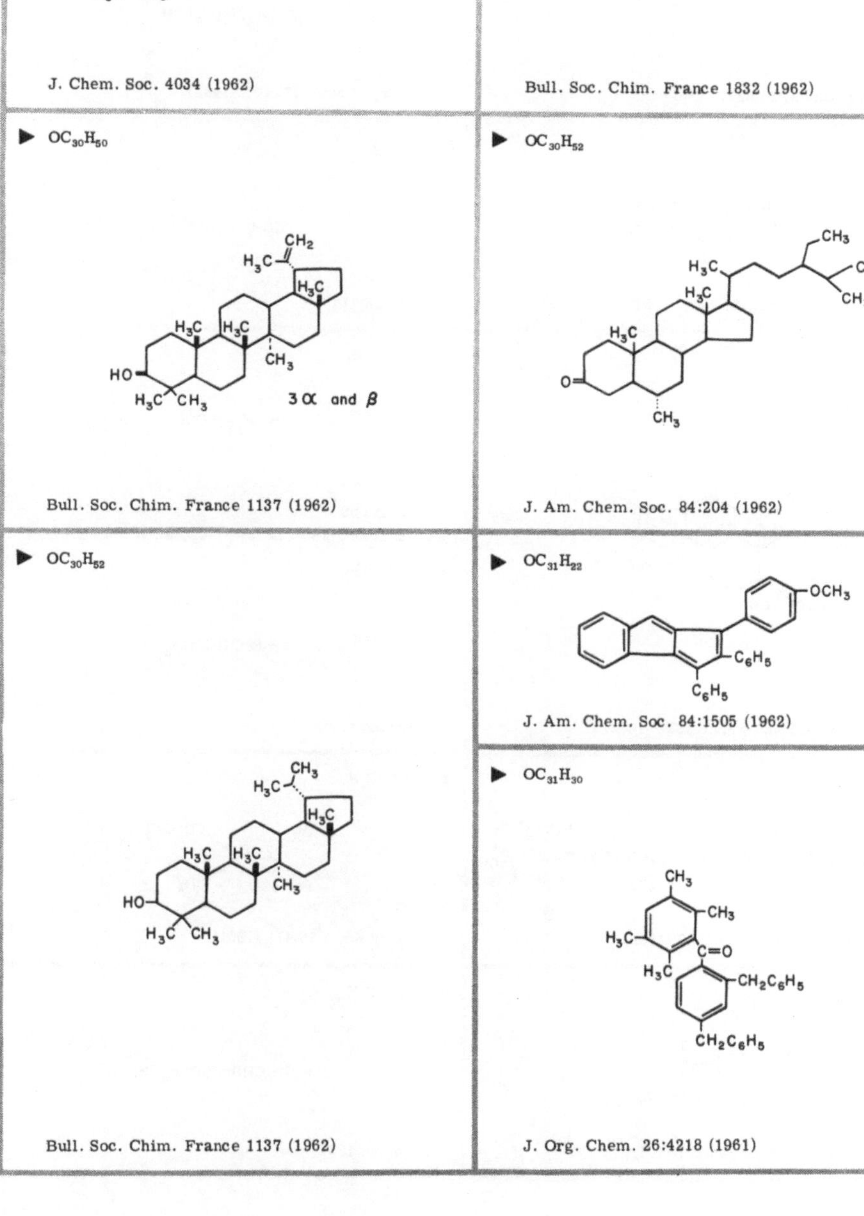

▶ $OC_{31}H_{34}$

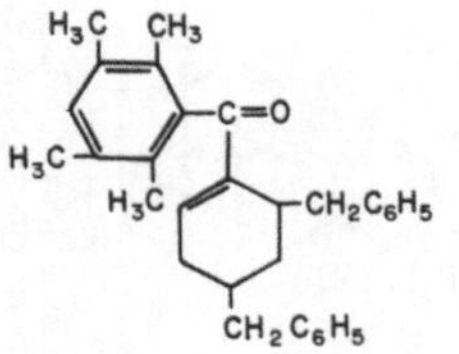

J. Org. Chem. 26:4217 (1961)

▶ $OC_{35}H_{26}$

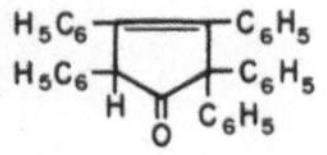

J. Am. Chem. Soc. 83:3727 (1961)

▶ $OC_{39}H_{52}$

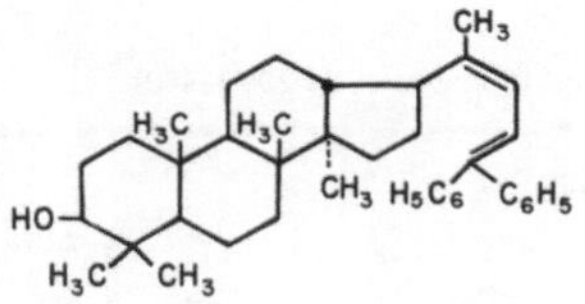

Bull. Soc. Chim. France 1832 (1962)

▶ $OC_{40}H_{56}$

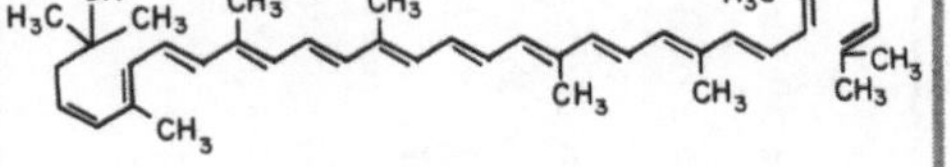

Acta Chem. Scand. 15:2058 (1961)

▶ $OC_{43}H_{30}$

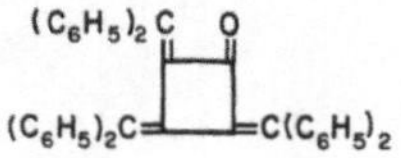

J. Am. Chem. Soc. 84:3397 (1962)

▶ $OC_{45}H_{74}$

$(CH_3)_2C{=}CHCH_2(CH_2C(CH_3){=}CHCH_2)_7CH_2C(CH_3){=}CHCH_2OH$

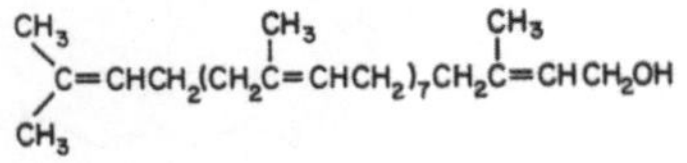

Varian 367

▶ $O_2IF_7C_7H_6$

$C_3F_7CH_2CHIO_2CCH_3$

J. Org. Chem. 27:3036 (1962)

▶ $O_2IC_3H_5$

$ICH_2CH_2COOH$

Varian 27

▶ $O_2BrC_4H_7$

$CH_3CH_2CH(Br)COOH$

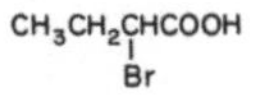

Varian 66

▶ $O_2BrC_4H_9$

$CH_2BrCH(OCH_3)_2$

Varian 419

▶ $O_2BrC_6H_9$

$(CH_3)(Br)C{=}C(CO_2CH_3)(CH_3)$

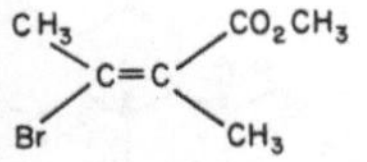

Arkiv Kemi 16:471 (1961)

▶ $O_2BrC_6H_{11}$

$CH_3CH_2CHBrCOOCH_2CH_3$

Varian 137

▶ $O_2BrC_{10}H_{15}$

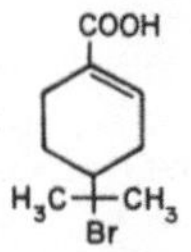


J. Org. Chem. 27:1033 (1962)

▶ $O_2BrC_{19}H_{26}D$

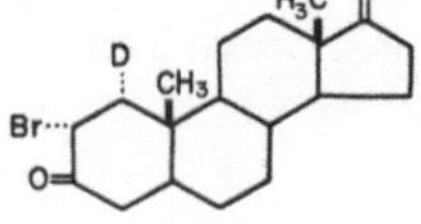


Tetrahedron Letters No. 18:835 (1962)

▶ $O_2BrC_{30}H_{43}$

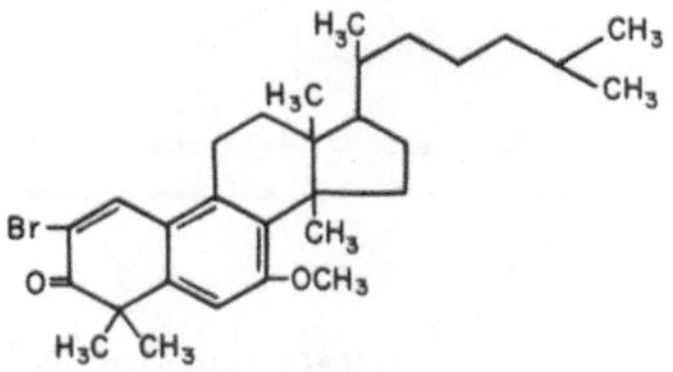


Chem. Pharm. Bull. 9:131 (1961)

▶ $O_2Br_2C_3H_4$

$CH_2Br-CHBr-COOH$

J. Chem. Phys. 37:2053, 3012 (1962)
Proc. Chem. Soc. 144 (1962)
Mol. Phys. 4:321 (1962)

▶ $O_2Br_2C_4H_6$

$CH_3CHBrCHBrCOOH$

Varian 403
Mol. Phys. 5:85 (1962)

▶ $O_2Br_2C_5H_8$

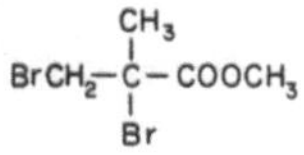


J. Am. Chem. Soc. 83:246 (1961)

▶ $O_2Br_2C_{15}H_{10}$

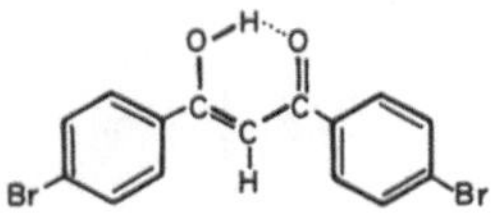


Varian 316

▶ $O_2Br_2C_{16}H_{12}$

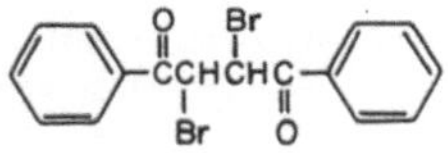


Varian 648

▶ $O_2Br_2C_{18}H_{14}$

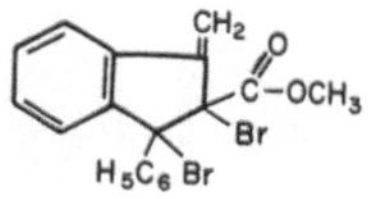


J. Org. Chem. 26:681 (1961)

▶ $O_2Br_4C_{15}H_{12}$

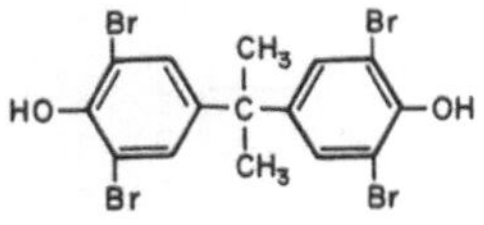


Varian 636

▶ $O_2ClC_2H_3$

$ClCH_2COOH$

J. Am. Chem. Soc. 84:3973 (1962)

| | |
|---|---|
| ▶ $O_2ClC_2H_3Hg$<br>$ClHgC(=O)OCH_3$<br>Chem. & Ind. 668 (1961) | ▶ $O_2ClC_6H_{11}$<br>$CH_3CH_2CH_2CH_2OC(=O)CH_2Cl$<br>Varian 138 |
| ▶ $O_2ClC_3H_5$<br>$H_3C-CH(Cl)-COOH$<br>Varian 25 | ▶ $O_2ClC_8H_7$<br>COCl / $OCH_3$<br>J. Chem. Phys. 37:2594 (1962) |
| ▶ $O_2ClC_4H_5$<br>$(Cl)(H_3C)C=C(COOH)(H)$<br>Varian 53<br>J. Am. Chem. Soc. 83:3897 (1961) | ▶ $O_2ClC_9H_7$<br>$(Cl)(H_5C_6)C=C(COOH)(H)$<br>J. Am. Chem. Soc. 83:3897 (1961) |
| ▶ $O_2ClC_4H_5$<br>$(CH_3)(Cl)C=C(COOH)(H)$<br>Varian 54<br>J. Am. Chem. Soc. 83:3897 (1961) | ▶ $O_2ClC_9H_7$<br>$(C_6H_5)(Cl)C=C(COOH)(H)$<br>J. Am. Chem. Soc. 83:3897 (1961) |
| ▶ $O_2ClC_4H_7$<br>$ClH_2C-C(=O)-OCH_2CH_3$<br>Varian 73 | ▶ $O_2ClC_{12}H_{11}$<br>$Cl-C_6H_4-CH=CH-C(CH_3)=CHCO_2H$<br>cis–cis<br>cis–trans<br>trans–cis<br>trans–trans<br>J. Org. Chem. 27:1990 (1962) |
| ▶ $O_2ClC_5H_3$<br>CCl (=O), O<br>Can. J. Chem. 39:909 (1961) | ▶ $O_2ClC_{13}H_9$<br>O, OH, C, Cl<br>J. Phys. Chem. 65:2023 (1961) |
| ▶ $O_2ClC_6H_5$<br>$ClH_2C$–, O, –CHO<br>Can. J. Chem. 39:909 (1961) | ▶ $O_2ClC_{14}H_{17}$<br>Cl, $OCOCH_3$<br>cis and trans<br>J. Org. Chem. 27:717 (1962) |

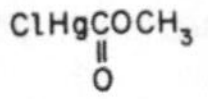

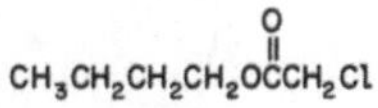

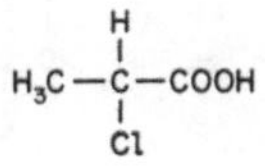

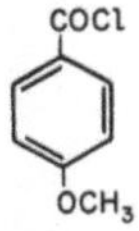

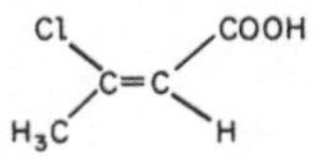

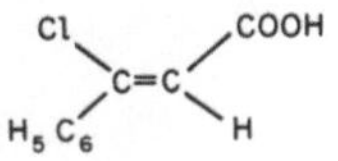

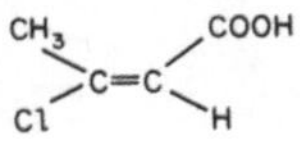

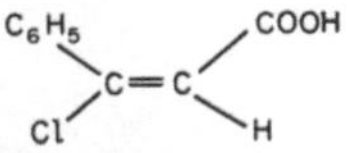

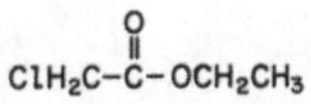

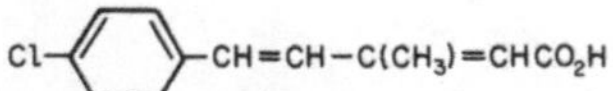

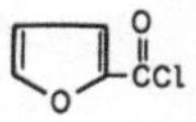

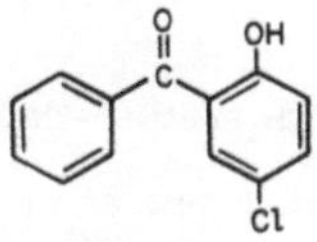

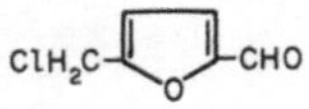

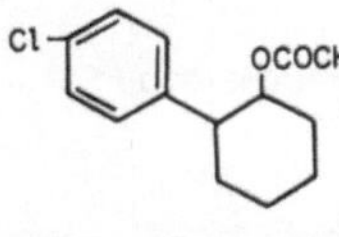

▶ $O_2ClC_{14}H_{17}$

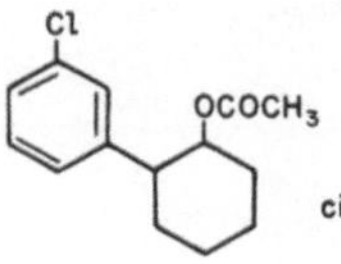


cis and trans

J. Org. Chem. 27:717 (1962)

▶ $O_2ClC_{14}H_{17}$

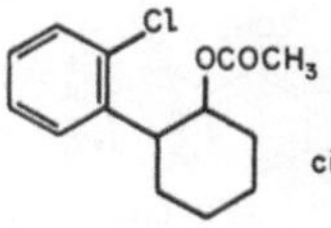


cis and trans

J. Org. Chem. 27:717 (1962)

▶ $O_2ClC_{16}H_{15}$

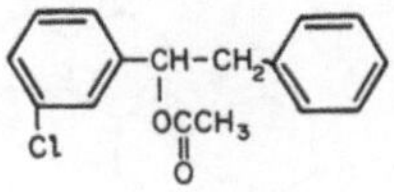


J. Am. Chem. Soc. 83:3647 (1961)

▶ $O_2ClC_{16}H_{15}$

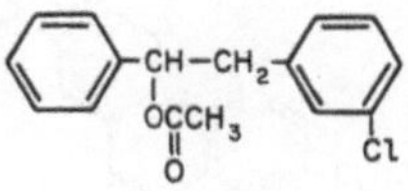


J. Am. Chem. Soc. 83:3647 (1961)

▶ $O_2ClC_{29}H_{47}$

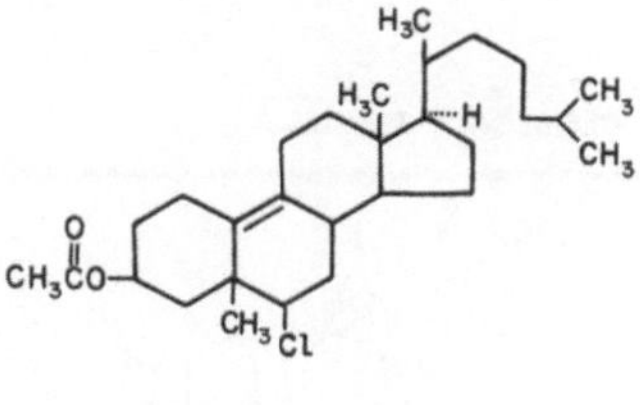


J. Org. Chem. 27:2811 (1962)

▶ $O_2Cl_2C_2H_2$

$Cl_2CHCOOH$

J. Am. Chem. Soc. 84:3973 (1962)

▶ $O_2Cl_2C_4H_6$

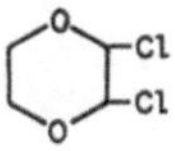


cis and trans

J. Org. Chem. 27:3183 (1962)

▶ $O_2Cl_2C_4H_6$

$CH_3CH_2OC(=O)CHCl_2$

Varian 59

▶ $O_2Cl_2C_5H_6$

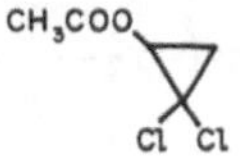


Rec. Trav. Chim. 80:740 (1961)

▶ $O_2Cl_2C_6H_8$

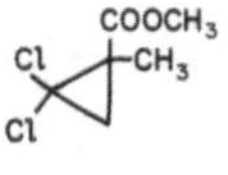


Chem. Ber. 95:2280 (1962)

▶ $O_2Cl_2C_7H_{12}$

$(CH_3)_3C-OCH_2CCl_2CHO$

J. Org. Chem. 26:4184 (1961)

▶ $O_2Cl_2C_8H_4$

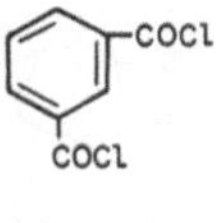


J. Chem. Phys. 37:2594 (1962)

▶ $O_2Cl_2C_8H_4$

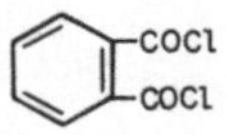


J. Chem. Phys. 37:2594 (1962)

▶ $O_2Cl_6C_{13}H_6$

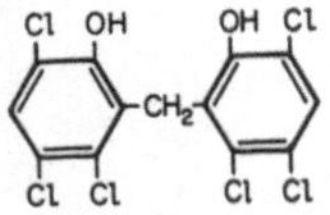


Varian 604

▶ $O_2Cl_2C_{17}H_{25}B$

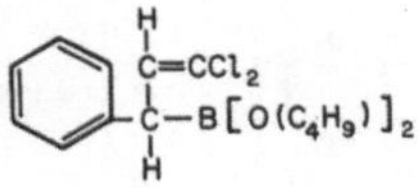


Varian 667

▶ $O_2FC_4H_5$

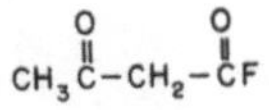


J. Org. Chem. 26:226 (1961)

▶ $O_2Cl_3C_7H_9$

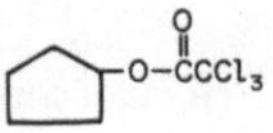


J. Chem. Soc. 1939 (1961)

▶ $O_2FC_{16}H_{17}$

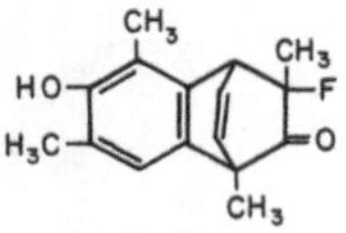


J. Am. Chem. Soc. 83:4197 (1961)

▶ $O_2Cl_3C_{12}H_{13}$

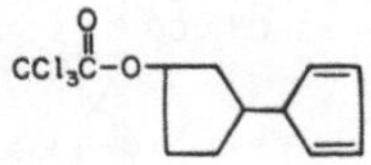


J. Chem. Soc. 1939 (1961)

▶ $O_2FC_{19}H_{25}$

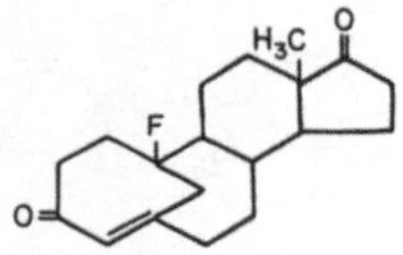


Tetrahedron Letters No. 26:1254 (1962)

▶ $O_2Cl_3C_{12}H_{17}$

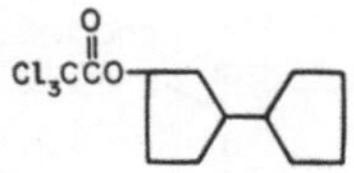


J. Chem. Soc. 1939 (1961)

▶ $O_2FC_{20}H_{25}$

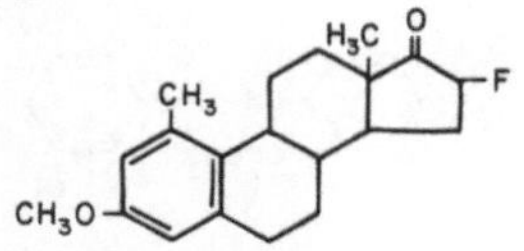


J. Am. Chem. Soc. 84:3784 (1962)

▶ $O_2Cl_3C_{17}H_{26}B$

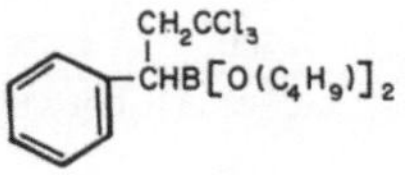


Varian 668

▶ $O_2F_2C_3H_4$

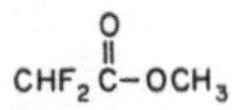


J. Chem. Soc. 3829 (1962)

▶ $O_2Cl_4C_4H_4$

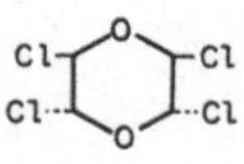


J. Org. Chem. 27:3183 (1962)

▶ $O_2F_2C_{16}H_{18}$

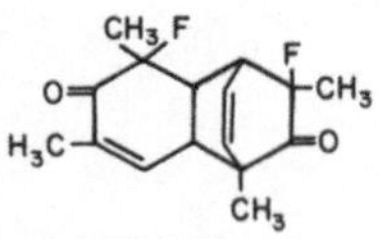


J. Am. Chem. Soc. 83:4197 (1961)

▶ $O_2F_2C_{19}H_{21}$

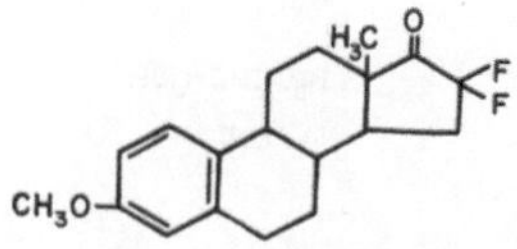

J. Am. Chem. Soc. 84:3784 (1962)

▶ $O_2F_2C_{20}H_{24}$

J. Am. Chem. Soc. 84:3784 (1962)

▶ $O_2F_2C_{21}H_{28}$

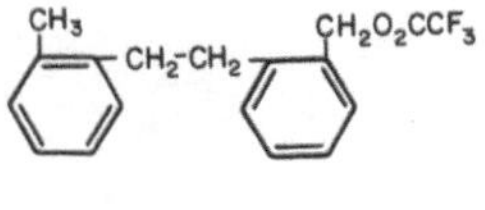

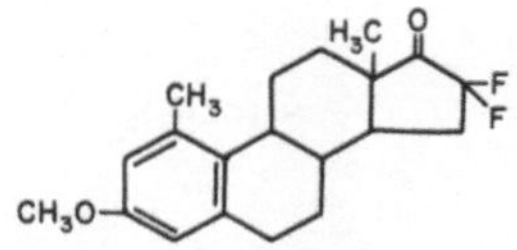

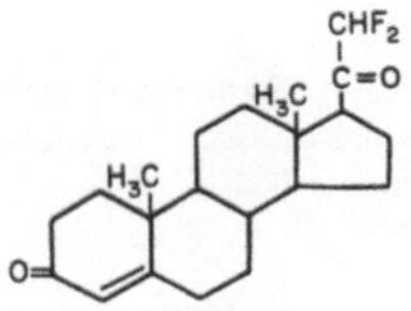

J. Am. Chem. Soc. 84:673 (1962)

▶ $O_2F_2C_{23}H_{36}$

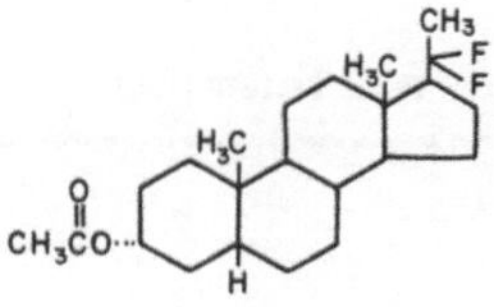

J. Org. Chem. 27:3167 (1962)

▶ $O_2F_3C_6H_3$

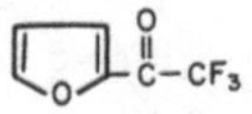

Can. J. Chem. 39:909 (1961)

▶ $O_2F_3C_{18}H_{17}$

J. Am. Chem. Soc. 83:960 (1961)

▶ $O_2F_5C_4H_3$

$CH_3OCF_2CF_2\overset{O}{\overset{\|}{C}}F$

J. Am. Chem. Soc. 84:4275 (1962)

▶ $O_2CH_2$

$Cu(HCO_2)_2 \cdot 4H_2O$

Bull. Chem. Soc. Japan 35:1205 (1962)

▶ $O_2C_2H_3D$

$CH_2DCOOH$

J. Chem. Phys. 37:3012 (1962)

▶ $O_2C_2H_4$

$H\overset{O}{\overset{\|}{C}}{}^{13}OCH_3$

Varian 9
J. Am. Chem. Soc. 84:2649 (1962)

▶ $O_2C_2H_4$

$CH_3COOH$

Varian 8
J. Chem. Phys. 37:2198 (1962)

▶ $O_2C_3H_4$

J. Am. Chem. Soc. 83:3504 (1961)

| | |
|---|---|
| ▶ $O_2C_3H_4$<br><br>Varian 409 | ▶ $O_2C_3H_8$<br>$H_3CCH(OH)CH_2OH$<br>Varian 45 |
| ▶ $O_2C_3H_4$<br>$CH_2{=}CH{-}COOH$<br>J. Chem. Phys. 36:1951 (1962) | ▶ $O_2C_3H_8$<br>$CH_2(OCH_3)_2$<br>J. Am. Chem. Soc. 84:3972 (1962) |
| ▶ $O_2C_3H_4$<br>$H_2C{=}CH{-}O{-}C(=O){-}H$<br>J. Chem. Phys. 36:2235 (1962) | ▶ $O_2C_4H_4$<br><br>Can. J. Chem. 39:220, 909 (1961) |
| ▶ $O_2C_3H_4$<br><br>J. Chem. Phys. 34:980 (1961) | ▶ $O_2C_4H_4$<br><br>Varian 51 |
| ▶ $O_2C_3H_6$<br>$C_2H_5CO_2H$<br>J. Chem. Phys. 34:1099 (1961)<br>Can. J. Chem. 40:1522 (1962)<br>J. Am. Chem. Soc. 84:34 (1962) | ▶ $O_2C_4H_4$<br><br>J. Chem. Phys. 34:1470 (1961) |
| ▶ $O_2C_3H_6$<br><br>Can. J. Chem. 40:1956 (1962) | ▶ $O_2C_4H_4$<br><br>J. Am. Chem. Soc. 84:3786 (1962) |
| ▶ $O_2C_3H_6$<br>$H\overset{O}{\overset{\Vert}{C}}{}^{13}OCH_2CH_3$<br>J. Am. Chem. Soc. 84:2649 (1962) | ▶ $O_2C_4H_6$<br><br>Varian 61<br>J. Mol. Phys. 4:385 (1961) |

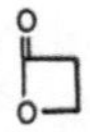

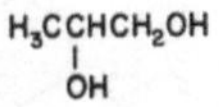

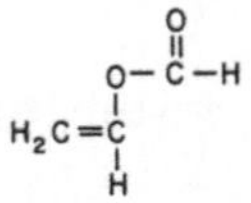

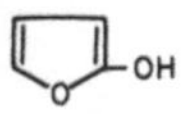

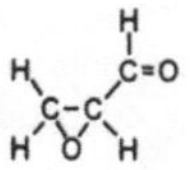

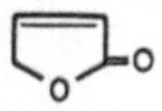

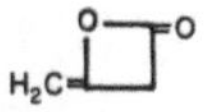

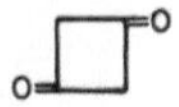

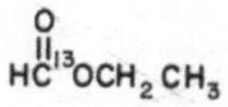

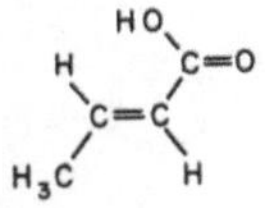

▶ $O_2C_4H_6$

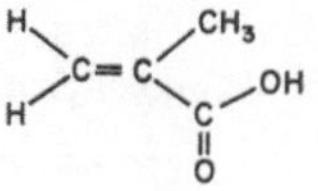

Varian 62

▶ $O_2C_4H_8$

$(CH_3)_2CHCOOH$

Varian 415
J. Chem. Phys. 34:1099 (1961)
J. Am. Chem. Soc. 83:1230 (1961)

▶ $O_2C_4H_6$

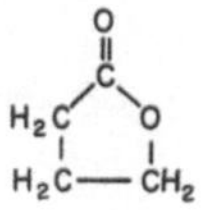

Varian 63

▶ $O_2C_4H_8$

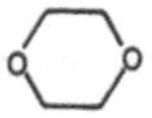

J. Chem. Soc. 2576 (1962)
J. Org. Chem. 27:3183 (1962)

▶ $O_2C_4H_6$

$H_2C=CH-COOCH_3$

Varian 64
J. Chem. Phys. 34:295 (1961)

▶ $O_2C_4H_8$

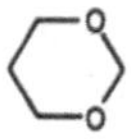

Tetrahedron Letters No. 16:686 (1962)

▶ $O_2C_4H_6$

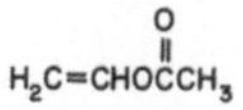

Varian 65

▶ $O_2C_4H_6$

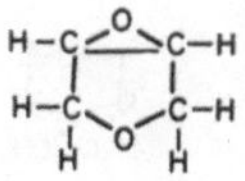

Varian 405

▶ $O_2C_4H_8$

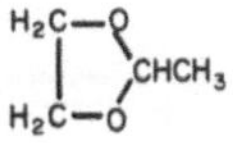

J. Am. Chem. Soc. 84:4876 (1962)

▶ $O_2C_4H_8$

$(CH_3)_2CHC^{13}OOH$

J. Am. Chem. Soc. 83:1230 (1961)

▶ $O_2C_4H_8$

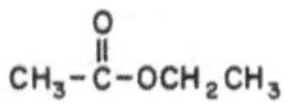

Varian 79
Bull. Soc. Chim. France 1679 (1961)
J. Am. Chem. Soc. 84:37 (1962)

▶ $O_2C_4H_8$

$CH_3CH_2CH_2CO_2H$

J. Chem. Phys. 34:1094, 1099 (1961)

▶ $O_2C_4H_8$

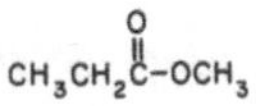

J. Am. Chem. Soc. 84:37 (1962)

▶ $O_2C_4H_{10}$

$CH_2OH(CH_2)_2CH_2OH$

Tetrahedron 18:791 (1962)

▶ $O_2C_5H_8$

$CH_3\overset{O}{\overset{\|}{C}}CH_2\overset{O}{\overset{\|}{C}}CH_3$

KETO and ENOL

J. Am. Chem. Soc. 83:2099 (1961)
Bull. Soc. Chim. France 1 (1962)

▶ $O_2C_4H_{10}$

$H_3CCH(OCH_3)_2$

Can. J. Chem. 40:1956 (1962)

▶ $O_2C_5H_8$

$CH_2{=}CH{-}COOC_2H_5$

J. Chem. Phys. 36:1951 (1962)

▶ $O_2C_4H_{10}$

$H_3C\underset{}{\overset{OH}{\overset{|}{C}}}HCH_2CH_2OH$

Varian 86

▶ $O_2C_5H_8$

$H_3CCH{=}\underset{CH_3}{\underset{|}{C}}{-}CO_2H$

cis and trans

Can. J. Chem. 39:505 (1961)
J. Am. Chem. Soc. 83:922 (1961)

▶ $O_2C_4H_{10}$

$H_3C{-}\overset{OH}{\overset{|}{C}}H{-}\overset{OH}{\overset{|}{C}}H{-}CH_3$

Varian 87

▶ $O_2C_5H_4$

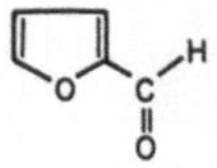

Varian 95

▶ $O_2C_5H_8$

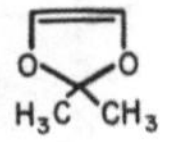

J. Am. Chem. Soc. 83:3504 (1961)

▶ $O_2C_5H_6$

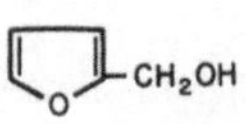

Varian 102

▶ $O_2C_5H_8$

$CH_3\underset{CH_3}{\underset{|}{C}}{=}CHCO_2H$

Varian 114
Bull. Chem. Soc. Japan 35:1194 (1962)

▶ $O_2C_5H_6$

$H_2C{=}CH{-}COOCH{=}CH_2$

J. Chem. Phys. 34:295 (1961)

▶ $O_2C_5H_8$

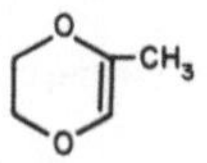

J. Org. Chem. 27:4461 (1962)

▶ $O_2C_5H_8$

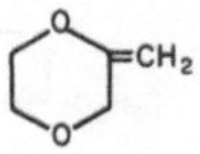

J. Org. Chem. 27:4461 (1962)

▶ $O_2C_5H_8$

$CH_2=C(CH_3)OCCH_3$ (O)

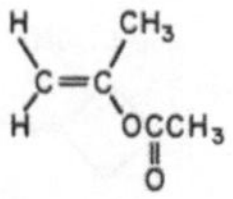

Varian 440
Arkiv Kemi 17:1 (1961)

▶ $O_2C_5H_8$

$CH_2=C(CH_3)CO_2CH_3$

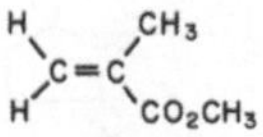

Varian 113
Arkiv Kemi 17:1 (1961)
J. Am. Chem. Soc. 83:922 (1961)

▶ $O_2C_5H_8$

$\triangleright$-$COOCH_3$

Varian 112

▶ $O_2C_5H_8$

Varian 437

▶ $O_2C_5H_8$

Varian 438

▶ $O_2C_5H_8$

Varian 439

▶ $O_2C_5H_{10}$

$(CH_3)_3CC^{13}OOH$

J. Am. Chem. Soc. 83:1230 (1961)

▶ $O_2C_5H_{10}$

$(CH_3)_2CHCOOCH_3$

J. Am. Chem. Soc. 83:1230 (1961)

▶ $O_2C_5H_{10}$

$(CH_3)_2CHC^{13}OOCH_3$

J. Am. Chem. Soc. 83:1230 (1961)

▶ $O_2C_5H_{10}$

$(CH_3)_2CH-OCOCH_3$

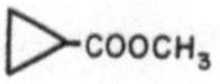

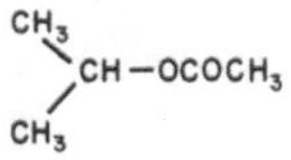

Tetrahedron 18:1147 (1962)

▶ $O_2C_5H_{10}$

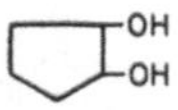

cis and trans

Bull. Chem. Soc. Japan 35:1742 (1962)

▶ $O_2C_5H_{10}$

Varian 445

▶ $O_2C_5H_{12}$

$CH_3CH(OCH_3)CH_2CH_2OH$

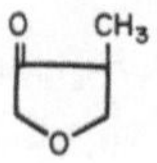

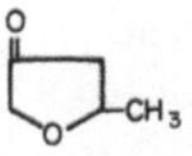

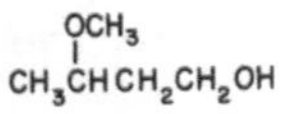

Varian 120

▶ $O_2C_6H_6$

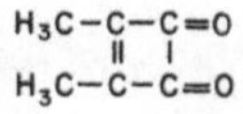

Tetrahedron Letters 657 (1961)

▶ $O_2C_6H_6$

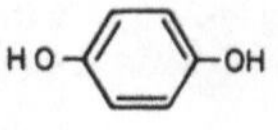

Can. J. Chem. 40:2124 (1962)
J. Chem. Phys. 37:2618 (1962)
Helv. Chim. Acta 45:568 (1962)

▶ $O_2C_6H_6$

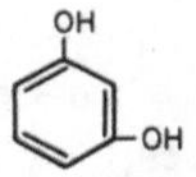

J. Chem. Phys. 37:2618 (1962)
Helv. Chim. Acta 45:568 (1962)

▶ $O_2C_6H_6$

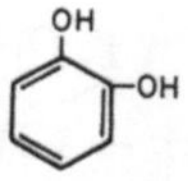

Varian 124
J. Chem. Phys. 37:2618 (1962)

▶ $O_2C_6H_8$

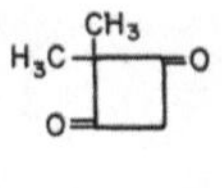

J. Org. Chem. 27:3736 (1962)

▶ $O_2C_6H_8$

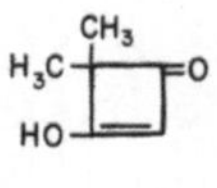

J. Org. Chem. 27:3736 (1962)

▶ $O_2C_6H_8$

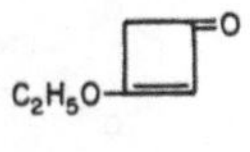

Tetrahedron Letters No. 23:1034 (1962)

▶ $O_2C_6H_8$

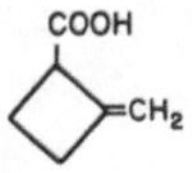

Varian 128

▶ $O_2C_6H_8$

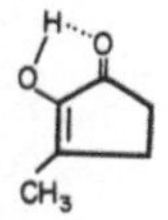

Varian 129

▶ $O_2C_6H_8$

$CH_3CH{=}CHCH{=}CHCOOH$

Varian 462

▶ $O_2C_6H_{10}$

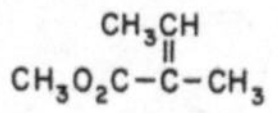

cis and trans

J. Chem. Soc. 1503 (1962)
J. Am. Chem. Soc. 83:922 (1961)

▶ $O_2C_6H_{10}$

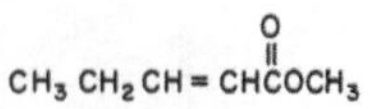

cis and trans

Can. J. Chem. 39:505 (1961)

▶ $O_2C_6H_{10}$

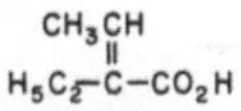

cis and trans

J. Am. Chem. Soc. 83:922 (1961)

▶ $O_2C_6H_{10}$

$H_2C{=}C(CH_3)COOCH_2CH_3$

Varian 135

▶ $O_2C_6H_{10}$

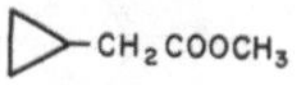


Chem. Ber. 95:2280 (1962)

▶ $O_2C_6H_{12}$

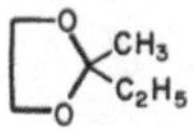


J. Org. Chem. 27:3183 (1962)

▶ $O_2C_6H_{12}$

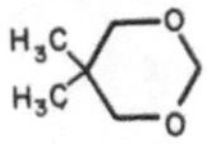


Tetrahedron Letters No. 16:686 (1962)

▶ $O_2C_6H_{12}$

$(CH_3)_3CC^{13}OOCH_3$

J. Am. Chem. Soc. 83:1230 (1961)

▶ $O_2C_6H_{12}$

$(CH_3)_3COC(=O)CH_3$

Varian 141

▶ $O_2C_6H_{12}$

$CH_3(CH_2)_3OC(=O)CH_3$

Varian 140

▶ $O_2C_6H_{12}$

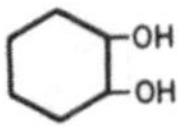


cis and trans

Bull. Chem. Soc. Japan 35:1742 (1962)

▶ $O_2C_6H_{12}$

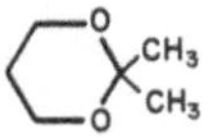


Tetrahedron Letters No. 16:686 (1962)

▶ $O_2C_6H_{12}$

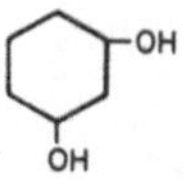


cis and trans

J. Org. Chem. 27:2361 (1962)

▶ $O_2C_6H_{14}$

$(CH_3CH_2O)_2CHCH_3$

Varian 143
J. Am. Chem. Soc. 83:4666 (1961)
J. Chem. Phys. 37:3012 (1962)

▶ $O_2C_6H_{15}B$

$CH_3CH(CH_3)CH_2B(OCH_3)_2$

J. Am. Chem. Soc. 84:4362 (1962)

▶ $O_2C_7H_6$

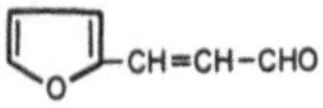


Varian 152
Can. J. Chem. 39:907 (1961); 40:1678 (1962)

▶ $O_2C_7H_6$

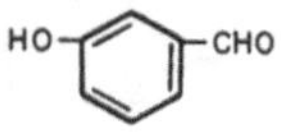


Can. J. Chem. 40:1073 (1962)

► $O_2C_7H_6$

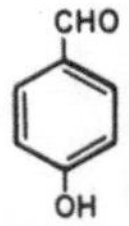

Can. J. Chem. 40:1073, 2124 (1962)

► $O_2C_7H_8$

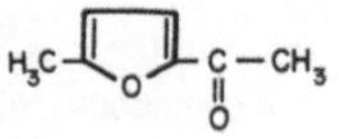

Varian 165

► $O_2C_7H_6$

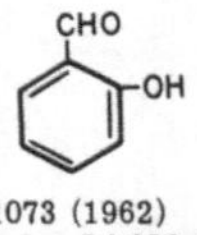

Can. J. Chem. 40:1073 (1962)
Bull. Chem. Soc. Japan 34:353 (1961)

► $O_2C_7H_8$

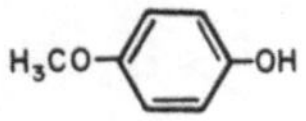

Can. J. Chem. 40:2122 (1962)

► $O_2C_7H_6$

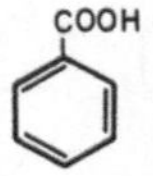

Tetrahedron Letters 519 (1961)

► $O_2C_7H_8$

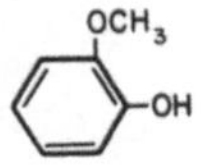

J. Chem. Phys. 37:1565 (1962)

► $O_2C_7H_8$

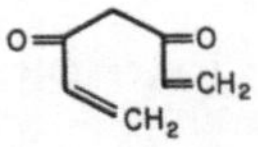

J. Org. Chem. 27:3328 (1962)

► $O_2C_7H_8$

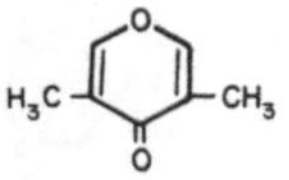

Varian 166

► $O_2C_7H_8$

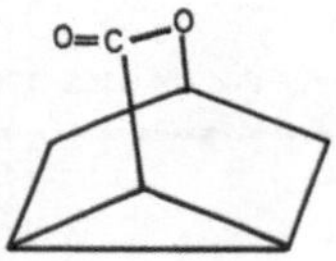

Tetrahedron Letters No. 22:1015 (1962)

► $O_2C_7H_9 \cdot HSO_4$

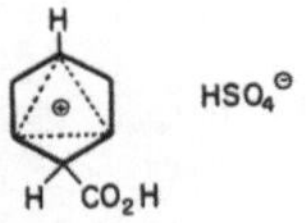

Tetrahedron Letters No. 22:1016 (1962)

► $O_2C_7H_{10}$

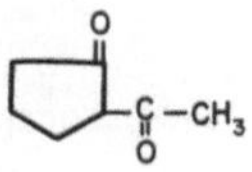

Tetrahedron 18:825 (1962)

► $O_2C_7H_8$

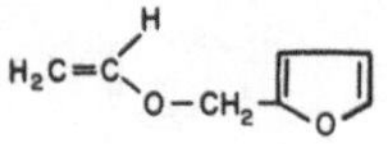

J. Chem. Soc. 2023 (1962)

► $O_2C_7H_{10}$

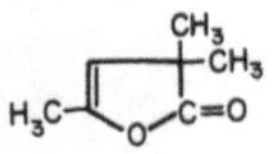

Varian 175

▶ $O_2C_7H_{10}$

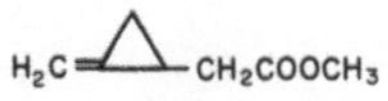

J. Am. Chem. Soc. 83:2379 (1961)

▶ $O_2C_7H_{10}$

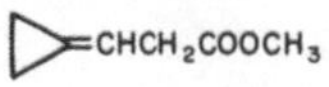

J. Am. Chem. Soc. 83:2379 (1961)

▶ $O_2C_7H_{10}$

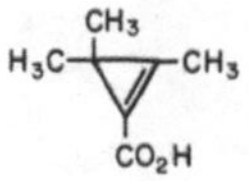

J. Am. Chem. Soc. 83:1003 (1961)

▶ $O_2C_7H_{12}$

$CH_2{=}CHCH_2O{-}CH_2O{-}CH_2CH{=}CH_2$

J. Chem. Soc. 4313 (1962)

▶ $O_2C_7H_{12}$

$HC{\equiv}C{-}CH(OC_2H_5)_2$

J. Am. Chem. Soc. 83:1978 (1961)

▶ $O_2C_7H_{12}$

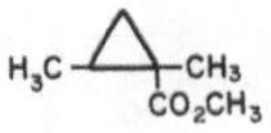

J. Am. Chem. Soc. 84:3736 (1962)

▶ $O_2C_7H_{13}B$

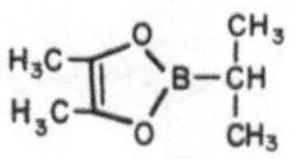

J. Org. Chem. 26:4915 (1961)

▶ $O_2C_7H_{14}$

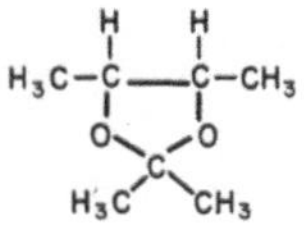

J. Am. Chem. Soc. 84:747 (1962)

▶ $O_2C_7H_{16}$

$CH_3CH_2CH_2OCH_2OCH_2CH_2CH_3$

J. Chem. Soc. 4313 (1962)

▶ $O_2C_8H_6$

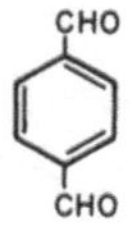

Can. J. Chem. 40:2333 (1962)

▶ $O_2C_8H_6$

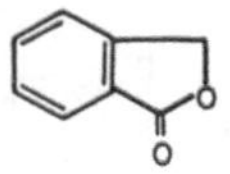

Varian 496

▶ $O_2C_8H_8$

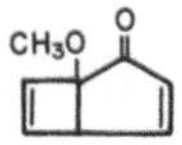

J. Am. Chem. Soc. 83:1768, 3919 (1961

▶ $O_2C_8H_8$

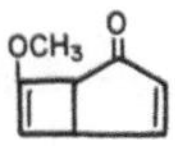

J. Am. Chem. Soc. 83:1768 (1961)

▶ $O_2C_8H_8$

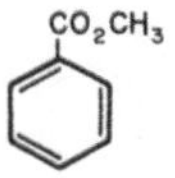

Tetrahedron Letters 519 (1961)

▶ $O_2C_8H_8$

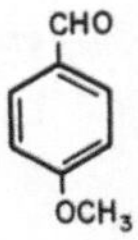

Varian 194
Can. J. Chem. 40:1073, 1866, 2331 (1962)
J. Chem. Phys. 37:2594 (1962)

▶ $O_2C_8H_{10}$

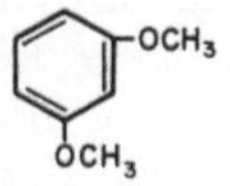

Can. J. Chem. 40:1866 (1962)
Helv. Chim. Acta 45:568 (1962)

▶ $O_2C_8H_8$

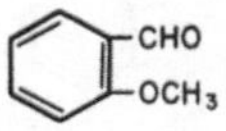

Can. J. Chem. 40:1073 (1962)

▶ $O_2C_8H_{10}$

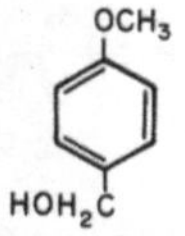

Can. J. Chem. 40:1866 (1962)

▶ $O_2C_8H_8$

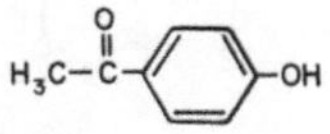

Can. J. Chem. 40:2124 (1962)

▶ $O_2C_8H_{10}$

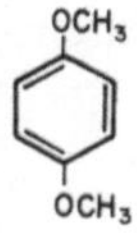

J. Org. Chem. 26:3224 (1961)
Can. J. Chem. 40:1759, 1866 (1962)
Helv. Chim. Acta 45:568 (1962)

▶ $O_2C_8H_8D_6$

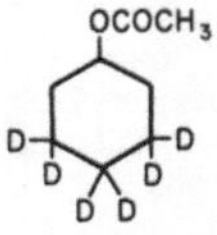

J. Am. Chem. Soc. 84:1053 (1962)

▶ $O_2C_8H_{10}$

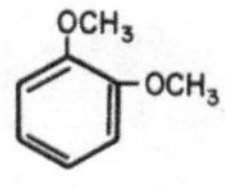

J. Chem. Phys. 37:2594 (1962)

▶ $O_2C_8H_{10}$

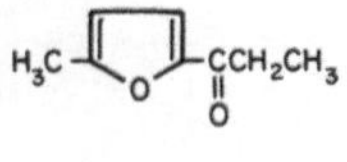

Varian 206

▶ $O_2C_8H_{10}$

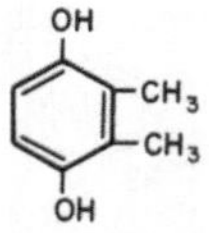

J. Chem. Soc. 2785 (1962)

▶ $O_2C_8H_{10}$

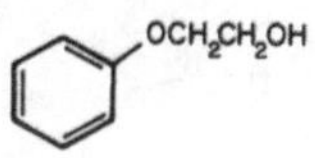

Varian 506

▶ $O_2C_8H_{11}D$

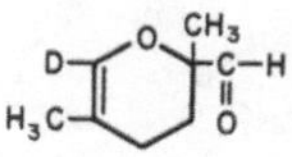

J. Am. Chem. Soc. 83:2198 (1961)

▶ $O_2C_8H_{11}D$

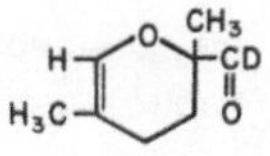

J. Am. Chem. Soc. 83:2198 (1961)

▶ $O_2C_8H_{12}$

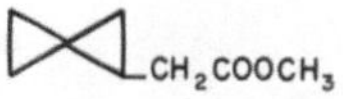

J. Am. Chem. Soc. 83:2379 (1961)

▶ $O_2C_8H_{12}$

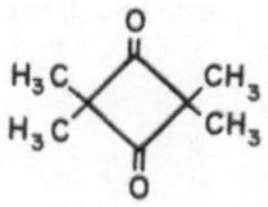

J. Am. Chem. Soc. 83:4867 (1961)

▶ $O_2C_8H_{12}$

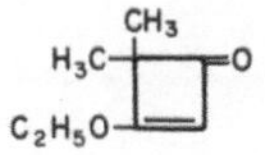

Tetrahedron Letters No. 23:1032 (1962)
J. Org. Chem. 27:3743 (1962)

▶ $O_2C_8H_{12}$

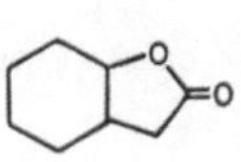

J. Org. Chem. 27:4141 (1962)

▶ $O_2C_8H_{12}$

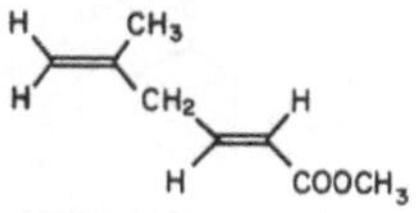

J. Org. Chem. 27:2732 (1962)

▶ $O_2C_8H_{12}$

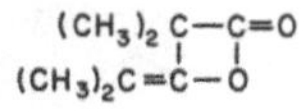

Varian 513
J. Org. Chem. 27:61 (1962)

▶ $O_2C_8H_{12}$

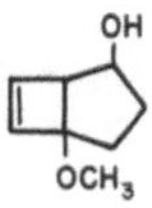

J. Am. Chem. Soc. 84:1213 (1962)

▶ $O_2C_8H_{12}$

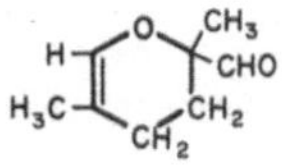

J. Am. Chem. Soc. 83:246 (1961)

▶ $O_2C_8H_{12}$

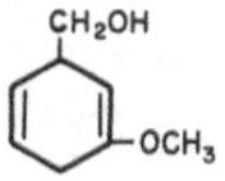

J. Am. Chem. Soc. 83:1005 (1961)

▶ $O_2C_8H_{12}$

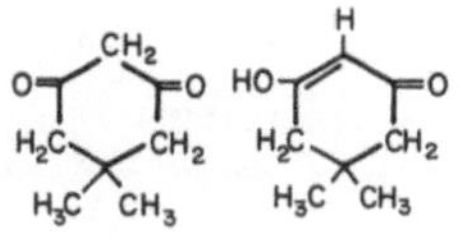

keto-enol MIXTURE

Varian 512

► $O_2C_8H_{14}$

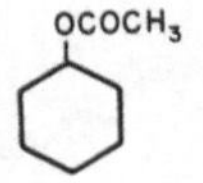

J. Am. Chem. Soc. 83:1146 (1961)
Tetrahedron 18:1147 (1962)

► $O_2C_8H_{16}$

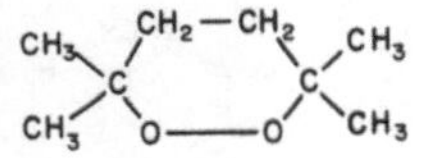

J. Am. Chem. Soc. 83:4357 (1961)

► $O_2C_8H_{14}$

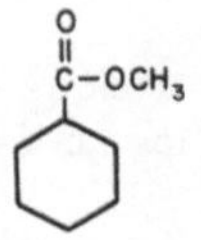

J. Org. Chem. 26:4727 (1961)

► $O_2C_8H_{16}$

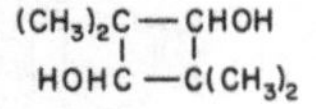

J. Org. Chem. 26:702 (1961)

► $O_2C_8H_{14}$

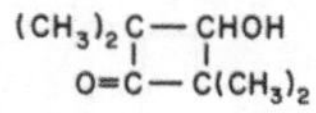

J. Org. Chem. 26:702 (1961)

► $O_2C_8H_{16}$

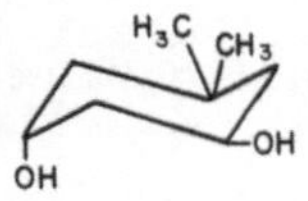

cis and trans

J. Org. Chem. 27:2361 (1962)

► $O_2C_8H_{14}$

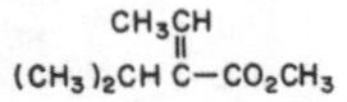

J. Am. Chem. Soc. 83:922 (1961)

► $O_2C_8H_{14}$

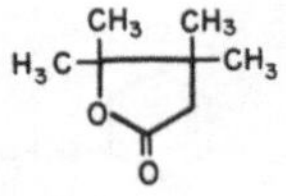

Varian 214
J. Org. Chem. 26:3517 (1961)

► $O_2C_8H_{17}B$

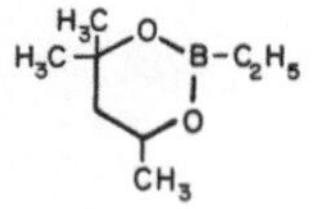

Varian 518

► $O_2C_8H_{14}$

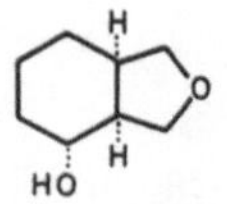

J. Org. Chem. 27:2972 (1962)

► $O_2C_8H_{18}$

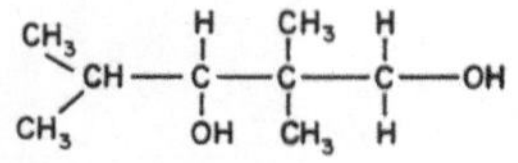

Varian 217

► $O_2C_8H_{15}B$

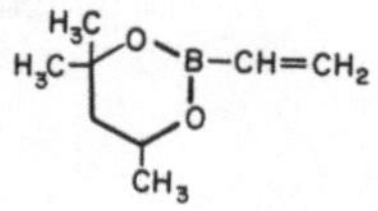

Varian 517

► $O_2C_8H_{19}B$

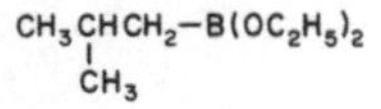

J. Am. Chem. Soc. Soc. 84:4362 (1962)

▶ $O_2C_9H_6$

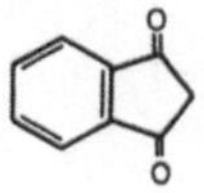


Varian 224

▶ $O_2C_9H_6$

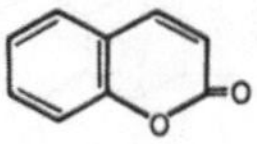


Varian 225

▶ $O_2C_9H_{10}$

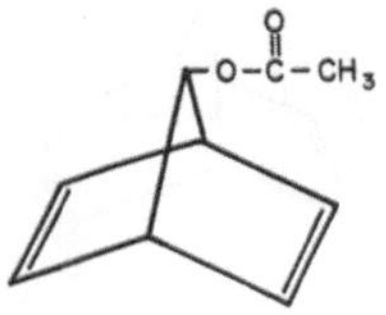


J. Org. Chem. 26:287 (1961)

▶ $O_2C_9H_8$

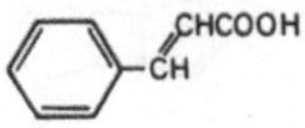


Varian 230

▶ $O_2C_9H_{10}$

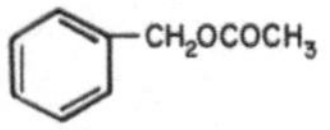


Varian 530

▶ $O_2C_9H_{10}$

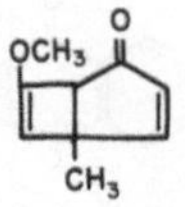


J. Am. Chem. Soc. 83:1768 (1961)

▶ $O_2C_9H_{10}$

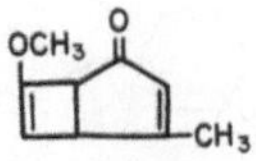


J. Am. Chem. Soc. 83:1768 (1961)

▶ $O_2C_9H_{10}$

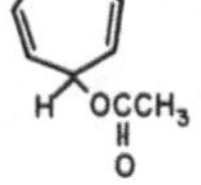


J. Org. Chem. 27:4714 (1962)

▶ $O_2C_9H_{10}$

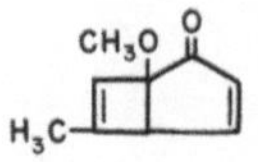


Varian 531
J. Am. Chem. Soc. 83:1768, 3914 (1961)

▶ $O_2C_9H_{10}$

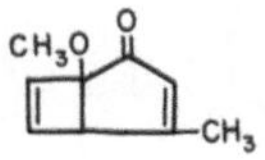


J. Am. Chem. Soc. 83:1768, 3919 (1961)

▶ $O_2C_9H_{10}$

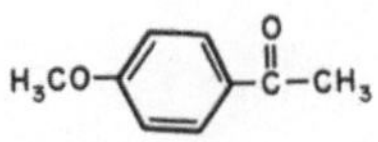


Can. J. Chem. 40:2331 (1962)
J. Chem. Phys. 37:2594 (1962)

▶ $O_2C_9H_{12}$

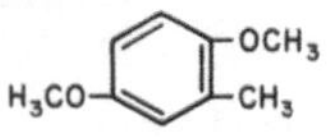


J. Org. Chem. 26:3224 (1961)
Can. J. Chem. 40:1866 (1962)

► $O_2C_9H_{12}$

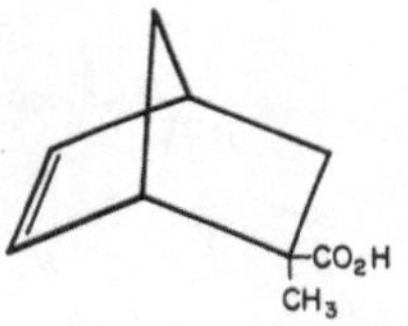


Can. J. Chem. 40:78 (1962)

► $O_2C_9H_{12}$

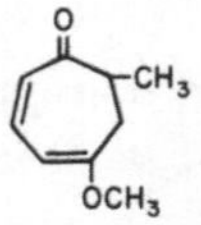


J. Am. Chem. Soc. 84:1213 (1962)

► $O_2C_9H_{14}$

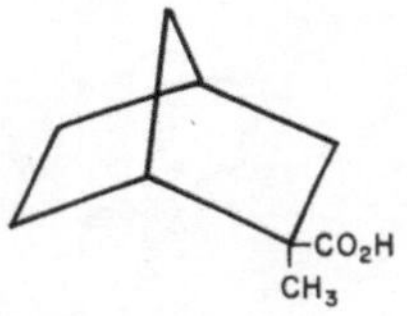


Can. J. Chem. 40:78 (1962)

► $O_2C_9H_{14}$

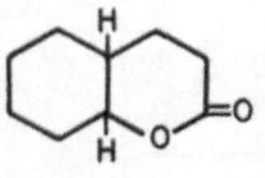


cis and trans

J. Org. Chem. 27:4141 (1962)

► $O_2C_9H_{16}$

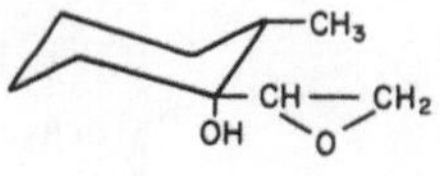


cis and trans

Mol. Phys. 4:311 (1961)

► $O_2C_9H_{18}$

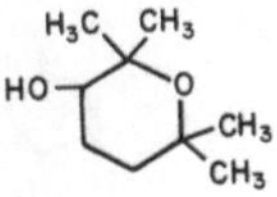


Tetrahedron Letters No. 17:770 (1962)

► $O_2C_{10}H_6$

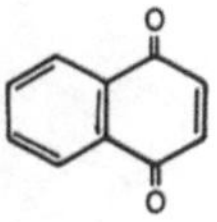


Varian 550

► $O_2C_{10}H_8$

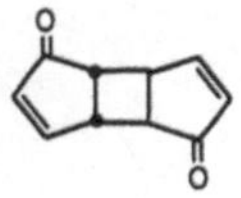


J. Am. Chem. Soc. 84:2344 (1962)

► $O_2C_{10}H_8$

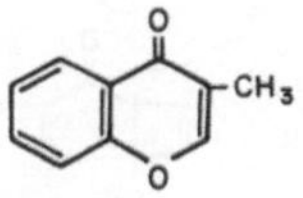


Varian 552

► $O_2C_{10}H_{10}$

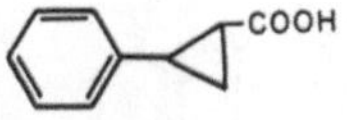


Chem. Ber. 95:2535 (1962)

► $O_2C_{10}H_{10}$

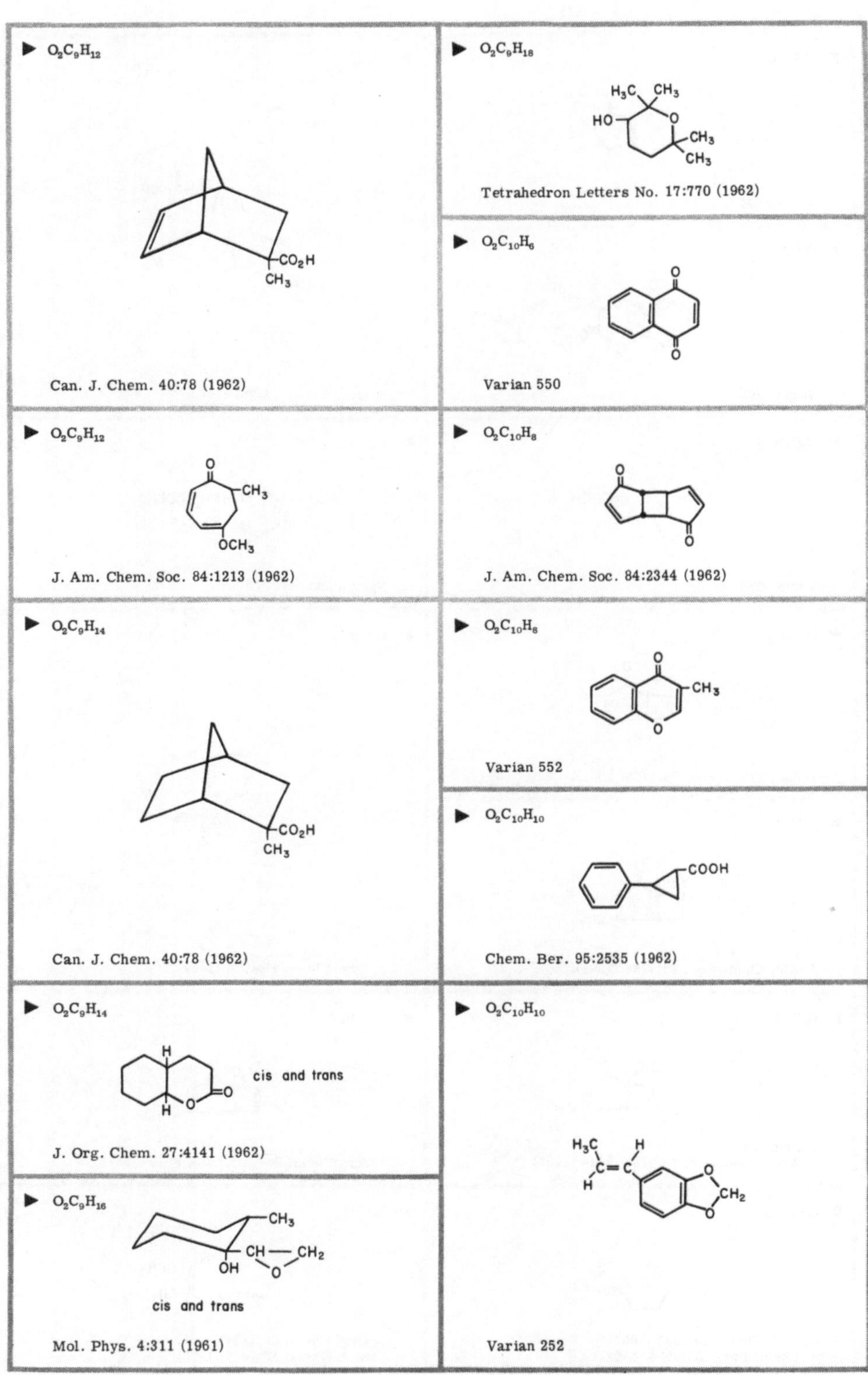

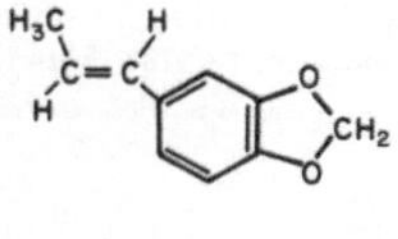


Varian 252

▶ $O_2C_{10}H_{10}$

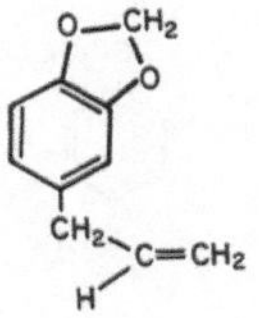

Varian 253

▶ $O_2C_{10}H_{12}$

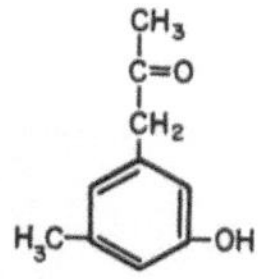

J. Chem. Soc. 3891 (1962)

▶ $O_2C_{10}H_{11}B$

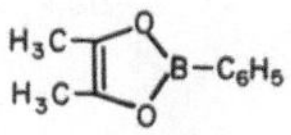

J. Org. Chem. 26:4915 (1961)

▶ $O_2C_{10}H_{12}$

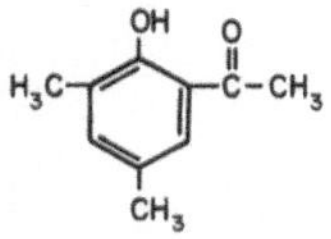

J. Org. Chem. 27:3658 (1962)

▶ $O_2C_{10}H_{12}$

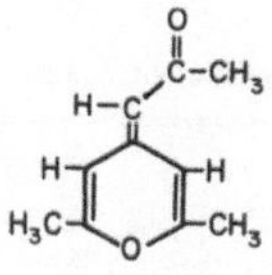

J. Chem. Soc. 3891 (1962)

▶ $O_2C_{10}H_{12}$

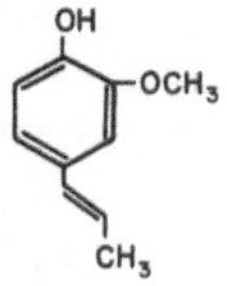

Can. J. Chem. 40:1866 (1962)

▶ $O_2C_{10}H_{12}$

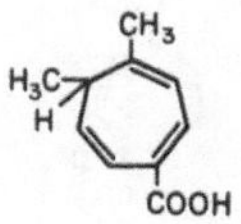

J. Org. Chem. 27:2292 (1962)

▶ $O_2C_{10}H_{12}$

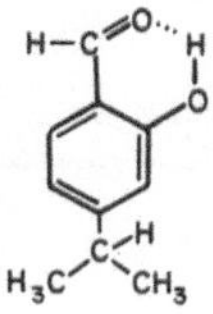

Varian 259

▶ $O_2C_{10}H_{12}$

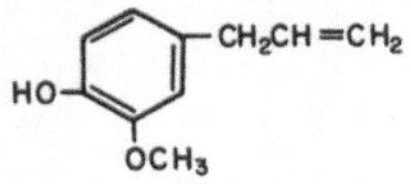


Varian 260

▶ $O_2C_{10}H_{12}$

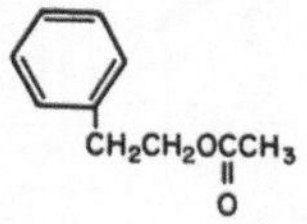


Varian 261

▶ $O_2C_{10}H_{12}$

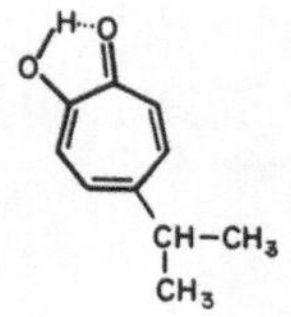


Varian 560

▶ $O_2C_{10}H_{14}$

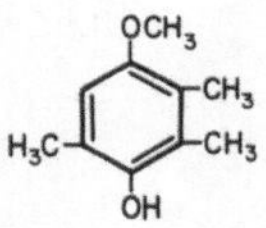


J. Org. Chem. 27:845 (1962)

▶ $O_2C_{10}H_{14}$

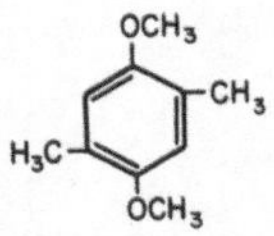


J. Org. Chem. 26:3224 (1961)

▶ $O_2C_{10}H_{14}$

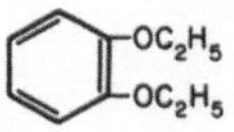


Can. J. Chem. 40:1866 (1962)

▶ $O_2C_{10}H_{14}$

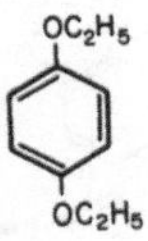


Can. J. Chem. 40:1866 (1962)

▶ $O_2C_{10}H_{14}$

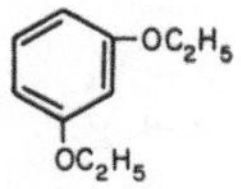


Can. J. Chem. 40:1866 (1962)

▶ $O_2C_{10}H_{14}$

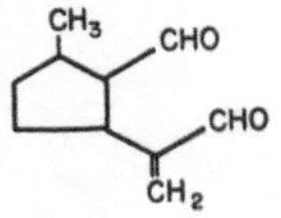


Tetrahedron Letters No. 1:32 (1962)

▶ $O_2C_{10}H_{14}$

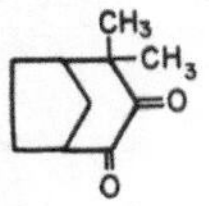


J. Org. Chem. 26:707 (1961)

▶ $O_2C_{10}H_{14}$

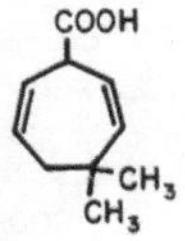


J. Org. Chem. 27:2787 (1962)

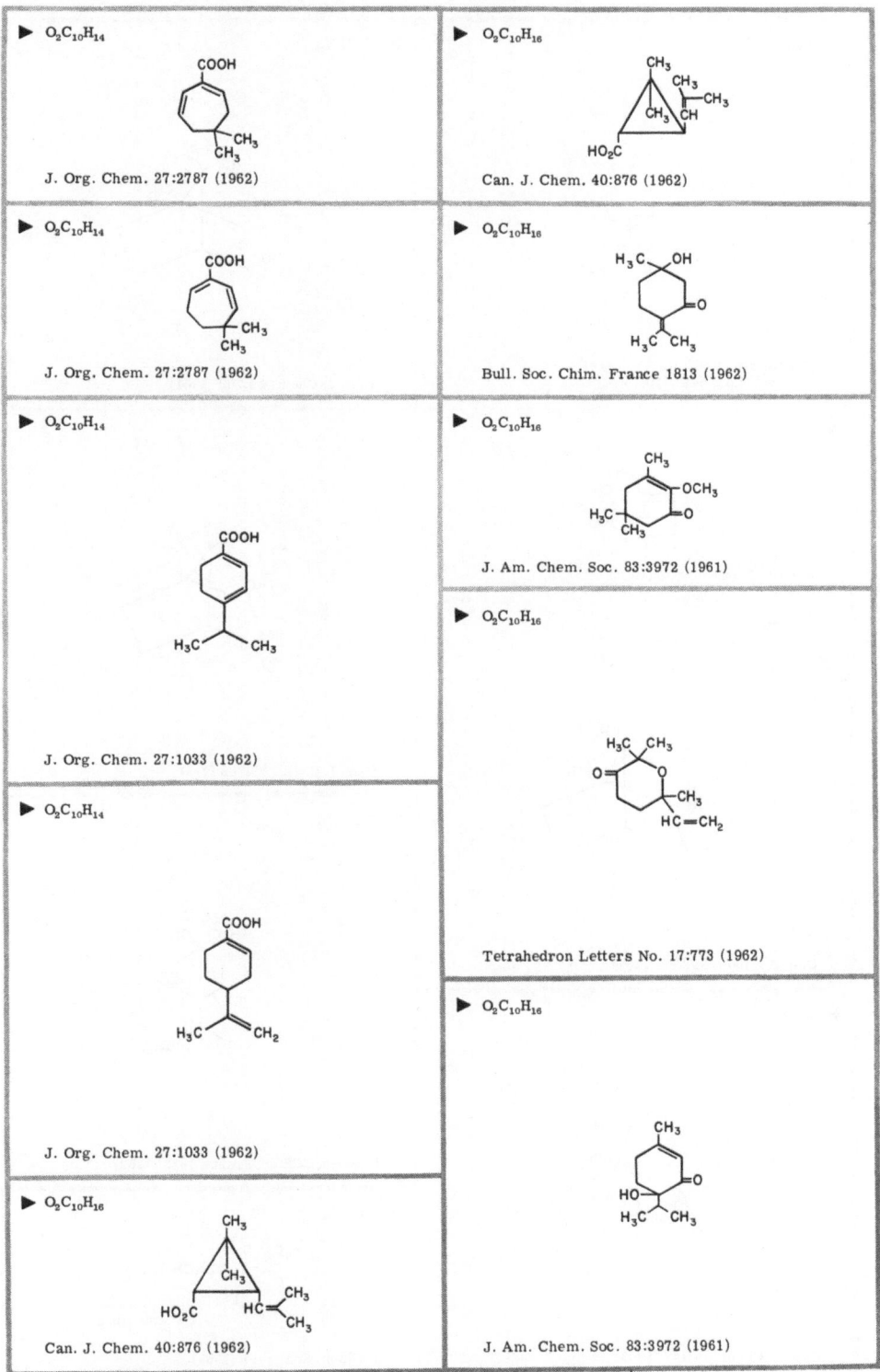

▶ $O_2C_{10}H_{14}$

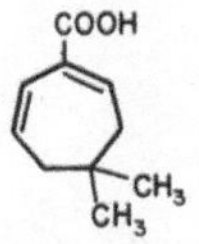

J. Org. Chem. 27:2787 (1962)

▶ $O_2C_{10}H_{14}$

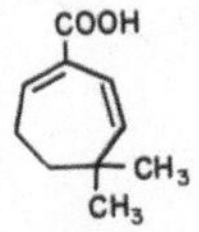

J. Org. Chem. 27:2787 (1962)

▶ $O_2C_{10}H_{14}$

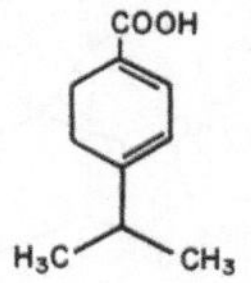

J. Org. Chem. 27:1033 (1962)

▶ $O_2C_{10}H_{14}$

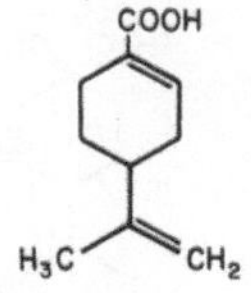

J. Org. Chem. 27:1033 (1962)

▶ $O_2C_{10}H_{16}$

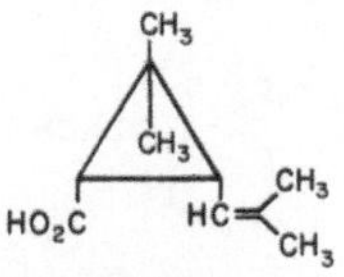

Can. J. Chem. 40:876 (1962)

▶ $O_2C_{10}H_{16}$

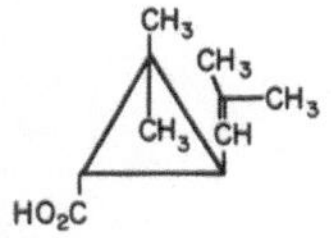

Can. J. Chem. 40:876 (1962)

▶ $O_2C_{10}H_{16}$

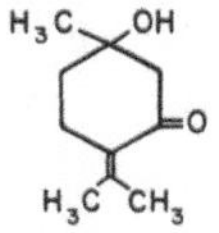

Bull. Soc. Chim. France 1813 (1962)

▶ $O_2C_{10}H_{16}$

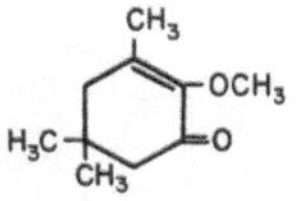

J. Am. Chem. Soc. 83:3972 (1961)

▶ $O_2C_{10}H_{16}$

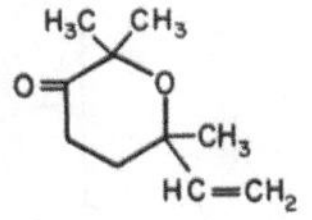

Tetrahedron Letters No. 17:773 (1962)

▶ $O_2C_{10}H_{16}$

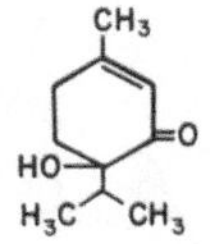

J. Am. Chem. Soc. 83:3972 (1961)

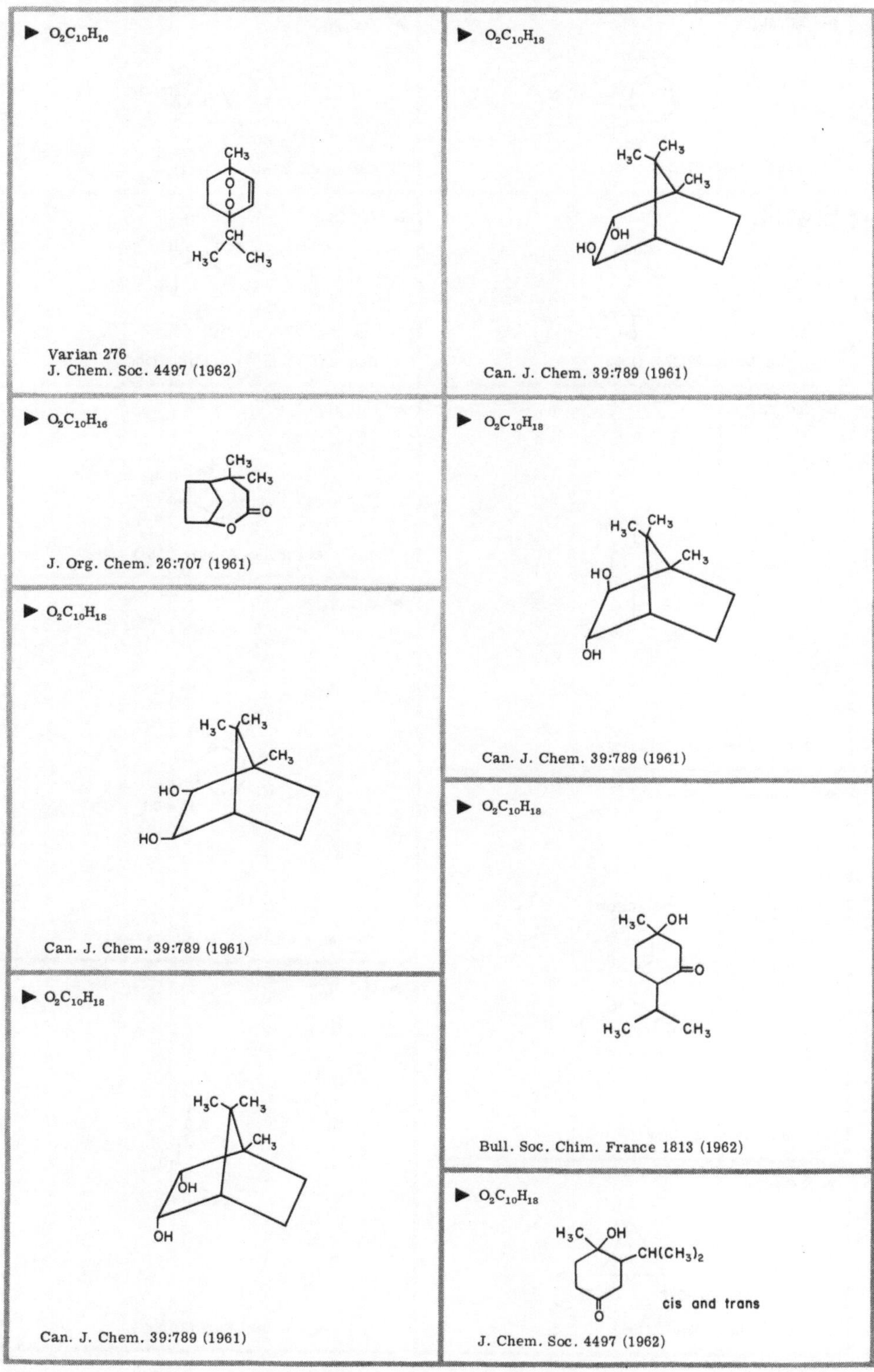

► $O_2C_{10}H_{16}$

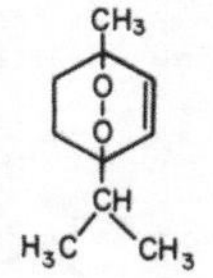


Varian 276
J. Chem. Soc. 4497 (1962)

► $O_2C_{10}H_{16}$

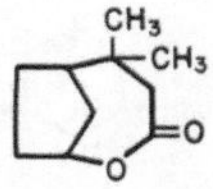


J. Org. Chem. 26:707 (1961)

► $O_2C_{10}H_{18}$

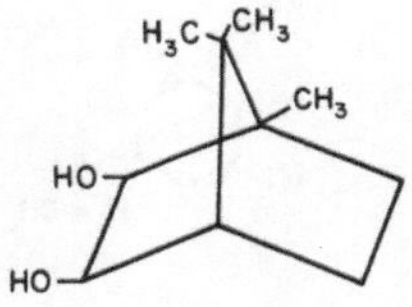


Can. J. Chem. 39:789 (1961)

► $O_2C_{10}H_{18}$

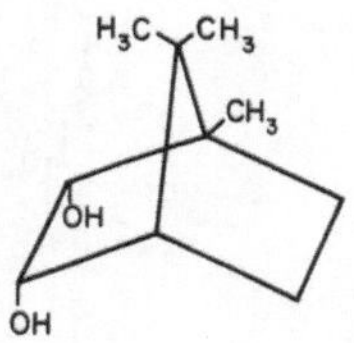


Can. J. Chem. 39:789 (1961)

► $O_2C_{10}H_{18}$

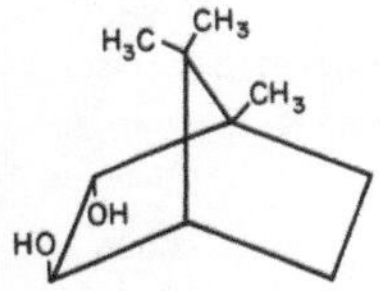


Can. J. Chem. 39:789 (1961)

► $O_2C_{10}H_{18}$

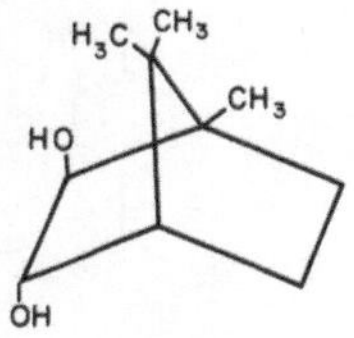


Can. J. Chem. 39:789 (1961)

► $O_2C_{10}H_{18}$

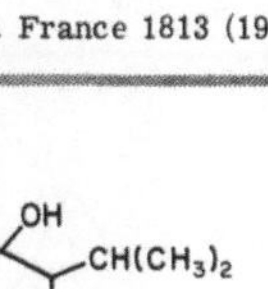


Bull. Soc. Chim. France 1813 (1962)

► $O_2C_{10}H_{18}$

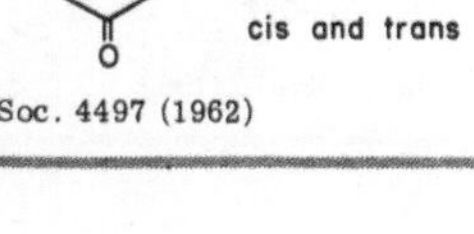


J. Chem. Soc. 4497 (1962)

▶ $O_2C_{10}H_{18}$

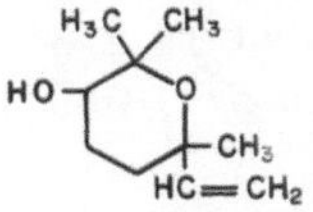

Tetrahedron Letters No. 17:770 (1962)

▶ $O_2C_{11}H_{12}$

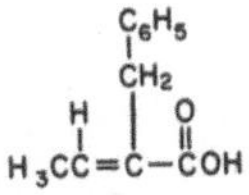

J. Org. Chem. 27:2682 (1962)

▶ $O_2C_{10}H_{18} \cdot Pd_2Cl_2$

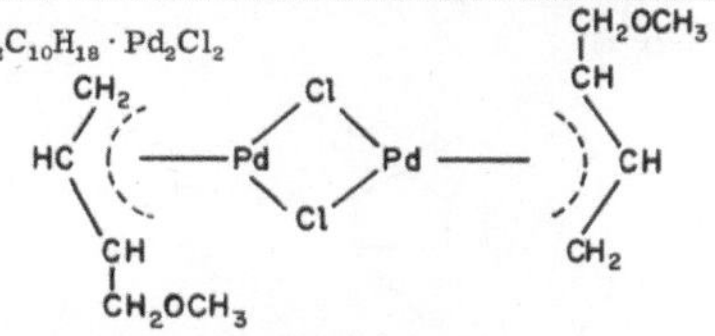

Chem. & Ind. 1190 (1962)

▶ $O_2C_{11}H_{12}$

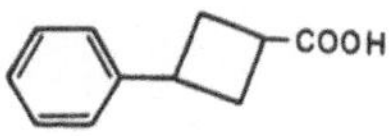

Chem. Ber. 95:2535 (1962)

▶ $O_2C_{10}H_{20}$

$CH_3(CH_2)_8-\overset{O}{\overset{\|}{C}}-OK$

Can. J. Chem. 39:359 (1961)

▶ $O_2C_{11}H_{14}$

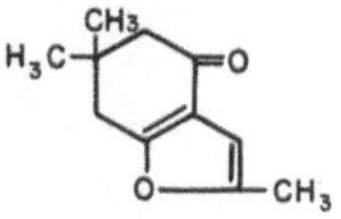

J. Org. Chem. 27:4502 (1962)

▶ $O_2C_{10}H_{20}$

$H_3C(CH_2)_7COOCH_3$

MCA Serial No. 56

▶ $O_2C_{11}H_{14}$

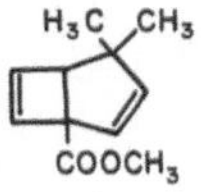

J. Org. Chem. 27:2292 (1962)

▶ $O_2C_{10}H_{20}$

$CH_3(CH_2)_4-\underset{\underset{CH_2-CH_3}{|}}{CH}-COOCH_3$

MCA Serial No. 57

▶ $O_2C_{11}H_{14}$

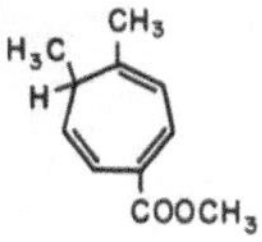

J. Org. Chem. 27:2292 (1962)

▶ $O_2C_{11}H_9D_5$

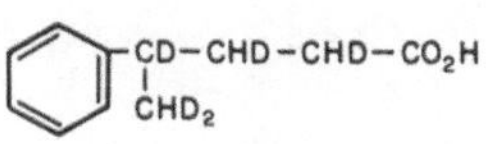

Bull. Soc. Chim. France 1747 (1962)

▶ $O_2C_{11}H_{14}$

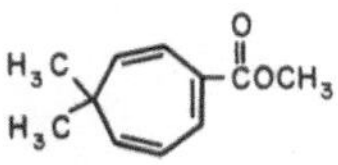

Tetrahedron Letters No. 19:839 (1962)

▶ $O_2C_{11}H_{10}$

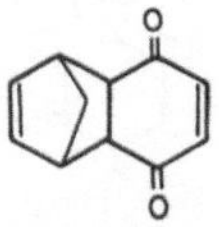

Helv. Chim. Acta 45:1870 (1962)

▶ $O_2C_{11}H_{15}D$

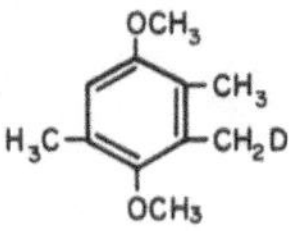

J. Org. Chem. 27:843 (1962)

▶ $O_2C_{11}H_{16}$

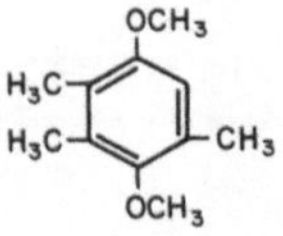


J. Org. Chem. 26:3224 (1961); 27:842 (1962)

▶ $O_2C_{11}H_{16}$

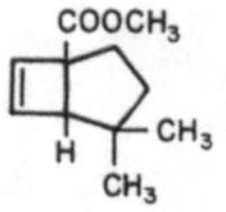


J. Org. Chem. 27:2788 (1962)

▶ $O_2C_{11}H_{16}$

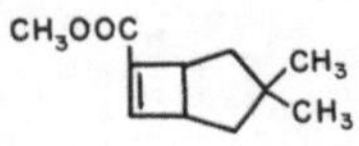


J. Org. Chem. 27:2788 (1962)

▶ $O_2C_{11}H_{16}$

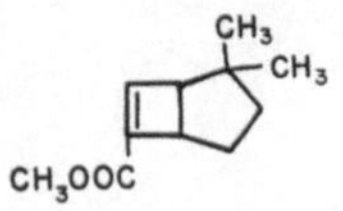


J. Org. Chem. 27:2788 (1962)

▶ $O_2C_{11}H_{16}$

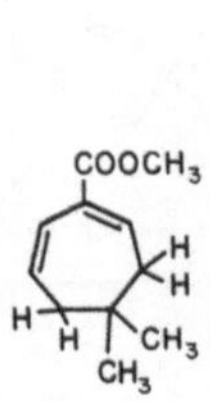


J. Org. Chem. 27:2790 (1962)

▶ $O_2C_{11}H_{18}$

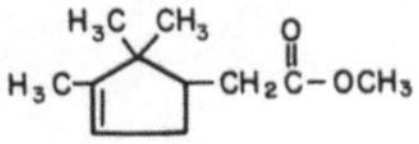


J. Org. Chem. 27:900 (1962)

▶ $O_2C_{11}H_{18}$

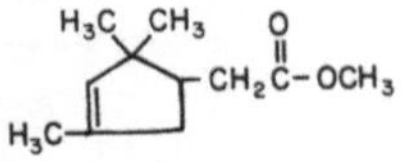


J. Org. Chem. 27:900 (1962)

▶ $O_2C_{11}H_{18}$

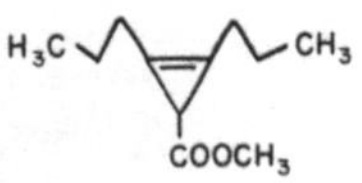


J. Am. Chem. Soc. 84:3168 (1962)

▶ $O_2C_{11}H_{18}$

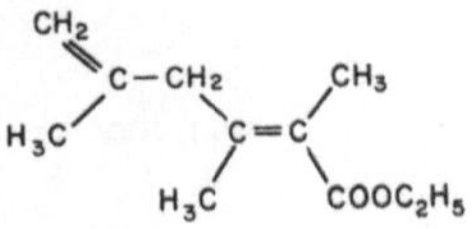


Chem. Ber. 95:443 (1962)

▶ $O_2C_{11}H_{18}$

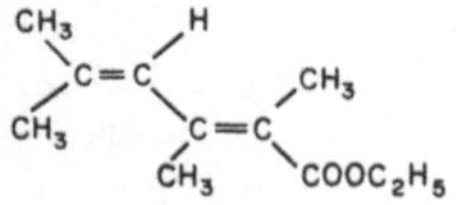


Chem. Ber. 95:443 (1962)

▶ $O_2C_{11}H_{18}$

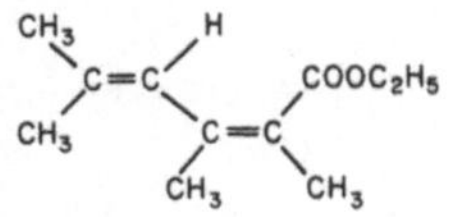


Chem. Ber. 95:443 (1962)

▶ $O_2C_{11}H_{18}$

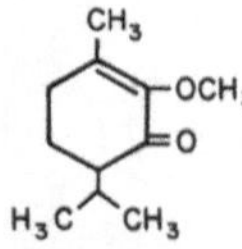


J. Am. Chem. Soc. 83:3972 (1961)

▶ $O_2C_{11}H_{18}$

$CH_3$ $CH_3O$ O $H_3C$ $CH_3$

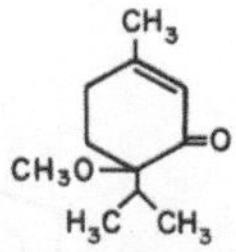

J. Am. Chem. Soc. 83:3972 (1961)

▶ $O_2C_{11}H_{20}$

$CH_2{=}CH(CH_2)_8CO_2H$

J. Am. Oil Chemists' Soc. 38:664 (1961)

▶ $O_2C_{12}H_{10}$

H H COOH H

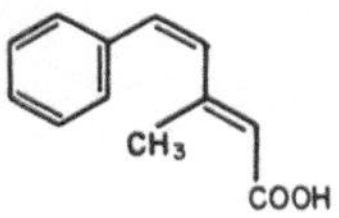

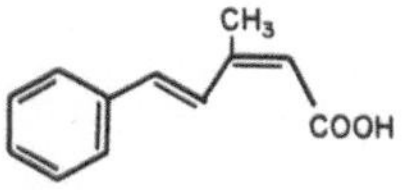

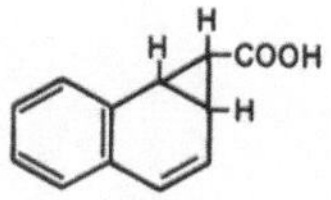

Varian 592
Chem. Ber. 94:2332 (1961)

▶ $O_2C_{12}H_{10}$

COOH

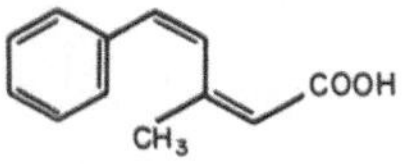

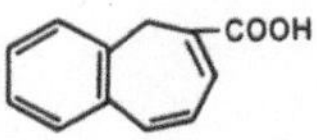

Varian 591
Chem. Ber. 94:2332 (1961)

▶ $O_2C_{12}H_{10}$

O O

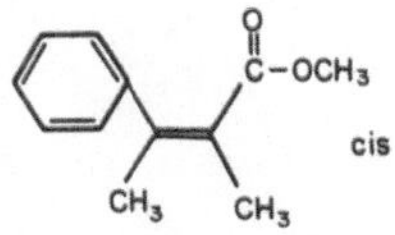

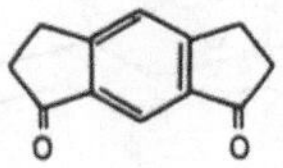

J. Am. Chem. Soc. 84:4611 (1962)

▶ $O_2C_{12}H_{12}$

cis-cis
cis-trans
trans-cis
trans-trans

$C_6H_5{-}CH{=}CH{-}C(CH_3){=}CHCO_2H$

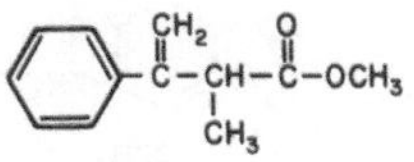

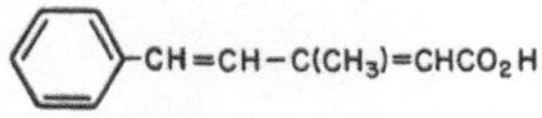

J. Org. Chem. 27:1989 (1962)

▶ $O_2C_{12}H_{12}$

$CH_3$ COOH

J. Org. Chem. 27:1536 (1962)

▶ $O_2C_{12}H_{12}$

$CH_3$ COOH

J. Org. Chem. 27:1536 (1962)

▶ $O_2C_{12}H_{12}$

$CH_3$ COOH

J. Org. Chem. 27:1536 (1962)

▶ $O_2C_{12}H_{12}$

COOH $CH_3$

J. Org. Chem. 27:1536 (1962)

▶ $O_2C_{12}H_{14}$

O C–$OCH_3$ $CH_3$ $CH_3$

cis and trans

J. Chem. Soc. 3778 (1962)

▶ $O_2C_{12}H_{14}$

$C_6H_5{-}C({=}CH_2){-}CH(CH_3){-}C({=}O){-}OCH_3$

J. Chem. Soc. 3778 (1962)

▶ $O_2C_{12}H_{16}$

$CH_3$ $CH_3O$ O $CH_3$ $CH_3$

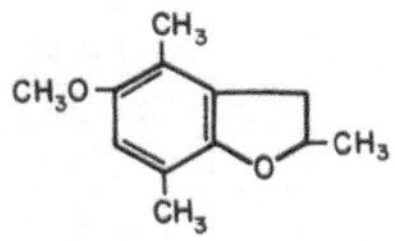

J. Org. Chem. 27:843 (1962)

▶ $O_2C_{12}H_{16}$

$OCH_3$ $H_3C$ $CH{=}CH_2$ $CH_3$ $OCH_3$

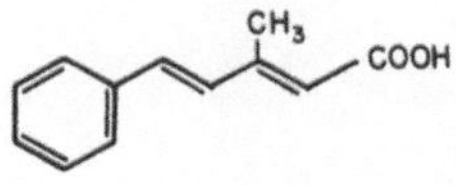

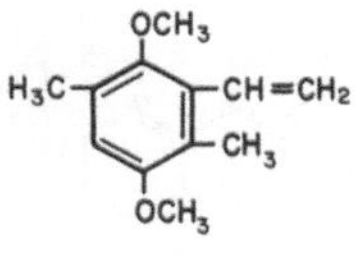

J. Org. Chem. 26:3224 (1961)

▶ $O_2C_{12}H_{18}$

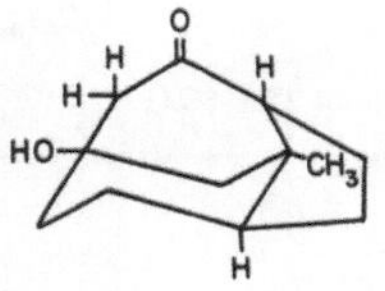

J. Org. Chem. 27:2018 (1962)

▶ $O_2C_{12}H_{18}$

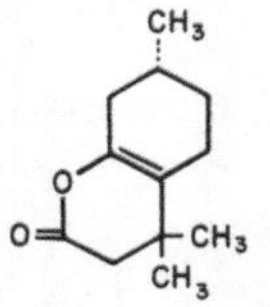

Acta Chem. Scand. 16:205 (1962)

▶ $O_2C_{12}H_{18}$

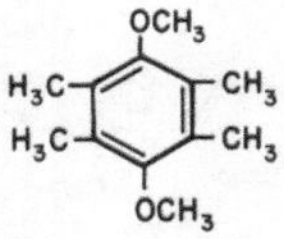

J. Org. Chem. 26:3224 (1961)

▶ $O_2C_{12}H_{18}$

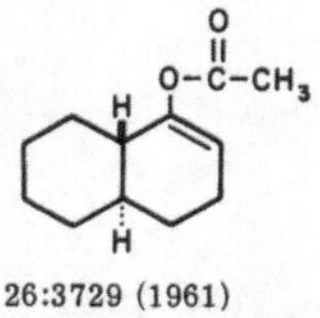

J. Org. Chem. 26:3729 (1961)

▶ $O_2C_{12}H_{18}$

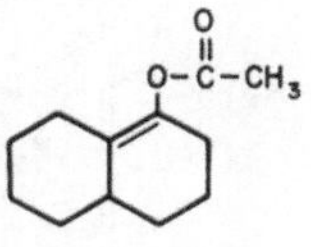

J. Org. Chem. 26:3729 (1961)

▶ $O_2C_{12}H_{20}$

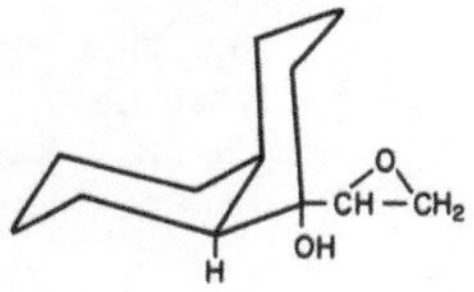

Mol. Phys. 4:311 (1961)

▶ $O_2C_{12}H_{20}$

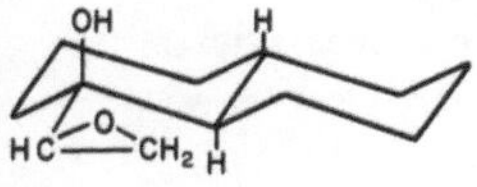

Mol. Phys. 4:311 (1961)

▶ $O_2C_{12}H_{20}$

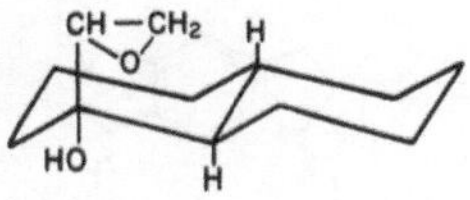

Mol. Phys. 4:311 (1961)

▶ $O_2C_{12}H_{20}$

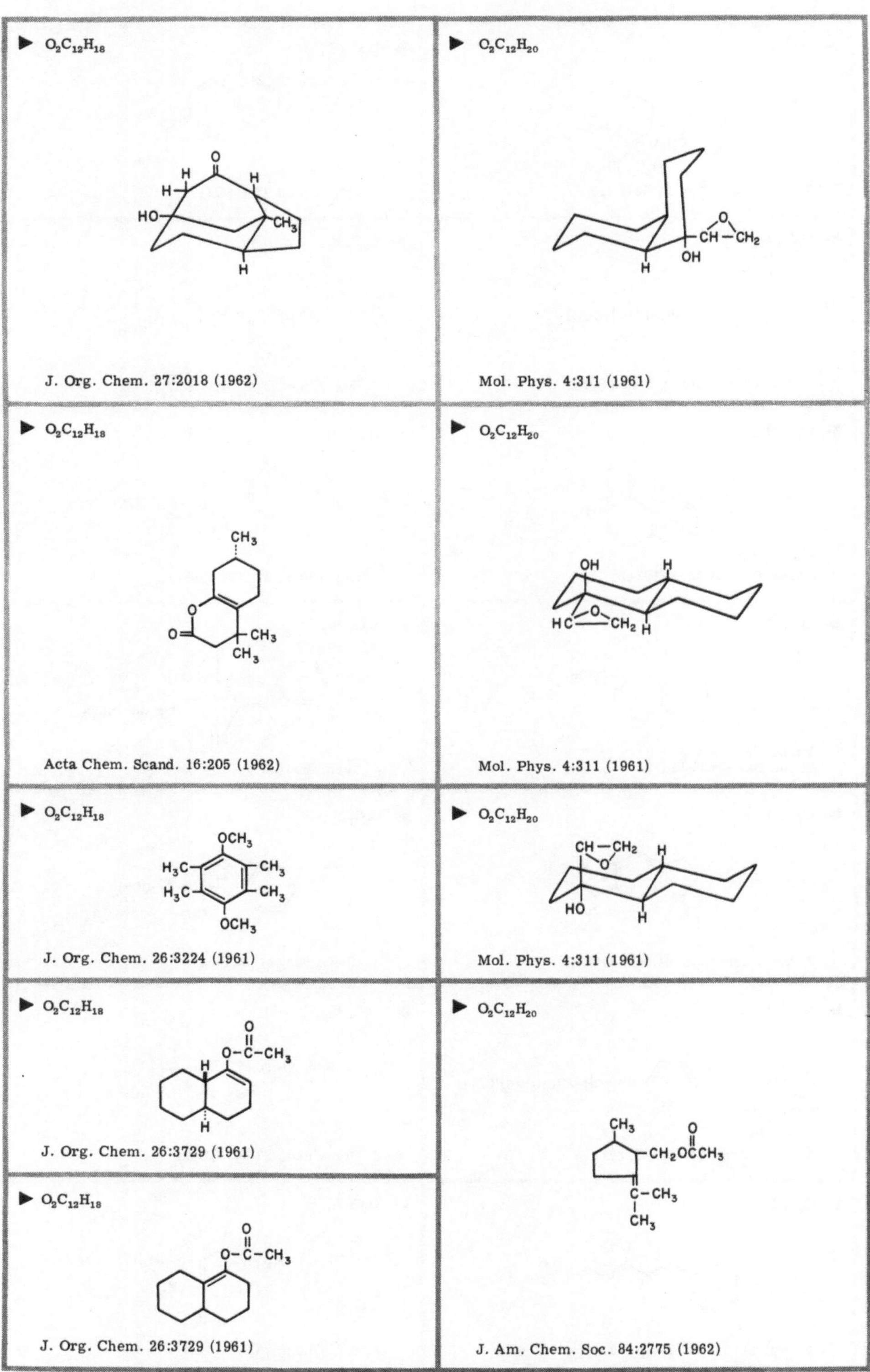

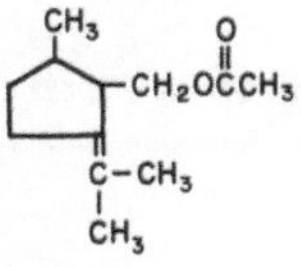

J. Am. Chem. Soc. 84:2775 (1962)

▶ $O_2C_{12}H_{20}$

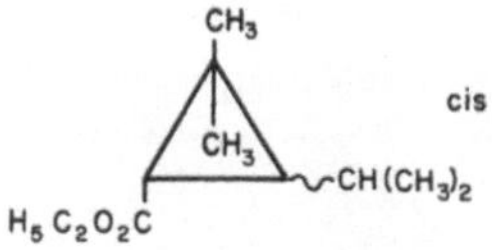

cis and trans

Can. J. Chem. 40:878 (1962)

▶ $O_2C_{12}H_{22}$

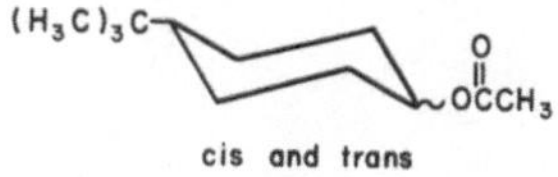

cis and trans

Tetrahedron Letters No. 3:97 (1962)

▶ $O_2C_{12}H_{22}$

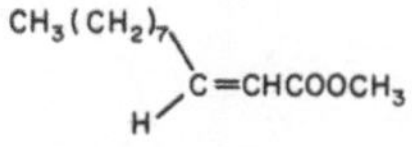

cis and trans

Tetrahedron 18:177 (1962)

▶ $O_2C_{12}H_{22}$

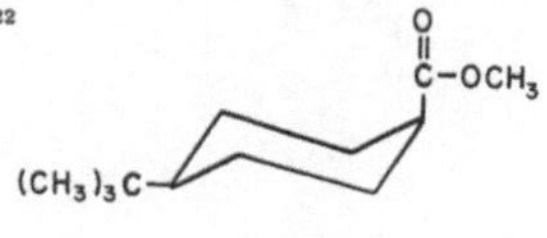

cis and trans

J. Org. Chem. 26:4727 (1961)

▶ $O_2C_{12}H_{24}$

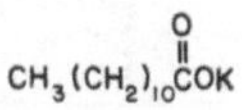

Can. J. Chem. 39:359 (1961)

▶ $O_2C_{12}H_{30}B_2$

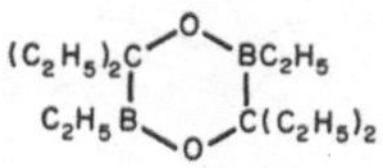

J. Am. Chem. Soc. 84:4715 (1962)

▶ $O_2C_{13}H_{10}$

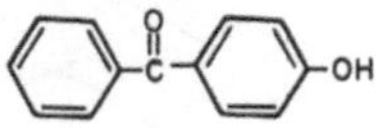

Can. J. Chem. 40:2124 (1962)

▶ $O_2C_{13}H_{10}$

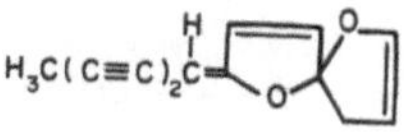

Chem. Ber. 94:3193 (1961)

▶ $O_2C_{13}H_{10}$

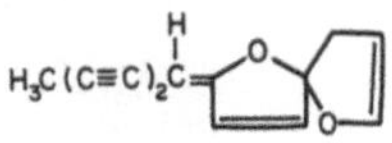

Chem. Ber. 94:3193 (1961)

▶ $O_2C_{13}H_{12}$

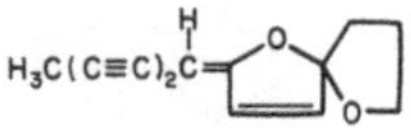

Chem. Ber. 94:3193 (1961)

▶ $O_2C_{13}H_{12}$

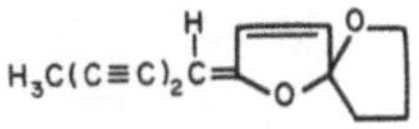

Chem. Ber. 94:3193 (1961)

▶ $O_2C_{13}H_{12}$

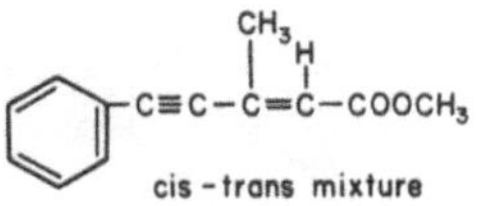

cis - trans mixture

J. Org. Chem. 27:1536 (1962)

▶ $O_2C_{13}H_{12}$

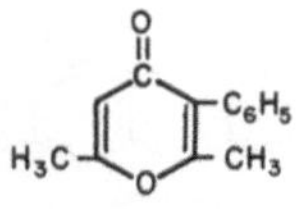

J. Am. Chem. Soc. 83:193 (1961)

► $O_2C_{13}H_{12}$

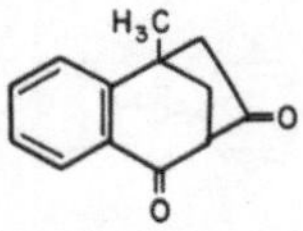


J. Am. Chem. Soc. 84:3517 (1962)

► $O_2C_{13}H_{14}$

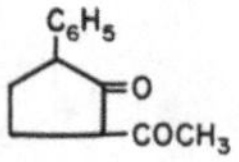


and ENOL

Chem. & Ind. 1146 (1962)

► $O_2C_{13}H_{14}$

$C_6H_5CH{=}CH{-}C(CH_3){-}CHCO_2CH_3$

cis–cis
cis–trans
trans–cis
trans–trans

J. Org. Chem. 27:1989 (1962)

► $O_2C_{13}H_{14}$

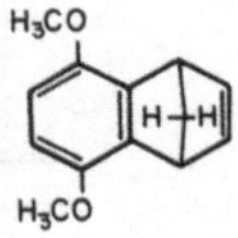


Varian 608

► $O_2C_{13}H_{16}$

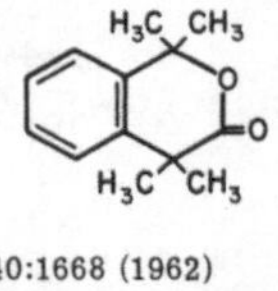


Can. J. Chem. 40:1668 (1962)

► $O_2C_{13}H_{18}$

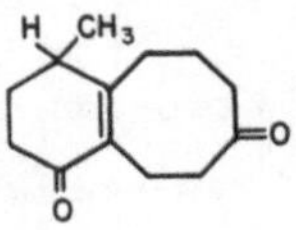


Tetrahedron Letters No. 16:730 (1962)

► $O_2C_{13}H_{18}$

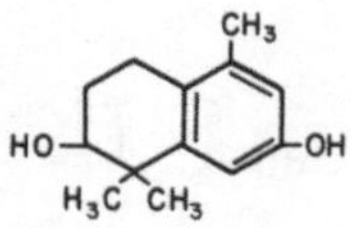


J. Chem. Soc. 1836 (1962)

► $O_2C_{13}H_{18}$

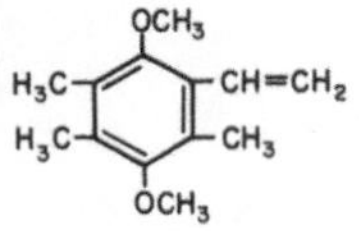


J. Org. Chem. 26:3224 (1961)

► $O_2C_{13}H_{18}$

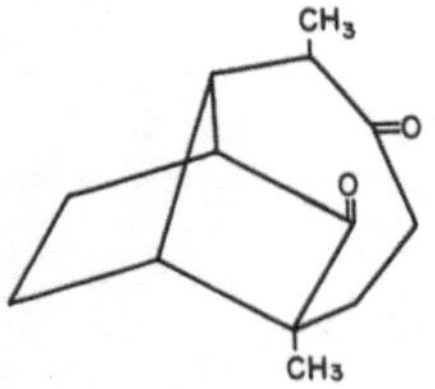


J. Am. Chem. Soc. 83:1251 (1961)

► $O_2C_{13}H_{20}$

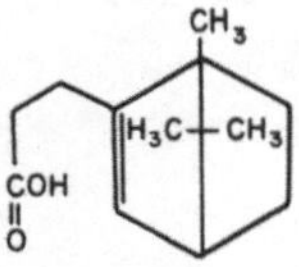


J. Am. Chem. Soc. 84:3205 (1962)

► $O_2C_{13}H_{22}$

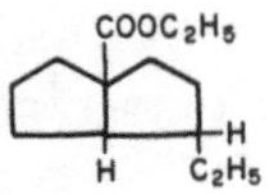


Tetrahedron 18:55 (1962)

► $O_2C_{13}H_{24}$

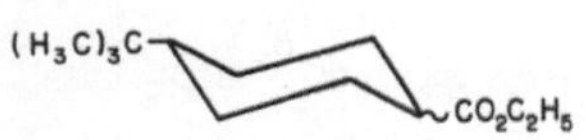


cis and trans

Tetrahedron Letters No. 3:97 (1962)

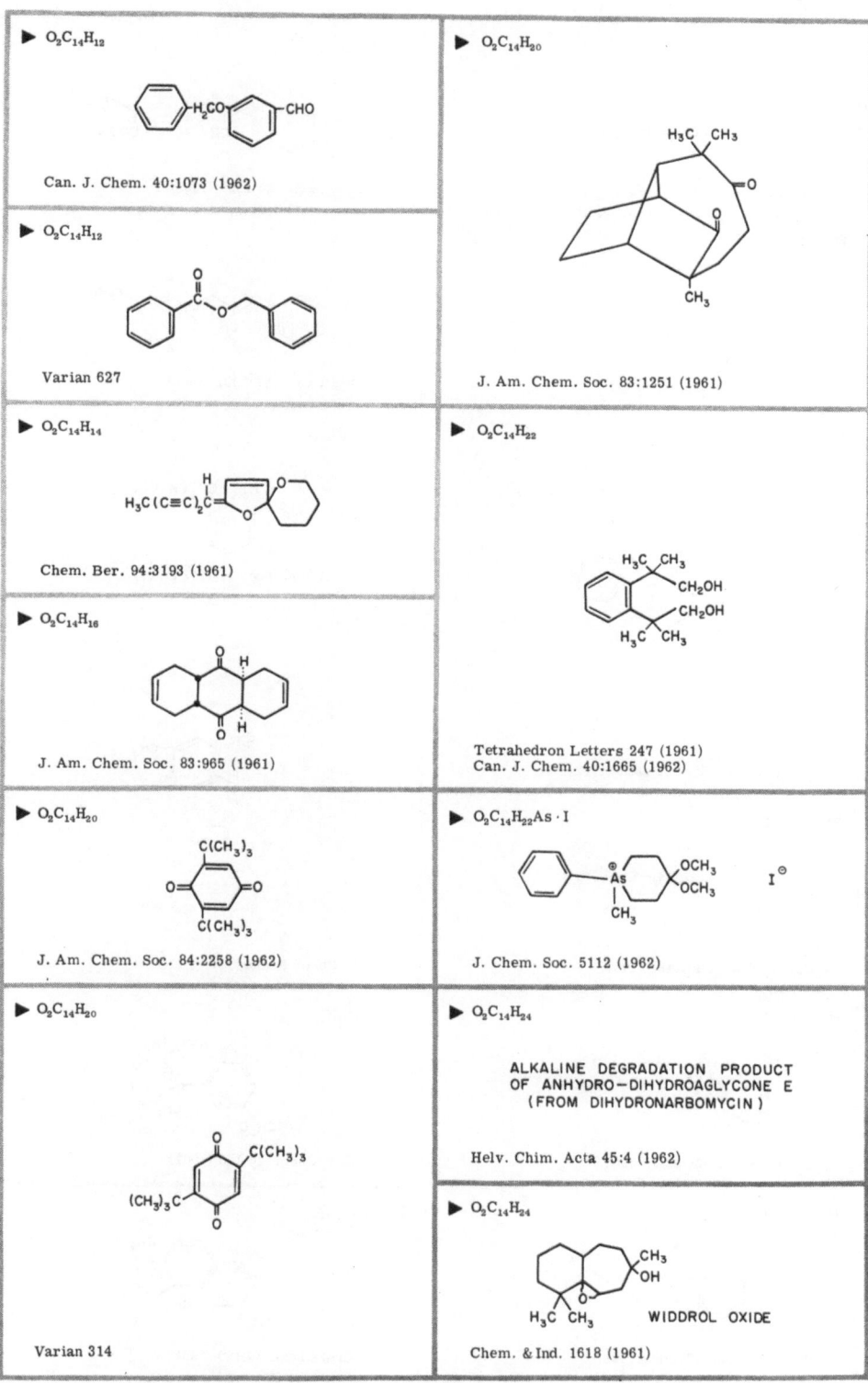

▶ $O_2C_{14}H_{12}$

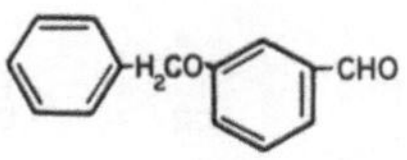


Can. J. Chem. 40:1073 (1962)

▶ $O_2C_{14}H_{12}$

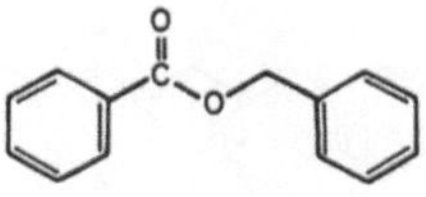


Varian 627

▶ $O_2C_{14}H_{14}$

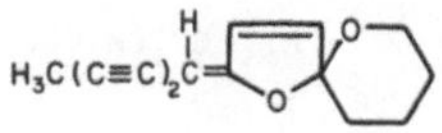


Chem. Ber. 94:3193 (1961)

▶ $O_2C_{14}H_{16}$

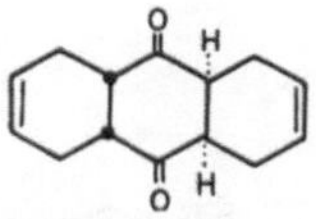


J. Am. Chem. Soc. 83:965 (1961)

▶ $O_2C_{14}H_{20}$

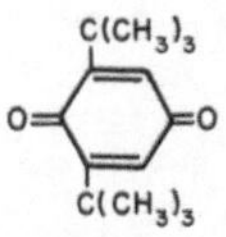


J. Am. Chem. Soc. 84:2258 (1962)

▶ $O_2C_{14}H_{20}$

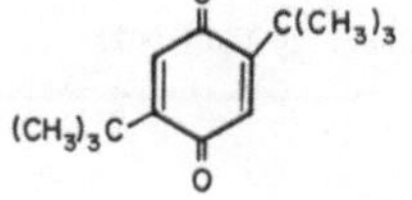


Varian 314

▶ $O_2C_{14}H_{20}$

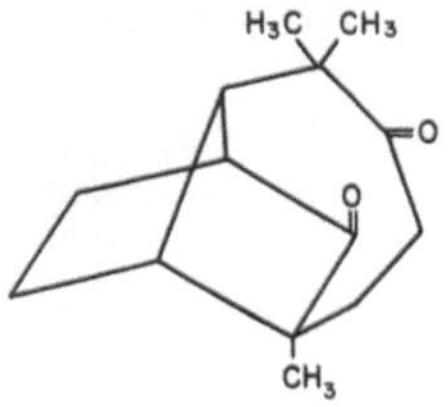


J. Am. Chem. Soc. 83:1251 (1961)

▶ $O_2C_{14}H_{22}$

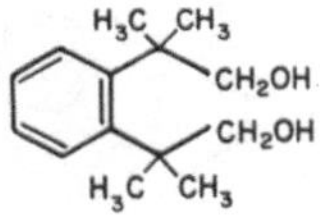


Tetrahedron Letters 247 (1961)
Can. J. Chem. 40:1665 (1962)

▶ $O_2C_{14}H_{22}As \cdot I$

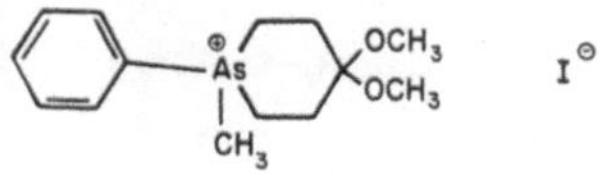


J. Chem. Soc. 5112 (1962)

▶ $O_2C_{14}H_{24}$

ALKALINE DEGRADATION PRODUCT
OF ANHYDRO-DIHYDROAGLYCONE E
(FROM DIHYDRONARBOMYCIN)

Helv. Chim. Acta 45:4 (1962)

▶ $O_2C_{14}H_{24}$

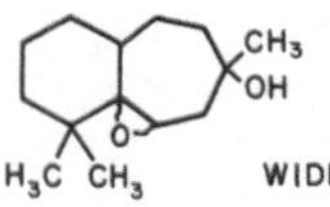


WIDDROL OXIDE

Chem. & Ind. 1618 (1961)

▶ $O_2C_{14}H_{28}$

$CH_3(CH_2)_{12}\overset{O}{\overset{\|}{C}}OK$

Can. J. Chem. 39:359 (1961)

▶ $O_2C_{14}H_{28}$

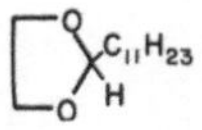

J. Org. Chem. 27:3183 (1962)

▶ $O_2C_{15}H_{12}$

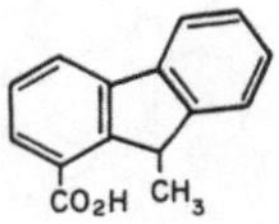

J. Am. Chem. Soc. 84:3531 (1962)

▶ $O_2C_{15}H_{14}$

$C_6H_5CH_2-\underset{C_6H_5}{\underset{|}{CH}}-CO_2H$

Can. J. Chem. 40:1483 (1962)

▶ $O_2C_{15}H_{16}$

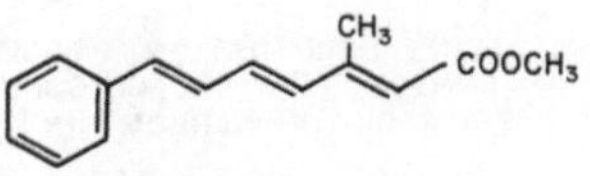

J. Org. Chem. 26:4286 (1961)

▶ $O_2C_{15}H_{16}$

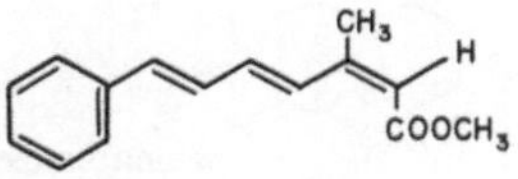

J. Org. Chem. 26:4286 (1961)

▶ $O_2C_{15}H_{16}$

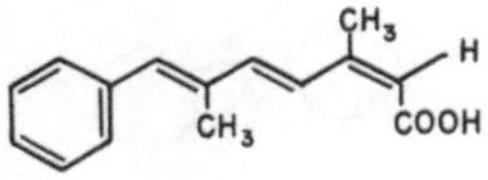

J. Org. Chem. 27:1991 (1962)

▶ $O_2C_{15}H_{16}$

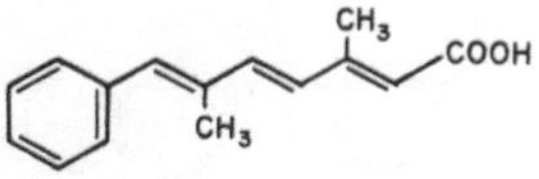

J. Org. Chem. 27:1991 (1962)

▶ $O_2C_{15}H_{18}$

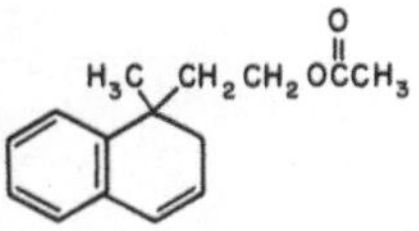

J. Am. Chem. Soc. 84:3517 (1962)

▶ $O_2C_{15}H_{18}$

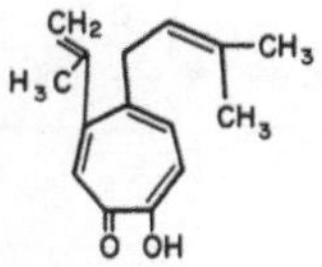

Acta Chem. Scand. 15:645 (1961)

▶ $O_2C_{15}H_{18}$

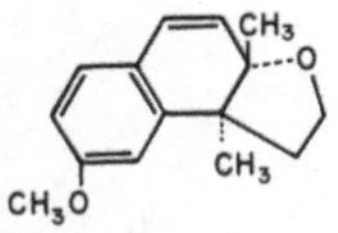

J. Org. Chem. 26:190 (1961)

▶ $O_2C_{15}H_{18}$

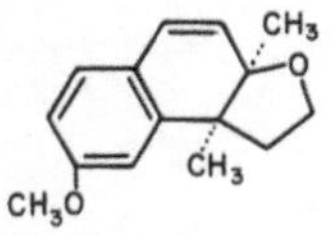

J. Org. Chem. 26:190 (1961)

▶ $O_2C_{15}H_{20}$

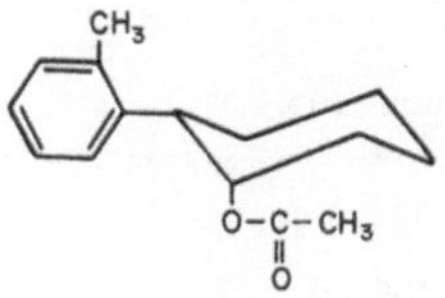

J. Org. Chem. 26:2648 (1961); 27:717 (1962)

▶ $O_2C_{15}H_{20}$

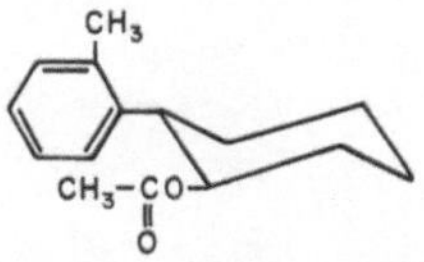

J. Org. Chem. 26:2648 (1961); 27:717 (1962)

▶ $O_2C_{15}H_{20}$

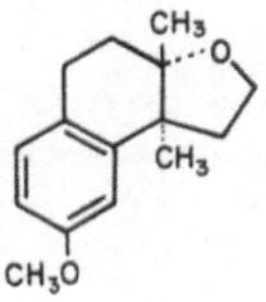

J. Org. Chem. 26:190 (1961)

▶ $O_2C_{15}H_{20}$

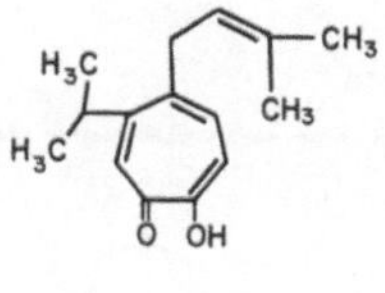

Acta Chem. Scand. 15:645 (1961)

▶ $O_2C_{15}H_{22}$

HELMINTHOSPORAL

Can. J. Chem. 39:1608 (1961)

▶ $O_2C_{15}H_{22}$

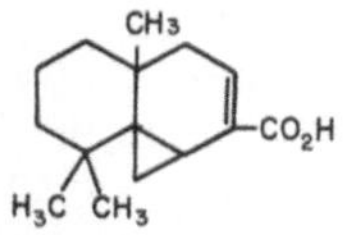

Acta Chem. Scand. 15:592 (1961)

▶ $O_2C_{15}H_{22}$

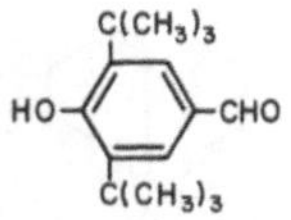

J. Am. Chem. Soc. 84:2258 (1962)

▶ $O_2C_{15}H_{22}$

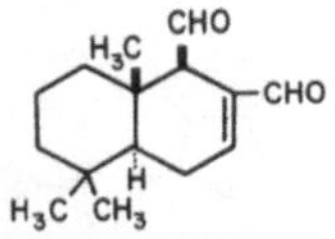

Australian J. Chem. 15:322 (1962)

▶ $O_2C_{15}H_{22}$

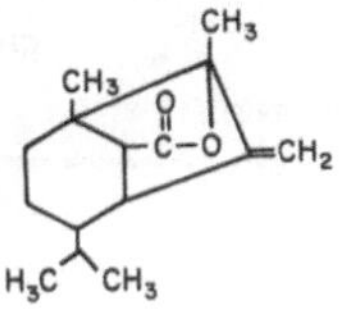

J. Am. Chem. Soc. 84:494 (1962)

▶ $O_2C_{15}H_{24}$

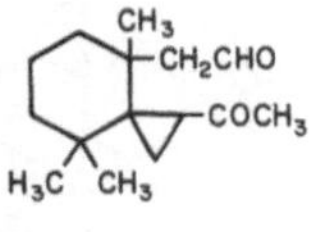

J. Org. Chem. 26:1964 (1961)

$O_2C_{15}H_{24}$
Acta Chem. Scand. 15:592 (1961)
$O_2C_{15}H_{24}$
Acta Chem. Scand. 15:592 (1961)
$O_2C_{15}H_{24}$
Chem. & Ind. 1424 (1962)
$O_2C_{15}H_{24}$
Chem. & Ind. 1424 (1962)
$O_2C_{15}H_{26}$
J. Org. Chem. 26:981 (1961)
$O_2C_{15}H_{26}$
Tetrahedron 18:365 (1962)
$O_2C_{15}H_{26}$
Tetrahedron Letters 281 (1961)
$O_2C_{15}H_{28}$
J. Am. Chem. Soc. 83:3096 (1961)
$O_2C_{16}H_{10}$
J. Am. Chem. Soc. 83:1387 (1961)
$O_2C_{16}H_{12}$
Tetrahedron 18:1008 (1962)
$O_2C_{16}H_{12}$
Varian 650
$O_2C_{16}H_{14}$
J. Am. Chem. Soc. 83:2780 (1961)

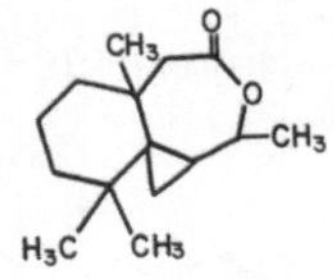

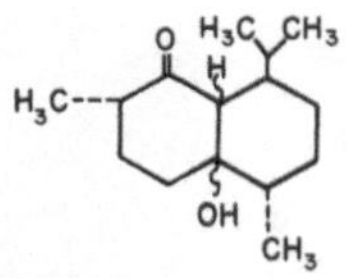

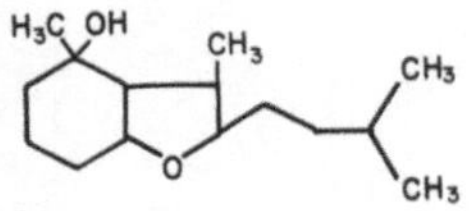

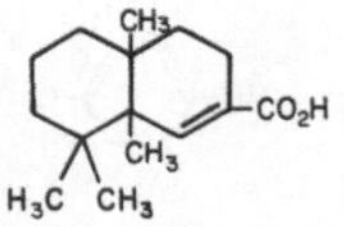

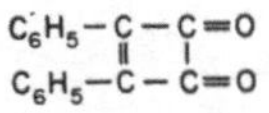

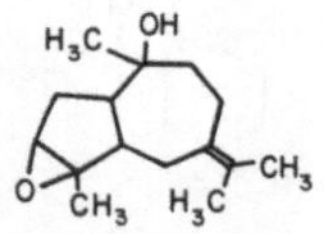

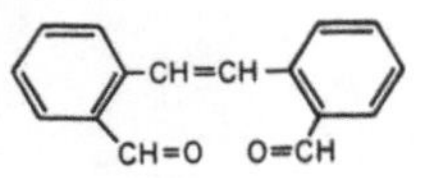

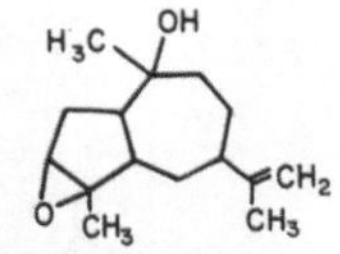

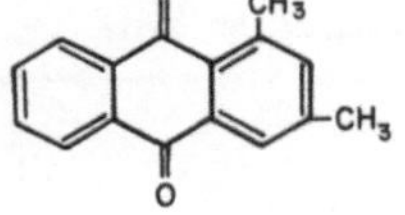

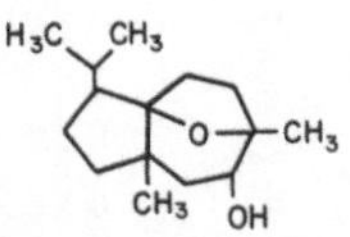

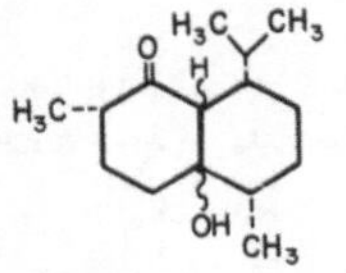

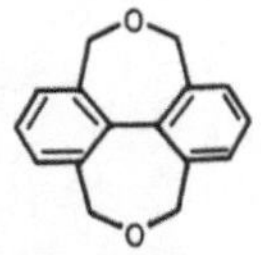

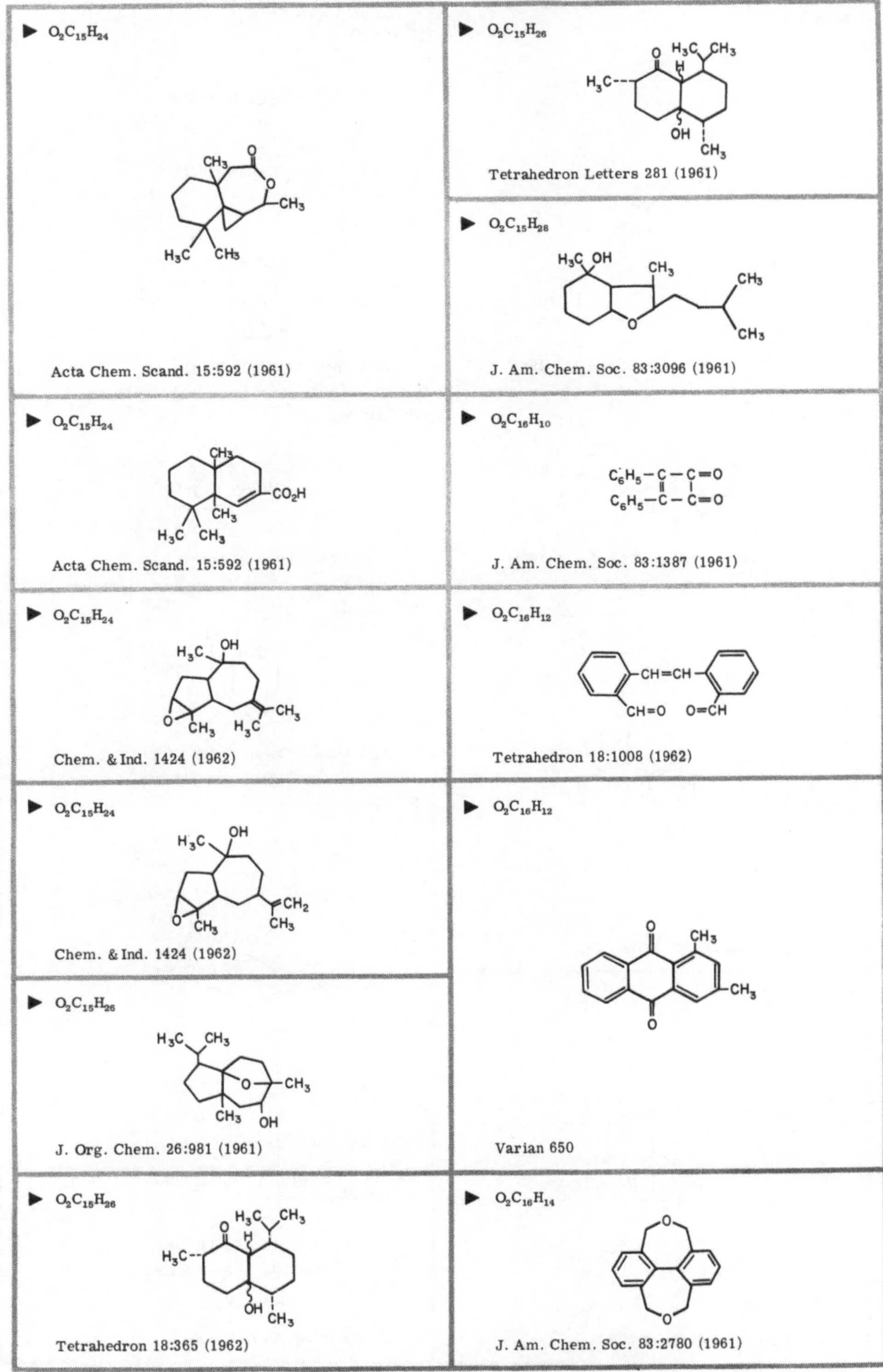

▶ $O_2C_{16}H_{16}$

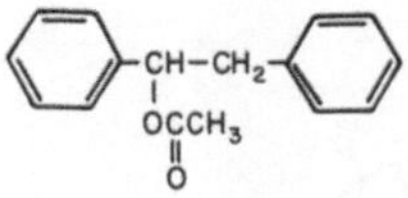

J. Am. Chem. Soc. 83:3647 (1961)

▶ $O_2C_{16}H_{18}$

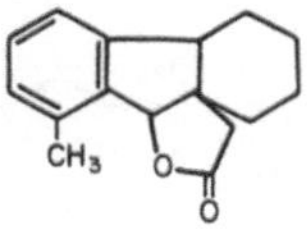

J. Am. Chem. Soc. 84:2614 (1962)

▶ $O_2C_{16}H_{16}$

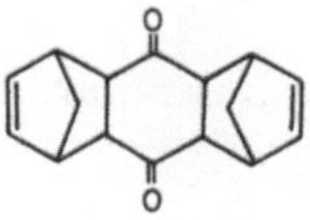

TWO ISOMERS

Helv. Chim. Acta 45:1870 (1962)

▶ $O_2C_{16}H_{18}$

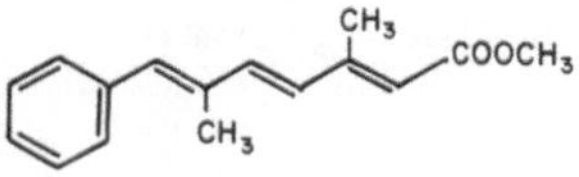

J. Org. Chem. 27:1991 (1962)

▶ $O_2C_{16}H_{16}$

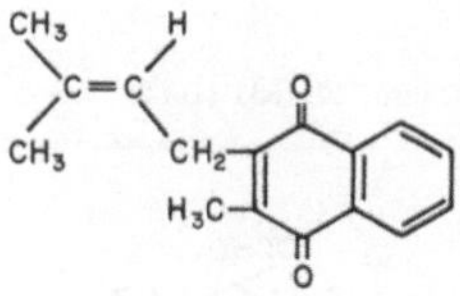

Chem. & Ind. 1020 (1962)

▶ $O_2C_{16}H_{18}$

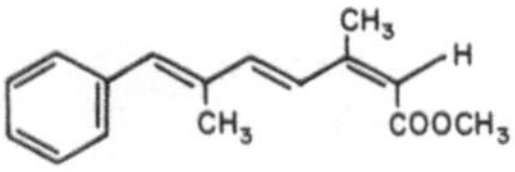

J. Org. Chem. 27:1991 (1962)

▶ $O_2C_{16}H_{22}$

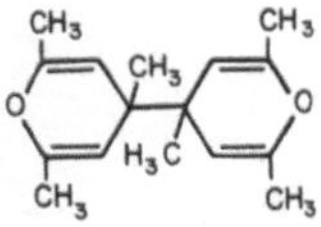

J. Org. Chem. 26:2260 (1961)

▶ $O_2C_{16}H_{16}$

$C_6H_5CH_2CH(C_6H_5)-CO_2CH_3$

Can. J. Chem. 40:1483 (1962)

▶ $O_2C_{16}H_{22}$

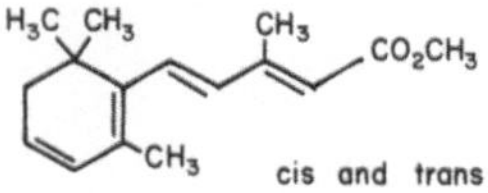

cis and trans

Helv. Chim. Acta 45:528 (1962)

▶ $O_2C_{16}H_{16}$

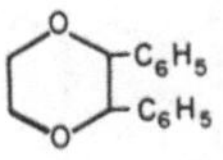

cis and trans

J. Org. Chem. 27:3184 (1962)

▶ $O_2C_{16}H_{24}$

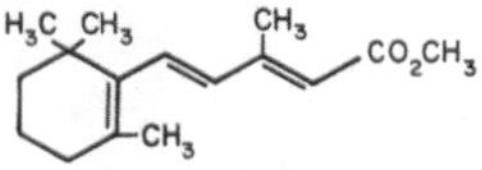

cis and trans

Helv. Chim. Acta 45:528 (1962)

▶ $O_2C_{16}H_{30}$

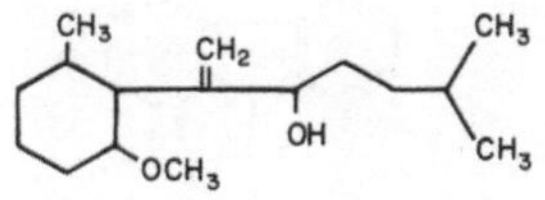


J. Am. Chem. Soc. 83:3096 (1961)

▶ $O_2C_{16}H_{30}$

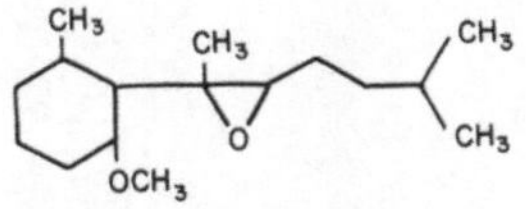


J. Am. Chem. Soc. 83:3096 (1961)

▶ $O_2C_{16}H_{32}$

$$CH_3(CH_2)_{14}\overset{\overset{O}{\|}}{C}OK$$

POTASSIUM PALMITATE

Can. J. Chem. 39:359 (1961)

▶ $O_2C_{17}H_{17}B$

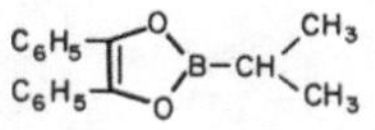


J. Org. Chem. 26:4915 (1961)

▶ $O_2C_{17}H_{18}$

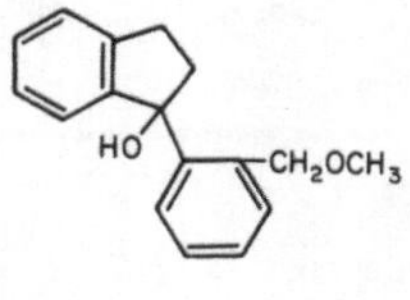


J. Am. Chem. Soc. 84:3531 (1962)

▶ $O_2C_{17}H_{20}$

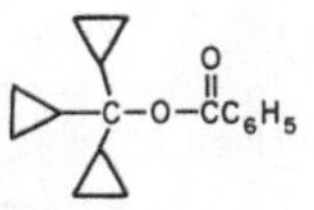


J. Am. Chem. Soc. 84:2462 (1962)

▶ $O_2C_{17}H_{22}$

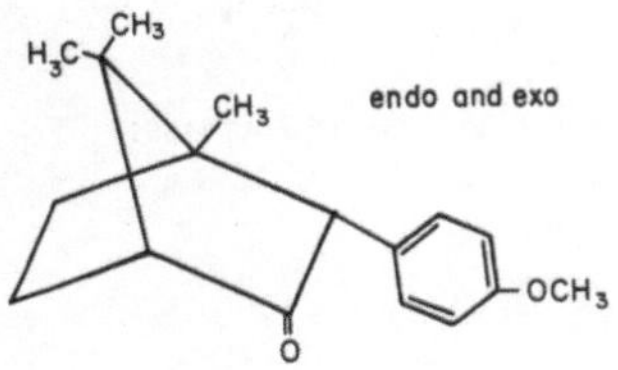


endo and exo

J. Org. Chem. 27:1532 (1962)

▶ $O_2C_{17}H_{24}$

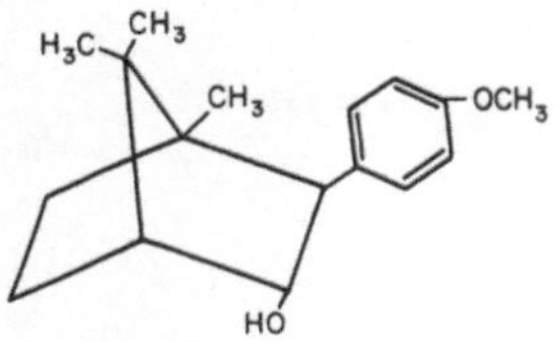


J. Org. Chem. 27:1531 (1962)

▶ $O_2C_{17}H_{26}$

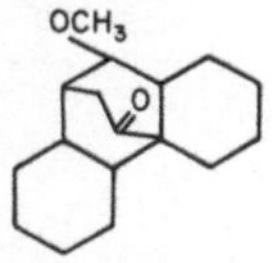


J. Am. Chem. Soc. 84:3779 (1962)

▶ $O_2C_{17}H_{26}$

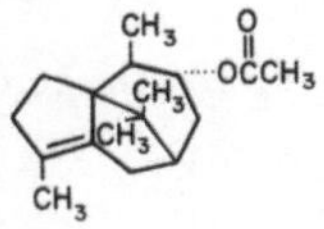


J. Am. Chem. Soc. 84:3205 (1962)

▶ $O_2C_{17}H_{28}$

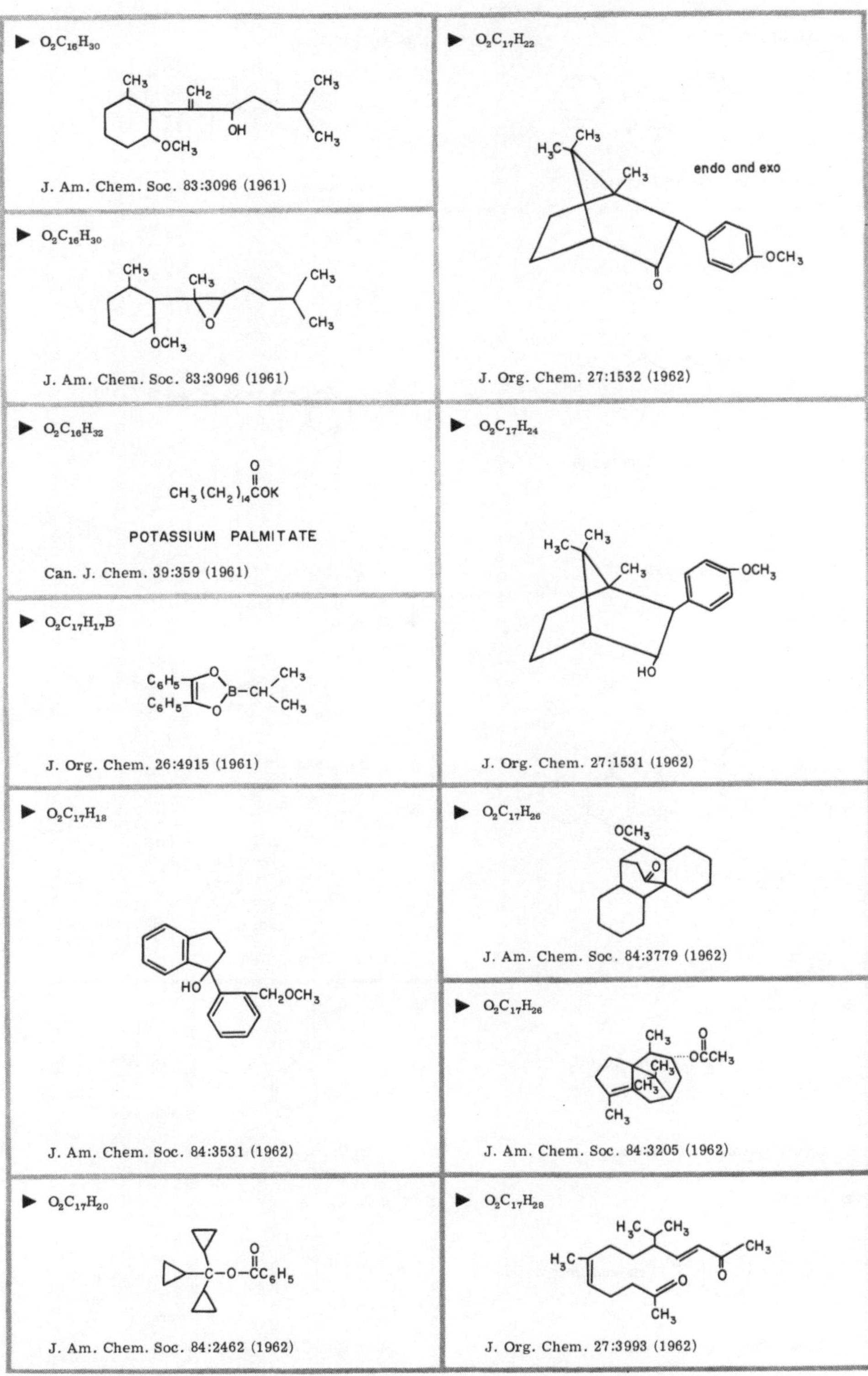

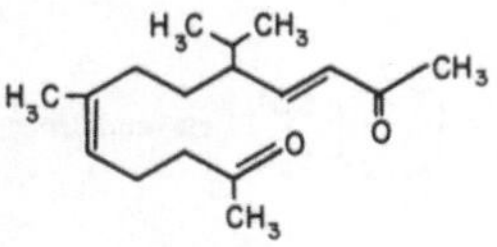


J. Org. Chem. 27:3993 (1962)

▶ $O_2C_{17}H_{28}$

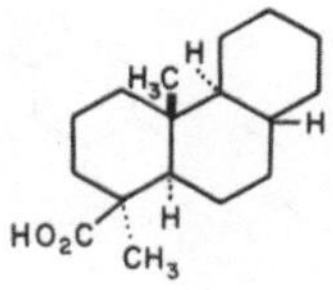


Can. J. Chem. 40:634 (1962)

▶ $O_2C_{18}H_{10}$

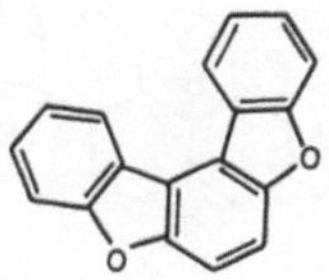


Acta Chem. Scand. 16:553 (1962)

▶ $O_2C_{18}H_{14}$

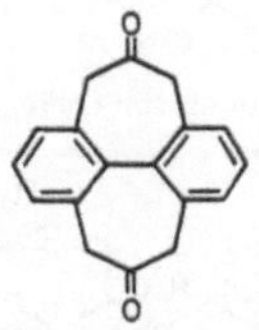


Tetrahedron Letters No. 22:1012 (1962)

▶ $O_2C_{18}H_{16}$

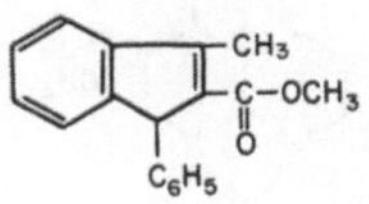


J. Org. Chem. 26:681 (1961)

▶ $O_2C_{18}H_{16}$

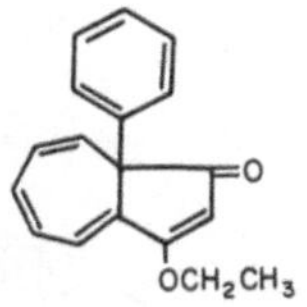


Proc. Chem. Soc. 21 (1962)
J. Chem. Soc. 2708 (1962)
Helv. Chim. Acta 45:600 (1962)

▶ $O_2C_{18}H_{20}$

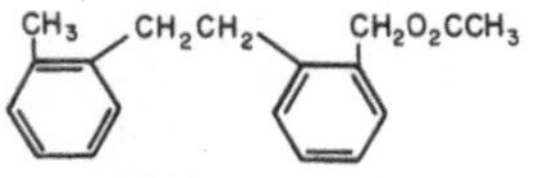


J. Am. Chem. Soc. 83:960 (1961)

▶ $O_2C_{18}H_{20}$

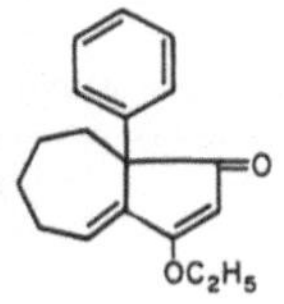


Proc. Chem. Soc. 21 (1962)
J. Chem. Soc. 2709 (1962)
Helv. Chim. Acta 45:600 (1962)

▶ $O_2C_{18}H_{22}$

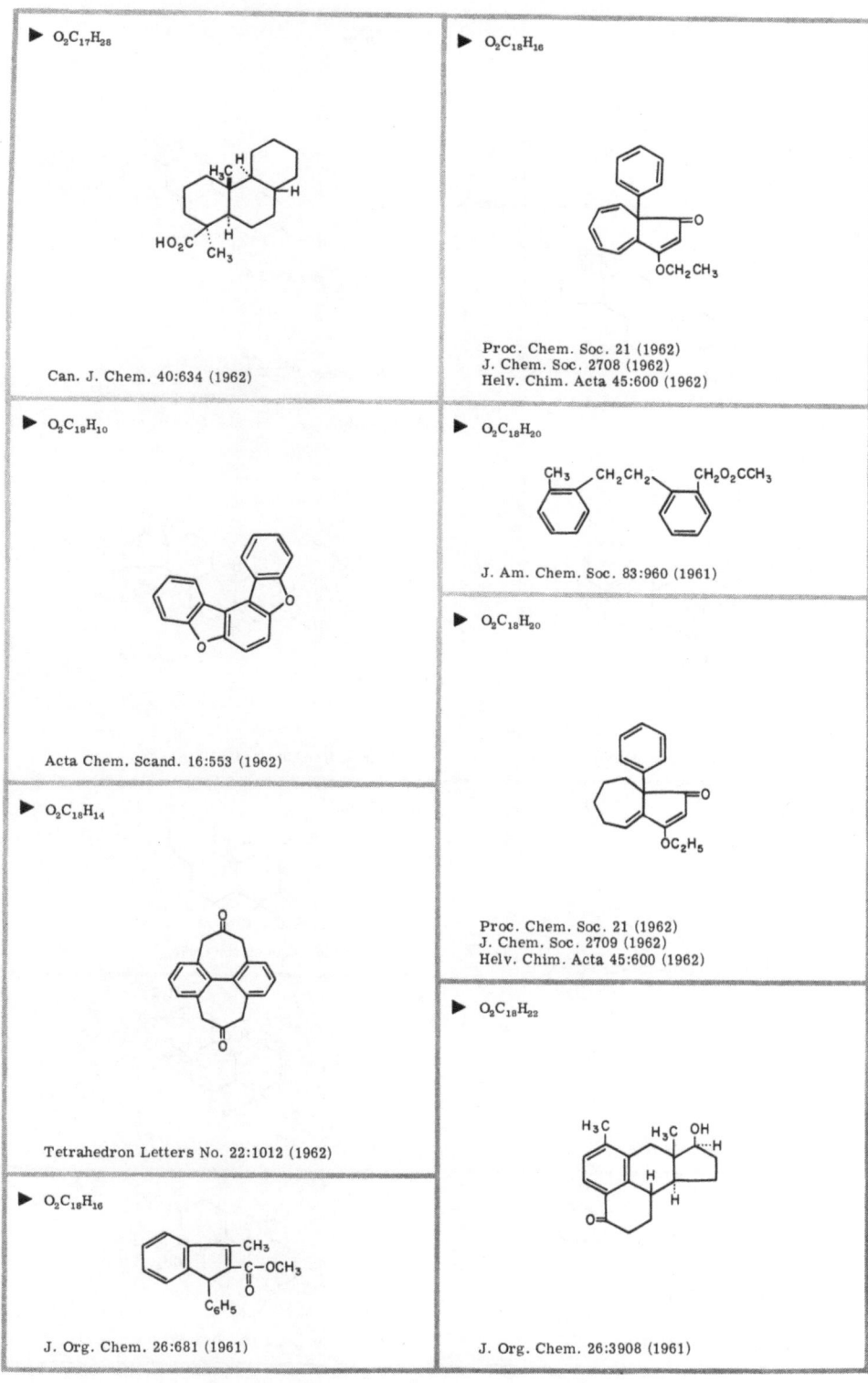

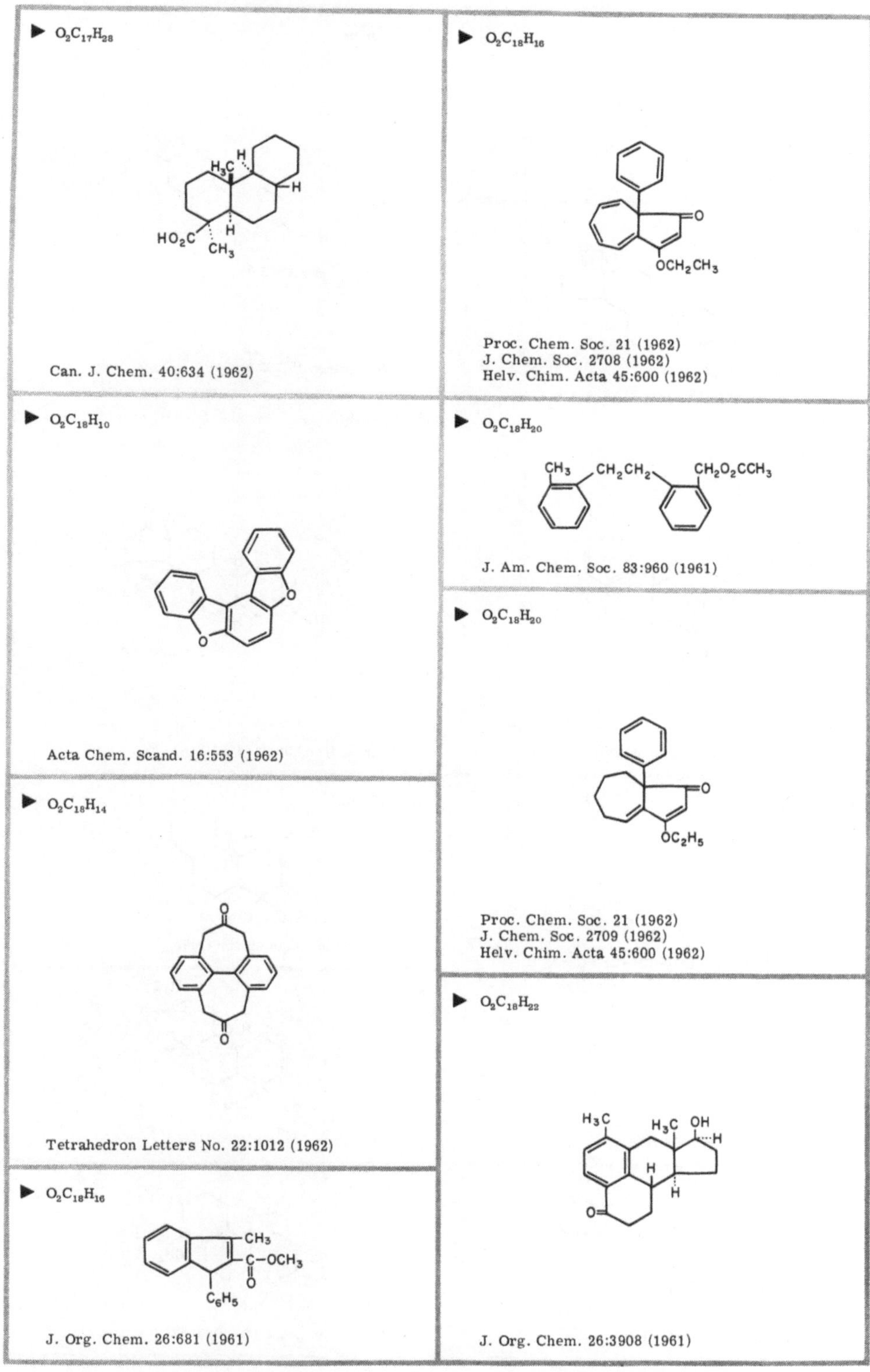

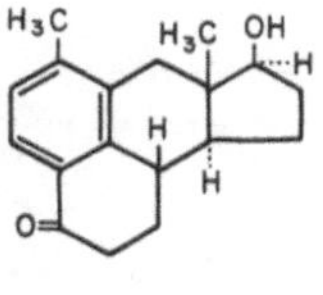


J. Org. Chem. 26:3908 (1961)

▶ $O_2C_{18}H_{22}$

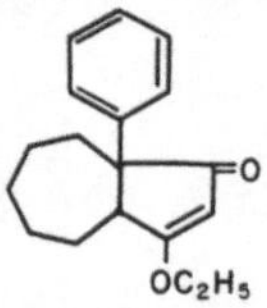

Helv. Chim. Acta 45:600 (1962)

▶ $O_2C_{18}H_{22}$

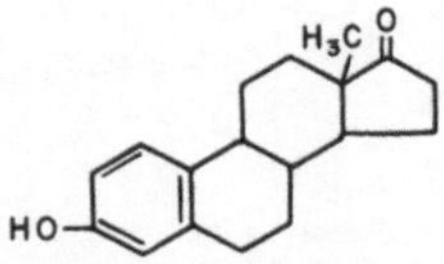

Chem. & Ind. 1716 (1962)

▶ $O_2C_{18}H_{26}$

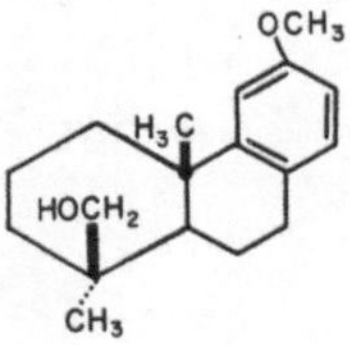

Tetrahedron Letters 360 (1961)

▶ $O_2C_{18}H_{28}$

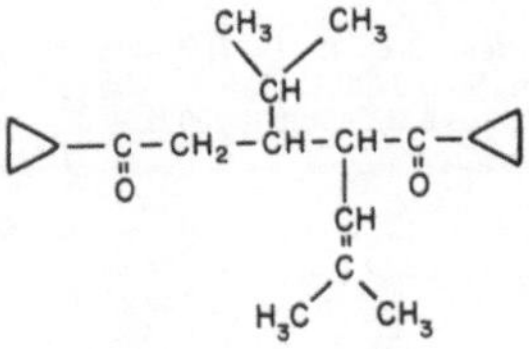

J. Org. Chem. 26:3693 (1961)

▶ $O_2C_{18}H_{28}$

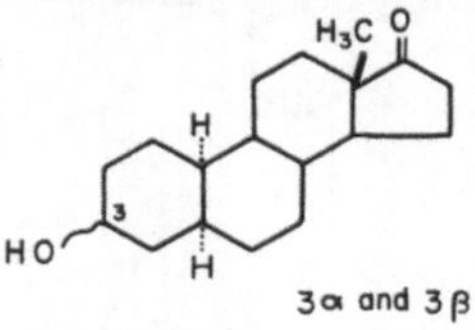

3α and 3β

Tetrahedron 15:202 (1961)

▶ $O_2C_{18}H_{28}$

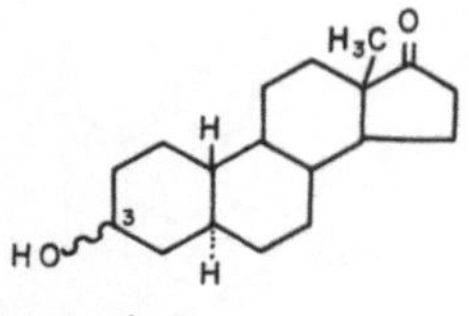

3α and 3β

Tetrahedron 15:202 (1961)

▶ $O_2C_{18}H_{28}$

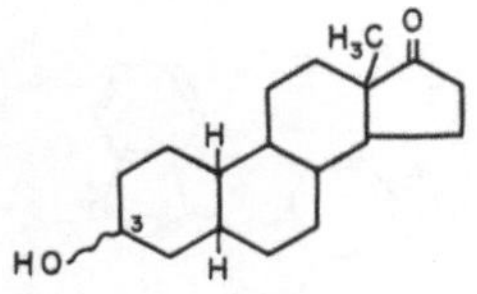

3α and 3β

Tetrahedron 15:202 (1961)

▶ $O_2C_{18}H_{30}$

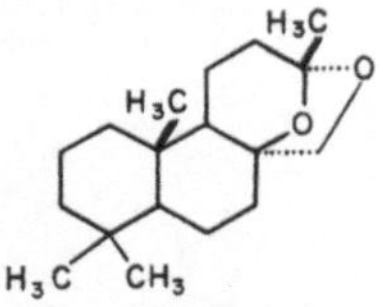

Helv. Chim. Acta 45:400 (1962)

▶ $O_2C_{18}H_{30}$

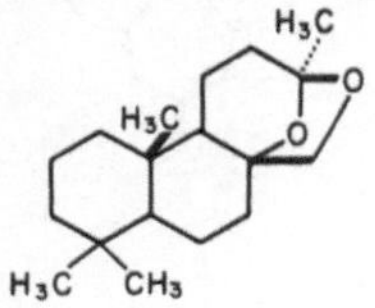

Helv. Chim. Acta 45:400 (1962)

▶ $O_2C_{18}H_{30}$

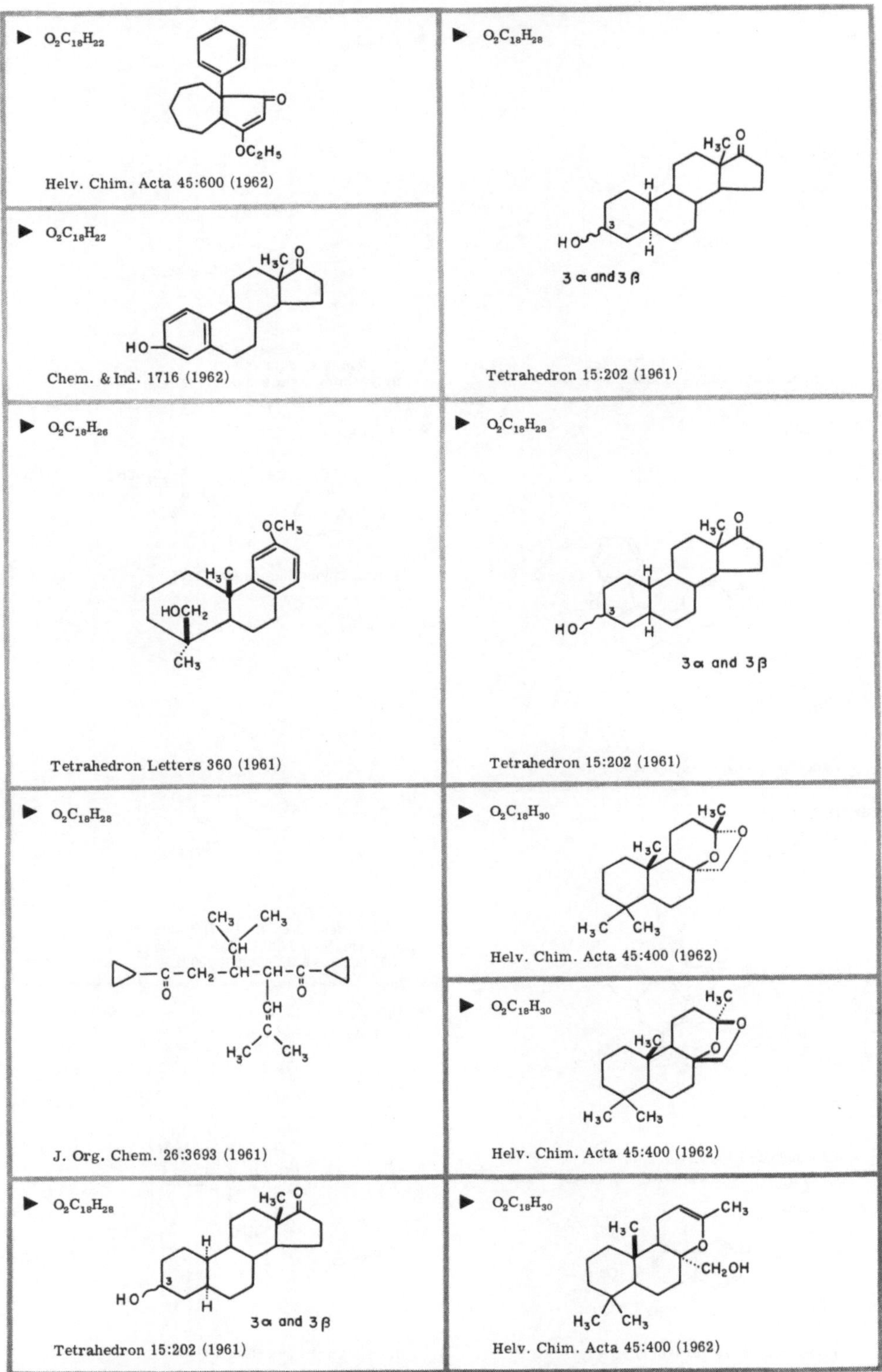

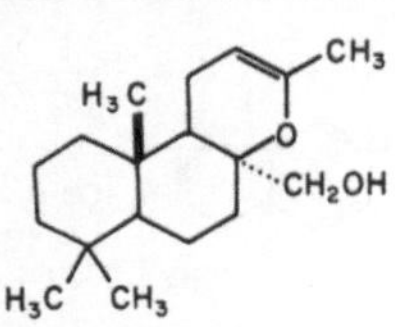

Helv. Chim. Acta 45:400 (1962)

▶ $O_2C_{18}H_{30}$

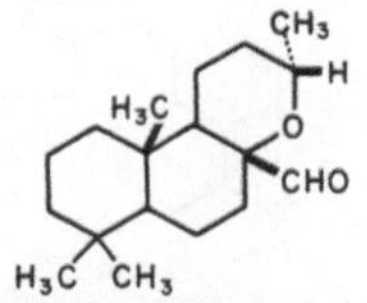

Helv. Chim. Acta 45:400 (1962)

▶ $O_2C_{18}H_{30}$

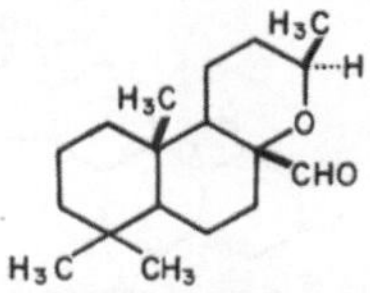

Helv. Chim. Acta 45:400 (1962)

▶ $O_2C_{18}H_{32}$

$CH_3(CH_2)_7C{\equiv}C(CH_2)_7CO_2H$

J. Am. Oil Chemists' Soc. 38:664 (1961)

▶ $O_2C_{18}H_{32}$

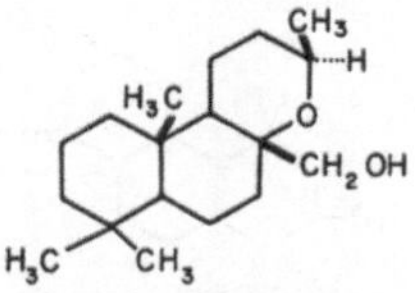

Helv. Chim. Acta 45:400 (1962)

▶ $O_2C_{18}H_{32}$

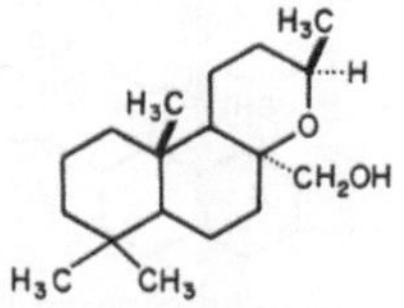

Helv. Chim. Acta 45:400 (1962)

▶ $O_2C_{18}H_{32}$

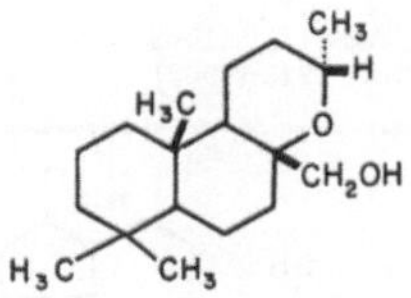

Helv. Chim. Acta 45:400 (1962)

▶ $O_2C_{19}H_{14}$

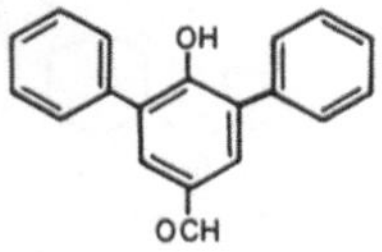

Varian 676

▶ $O_2C_{19}H_{16}$

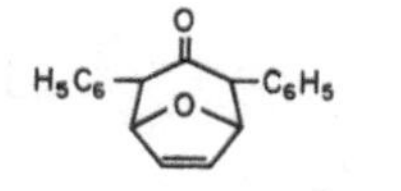

two isomers

J. Am. Chem. Soc. 84:4979 (1962)

▶ $O_2C_{19}H_{16}$

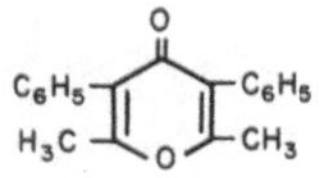

J. Am. Chem. Soc. 83:193 (1961)

▶ $O_2C_{19}H_{22}$

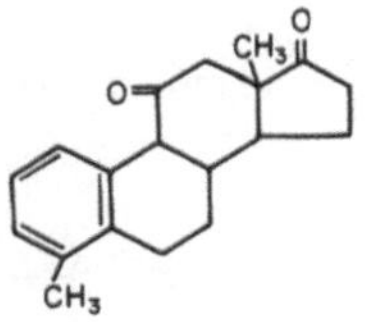

Chem. & Ind. 1716 (1962)

▶ $O_2C_{19}H_{22}$

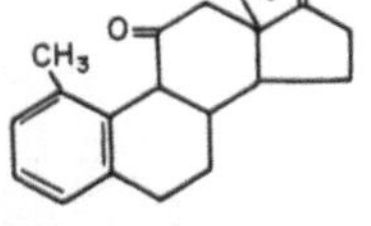

Chem. & Ind. 1716 (1962)

▶ $O_2C_{19}H_{22}$

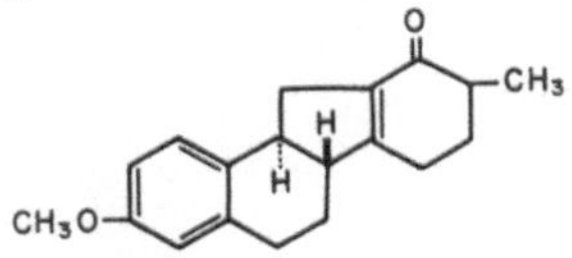

J. Org. Chem. 26:4588 (1961)

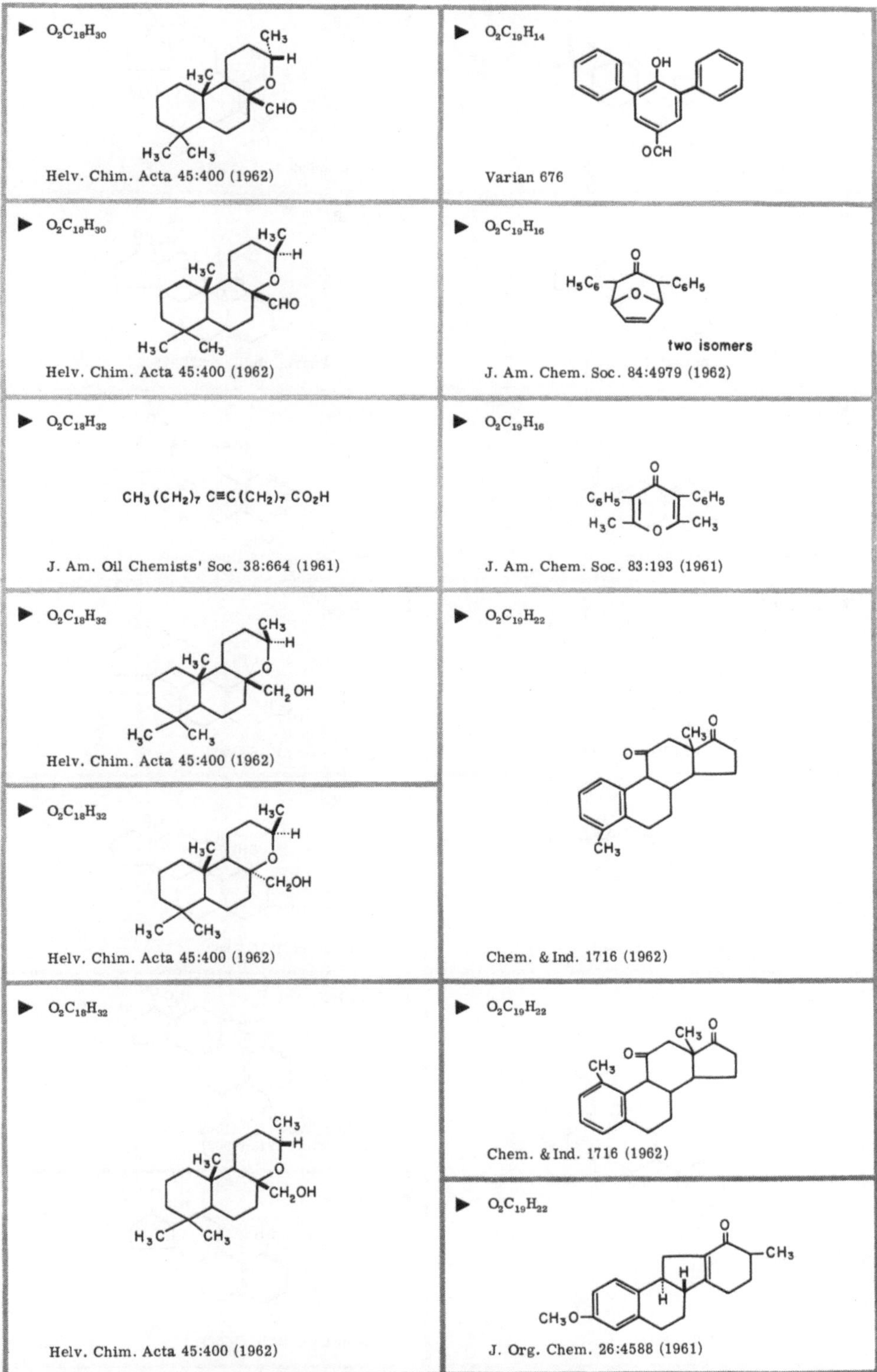

▶ $O_2C_{19}H_{22}$

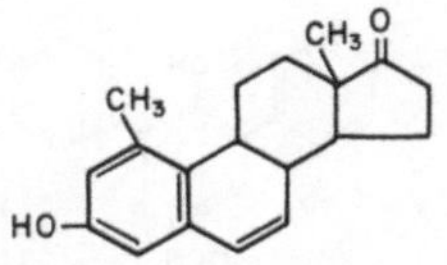


Chem. & Ind. 1716 (1962)

▶ $O_2C_{19}H_{24}$

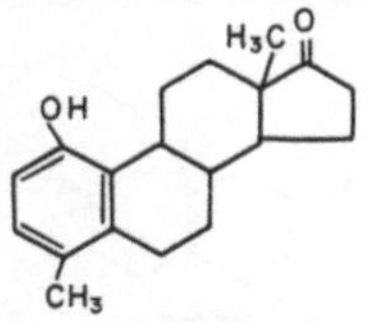


Chem. & Ind. 1716 (1962)

▶ $O_2C_{19}H_{22}$

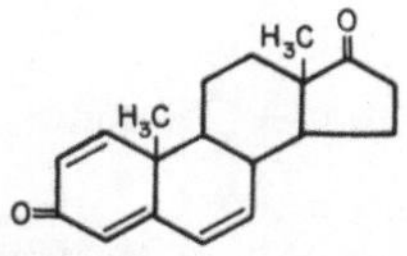


Varian 335
Helv. Chim. Acta 44:1380 (1961)

▶ $O_2C_{19}H_{24}$

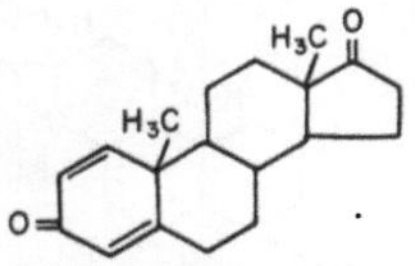


Chem. Pharm. Bull. 10:338 (1962)

▶ $O_2C_{19}H_{22}$

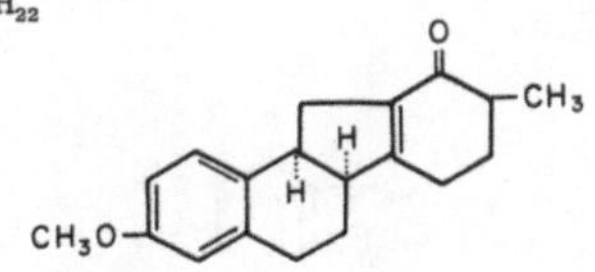


J. Org. Chem. 26:4588 (1961)

▶ $O_2C_{19}H_{24}$

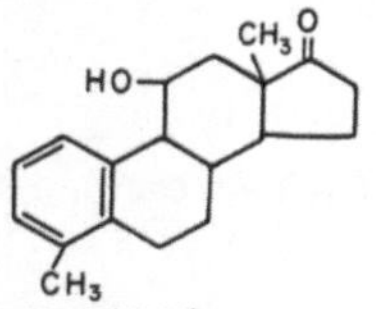


Chem. & Ind. 1716 (1962)

▶ $O_2C_{19}H_{22}$

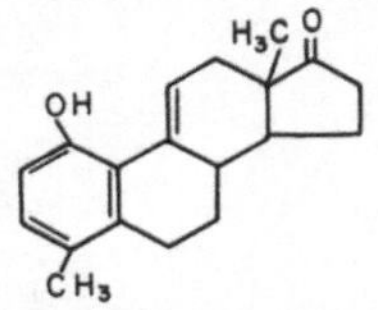


J. Am. Chem. Soc. 83:4627 (1961)

▶ $O_2C_{19}H_{24}$

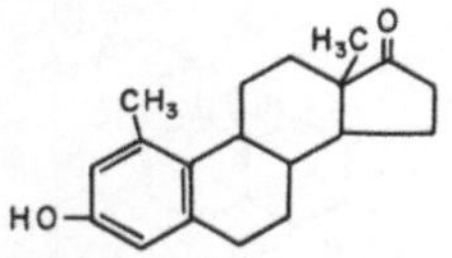


Chem. & Ind. 1716 (1962)

▶ $O_2C_{19}H_{24}$

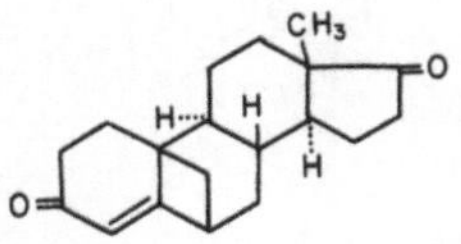


Helv. Chim. Acta 45:2615 (1962)

▶ $O_2C_{19}H_{24}$

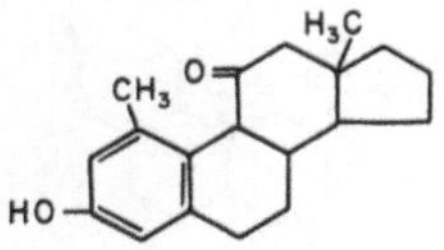


Chem. & Ind. 1716 (1962)

▶ $O_2C_{19}H_{24}$

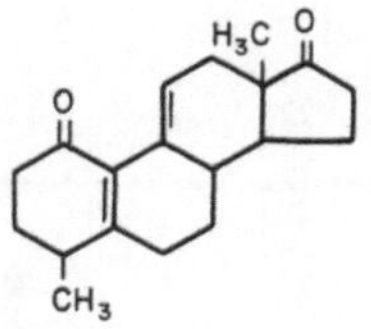


Chem. & Ind. 1530 (1961)

▶ $O_2C_{19}H_{26}$

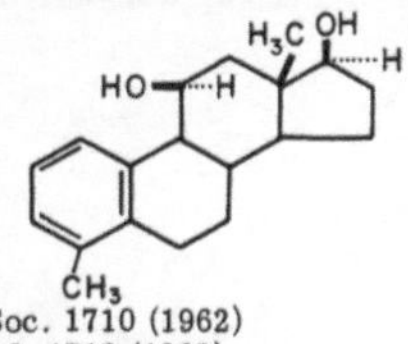


J. Chem. Soc. 1710 (1962)
Chem. & Ind. 1716 (1962)

▶ $O_2C_{19}H_{26}$

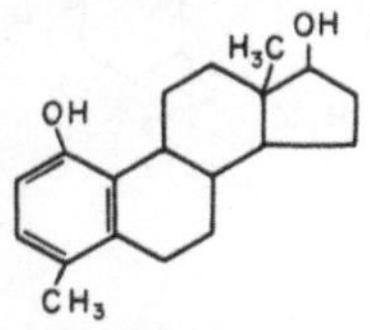


Chem. & Ind. 1716 (1962)

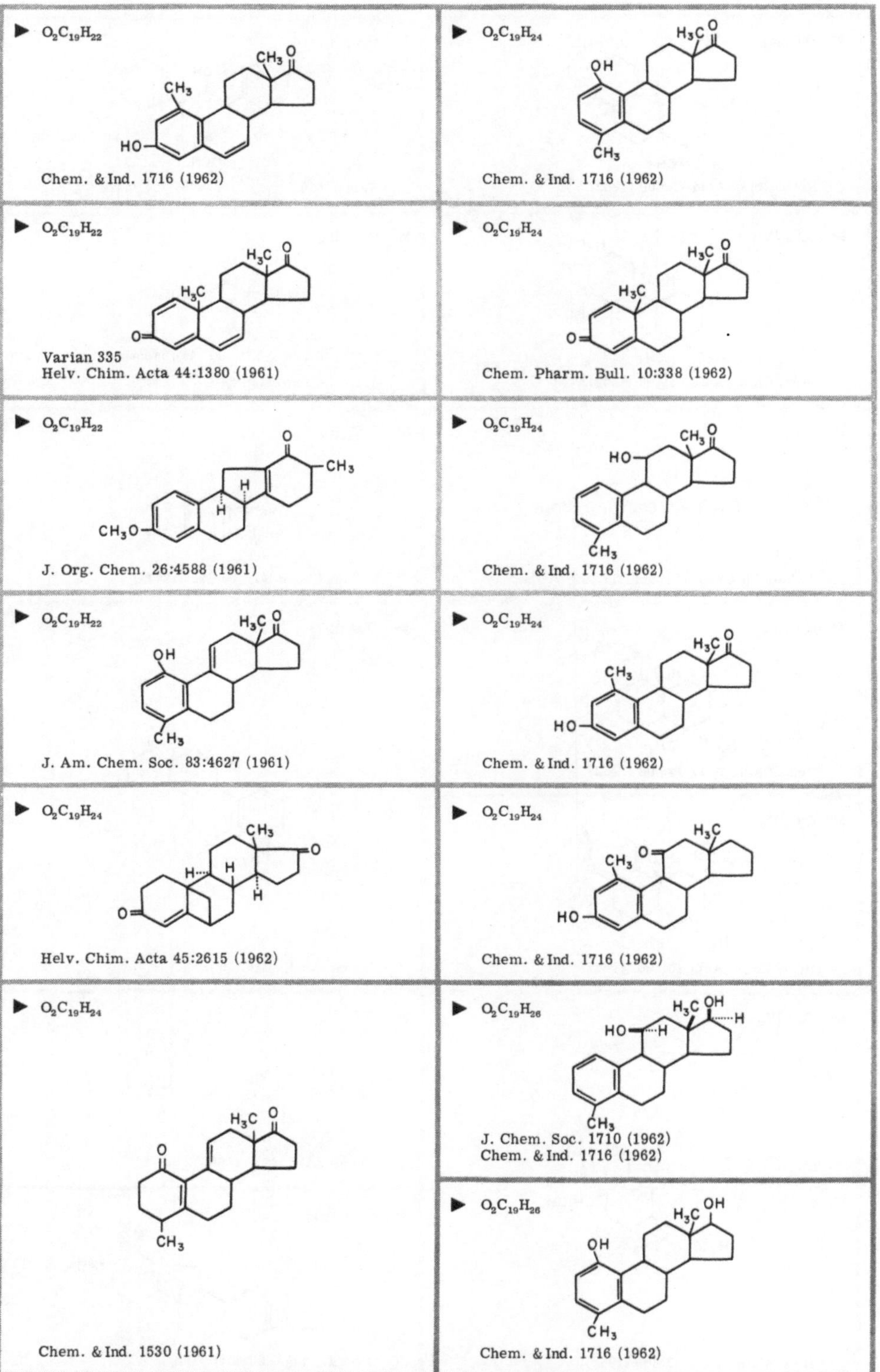

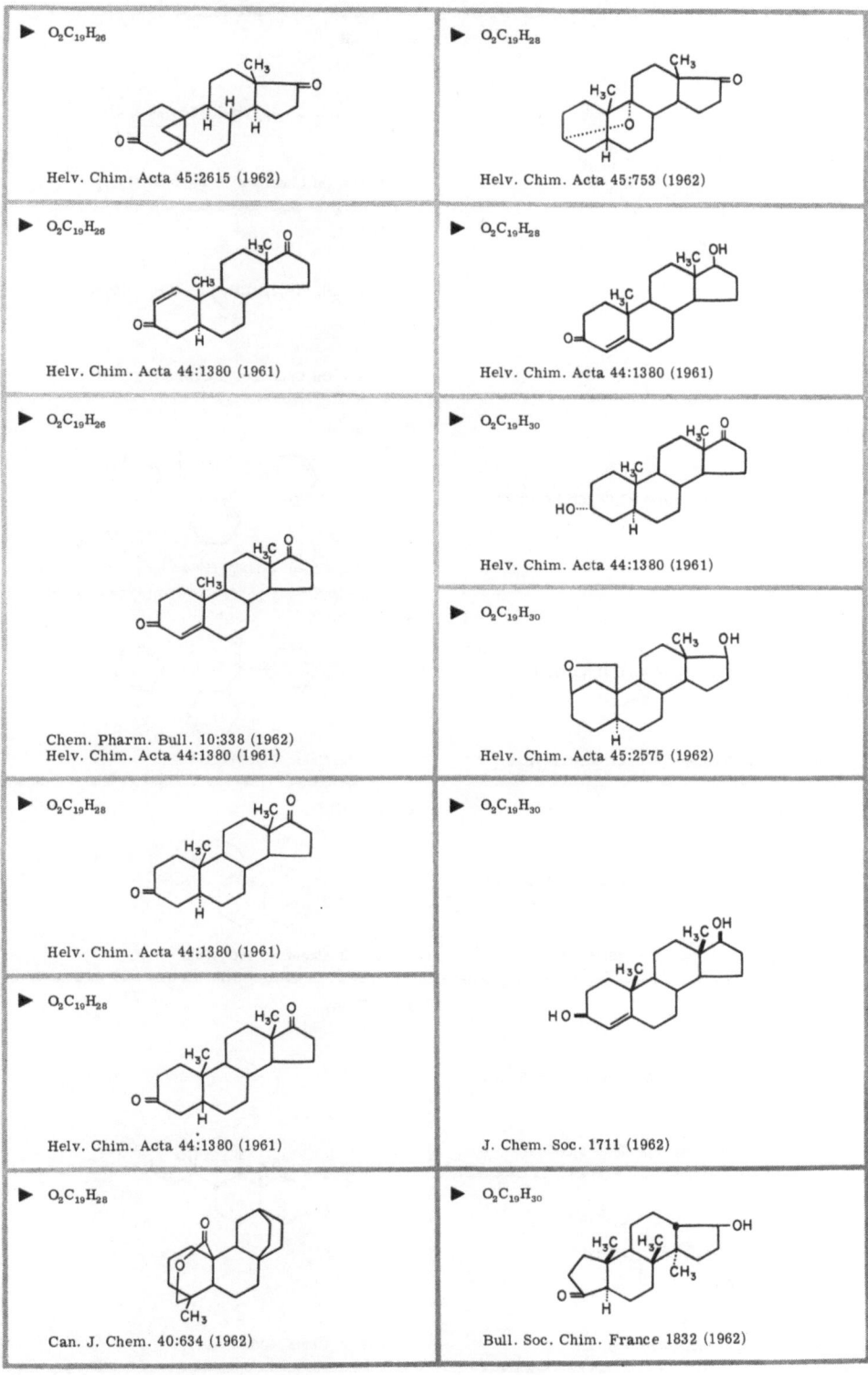

► $O_2C_{19}H_{26}$

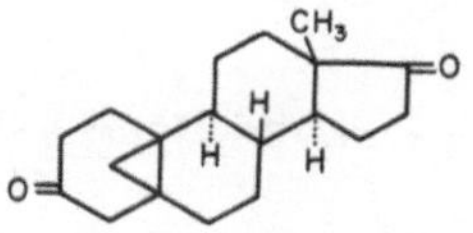

Helv. Chim. Acta 45:2615 (1962)

► $O_2C_{19}H_{26}$

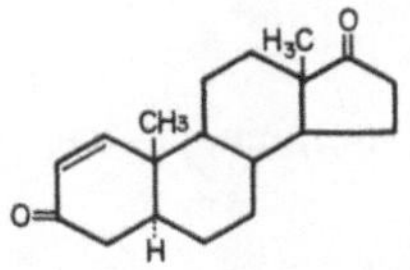

Helv. Chim. Acta 44:1380 (1961)

► $O_2C_{19}H_{26}$

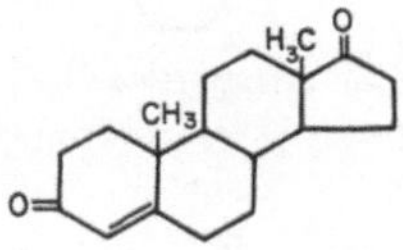

Chem. Pharm. Bull. 10:338 (1962)
Helv. Chim. Acta 44:1380 (1961)

► $O_2C_{19}H_{28}$

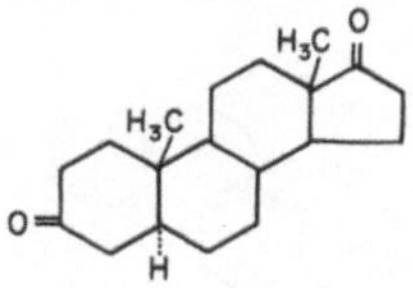

Helv. Chim. Acta 44:1380 (1961)

► $O_2C_{19}H_{28}$

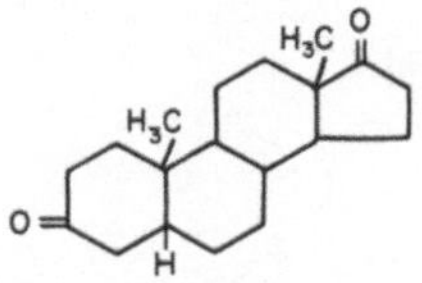

Helv. Chim. Acta 44:1380 (1961)

► $O_2C_{19}H_{28}$

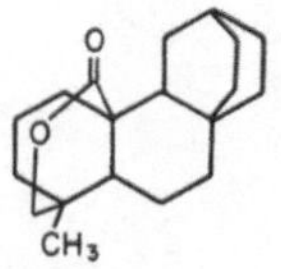

Can. J. Chem. 40:634 (1962)

► $O_2C_{19}H_{28}$

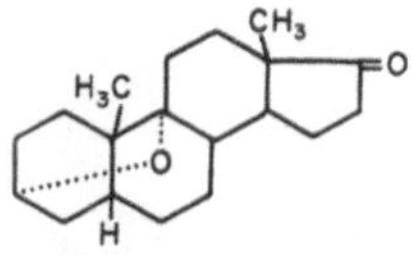

Helv. Chim. Acta 45:753 (1962)

► $O_2C_{19}H_{28}$

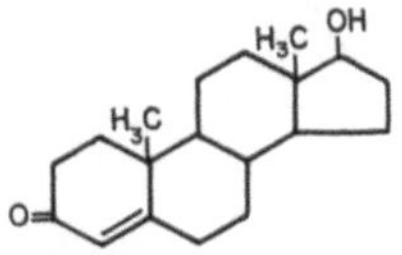

Helv. Chim. Acta 44:1380 (1961)

► $O_2C_{19}H_{30}$

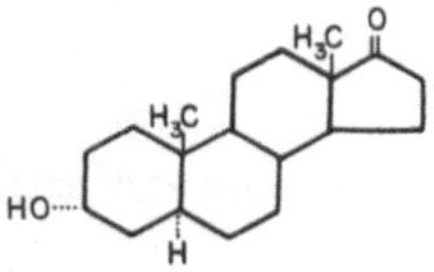

Helv. Chim. Acta 44:1380 (1961)

► $O_2C_{19}H_{30}$

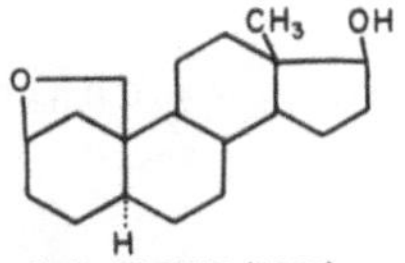

Helv. Chim. Acta 45:2575 (1962)

► $O_2C_{19}H_{30}$

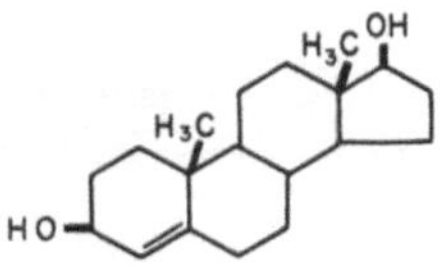

J. Chem. Soc. 1711 (1962)

► $O_2C_{19}H_{30}$

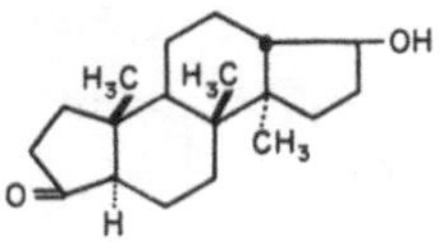

Bull. Soc. Chim. France 1832 (1962)

▶ $O_2C_{19}H_{30}$

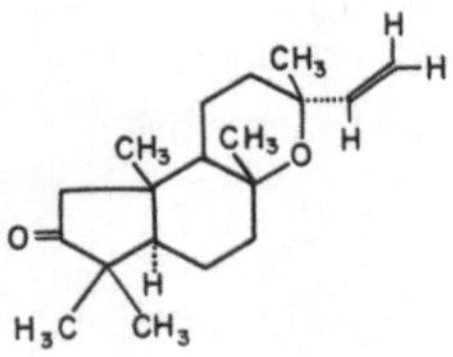

J. Chem. Soc. 3741 (1962)

▶ $O_2C_{19}H_{32}$

$CH_3CH_2CH{=}CH(CH_2CH{=}CH)_2CH_2(CH_2)_5CH_2\overset{O}{\overset{\|}{C}}OCH_3$

Varian 337

▶ $O_2C_{19}H_{34}$

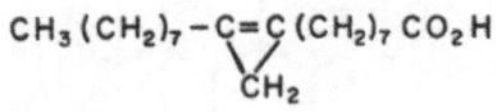

J. Am. Oil Chemists' Soc. 38:664 (1961)

▶ $O_2C_{19}H_{34}$

$CH_3(CH_2)_4\ CH{=}CHCH_2\ CH{=}CH(CH_2)_7\ CO_2CH_3$

J. Am. Oil Chemists' Soc. 38:664 (1961)

▶ $O_2C_{19}H_{34}$

$CH_3(CH_2)_4{-}CH{=}CH{-}CH{=}CH{-}(CH_2)_8\ CO_2CH_3$

J. Am. Oil Chemists' Soc. 38:664 (1961)

▶ $O_2C_{19}H_{34}$

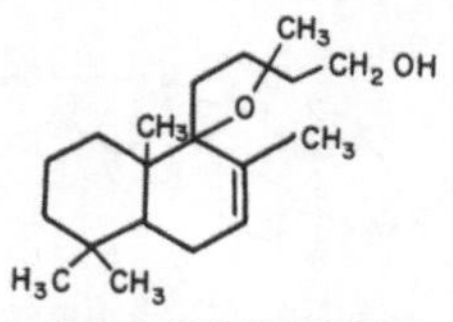

Acta Chem. Scand. 16:1675 (1962)

▶ $O_2C_{19}H_{36}$

$CH_3(CH_2)_7\ CH{=}CH(CH_2)_7\ CO_2\ CH_3$

J. Am. Oil Chemists' Soc. 38:664 (1961)

▶ $O_2C_{19}H_{36}$

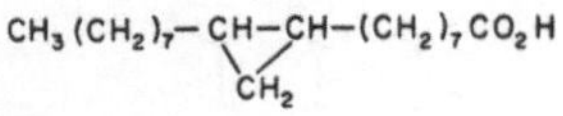

J. Am. Oil Chemists' Soc. 38:664 (1961)

▶ $O_2C_{20}H_{18}$

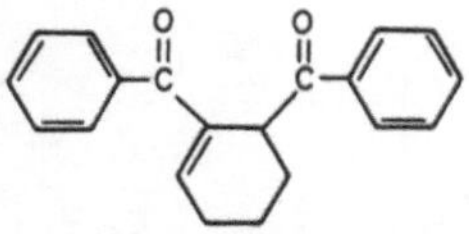

J. Org. Chem. 27:1960 (1962)

▶ $O_2C_{20}H_{18}$

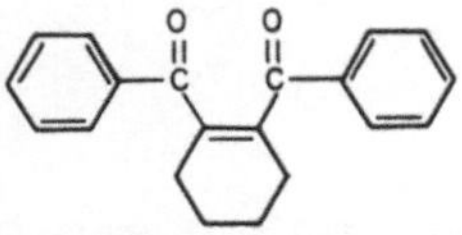

J. Org. Chem. 27:1960 (1962)

▶ $O_2C_{20}H_{22}$

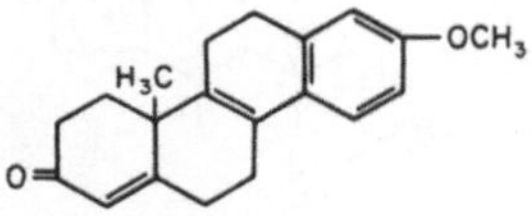

Can. J. Chem. 40:986 (1962)

▶ $O_2C_{20}H_{24}$

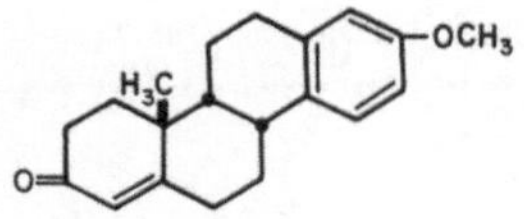

Can. J. Chem. 40:984 (1962)

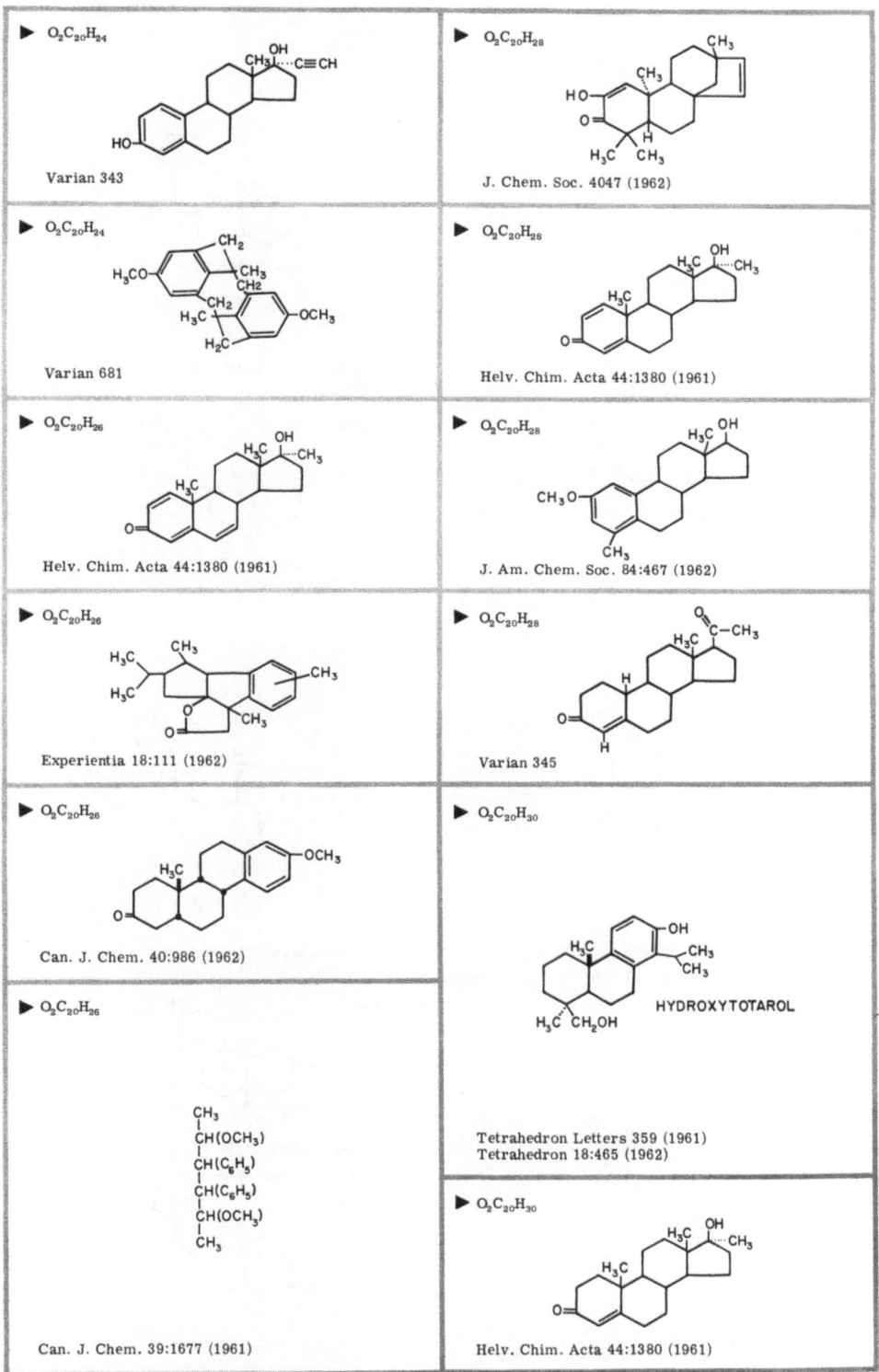

▶ $O_2C_{20}H_{24}$

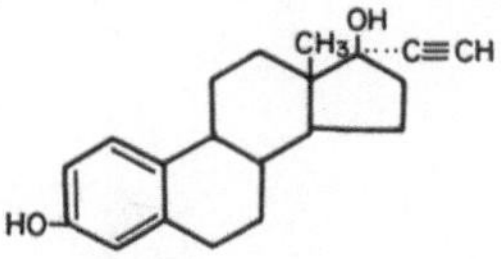

Varian 343

▶ $O_2C_{20}H_{28}$

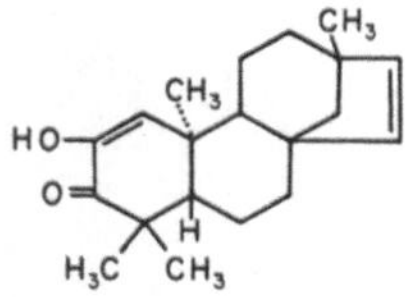

J. Chem. Soc. 4047 (1962)

▶ $O_2C_{20}H_{24}$

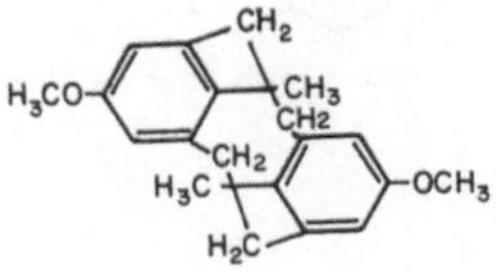

Varian 681

▶ $O_2C_{20}H_{28}$

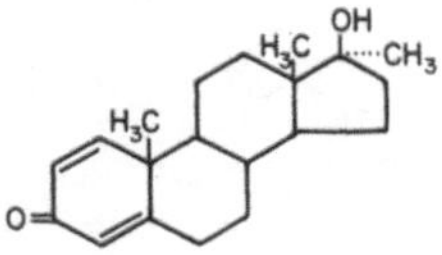

Helv. Chim. Acta 44:1380 (1961)

▶ $O_2C_{20}H_{26}$

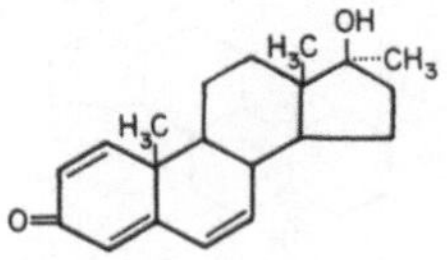

Helv. Chim. Acta 44:1380 (1961)

▶ $O_2C_{20}H_{28}$

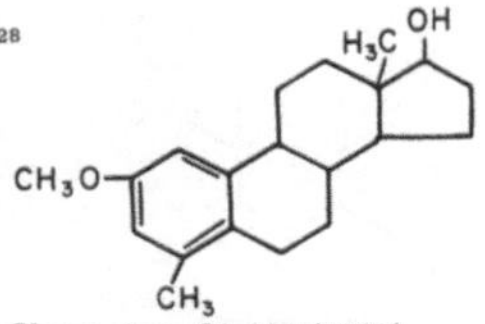

J. Am. Chem. Soc. 84:467 (1962)

▶ $O_2C_{20}H_{26}$

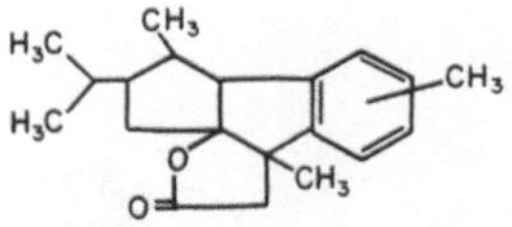

Experientia 18:111 (1962)

▶ $O_2C_{20}H_{28}$

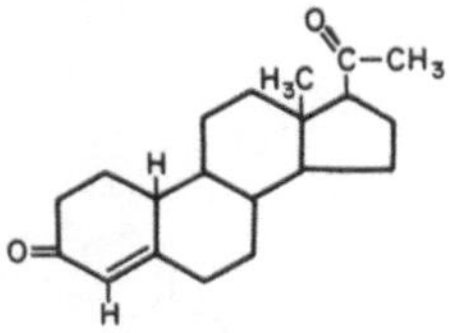

Varian 345

▶ $O_2C_{20}H_{26}$

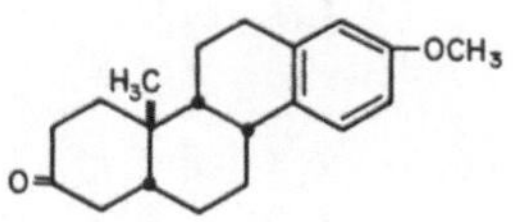

Can. J. Chem. 40:986 (1962)

▶ $O_2C_{20}H_{30}$

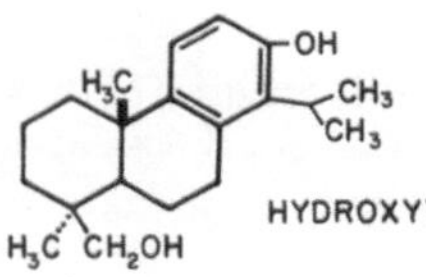

HYDROXYTOTAROL

Tetrahedron Letters 359 (1961)
Tetrahedron 18:465 (1962)

▶ $O_2C_{20}H_{26}$

$CH_3$
|
$CH(OCH_3)$
|
$CH(C_6H_5)$
|
$CH(C_6H_5)$
|
$CH(OCH_3)$
|
$CH_3$

Can. J. Chem. 39:1677 (1961)

▶ $O_2C_{20}H_{30}$

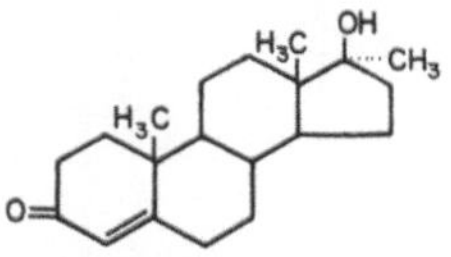

Helv. Chim. Acta 44:1380 (1961)

▶ $O_2C_{20}H_{30}$

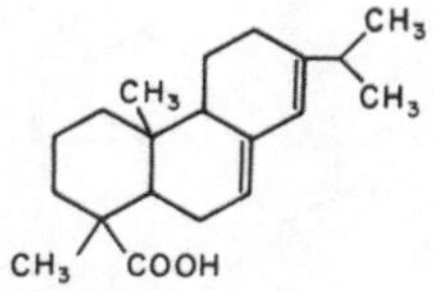

J. Org. Chem. 27:3535 (1962)

▶ $O_2C_{20}H_{30}$

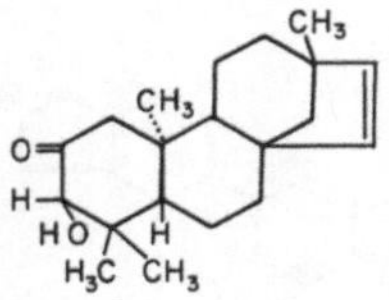

J. Chem. Soc. 4048 (1962)

▶ $O_2C_{20}H_{30}$

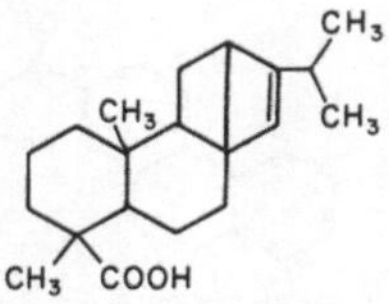

J. Org. Chem. 27:1178 (1962)

▶ $O_2C_{20}H_{30}$

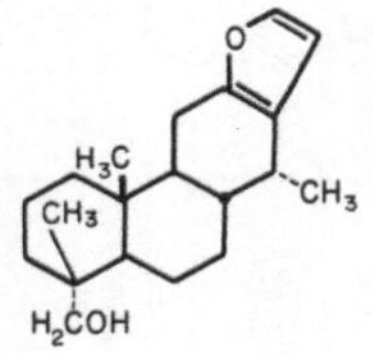

Tetrahedron Letters 361 (1961)

▶ $O_2C_{20}H_{30}$

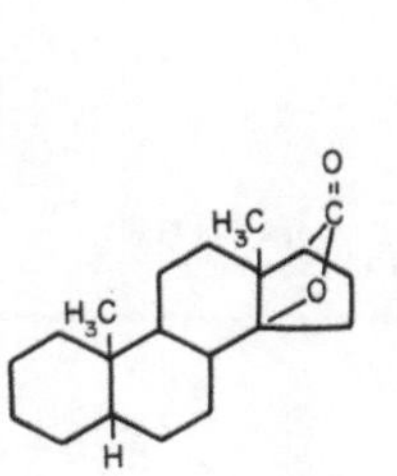

Helv. Chim. Acta 44:1380 (1961)

▶ $O_2C_{20}H_{30}$

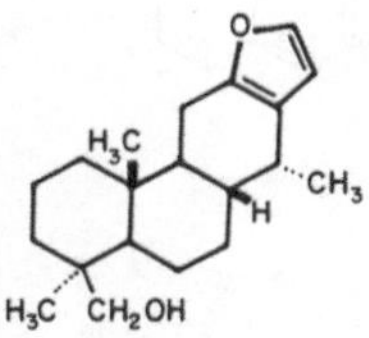

Tetrahedron Letters 361 (1961)

▶ $O_2C_{20}H_{32}$

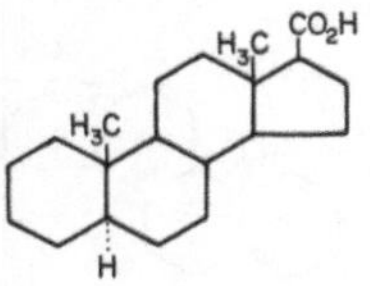

Helv. Chim. Acta 44:1380 (1961)

▶ $O_2C_{20}H_{32}$

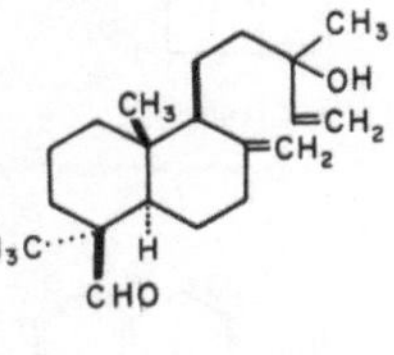

Acta Chem. Scand. 15:1303 (1961)

▶ $O_2C_{20}H_{32}$

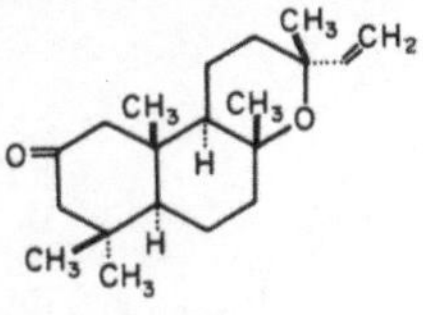

Tetrahedron 18:285 (1962)

▶ $O_2C_{20}H_{32}$

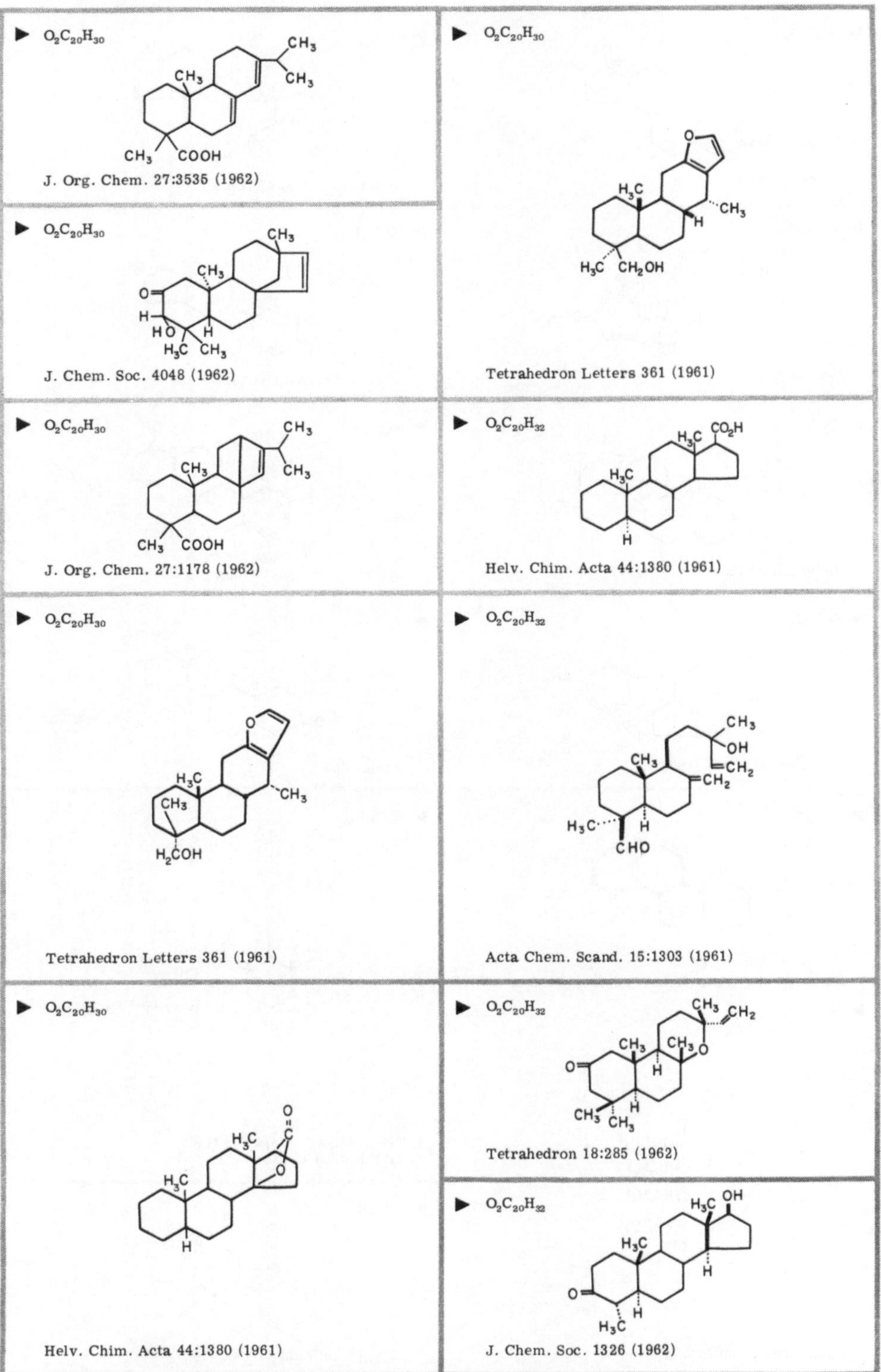

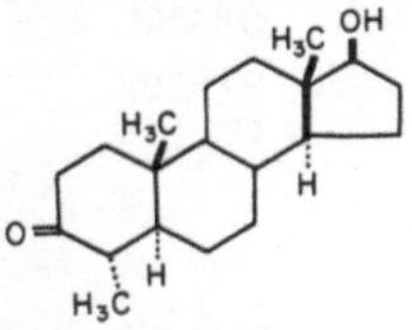

J. Chem. Soc. 1326 (1962)

$O_2C_{20}H_{32}$
J. Org. Chem. 27:3993 (1962)
$O_2C_{20}H_{32}$
J. Org. Chem. 27:3993 (1962)
$O_2C_{20}H_{34}$
Acta Chem. Scand. 15:1303 (1961)
$O_2C_{20}H_{34}$
Tetrahedron 18:169 (1962)
$O_2C_{20}H_{34}$
J. Org. Chem. 27:3993 (1962)
$O_2C_{20}H_{34}$
Helv. Chim. Acta 45:400 (1962)
$O_2C_{20}H_{36}$
Helv. Chim. Acta 45:400 (1962)
$O_2C_{20}H_{36}$
Helv. Chim. Acta 45:400 (1962)
$O_2C_{20}H_{36}$
Helv. Chim. Acta 45:400 (1962)
$O_2C_{20}H_{38}$
CH3(CH2)7 CH–CH(CH2)7 CO2CH3
CH2
J. Am. Oil Chemists' Soc. 38:664 (1961)

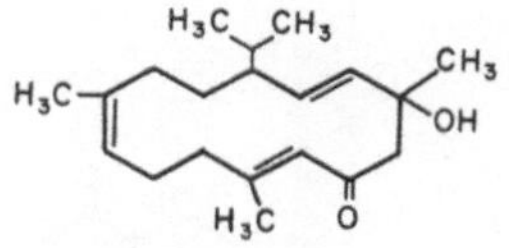

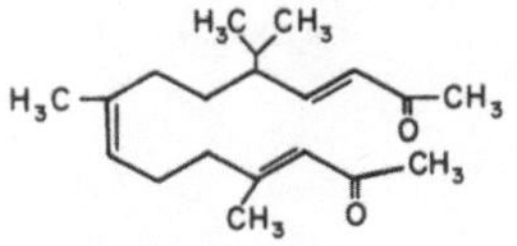

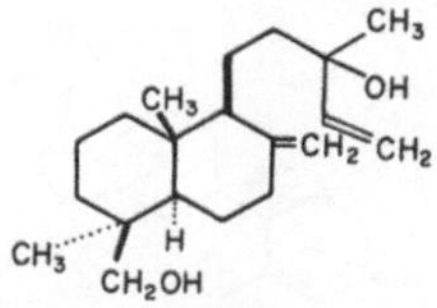

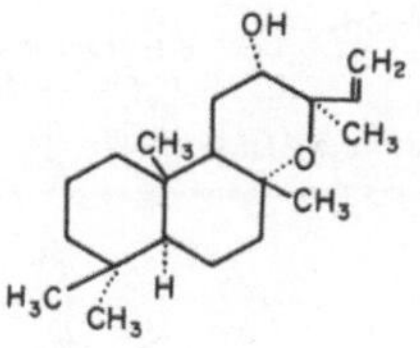

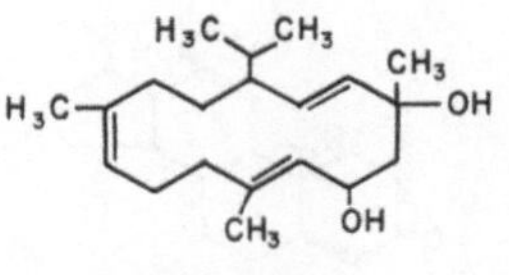

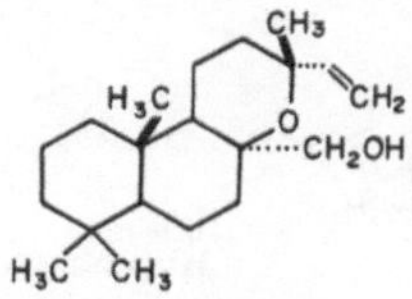

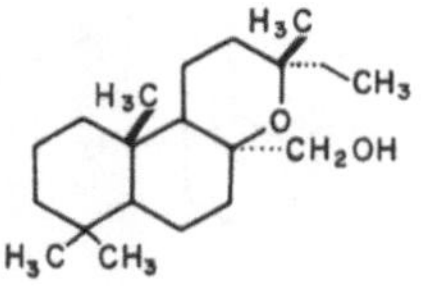

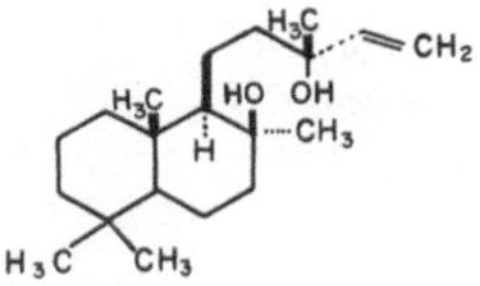

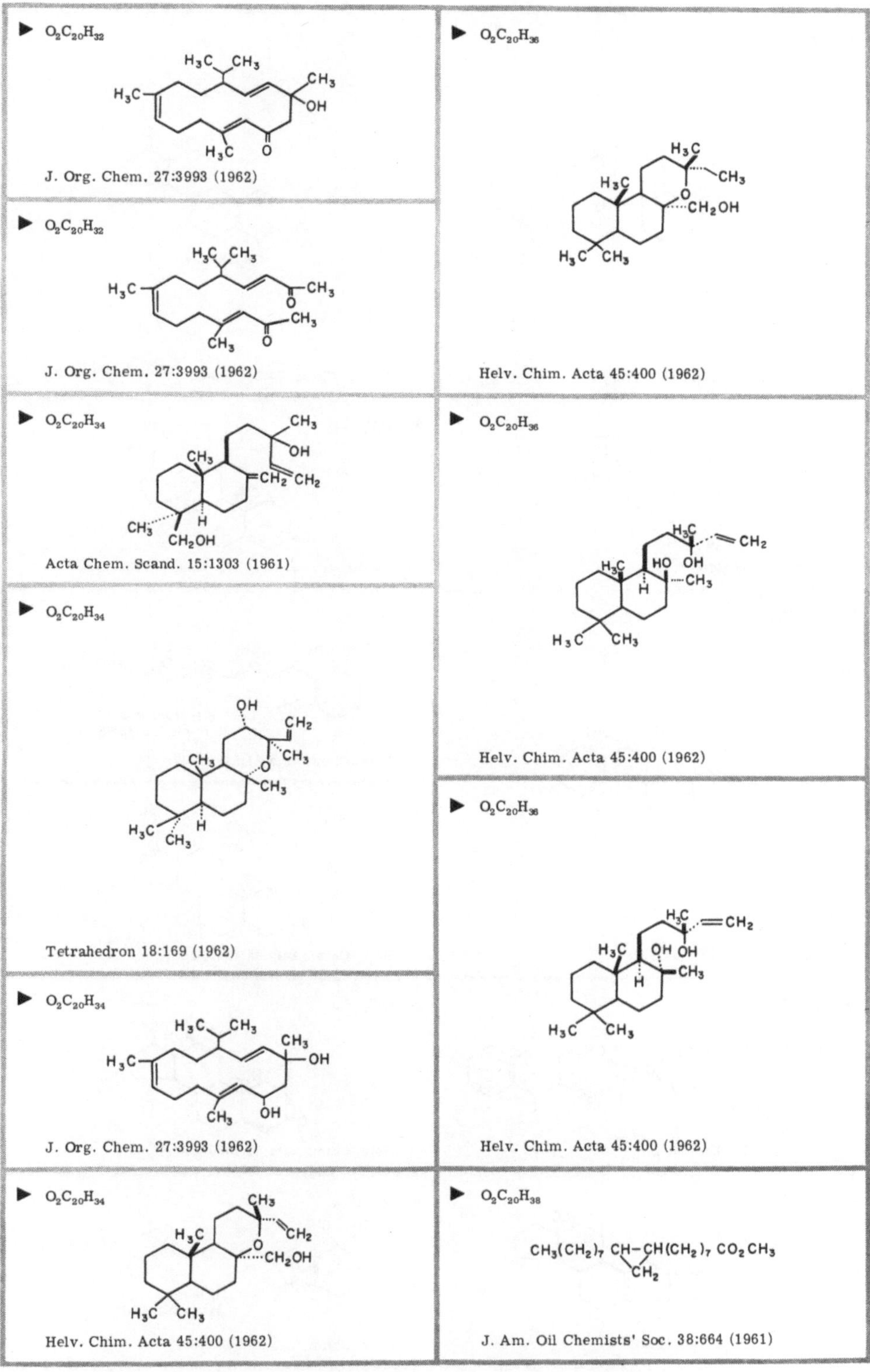

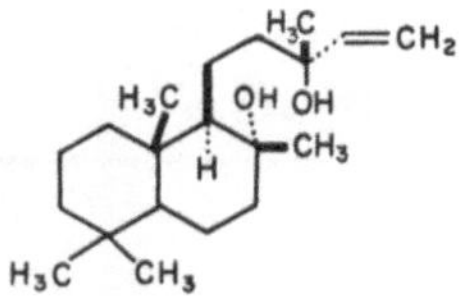

▶ $O_2C_{20}H_{38}$

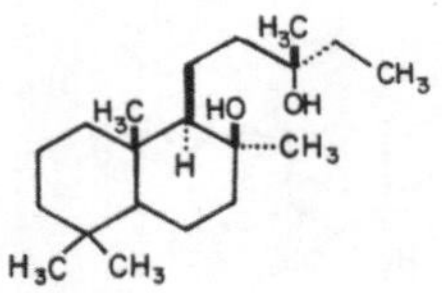

Helv. Chim. Acta 45:400 (1962)

▶ $O_2C_{20}H_{38}$

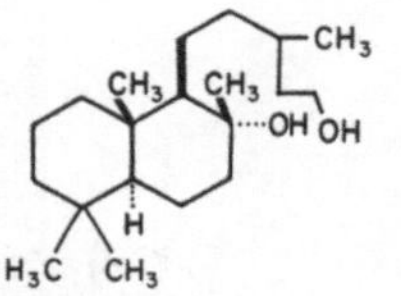

J. Chem. Soc. 4708 (1962)

▶ $O_2C_{21}H_{24}$

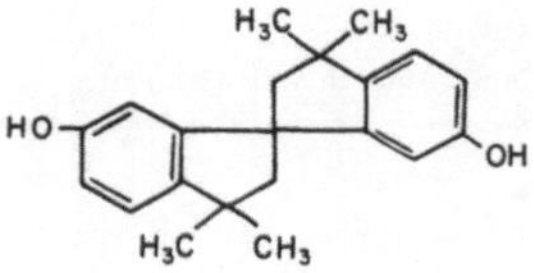

J. Chem. Soc. 417 (1962)

▶ $O_2C_{21}H_{24}$

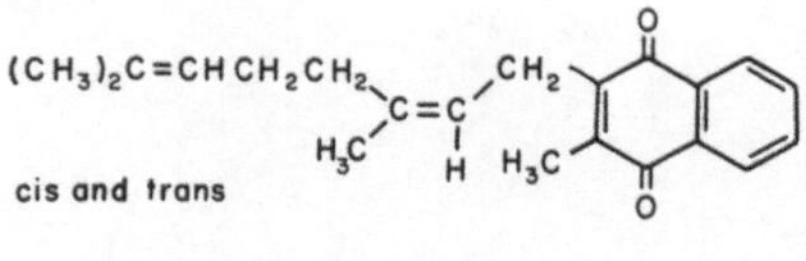


Chem. & Ind. 1020 (1962)

▶ $O_2C_{21}H_{26}$

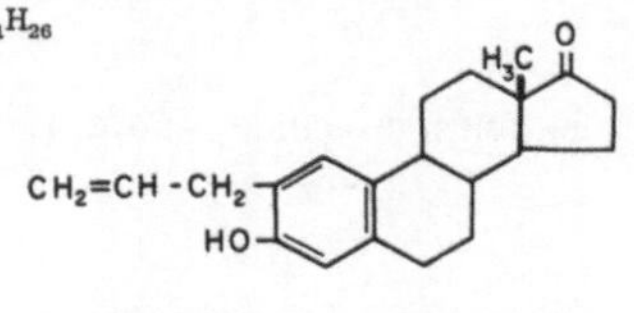


J. Org. Chem. 27:357 (1962)

▶ $O_2C_{21}H_{26}$

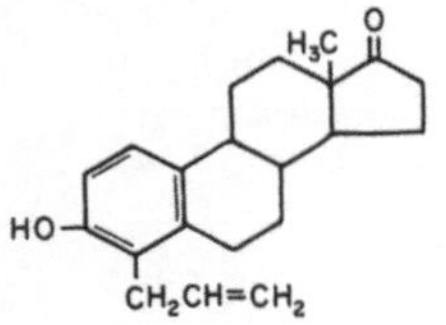


J. Org. Chem. 27:357 (1962)

▶ $O_2C_{21}H_{28}$

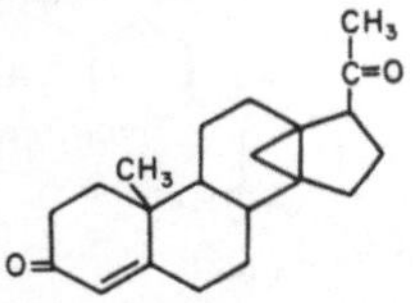

Tetrahedron Letters No. 11:455 (1962)

▶ $O_2C_{21}H_{28}$

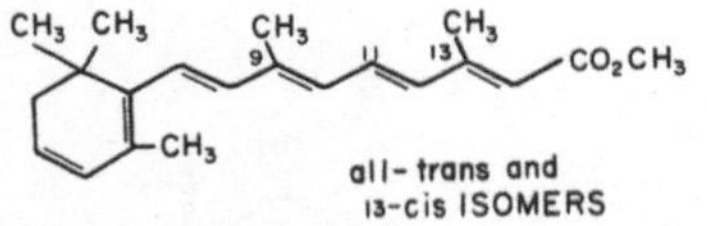


Helv. Chim. Acta 45:548 (1962)

▶ $O_2C_{21}H_{28}$

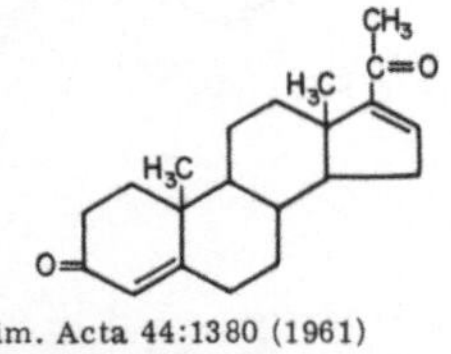

Helv. Chim. Acta 44:1380 (1961)

▶ $O_2C_{21}H_{28}$

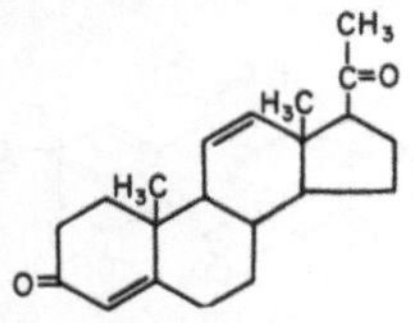

Helv. Chim. Acta 44:1380 (1961)

▶ $O_2C_{21}H_{30}$

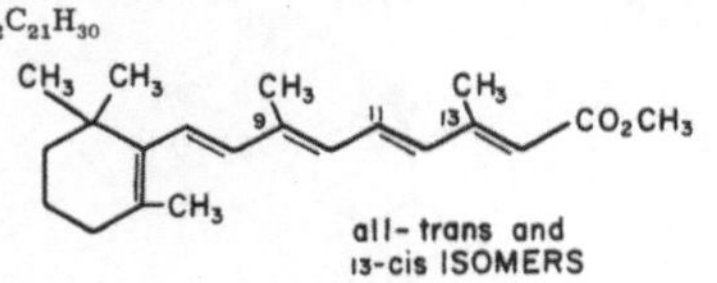


Helv. Chim. Acta 45:548 (1962)

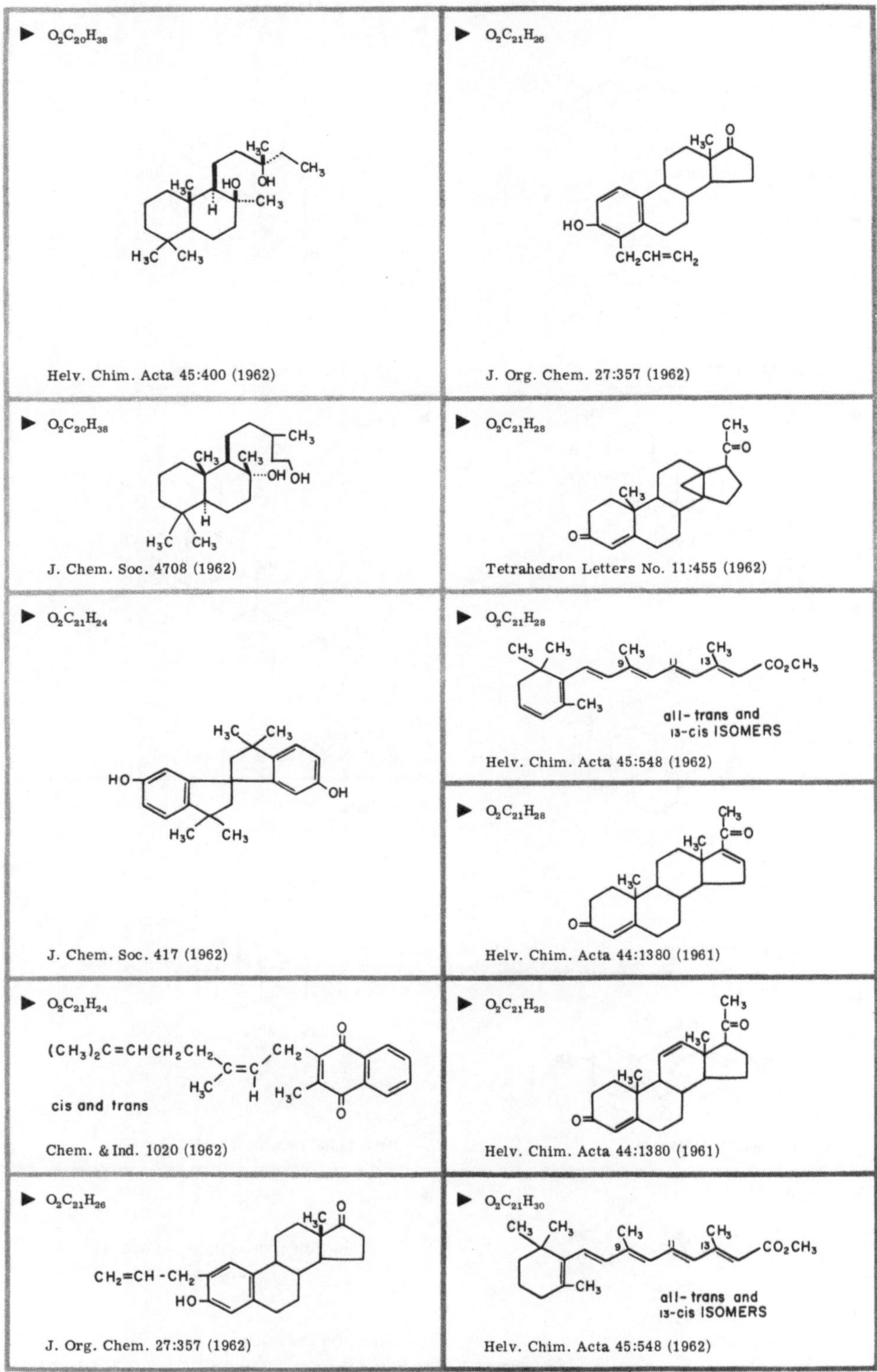

▶ $O_2C_{21}H_{30}$

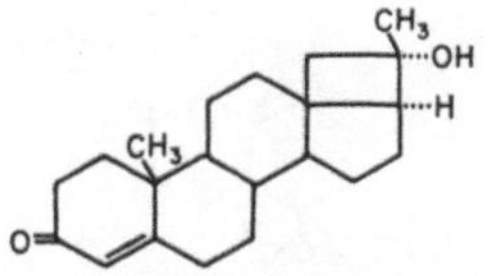


Helv. Chim. Acta 44:1927 (1961)

▶ $O_2C_{21}H_{30}$

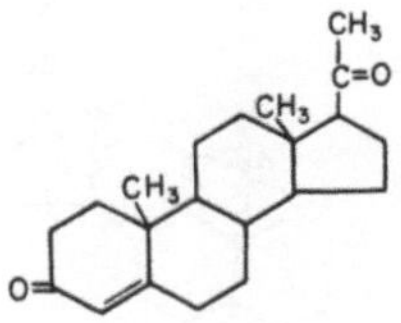

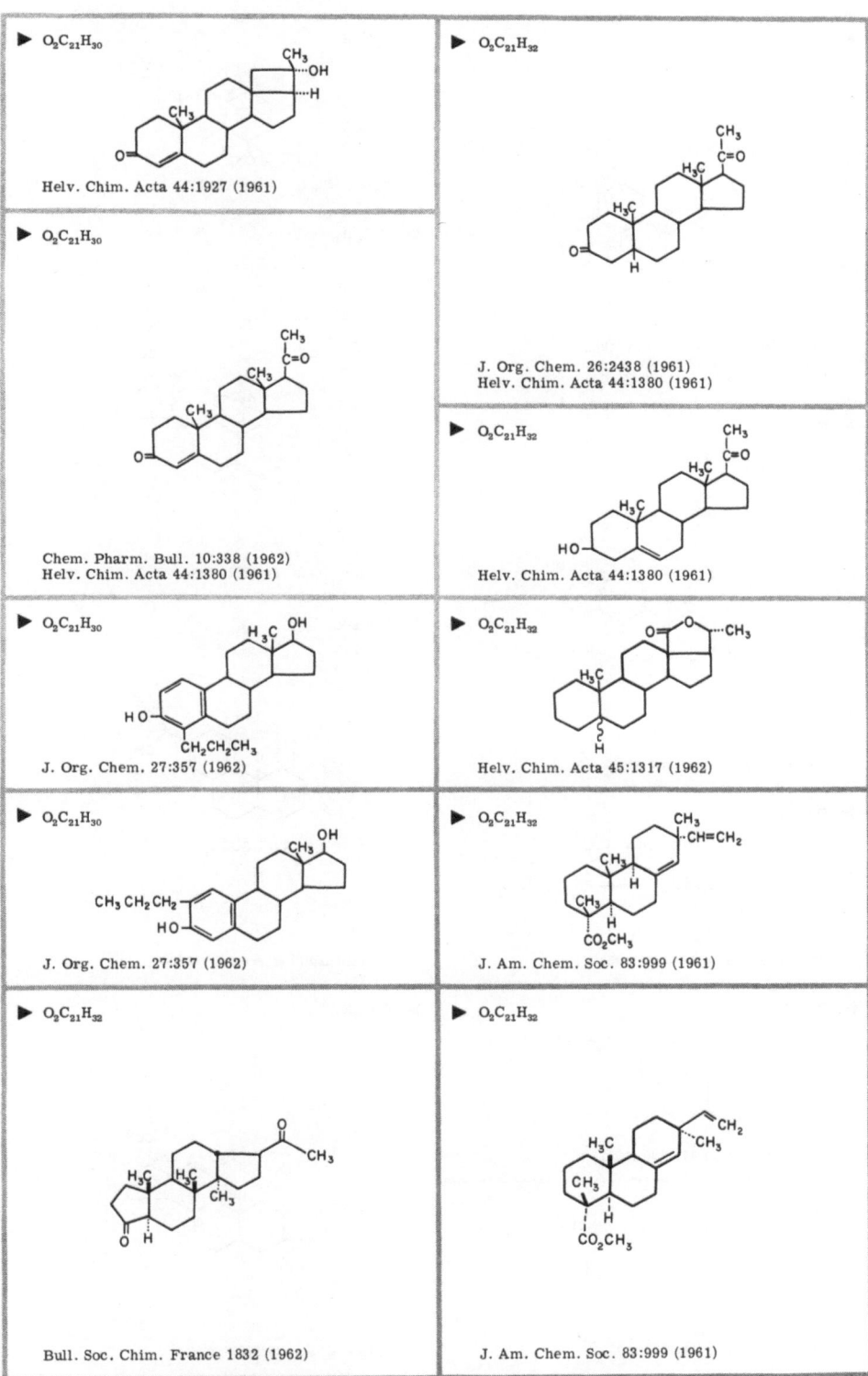

Chem. Pharm. Bull. 10:338 (1962)
Helv. Chim. Acta 44:1380 (1961)

▶ $O_2C_{21}H_{30}$

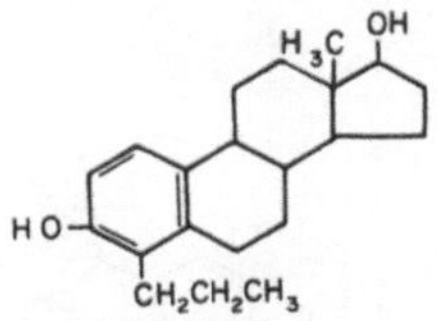


J. Org. Chem. 27:357 (1962)

▶ $O_2C_{21}H_{30}$

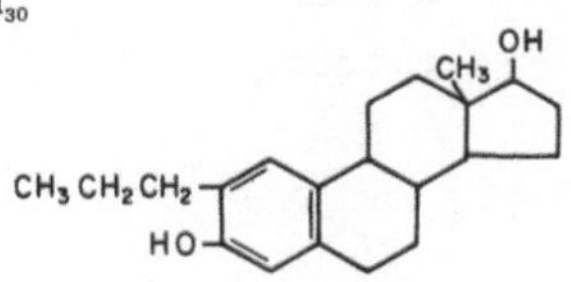


J. Org. Chem. 27:357 (1962)

▶ $O_2C_{21}H_{32}$

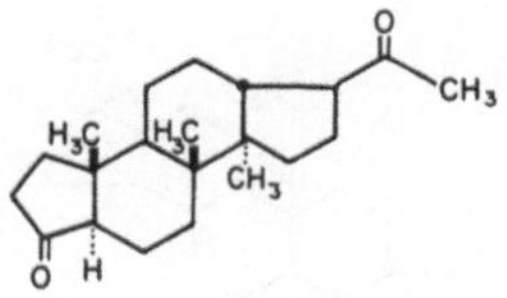


Bull. Soc. Chim. France 1832 (1962)

▶ $O_2C_{21}H_{32}$

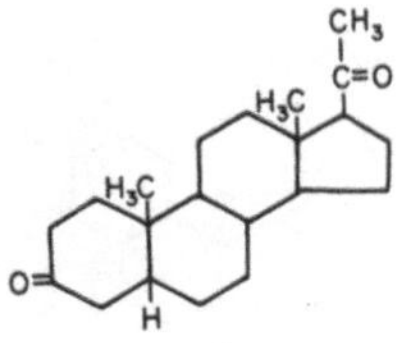


J. Org. Chem. 26:2438 (1961)
Helv. Chim. Acta 44:1380 (1961)

▶ $O_2C_{21}H_{32}$

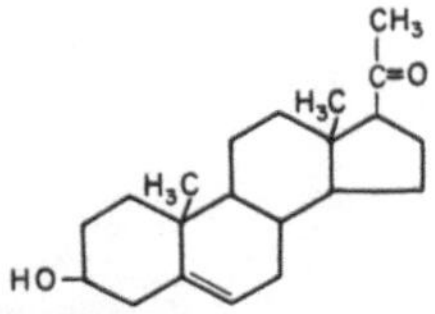


Helv. Chim. Acta 44:1380 (1961)

▶ $O_2C_{21}H_{32}$

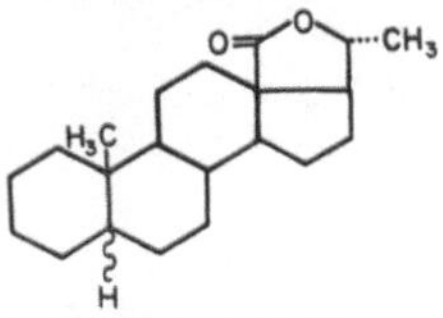


Helv. Chim. Acta 45:1317 (1962)

▶ $O_2C_{21}H_{32}$

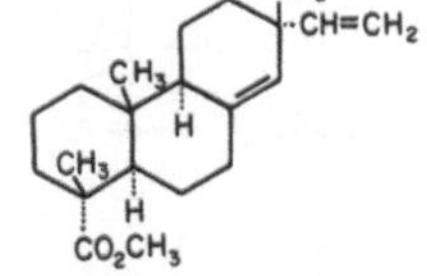


J. Am. Chem. Soc. 83:999 (1961)

▶ $O_2C_{21}H_{32}$

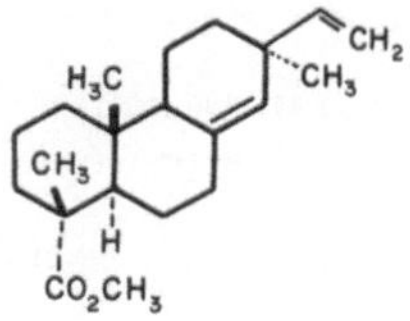


J. Am. Chem. Soc. 83:999 (1961)

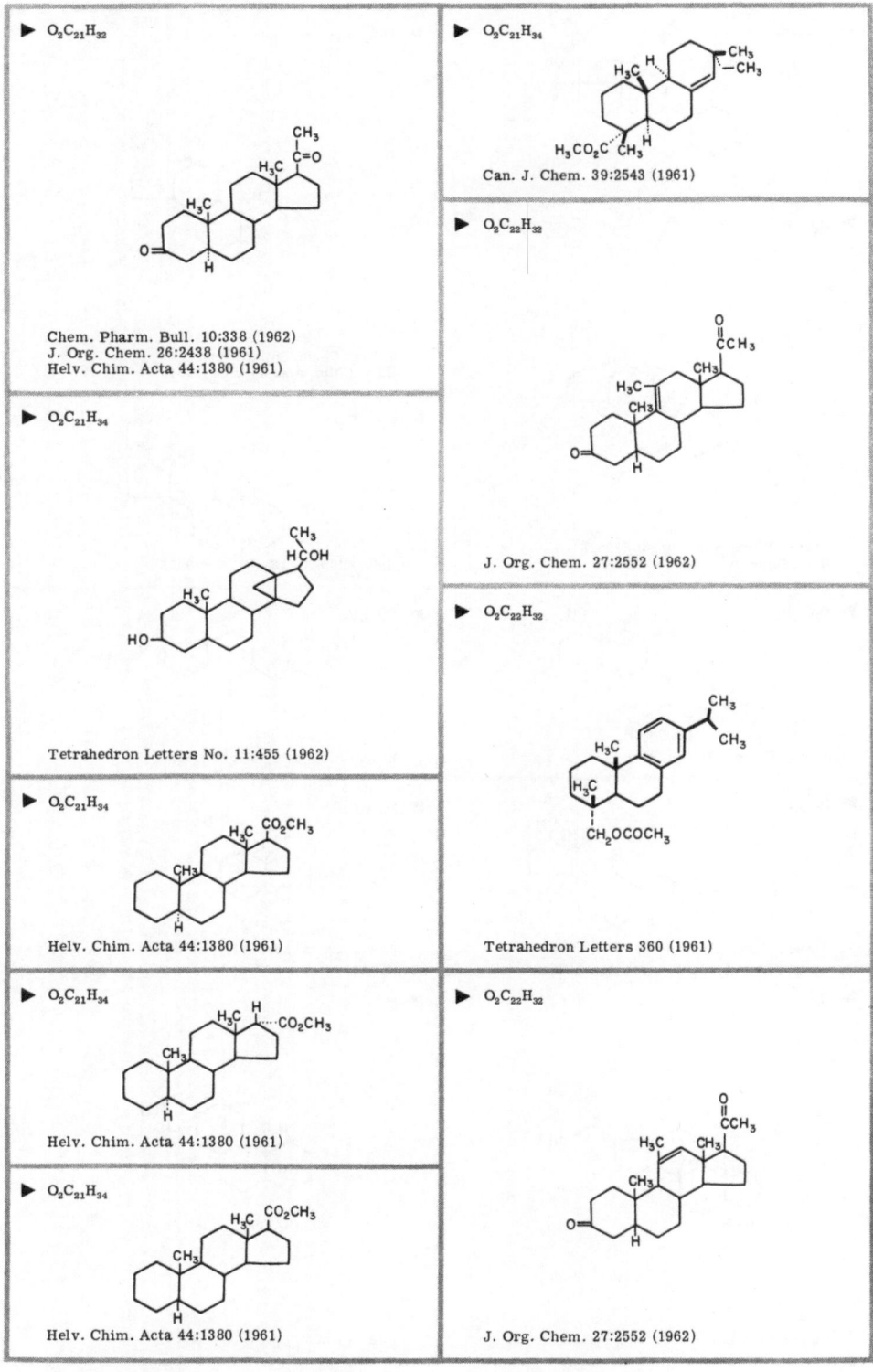

▶ $O_2C_{21}H_{32}$

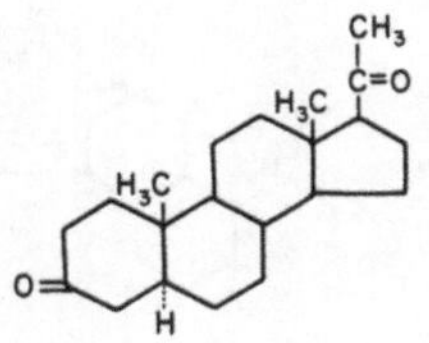

Chem. Pharm. Bull. 10:338 (1962)
J. Org. Chem. 26:2438 (1961)
Helv. Chim. Acta 44:1380 (1961)

▶ $O_2C_{21}H_{34}$

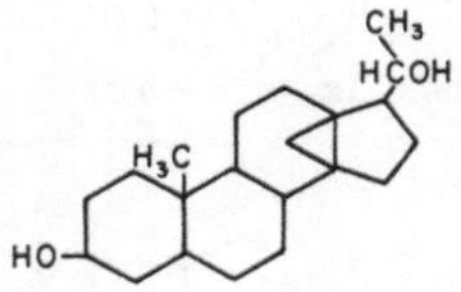

Tetrahedron Letters No. 11:455 (1962)

▶ $O_2C_{21}H_{34}$

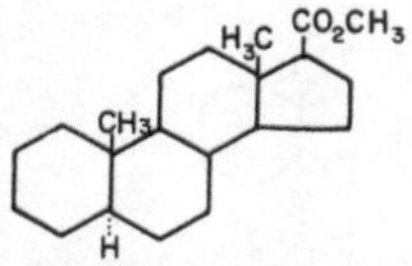

Helv. Chim. Acta 44:1380 (1961)

▶ $O_2C_{21}H_{34}$

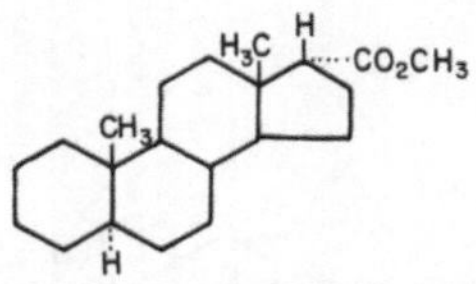

Helv. Chim. Acta 44:1380 (1961)

▶ $O_2C_{21}H_{34}$

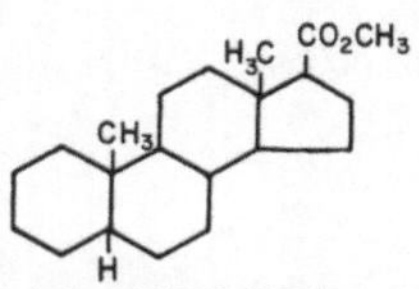

Helv. Chim. Acta 44:1380 (1961)

▶ $O_2C_{21}H_{34}$

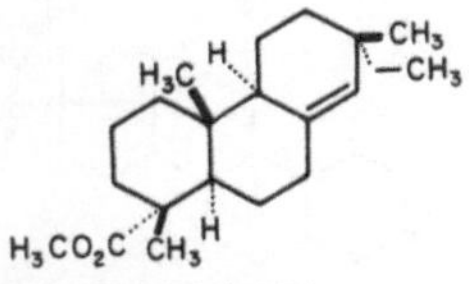

Can. J. Chem. 39:2543 (1961)

▶ $O_2C_{22}H_{32}$

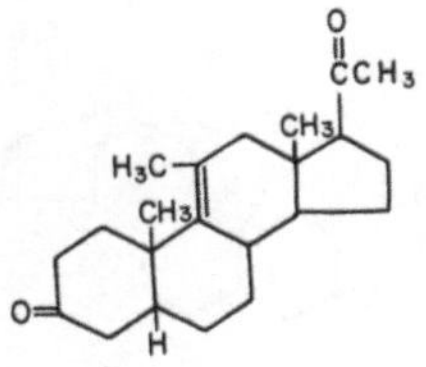

J. Org. Chem. 27:2552 (1962)

▶ $O_2C_{22}H_{32}$

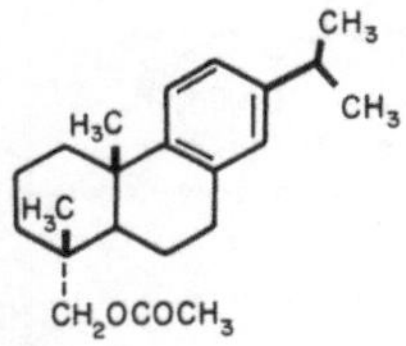

Tetrahedron Letters 360 (1961)

▶ $O_2C_{22}H_{32}$

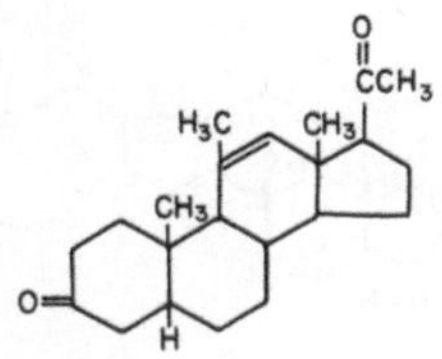

J. Org. Chem. 27:2552 (1962)

▶ $O_2C_{22}H_{38}$

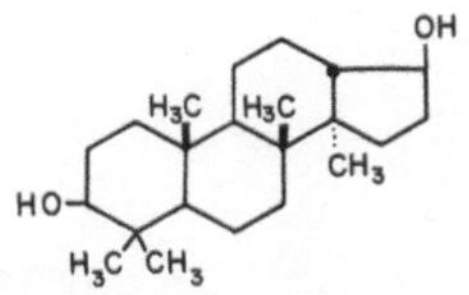

Bull. Soc. Chim. France 1832 (1962)

▶ $O_2C_{23}H_{22}$

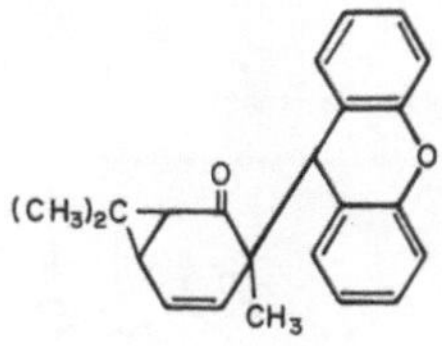

Varian 693

▶ $O_2C_{23}H_{22}$

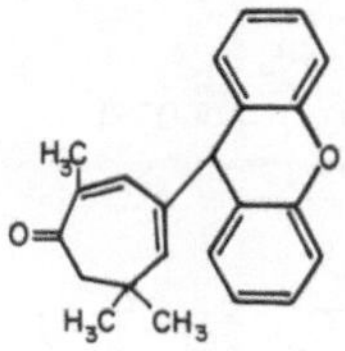

Varian 694

▶ $O_2C_{23}H_{30}$

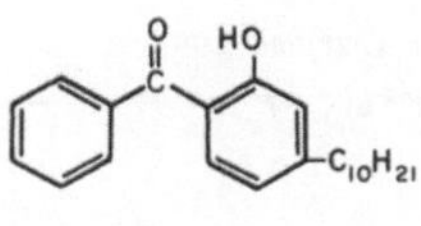

J. Phys. Chem. 65:2023 (1961)

▶ $O_2C_{23}H_{32}$

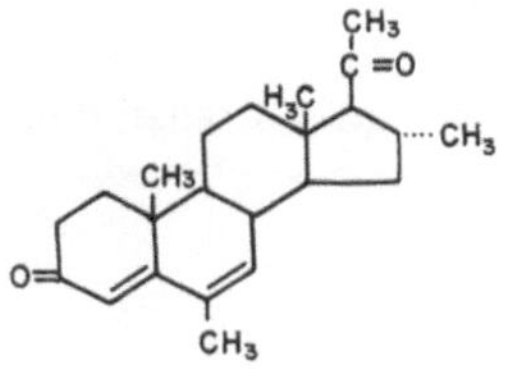

J. Org. Chem. 27:1139 (1962)

▶ $O_2C_{23}H_{32}$

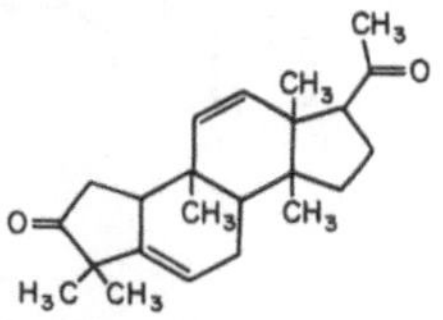

J. Org. Chem. 27:4546 (1962)

▶ $O_2C_{23}H_{34}$

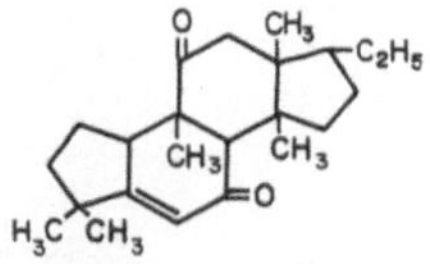

J. Org. Chem. 27:4546 (1962)

▶ $O_2C_{23}H_{34}$

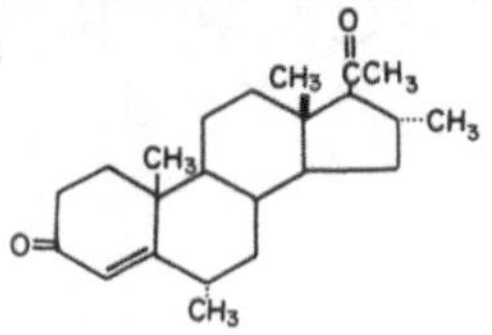

J. Org. Chem. 27:1139 (1962)

▶ $O_2C_{23}H_{36}$

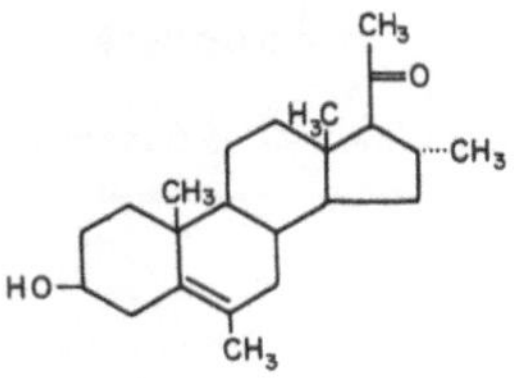

J. Org. Chem. 27:1139 (1962)

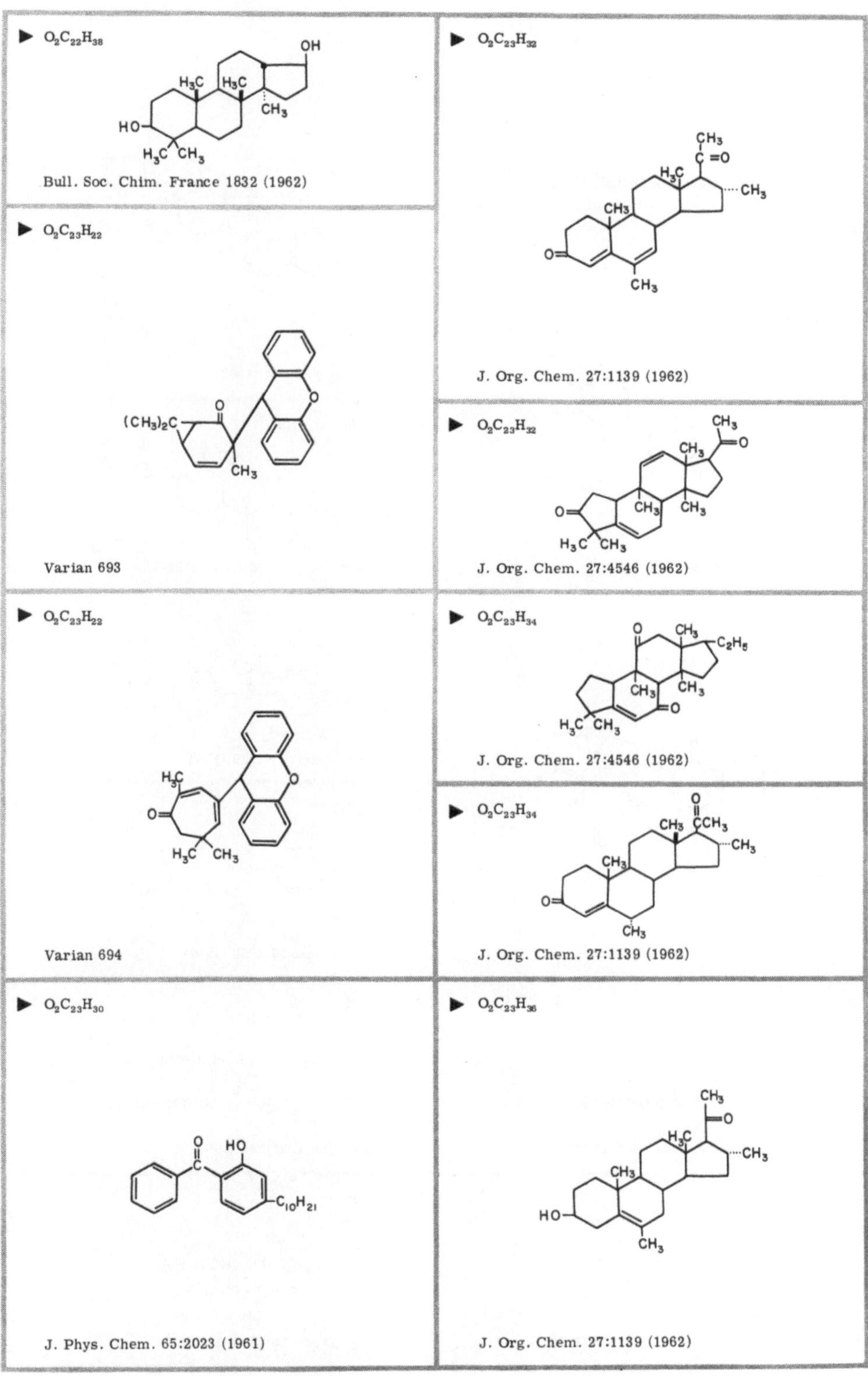

► $O_2C_{23}H_{36}$

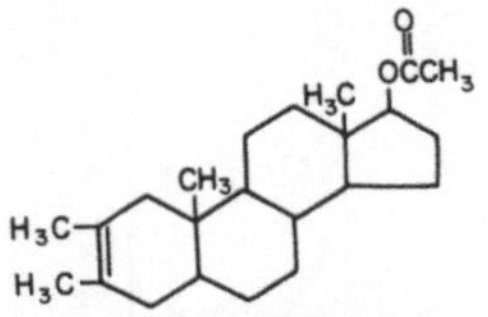

Tetrahedron Letters No. 26:1252 (1962)

► $O_2C_{23}H_{38}$

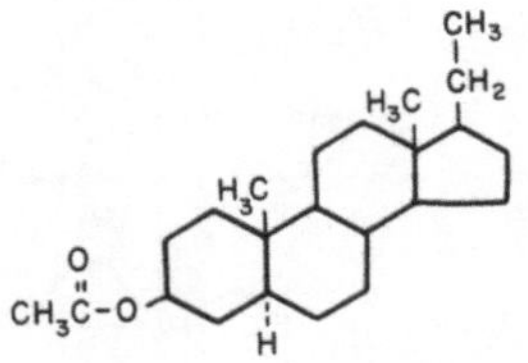

Helv. Chim. Acta 44:1380 (1961)

► $O_2C_{24}H_{20}$

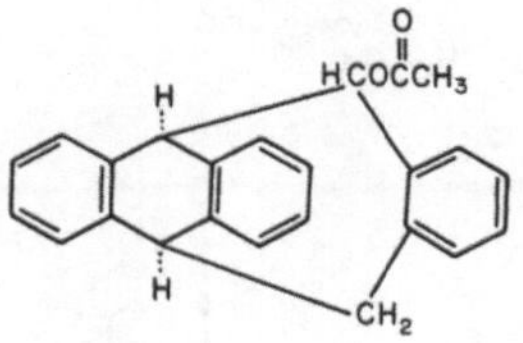

J. Am. Chem. Soc. 84:3562 (1962)

► $O_2C_{24}H_{20}$

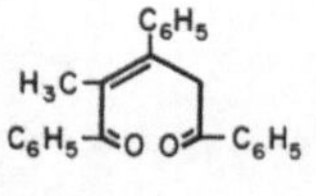

Tetrahedron Letters No. 25:1217 (1962)

► $O_2C_{24}H_{20}$

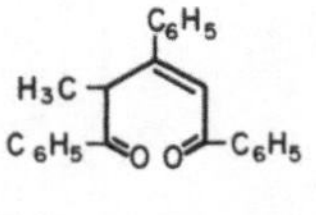

Tetrahedron Letters No. 25:1216 (1962)

► $O_2C_{24}H_{24}$

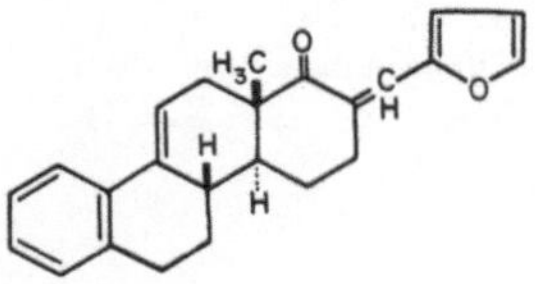

J. Chem. Soc. 274 (1962)

► $O_2C_{24}H_{24}$

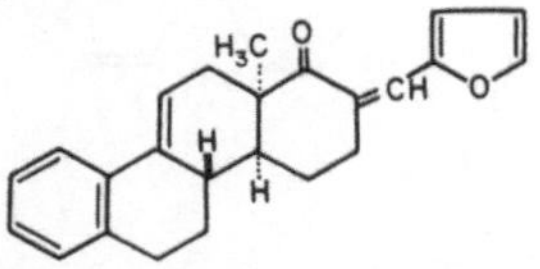

J. Chem. Soc. 274 January (1962)

► $O_2C_{24}H_{26}$

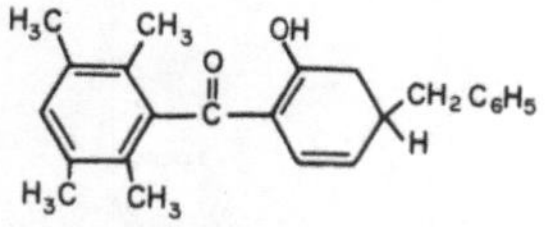

J. Org. Chem. 27:1219 (1962)

► $O_2C_{24}H_{26}$

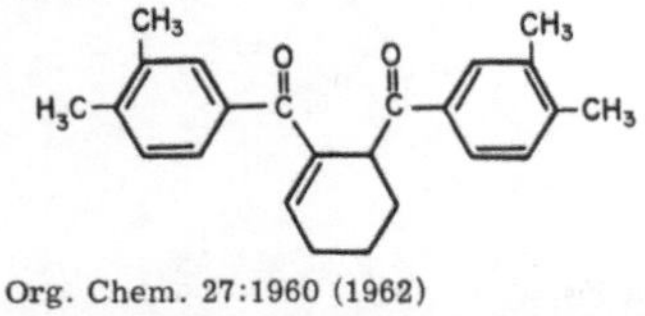

J. Org. Chem. 27:1960 (1962)

► $O_2C_{24}H_{28}$

$C_6H_5-\overset{O}{\overset{\|}{C}}-CH_2-CH-CH(CH_3)_2$

$C_6H_5-\overset{O}{\overset{\|}{C}}-CH-CH=C(CH_3)_2$

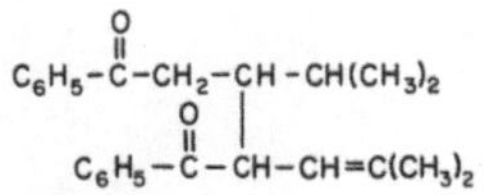

J. Org. Chem. 26:246 (1961)

► $O_2C_{24}H_{32}$

$C_6H_5CH(OH)CH_2CHCH(CH_3)_2$

$C_6H_5CH(OH)CHCH=C(CH_3)_2$

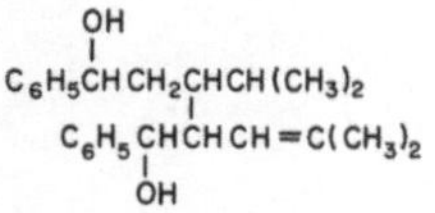

J. Org. Chem. 26:3559 (1961)

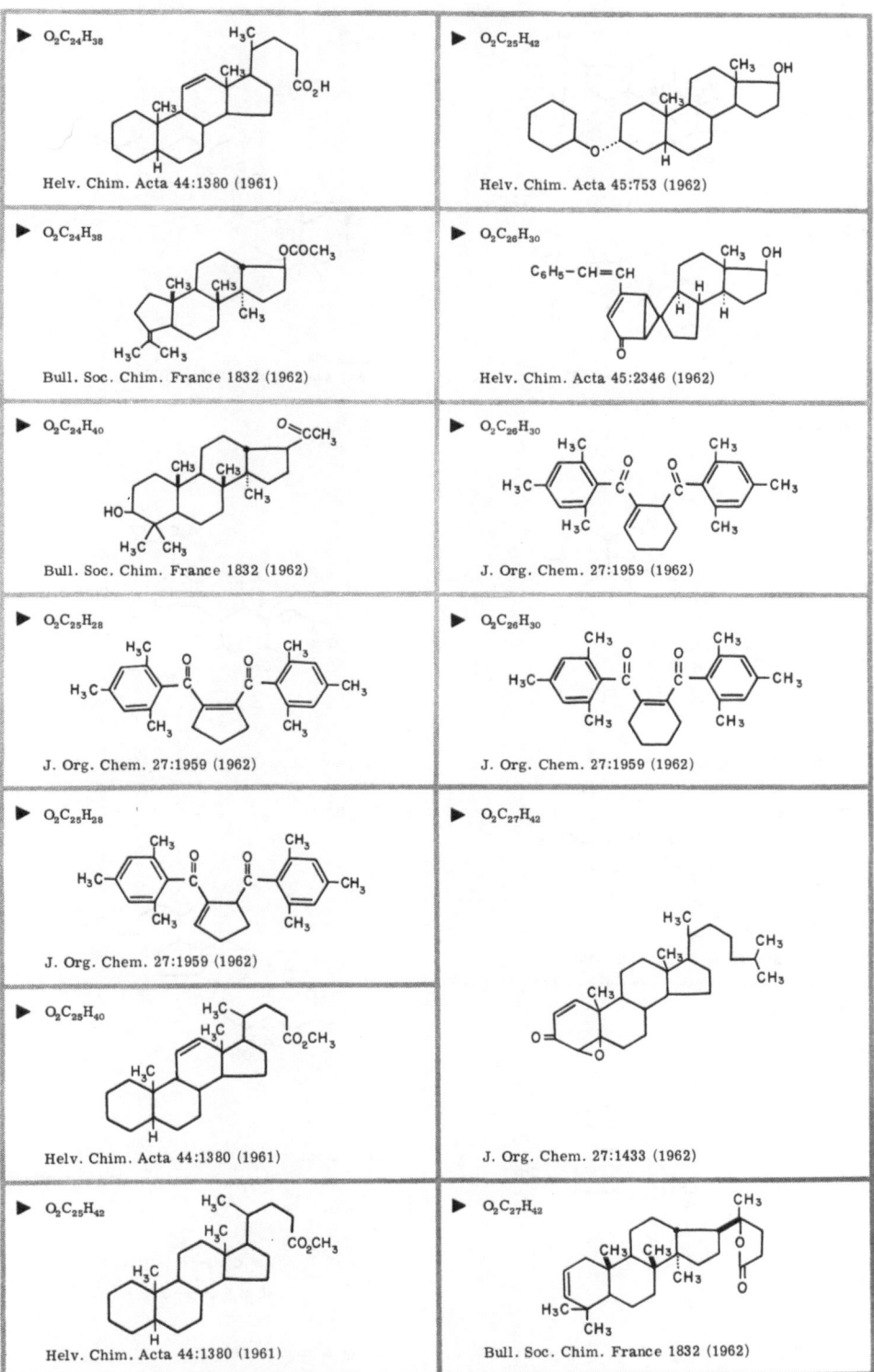

▶ $O_2C_{24}H_{38}$

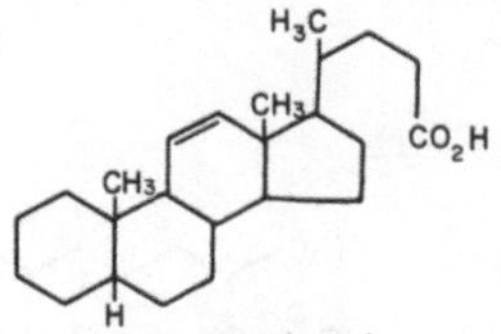

Helv. Chim. Acta 44:1380 (1961)

▶ $O_2C_{25}H_{42}$

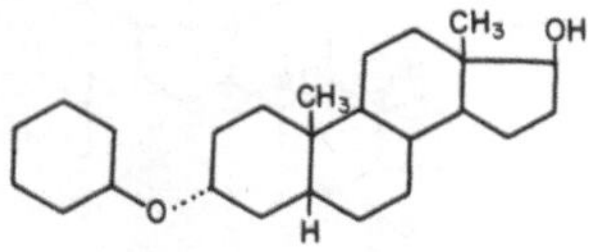

Helv. Chim. Acta 45:753 (1962)

▶ $O_2C_{24}H_{38}$

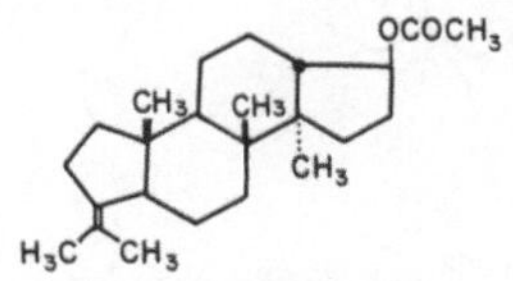

Bull. Soc. Chim. France 1832 (1962)

▶ $O_2C_{26}H_{30}$

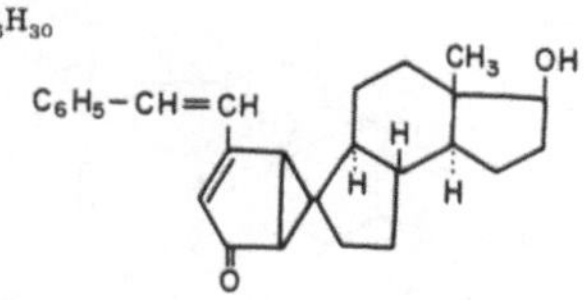

Helv. Chim. Acta 45:2346 (1962)

▶ $O_2C_{24}H_{40}$

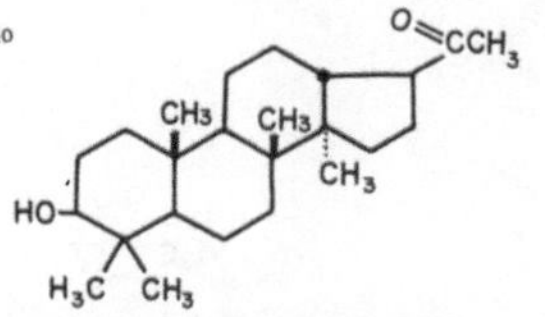

Bull. Soc. Chim. France 1832 (1962)

▶ $O_2C_{26}H_{30}$

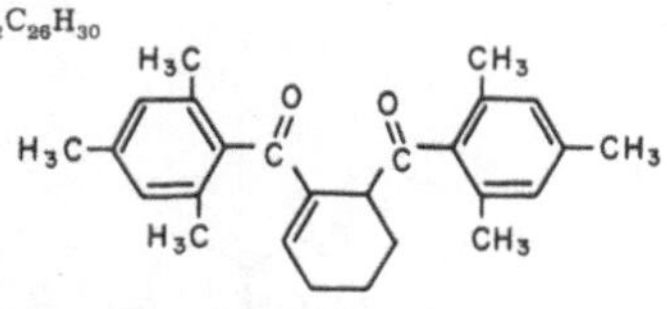

J. Org. Chem. 27:1959 (1962)

▶ $O_2C_{25}H_{28}$

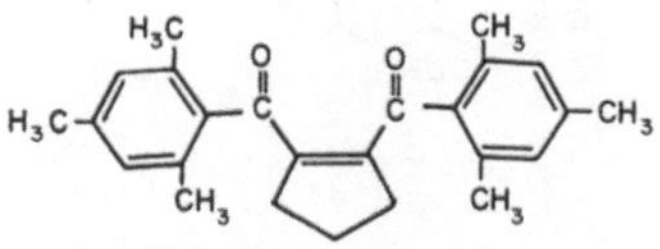

J. Org. Chem. 27:1959 (1962)

▶ $O_2C_{26}H_{30}$

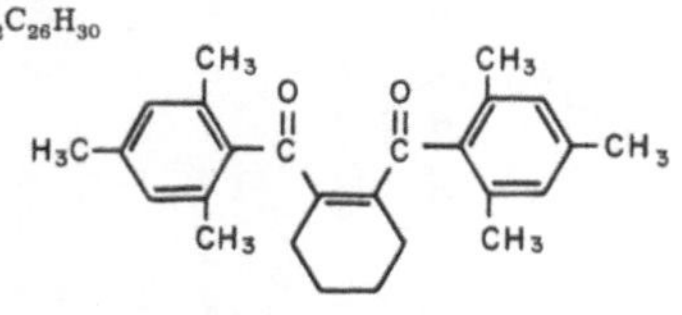

J. Org. Chem. 27:1959 (1962)

▶ $O_2C_{25}H_{28}$

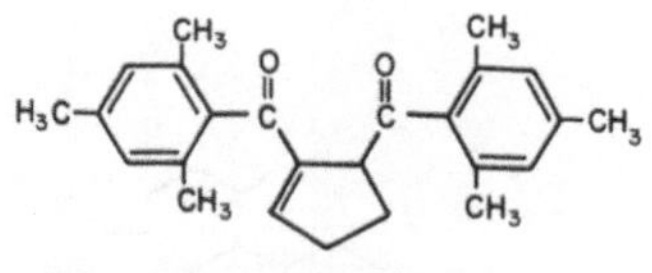

J. Org. Chem. 27:1959 (1962)

▶ $O_2C_{27}H_{42}$

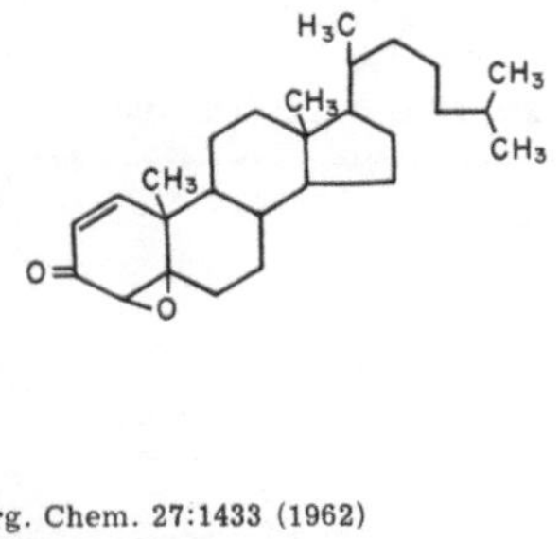

J. Org. Chem. 27:1433 (1962)

▶ $O_2C_{25}H_{40}$

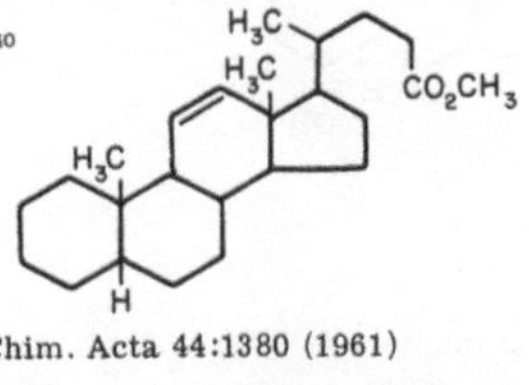

Helv. Chim. Acta 44:1380 (1961)

▶ $O_2C_{25}H_{42}$

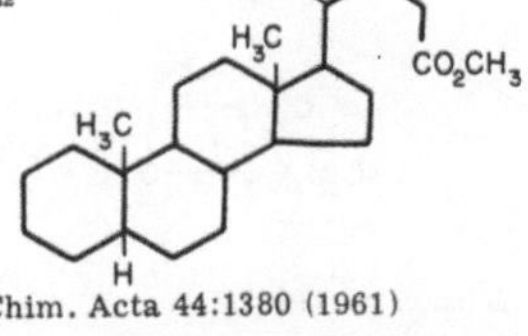

Helv. Chim. Acta 44:1380 (1961)

▶ $O_2C_{27}H_{42}$

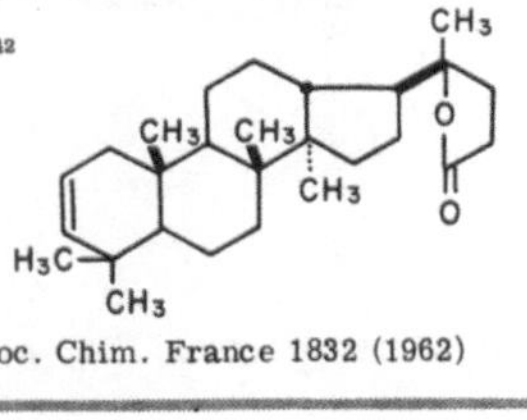

Bull. Soc. Chim. France 1832 (1962)

▶ $O_2C_{27}H_{42}$

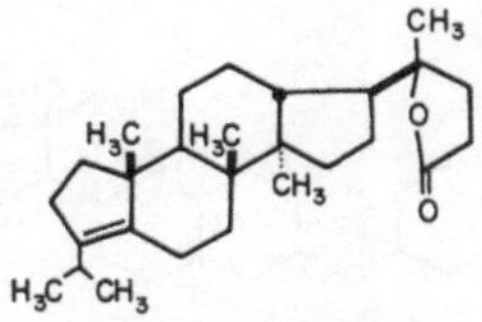

Bull. Soc. Chim. France 1832 (1962)

▶ $O_2C_{27}H_{42}$

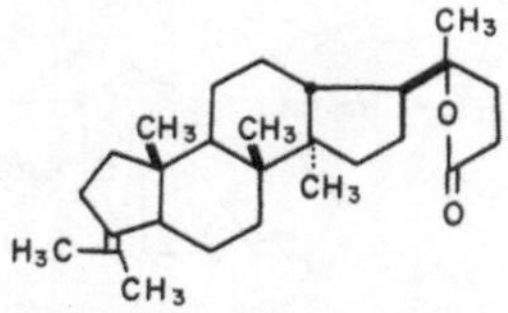

Bull. Soc. Chim. France 1832 (1962)

▶ $O_2C_{27}H_{44}$

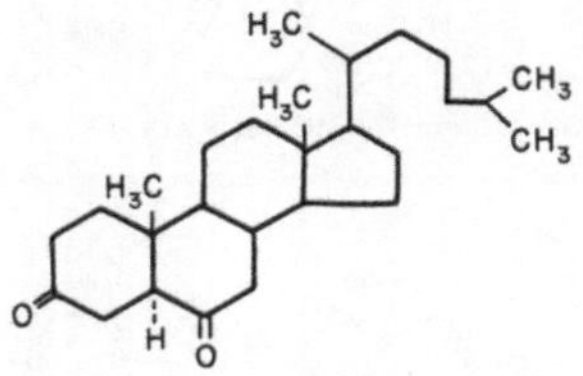

Helv. Chim. Acta 44:1380 (1961)

▶ $O_2C_{27}H_{44}$

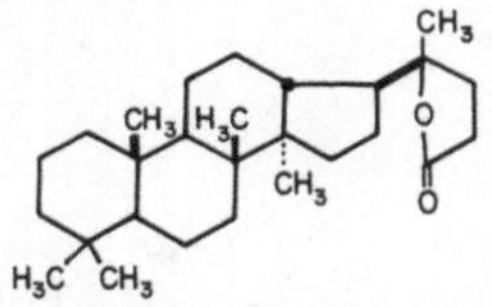

Bull. Soc. Chim. France 1832 (1962)

▶ $O_2C_{27}H_{44}$

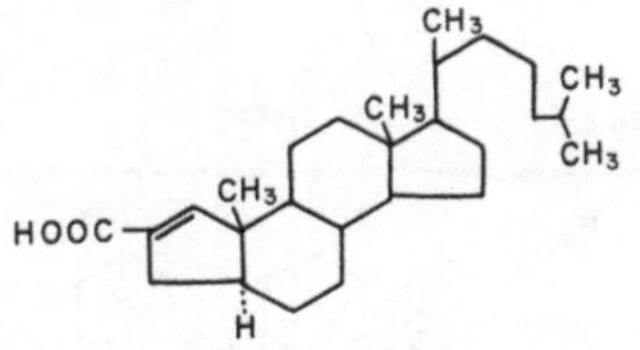

J. Org. Chem. 27:4026 (1962)

▶ $O_2C_{27}H_{44}$

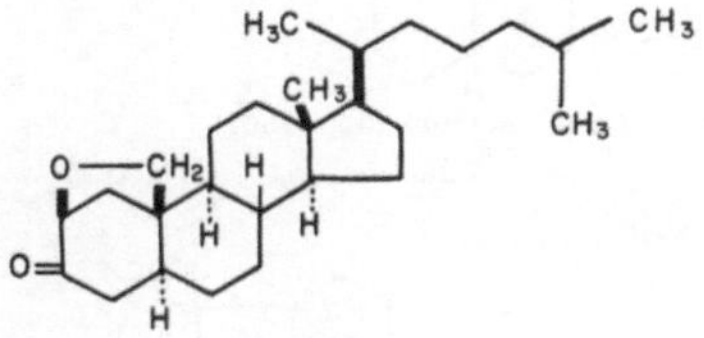

Varian 698
J. Am. Chem. Soc. 84:4976 (1962)

▶ $O_2C_{27}H_{46}$

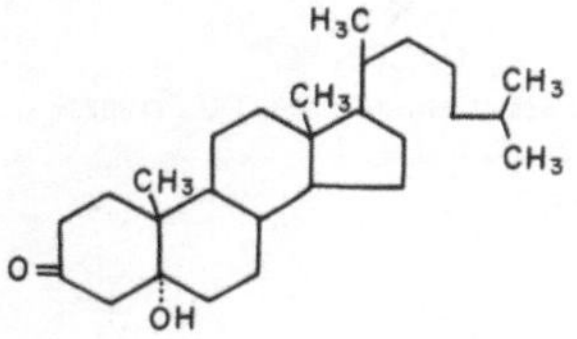

Bull. Soc. Chim. France 2444 (1961)

▶ $O_2C_{27}H_{48}$

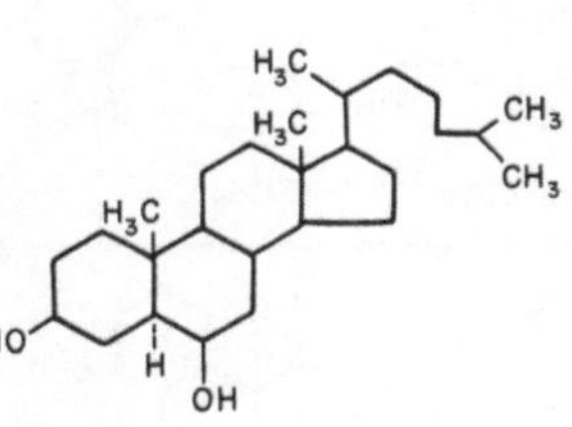

Helv. Chim. Acta 44:1380 (1961)

▶ $O_2C_{28}H_{20}$

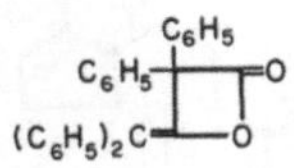

Chem. & Ind. 1313 (1961)

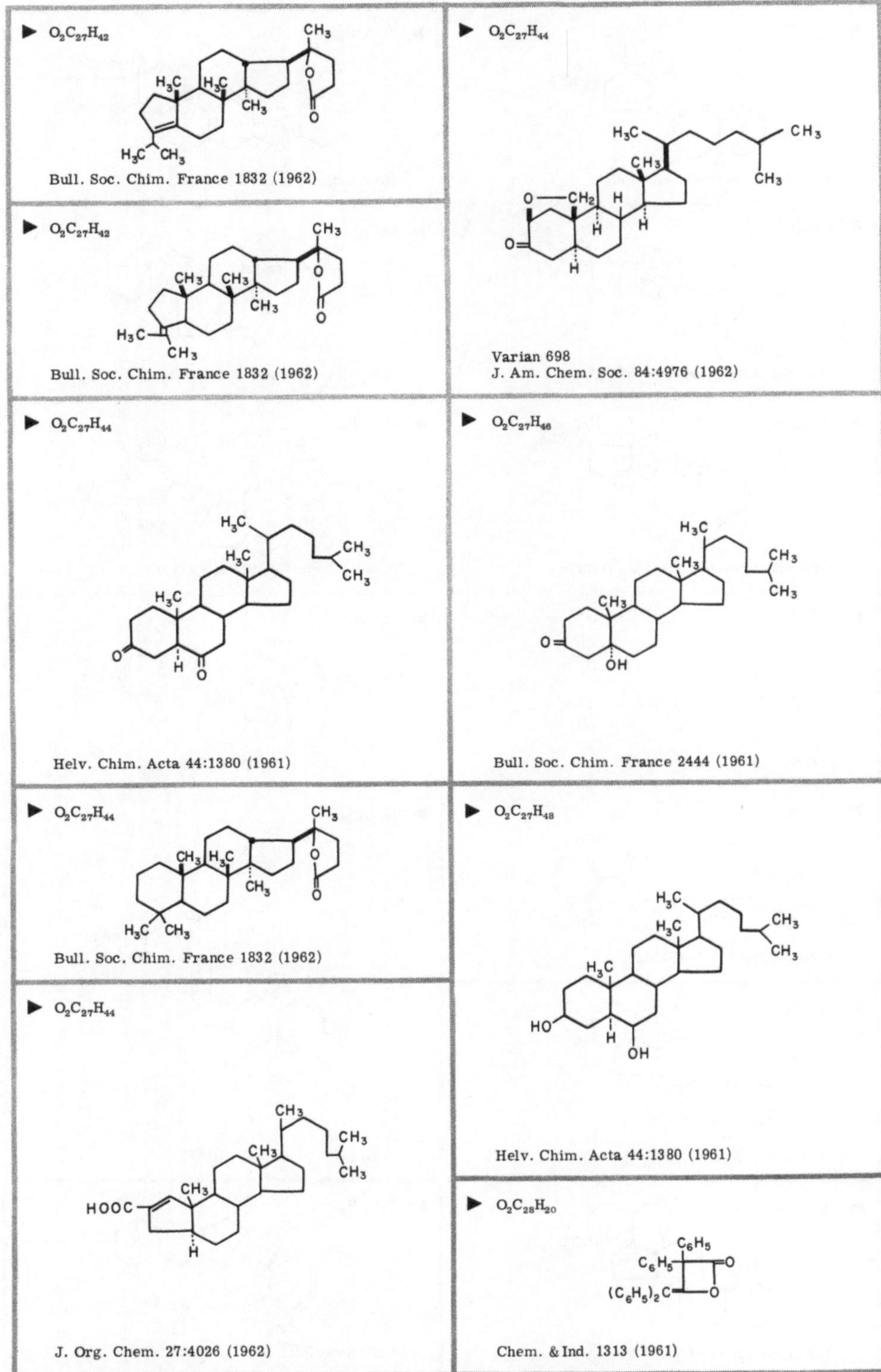

► $O_2C_{28}H_{32}$

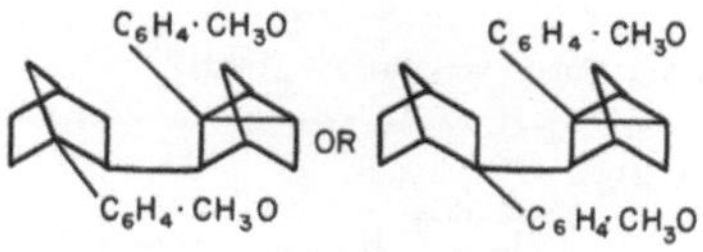

2-p-ANISYLNORBORNENE DIMER

J. Org. Chem. 26:3749 (1961)

► $O_2C_{28}H_{48}$

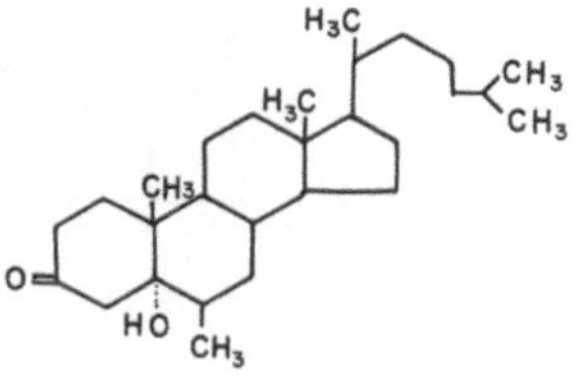

Bull. Soc. Chim. France 2444 (1961)

► $O_2C_{28}H_{34}$

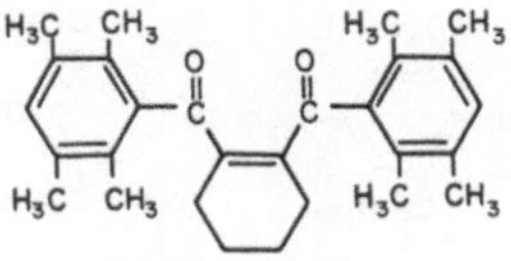

J. Org. Chem. 27:1959 (1962)

► $O_2C_{29}H_{22}$

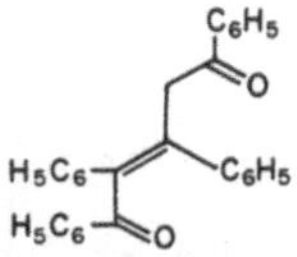

Tetrahedron Letters No. 25:1214 (1962)

► $O_2C_{28}H_{34}$

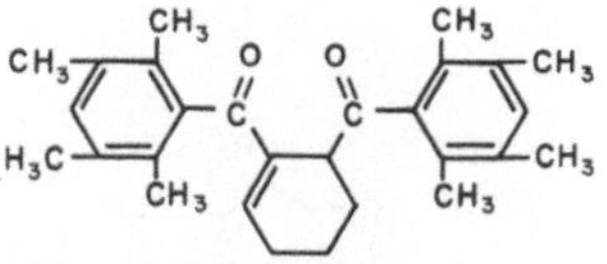

J. Org. Chem. 27:1959 (1962)

► $O_2C_{29}H_{22}$

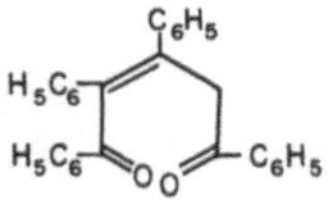

Tetrahedron Letters No. 25:1214 (1962)

► $O_2C_{28}H_{38}$

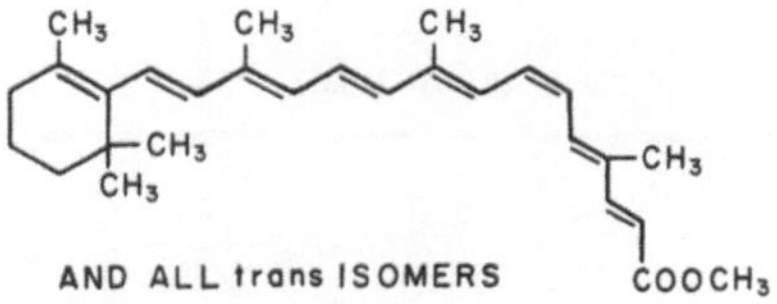

AND ALL trans ISOMERS

J. Chem. Soc. 1625 (1961)

► $O_2C_{29}H_{44}$

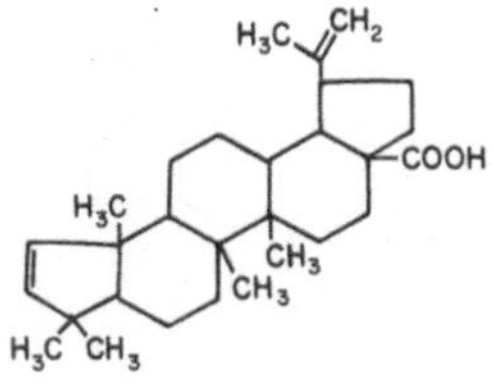

J. Org. Chem. 27:4072 (1962)

► $O_2C_{28}H_{40}$

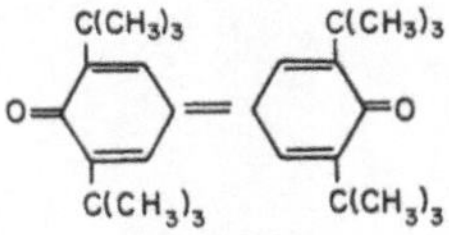

J. Am. Chem. Soc. 84:2258 (1962)

► $O_2C_{29}H_{46}$

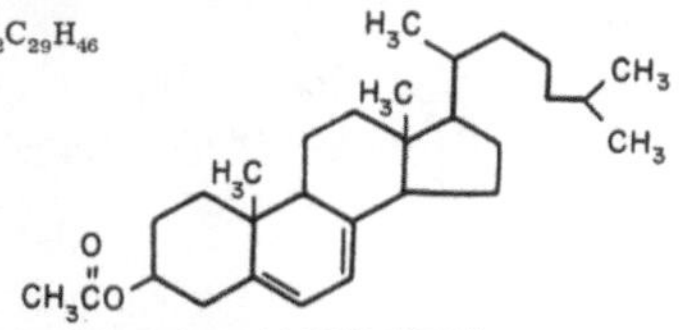

Helv. Chim. Acta 44:1380 (1961)

▶ $O_2C_{29}H_{46}$

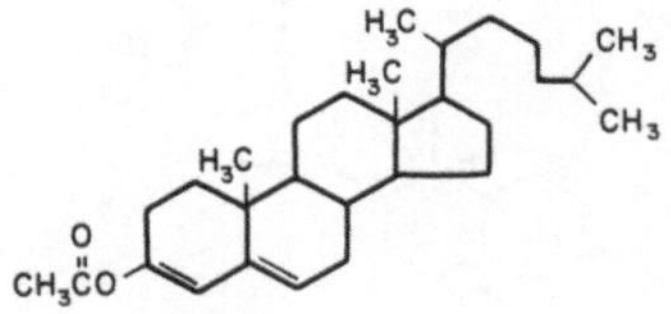


Helv. Chim. Acta 44:1380 (1961)

▶ $O_2C_{29}H_{48}$

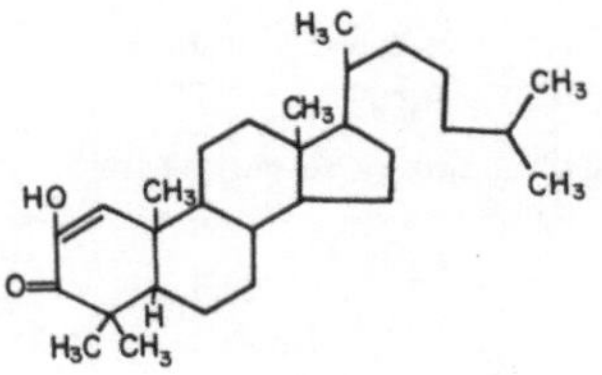


J. Chem. Soc. 2725 (1961)

▶ $O_2C_{29}H_{48}$

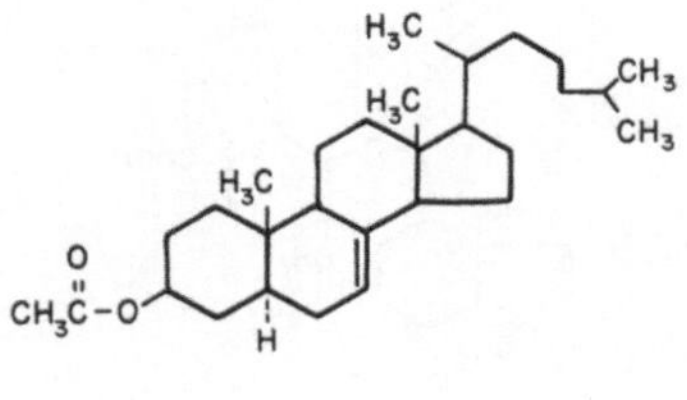


Helv. Chim. Acta 44:1380 (1961)

▶ $O_2C_{29}H_{50}$

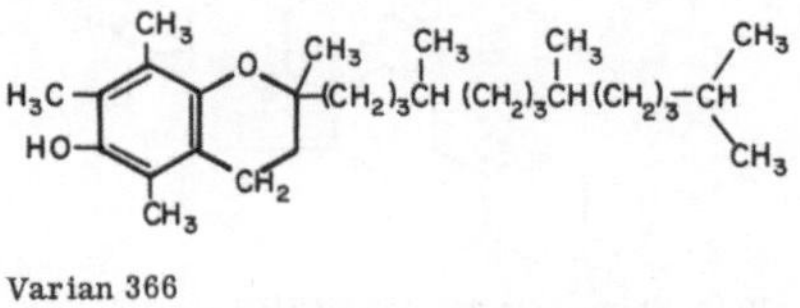


Varian 366

▶ $O_2C_{30}H_{20}$

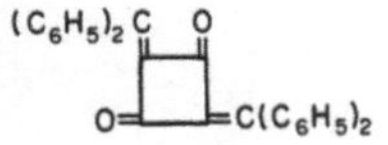


J. Am. Chem. Soc. 84:3397 (1962)

▶ $O_2C_{30}H_{42}$

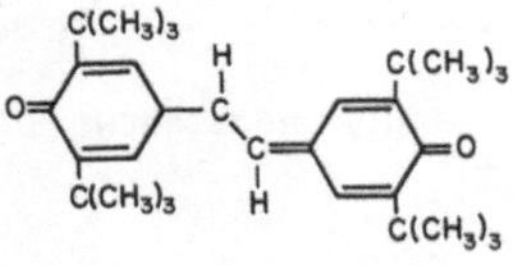


J. Am. Chem. Soc. 84:2258 (1962)

▶ $O_2C_{30}H_{44}$

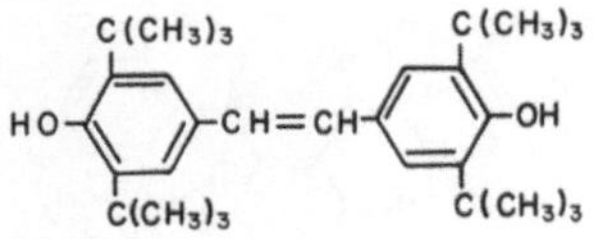


J. Am. Chem. Soc. 84:2258 (1962)

▶ $O_2C_{30}H_{46}$

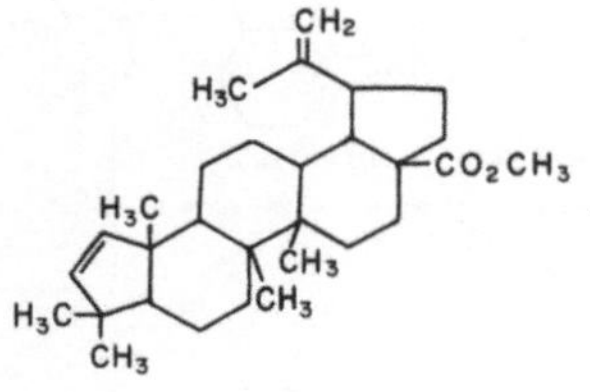


Can. J. Chem. 40:1634 (1962)

▶ $O_2C_{30}H_{46}$

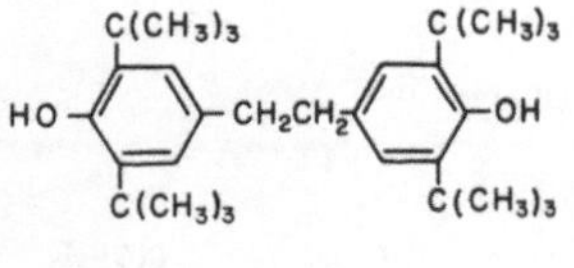


J. Am. Chem. Soc. 84:2258 (1962)

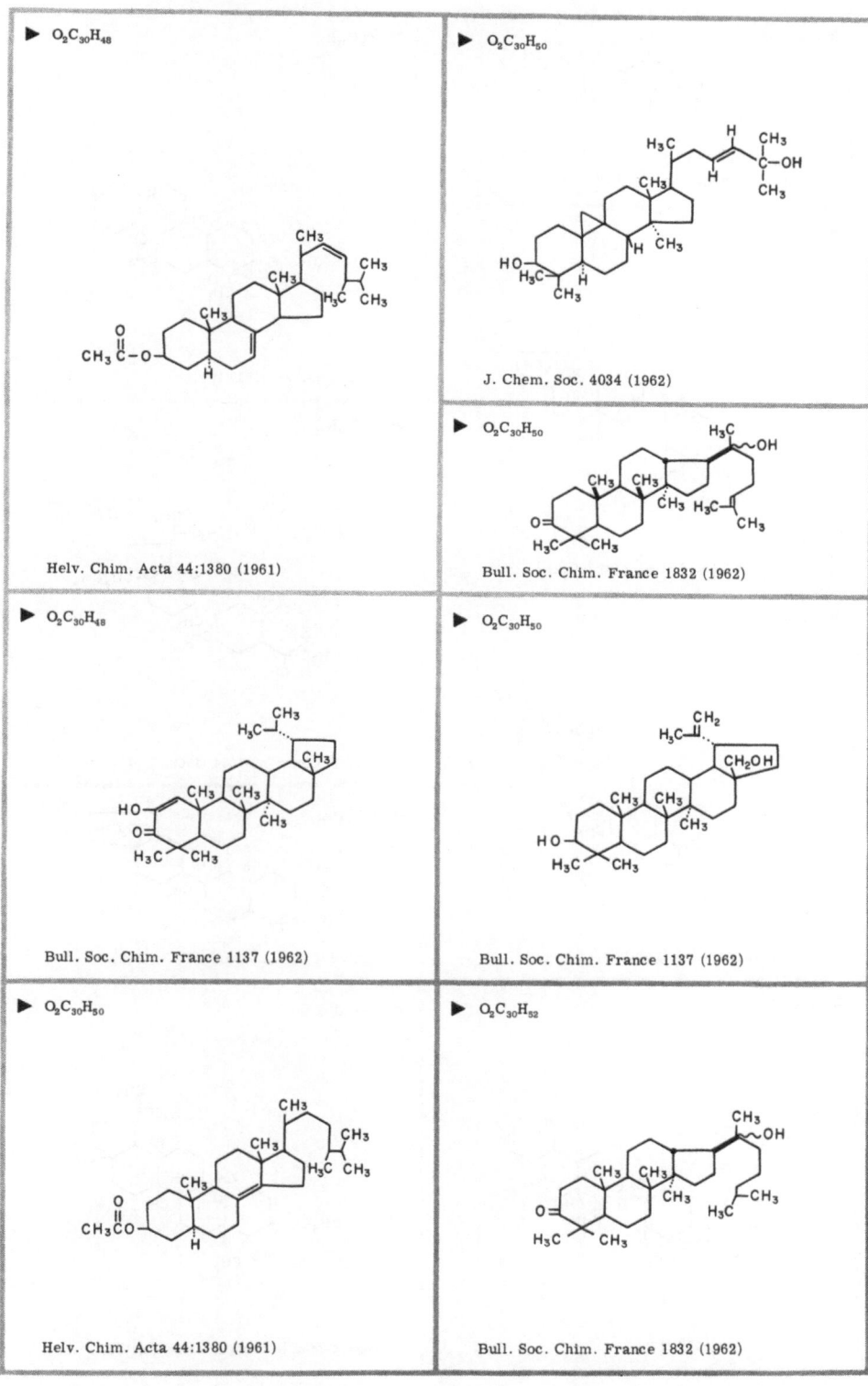

▶ $O_2C_{30}H_{48}$

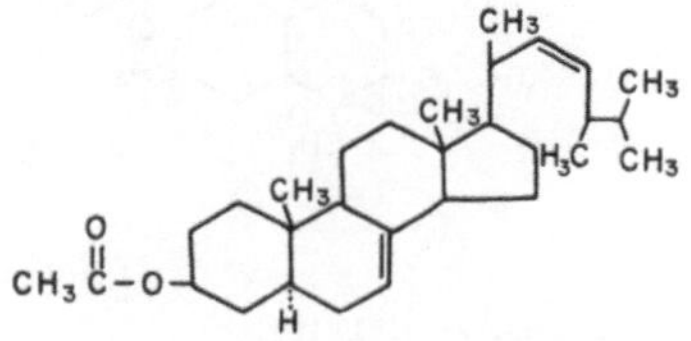

Helv. Chim. Acta 44:1380 (1961)

▶ $O_2C_{30}H_{50}$

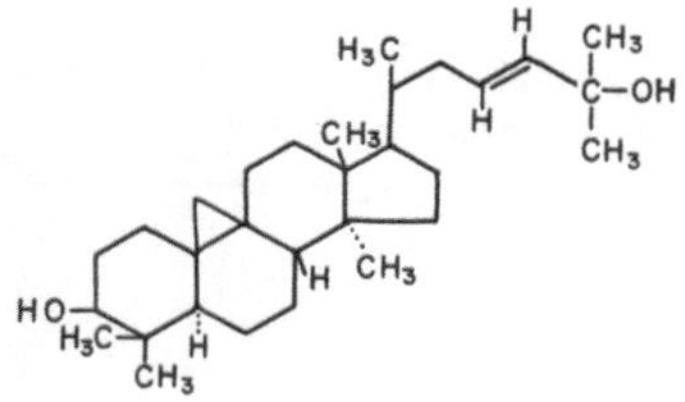

J. Chem. Soc. 4034 (1962)

▶ $O_2C_{30}H_{50}$

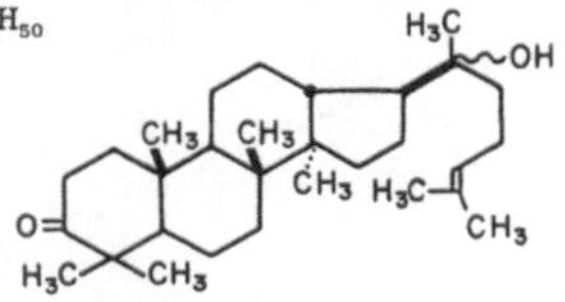

Bull. Soc. Chim. France 1832 (1962)

▶ $O_2C_{30}H_{48}$

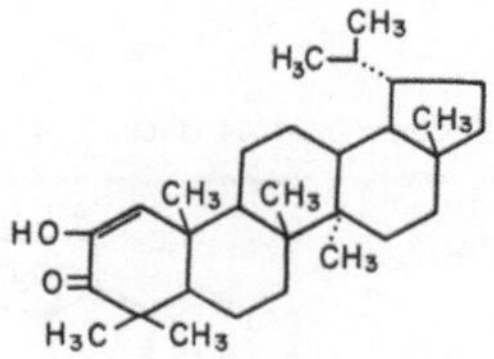

Bull. Soc. Chim. France 1137 (1962)

▶ $O_2C_{30}H_{50}$

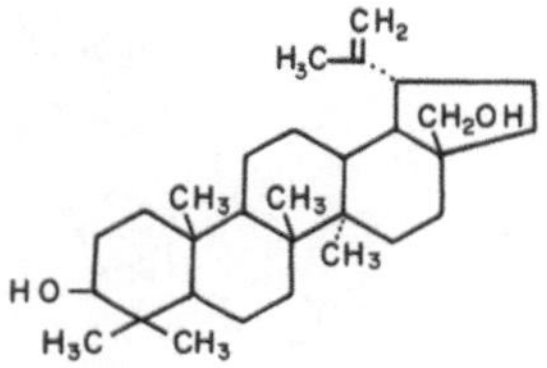

Bull. Soc. Chim. France 1137 (1962)

▶ $O_2C_{30}H_{50}$

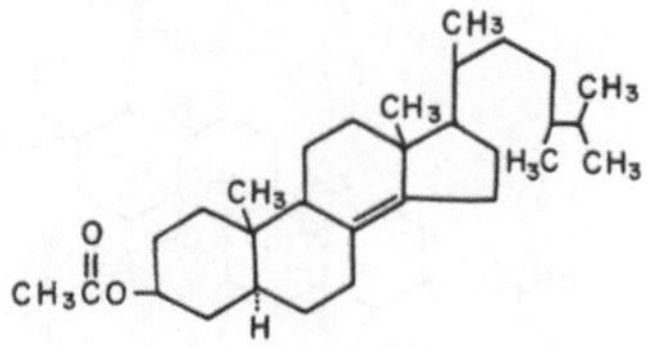

Helv. Chim. Acta 44:1380 (1961)

▶ $O_2C_{30}H_{52}$

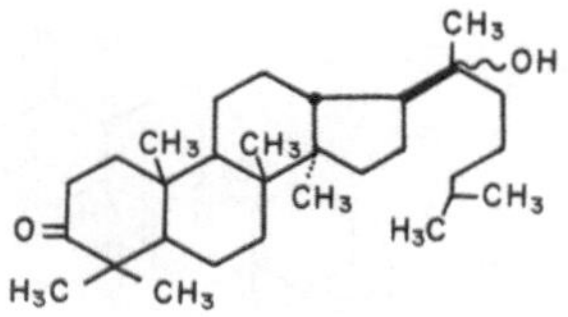

Bull. Soc. Chim. France 1832 (1962)

► $O_2C_{30}H_{52}$

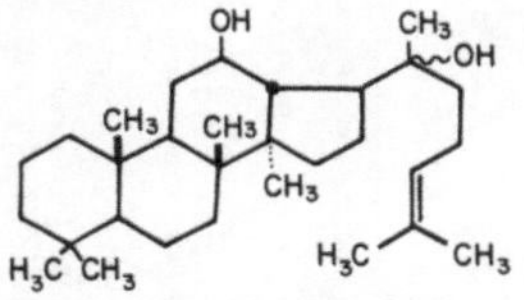

Bull. Soc. Chim. France 1832 (1962)

► $O_2C_{30}H_{52}$

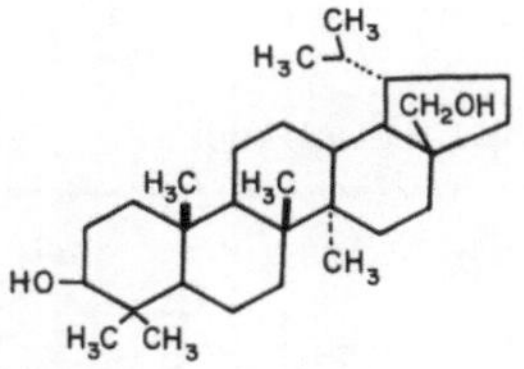

Bull. Soc. Chim. France 1137 (1962)

► $O_2C_{31}H_{44}$

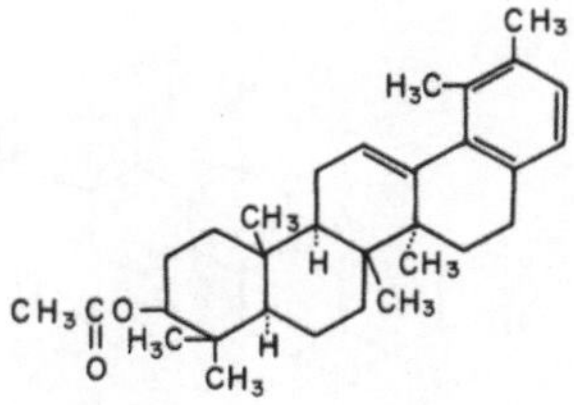

J. Chem. Soc. 5172 (1962)

► $O_2C_{31}H_{46}$

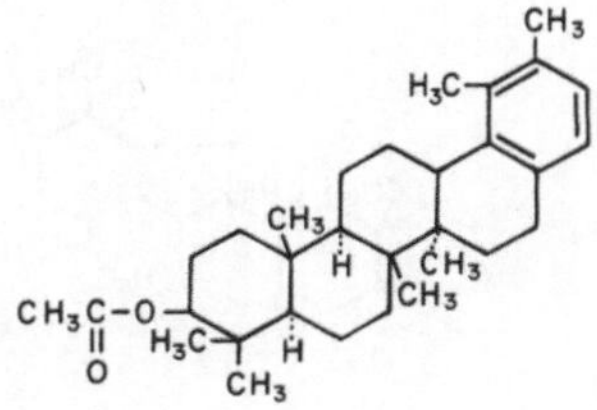

J. Chem. Soc. 5172 (1962)

► $O_2C_{31}H_{48}$

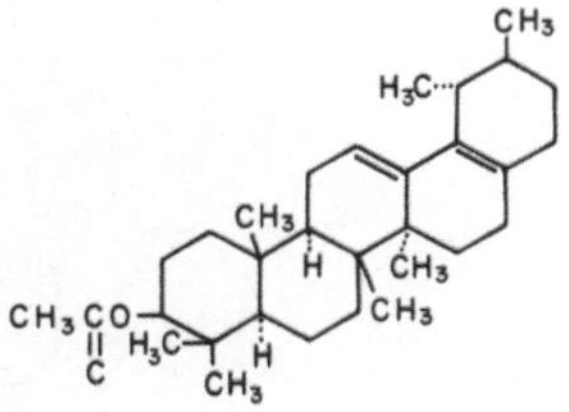

J. Chem. Soc. 5171 (1962)

► $O_2C_{31}H_{48}$

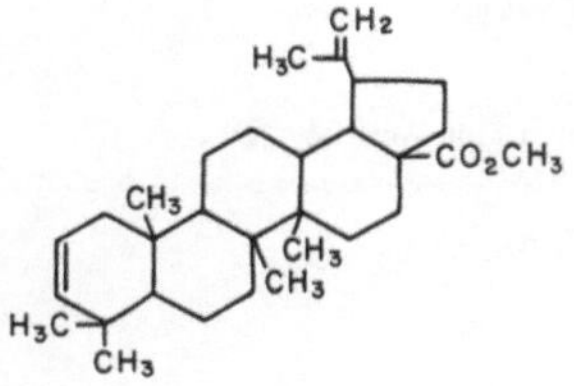

Can. J. Chem. 40:1634 (1962)

► $O_2C_{31}H_{52}$

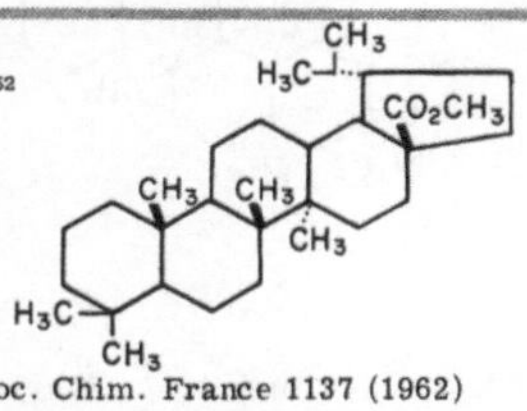

Bull. Soc. Chim. France 1137 (1962)

► $O_2C_{32}H_{52}$

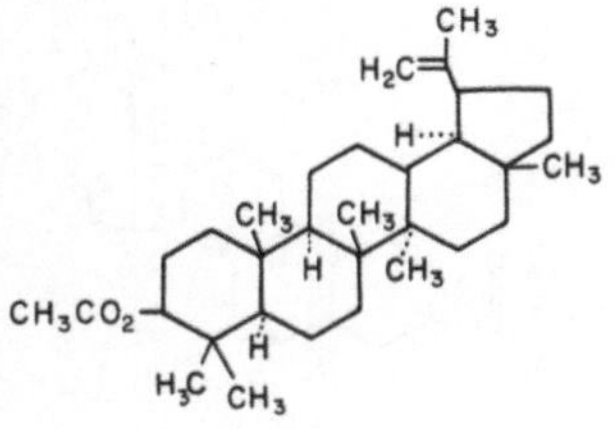

Bull. Chem. Soc. Japan 35:1899 (1962)

▶ $O_2C_{32}H_{52}$

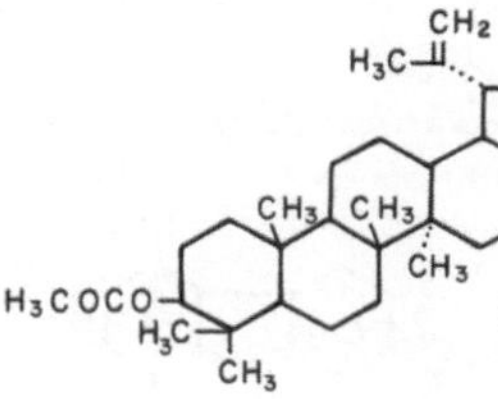

3 α and β

Bull. Soc. Chim. France 1137 (1962)

▶ $O_2C_{32}H_{54}$

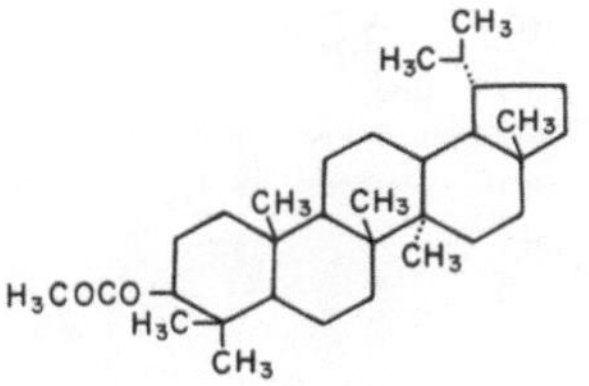

Bull. Soc. Chim. France 1137 (1962)

▶ $O_2C_{33}H_{44}$

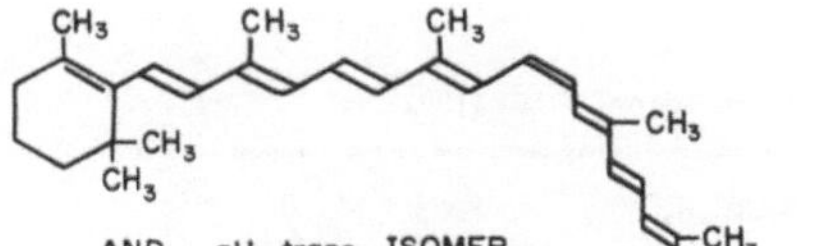

AND all-trans ISOMER

J. Chem. Soc. 1625 (1961)

▶ $O_2C_{34}H_{50}$

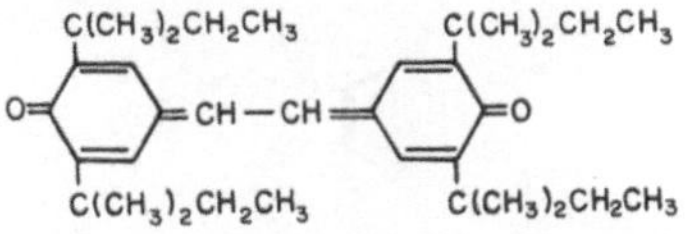

J. Am. Chem. Soc. 84:2258 (1962)

▶ $O_2C_{36}H_{28}$

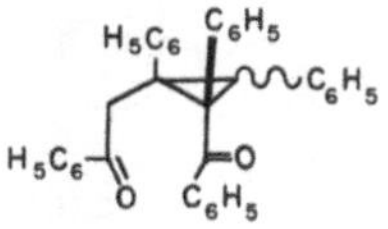

J. Org. Chem. 27:19 (1962)

▶ $O_2C_{36}H_{28}$

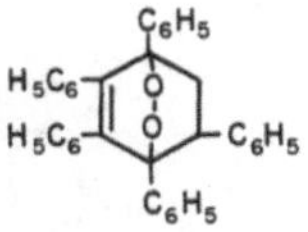

J. Org. Chem. 27:17 (1962)

▶ $O_2C_{36}H_{30}$

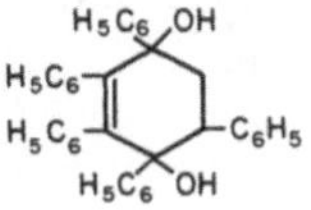

J. Org. Chem. 27:17 (1962)

▶ $O_2C_{36}H_{38}$

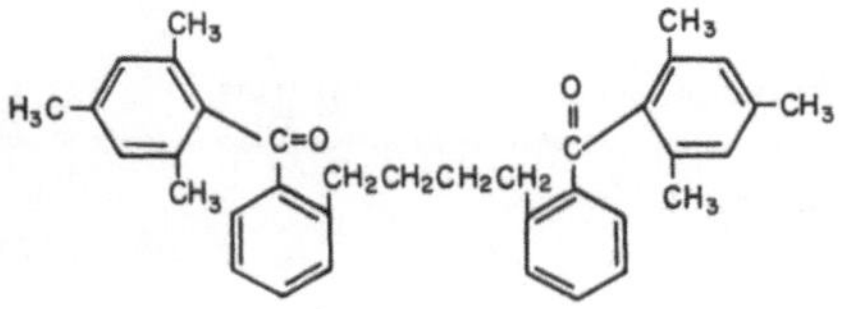

J. Org. Chem. 26:1043 (1961)

▶ $O_2C_{37}H_{54}$

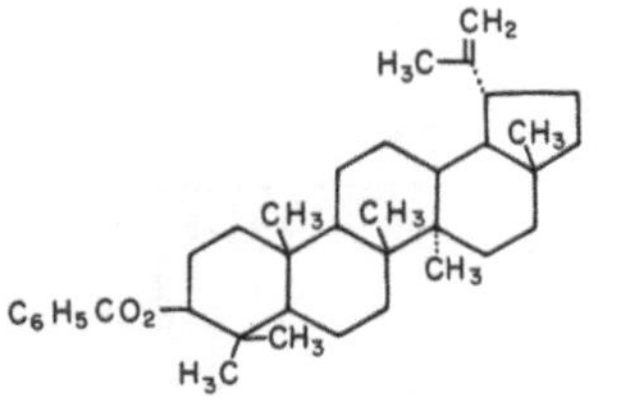

Bull. Soc. Chim. France 1137 (1962)

▶ $O_2C_{40}H_{36}$

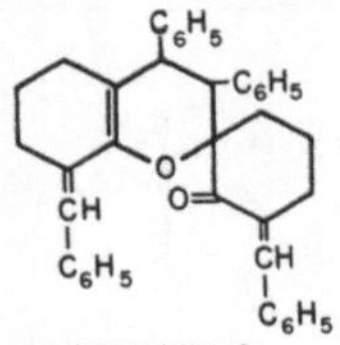

J. Org. Chem. 26:2192 (1961)

▶ $O_2C_{40}H_{58}$

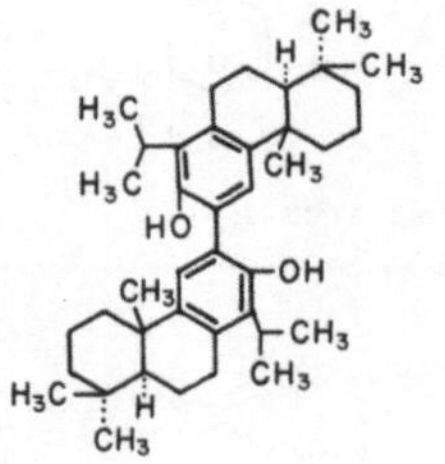

Chem. & Ind. 1757 (1962)

▶ $O_3BrC_{18}H_{17}$

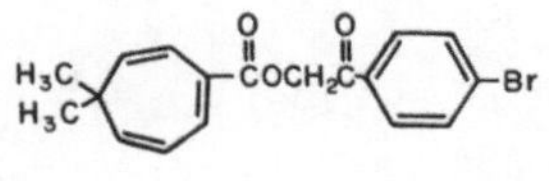

Tetrahedron Letters No. 19:841 (1962)

▶ $O_3BrC_{23}H_{31}$

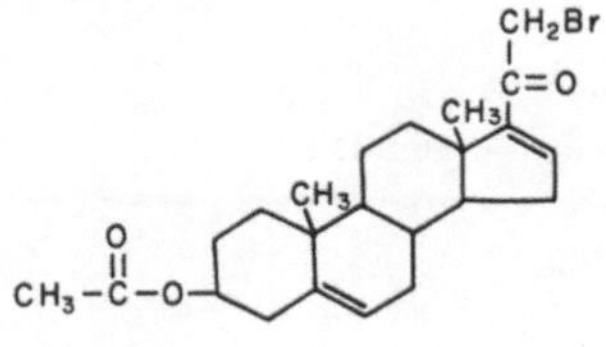

J. Org. Chem. 27:4397 (1962)

▶ $O_3BrC_{24}H_{37}$

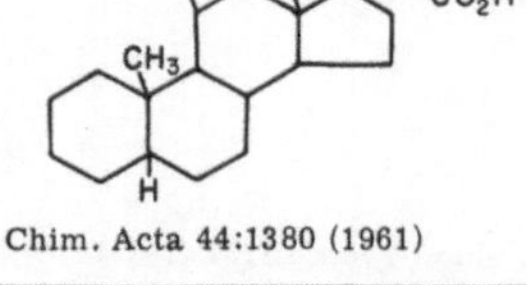

Helv. Chim. Acta 44:1380 (1961)

▶ $O_3BrC_{24}H_{37}$

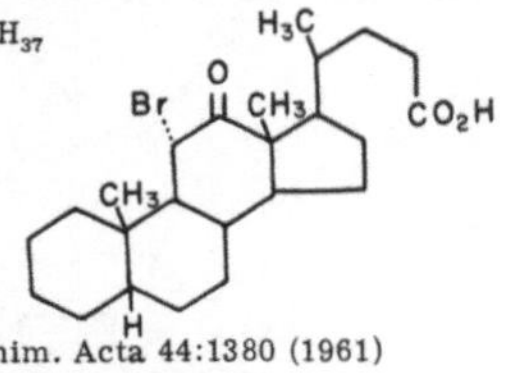

Helv. Chim. Acta 44:1380 (1961)

▶ $O_3BrC_{27}H_{41}$

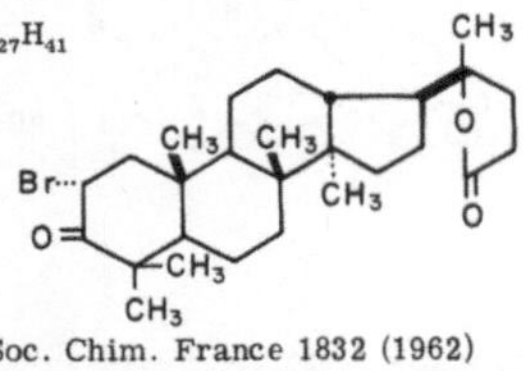

Bull. Soc. Chim. France 1832 (1962)

▶ $O_3BrC_{27}H_{43}$

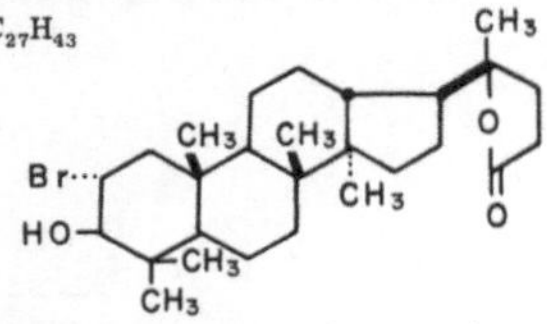

Bull. Soc. Chim. France 1832 (1962)

▶ $O_3BrC_{29}H_{47}$

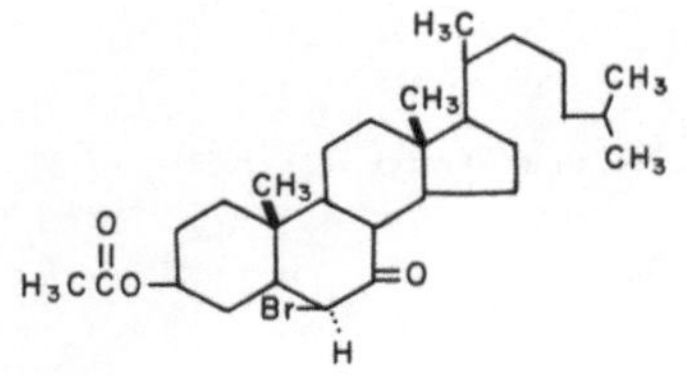

Tetrahedron 18:791 (1962)

▶ $O_3ClC_8H_7$

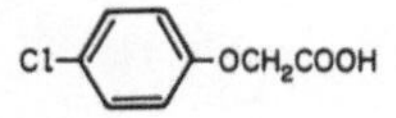

Varian 500

▶ $O_3ClC_8H_7$

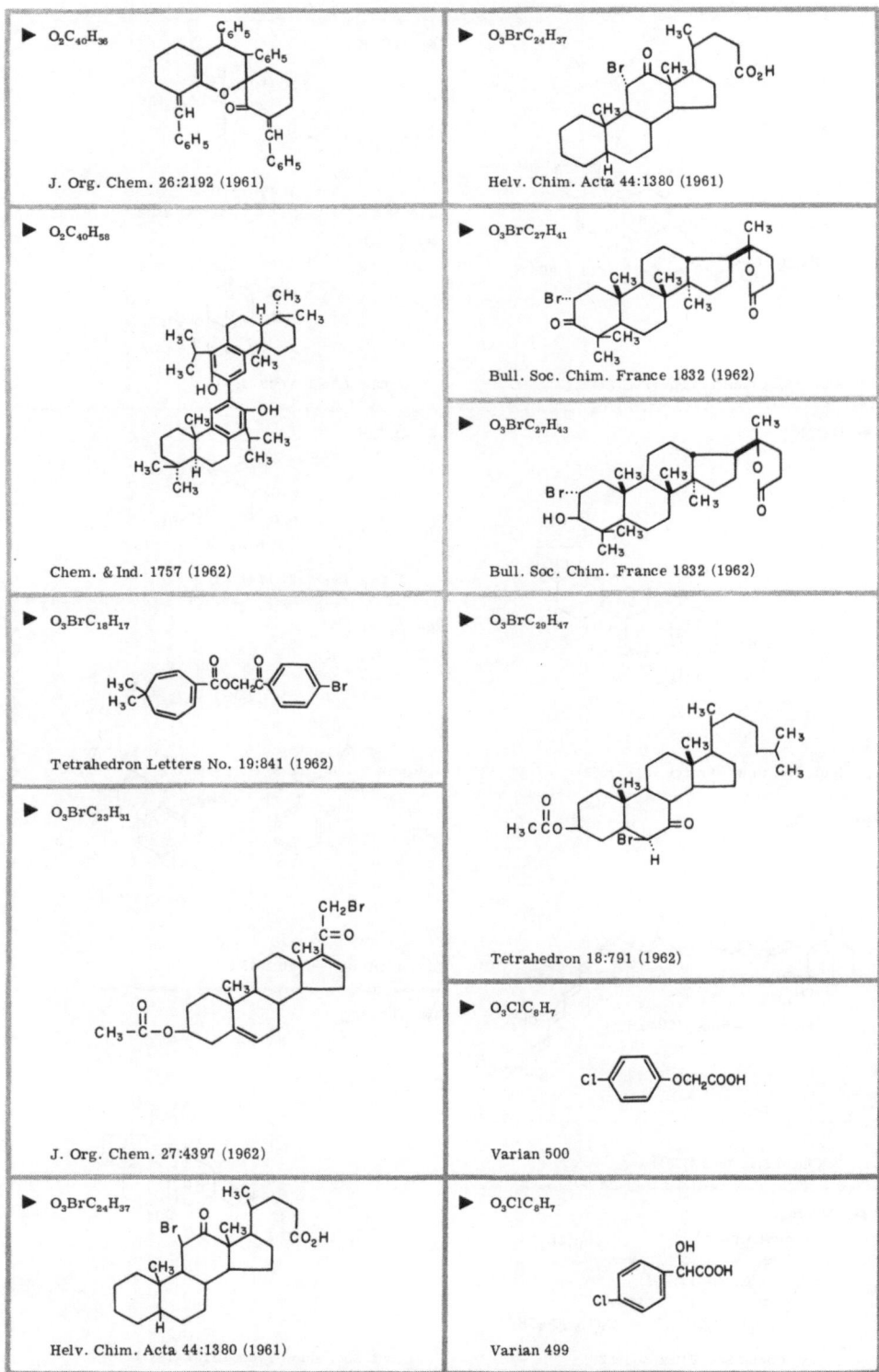

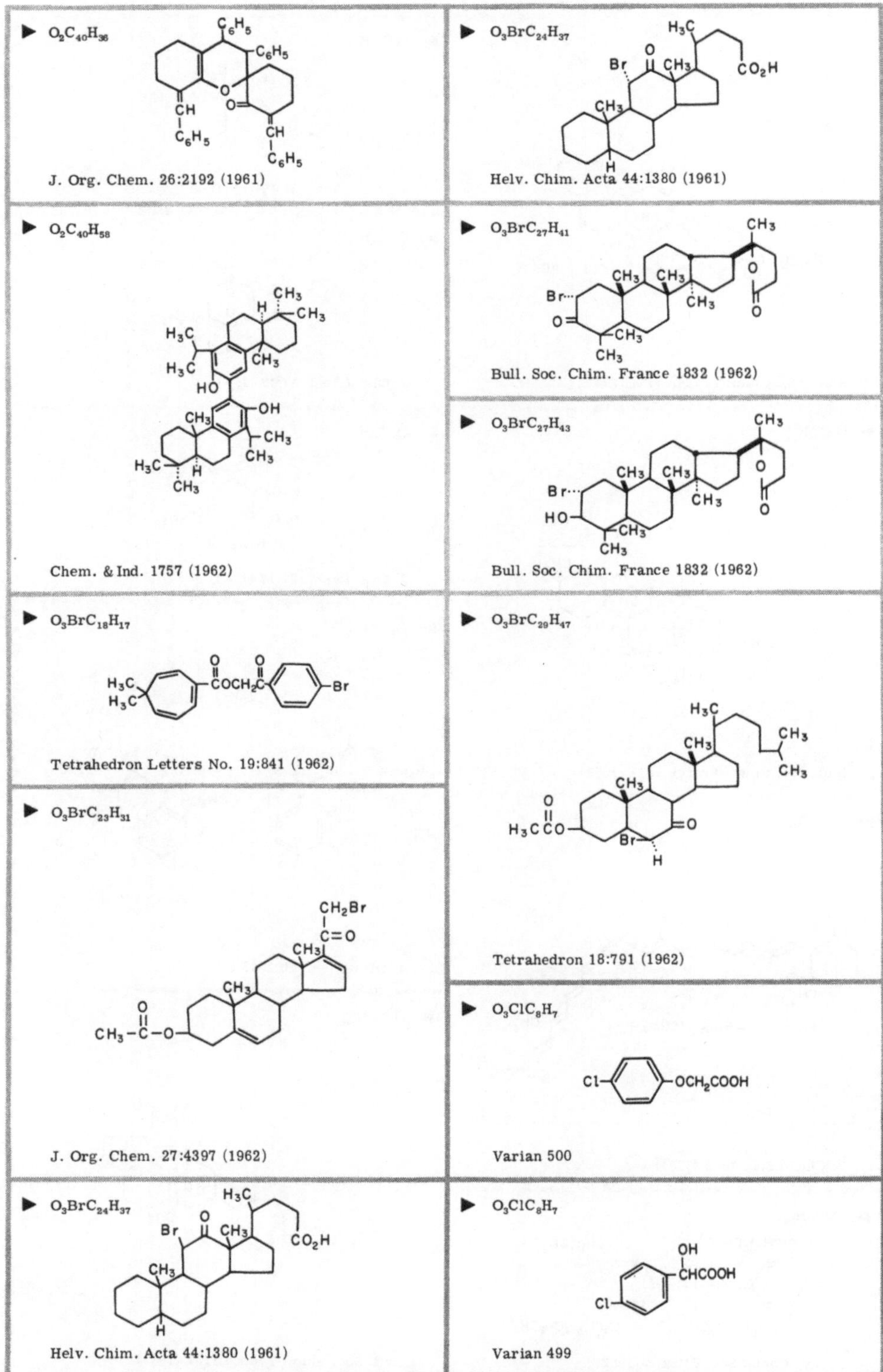

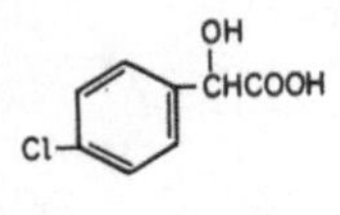

Varian 499

▶ $O_3ClC_9H_{13}$

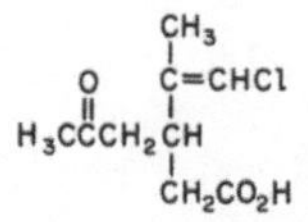

Tetrahedron Letters No. 21:941 (1962)

▶ $O_3ClC_{15}H_{17}$

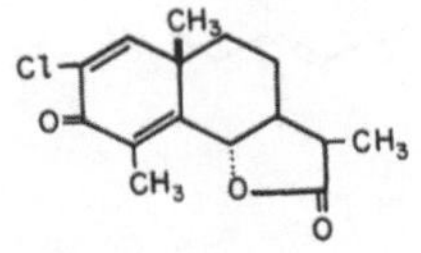

J. Chem. Soc. 1680 (1962)

▶ $O_3ClC_{15}H_{19}$

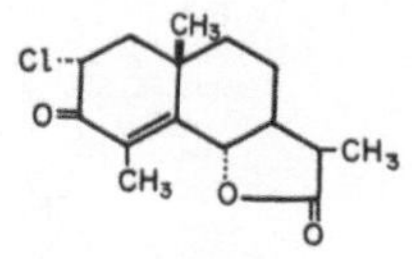

J. Chem. Soc. 1680 (1962)

▶ $O_3ClC_{21}H_{29}$

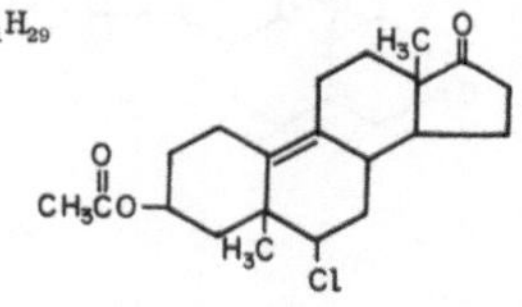

J. Org. Chem. 27:2811 (1962)

▶ $O_3ClC_{22}H_{29}$

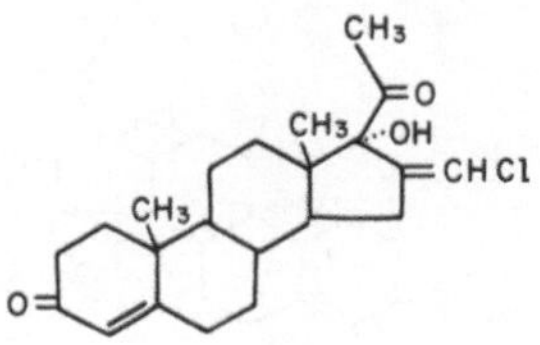

Chem. Ber. 95:2110 (1962)

▶ $O_3ClC_{22}H_{29}$

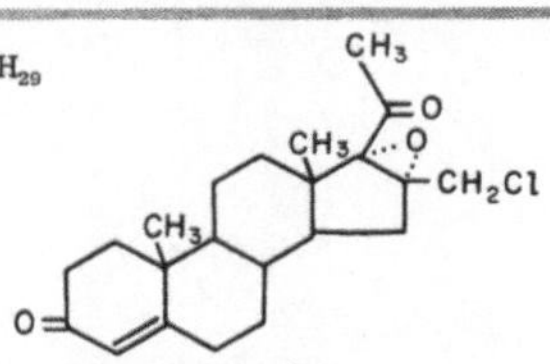

Chem. Ber. 95:2110 (1962)

▶ $O_3ClC_{23}H_{33}$

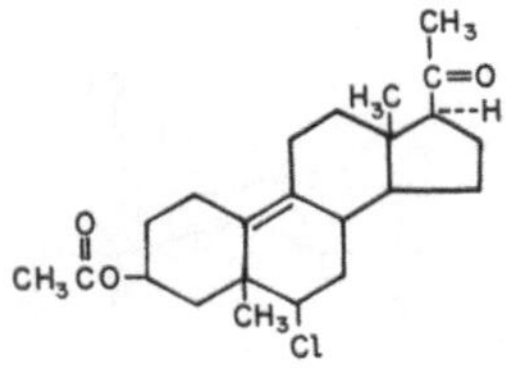

J. Org. Chem. 27:2811 (1962)

▶ $O_3ClC_{25}H_{29}$

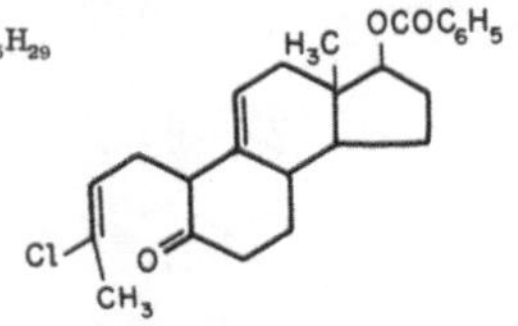

Bull. Soc. Chim. France 1677 (1961)

▶ $O_3ClC_{25}H_{37}$

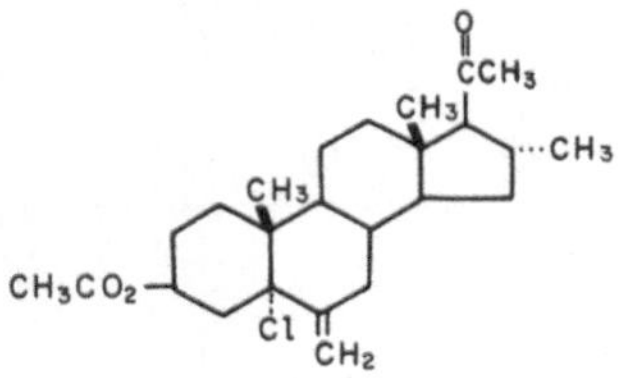

J. Org. Chem. 27:1139 (1962)

▶ $O_3Cl_2C_{15}H_{18}$

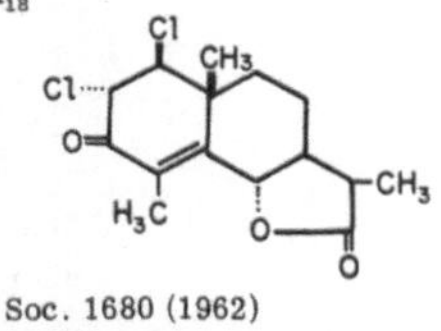

J. Chem. Soc. 1680 (1962)

▶ $O_3Cl_6C_5H_2$

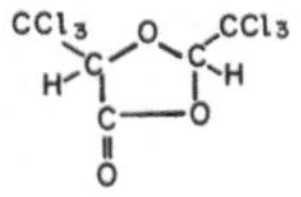

Varian 425

▶ $O_3Cl_6C_9H_{12}$

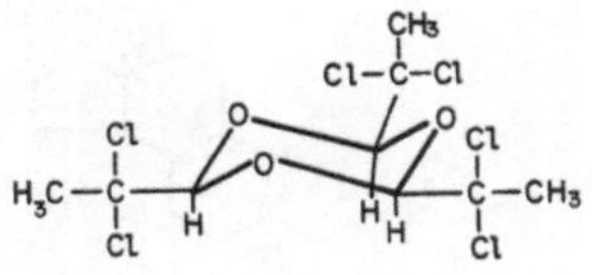

Varian 536

▶ $O_3Cl_9C_6H_3$

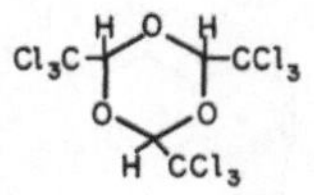

J. Org. Chem. 26:1458 (1961)

▶ $O_3Cl_9C_6H_3$

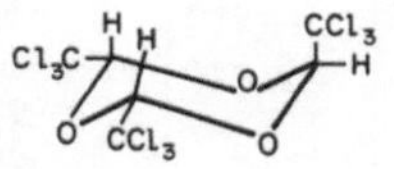

Varian 449

▶ $O_3Cl_9C_6H_3$

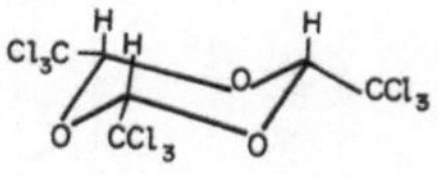

Varian 450

▶ $O_3FC_{21}H_{27}$

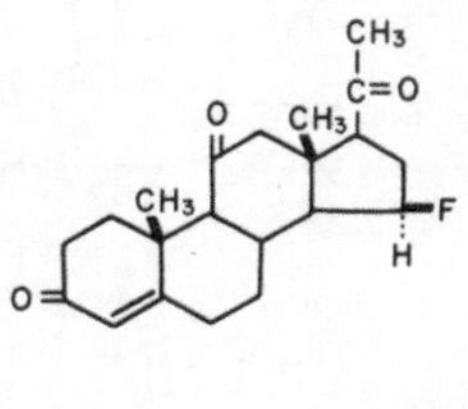

Tetrahedron Letters No. 23:1067 (1962)

▶ $O_3FC_{22}H_{29}$

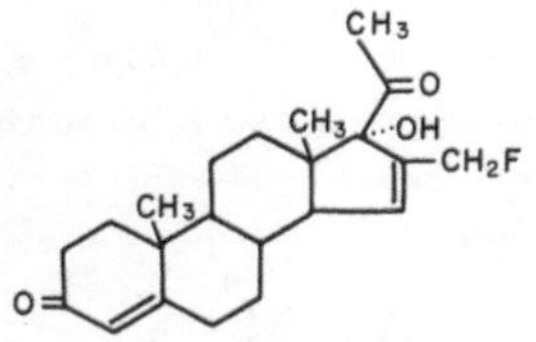

Chem. Ber. 95:2110 (1962)

▶ $O_3FC_{22}H_{29}$

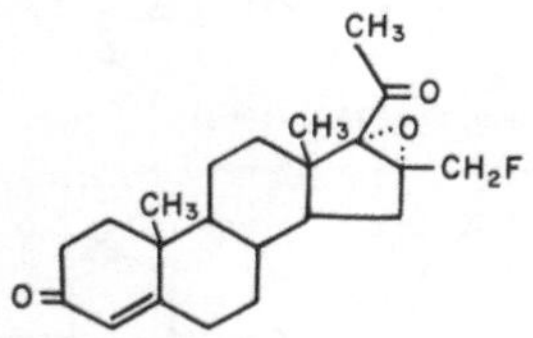

Chem. Ber. 95:2110 (1962)

▶ $O_3FC_{22}H_{29}$

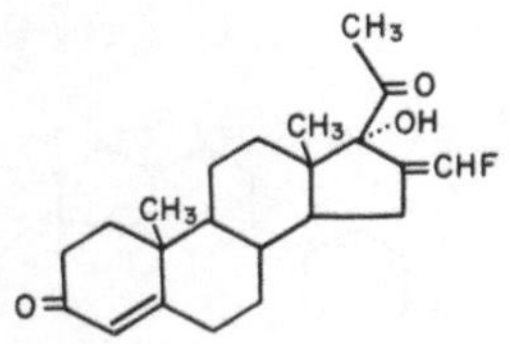

Chem. Ber. 95:2110 (1962)

▶ $O_3F_2C_4H_4$

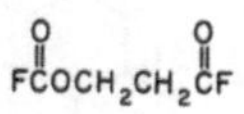

J. Am. Chem. Soc. 84:4275 (1962)

▶ $O_3F_2C_5H_6$

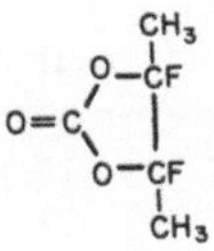

J. Am. Chem. Soc. 84:4275 (1962)

▶ $O_3F_2C_6H_8$

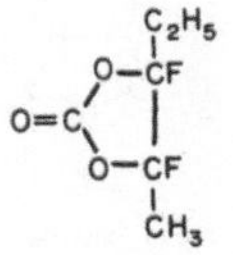

J. Am. Chem. Soc. 84:4275 (1962)

▶ $O_3F_2C_{10}H_8$

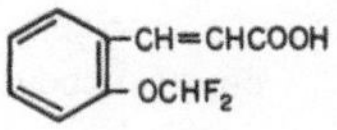

J. Chem. Soc. 3829 (1962)

▶ $O_3F_4C_5H_6$

$CH_3OCF_2CF_2\overset{O}{\overset{\|}{C}}OCH_3$

J. Am. Chem. Soc. 84:4275 (1962)

▶ $O_3C_3H_2$

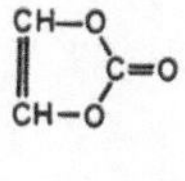

J. Chem. Phys. 36:2644 (1962)

▶ $O_3C_3H_6$

$(CH_3O)_2CO$

J. Chem. Phys. 37:2198 (1962)
J. Chem. Soc. 1372 (1962)

▶ $O_3C_3H_8$

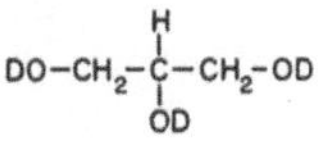

Varian 395

▶ $O_3C_3H_9B$

$(H_3C)_3BO_3$

J. Chem. Soc. 1372 (1962)

▶ $O_3C_4H_2$

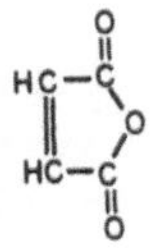

Varian 48
Tetrahedron Letters No. 23:1063 (1962)

▶ $O_3C_4H_2 \cdot O_4C_4Fe$

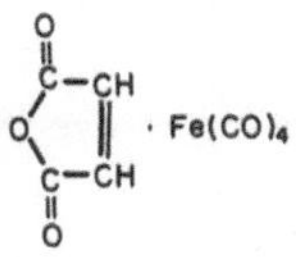

Tetrahedron Letters No. 23:1063 (1962)

▶ $O_3C_4H_6$

$CH_3\overset{O}{\overset{\|}{C}}-O-\overset{O}{\overset{\|}{C}}-CH_3$

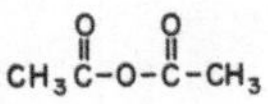

Bull. Chem. Soc. Japan 34:143 (1961)

▶ $O_3C_5H_4$

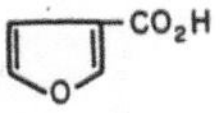

$CO_2H$

Can. J. Chem. 39:909 (1961)

▶ $O_3C_5H_4$

$CH_3$ H

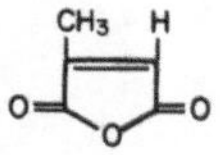

Varian 427

▶ $O_3C_4H_8$

$CH_3CH_2OCH_2COOH$

Varian 417

▶ $O_3C_5H_8$

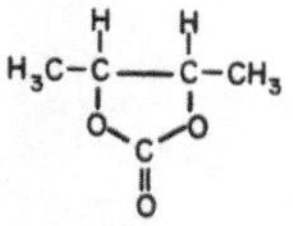

meso and DL

J. Am. Chem. Soc. 84:747 (1962)

▶ $O_3C_4H_{10}$

$CH(OCH_3)_3$

J. Am. Chem. Soc. 84:3973 (1962)

▶ $O_3C_5H_8$

$CH_3CH=C(CH_2OH)COOH$

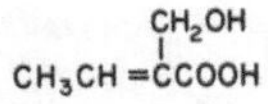

J. Org. Chem. 26:3047 (1961)

▶ $O_3C_4H_{12}Si$

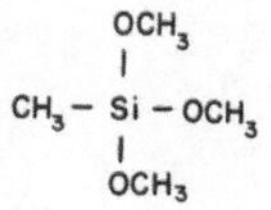

$CH_3-Si(OCH_3)_3$

Varian 93

▶ $O_3C_5H_4$

$CO_2H$

Can. J. Chem. 39:909 (1961)

▶ $O_3C_5H_{10}$

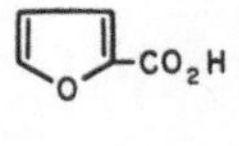

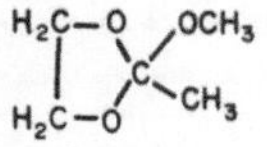

J. Am. Chem. Soc. 84:4876 (1962)

▶ $O_3C_6H_6$

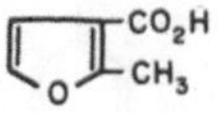


Can. J. Chem. 39:909 (1961)

▶ $O_3C_6H_6$

Varian 125

▶ $O_3C_6H_6$

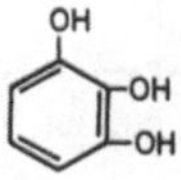


J. Chem. Phys. 37:2618 (1962)

▶ $O_3C_6H_6$

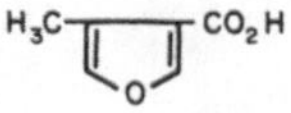


Can. J. Chem. 39:909 (1961)

▶ $O_3C_6H_6 \cdot HClO_4$

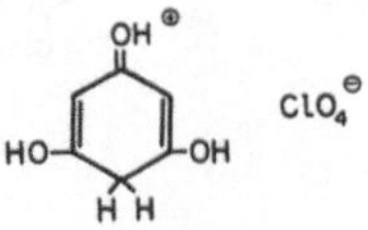


J. Am. Chem. Soc. 84:4343 (1962)

▶ $O_3C_6H_8$

$CH_3\overset{O}{\overset{\|}{C}}OCH_2C\equiv CCH_2OH$

Varian 463

▶ $O_3C_6H_{10}$

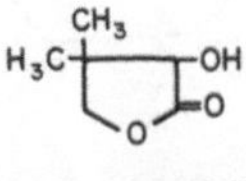


Angew. Chem. 74:751, 995 (1962)
Angew. Chem. Intern. Ed. 1:592 (1962)

▶ $O_3C_6H_{10}$

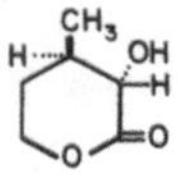


VERRUCARIN ACID LACTONE

Varian 467
Helv. Chim. Acta 45:1726 (1962)

▶ $O_3C_6H_{10}$

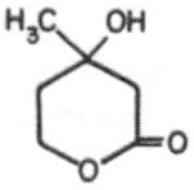


Varian 466

▶ $O_3C_6H_{12}$

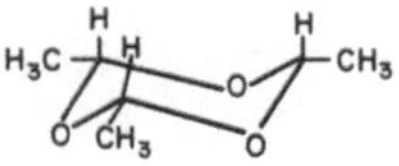


Varian 474

▶ $O_3C_6H_{14}$

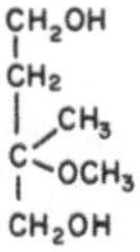


Helv. Chim. Acta 45:138 (1962)

▶ $O_3C_6H_{14}$

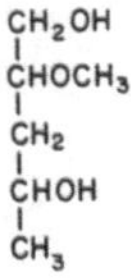


Helv. Chim. Acta 45:138 (1962)

▶ $O_3C_7H_6$

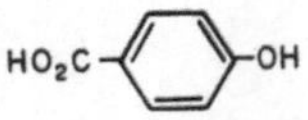

Can. J. Chem. 40:2124 (1962)

▶ $O_3C_7H_8$

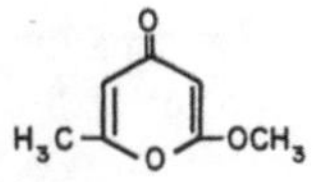

J. Org. Chem. 27:3715 (1962)

▶ $O_3C_7H_6$

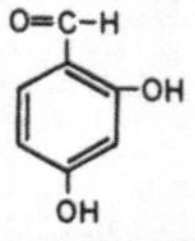

J. Chem. Phys. 37:1009 (1962)

▶ $O_3C_7H_8$

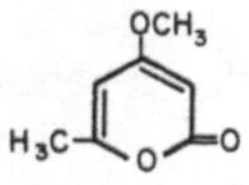

J. Org. Chem. 27:3715 (1962)

▶ $O_3C_7H_6$

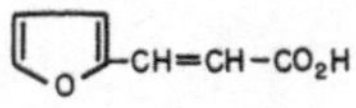

Can. J. Chem. 39:909 (1961)

▶ $O_3C_7H_{10}$

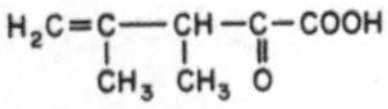

Chem. Ber. 95:500 (1962)

▶ $O_3C_7H_8$

Can. J. Chem. 39:909 (1961)

▶ $O_3C_7H_{10}$

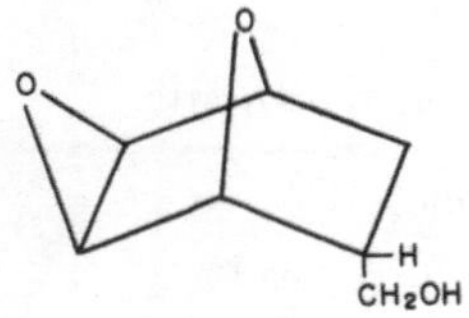

J. Am. Chem. Soc. 84:4122 (1962)

▶ $O_3C_7H_8$

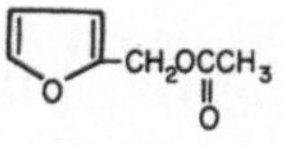

Varian 167

▶ $O_3C_7H_8$

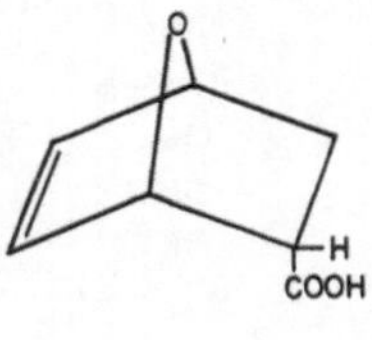

J. Am. Chem. Soc. 84:4121 (1962)

▶ $O_3C_7H_{10}$

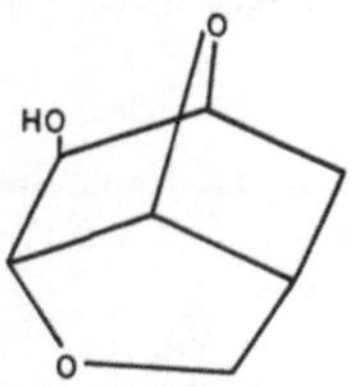

J. Am. Chem. Soc. 84:4122 (1962)

▶ $O_3C_7H_{12}$

$CH_2CH_3$, $H_2C$, $CH_2$, $CH_2$, O, O, O, C, H

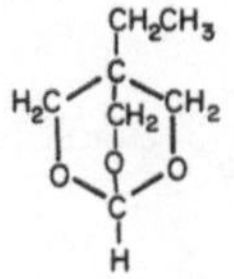

Varian 493

▶ $O_3C_8H_{12}$

$CO_2C_2H_5$, H, O

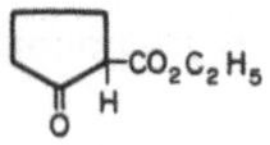

Can. J. Chem. 40:363 (1962)

▶ $O_3C_7H_{16}$

$HOH_2CCH(CH_3)CH(OH)CH(CH_3)CH_2OH$

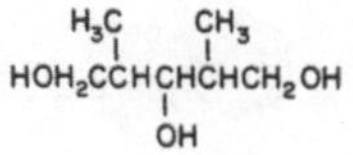

J. Am. Chem. Soc. 84:1512 (1962)

▶ $O_3C_8H_{12}$

O, O, O

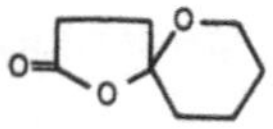

Chem. Ber. 95:1733 (1962)

▶ $O_3C_7H_{18}Si$

$CH_3Si(OCH_2CH_3)_3$

Varian 184

▶ $O_3C_8H_{14}$

$CH_2{=}CHCH_2O(CH_2O)_2CH_2CH{=}CH_2$

J. Chem. Soc. 4313 (1962)

▶ $O_3C_8H_6$

O, O, CHO

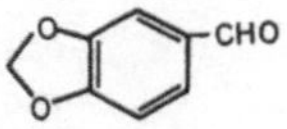

Varian 187
Can. J. Chem. 40:1073 (1962)

▶ $O_3C_8H_{14}$

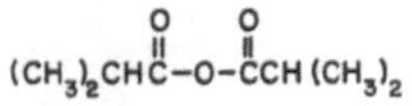

$(CH_3)_2CHC(=O)-O-C(=O)CH(CH_3)_2$

Varian 516

▶ $O_3C_8H_8$

$OCH_3$, $CO_2H$

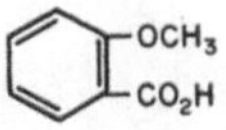

Varian 195
Can. J. Chem. 40:1866 (1962)

▶ $O_3C_8H_{14}Hg$

$CH_3C(=O)OHg$, H, HO, H

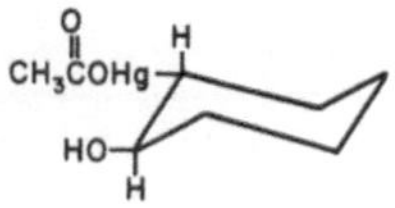

Chem. & Ind. 2053 (1961)

▶ $O_3C_8H_8$

$OCH_3$, $HO_2C$

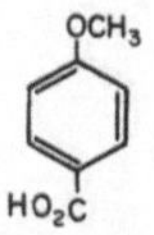

Varian 196
Can. J. Chem. 40:1866 (1962)

▶ $O_3C_8H_8$

OH, $OCH_3$, CHO

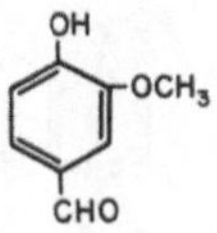

Varian 197
Can. J. Chem. 40:1073, 1866 (1962)

▶ $O_3C_8H_{18}$

$CH_3(CH_2)_2(CH_2O)_2(CH_2)_2CH_3$

J. Chem. Soc. 4313 (1962)

▶ $O_3C_8H_{18}$

$CH_3(CH_2)_3C(OCH_3)_3$

Varian 218

▶ $O_3C_9H_{10}$

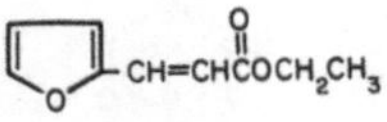


Varian 235

▶ $O_3C_8H_{18}Si$

$H_2C{=}CHSi(OC_2H_5)_3$

Varian 219

▶ $O_3C_9H_{10}$

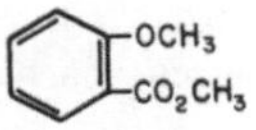


Can. J. Chem. 40:1866 (1962)

▶ $O_3C_8H_{20}Si$

$C_2H_5Si(OC_2H_5)_3$

Varian 222

▶ $O_3C_9H_{10}$

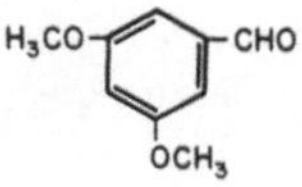


Can. J. Chem. 40:1073 (1962)

▶ $O_3C_9H_8$

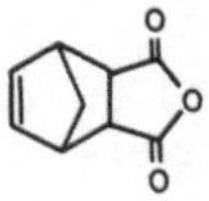


endo

Helv. Chim. Acta 45:1870 (1962)

▶ $O_3C_9H_{10}$

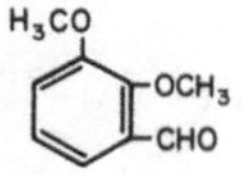


Can. J. Chem. 40:1866 (1962)

▶ $O_3C_9H_8$

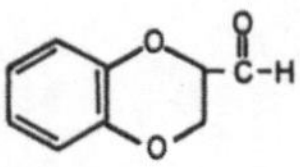


Tetrahedron 18:289 (1962)

▶ $O_3C_9H_{10}$

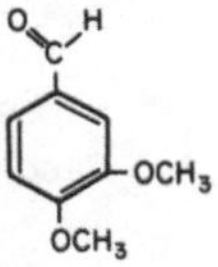


Varian 236

▶ $O_3C_9H_8$

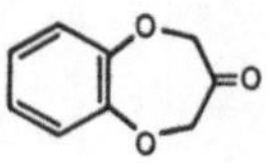


Tetrahedron 18:289 (1962)

▶ $O_3C_9H_{10}$

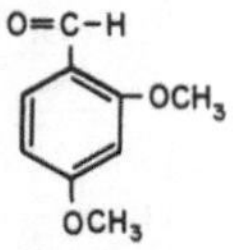


J. Chem. Phys. 37:1009 (1962)

▶ $O_3C_9H_{10}$

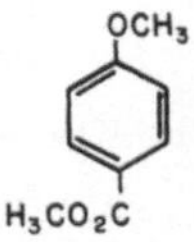


Can. J. Chem. 40:1866 (1962)

▶ $O_3C_9H_{10}$

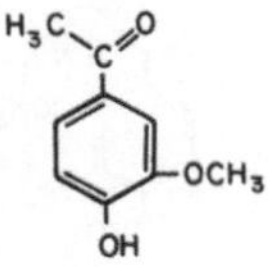


Varian 234

▶ $O_3C_9H_{12}$

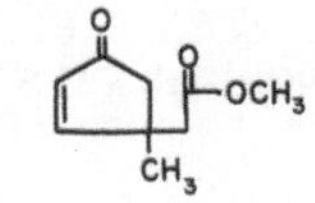


J. Am. Chem. Soc. 83:1768 (1961)

▶ $O_3C_9H_{14}$

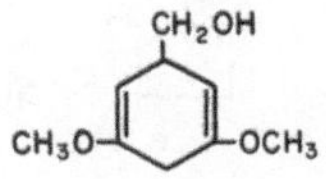


J. Am. Chem. Soc. 83:1005 (1961)

▶ $O_3C_9H_{14}Hg$

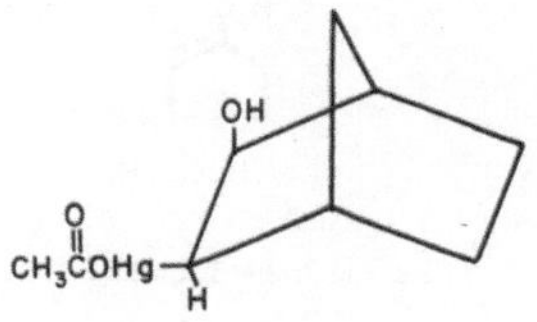


Chem. & Ind. 2053 (1961)

▶ $O_3C_9H_{18}$

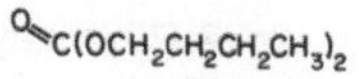


Varian 243

▶ $O_3C_{10}H_8$

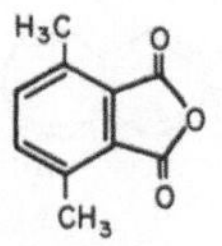


Bull. Soc. Chim. France 1255 (1962)

▶ $O_3C_{10}H_{10}$

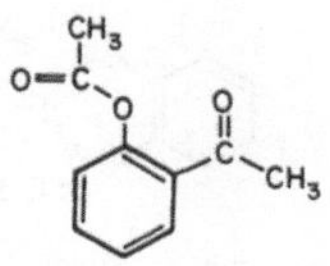


Varian 254

▶ $O_3C_{10}H_{12}$

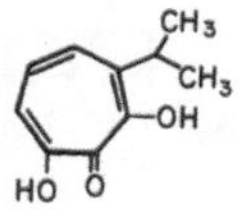


J. Org. Chem. 26:174 (1961)

▶ $O_3C_{10}H_{14}$

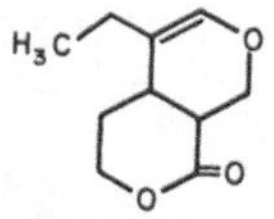


Tetrahedron 16:192 (1961)

▶ $O_3C_{10}H_{14}$

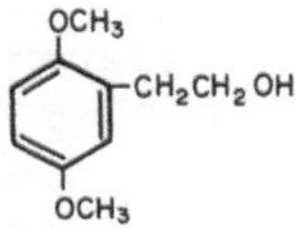


J. Org. Chem. 26:3225 (1961)

▶ $O_3C_{10}H_{14}$

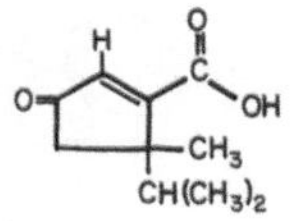


Varian 567

▶ $O_3C_{10}H_{16}$

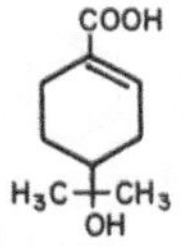


J. Org. Chem. 27:1033 (1962)

▶ $O_3C_{10}H_{18}$

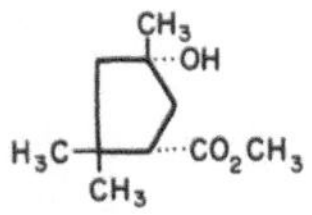


J. Am. Chem. Soc. 83:3972 (1961)

▶ $O_3C_{10}H_{18}$

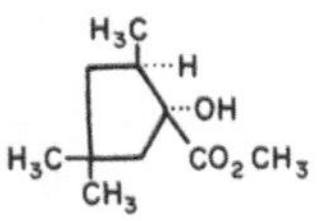


J. Am. Chem. Soc. 83:3972 (1961)

▶ $O_3C_{10}H_{18}$

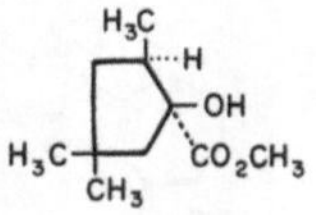

J. Am. Chem. Soc. 83:3972 (1961)

▶ $O_3C_{10}H_{20}$

Helv. Chim. Acta 45:129 (1962)

▶ $O_3C_{11}H_8$

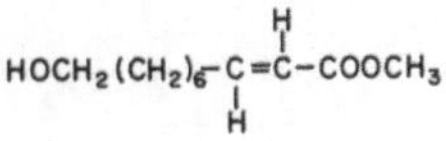

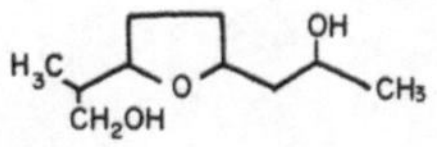

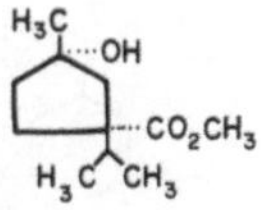

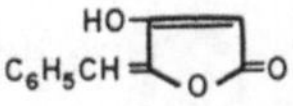

Chem. & Ind. 561 (1962)

▶ $O_3C_{11}H_{14}$

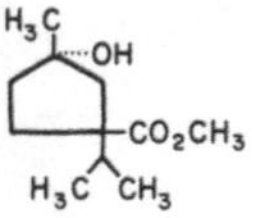

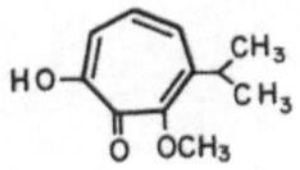

J. Org. Chem. 26:1679 (1961)

▶ $O_3C_{11}H_{14}$

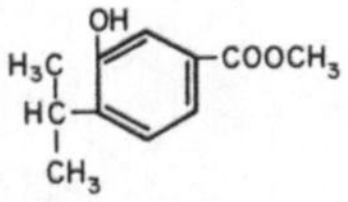

Proc. Chem. Soc. 219 (1962)

▶ $O_3C_{11}H_{18}$

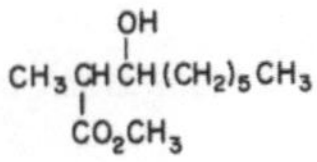

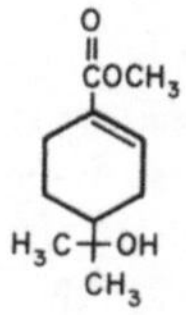

Tetrahedron Letters No. 16:710 (1962)

▶ $O_3C_{11}H_{20}$

Tetrahedron 18:177 (1962)

▶ $O_3C_{11}H_{20}$

J. Am. Chem. Soc. 83:3972 (1961)

▶ $O_3C_{11}H_{20}$

cis and trans ISOMERS

J. Am. Chem. Soc. 83:3972 (1961)

▶ $O_3C_{11}H_{22}$

Helv. Chim. Acta 45:129 (1962)

▶ $O_3C_{12}H_{10}$

J. Org. Chem. 27:3225 (1962)

▶ $O_3C_{12}H_{12}$

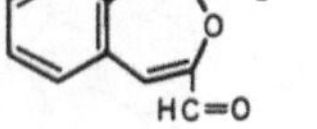

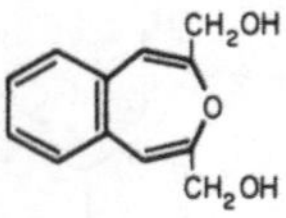

J. Org. Chem. 27:3225 (1962)

► $O_3C_{12}H_{12}$

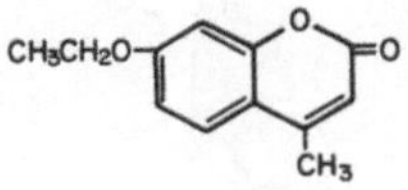

Varian 294

► $O_3C_{12}H_{14}$

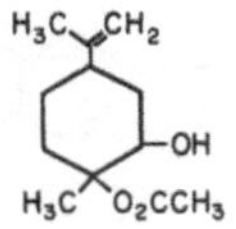

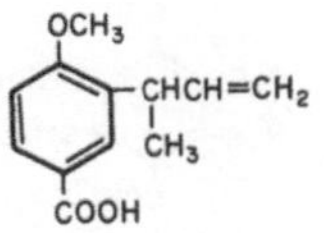

J. Org. Chem. 26:4787 (1961)

► $O_3C_{12}H_{14}$

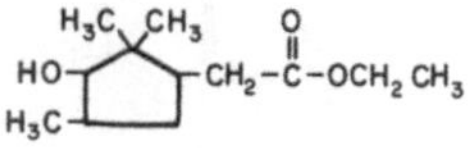

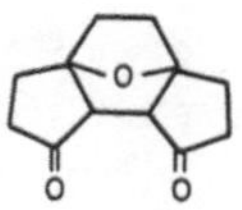

J. Am. Chem. Soc. 84:4611 (1962)

► $O_3C_{12}H_{14}$

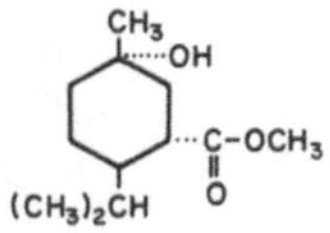

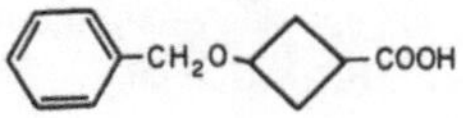

Chem. Ber. 95:2535 (1962)

► $O_3C_{12}H_{16}$

J. Org. Chem. 26:3729 (1961)

► $O_3C_{12}H_{18}$

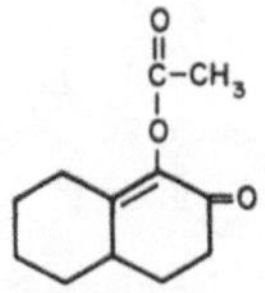

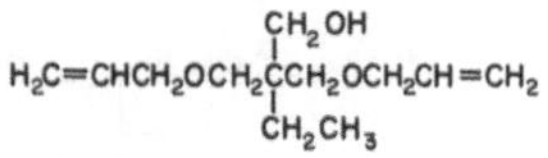

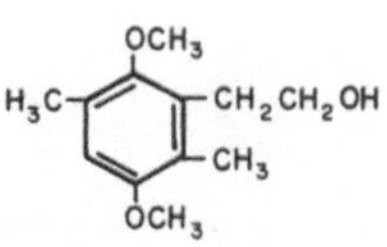

J. Org. Chem. 26:3225 (1961)

► $O_3C_{12}H_{18}$

J. Org. Chem. 26:3729 (1961)

► $O_3C_{12}H_{20}$

Bull. Chem. Soc. Japan 35:1899 (1962)

► $O_3C_{12}H_{22}$

J. Org. Chem. 27:900 (1962)

► $O_3C_{12}H_{22}$

Varian 297

► $O_3C_{12}H_{22}$

$CH_3OCH_2(CH_2)_6CH{=}CHCOOCH_3$

cis and trans

Tetrahedron 18:177 (1962)

► $O_3C_{12}H_{22}$

Varian 603

► $O_3C_{13}H_8$

Tetrahedron 18:841 (1962)

► $O_3C_{13}H_8$

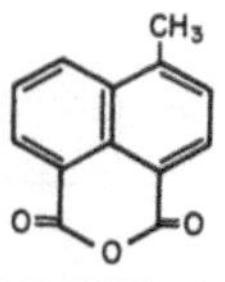

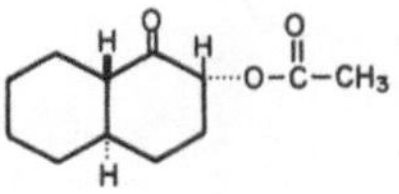

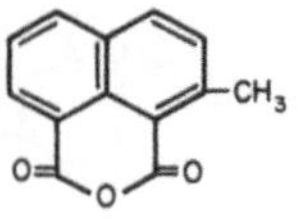

Tetrahedron 18:841 (1962)

▶ $O_3C_{13}H_{10}$

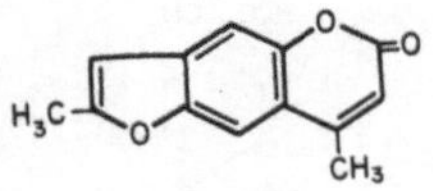


J. Org. Chem. 27:2569 (1962)

▶ $O_3C_{13}H_{20}$

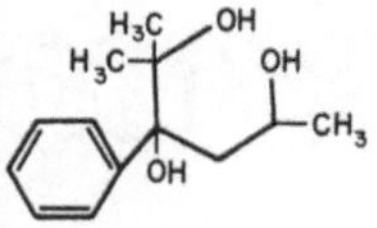


J. Chem. Soc. 4743 (1962)

▶ $O_3C_{13}H_{10}$

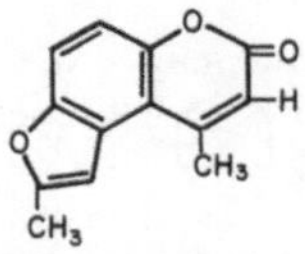


J. Org. Chem. 27:2569 (1962)

▶ $O_3C_{13}H_{20}$

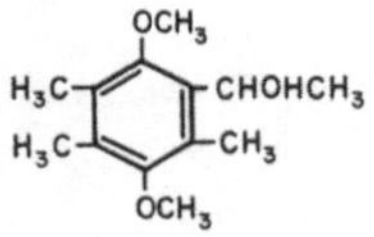


J. Org. Chem. 26:3225 (1961)

▶ $O_3C_{13}H_{10}$

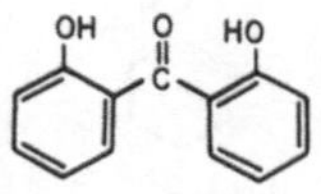


J. Phys. Chem. 65:2023 (1961)

▶ $O_3C_{13}H_{20}$

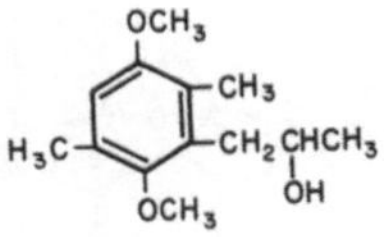


J. Org. Chem. 27:842 (1962)

▶ $O_3C_{13}H_{14}$

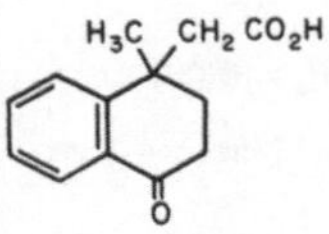


J. Am. Chem. Soc. 84:3517 (1962)

▶ $O_3C_{13}H_{20}$

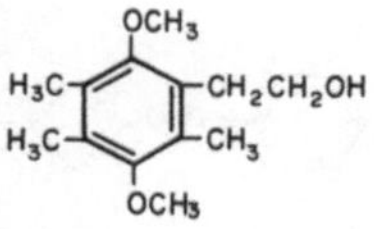


J. Org. Chem. 26:3225 (1961)

▶ $O_3C_{13}H_{16}$

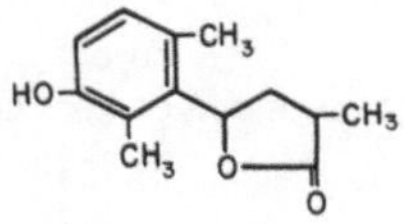


J. Org. Chem. 26:2615 (1961)

▶ $O_3C_{14}H_{12}$

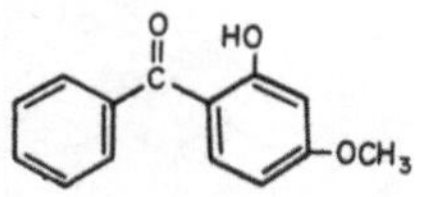


J. Phys. Chem. 65:2023 (1961)

▶ $O_3C_{13}H_{18}$

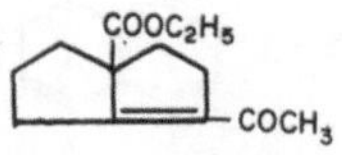


Tetrahedron 18:55 (1962)

▶ $O_3C_{13}H_{18}$

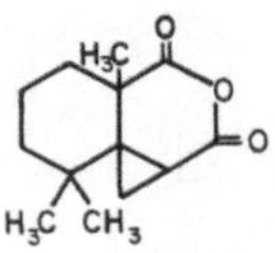


Acta Chem. Scand. 15:592 (1961)

▶ $O_3C_{14}H_{12}$

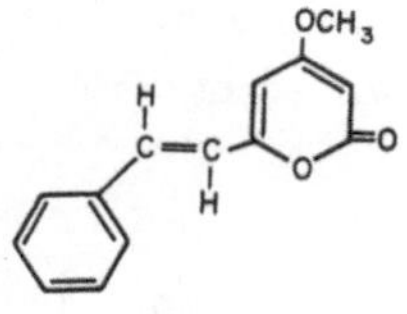


J. Org. Chem. 27:3715 (1962)

▶ $O_3C_{14}H_{14}$

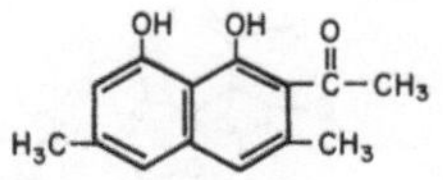

J. Chem. Soc. 3753 (1962)

▶ $O_3C_{14}H_{14}$

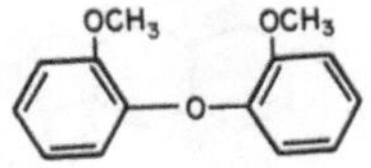

J. Chem. Phys. 37:1565 (1962)

▶ $O_3C_{14}H_{14}$

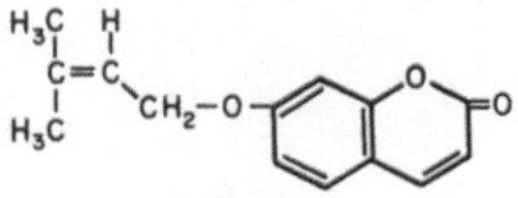

Chem. & Ind. 1020 (1962)

▶ $O_3C_{14}H_{14}$

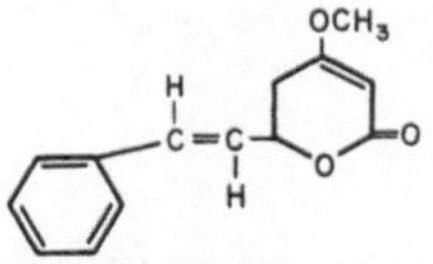

J. Org. Chem. 27:3715 (1962)

▶ $O_3C_{14}H_{16}$

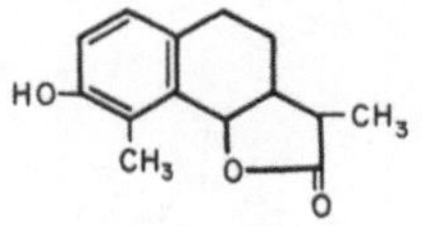

J. Org. Chem. 26:2614 (1961)

▶ $O_3C_{14}H_{18}$

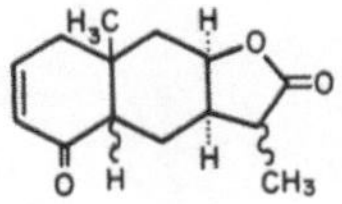

J. Org. Chem. 27:906 (1962)

▶ $O_3C_{14}H_{20}$

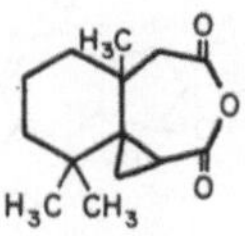

Acta Chem. Scand. 15:592 (1961)

▶ $O_3C_{14}H_{22}$

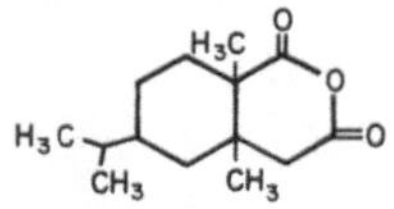

Tetrahedron 12:107 (1961)

▶ $O_3C_{15}H_{12}$

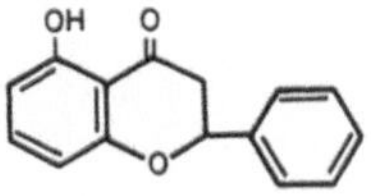

Bull. Soc. Chim. France 1962 (1962)

▶ $O_3C_{15}H_{14}$

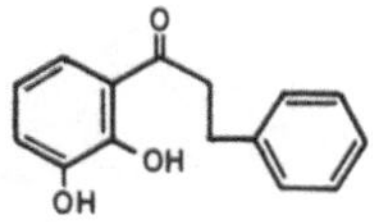

Bull. Soc. Chim. France 1962 (1962)

▶ $O_3C_{15}H_{16}$

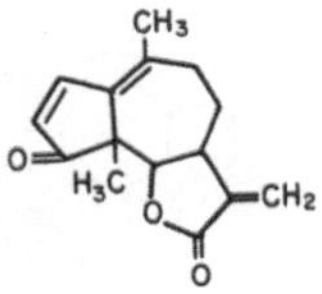

J. Am. Chem. Soc. 84:2601 (1962)
Tetrahedron Letters No. 2:86 (1961)

▶ $O_3C_{15}H_{16}$

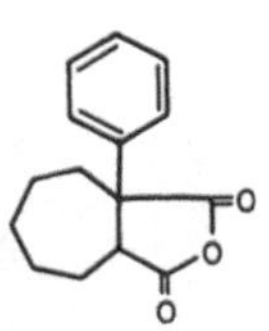

Helv. Chim. Acta 45:600 (1962)

▶ $O_3C_{15}H_{16}$

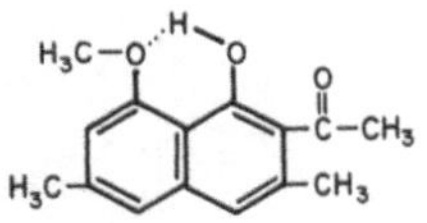

J. Chem. Soc. 3755 (1962)

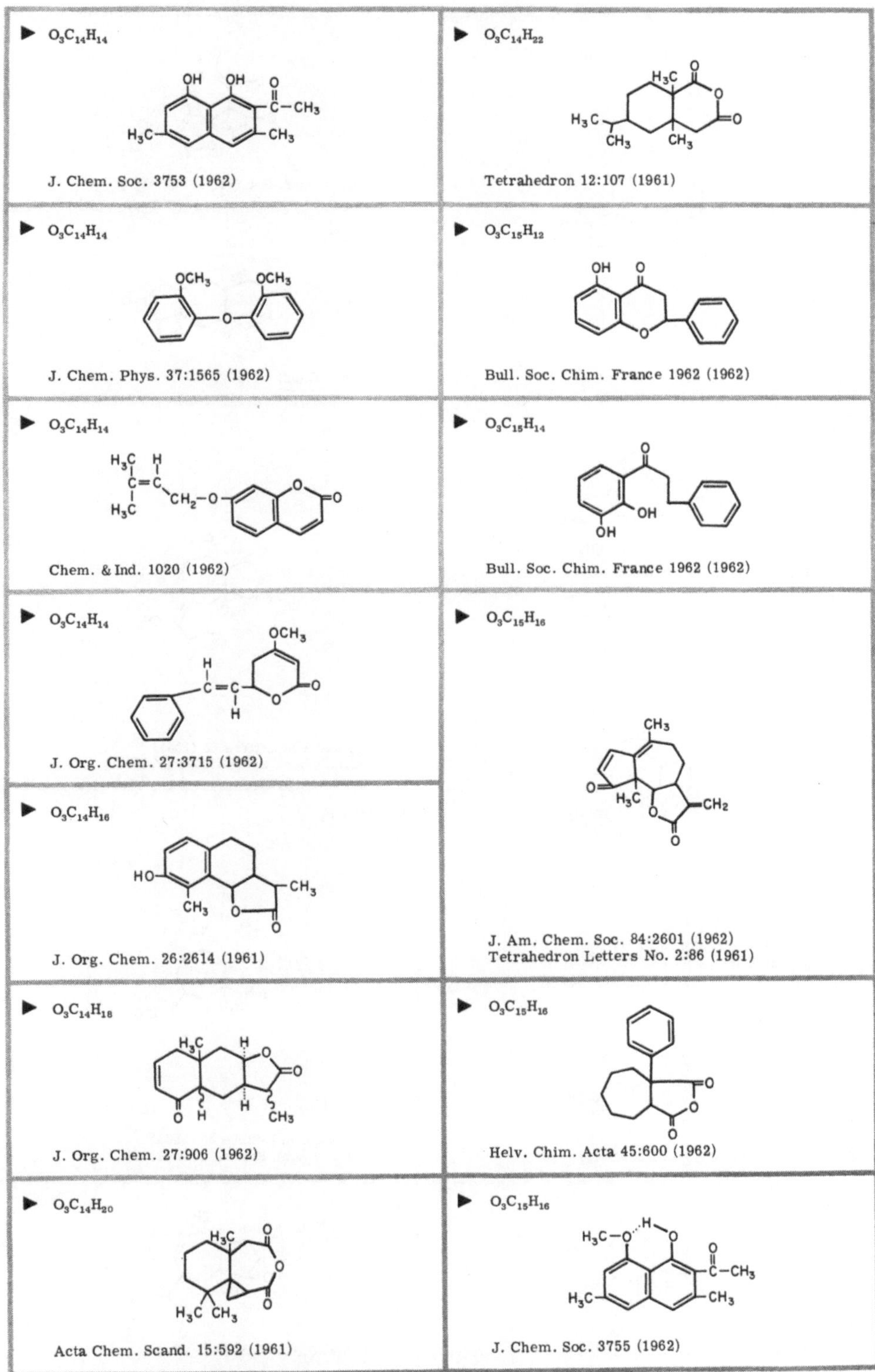

► $O_3C_{15}H_{18}$

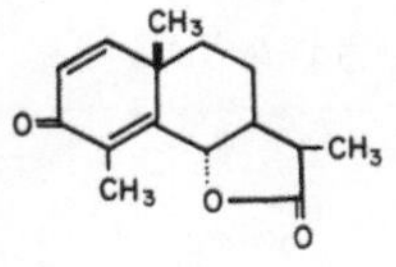

J. Chem. Soc. 1680 (1962)

► $O_3C_{15}H_{18}$

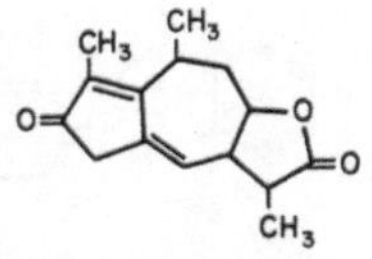

J. Am. Chem. Soc. 84:3857 (1962)

► $O_3C_{15}H_{18}$

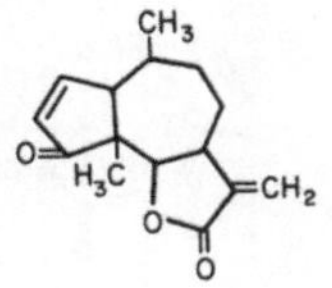

Tetrahedron Letters 86 (1961)
J. Am. Chem. Soc. 84:2601 (1962)

► $O_3C_{15}H_{18}$

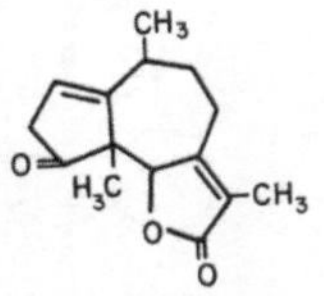

J. Am. Chem. Soc. 84:2601 (1962)

► $O_3C_{15}H_{18}$

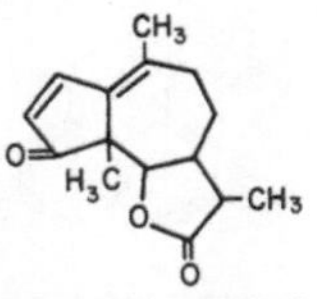

J. Am. Chem. Soc. 84:2601 (1962)

► $O_3C_{15}H_{18}$

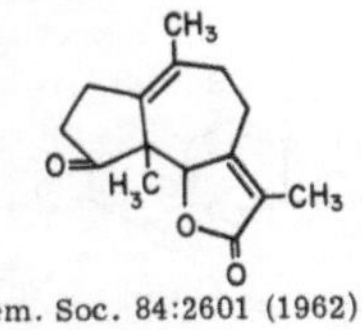

J. Am. Chem. Soc. 84:2601 (1962)

► $O_3C_{15}H_{18}$

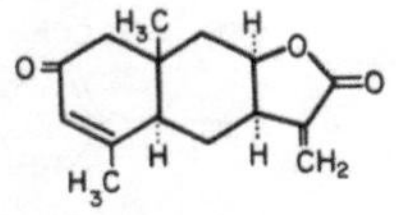

J. Org. Chem. 27:4041 (1962)

► $O_3C_{15}H_{20}$

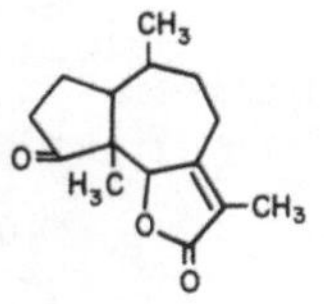

Tetrahedron Letters 86 (1961)
J. Am. Chem. Soc. 84:2601 (1962)

► $O_3C_{15}H_{20}$

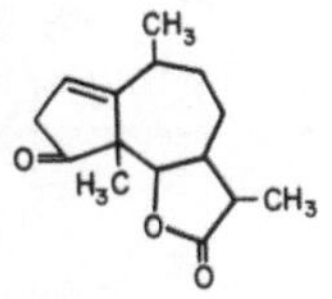

Tetrahedron Letters 86 (1961)
J. Am. Chem. Soc. 84:2601 (1962)

► $O_3C_{15}H_{20}$

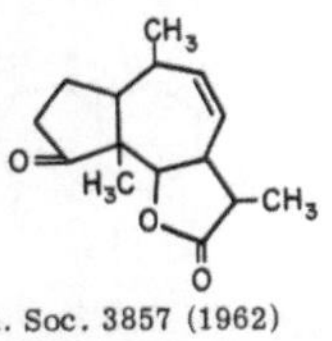

J. Am. Chem. Soc. 3857 (1962)

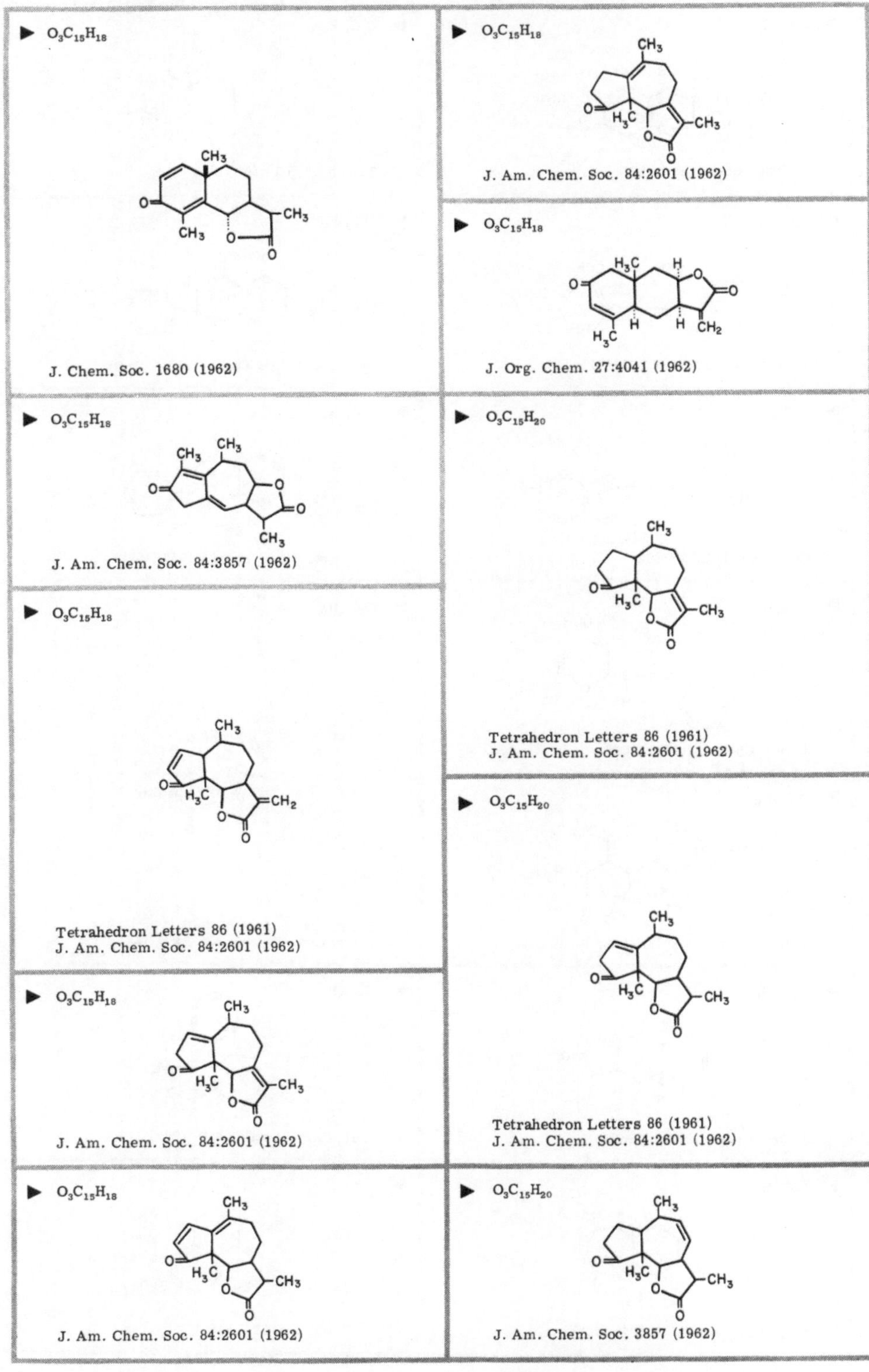

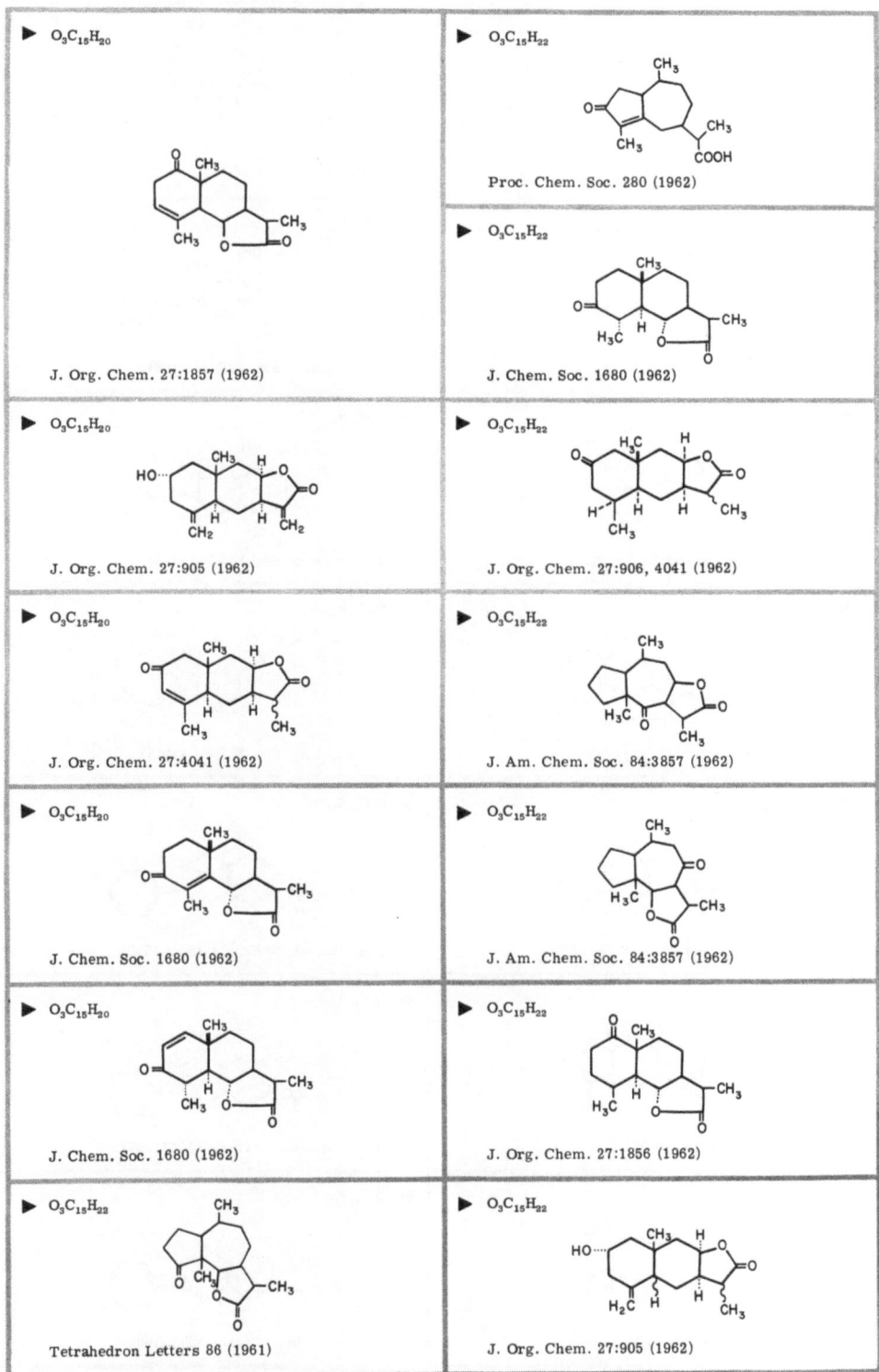

► $O_3C_{15}H_{20}$

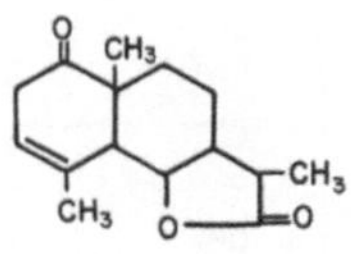

J. Org. Chem. 27:1857 (1962)

► $O_3C_{15}H_{22}$

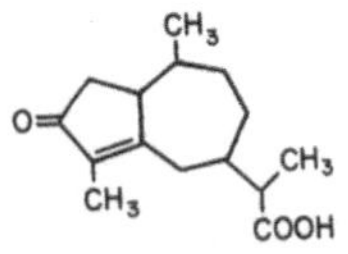

Proc. Chem. Soc. 280 (1962)

► $O_3C_{15}H_{22}$

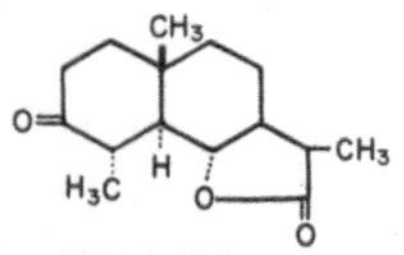

J. Chem. Soc. 1680 (1962)

► $O_3C_{15}H_{20}$

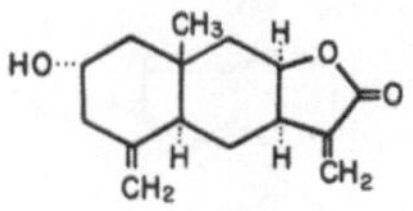

J. Org. Chem. 27:905 (1962)

► $O_3C_{15}H_{22}$

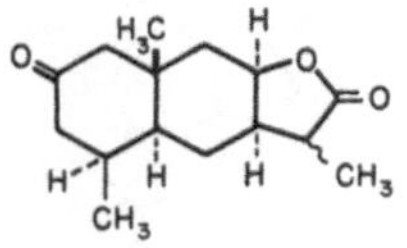

J. Org. Chem. 27:906, 4041 (1962)

► $O_3C_{15}H_{20}$

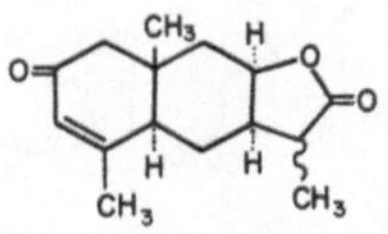

J. Org. Chem. 27:4041 (1962)

► $O_3C_{15}H_{22}$

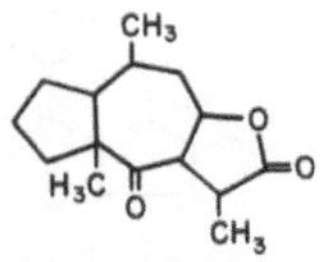

J. Am. Chem. Soc. 84:3857 (1962)

► $O_3C_{15}H_{20}$

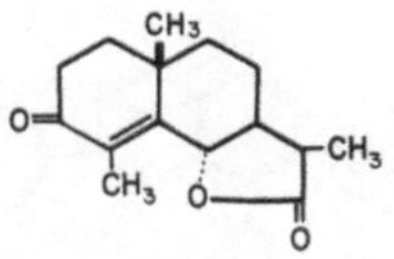

J. Chem. Soc. 1680 (1962)

► $O_3C_{15}H_{22}$

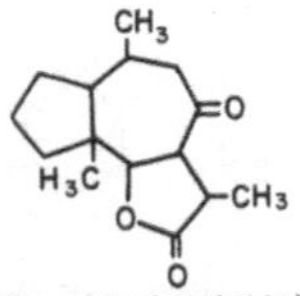

J. Am. Chem. Soc. 84:3857 (1962)

► $O_3C_{15}H_{20}$

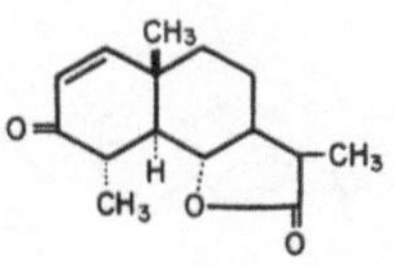

J. Chem. Soc. 1680 (1962)

► $O_3C_{15}H_{22}$

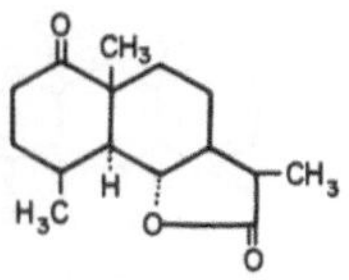

J. Org. Chem. 27:1856 (1962)

► $O_3C_{15}H_{22}$

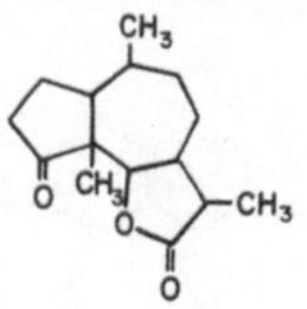

Tetrahedron Letters 86 (1961)

► $O_3C_{15}H_{22}$

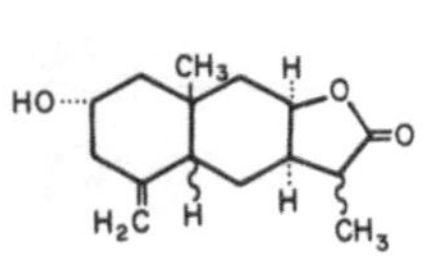

J. Org. Chem. 27:905 (1962)

▶ $O_3C_{15}H_{22}$

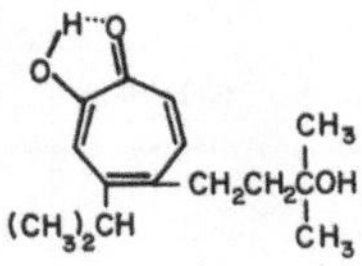


Varian 647

▶ $O_3C_{15}H_{26}$

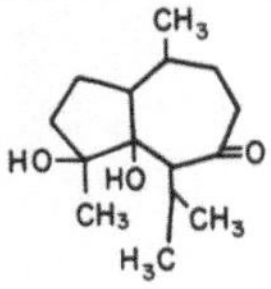


J. Chem. Soc. 1963 (1962)

▶ $O_3C_{15}H_{26}$

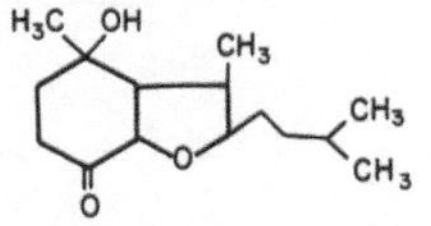


J. Am. Chem. Soc. 83:3096 (1961)

▶ $O_3C_{15}H_{24}$

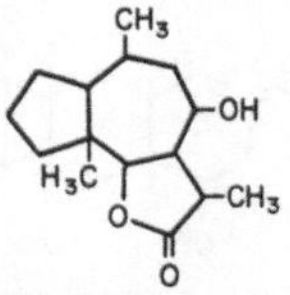


J. Am. Chem. Soc. 84:3857 (1962)

▶ $O_3C_{15}H_{28}$

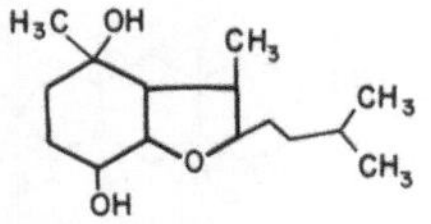


J. Am. Chem. Soc. 83:3096 (1961)

▶ $O_3C_{15}H_{24}$

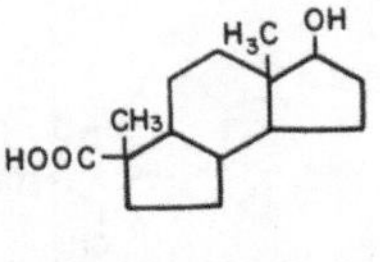


Tetrahedron 18:1015 (1962)

▶ $O_3C_{15}H_{28}$

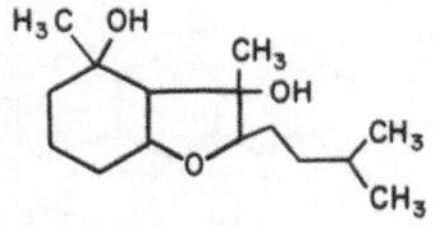


J. Am. Chem. Soc. 83:3096 (1961)

▶ $O_3C_{15}H_{24}$

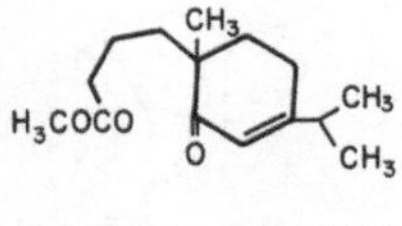


Tetrahedron Letters No. 5:172 (1962)

▶ $O_3C_{16}H_{12}$

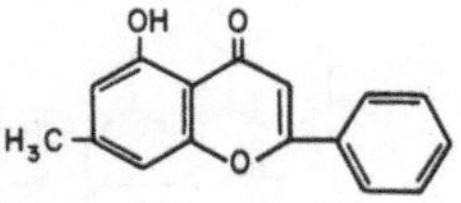


Bull. Soc. Chim. France 1962 (1962)

▶ $O_3C_{15}H_{24}$

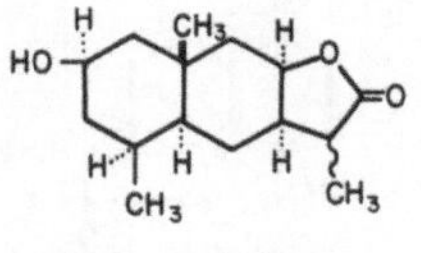


J. Org. Chem. 27:907 (1962)

▶ $O_3C_{16}H_{12}$

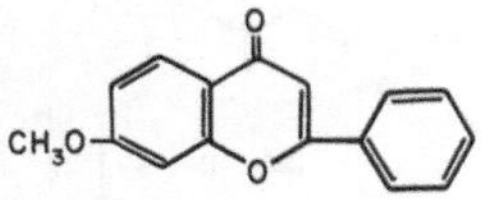


Bull. Soc. Chim. France 1962 (1962)

▶ $O_3C_{15}H_{24}$

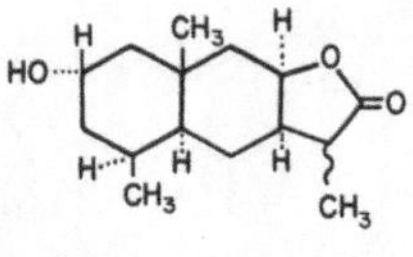


J. Org. Chem. 27:905 (1962)

▶ $O_3C_{16}H_{12}$

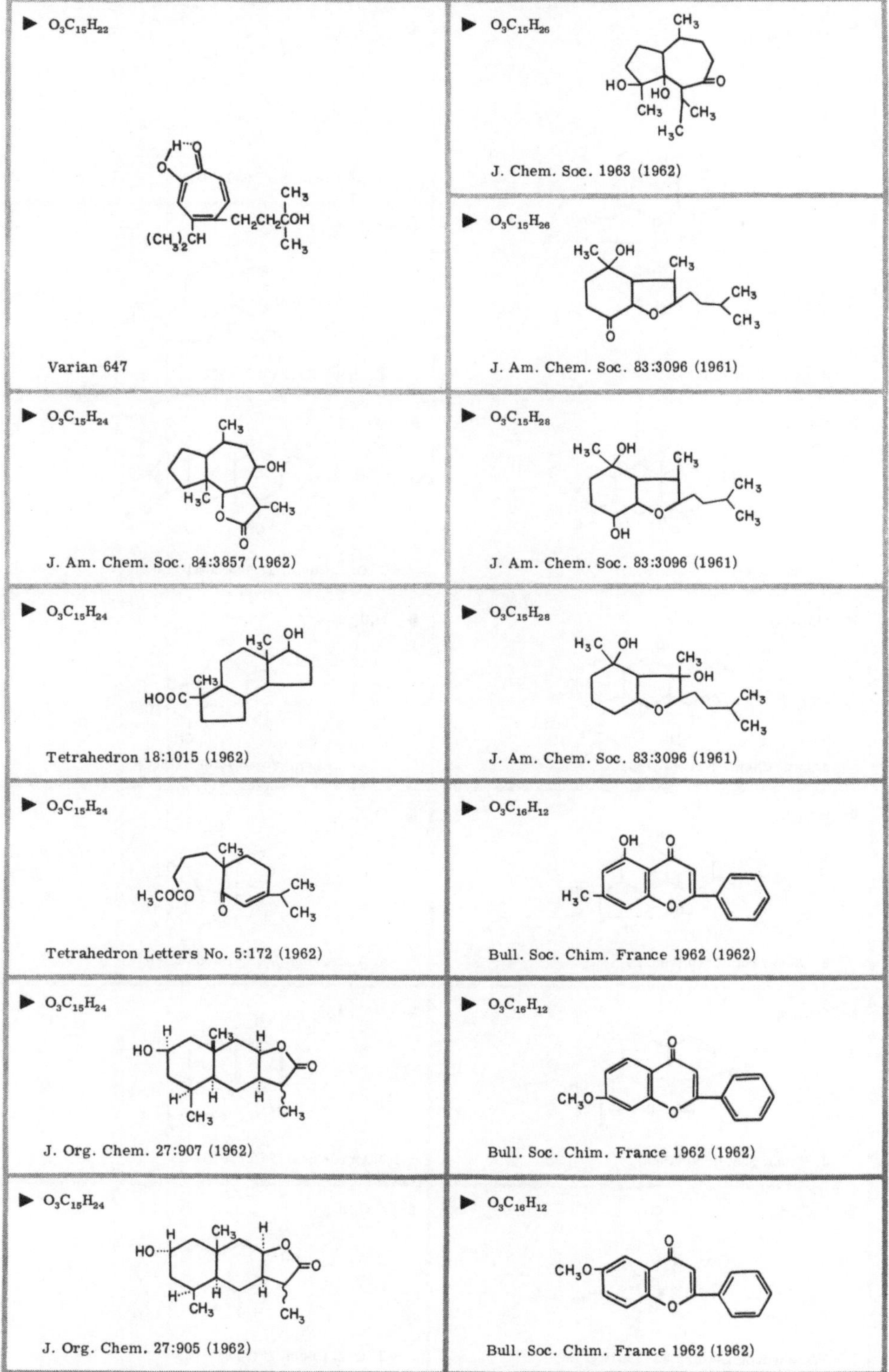

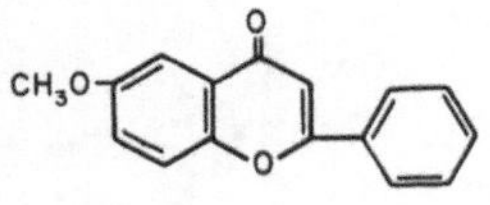


Bull. Soc. Chim. France 1962 (1962)

▶ $O_3C_{16}H_{14}$

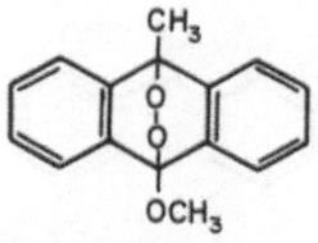


Tetrahedron 18:388 (1962)

▶ $O_3C_{16}H_{14}$

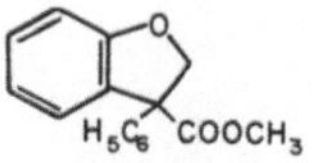


J. Org. Chem. 26:4826 (1961)

▶ $O_3C_{16}H_{14}$

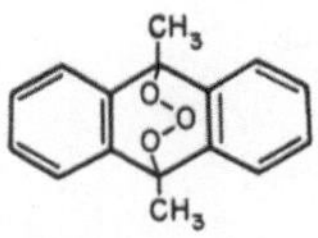


Tetrahedron 18:388 (1962)

▶ $O_3C_{17}H_{14}$

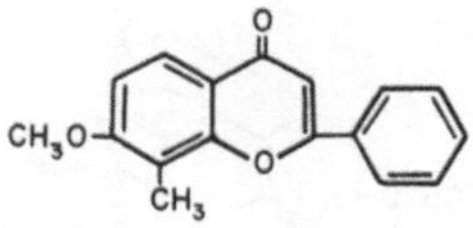


Bull. Soc. Chim. France 1962 (1962)

▶ $O_3C_{17}H_{16}$

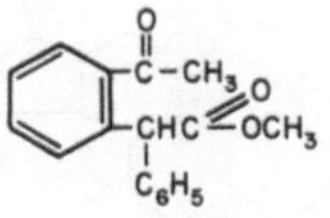


J. Org. Chem. 26:681 (1961)

▶ $O_3C_{17}H_{16}$

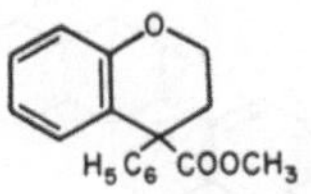


J. Org. Chem. 26:4826 (1961)

▶ $O_3C_{17}H_{18}$

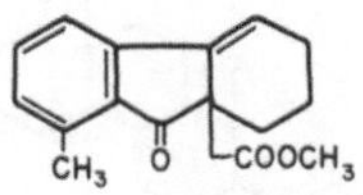


J. Am. Chem. Soc. 84:2614 (1962)

▶ $O_3C_{17}H_{18}$

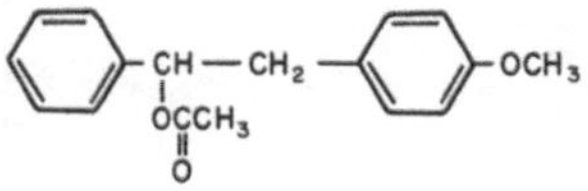


J. Am. Chem. Soc. 83:3647 (1961)

▶ $O_3C_{17}H_{18}$

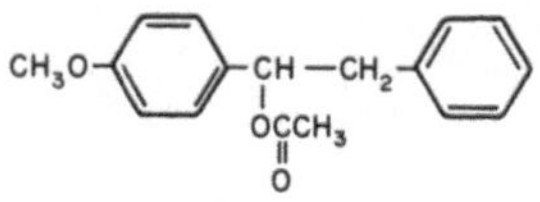


J. Am. Chem. Soc. 83:3647 (1961)

▶ $O_3C_{17}H_{28}$

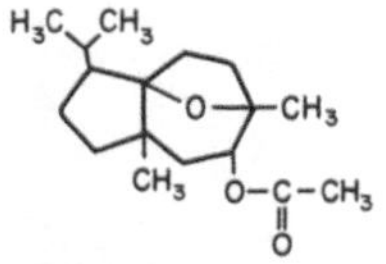


J. Org. Chem. 26:981 (1961)

▶ $O_3C_{18}H_{18}$

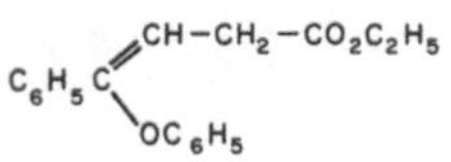


J. Am. Chem. Soc. 84:4149 (1962)

▶ $O_3C_{18}H_{18}$

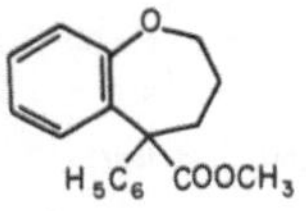


J. Org. Chem. 26:4826 (1961)

▶ $O_3C_{18}H_{22}$

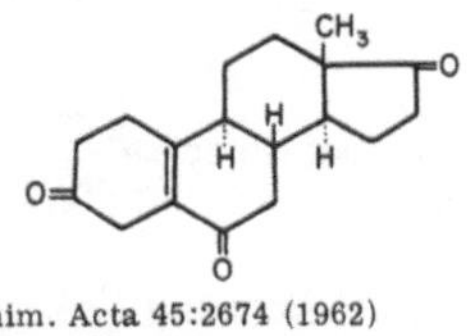


Helv. Chim. Acta 45:2674 (1962)

▶ $O_3C_{18}H_{22}$

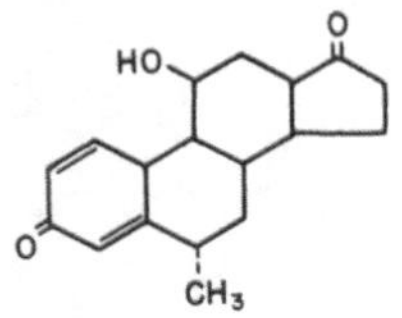

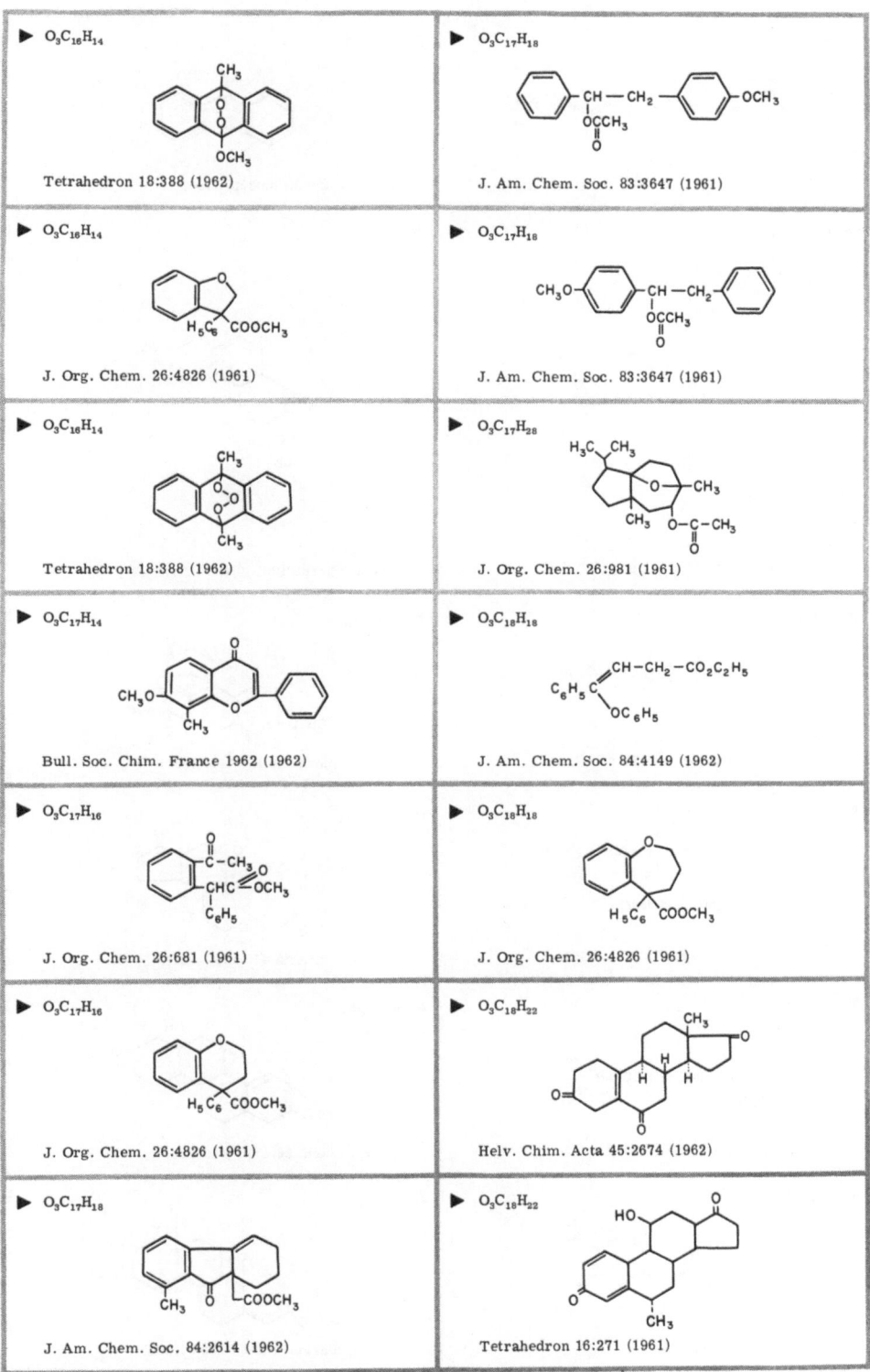

Tetrahedron 16:271 (1961)

▶ $O_3C_{18}H_{22}$

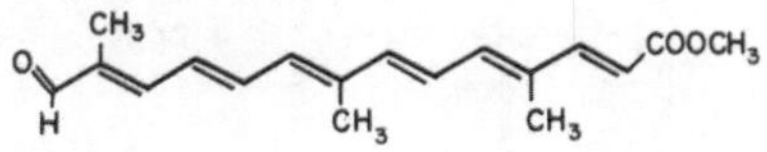

J. Chem. Soc. 1625 (1961)

▶ $O_3C_{18}H_{28}$

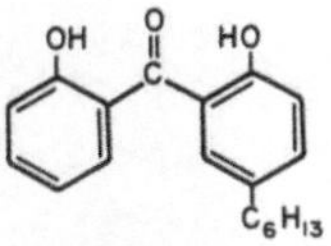

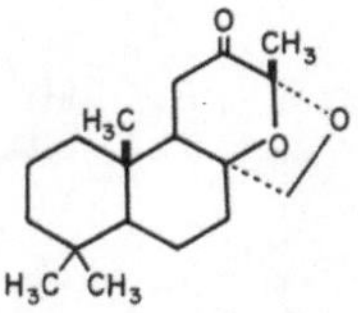

Helv. Chim. Acta 45:400 (1962)

▶ $O_3C_{18}H_{36}$

$CH_3(CH_2)_5CH(OH){-}(CH_2)_{10}CO_2H$

J. Am. Oil Chemists' Soc. 38:664 (1961)

▶ $O_3C_{18}H_{38}$

$CH_3(CH_2)_7CH(OH){-}CH(OH){-}(CH_2)_7CH_2OH$

erythro and threo

Bull. Chem. Soc. Japan 35:1742 (1962)

▶ $O_3C_{19}H_{20}$

$(C_6H_5)(C_6H_5O)C{=}CHCH_2CO_2CH(CH_3)_2$

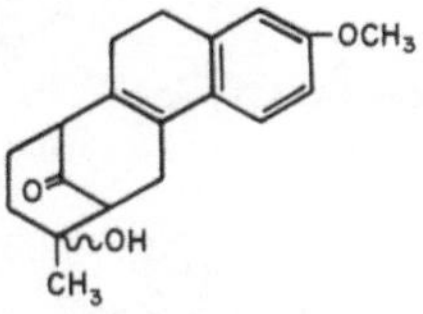

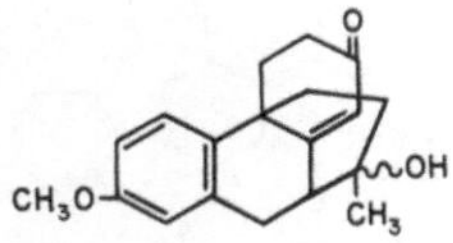

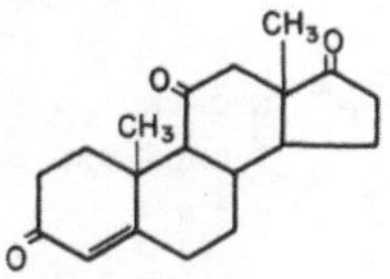

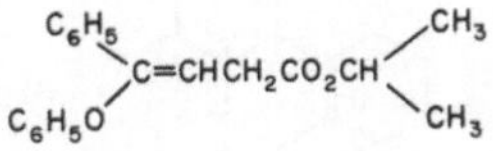

J. Am. Chem. Soc. 84:4148 (1962)

▶ $O_3C_{19}H_{22}$

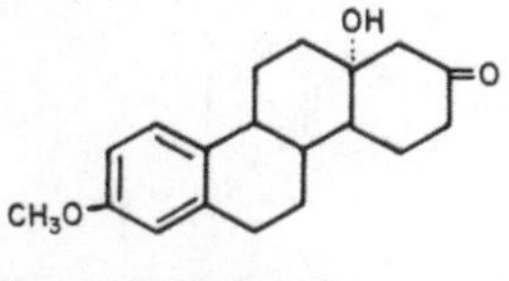

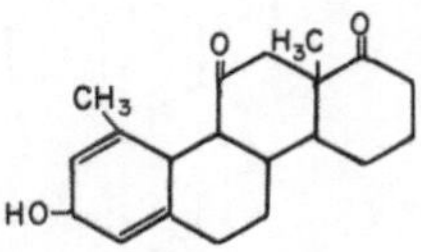

Chem. & Ind. 1716 (1962)

▶ $O_3C_{19}H_{22}$

J. Phys. Chem. 65:2023 (1961)

▶ $O_3C_{19}H_{22}$

Chem. Pharm. Bull. 9:756 (1961)

▶ $O_3C_{19}H_{22}$

Chem. Pharm. Bull. 9:756 (1961)

▶ $O_3C_{19}H_{24}$

Helv. Chim. Acta 44:1380 (1961)

▶ $O_3C_{19}H_{24}$

J. Org. Chem. 26:4588 (1961)

▶ $O_3C_{19}H_{24}$

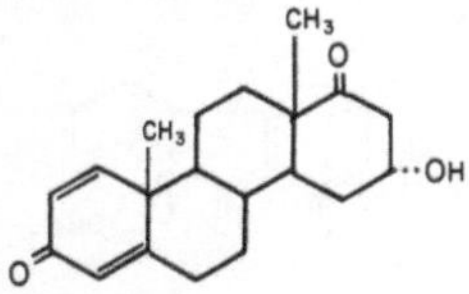

Chem. Pharm. Bull. 10:338 (1962)

► $O_3C_{19}H_{24}$

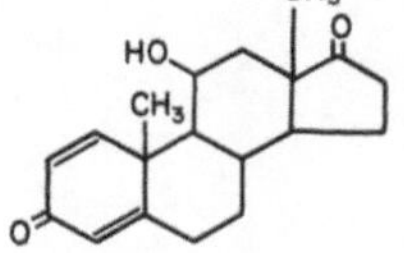

Chem. Pharm. Bull. 10:338 (1962)

► $O_3C_{19}H_{24}$

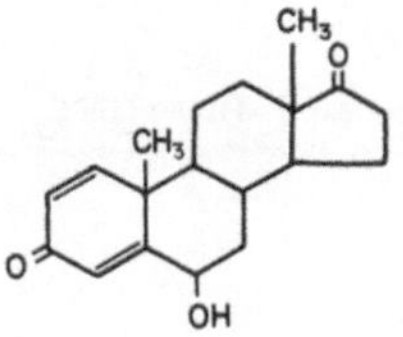

J. Org. Chem. 27:3161 (1962)

► $O_3C_{19}H_{26}$

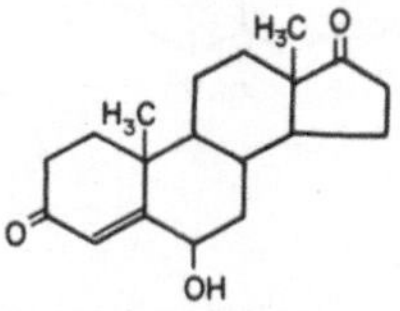

Chem. Pharm. Bull. 10:338 (1962)

► $O_3C_{19}H_{26}$

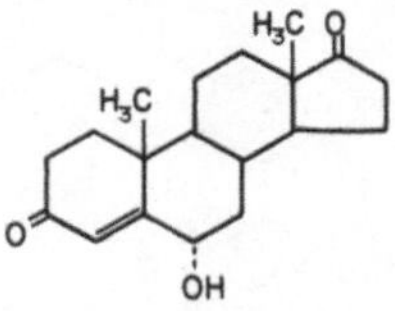

Chem. Pharm. Bull. 10:338 (1962)

► $O_3C_{19}H_{26}$

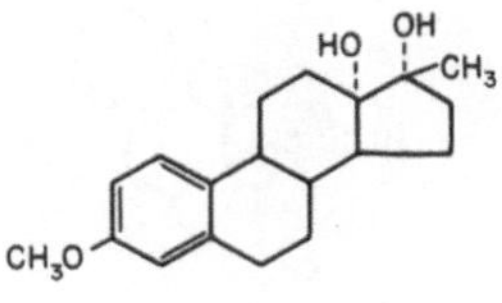

J. Org. Chem. 26:4587 (1961)

► $O_3C_{19}H_{26}$

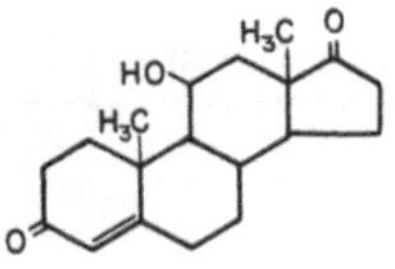

Chem. Pharm. Bull. 10:338 (1962)
Helv. Chim. Acta 44:1380 (1961)

► $O_3C_{19}H_{26}$

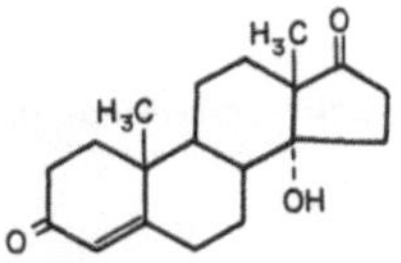

Chem. Pharm. Bull. 10:338 (1962)

► $O_3C_{19}H_{26}$

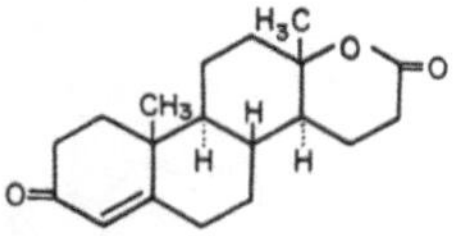

Helv. Chim. Acta 45:2674 (1962)

► $O_3C_{19}H_{26}$

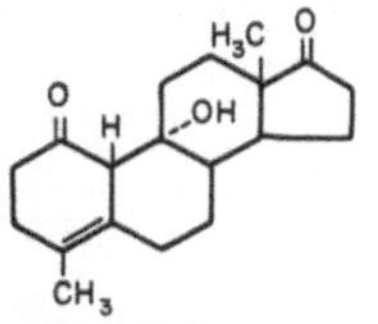

Chem. & Ind. 1530 (1961)

► $O_3C_{19}H_{26}$

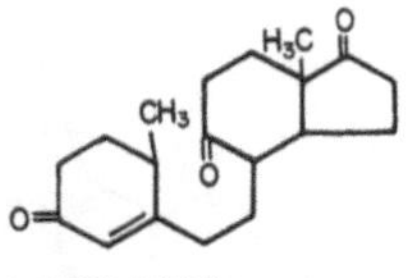

Chem. & Ind. 1530 (1961)

► $O_3C_{19}H_{28}$

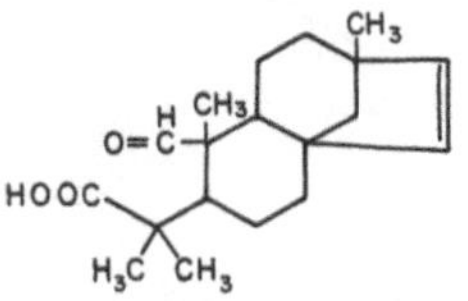

J. Chem. Soc. 4050 (1962)

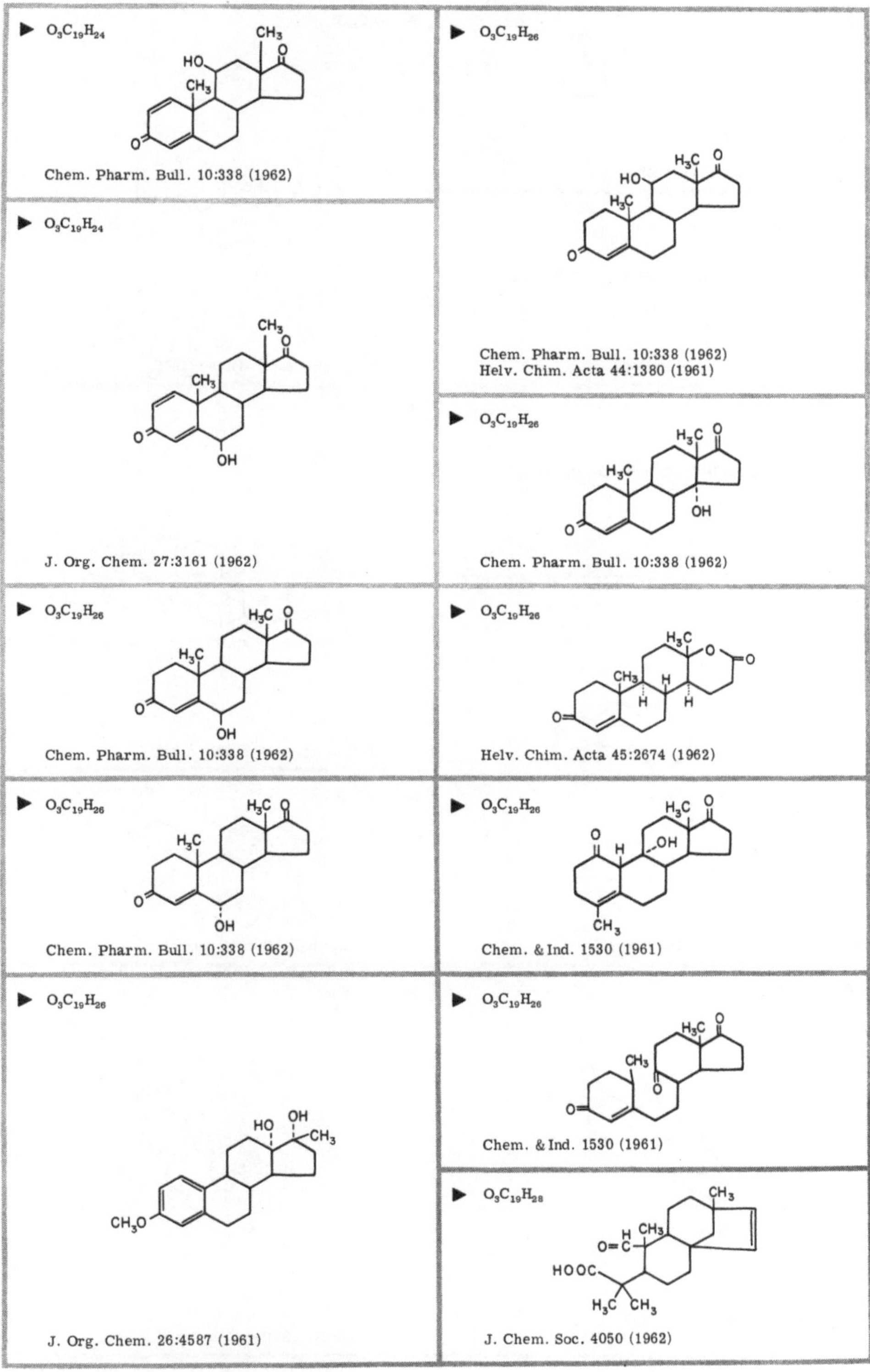

▶ $O_3C_{19}H_{30}$

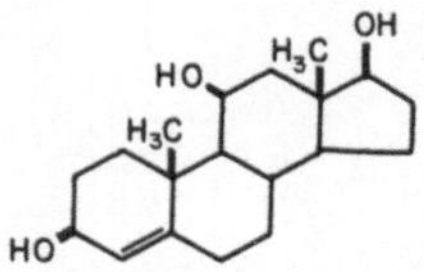

J. Chem. Soc. 1711 (1962)

▶ $O_3C_{19}H_{34}$

$CH_3\,CH_2\,CH=CH(CH_2)_2\,CH(OH)CH_2\,CH=CH(CH_2)_7\,COOCH_3$

J. Org. Chem. 27:3113 (1962)

▶ $O_3C_{19}H_{36}$

$CH_3(CH_2)_5CH(OH)CH_2CH=CH(CH_2)_7COOCH_3$

J. Org. Chem. 27:3116 (1962)

▶ $O_3C_{20}H_{18}$

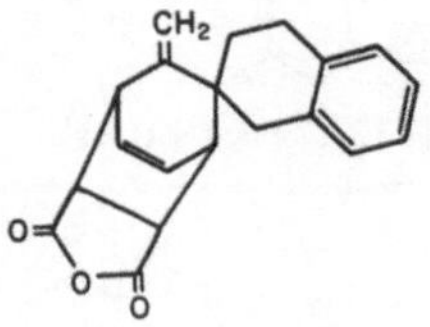

J. Org. Chem. 26:2607 (1961)

▶ $O_3C_{20}H_{24}$

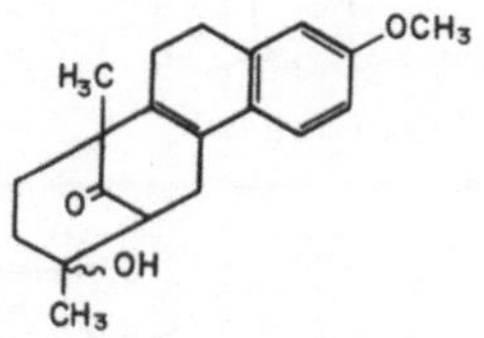

Can. J. Chem. 40:984 (1962)
Chem. Pharm. Bull. 9:756 (1961)

▶ $O_3C_{20}H_{28}$

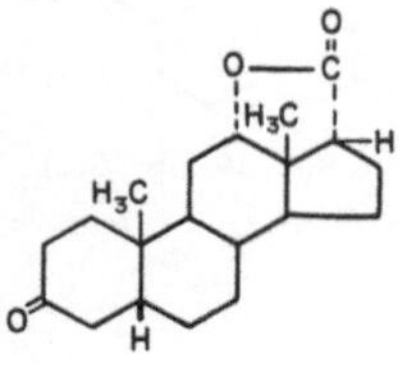

Helv. Chim. Acta 44:1380 (1961)

▶ $O_3C_{20}H_{28}$

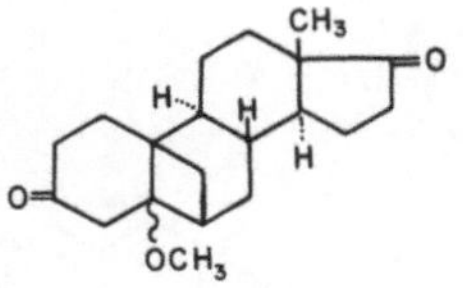

Helv. Chim. Acta 45:2615 (1962)

▶ $O_3C_{20}H_{28}$

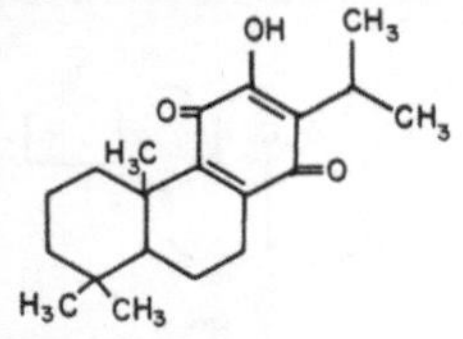

Can. J. Chem. 40:1541 (1962)

▶ $O_3C_{20}H_{28}$

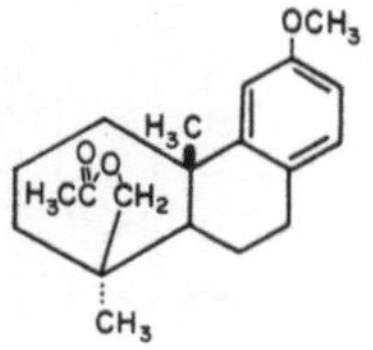

Tetrahedron Letters 360 (1961)

▶ $O_3C_{20}H_{28}$

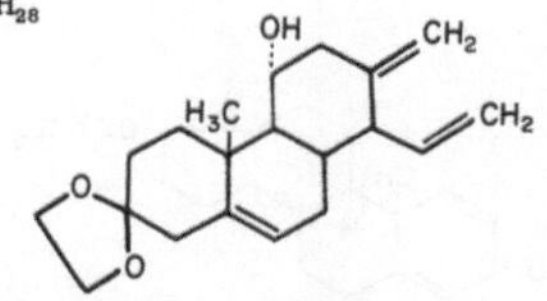

Helv. Chim. Acta 44:1927 (1961)

▶ $O_3C_{20}H_{32}$

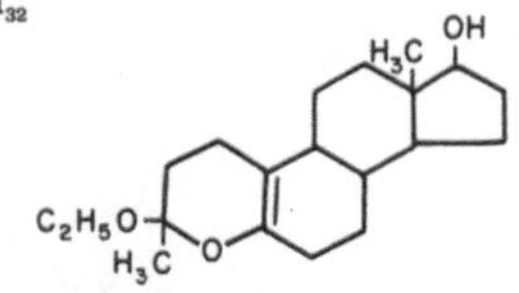

Bull. Soc. Chim. France 1679 (1961)

▶ $O_3C_{20}H_{30}$

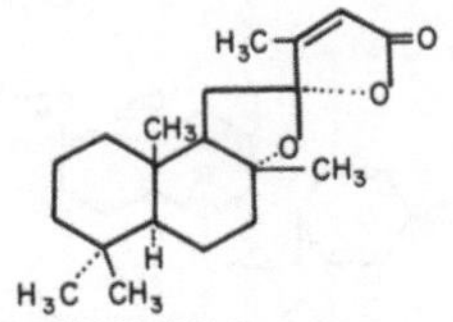

Tetrahedron 14:246 (1961)

▶ $O_3C_{20}H_{32}$

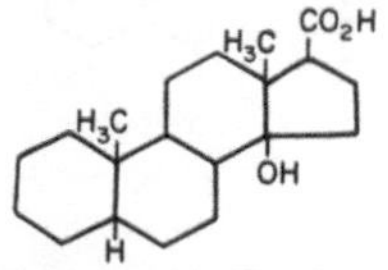

Helv. Chim. Acta 44:1380 (1961)

▶ $O_3C_{20}H_{30}$

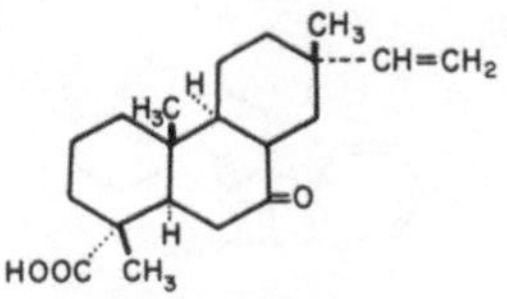

J. Org. Chem. 27:1931 (1962)

▶ $O_3C_{20}H_{34}$

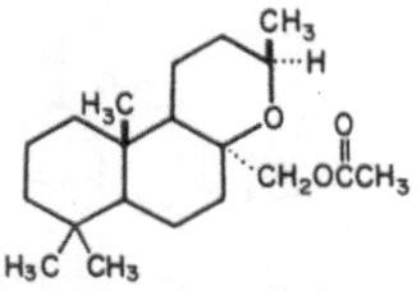

Helv. Chim. Acta 45:400 (1962)

▶ $O_3C_{20}H_{30}$

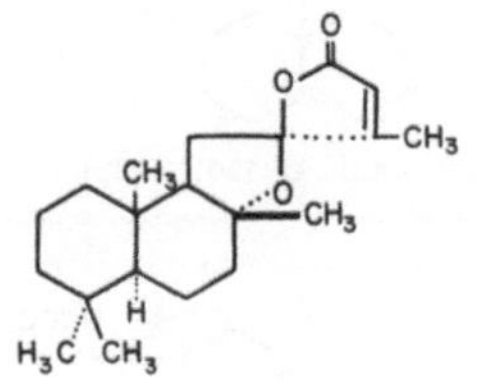

Tetrahedron 14:246 (1961)

▶ $O_3C_{20}H_{34}$

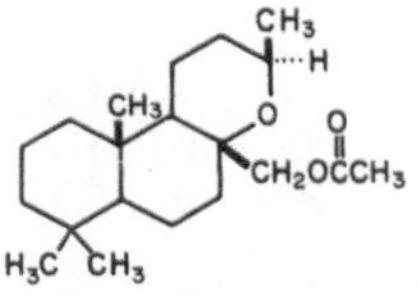

Helv. Chim. Acta 45:400 (1962)

▶ $O_3C_{20}H_{34}$

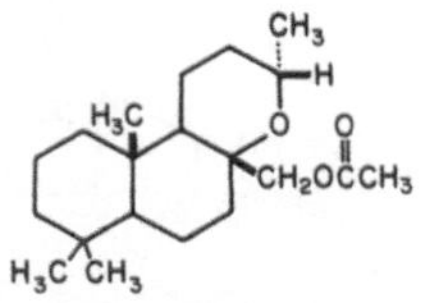

Helv. Chim. Acta 45:400 (1962)

▶ $O_3C_{20}H_{32}$

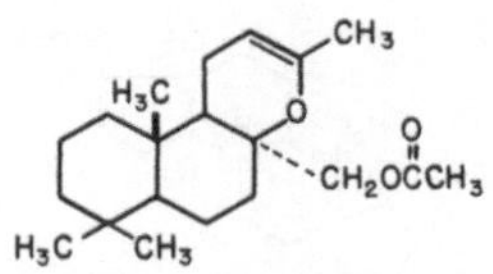

Helv. Chim. Acta 45:400 (1962)

▶ $O_3C_{21}H_{24}$

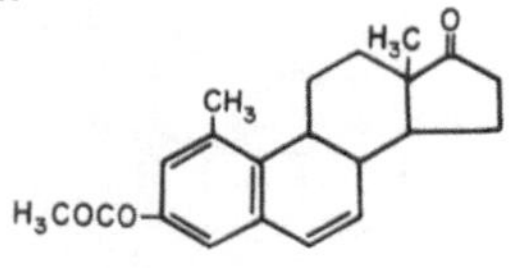

Chem. & Ind. 1716 (1962)

▶ $O_3C_{20}H_{32}$

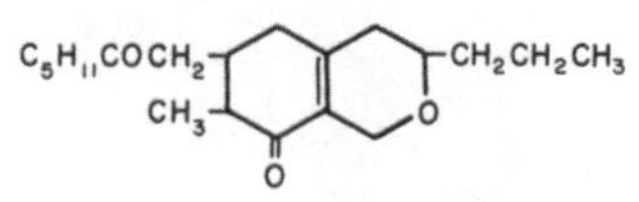

J. Chem. Soc. 4579 (1961)

▶ $O_3C_{21}H_{26}$

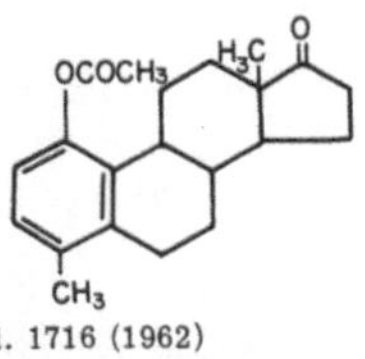

Chem. & Ind. 1716 (1962)

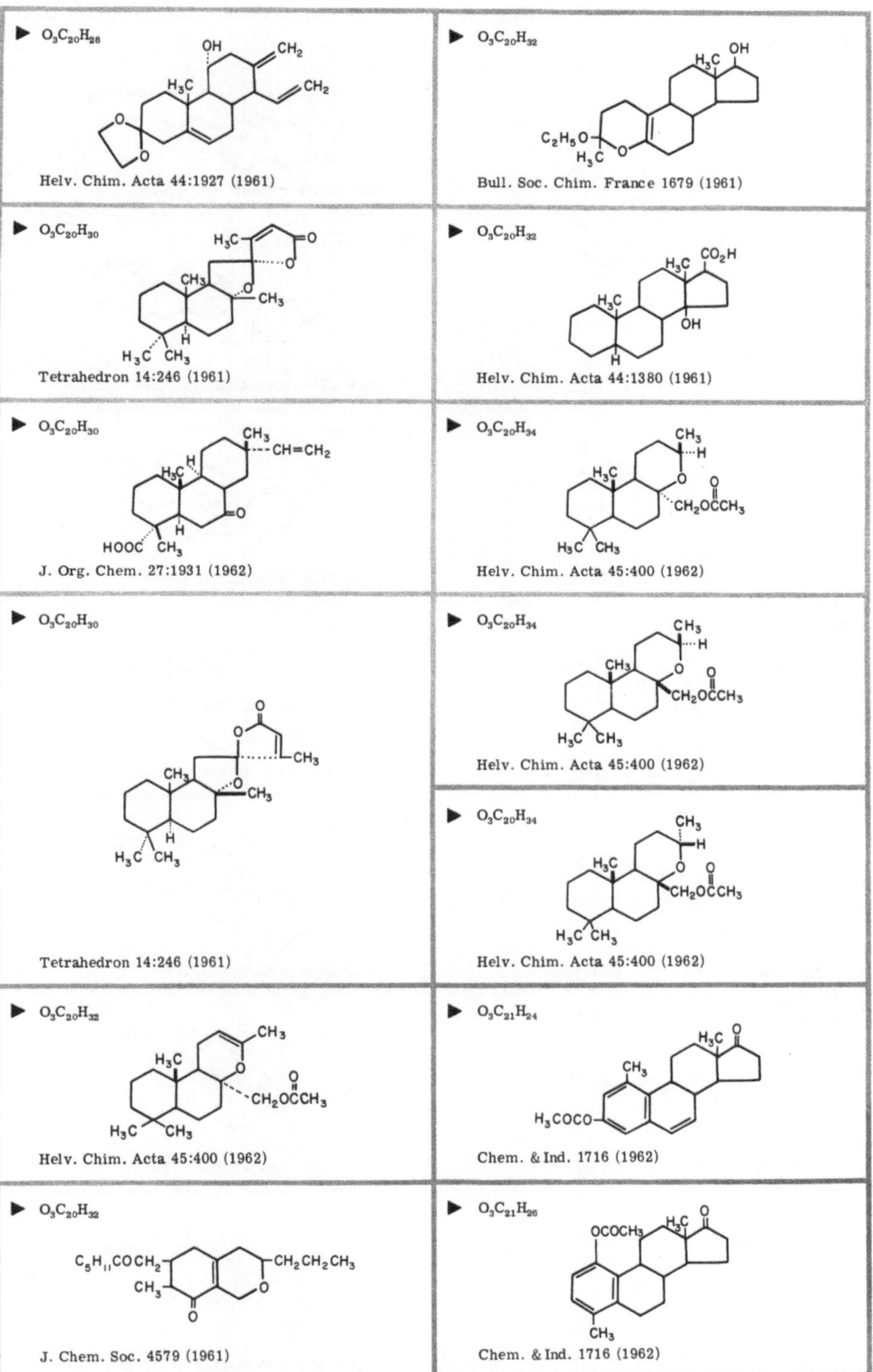

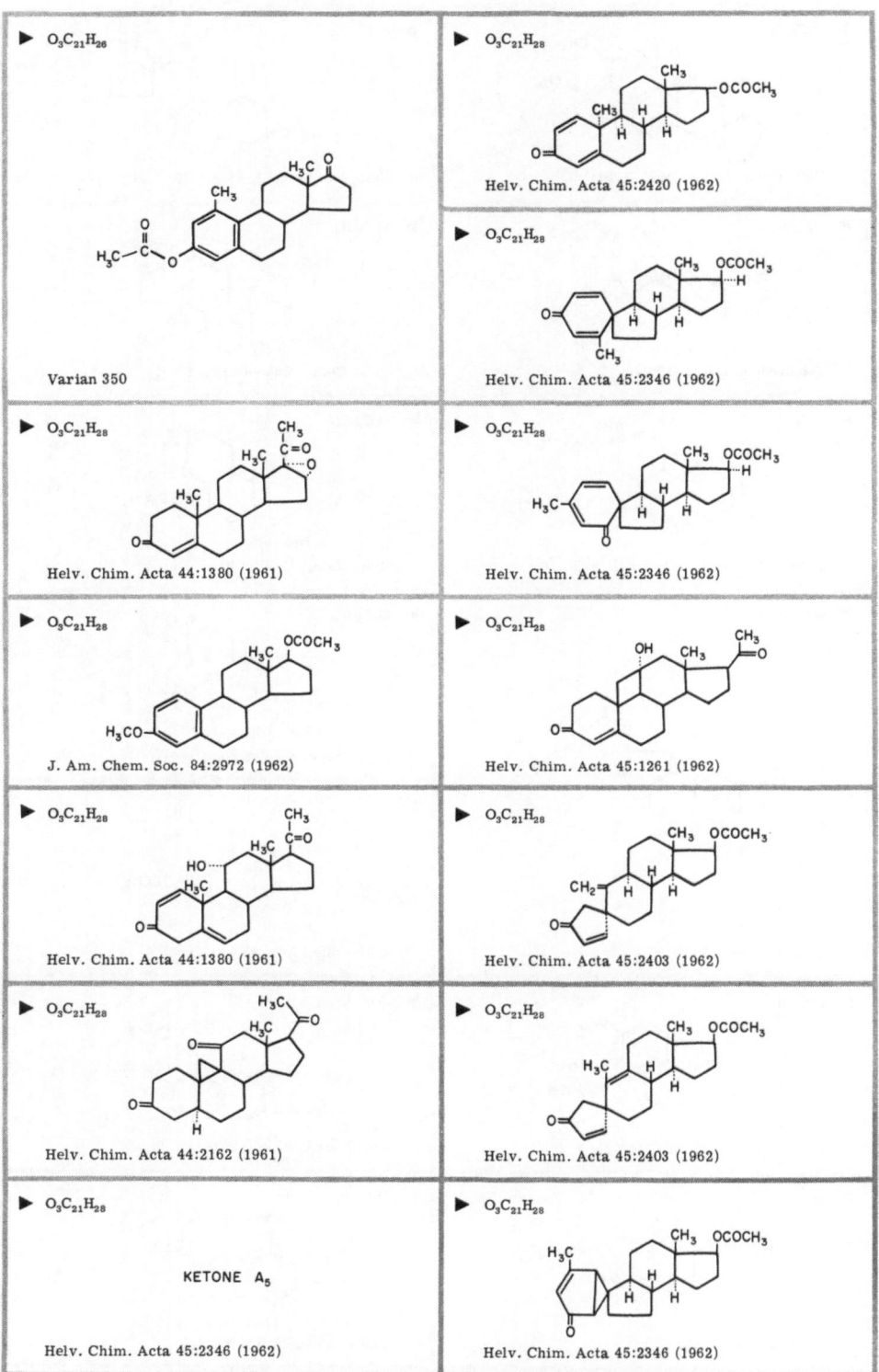

▶ $O_3C_{21}H_{26}$

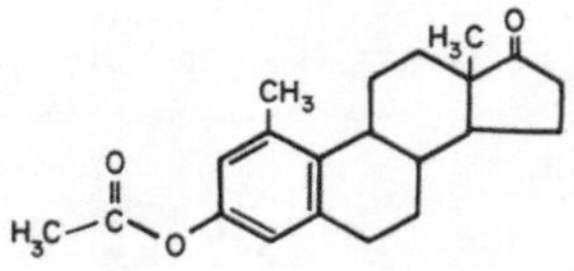

Varian 350

▶ $O_3C_{21}H_{28}$

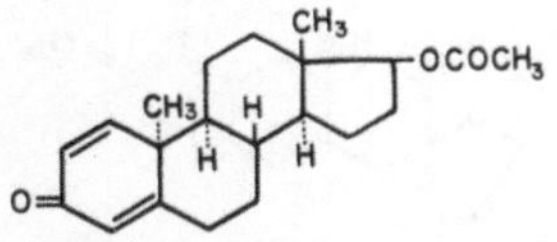

Helv. Chim. Acta 45:2420 (1962)

▶ $O_3C_{21}H_{28}$

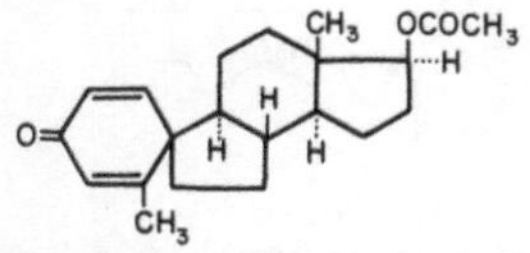

Helv. Chim. Acta 45:2346 (1962)

▶ $O_3C_{21}H_{28}$

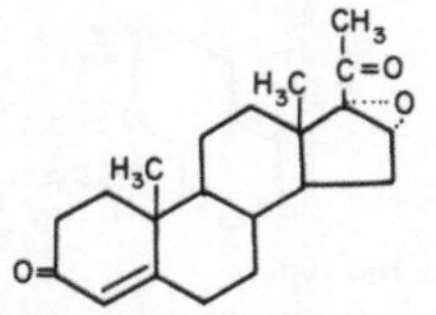

Helv. Chim. Acta 44:1380 (1961)

▶ $O_3C_{21}H_{28}$

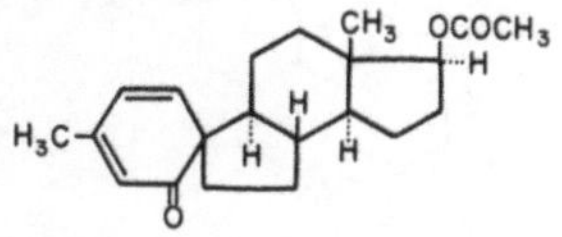

Helv. Chim. Acta 45:2346 (1962)

▶ $O_3C_{21}H_{28}$

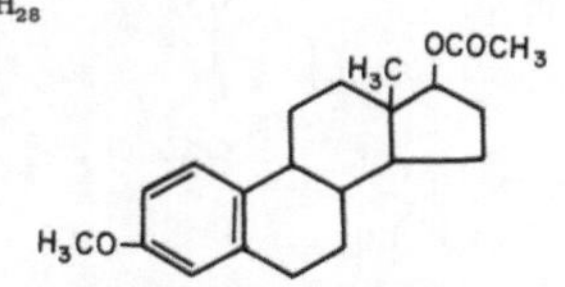

J. Am. Chem. Soc. 84:2972 (1962)

▶ $O_3C_{21}H_{28}$

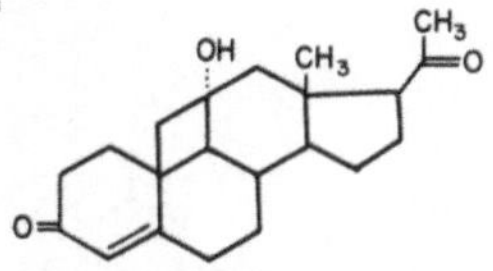

Helv. Chim. Acta 45:1261 (1962)

▶ $O_3C_{21}H_{28}$

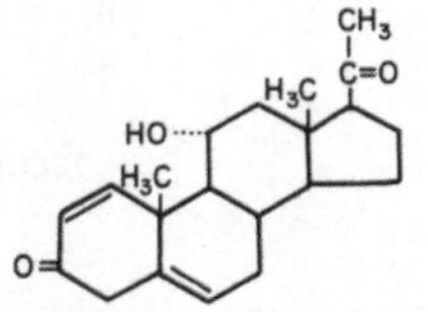

Helv. Chim. Acta 44:1380 (1961)

▶ $O_3C_{21}H_{28}$

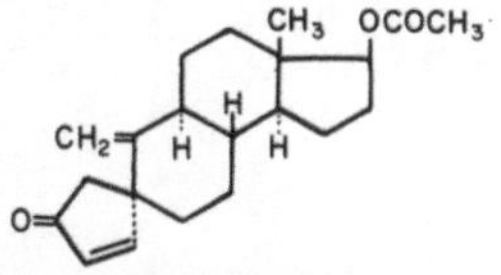

Helv. Chim. Acta 45:2403 (1962)

▶ $O_3C_{21}H_{28}$

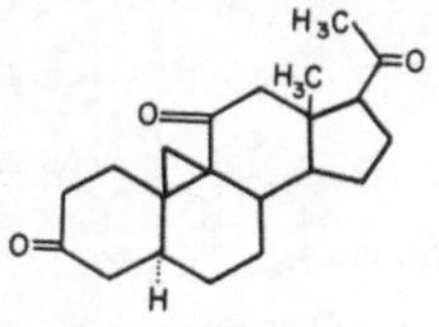

Helv. Chim. Acta 44:2162 (1961)

▶ $O_3C_{21}H_{28}$

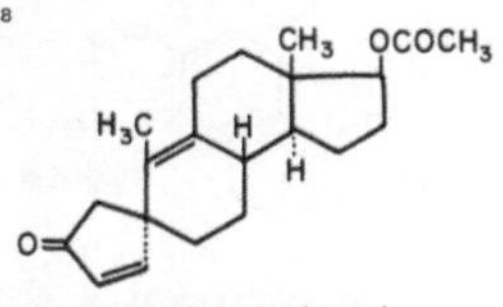

Helv. Chim. Acta 45:2403 (1962)

▶ $O_3C_{21}H_{28}$

KETONE $A_5$

Helv. Chim. Acta 45:2346 (1962)

▶ $O_3C_{21}H_{28}$

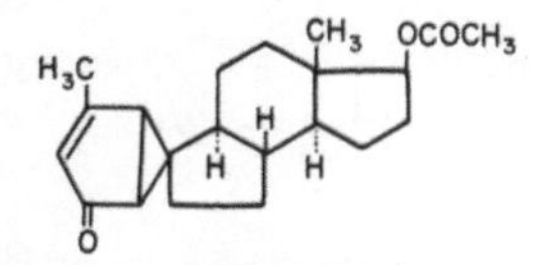

Helv. Chim. Acta 45:2346 (1962)

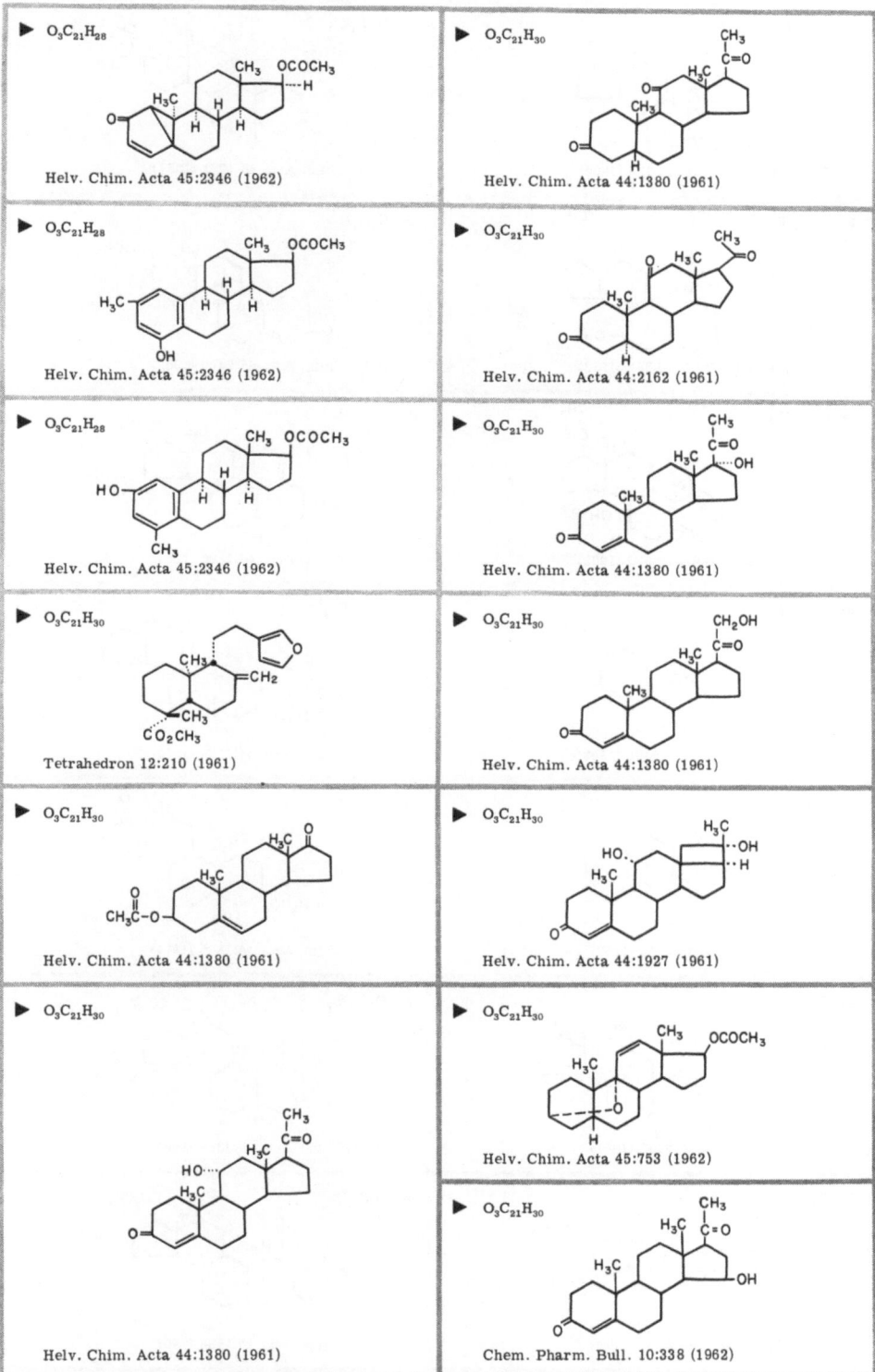

▶ $O_3C_{21}H_{28}$

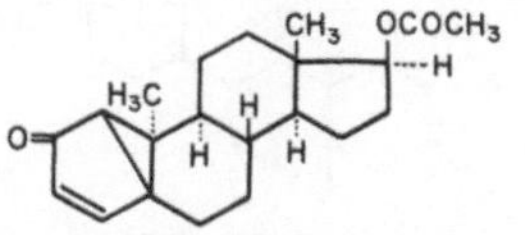

Helv. Chim. Acta 45:2346 (1962)

▶ $O_3C_{21}H_{28}$

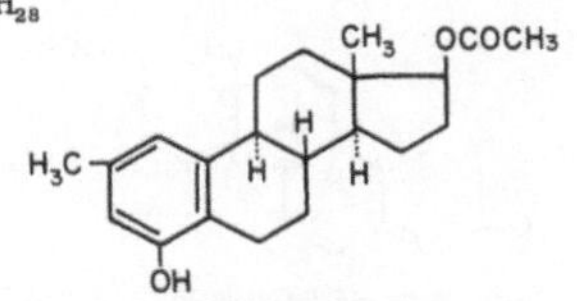

Helv. Chim. Acta 45:2346 (1962)

▶ $O_3C_{21}H_{28}$

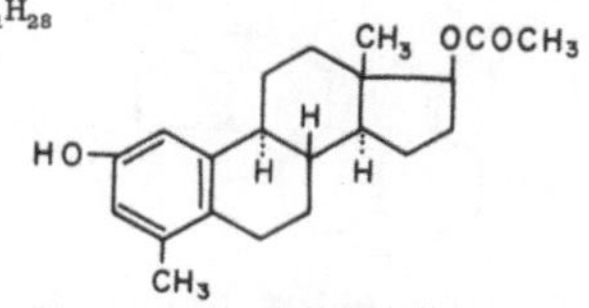

Helv. Chim. Acta 45:2346 (1962)

▶ $O_3C_{21}H_{30}$

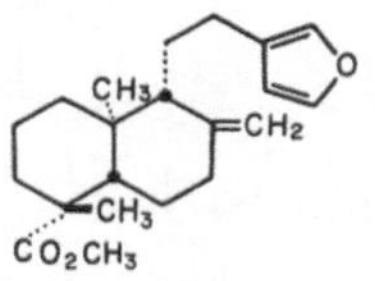

Tetrahedron 12:210 (1961)

▶ $O_3C_{21}H_{30}$

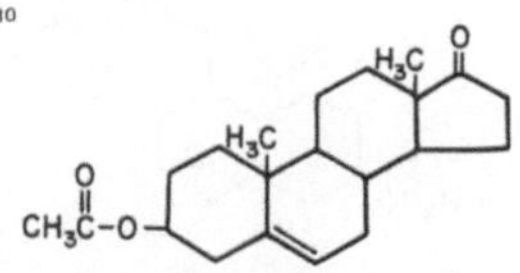

Helv. Chim. Acta 44:1380 (1961)

▶ $O_3C_{21}H_{30}$

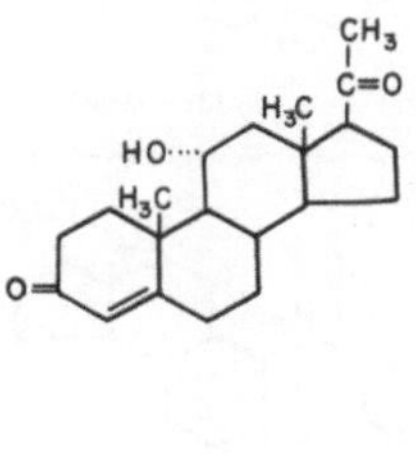

Helv. Chim. Acta 44:1380 (1961)

▶ $O_3C_{21}H_{30}$

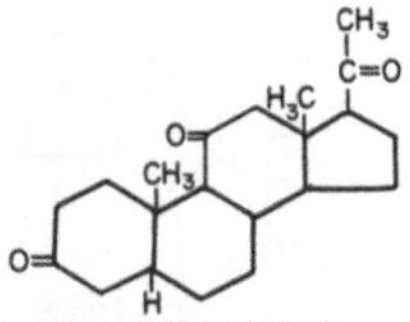

Helv. Chim. Acta 44:1380 (1961)

▶ $O_3C_{21}H_{30}$

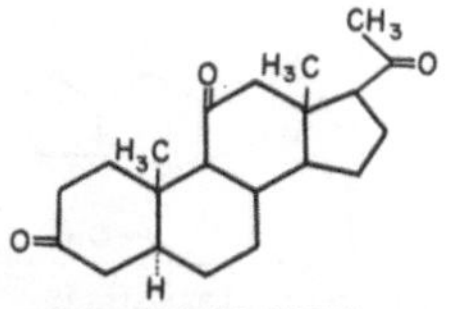

Helv. Chim. Acta 44:2162 (1961)

▶ $O_3C_{21}H_{30}$

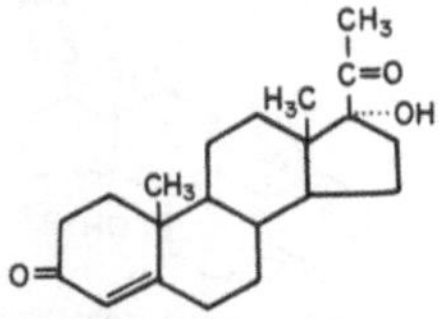

Helv. Chim. Acta 44:1380 (1961)

▶ $O_3C_{21}H_{30}$

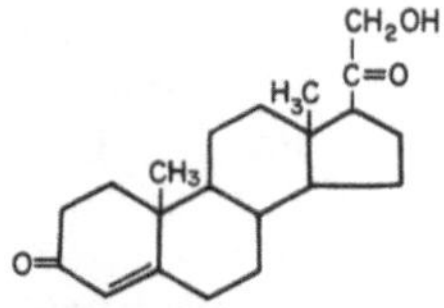

Helv. Chim. Acta 44:1380 (1961)

▶ $O_3C_{21}H_{30}$

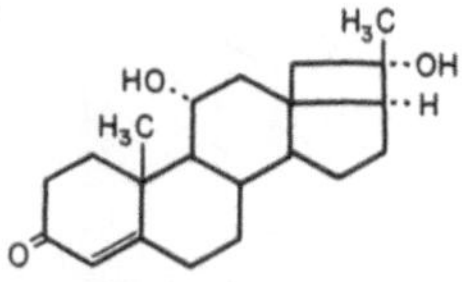

Helv. Chim. Acta 44:1927 (1961)

▶ $O_3C_{21}H_{30}$

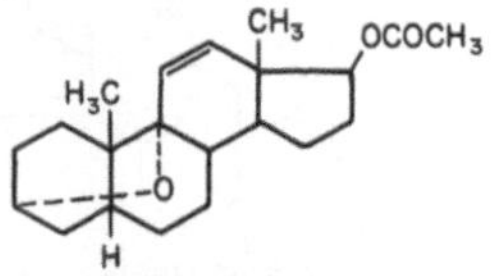

Helv. Chim. Acta 45:753 (1962)

▶ $O_3C_{21}H_{30}$

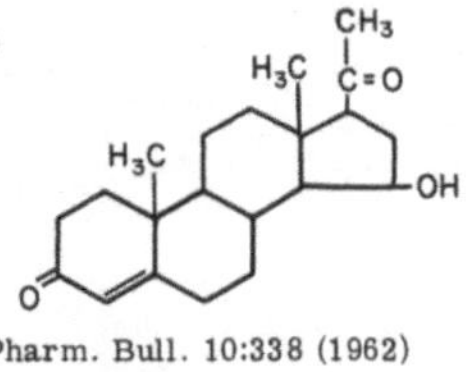

Chem. Pharm. Bull. 10:338 (1962)

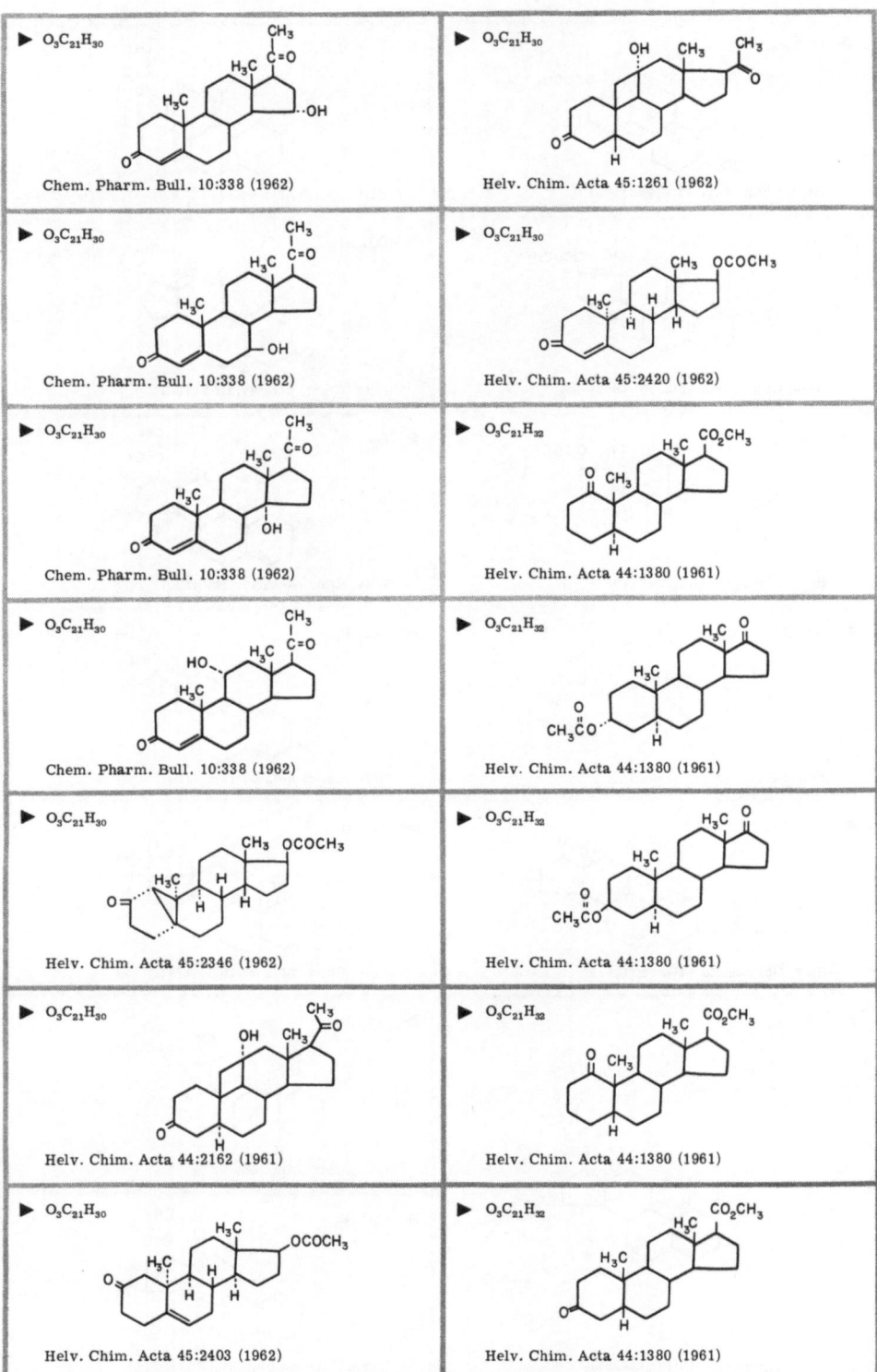

▶ $O_3C_{21}H_{30}$

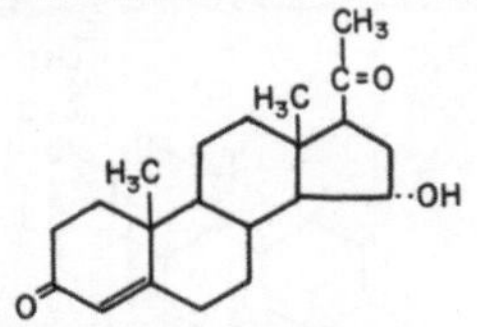

Chem. Pharm. Bull. 10:338 (1962)

▶ $O_3C_{21}H_{30}$

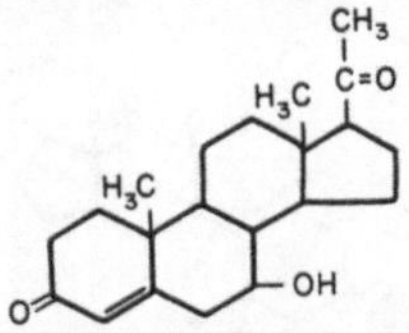

Chem. Pharm. Bull. 10:338 (1962)

▶ $O_3C_{21}H_{30}$

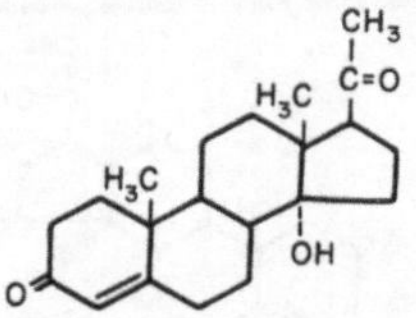

Chem. Pharm. Bull. 10:338 (1962)

▶ $O_3C_{21}H_{30}$

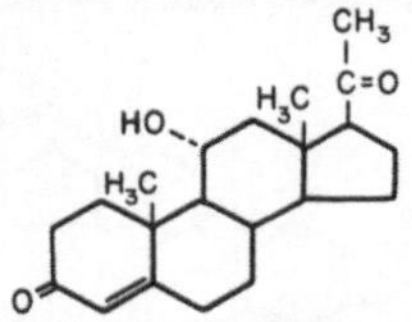

Chem. Pharm. Bull. 10:338 (1962)

▶ $O_3C_{21}H_{30}$

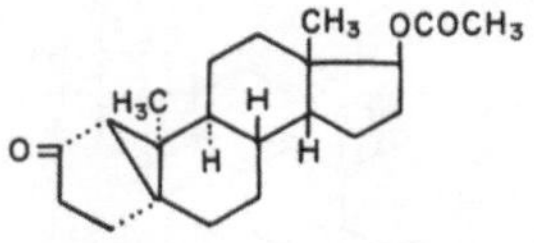

Helv. Chim. Acta 45:2346 (1962)

▶ $O_3C_{21}H_{30}$

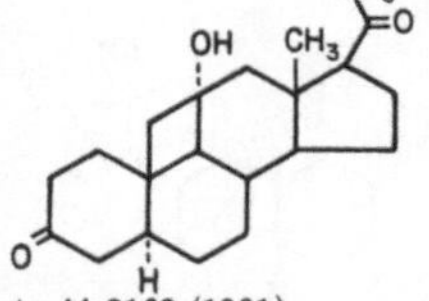

Helv. Chim. Acta 44:2162 (1961)

▶ $O_3C_{21}H_{30}$

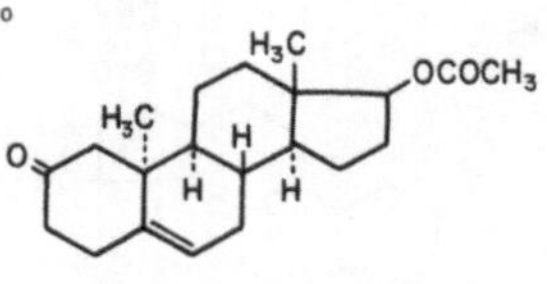

Helv. Chim. Acta 45:2403 (1962)

▶ $O_3C_{21}H_{30}$

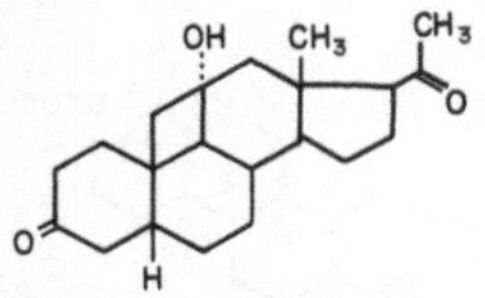

Helv. Chim. Acta 45:1261 (1962)

▶ $O_3C_{21}H_{30}$

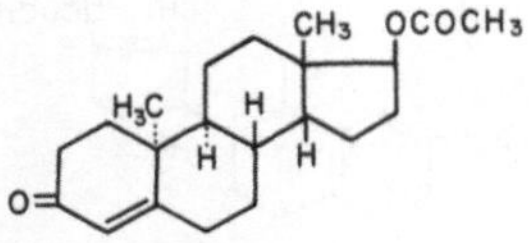

Helv. Chim. Acta 45:2420 (1962)

▶ $O_3C_{21}H_{32}$

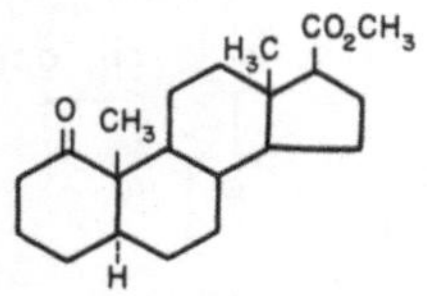

Helv. Chim. Acta 44:1380 (1961)

▶ $O_3C_{21}H_{32}$

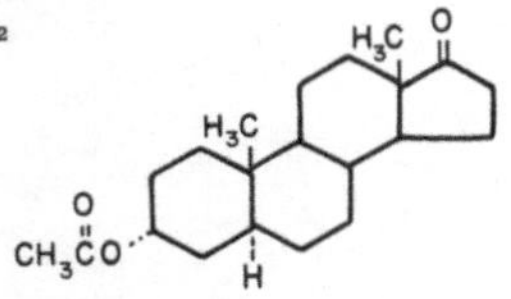

Helv. Chim. Acta 44:1380 (1961)

▶ $O_3C_{21}H_{32}$

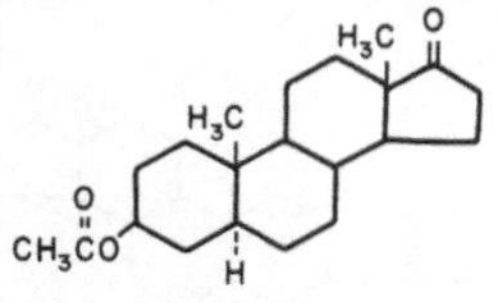

Helv. Chim. Acta 44:1380 (1961)

▶ $O_3C_{21}H_{32}$

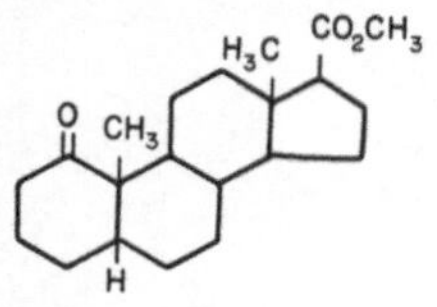

Helv. Chim. Acta 44:1380 (1961)

▶ $O_3C_{21}H_{32}$

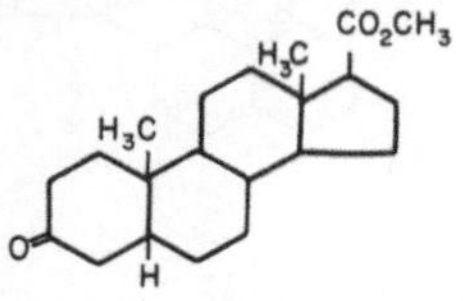

Helv. Chim. Acta 44:1380 (1961)

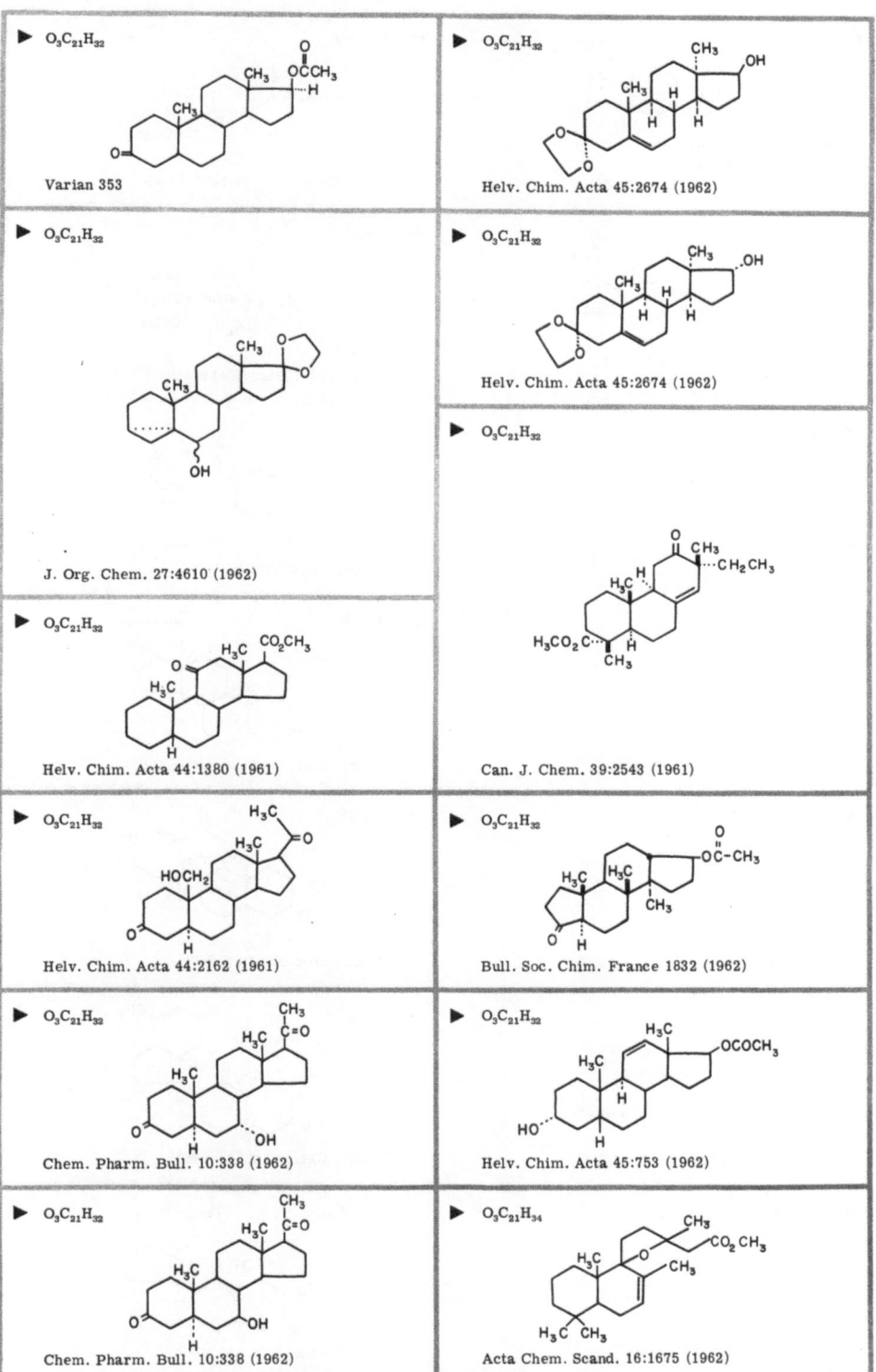

▶ $O_3C_{21}H_{32}$

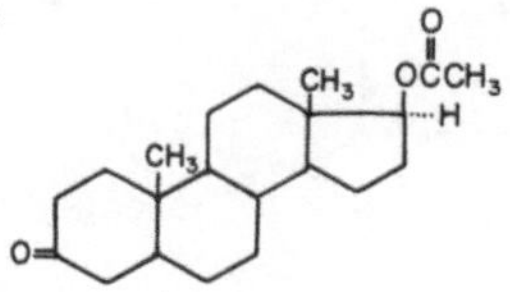

Varian 353

▶ $O_3C_{21}H_{32}$

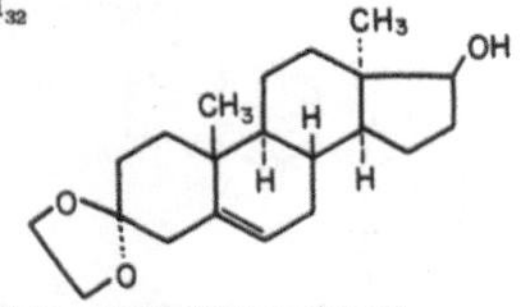

Helv. Chim. Acta 45:2674 (1962)

▶ $O_3C_{21}H_{32}$

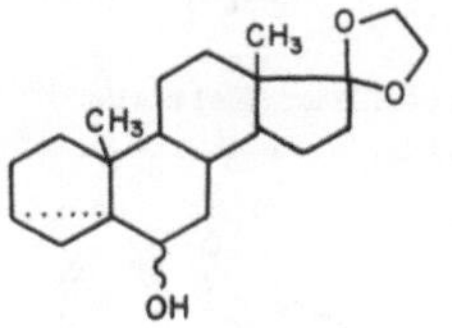

J. Org. Chem. 27:4610 (1962)

▶ $O_3C_{21}H_{32}$

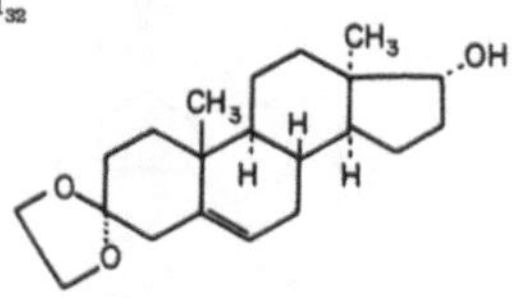

Helv. Chim. Acta 45:2674 (1962)

▶ $O_3C_{21}H_{32}$

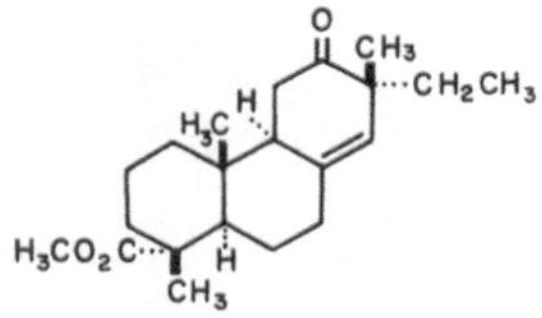

Can. J. Chem. 39:2543 (1961)

▶ $O_3C_{21}H_{32}$

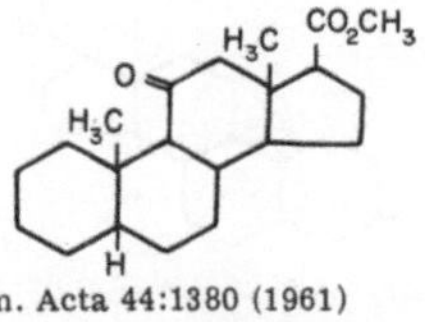

Helv. Chim. Acta 44:1380 (1961)

▶ $O_3C_{21}H_{32}$

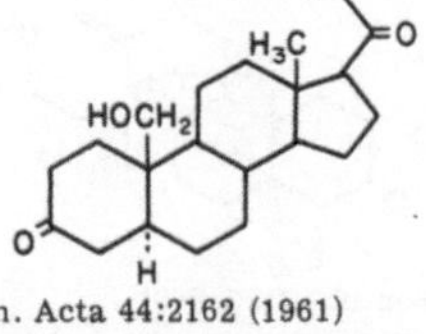

Helv. Chim. Acta 44:2162 (1961)

▶ $O_3C_{21}H_{32}$

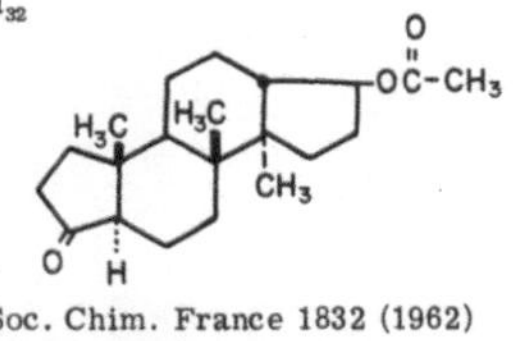

Bull. Soc. Chim. France 1832 (1962)

▶ $O_3C_{21}H_{32}$

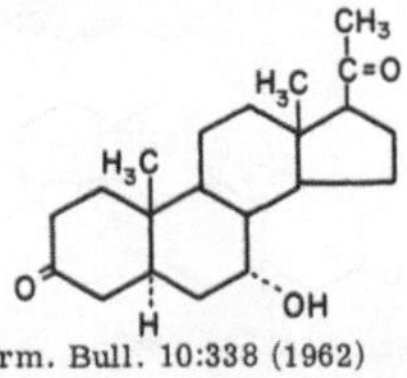

Chem. Pharm. Bull. 10:338 (1962)

▶ $O_3C_{21}H_{32}$

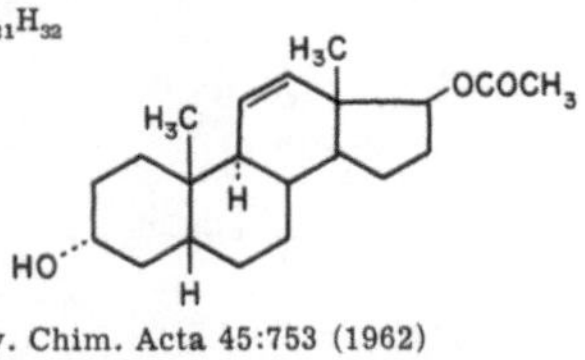

Helv. Chim. Acta 45:753 (1962)

▶ $O_3C_{21}H_{32}$

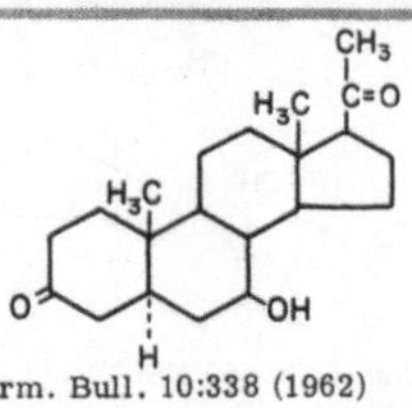

Chem. Pharm. Bull. 10:338 (1962)

▶ $O_3C_{21}H_{34}$

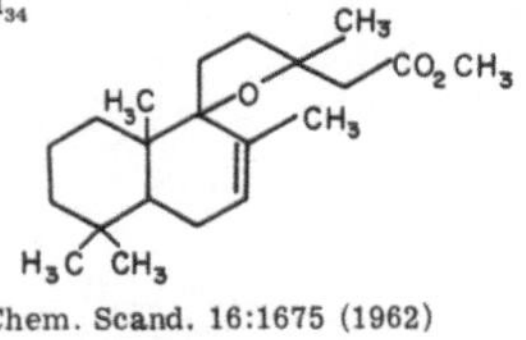

Acta Chem. Scand. 16:1675 (1962)

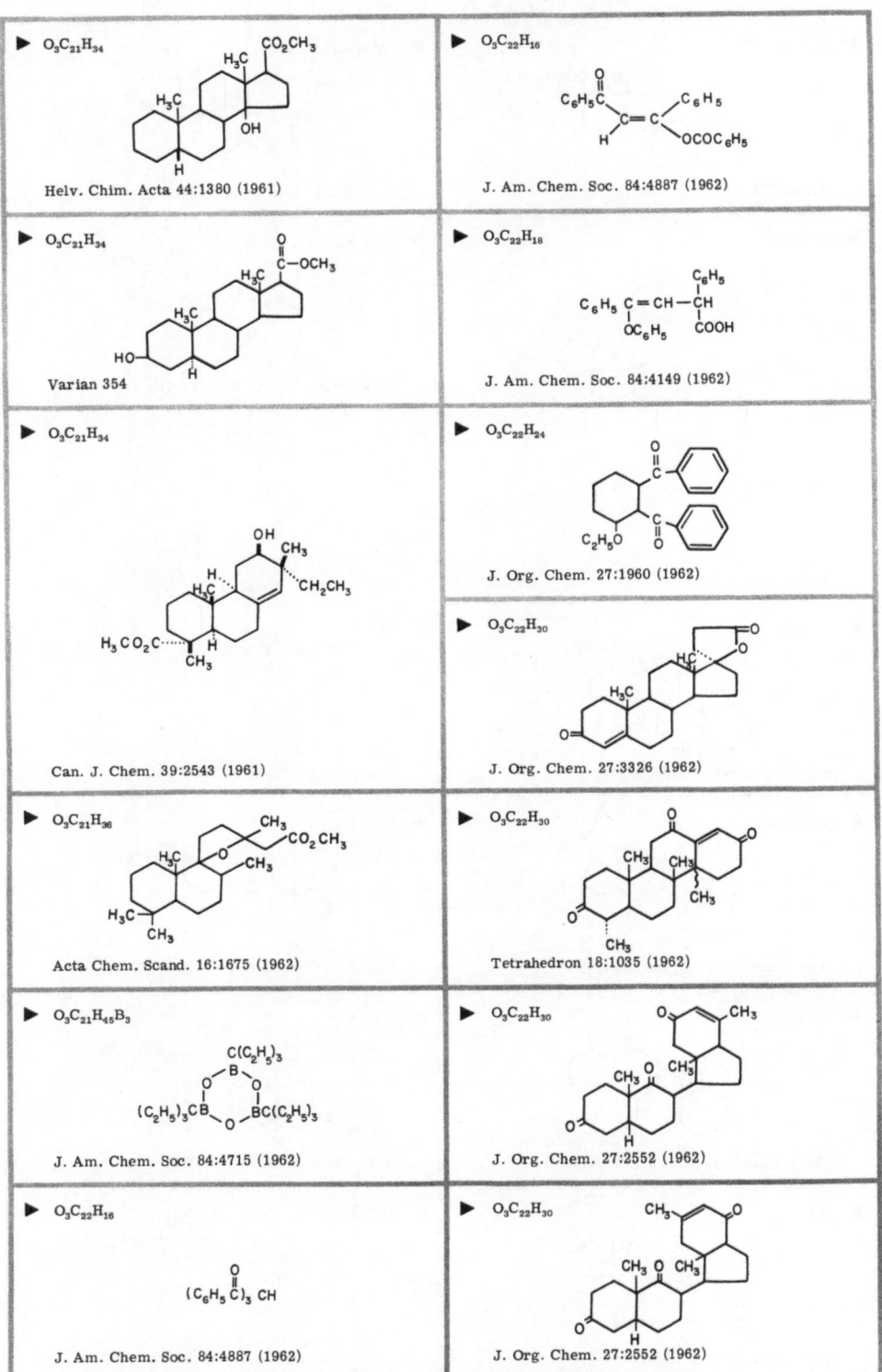

▶ $O_3C_{21}H_{34}$

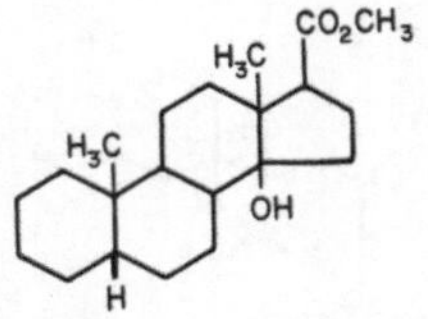

Helv. Chim. Acta 44:1380 (1961)

▶ $O_3C_{21}H_{34}$

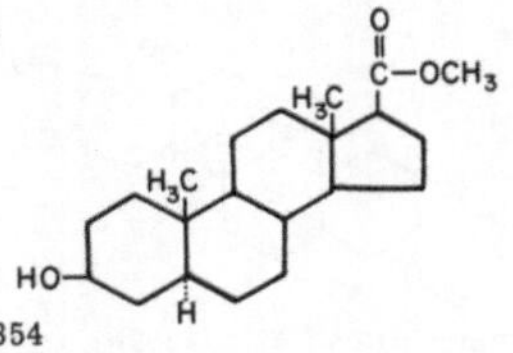

Varian 354

▶ $O_3C_{21}H_{34}$

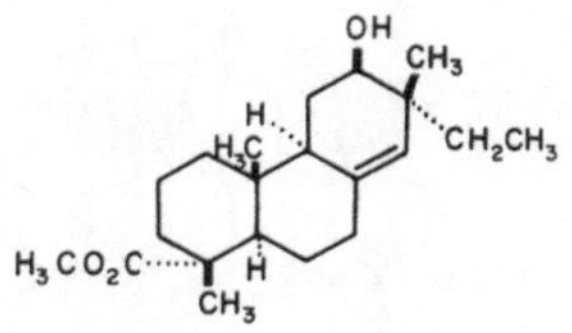

Can. J. Chem. 39:2543 (1961)

▶ $O_3C_{21}H_{36}$

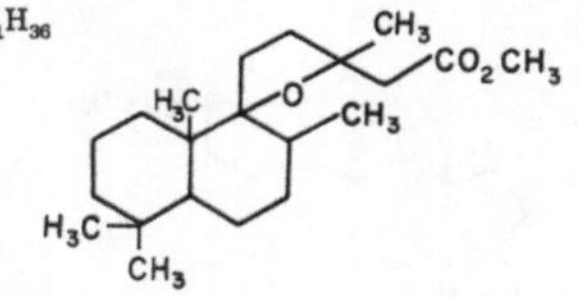

Acta Chem. Scand. 16:1675 (1962)

▶ $O_3C_{21}H_{45}B_3$

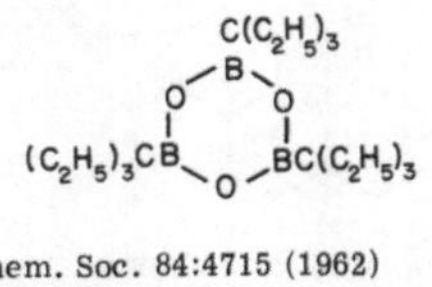

J. Am. Chem. Soc. 84:4715 (1962)

▶ $O_3C_{22}H_{16}$

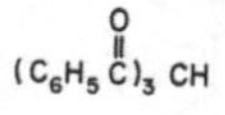

J. Am. Chem. Soc. 84:4887 (1962)

▶ $O_3C_{22}H_{16}$

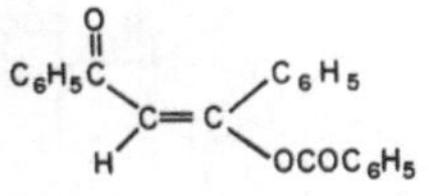

J. Am. Chem. Soc. 84:4887 (1962)

▶ $O_3C_{22}H_{18}$

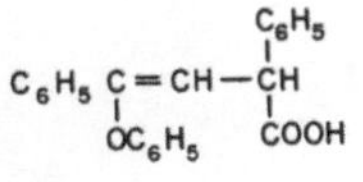

J. Am. Chem. Soc. 84:4149 (1962)

▶ $O_3C_{22}H_{24}$

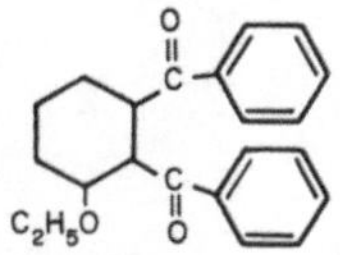

J. Org. Chem. 27:1960 (1962)

▶ $O_3C_{22}H_{30}$

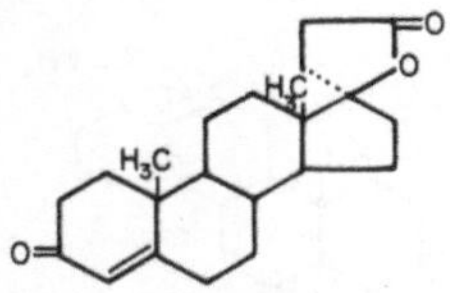

J. Org. Chem. 27:3326 (1962)

▶ $O_3C_{22}H_{30}$

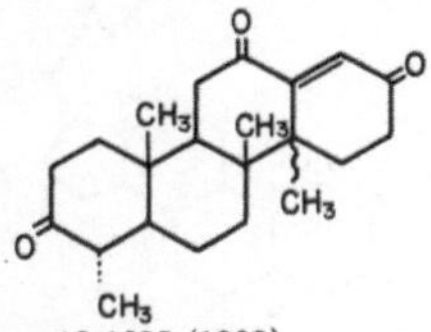

Tetrahedron 18:1035 (1962)

▶ $O_3C_{22}H_{30}$

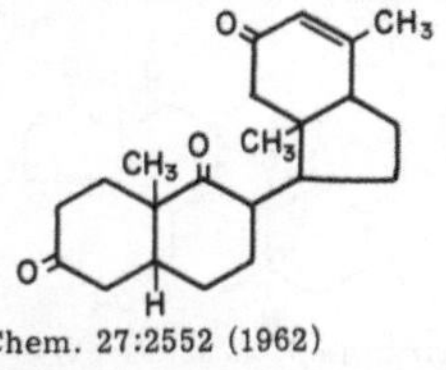

J. Org. Chem. 27:2552 (1962)

▶ $O_3C_{22}H_{30}$

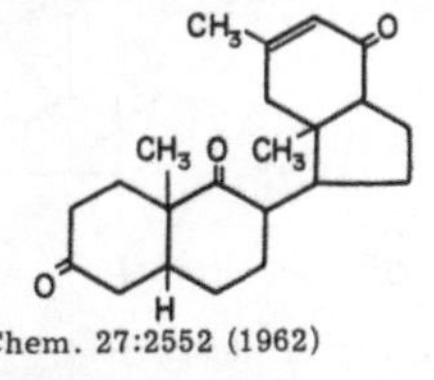

J. Org. Chem. 27:2552 (1962)

▶ $O_3C_{22}H_{32}$

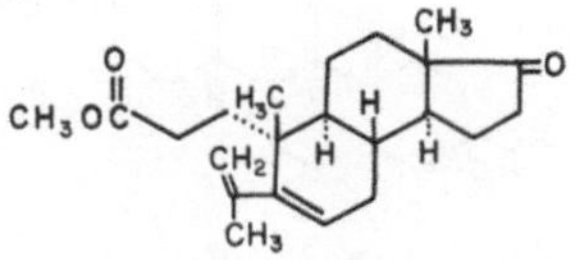

Helv. Chim. Acta 45:2674 (1962)

▶ $O_3C_{22}H_{32}$

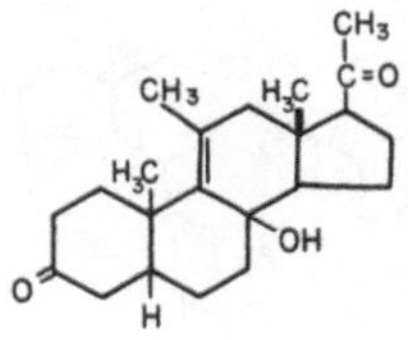

J. Org. Chem. 27:2553 (1962)

▶ $O_3C_{22}H_{32}$

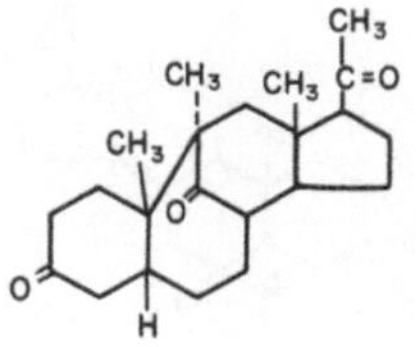

J. Org. Chem. 27:2553 (1962)

▶ $O_3C_{23}H_{18}$

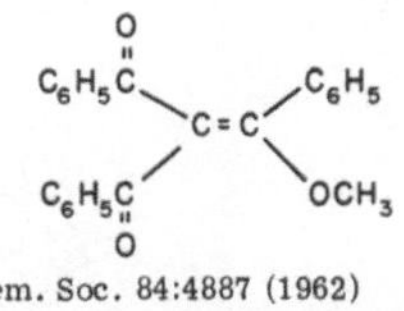

J. Am. Chem. Soc. 84:4887 (1962)

▶ $O_3C_{23}H_{30}$

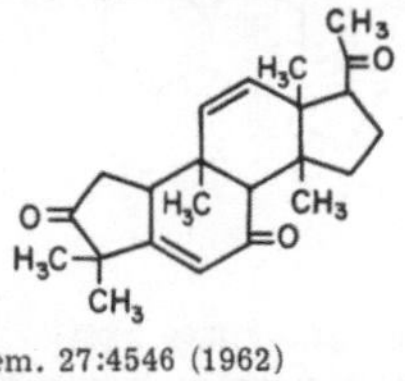

J. Org. Chem. 27:4546 (1962)

▶ $O_3C_{23}H_{30}$

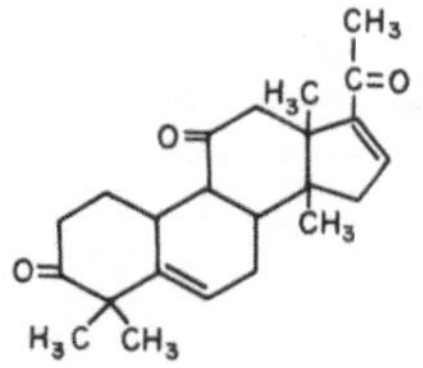

Tetrahedron Letters No. 8:316 (1962)

▶ $O_3C_{23}H_{30}$

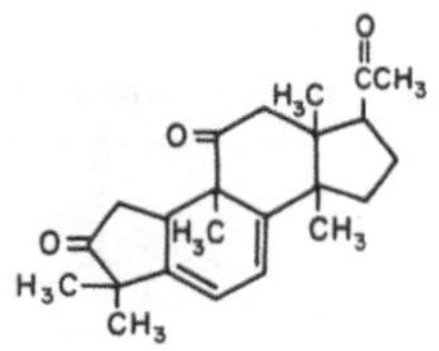

J. Org. Chem. 27:4546 (1962)

▶ $O_3C_{23}H_{32}$

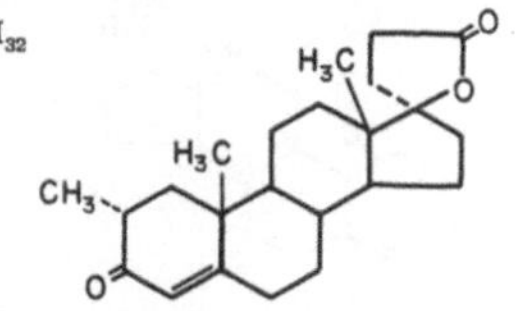

J. Org. Chem. 26:3080 (1961)

▶ $O_3C_{23}H_{32}$

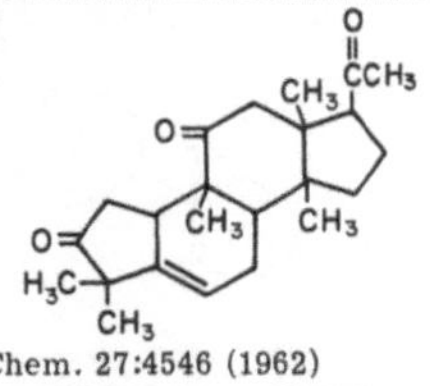

J. Org. Chem. 27:4546 (1962)

▶ $O_3C_{23}H_{32}$

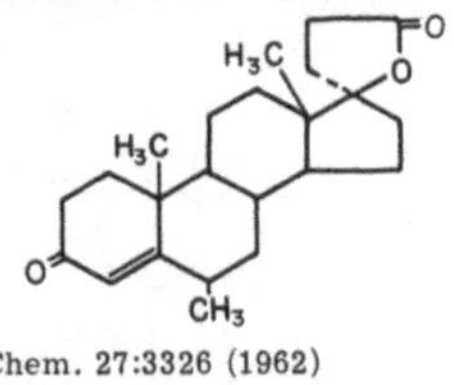

J. Org. Chem. 27:3326 (1962)

▶ $O_3C_{23}H_{32}$

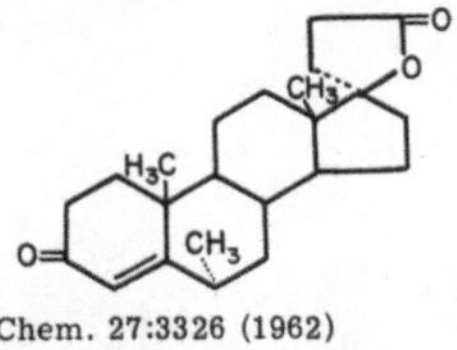

J. Org. Chem. 27:3326 (1962)

▶ $O_3C_{23}H_{36}$

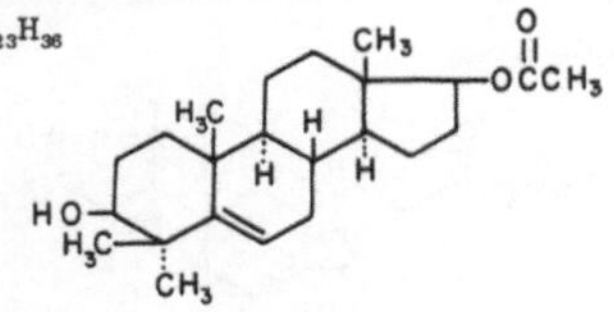

Helv. Chim. Acta 45:2674 (1962)

▶ $O_3C_{23}H_{32}$

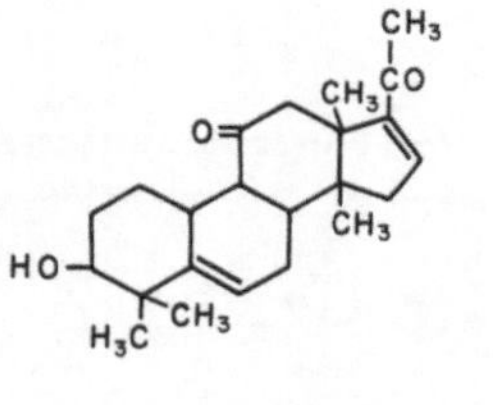

Tetrahedron Letters No. 8:316 (1962)

▶ $O_3C_{23}H_{36}$

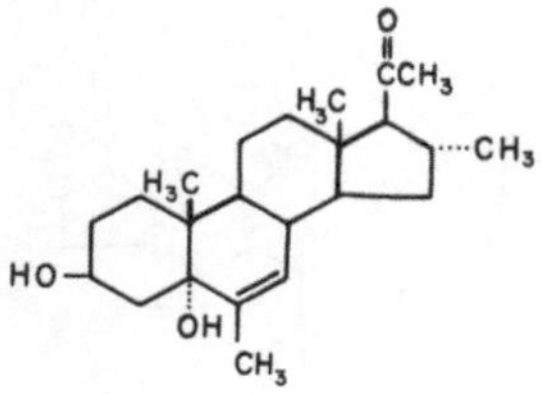

J. Org. Chem. 27:1139 (1962)

▶ $O_3C_{23}H_{34}$

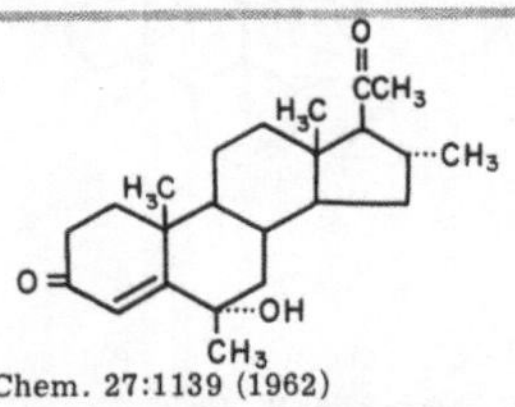

J. Org. Chem. 27:1139 (1962)

▶ $O_3C_{23}H_{34}$

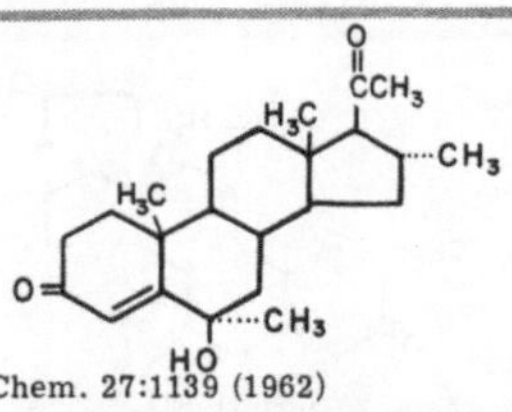

J. Org. Chem. 27:1139 (1962)

▶ $O_3C_{23}H_{36}$

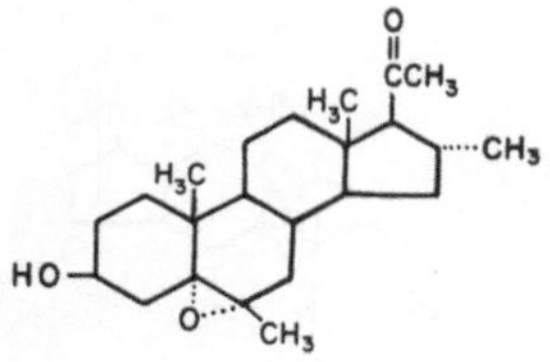

J. Org. Chem. 27:1139 (1962)

▶ $O_3C_{23}H_{36}$

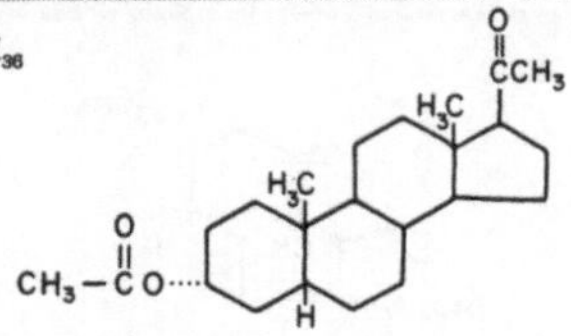

J. Org. Chem. 27:3166 (1962)

▶ $O_3C_{23}H_{36}$

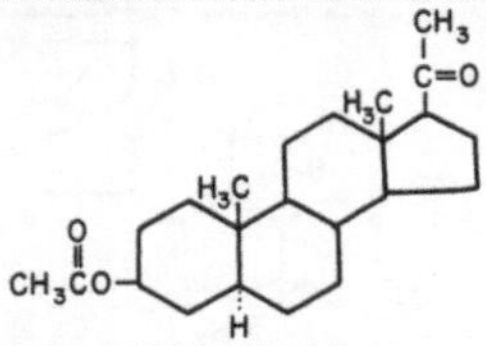

Helv. Chim. Acta 44:1380 (1961)

▶ $O_3C_{23}H_{36}$

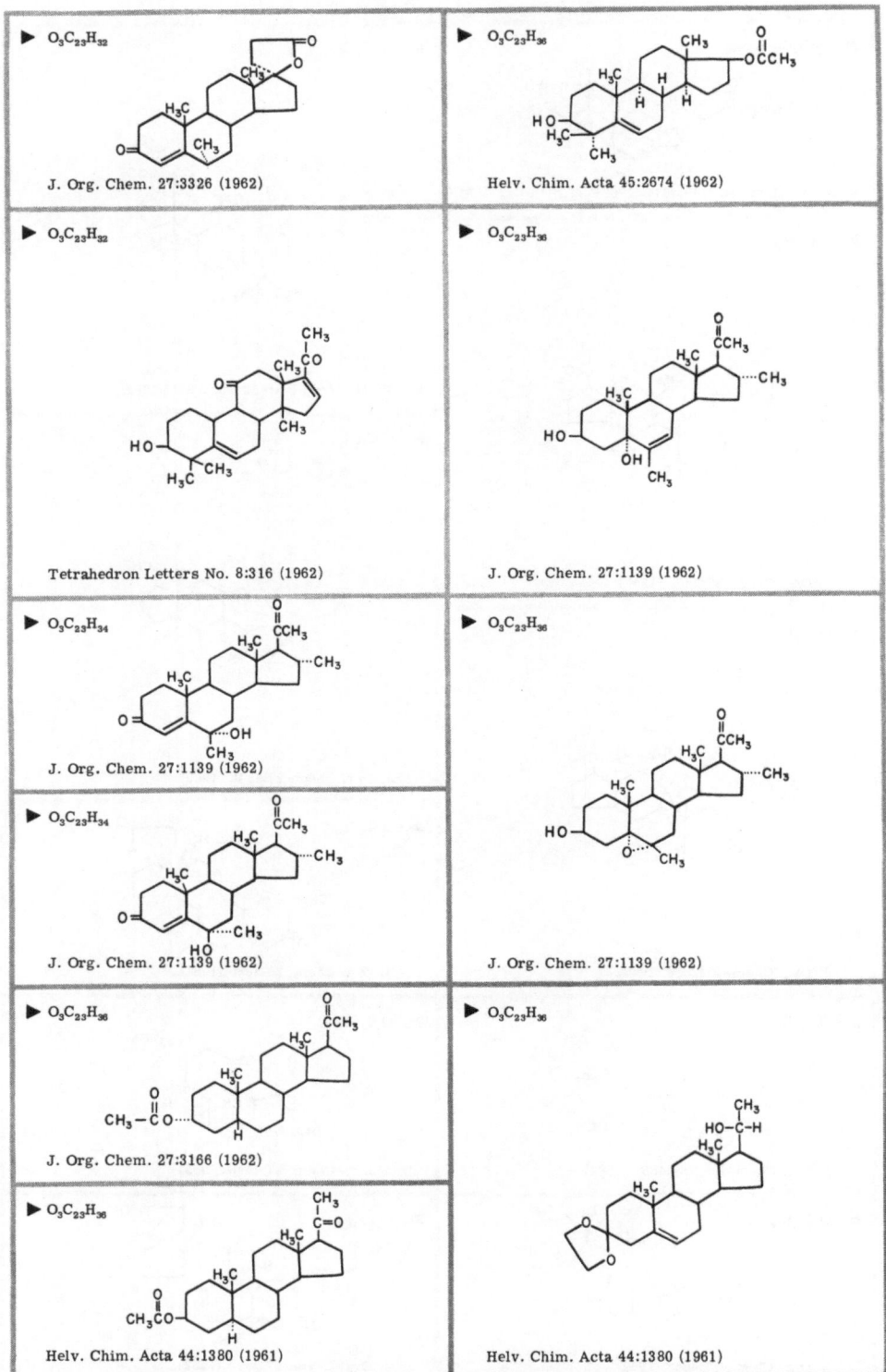

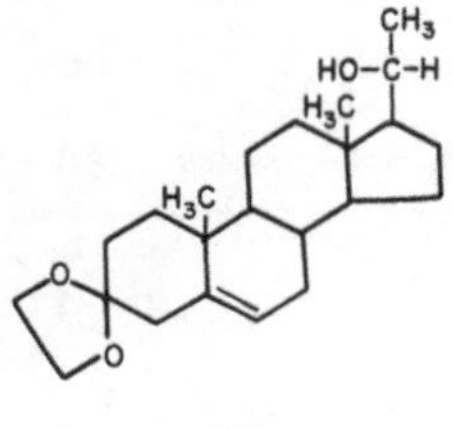

Helv. Chim. Acta 44:1380 (1961)

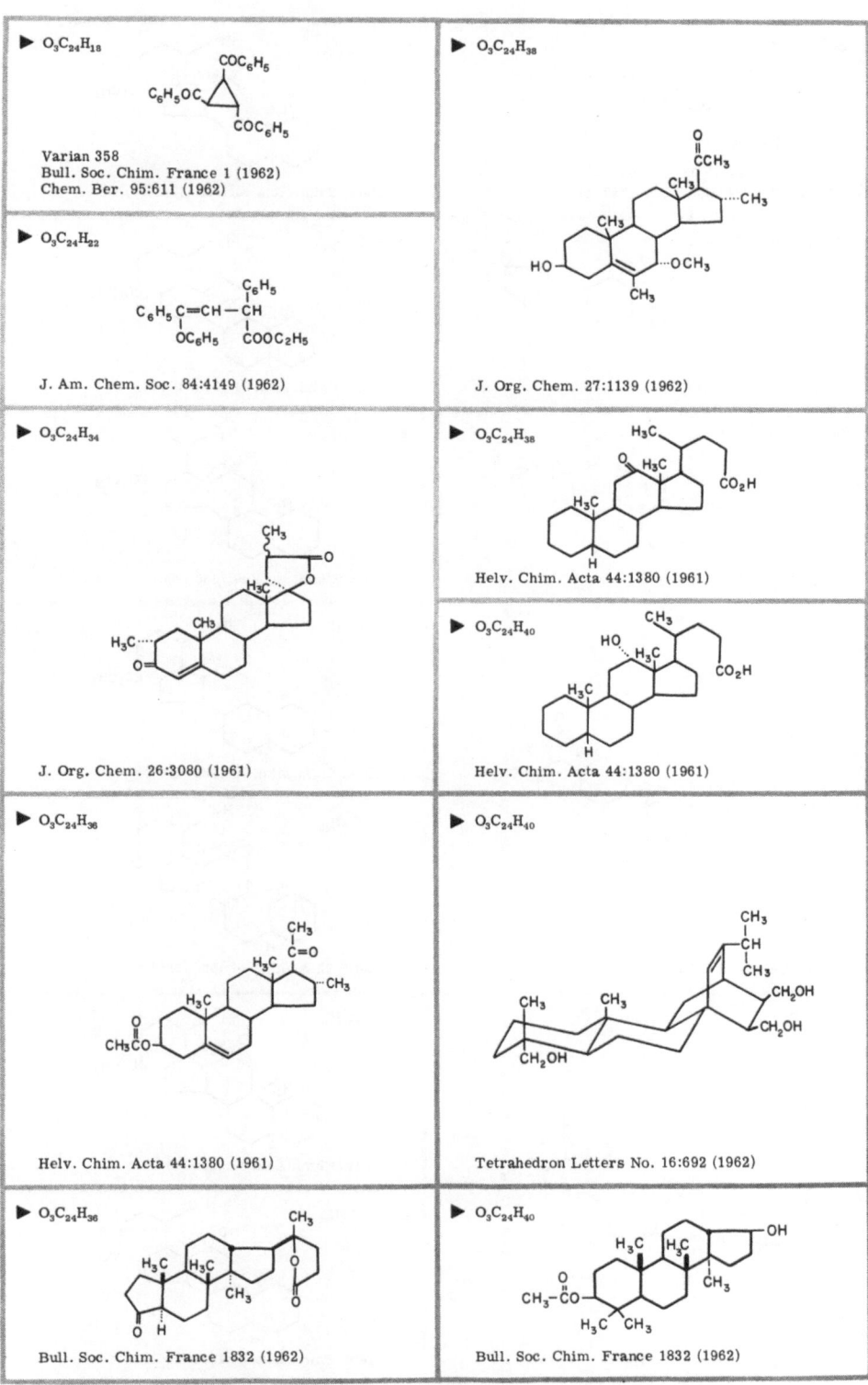

▶ $O_3C_{24}H_{18}$

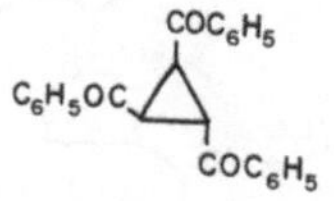

Varian 358
Bull. Soc. Chim. France 1 (1962)
Chem. Ber. 95:611 (1962)

▶ $O_3C_{24}H_{22}$

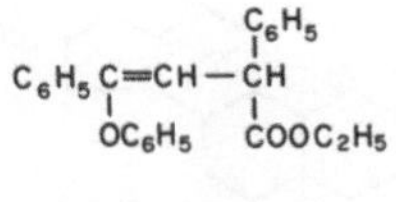

J. Am. Chem. Soc. 84:4149 (1962)

▶ $O_3C_{24}H_{38}$

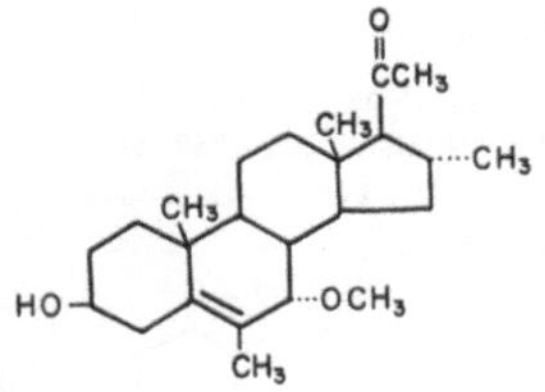

J. Org. Chem. 27:1139 (1962)

▶ $O_3C_{24}H_{34}$

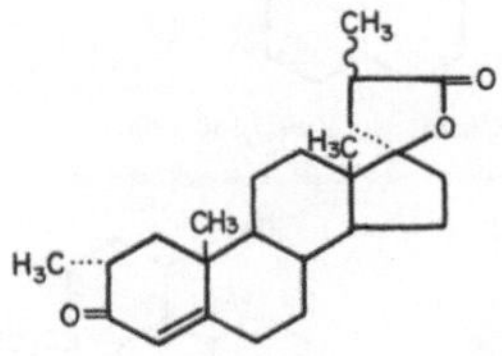

J. Org. Chem. 26:3080 (1961)

▶ $O_3C_{24}H_{38}$

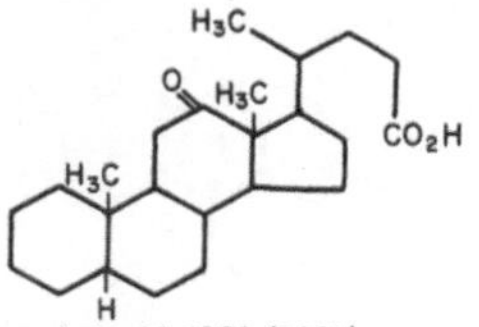

Helv. Chim. Acta 44:1380 (1961)

▶ $O_3C_{24}H_{40}$

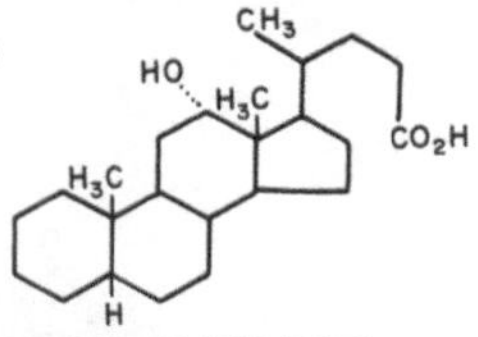

Helv. Chim. Acta 44:1380 (1961)

▶ $O_3C_{24}H_{36}$

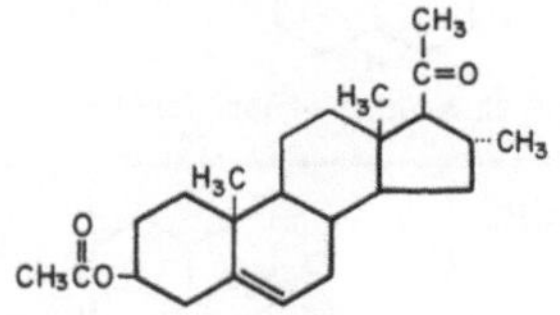

Helv. Chim. Acta 44:1380 (1961)

▶ $O_3C_{24}H_{40}$

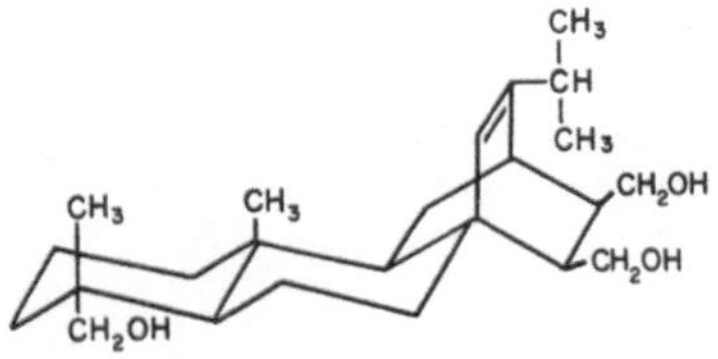

Tetrahedron Letters No. 16:692 (1962)

▶ $O_3C_{24}H_{36}$

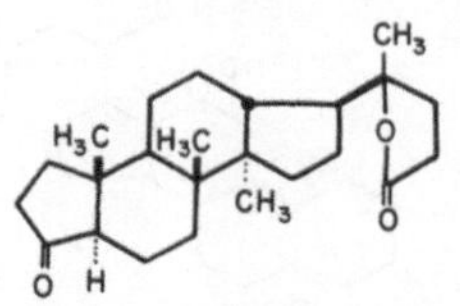

Bull. Soc. Chim. France 1832 (1962)

▶ $O_3C_{24}H_{40}$

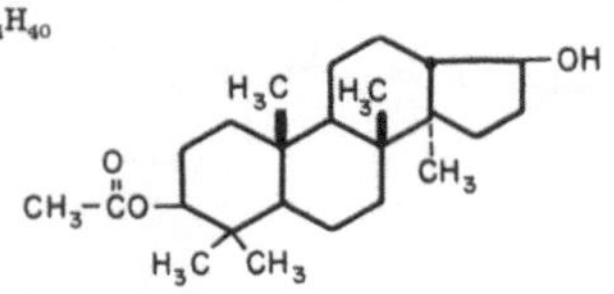

Bull. Soc. Chim. France 1832 (1962)

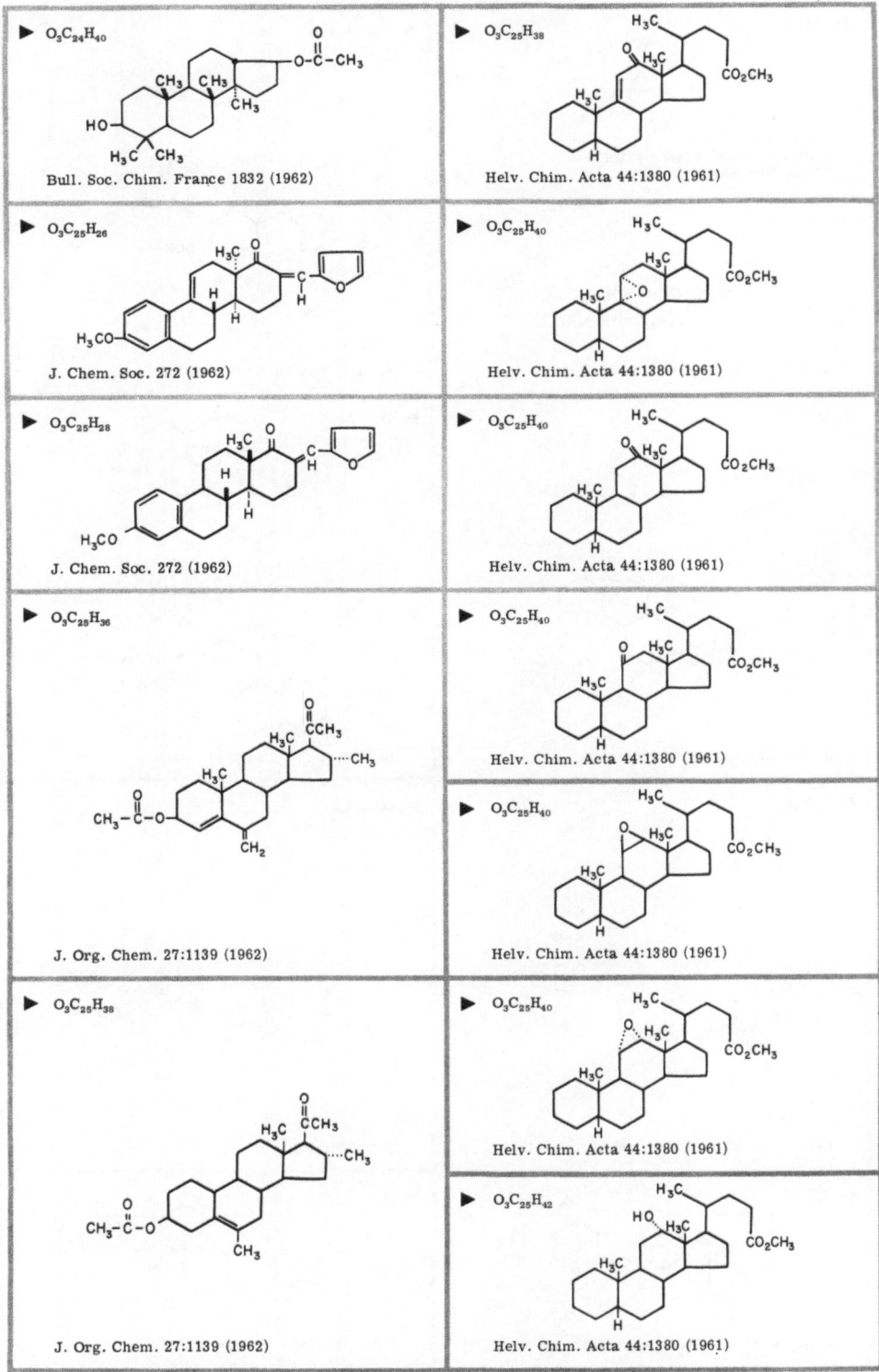

▶ $O_3C_{24}H_{40}$

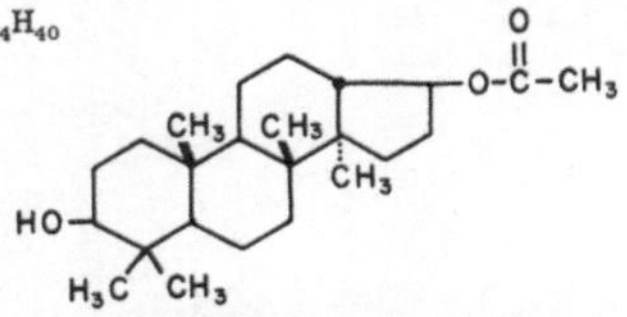

Bull. Soc. Chim. France 1832 (1962)

▶ $O_3C_{25}H_{26}$

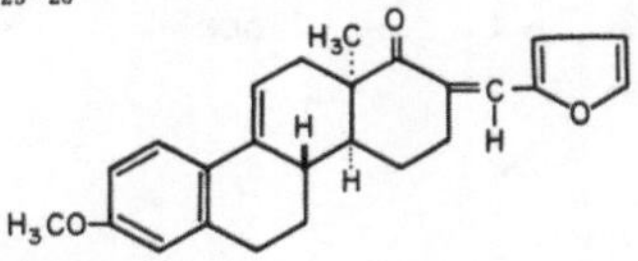

J. Chem. Soc. 272 (1962)

▶ $O_3C_{25}H_{28}$

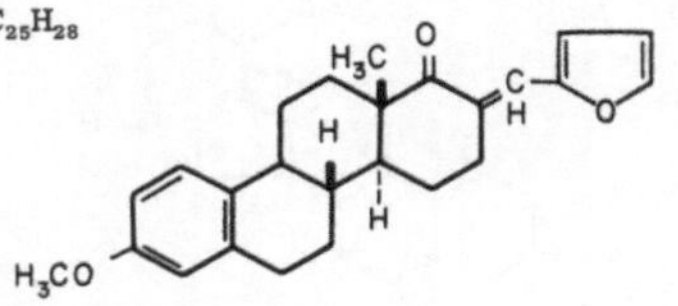

J. Chem. Soc. 272 (1962)

▶ $O_3C_{25}H_{36}$

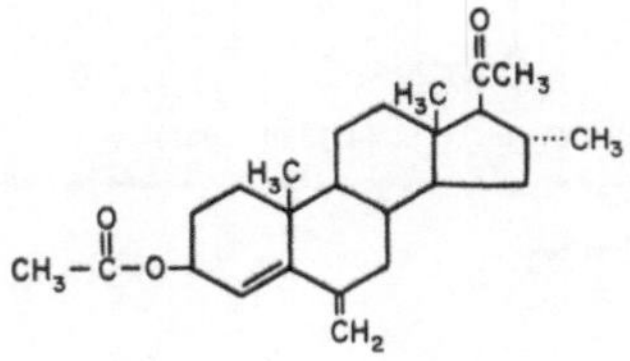

J. Org. Chem. 27:1139 (1962)

▶ $O_3C_{25}H_{38}$

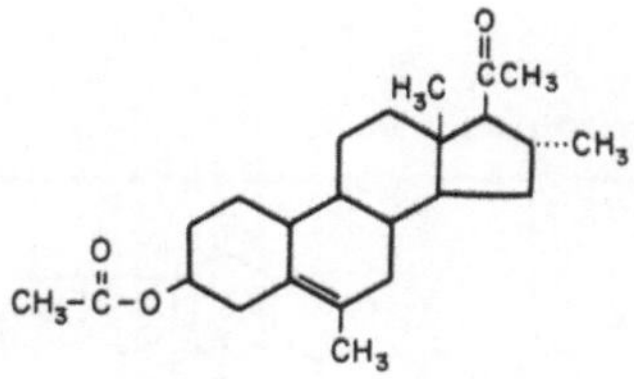

J. Org. Chem. 27:1139 (1962)

▶ $O_3C_{25}H_{38}$

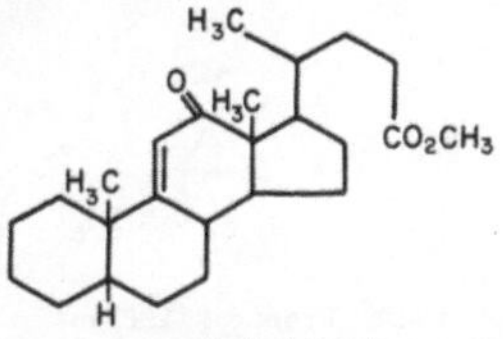

Helv. Chim. Acta 44:1380 (1961)

▶ $O_3C_{25}H_{40}$

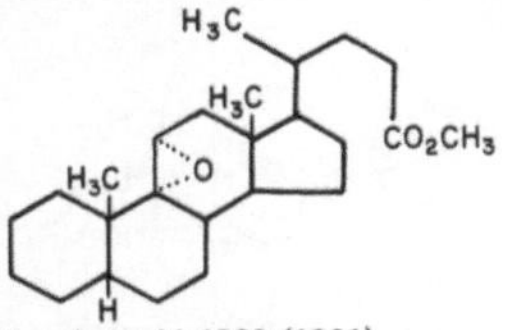

Helv. Chim. Acta 44:1380 (1961)

▶ $O_3C_{25}H_{40}$

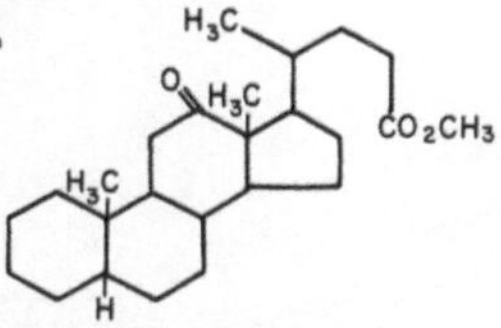

Helv. Chim. Acta 44:1380 (1961)

▶ $O_3C_{25}H_{40}$

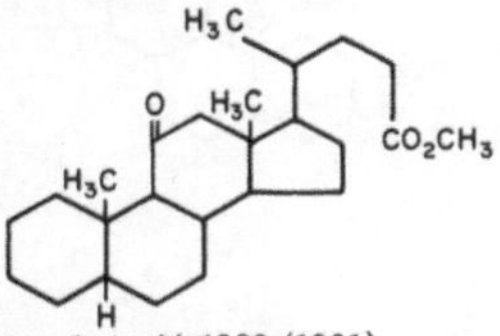

Helv. Chim. Acta 44:1380 (1961)

▶ $O_3C_{25}H_{40}$

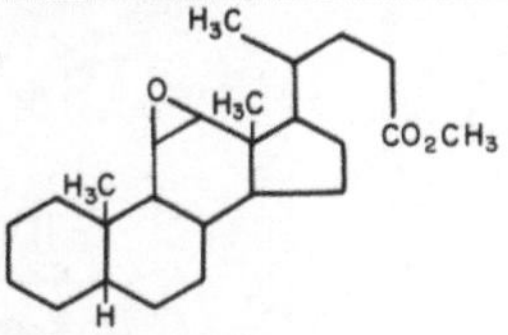

Helv. Chim. Acta 44:1380 (1961)

▶ $O_3C_{25}H_{40}$

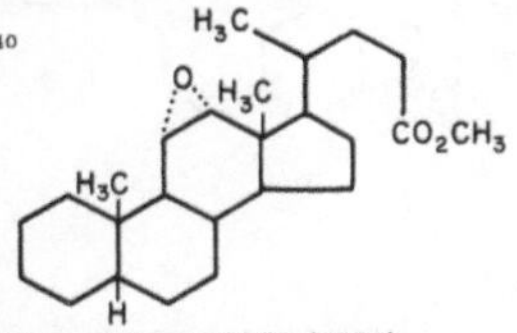

Helv. Chim. Acta 44:1380 (1961)

▶ $O_3C_{25}H_{42}$

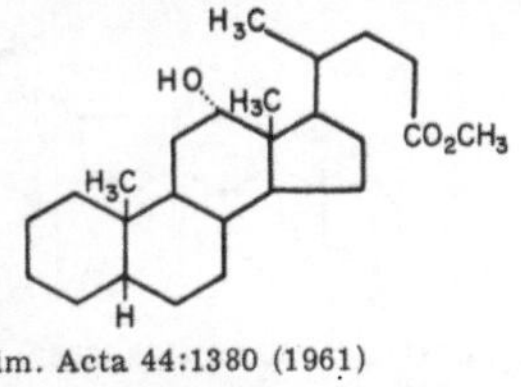

Helv. Chim. Acta 44:1380 (1961)

► $O_3C_{25}H_{42}$

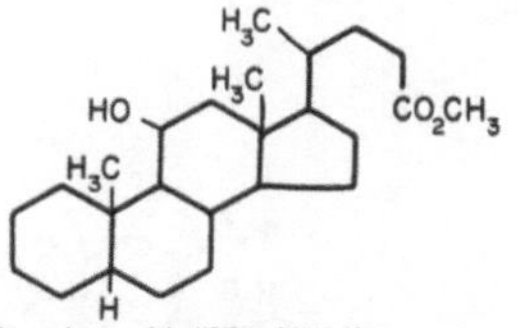

Helv. Chim. Acta 44:1380 (1961)

► $O_3C_{26}H_{40}$

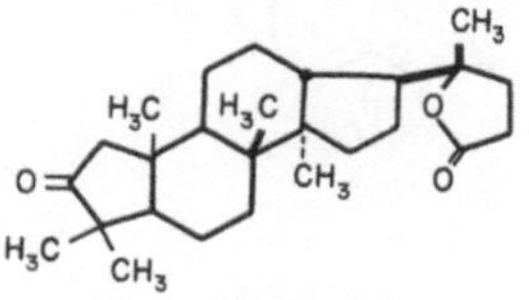

Bull. Soc. Chim. France 1832 (1962)

► $O_3C_{26}H_{42}$

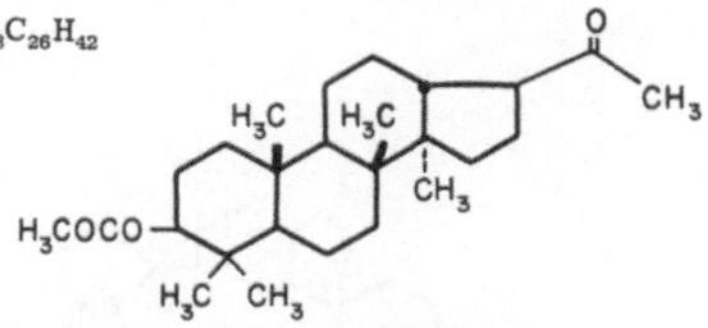

Bull. Soc. Chim. France 1832 (1962)

► $O_3C_{27}H_{40}$

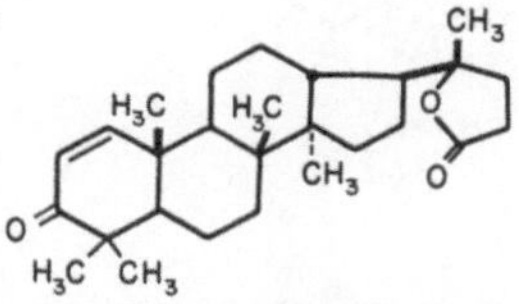

Bull. Soc. Chim. France 1832 (1962)

► $O_3C_{27}H_{42}$

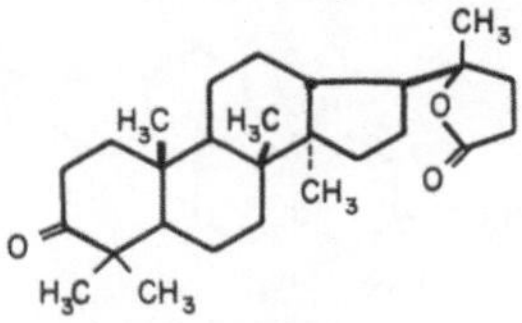

Bull. Soc. Chim. France 1832 (1962)

► $O_3C_{27}H_{42}$

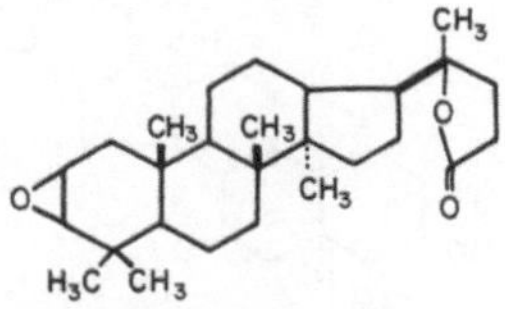

Bull. Soc. Chim. France 1832 (1962)

► $O_3C_{27}H_{44}$

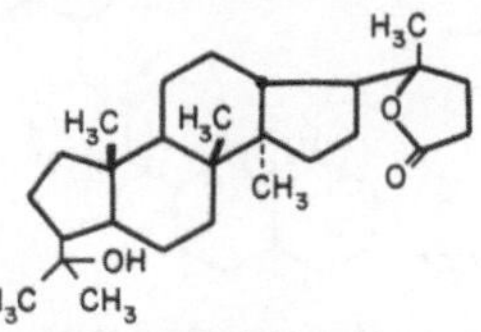

Bull. Soc. Chim. France 1832 (1962)

► $O_3C_{27}H_{44}$

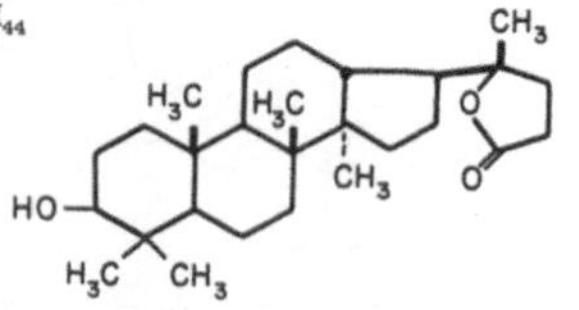

Bull. Soc. Chim. France 1832 (1962)

► $O_3C_{29}H_{46}$

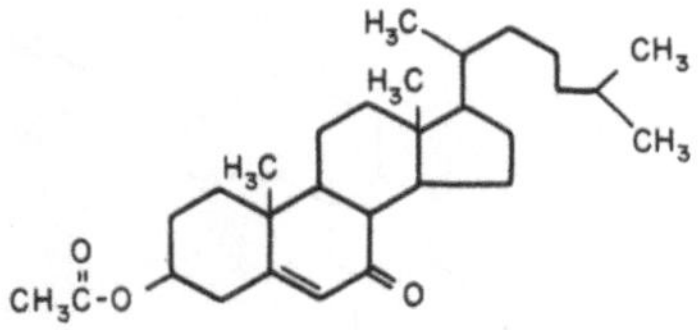

Helv. Chim. Acta 44:1380 (1961)

► $O_3C_{29}H_{47}D$

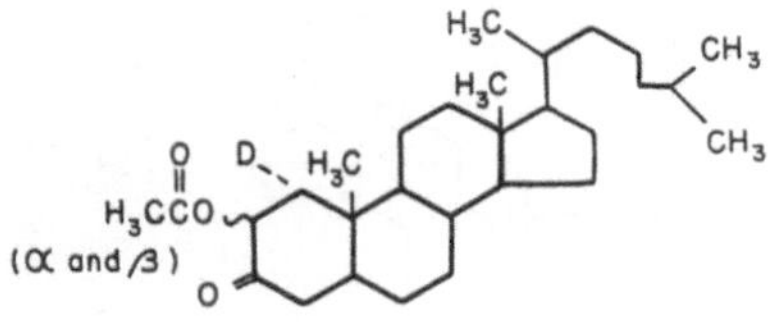

Tetrahedron Letters No. 15:650 (1962)

► $O_3C_{29}H_{48}$

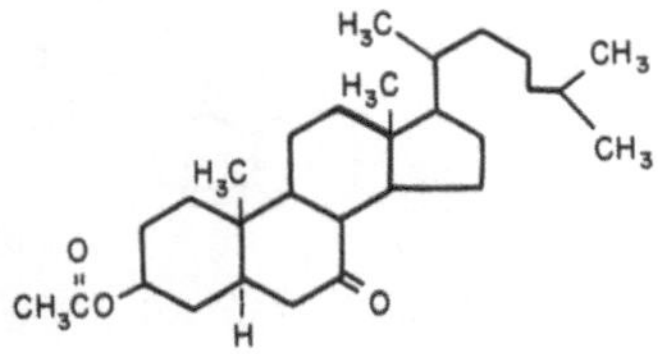

Helv. Chim. Acta 44:1380 (1961)

▶ $O_3C_{29}H_{48}$

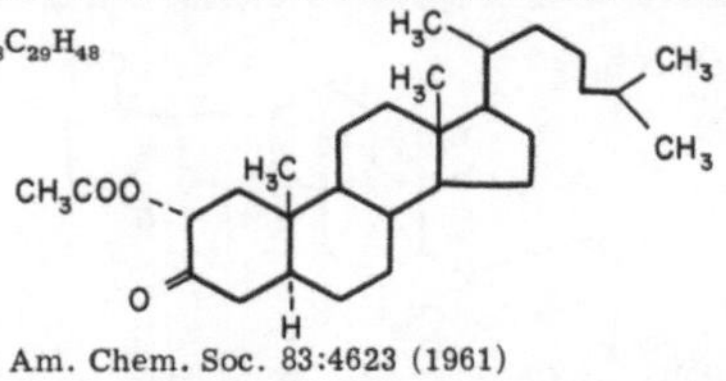

J. Am. Chem. Soc. 83:4623 (1961)

▶ $O_3C_{29}H_{48}$

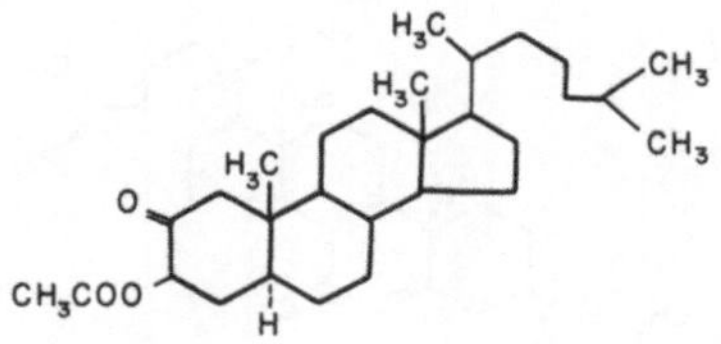

J. Am. Chem. Soc. 83:4623 (1961)

▶ $O_3C_{29}H_{48}$

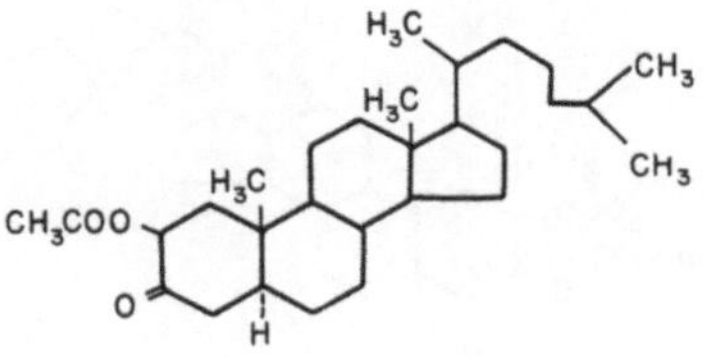

J. Am. Chem. Soc. 83:4623 (1961)

▶ $O_3C_{29}H_{48}$

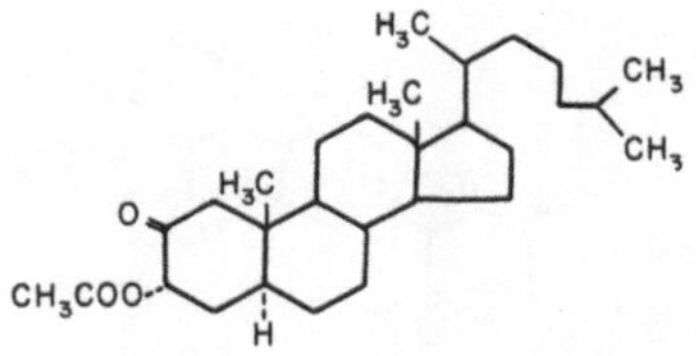

J. Am. Chem. Soc. 83:4623 (1961)

▶ $O_3C_{29}H_{48}$

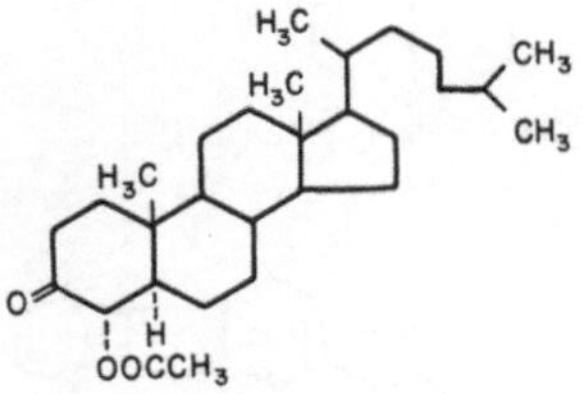

J. Am. Chem. Soc. 83:4623 (1961)

▶ $O_3C_{29}H_{48}$

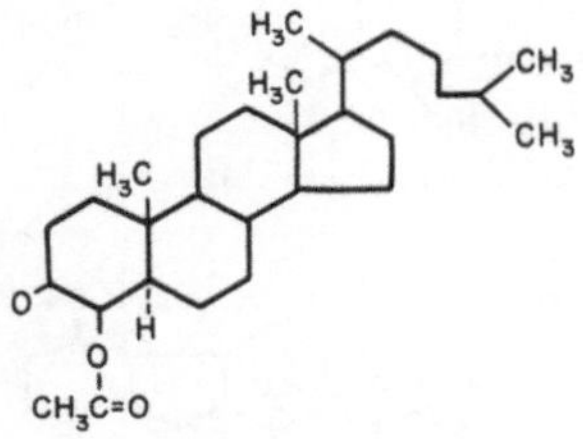

J. Am. Chem. Soc. 83:4623 (1961)

▶ $O_3C_{29}H_{48}$

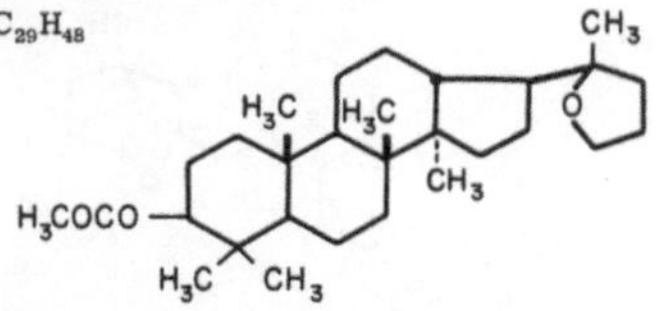

Bull. Soc. Chim. France 1832 (1962)

▶ $O_3C_{29}H_{50}$

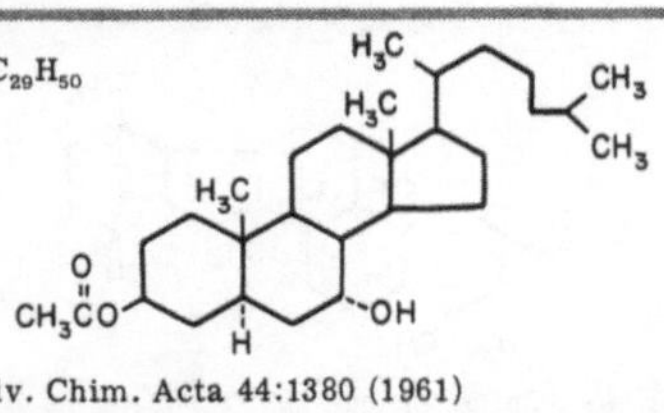

Helv. Chim. Acta 44:1380 (1961)

▶ $O_3C_{29}H_{50}$

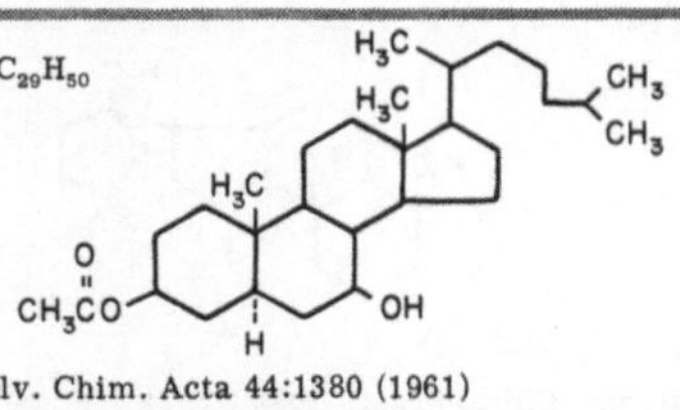

Helv. Chim. Acta 44:1380 (1961)

▶ $O_3C_{29}H_{50}$

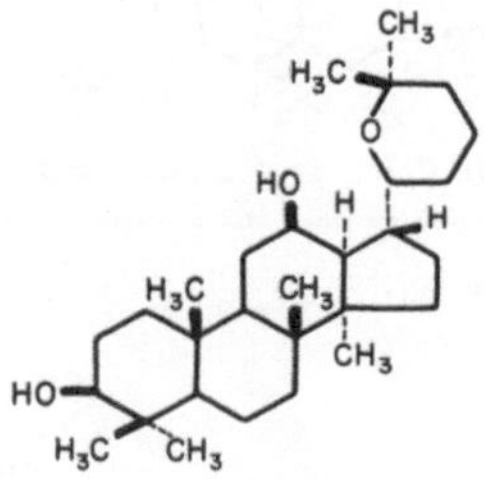

Tetrahedron Letters No. 26:1239 (1962)

▶ $O_3C_{30}H_{46}$

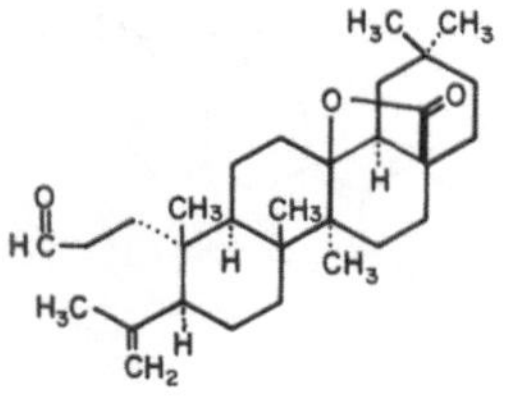

Helv. Chim. Acta 45:2674 (1962)

▶ $O_3C_{30}H_{26}$

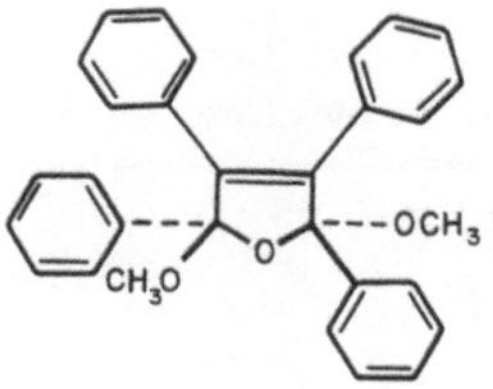

J. Org. Chem. 27:2764 (1962)

▶ $O_3C_{30}H_{48}$

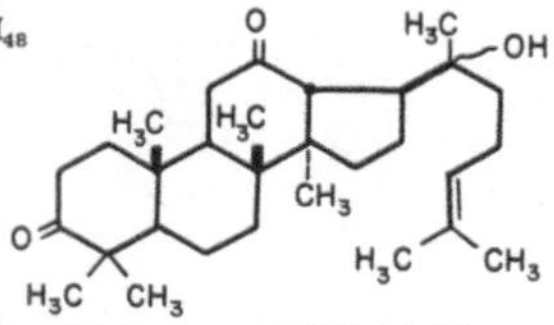

Bull. Soc. Chim. France 1832 (1962)

▶ $O_3C_{30}H_{50}$

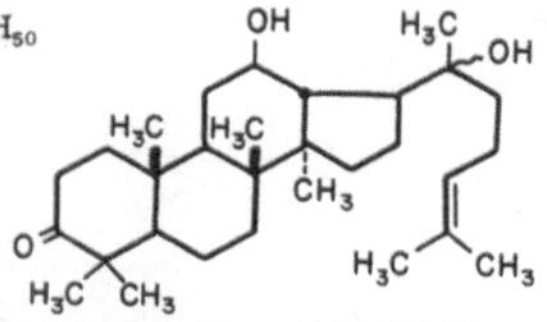

Bull. Soc. Chim. France 1832 (1962)

▶ $O_3C_{30}H_{26}$

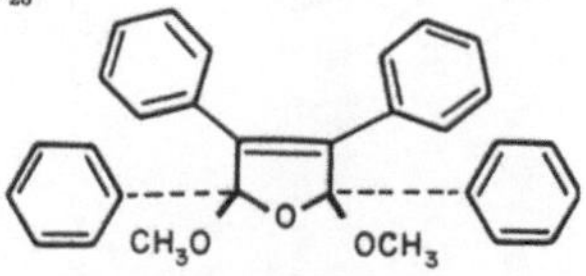

J. Org. Chem. 27:2764 (1962)

▶ $O_3C_{30}H_{52}$

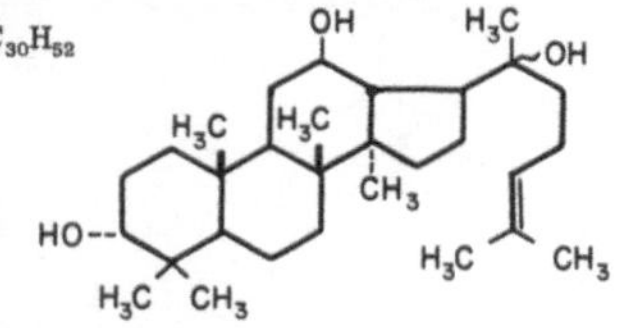

Bull. Soc. Chim. France 1832 (1962)

▶ $O_3C_{30}H_{42}$

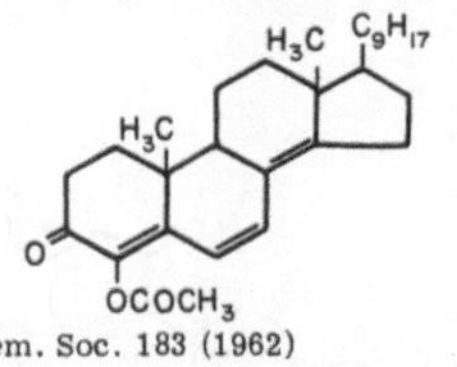

Proc. Chem. Soc. 183 (1962)

▶ $O_3C_{30}H_{46}$

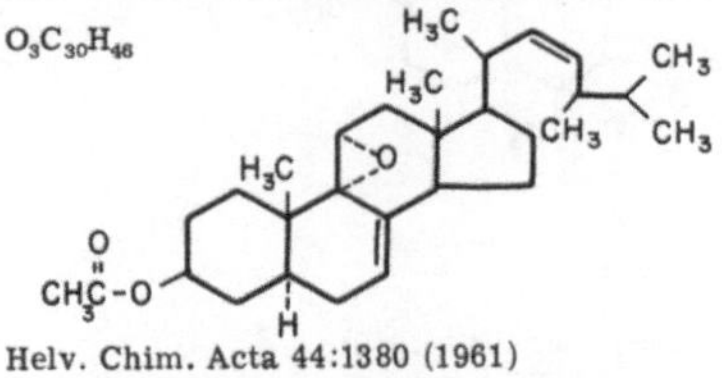

Helv. Chim. Acta 44:1380 (1961)

▶ $O_3C_{31}H_{46}$

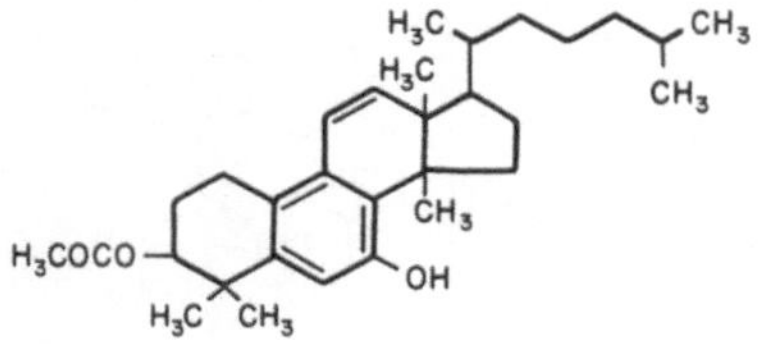

Chem. Pharm. Bull. 9:131 (1961)

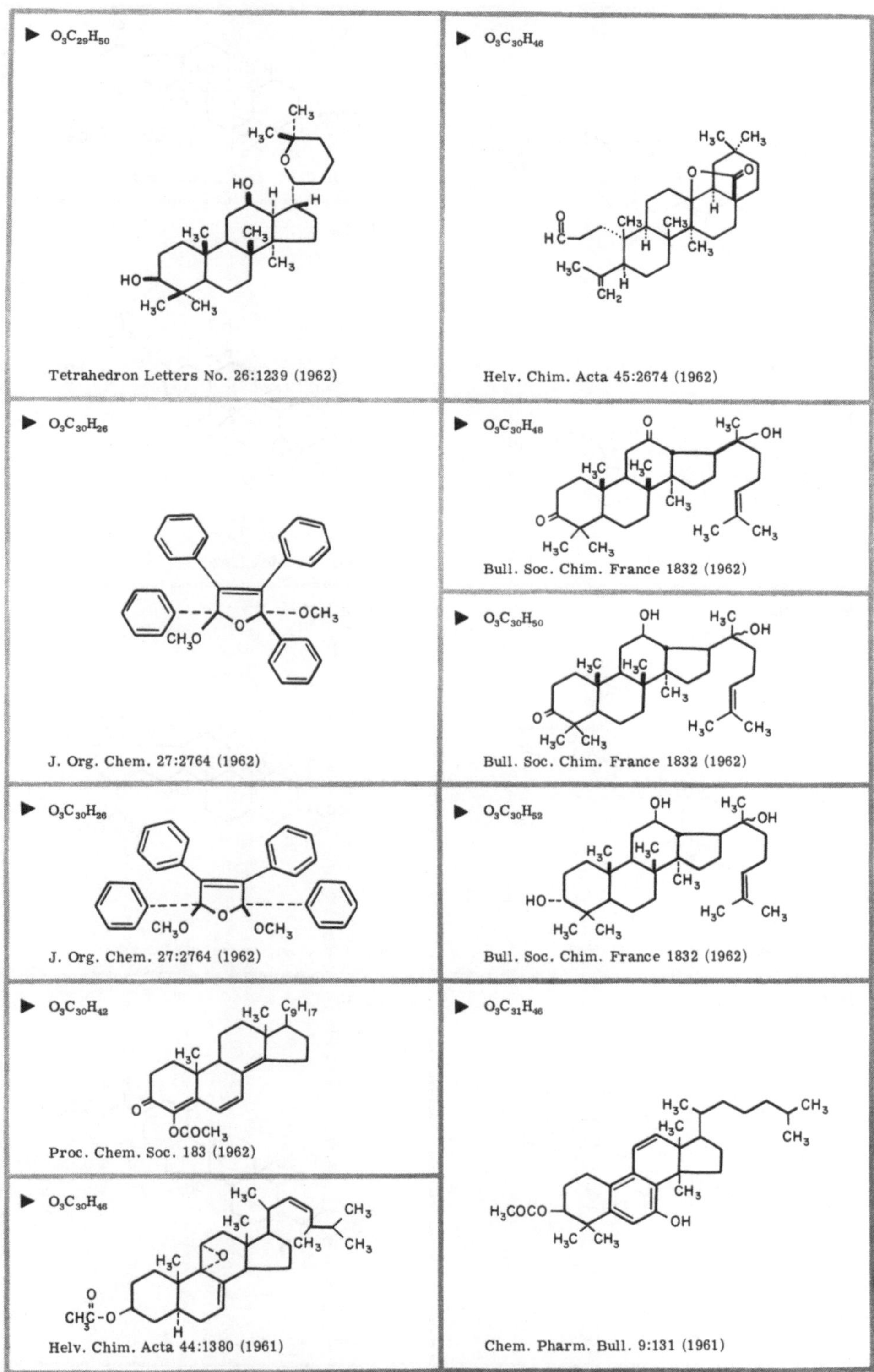

▶ $O_3C_{31}H_{48}$

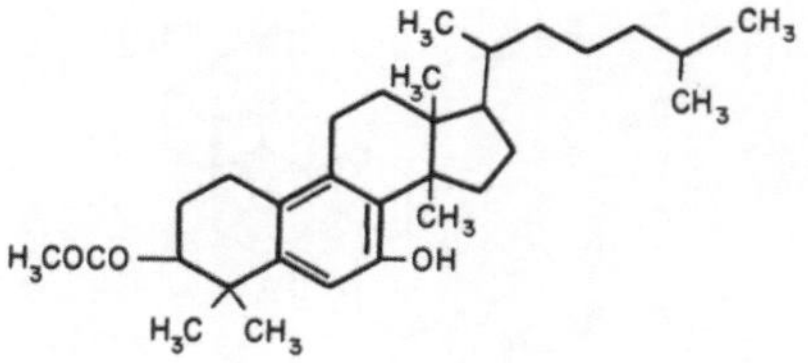

Chem. Pharm. Bull. 9:131 (1961)

▶ $O_3C_{31}H_{48}$

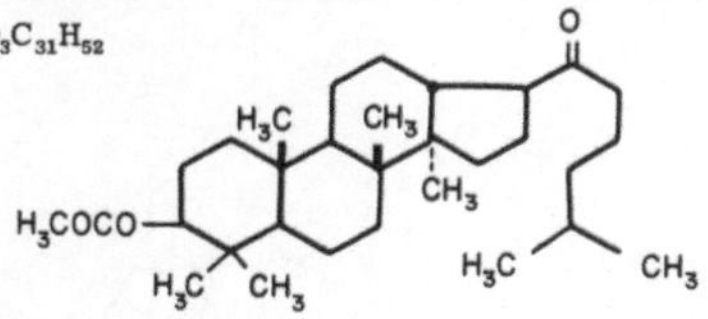

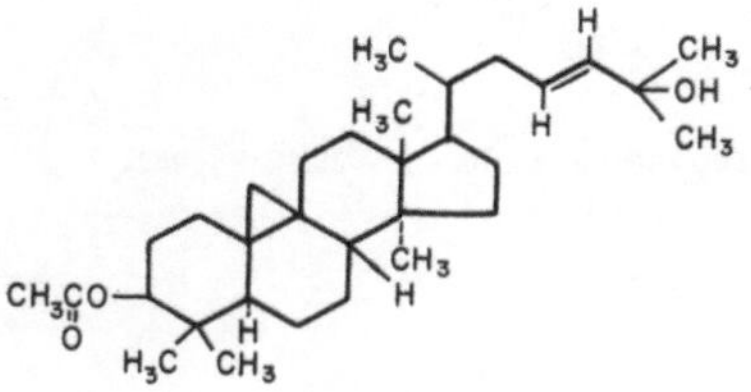

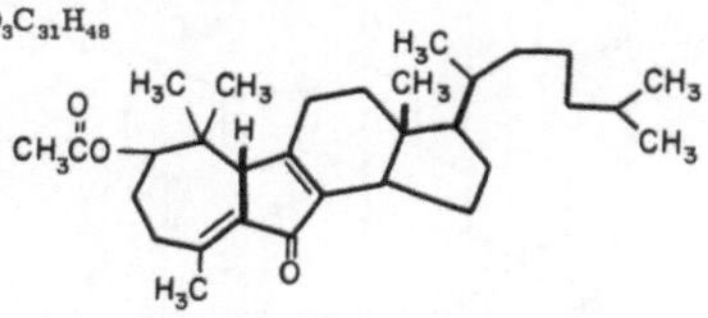

J. Chem. Soc. 1217 (1961)

▶ $O_3C_{31}H_{50}$

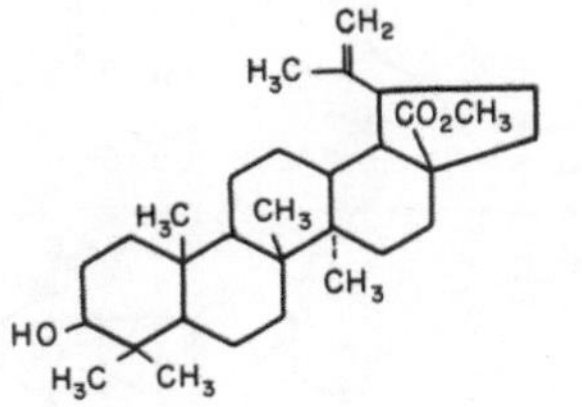

Bull. Soc. Chim. France 1137 (1962)

▶ $O_3C_{31}H_{50}$

Bull. Soc. Chim. France 1137 (1962)

▶ $O_3C_{31}H_{52}$

Bull. Soc. Chim. France 1832 (1962)

▶ $O_3C_{32}H_{52}$

J. Chem. Soc. 4034 (1962)

▶ $O_3C_{32}H_{54}$

Bull. Soc. Chim. France 1832 (1962)

▶ $O_3C_{33}H_{54}$

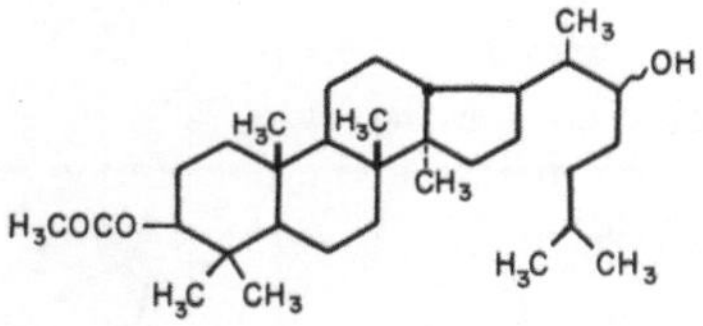

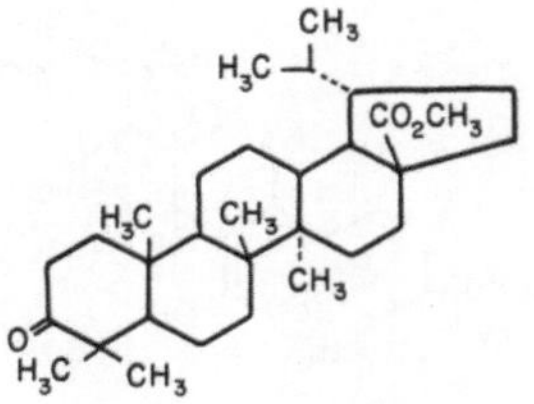

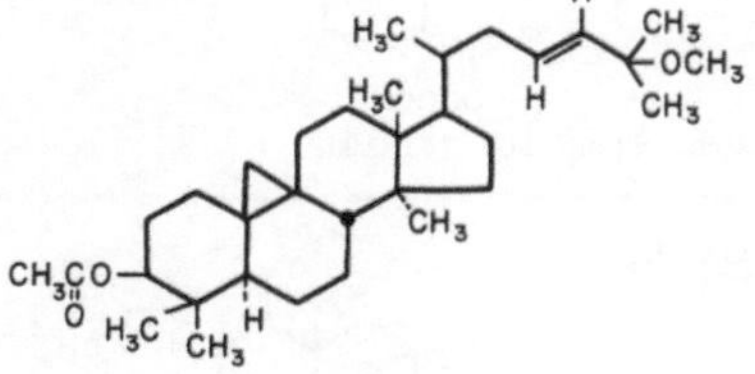

J. Chem. Soc. 4035 (1962)

▶ O4BrC15H19
J. Org. Chem. 27:1898 (1962)
▶ O4BrC21H29
J. Am. Chem. Soc. 84:4356 (1962)
▶ O4BrC21H29
J. Am. Chem. Soc. 84:4356 (1962)
▶ O4BrC27H39
Bull. Soc. Chim. France 1832 (1962)
▶ O4BrC29H45
Bull. Soc. Chim. France 1832 (1962)
▶ O4Br2C8H10
J. Chem. Phys. 37:2725 (1962)
▶ O4ClCH3
CH3OClO3
J. Chem. Soc. 1372 (1962)
▶ O4ClC7H5
J. Org. Chem. 26:1029 (1961)
▶ O4ClC24H31
Chem. Ber. 95:2110 (1962)
▶ O4FC23H31
J. Org. Chem. 26:3080 (1961)
▶ O4FC23H33
J. Am. Chem. Soc. 84:3784 (1962)
▶ O4FC24H31
Chem. Ber. 95:2110 (1962)
▶ O4F2C16H14
J. Chem. Soc. 3829 (1962)

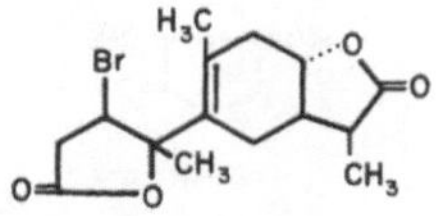

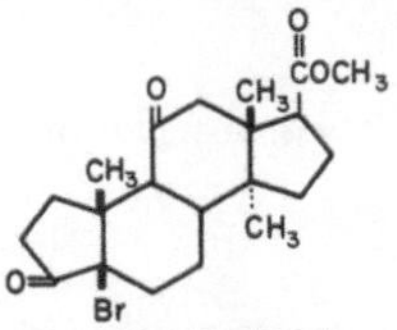

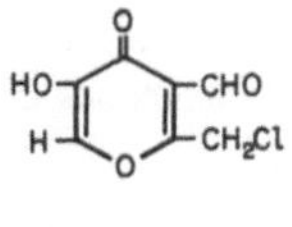

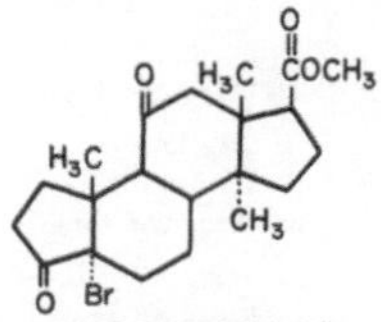

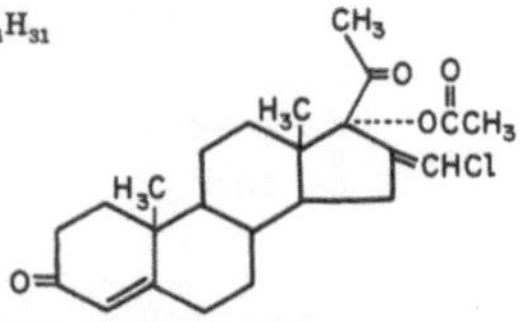

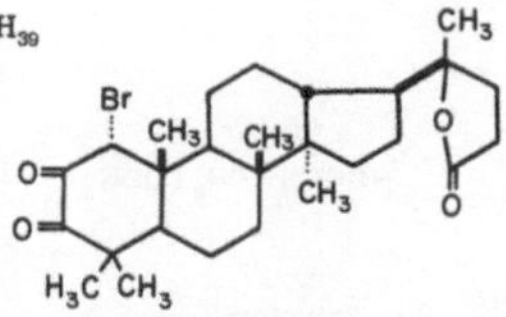

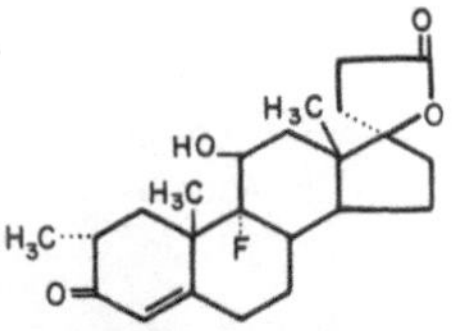

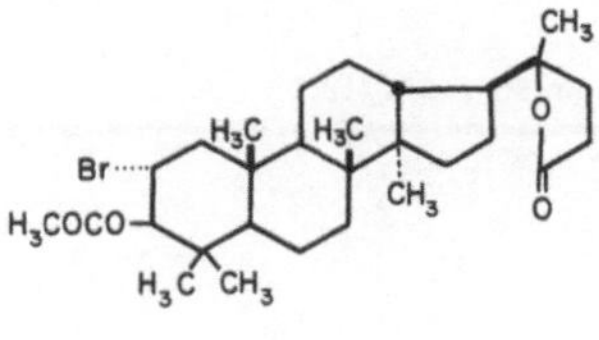

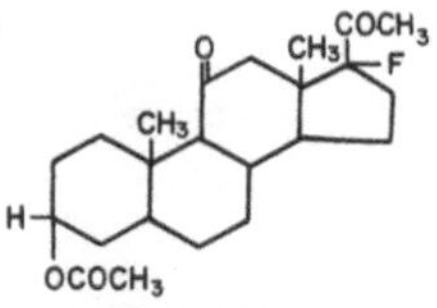

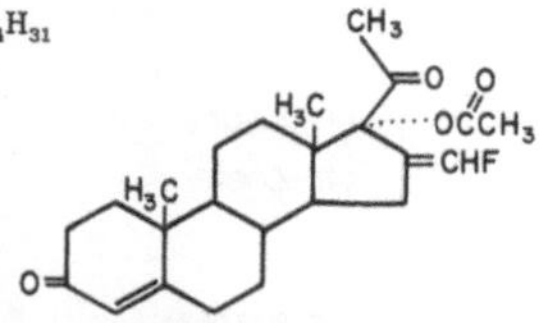

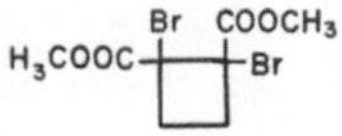

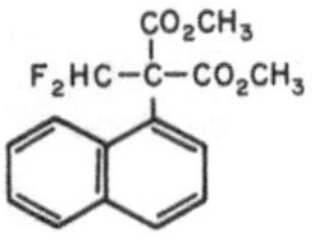

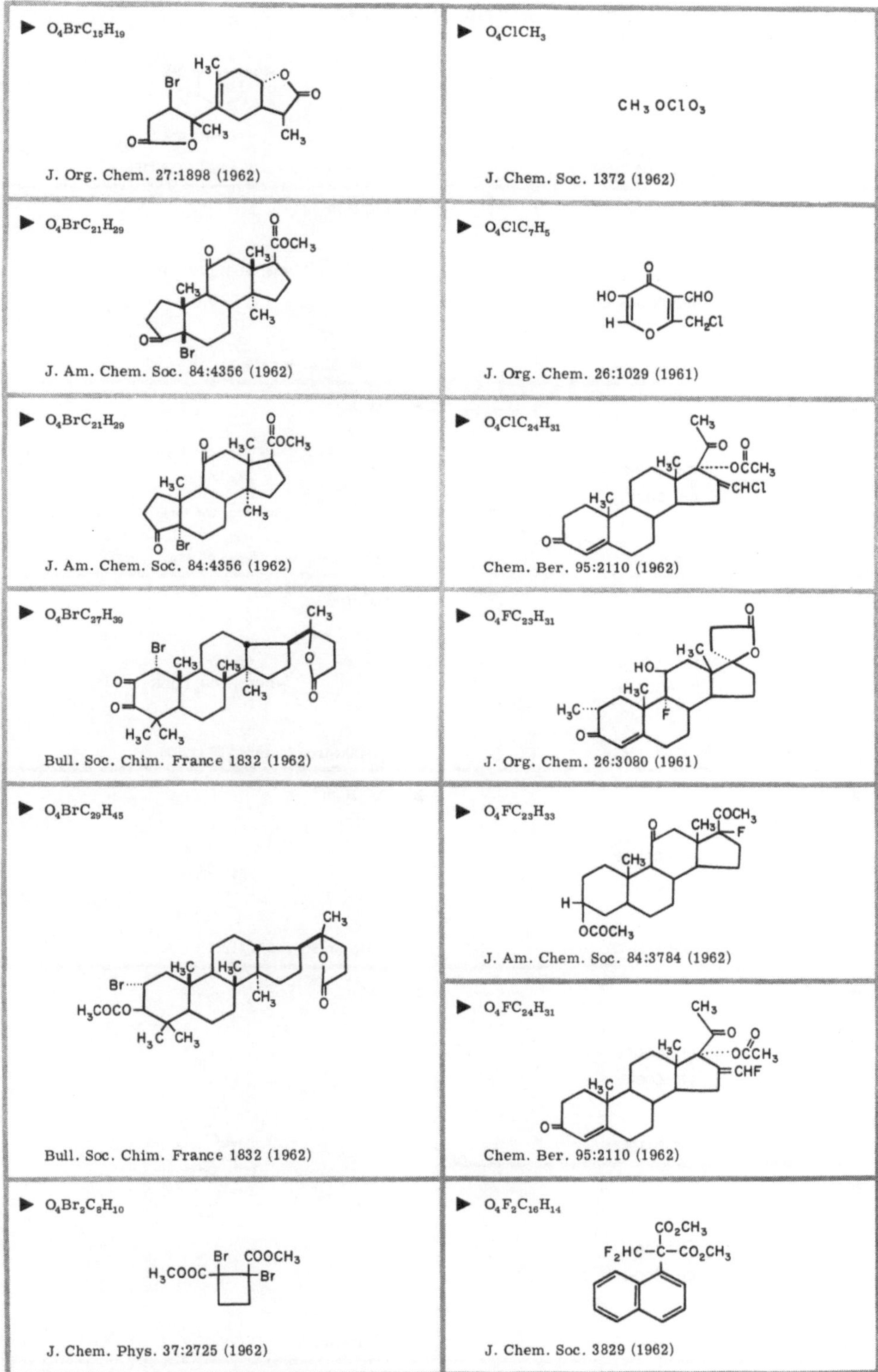

▶ $O_4C_3H_4$

$CH_2(COOH)_2$

J. Chem. Phys. 37:2198 (1962)
J. Am. Chem. Soc. 84:3973 (1962)

▶ $O_4C_4H_4$

$CH{-}COOCs$ ‖ $CH{-}COOCs$

J. Phys. Chem. 66:1702 (1962)

▶ $O_4C_4H_4$

$CH{-}COOK$ ‖ $CH{-}COOK$

J. Phys. Chem. 66:1702 (1962)

▶ $O_4C_4H_4$

$CH{-}COOLi$ ‖ $CH{-}COOLi$

J. Phys. Chem. 66:1702 (1962)

▶ $O_4C_4H_4$

$CH{-}COONa$ ‖ $CH{-}COONa$

J. Phys. Chem. 66:1702 (1962)

▶ $O_4C_4H_4$

$CH{-}COORb$ ‖ $CH{-}COORb$

J. Phys. Chem. 66:1702 (1962)

▶ $O_4C_4H_4$

$CH{-}COOH$ ‖ $CH{-}COOH$

J. Phys. Chem. 66:1702 (1962)

▶ $O_4C_4H_6$

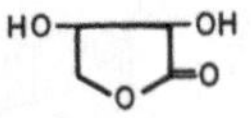


cis and trans

Bull. Chem. Soc. Japan 35:1742 (1962)

▶ $O_4C_4H_6Hg$

$CH_3C(=O)OHgC(=O)CH_3$

Chem. &Ind. 668 (1961)

▶ $O_4C_4H_8$

$CH_3CH(OH){-}CH(OH){-}CO_2H$

erythro and threo

Bull. Chem. Soc. Japan 35:1742 (1962)

▶ $O_4C_4H_8$

$HOOCH(CH_3){-}CH_2COOH$

Tetrahedron Letters 708 (1961)

▶ $O_4C_4H_{12}Si$

$(H_3CO)_4Si$

J. Chem. Soc. 1372 (1962)

▶ $O_4C_5H_6$

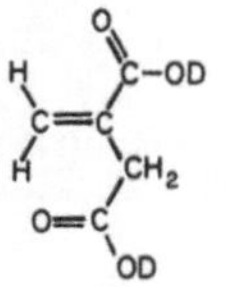


Varian 433

▶ $O_4C_5H_8$

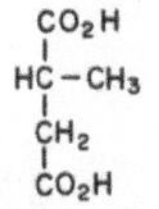

$HO_2C-CH(CH_3)-CH_2-CO_2H$

Can. J. Chem. 39:2267 (1961)

▶ $O_4C_5H_8$

$CH_3C(=O)-O-CH_2-O-C(=O)CH_3$

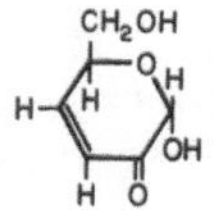

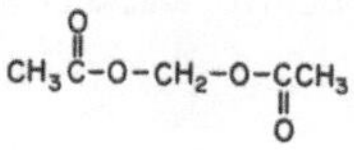

Chem. Ber. 94:3317 (1961)

▶ $O_4C_5H_{10}$

$CH_3CH(OH)-CH(OH)-CO_2CH_3$

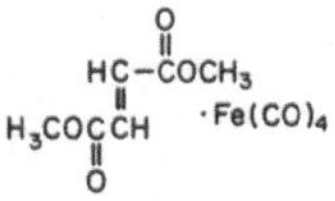

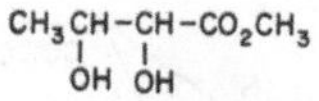

erythro and threo

Bull. Chem. Soc. Japan 35:1742 (1962)

▶ $O_4C_5H_{10}$

$(CH_3O)_2CHCOOCH_3$

Varian 446

▶ $O_4C_6H_6$

Varian 455

▶ $O_4C_6H_6$

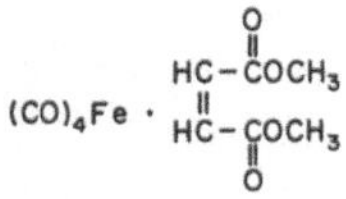

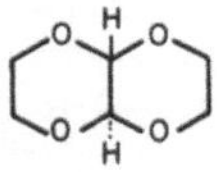

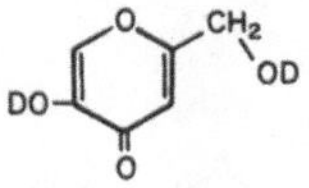

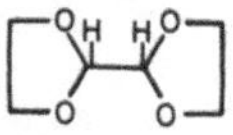

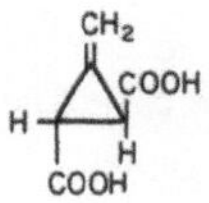

J. Am. Chem. Soc. 84:2249 (1962)

▶ $O_4C_6H_8$

$CH_3O_2CCH=CHCO_2CH_3$

cis and trans

Tetrahedron Letters No. 23:1063 (1962)
J. Chem. Soc. 2881 (1961)

▶ $O_4C_6H_8$

cis and trans

Australian J. Chem. 15:503 (1962)

▶ $O_4C_6H_8 \cdot Fe(CO)_4$

$HC(-COCH_3{=}O)=CH-C(=O)OCH_3 \cdot Fe(CO)_4$

Tetrahedron Letters No. 23:1063 (1962)

▶ $O_4C_6H_8 \cdot Fe(CO)_4$

Tetrahedron Letters No. 23:1063 (1962)

▶ $O_4C_6H_{10}$

J. Org. Chem. 27:3183 (1962)

▶ $O_4C_6H_{10}$

J. Org. Chem. 27:3183 (1962)

▶ $O_4C_6H_{10}$

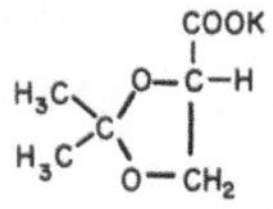

J. Am. Chem. Soc. 83:3901 (1961)

▶ $O_4C_6H_{10}$

$HO_2C-CH_2-CH(CH_3)-CH_2-CO_2H$

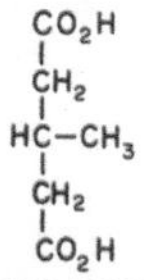

Can. J. Chem. 39:2267 (1961)

► $O_4C_7H_8$

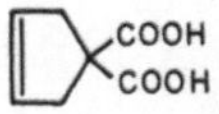

J. Org. Chem. 27:3367 (1962)

► $O_4C_7H_8$

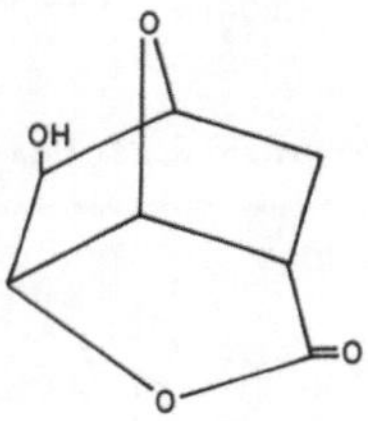

J. Am. Chem. Soc. 84:4123 (1962)

► $O_4C_7H_{10}$

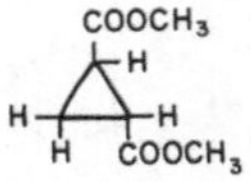

J. Org. Chem. 27:4312 (1962)

► $O_4C_7H_{10}$

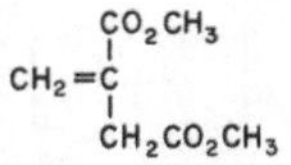

J. Am. Chem. Soc. 83:922 (1961)

► $O_4C_7H_{10}$

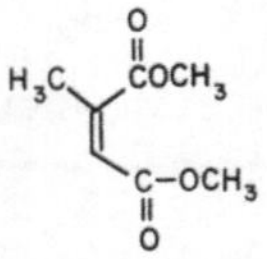

METHYL CITRACONATE

Can. J. Chem. 39:505 (1961)

► $O_4C_7H_{10}$

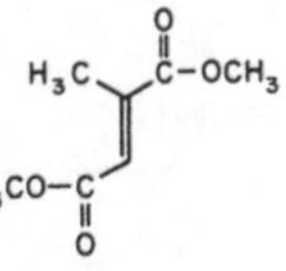

METHYL MESACONATE

Can. J. Chem. 39:505 (1961)

► $O_4C_7H_{10}$

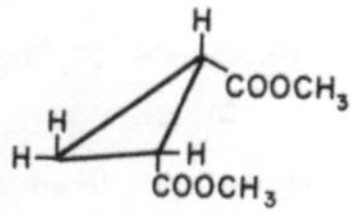

J. Org. Chem. 27:4312 (1962)

► $O_4C_7H_{10}$

$(CH_3COO)_2CHCH{=}CH_2$

Varian 492

► $O_4C_7H_{12}$

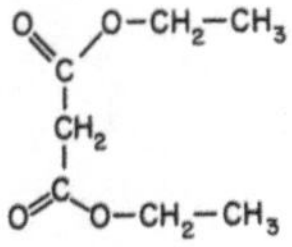

Varian 181

► $O_4C_7H_{12}$

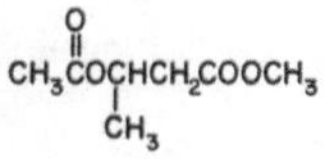

Varian 182

► $O_4C_7H_{14}$

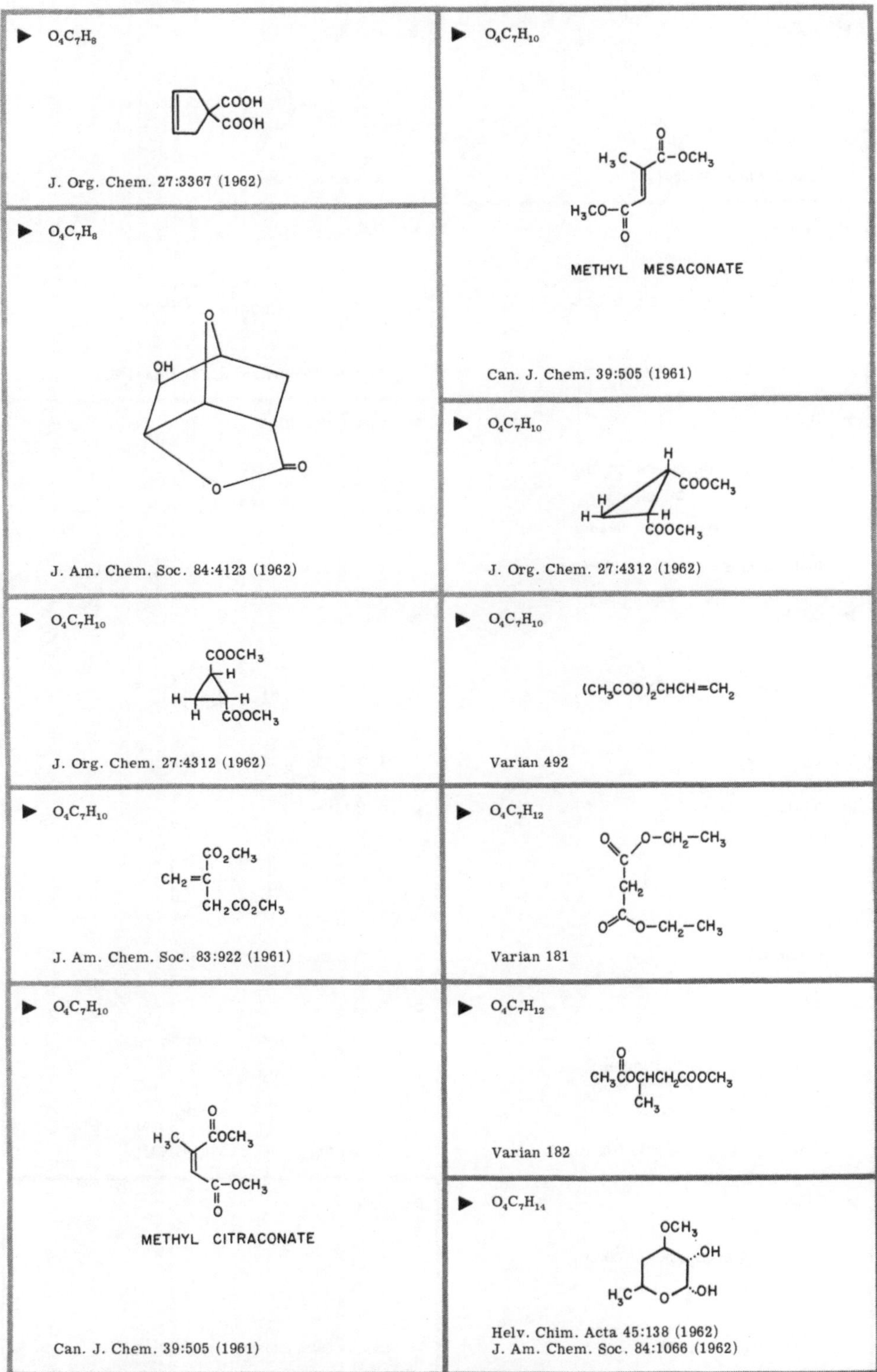

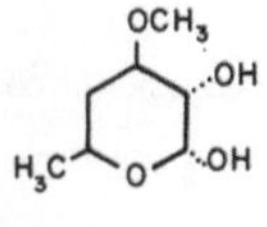

Helv. Chim. Acta 45:138 (1962)
J. Am. Chem. Soc. 84:1066 (1962)

▶ $O_4C_7H_{14}$

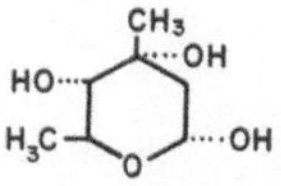

Tetrahedron 18:1265 (1962)

▶ $O_4C_8H_{10}$

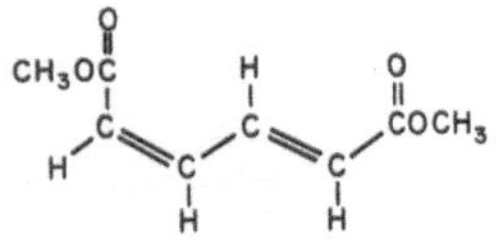

J. Am. Chem. Soc. 84:4307 (1962)

▶ $O_4C_7H_{14}$

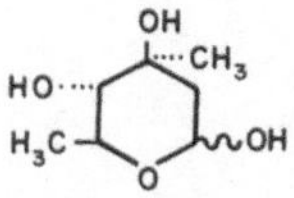

Tetrahedron 18:1265 (1962)

▶ $O_4C_8H_{10}$

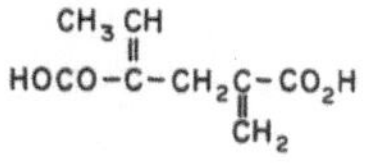

J. Am. Chem. Soc. 83:922 (1961)

▶ $O_4C_7H_{14}$

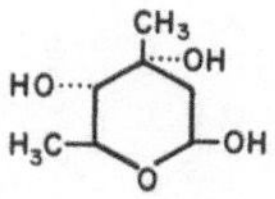

Tetrahedron 18:1265 (1962)

▶ $O_4C_8H_{12}$

$H_5C_2OOC-CH=CH-COOC_2H_5$

cis and trans

Varian 212
J. Chem. Phys. 37:2729 (1962)
Tetrahedron Letters No. 23:1063 (1962)

▶ $O_4C_8H_6$

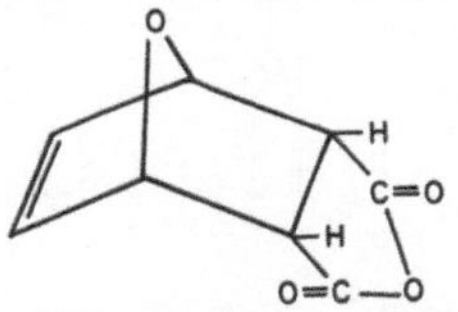

Tetrahedron Letters No. 25:1220 (1962)

▶ $O_4C_8H_{12}$

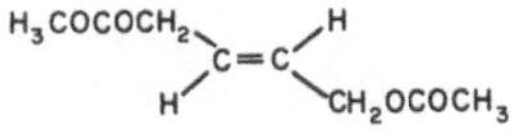

Tetrahedron 18:791 (1962)

▶ $O_4C_8H_6$

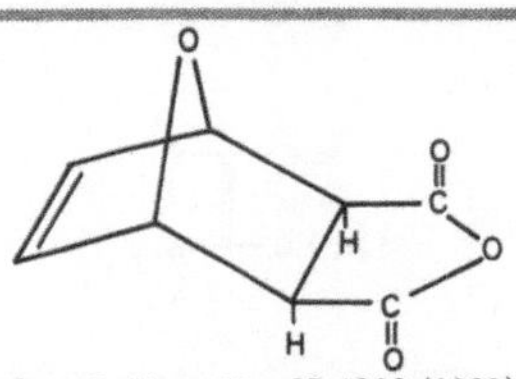

Tetrahedron Letters No. 25:1220 (1962)

▶ $O_4C_8H_{12}$

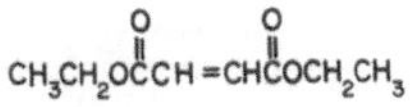

Varian 213

▶ $O_4C_8H_8$

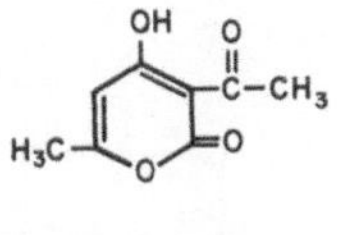

Arkiv Kemi 17:523 (1961)

▶ $O_4C_8H_{12}$

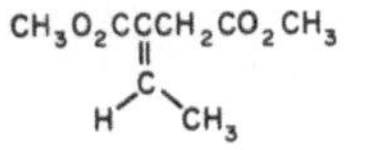

J. Am. Chem. Soc. 83:922 (1961)

▶ $O_4C_8H_8$

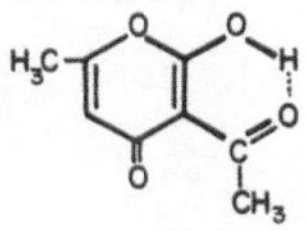

Varian 504

▶ $O_4C_8H_{12} \cdot Fe(CO)_4$

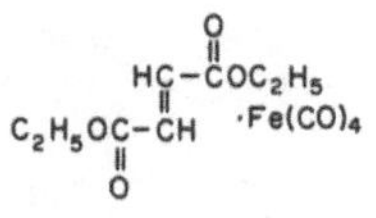

Tetrahedron Letters No. 23:1063 (1962)

▶ $O_4C_8H_{14}$

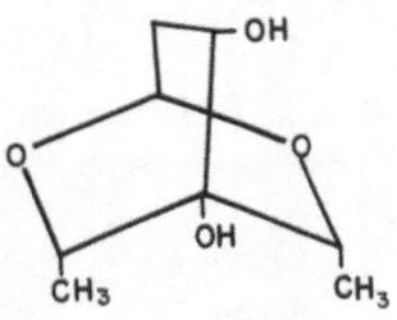

J. Am. Chem. Soc. 84:3183 (1962)

▶ $O_4C_8H_{18}$

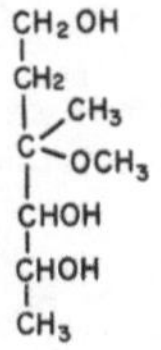

Helv. Chim. Acta 45:138 (1962)

▶ $O_4C_8H_{14}$

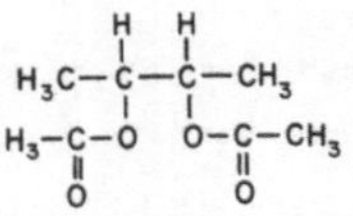

meso and dl

J. Am. Chem. Soc. 84:743 (1962)

▶ $O_4C_9H_{10}$

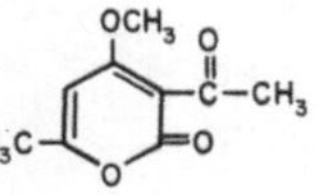

Arkiv Kemi 17:523 (1961)

▶ $O_4C_8H_{14}$

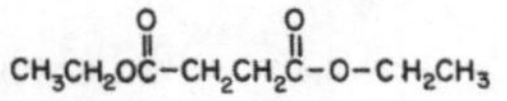

Varian 215

▶ $O_4C_9H_{10}$

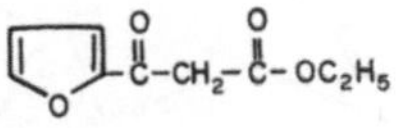

Can. J. Chem. 39:909 (1961)

▶ $O_4C_8H_{16}$

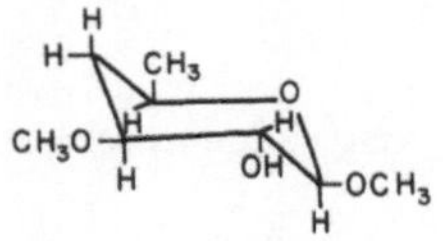

J. Am. Chem. Soc. 84:1066 (1962)

▶ $O_4C_9H_{12}$

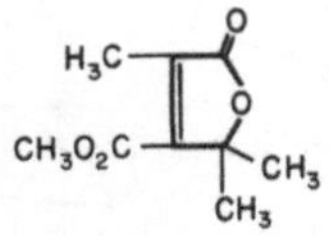

Bull. Chem. Soc. Japan 35:1194 (1962)

▶ $O_4C_8H_{16}$

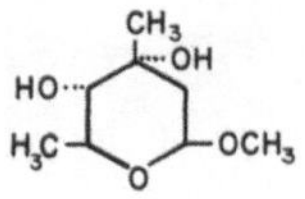

Tetrahedron 18:1265 (1962)

▶ $O_4C_8H_{16}$

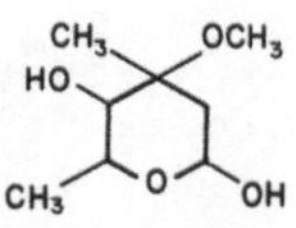

Tetrahedron 18:1265 (1962)

▶ $O_4C_9H_{16}$

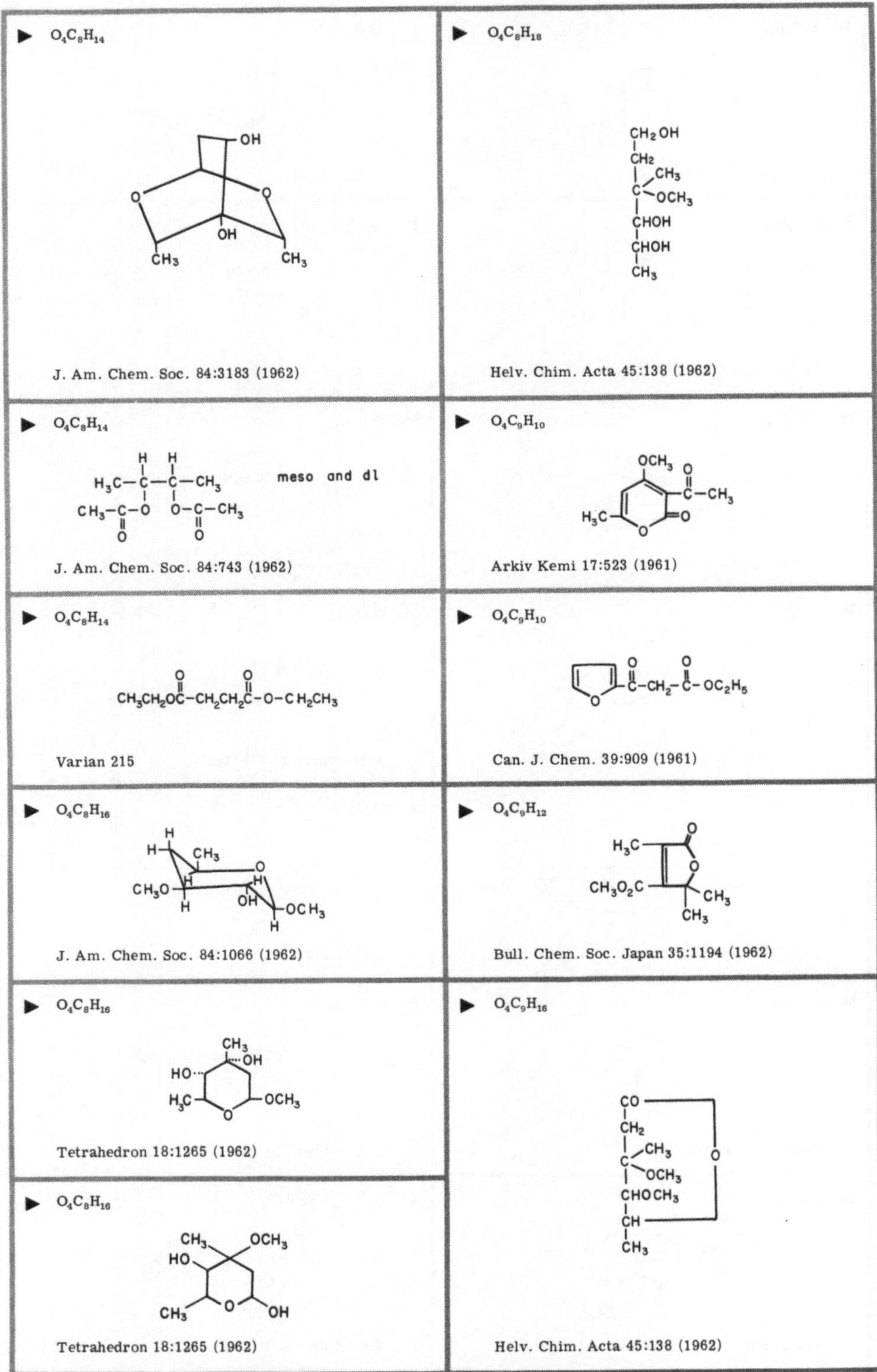

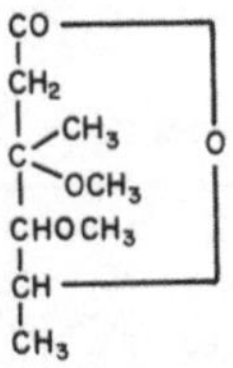

Helv. Chim. Acta 45:138 (1962)

▶ $O_4C_9H_{16}$

$CH_2{=}CHCH_2O{-}(CH_2O)_3{-}CH_2CH{=}CH_2$

J. Chem. Soc. 4313 (1962)

▶ $O_4C_9H_{20}$

$CH_3CH_2CH_2O(CH_2O)_3CH_2CH_2CH_3$

J. Chem. Soc. 4313 (1962)

▶ $O_4C_{10}H_8$

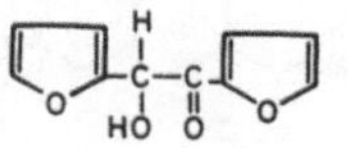

Varian 247

▶ $O_4C_{10}H_{10}$

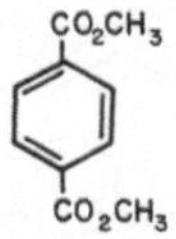

Helv. Chim. Acta 45:568 (1962)

▶ $O_4C_{10}H_{10}$

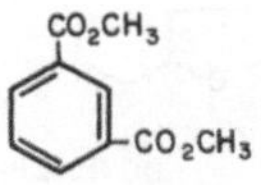

Helv. Chim. Acta 45:568 (1962)

▶ $O_4C_{10}H_{10}$

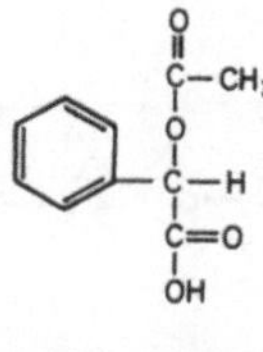

Varian 554

▶ $O_4C_{10}H_{12}$

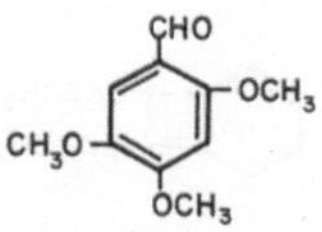

J. Org. Chem. 27:3733 (1962)

▶ $O_4C_{10}H_{12}$

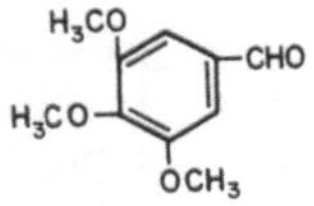

Can. J. Chem. 40:1073 (1962)

▶ $O_4C_{10}H_{12}$

$H_2C{=}CHCH_2OC(=O)CH{=}CHC(=O)OCH_2CH{=}CH_2$

Varian 262

▶ $O_4C_{10}H_{12}$

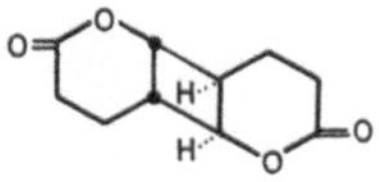

J. Am. Chem. Soc. 84:2344 (1962)

▶ $O_4C_{10}H_{12}$

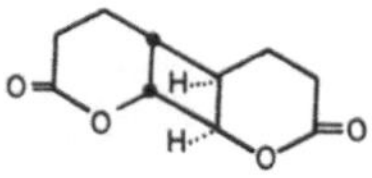

J. Am. Chem. Soc. 84:2344 (1962)

▶ $O_4C_{10}H_{14}$

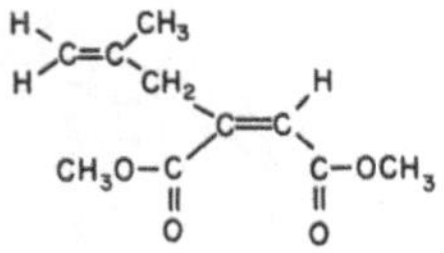

J. Org. Chem. 27:2731 (1962)

▶ $O_4C_{10}H_{14}$

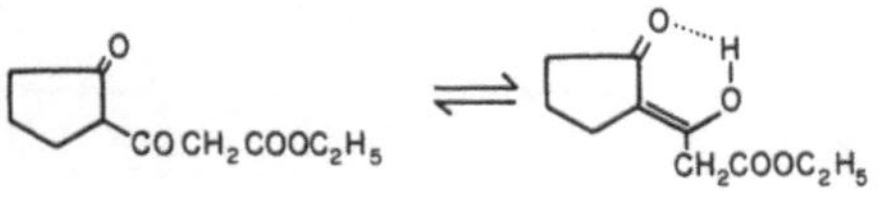

Tetrahedron 18:823 (1962)

▶ $O_4C_{10}H_{14}$

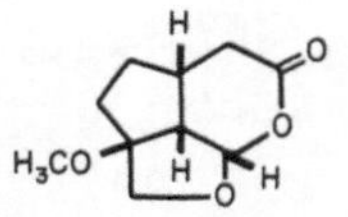

Tetrahedron Letters 110 (1961)

▶ $O_4C_{10}H_{14}$

$CH_2{=}C(CO_2CH_3)-CH_2-C(CO_2CH_3){=}CH-CH_3$

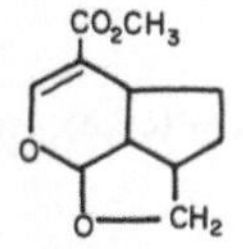

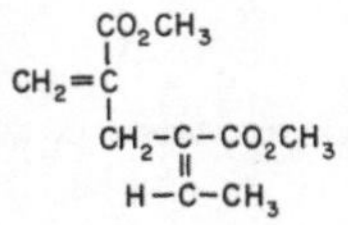

J. Am. Chem. Soc. 83:922 (1961)

▶ $O_4C_{10}H_{16}$

Tetrahedron Letters No. 14:612 (1962)

▶ $O_4C_{10}H_{16}$

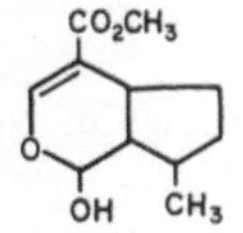

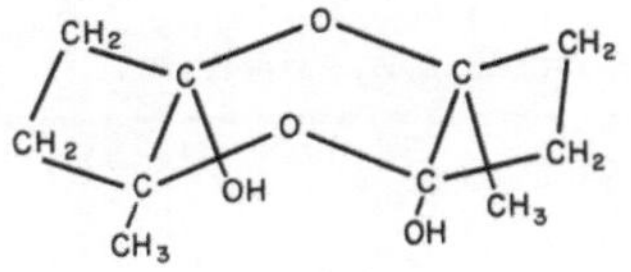

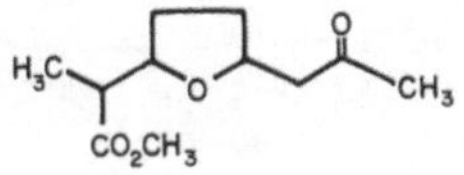

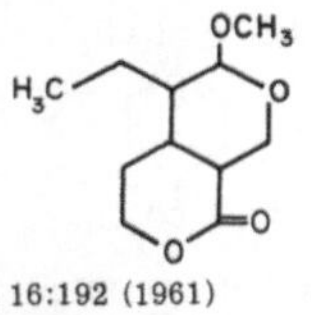

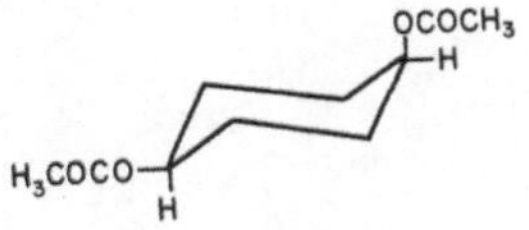

Tetrahedron 18:791 (1962)

▶ $O_4C_{10}H_{18}$

$H_5C_2O_2C-(CH_2)_4-CO_2C_2H_5$

Can. J. Chem. 40:363 (1962)

▶ $O_4C_{11}H_{14}$

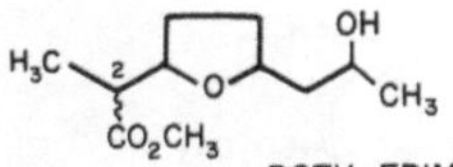

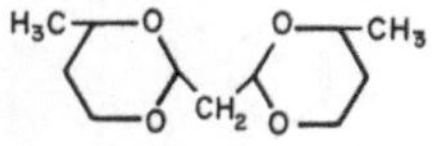

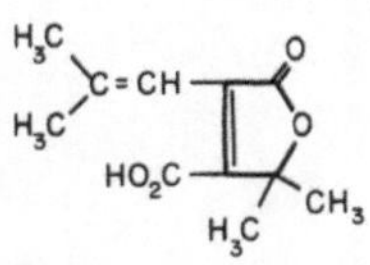

Bull. Chem. Soc. Japan 35:1194 (1962)

▶ $O_4C_{11}H_{14}$

J. Org. Chem. 26:1192 (1961)

▶ $O_4C_{11}H_{16}$

J. Org. Chem. 26:1192 (1961)

▶ $O_4C_{11}H_{18}$

Helv. Chim. Acta 45:129 (1962)

▶ $O_4C_{11}H_{18}$

Tetrahedron 16:192 (1961)

▶ $O_4C_{11}H_{20}$

BOTH EPIMERS (C-2)

Helv. Chim. Acta 45:129, 620 (1962)

▶ $O_4C_{11}H_{20}$

Varian 588

▶ $O_4C_{12}H_{14}$

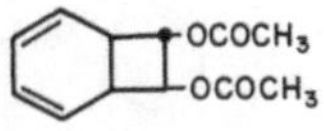

Tetrahedron Letters 721 (1961)

▶ $O_4C_{12}H_{14}$

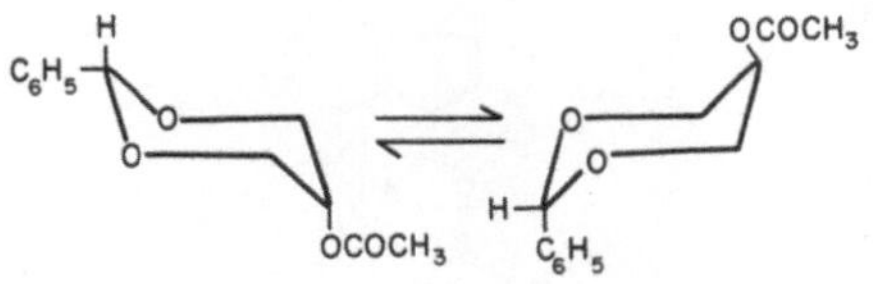

J. Chem. Soc. 5005 (1961)

▶ $O_4C_{12}H_{14}$

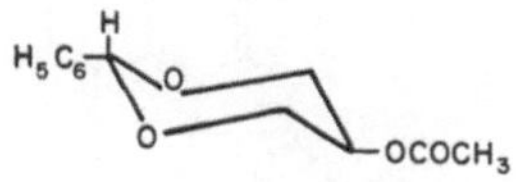

J. Chem. Soc. 5005 (1961)

▶ $O_4C_{12}H_{14}$

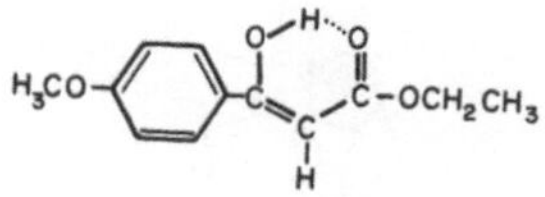

Varian 295

▶ $O_4C_{12}H_{16}$

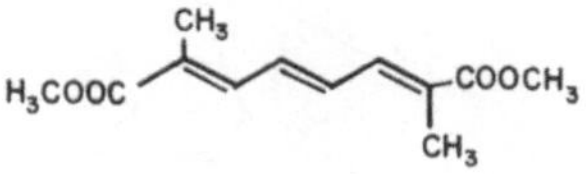

J. Chem. Soc. 1625 (1961)

▶ $O_4C_{12}H_{16}$

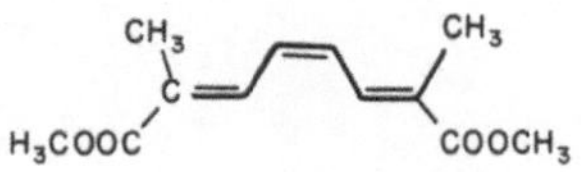

J. Chem. Soc. 1625 (1961)

▶ $O_4C_{12}H_{18}$

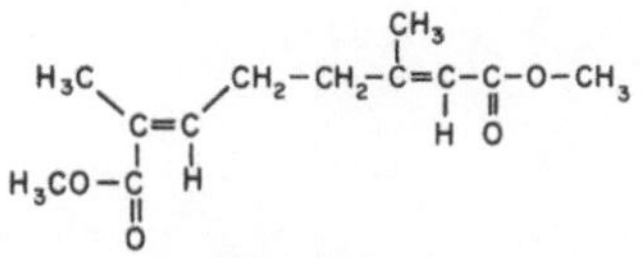

J. Chem. Soc. 1503 (1962)

▶ $O_4C_{12}H_{20}$

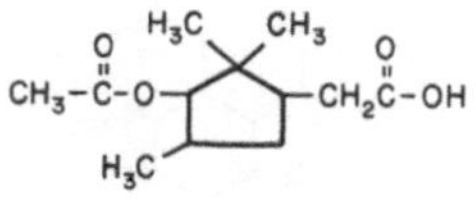

J. Org. Chem. 27:900 (1962)

▶ $O_4C_{12}H_{22}$

$CH_3OOC-(CH_2)_8-COOCH_3$

MCA Serial No. 58

▶ $O_4C_{12}H_{22}$

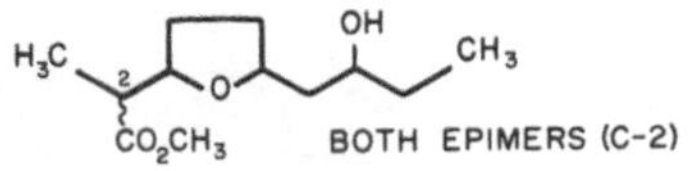


Helv. Chim. Acta 45:620 (1962)

▶ $O_4C_{12}H_{22}$

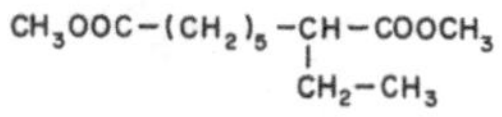

MCA Serial No. 59

▶ $O_4C_{12}H_{22}$

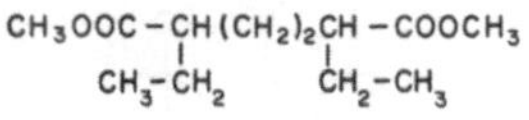

MCA Serial No. 60

▶ $O_4C_{13}H_8$

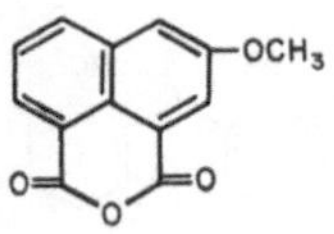

Tetrahedron 18:841 (1962)

▶ $O_4C_{13}H_8$

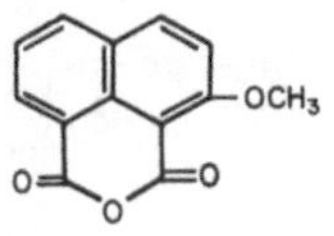

Tetrahedron 18:841 (1962)

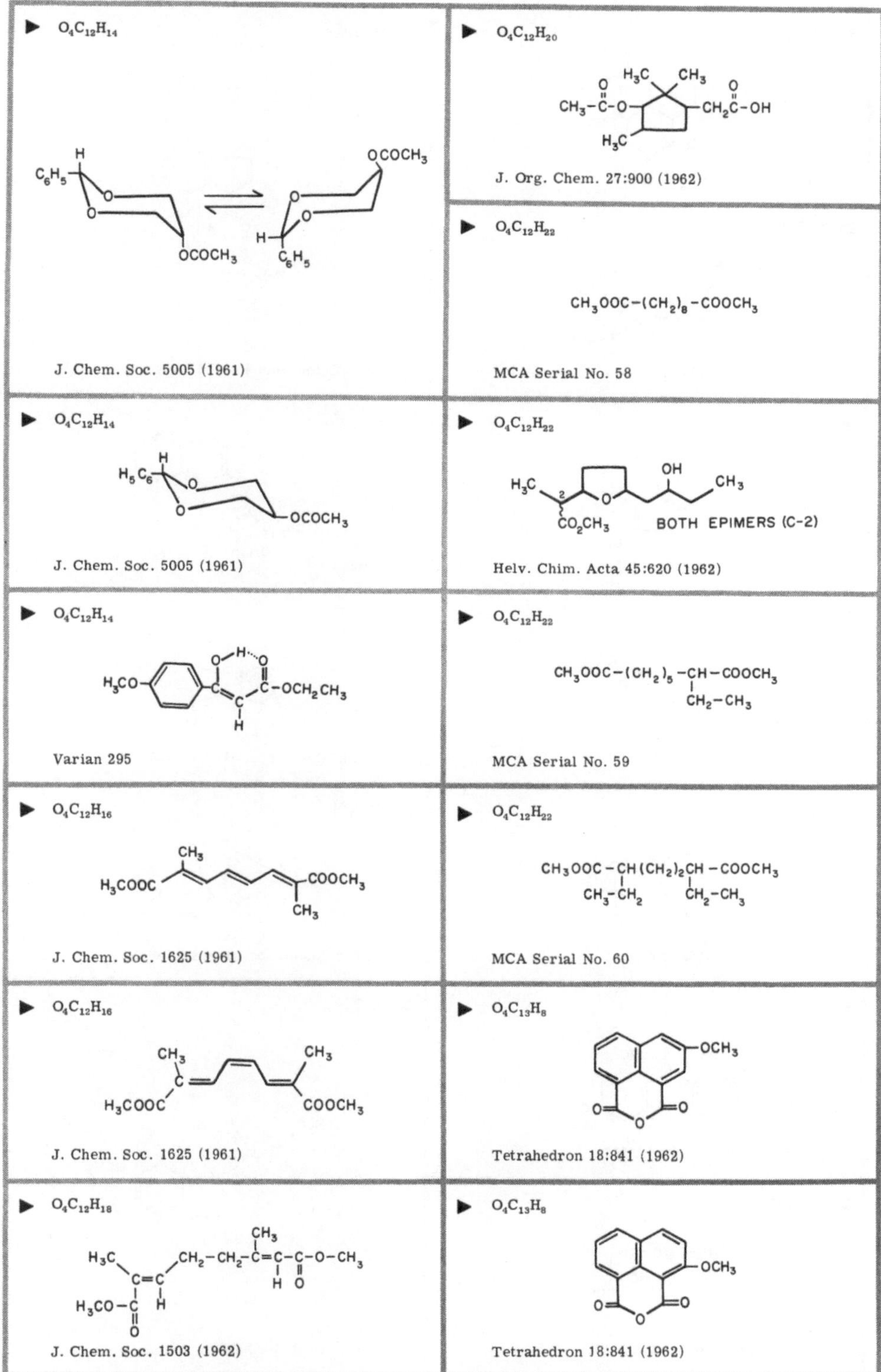

► $O_4C_{13}H_8$

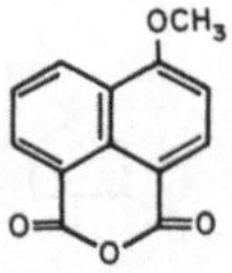

Tetrahedron 18:841 (1962)

► $O_4C_{13}H_{16}$

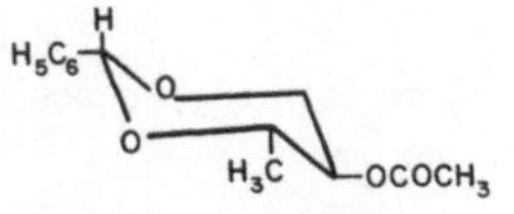

J. Chem. Soc. 5005 (1961)

► $O_4C_{13}H_{16}$

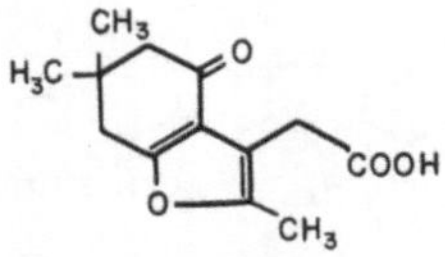

J. Org. Chem. 27:4502 (1962)

► $O_4C_{13}H_{16}$

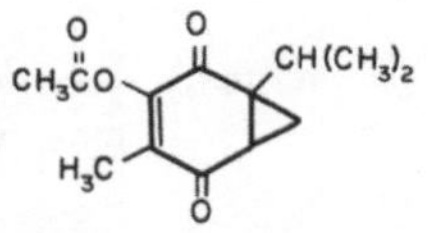

J. Chem. Soc. 5188 (1962)

► $O_4C_{13}H_{18}$

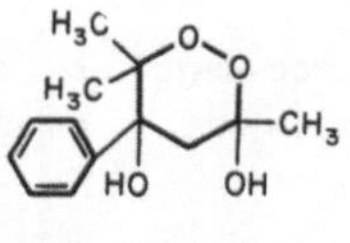

J. Chem. Soc. 4743 (1962)

► $O_4C_{14}H_{12}$

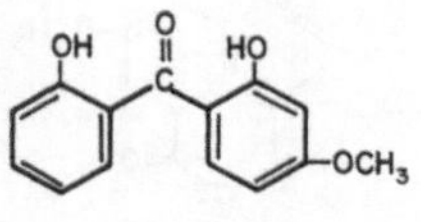

J. Phys. Chem. 65:2023 (1961)

► $O_4C_{14}H_{12}$

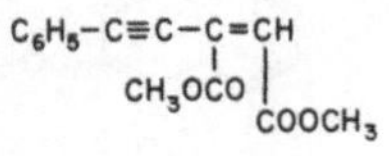

J. Am. Chem. Soc. 83:2018 (1961)

► $O_4C_{14}H_{12}$

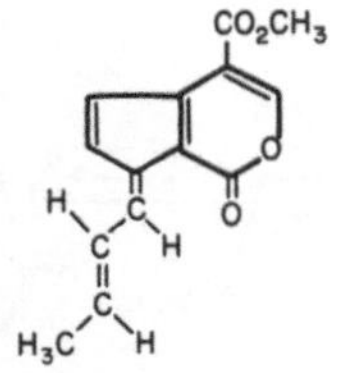

Helv. Chim. Acta 45:1406 (1962)

► $O_4C_{14}H_{14}$

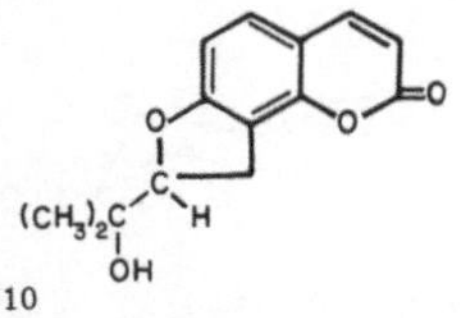

Varian 310

► $O_4C_{14}H_{16}$

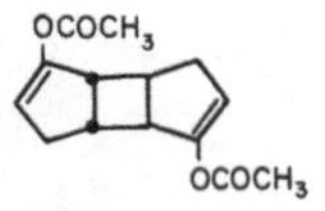

J. Am. Chem. Soc. 84:2344 (1962)

► $O_4C_{14}H_{16}$

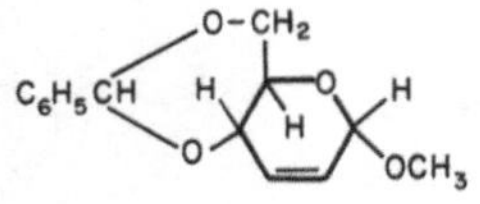

J. Am. Chem. Soc. 83:3827 (1961)

► $O_4C_{14}H_{18}$

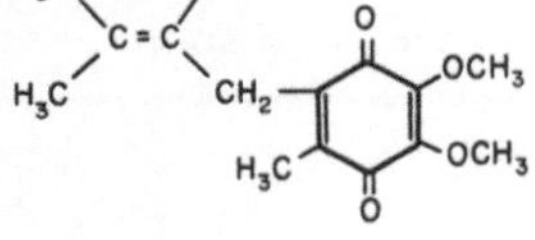

Chem. & Ind. 1020 (1962)

▶ $O_4C_{14}H_{22}$

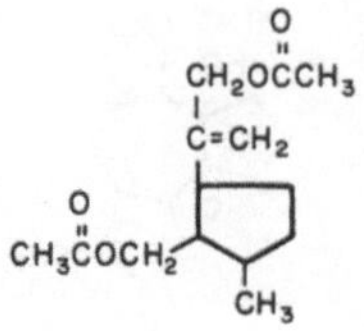

J. Org. Chem. 26:1192 (1961)

▶ $O_4C_{14}H_{24}$

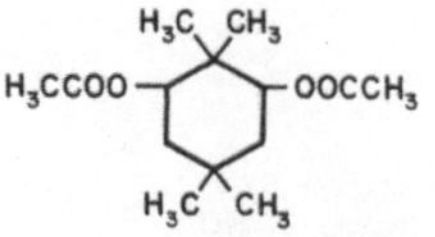

J. Chem. Phys. 34:594 (1961); 37:2473 (1962)

▶ $O_4C_{14}H_{24}$

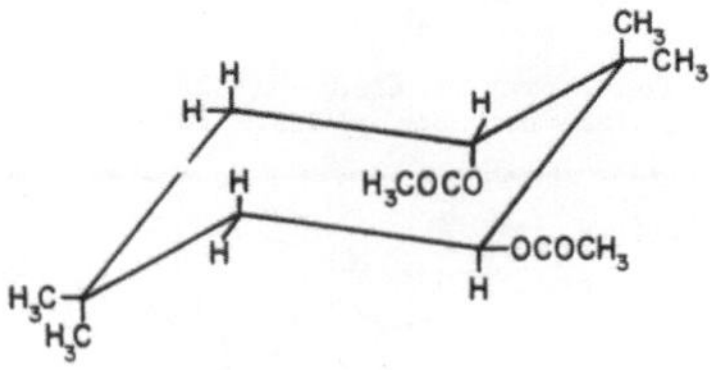

J. Chem. Phys. 34:594 (1961)

▶ $O_4C_{14}H_{24}$

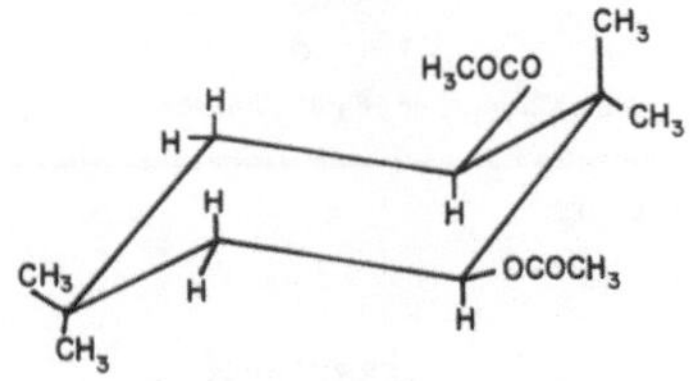

J. Chem. Phys. 34:594 (1961)

▶ $O_4C_{15}H_{12}$

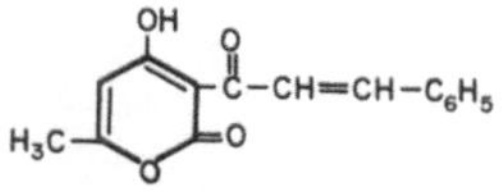

Arkiv Kemi 17:523 (1961)

▶ $O_4C_{15}H_{14}$

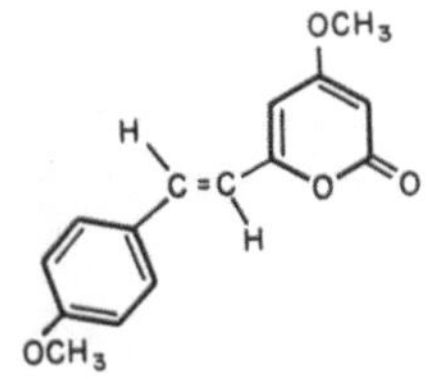

J. Org. Chem. 27:3715 (1962)

▶ $O_4C_{15}H_{14}$

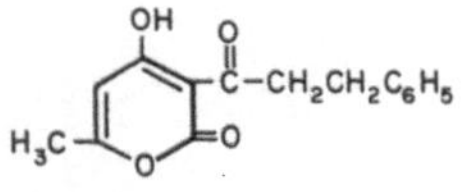

Arkiv Kemi 17:523 (1961)

▶ $O_4C_{15}H_{14}$

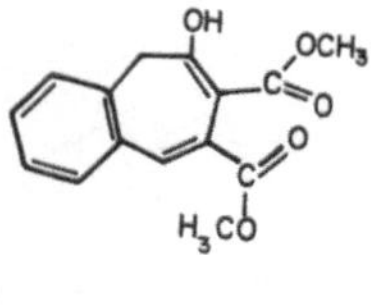

Varian 639

▶ $O_4C_{15}H_{16}$

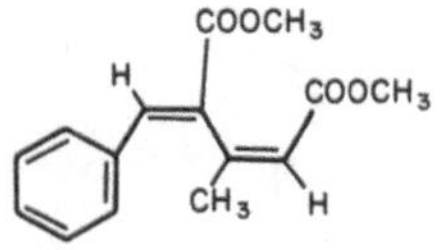

J. Org. Chem. 26:4286 (1961)

▶ $O_4C_{15}H_{18}$

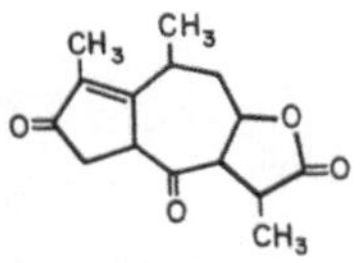

J. Am. Chem. Soc. 84:3857 (1962)

▶ $O_4C_{15}H_{18}$

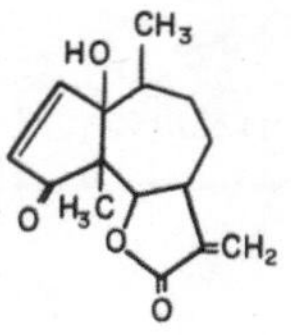

Tetrahedron Letters 86 (1961)

▶ $O_4C_{15}H_{20}$

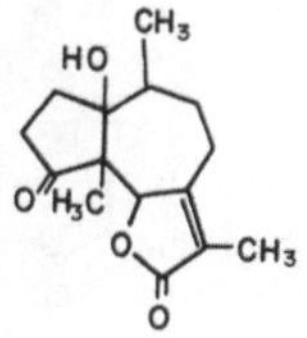

Tetrahedron Letters 86 (1961)
J. Am. Chem. Soc. 84:2601 (1962)

▶ $O_4C_{15}H_{18}$

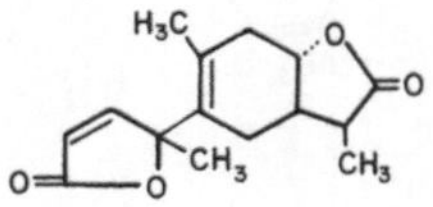

J. Org. Chem. 27:1898 (1962)

▶ $O_4C_{15}H_{18}$

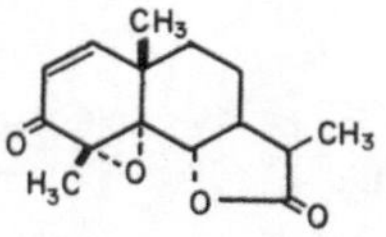

J. Chem. Soc. 1680 (1962)

▶ $O_4C_{15}H_{20}$

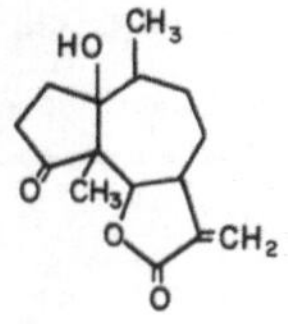

Tetrahedron Letters 86 (1961)
J. Am. Chem. Soc. 84:2601 (1962)

▶ $O_4C_{15}H_{20}$

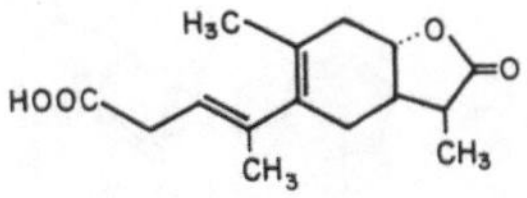

J. Org. Chem. 27:1898 (1962)

▶ $O_4C_{15}H_{20}$

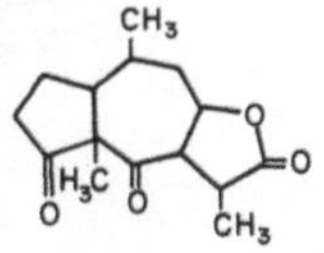

J. Am. Chem. Soc. 84:3857 (1962)

▶ $O_4C_{15}H_{20}$

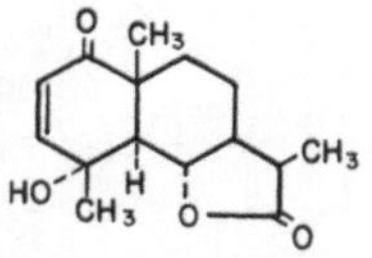

J. Org. Chem. 27:1856 (1962)

▶ $O_4C_{15}H_{20}$

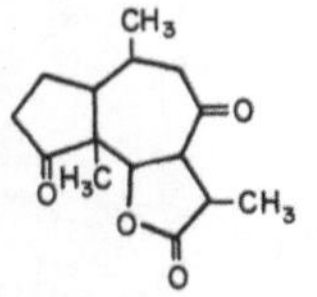

J. Am. Chem. Soc. 84:3857 (1962)

▶ $O_4C_{15}H_{20}$

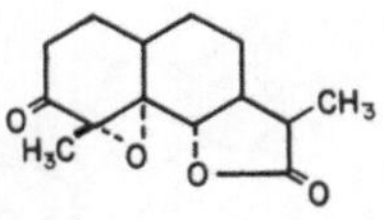

J. Chem. Soc. 1680 (1962)

▶ $O_4C_{15}H_{22}$

VERRUCAROL

Helv. Chim. Acta 45:1726 (1962)

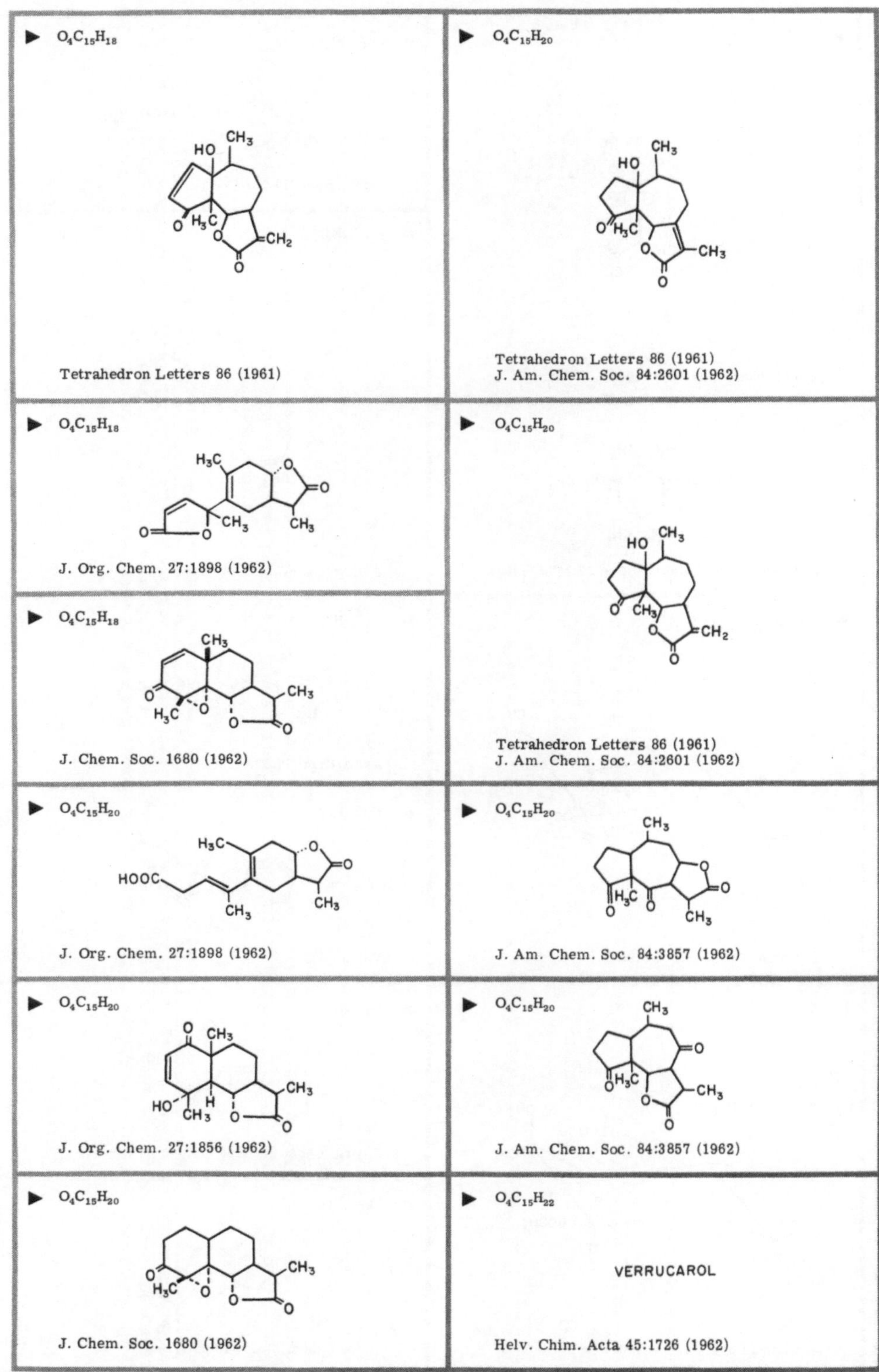

▶ $O_4C_{15}H_{22}$

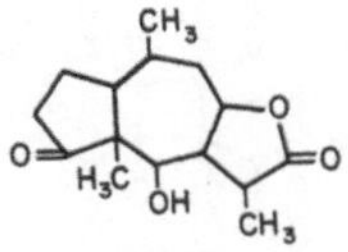

J. Am. Chem. Soc. 84:3857 (1962)

▶ $O_4C_{15}H_{22}$

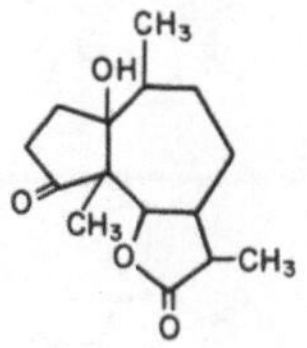

Tetrahedron Letters 86 (1961)
J. Am. Chem. Soc. 84:2601 (1962)

▶ $O_4C_{15}H_{22}$

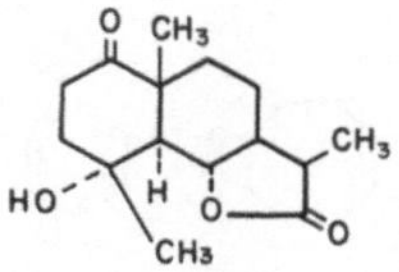

J. Org. Chem. 27:1856 (1962)

▶ $O_4C_{15}H_{22}$

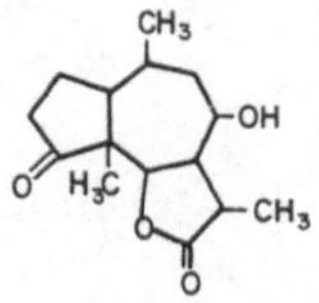

J. Am. Chem. Soc. 84:3857 (1962)

▶ $O_4C_{15}H_{22}$

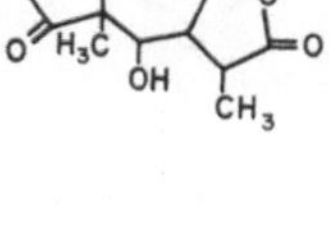

J. Am. Chem. Soc. 84:3857 (1962)

▶ $O_4C_{15}H_{24}$

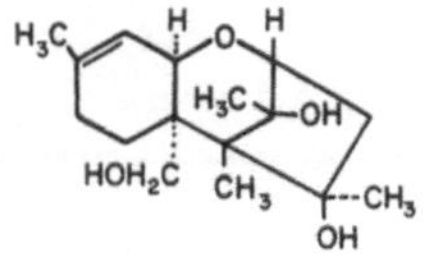

DIHYDROVERRUCAROL-B

Helv. Chim. Acta 45:1726 (1962)

▶ $O_4C_{15}H_{26}$

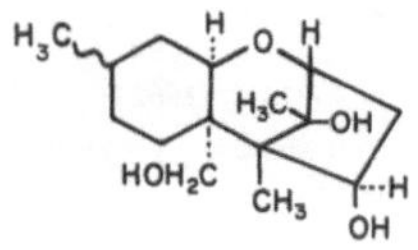

TETRAHYDROVERRUCAROL

Helv. Chim. Acta 45:1726 (1962)

▶ $O_4C_{16}H_{12}$

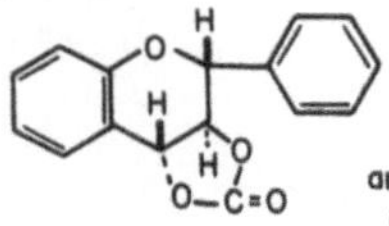

and cis isomer

Tetrahedron Letters 430 (1961)

▶ $O_4C_{16}H_{12}$

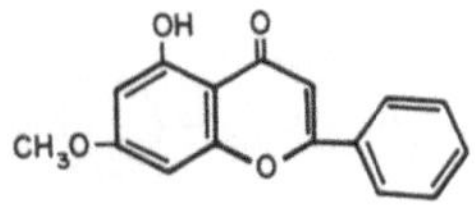

Bull. Soc. Chim. France 1962 (1962)

▶ $O_4C_{16}H_{14}$

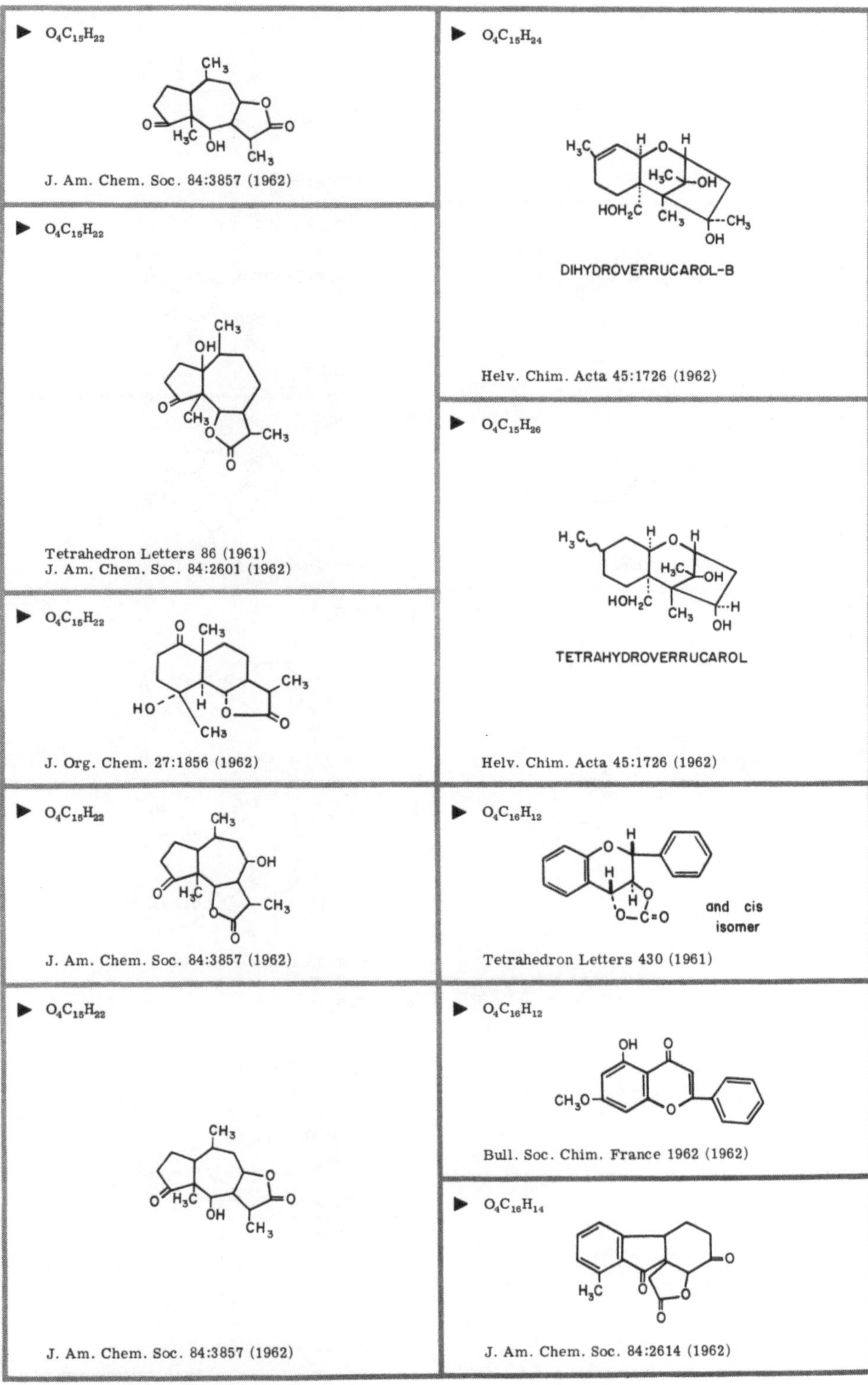

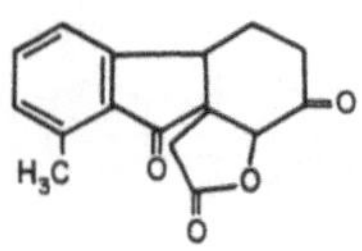

J. Am. Chem. Soc. 84:2614 (1962)

▶ $O_4C_{16}H_{14}$

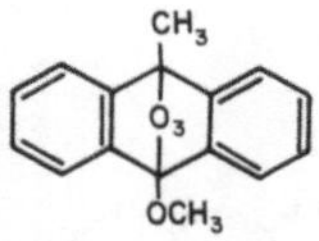

Tetrahedron 18:388 (1962)

▶ $O_4C_{16}H_{16}$

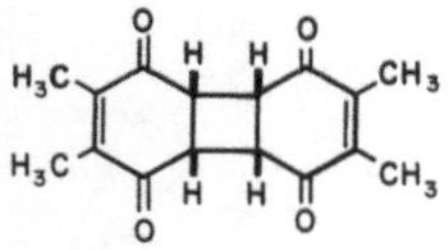

J. Chem. Soc. 1718 (1962)

▶ $O_4C_{16}H_{14}$

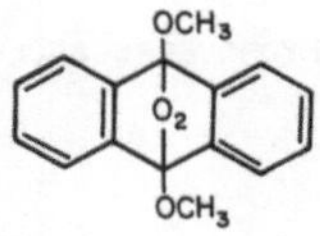

Tetrahedron 18:388 (1962)

▶ $O_4C_{16}H_{16}$

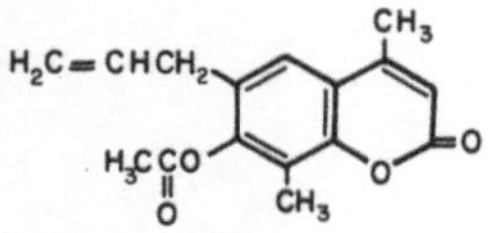

Varian 323

▶ $O_4C_{16}H_{16}$

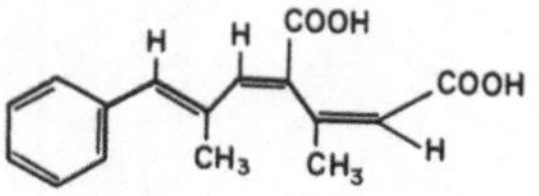

J. Org. Chem. 27:1991 (1962)

▶ $O_4C_{16}H_{18}$

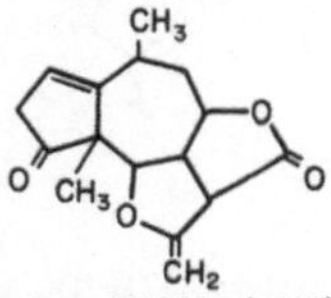

J. Am. Chem. Soc. 84:3857 (1962)

▶ $O_4C_{16}H_{16}$

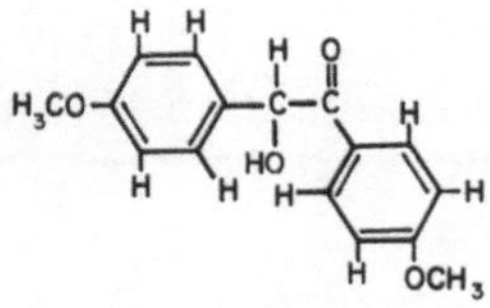

Varian 655

▶ $O_4C_{16}H_{20}$

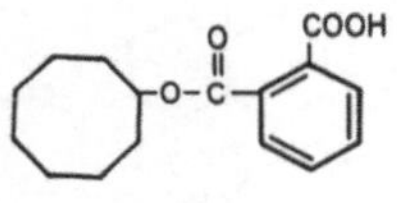

J. Am. Chem. Soc. 84:2394 (1962)

▶ $O_4C_{16}H_{22}$

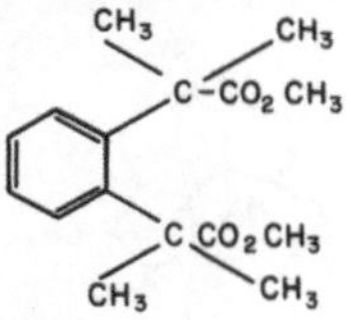

Tetrahedron Letters 245 (1961)

▶ $O_4C_{16}H_{16}$

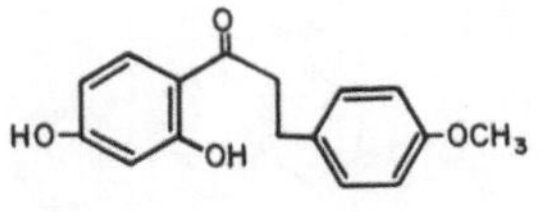

Bull. Soc. Chim. France 1962 (1962)

▶ $O_4C_{16}H_{26}$

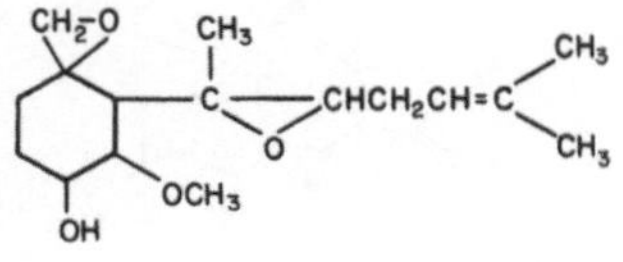

J. Am. Chem. Soc. 83:3096 (1961)

▶ $O_4C_{16}H_{16}$

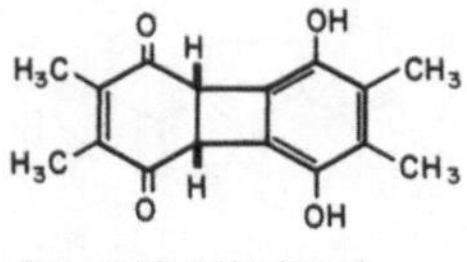

J. Chem. Soc. 1718, 2783 (1962)

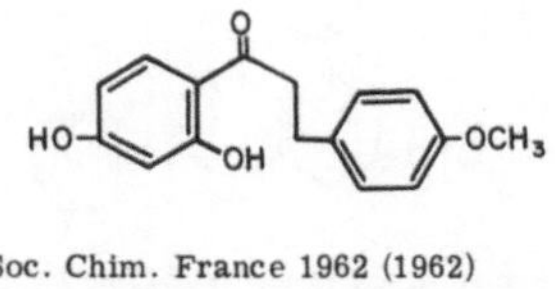

▶ $O_4C_{16}H_{30}$

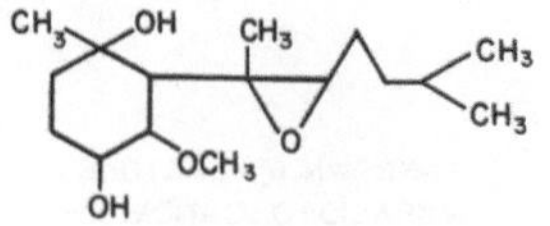


J. Am. Chem. Soc. 83:3096 (1961)

▶ $O_4C_{17}H_{14}$

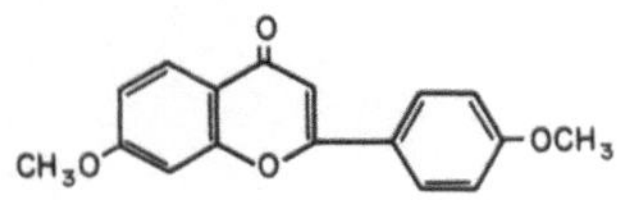


Bull. Soc. Chim. France 1962 (1962)

▶ $O_4C_{16}H_{32}$

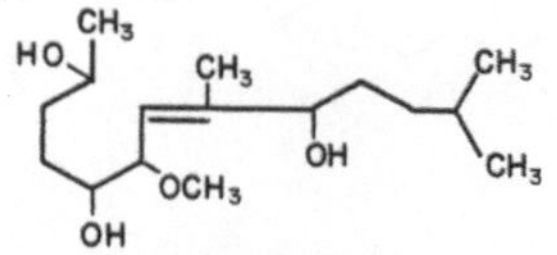


J. Am. Chem. Soc. 83:3096 (1961)

▶ $O_4C_{17}H_{16}$

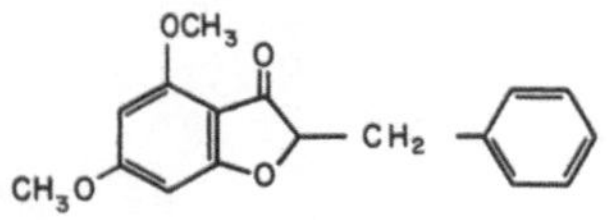


Bull. Soc. Chim. France 1962 (1962)

▶ $O_4C_{16}H_{32}Pt_2$

$$\left[(CH_3)_3Pt-CH(\overset{O}{\overset{\|}{C}}-CH_3)_2\right]_2$$

J. Chem. Soc. 4736 (1962)

▶ $O_4C_{17}H_{16}$

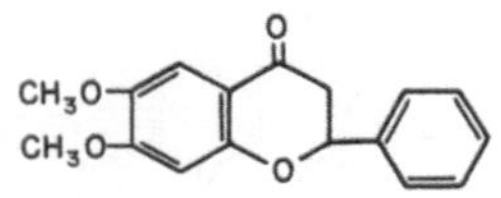


Bull. Soc. Chim. France 1962 (1962)

▶ $O_4C_{17}H_{14}$

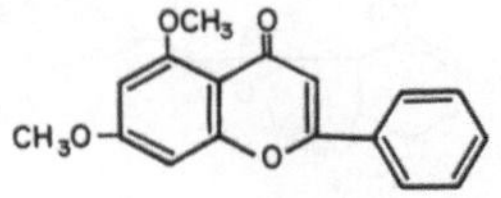


Bull. Soc. Chim. France 1962 (1962)

▶ $O_4C_{17}H_{16}$

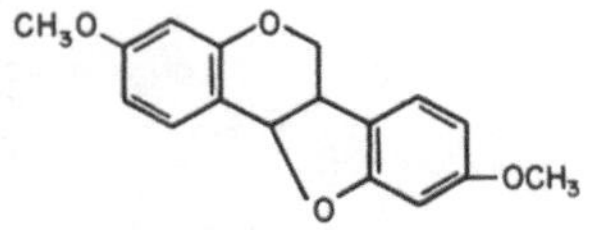


Experientia 18:163 (1962)

▶ $O_4C_{17}H_{14}$

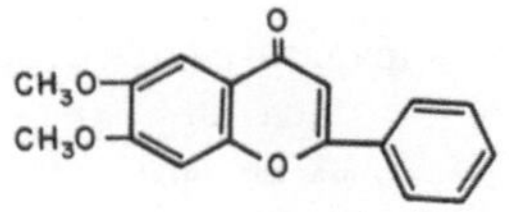


Bull. Soc. Chim. France 1962 (1962)

▶ $O_4C_{17}H_{18}$

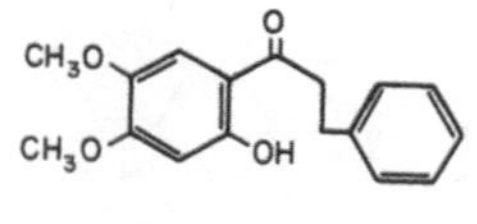


Bull. Soc. Chim. France 1962 (1962)

▶ $O_4C_{17}H_{14}$

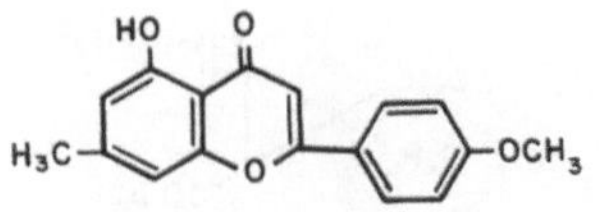


Bull. Soc. Chim. France 1962 (1962)

▶ $O_4C_{17}H_{18}$

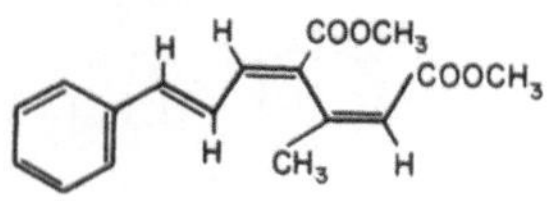


J. Org. Chem. 26:4286 (1961)

▶ $O_4C_{17}H_{14}$

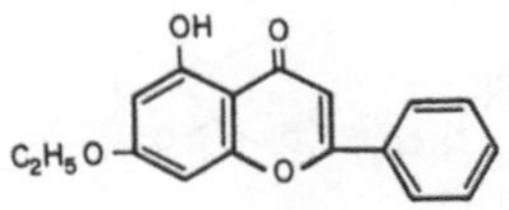


Bull. Soc. Chim. France 1962 (1962)

▶ $O_4C_{17}H_{24}$

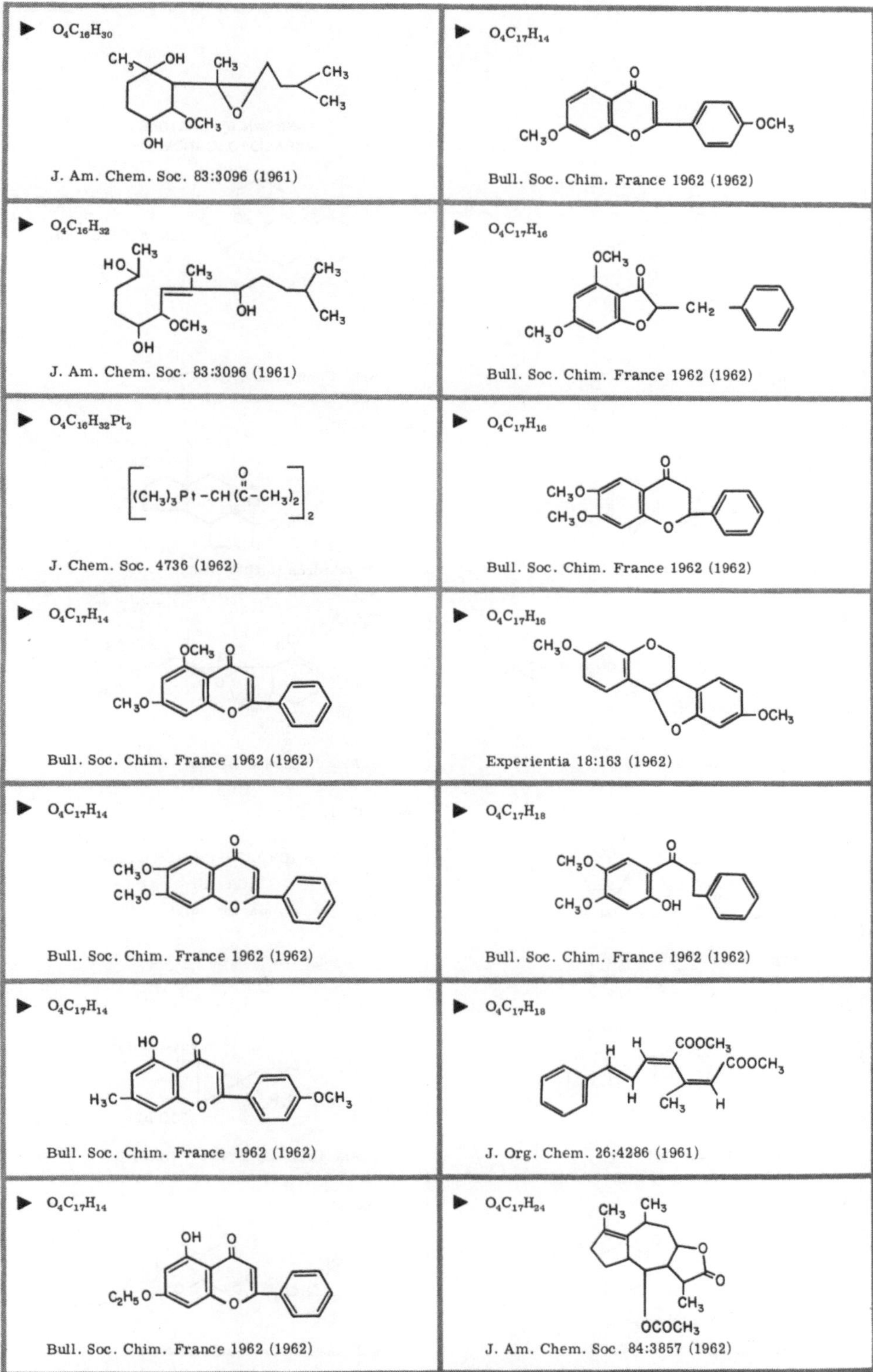

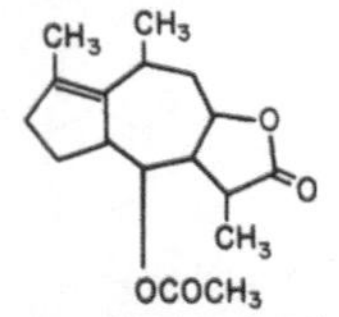


J. Am. Chem. Soc. 84:3857 (1962)

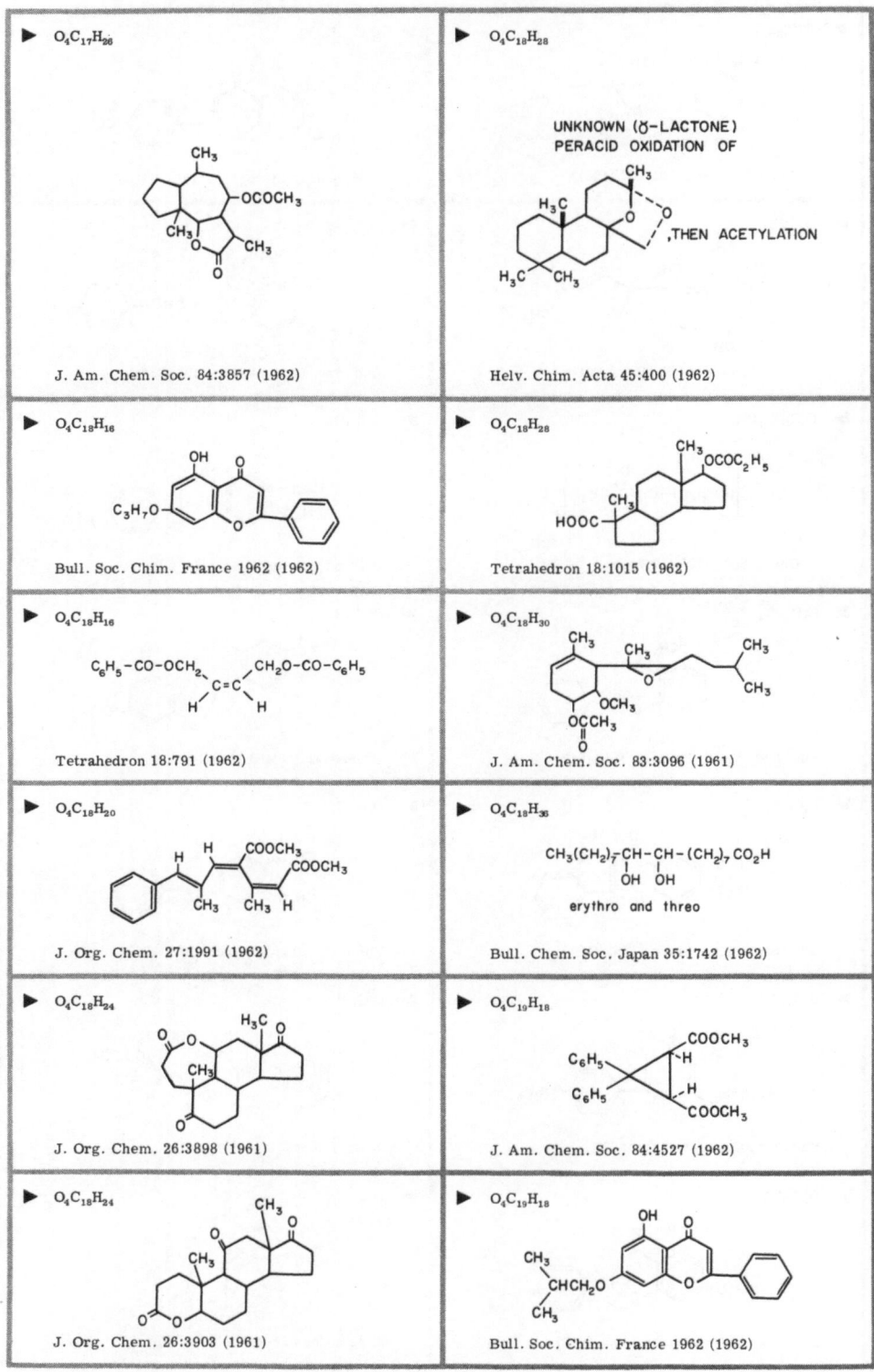

▶ $O_4C_{17}H_{26}$

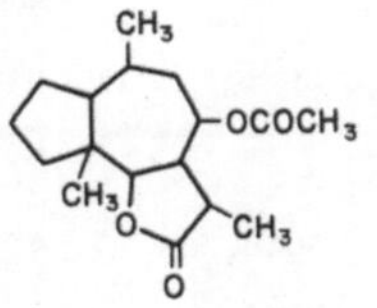

J. Am. Chem. Soc. 84:3857 (1962)

▶ $O_4C_{18}H_{28}$

UNKNOWN (δ-LACTONE)
PERACID OXIDATION OF

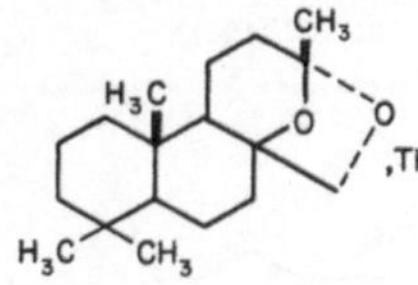

,THEN ACETYLATION

Helv. Chim. Acta 45:400 (1962)

▶ $O_4C_{18}H_{16}$

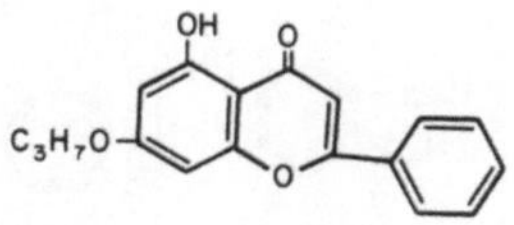

Bull. Soc. Chim. France 1962 (1962)

▶ $O_4C_{18}H_{28}$

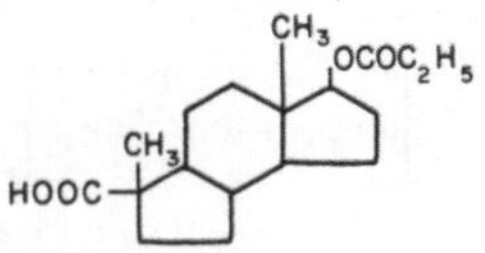

Tetrahedron 18:1015 (1962)

▶ $O_4C_{18}H_{16}$

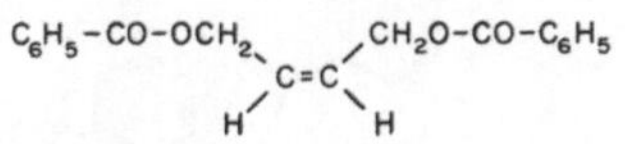

Tetrahedron 18:791 (1962)

▶ $O_4C_{18}H_{30}$

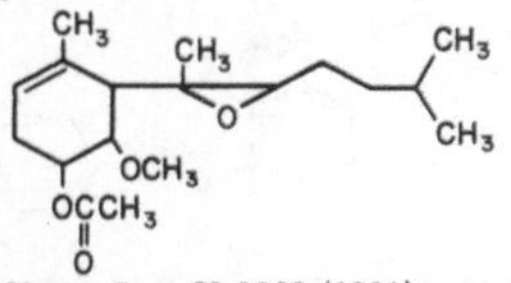

J. Am. Chem. Soc. 83:3096 (1961)

▶ $O_4C_{18}H_{20}$

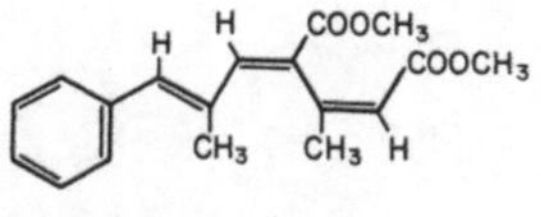

J. Org. Chem. 27:1991 (1962)

▶ $O_4C_{18}H_{36}$

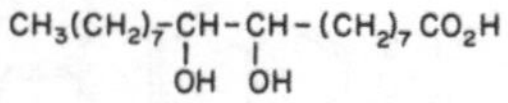

erythro and threo

Bull. Chem. Soc. Japan 35:1742 (1962)

▶ $O_4C_{18}H_{24}$

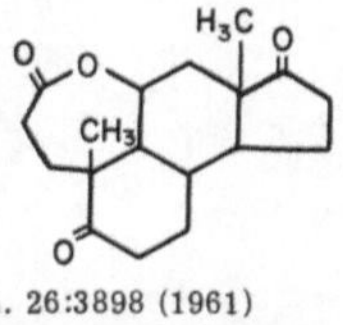

J. Org. Chem. 26:3898 (1961)

▶ $O_4C_{19}H_{18}$

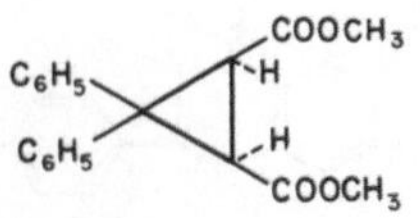

J. Am. Chem. Soc. 84:4527 (1962)

▶ $O_4C_{18}H_{24}$

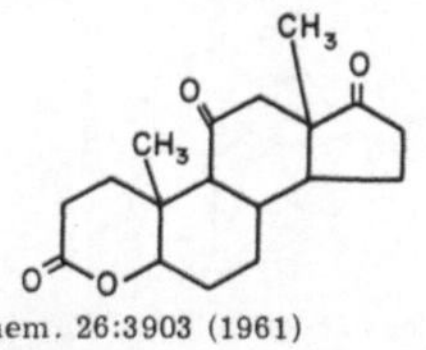

J. Org. Chem. 26:3903 (1961)

▶ $O_4C_{19}H_{18}$

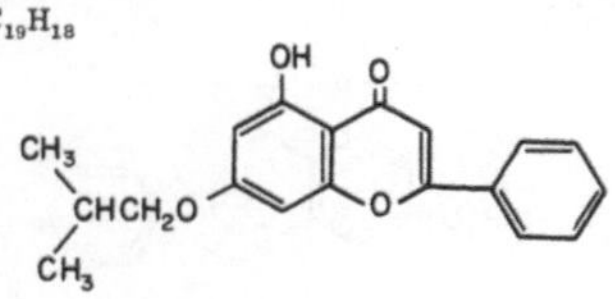

Bull. Soc. Chim. France 1962 (1962)

▶ $O_4C_{19}H_{22}$

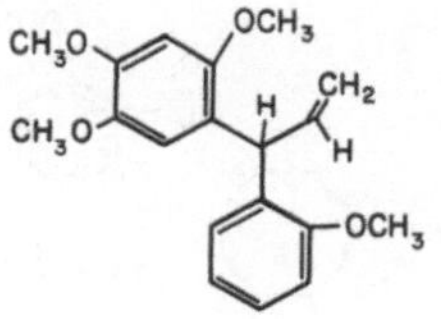

Varian 678
Tetrahedron 18:1504 (1962)

▶ $O_4C_{20}H_{20}$

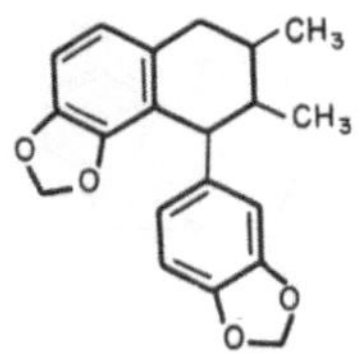

J. Chem. Soc. 1782 (1962)

▶ $O_4C_{19}H_{22}$

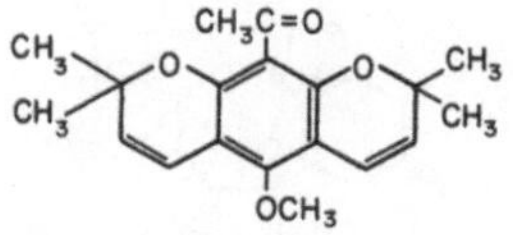

Tetrahedron 16:206 (1961)

▶ $O_4C_{19}H_{26}$

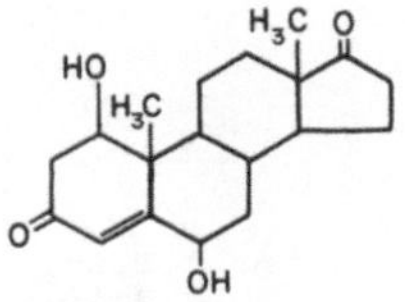

J. Org. Chem. 27:3160 (1962)

▶ $O_4C_{20}H_{20}$

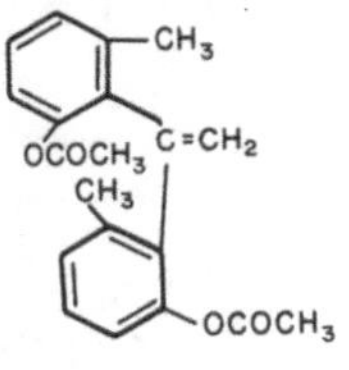

J. Org. Chem. 26:3656 (1961)

▶ $O_4C_{19}H_{26}$

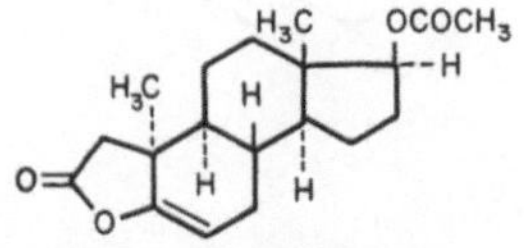

Helv. Chim. Acta 45:2346 (1962)

▶ $O_4C_{20}H_{22}$

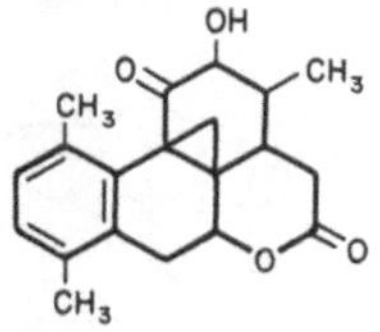

Tetrahedron Letters No. 23:1087 (1962)

▶ $O_4C_{20}H_{16}$

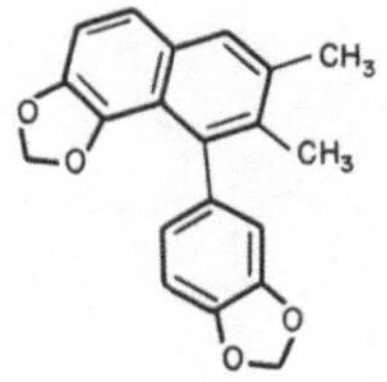

J. Chem. Soc. 1782 (1962)

▶ $O_4C_{20}H_{22}$

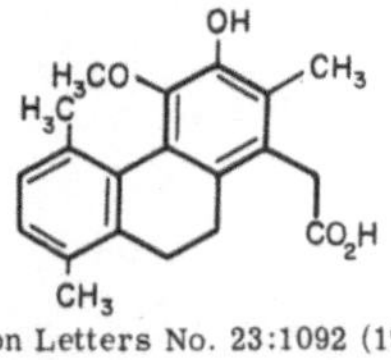

Tetrahedron Letters No. 23:1092 (1962)

▶ $O_4C_{20}H_{24}$

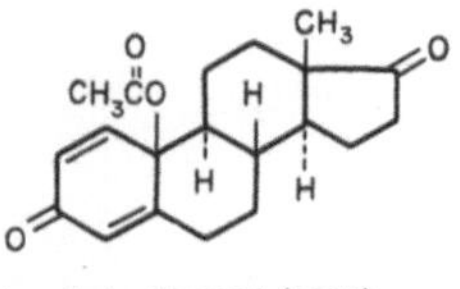

Helv. Chim. Acta 45:2674 (1962)

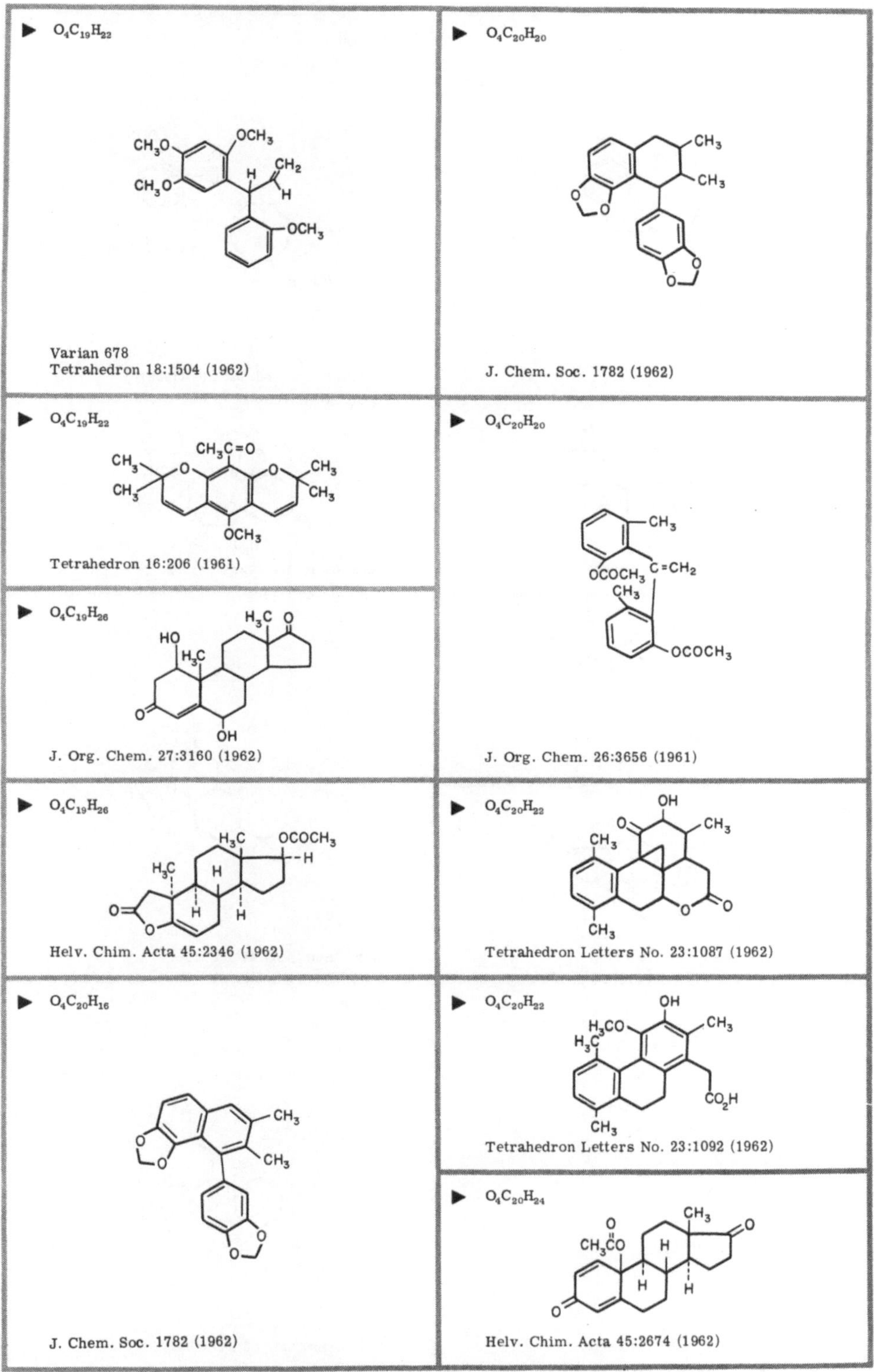

▶ $O_4C_{20}H_{24}$

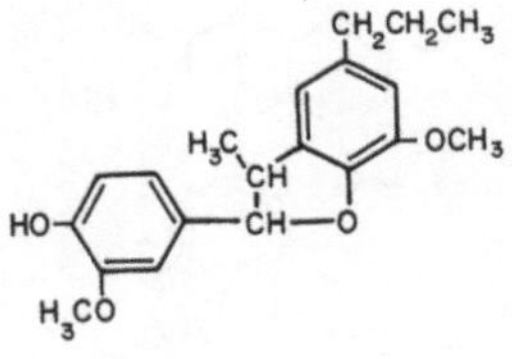


Varian 682

▶ $O_4C_{20}H_{28}$

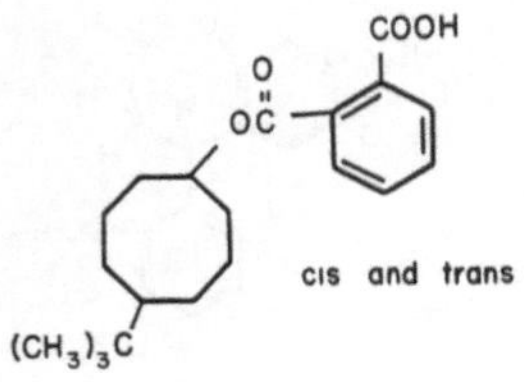


J. Am. Chem. Soc. 84:2394 (1962)

▶ $O_4C_{20}H_{26}$

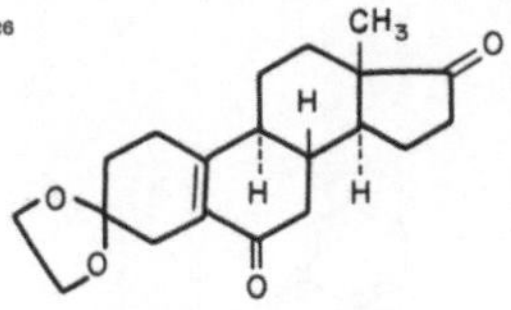


Helv. Chim. Acta 45:2674 (1962)

▶ $O_4C_{20}H_{30}$

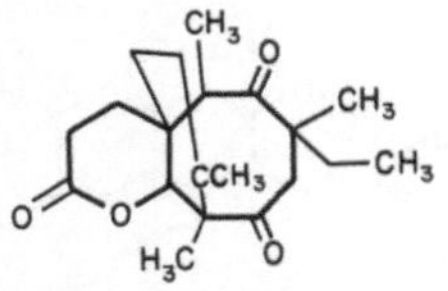


Gazz. Chim. Ital. 92:884 (1962)

▶ $O_4C_{20}H_{26}$

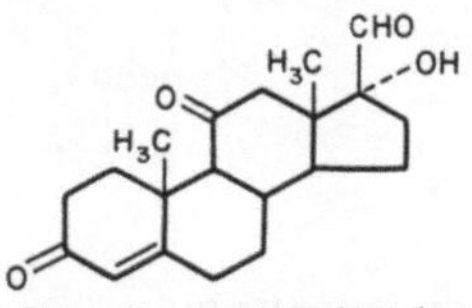


Bull. Soc. Chim. France 1679 (1961)

▶ $O_4C_{20}H_{26}$

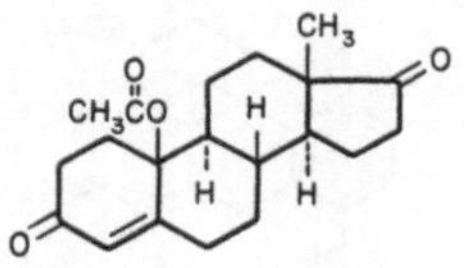


Helv. Chim. Acta 45:2674 (1962)

▶ $O_4C_{20}H_{32}$

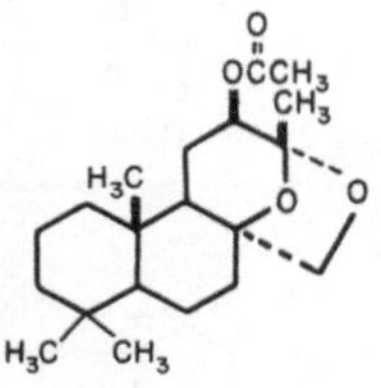


Helv. Chim. Acta 45:400 (1962)

▶ $O_4C_{20}H_{28}$

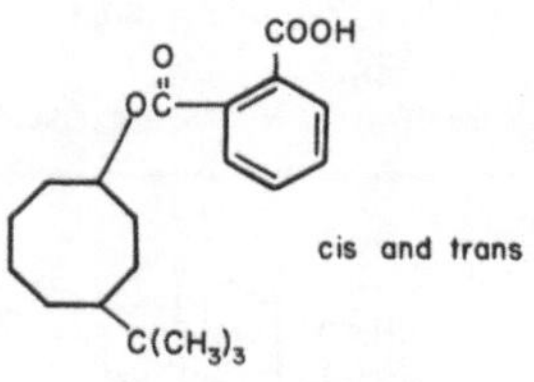


J. Am. Chem. Soc. 84:2394 (1962)

▶ $O_4C_{20}H_{32}$

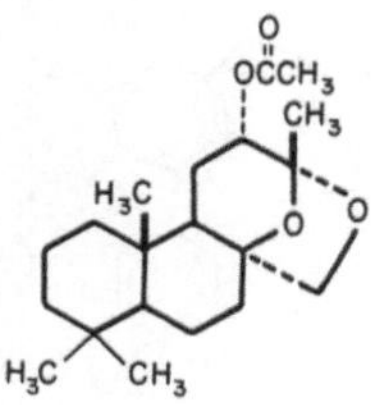


Helv. Chim. Acta 45:400 (1962)

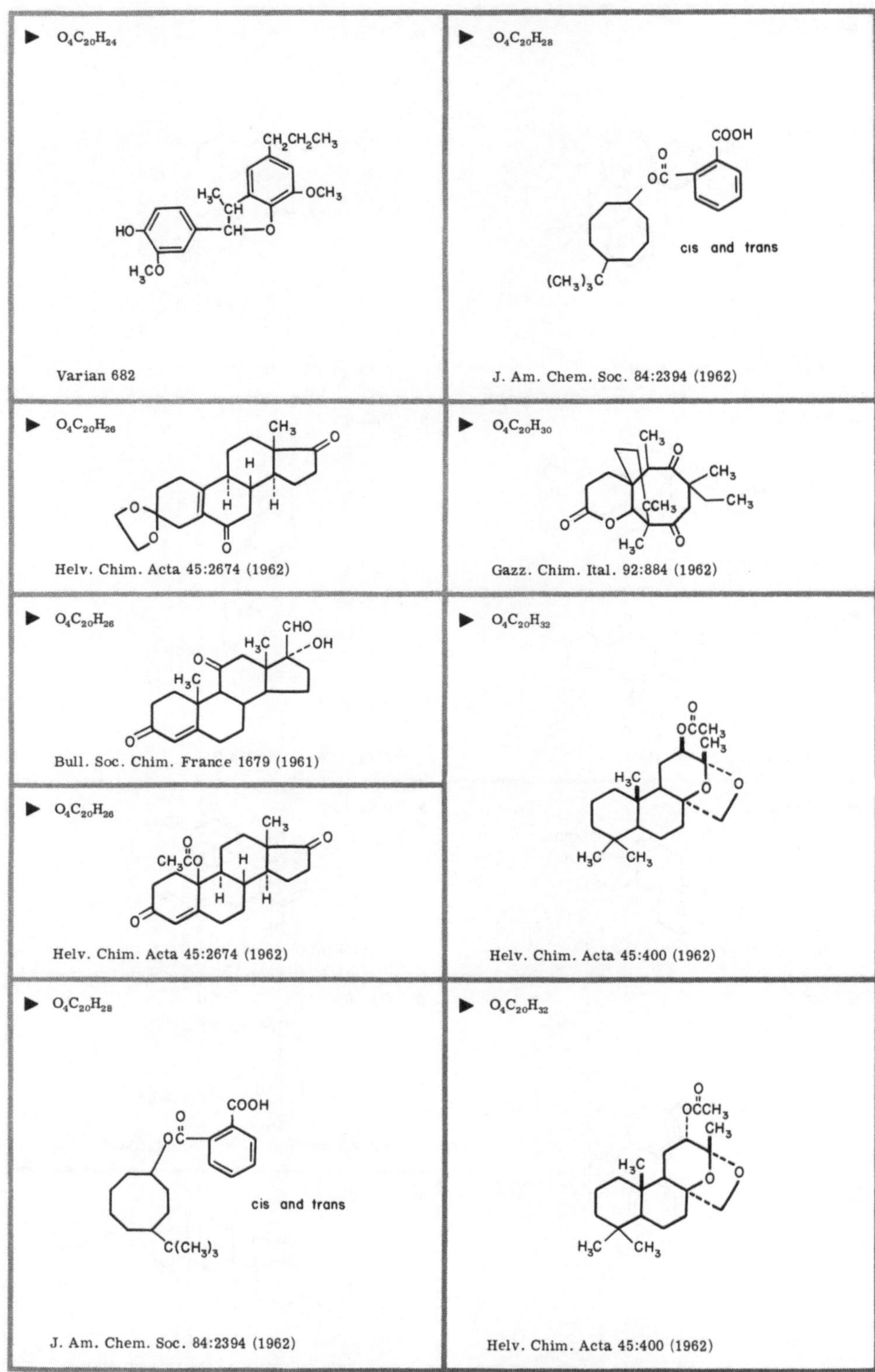

▶ $O_4C_{20}H_{32}$

ANHYDRO-DIHYDROAGLYCONE B
AND
ANHYDRO-DIHYDROAGLYCONE D
(BOTH FROM DIHYDRONARBOMYCIN)

Helv. Chim. Acta 45:4 (1962)

▶ $O_4C_{20}H_{32}$

ANHYDRO-DIHYDROAGLYCONE A
AND
ANHYDRO-DIHYDROAGLYCONE E
(BOTH FROM DIHYDRONARBOMYCIN)

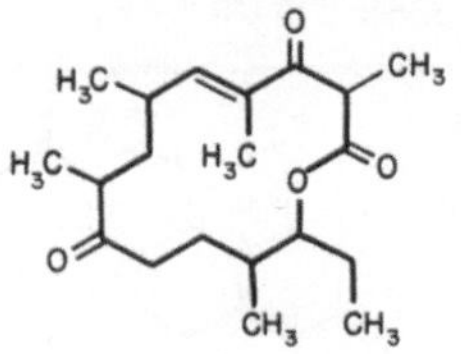

Helv. Chim. Acta 45:4 (1962)

▶ $O_4C_{20}H_{40}Pt_2$

$$\left[(CH_3)_3Pt-CH(\overset{O}{\overset{\|}{C}}-C_2H_5)_2\right]_2$$

J. Chem. Soc. 4736 (1962)

▶ $O_4C_{21}H_{20}$

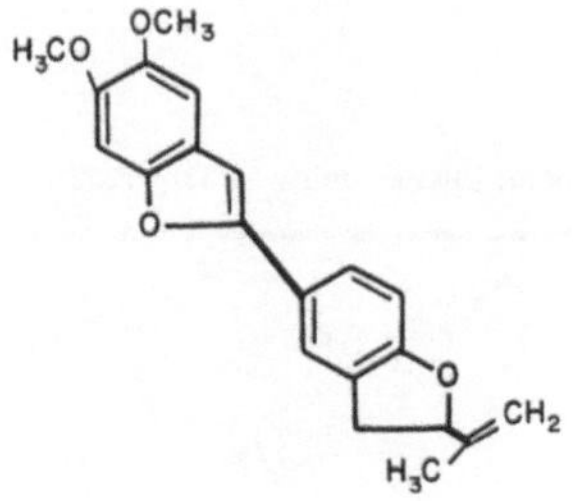

J. Chem. Soc. 775 (1962)

▶ $O_4C_{21}H_{26}$

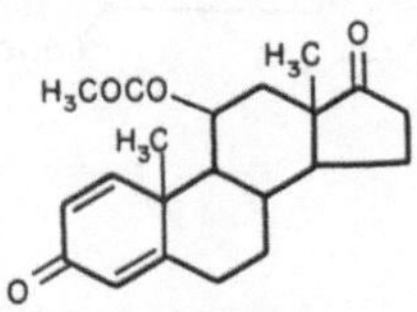

Chem. Pharm. Bull. 10:338 (1962)

▶ $O_4C_{21}H_{26}$

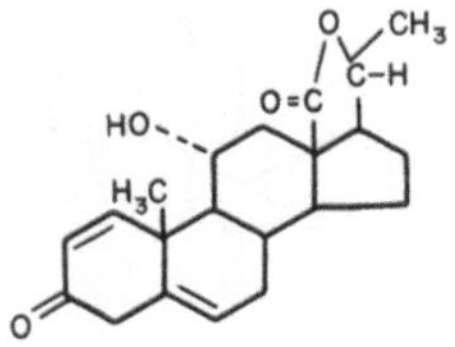

Helv. Chim. Acta 44:1380 (1961)

▶ $O_4C_{21}H_{26}$

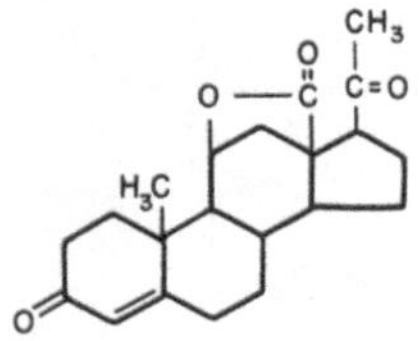

Helv. Chim. Acta 44:1380 (1961)

▶ $O_4C_{21}H_{28}$

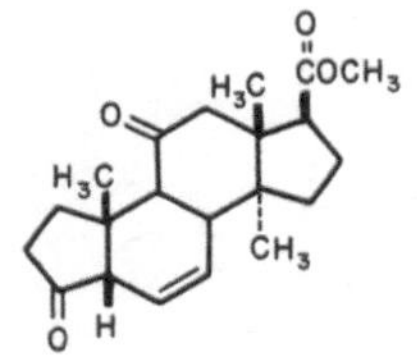

J. Am. Chem. Soc. 84:4356 (1962)

▶ $O_4C_{21}H_{28}$

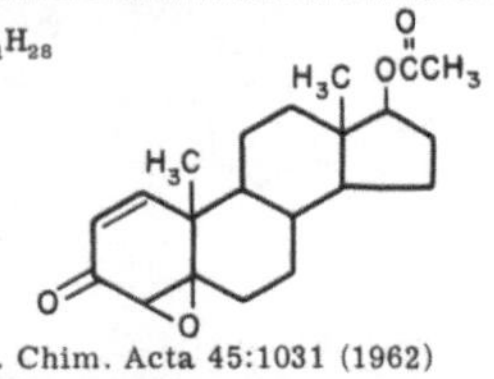

Helv. Chim. Acta 45:1031 (1962)

▶ $O_4C_{21}H_{28}$

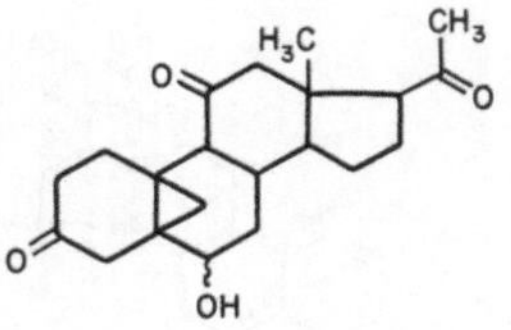

Helv. Chim. Acta 45:1261 (1962)

▶ $O_4C_{21}H_{28}$

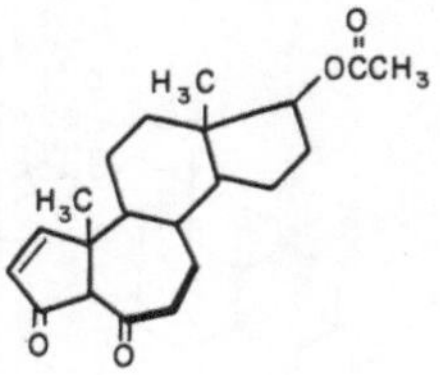

Helv. Chim. Acta 45:1031 (1962)

▶ $O_4C_{21}H_{28}$

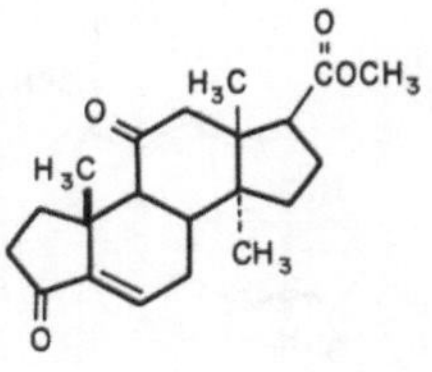

J. Am. Chem. Soc. 84:4356 (1962)

▶ $O_4C_{21}H_{28}$

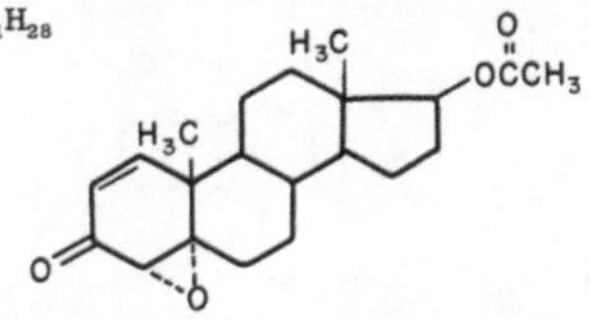

Helv. Chim. Acta 45:1031 (1962)

▶ $O_4C_{21}H_{28}$

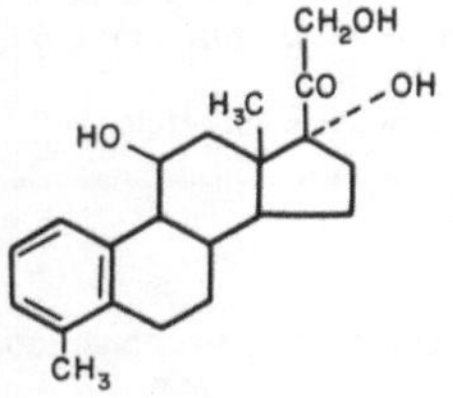

Chem. & Ind. 1716 (1962)

▶ $O_4C_{21}H_{28}$

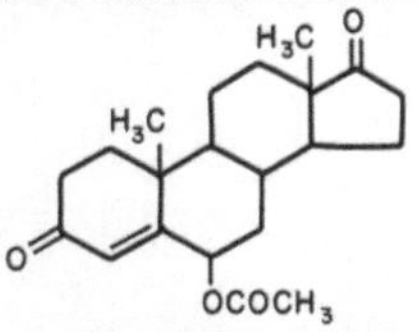

Chem. Pharm. Bull. 10:338 (1962)

▶ $O_4C_{21}H_{28}$

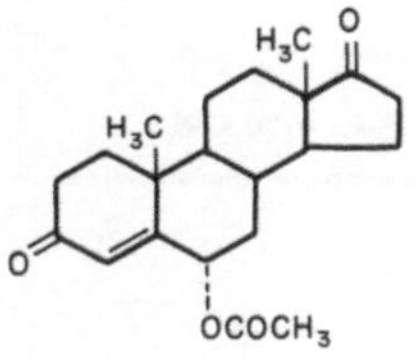

Chem. Pharm. Bull. 10:338 (1962)

▶ $O_4C_{21}H_{28}$

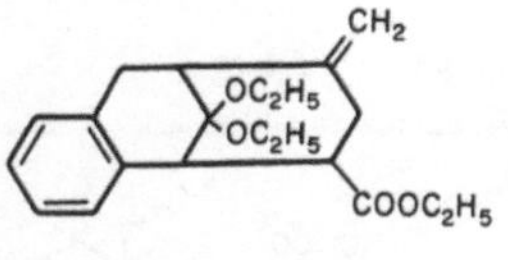

Tetrahedron 14:25 (1961)

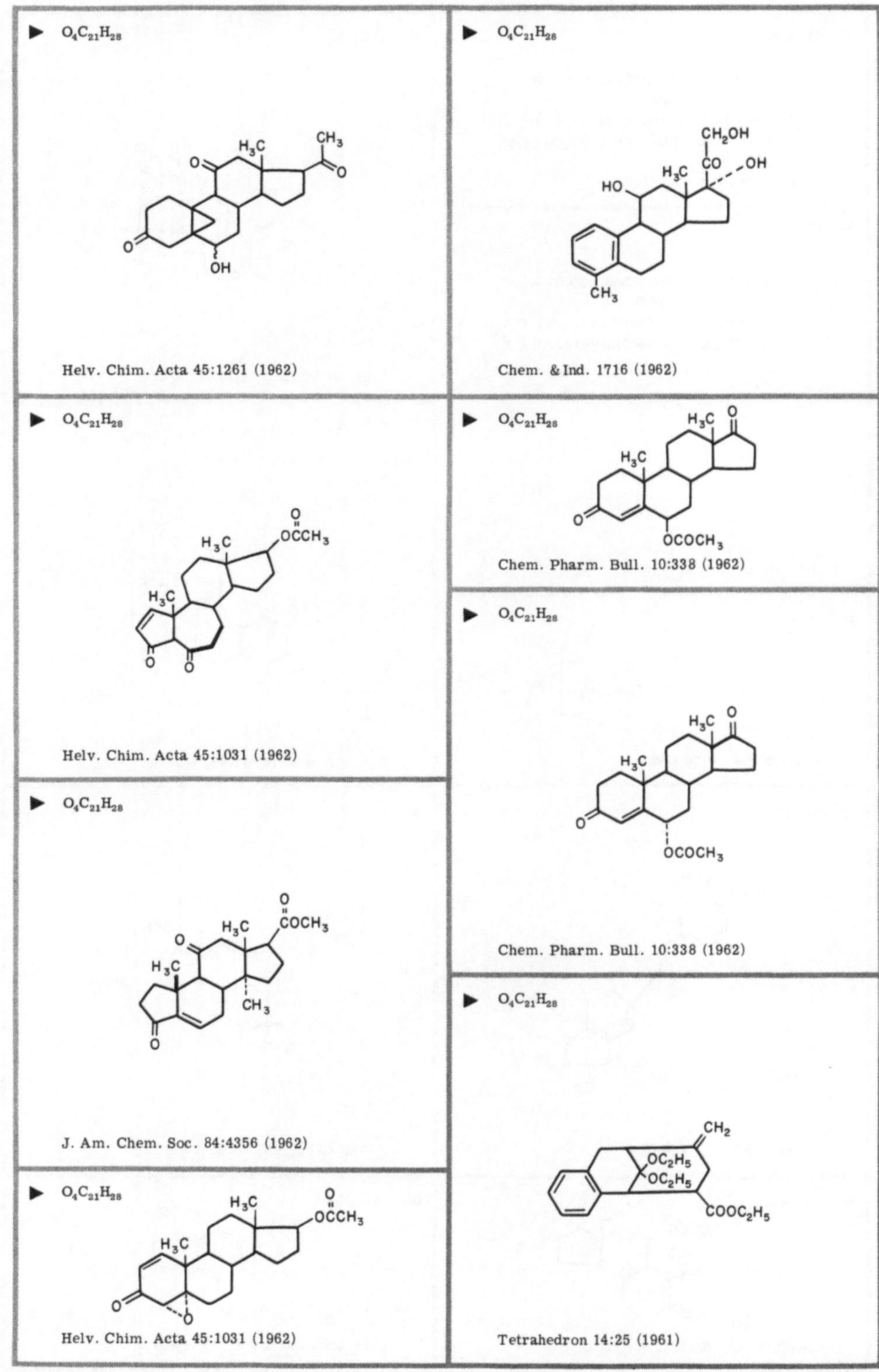

▶ $O_4C_{21}H_{28}$

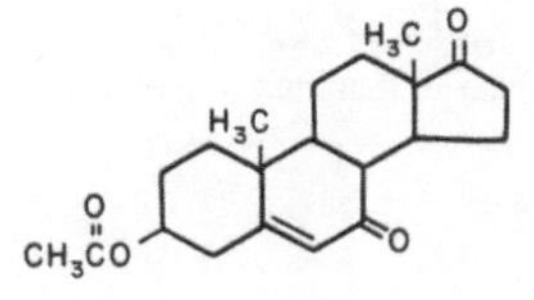


Helv. Chim. Acta 44:1380 (1961)

▶ $O_4C_{21}H_{28}$

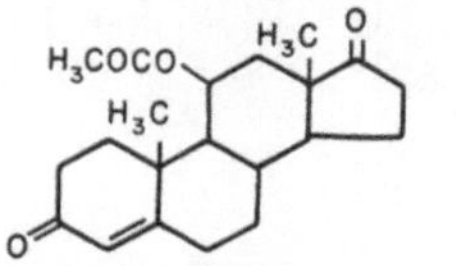


Chem. Pharm. Bull. 10:338 (1962)

▶ $O_4C_{21}H_{28}$

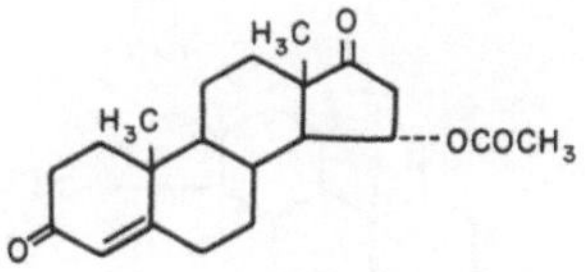


Chem. Pharm. Bull. 10:338 (1962)

▶ $O_4C_{21}H_{30}$

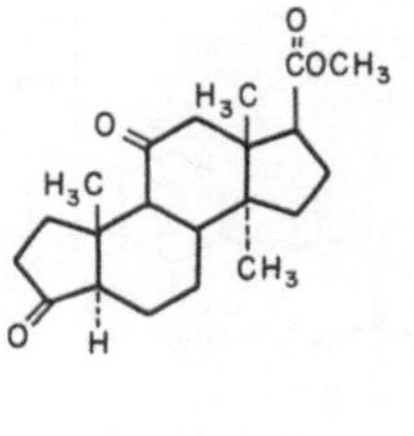


J. Am. Chem. Soc. 84:4356 (1962)

▶ $O_4C_{21}H_{30}$

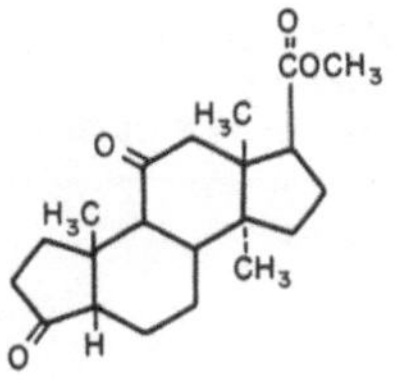


J. Am. Chem. Soc. 84:4356 (1962)

▶ $O_4C_{21}H_{30}$

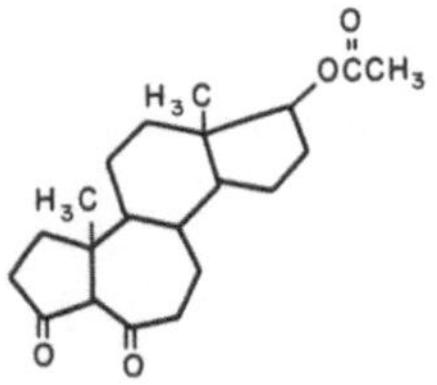


Helv. Chim. Acta 45:1031 (1962)

▶ $O_4C_{21}H_{30}$

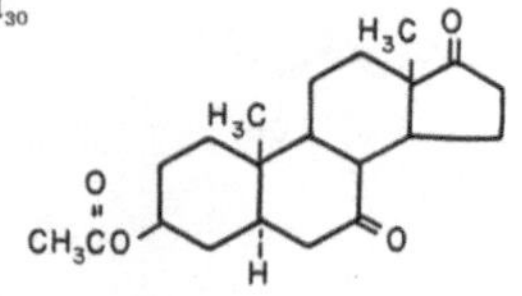


Helv. Chim. Acta 44:1380 (1961)

▶ $O_4C_{21}H_{30}$

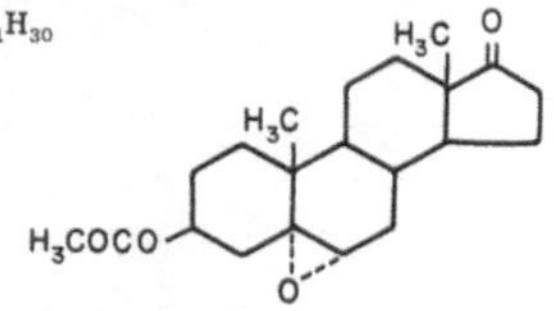


J. Am. Chem. Soc. 84:3206 (1962)

▶ $O_4C_{21}H_{30}$

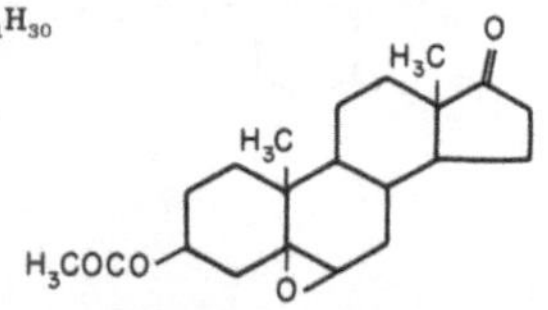


J. Am. Chem. Soc. 84:3206 (1962)

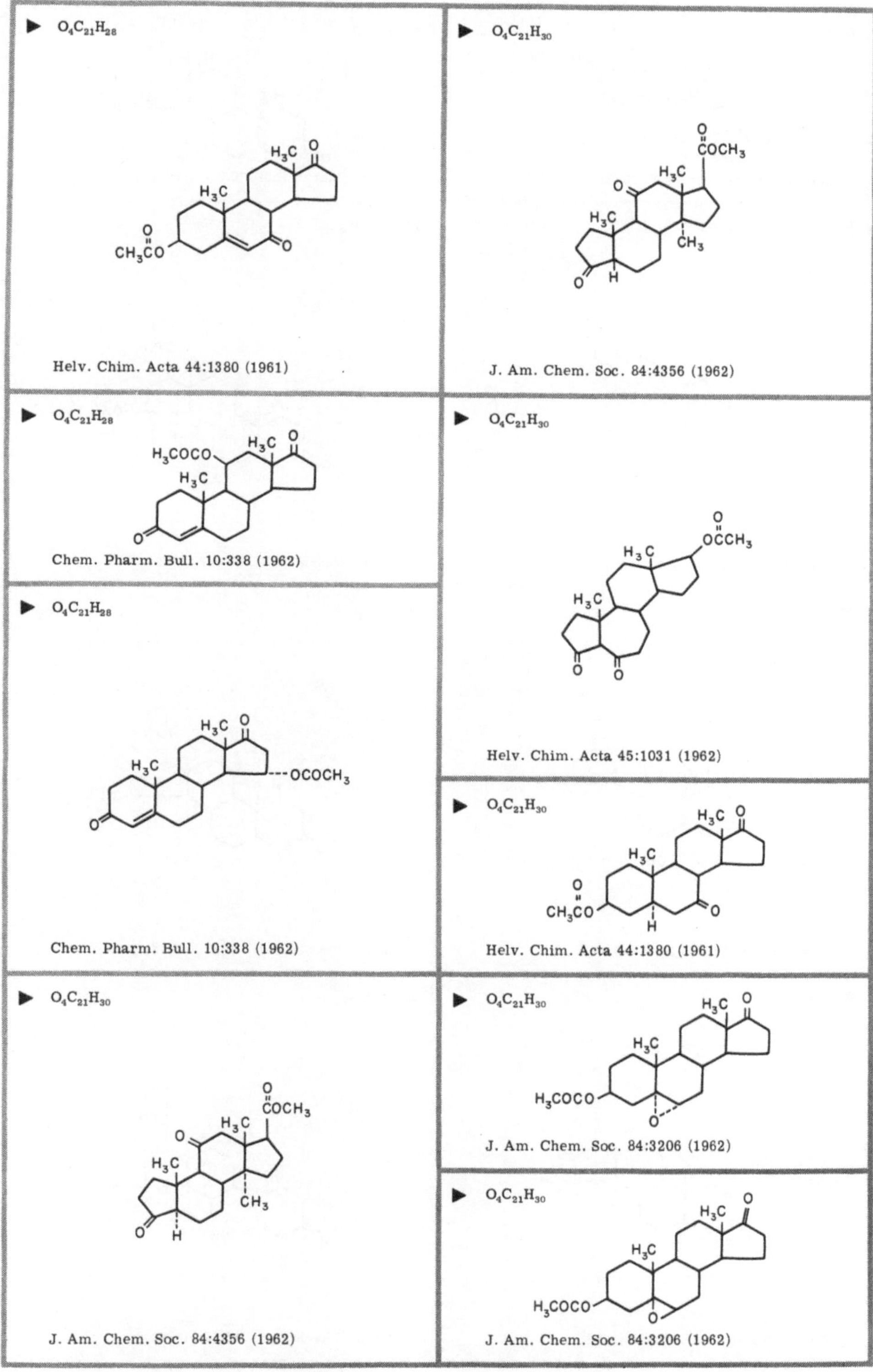

▶ $O_4C_{21}H_{30}$

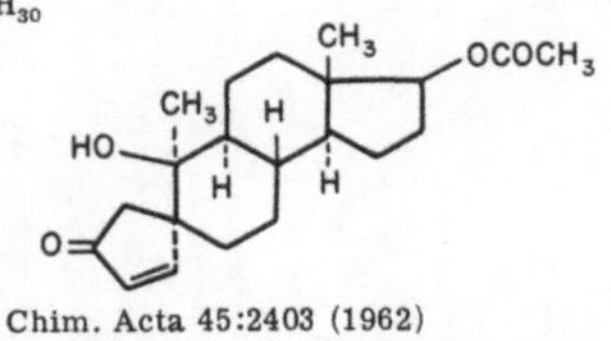


Helv. Chim. Acta 45:2403 (1962)

▶ $O_4C_{21}H_{30}$

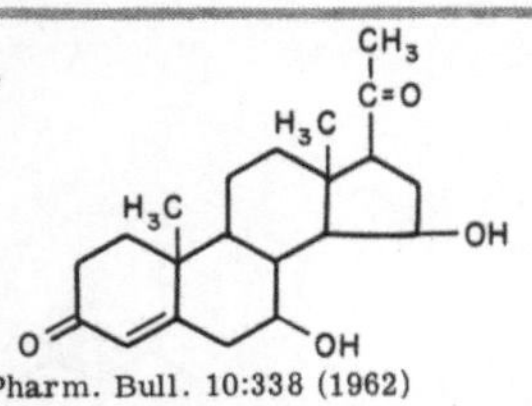


Chem. Pharm. Bull. 10:338 (1962)

▶ $O_4C_{21}H_{30}$

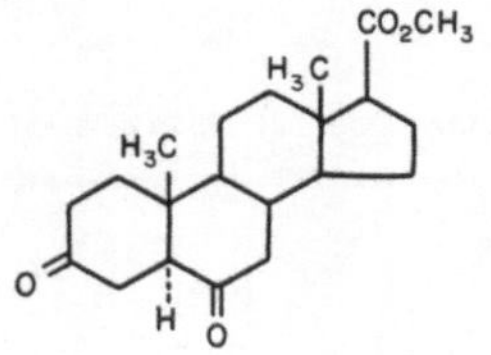


Helv. Chim. Acta 44:1380 (1961)

▶ $O_4C_{21}H_{30}$

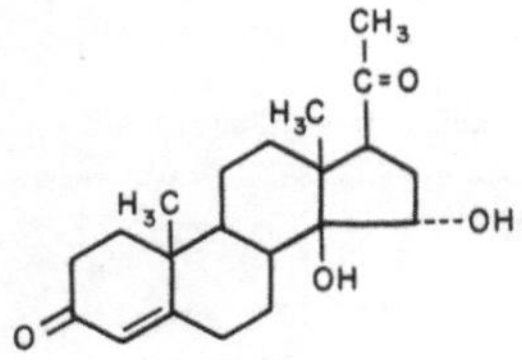


Chem. Pharm. Bull. 10:338 (1962)

▶ $O_4C_{21}H_{30}$

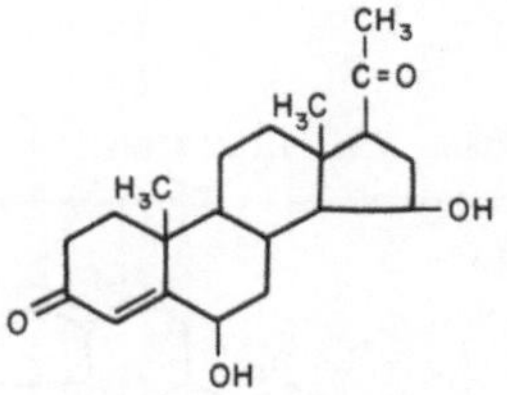


Chem. Pharm. Bull. 10:338 (1962)

▶ $O_4C_{21}H_{30}$

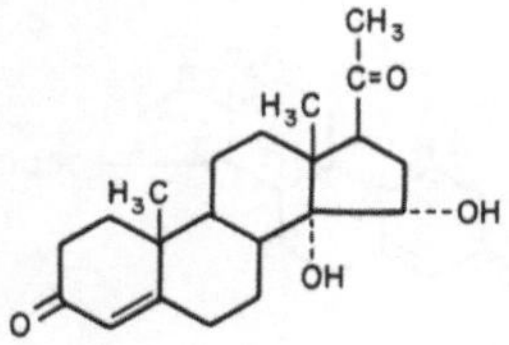


Chem. Pharm. Bull. 10:338 (1962)

▶ $O_4C_{21}H_{30}$

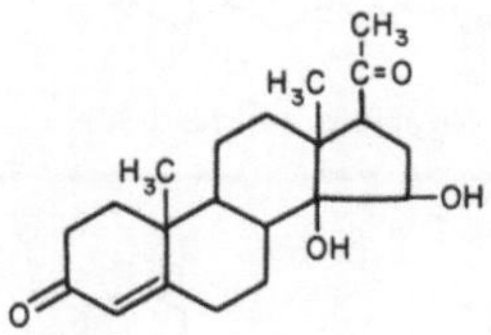


Chem. Pharm. Bull. 10:338 (1962)

▶ $O_4C_{21}H_{30}$

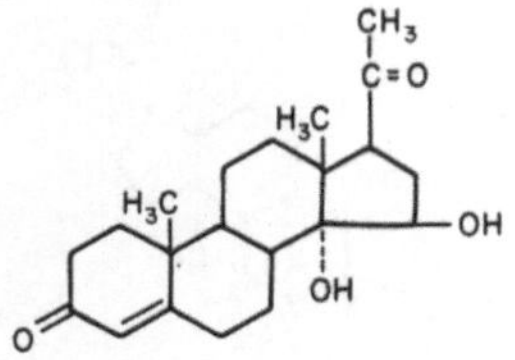


Chem. Pharm. Bull. 10:338 (1962)

► $O_4C_{21}H_{30}$

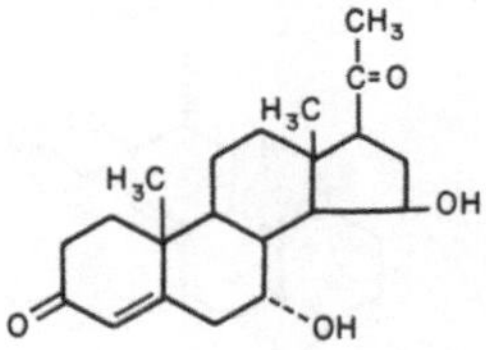

Chem. Pharm. Bull. 10:338 (1962)

► $O_4C_{21}H_{34}$

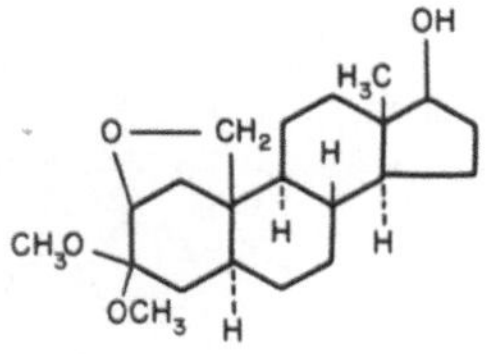

J. Am. Chem. Soc. 84:4976 (1962)

► $O_4C_{21}H_{32}$

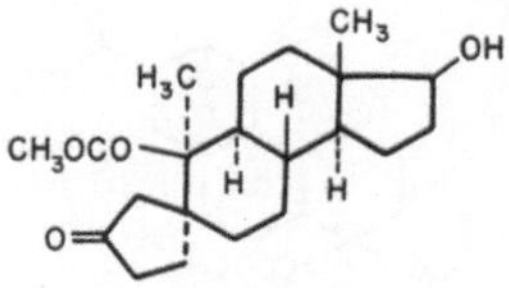

Helv. Chim. Acta 45:2403 (1962)

► $O_4C_{21}H_{34}$

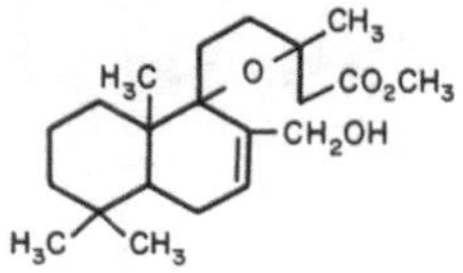

Acta Chem. Scand. 16:1675 (1962)

► $O_4C_{21}H_{32}$

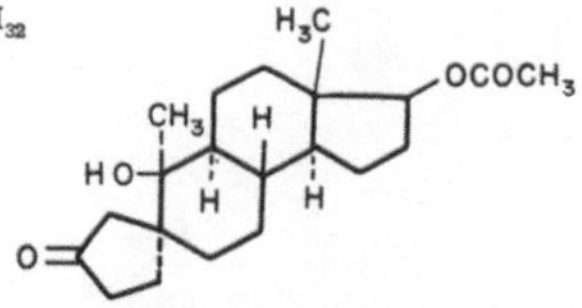

Helv. Chim. Acta 45:2403 (1962)

► $O_4C_{21}H_{34}$

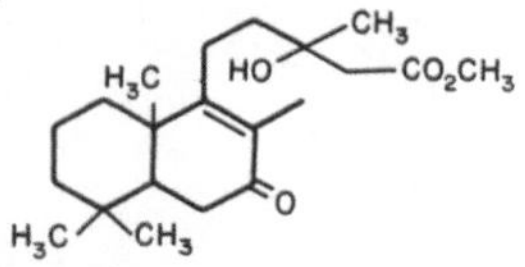

Acta Chem. Scand. 16:1675 (1962)

► $O_4C_{21}H_{32}$

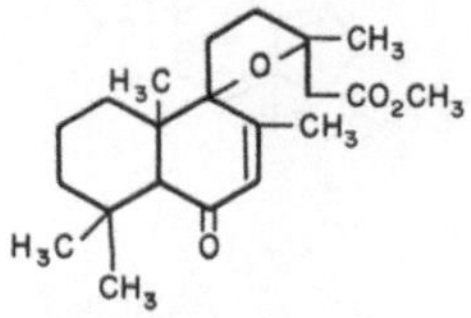

Gazz. Chim. Ital. 92:983 (1962)

► $O_4C_{21}H_{34}$

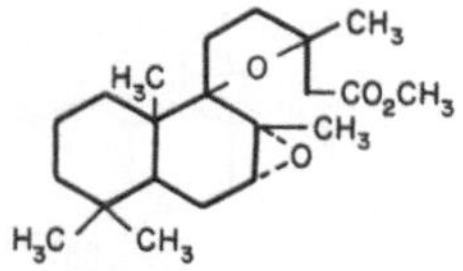

Gazz. Chim. Ital. 92:995 (1962)

► $O_4C_{21}H_{32}$

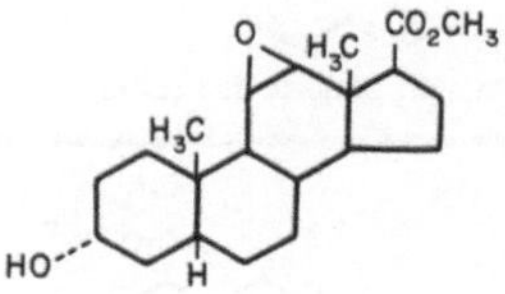

Helv. Chim. Acta 44:1380 (1961)

► $O_4C_{22}H_{22}$

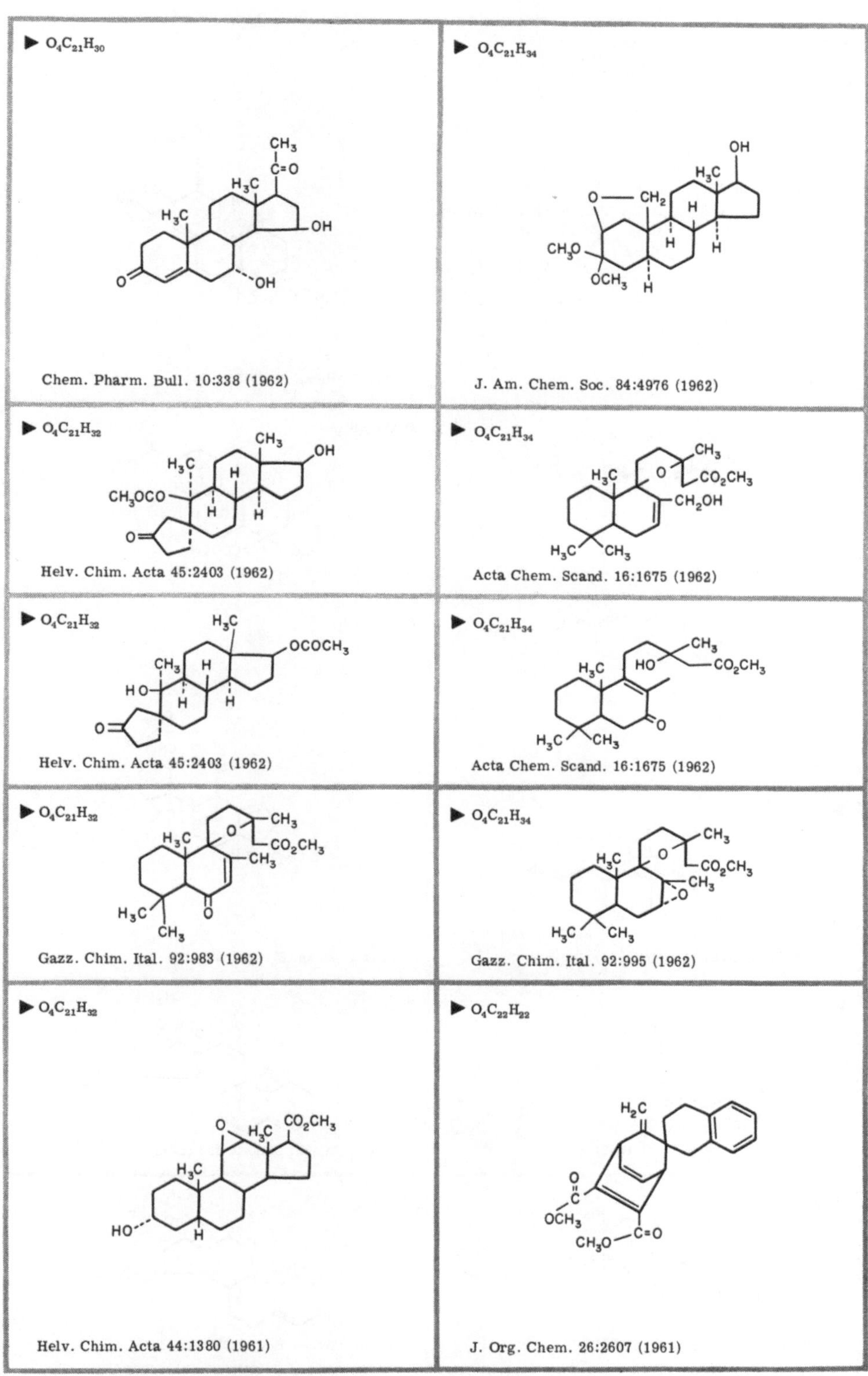

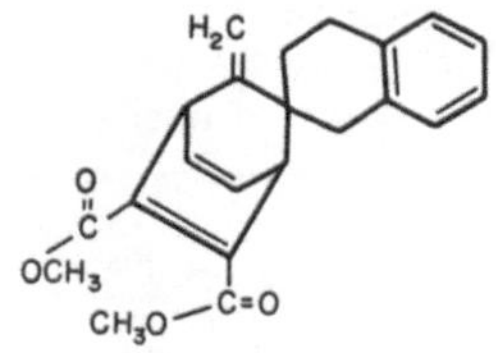

J. Org. Chem. 26:2607 (1961)

▶ $O_4C_{22}H_{24}$

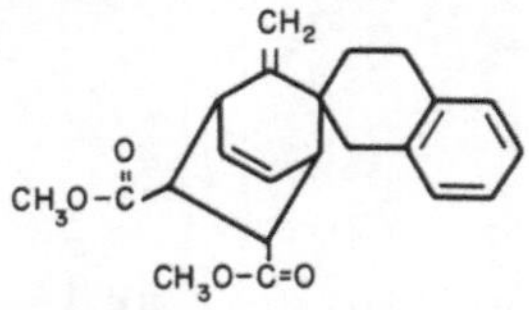

J. Org. Chem. 26:2607 (1961)

▶ $O_4C_{22}H_{26}$

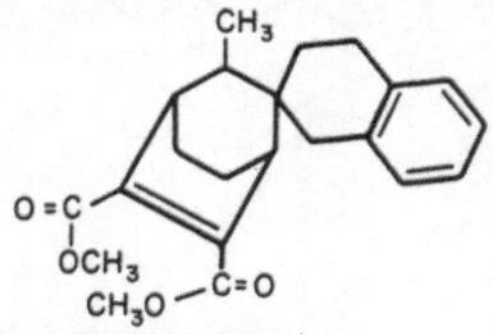

J. Org. Chem. 26:2607 (1961)

▶ $O_4C_{22}H_{26}$

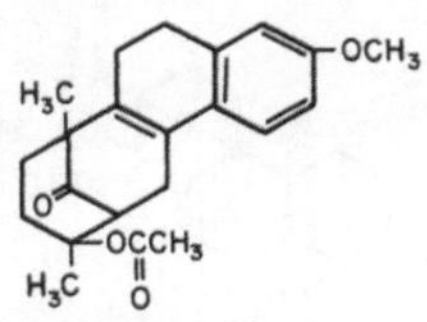

Can. J. Chem. 40:984 (1962)

▶ $O_4C_{22}H_{28}$

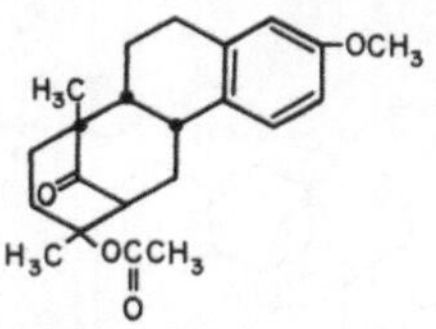

Can. J. Chem. 40:984 (1962)

▶ $O_4C_{22}H_{32}$

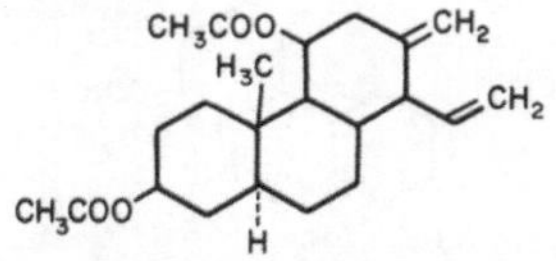

Helv. Chim. Acta 44:1927 (1961)

▶ $O_4C_{22}H_{32}$

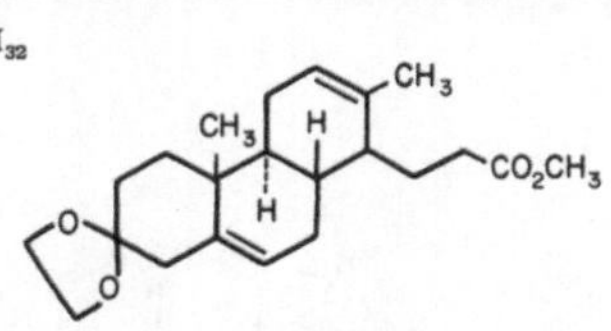

Helv. Chim. Acta 45:2674 (1962)

▶ $O_4C_{22}H_{32}$

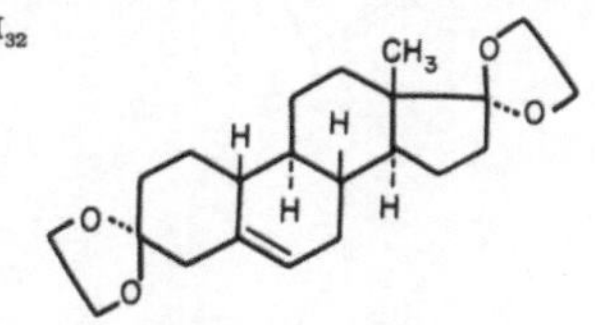

Helv. Chim. Acta 45:2674 (1962)

▶ $O_4C_{22}H_{34}$

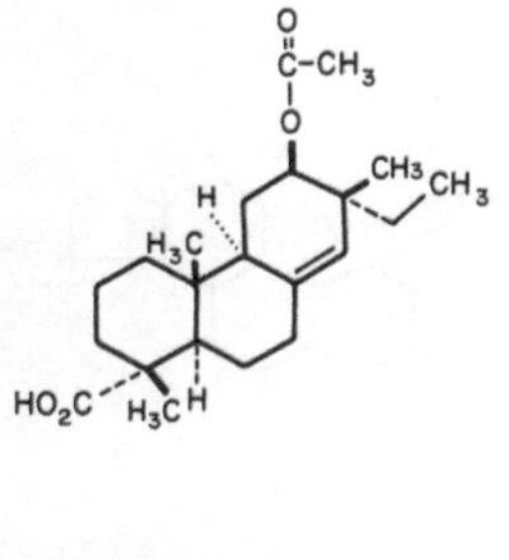

Can. J. Chem. 39:2543 (1961)

▶ $O_4C_{22}H_{34}$

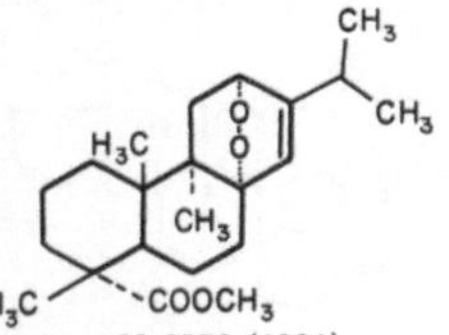

J. Am. Chem. Soc. 83:2570 (1961)

▶ $O_4C_{23}H_{28}$

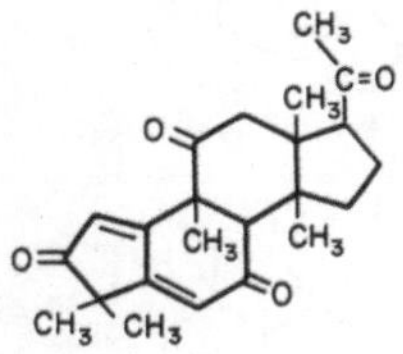

J. Org. Chem. 27:4546 (1962)

▶ $O_4C_{23}H_{32}$

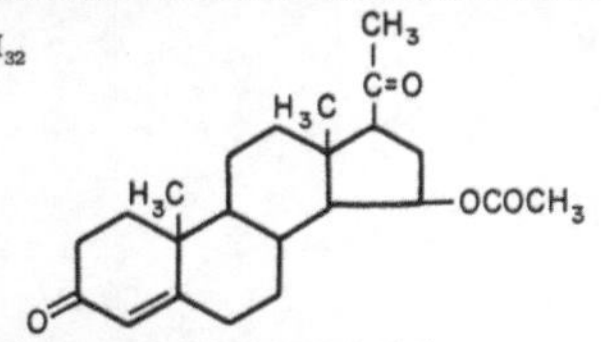

Chem. Pharm. Bull. 10:338 (1962)

▶ $O_4C_{23}H_{32}$

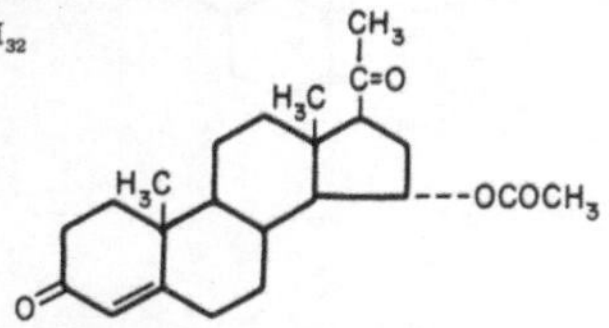

Chem. Pharm. Bull. 10:338 (1962)

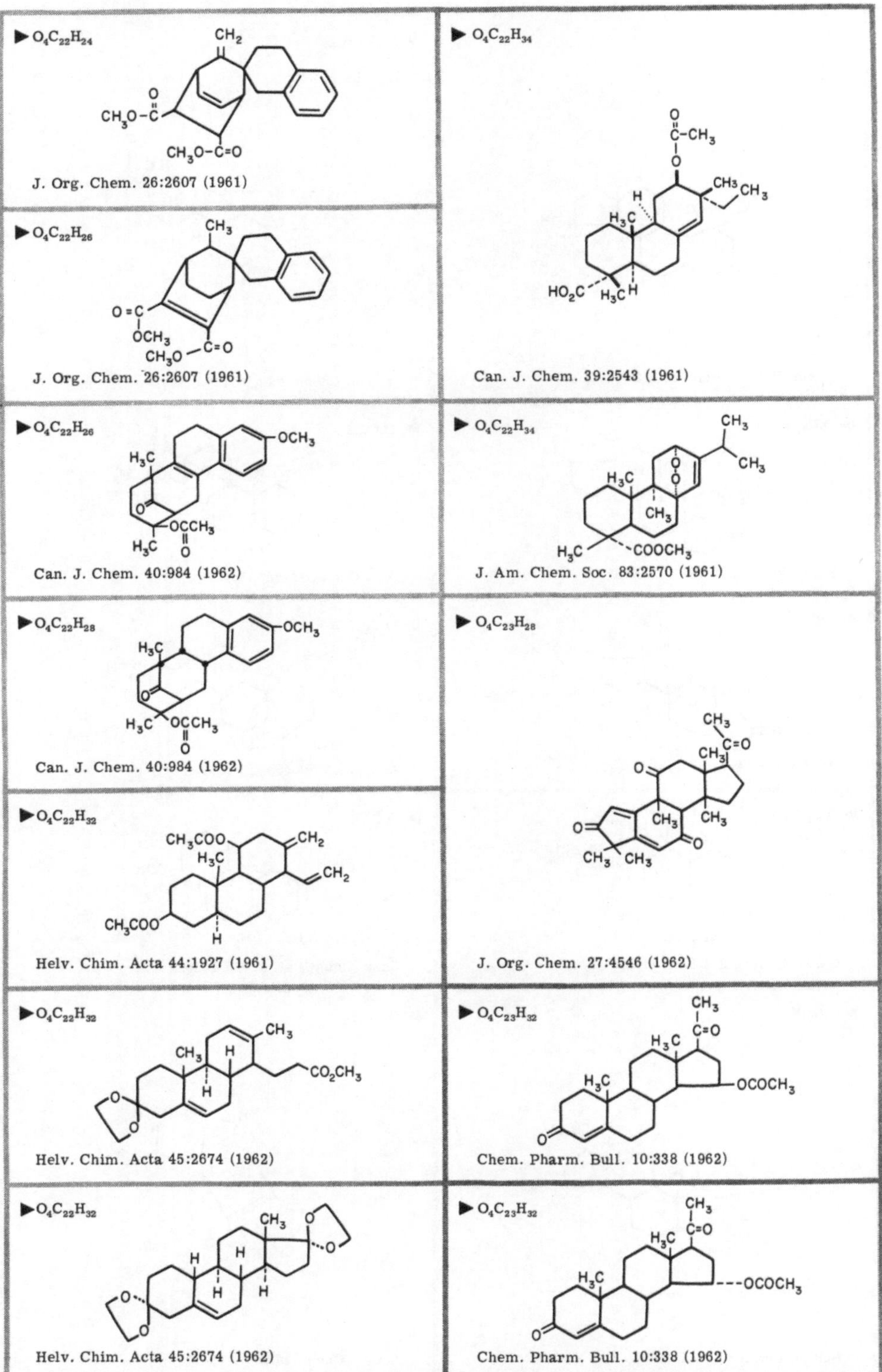

▶ $O_4C_{23}H_{32}$

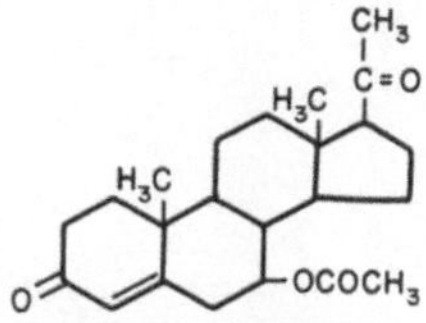

Chem. Pharm. Bull. 10:338 (1962)

▶ $O_4C_{23}H_{32}$

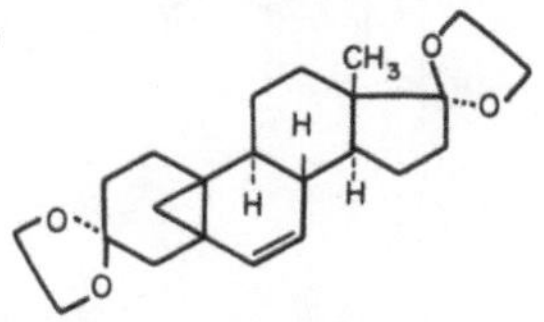

Helv. Chim. Acta 45:2615 (1962)

▶ $O_4C_{23}H_{32}$

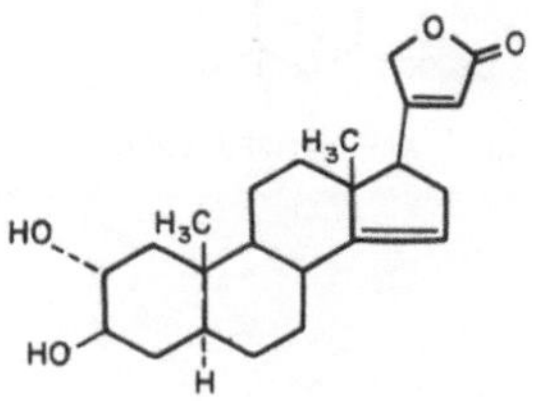

Chem. & Ind. 1724 (1962)

▶ $O_4C_{23}H_{32}$

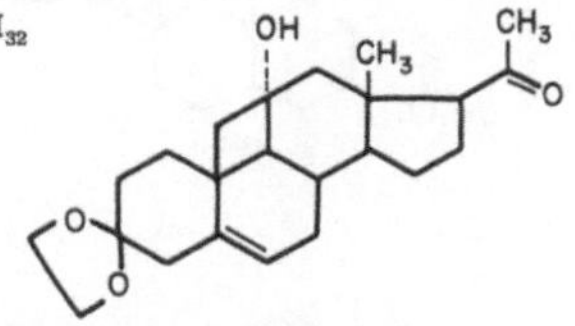

Helv. Chim. Acta 45:1261 (1962)

▶ $O_4C_{23}H_{32}$

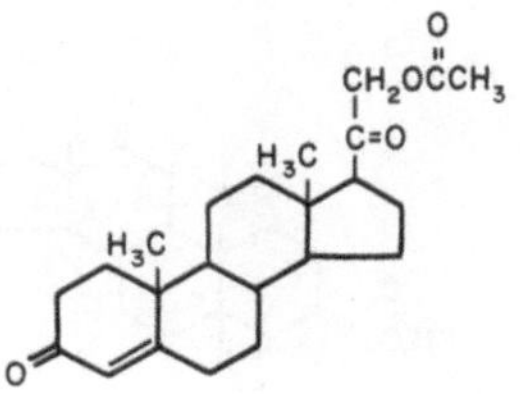

Helv. Chim. Acta 44:1380 (1961)

▶ $O_4C_{23}H_{32}$

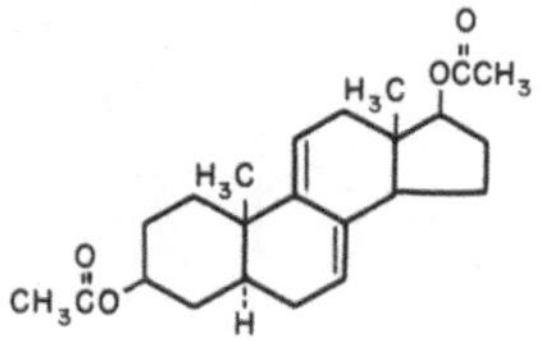

Helv. Chim. Acta 44:1380 (1961)

▶ $O_4C_{23}H_{32}$

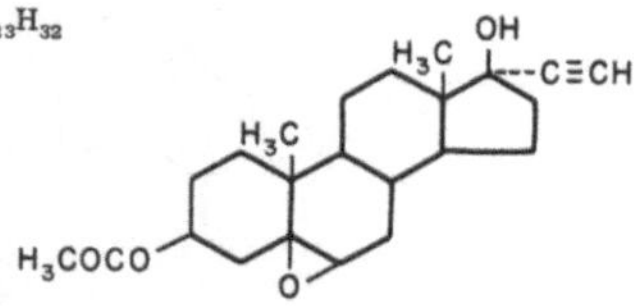

J. Am. Chem. Soc. 84:3206 (1962)

▶ $O_4C_{23}H_{32}$

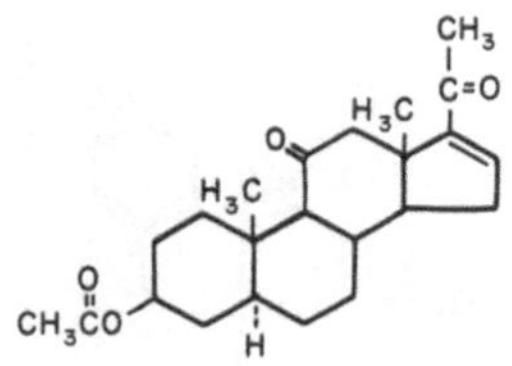

Helv. Chim. Acta 44:1380 (1961)

▶ $O_4C_{23}H_{32}$

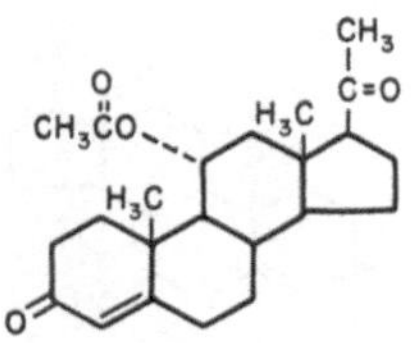

Chem. Pharm. Bull. 10:338 (1962)
Helv. Chim. Acta 44:1380 (1961)

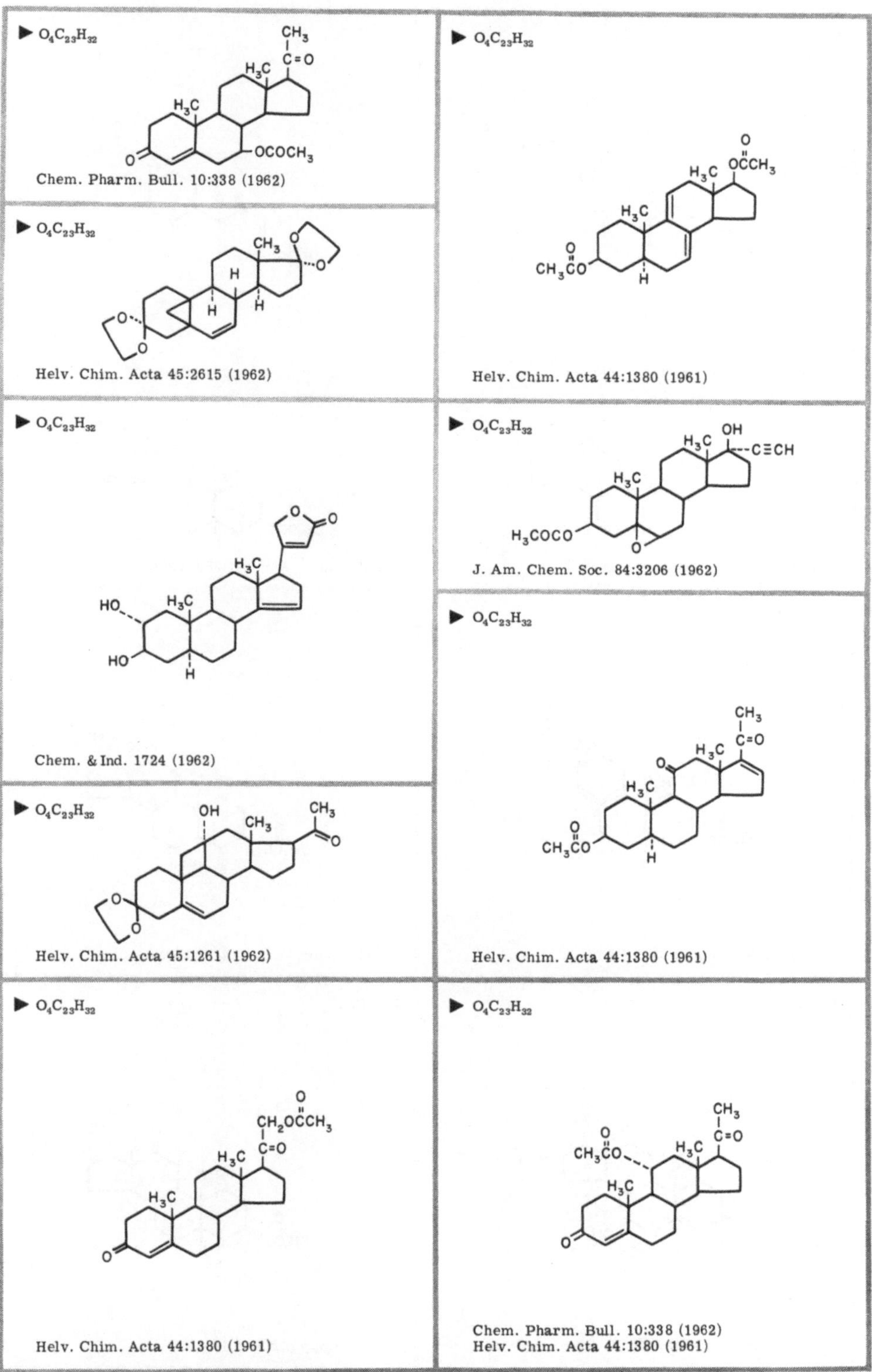

▶ $O_4C_{23}H_{34}$

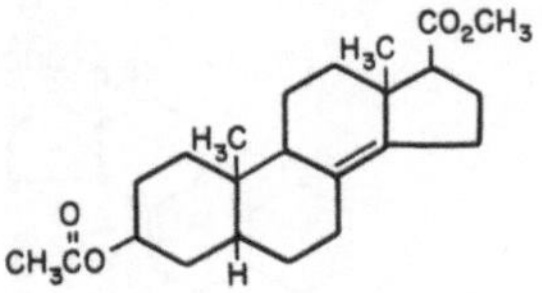

Helv. Chim. Acta 44:1380 (1961)

▶ $O_4C_{23}H_{34}$

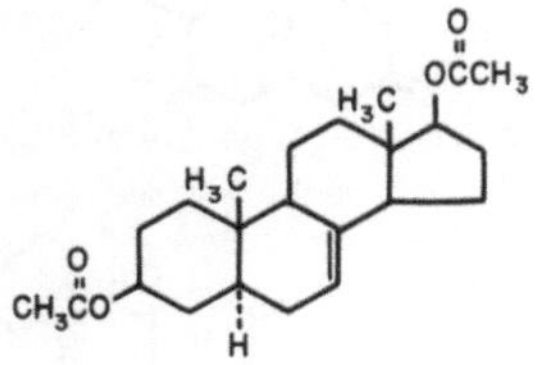

Helv. Chim. Acta 44:1380 (1961)

▶ $O_4C_{23}H_{34}$

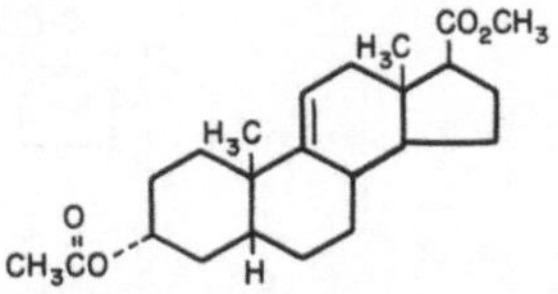

Helv. Chim. Acta 44:1380 (1961)

▶ $O_4C_{23}H_{34}$

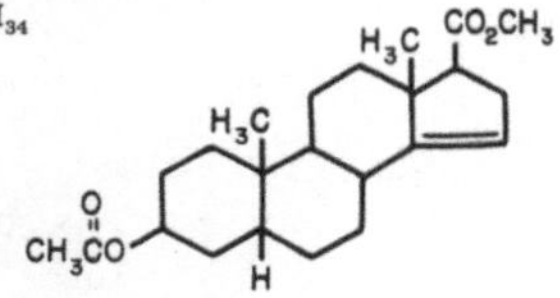

Helv. Chim. Acta 44:1380 (1961)

▶ $O_4C_{23}H_{34}$

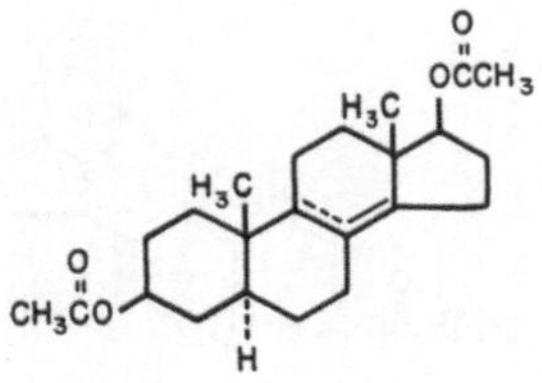

Helv. Chim. Acta 44:1380 (1961)

▶ $O_4C_{23}H_{34}$

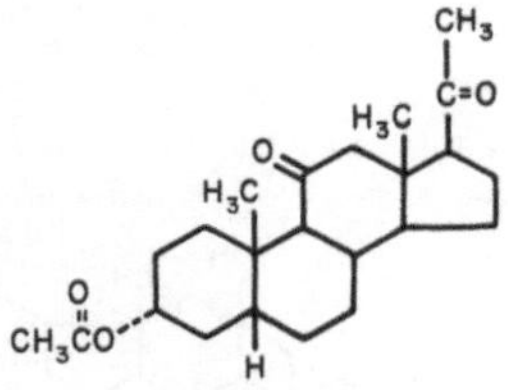

Helv. Chim. Acta 44:1380 (1961)

▶ $O_4C_{23}H_{34}$

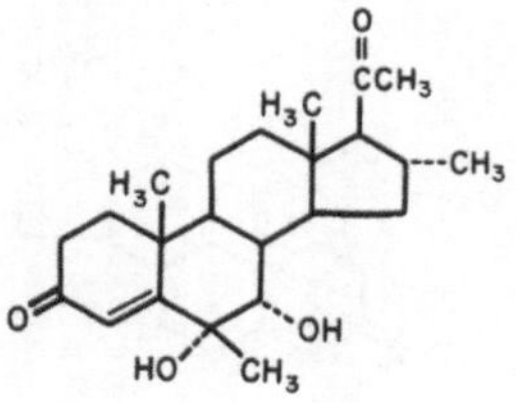

J. Org. Chem. 27:1139 (1962)

▶ $O_4C_{23}H_{34}$

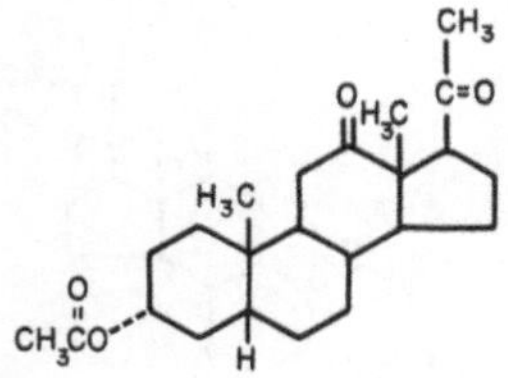

Helv. Chim. Acta 44:1380 (1961)

▶ $O_4C_{23}H_{34}$

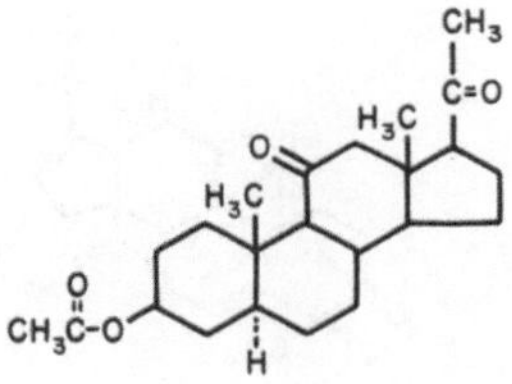


Helv. Chim. Acta 44:1380 (1961)

▶ $O_4C_{23}H_{34}$

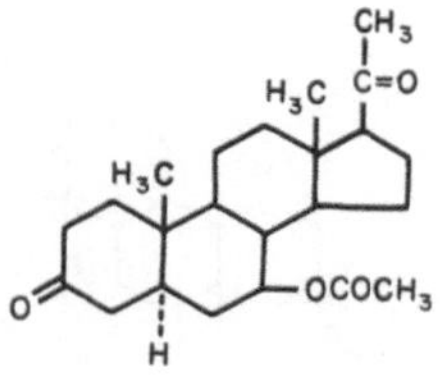


Chem. Pharm. Bull. 10:338 (1962)

▶ $O_4C_{23}H_{34}$

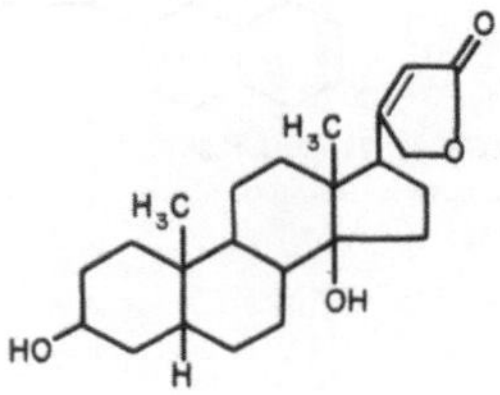


Helv. Chim. Acta 44:1380 (1961)

▶ $O_4C_{23}H_{34}$

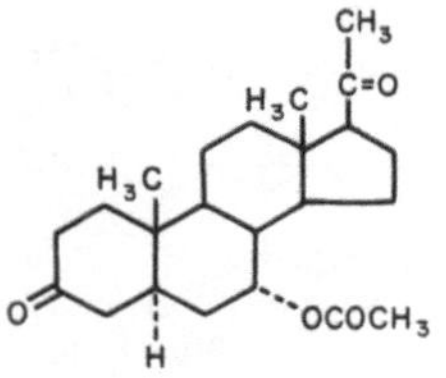


Chem. Pharm. Bull. 10:338 (1962)

▶ $O_4C_{23}H_{34}$

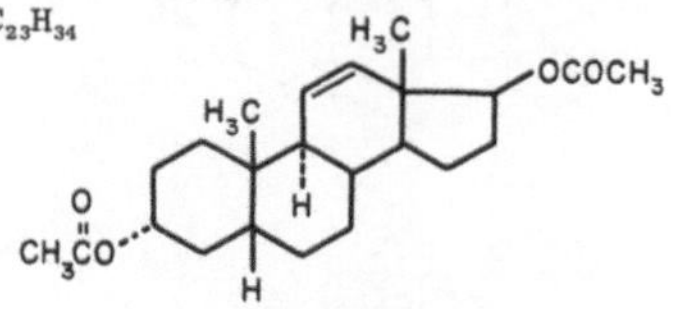


Helv. Chim. Acta 45:753 (1962)

▶ $O_4C_{23}H_{34}$

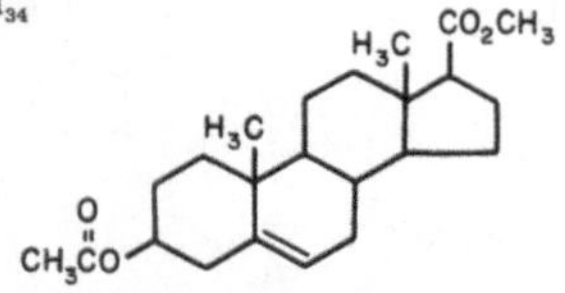


Helv. Chim. Acta 44:1380 (1961)

▶ $O_4C_{23}H_{34}$

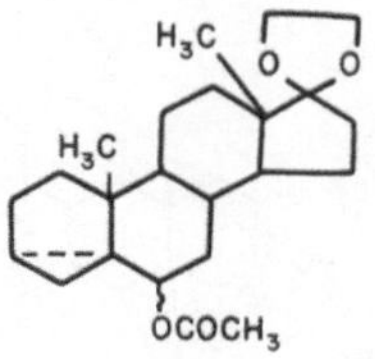


J. Org. Chem. 27:4610 (1962)

▶ $O_4C_{23}H_{34}$

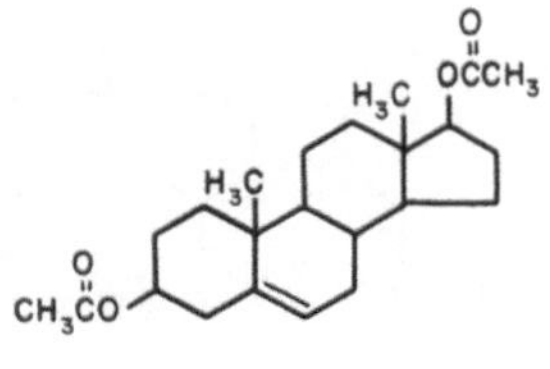


Helv. Chim. Acta 44:1380 (1961)

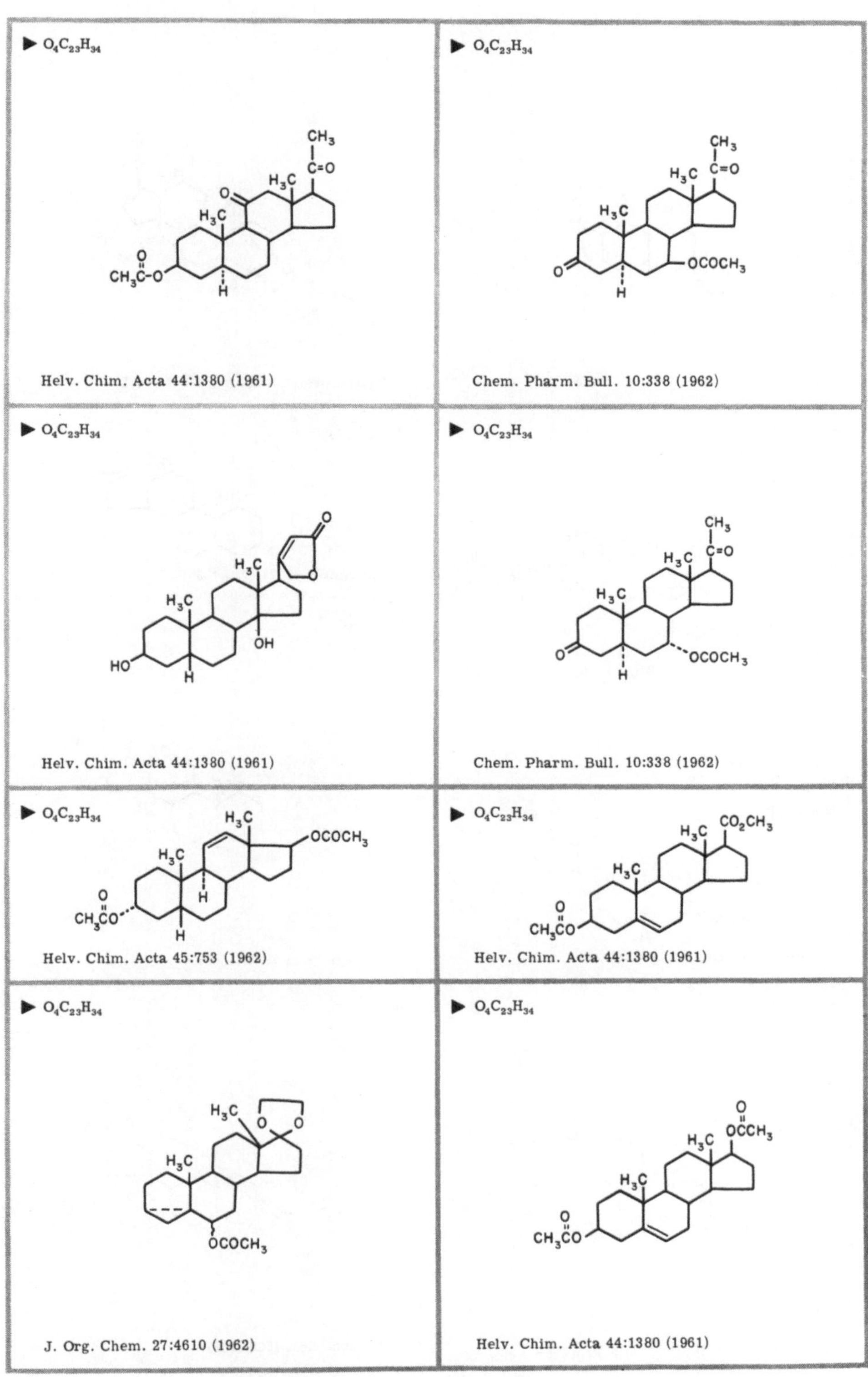

▶ $O_4C_{23}H_{34}$

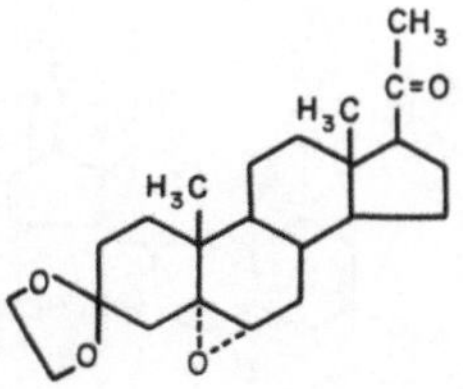

J. Am. Chem. Soc. 84:3206 (1962)

▶ $O_4C_{24}H_{30}$

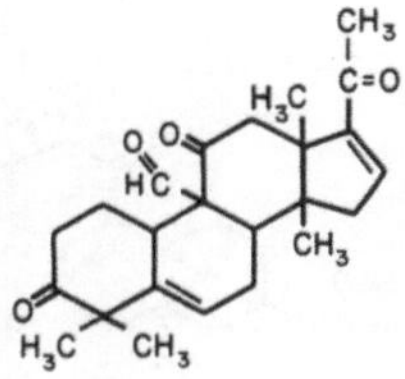

Tetrahedron Letters No. 8:316 (1962)

▶ $O_4C_{23}H_{36}$

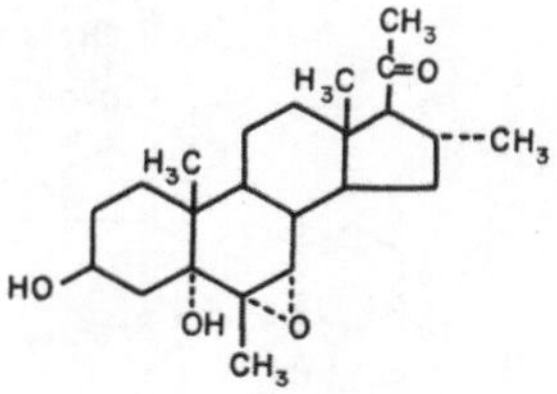

J. Org. Chem. 27:1139 (1962)

▶ $O_4C_{24}H_{36}$

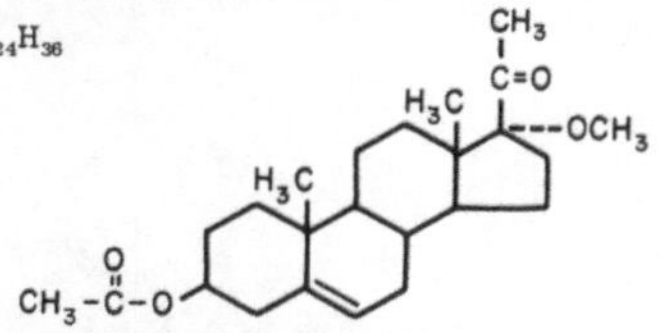

J. Org. Chem. 27:4397 (1962)

▶ $O_4C_{24}H_{36}$

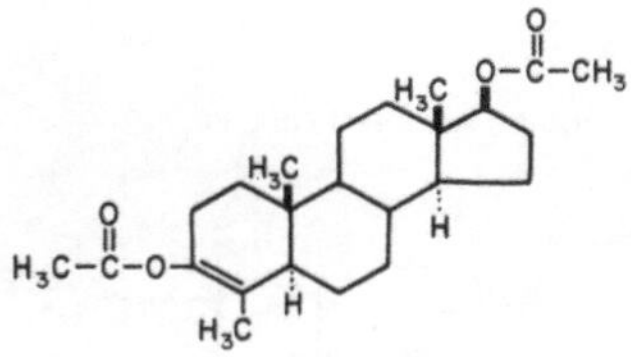

J. Chem. Soc. 1326 (1962)

▶ $O_4C_{23}H_{36}$

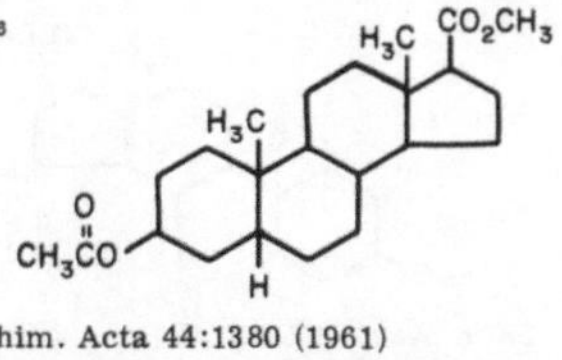

Helv. Chim. Acta 44:1380 (1961)

▶ $O_4C_{23}H_{36}$

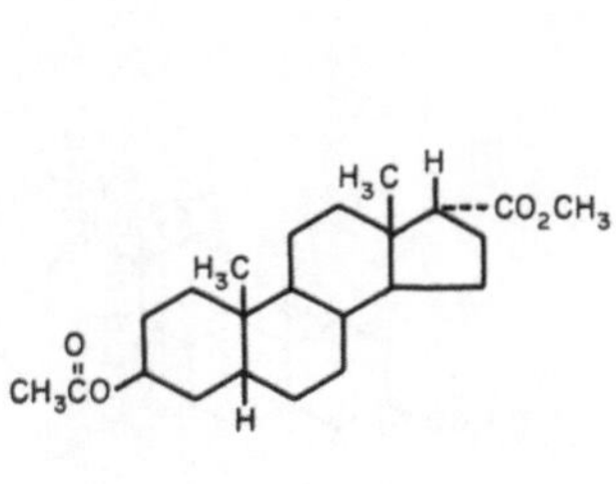

Helv. Chim. Acta 44:1380 (1961)

▶ $O_4C_{24}H_{36}$

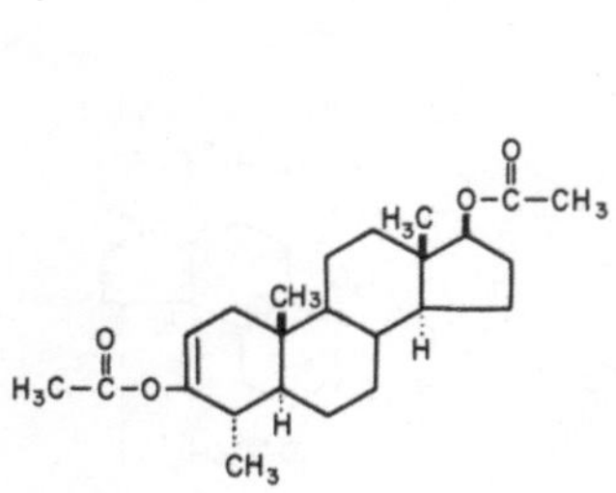

J. Chem. Soc. 1326 (1962)

▶ $O_4C_{24}H_{36}$

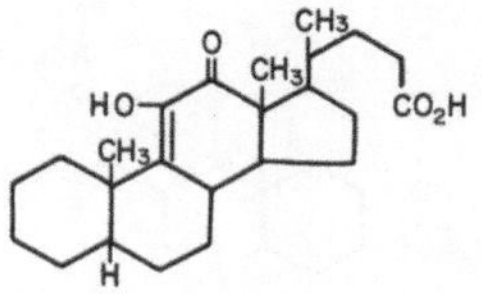

Helv. Chim. Acta 44:1380 (1961)

▶ $O_4C_{24}H_{36}$

UNKNOWN COMPOUND 'A' PRODUCED BY UV.-IRRADIATION OF 3,20-DI-ETHYLENEDIOXY 11-KETO-Δ5-PREGNENE

Helv. Chim. Acta 45:1261 (1962)

▶ $O_4C_{24}H_{40}$

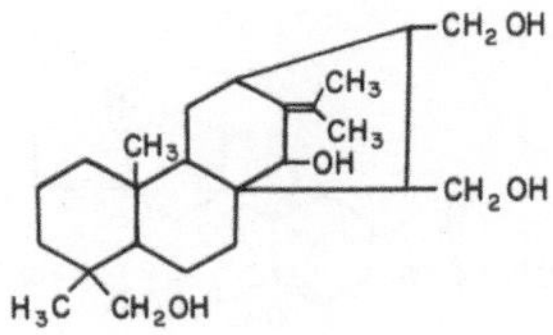

J. Org. Chem. 27:3535 (1962)

▶ $O_4C_{24}H_{48}Pt_2$

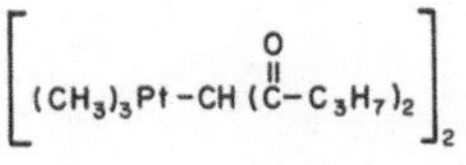

J. Chem. Soc. 4736 (1962)

▶ $O_4C_{25}H_{26}$

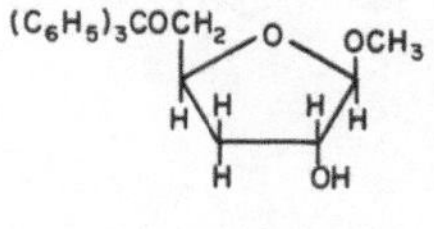

J. Am. Chem. Soc. 83:1900 (1961)

▶ $O_4C_{25}H_{30}$

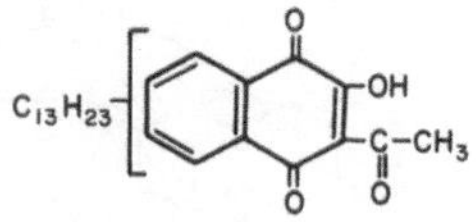

Tetrahedron Letters No. 23:1049 (1962)

▶ $O_4C_{25}H_{34}$

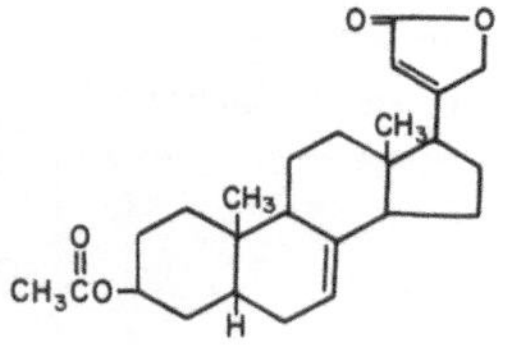

Helv. Chim. Acta 44:1380 (1961)

▶ $O_4C_{25}H_{34}$

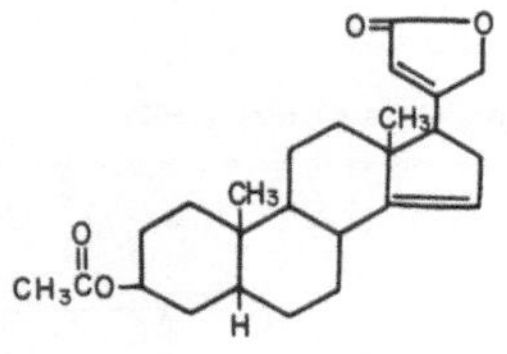

Helv. Chim. Acta 44:1380 (1961)

▶ $O_4C_{25}H_{34}$

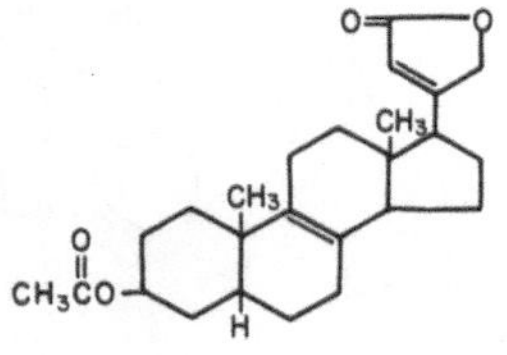

Helv. Chem. Acta 44:1380 (1961)

▶ $O_4C_{25}H_{34}$

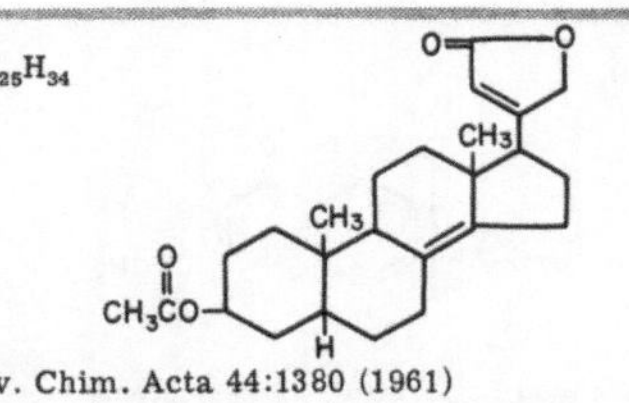

Helv. Chim. Acta 44:1380 (1961)

▶ $O_4C_{25}H_{36}$

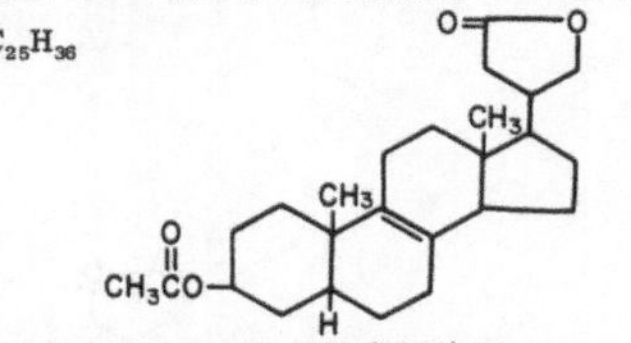

Helv. Chim. Acta 44:1380 (1961)

▶ $O_4C_{25}H_{36}$

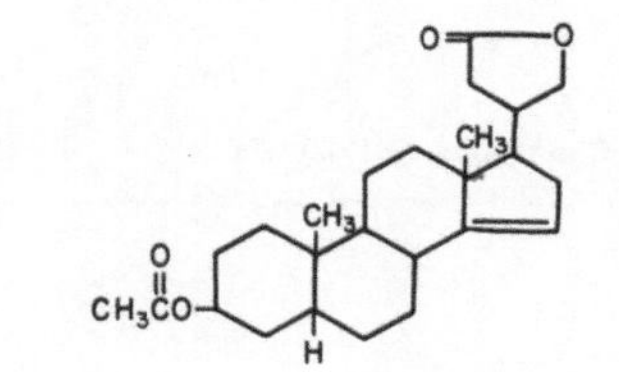

Helv. Chim. Acta 44:1380 (1961)

▶ $O_4C_{25}H_{36}$

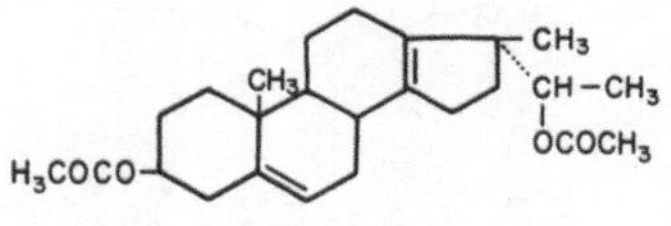

Gazz. Chim. Ital. 92:1118 (1962)

▶ $O_4C_{25}H_{36}$

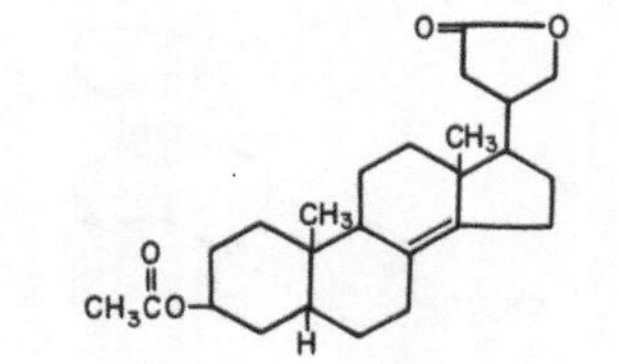

Helv. Chim. Acta 44:1380 (1961)

▶ $O_4C_{25}H_{38}$

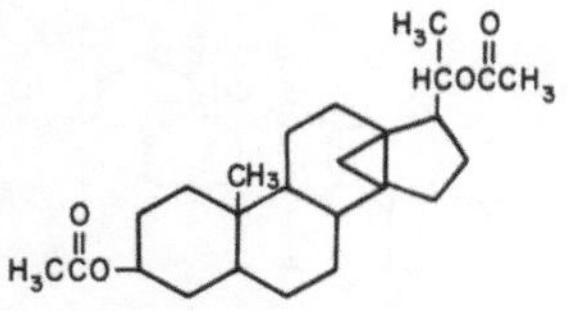

Collection Czech. Chem. Commun. 27:2771 (1962)
Tetrahedron Letters No. 12:501 (1962)

▶ $O_4C_{25}H_{38}$

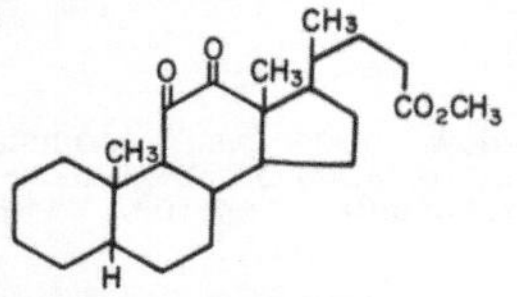

Helv. Chim. Acta 44:1380 (1961)

▶ $O_4C_{25}H_{38}$

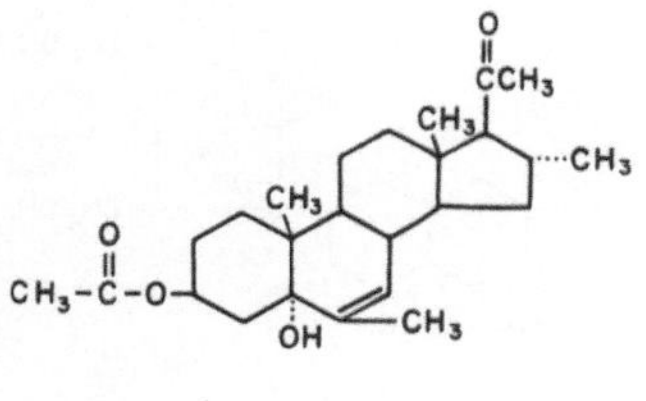

J. Org. Chem. 27:1139 (1962)

▶ $O_4C_{25}H_{38}$

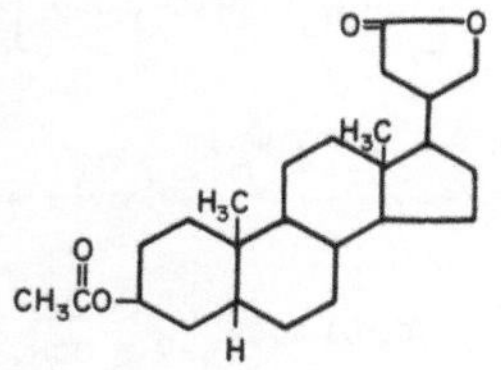

Helv. Chim. Acta 44:1380 (1961)

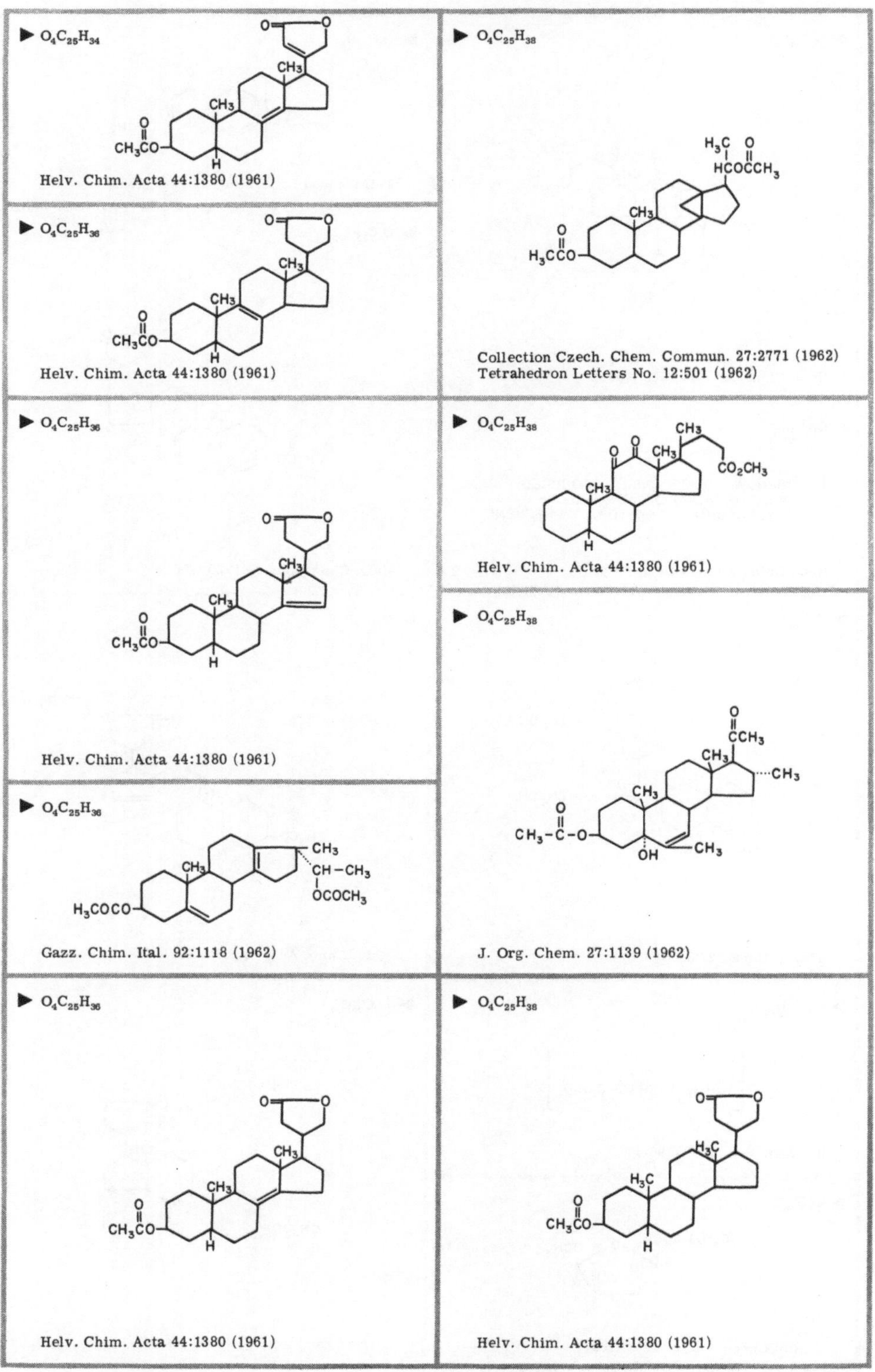

▶ $O_4C_{25}H_{38}$

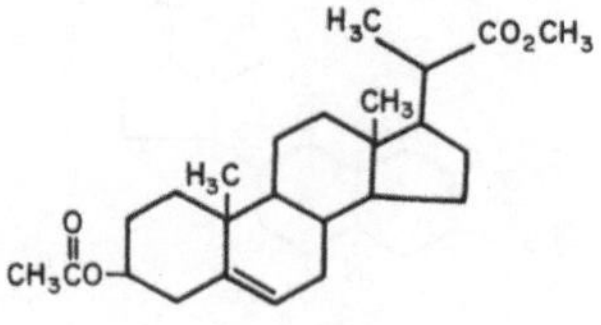

Helv. Chim. Acta 44:1380 (1961)

▶ $O_4C_{25}H_{40}$

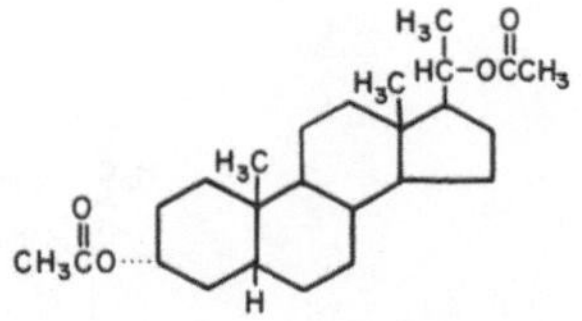

Helv. Chim. Acta 44:1380 (1961)

▶ $O_4C_{26}H_{32}$

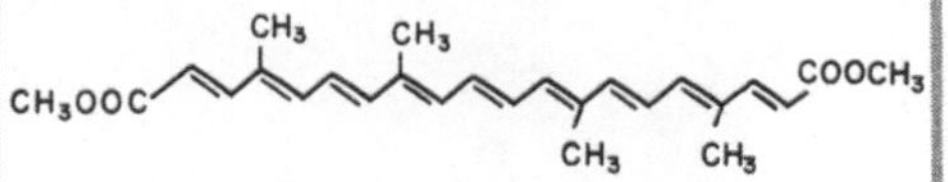

and other cis-trans isomers

J. Chem. Soc. 1625 (1961)

▶ $O_4C_{26}H_{40}$

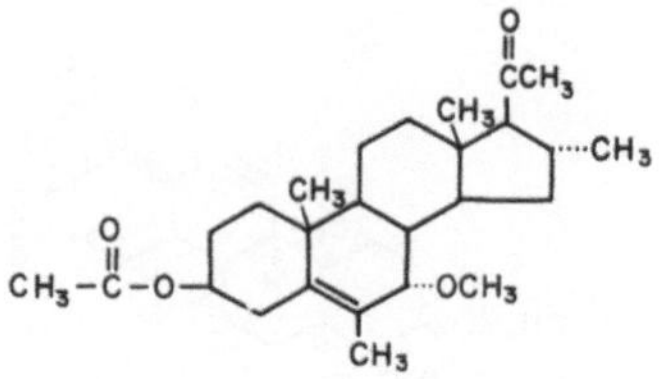

J. Org. Chem. 27:1139 (1962)

▶ $O_4C_{26}H_{40}$

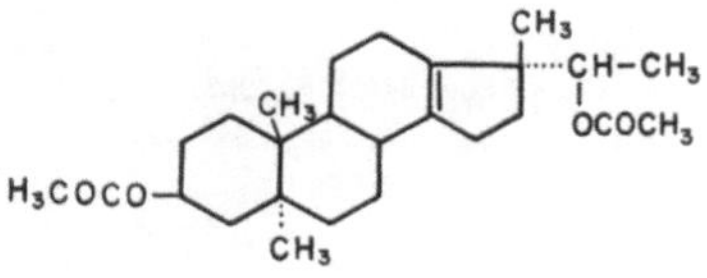

Gazz. Chim. Ital. 92:1118 (1962)

▶ $O_4C_{26}H_{42}$

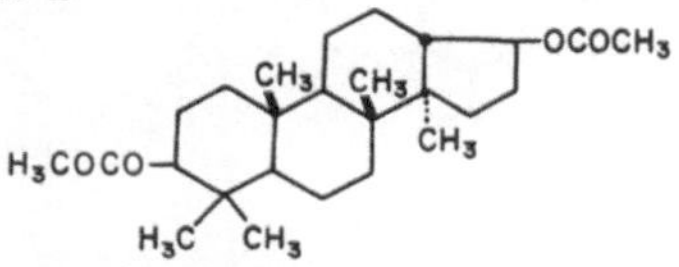

Bull. Soc. Chim. France 1832 (1962)

▶ $O_4C_{26}H_{42}$

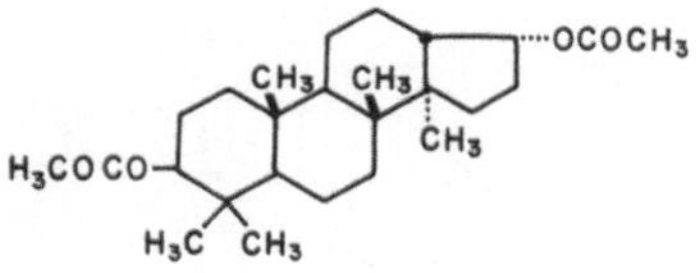

Bull. Soc. Chim. France 1832 (1962)

▶ $O_4C_{29}H_{46}$

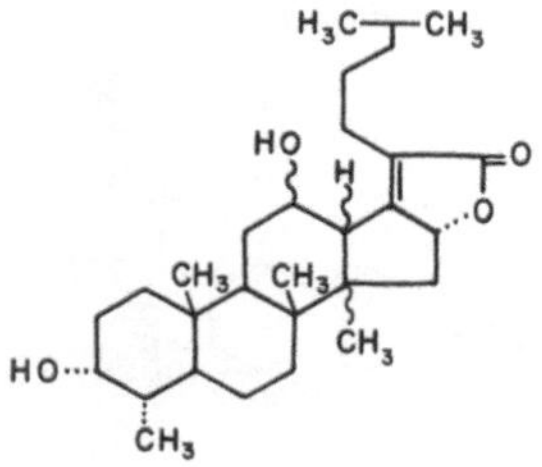

Tetrahedron 18:1037 (1962)

▶ $O_4C_{30}H_{22}$

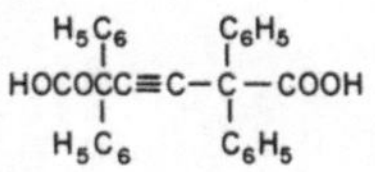

J. Am. Chem. Soc. 84:3278 (1962)

▶ $O_4C_{30}H_{40}$

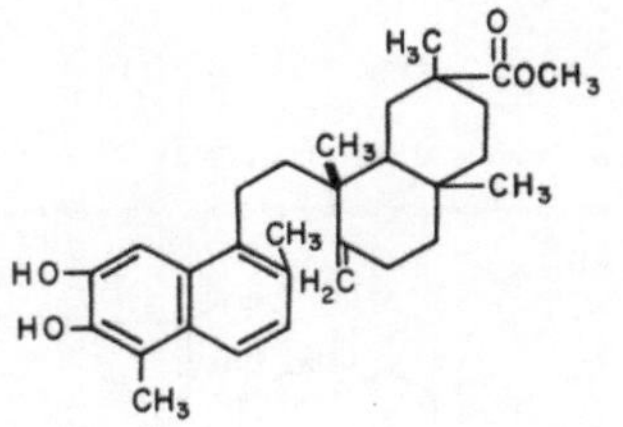

Tetrahedron Letters No. 14:604 (1962)

▶ $O_4C_{31}H_{46}$

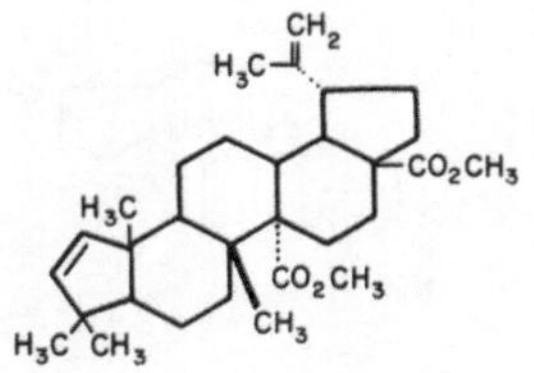

Can. J. Chem. 40:1634 (1962)

▶ $O_4C_{31}H_{50}$

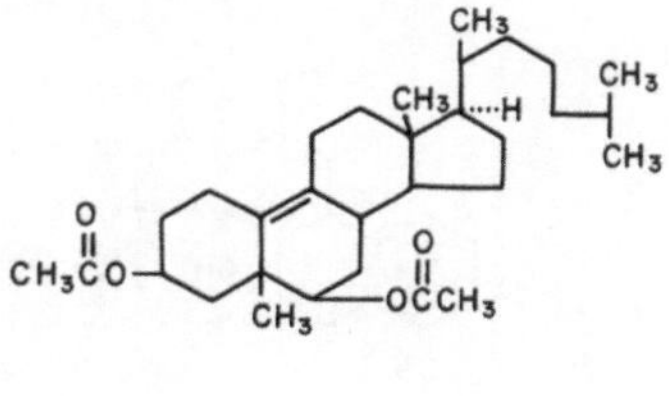

J. Org. Chem. 27:2811 (1962)

▶ $O_4C_{31}H_{52}$

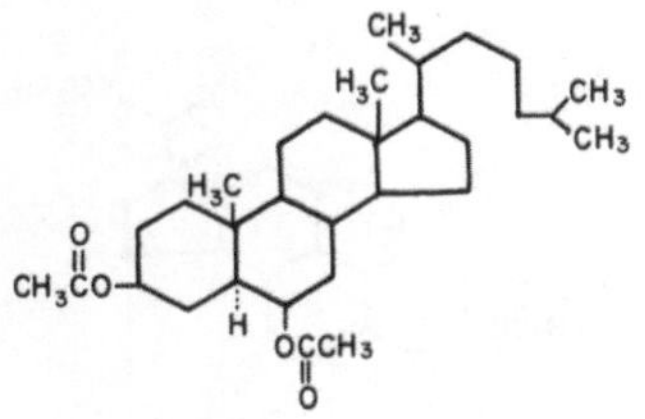

Helv. Chim. Acta 44:1380 (1961)

▶ $O_4C_{32}H_{46}$

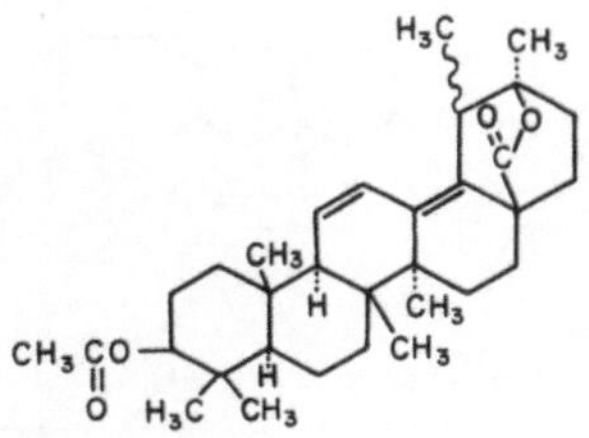

J. Chem. Soc. 5175 (1962)

▶ $O_4C_{32}H_{46}$

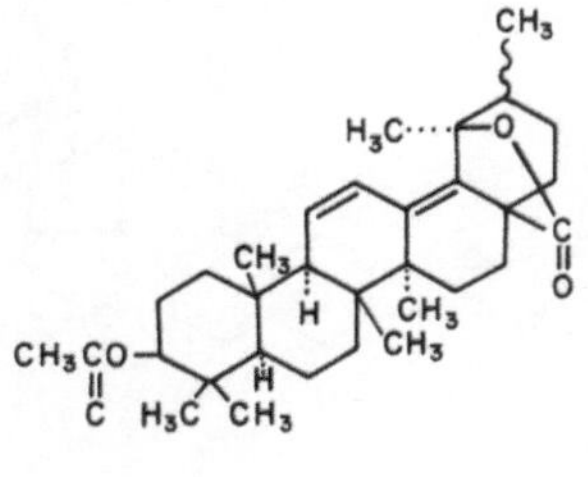

J. Chem. Soc. 5174 (1962)

▶ $O_4C_{32}H_{46}$

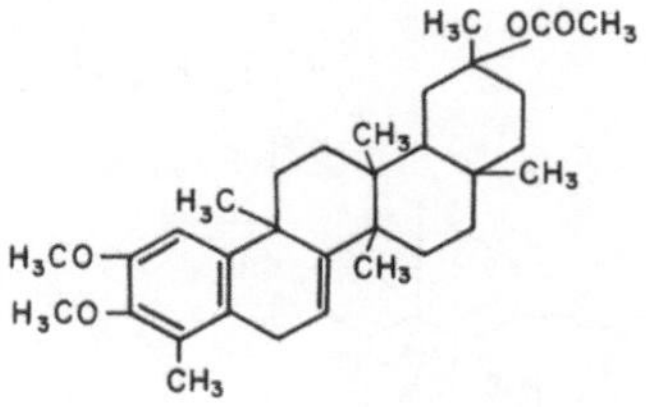

Tetrahedron Letters No. 14:604 (1962)

▶ $O_4C_{32}H_{48}$

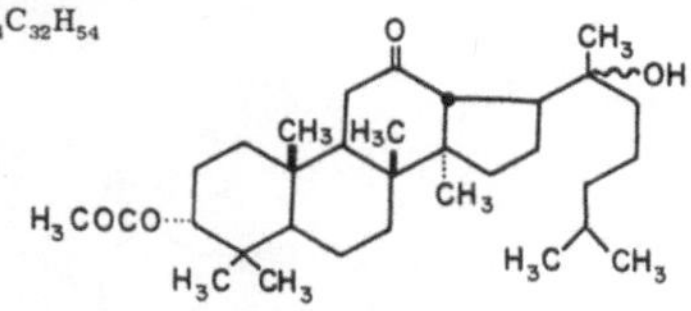

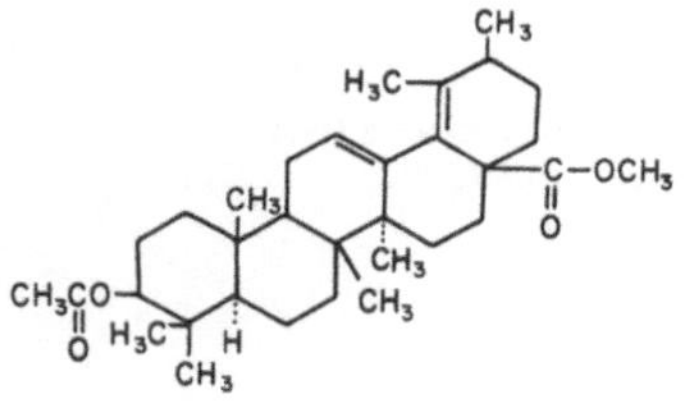

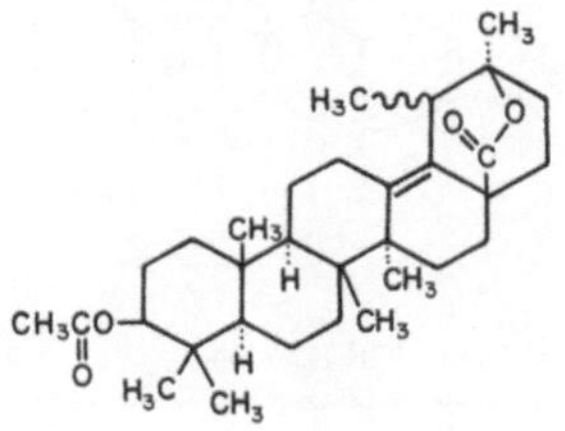

J. Chem. Soc. 5170 (1962)

▶ $O_4C_{32}H_{50}$

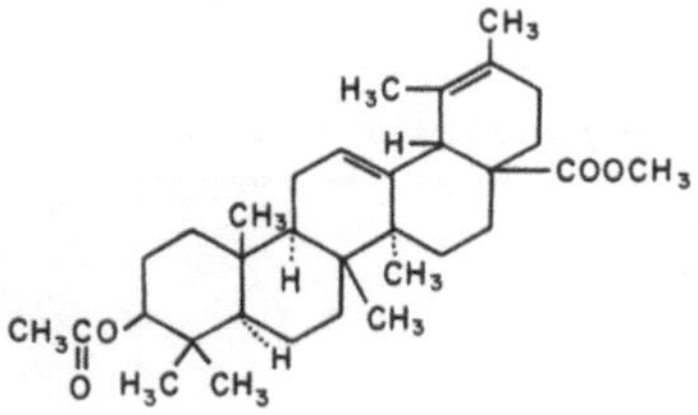

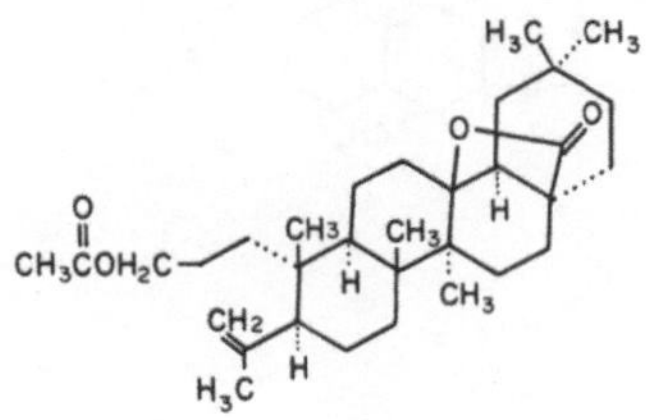

Helv. Chim. Acta 45:2674 (1962)

▶ $O_4C_{32}H_{54}$

Bull. Soc. Chim. France 1832 (1962)

▶ $O_4C_{33}H_{50}$

J. Chem. Soc. 5169 (1962)

▶ $O_4C_{33}H_{50}$

J. Chem. Soc. 5169 (1962)

▶ $O_4C_{33}H_{52}$

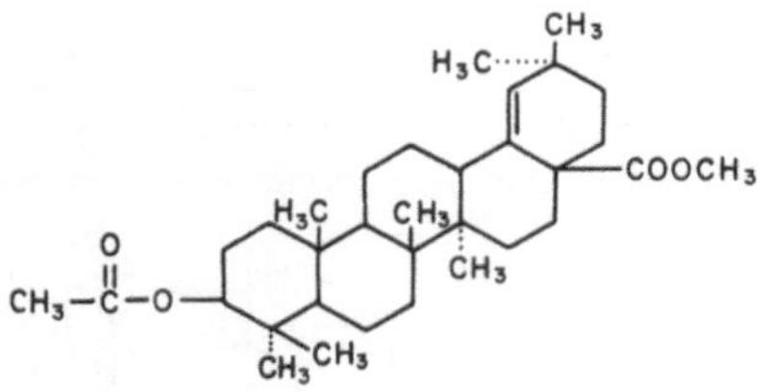

J. Org. Chem. 27:4512 (1962)

► $O_4C_{33}H_{52}$

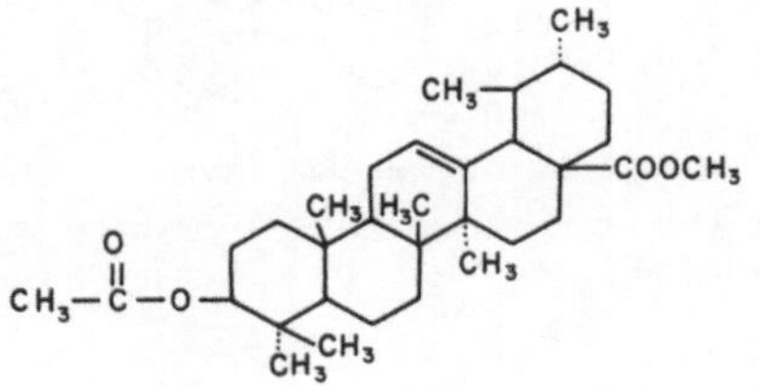

J. Org. Chem. 27:4512 (1962)

► $O_4C_{34}H_{54}$

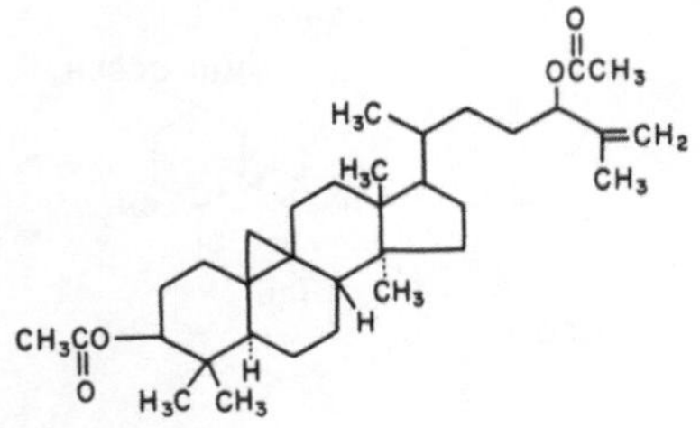

J. Chem. Soc. 4035 (1962)

► $O_4C_{33}H_{52}$

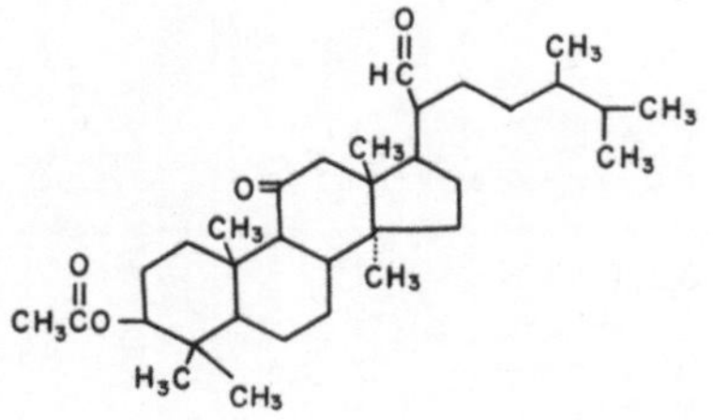

J. Org. Chem. 27:4710 (1962)

► $O_4C_{34}H_{54}$

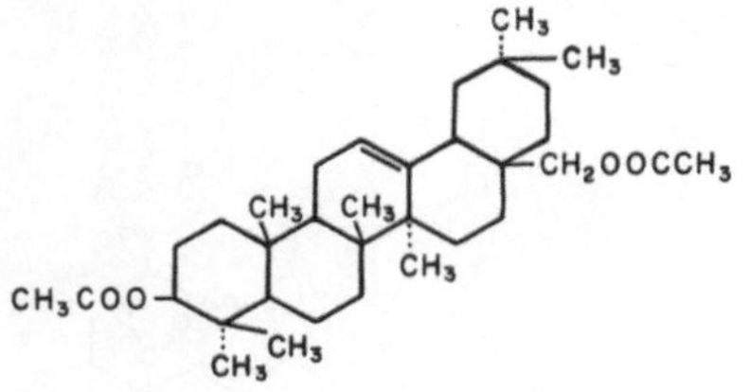

J. Org. Chem. 27:4512 (1962)

► $O_4C_{33}H_{54}$

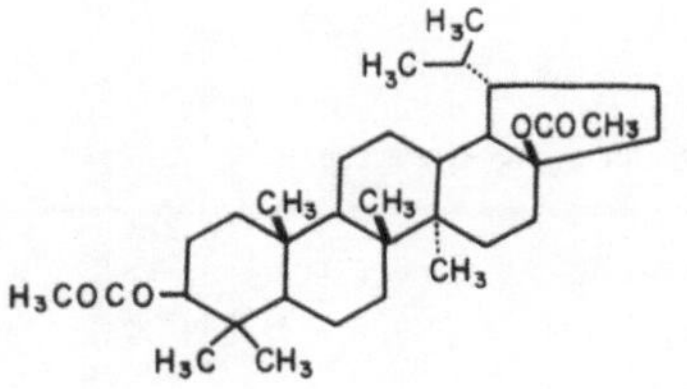

Bull. Soc. Chim. France 1137 (1962)

► $O_4C_{34}H_{54}$

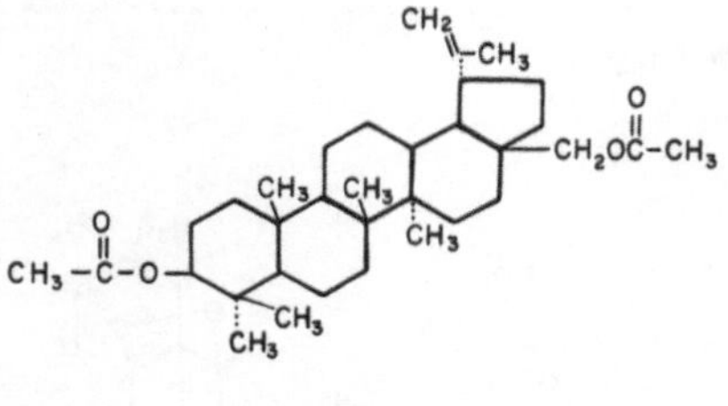

Bull. Soc. Chim. France 1137 (1962)
J. Org. Chem. 27:4512 (1962)

► $O_4C_{34}H_{48}$

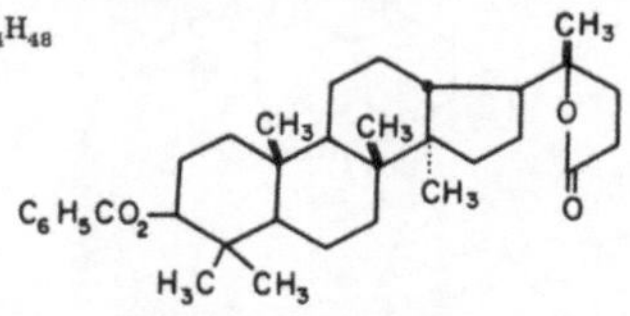

Bull. Soc. Chim. France 1832 (1962)

► $O_4C_{38}H_{56}$

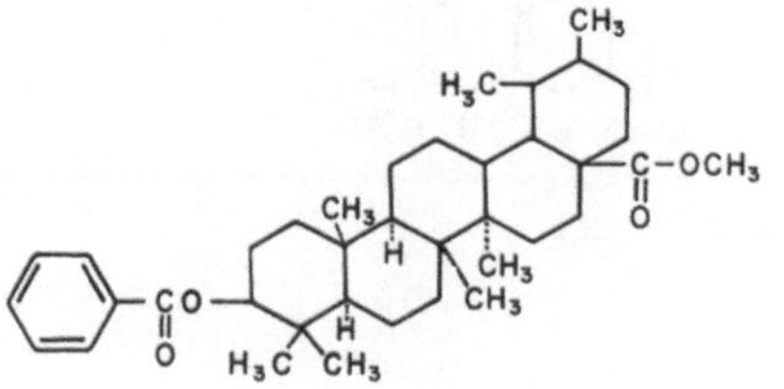

J. Chem. Soc. 5173 (1962)

► $O_5BrC_{15}H_{17}$

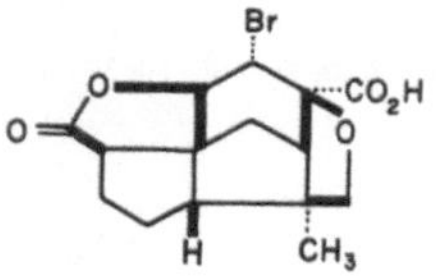

Tetrahedron 18:1321 (1962)

► $O_4C_{40}H_{56}$

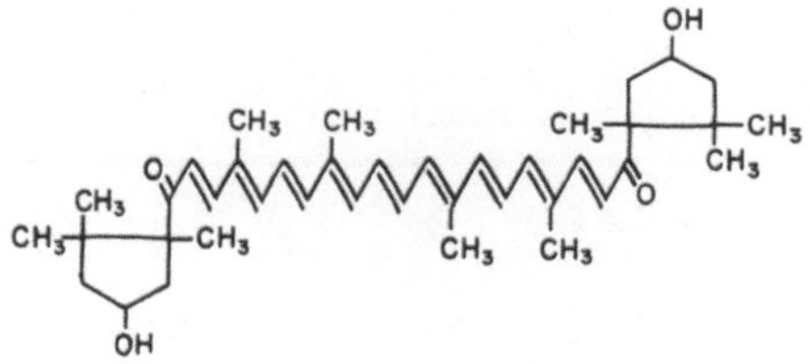

Proc. Chem. Soc. 215 (1962)

► $O_5BrC_{16}H_{19}$

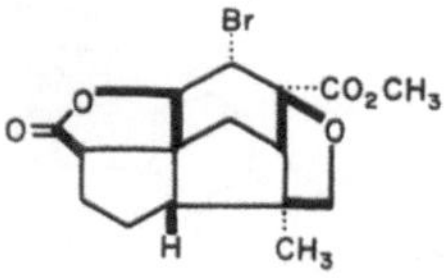

Tetrahedron 18:1321 (1962)

► $O_5IC_{10}H_{11}$

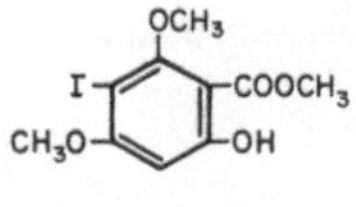

Helv. Chim. Acta 45:2241 (1962)

► $O_5ClC_8H_7$

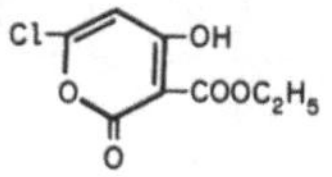

J. Chem. Soc. 2608 (1962)

► $O_5IC_{10}H_{11}$

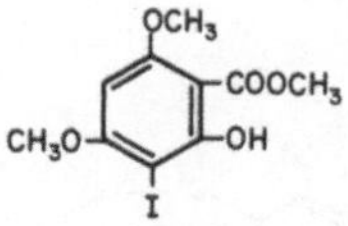

Helv. Chim. Acta 45:2241 (1962)

► $O_5Cl_{10}H_{11}$

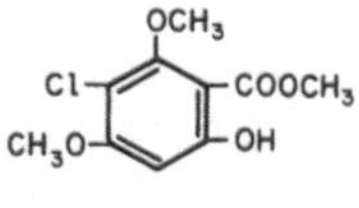

Helv. Chim. Acta 45:2241 (1962)

► $O_5BrC_6H_{11}$

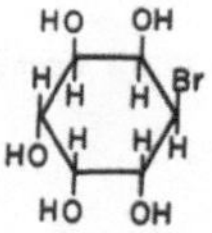

J. Am. Chem. Soc. 83:2335 (1961)

► $O_5ClC_{10}H_{11}$

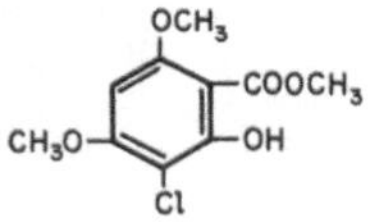

Helv. Chim. Acta 45:2241 (1962)

▶ $O_5ClC_{16}H_{17}$

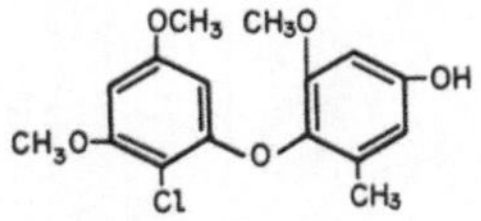

Helv. Chim. Acta 45:813 (1962)

▶ $O_5ClC_{17}H_{19}$

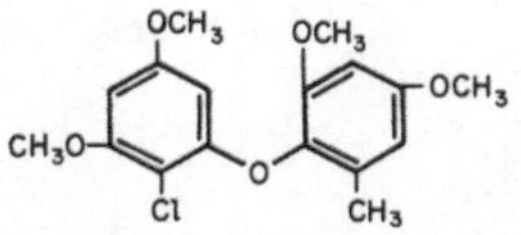

Helv. Chim. Acta 45:813 (1962)

▶ $O_5ClC_{24}H_{31}$

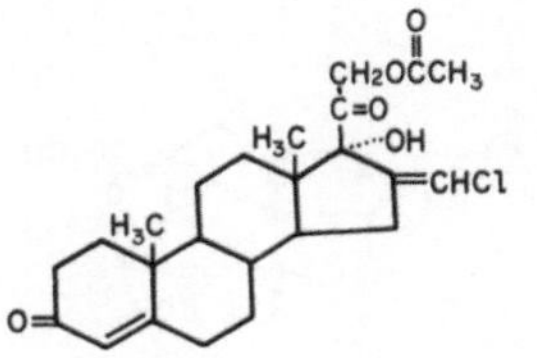

Chem. Ber. 95:2110 (1962)

▶ $O_5ClC_{24}H_{35}$

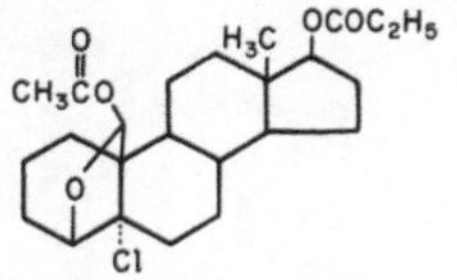

Helv. Chim. Acta 45:2575 (1962)

▶ $O_5ClC_{25}H_{35}$

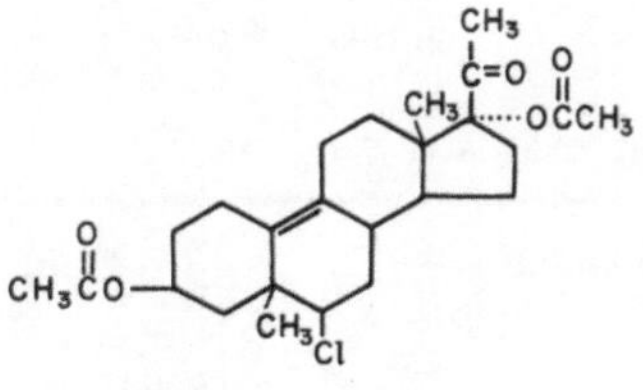

J. Org. Chem. 27:2811 (1962)

▶ $O_5ClC_{27}H_{47}$

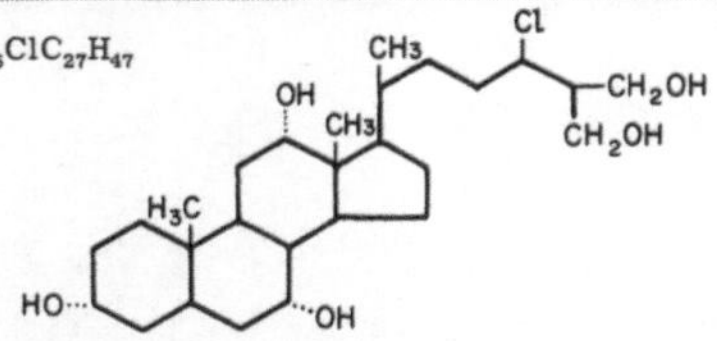

J. Chem. Soc. 2817 (1961)

▶ $O_5Cl_3C_{20}H_{17}$

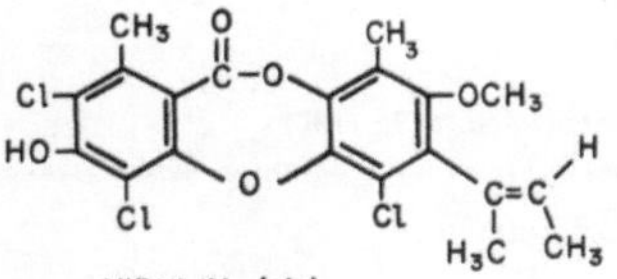

NIDULIN (cis)

J. Org. Chem. 26:3011 (1961)

▶ $O_5Cl_3C_{20}H_{19}$

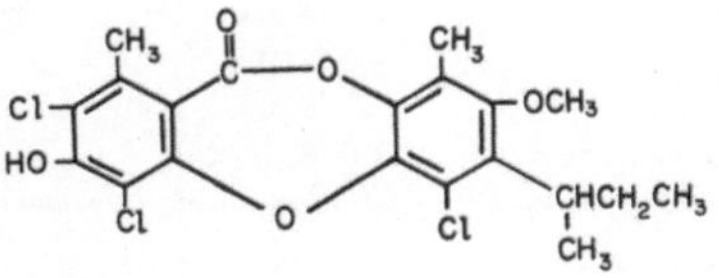

DIHYDRONIDULIN

J. Org. Chem. 26:1339, 3011 (1961)

▶ $O_5Cl_3C_{21}H_{19}$

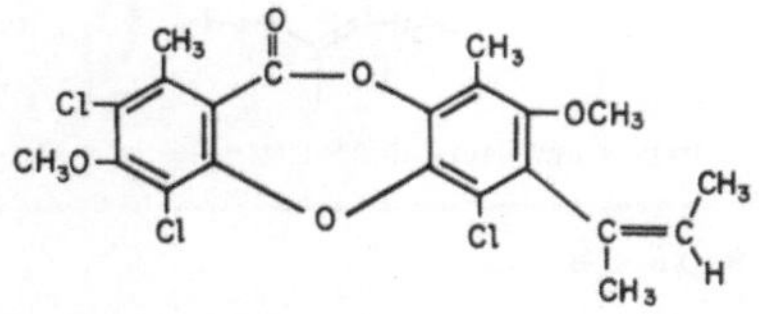

O-METHYLISONIDULIN (trans)

J. Org. Chem. 26:1339, 3011 (1961)

▶ $O_5FC_8H_{11}$

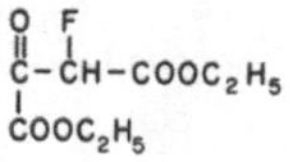

J. Org. Chem. 27:1165 (1962)

▶ $O_5FC_{23}H_{31}$

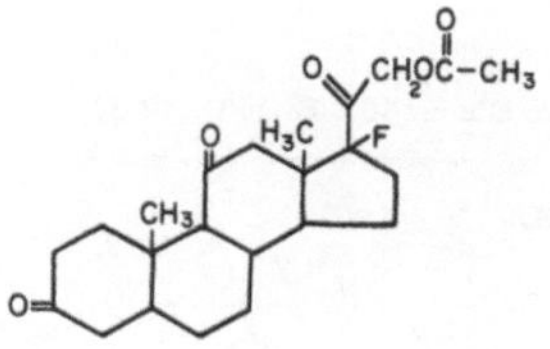

J. Am. Chem. Soc. 84:3784 (1962)

▶ $O_5FC_{24}H_{31}$

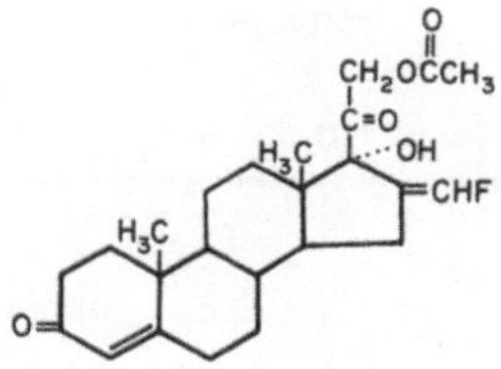

Chem. Ber. 95:2110 (1962)

▶ $O_5F_2C_{17}H_{14}$

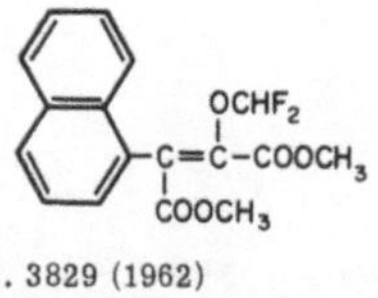

J. Chem. Soc. 3829 (1962)

▶ $O_5C_4H_2D_4$

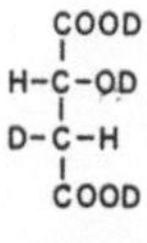

J. Am. Chem. Soc. 83:3634 (1961)

▶ $O_5C_4H_2D_4$

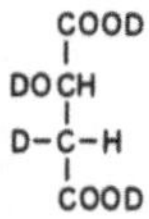

J. Am. Chem. Soc. 83:3634 (1961)

▶ $O_5C_4H_4$

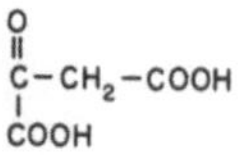

J. Org. Chem. 27:1165 (1962)

▶ $O_5C_5H_{10}$

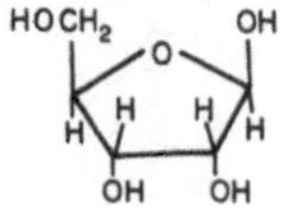

RIBOSE

J. Am. Chem. Soc. 84:4464 (1962)

▶ $O_5C_6H_4$

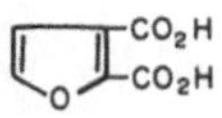

Can. J. Chem. 39:909 (1961)

▶ $O_5C_6H_{10}$

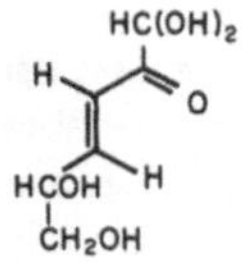

Australian J. Chem. 15:503 (1962)

▶ $O_5C_6H_{10}$

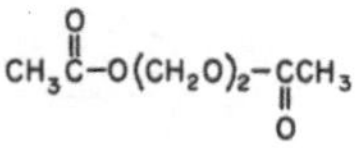

Chem. Ber. 94:3317 (1961)

▶ $O_5C_6H_{12}$

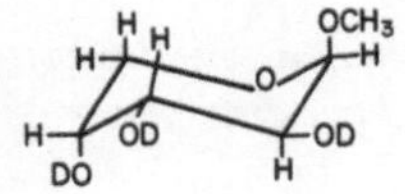

Varian 475

▶ $O_5C_6H_{12}$

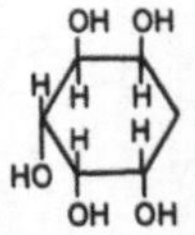

J. Am. Chem. Soc. 83:2335 (1961)

▶ $O_5C_6H_{12}$

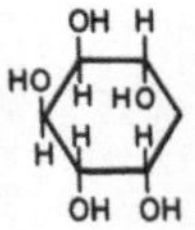

J. Am. Chem. Soc. 83:2335 (1961)

▶ $O_5C_6H_{12}$

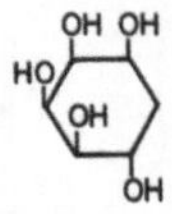

J. Am. Chem. Soc. 83:4243 (1961)

▶ $O_5C_7H_6$

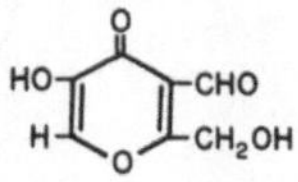

J. Org. Chem. 26:1029 (1961)

▶ $O_5C_7H_{10}$

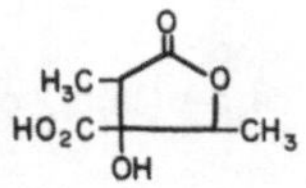

Australian J. Chem. 15:139 (1962)

▶ $O_5C_7H_{12}$

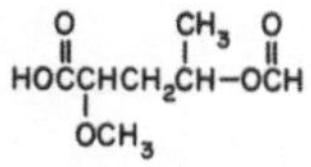

J. Am. Chem. Soc. 83:3352 (1961)

▶ $O_5C_8H_5Mn$

$CH_2{=}CH{-}CH_2{-}Mn(CO)_5$

J. Am. Chem. Soc. 83:1601 (1961)

▶ $O_5C_8H_8$

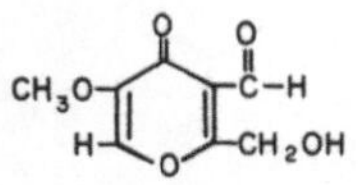

J. Org. Chem. 26:1029 (1961)

▶ $O_5C_8H_{12}$

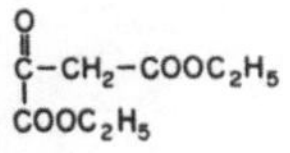

J. Org. Chem. 27:1165 (1962)

▶ $O_5C_8H_{16}$

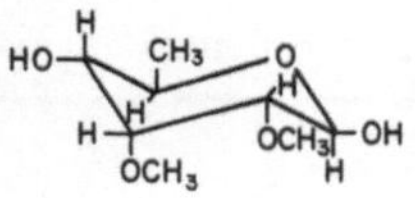

J. Am. Chem. Soc. 84:880 (1962)

▶ $O_5C_9H_{10}$

$O{=}C(CH{=}CH{-}CO_2CH_3)_2$

Tetrahedron Letters No. 1:1 (1961)

► $O_5C_9H_{10}$

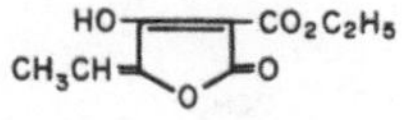

Chem. & Ind. 561 (1962)

► $O_5C_9H_{14}$

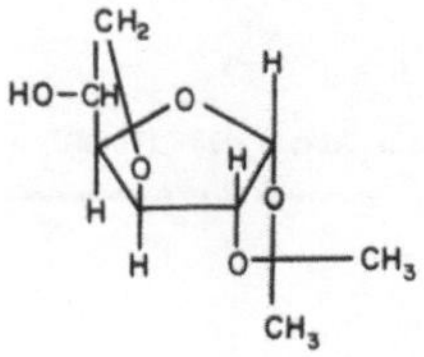

J. Chem. Soc. 3701 (1962)

► $O_5C_9H_{14}$

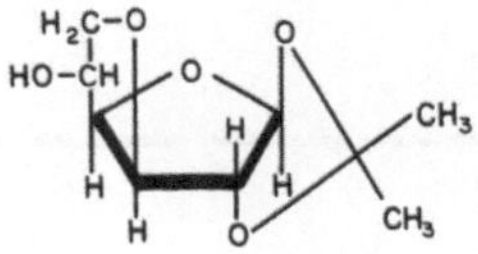

J. Chem. Soc. 2081 (1962)

► $O_5C_9H_{14}$

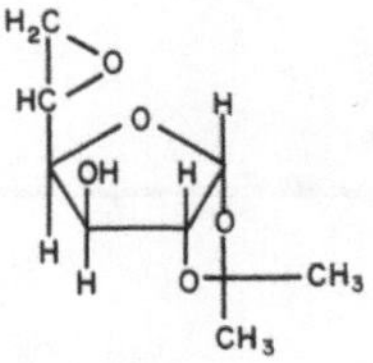

J. Chem. Soc. 3702 (1962)
Chem. & Ind. 213 (1962)

► $O_5C_9H_{18}$

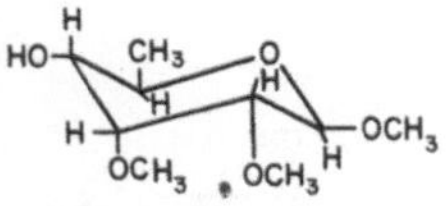

J. Am. Chem. Soc. 84:880 (1962)

► $O_5C_{10}H_{14}$

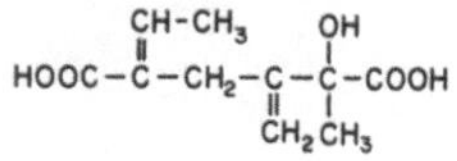

J. Org. Chem. 27:4132 (1962)

► $O_5C_{10}H_{16}$

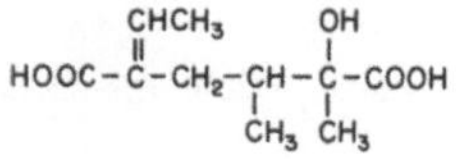

J. Org. Chem. 27:4132 (1962)

► $O_5C_{10}H_{16}$

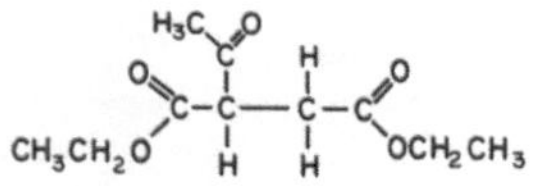

Varian 277

► $O_5C_{10}H_{16}$

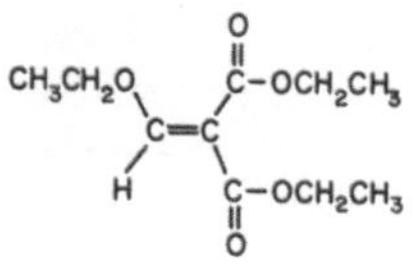

Varian 573

► $O_5C_{10}H_{18}$

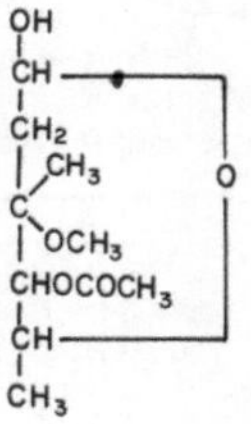

Helv. Chim. Acta 45:138 (1962)

► $O_5C_{10}H_{18}$

$CH_2{=}CHCH_2O{-}(CH_2O)_4{-}CH_2CH{=}CH_2$

J. Chem. Soc. 4313 (1962)

► $O_5C_{10}H_{22}$

$CH_3CH_2CH_2O{-}(CH_2O)_4{-}CH_2CH_2CH_3$

J. Chem. Soc. 4313 (1962)

► $O_5C_{11}H_{14}$

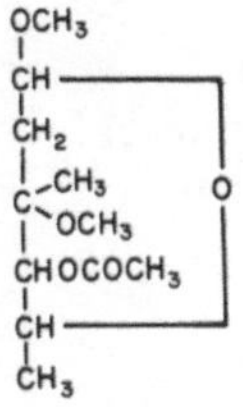

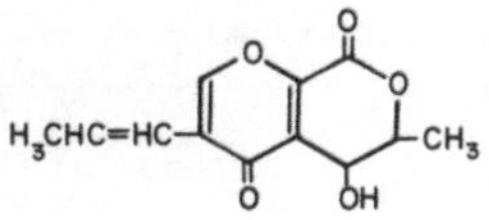

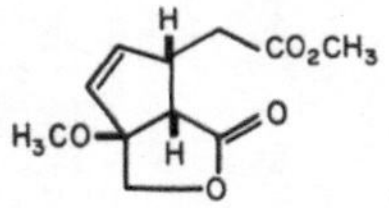

Tetrahedron Letters 110 (1961)

► $O_5C_{11}H_{14}$

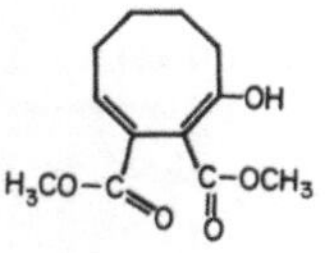

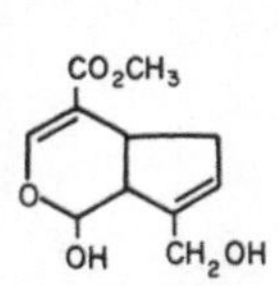

J. Org. Chem. 26:1192 (1961)

► $O_5C_{11}H_{14}$

Varian 583

► $O_5C_{11}H_{20}$

Helv. Chim. Acta 45:138 (1962)

► $O_5C_{12}H_{12}$

Varian 595

► $O_5C_{12}H_{16}$

Varian 600

► $O_5C_{12}H_{18}$

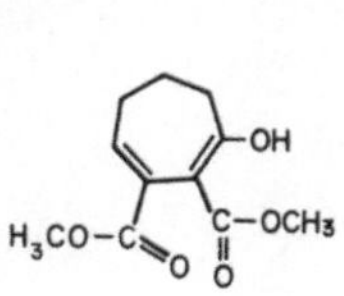

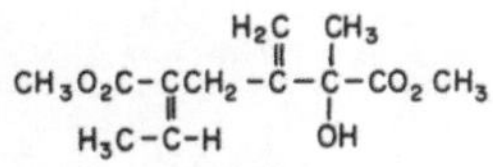

J. Am. Chem. Soc. 83:922 (1961)

▶ $O_5C_{12}H_{18}$

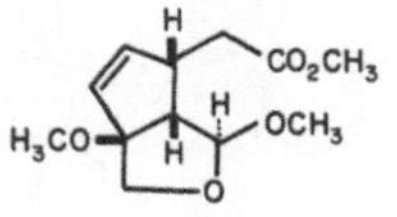

Tetrahedron Letters 110 (1961)

▶ $O_5C_{12}H_{20}$

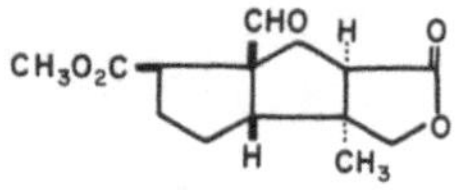

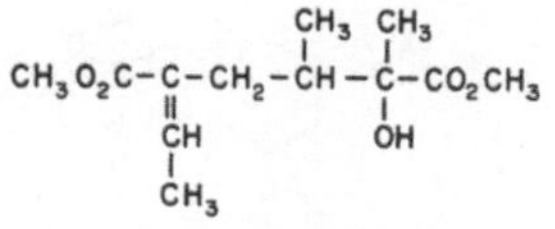

J. Am. Chem. Soc. 83:922 (1961)

▶ $O_5C_{13}H_{22}$

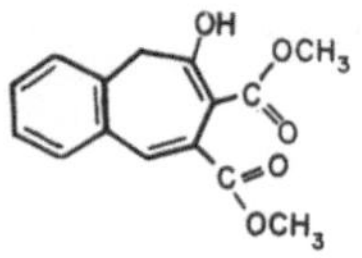

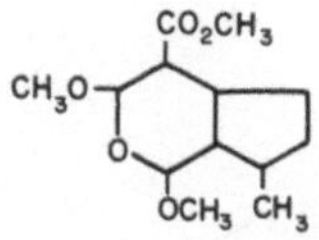

J. Org. Chem. 26:1192 (1961)

▶ $O_5C_{14}H_{10}$

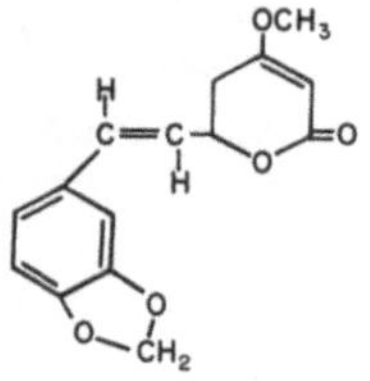

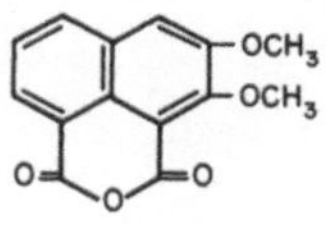

Tetrahedron 18:841 (1962)

▶ $O_5C_{14}H_{12}$

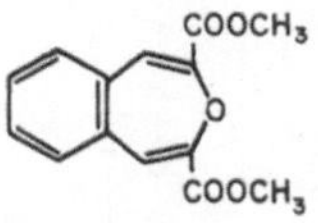

J. Org. Chem. 27:3225 (1962)

▶ $O_5C_{14}H_{12}$

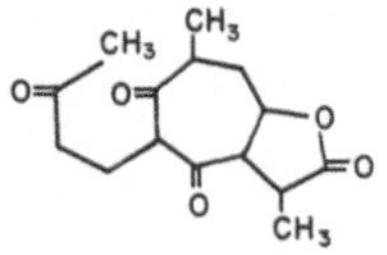

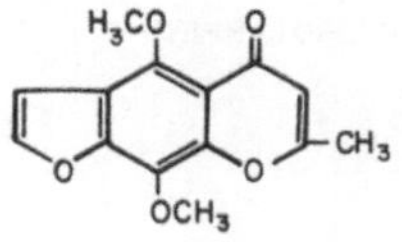

Varian 628

▶ $O_5C_{14}H_{14}$

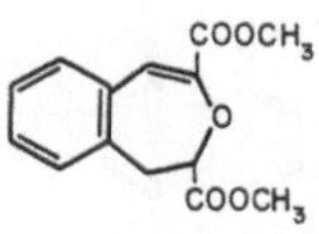

J. Org. Chem. 27:3225 (1962)

▶ $O_5C_{14}H_{18}$

Tetrahedron 18:1321 (1962)

▶ $O_5C_{15}H_{14}$

Varian 639

▶ $O_5C_{15}H_{14}$

J. Org. Chem. 27:3715 (1962)

▶ $O_5C_{15}H_{20}$

J. Am. Chem. Soc. 84:3857 (1962)

▶ $O_5C_{16}H_{12}$

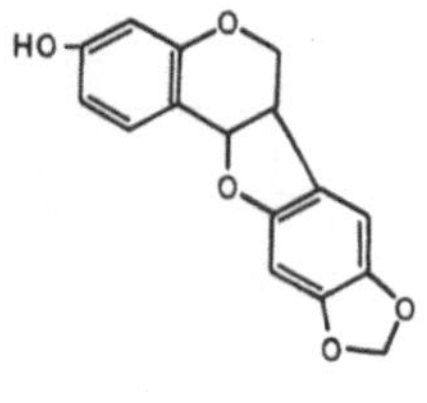

Experientia 18:161 (1962)

▶ $O_5C_{16}H_{14}$

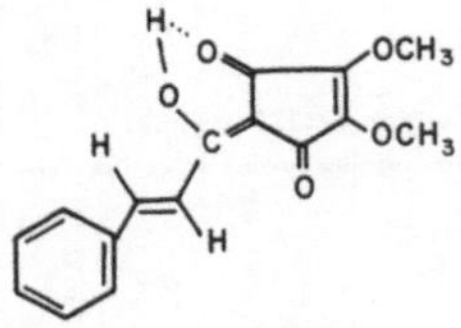

J. Chem. Soc. 4341 (1962)

▶ $O_5C_{16}H_{14}$

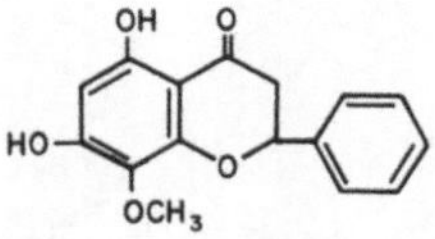

Bull. Soc. Chim. France 1962 (1962)

▶ $O_5C_{16}H_{16}$

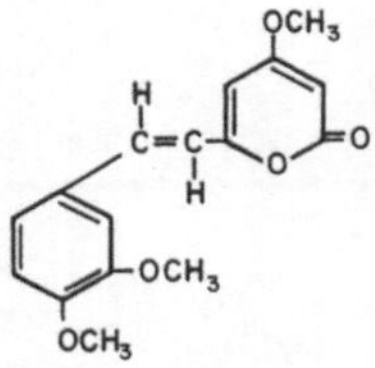

J. Org. Chem. 27:3715 (1962)

▶ $O_5C_{16}H_{16}$

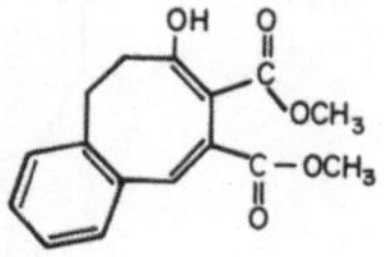

Varian 656

▶ $O_5C_{16}H_{16}$

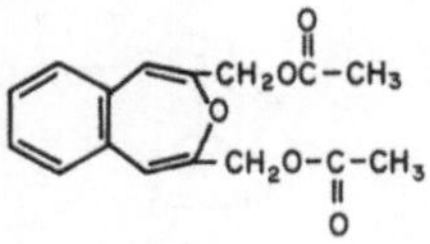

J. Org. Chem. 27:3225 (1962)

▶ $O_5C_{16}H_{22}$

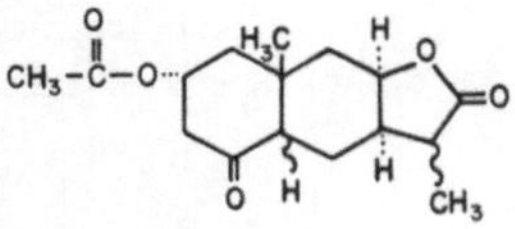

J. Org. Chem. 27:909 (1962)

▶ $O_5C_{17}H_{12}$

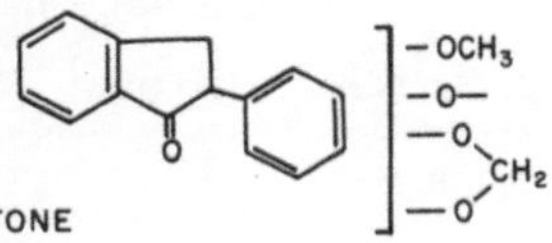

NEORAUTONE

J. Org. Chem. 26:5015 (1961)

▶ $O_5C_{17}H_{14}$

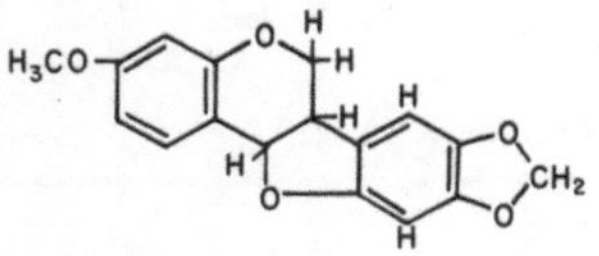

PTEROCARPIN

Tetrahedron Letters 287 (1961)

▶ $O_5C_{17}H_{14}$

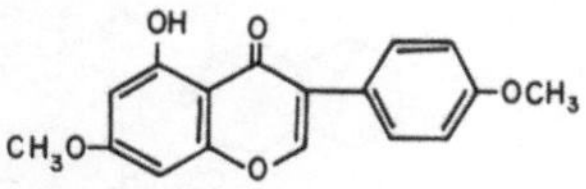

Bull. Soc. Chim. France 1962 (1962)

▶ $O_5C_{17}H_{16}$

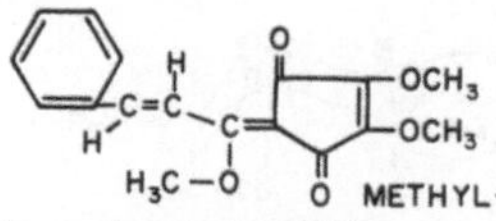

METHYL-LINDERONE

J. Chem. Soc. 4340 (1962)
Proc. Chem. Soc. 455 (1961)

▶ $O_5C_{17}H_{20}$

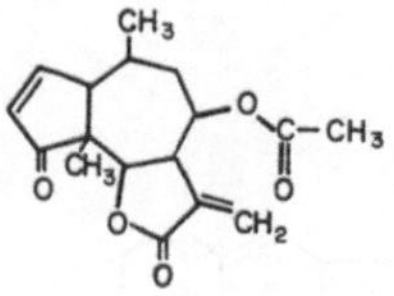

Varian 666

▶ $O_5C_{17}H_{20}$

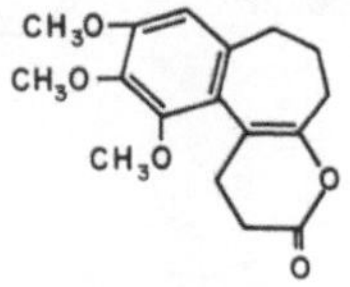

Tetrahedron 14:13 (1961)

▶ $O_5C_{17}H_{20}$

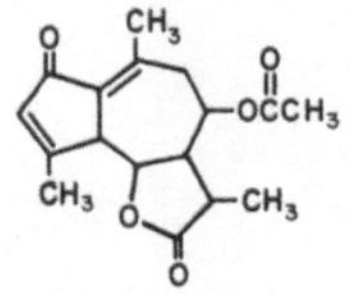

J. Am. Chem. Soc. 83:1139 (1961)

▶ $O_5C_{17}H_{20}$

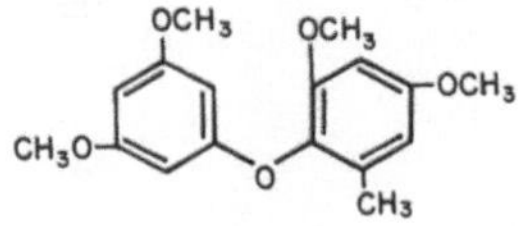

Helv. Chim. Acta 45:813 (1962)

▶ $O_5C_{17}H_{20}$

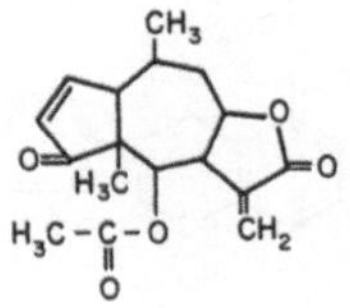

J. Org. Chem. 27:4044, 4127 (1962)

▶ $O_5C_{17}H_{22}$

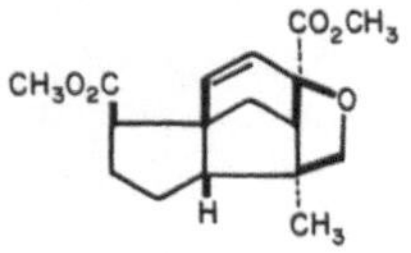

Tetrahedron 18:547, 1321 (1962)

▶ $O_5C_{17}H_{22}$

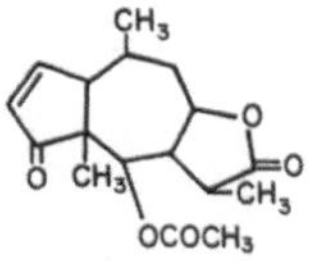

J. Am. Chem. Soc. 84:3857 (1962)

▶ $O_5C_{17}H_{22}$

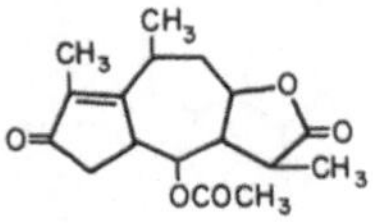

J. Am. Chem. Soc. 84:3857 (1962)

▶ $O_5C_{17}H_{22}$

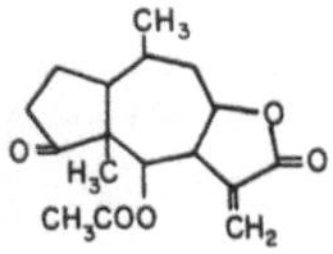

J. Org. Chem. 27:4127 (1962)

▶ $O_5C_{17}H_{22}$

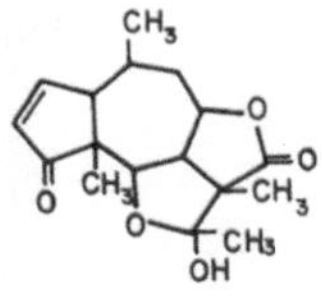

J. Am. Chem. Soc. 84:3857 (1962)

► $O_5C_{17}H_{22}$

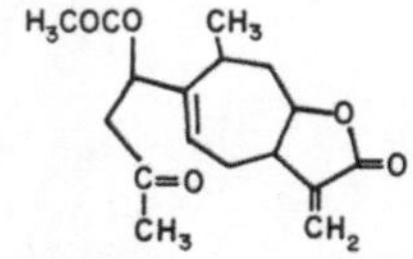


J. Org. Chem. 27:2692 (1962)

► $O_5C_{17}H_{42}Si_4$

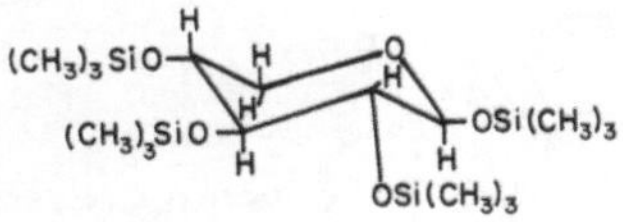


Tetrahedron 18:1145 (1962)

► $O_5C_{17}H_{24}$

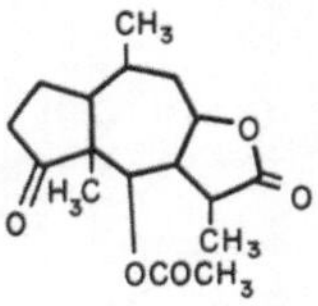


J. Org. Chem. 27:4127 (1962)
J. Am. Chem. Soc. 84:3857 (1962)

► $O_5C_{18}H_{12}$

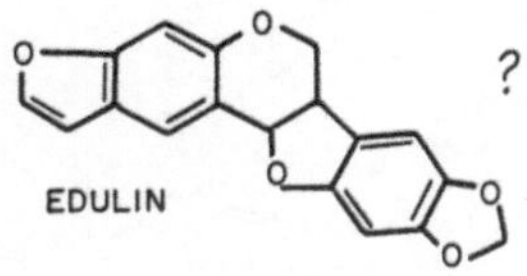


J. Org. Chem. 26:5016 (1961)

► $O_5C_{17}H_{24}$

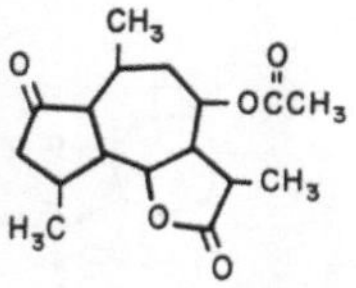


J. Am. Chem. Soc. 83:1139 (1961)

► $O_5C_{17}H_{24}$

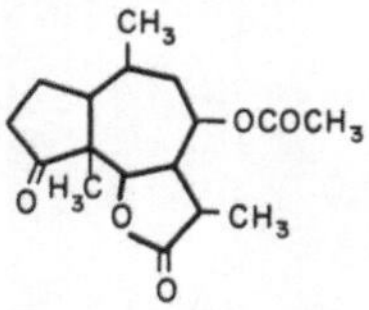


J. Am. Chem. Soc. 84:3857 (1962)

► $O_5C_{18}H_{16}$

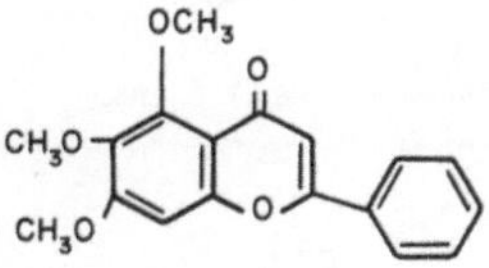


Bull. Soc. Chim. France 1962 (1962)

► $O_5C_{17}H_{42}Si_4$




Tetrahedron 18:1145 (1962)

► $O_5C_{18}H_{16}$

Bull. Soc. Chim. France 1962 (1962)

▶ $O_5C_{18}H_{20}$

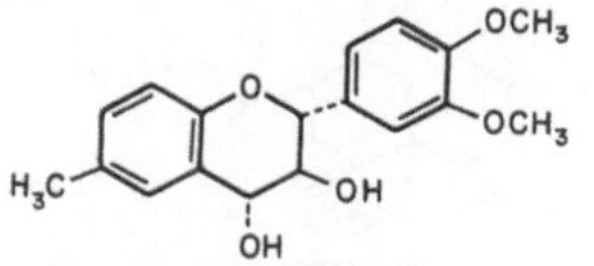

J. Chem. Soc. 3861 (1962)

▶ $O_5C_{18}H_{20}$

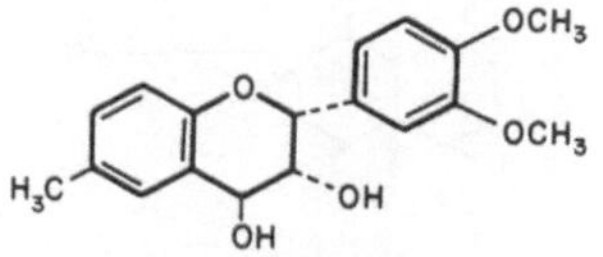

J. Chem. Soc. 3861 (1962)

▶ $O_5C_{18}H_{20}$

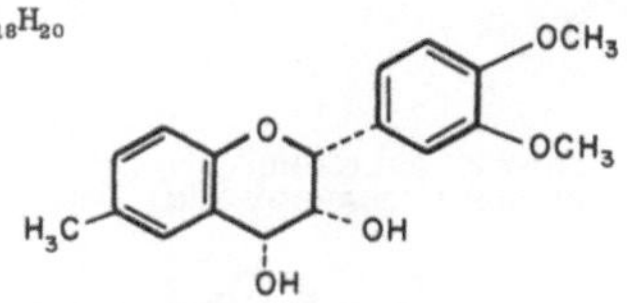

J. Chem. Soc. 3861 (1962)

▶ $O_5C_{18}H_{20}$

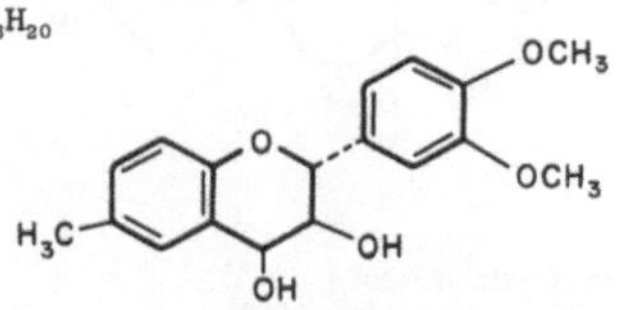

J. Chem. Soc. 3861 (1962)

▶ $O_5C_{18}H_{22}$

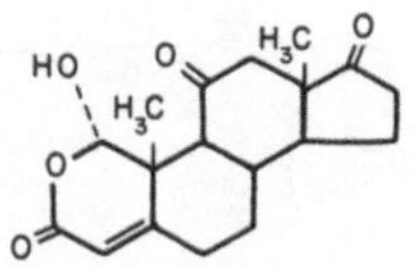

Tetrahedron 18:767 (1962)

▶ $O_5C_{19}H_{18}$

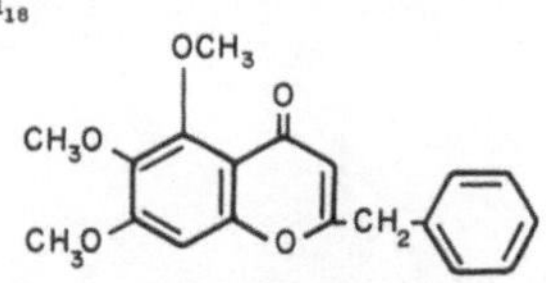

Bull. Soc. Chim. France 1962 (1962)

▶ $O_5C_{19}H_{18}$

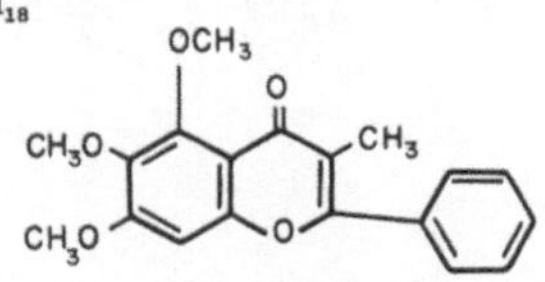

Bull. Soc. Chim. France 1962 (1962)

▶ $O_5C_{18}H_{20}$

SAMADEROL

Bull. Soc. Chim. France 1255, 1715 (1962)

▶ $O_5C_{19}H_{22}$

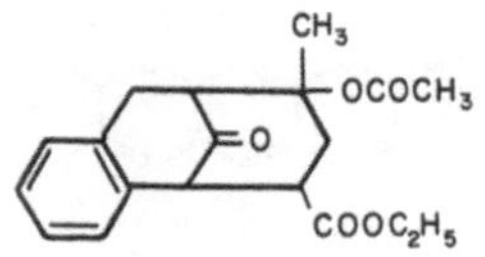

Tetrahedron 14:26 (1961)

▶ $O_5C_{19}H_{24}$

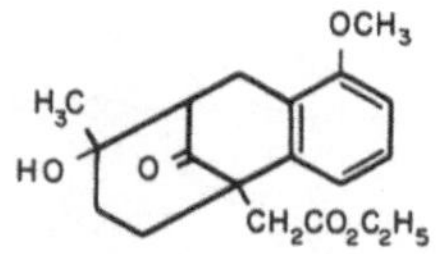

Bull. Chem. Soc. Japan 34:274 (1961)

▶ $O_5C_{19}H_{26}$

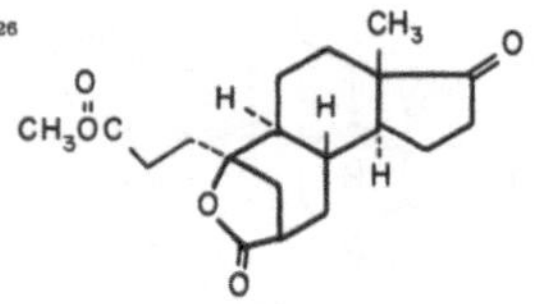

Helv. Chim. Acta 45:2615 (1962)

▶ $O_5C_{19}H_{28}$

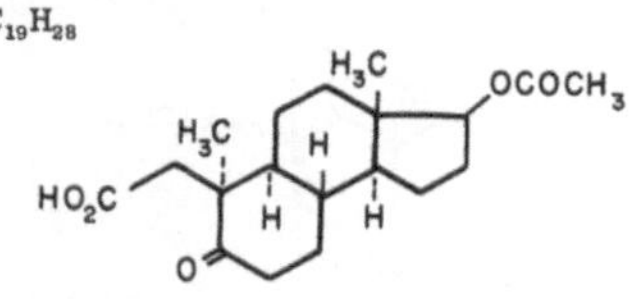

Helv. Chim. Acta 45:2346 (1962)

▶ $O_5C_{20}H_{20}$

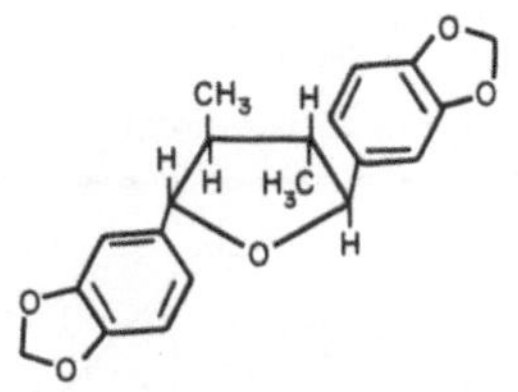

Australian J. Chem. 15:305 (1962)

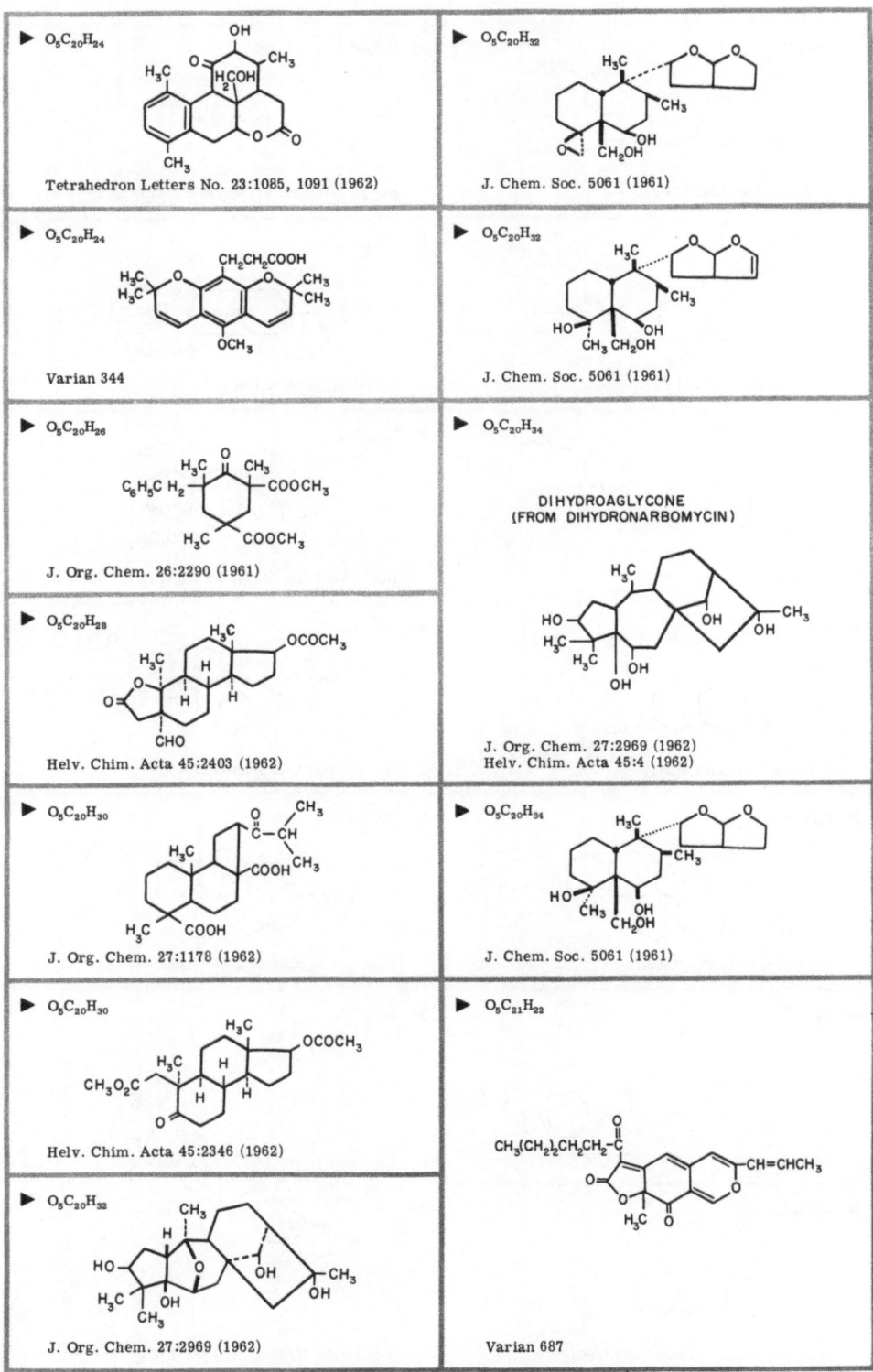

$O_5C_{20}H_{24}$
Tetrahedron Letters No. 23:1085, 1091 (1962)
$O_5C_{20}H_{24}$
Varian 344
$O_5C_{20}H_{26}$
J. Org. Chem. 26:2290 (1961)
$O_5C_{20}H_{28}$
Helv. Chim. Acta 45:2403 (1962)
$O_5C_{20}H_{30}$
J. Org. Chem. 27:1178 (1962)
$O_5C_{20}H_{30}$
Helv. Chim. Acta 45:2346 (1962)
$O_5C_{20}H_{32}$
J. Org. Chem. 27:2969 (1962)
$O_5C_{20}H_{32}$
J. Chem. Soc. 5061 (1961)
$O_5C_{20}H_{32}$
J. Chem. Soc. 5061 (1961)
$O_5C_{20}H_{34}$
DIHYDROAGLYCONE
(FROM DIHYDRONARBOMYCIN)
J. Org. Chem. 27:2969 (1962)
Helv. Chim. Acta 45:4 (1962)
$O_5C_{20}H_{34}$
J. Chem. Soc. 5061 (1961)
$O_5C_{21}H_{22}$
Varian 687

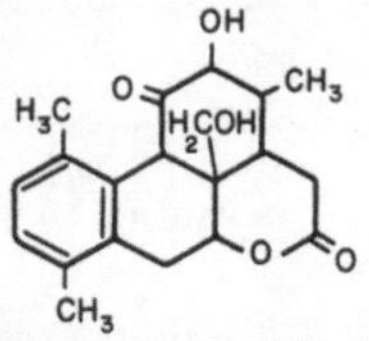

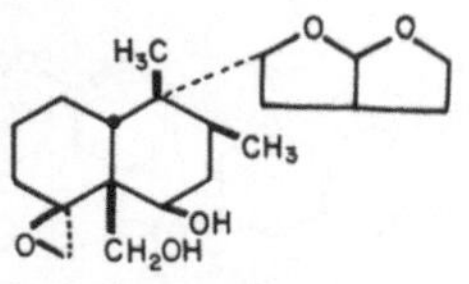

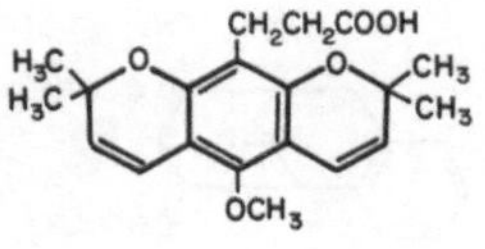

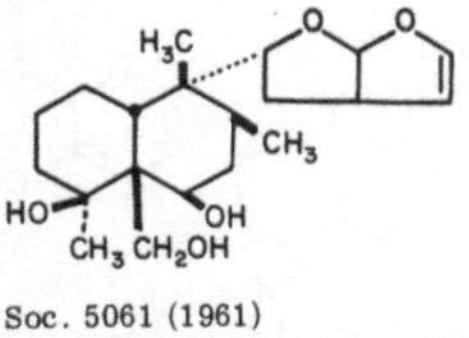

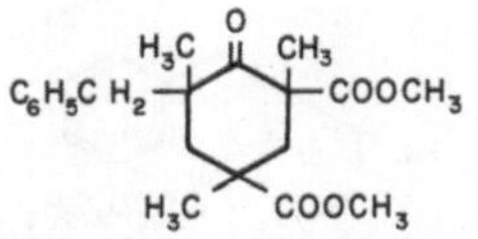

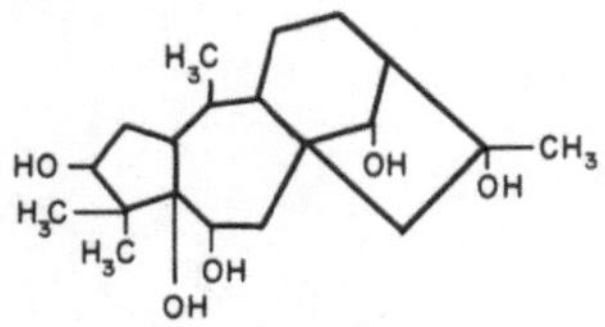

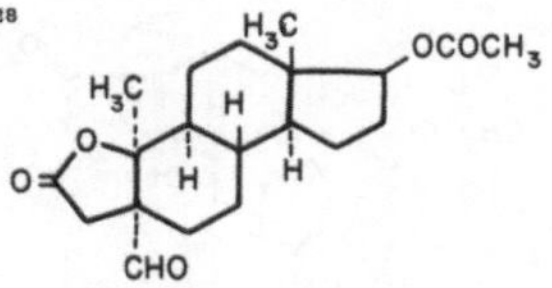

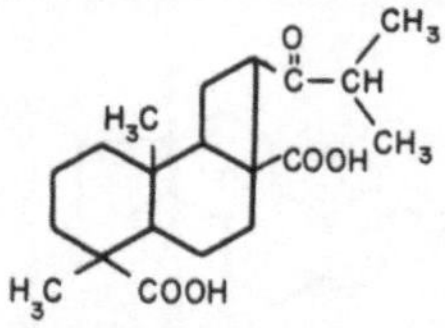

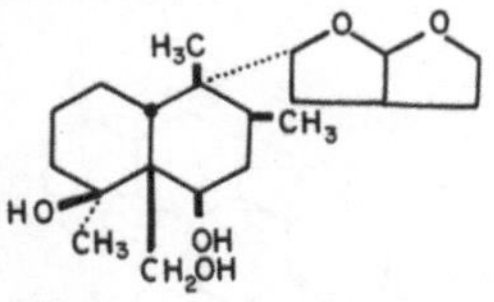

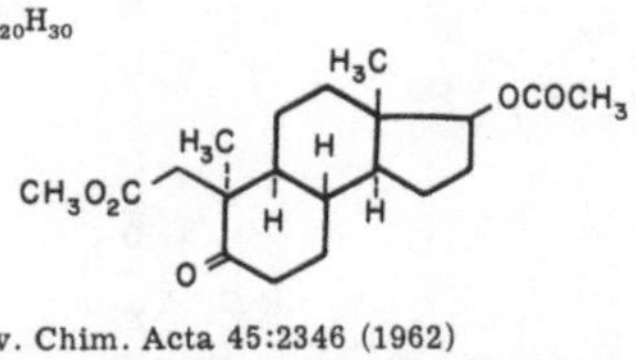

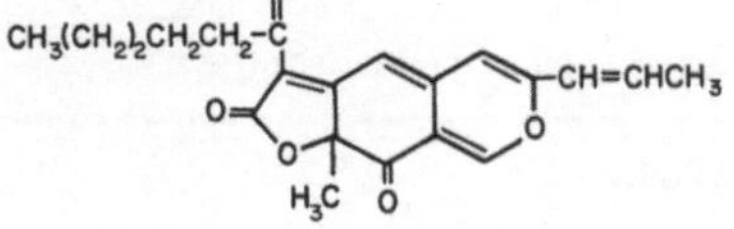

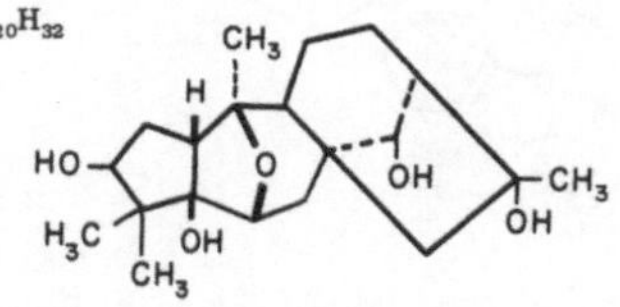

▶ $O_5C_{21}H_{24}$

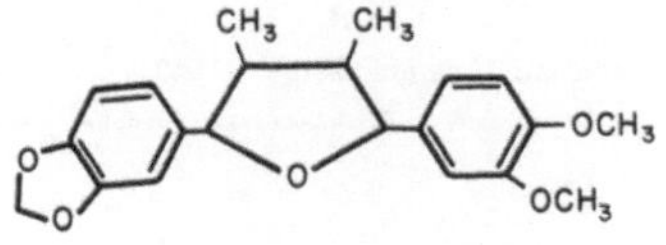

Australian J. Chem. 15:305 (1962)

▶ $O_5C_{21}H_{28}$

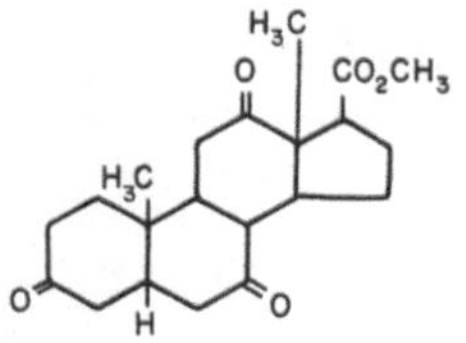

Helv. Chim. Acta 44:1380 (1961)

▶ $O_5C_{21}H_{26}$

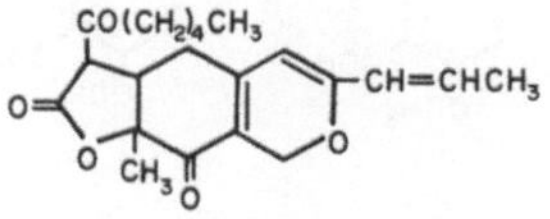

J. Chem. Soc. 4579 (1961)

▶ $O_5C_{21}H_{30}$

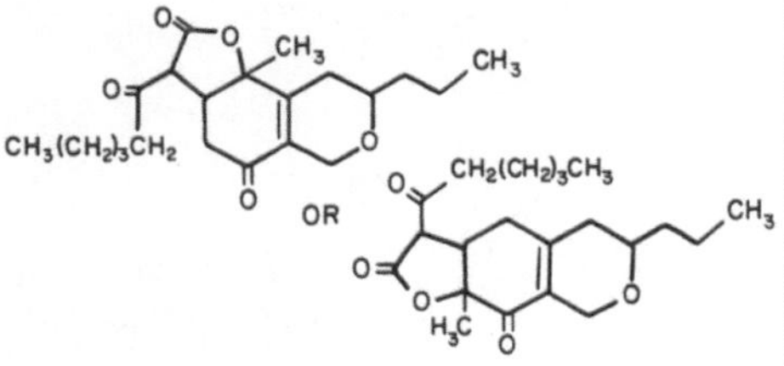

J. Chem. Soc. 4579 (1961)
Tetrahedron 18:1195 (1962)

▶ $O_5C_{21}H_{26}$

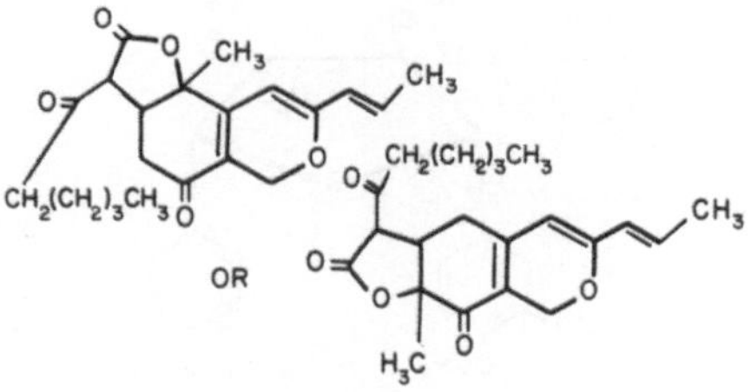

Tetrahedron 18:1195 (1962)

▶ $O_5C_{21}H_{30}$

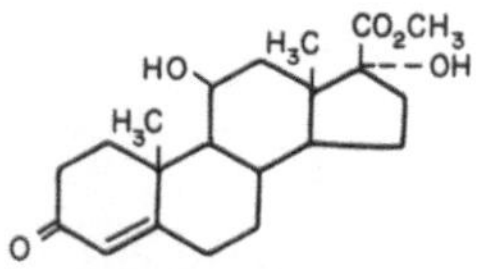

Helv. Chim. Acta 44:1380 (1961)

▶ $O_5C_{21}H_{28}$

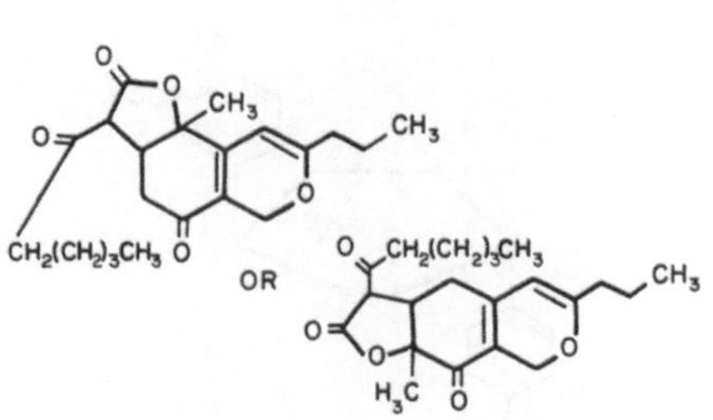

J. Chem. Soc. 4579 (1961)
Tetrahedron 18:1195 (1962)

▶ $O_5C_{21}H_{30}$

WITHAFERIN
(STRUCTURE UNKNOWN)

J. Chem. Soc. 2925 (1962)

▶ $O_5C_{21}H_{32}$

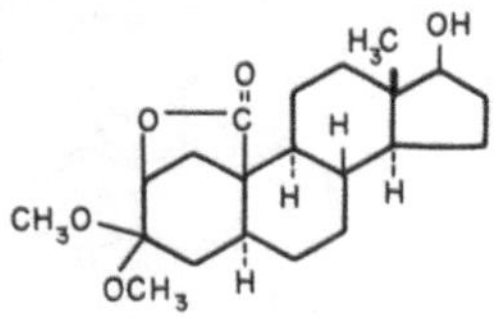

J. Am. Chem. Soc. 84:4976 (1962)

► $O_5C_{21}H_{32}$

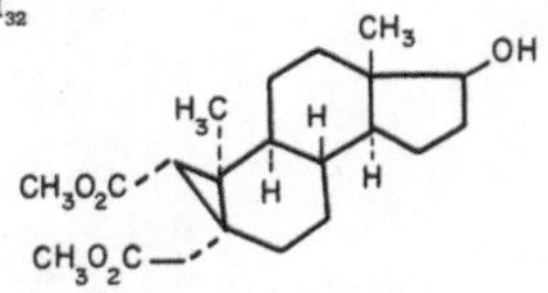

Helv. Chim. Acta 45:2346 (1962)

► $O_5C_{21}H_{32}$

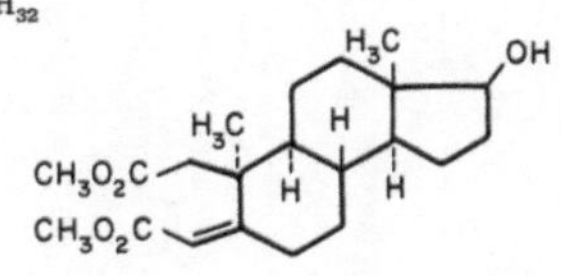

Helv. Chim. Acta 45:2346 (1962)

► $O_5C_{21}H_{32}$

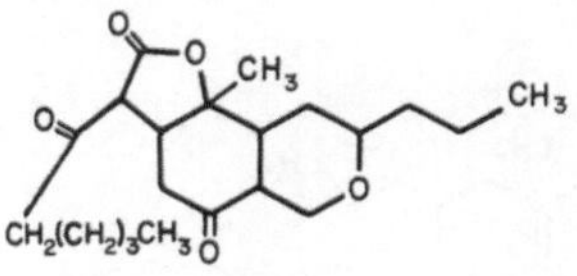

OR

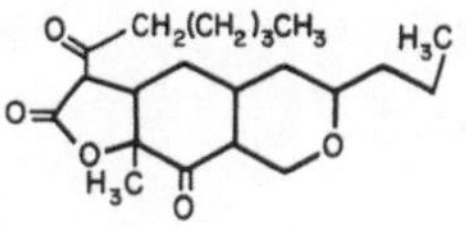

Tetrahedron 18:1195 (1962)

► $O_5C_{21}H_{36}$

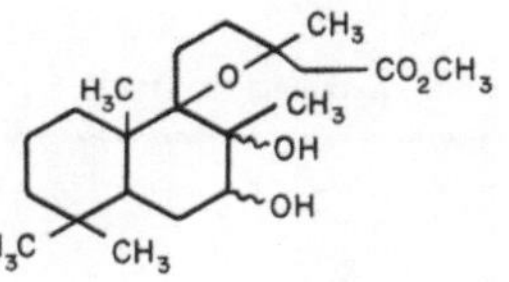

Acta Chem. Scand. 16:1675 (1962)

► $O_5C_{22}H_{28}$

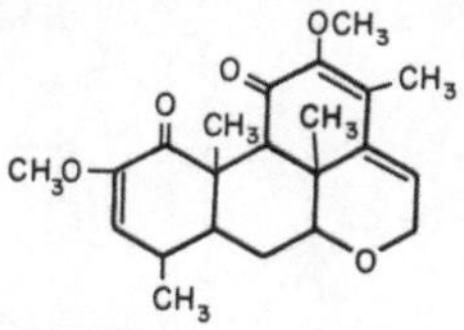

Australian J. Chem. 15:807 (1962)

► $O_5C_{22}H_{28}$

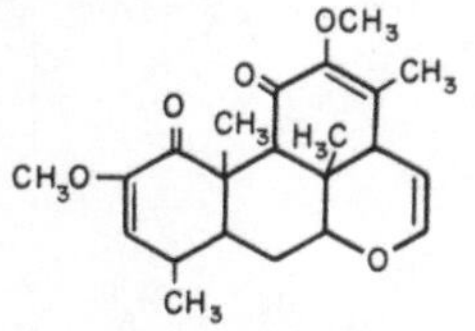

Australian J. Chem. 15:807 (1962)

► $O_5C_{22}H_{28}$

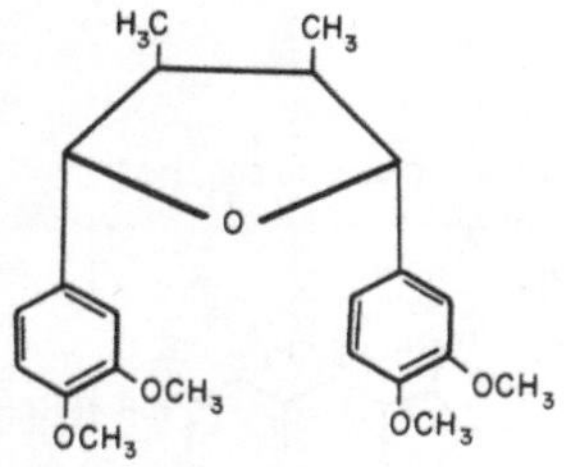

J. Chem. Soc. 1460 (1962)
Australian J. Chem. 15:305 (1962)

► $O_5C_{22}H_{28}$

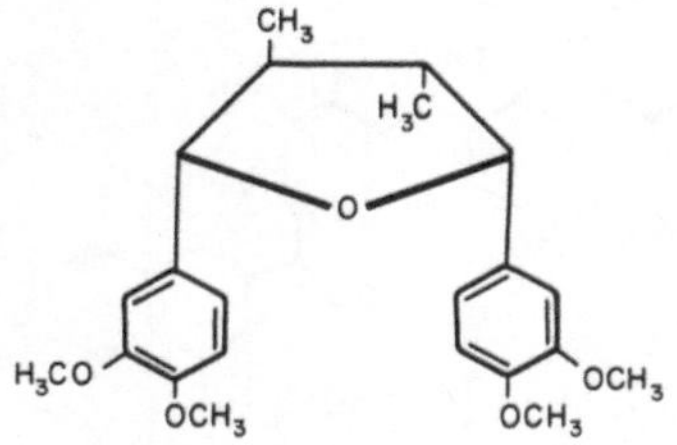

J. Chem. Soc. 1460 (1962)

► $O_5C_{22}H_{28}$

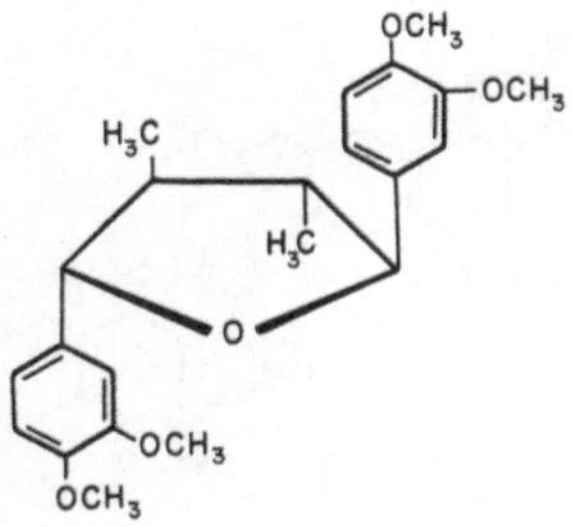

J. Chem. Soc. 1460 (1962)

► $O_5C_{22}H_{28}$

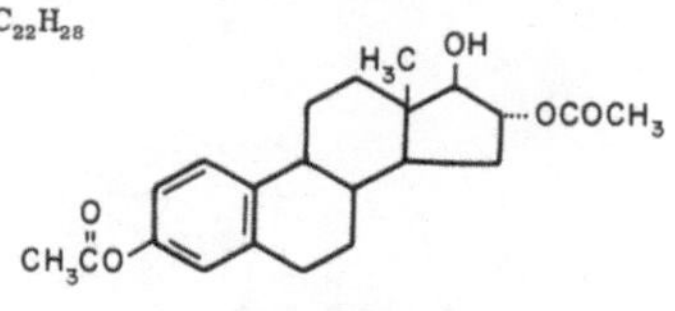

J. Am. Chem. Soc. 84:2972 (1962)

► $O_5C_{22}H_{30}$

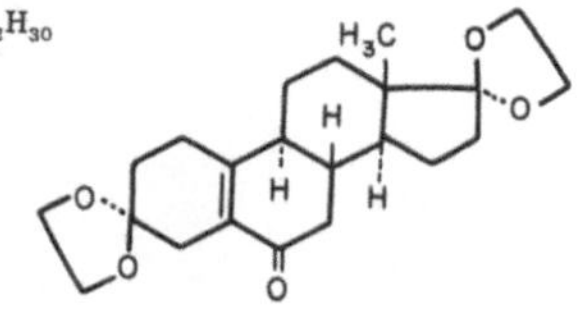

Helv. Chim. Acta 45:2674 (1962)

► $O_5C_{22}H_{30}$

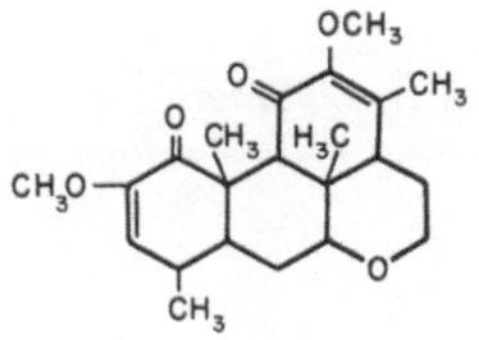

Australian J. Chem. 15:807 (1962)

► $O_5C_{22}H_{30}$

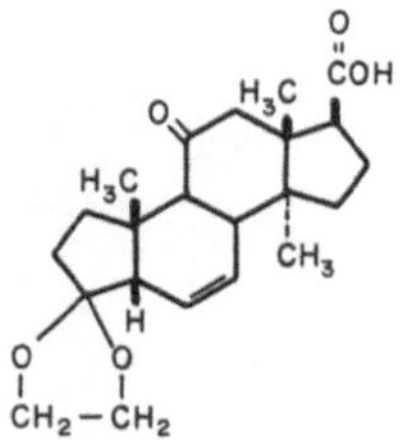

J. Am. Chem. Soc. 84:4356 (1962)

► $O_5C_{22}H_{30}$

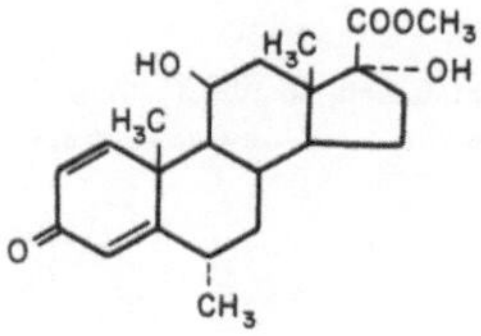

Tetrahedron 16:271 (1961)

► $O_5C_{22}H_{30}$

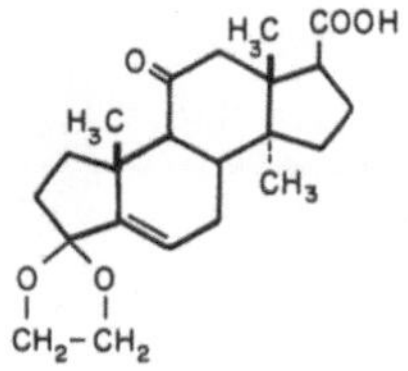

J. Am. Chem. Soc. 84:4356 (1962)

► $O_5C_{22}H_{30}$

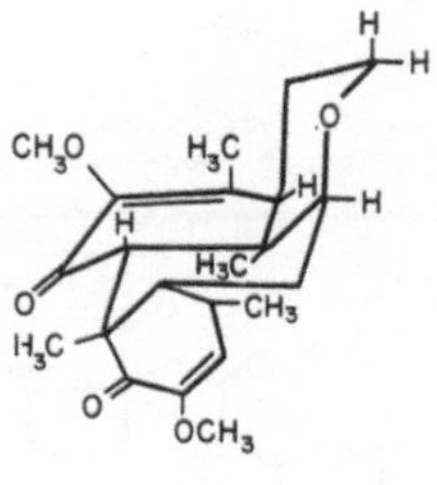

Tetrahedron 18:1433 (1962)

► $O_5C_{22}H_{32}$

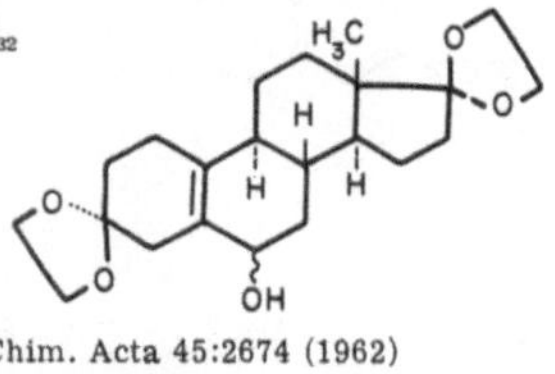

Helv. Chim. Acta 45:2674 (1962)

► $O_5C_{22}H_{34}$

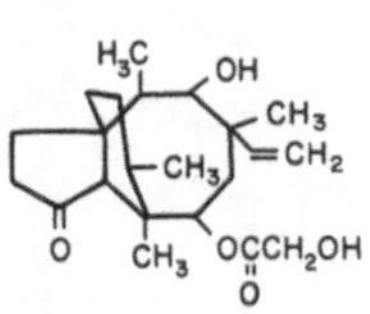

Gazz. Chim. Ital. 92:884 (1962)

▶ $O_5C_{23}H_{26}$

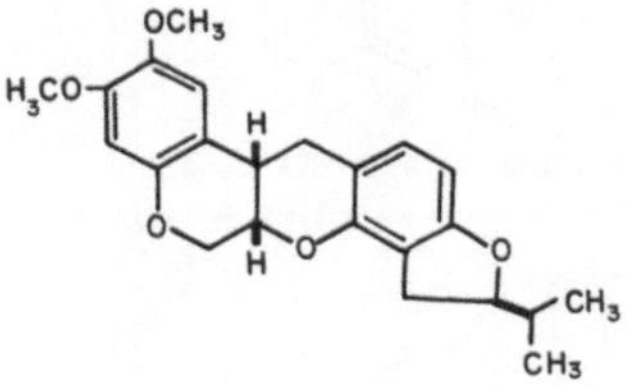

J. Chem. Soc. 775 (1962)

▶ $O_5C_{23}H_{30}$

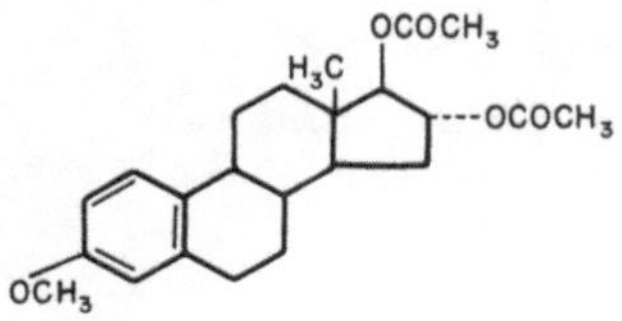

J. Am. Chem. Soc. 84:2972 (1962)

▶ $O_5C_{23}H_{26}$

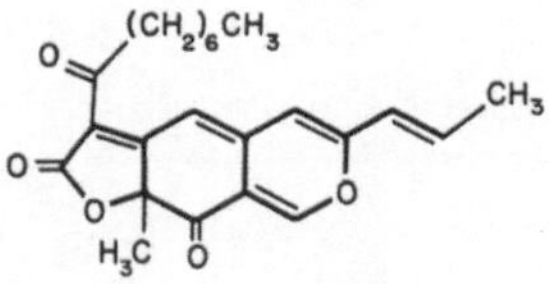

Tetrahedron 18:1176 (1962)

▶ $O_5C_{23}H_{30}$

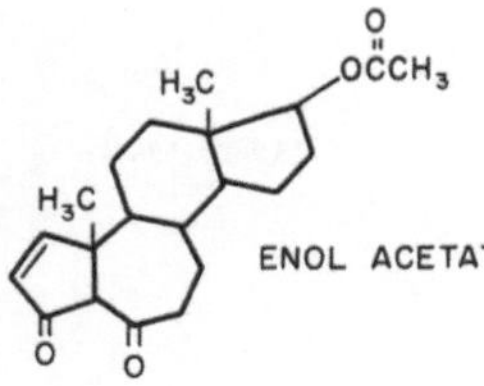

ENOL ACETATE

Helv. Chim. Acta 45:1031 (1962)

▶ $O_5C_{23}H_{28}$

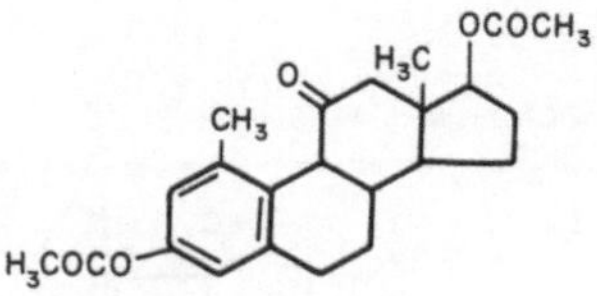

Chem. & Ind. 1716 (1962)

▶ $O_5C_{23}H_{30}$

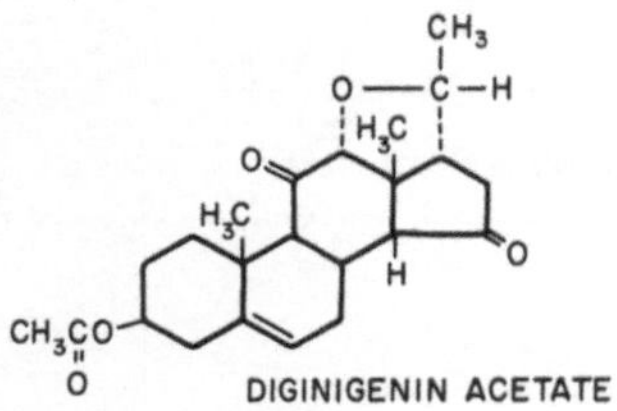

DIGINIGENIN ACETATE

J. Chem. Soc. 3615, 3620 (1962)
Proc. Chem. Soc. 65 (1962)

▶ $O_5C_{23}H_{28}$

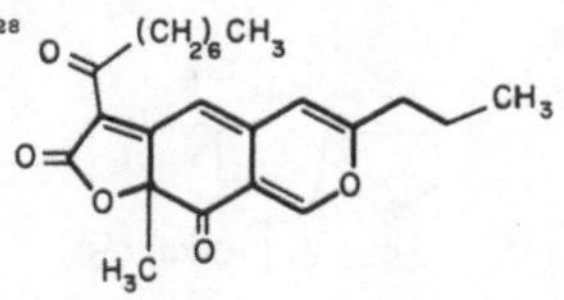

Tetrahedron 18:1176 (1962)

▶ $O_5C_{23}H_{30}$

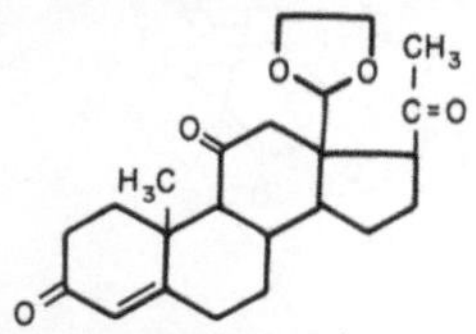

Helv. Chim. Acta 45:331 (1962)

▶ $O_5C_{23}H_{30}$

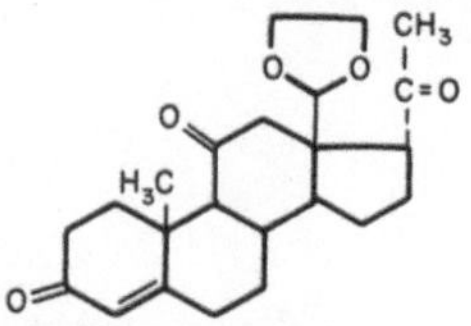


Helv. Chim. Acta 45:331 (1962)

▶ $O_5C_{23}H_{32}$

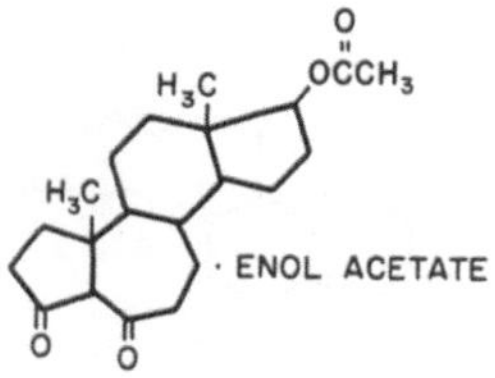


Helv. Chim. Acta 45:1031 (1962)

▶ $O_5C_{23}H_{32}$

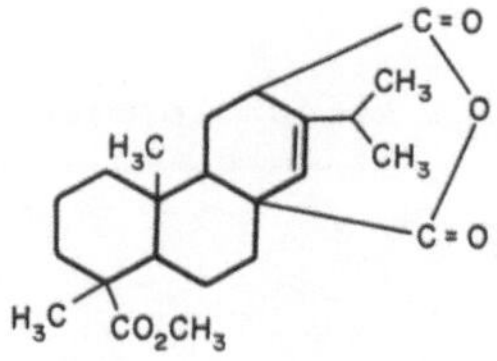


J. Org. Chem. 27:3535 (1962)

▶ $O_5C_{23}H_{32}$

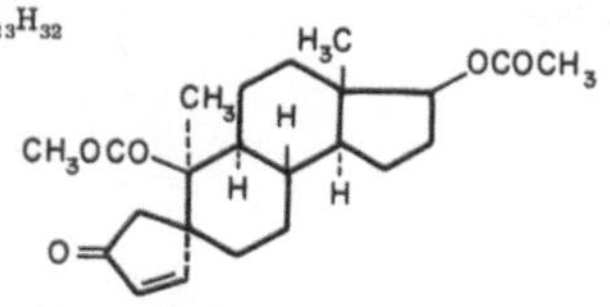


Helv. Chim. Acta 45:2403 (1962)

▶ $O_5C_{23}H_{32}$

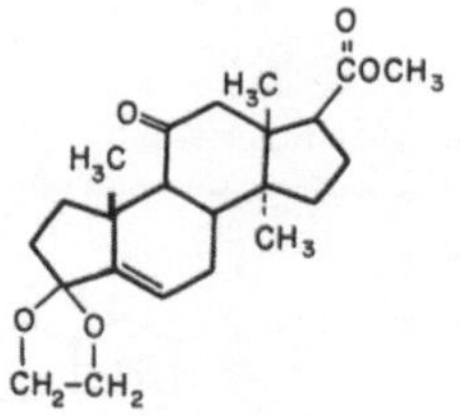


J. Am. Chem. Soc. 84:4356 (1962)

▶ $O_5C_{23}H_{32}$

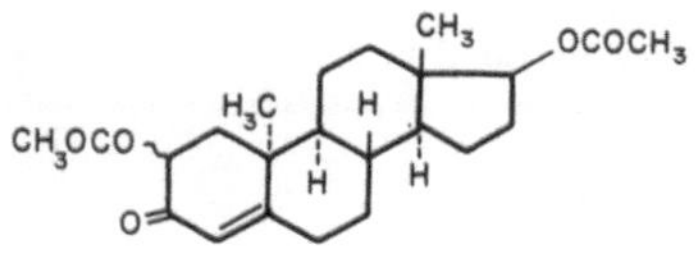


Helv. Chim. Acta 45:2420 (1962)

▶ $O_5C_{23}H_{32}$

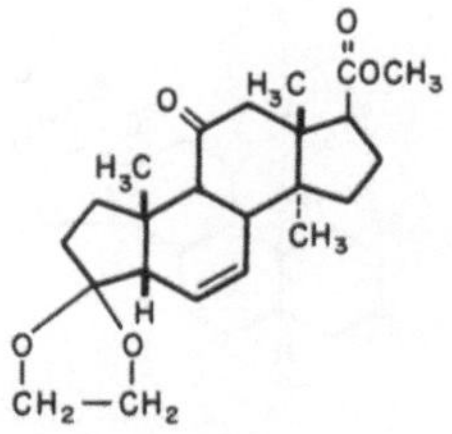


J. Am. Chem. Soc. 84:4356 (1962)

▶ $O_5C_{23}H_{32}$

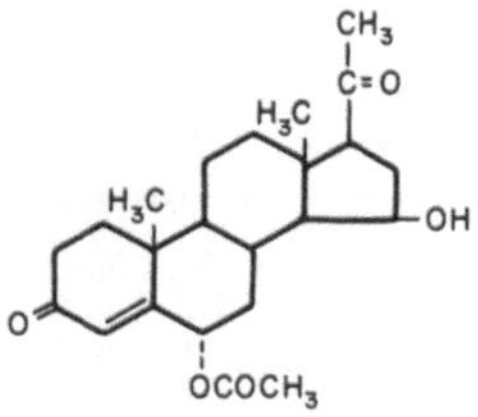


Chem. Pharm. Bull. 10:338 (1962)

► $O_5C_{23}H_{32}$

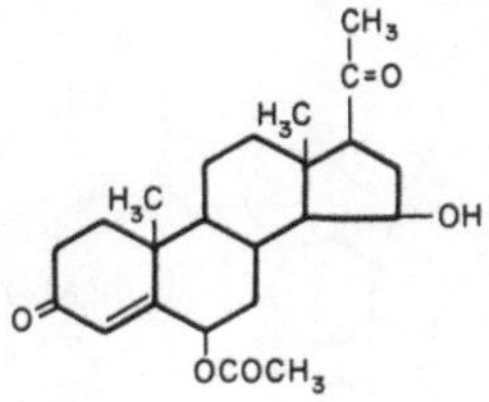

Chem. Pharm. Bull. 10:338 (1962)

► $O_5C_{23}H_{32}$

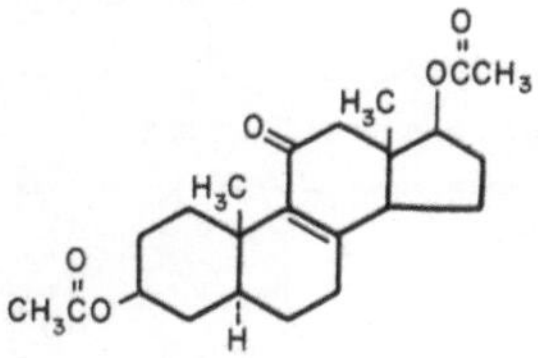

Helv. Chim. Acta 44:1380 (1961)

► $O_5C_{23}H_{32}$

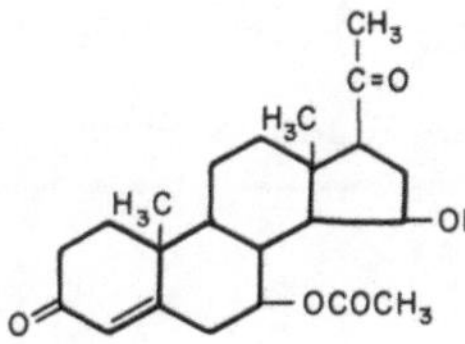

Chem. Pharm. Bull. 10:338 (1962)

► $O_5C_{23}H_{32}$

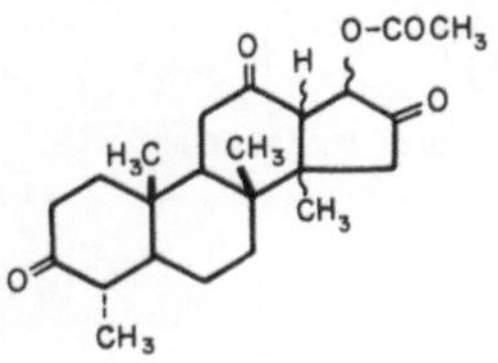

Tetrahedron 18:1035 (1962)

► $O_5C_{23}H_{32}$

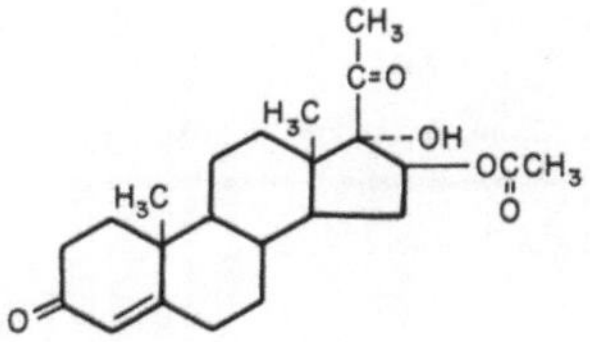

Helv. Chim. Acta 44:1380 (1961)

► $O_5C_{23}H_{32}$

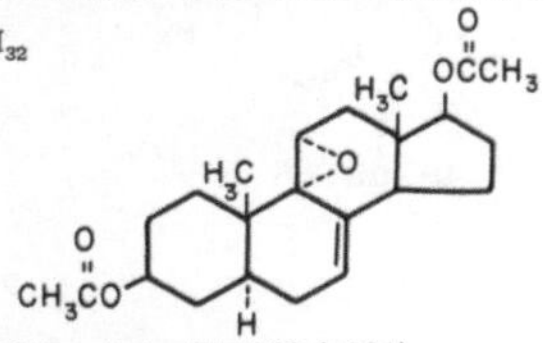

Helv. Chim. Acta 44:1380 (1961)

► $O_5C_{23}H_{32}$

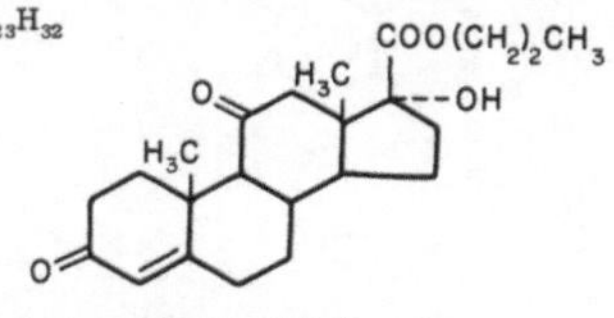

Tetrahedron 16:271 (1961)

► $O_5C_{23}H_{32}$

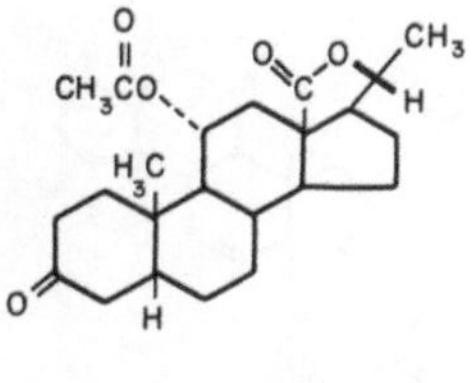

Helv. Chim. Acta 44:1380 (1961)

▶ $O_5C_{23}H_{34}$

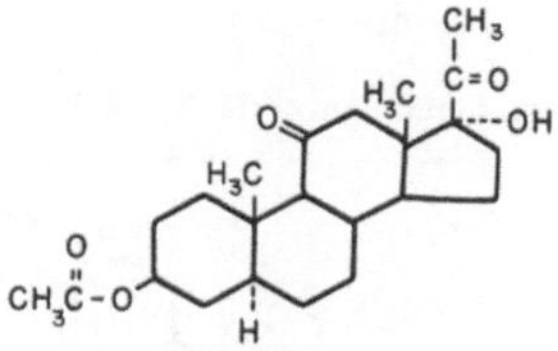


Helv. Chim. Acta 44:1380 (1961)

▶ $O_5C_{23}H_{34}$

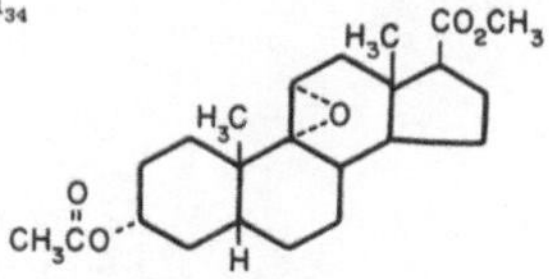


Helv. Chim. Acta 44:1380 (1961)

▶ $O_5C_{23}H_{34}$

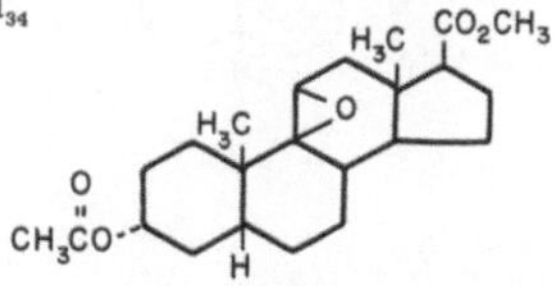


Helv. Chim. Acta 44:1380 (1961)

▶ $O_5C_{23}H_{34}$

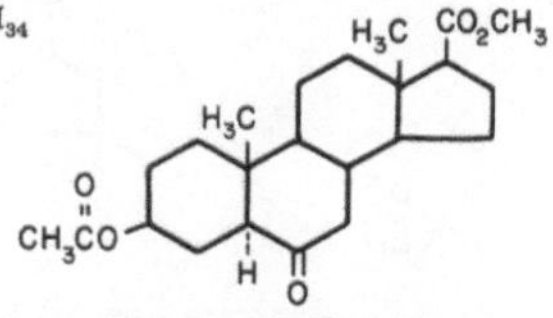


Helv. Chim. Acta 44:1380 (1961)

▶ $O_5C_{23}H_{34}$

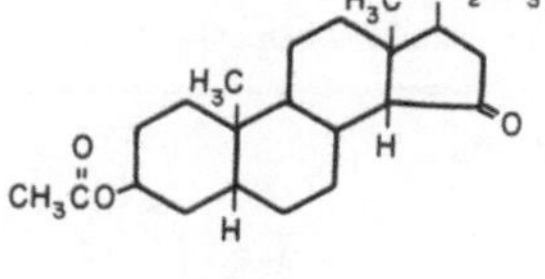


Helv. Chim. Acta 44:1380 (1961)

▶ $O_5C_{23}H_{34}$

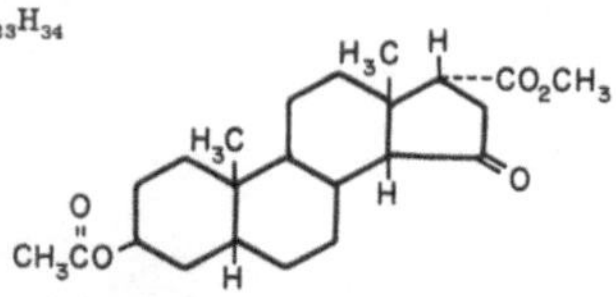


Helv. Chim. Acta 44:1380 (1961)

▶ $O_5C_{23}H_{34}$

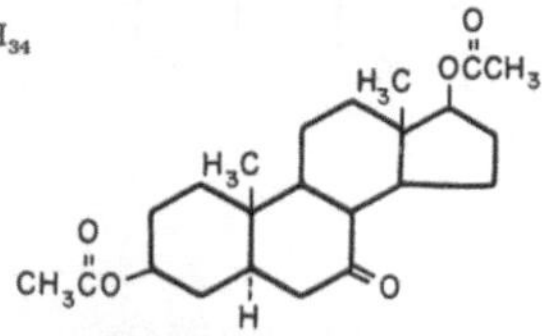


Helv. Chim. Acta 44:1380 (1961)

▶ $O_5C_{23}H_{34}$

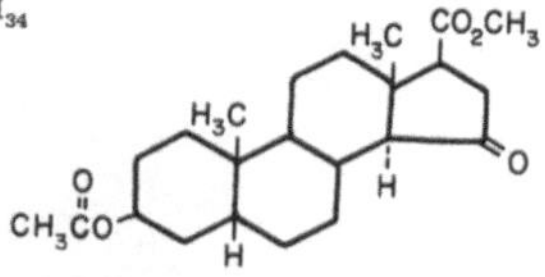


Helv. Chim. Acta 44:1380 (1961)

▶ $O_5C_{23}H_{34}$

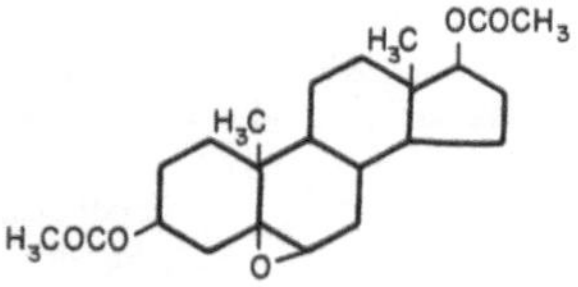


J. Am. Chem. Soc. 84:3206 (1962)

▶ $O_5C_{23}H_{34}$

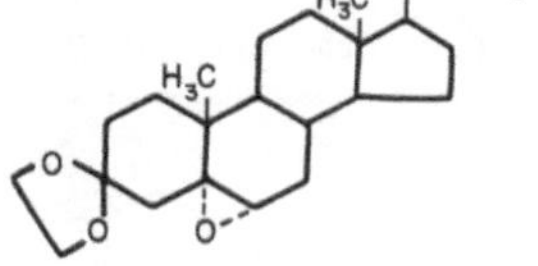


J. Am. Chem. Soc. 84:3206 (1962)

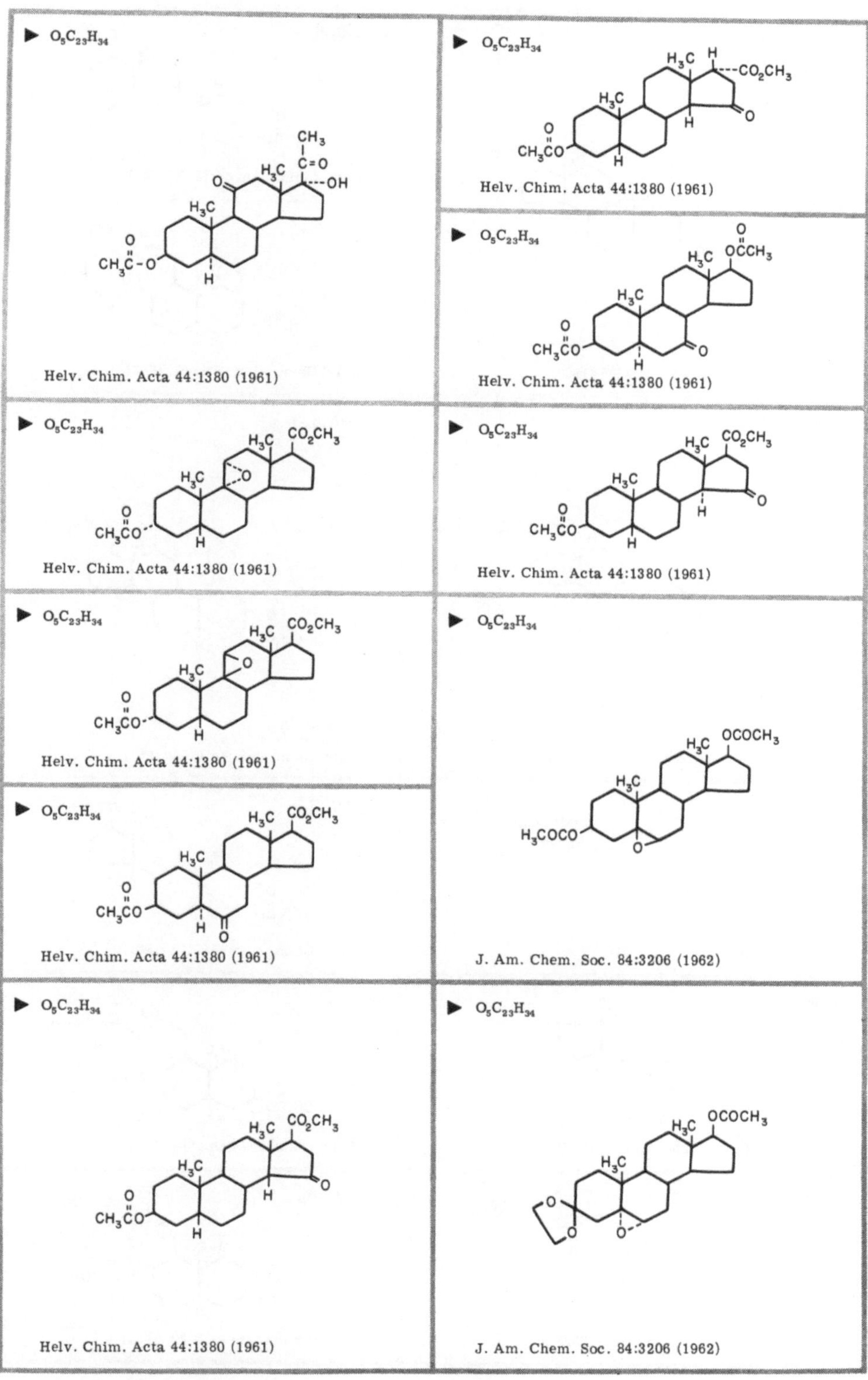

▶ $O_5C_{23}H_{34}$

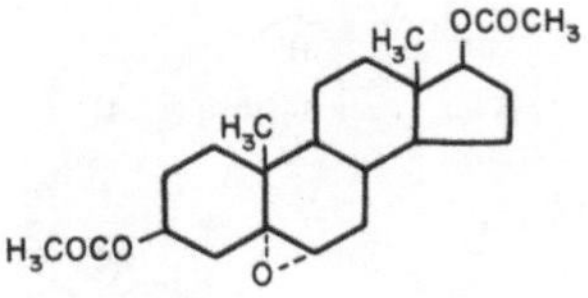


J. Am. Chem. Soc. 84:3206 (1962)

▶ $O_5C_{23}H_{34}$

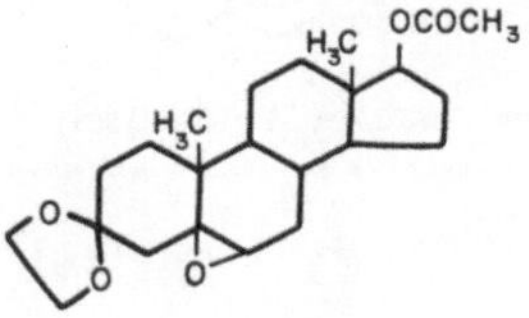


J. Am. Chem. Soc. 84:3206 (1962)

▶ $O_5C_{23}H_{34}$

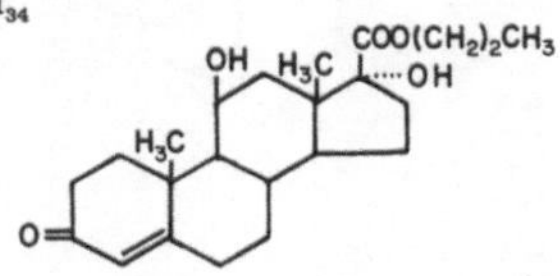


Tetrahedron 16:271 (1961)

▶ $O_5C_{23}H_{34}$

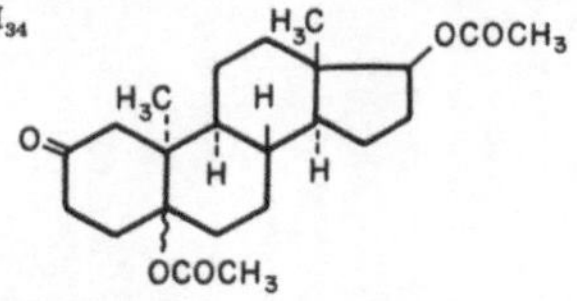


Helv. Chim. Acta 45:2420 (1962)

▶ $O_5C_{23}H_{34}$

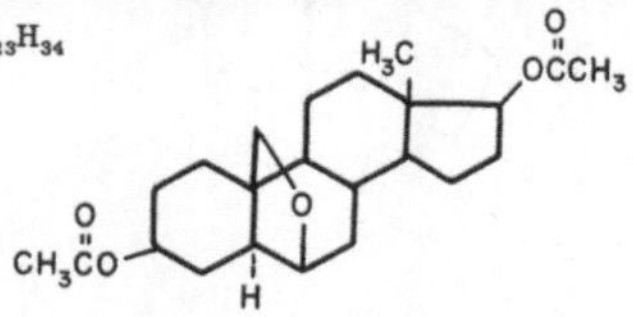


Helv. Chim. Acta 45:2161 (1962)

▶ $O_5C_{23}H_{34}$

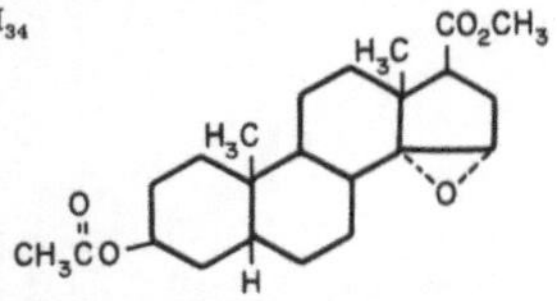


Helv. Chim. Acta 44:1380 (1961)

▶ $O_5C_{23}H_{34}$

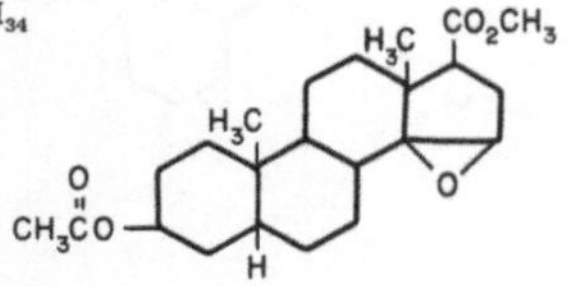


Helv. Chim. Acta 44:1380 (1961)

▶ $O_5C_{23}H_{36}$

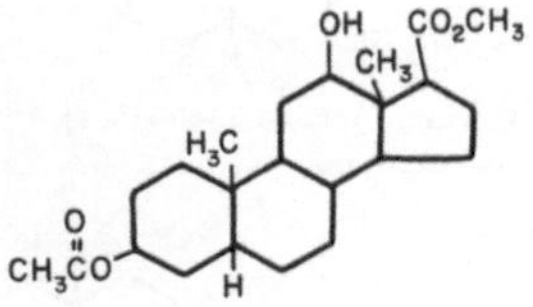


Helv. Chim. Acta 44:1380 (1961)

▶ $O_5C_{23}H_{36}$

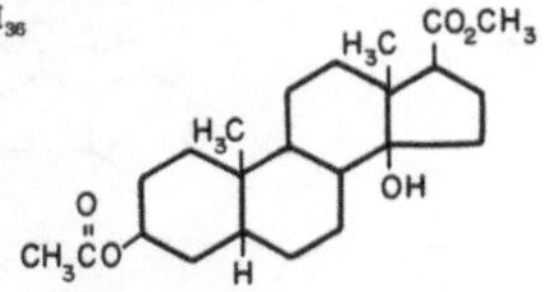


Helv. Chim. Acta 44:1380 (1961)

▶ $O_5C_{23}H_{36}$

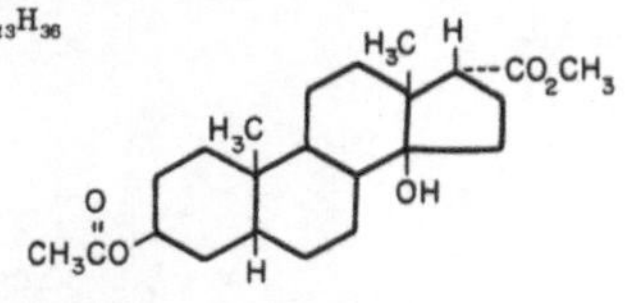


Helv. Chim. Acta 44:1380 (1961)

▶ $O_5C_{23}H_{36}$

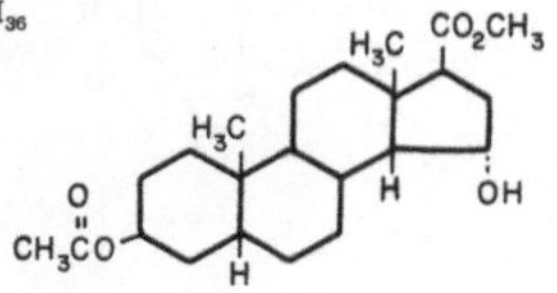


Helv. Chim. Acta 44:1380 (1961)

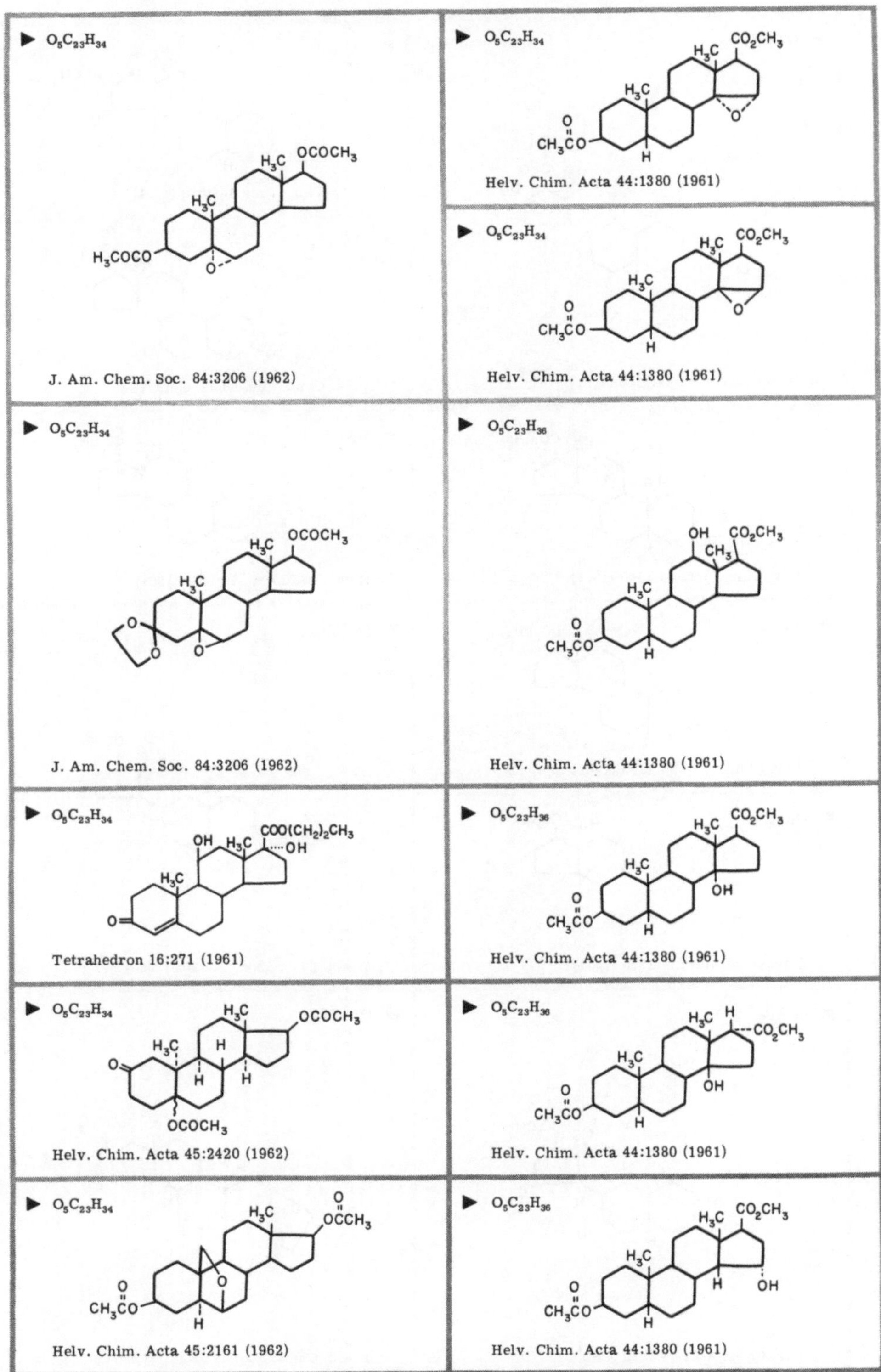

► $O_5C_{24}H_{30}$

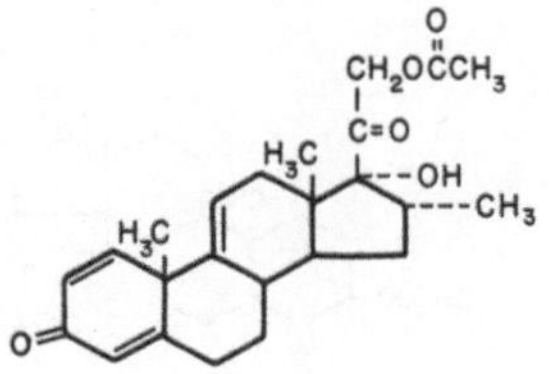

Helv. Chim. Acta 44:1380 (1961)

► $O_5C_{24}H_{32}$

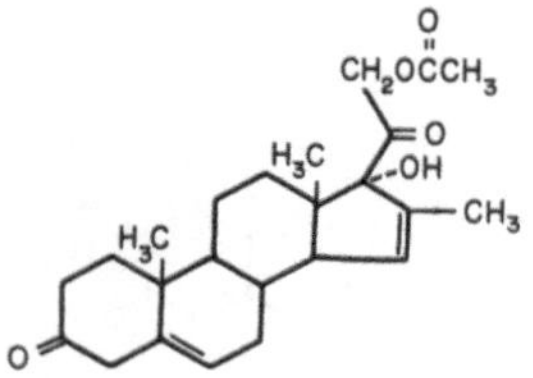

Chem. Ber. 95:2110 (1962)

► $O_5C_{24}H_{32}$

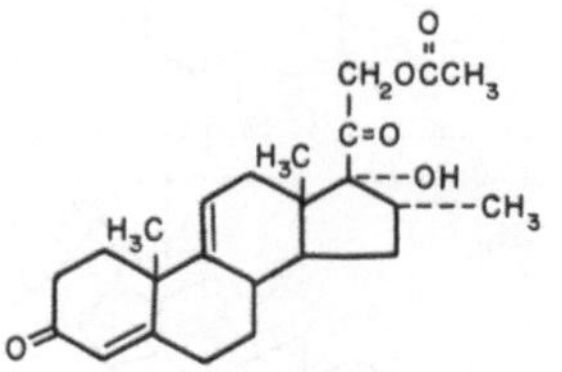

Helv. Chim. Acta 44:1380 (1961)

► $O_5C_{24}H_{32}$

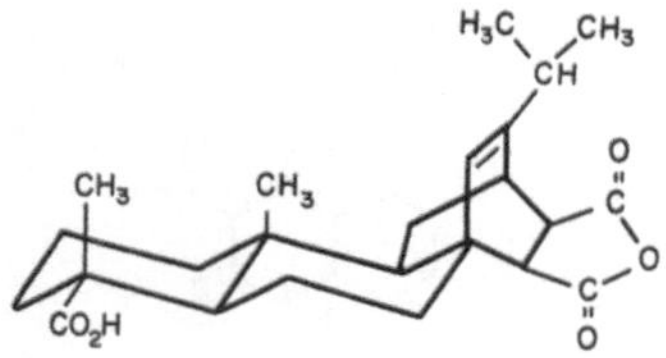

Tetrahedron Letters No. 16:692 (1962)

► $O_5C_{24}H_{32}$

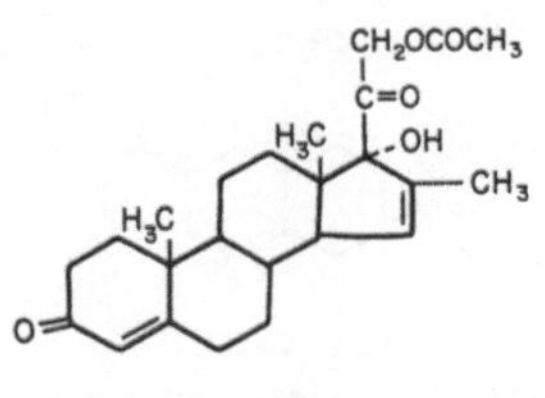

Chem. Ber. 95:2110 (1962)

► $O_5C_{24}H_{34}$

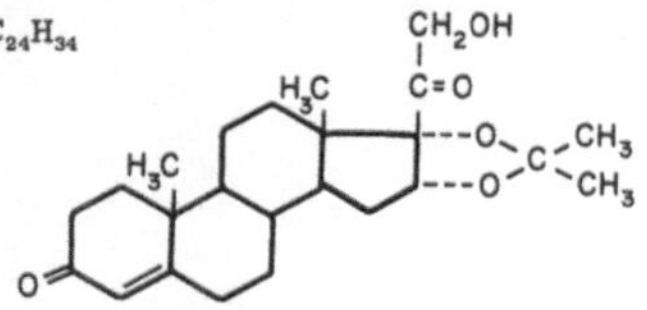

J. Org. Chem. 26:5036 (1961)

► $O_5C_{24}H_{34}$

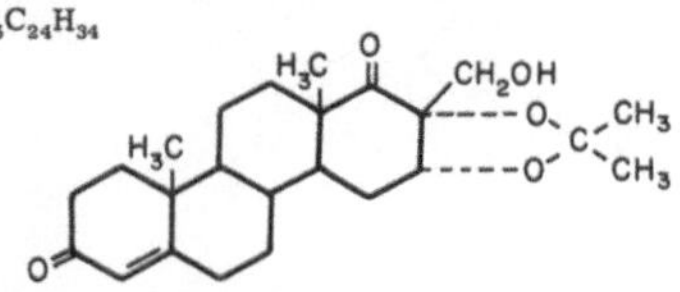

J. Org. Chem. 26:5036 (1961)

► $O_5C_{24}H_{34}$

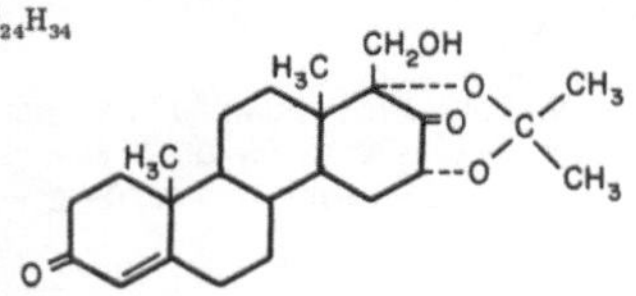

J. Org. Chem. 26:5036 (1961)

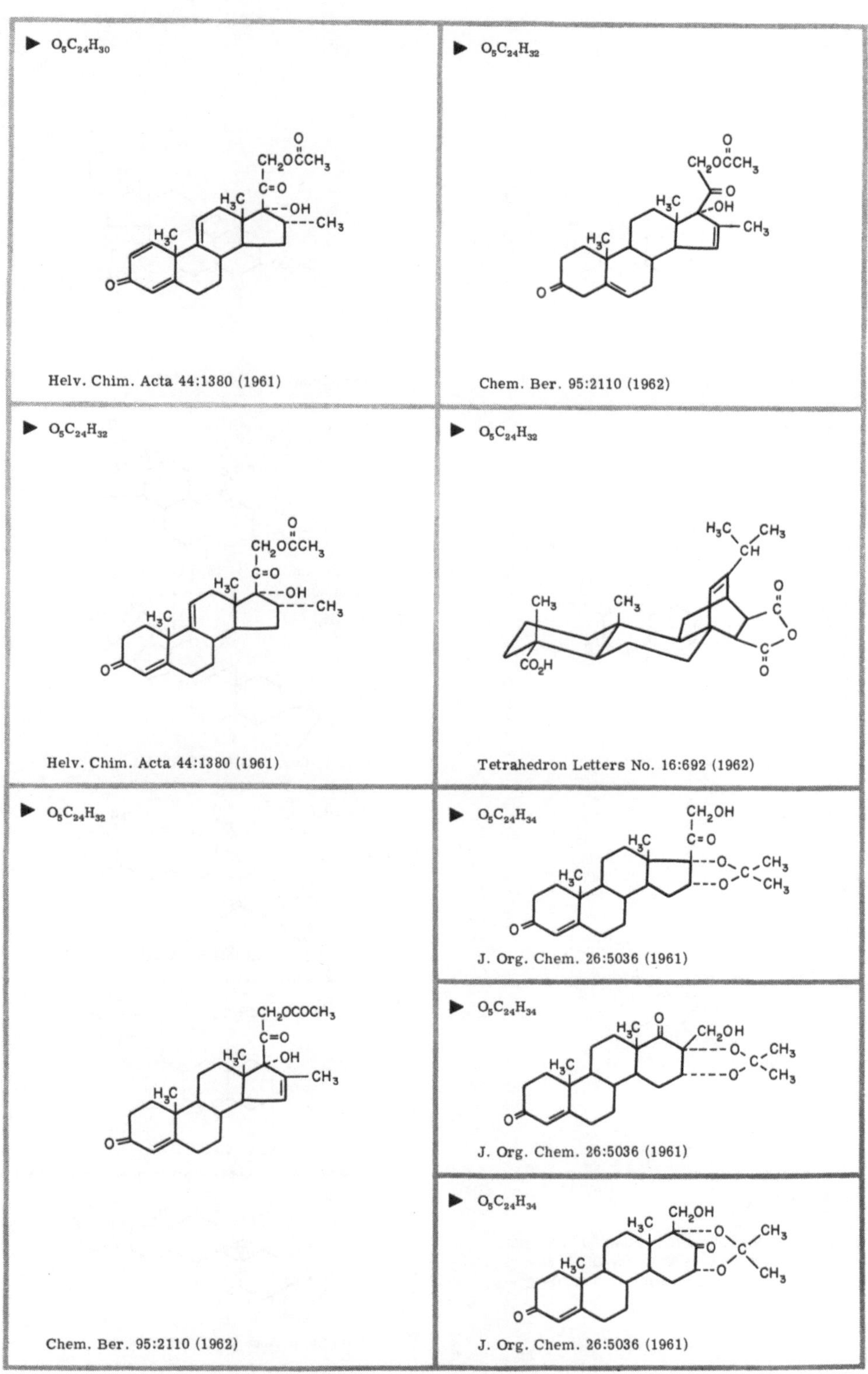

► $O_5C_{24}H_{34}$

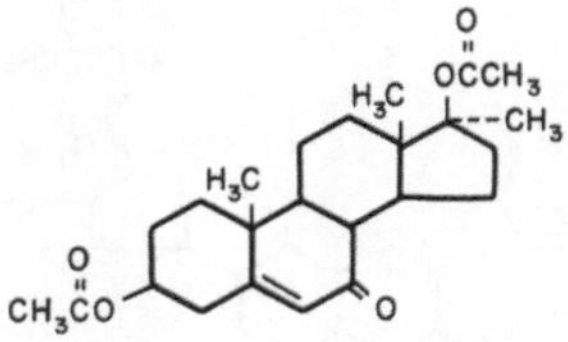

Helv. Chim. Acta 44:1380 (1961)

► $O_5C_{24}H_{36}$

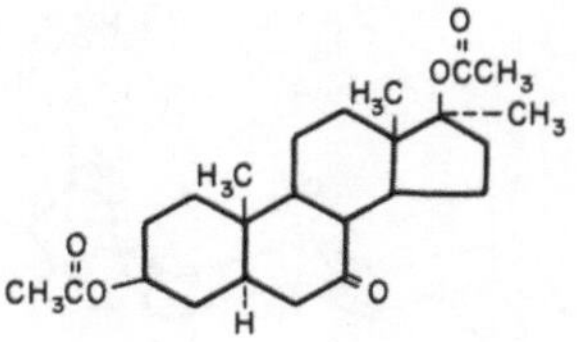

Helv. Chim. Acta 44:1380 (1961)

► $O_5C_{25}H_{32}$

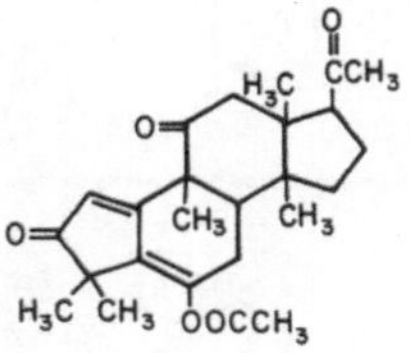

J. Org. Chem. 27:4546 (1962)

► $O_5C_{25}H_{36}$

UNKNOWN COMPOUND "D" PRODUCED BY UV.-IRRADIATION OF 3,20-DIETHYLENEDIOXY-11-KETO-$\Delta^5$-PREGNENE

Helv. Chim. Acta 45:1261 (1962)

► $O_5C_{25}H_{36}$

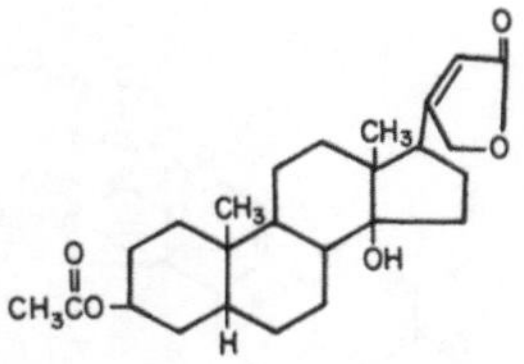

Helv. Chim. Acta 44:1380 (1961)

► $O_5C_{25}H_{36}$

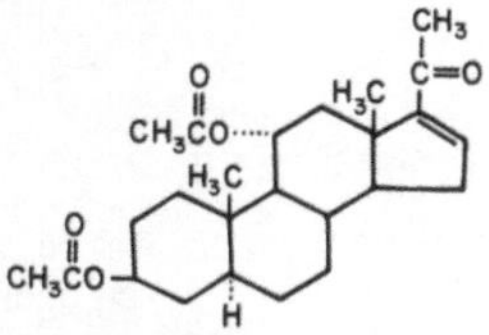

Helv. Chim. Acta 44:1380 (1961)

► $O_5C_{25}H_{36}$

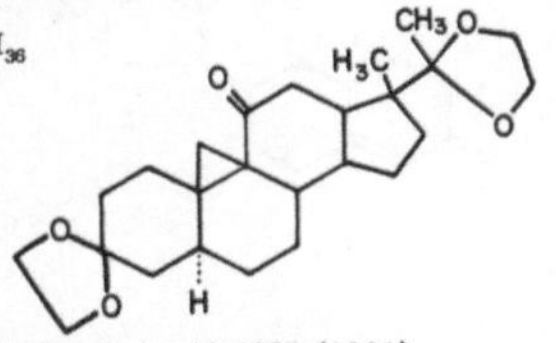

Helv. Chim. Acta 44:2162 (1961)

► $O_5C_{25}H_{36}$

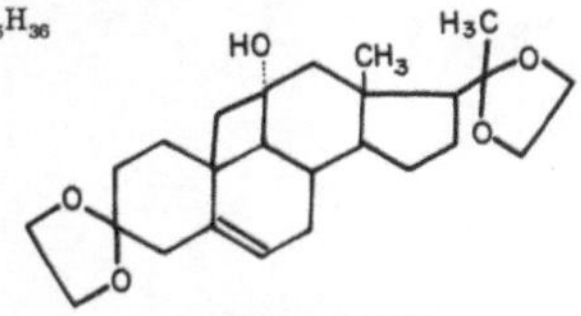

Helv. Chim. Acta 45:1261 (1962)

► $O_5C_{25}H_{36}$

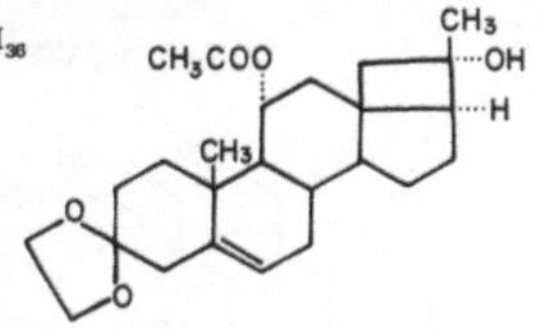

Helv. Chim. Acta 44:1927 (1961)

► $O_5C_{25}H_{36}$

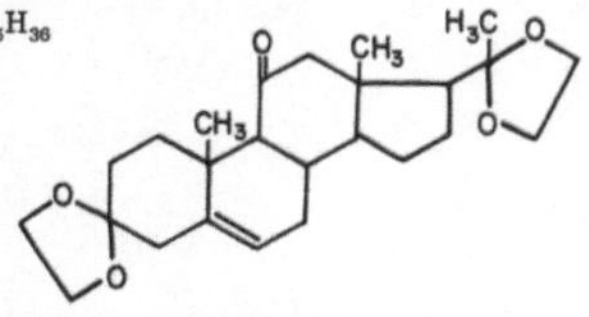

Helv. Chim. Acta 45:1261 (1962)

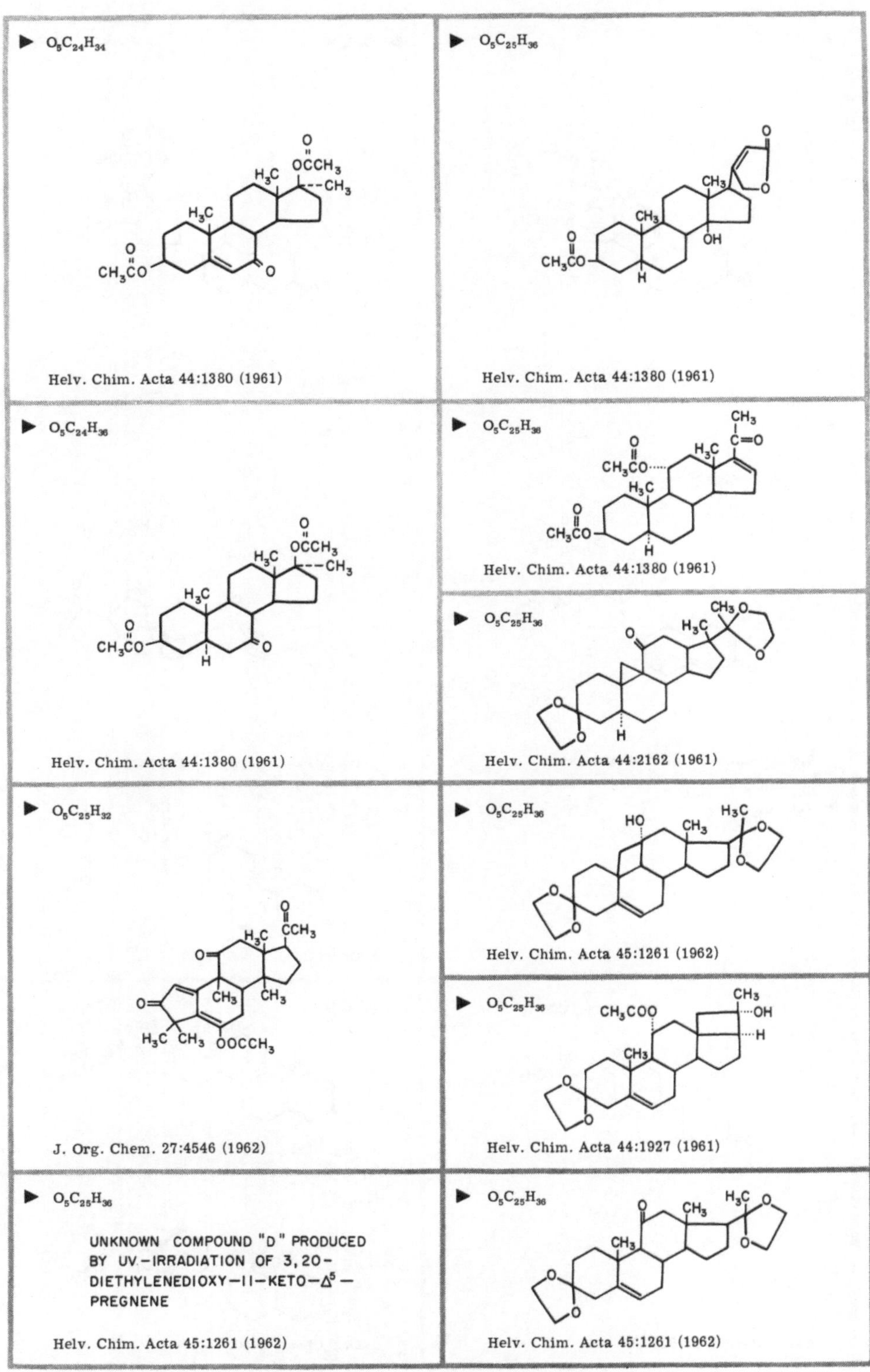

▶ $O_5C_{25}H_{38}$

5α, 6α-EPOXYPREGNAN-3,20-DIONE-3,20-DIETHYLENE KETAL

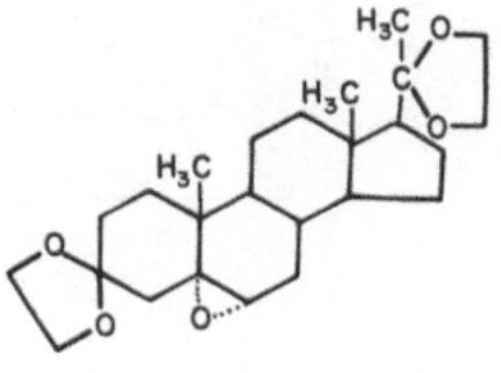

J. Am. Chem. Soc. 84:3206 (1962)

▶ $O_5C_{25}H_{38}$

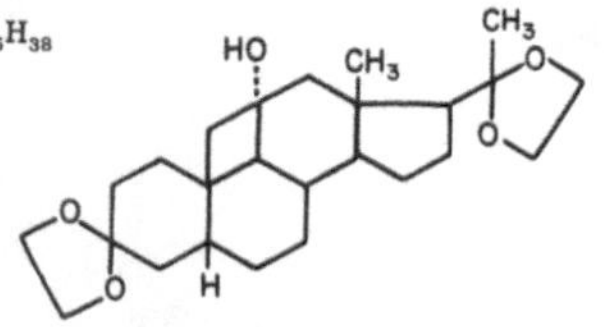

Helv. Chim. Acta 45:1261 (1962)

▶ $O_5C_{25}H_{38}$

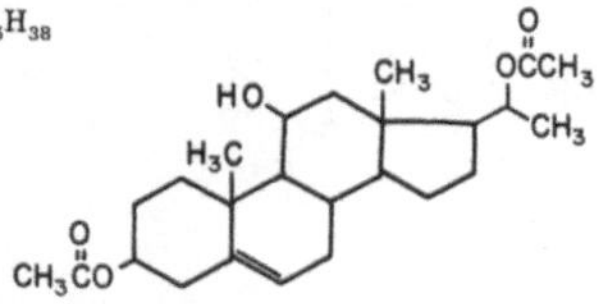

Helv. Chim. Acta 45:2161 (1962)

▶ $O_5C_{25}H_{38}$

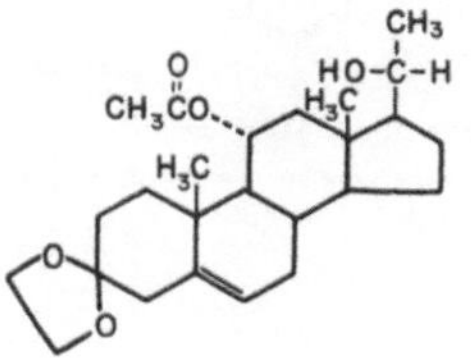

Helv. Chim. Acta 44:1380 (1961)

▶ $O_5C_{25}H_{38}$

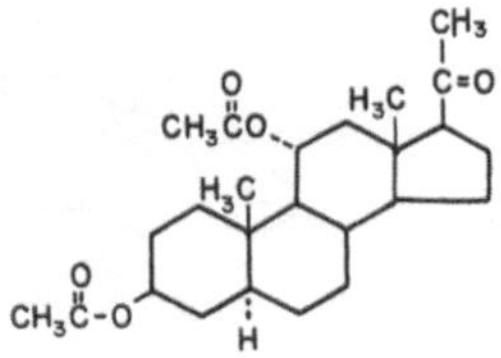

Helv. Chim. Acta 44:1380 (1961)

▶ $O_5C_{25}H_{38}$

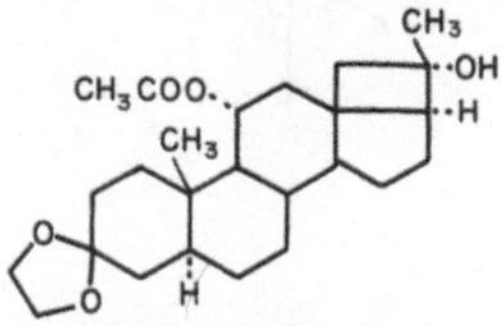

Helv. Chim. Acta 44:1927 (1961)

▶ $O_5C_{25}H_{38}$

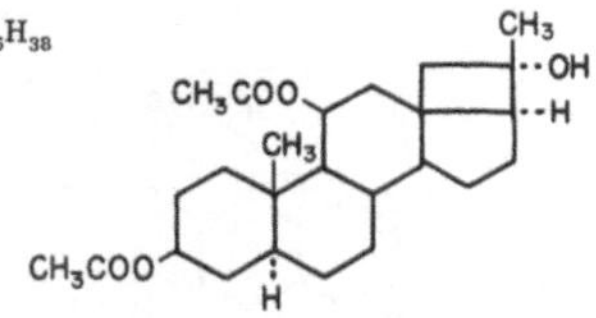

Helv. Chim. Acta 44:1927 (1961)

▶ $O_5C_{25}H_{38}$

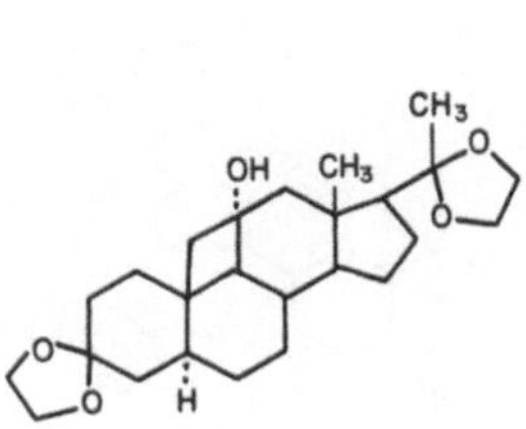

Helv. Chim. Acta 44:2162 (1961)

▶ $O_5C_{25}H_{38}$

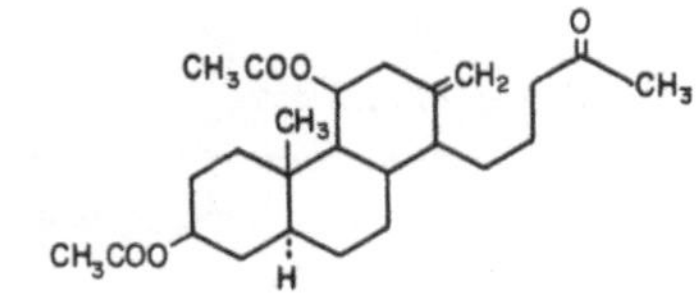

Helv. Chim. Acta 44:1927 (1961)

▶ $O_5C_{25}H_{38}$

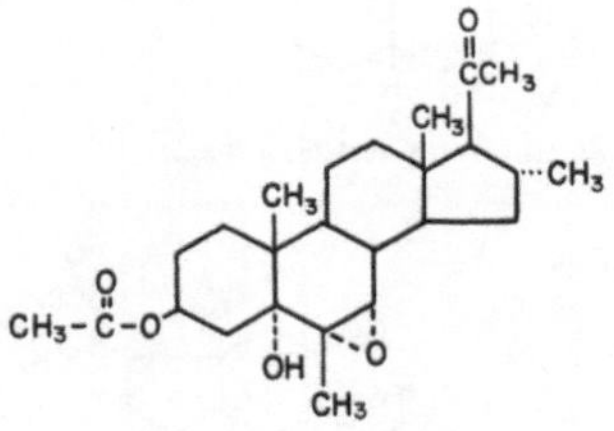

J. Org. Chem. 27:1139 (1962)

▶ $O_5C_{25}H_{38}$

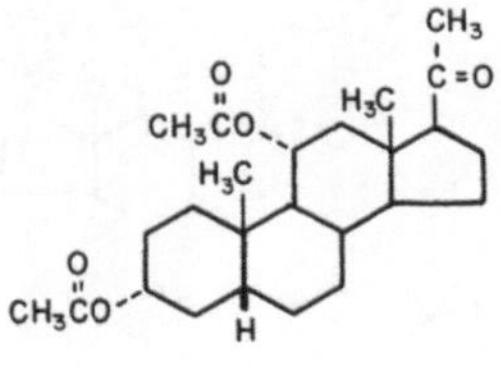

Helv. Chim. Acta 44:1380 (1961)

▶ $O_5C_{25}H_{38}$

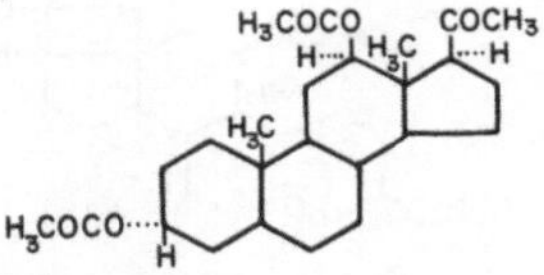

Varian 361

▶ $O_5C_{25}H_{40}$

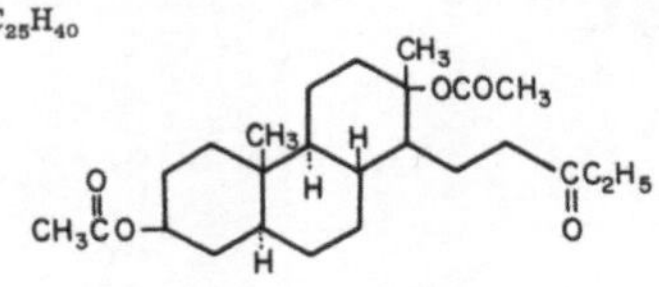

Helv. Chim. Acta 45:2674 (1962)

▶ $O_5C_{25}H_{42}$

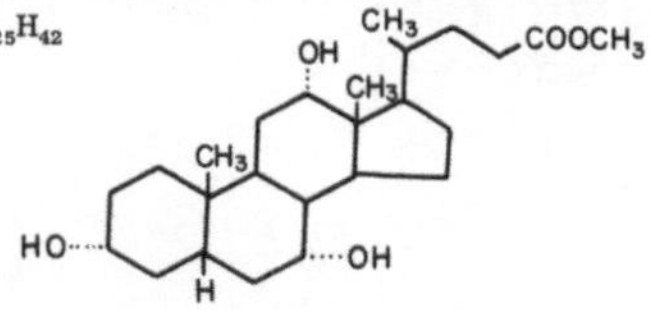

J. Chem. Soc. 2817 (1961)

▶ $O_5C_{26}H_{24}$

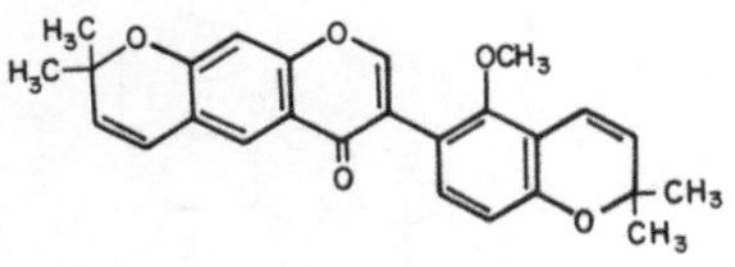

Varian 696

▶ $O_5C_{26}H_{30}$

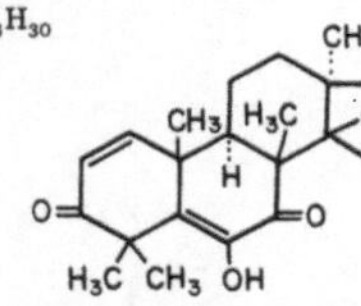

CEDRELONE

Proc. Chem. Soc. 446 (1961)

▶ $O_5C_{26}H_{36}$

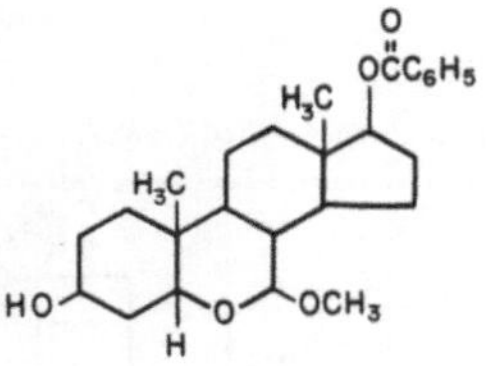

J. Am. Chem. Soc. 83:3071 (1961)

▶ $O_5C_{26}H_{36}$

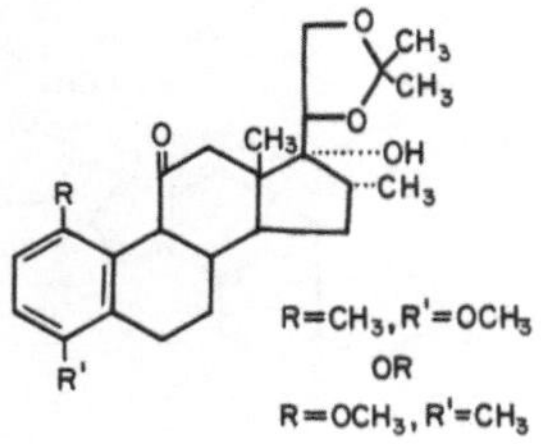

J. Org. Chem. 27:915 (1962)

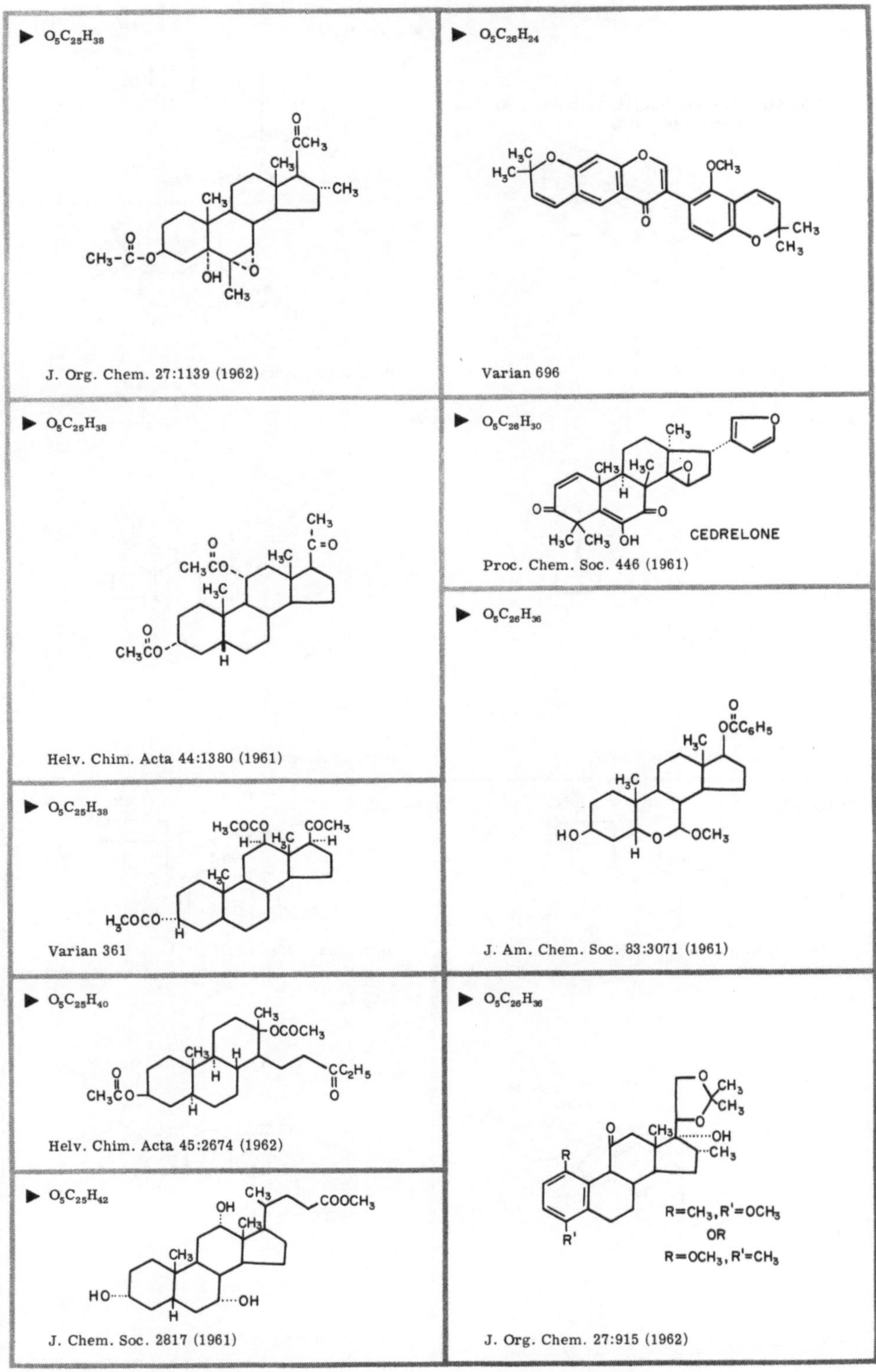

▶ $O_5C_{26}H_{38}$

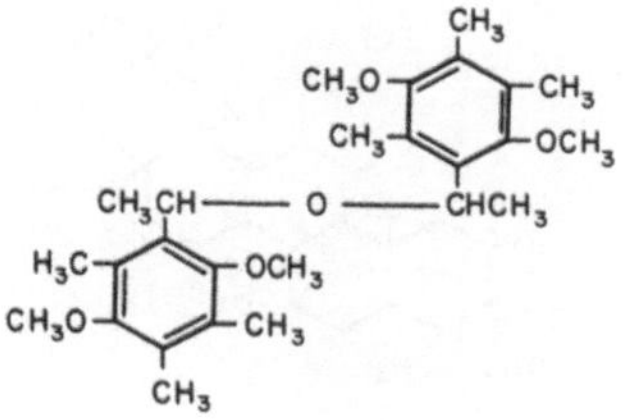

J. Org. Chem. 26:3225 (1961)

▶ $O_5C_{27}H_{44}$

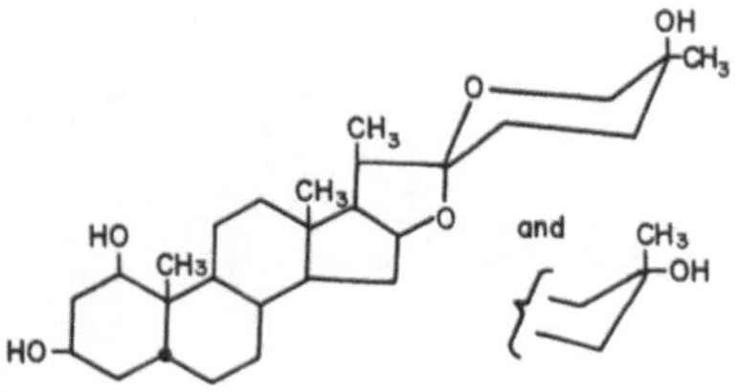

Tetrahedron Letters No. 24:1111 (1962)

▶ $O_5C_{26}H_{40}$

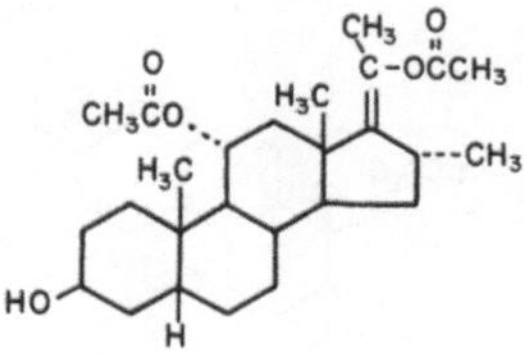

Helv. Chim. Acta 44:1380 (1961)

▶ $O_5C_{27}H_{46}$

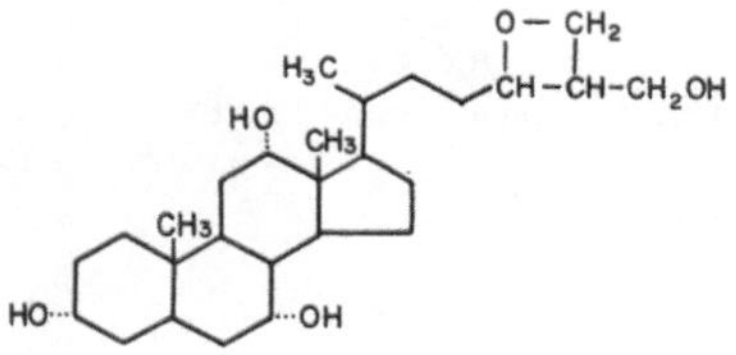

J. Chem. Soc. 2817 (1961)

▶ $O_5C_{27}H_{40}$

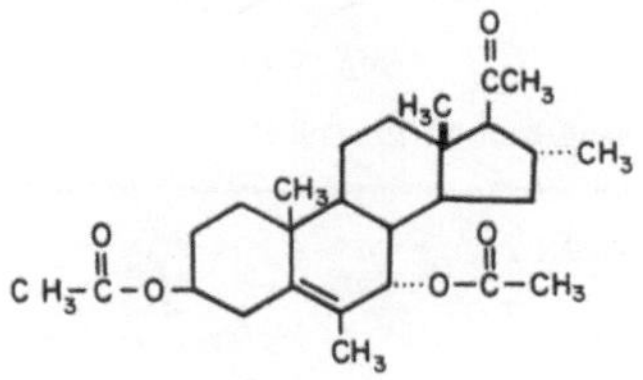

J. Org. Chem. 27:1139 (1962)

▶ $O_5C_{27}H_{48}$

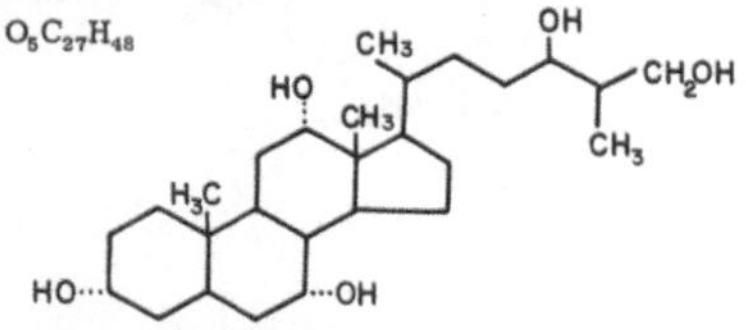

J. Chem. Soc. 2817 (1961)

▶ $O_5C_{27}H_{42}$

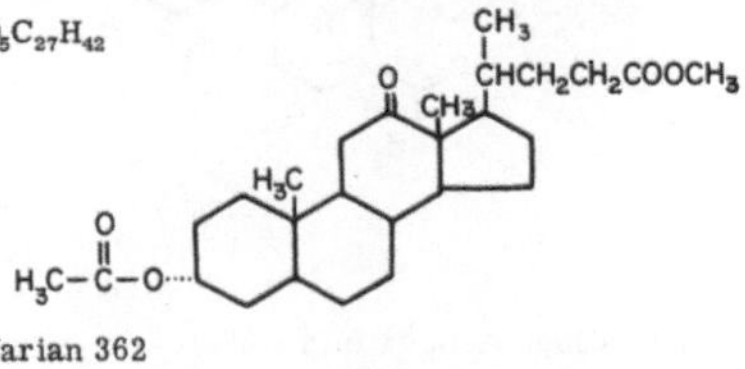

Varian 362

▶ $O_5C_{28}H_{36}$

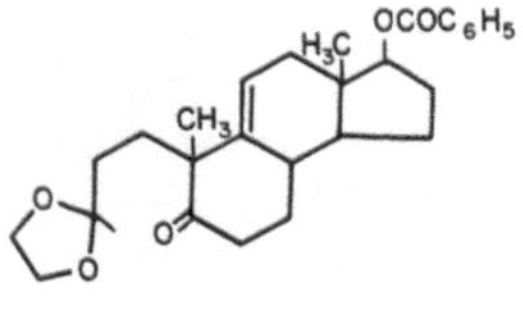

Bull. Soc. Chim. France 1679 (1961)

▶ $O_5C_{28}H_{44}$

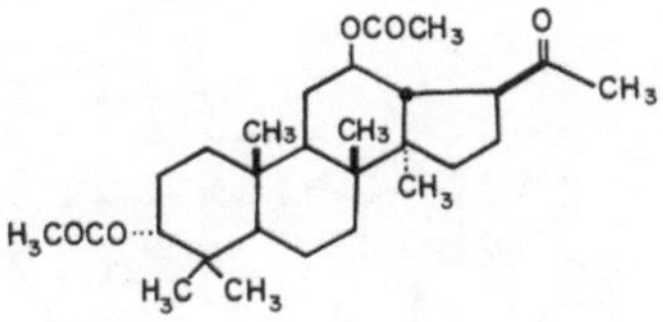


Bull. Soc. Chim. France 1832 (1962)

▶ $O_5C_{29}H_{42}$

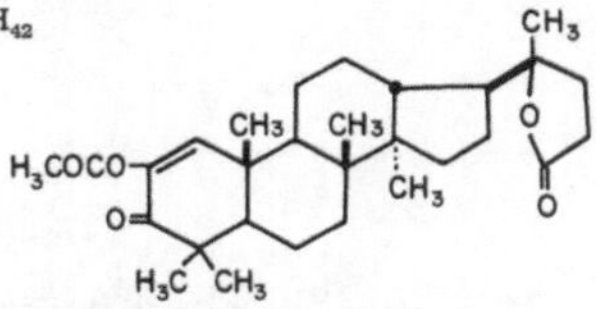


Bull. Soc. Chim. France 1832 (1962)

▶ $O_5C_{29}H_{44}$

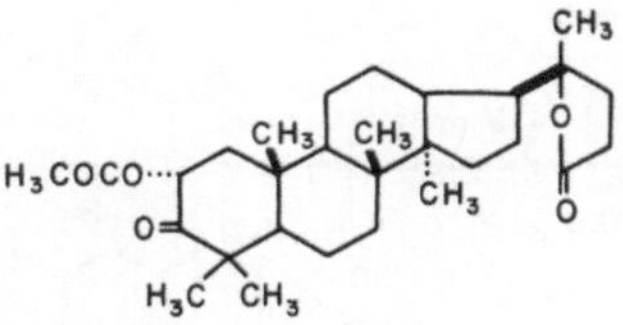


Bull. Soc. Chim. France 1832 (1962)

▶ $O_5C_{30}H_{44}$

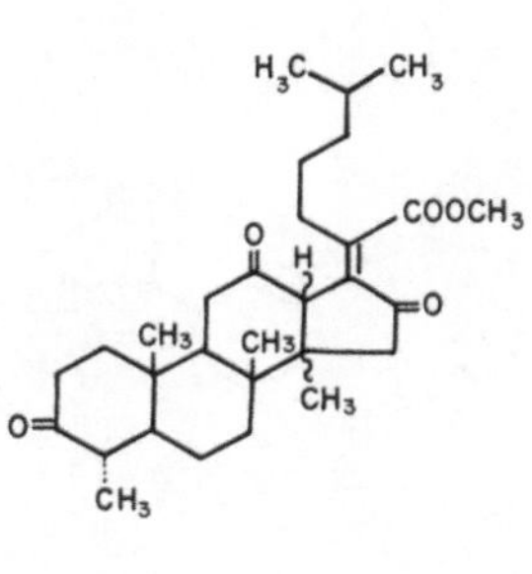


Tetrahedron 18:1035 (1962)

▶ $O_5C_{31}H_{46}$

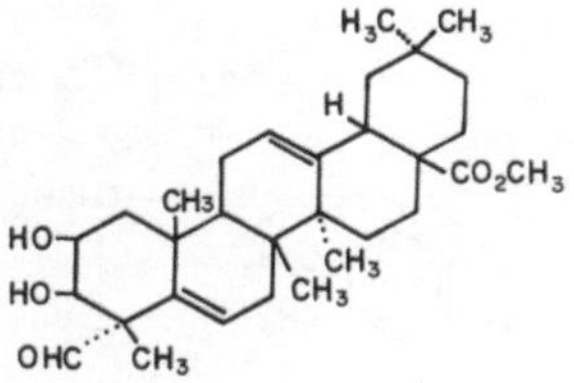


J. Chem. Soc. 4308 (1961)

▶ $O_5C_{31}H_{46}$

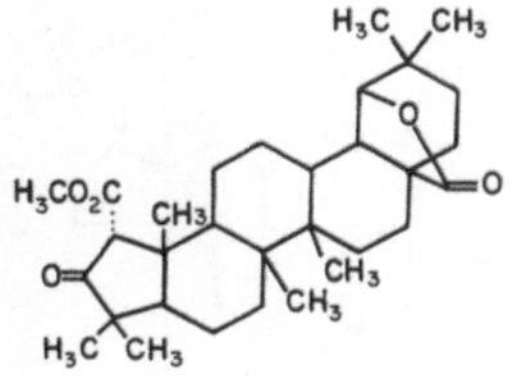


Can. J. Chem. 40:791 (1962)

▶ $O_5C_{31}H_{46}$

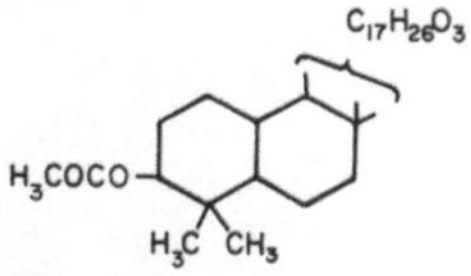


Tetrahedron 18:791 (1962)

▶ $O_5C_{31}H_{48}$

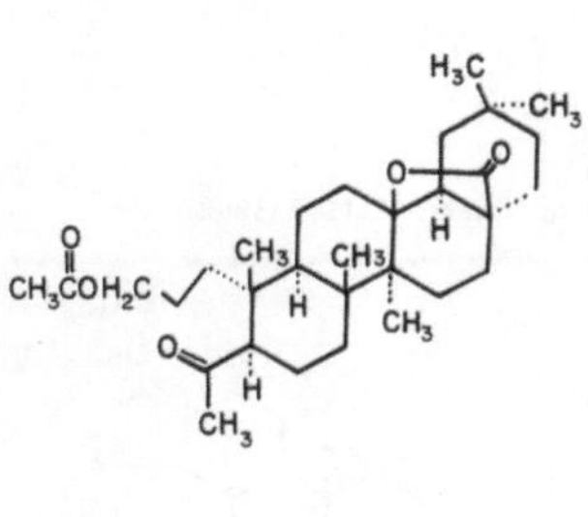


Helv. Chim. Acta 45:2674 (1962)

▶ $O_5C_{31}H_{48}$

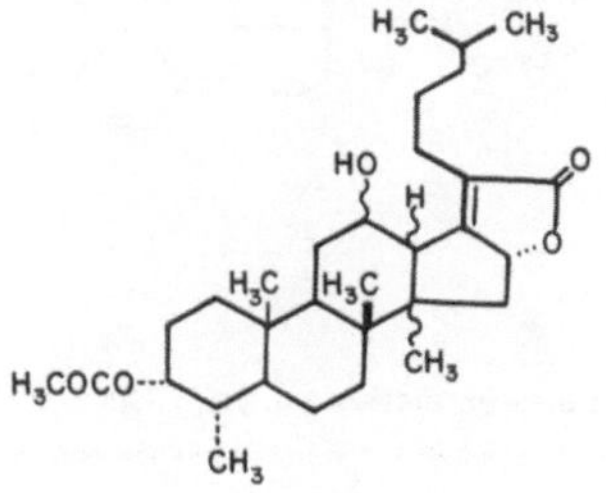

Tetrahedron 18:1047 (1962)

▶ $O_5C_{32}H_{44}$

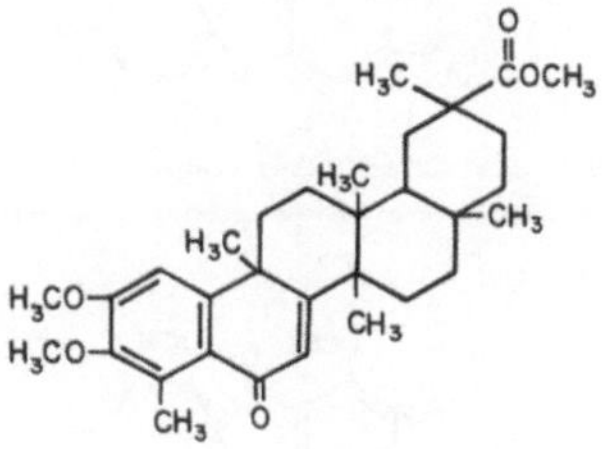

Tetrahedron Letters No. 14:604 (1962)

▶ $O_5C_{32}H_{50}$

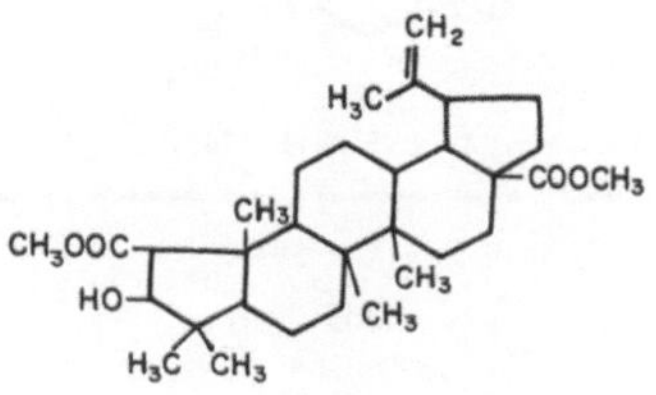

J. Org. Chem. 27:4072 (1962)

▶ $O_5C_{33}H_{50}$

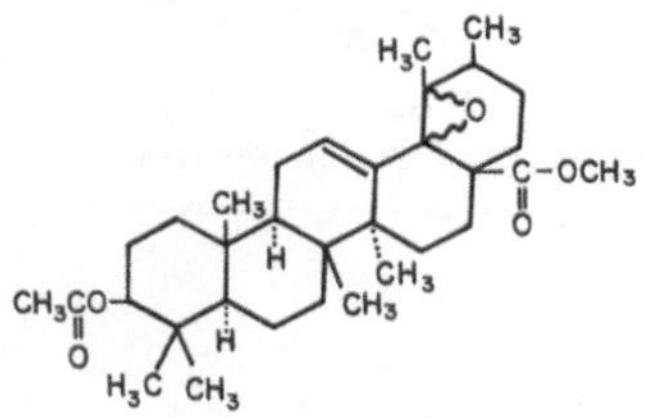

J. Chem. Soc. 5170 (1962)

▶ $O_5C_{33}H_{50}$

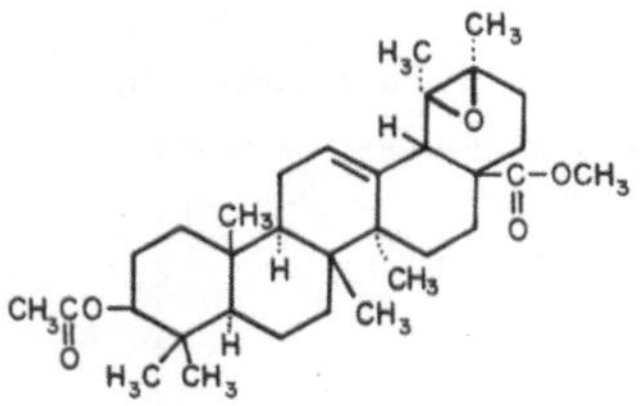

J. Chem. Soc. 5170 (1962)

▶ $O_5C_{33}H_{50}$

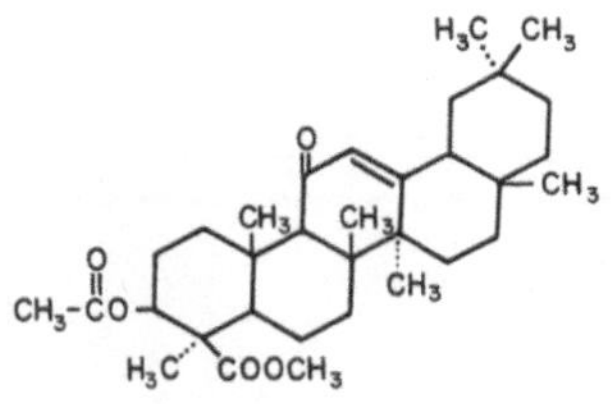

J. Org. Chem. 27:4512 (1962)

▶ $O_5C_{34}H_{56}$

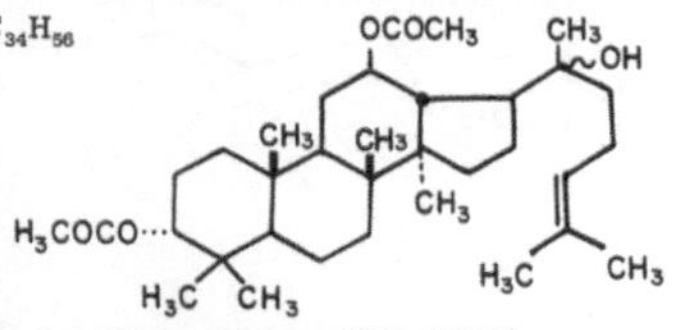

Bull. Soc. Chim. France 1832 (1962)

▶ $O_5C_{36}H_{42}$

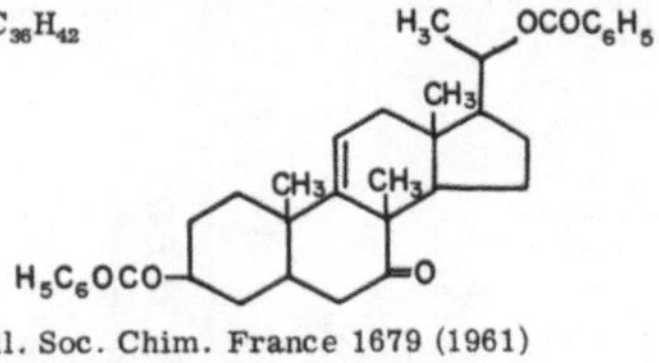

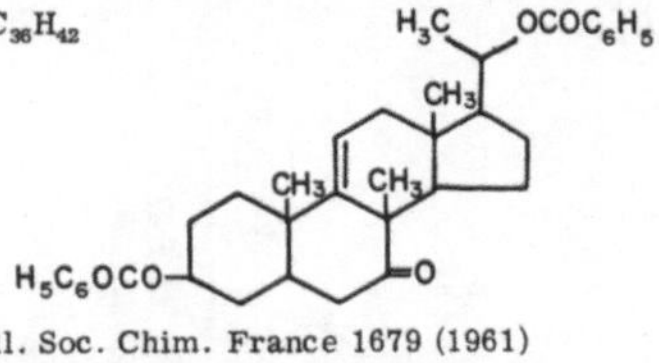

Bull. Soc. Chim. France 1679 (1961)

▶ $O_5C_{37}H_{58}$

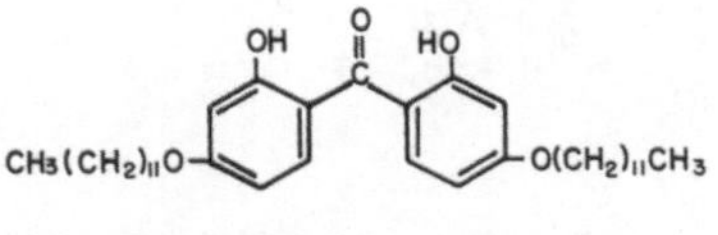

J. Phys. Chem. 65:2023 (1961)

▶ $O_5C_{39}H_{76}$

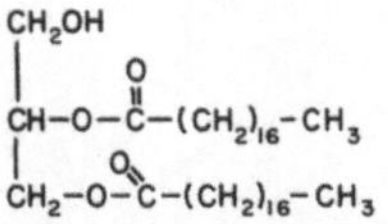

J. Am. Oil Chemists' Soc. 38:664 (1961)

▶ $O_6BrC_{24}H_{33}$

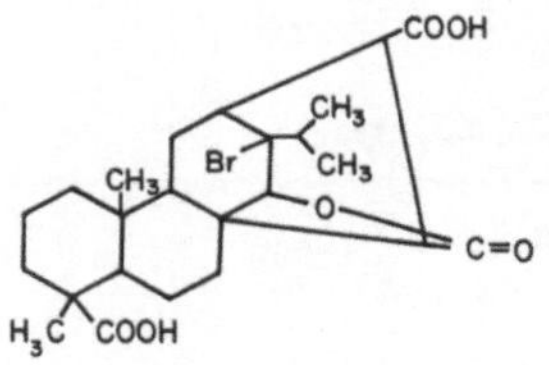

J. Org. Chem. 27:3535 (1962)

▶ $O_6BrC_{26}H_{37}$

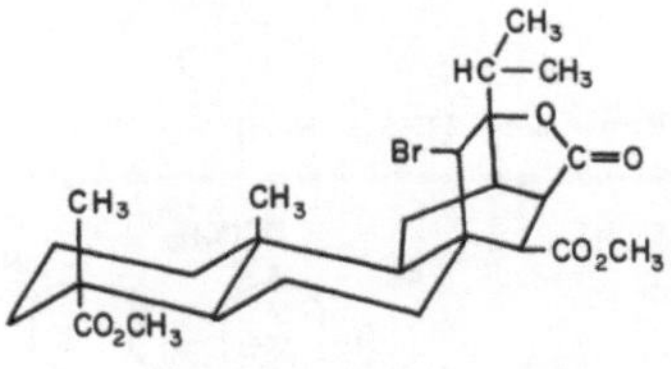

Tetrahedron Letters No. 16:693 (1962)

▶ $O_6Br_2C_{20}H_{16}$

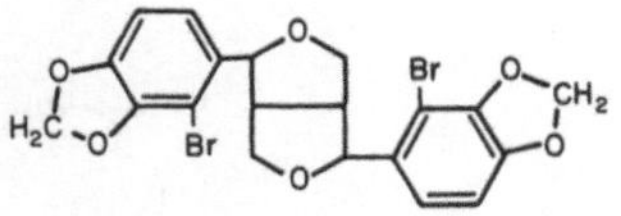

Tetrahedron Letters No. 4:157 (1962)

▶ $O_6ClC_{16}H_{15}$

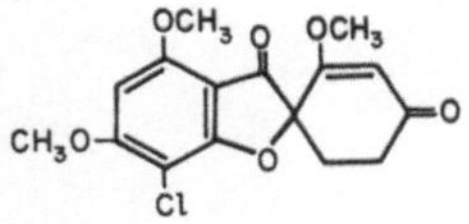

Helv. Chim. Acta 45:2241 (1962)

▶ $O_6ClC_{16}H_{15}$

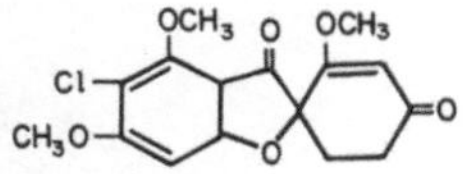

Helv. Chim. Acta 45:2241 (1962)

▶ $O_6ClC_{17}H_{17}$

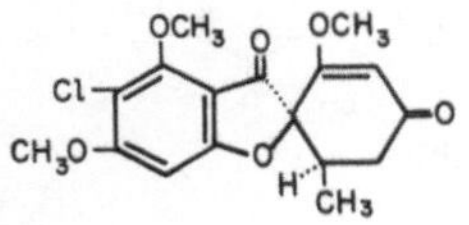

Helv. Chim. Acta 45:2241 (1962)

▶ $O_6ClC_{17}H_{17}$

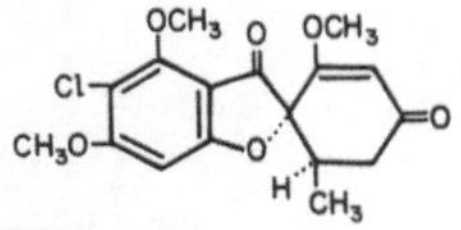

Helv. Chim. Acta 45:2241 (1962)

▶ $O_6ClC_{17}H_{17}$

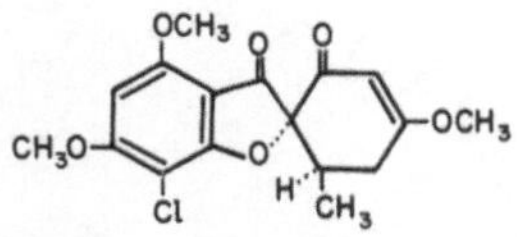

Helv. Chim. Acta 45:2241 (1962)

► $O_6ClC_{17}H_{17}$

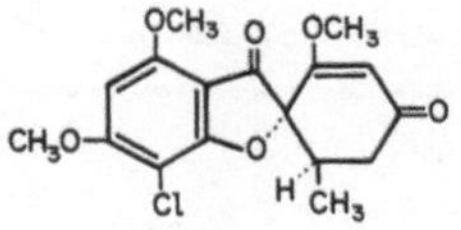

Helv. Chim. Acta 45:2241 (1962)

► $O_6ClC_{25}H_{31}$

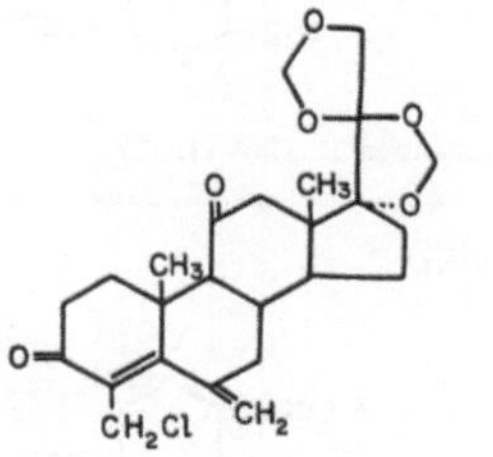

J. Org. Chem. 26:977 (1961)

► $O_6Cl_3C_{11}H_{11}$

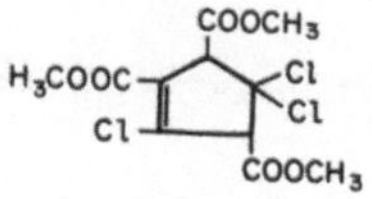

Varian 284

► $O_6FC_{24}H_{29}$

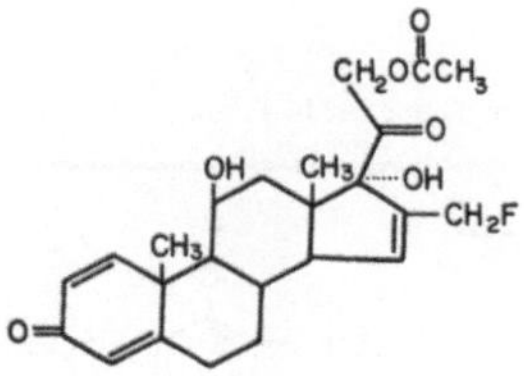

Chem. Ber. 95:2110 (1962)

► $O_6C_4H_6$

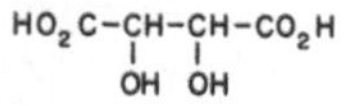

meso and d,l

Bull. Chem. Soc. Japan 35:1742 (1962)

► $O_6C_6H_6$

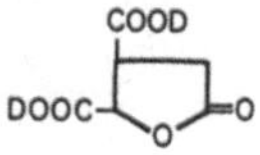

Varian 456

► $O_6C_6H_6$

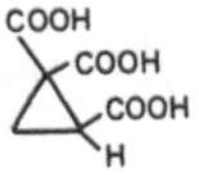

J. Am. Chem. Soc. 84:2249 (1962)

► $O_6C_6H_8$

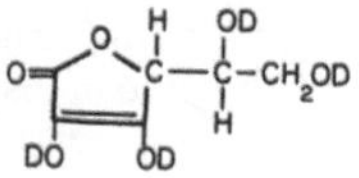

Varian 464

► $O_6C_6H_{10}$

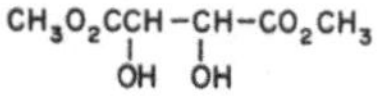

meso and d,l

Bull. Chem. Soc. Japan 35:1742 (1962)

► $O_6C_6H_{12}$

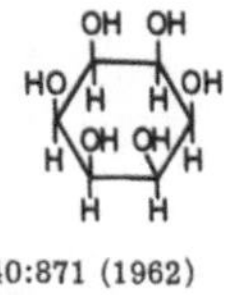

Can. J. Chem. 40:871 (1962)

► $O_6C_6H_{12}$

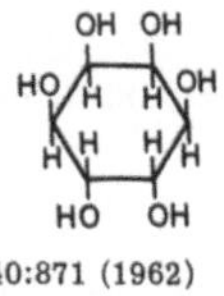

Can. J. Chem. 40:871 (1962)

▶ $O_6C_6H_{12}$

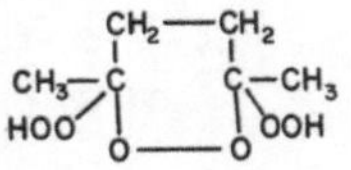

J. Org. Chem. 27:4319 (1962)

▶ $O_6C_6H_{12}$

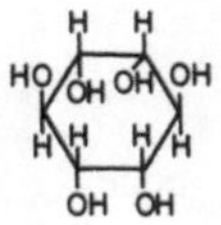

Can. J. Chem. 40:871 (1962)

▶ $O_6C_7H_{12}$

$CH_3C(=O)-O-(CH_2O)_3C(=O)CH_3$

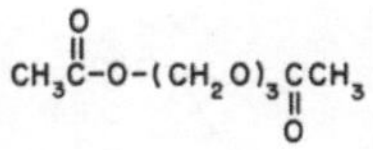

Chem. Ber. 94:3317 (1961)

▶ $O_6C_8H_{12}$

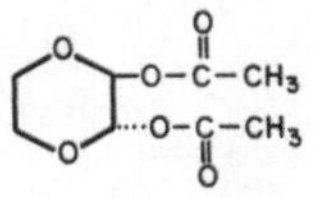

J. Org. Chem. 27:3183 (1962)

▶ $O_6C_9H_{14}$

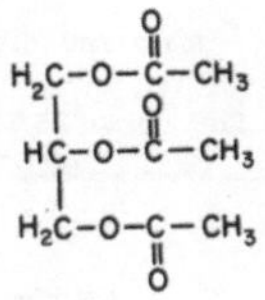

Varian 242

▶ $O_6C_9H_{14}$

$H_3COCOCH(COOC_2H_5)_2$

Varian 546

▶ $O_6C_{11}H_{14}$

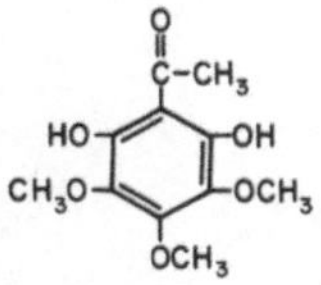

Tetrahedron 14:299 (1961)

▶ $O_6C_{11}H_{18}$

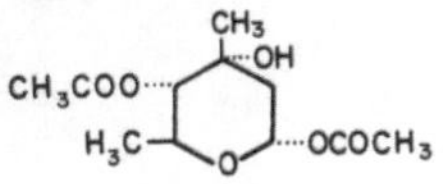

Tetrahedron 18:1265 (1962)

▶ $O_6C_{11}H_{18}$

Tetrahedron 18:1265 (1962)

▶ $O_6C_{11}H_{18}$

Tetrahedron 18:1265 (1962)

▶ $O_6C_{11}H_{20}$

$CH_2{=}CHCH_2O-(CH_2O)_5-CH_2CH{=}CH_2$

J. Chem. Soc. 4313 (1962)

▶ $O_6C_{11}H_{24}$

$CH_3CH_2CH_2O(CH_2O)_5CH_2CH_2CH_3$

J. Chem. Soc. 4313 (1962)

▶ $O_6C_{12}H_{16}$

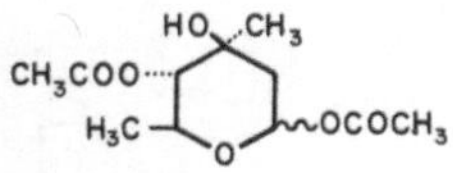

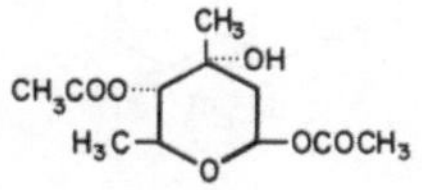

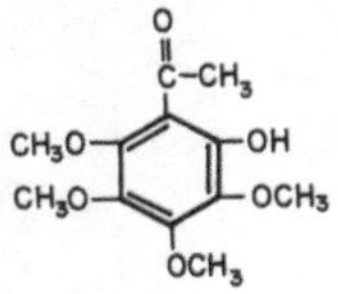

Tetrahedron 14:298 (1961)

▶ $O_6C_{12}H_{18}$

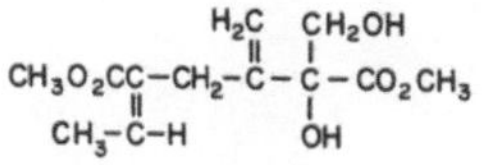

J. Am. Chem. Soc. 83:922 (1961)

▶ $O_6C_{12}H_{18}$

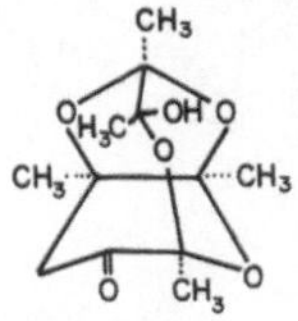

J. Chem. Soc. 4882 (1961)

▶ $O_6C_{12}H_{18}$

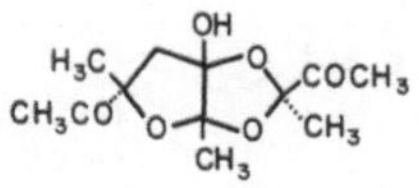

J. Chem. Soc. 4882 (1961)

▶ $O_6C_{12}H_{20}$

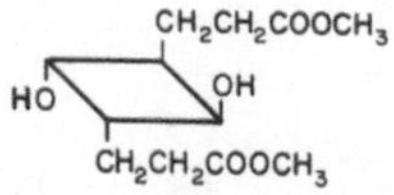

J. Am. Chem. Soc. 84:2344 (1962)

▶ $O_6C_{12}H_{20}$

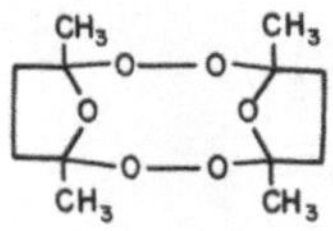

J. Org. Chem. 27:4319 (1962)

▶ $O_6C_{12}H_{20}$

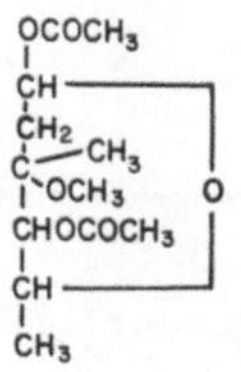

Tetrahedron 18:1265 (1962)
Helv. Chim. Acta 45:138 (1962)

▶ $O_6C_{14}H_{18}$

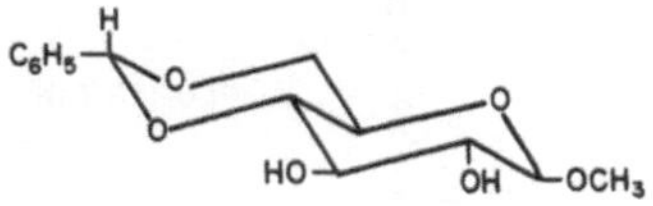

J. Chem. Soc. 5005 (1961)

▶ $O_6C_{14}H_{18}$

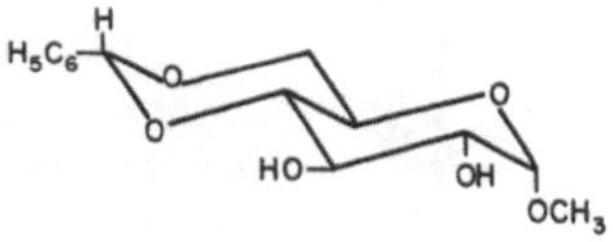

J. Chem. Soc. 5005 (1961)

▶ $O_6C_{15}H_{12}$

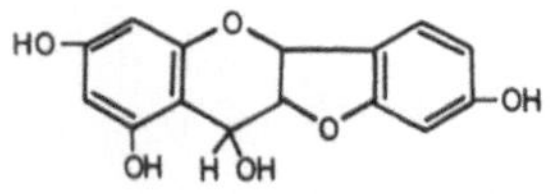

Tetrahedron Letters No. 18:790 (1962)

▶ $O_6C_{15}H_{14}$

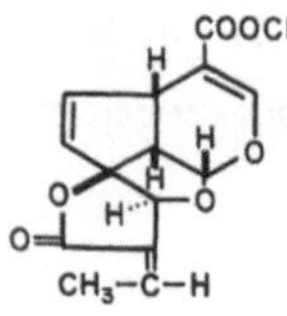

ISOPLUMERICIN

Helv. Chim. Acta 44:1447 (1961)

▶ $O_6C_{15}H_{14}$

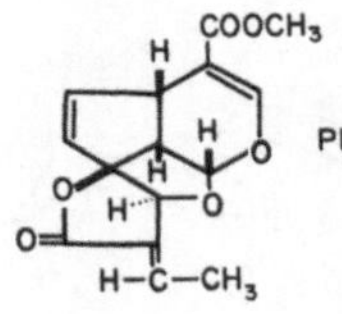

PLUMERICIN

Varian 640
Helv. Chim. Acta 44:1447 (1961)

▶ $O_6C_{15}H_{16}$

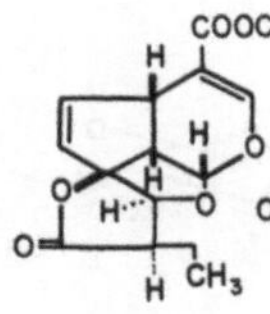

α-DIHYDROPLUMERICIN

Varian 641
Helv. Chim. Acta 44:1447 (1961)

▶ $O_6C_{15}H_{16}$

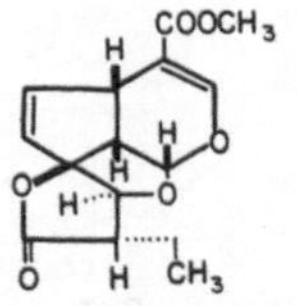

β-DIHYDROPLUMERICIN

Varian 641
Helv. Chim. Acta 44:1447 (1961)

▶ $O_6C_{15}H_{18}$

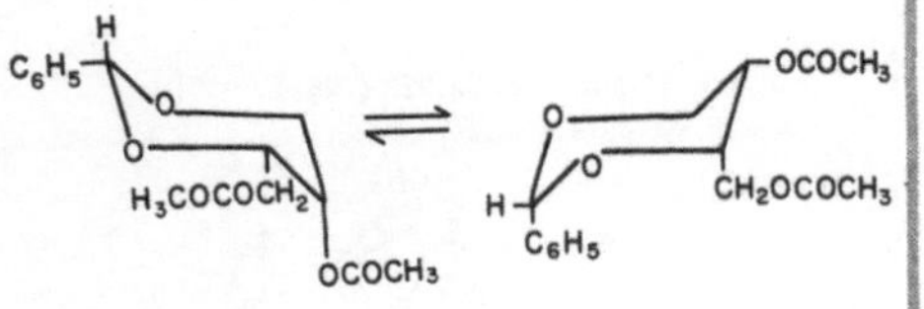

J. Chem. Soc. 5005 (1961)

▶ $O_6C_{15}H_{18}$

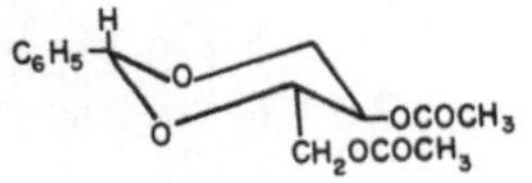

J. Chem. Soc. 5005 (1961)

▶ $O_6C_{15}H_{18}$

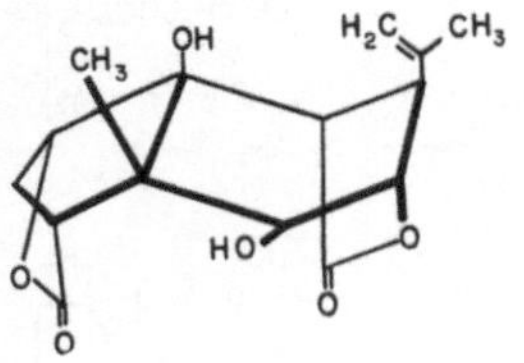

J. Chem. Soc. 3006 (1961)

▶ $O_6C_{15}H_{21}Rh$

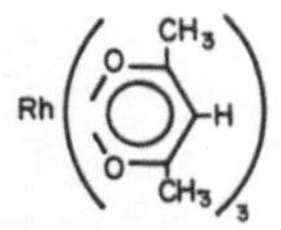

Varian 644

▶ $O_6C_{15}H_{22}$

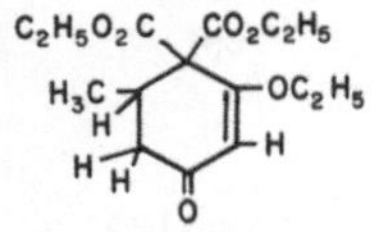

J. Am. Chem. Soc. 84:310 (1962)

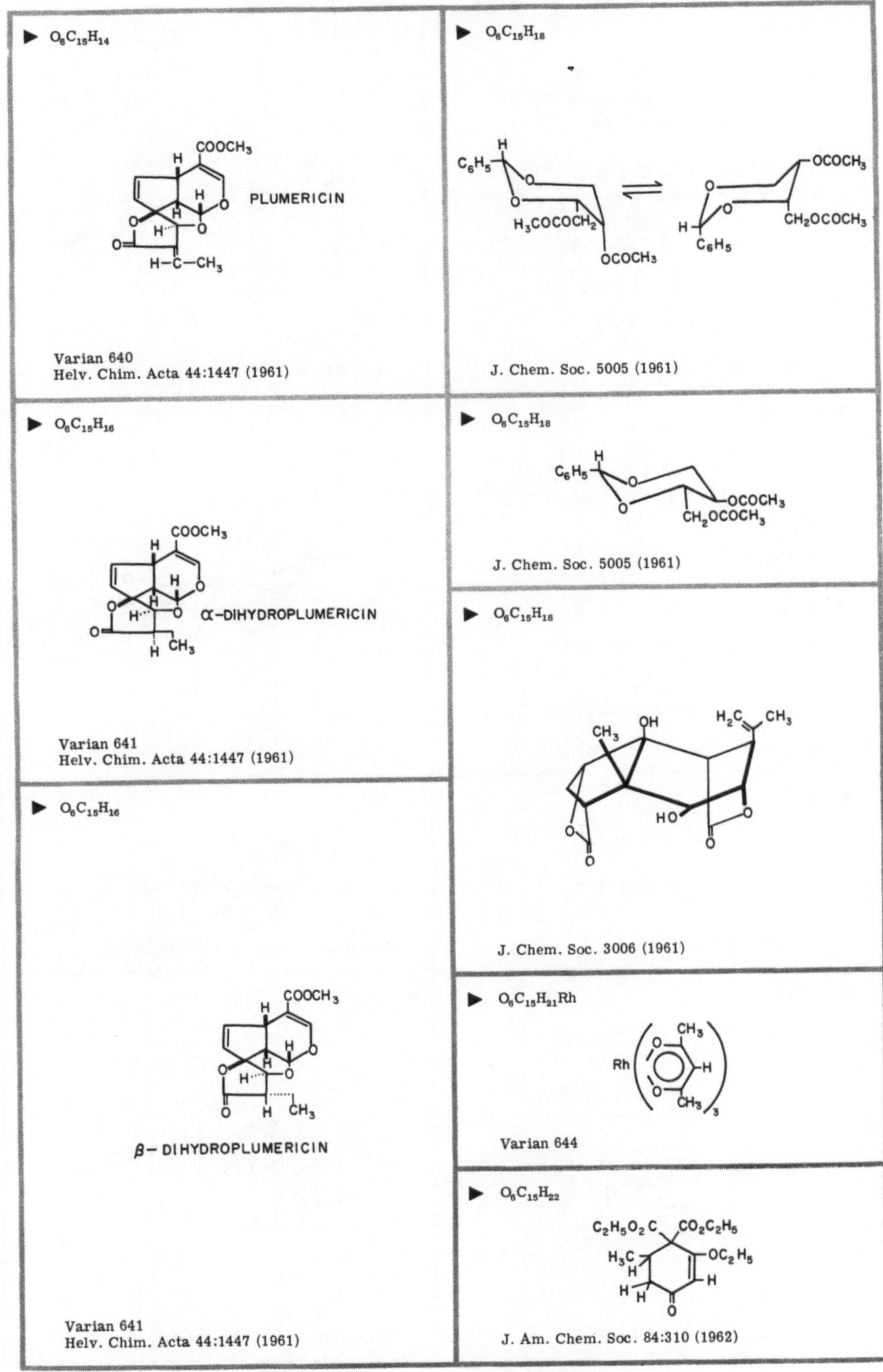

▶ $O_6C_{15}H_{24}$

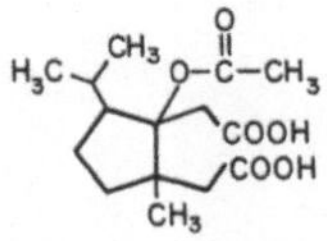

J. Org. Chem. 26:982 (1961)

▶ $O_6C_{16}H_{20}$

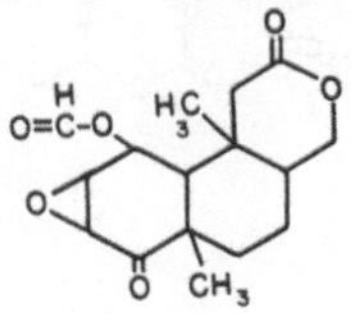

J. Chem. Soc. 4817 (1962)

▶ $O_6C_{16}H_{28}$

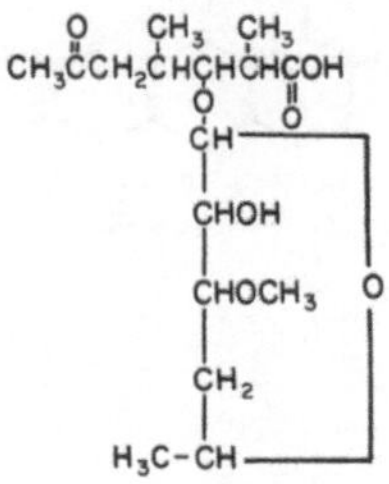

J. Am. Chem. Soc. 84:1512 (1962)

▶ $O_6C_{17}H_{18}$

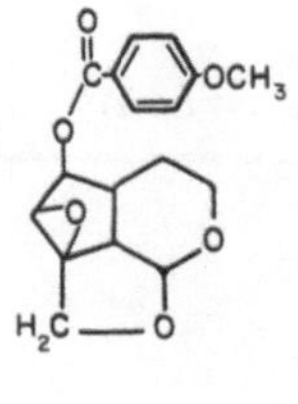

Tetrahedron Letters No. 8:328 (1962)

▶ $O_6C_{17}H_{18}$

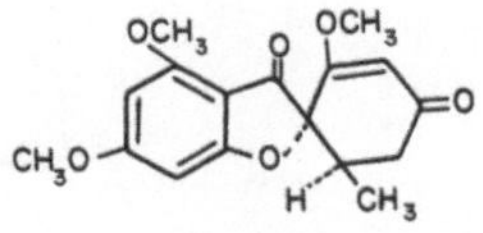

Helv. Chim. Acta 45:2241 (1962)

▶ $O_6C_{17}H_{18}$

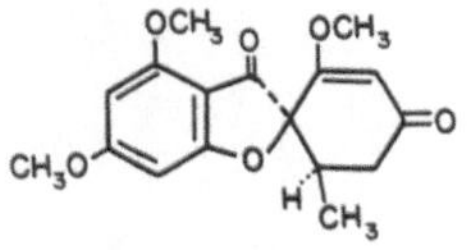

Helv. Chim. Acta 45:2241 (1962)

▶ $O_6C_{17}H_{24}$

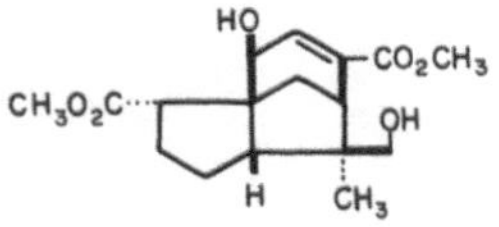

Tetrahedron 18:1321 (1962)

▶ $O_6C_{17}H_{24}$

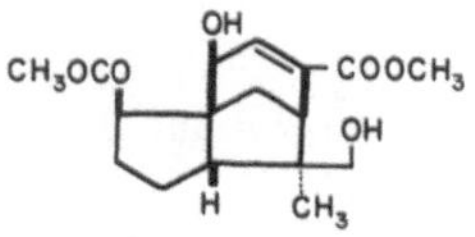

Tetrahedron 18:547, 1321 (1962)

▶ $O_6C_{17}H_{24}$

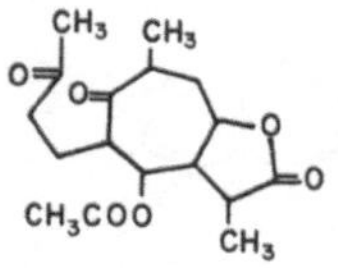

J. Am. Chem. Soc. 84:3857 (1962)

▶ $O_6C_{18}H_{12}$

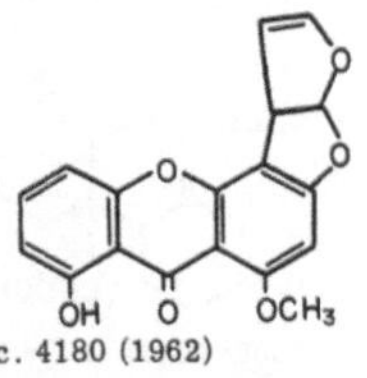

J. Chem. Soc. 4180 (1962)

▶ $O_6C_{18}H_{14}$

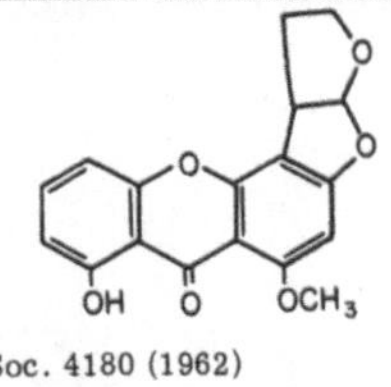

J. Chem. Soc. 4180 (1962)

▶ $O_6C_{18}H_{14}$

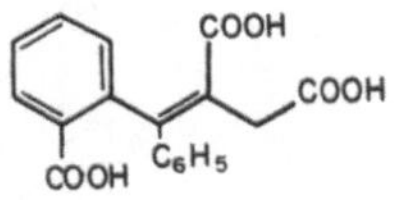

J. Am. Chem. Soc. 84:1315 (1962)

▶ $O_6C_{18}H_{16}$

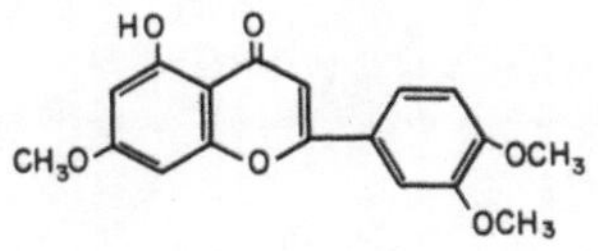


Bull. Soc. Chim. France 1962 (1962)

▶ $O_6C_{18}H_{18}$

SAMADERINE A

Bull. Soc. Chim. France 1715 (1962)

▶ $O_6C_{18}H_{18}$

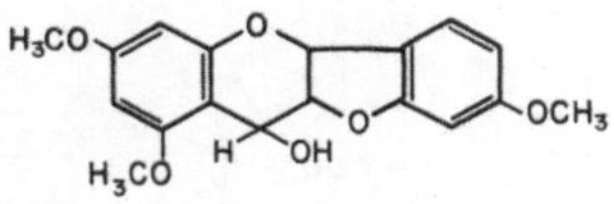


Tetrahedron Letters No. 18:791 (1962)

▶ $O_6C_{18}H_{20}$

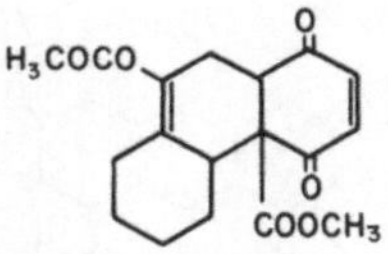


J. Chem. Soc. 2908 (1961)

▶ $O_6C_{18}H_{20}$

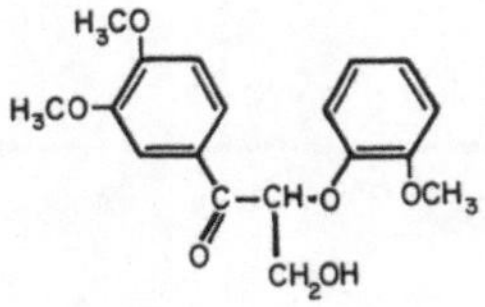


Varian 675

▶ $O_6C_{19}H_{12}$

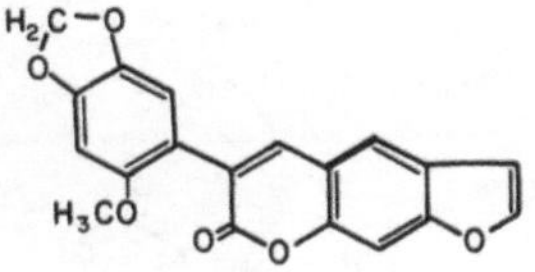


J. Chem. Soc. 775 (1962)

▶ $O_6C_{19}H_{12}$

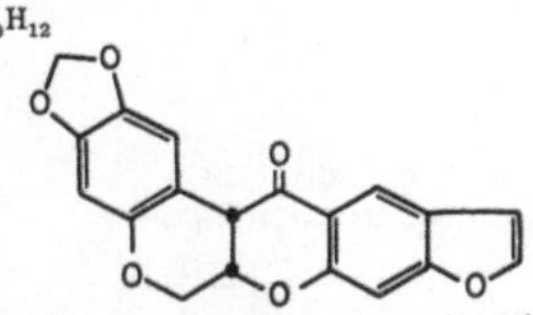

Tetrahedron Letters No. 18:802 (1962)

▶ $O_6C_{19}H_{14}$

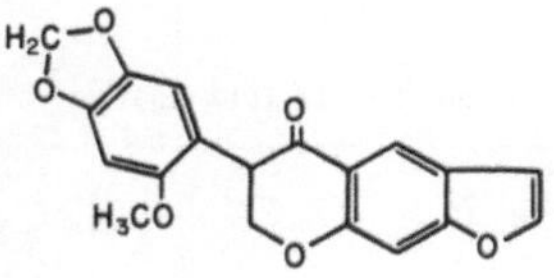


J. Chem. Soc. 775 (1962)

▶ $O_6C_{19}H_{18}$

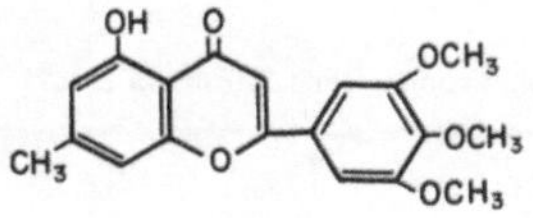


Bull. Soc. Chim. France 1962 (1962)

► $O_6C_{19}H_{22}$

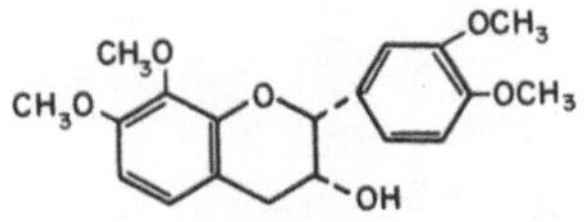


J. Chem. Soc. 4504 (1962)

► $O_6C_{20}H_{16}$

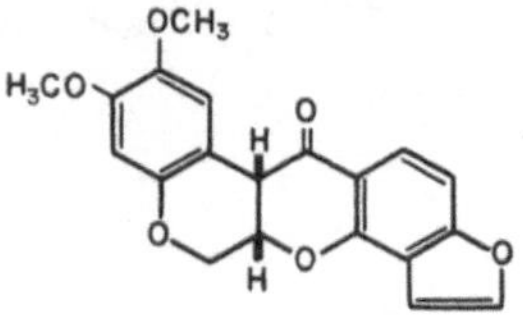


J. Chem. Soc. 775 (1962)

► $O_6C_{19}H_{22}$

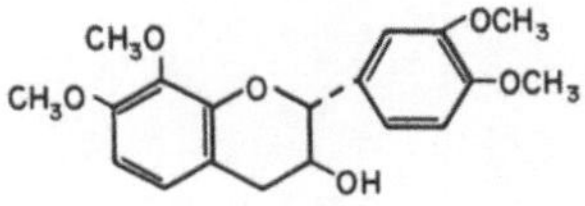


J. Chem. Soc. 4504 (1962)

► $O_6C_{19}H_{22}$

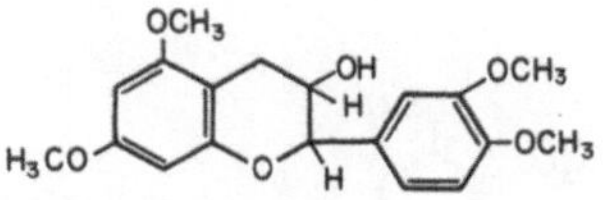


Varian 679
Tetrahedron Letters No. 18:792 (1962)

► $O_6C_{20}H_{16}$

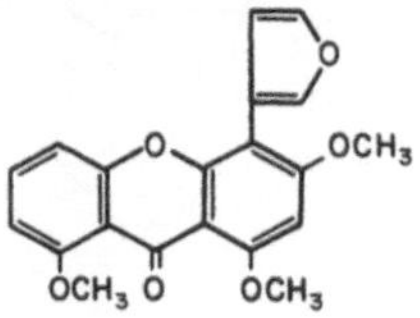


J. Chem. Soc. 4180 (1962)

► $O_6C_{19}H_{26}$

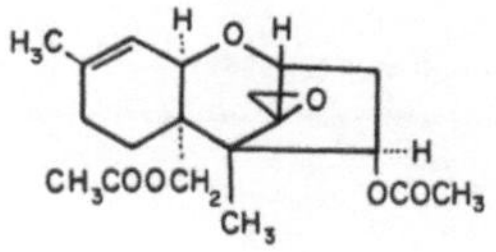


Helv. Chim. Acta 45:1726 (1962)

► $O_6C_{20}H_{18}$

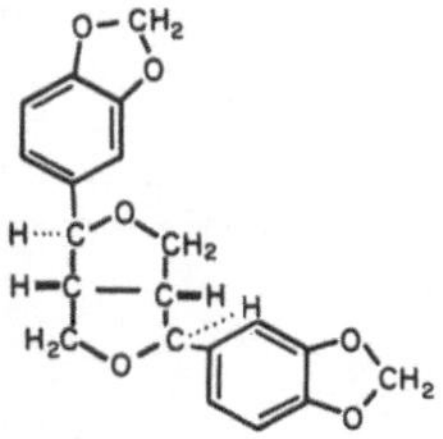


(+)-SESAMIN

Chem. Ber. 94:2522 (1961)
Tetrahedron Letters No. 4:157 (1962)

► $O_6C_{19}H_{30}$

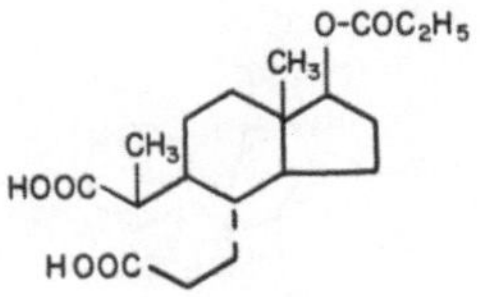


Tetrahedron 18:1015 (1962)

► $O_6C_{20}H_{18}$

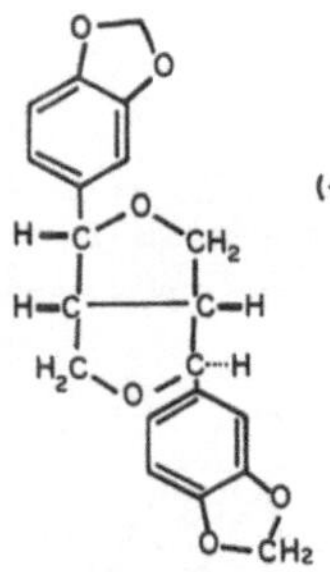


(+)-ASARININ

Chem. Ber. 94:2522 (1961)
Tetrahedron Letters No. 4:157 (1962)

▶ $O_6C_{20}H_{18}$

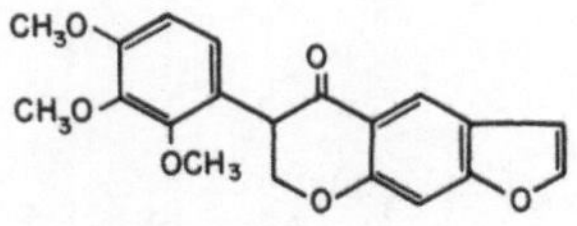

Chem. & Ind. 1946 (1962)

▶ $O_6C_{20}H_{20}$

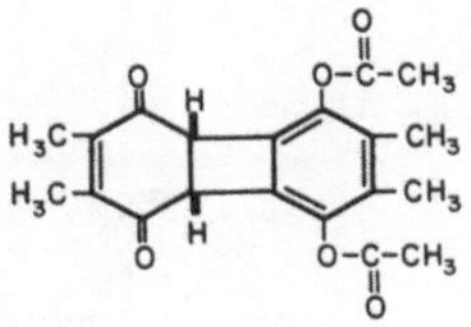

J. Chem. Soc. 1718 (1962)

▶ $O_6C_{20}H_{24}$

GLAUCANOL

Bull. Soc. Chim. France 1255 (1962)

▶ $O_6C_{20}H_{26}$

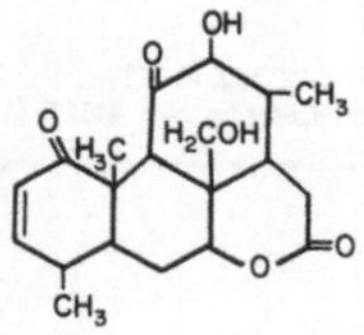

Tetrahedron Letters No. 23:1085 (1962)

▶ $O_6C_{20}H_{34}$

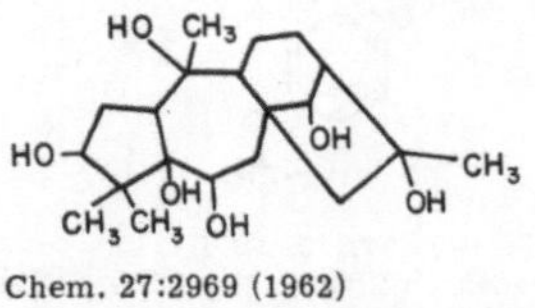

J. Org. Chem. 27:2969 (1962)

▶ $O_6C_{20}H_{36}$

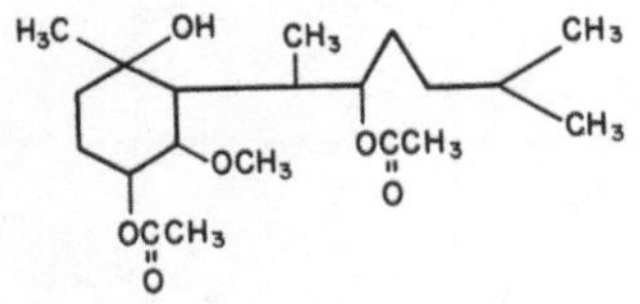

J. Am. Chem. Soc. 83:3096 (1961)

▶ $O_6C_{21}H_{22}$

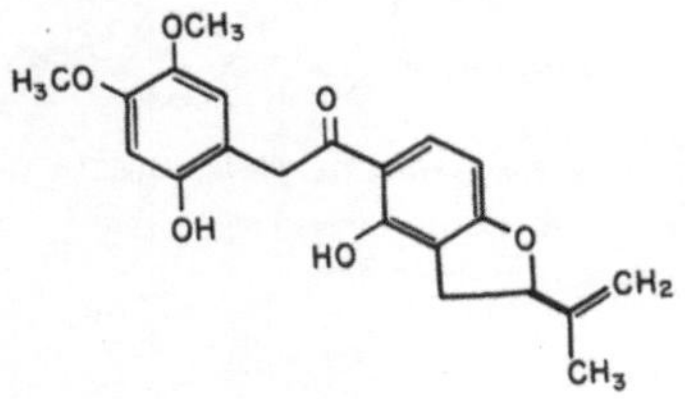

J. Chem. Soc. 775 (1962)

▶ $O_6C_{21}H_{22}$

SAMADEROL ACETATE

Bull. Soc. Chim. France 1715 (1962)

▶ $O_6C_{21}H_{26}$

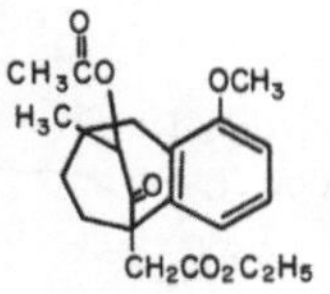

Bull. Chem. Soc. Japan 34:274 (1961)

▶ $O_6C_{21}H_{30}$

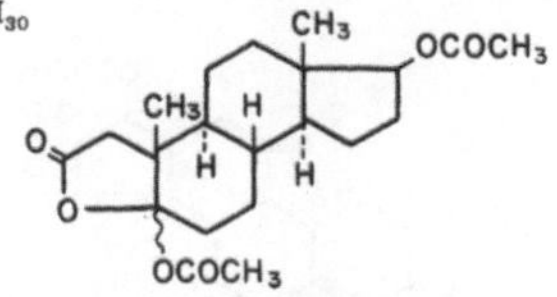

Helv. Chim. Acta 45:2346 (1962)

▶ $O_6C_{22}H_{20}$

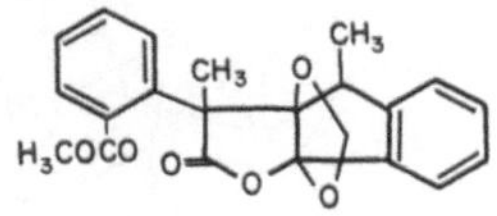

Tetrahedron Letters No. 19:848 (1962)

▶ $O_6C_{21}H_{30}$

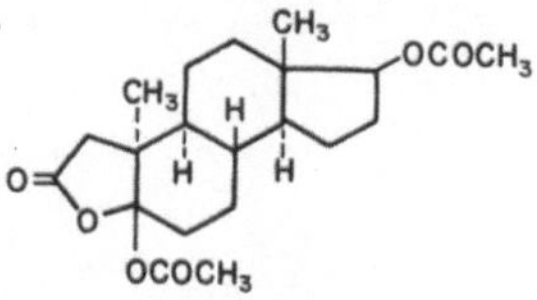

Helv. Chim. Acta 45:2346 (1962)

▶ $O_6C_{22}H_{22}$

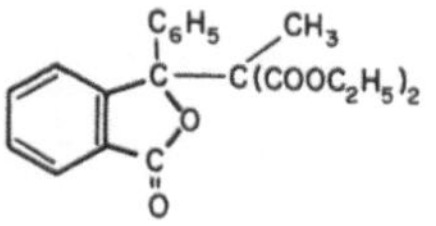

J. Org. Chem. 27:324 (1962)

▶ $O_6C_{21}H_{32}$

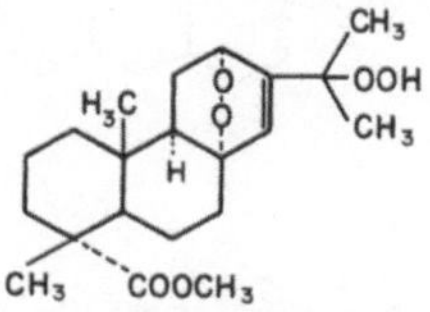

J. Am. Chem. Soc. 83:2570 (1961)

▶ $O_6C_{22}H_{24}$

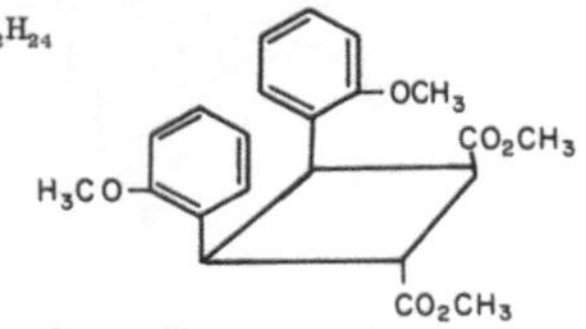

Can. J. Chem. 40:1251 (1962)

▶ $O_6C_{21}H_{34}$

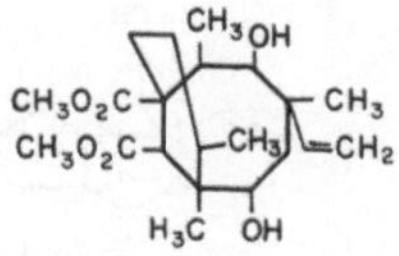

Gazz. Chim. Ital. 92:884 (1962)

▶ $O_6C_{22}H_{26}$

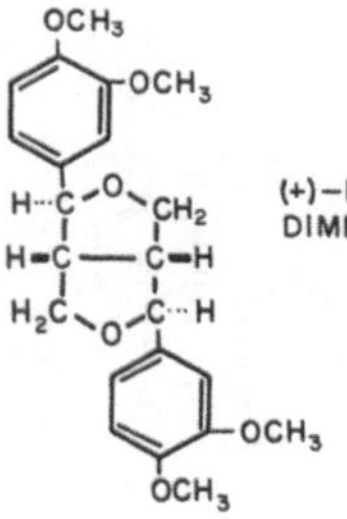

(+)-PINORESINOL DIMETHYL ETHER

Tetrahedron Letters No. 4:157 (1962)
Chem. Ber. 94:2522 (1961)

▶ $O_6C_{22}H_{20}$

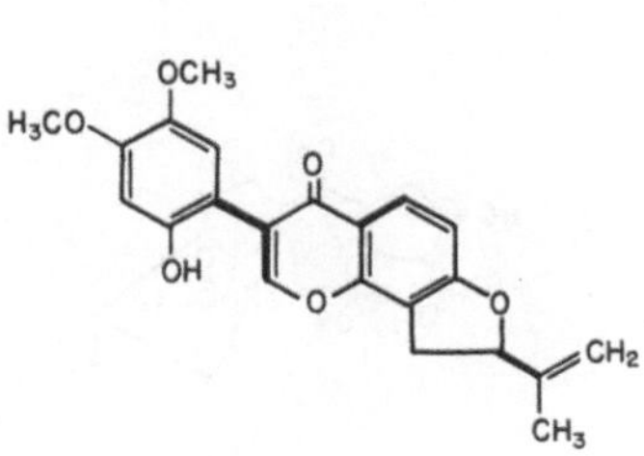

J. Chem. Soc. 775 (1962)

▶ $O_6C_{22}H_{26}$

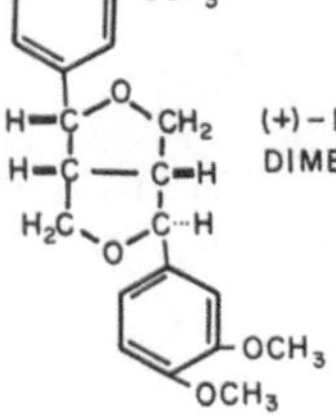

(+)-EPI-PINORESINOL DIMETHYL ETHER

Tetrahedron Letters No. 4:157 (1962)
Chem. Ber. 94:2522 (1961)

► $O_6C_{22}H_{28}$

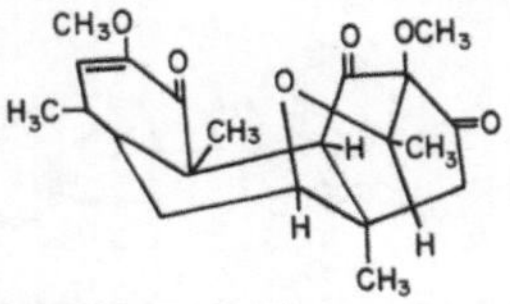

Tetrahedron 18:1433 (1962)

► $O_6C_{22}H_{28}$

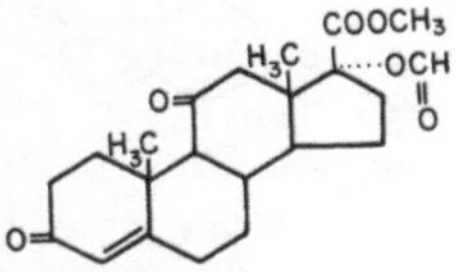

Tetrahedron 16:271 (1961)

► $O_6C_{22}H_{28}$

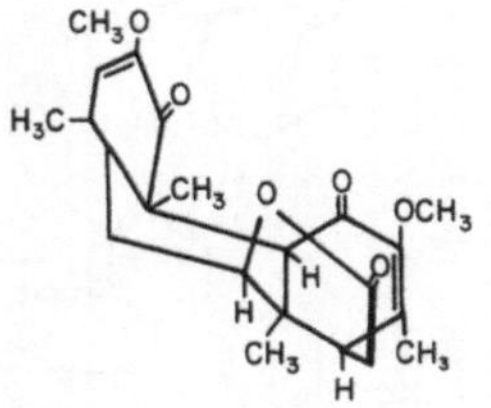

Tetrahedron 18:1433 (1962)

► $O_6C_{22}H_{30}$

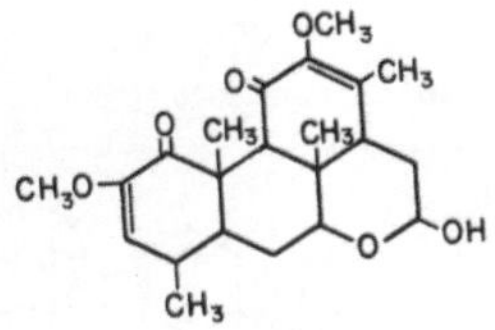

Australian J. Chem. 15:807 (1962)
Tetrahedron 18:1433 (1962)

► $O_6C_{22}H_{28}$

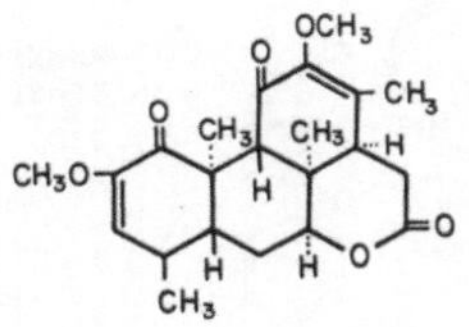

Australian J. Chem. 15:807 (1962)
Tetrahedron 18:1433 (1962)

► $O_6C_{22}H_{32}$

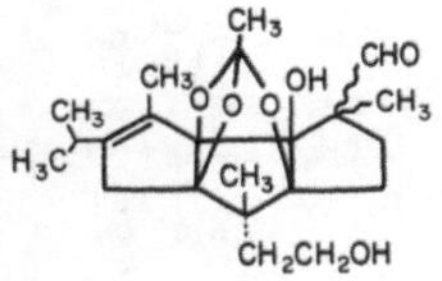

Experientia 18:549 (1962)

► $O_6C_{22}H_{28}$

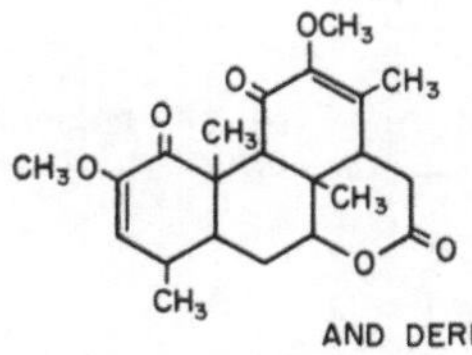

AND DERIVATIVES

Tetrahedron 15:100 (1961)

► $O_6C_{22}H_{36}$

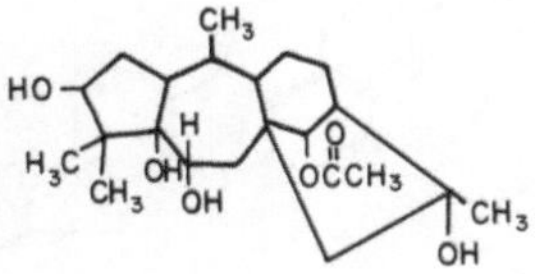

J. Org. Chem. 27:2969 (1962)

▶ $O_6C_{22}H_{36}$

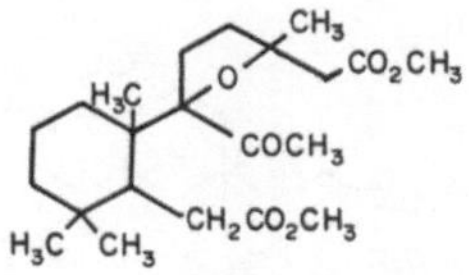

Acta Chem. Scand. 16:1675 (1962)

▶ $O_6C_{23}H_{20}$

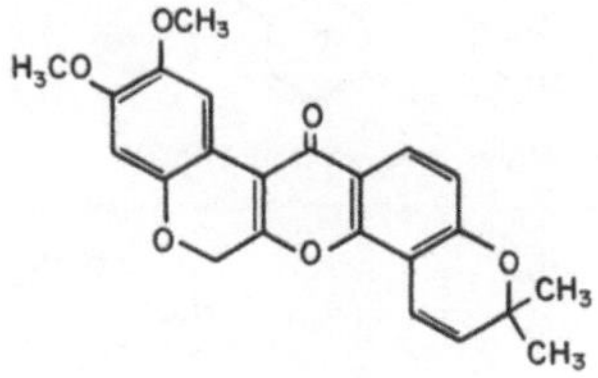

J. Chem. Soc. 775 (1962)

▶ $O_6C_{23}H_{20}$

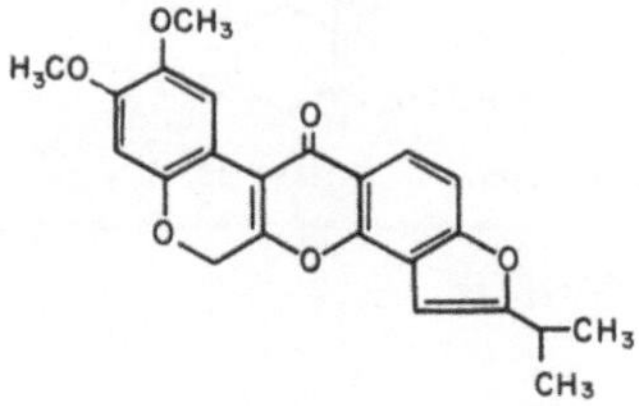

J. Chem. Soc. 775 (1962)

▶ $O_6C_{23}H_{22}$

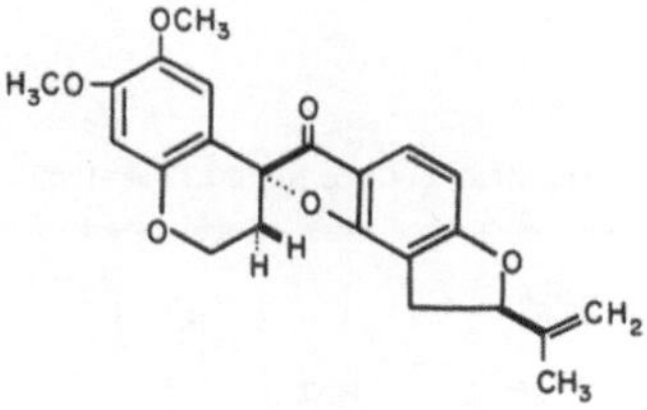

J. Chem. Soc. 775 (1962)

▶ $O_6C_{23}H_{22}$

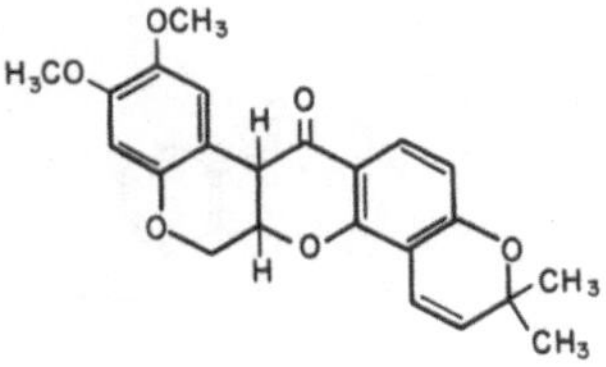

J. Chem. Soc. 775 (1962)

▶ $O_6C_{23}H_{22}$

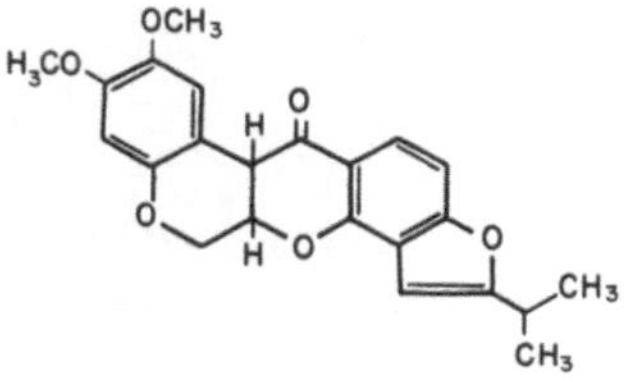

J. Chem. Soc. 775 (1962)

▶ $O_6C_{23}H_{22}$

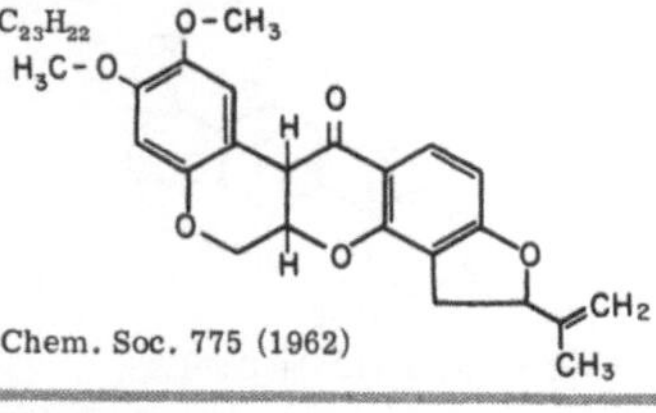

J. Chem. Soc. 775 (1962)

▶ $O_6C_{23}H_{24}$

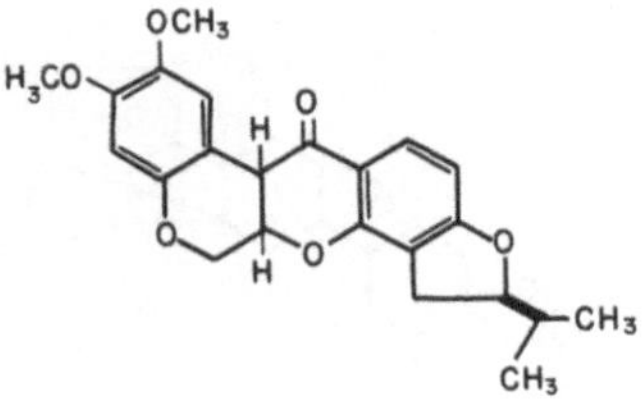

J. Chem. Soc. 775 (1962)

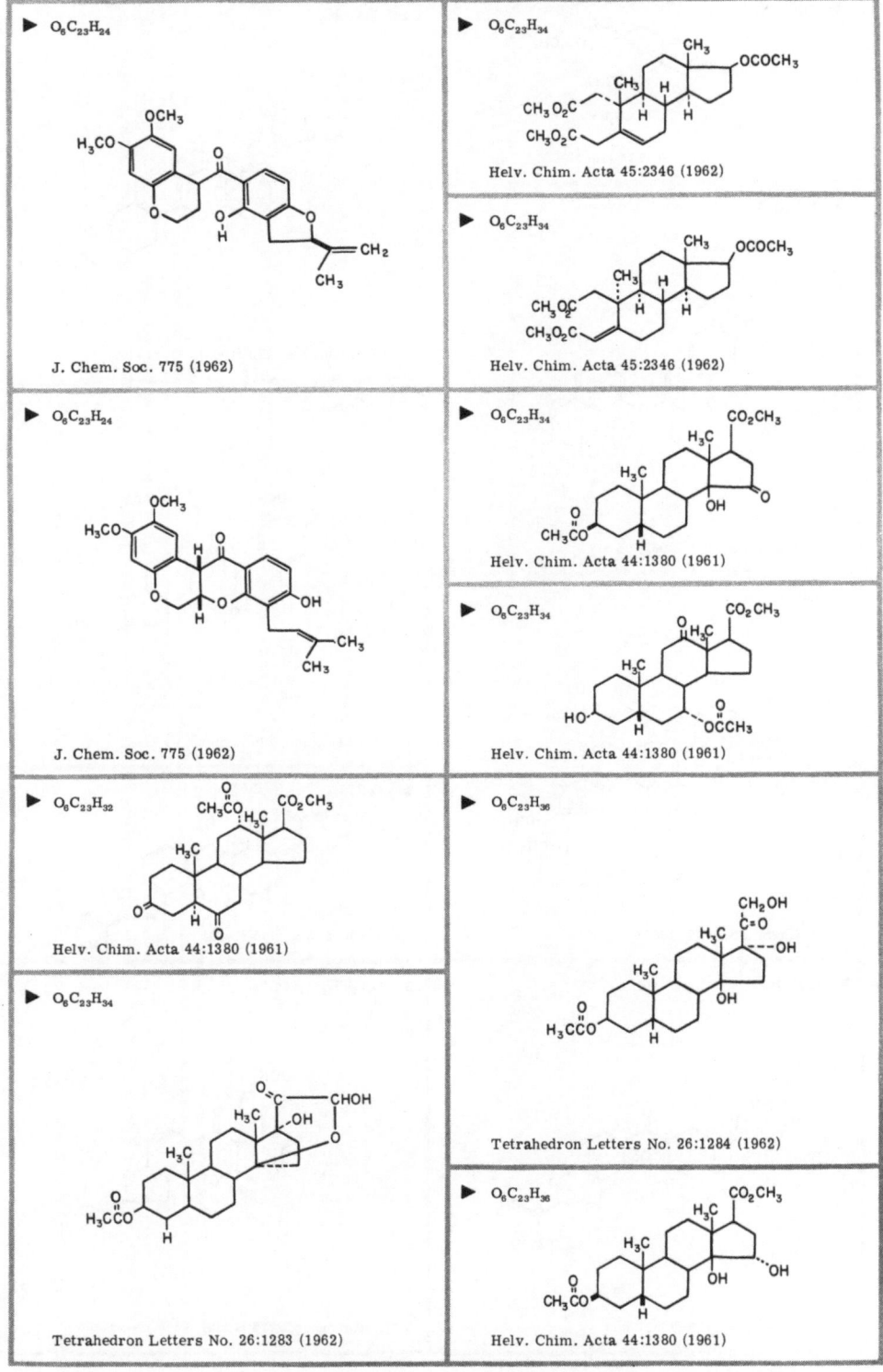

▶ $O_6C_{23}H_{24}$

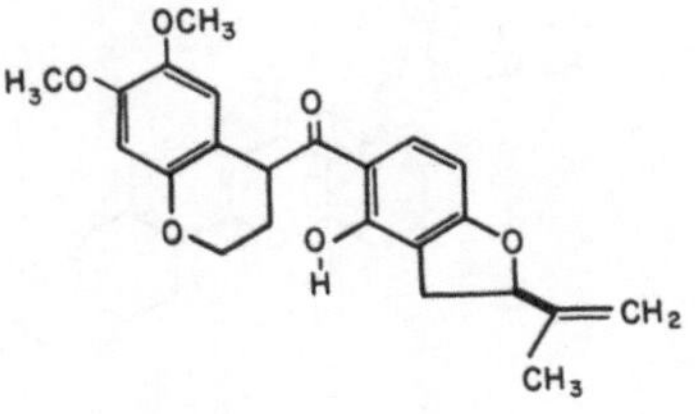


J. Chem. Soc. 775 (1962)

▶ $O_6C_{23}H_{24}$

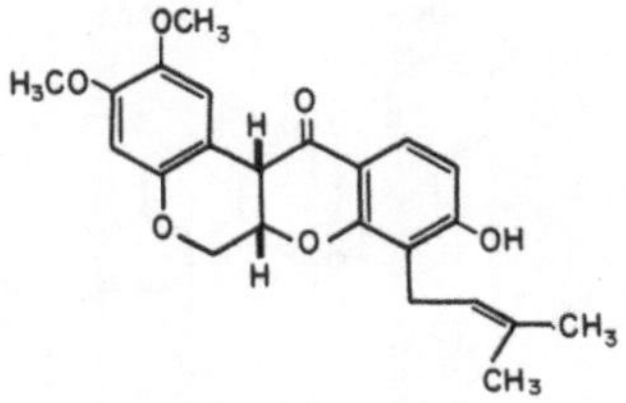


J. Chem. Soc. 775 (1962)

▶ $O_6C_{23}H_{32}$

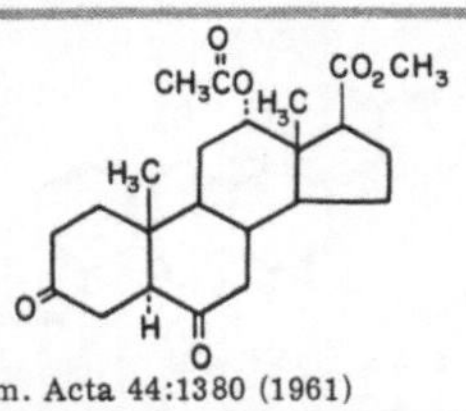


Helv. Chim. Acta 44:1380 (1961)

▶ $O_6C_{23}H_{34}$

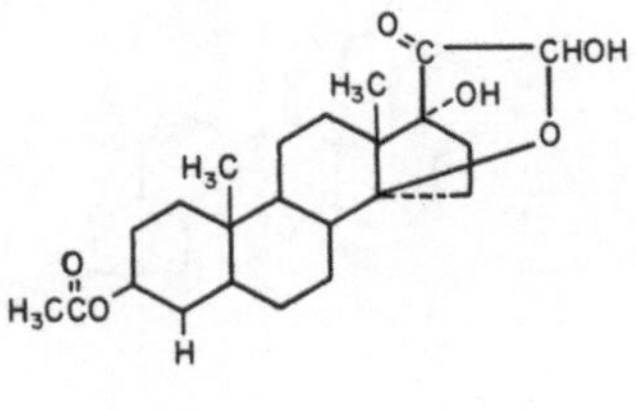


Tetrahedron Letters No. 26:1283 (1962)

▶ $O_6C_{23}H_{34}$

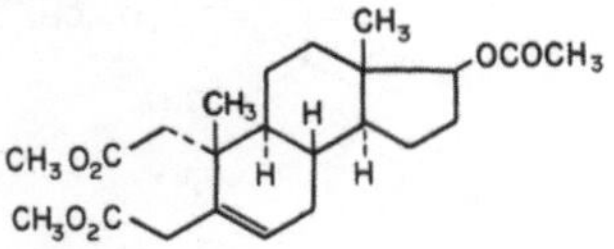


Helv. Chim. Acta 45:2346 (1962)

▶ $O_6C_{23}H_{34}$

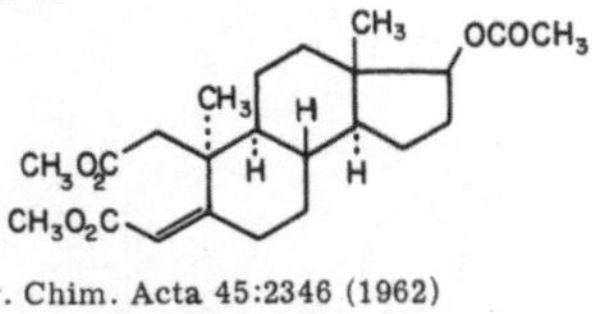


Helv. Chim. Acta 45:2346 (1962)

▶ $O_6C_{23}H_{34}$

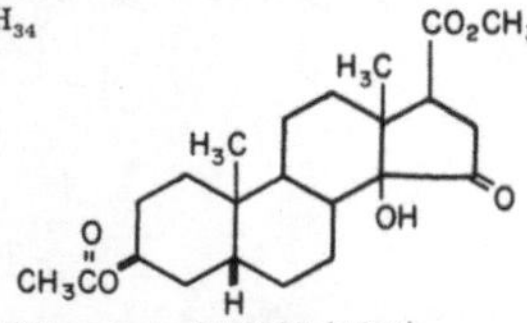


Helv. Chim. Acta 44:1380 (1961)

▶ $O_6C_{23}H_{34}$

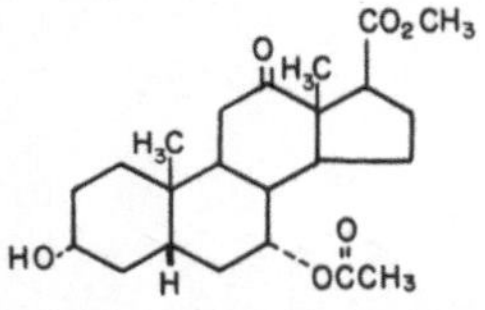


Helv. Chim. Acta 44:1380 (1961)

▶ $O_6C_{23}H_{36}$

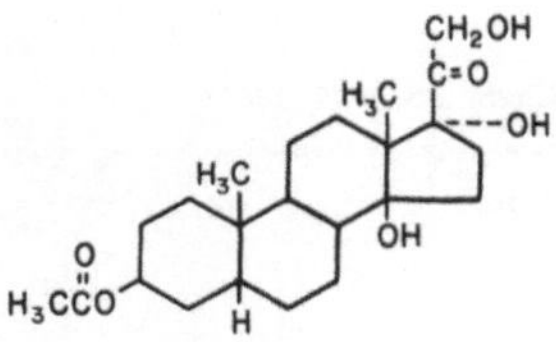


Tetrahedron Letters No. 26:1284 (1962)

▶ $O_6C_{23}H_{36}$

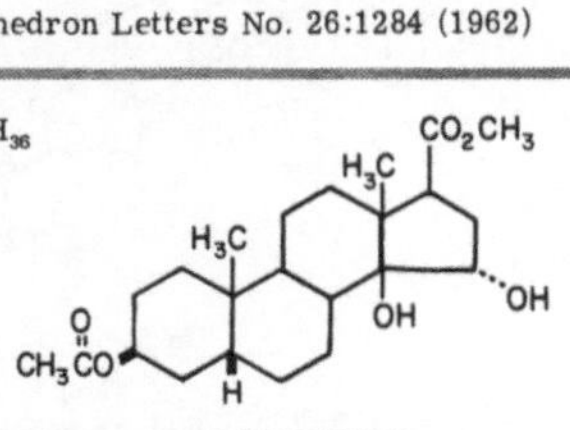


Helv. Chim. Acta 44:1380 (1961)

▶ $O_6C_{24}H_{25}$

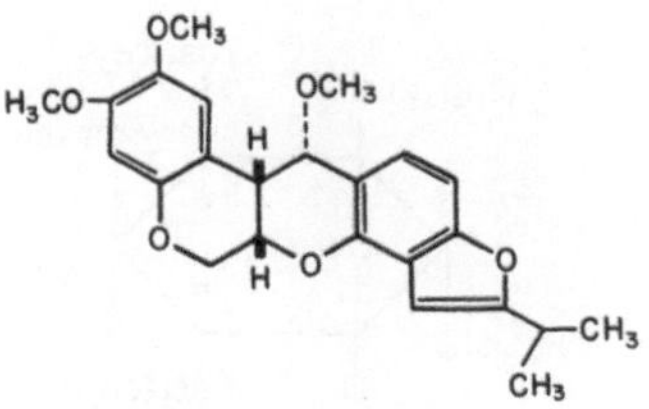

J. Chem. Soc. 775 (1962)

▶ $O_6C_{24}H_{30}$

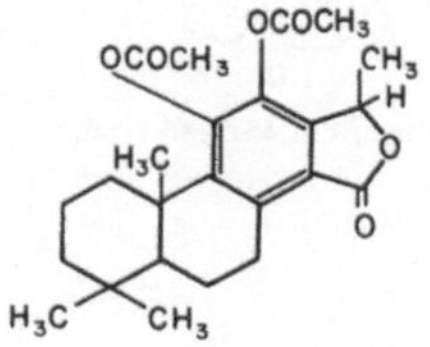

Chem. Ber. 95:3034 (1962)

▶ $O_6C_{24}H_{30}$

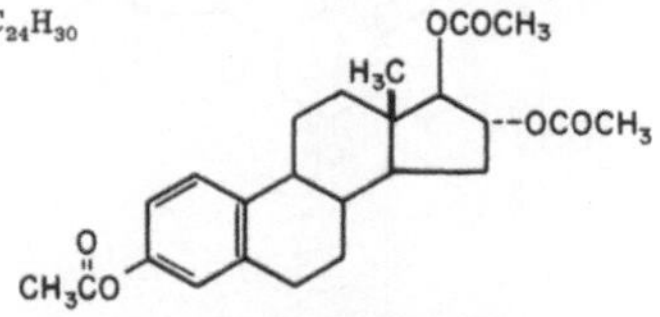

J. Am. Chem. Soc. 84:2972 (1962)

▶ $O_6C_{24}H_{30}$

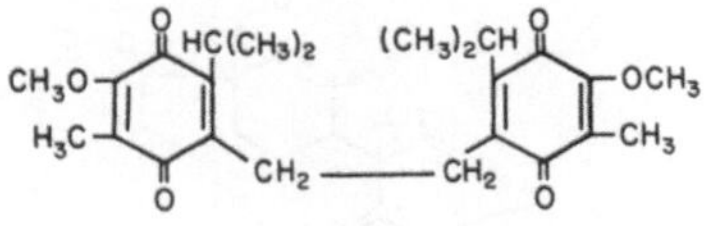

J. Chem. Soc. 5190 (1962)

▶ $O_6C_{24}H_{32}$

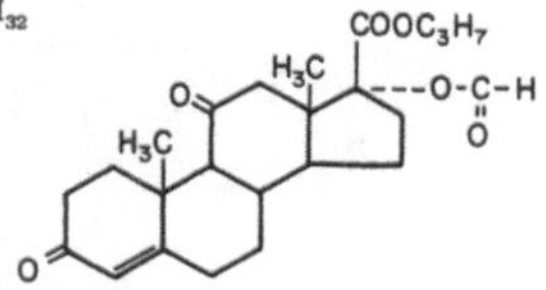

Tetrahedron 16:281 (1961)

▶ $O_6C_{24}H_{32}$

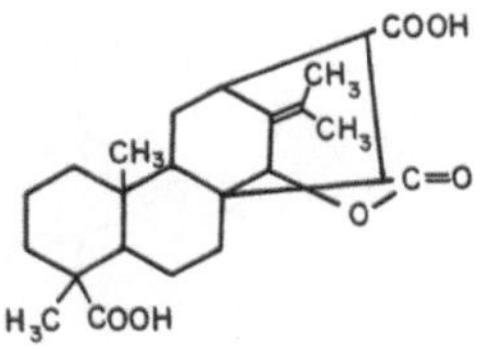

J. Org. Chem. 27:3535 (1962)

▶ $O_6C_{24}H_{32}$

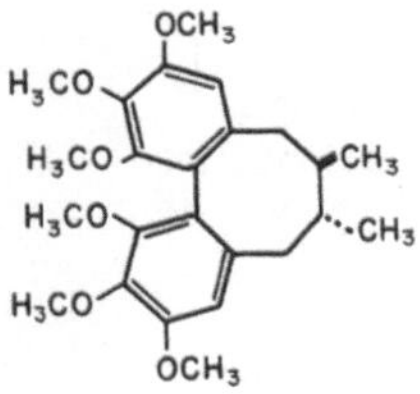

Tetrahedron Letters No. 9:361 (1962)

▶ $O_6C_{24}H_{32}$

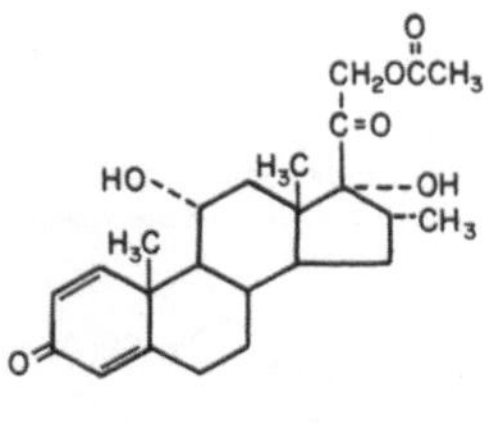

Helv. Chim. Acta 44:1380 (1961)

► $O_6C_{24}H_{34}$

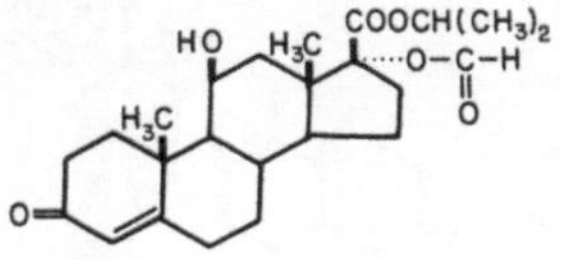

Tetrahedron 16:271 (1961)

► $O_6C_{24}H_{34}$

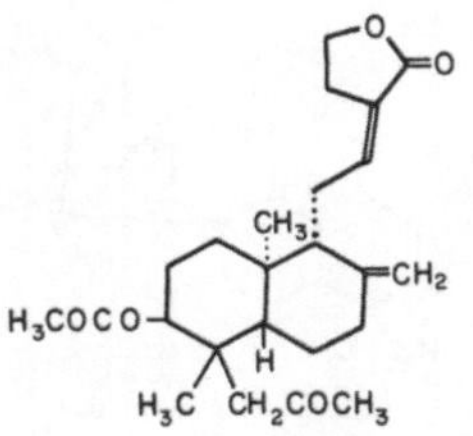

Tetrahedron 18:397 (1962)

► $O_6C_{24}H_{34}$

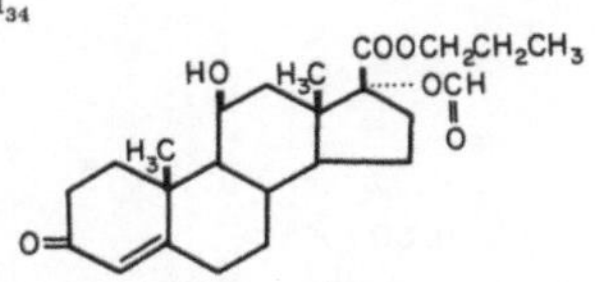

Tetrahedron 16:271 (1961)

► $O_6C_{24}H_{34}$

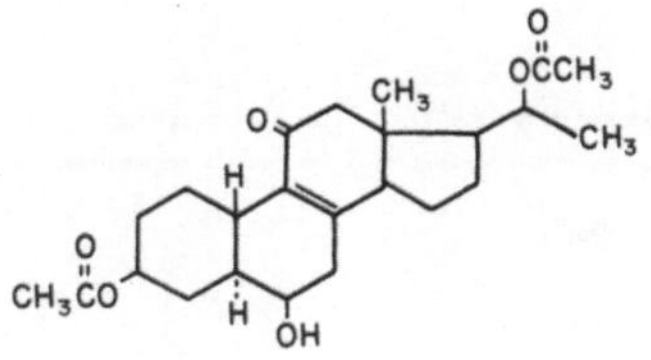

Helv. Chim. Acta 45:2161 (1962)

► $O_6C_{24}H_{36}$

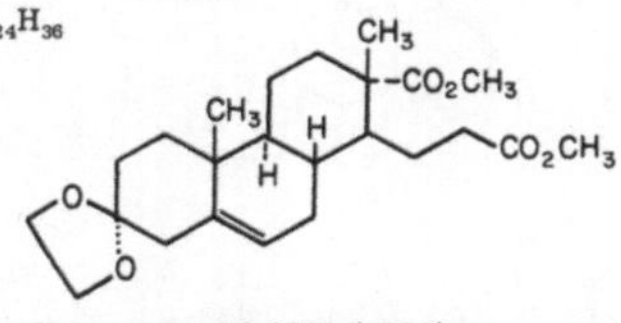

Helv. Chim. Acta 45:2674 (1962)

► $O_6C_{24}H_{60}Si_6$

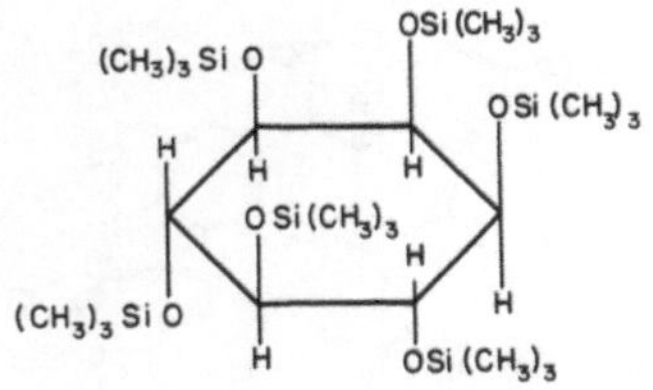

Tetrahedron 18:1147 (1962)

► $O_6C_{25}H_{30}$

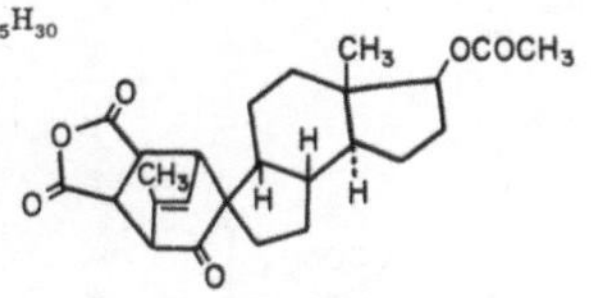

Helv. Chim. Acta 45:2346 (1962)

► $O_6C_{25}H_{32}$

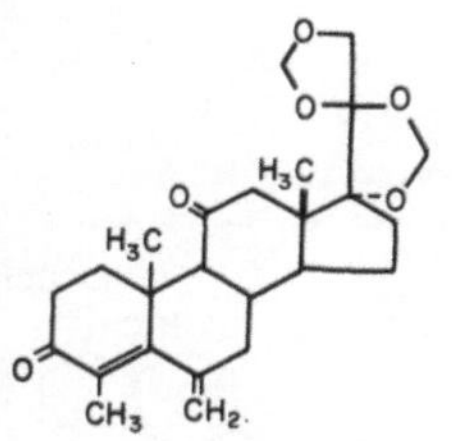

J. Org. Chem. 26:977 (1961)

► $O_6C_{25}H_{34}$

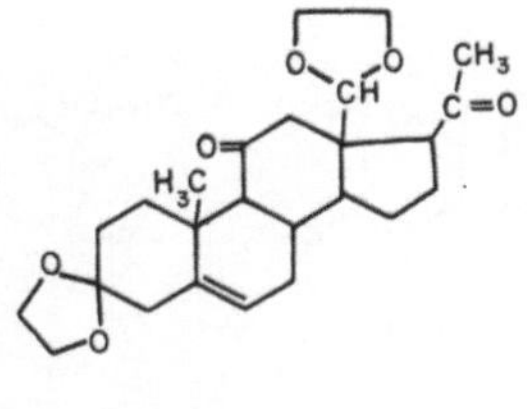

Helv. Chim. Acta 45:331 (1962)

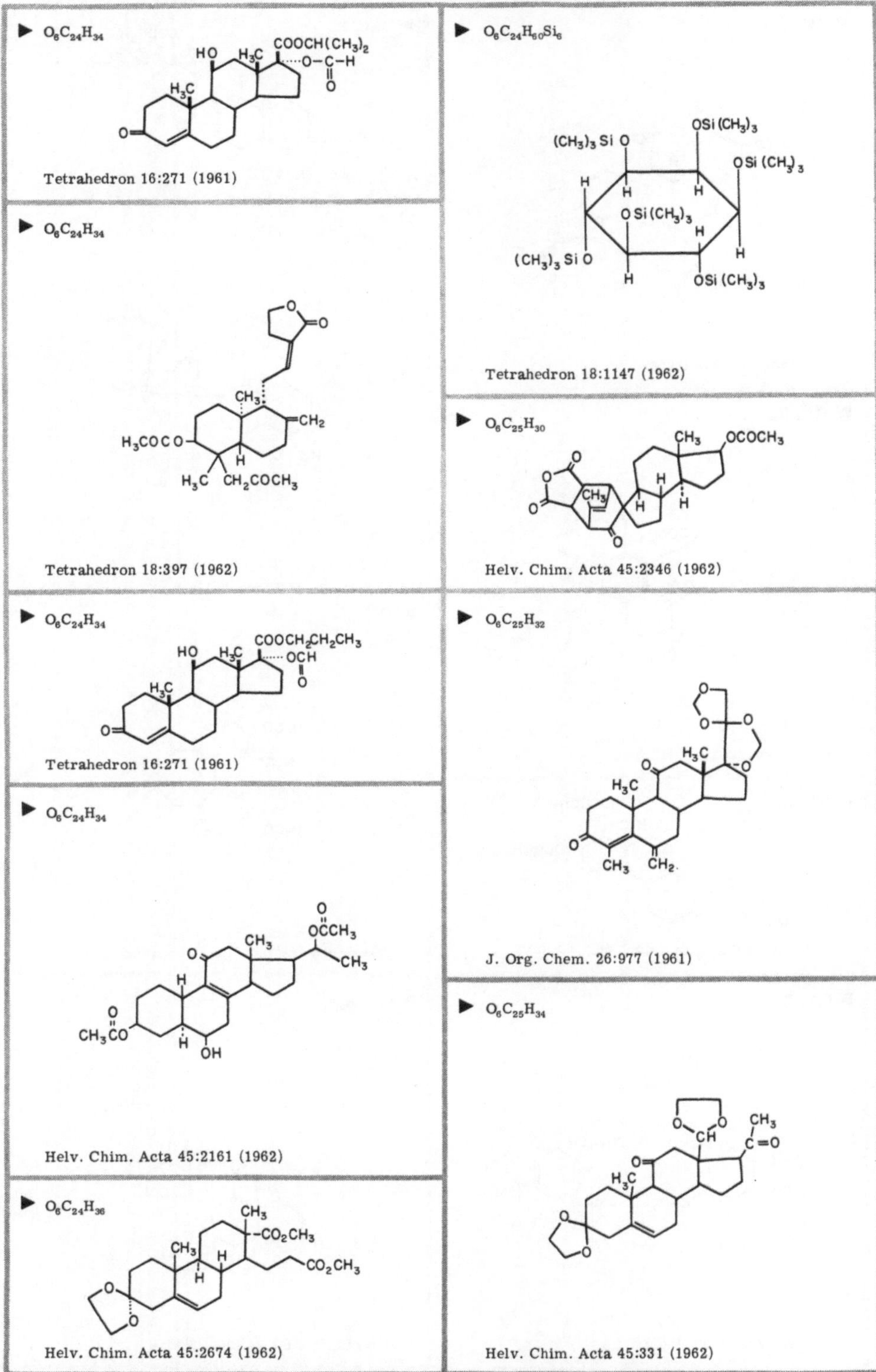

▶ $O_6C_{25}H_{34}$

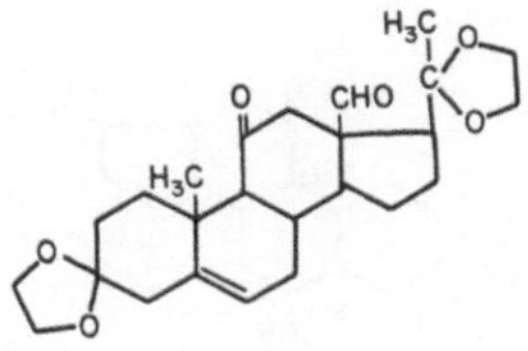

Helv. Chim. Acta 45:331 (1962)

▶ $O_6C_{25}H_{34}$

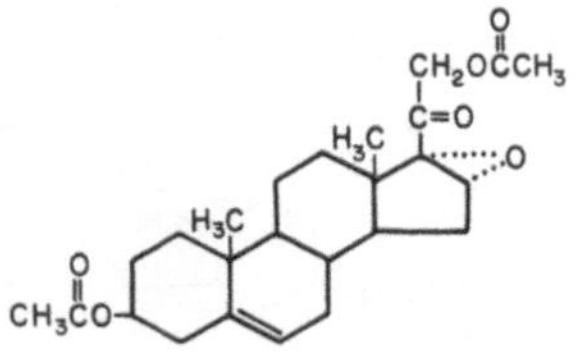

Helv. Chim. Acta 44:1380 (1961)

▶ $O_6C_{25}H_{34}$

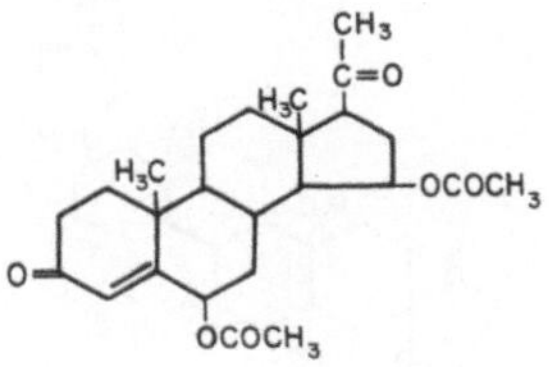

Chem. Pharm. Bull. 10:338 (1962)

▶ $O_6C_{25}H_{34}$

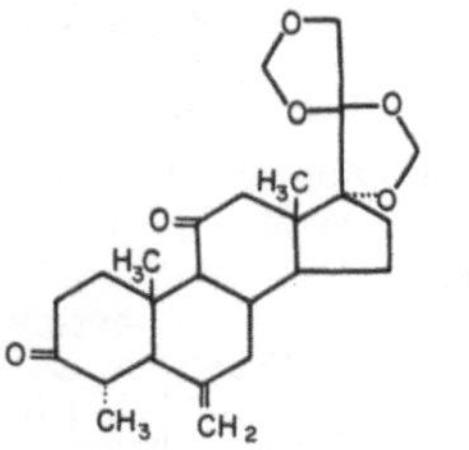

J. Org. Chem. 26:977 (1961)

▶ $O_6C_{25}H_{34}$

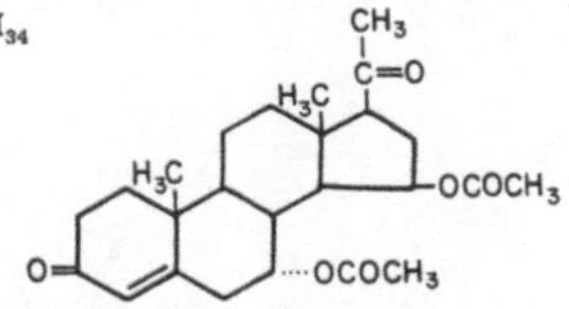

Chem. Pharm. Bull. 10:338 (1962)

▶ $O_6C_{25}H_{36}$

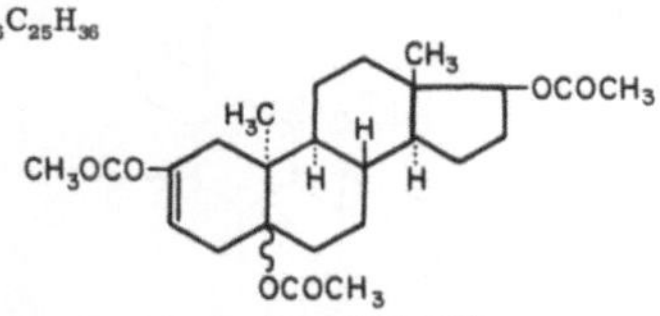

Helv. Chim. Acta 45:2420 (1962)

▶ $O_6C_{25}H_{34}$

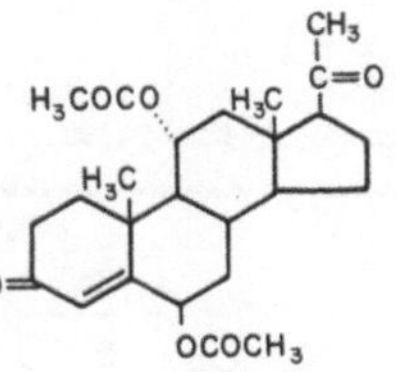

Chem. Pharm. Bull. 10:338 (1962)

▶ $O_6C_{25}H_{36}$

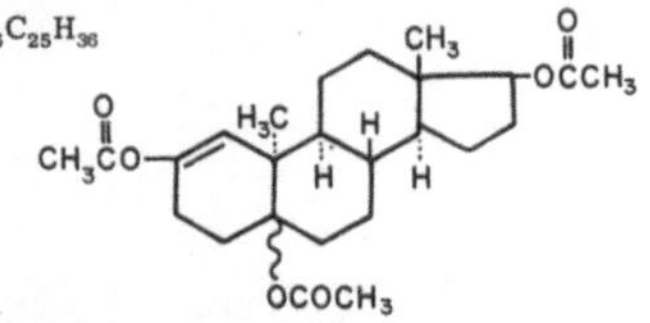

Helv. Chim. Acta 45:2420 (1962)

▶ $O_6C_{25}H_{36}$

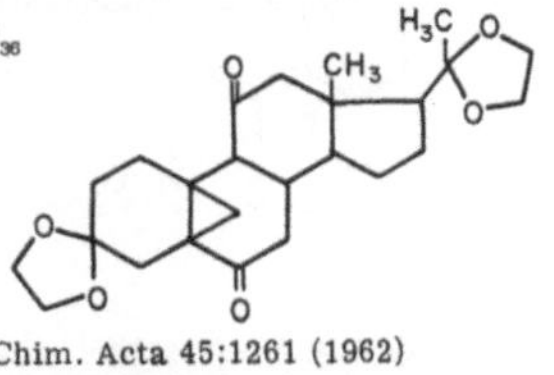

Helv. Chim. Acta 45:1261 (1962)

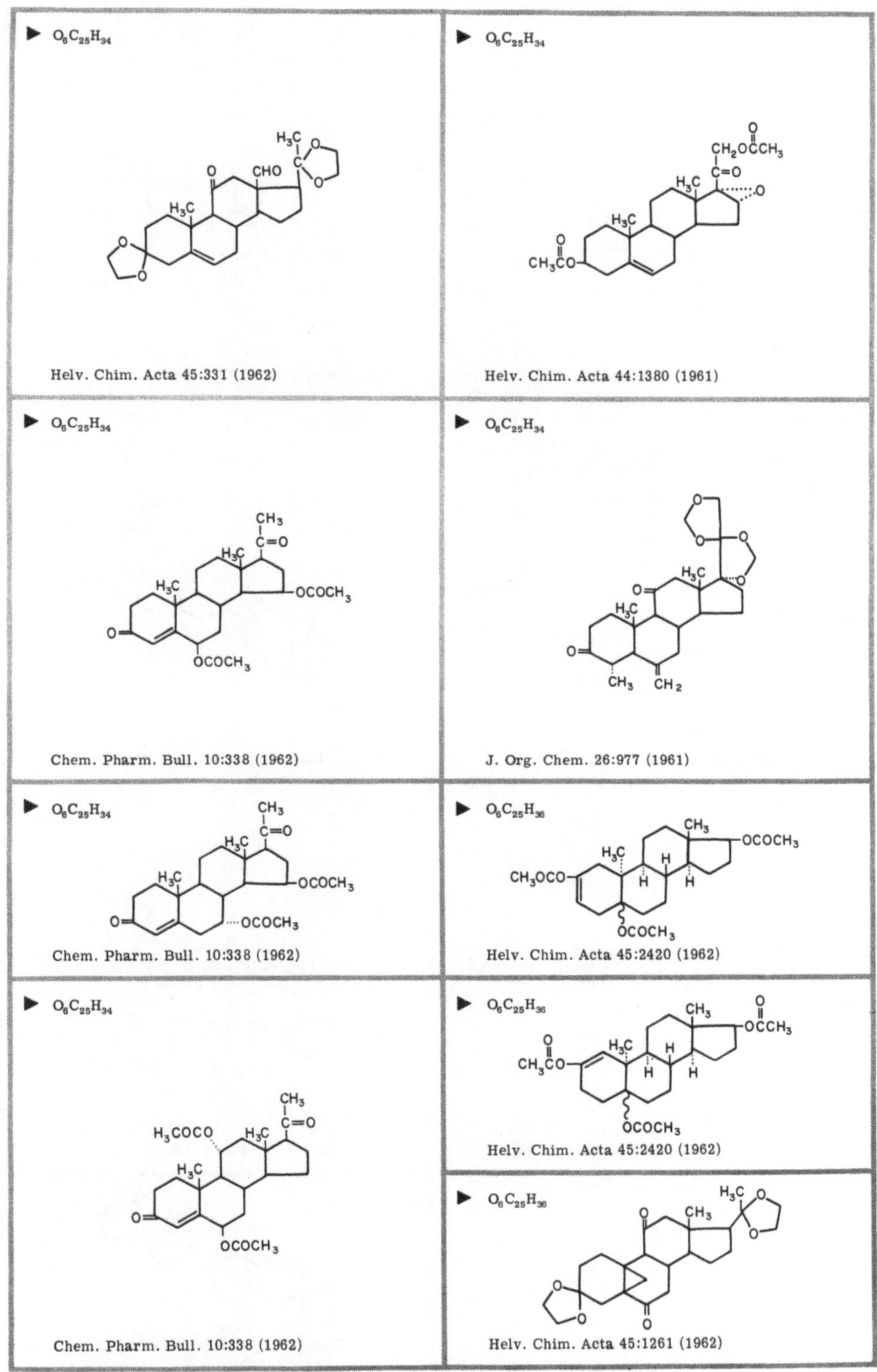

▶ $O_6C_{25}H_{36}$

5α,6α-EPOXYPREGNAN-3,20-DIONE-21-OL 3-ETHYLENE KETAL 21-ACETATE

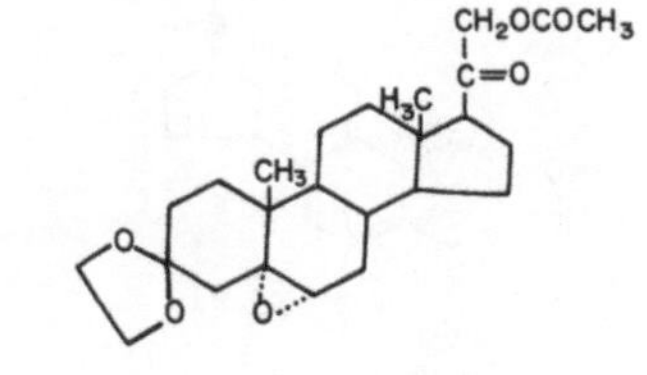

J. Am. Chem. Soc. 84:3206 (1962)

▶ $O_6C_{25}H_{36}$

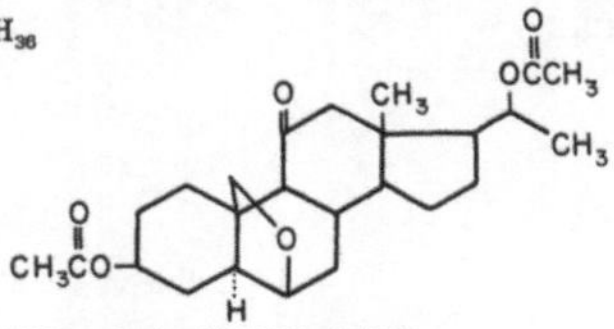

Helv. Chim. Acta 45:2161 (1962)

▶ $O_6C_{25}H_{36}$

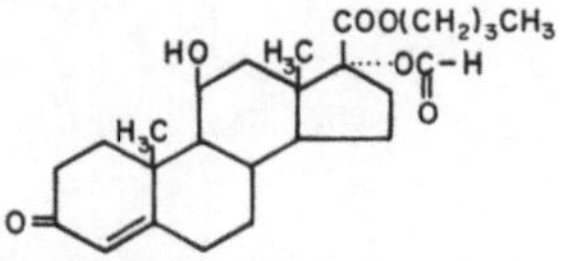

Tetrahedron 16:271 (1961)

▶ $O_6C_{25}H_{36}$

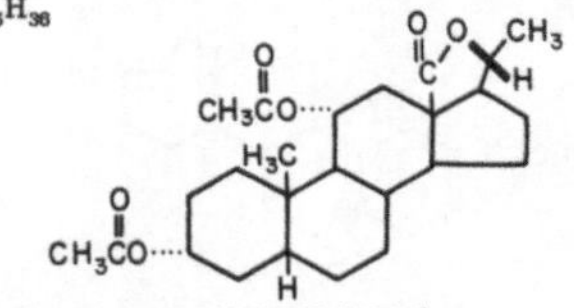

Helv. Chim. Acta 44:1380 (1961)

▶ $O_6C_{25}H_{36}$

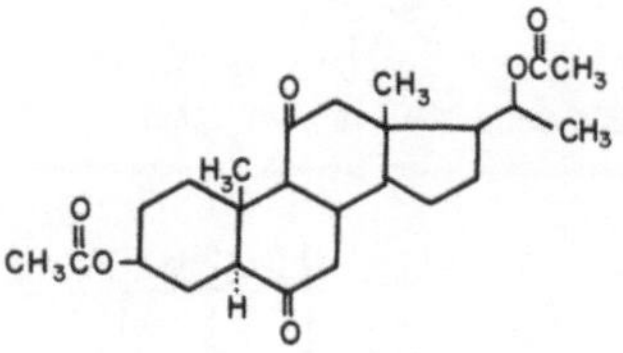

Helv. Chim. Acta 45:2161 (1962)

▶ $O_6C_{25}H_{38}$

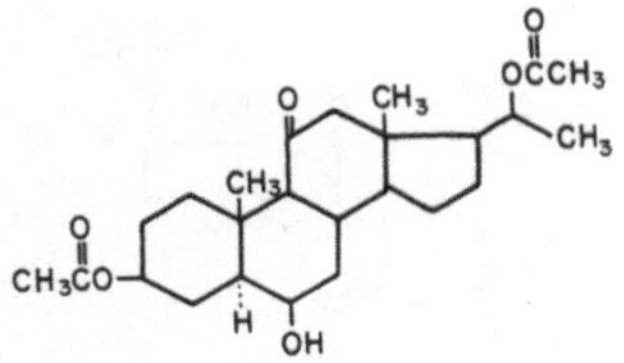

Helv. Chim. Acta 45:2161 (1962)

▶ $O_6C_{25}H_{38}$

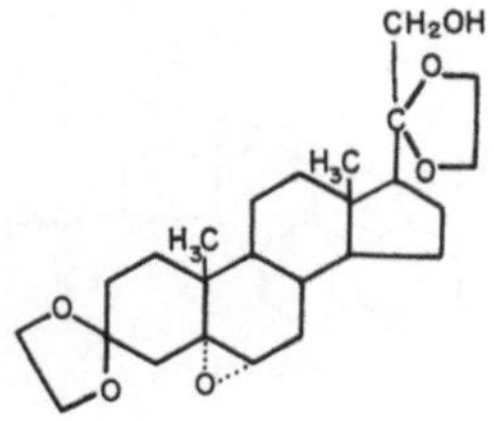

J. Am. Chem. Soc. 84:3206 (1962)

▶ $O_6C_{25}H_{38}$

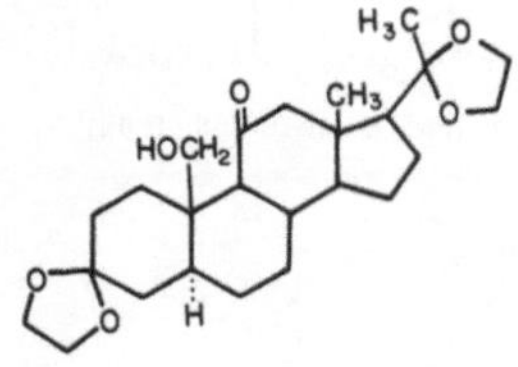

Helv. Chim. Acta 44:2162 (1961)

▶ $O_6C_{25}H_{38}$

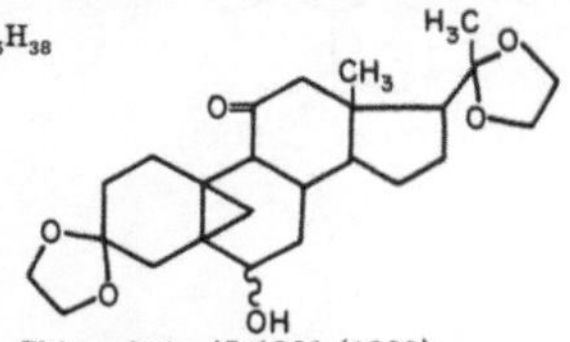

Helv. Chim. Acta 45:1261 (1962)

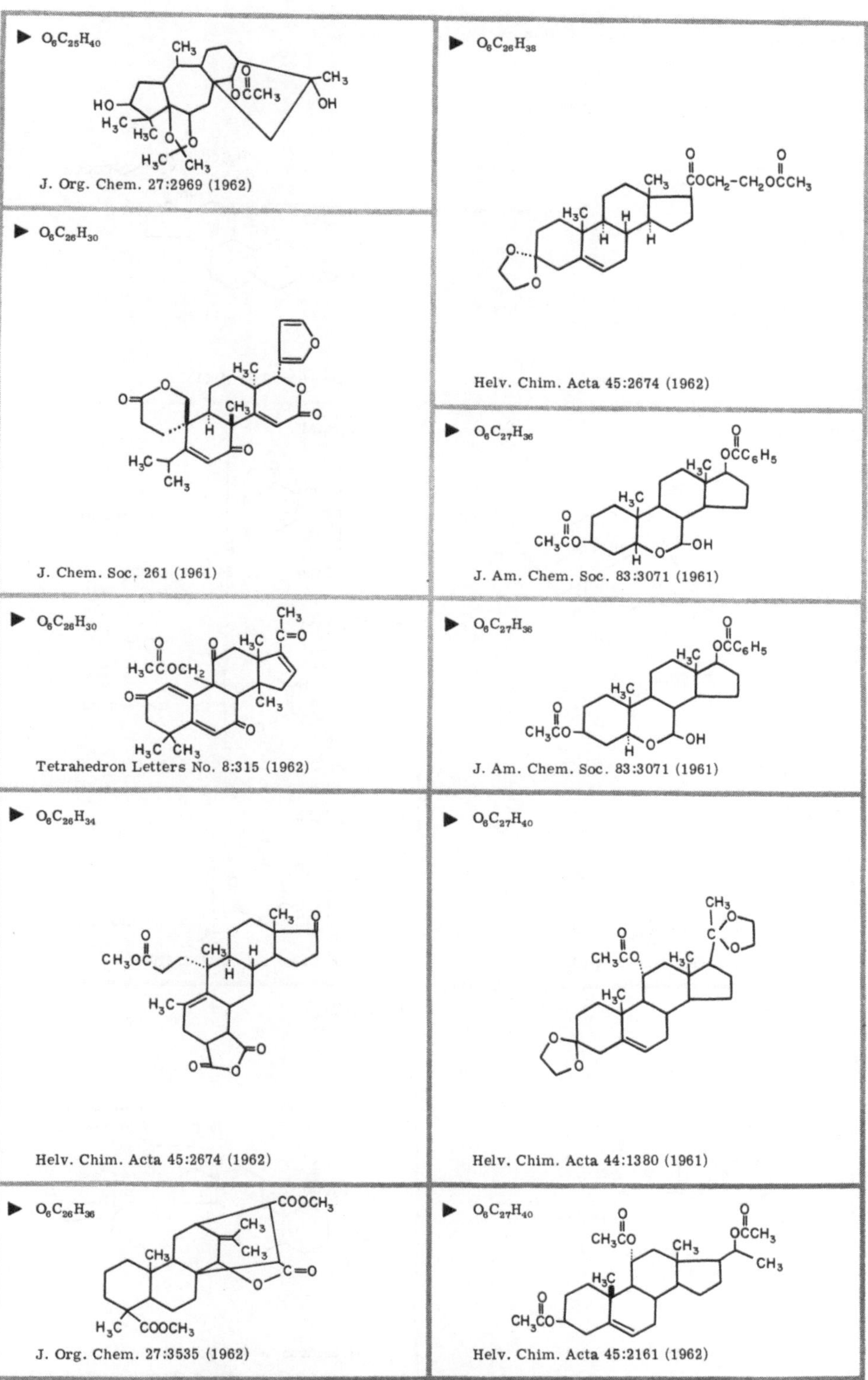

▶ $O_6C_{25}H_{40}$

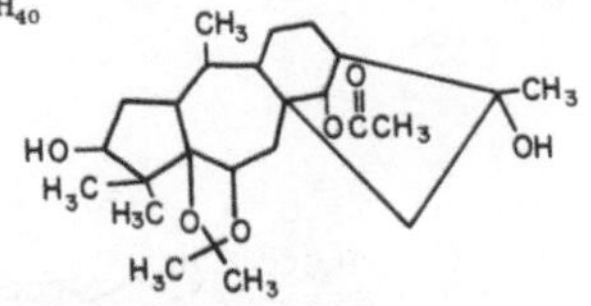

J. Org. Chem. 27:2969 (1962)

▶ $O_6C_{26}H_{30}$

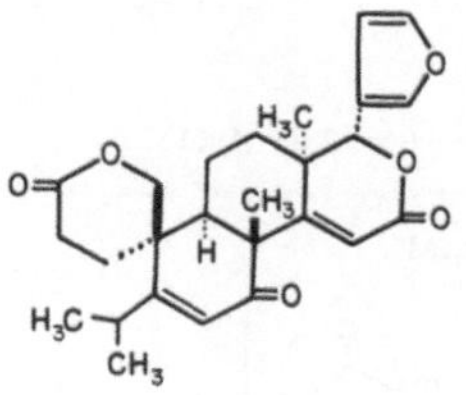

J. Chem. Soc. 261 (1961)

▶ $O_6C_{26}H_{30}$

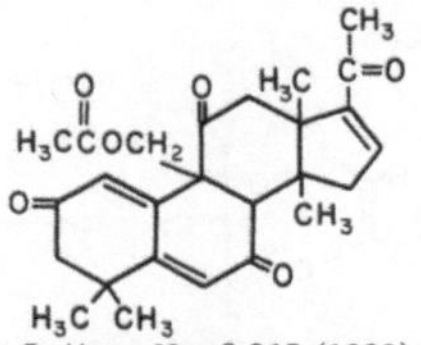

Tetrahedron Letters No. 8:315 (1962)

▶ $O_6C_{26}H_{34}$

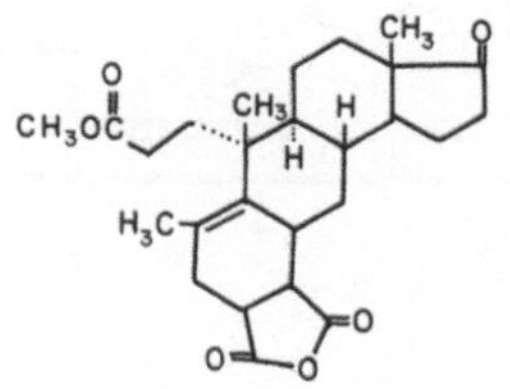

Helv. Chim. Acta 45:2674 (1962)

▶ $O_6C_{26}H_{36}$

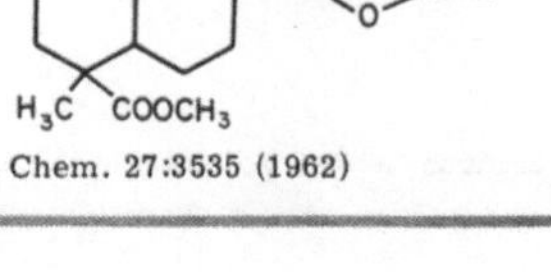

J. Org. Chem. 27:3535 (1962)

▶ $O_6C_{26}H_{38}$

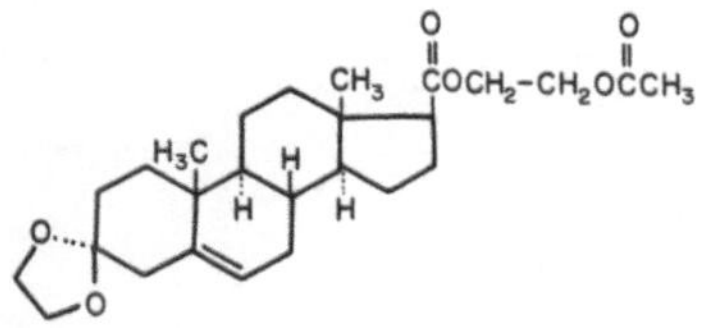

Helv. Chim. Acta 45:2674 (1962)

▶ $O_6C_{27}H_{36}$

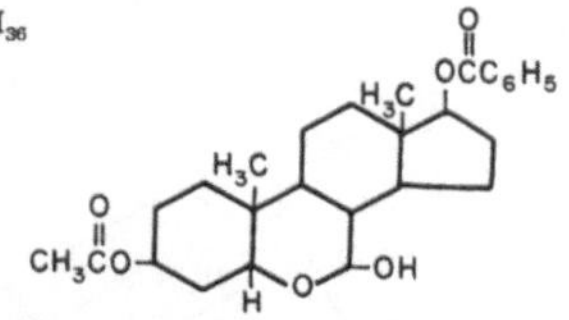

J. Am. Chem. Soc. 83:3071 (1961)

▶ $O_6C_{27}H_{36}$

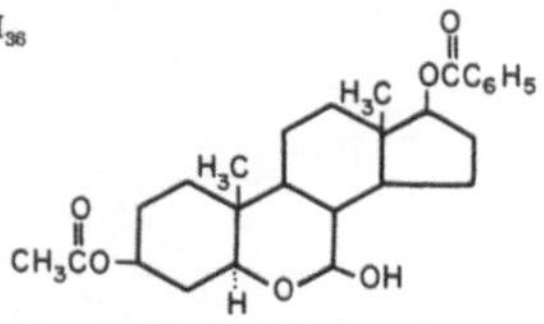

J. Am. Chem. Soc. 83:3071 (1961)

▶ $O_6C_{27}H_{40}$

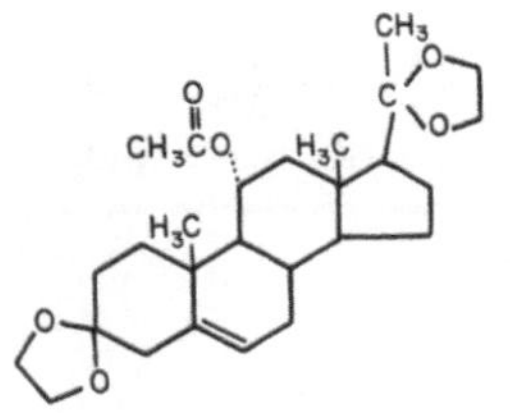

Helv. Chim. Acta 44:1380 (1961)

▶ $O_6C_{27}H_{40}$

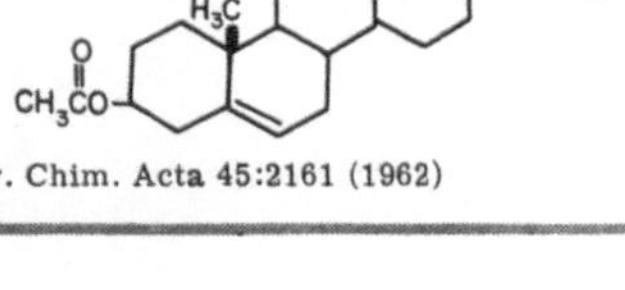

Helv. Chim. Acta 45:2161 (1962)

▶ $O_6C_{27}H_{40}$

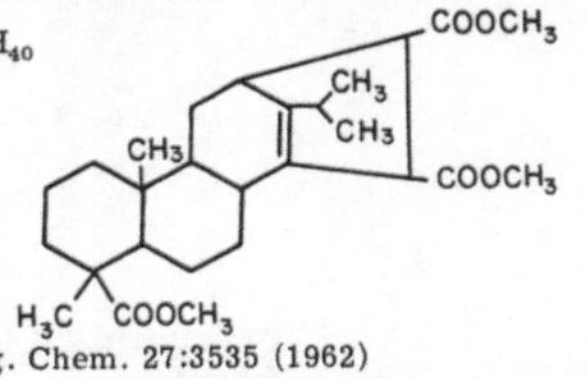


J. Org. Chem. 27:3535 (1962)

▶ $O_6C_{27}H_{42}$

UNKNOWN COMPOUND 'B' PRODUCED BY UV.-IRRADIATION OF 3,20-DIETHYLENE-DIOXY-11-KETO-$\Delta^5$-PREGNENE

Helv. Chim. Acta 45:1261 (1962)

▶ $O_6C_{27}H_{42}$

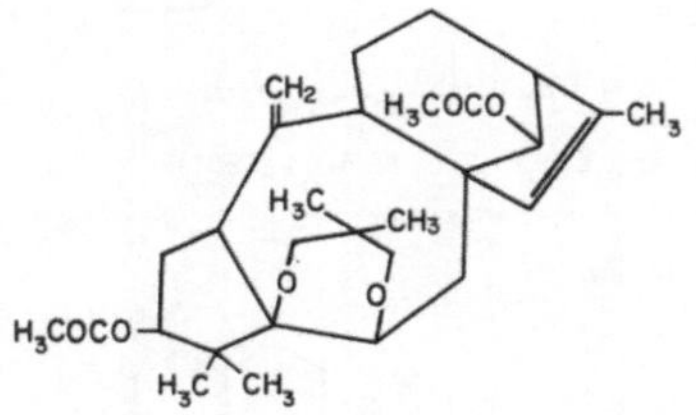


Tetrahedron Letters No. 2:61 (1961)

▶ $O_6C_{28}H_{38}$

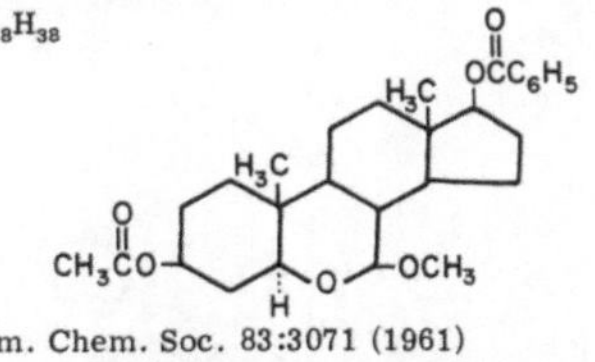


J. Am. Chem. Soc. 83:3071 (1961)

▶ $O_6C_{29}H_{22}$

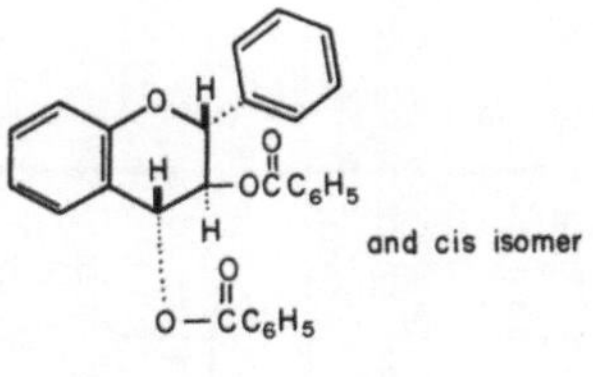


Tetrahedron Letters 431 (1961)

▶ $O_6C_{29}H_{46}$

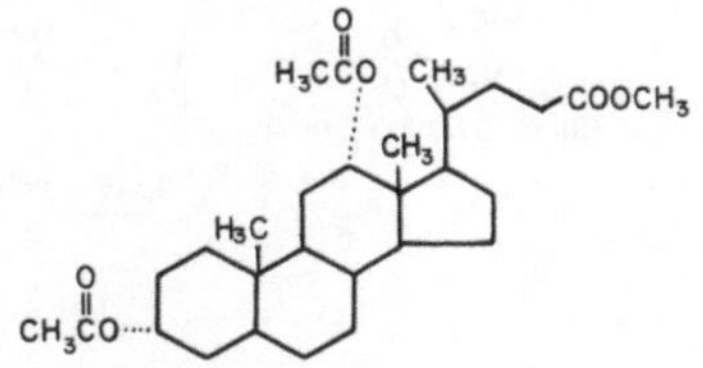


J. Chem. Soc. 2817 (1961)

▶ $O_6C_{30}H_{27}Al$

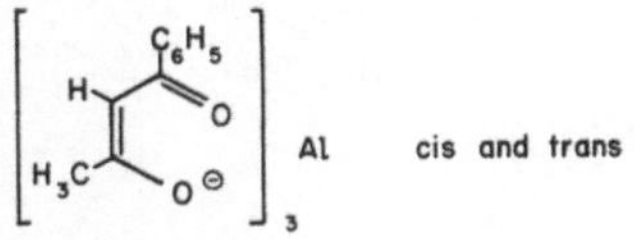


J. Am. Chem. Soc. 84:2303 (1962)

▶ $O_6C_{30}H_{27}Co$

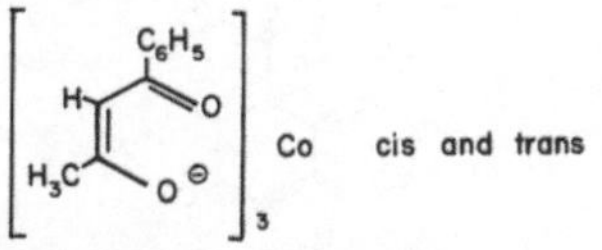


J. Am. Chem. Soc. 84:2303 (1962)

▶ $O_6C_{30}H_{27}Rh$

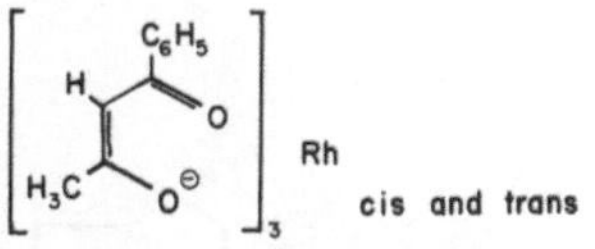


J. Am. Chem. Soc. 84:2303 (1962)

▶ $O_6C_{31}H_{46}$

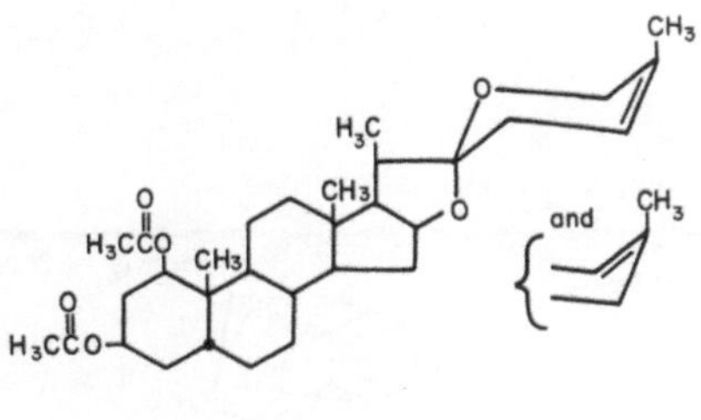


Tetrahedron Letters No. 24:1109 (1962)

► $O_6C_{32}H_{42}$

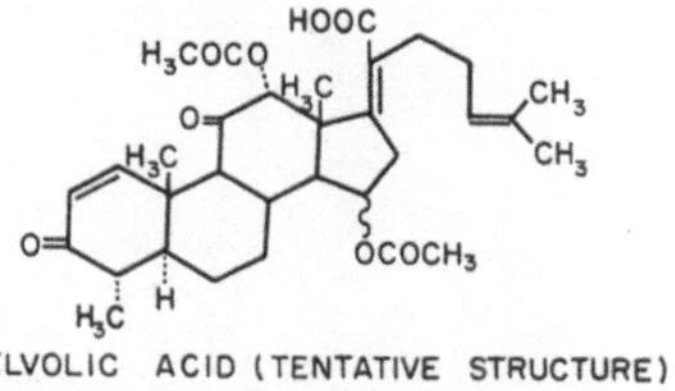

HELVOLIC ACID (TENTATIVE STRUCTURE)

J. Org. Chem. 26:4529 (1961)

► $O_6C_{33}H_{44}$

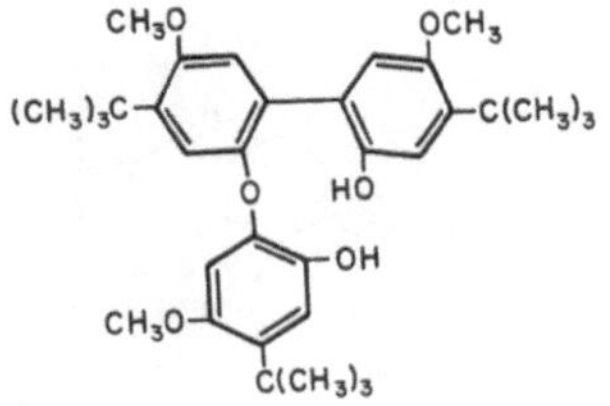

J. Chem. Soc. 4989 (1962)

► $O_6C_{32}H_{50}$

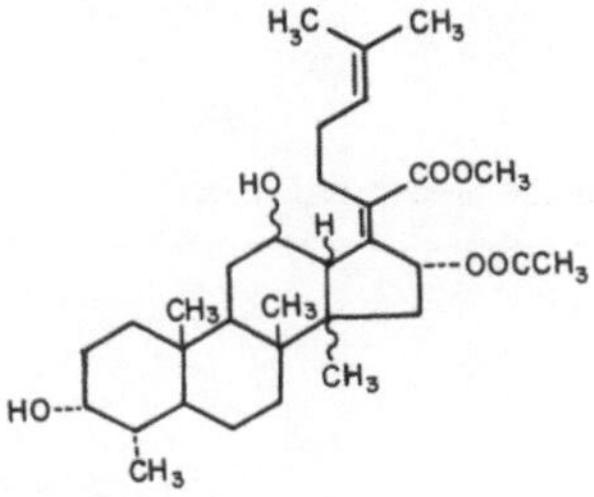

Tetrahedron 18:1035 (1962)

► $O_6C_{33}H_{48}$

METHYL TETRAHYDROHELVOLATE (TENTATIVE STRUCTURE)

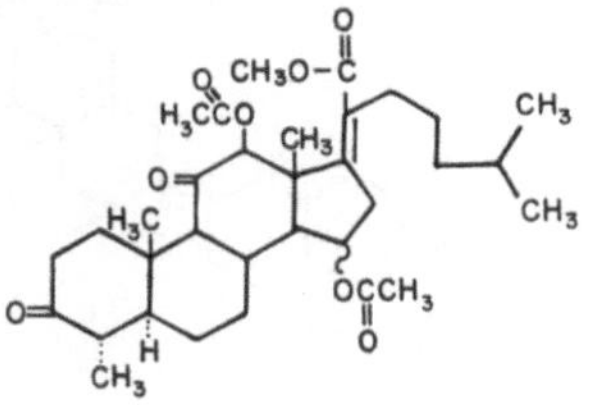

J. Org. Chem. 26:4529 (1961)

► $O_6C_{33}H_{42}$

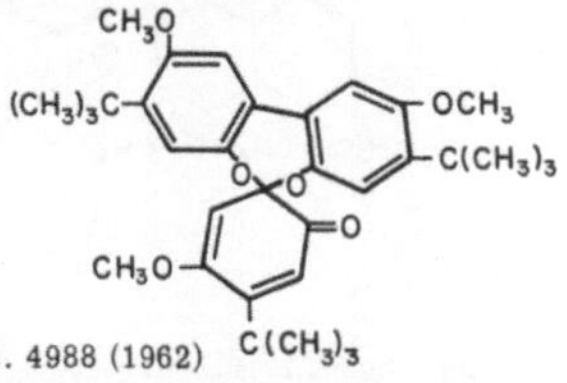

J. Chem. Soc. 4988 (1962)

► $O_6C_{33}H_{44}$

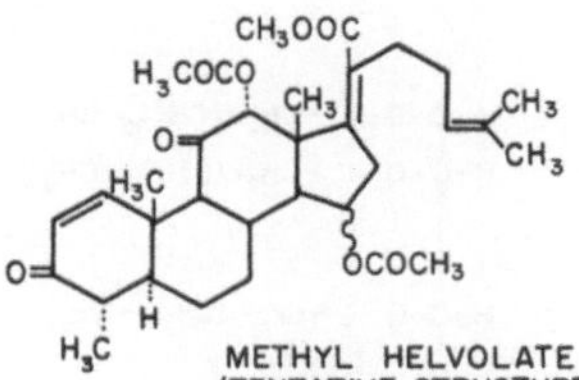

METHYL HELVOLATE (TENTATIVE STRUCTURE)

J. Org. Chem. 26:4529 (1961)

► $O_6C_{33}H_{50}$

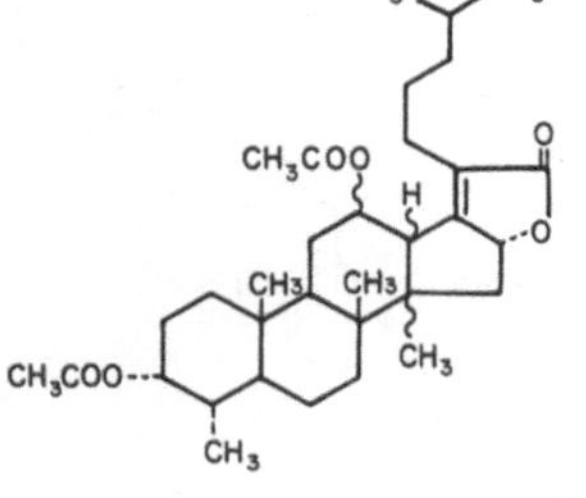

Tetrahedron 18:1047 (1962)

▶ $O_6C_{33}H_{52}$

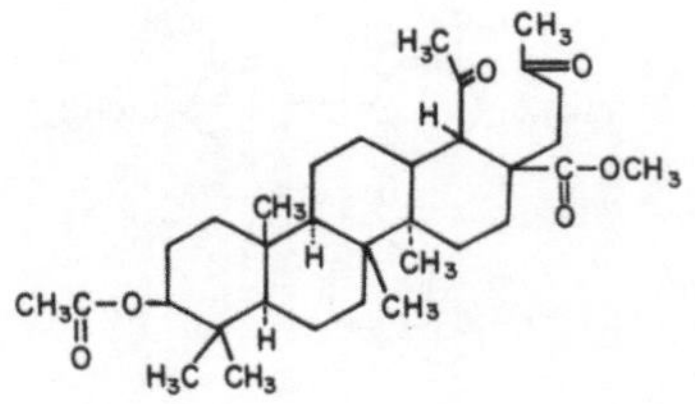


J. Chem. Soc. 5173 (1962)

▶ $O_6C_{34}H_{50}$

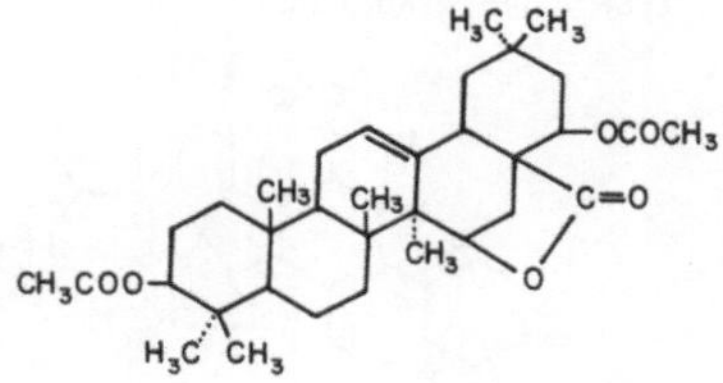


J. Org. Chem. 27:4512 (1962)

▶ $O_6C_{35}H_{48}$

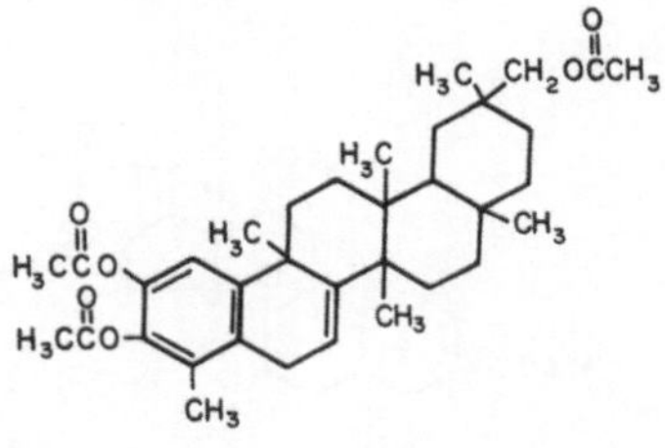


Tetrahedron Letters No. 14:606 (1962)

▶ $O_6C_{36}H_{56}$

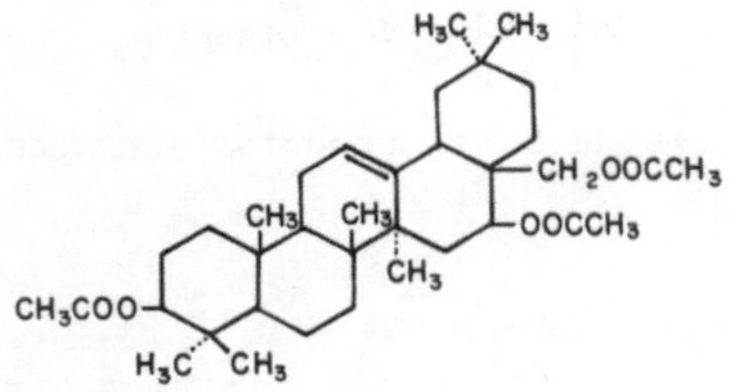


J. Org. Chem. 27:4512 (1962)

▶ $O_6C_{36}H_{58}$

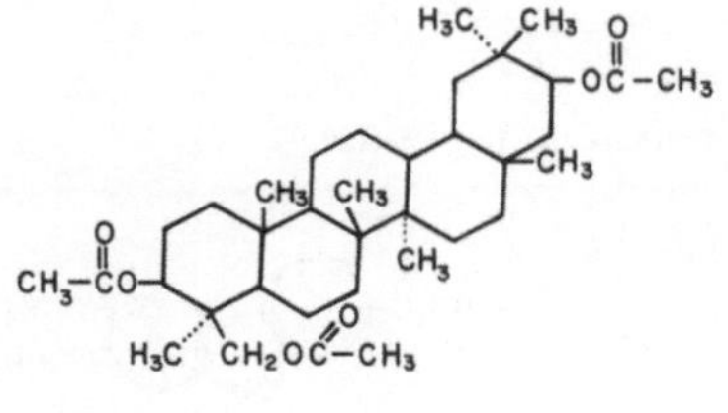


J. Org. Chem. 27:4512 (1962)

▶ $O_6C_{57}H_{110}$

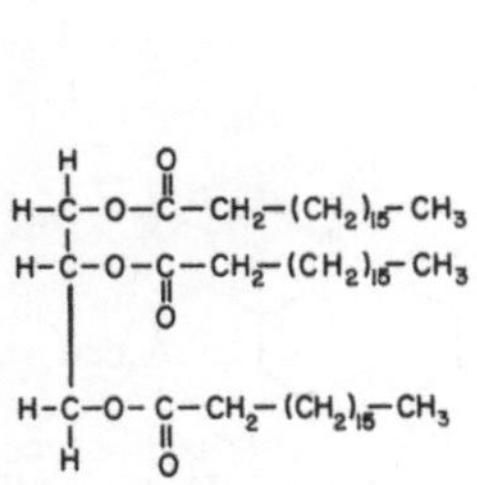


Varian 368

▶ $O_7C_8H_{14}$

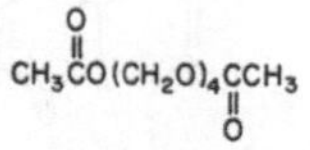

Chem. Ber. 94:3317 (1961)

▶ $O_7C_{11}H_{18}$

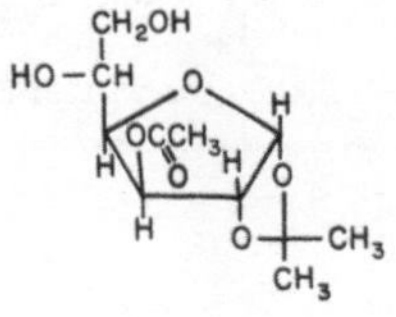

J. Chem. Soc. 3699 (1962)
Chem. & Ind. 213 (1962)

▶ $O_7C_{12}H_{22}$

$CH_2=CHCH_2O-(CH_2O)_6-CH_2CH=CH_2$

J. Chem. Soc. 4313 (1962)

▶ $O_7C_{13}H_{16}$

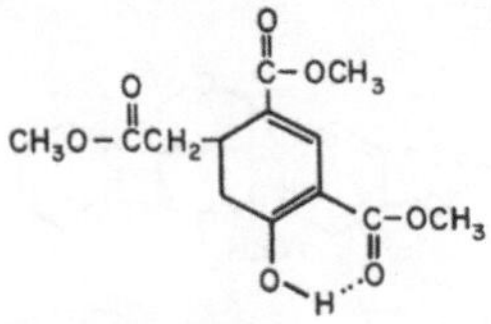

J. Chem. Soc. 4720 (1962)

▶ $O_7C_{16}H_{22}$

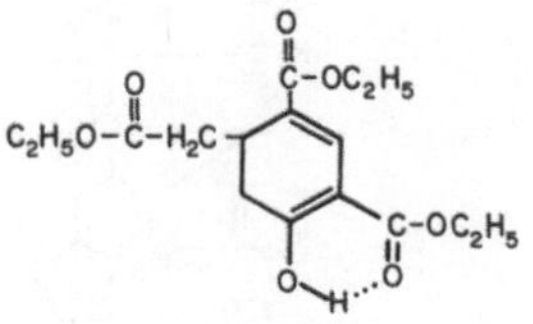

J. Chem. Soc. 4721 (1962)

▶ $O_7C_{17}H_{20}$

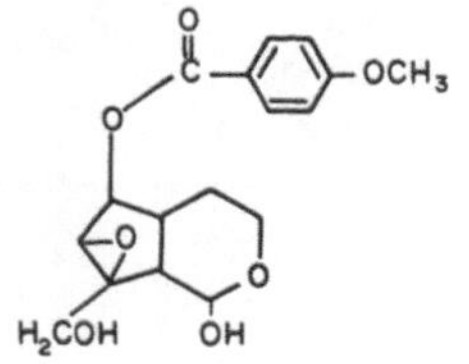

Tetrahedron Letters No. 8:327 (1962)

▶ $O_7C_{17}H_{20}$

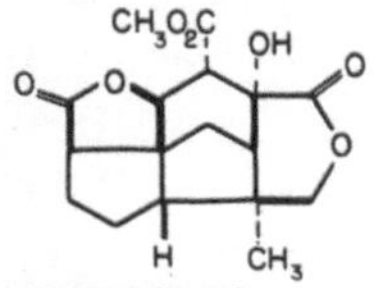

Tetrahedron 18:1321 (1962)

▶ $O_7C_{17}H_{22}$

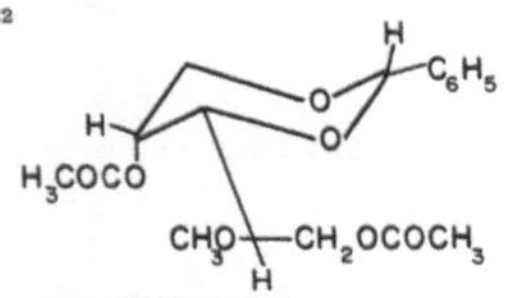

J. Chem. Soc. 5005 (1961)

▶ $O_7C_{18}H_{16}$

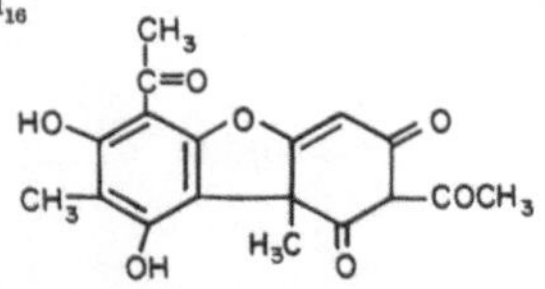

Acta Chem. Scand. 16:583 (1962)

▶ $O_7C_{18}H_{16}$

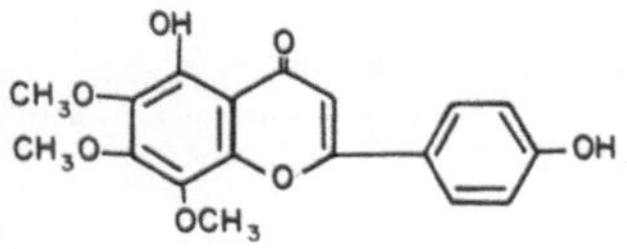

Tetrahedron 14:296 (1961)

▶ $O_7C_{19}H_{18}$

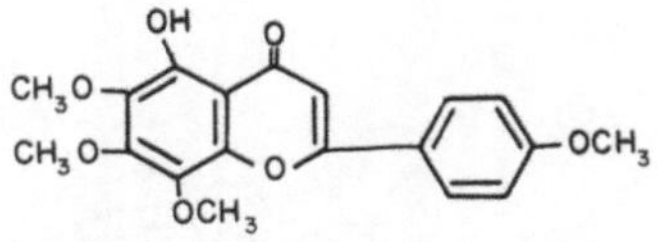


Tetrahedron 14:300 (1961)

▶ $O_7C_{19}H_{22}$

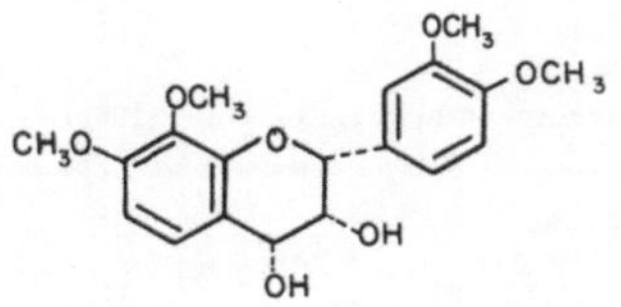


J. Chem. Soc. 4503 (1962)

▶ $O_7C_{19}H_{22}$

SAMADERINE B

Bull. Soc. Chim. France 1715 (1962)

▶ $O_7C_{19}H_{24}$

SAMADERINE C

Bull. Soc. Chim. France 1715 (1962)

▶ $O_7C_{20}H_{20}$

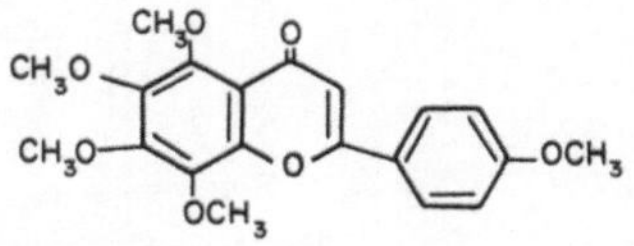


Tetrahedron 14:300 (1961)

▶ $O_7C_{20}H_{20}$

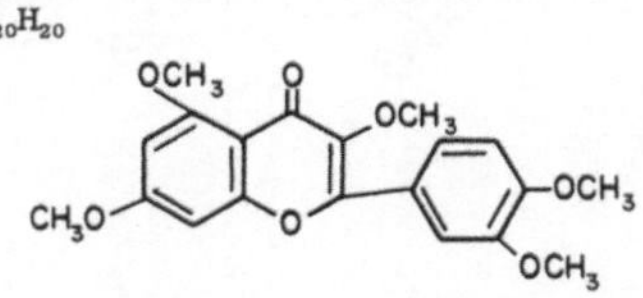


Bull. Soc. Chim. France 1962 (1962)

▶ $O_7C_{20}H_{24}$

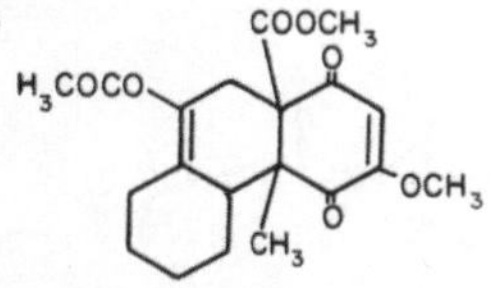


J. Chem. Soc. 2908 (1961)

▶ $O_7C_{20}H_{28}$

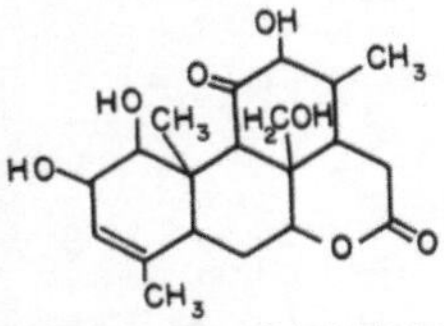


Tetrahedron Letters No. 23:1091 (1962)

▶ $O_7C_{21}H_{16}$

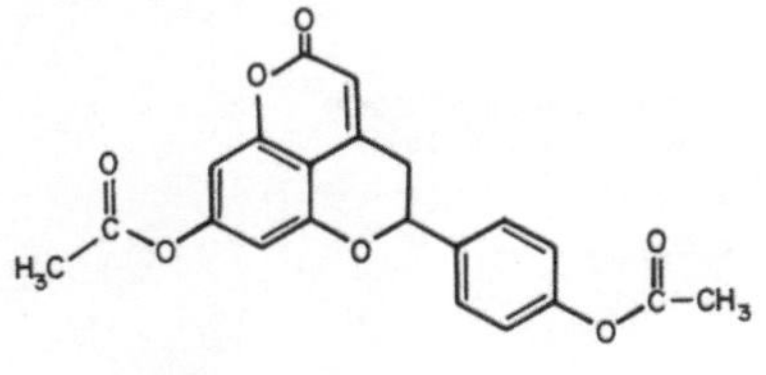


Varian 686

▶ $O_7C_{21}H_{20}$

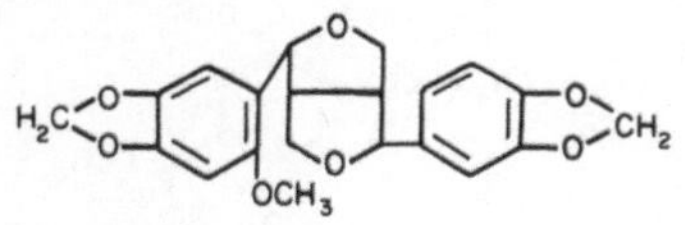


J. Org. Chem. 27:3234 (1962)

▶ $O_7C_{22}H_{22}$

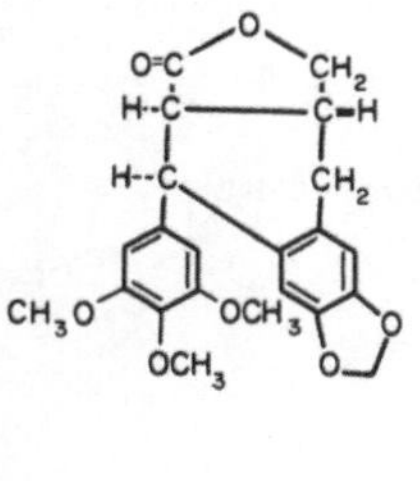


Bull. Soc. Chim. France 1722 (1962)

▶ $O_7C_{22}H_{24}$

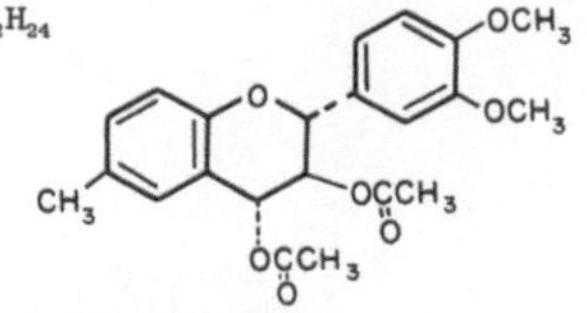

J. Chem. Soc. 3861 (1962)

▶ $O_7C_{22}H_{24}$

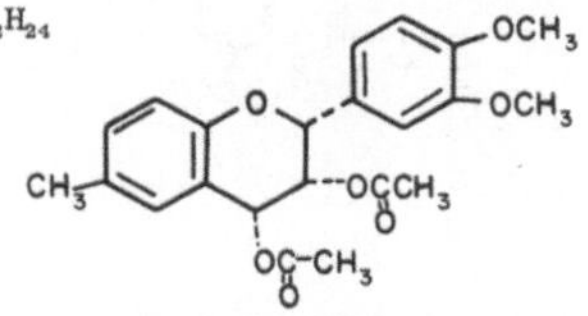

J. Chem. Soc. 3861 (1962)

▶ $O_7C_{22}H_{24}$

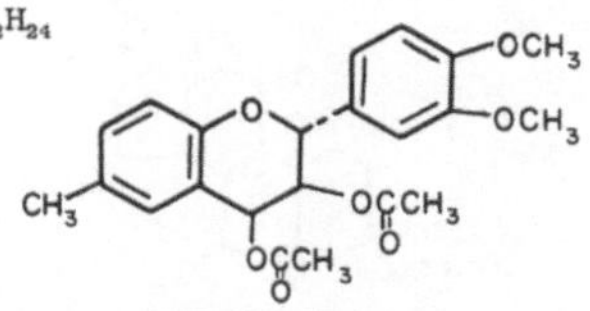

J. Chem. Soc. 3861 (1962)

▶ $O_7C_{22}H_{24}$

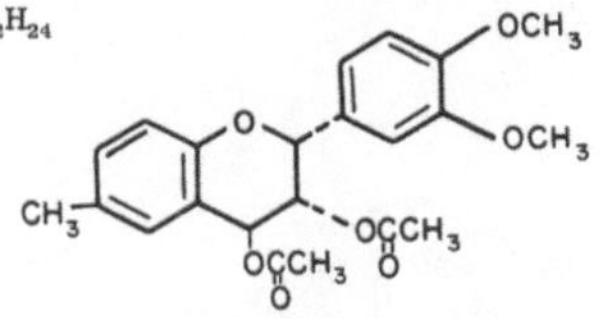

J. Chem. Soc. 3861 (1962)

▶ $O_7C_{22}H_{30}$

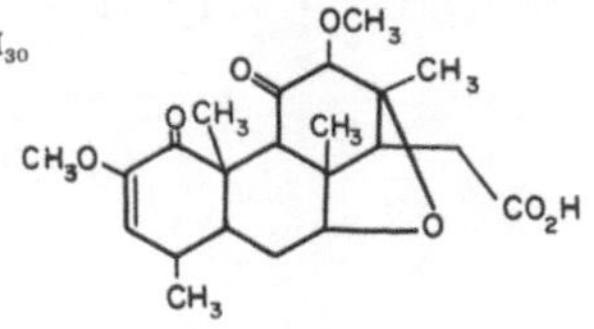

Tetrahedron 18:1433 (1962)

▶ $O_7C_{22}H_{34}$

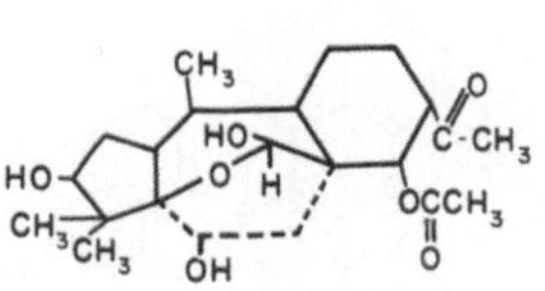

J. Org. Chem. 27:2969 (1962)

▶ $O_7C_{22}H_{36}$

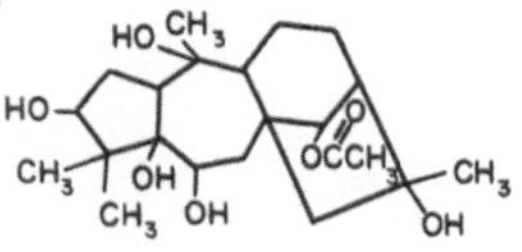

J. Org. Chem. 27:2969 (1962)

▶ $O_7C_{23}H_{22}$

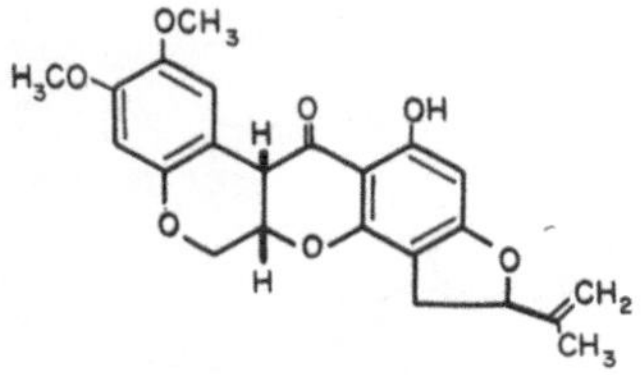

J. Chem. Soc. 775 (1962)

▶ $O_7C_{23}H_{22}$

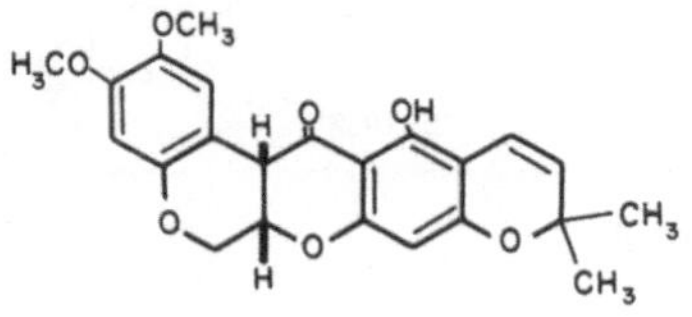

J. Chem. Soc. 775 (1962)

▶ $O_7C_{23}H_{22}$

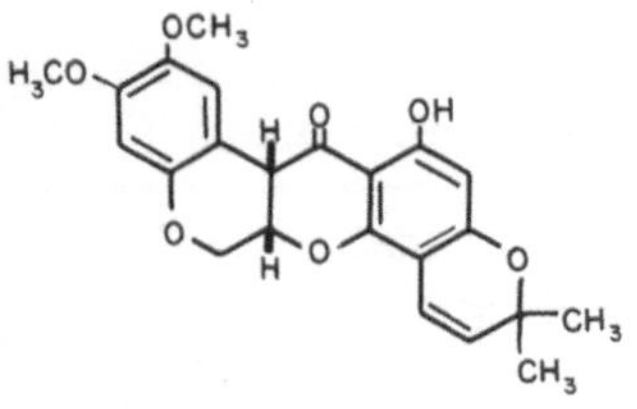

J. Chem. Soc. 775 (1962)

▶ $O_7C_{23}H_{22}$

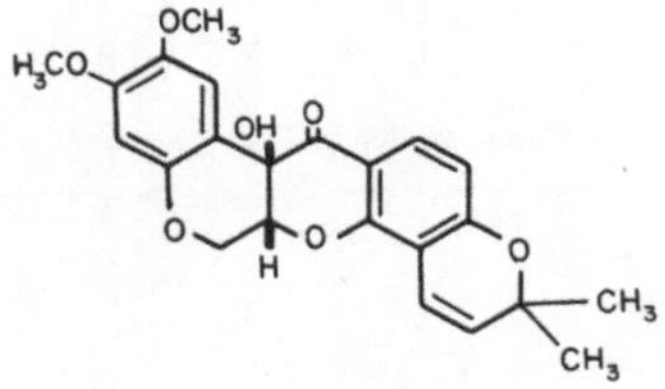

J. Chem. Soc. 775 (1962)

▶ $O_7C_{23}H_{22}$

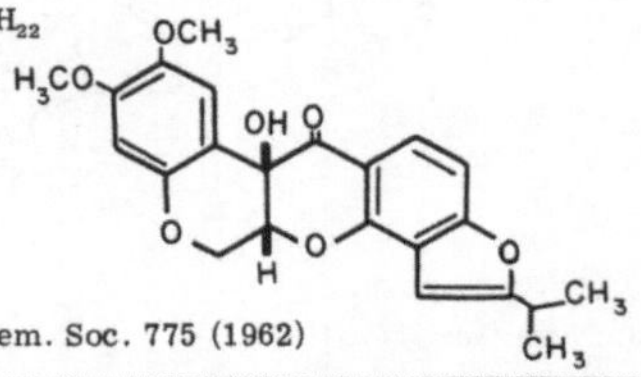

J. Chem. Soc. 775 (1962)

▶ $O_7C_{23}H_{22}$

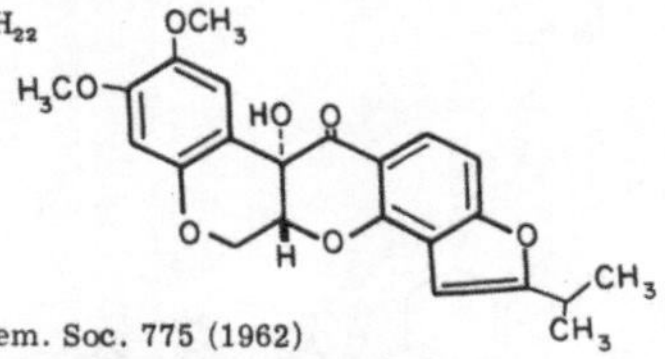

J. Chem. Soc. 775 (1962)

▶ $O_7C_{23}H_{22}$

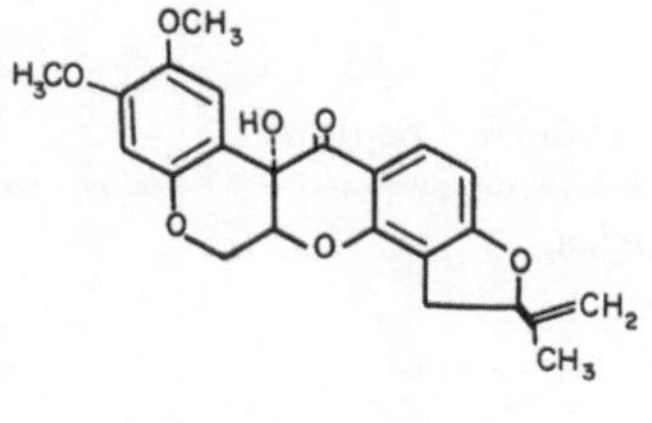

J. Chem. Soc. 775 (1962)

▶ $O_7C_{23}H_{22}$

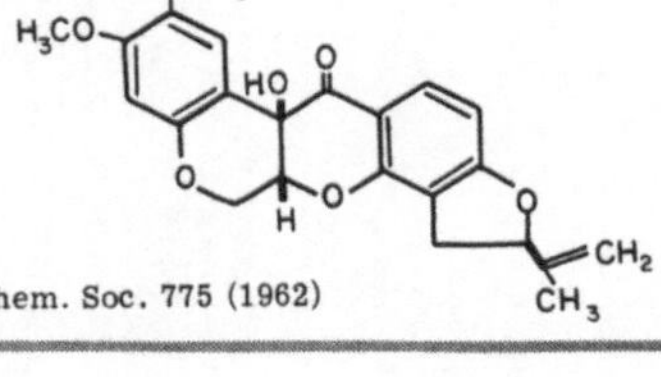

J. Chem. Soc. 775 (1962)

▶ $O_7C_{23}H_{26}$

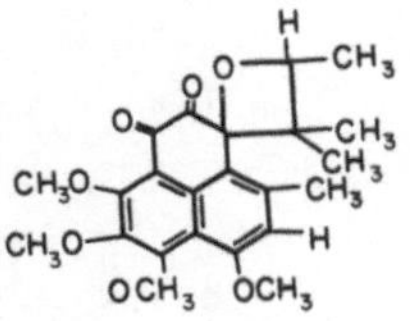

Tetrahedron 18:845 (1962)

▶ $O_7C_{23}H_{30}$

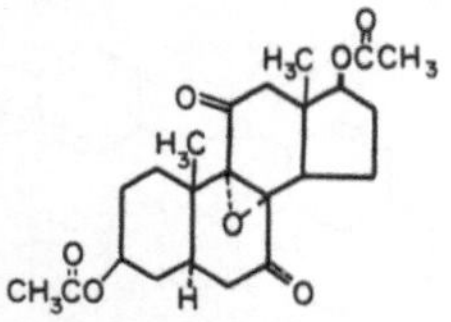

Helv. Chim. Acta 44:1380 (1961)

▶ $O_7C_{24}H_{24}$

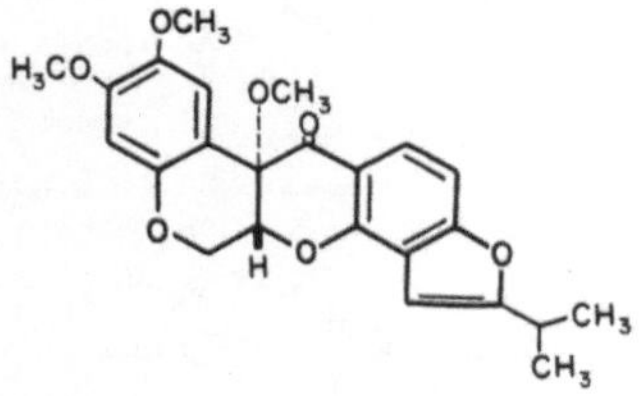

J. Chem. Soc. 775 (1962)

▶ $O_7C_{24}H_{24}$

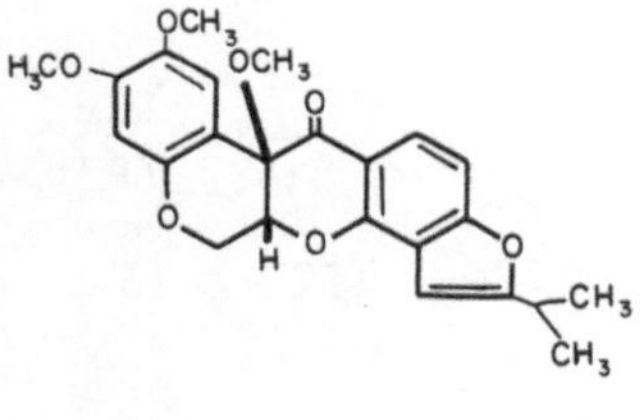

J. Chem. Soc. 775 (1962)

▶ $O_7C_{24}H_{24}$

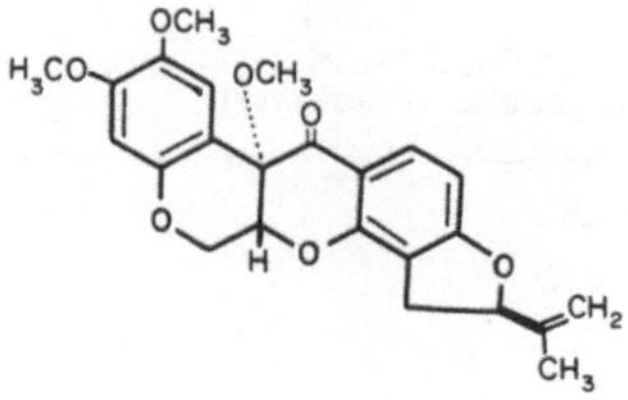


J. Chem. Soc. 775 (1962)

▶ $O_7C_{24}H_{24}$

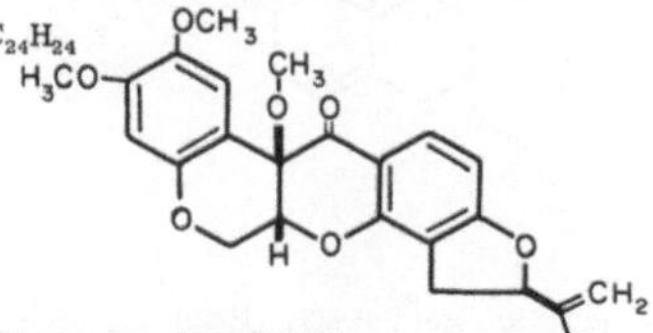


J. Chem. Soc. 775 (1962)

▶ $O_7C_{24}H_{32}$

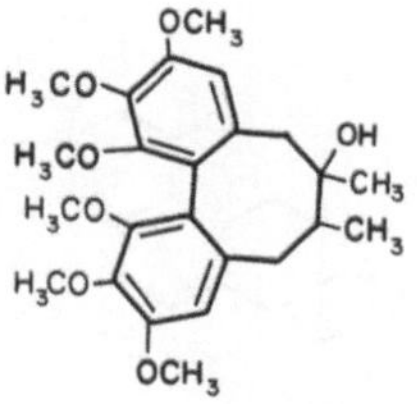


Tetrahedron Letters 730 (1961)

▶ $O_7C_{24}H_{32}$

(— position uncertain)

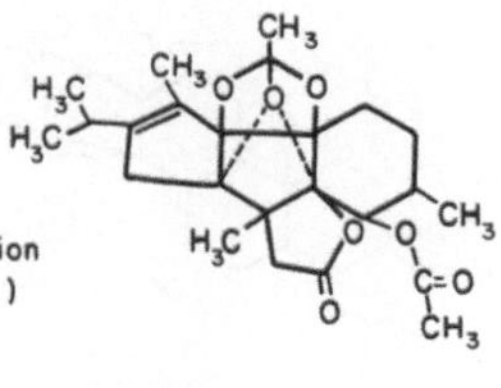


Experientia 18:111 (1962)

▶ $O_7C_{24}H_{34}$

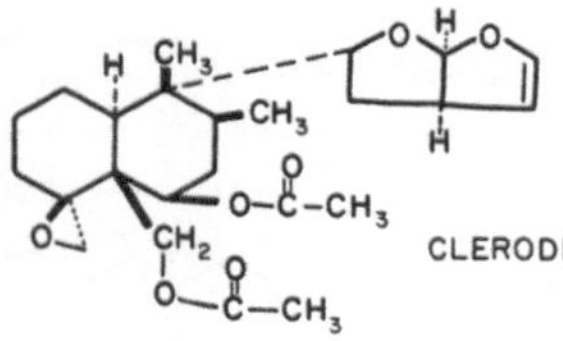


CLERODIN

Proc. Chem. Soc. 76 (1961)
J. Chem. Soc. 5061 (1961)

▶ $O_7C_{24}H_{36}$

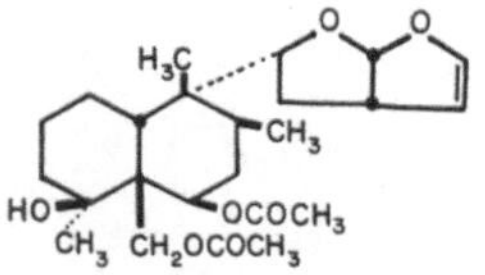


J. Chem. Soc. 5061 (1961)

▶ $O_7C_{24}H_{36}$

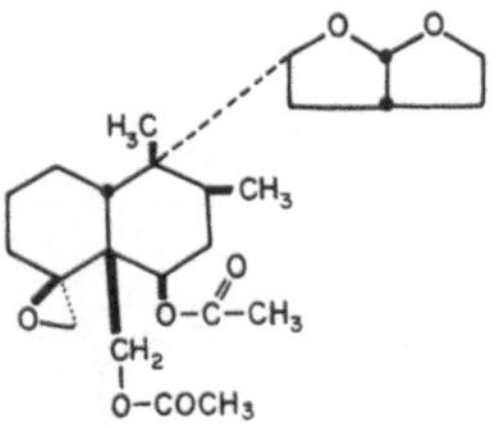


J. Chem. Soc. 5061 (1961)

▶ $O_7C_{24}H_{38}$

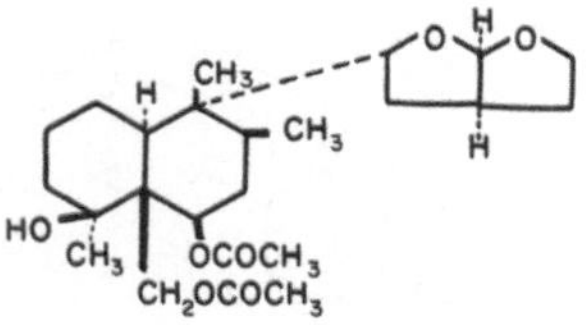


J. Chem. Soc. 5061 (1961)

▶ $O_7C_{25}H_{24}$

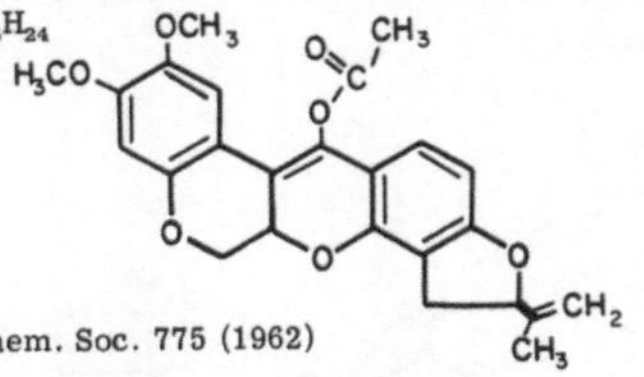

J. Chem. Soc. 775 (1962)

▶ $O_7C_{25}H_{30}$

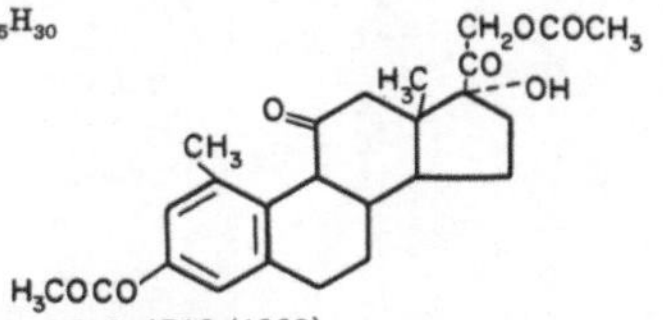

Chem. & Ind. 1716 (1962)

▶ $O_7C_{25}H_{30}$

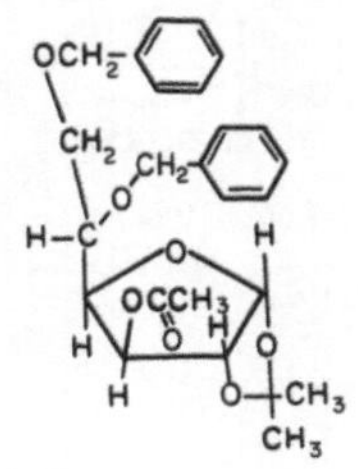

J. Chem. Soc. 3702 (1962)

▶ $O_7C_{25}H_{34}$

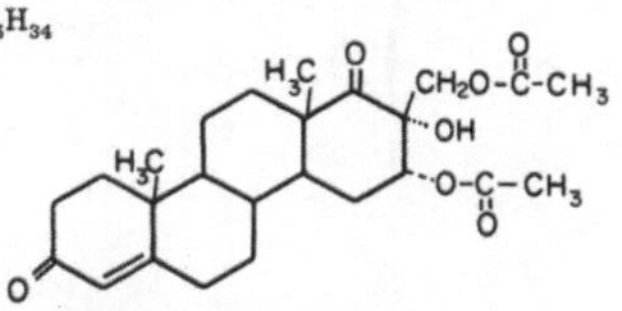

J. Org. Chem. 26:5036 (1961)

▶ $O_7C_{25}H_{34}$

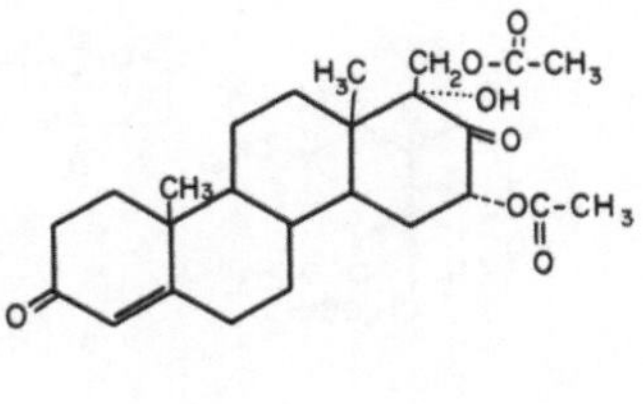

J. Org. Chem. 26:5036 (1961)

▶ $O_7C_{25}H_{34}$

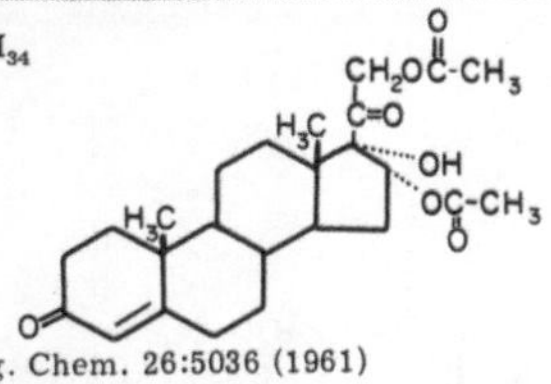

J. Org. Chem. 26:5036 (1961)

▶ $O_7C_{25}H_{34}$

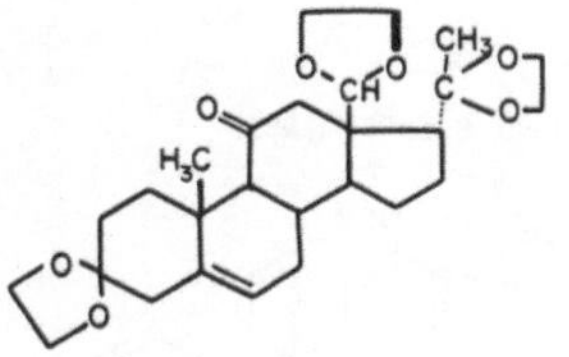

Helv. Chim. Acta 45:331 (1962)

▶ $O_7C_{25}H_{34}$

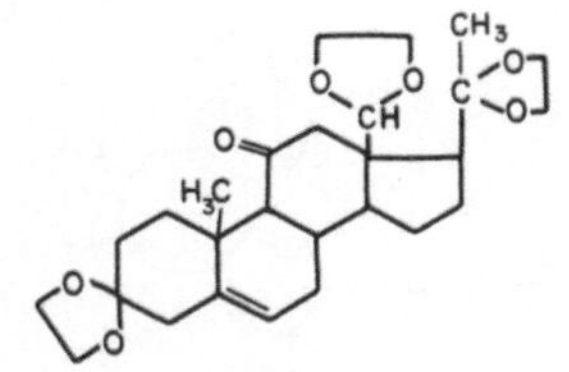

Helv. Chim. Acta 45:331 (1962)

▶ $O_7C_{25}H_{36}$

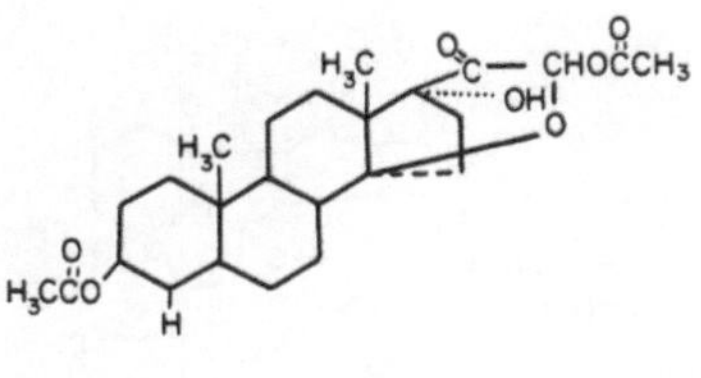

Tetrahedron Letters No. 26:1284 (1962)

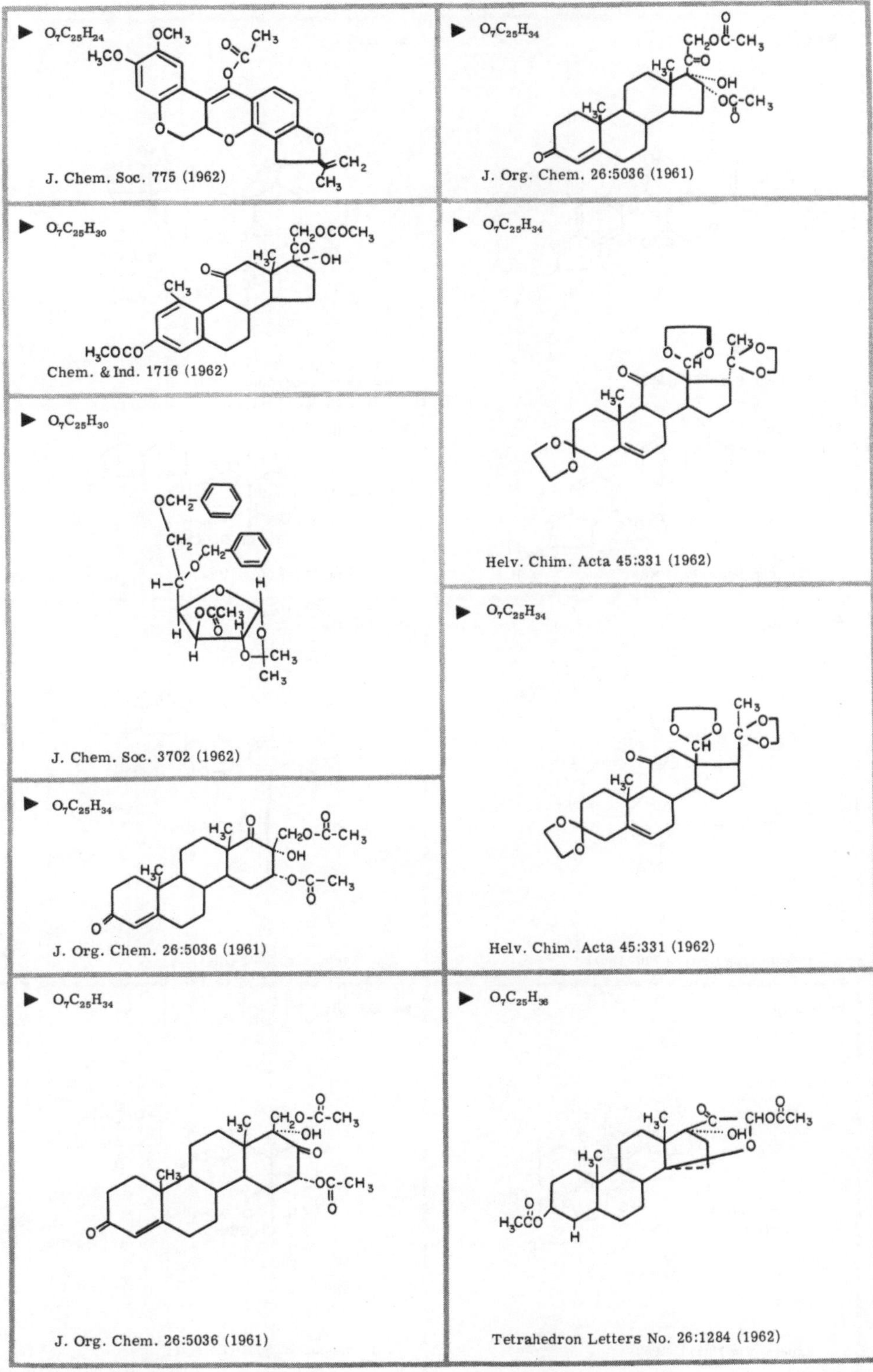

▶ $O_7C_{25}H_{40}$

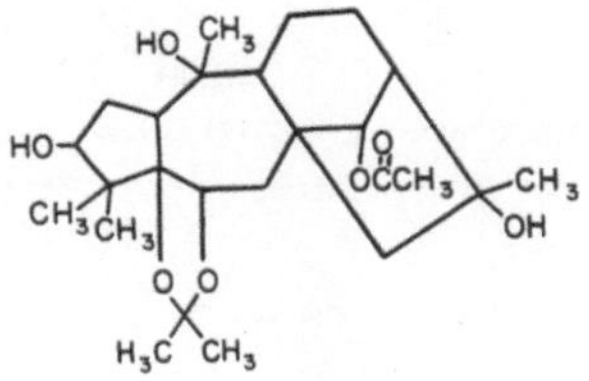

J. Org. Chem. 27:2969 (1962)

▶ $O_7C_{28}H_{40}$

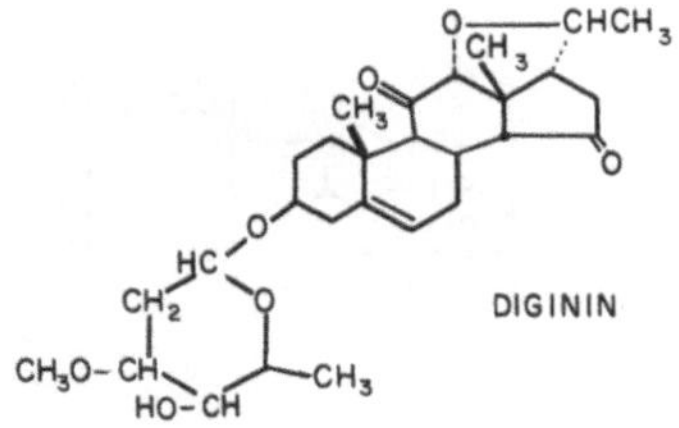

DIGININ

Proc. Chem. Soc. 65 (1962)

▶ $O_7C_{26}H_{20}$

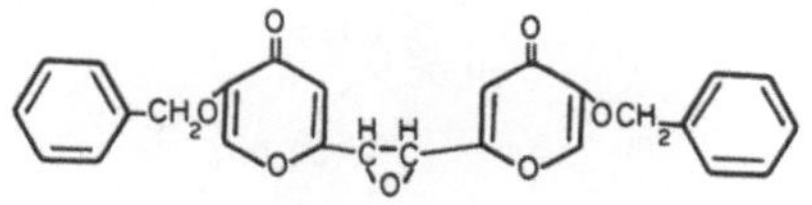

J. Chem. Soc. 441 (1962)

▶ $O_7C_{28}H_{40}$

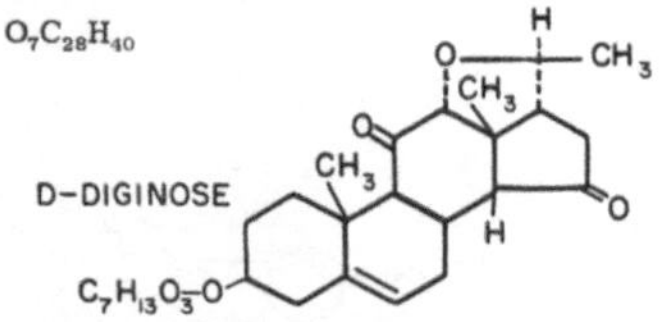


J. Chem. Soc. 3615 (1962)

▶ $O_7C_{27}H_{40}$

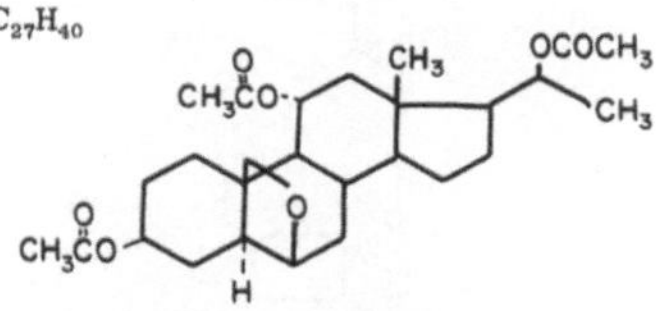

Helv. Chim. Acta 45:2161 (1962)

▶ $O_7C_{30}H_{46}$

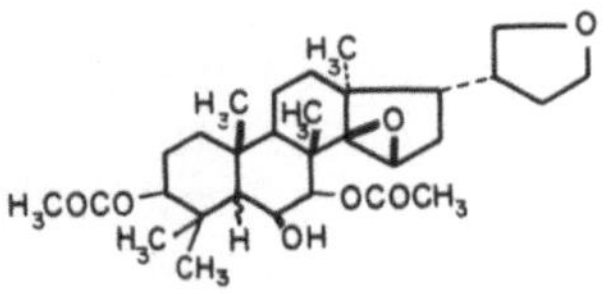

FROM CEDRELONE VIA $NaBH_4$ AND ACETYLATION

Proc. Chem. Soc. 446 (1961)

▶ $O_7C_{27}H_{40}$

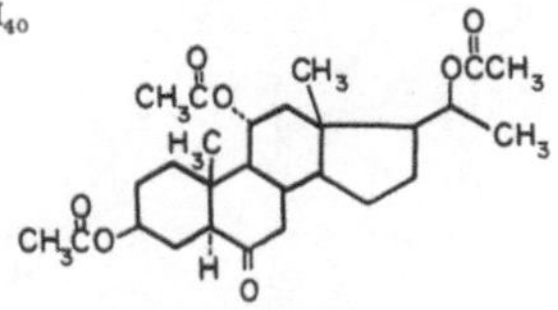

Helv. Chim. Acta 45:2161 (1962)

▶ $O_7C_{28}H_{32}$

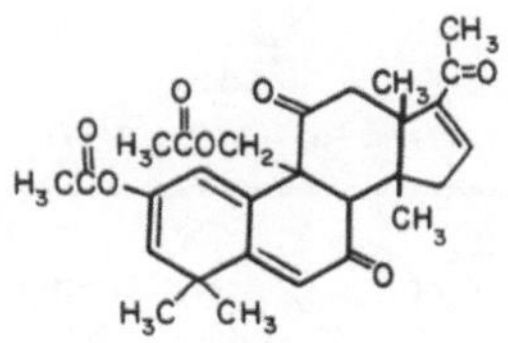

Tetrahedron Letters No. 8 :315 (1962)

▶ $O_7C_{31}H_{40}$

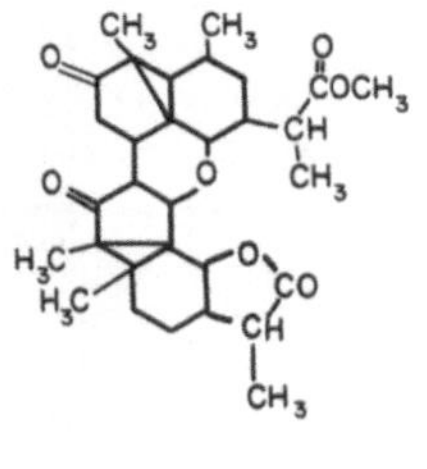

Tetrahedron Letters No. 8:335 (1962)

▶ $O_7C_{44}H_{66}$

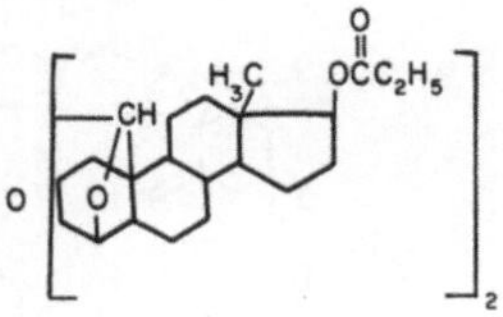

Helv. Chim. Acta 45:2575 (1962)

▶ $O_8C_{12}H_{16}$

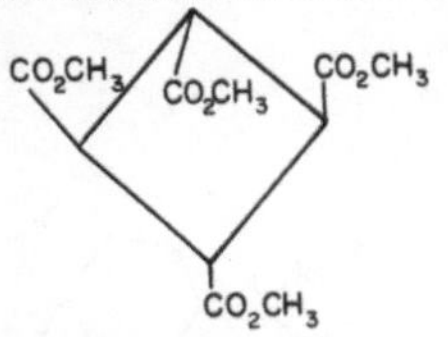

J. Am. Chem. Soc. 83:2725 (1961)

▶ $O_8C_{12}H_{16}$

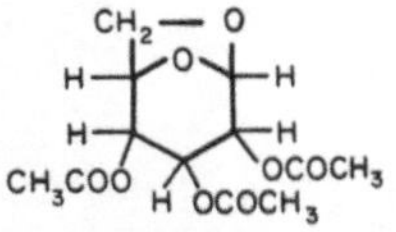

Proc. Chem. Soc. 382 (1962)

▶ $O_7C_{55}H_{88}$

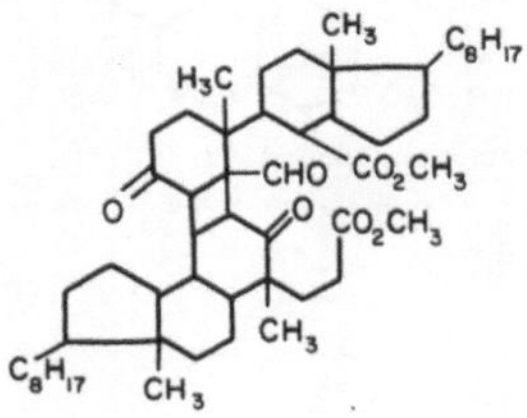

Helv. Chim. Acta 45:2342 (1962)

▶ $O_8C_{12}H_{16}$

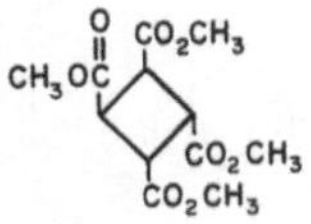

J. Am. Chem. Soc. 83:2725, 2785 (1961)

▶ $O_8C_{12}H_{22}$

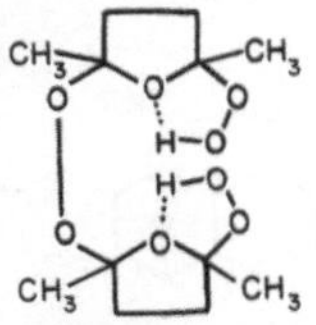

J. Org. Chem. 27:4319 (1962)

▶ $O_8FC_{24}H_{31}$

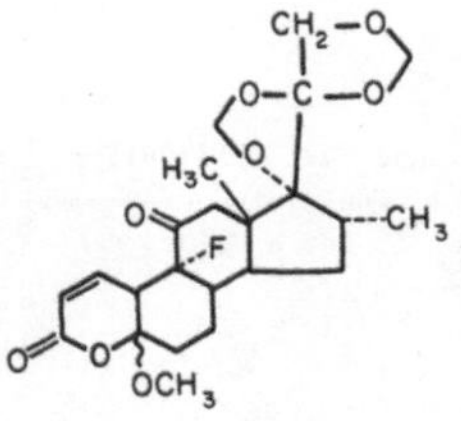

J. Am. Chem. Soc. 84:1270 (1962)

▶ $O_8C_{13}H_{20}$

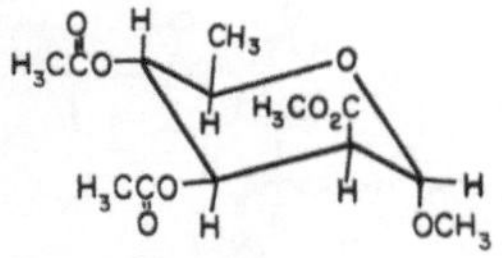

J. Chem. Soc. 2503 June (1962)

▶ $O_8C_{16}H_{16}$

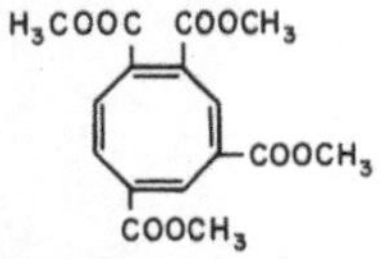

J. Am. Chem. Soc. 83:2944 (1961)

▶ $O_8C_9H_{16}$

$CH_3C(=O)O(CH_2O)_5C(=O)CH_3$

Chem. Ber. 94:3317 (1961)

▶ $O_8C_{16}H_{24}$

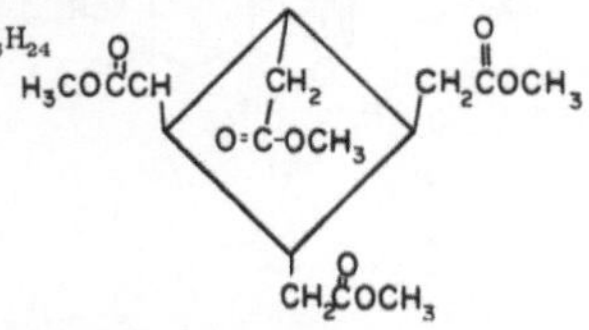

J. Am. Chem. Soc. 84:1008 (1962)

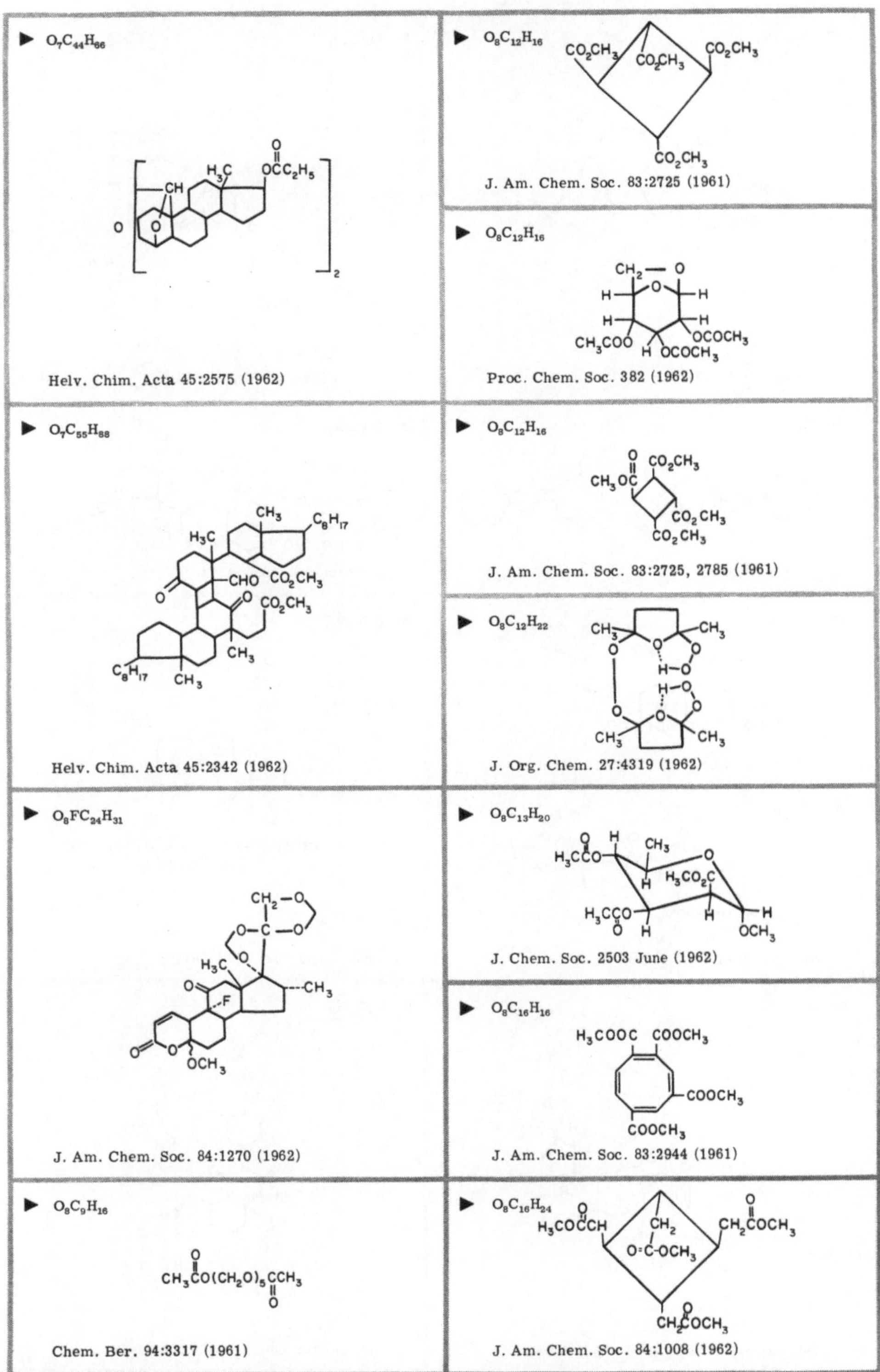

▶ $O_8C_{16}H_{24}$

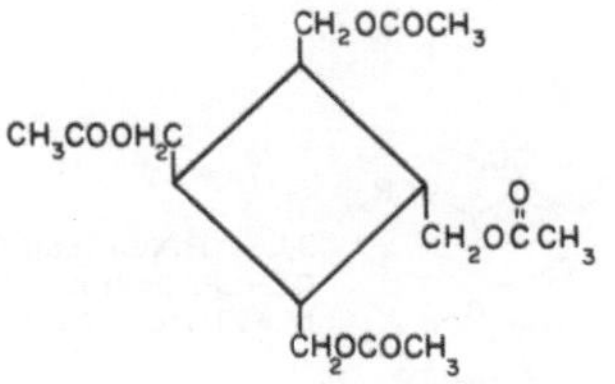

J. Am. Chem. Soc. 83:2725 (1961)

▶ $O_8C_{20}H_{18}$

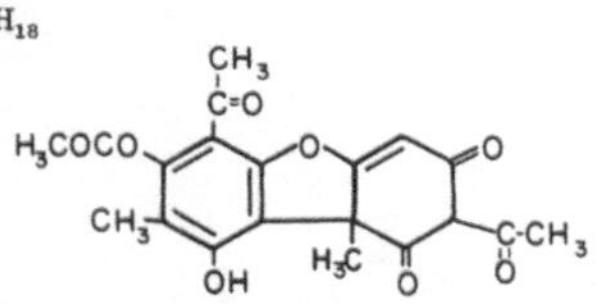

Acta Chem. Scand. 16:583 (1962)

▶ $O_8C_{20}H_{24}$

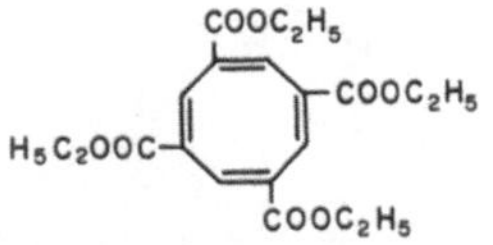

J. Am. Chem. Soc. 83:2944 (1961)

▶ $O_8C_{18}H_{16}$

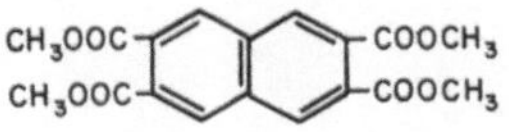

J. Org. Chem. 27:3354 (1962)

▶ $O_8C_{20}H_{24}$

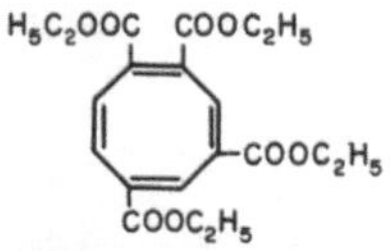

J. Am. Chem. Soc. 83:2944 (1961)

▶ $O_8C_{18}H_{22}$

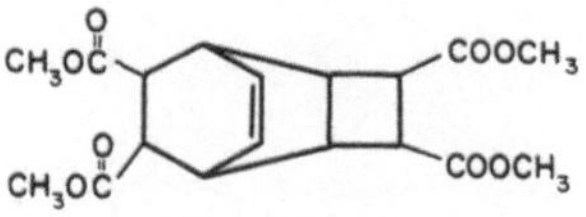

J. Am. Chem. Soc. 83:1705 (1961)

▶ $O_8C_{20}H_{30}$

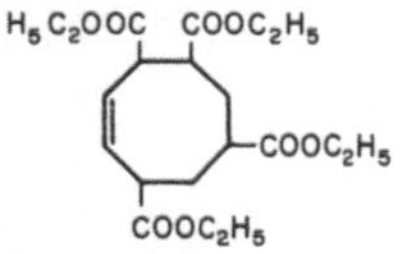

J. Am. Chem. Soc. 83:2944 (1961)

▶ $O_8C_{20}H_{18}$

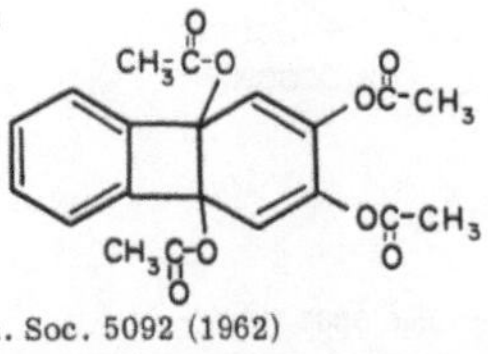

J. Chem. Soc. 5092 (1962)

▶ $O_8C_{20}H_{32}$

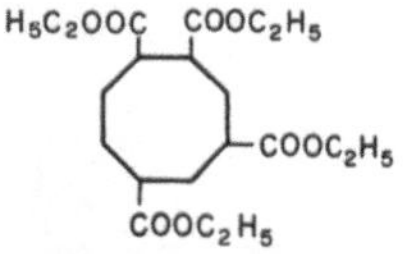

J. Am. Chem. Soc. 83:2944 (1961)

▶ $O_8C_{20}H_{18}$

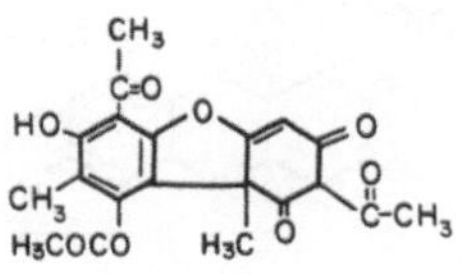

Acta Chem. Scand. 16:583 (1962)

▶ $O_8C_{22}H_{28}$

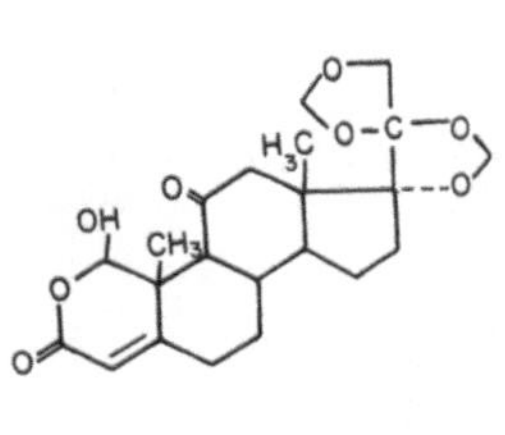

J. Am. Chem. Soc. 84:1270 (1962)

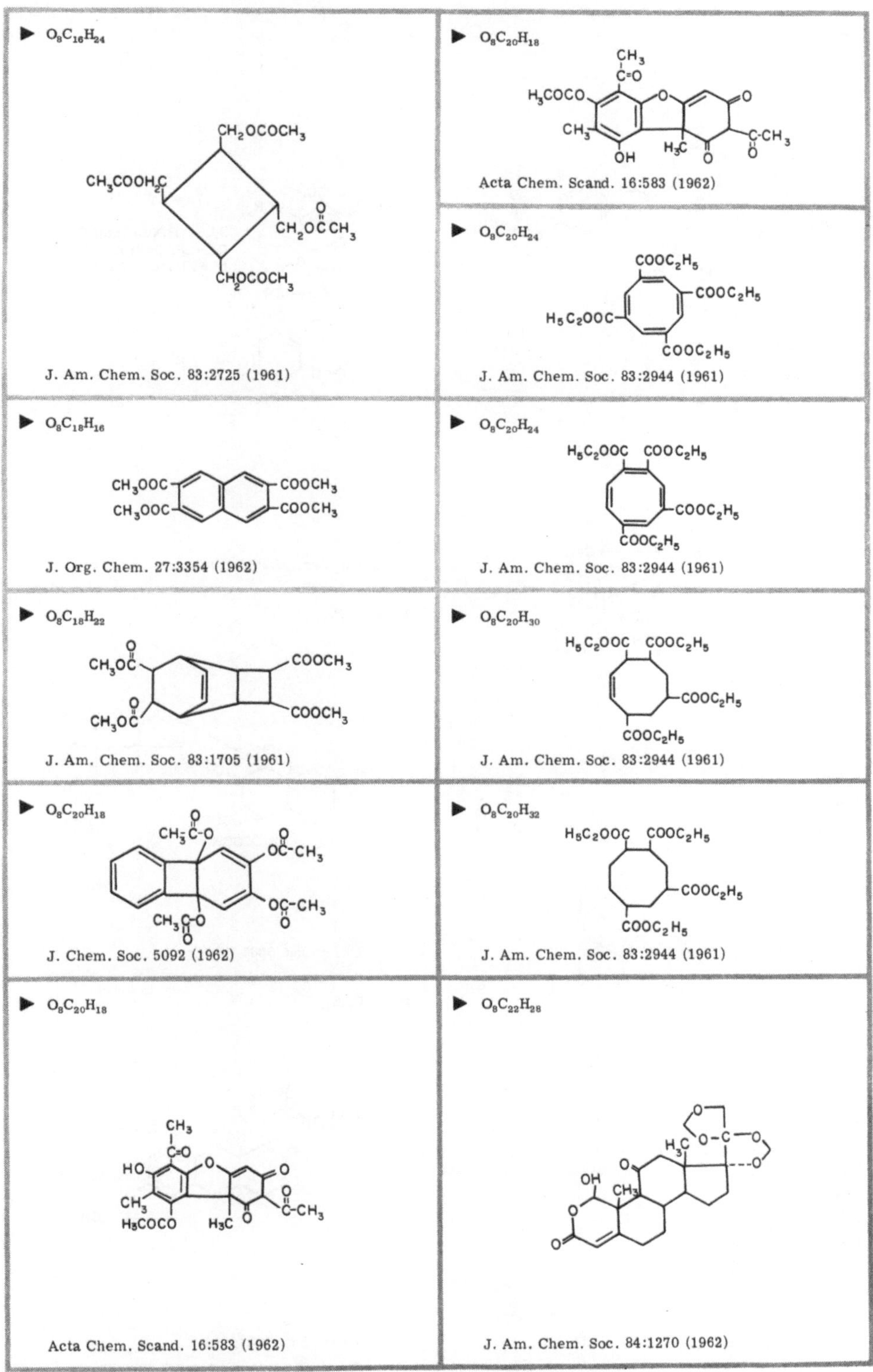

▶ $O_8C_{23}H_{18}$

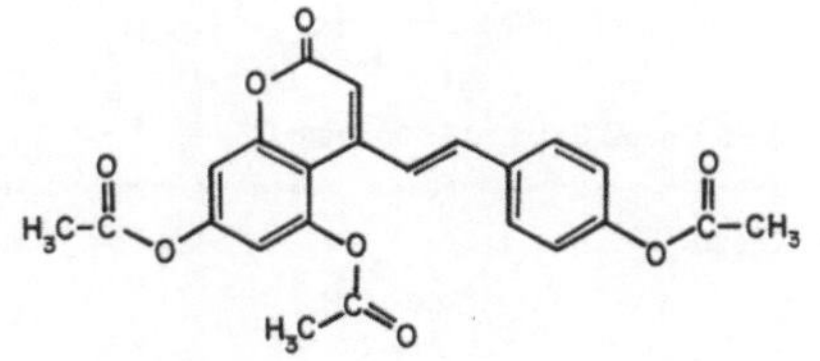

Varian 692

▶ $O_8C_{23}H_{22}$

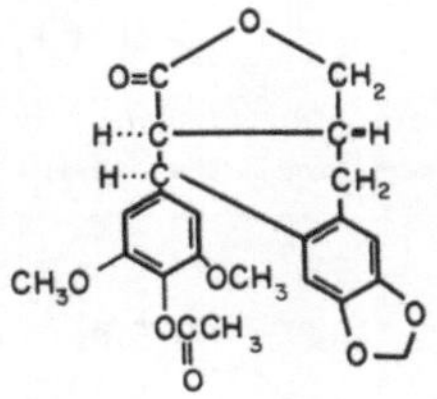

Bull. Soc. Chim. France 1722 (1962)

▶ $O_8C_{23}H_{30}$

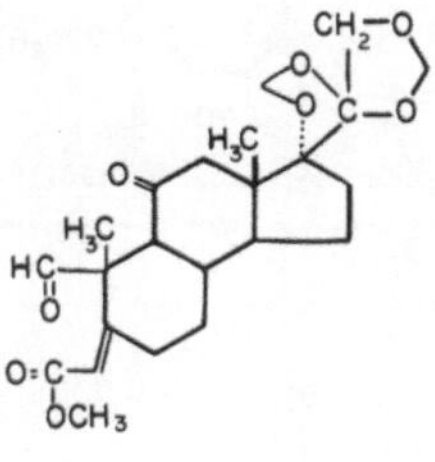

J. Am. Chem. Soc. 84:1270 (1962)

▶ $O_8C_{24}H_{26}$

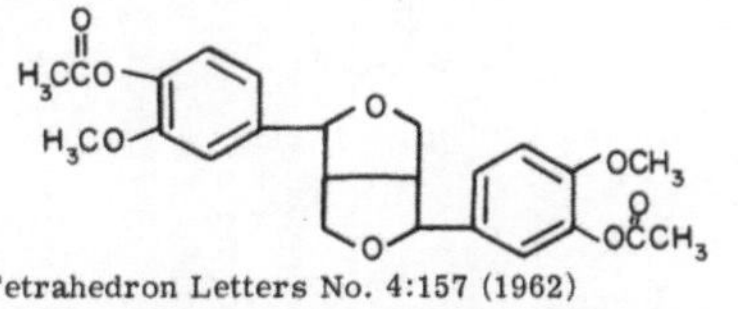

Tetrahedron Letters No. 4:157 (1962)

▶ $O_8C_{24}H_{30}$

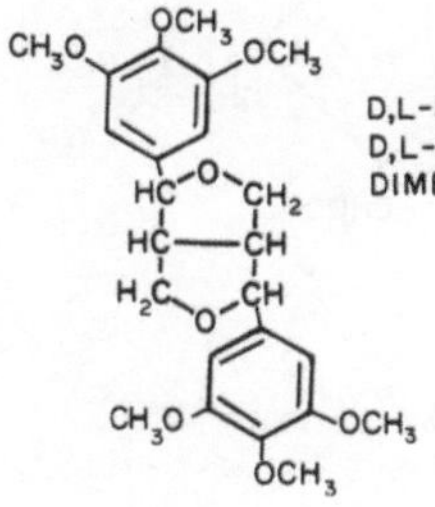

D,L-SYRINGARESINOL
D,L-EPI SYRINGARESINOL
DIMETHYL ETHERS

Chem. Ber. 94:2522 (1961)

▶ $O_8C_{24}H_{34}$

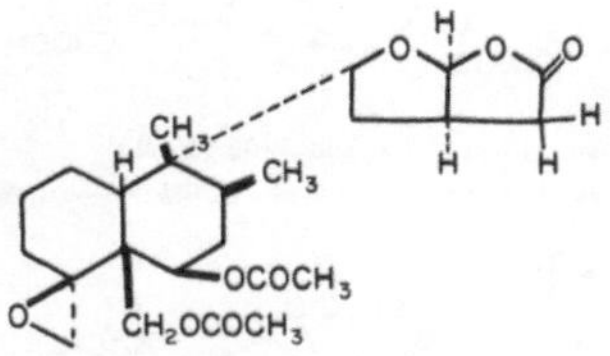

J. Chem. Soc. 5061 (1961)

▶ $O_8C_{25}H_{24}$

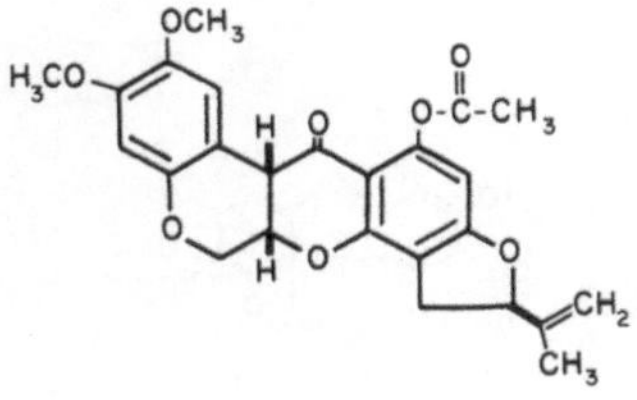

J. Chem. Soc. 775 (1962)

▶ $O_8C_{25}H_{24}$

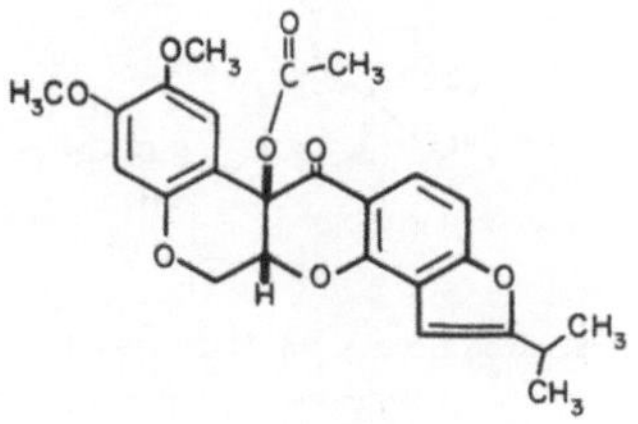

J. Chem. Soc. 775 (1962)

▶ $O_9C_{26}H_{36}$

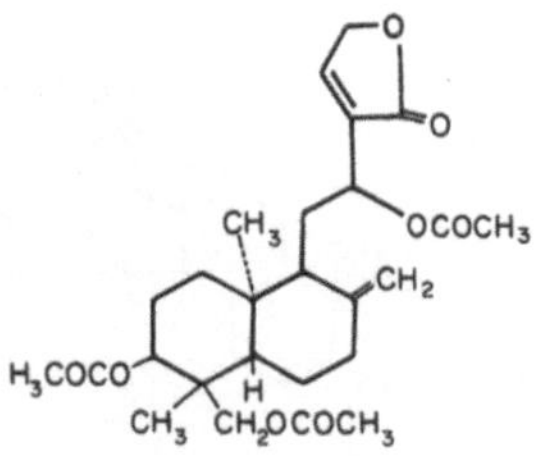

Tetrahedron 18:397 (1962)

▶ $O_8C_{25}H_{24}$

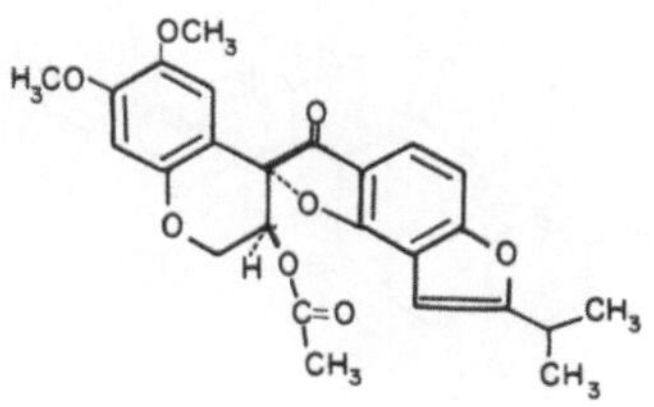

J. Chem. Soc. 775 (1962)

▶ $O_8C_{27}H_{34}$

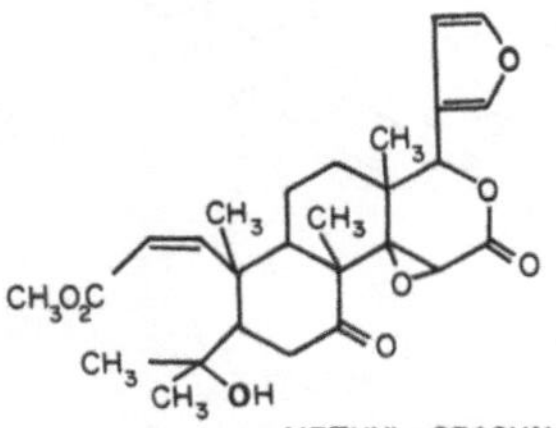

METHYL OBACUNOATE

J. Chem. Soc. 255 (1961)

▶ $O_8C_{26}H_{34}$

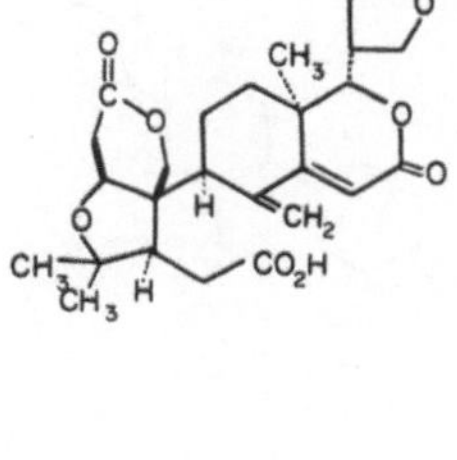

J. Chem. Soc. 255 (1961)

▶ $O_8C_{27}H_{38}$

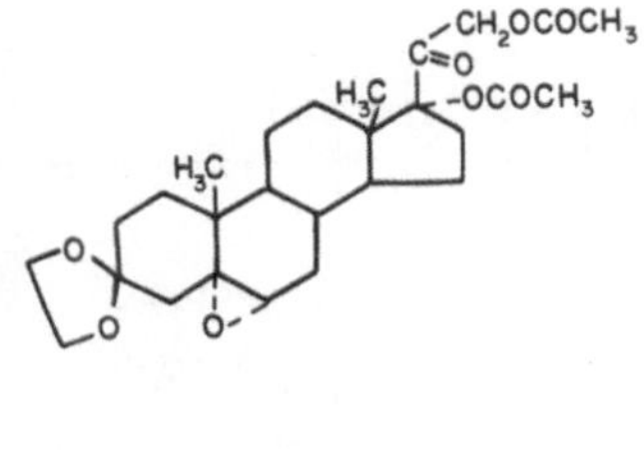

J. Am. Chem. Soc. 84:3206 (1962)

▶ $O_8C_{27}H_{38}$

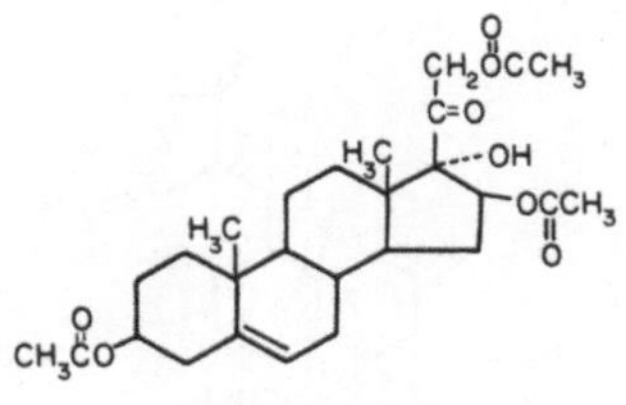

Helv. Chim. Acta 44:1380 (1961)

▶ $O_8C_{28}H_{36}$

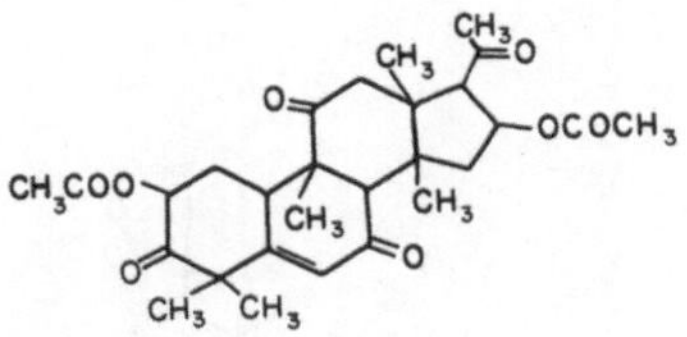

J. Org. Chem. 27:4546 (1962)

▶ $O_8C_{28}H_{40}$

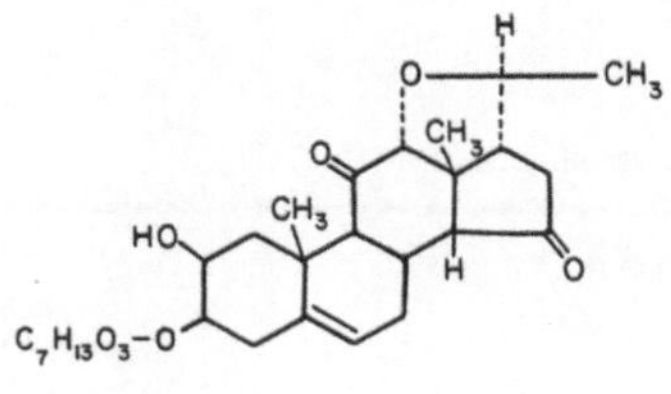

J. Chem. Soc. 3615 (1962)

▶ $O_8C_{30}H_{27}As$

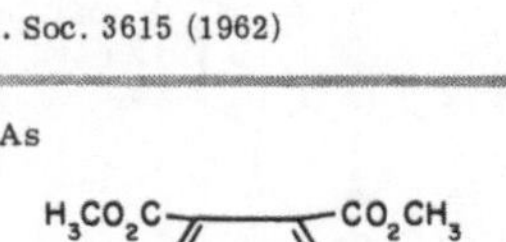

Tetrahedron Letters 477 (1961)

▶ $O_8C_{30}H_{40}$

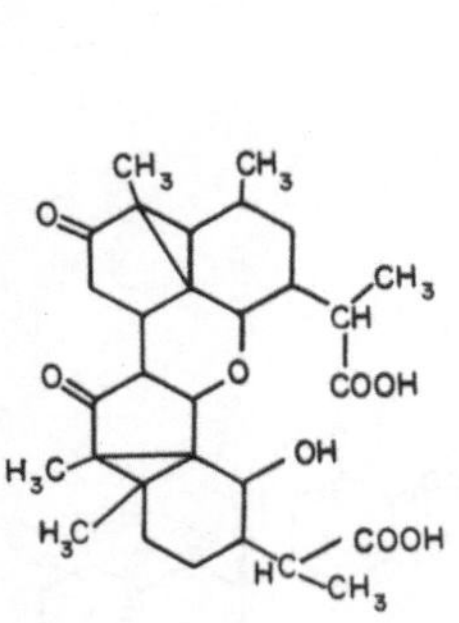

Tetrahedron Letters No. 8:335 (1962)

▶ $O_8C_{31}H_{48}$

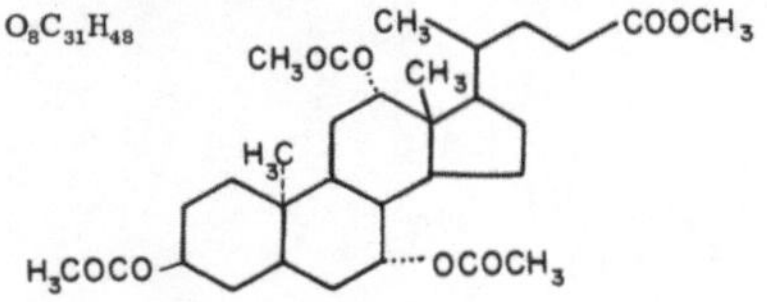

J. Chem. Soc. 2817 (1961)

▶ $O_8C_{32}H_{46}$

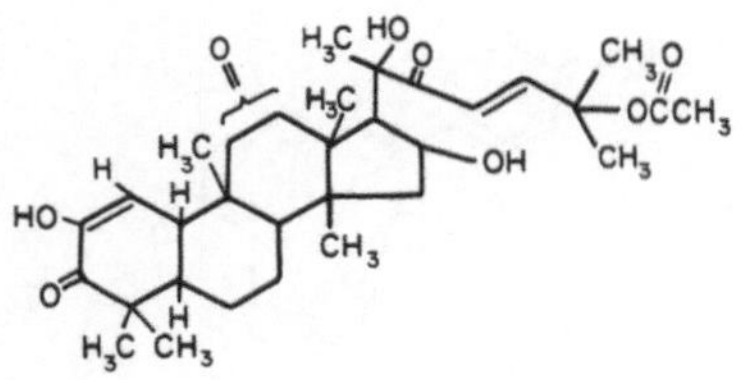

Tetrahedron Letters No. 18:616 (1961)

▶ $O_8C_{33}H_{52}$

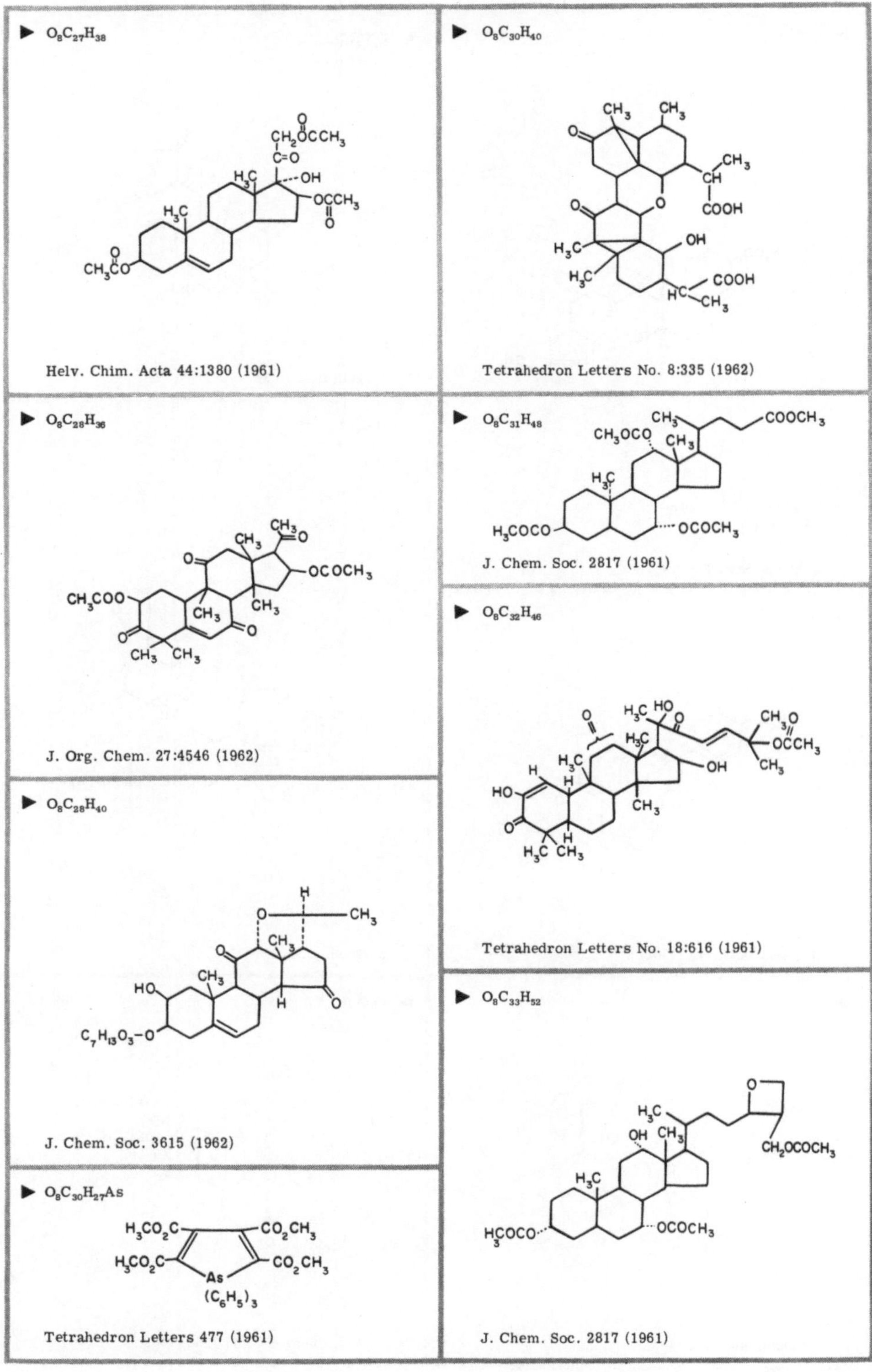

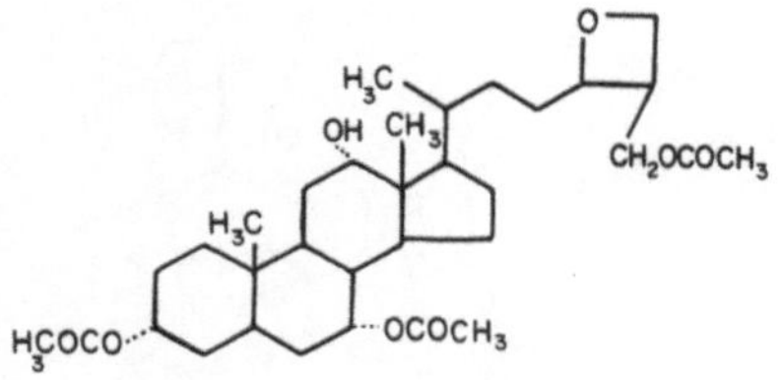

J. Chem. Soc. 2817 (1961)

► $O_8C_{34}H_{50}$

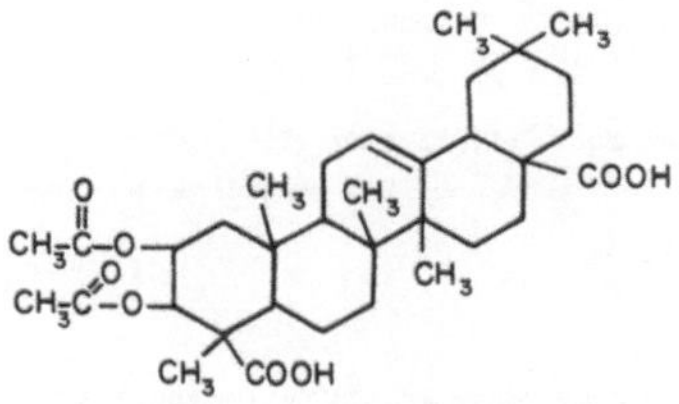

J. Org. Chem. 26:1241 (1961)

► $O_8C_{36}H_{32}$

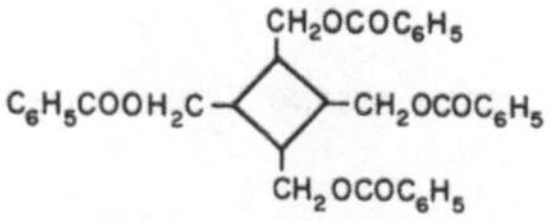

J. Am. Chem. Soc. 83:2725 (1961)

► $O_8C_{37}H_{56}$

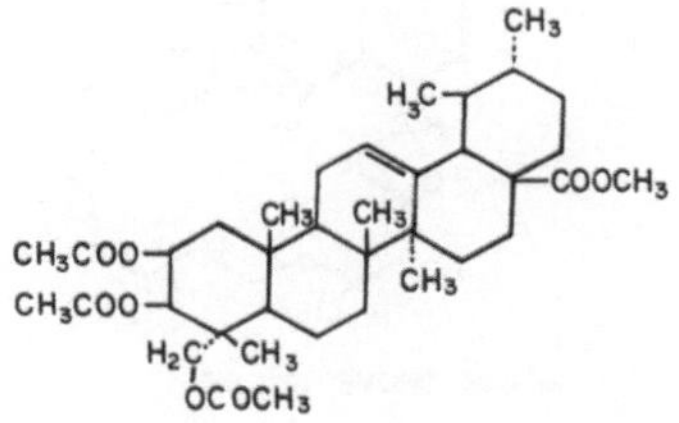

J. Org. Chem. 27:4512 (1962)

► $O_9IC_{19}H_{21}$

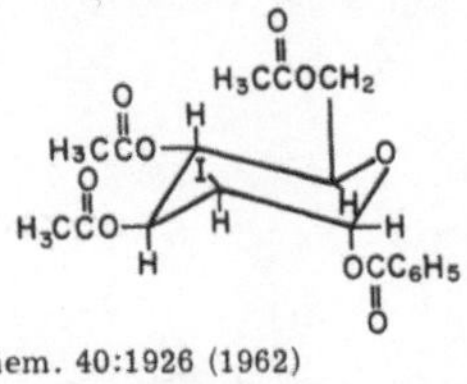

Can. J. Chem. 40:1926 (1962)

► $O_9IC_{19}H_{21}$

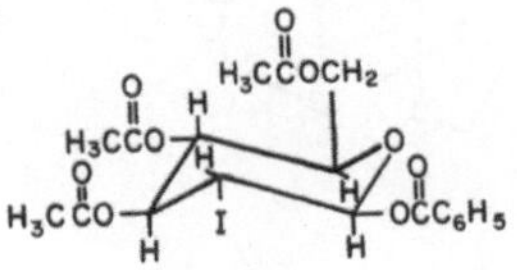

Can. J. Chem. 40:1927 (1962)

► $O_9C_{10}H_{18}$

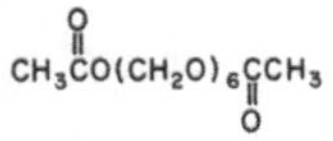

Chem. Ber. 94:3317 (1961)

► $O_9C_{14}H_{20}$

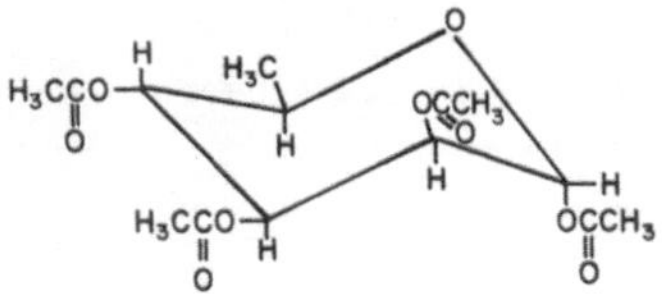

J. Chem. Soc. 2503 (1962)

► $O_9C_{15}H_{22}$

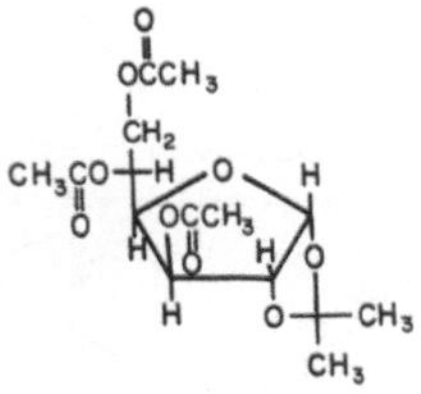

J. Chem. Soc. 3699 (1962)
Chem. & Ind. 213 (1962)

► $O_9C_{15}H_{22}$

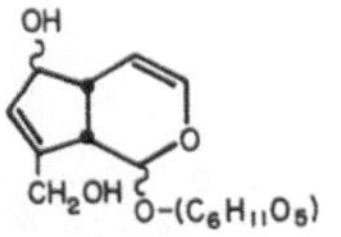

Tetrahedron Letters No. 8:325 (1962)

► $O_9C_{21}H_{24}$

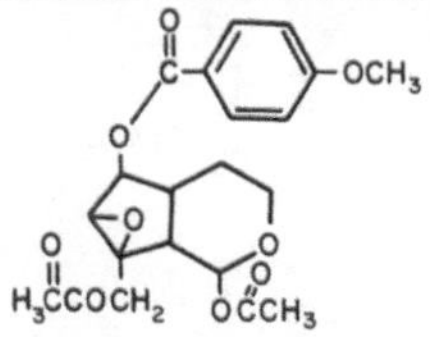

Tetrahedron Letters No. 8:327 (1962)

▶ $O_9C_{22}H_{20}$

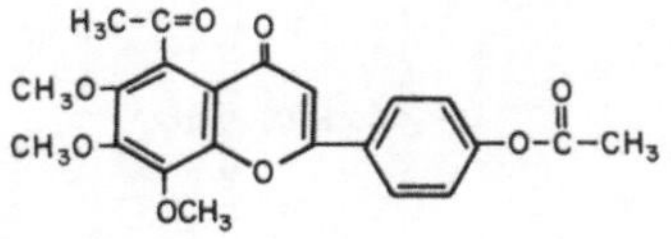

Tetrahedron 14:301 (1961)

▶ $O_9C_{22}H_{20}$

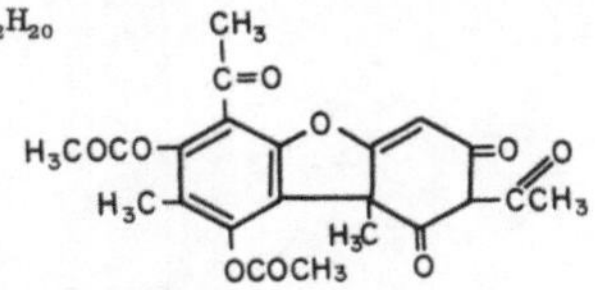

Acta Chem. Scand. 16:583 (1962)

▶ $O_9C_{22}H_{26}$

Tetrahedron 18:1208 (1962)

▶ $O_9C_{23}H_{26}$

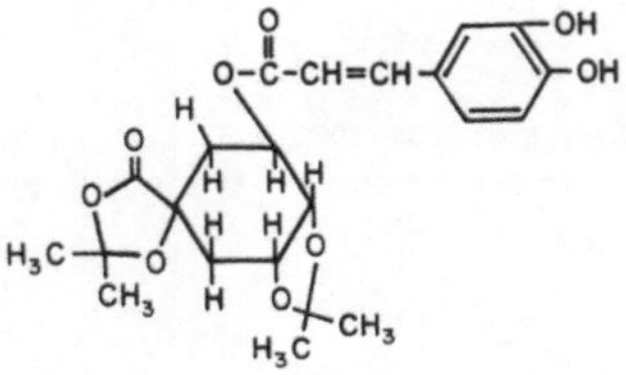

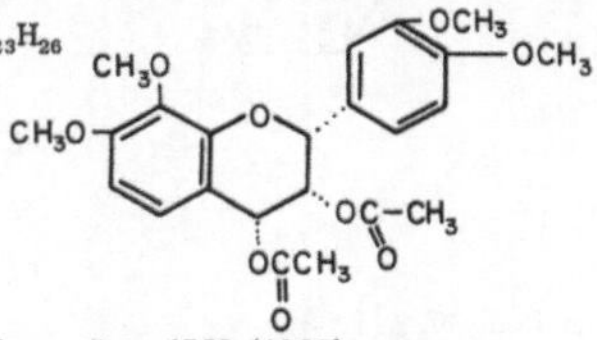

J. Chem. Soc. 4503 (1962)

▶ $O_9C_{23}H_{28}$

Tetrahedron 18:1208 (1962)

▶ $O_9C_{23}H_{28}$

DIACETYL-SAMADERINE C

Bull. Soc. Chim. France 1715 (1962)

▶ $O_9C_{23}H_{32}$

TETRAHYDRO-DIACETYLSAMADERINE C

Bull. Soc. Chim. France 1715 (1962)

▶ $O_9C_{26}H_{30}$

ISOGLAUCANOL TRIACETATE

Bull. Soc. Chim. France 1432 (1961)

▶ $O_9C_{27}H_{40}$

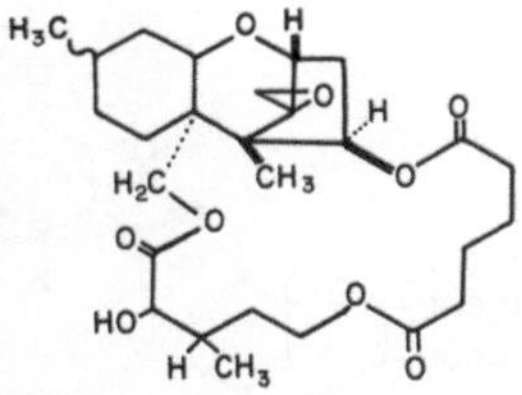

HEXAHYDROVERRUCARIN A

Helv. Chim. Acta 45:1726 (1962)

▶ $O_9C_{29}H_{42}$

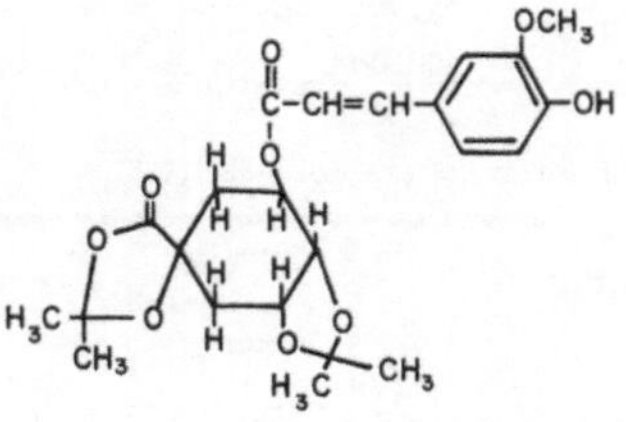

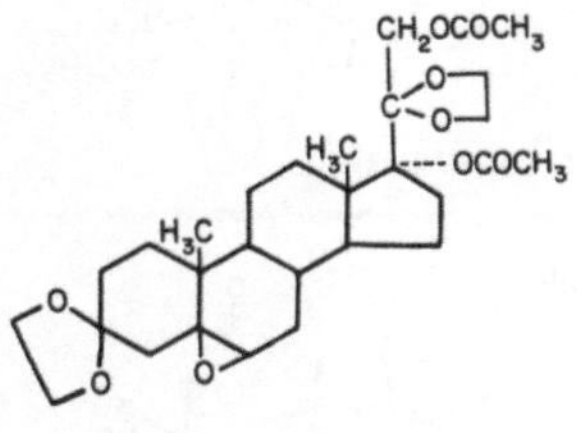

J. Am. Chem. Soc. 84:3206 (1962)

▶ $O_{9}C_{35}H_{54}$

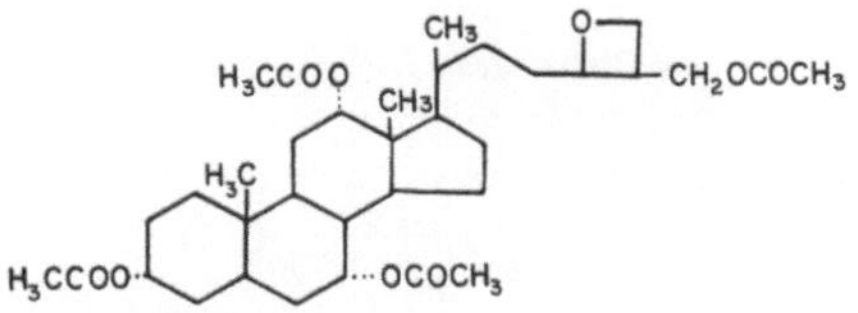

J. Chem. Soc. 2817 (1961)

▶ $O_{10}C_{15}H_{22}$

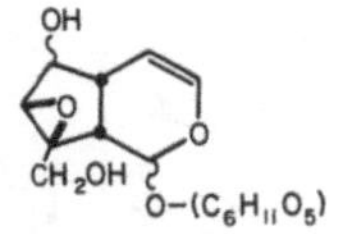

Tetrahedron Letters No. 8:326 (1962)

▶ $O_{10}C_{15}H_{24}$

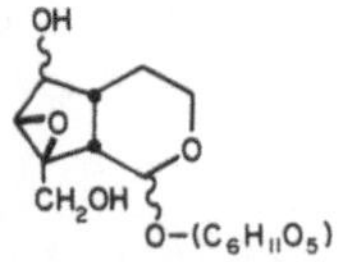

Tetrahedron Letters No. 8:326 (1962)

▶ $O_{10}C_{15}H_{20}D_{2}$

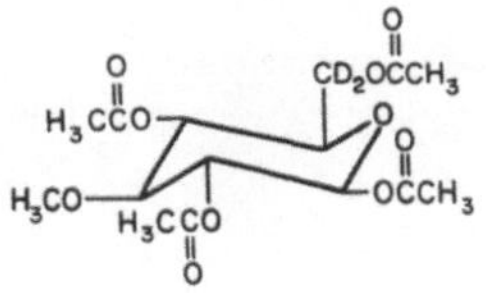

Can. J. Chem. 40:1957 (1962)

▶ $O_{10}C_{17}H_{26}$

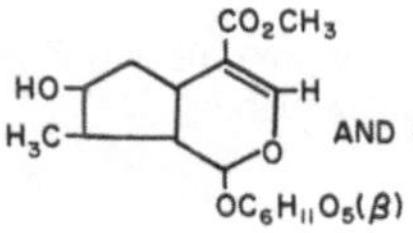

AND PENTA-ACETATE

Tetrahedron Letters 395 (1961)

▶ $O_{10}C_{18}H_{20}$

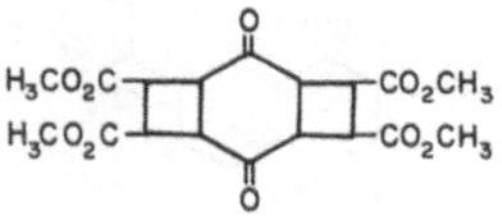

Tetrahedron Letters No. 1:1 (1961)

▶ $O_{10}C_{15}H_{22}$

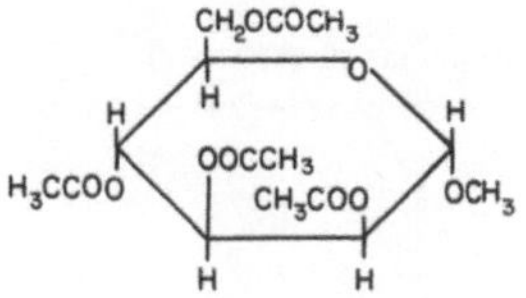

Chem. & Ind. 1827 (1962)

▶ $O_{10}C_{23}H_{34}$

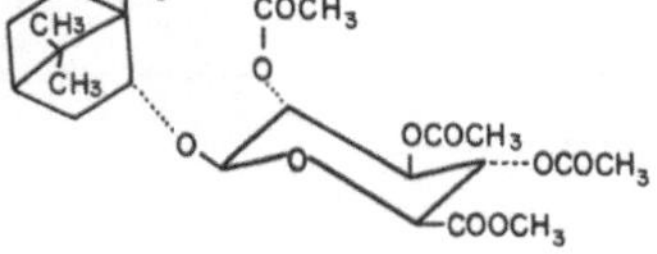

J. Am. Chem. Soc. 84:2972 (1962)

▶ $O_{10}C_{15}H_{22}$

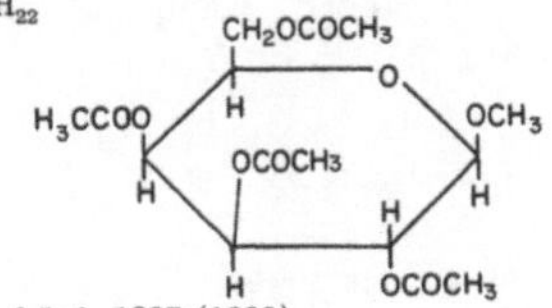

Chem. & Ind. 1827 (1962)

▶ $O_{10}C_{28}H_{32}$

GLAUCANOL TETRAACETATE

Bull. Soc. Chim. France 1432 (1961)

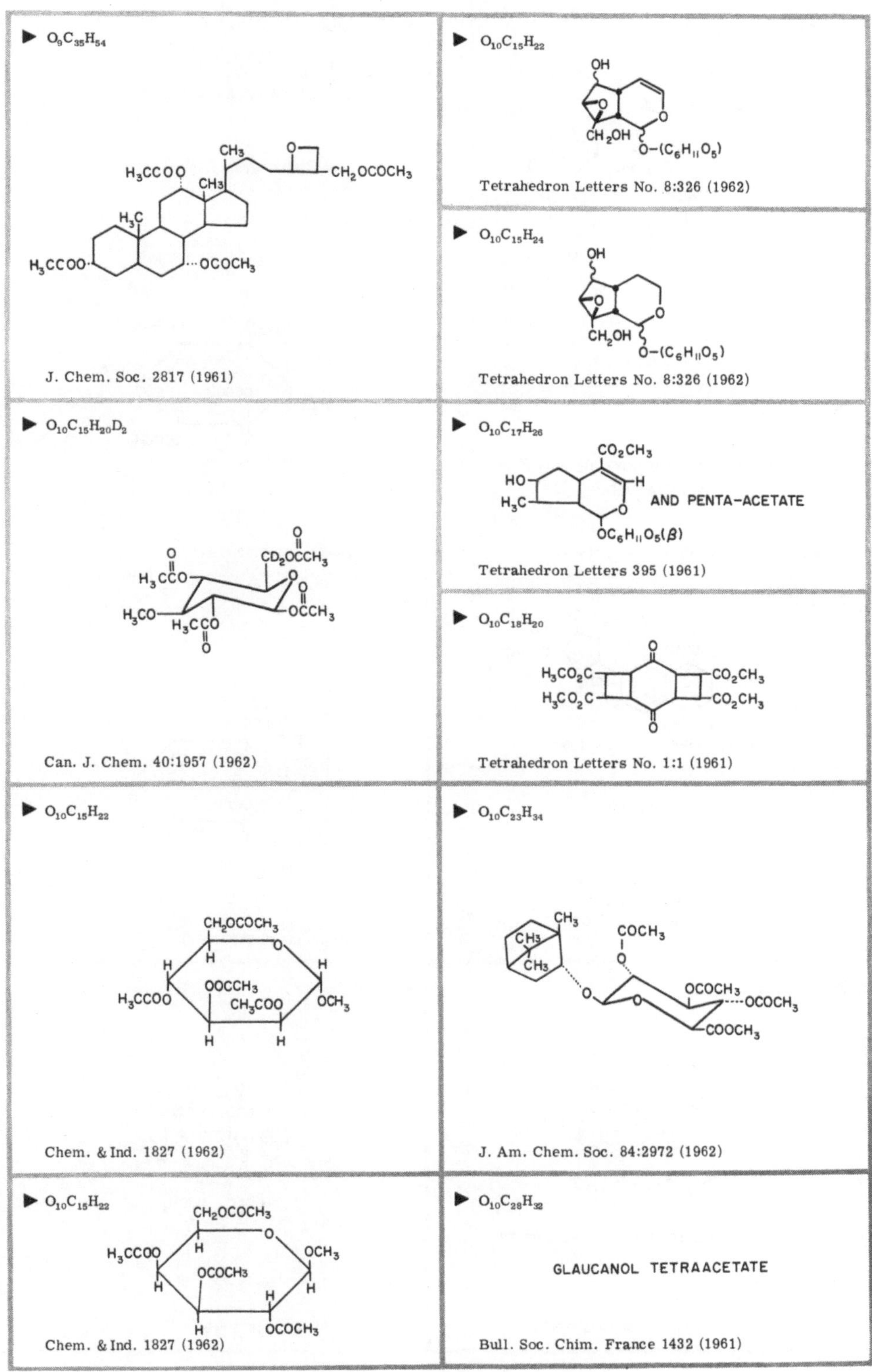

▶ $O_{10}C_{29}H_{36}$

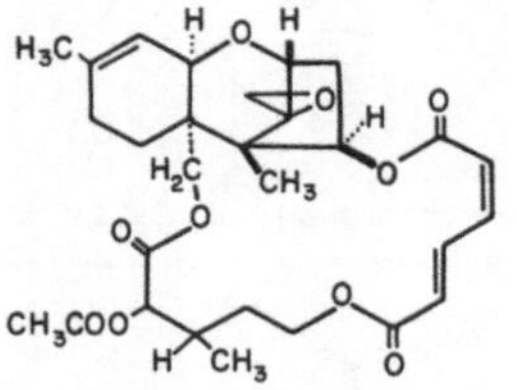

VERRUCARIN A, MONO-O-ACETYL DERIVATIVE

Helv. Chim. Acta 45:1726 (1962)

▶ $O_{10}C_{29}H_{40}$

TRIACETATE OF DESOXY-HEXAHYDRO-GLAUCARUBOL

Bull. Soc. Chim. France 1432 (1961)

▶ $O_{10}C_{30}H_{38}$

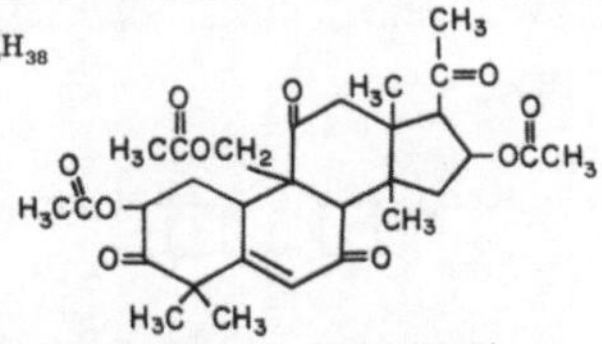

Tetrahedron Letters No. 8:315 (1962)

▶ $O_{11}C_{16}H_{22}$

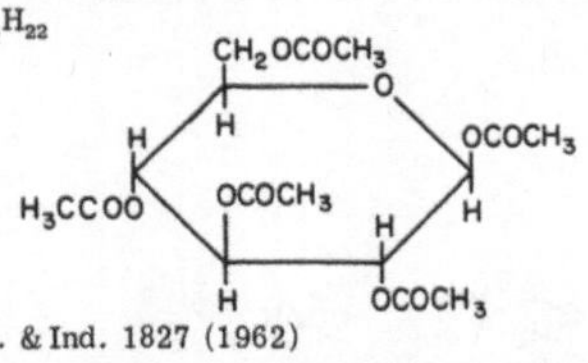

Chem. & Ind. 1827 (1962)

▶ $O_{11}C_{28}H_{36}$

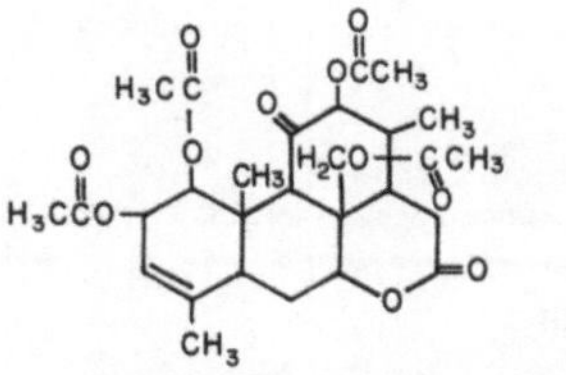

CHAPARRIN TETRAACETATE

Tetrahedron Letters No. 23:1085 (1962)

▶ $O_{11}C_{30}H_{36}$

DIHYDROGLAUCANOL PENTAACETATE

Bull. Soc. Chim. France 1432 (1961)

▶ $O_{12}C_{18}H_{24}$

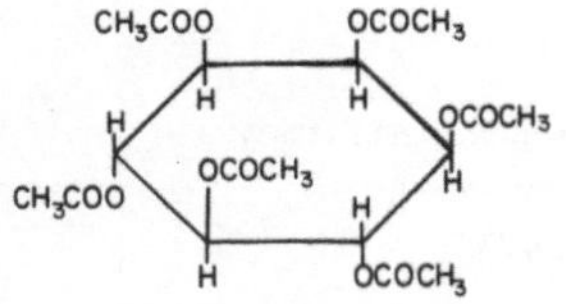

Tetrahedron 18:1147 (1962)

▶ $O_{12}C_{18}H_{24}$

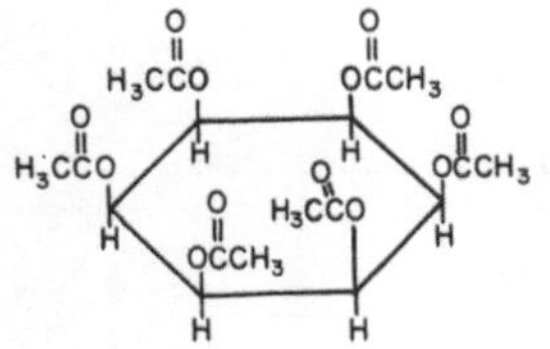

Can. J. Chem. 40:871 (1962)

▶ $O_{12}H_{18}H_{24}$

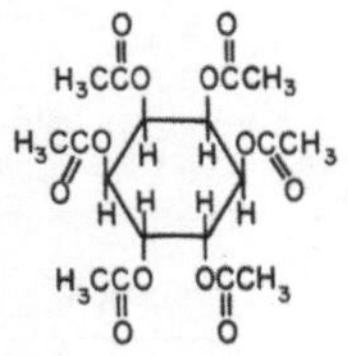

Can. J. Chem. 40:871 (1962)

► $O_{12}C_{18}H_{24}$

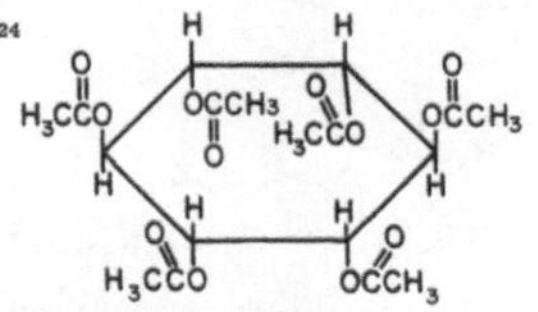

Can. J. Chem. 40:871 (1962)

► $O_{12}C_{43}H_{70}$

TRINACTIN

Helv. Chim. Acta 45:620 (1962)

► $O_{12}C_{22}H_{26}$

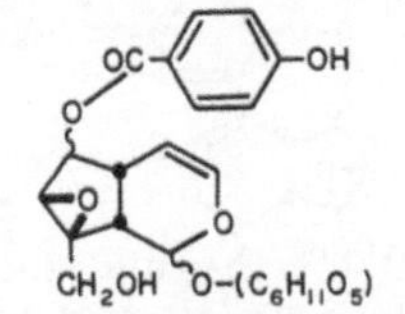

Tetrahedron Letters No. 8:325 (1962)

► $O_{13}C_{24}H_{32}$

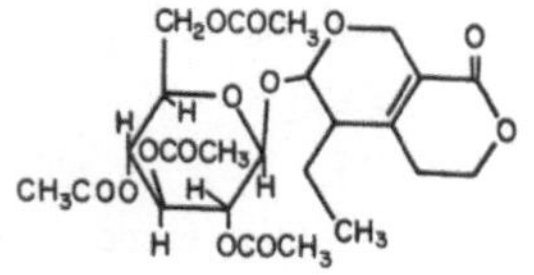

Tetrahedron 16:192 (1961)

► $O_{12}C_{22}H_{28}$

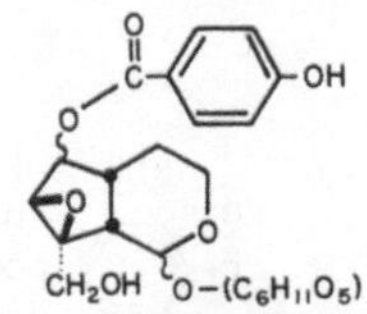

Tetrahedron Letters No. 8:325 (1962)

► $O_{13}C_{24}H_{32}$

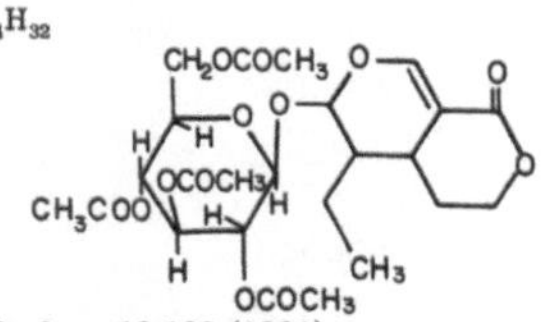

Tetrahedron 16:192 (1961)

► $O_{12}C_{40}H_{64}$

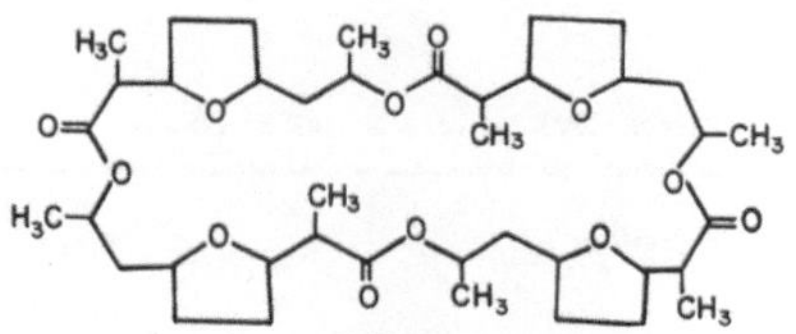

NONACTIN

Helv. Chim. Acta 45:129, 620 (1962)

► $O_{13}C_{30}H_{38}$

GLAUCARUBOL PENTAACETATE

Bull. Soc. Chim. France 1432 (1961)

► $O_{12}C_{41}H_{66}$

MONACTIN

Helv. Chim. Acta 45:620 (1962)

► $O_{13}C_{34}H_{44}$

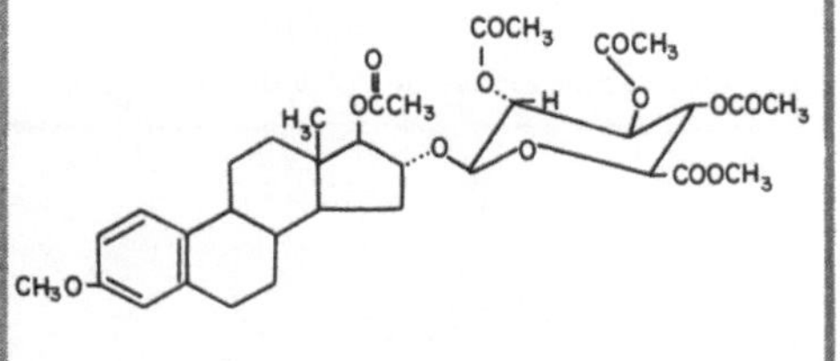

J. Am. Chem. Soc. 84:2972 (1962)

► $O_{12}C_{42}H_{68}$

DINACTIN

Helv. Chim. Acta 45:620 (1962)

► $O_{14}C_{32}H_{30-34}$

SECALONIC ACID

Chem. Ber. 95:1328 (1962)

► $O_{15}C_{27}H_{34}$

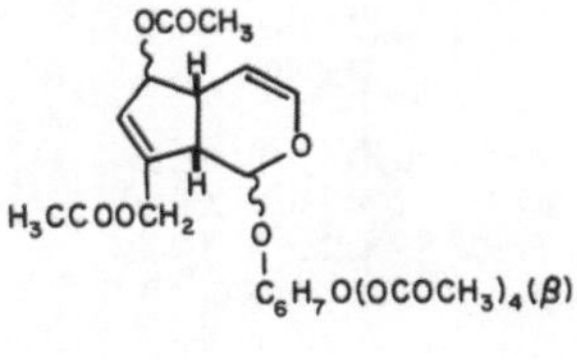


Tetrahedron Letters 110 (1961)

► $O_{16}C_{27}H_{34}$

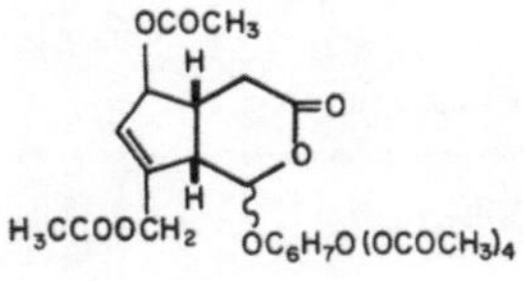


Tetrahedron Letters 110 (1961)

► $O_{16}C_{27}H_{34}$

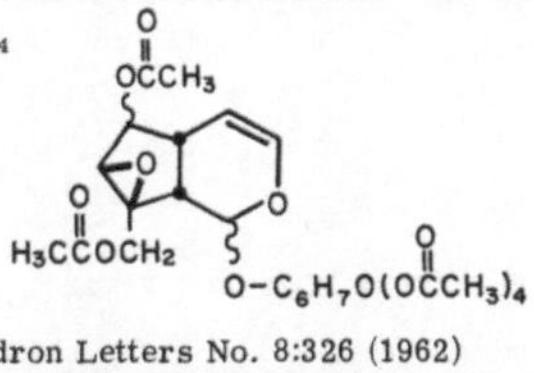


Tetrahedron Letters No. 8:326 (1962)

► $O_{16}C_{27}H_{36}$

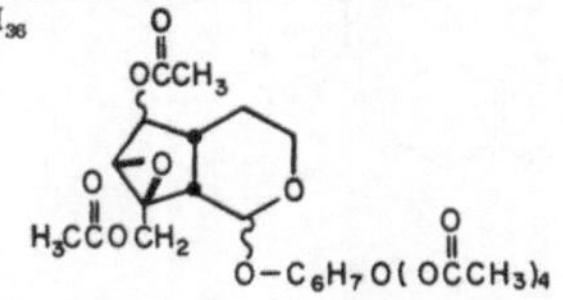


Tetrahedron Letters No. 8:326 (1962)

► $O_{18}C_{29}H_{38}$

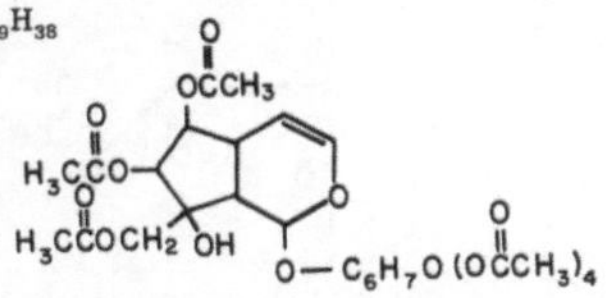


Tetrahedron Letters No. 8:327 (1962)

► $O_{18}C_{34}H_{38}$

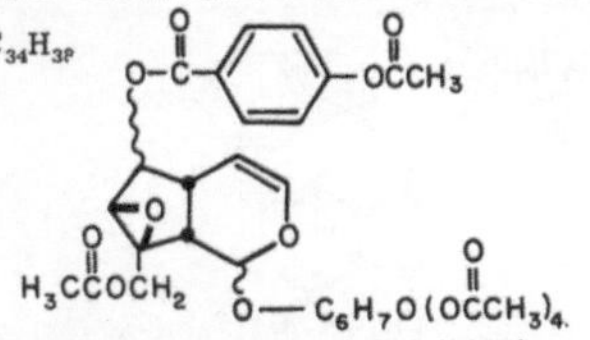


Tetrahedron Letters No. 8:326 (1962)

► $O_{18}C_{34}H_{40}$

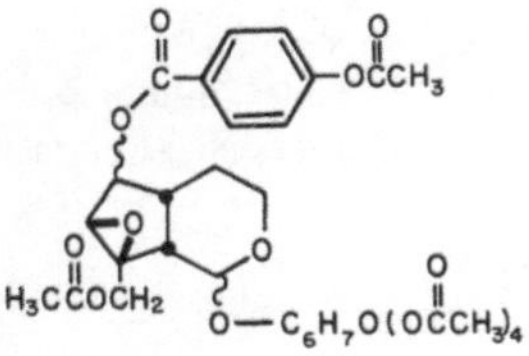


Tetrahedron Letters No. 8:326 (1962)

► $O_{22}C_{57}H_{82}$

FILIPIN DECA-ACETATE

Tetrahedron Letters 383 (1961)

► $IBrC_6H_4$

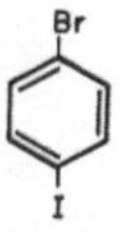


J. Chem. Phys. 37:2594 (1962)
Helv. Chim. Acta 45:568 (1962)

► $IClC_6H_4$

J. Chem. Phys. 37:2594 (1962)
Helv. Chim. Acta 45:568 (1962)

▶ $IClC_6H_7D_3$

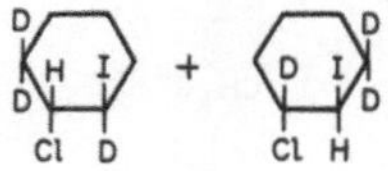


Can. J. Chem. 40:1872 (1962)

▶ $IClC_6H_{10}$

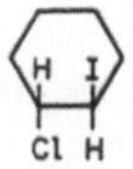


Can. J. Chem. 40:1872 (1962)

▶ $ICl_6F_7C_{14}H_8$

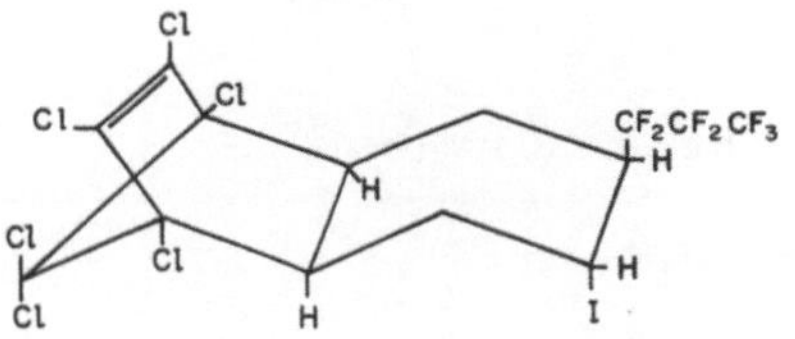


J. Org. Chem. 27:3029 (1962)

▶ $ICl_6F_7C_{15}H_8$

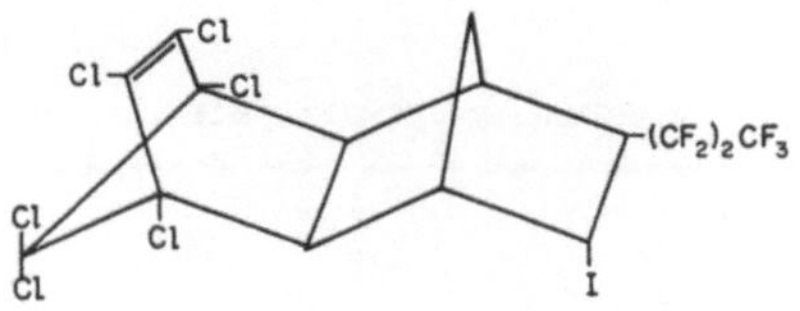


J. Org. Chem. 27:3029 (1962)

▶ $IFC_6H_4$

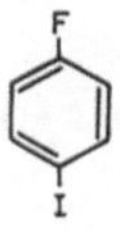


Helv. Chim. Acta 45:568 (1962)

▶ $IF_7C_7H_6$

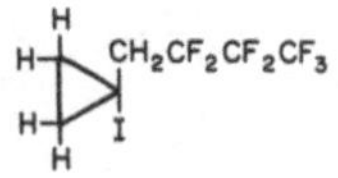


J. Org. Chem. 27:2722 (1962)

▶ $IF_7C_9H_{10}$

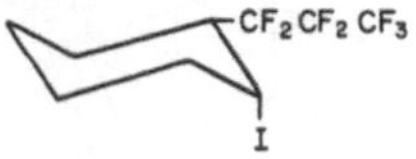


J. Am. Chem. Soc. 84:3020 (1962)

▶ $IF_7C_9H_{10}$

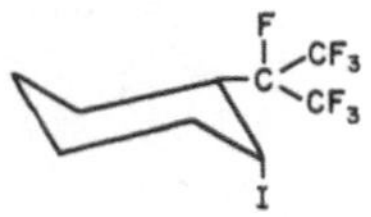


J. Am. Chem. Soc. 84:3020 (1962)

▶ $IF_7C_9H_{10}$

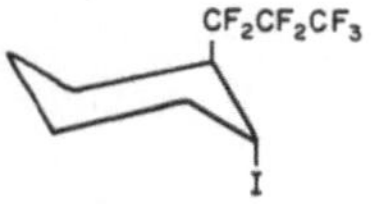


J. Am. Chem. Soc. 84:3020 (1962)

▶ $IF_7C_9H_{10}$

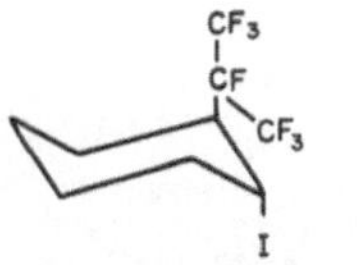


J. Am. Chem. Soc. 84:3020 (1962)

▶ $IF_7C_{10}H_8$

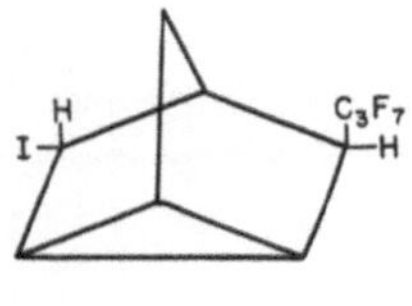


J. Org. Chem. 27:3032 (1962)

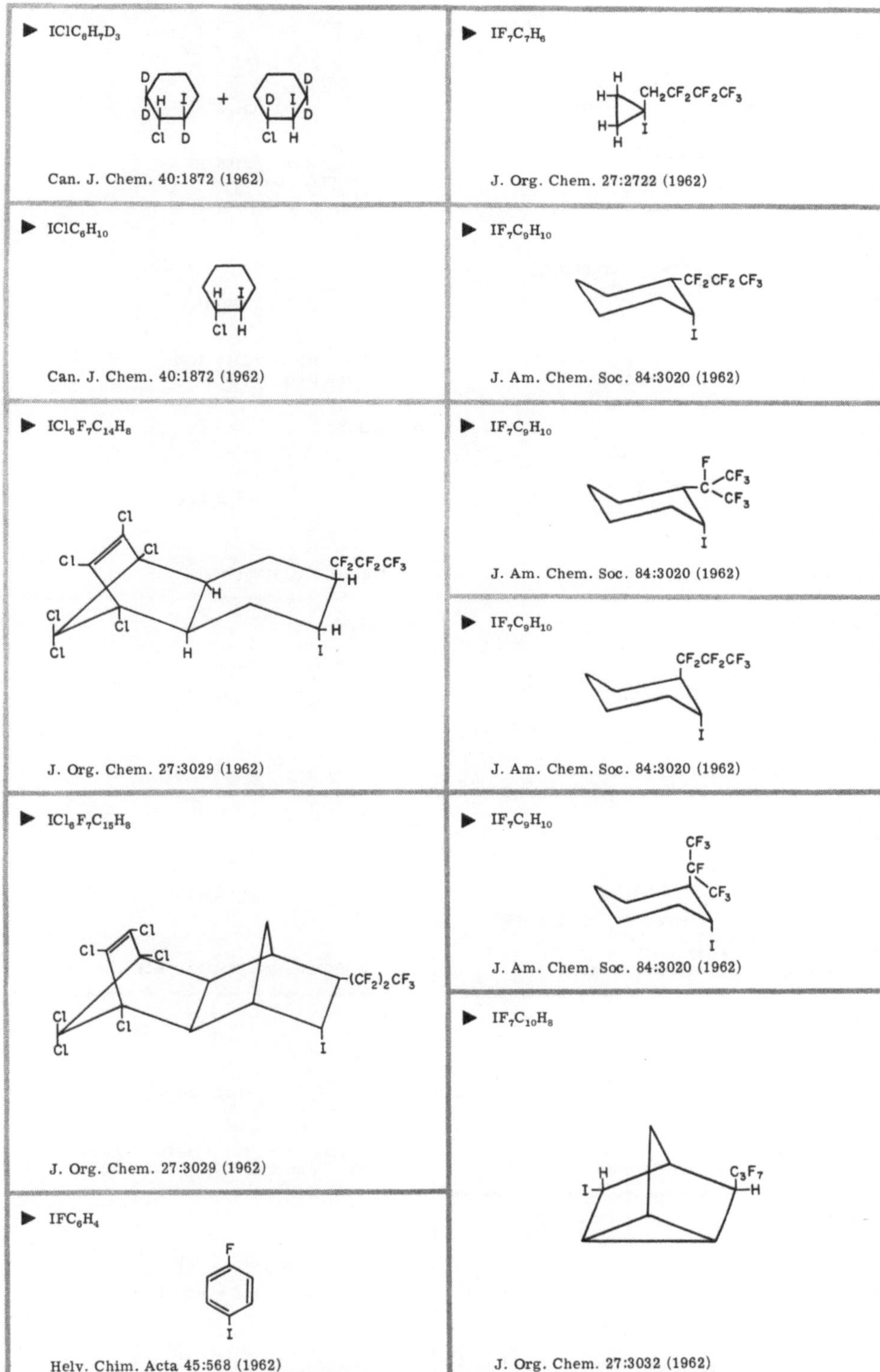

▶ $IF_7C_{10}H_8$

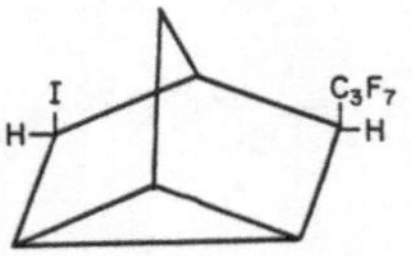

J. Org. Chem. 27:3032 (1962)

▶ $IF_7C_{10}H_{10}$

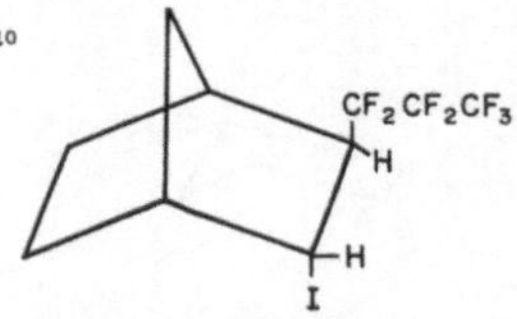

J. Org. Chem. 27:3028 (1962)

▶ $IF_7C_{13}H_{16}$

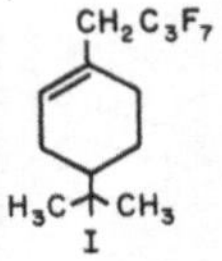

J. Org. Chem. 27:3032 (1962)

▶ $ICH_2D$

$CH_2DI$

J. Chem. Phys. 37:3012 (1962)

▶ $ICH_3$

$CH_3I$

J. Chem. Phys. 34:1099 (1961); 37:2198 (1962)
J. Phys. Chem. 66:2653 (1962)
J. Mol. Phys. 4:369 (1961)
Bull. Chem. Soc. Japan 34:143 (1961)

▶ $ICH_3Mg$

$CH_3MgI$

J. Am. Chem. Soc. 84:912 (1962)

▶ $IC_2H_5$

$C^{13}H_3CH_2I$

J. Am. Chem. Soc. 83:4479 (1961)

▶ $IC_2H_5$

$CH_3C^{13}H_2I$

J. Chem. Phys. 37:2198 (1962)
J. Am. Chem. Soc. 83:4479 (1961)

▶ $IC_2H_5$

$CH_3CH_2I$

Varian 13
J. Chem. Phys. 37:2198 (1962)
J. Chem. Phys 34:1099 (1961)

▶ $I_2C_2H_5Hg$

$C_2H_5HgI$

J. Chem. Soc. 5127 (1962)

▶ $IC_2H_5Zn$

$C_2H_5ZnI$

J. Chem. Soc. 5128 (1962)

▶ $IC_3H_3$

$CH_2{=}C{=}CHI$

J. Am. Chem. Soc. 83:4729 (1961)

▶ $IC_3H_3$

$HC{\equiv}C{-}CH_2I$

Rec. Trav. Chim. 81:635 (1962)
J. Am. Chem. Soc. 83:1978, 4729 (1961)

▶ $IC_3H_5$

$H_2C{=}CHCH_2I$

Varian 26

▶ $IC_3H_7$

$(CH_3)_2CHI$

Varian 392
J. Chem. Phys. 34:1099 (1961)

▶ $IC_3H_7$

$CH_3CH_2CH_2I$

J. Chem. Phys. 34:1094, 1099 (1961)

▶ $IC_3H_7Mg$

$CH_3CH_2CH_2MgI$

J. Chem. Soc. 5127 (1962)

▶ $IC_4H_9$

$CH_3CH_2CH_2CH_2I$

Varian 81

▶ $IC_4H_9$

$CH_3CH_2CHICH_3$

Varian 82

▶ $IC_7H_7$

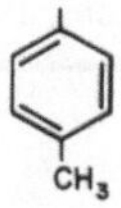


J. Chem. Phys. 37:2594 (1962)
Helv. Chim. Acta 45:568 (1962)

▶ $IH_3Si$

$SiH_3I$

J. Chem. Phys. 37:2198 (1962)

▶ $I_2CH_2$

$CH_2I_2$

J. Chem. Phys. 37:2198 (1962)

▶ $I_2CH_2$

$C^{13}H_2I_2$

J. Am. Chem. Soc. 83:4479 (1961)

▶ $I_2C_5H_{10}$

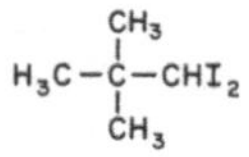


J. Chem. Soc. 475 (1962)

▶ $I_2C_6H_4$

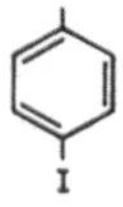


J. Chem. Phys. 37:2594 (1962)
Helv. Chim. Acta 45:568 (1962)

▶ $I_2C_6H_4$

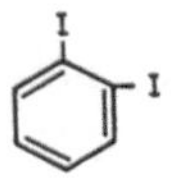


J. Chem. Phys. 37:2594 (1962)

▶ $I_2C_6H_4$

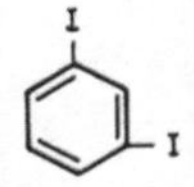

J. Chem. Phys. 37:2594 (1962)

▶ $I_2C_{16}H_{16}$

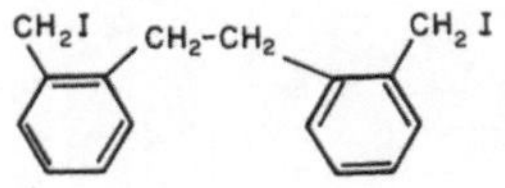


J. Am. Chem. Soc. 83:960 (1961)

▶ $I_2H_2Si$

$SiH_2I_2$

J. Chem. Phys. 37:2198 (1962)

▶ $I_3CH$

$CHI_3$

J. Mol. Phys. 4:369 (1961)

▶ $BrClC_2H_4$

$ClCH_2CH_2Br$

Varian 5

▶ $BrClC_3H_4Pd$

$CH_2{=}C(Br){-}CH_2PdCl$

Chem. & Ind. 745 (1961)

▶ $BrClC_3H_6$

$Cl-CH_2CH_2CH_2Br$

Varian 29
J. Org. Chem. 26:4846 (1961)

▶ $BrClC_4H_8$

$BrCH_2CH(CH_3)CH_2Cl$

J. Am. Chem. Soc. 83:3671 (1961)

▶ $BrClC_6H_4$

J. Chem. Phys. 37:2594 (1962)
Helv. Chim. Acta 45:568 (1962)

▶ $BrFC_2H_2$

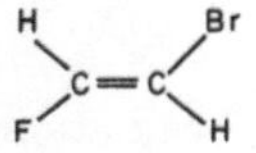


trans and cis

J. Org. Chem. 27:3500 (1962)

▶ $BrFC_6H_4$

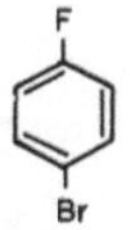


Varian 121
Helv. Chim. Acta 45:568 (1962)

▶ $BrF_2C_2H_3$

$CF_2BrCH_3$

J. Mol. Spectr. 7:307, 393 (1961)

▶ $BrF_3C_2H_2$

$CF_3CH_2Br$

J. Mol. Spectr. 7:307, 393 (1961)

▶ $BrF_5C_3H_2$

$CHF_2CF_2CHFBr$

J. Mol. Spectr. 7:322, 393 (1961)

▶ $BrF_5C_3H_2$

$CF_2BrCF_2CH_2F$

J. Mol. Spectr. 7:322, 393 (1961)

▶ $BrC_2H_5Hg$

$C_2H_5HgBr$

J. Chem. Soc. 5127 (1962)

▶ $BrCH_2D$

$CH_2DBr$

J. Chem. Phys. 37:3012 (1962)

▶ $BrC_2H_5Mg$

$C_2H_5MgBr$

J. Chem. Soc. 5127 (1962)

▶ $BrCH_3$

$CH_3Br$

J. Chem. Phys. 34:1099 (1961); 37:2198 (1962)
J. Phys. Chem. 66:2653 (1962)

▶ $BrC_3H_3$

$H_2C{=}C{=}CHBr$

J. Am. Chem. Soc. 83:4729 (1961)

▶ $BrC_2H_3$

$CH_2{=}CHBr$

Can. J. Chem. 40:2 (1962)
J. Phys. Chem. 66:2653 (1962)
J. Am. Chem. Soc. 83:2045 (1961)

▶ $BrC_3H_3$

$H{-}C{\equiv}C{-}CH_2Br$

Rec. Trav. Chim. 81:635 (1962)
J. Am. Chem. Soc. 83:1729, 1978 (1961)

▶ $BrC_2H_5$

$CH_3CH_2Br$

Varian 10
Can. J. Chem. 40:1956 (1962)
J. Mol. Spectr. 7:307, 393 (1961)
J. Chem. Phys. 34:1099 (1961); 37:2198 (1962)

▶ $BrC_2H_5$

$CH_3C^{13}H_2Br$

J. Chem. Phys. 37:2198 (1962)
J. Am. Chem. Soc. 83:4479 (1961)

▶ $BrC_3H_5$

$CH_3{-}C(Br){=}CH_2$

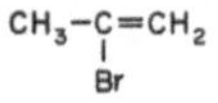

Varian 23
Arkiv Kemi 17:1 (1961)
J. Am. Chem. Soc. 83:2045 (1961)

▶ $BrC_2H_5$

$C^{13}H_3CH_2Br$

J. Am. Chem. Soc. 83:4479 (1961)

▶ $BrC_3H_5$

$(CH_3)(H)C{=}CHBr$ cis and trans

J. Am. Chem. Soc. 83:2045 (1961)

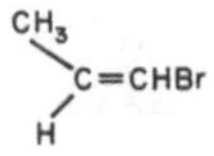

▶ $BrC_3H_5$

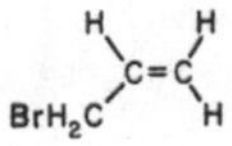

Varian 24

▶ $BrC_3H_7$

$(CH_3)_2\,C^{13}HBr$

J. Am. Chem. Soc. 83:4479 (1961)
J. Chem. Phys. 37:2198 (1962)

▶ $BrC_3H_7$

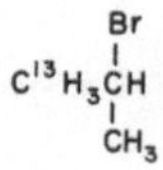

J. Am. Chem. Soc. 83:4479 (1961)

▶ $BrC_3H_7$

$CH_3CH_2CH_2Br$

J. Chem. Phys. 34:1094, 1099 (1961)

▶ $BrC_3H_7$

$(CH_3)_2CHBr$

Varian 391
Can. J. Chem. 40:1956 (1962)
J. Chem. Phys. 34:1099 (1961); 37:2198 (1962)
J. Am. Chem. Soc. 83:1230 (1961)

▶ $BrC_3H_7Mg$

$CH_3CH_2CH_2MgBr$

J. Chem. Soc. 5127 (1962)

▶ $BrC_4H_7$

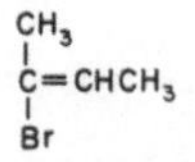

cis and trans

J. Org. Chem. 26:623 (1961)

▶ $BrC_4H_7$

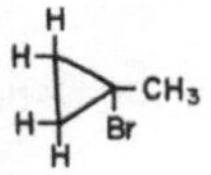

J. Org. Chem. 27:2722 (1962)

▶ $BrC_4H_7Mg$

$CH_3CH{=}CHCH_2MgBr$

J. Am. Chem. Soc. 83:494 (1961)

▶ $BrC_4H_9$

$CH_3CH_2CHBrCH_3$

Varian 418

▶ $BrC_5H_7$

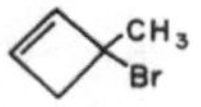

J. Am. Chem. Soc. 84:784 (1962)

▶ $BrC_5H_9$

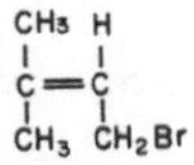

Chem. & Ind. 1020 (1962)

▶ $BrC_7H_7$

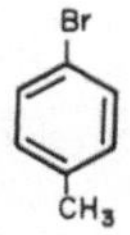

J. Chem. Phys. 37:2594 (1962)
Helv. Chim. Acta 45:568 (1962)

▶ $BrC_7H_7$

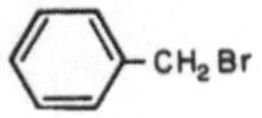

J. Chem. Phys. 37:2198 (1962)

▶ $BrC_7H_7$

$C_6H_5C^{13}H_2Br$

J. Am. Chem. Soc. 83:4479 (1961)

▶ $BrC_8H_7$

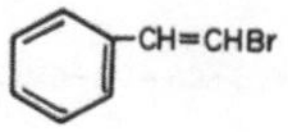


Varian 497

▶ $BrC_8H_{11}$

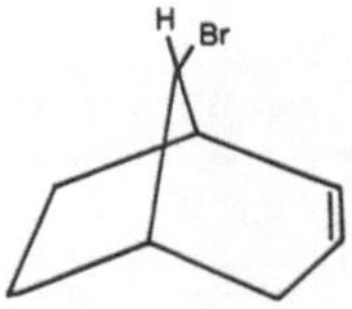


J. Am. Chem. Soc. 84:2226 (1962)

▶ $BrC_8H_{13}$

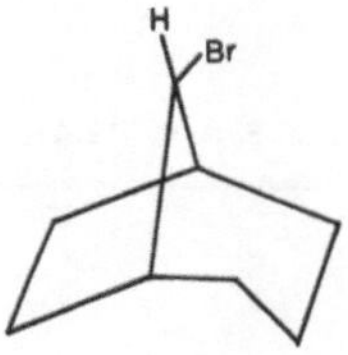


J. Am. Chem. Soc. 84:2226 (1962)

▶ $BrC_9H_9$

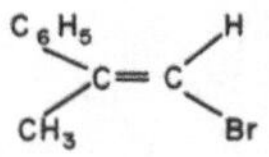


cis and trans

J. Am. Chem. Soc. 84:2252 (1962)

▶ $BrC_9H_{11}$

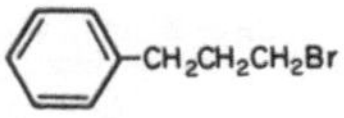


Varian 237

▶ $BrC_{10}H_{15}$

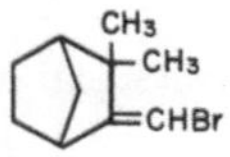


J. Org. Chem. 26:707 (1961)

▶ $BrC_{10}H_{17}$

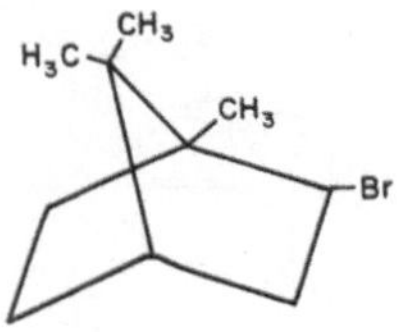


J. Org. Chem. 26:707 (1961)

▶ $BrC_{10}H_{19}$

cis and trans

Tetrahedron Letters No. 3:97 (1962)

▶ $BrC_{13}H_{11}$

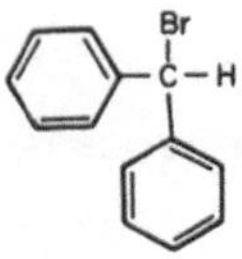


Varian 606

▶ $BrC_{16}H_{15}$

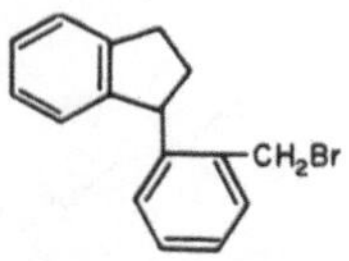


J. Am. Chem. Soc. 84:3531 (1962)

▶ $BrC_{16}H_{17}$

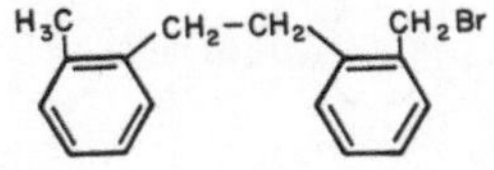

J. Am. Chem. Soc. 83:960 (1961)

▶ $BrC_{16}H_{18}As$

$[C_{12}H_9\,As(CH_3)_2CH{=}CH_2]\,Br$

J. Chem. Soc. 2762 (1961)

▶ $BrC_{25}H_{39}$

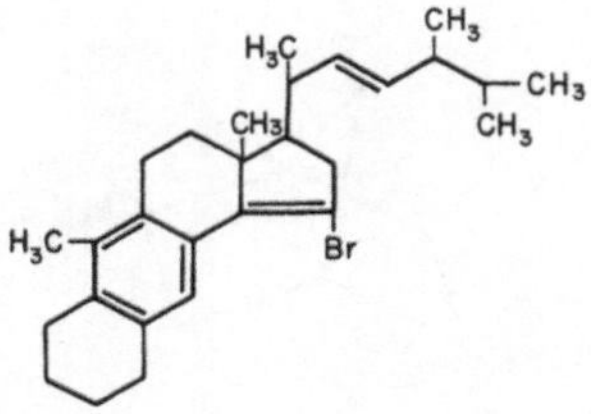

J. Am. Chem. Soc. 83:4811 (1961)

▶ $BrH_3Si$

$SiH_3Br$

J. Chem. Phys. 37:2198 (1962)

▶ $Br_2Cl_2F_2C_2$

CFClBr–CFClBr

J. Chem. Phys. 37:411 (1962)

▶ $Br_2FC_2H$

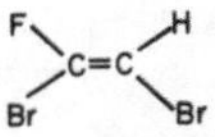

J. Org. Chem. 27:3500 (1962)

▶ $Br_2FC_2H$

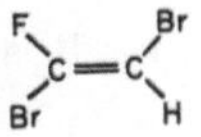

J. Org. Chem. 27:3500 (1962)

▶ $Br_2F_4C_4H_2$

$BrF_2C{-}CH{=}CH{-}CBrF_2$

J. Am. Chem. Soc. 83:382 (1961)

▶ $Br_2CH_2$

$C^{13}H_2Br_2$

J. Am. Chem. Soc. 83:4479 (1961)

▶ $Br_2CH_2$

$CH_2Br_2$

J. Chem. Phys. 37:2198 (1962)

▶ $Br_2C_2H_4$

$BrC^{13}H_2C^{13}H_2Br$

J. Am. Chem. Soc. 83:4479 (1961)

▶ $Br_2C_2H_4$

$BrCH_2CH_2Br$

J. Chem. Phys. 37:2198 (1962)

▶ $Br_2C_2H_4$

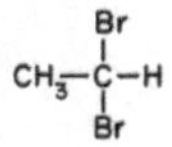

Varian 370

▶ $Br_2C_3H_4$

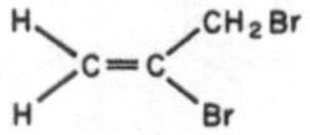


Varian 17
J. Am. Chem. Soc. 83:231 (1961)

▶ $Br_2C_3H_6$

$BrCH_2CH_2CH_2Br$

J. Org. Chem. 26:4846 (1961)

▶ $Br_2C_3H_6$

$CH_3CHBr\,CH_2Br$

J. Chem. Phys. 37:3012 (1962)
Proc. Chem. Soc. 213 (1962)

▶ $Br_2C_4H_4$

$BrCH_2{-}C{\equiv}C{-}CH_2Br$

J. Chem. Phys. 36:2644 (1962)

▶ $Br_2C_4H_6$

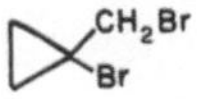


J. Org. Chem. 27:2722 (1962)

▶ $Br_2C_4H_8$

$CH_3CHBrCHBrCH_3$

meso and D,L

J. Am. Chem. Soc. 84:743, 747 (1962)
Can. J. Chem. 39:2266 (1961)

▶ $Br_2C_4H_8$

$CH_3CH_2CHBrCH_2Br$

Varian 74

▶ $Br_2C_4H_8$

$(CH_3)_2CBrCH_2Br$

Varian 412

▶ $Br_2C_6H_4$

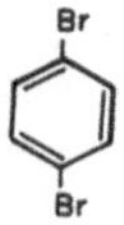


J. Chem. Phys. 37:2594 (1962)
Helv. Chim. Acta 45:568 (1962)

▶ $Br_2C_6H_4$

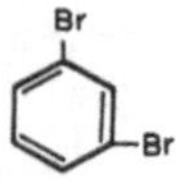


J. Chem. Phys. 37:2594 (1962)
Helv. Chim. Acta 45:568 (1962)

▶ $Br_2C_6H_4$

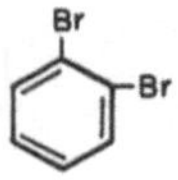


J. Chem. Phys. 37:2594 (1962)

▶ $Br_2C_6H_8$

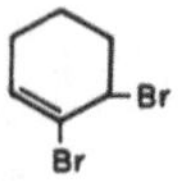


J. Org. Chem. 27:748 (1962)

▶ $Br_2C_6H_{10}$

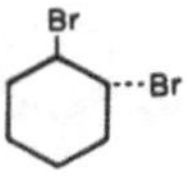


Trans. Faraday Soc. 57:390 (1961)

▶ $Br_2C_8H_8$

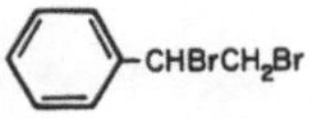


Varian 503

▶ $Br_2C_9H_9D$

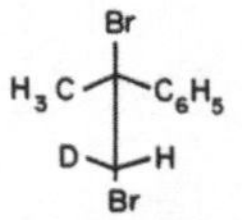


J. Am. Chem. Soc. 84:2252 (1962)

▶ $Br_2C_9H_9D$

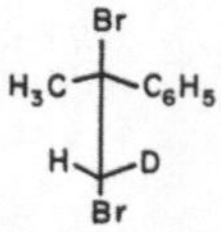


J. Am. Chem. Soc. 84:2252 (1962)

▶ $Br_2C_9H_{10}$

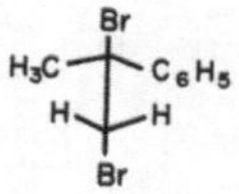


J. Am. Chem. Soc. 84:2252 (1962)

▶ $Br_2C_{10}H_6$

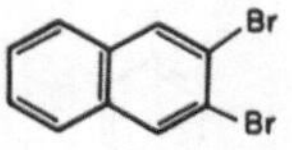


Varian 244

▶ $Br_2C_{12}H_8$

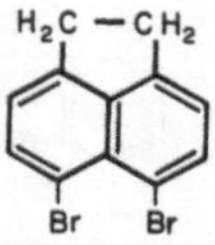


J. Org. Chem. 27:674 (1962)

▶ $Br_2C_{14}H_8$

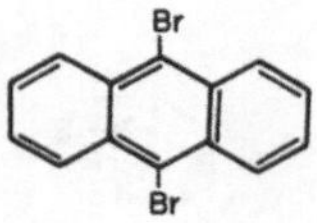


Varian 618

▶ $Br_2C_{16}H_{12}$

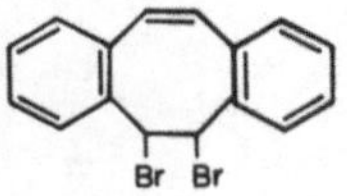


Tetrahedron 18:1005 (1962)

▶ $Br_2C_{16}H_{16}$

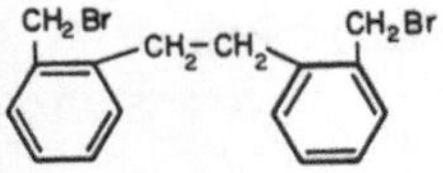


J. Am. Chem. Soc. 83:960 (1961)

▶ $Br_2C_{30}H_{20}$

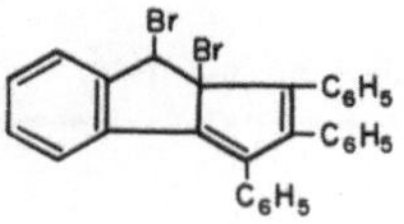


J. Am. Chem. Soc. 84:1505 (1962)

▶ $Br_3CH$

$CHBr_3$

J. Chem. Phys. 37:2198 (1962)

▶ $Br_3C_2H_3$

$CHBr_2-CH_2Br$

J. Chem. Phys. 36:3353 (1962)

▶ $Br_3C_6H_3$

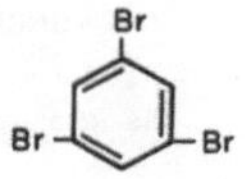


Can. J. Chem. 40:1759 (1962)

▶ $Br_4C_{10}H_8$

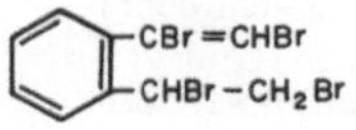


J. Org. Chem. 26:3584 (1961)

▶ $Br_4C_{10}H_{10}$

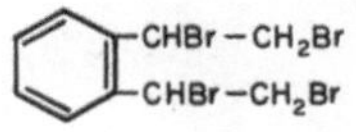


J. Org. Chem. 26:3584 (1961)

▶ $Br_4C_{16}H_{12}$

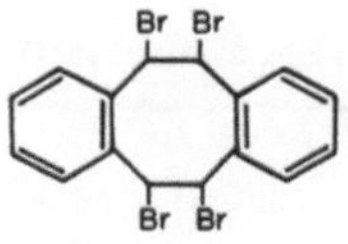


Tetrahedron 18:1006 (1962)

▶ $ClFC_3H_4$

$H_3C-CF=CHCl$

J. Chem. Phys. 37:456 (1962)

▶ $ClFC_6H_4$

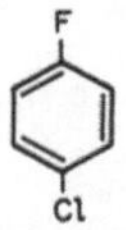


Helv. Chim. Acta 45:568 (1962)

▶ $ClF_2CH$

$CHF_2Cl$

J. Am. Chem. Soc. 84:3973 (1962)

▶ $ClF_2C_2H_3$

$CF_2ClCH_3$

J. Mol. Spectr. 7:307, 393 (1961)

▶ $ClF_2C_3H_5$

$CH_3CF_2CH_2Cl$

Varian 381

▶ $ClF_3C_2H_2$

$CF_3CH_2Cl$

J. Mol. Spectr. 7:307, 393 (1961)

▶ $ClF_3C_{10}H_{10}$

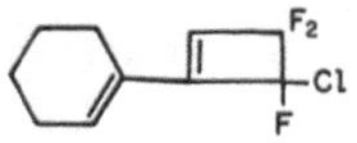


J. Am. Chem. Soc. 83:875 (1961)

▶ $ClF_3C_{12}H_{10}Ni$

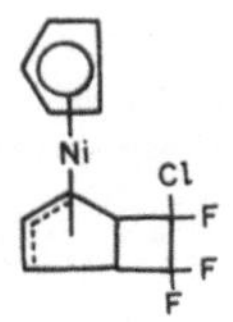


J. Am. Chem. Soc. 84:497 (1962)

▶ $ClF_4C_3H$

$CClF=CHCF_3$

J. Mol. Spectr. 7:385 (1961)

▶ $ClF_5C_3H_2$

$CHF_2CF_2CHFCl$

J. Mol. Spectr. 7:322, 393 (1961)

▶ $ClF_5C_3H_2$

$CF_2ClCF_2CH_2F$

J. Mol. Spectr. 7:322, 393 (1961)

▶ $ClCH_2D$

$CH_2DCl$

J. Chem. Phys 37:3012 (1962)

▶ $ClCH_3$

$CH_3Cl$

J. Chem. Phys. 34:1099 (1961); 37:2198 (1962)

▶ $ClCH_3Hg$

$CH_3HgCl$

J. Chem. Soc. 5127 (1962)

▶ $ClC_2HD_2$

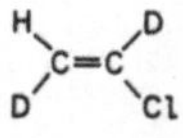


J. Chem. Phys. 34:2136 (1961)

▶ $ClC_2H_2D$

$CHD=CHCl$

cis and trans

J. Chem. Phys. 34:2136 (1961)

▶ $(ClC_2H_2D)_X$

$-(CH_2CDCl)_x-$

α-DEUTEROPOLYVINYLCHLORIDE

Chem. & Ind. 1826 (1962)

▶ $ClC_2H_3$

$CHCl=C^{13}H_2$

J. Chem. Phys. 37:2198 (1962)

▶ $ClC_2H_3$

$CH_2=C^{13}HCl$

J. Chem. Phys. 37:2198 (1962)

▶ $ClC_2H_3$

$CH_2=CHCl$

J. Am. Chem. Soc. 83:1300 (1961)
Can. J. Chem. 40:2 (1962)
J. Chem. Phys. 34:2136 (1961); 36:2644 (1962); 37:2198 (1962)

▶ $ClC_2H_5$

$CH_3CH_2Cl$

Varian 11
Can. J. Chem. 40:1956 (1962)
J. Chem. Phys. 34:1099 (1961); 37:2198 (1962)
J. Phys. Chem. 66:2653 (1962)
J. Mol. Spectr. 7:307, 393 (1961)

▶ $ClC_2H_5Hg$

$C_2H_5HgCl$

J. Chem. Soc. 5127 (1962)

▶ $ClC_3H_3$

$H_2C{=}C{=}CHCl$

J. Am. Chem. Soc. 83:4729 (1961)

▶ $ClC_3H_3$

$HC{\equiv}CCH_2Cl$

Trans. Faraday Soc. 57:28 (1961)
J. Am. Chem. Soc. 83:1978, 4729 (1961)
J. Chem. Phys. 36:2644 (1962)
Rec. Trav. Chim. 81:635 (1962)

▶ $ClC_3H_4 \cdot Co(CO)_3$

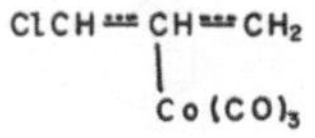


J. Am. Chem. Soc. 83:1601 (1961)

▶ $ClC_3H_4 \cdot Co(CO)_3$

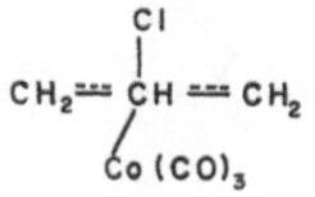


J. Am. Chem. Soc. 83:1601 (1961)

▶ $ClC_3H_5$

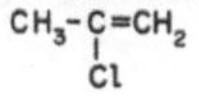


Arkiv Kemi 17:1 (1961)
J. Am. Chem. Soc. 83:2045 (1961)

▶ $ClC_3H_5$

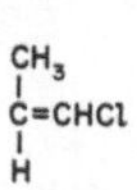


cis and trans

J. Am. Chem. Soc. 83:2045 (1961)

▶ $ClC_3H_5Pd$

$CH_2{=}CH-CH_2-PdCl$

Chem. & Ind. 745 (1961)

▶ $ClC_3H_7$

$CH_3CH_2CH_2Cl$

J. Chem. Phys. 34:1094, 1099 (1961)

▶ $ClC_3H_7$

$(H_3C)_2CHCl$

Can. J. Chem. 40:1956 (1962)
J. Chem. Phys. 34:1099 (1961)

▶ $ClC_3H_7Mg$

$CH_3CH_2CH_2MgCl$

J. Chem. Soc. 5127 (1962)

▶ $ClC_4H_5$

$CH_3C{\equiv}CCH_2Cl$

Arkiv Kemi 16:471 (1961)

▶ $ClC_4H_7$

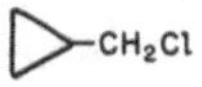


J. Am. Chem. Soc. 83:1987 (1961)

▶ $ClC_4H_7$

$CH_2{=}CHCH_2CH_2Cl$

J. Am. Chem. Soc. 83:1987 (1961)

► $ClC_4H_7$

$(CH_3)_2C{=}CHCl$

Varian 67

► $ClC_4H_7Pd$

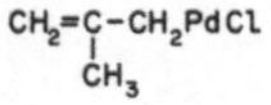

Chem. & Ind. 745 (1961)

► $ClC_4H_{10}Al$

$(C_2H_5)_2AlCl$

Rec. Trav. Chim. 81:729 (1962)

► $ClC_5H_9$

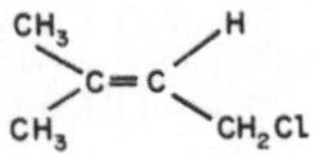

Chem. & Ind. 1020 (1962)

► $ClC_5H_{11}$

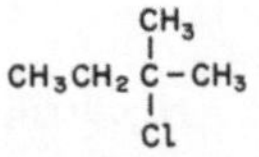

Varian 447
J. Am. Chem. Soc. 84:37 (1962)

► $ClC_6H_5$

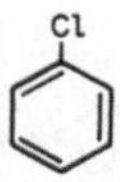

Helv. Chem. Acta 45:568 (1962)

► $ClC_6H_5 \cdot Cr(CO)_3$

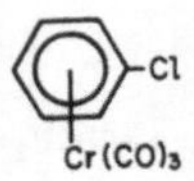

J. Chem. Soc. 3660 (1962)

► $ClC_6H_{11}$

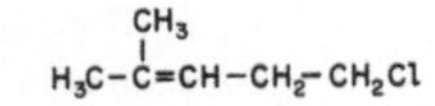

J. Am. Chem. Soc. 84:2611 (1962)

► $ClC_7H_7$

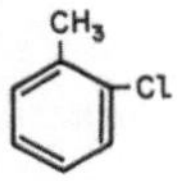

Bull. Chem. Soc. Japan 34:143 (1961)

► $ClC_7H_7$

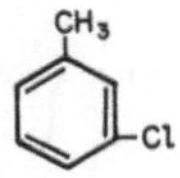

Bull. Chem. Soc. Japan 34:143 (1961)

► $ClC_7H_7$

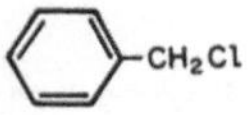

J. Chem. Phys. 37:2198 (1962)

► $ClC_7H_7$

$C_6H_5C^{13}H_2Cl$

J. Am. Chem. Soc. 83:4479 (1961)

► $ClC_7H_7$

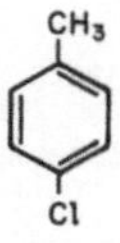

Bull. Chem. Soc. Japan 34:143 (1961)
J. Chem. Phys. 37:2594 (1962)
Can. J. Chem. 40:2331 (1962)
Helv. Chim. Acta 45:568 (1962)

▶ $ClC_7H_7 \cdot Cr(CO)_3$

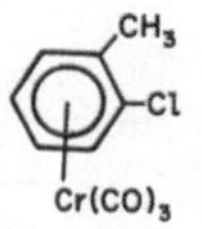

J. Chem. Soc. 3660 (1962)

▶ $ClC_7H_7 \cdot Cr(CO)_3$

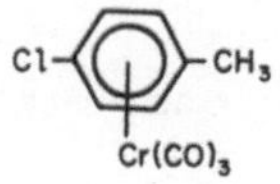

J. Chem. Soc. 3660 (1962)

▶ $ClC_8H_7$

Varian 498

▶ $ClC_8H_{12} \cdot SbF_6$

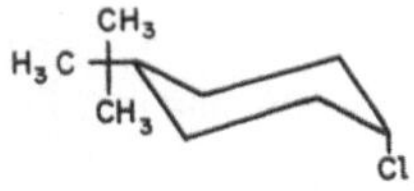

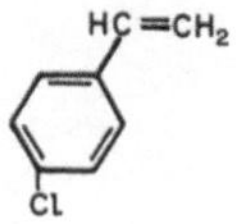

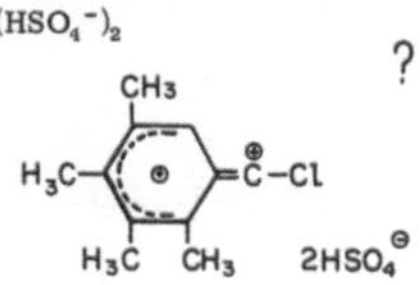

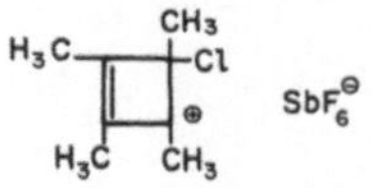

J. Am. Chem. Soc. 84:3199 (1962)

▶ $ClC_{10}H_{11} \cdot (HSO_4^-)_2$

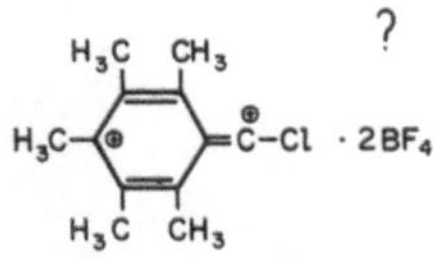

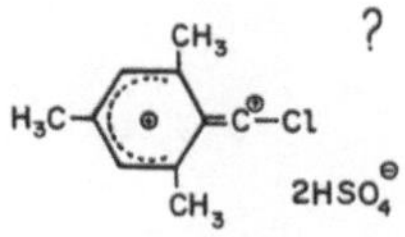

J. Am. Chem. Soc. 83:4460 (1961)

▶ $ClC_{10}H_{11} \cdot 2BF_4$

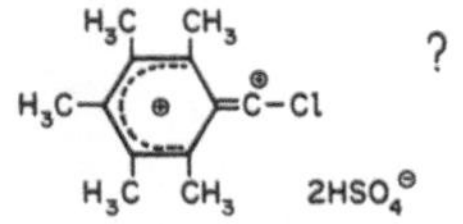

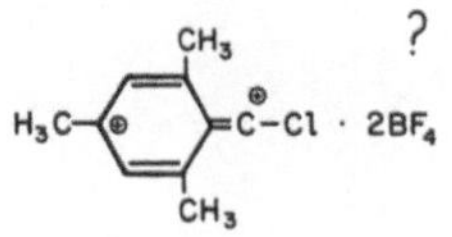

Tetrahedron Letters No. 22:985 (1962)

▶ $ClC_{10}H_{13}$

Varian 263

▶ $ClC_{10}H_{19}$

J. Am. Chem. Soc. 84:2463 (1962)

▶ $ClC_{10}H_{19}$

J. Am. Chem. Soc. 84:2463 (1962)

▶ $ClC_{11}H_{13} \cdot (HSO_4^-)_2$

J. Am. Chem. Soc. 83:4460 (1961)

▶ $ClC_{12}H_{15} \cdot 2BF_4$

Tetrahedron Letters No. 22:984 (1962)

▶ $ClC_{12}H_{15} \cdot (HSO_4^-)_2$

J. Am. Chem. Soc. 83:4460 (1961)

▶ $ClC_{13}H_{11}$

Varian 176

▶ $ClC_{16}H_{17}$

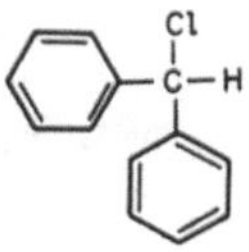

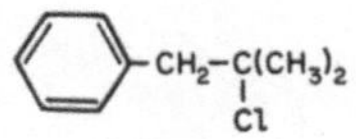

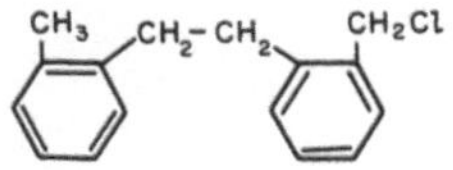

J. Am. Chem. Soc. 83:960 (1961)

► $ClH_3Si$

$SiH_3Cl$

J. Chem. Phys. 37:2198 (1962)

► $Cl_2F_2C_{11}H_{10}$

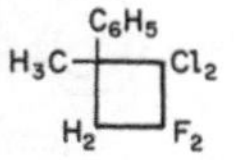


J. Am. Chem. Soc. 84:2935 (1962)

► $Cl_2FCH$

$CHFCl_2$

J. Am. Chem. Soc. 84:3973 (1962)

► $Cl_2F_3C_3H$

$CCl_2{=}CHCF_3$

J. Mol. Spectr. 7:385 (1961)

► $Cl_2FC_3H_5$

$CH_3CFClCH_2Cl$

Varian 382

► $Cl_2CHD$

$CHDCl_2$

J. Chem. Phys. 37:3012 (1962)

► $Cl_2FC_{10}H_9$

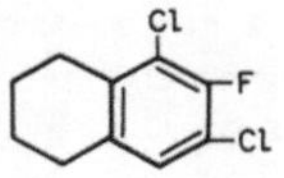


J. Am. Chem. Soc. 83:874 (1961)

► $Cl_2CH_2$

$CH_2Cl_2$

J. Chem. Phys. 37:2198 (1962)

► $Cl_2CH_2$

$C^{13}H_2Cl_2$

J. Am. Chem. Soc. 83:4479 (1961)

► $Cl_2F_2C_2H_2$

$CHCl_2CHF_2$

J. Chem. Phys. 36:3353 (1962)

► $Cl_2F_2C_2H_2$

$CH_2ClCF_2Cl$

Varian 369

► $Cl_2C_2H_2$

H H<br>
| |<br>
ClC=CCl

cis and trans

J. Chem. Phys. 34:2136 (1961); 36:2644 (1962); 37:2198, 2729 (1962)
J. Am. Chem. Soc. 84:3405 (1962)

▶ $Cl_2C_2H_2$

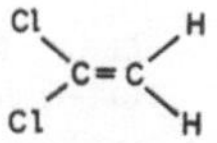

J. Chem. Phys. 34:2136 (1961); 36:2644 (1962); 37:2198 (1962)

▶ $Cl_2C_2H_4$

$CH_3CHCl_2$

J. Chem. Phys. 37:2198 (1962)

▶ $Cl_2C_2H_4$

$ClCH_2CH_2Cl$

J. Chem. Phys. 37:2198 (1962)

▶ $Cl_2C_2H_4$

$ClC^{13}H_2C^{13}H_2Cl$

J. Am. Chem. Soc. 83:4479 (1961)

▶ $Cl_2C_3H_4$

$HCCl=CHCH_2Cl$

cis and trans

J. Am. Chem. Soc. 83:231 (1961)

▶ $Cl_2C_3H_4$

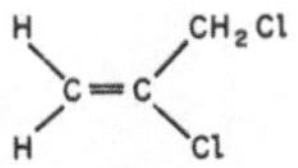

Varian 18
J. Am. Chem. Soc. 83:231 (1961)

▶ $Cl_2C_3H_4Pd$

$CH_2=C(Cl)-CH_2PdCl$

Chem. & Ind. 745 (1961)

▶ $Cl_2C_3H_4Pd$

$CHCl=CH-CH_2PdCl$

Chem. & Ind. 745 (1961)

▶ $Cl_2C_3H_6$

$ClCH_2CH_2CH_2Cl$

Varian 31
J. Org. Chem. 26:4845 (1961)

▶ $Cl_2C_3H_6$

$CH_3CHClCH_2Cl$

Varian 30
J. Chem. Phys. 37:3012 (1962)
Proc. Chem. Soc. 213 (1962)

▶ $Cl_2C_4H_4$

$ClCH_2-C\equiv C-CH_2Cl$

J. Chem. Phys. 36:2644 (1962)

▶ $Cl_2C_4H_4Si$

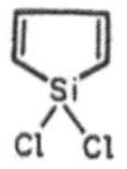

J. Am. Chem. Soc. 83:5029 (1961); 84:4727 (1962)

▶ $Cl_2C_4H_6$

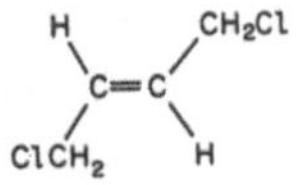

Varian 404

▶ $Cl_2C_4H_6Si$

$(H_2C=CH)_2SiCl_2$

J. Am. Chem. Soc. 84:4727 (1962)

▶ $Cl_2C_4H_8$

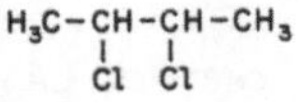

meso and D,L

J. Am. Chem. Soc. 84:743 (1962)

▶ $Cl_2C_6H_4$

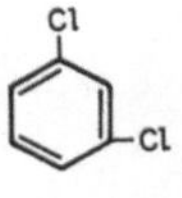

Helv. Chim. Acta 45:568 (1962)

▶ $Cl_2C_6H_4$

Can. J. Chem. 40:1759 (1962)
J. Chem. Phys. 37:2594 (1962)
Helv. Chim. Acta 45:568 (1962)

▶ $Cl_2C_6H_4$

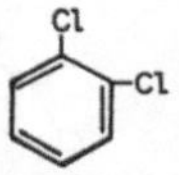

J. Chem. Phys. 37:2594 (1962)
Helv. Chim. Acta 45:568 (1962)

▶ $Cl_2C_6H_{10}$

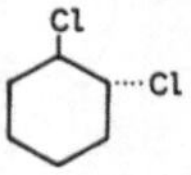

Trans. Faraday Soc. 57:390 (1961)

▶ $Cl_2C_6H_{10}$

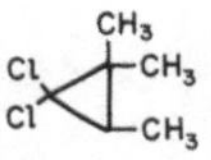

Chem. Ber. 95:2280 (1962)

▶ $Cl_2C_7H_{12}$

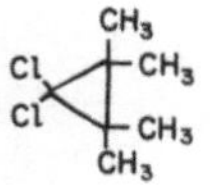

Chem. Ber. 95:2280 (1962)

▶ $Cl_2C_8H_{12} \cdot Pd_2Cl_2$

$[PdCl(C_4H_6Cl)]_2$

Chem. & Ind. 1190 (1962)

▶ $Cl_2C_{10}H_{10}$

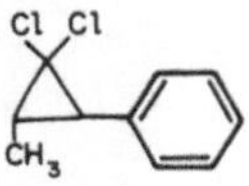

J. Am. Chem. Soc. 84:2249 (1962)

▶ $Cl_2C_{10}H_{10}$

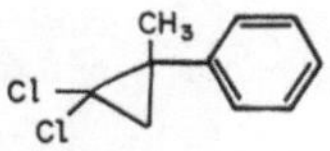

J. Am. Chem. Soc. 84:2249 (1962)
Chem. Ber. 95:2280 (1962)

▶ $Cl_2C_{14}H_8$

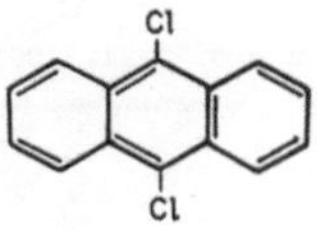

Varian 619

▶ $Cl_3FC_2H_2$

$CHCl_2CHFCl$

J. Chem. Phys. 36:3353 (1962)

▶ $Cl_3F_2C_2H$

$CHCl_2{-}CClF_2$

J. Chem. Phys. 37:1466 (1962)

▶ $Cl_3CH$

$CHCl_3$

J. Chem. Phys. 37:2198 (1962)

▶ $Cl_3CH$

$C^{13}HCl_3$

J. Am. Chem. Soc. 83:4479 (1961)

▶ $Cl_3C_2H$

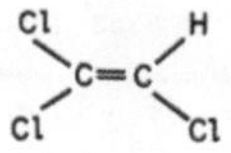

J. Chem. Phys. 34:2136 (1961); 36:2644 (1962); 37:2198 (1962)

▶ $Cl_3C_2HD_2$

$CHD_2CCl_3$

J. Chem. Phys. 37:3012 (1962)

▶ $Cl_3C_2H_2D$

$CH_2DCCl_3$

J. Chem. Phys. 37:3012 (1962)

▶ $Cl_3C_2H_3$

$CH_3CCl_3$

J. Chem. Phys. 37:2198 (1962)

▶ $Cl_3C_2H_3$

$Cl_2CHCH_2Cl$

Varian 2

▶ $Cl_3C_2H_3Si$

$H_2C{=}CHSiCl_3$

Varian 3
J. Am. Chem. Soc. 84:4727 (1962)

▶ $Cl_3C_3H_3$

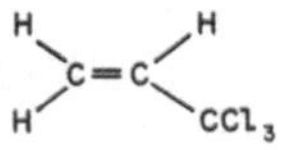

J. Chem. Phys. 36:566 (1962)

▶ $Cl_3C_3H_5$

$CH_3CCl_2CH_2Cl$

Varian 383

▶ $Cl_3C_6H_3$

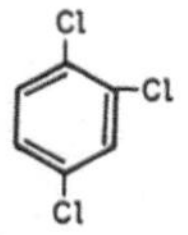

Mol. Phys. 4:265 (1961)

▶ $Cl_3C_6H_3$

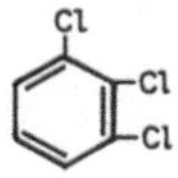

Helv. Chim. Acta 45:568 (1962)

▶ $Cl_3C_6H_3$

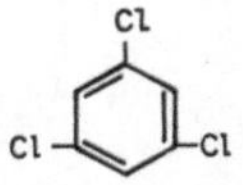

Can. J. Chem. 40:1759 (1962)

▶ $Cl_3HSi$

$SiHCl_3$

J. Chem. Phys. 37:2198 (1962)

▶ $Cl_3C_6H_6As$

$As(CH=CHCl)_3$

J. Chem. Soc. 2762 (1961)

▶ $Cl_4FC_2H$

$CFCl_2 \cdot CHCl_2$

Can. J. Chem. 39:39 (1961)
J. Chem. Phys. 36:3353 (1962)

▶ $Cl_3C_{10}H_{11}$

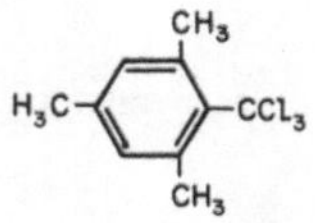

J. Am. Chem. Soc. 83:4460 (1961)

▶ $Cl_4C_2H_2$

$Cl_2C^{13}HC^{13}HCl_2$

J. Am. Chem. Soc. 83:4479 (1961)

▶ $Cl_3C_{11}H_{13}$

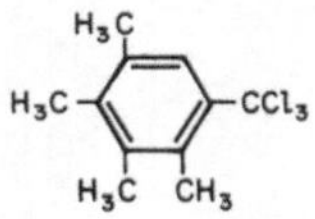

J. Am. Chem. Soc. 83:4460 (1961)

▶ $Cl_4C_2H_2$

$CHCl_2CHCl_2$

J. Chem. Phys. 36:3353 (1962); 37:2198 (1962)

▶ $Cl_3C_{12}H_{15}$

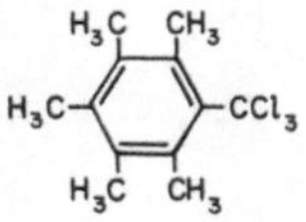

J. Am. Chem. Soc. 83:4460 (1961)

▶ $Cl_5C_3H_3$

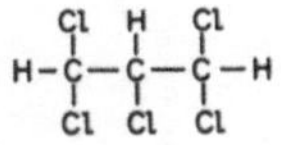

Varian 377

▶ $Cl_3C_{20}H_{35}$

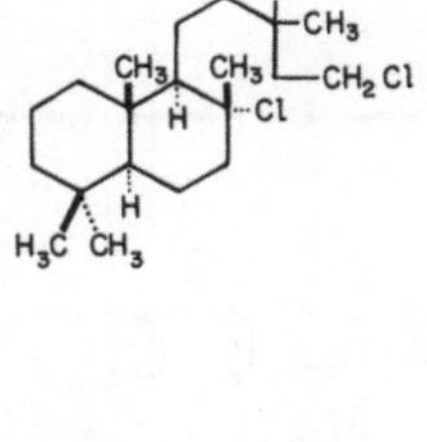

Tetrahedron 18:285 (1962)

▶ $Cl_5C_{14}H_9$

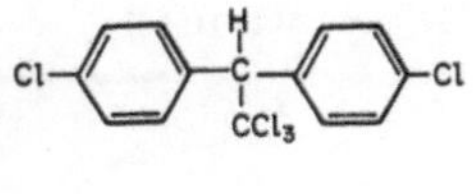

Varian 620

▶ $Cl_{12}C_{20}H_8$

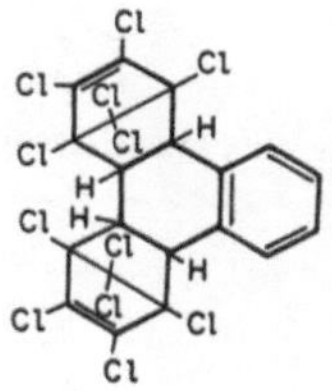

Varian 338

▶ $FCH_2D$

$CH_2DF$

J. Chem. Phys. 37:3012 (1962)

▶ $FCH_3$

$CH_3F$

J. Chem. Phys. 34:1099 (1961); 37:2198 (1962)

▶ $FC_2H_3$

$CH_2{=}CHF$

Can. J. Chem. 40:2 (1962)

▶ $FC_2H_5$

$CH_3CH_2F$

J. Chem. Phys. 34:1099 (1961)
J. Mol. Spectr. 7:307, 393 (1961)
J. Am. Chem. Soc. 83:4473 (1961)

▶ $FC_3H_5$

$CH_3-CH{=}CHF$

J. Mol. Spectr. 9:30 (1962)

▶ $FC_3H_7$

$CH_3CHFCH_3$

J. Chem. Phys. 34:1099 (1961)

▶ $FC_3H_7$

$CH_3CH_2CH_2F$

J. Chem. Phys. 34:1099 (1961)

▶ $FC_6H_5$

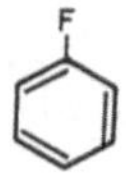

Tetrahedron Letters 519 (1961)

▶ $FC_6H_5 \cdot Cr(CO)_3$

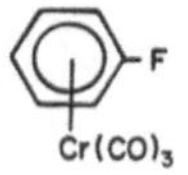

J. Chem. Soc. 3660 (1962)

▶ $FC_7H_7$

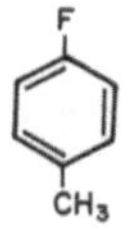

Helv. Chim. Acta 45:568 (1962)

▶ $FC_7H_7$

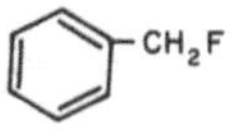

J. Am. Chem. Soc. 84:3972 (1962)

▶ $FC_7H_7 \cdot Cr(CO)_3$

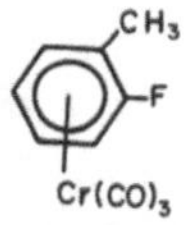

J. Chem. Soc. 3660 (1962)

▶ $FC_7H_7 \cdot Cr(CO)_3$

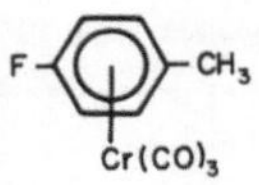


J. Chem. Soc. 3660 (1962)

▶ $FC_8H_7$

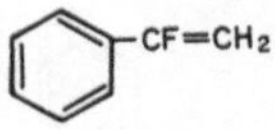


J. Org. Chem. 27:4019 (1962)

▶ $FH_3Si$

$SiH_3F$

J. Chem. Phys. 37:2198 (1962)

▶ $F_2CH_2$

$CH_2F_2$

J. Chem. Phys. 37:2198, 2907 (1962)
J. Am. Chem. Soc. 84:3973 (1962)

▶ $F_2C_2H_2$

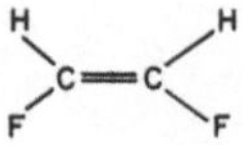


J. Chem. Soc. 743 (1961)

▶ $F_2C_2H_4$

$CHF_2CH_3$

J. Mol. Spectr. 7:307, 393 (1961)
J. Chem. Phys. 37:2907 (1962)

▶ $F_2C_6H_4$

Helv. Chim. Acta 45:568 (1962)

▶ $F_2H_2Si$

$SiH_2F_2$

J. Chem. Phys. 37:2198 (1962)

▶ $F_3CH$

$CHF_3$

J. Chem. Phys. 37:2731 (1962)
J. Am. Chem. Soc. 84:3973 (1962)

▶ $F_3CH_3SeSi$

$CF_3SeSiH_3$

J. Chem. Soc. 2293 (1962)

▶ $F_3C_2H_3$

$CF_3CH_3$

J. Mol. Spectr. 7:307, 393 (1961)

▶ $F_3C_3H$

$CF_3-C\equiv CH$

J. Mol. Spectr. 7:385, 393 (1961)

▶ $F_3C_3H_5$

$CF_3CH_2CH_3$

J. Mol. Spectr. 7:322, 393 (1961)

▶ $F_3C_7H_5$

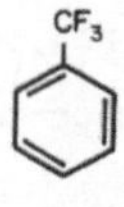


J. Am. Chem. Soc. 83:3428 (1961)

▶ $F_3C_7H_9$

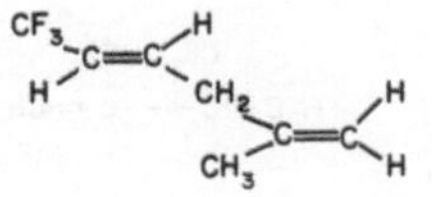


J. Org. Chem. 27:2731 (1962)

▶ $F_3C_8H_5$

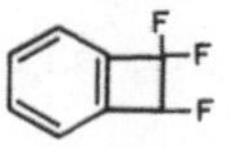


J. Org. Chem. 27:1560 (1962)

▶ $F_3C_8H_7$

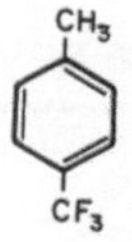


Varian 190

▶ $F_4C_2H_2$

$CF_3CFH_2$

J. Mol. Spectr. 7:307, 393 (1961)

▶ $F_4C_2H_2$

$CF_2HCF_2H$

J. Mol. Spectr. 7:307, 393 (1961)

▶ $F_4C_6H_6$

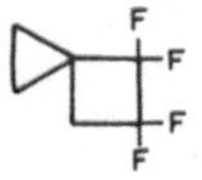


J. Org. Chem. 27:2722 (1962)

▶ $F_4C_6H_{10}$

$CH_3CH_2CF_2CF_2CH_2CH_3$

J. Mol. Spectr. 7:322, 393 (1961)

▶ $F_4C_{12}H_{10}Ni$

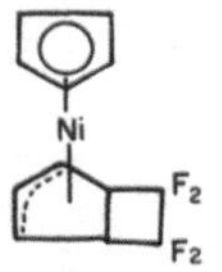


J. Am. Chem. Soc. 84:497 (1962)

▶ $F_4C_{16}H_{10}$

$C_6H_5CF=CFCF=CFC_6H_5$

J. Am. Chem. Soc. 83:1387 (1961)

▶ $F_4C_{21}H_{32}$

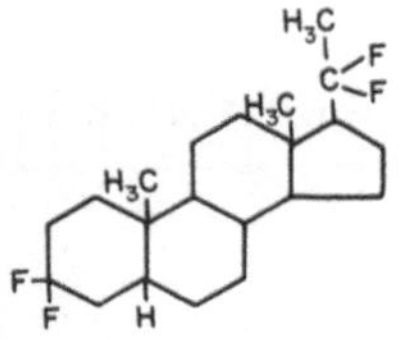


J. Org. Chem. 26:2438 (1961)

► $F_4C_{21}H_{32}$

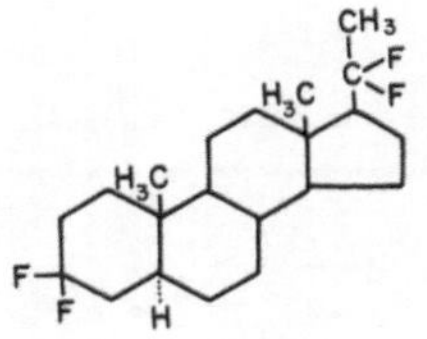

J. Org. Chem. 26:2438 (1961)

► $F_5C_2H$

$CF_3CF_2H$

J. Mol. Spectr. 7:307, 393 (1961)

► $F_5C_3H_3$

$CHF_2CF_2CH_2F$

J. Mol. Spectr. 7:322, 393 (1961)

► $F_6C_3H_2$

$CF_3CF_2CH_2F$

J. Mol. Spectr. 7:322, 393 (1961)

► $F_6C_3H_2$

$CF_3CH_2CF_3$

J. Mol. Spectr. 7:322, 393 (1961)

► $F_6C_3H_2$

$CF_2HCF_2CF_2H$

J. Mol. Spectr. 7:322, 393 (1961)

► $F_6C_5H_2$

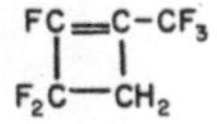

J. Am. Chem. Soc. 83:389 (1961)

► $F_6C_6H_4$

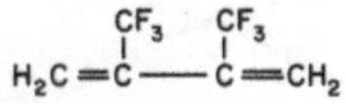

J. Am. Chem. Soc. 83:391 (1961)

► $F_6C_6H_8$

$CF_3(CH_2)_4CF_3$

J. Mol. Spectr. 7:322, 393 (1961)

► $F_6C_8H_4$

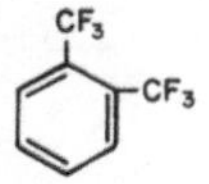

J. Am. Chem. Soc. 83:3428 (1961)

► $F_6C_8H_6$

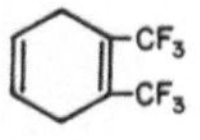

J. Am. Chem. Soc. 83:391 (1961)

► $F_6C_8H_8$

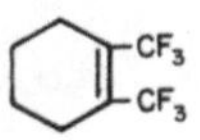

J. Am. Chem. Soc. 83:391 (1961)

▶ $F_6C_8H_8$

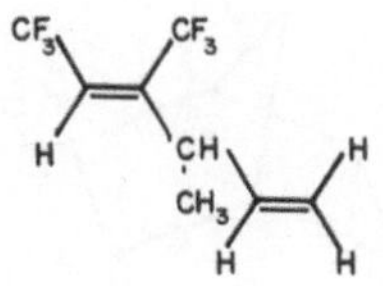

J. Org. Chem. 27:2731 (1962)

▶ $F_6C_8H_8$

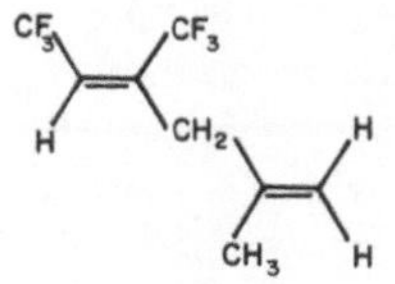

J. Org. Chem. 27:2731 (1962)

▶ $F_6C_{10}H_8$

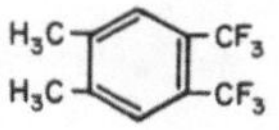

J. Am. Chem. Soc. 83:391 (1961)

▶ $F_6C_{10}H_{10}$

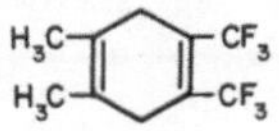

J. Am. Chem. Soc. 83:391 (1961)

▶ $F_6C_{14}H_8$

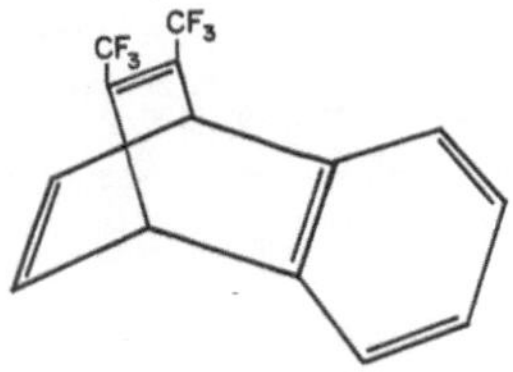

J. Am. Chem. Soc. 83:3428 (1961)

▶ $F_6C_{14}H_{14}$

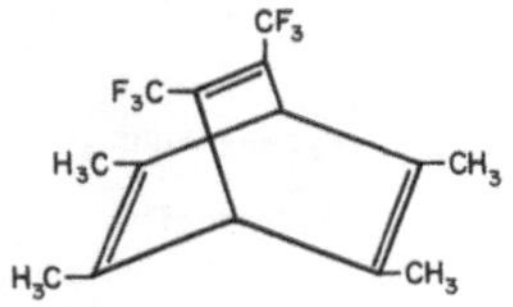

J. Am. Chem. Soc. 83:3428 (1961)

▶ $F_6C_{14}H_{14} \cdot Fe(CO)_3$

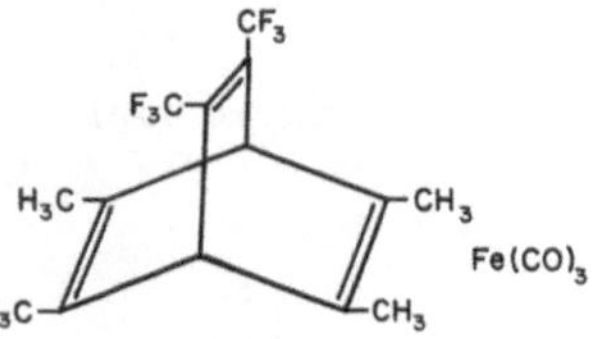

J. Am. Chem. Soc. 84:4705 (1962)

▶ $F_6C_{19}H_{19} \cdot Co$

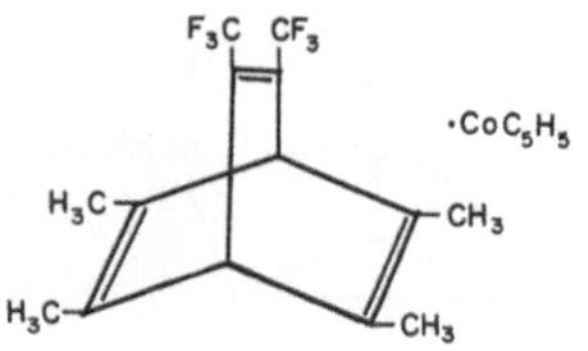

J. Am. Chem. Soc. 84:4705 (1962)

► $F_7C_3H$

$CF_3CF_2CF_2H$

J. Mol. Spectr. 7:322, 393 (1961)

► $F_7C_3H_3SeSi$

$CF_3(CF_2)_2SeSiH_3$

J. Chem. Soc. 2293 (1962)

► $F_7C_5H_3$

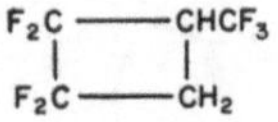


J. Am. Chem. Soc. 83:389 (1961)

► $F_7C_6H_7$

$CF_3CF_2CF_2CH_2CH_2CH_3$

J. Mol. Spectr. 7:322, 393 (1961)

► $F_7C_8H_5 \cdot Co(CO)I$

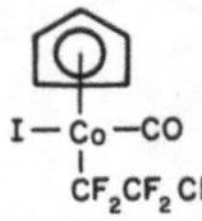


J. Am. Chem. Soc. 83:3593 (1961)

► $F_7C_{10}H_9$

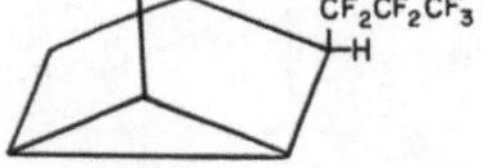


J. Org. Chem. 27:3032 (1962)

► $F_7C_{10}H_{11}$

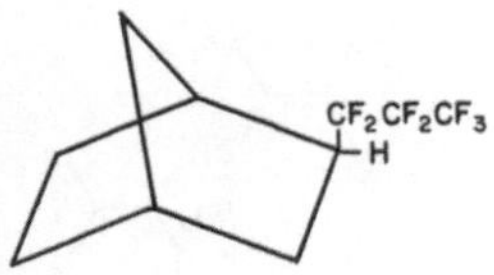


J. Org. Chem. 27:3028 (1962)

► $F_7C_{13}H_{13}$

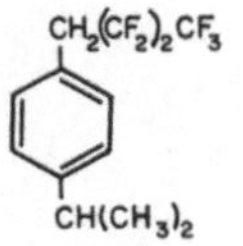


J. Org. Chem. 27:3032 (1962)

► $F_8C_4H_2$

$CF_2HCF_2CF_2CF_2H$

J. Mol. Spectr. 7:322, 393 (1961)

► $F_8C_4H_2$

$CF_3CF_2CF_2CH_2F$

J. Mol. Spectr. 7:322, 393 (1961)

► $F_9C_5H_3$

$CF_2HCF_2CF_2CF_2CH_2F$

J. Mol. Spectr. 7:322, 393 (1961)

► $F_9C_9H_3$

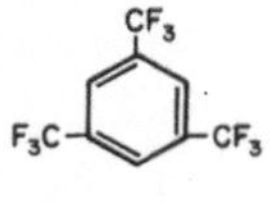


Mol. Phys. 5:1 (1962)

▶ $F_9C_9H_3$

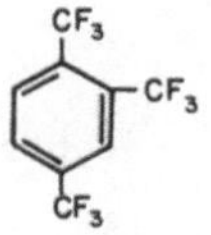


J. Am. Chem. Soc. 83:3428 (1961)

▶ $F_{18}C_{18}H_4$

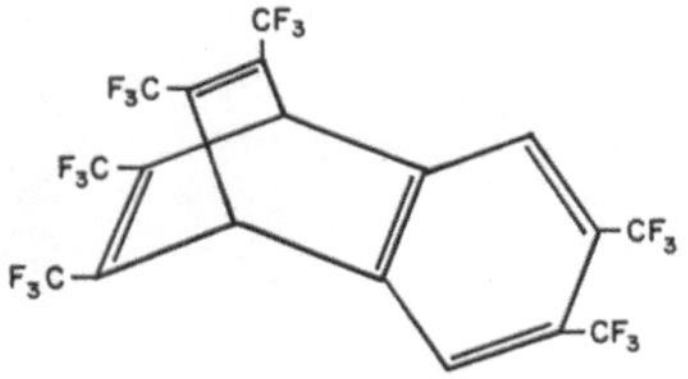


J. Am. Chem. Soc. 83:3428 (1961)

▶ $F_{12}C_{10}H_2$

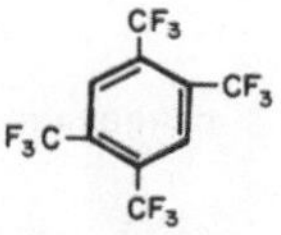


J. Am. Chem. Soc. 83:3428 (1961)

▶ $F_{12}C_{12}H_8$

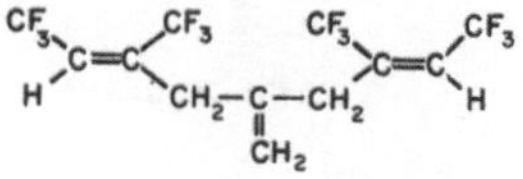


J. Org. Chem. 27:2731 (1962)

▶ $F_{20}C_{12}H_2$

$CHF_2(CF_2)_4C{\equiv}C(CF_2)_4CF_2H$

J. Am. Chem. Soc. 83:3424 (1961)

▶ $F_{12}C_{13}H_5Co$

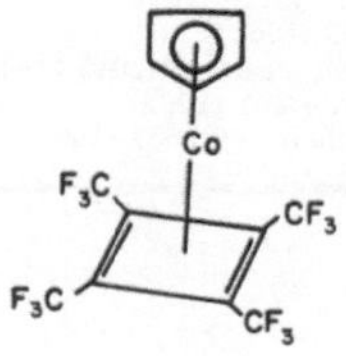


J. Chem. Soc. 3491 (1962)

▶ $F_{20}C_{22}H_{16}$

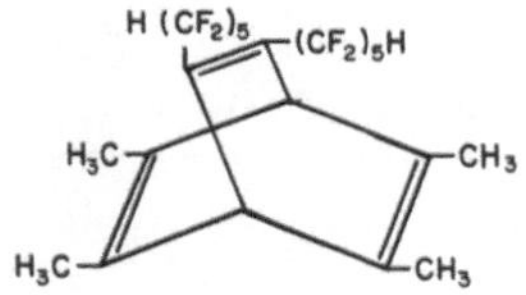


J. Am. Chem. Soc. 83:3428 (1961)

▶ $F_{12}C_{14}H_4$

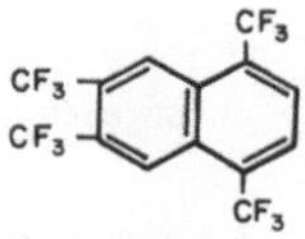


J. Am. Chem. Soc. 83:3428 (1961)

▶ $CH_3 \cdot HgClO_4$

$CH_3HgClO_4$

J. Chem. Soc. 5127 (1962)

▶ $F_{16}C_8H_2$

$CF_2H(CF_2)_6CF_2H$

J. Mol. Spectr. 7:322, 393 (1961)

▶ $CH_3D$

$CH_3D$

J. Chem. Phys. 37:3012 (1962)

► $CH_3Li$

$CH_3Li$

J. Am. Chem. Soc. 84:912 (1962)

► $CH_4$

$CH_4$

J. Chem. Phys. 34:1062, 1099 (1961); 35:1174 (1961); 37:2198 (1962)
J. Am. Chem. Soc. 83:2045, 4726 (1961)
Mol. Phys. 4:361 (1962)

► $C_2H_2$

$HC{\equiv}CH$

J. Chem. Phys. 35:1174 (1961)
J. Am. Chem. Soc. 83:1978 (1961)

► $C_2H_3D$

$CH_2{=}CHD$

J. Mol. Spectr. 8:475 (1962)

► $C_2H_3Li$

$CH_2{=}CHLi$

Can. J. Chem. 40:2 (1962)
J. Am. Chem. Soc. 83:1306 (1961)

► $C_2H_4$

$CH_2{=}CH_2$

J. Chem. Phys. 35:1174 (1961); 36:2441, 2644 (1962); 37:2198 (1962)
J. Phys. Chem. 66:2653 (1962)
J. Am. Chem. Soc. 83:2045 (1961)
J. Mol. Spectr. 8:475 (1962)

► $C_2H_4 \cdot PtCl_3K$

$K[C_2H_4PtCl_3]$

J. Chem. Soc. 4736 (1962)

► $C_2H_5 \cdot HgClO_4$

$C_2H_5HgClO_4$

J. Chem. Soc. 5127 (1962)

► $C_2H_5B_3$

CARBORANE-3

J. Am. Chem. Soc. 84:3837 (1962)

► $C_2H_5Li$

$CH_3CH_2Li$

J. Am. Chem. Soc. 84:1371 (1962)

► $C_2H_6$

$H_3C{-}CH_3$

J. Chem. Phys. 34:1099 (1961); 35:1174 (1961); 37:2198 (1962)
J. Am. Chem. Soc. 83:2045 (1961)
Mol. Phys. 4:361 (1962)
J. Phys. Chem. 66:2653 (1962)

► $C_2H_6Cd$

$(CH_3)_2Cd$

J. Am. Chem. Soc. 84:912 (1962)

► $C_2H_6Hg$

$CH_3HgCH_3$

J. Am. Chem. Soc. 84:912 (1962)
J. Chem. Soc. 5127 (1962)

► $C_2H_6Zn$

$(CH_3)_2Zn$

J. Am. Chem. Soc. 84:912 (1962)

▶ $C_3H_4$

$CH_3C{\equiv}CH$

Varian 16
Rec. Trav. Chim. 81:635 (1962)
J. Chem. Phys. 36:2644 (1962); 37:2198 (1962)
J. Mol. Spectr. 7:393 (1961)
J. Am. Chem. Soc. 83:4729 (1961); 84:37 (1962)

▶ $C_3H_4$

J. Am. Chem. Soc. 83:1226 (1961)

▶ $C_3H_4$

$CH_3C{\equiv}C^{13}{-}H$

J. Chem. Phys. 37:2198 (1962)

▶ $C_3H_4$

$CH_2{=}C{=}CH_2$

J. Phys. Chem. 66:2653 (1962)
J. Chem. Phys. 36:2644 (1962)
J. Am. Chem. Soc. 83:4729 (1961)

▶ $C_3H_5 \cdot Co(CO)_3$

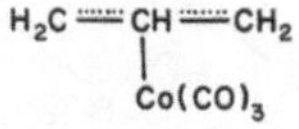

J. Am. Chem. Soc. 83:1097, 1601 (1961)

▶ $C_3H_5 \cdot Mn(CO)_4$

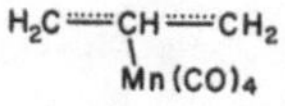

J. Am. Chem. Soc. 83:1601 (1961)

▶ $C_3H_5Li$

$CH_2{=}CHCH_2Li$

J. Am. Chem. Soc. 83:1306 (1961)

▶ $C_3H_6$

J. Am. Chem. Soc. 83:1226 (1961)

▶ $C_3H_6$

$CH_3{-}CH{=}CH_2$

J. Am. Chem. Soc. 83:231, 2045 (1961)

▶ $C_3H_8$

$(CH_3)_2CH_2$

Mol. Phys. 4:361 (1962)
J. Chem. Phys. 34:1099 (1961)
J. Am. Chem. Soc. 83:2045 (1961)

▶ $C_3H_9 \cdot NO_3Pt$

$[(CH_3)_3Pt]NO_3$

J. Chem. Soc. 4736 (1962)

▶ $C_3H_9Al$

$(CH_3)_3Al$

J. Am. Chem. Soc. 84:912 (1962)

▶ $C_3H_9B$

$(CH_3)_3B$

J. Am. Chem. Soc. 83:4138 (1961)

▶ $C_3H_{10}B_4$

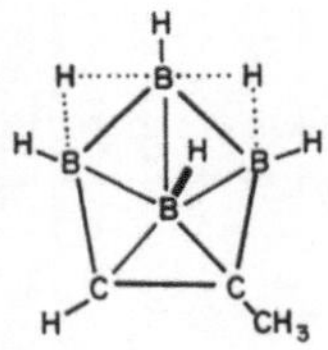

J. Am. Chem. Soc. 84:2830 (1962)

▶ $C_3H_{10}Pb$

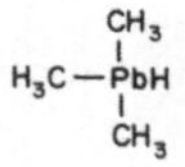

J. Chem. Soc. 1145 (1962)

▶ $C_4H_2$

$HC{\equiv}C{-}C{\equiv}CH$

J. Chem. Phys. 37:2198 (1962)

▶ $C_4H_4$

$HC{\equiv}CCH{=}CH_2$

J. Am. Chem. Soc. 83:1978 (1961)
Rec. Trav. Chim. 81:635 (1962)

▶ $C_4H_5Si$

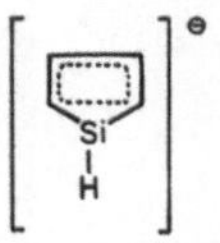

J. Am. Chem. Soc. 83:5029 (1961); 84:4727 (1962)

▶ $C_4H_6$

$CH_2{=}CH{-}CH{=}CH_2$

J. Chem. Soc. 594 (1961)

▶ $C_4H_6$

J. Am. Chem. Soc. 83:1226 (1961)

▶ $C_4H_6$

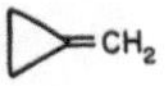

J. Org. Chem. 27:2722 (1962)

▶ $C_4H_6$

$CH_3{-}C{\equiv}C{-}CH_3$

J. Phys. Chem. 66:2653 (1962)
J. Chem. Phys. 35:1174 (1961)

▶ $C_4H_6 \cdot Fe(CO)_3$

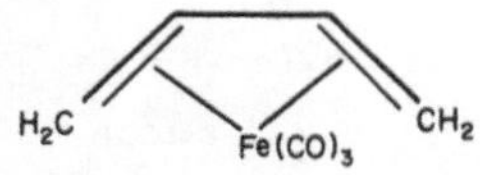

J. Chem. Soc. 594 (1961)

▶ $C_4H_6Hg$

$Hg(CH{=}CH_2)_2$

J. Phys. Chem. 65:224 (1961)

▶ $C_4H_6Si$

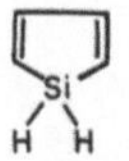

J. Am. Chem. Soc. 83:5029 (1961); 84:4727 (1962)

▶ $C_4H_7 \cdot ClFe(CO)_3$

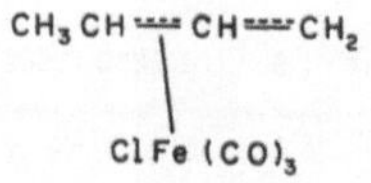

J. Am. Chem. Soc. 83:3726 (1961)

▶ $C_4H_7 \cdot Co(CO)_3$

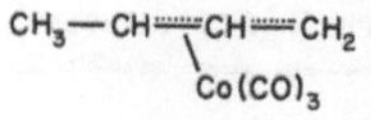

J. Am. Chem. Soc. 83:1601 (1961)

▶ $C_4H_7 \cdot Mn(CO)_4$

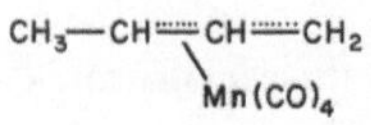

J. Am. Chem. Soc. 83:1601 (1961)

▶ $C_4H_7Si$

$(CH_2{=}CH)_2SiH$

J. Am. Chem. Soc. 84:4727 (1962)

▶ $C_4H_8$

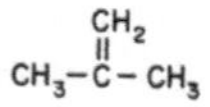

Bull. Chem. Soc. Japan 35:1899 (1962)
J. Chem. Phys. 36:2644 (1962)
J. Am. Chem. Soc. 83:2045 (1961)

▶ $C_4H_8$

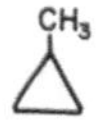

Chem. Ber. 95:2280 (1962)

▶ $C_4H_8$

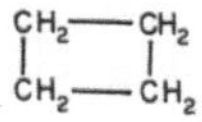

J. Chem. Phys. 35:1174 (1961)
J. Am. Chem. Soc. 83:1226 (1961)

▶ $C_4H_8$

$CH_3\,CH_2{-}CH{=}CH_2$

J. Am. Chem. Soc. 83:231 (1961)

▶ $C_4H_8$

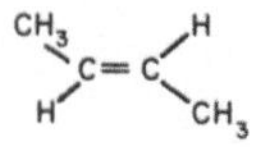

cis and trans

J. Am. Chem. Soc. 83:2045 (1961)

▶ $C_4H_{10}$

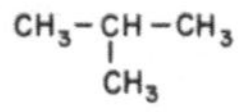

Mol. Phys. 4:361 (1962)
J. Am. Chem. Soc. 83:2045 (1961)

▶ $C_4H_{10}Cd$

$(C_2H_5)_2Cd$

J. Chem. Soc. 5128 (1962)

▶ $C_4H_{10}Hg$

$(CH_3CH_2)_2Hg^{199}$

J. Chem. Soc. 5127 (1962)
J. Am. Chem. Soc. 83:4473 (1961)

▶ $C_4H_{10}Mg$

$(C_2H_5)_2Mg$

J. Chem. Soc. 5127 (1962)

▶ $C_4H_{10}Tl$

$(C_2H_5)_2Tl$

J. Chem. Soc. 850 (1962)

▶ $C_4H_{10}Zn$

$(C_2H_5)_2Zn$

J. Chem. Soc. 5128 (1962)

▶ $C_4H_{12}B_2$

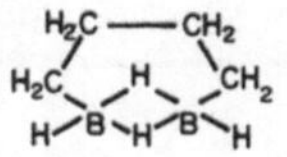


J. Am. Chem. Soc. 84:3840 (1962)

▶ $C_4H_{12}B_4$

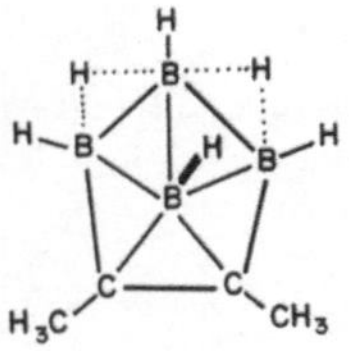


J. Am. Chem. Soc. 84:2830 (1962)

▶ $C_4H_{12}Si$

$Si(CH_3)_4$

J. Chem. Phys. 35:1174 (1961); 37:2198 (1962)

▶ $C_4H_{12}Sn$

$Sn(CH_3)_4$

Varian 424
J. Chem. Phys. 35:1174 (1961)

▶ $C_5H_5 \cdot HFeMn(CO)_7$

$[\pi-C_5H_5FeMn(CO)_7]H^{\oplus}$

J. Chem. Soc. 3654 (1962)

▶ $C_5H_5 \cdot FeMn(CO)_7HPF_6$

$[\pi-C_5H_5FeMn(CO)_7]H^{\oplus}PF_6^{\ominus}$

J. Chem. Soc. 3654 (1962)

▶ $C_5H_5 \cdot FeMn(CO)_7DPF_6$

$[\pi-C_5H_5FeMn(CO)_7]D^{\oplus}PF_6^{\ominus}$

J. Chem. Soc. 3654 (1962)

▶ $C_5H_5W \cdot (CO)_3H_2$

$\pi\text{-}C_5H_5W(CO)_3H_2^{\oplus}$

J. Chem. Soc. 3660 (1962)

▶ $C_5H_6$

HC—CH, HC—CH, $CH_2$

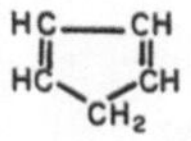

J. Am. Chem. Soc. 84:4727 (1962)

▶ $C_5H_6$

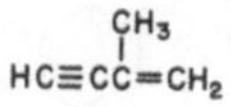

$HC{\equiv}C\text{-}C(CH_3){=}CH_2$

Varian 99

▶ $C_5H_8$

$H_3C$, $CH_3$

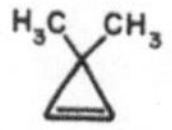

Proc. Chem. Soc. 152 (1962)
J. Am. Chem. Soc. 83:2015 (1961)

▶ $C_5H_8$

$=CH_2$

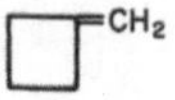

Varian 109

▶ $C_5H_8$

$(CH_3)_2C{=}C{=}C^{13}H_2$

J. Chem. Phys. 37:2198 (1962)

▶ $C_5H_8$

J. Am. Chem. Soc. 83:1226 (1961)

▶ $C_5H_8$

$CH_2{=}CH\text{-}C(CH_3){=}CH_2$

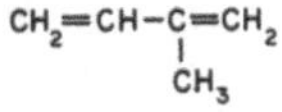

Bull. Chem. Soc. Japan 35:1899 (1962)

▶ $C_5H_8$

$HC{\equiv}C(CH_2)_2CH_3$

Rec. Trav. Chim. 81:635 (1962)
J. Am. Chem. Soc. 83:1978 (1961)

▶ $(C_5H_8)n$

$[CH_2{=}C(CH_3)\text{-}CH\text{-}CH_2\text{-}]_n$

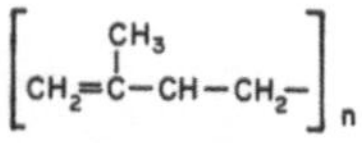

3,4-POLYISOPRENE

Anal. Chem. 34:1793 (1962)

▶ $(C_5H_8)n$

$[-CH_2(CH_3)C{=}C(H)CH_2-]_n$

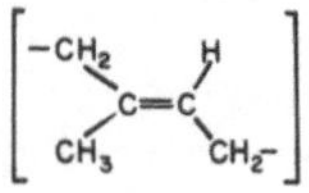

trans-1,4-POLYISOPRENE

Anal. Chem. 34:1793 (1962)

▶ $(C_5H_8)n$

$[(CH_3)(-CH_2)C{=}C(H)CH_2-]_n$

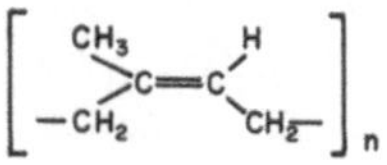

cis-1,4-POLYISOPRENE

Anal. Chem. 34:1793 (1962)

▶ $C_5H_9 \cdot Fe(CO)_3BF_4$

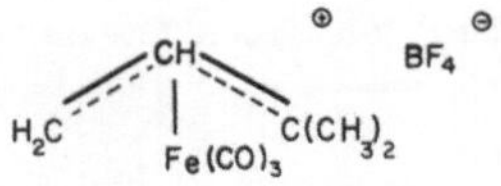


J. Am. Chem. Soc. 84:4591 (1962)

▶ $C_5H_{14}B_4$

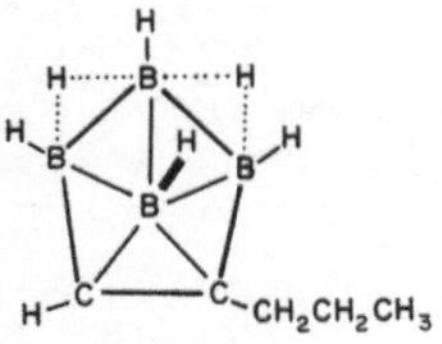


J. Am. Chem. Soc. 84:2830 (1962)

▶ $C_5H_{10}$

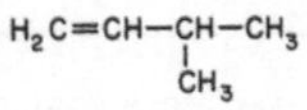


J. Am. Chem. Soc. 84:2748 (1962)

▶ $C_6H_5D$

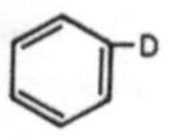


Can. J. Chem. 39:230 (1961)

▶ $C_5H_{10}$

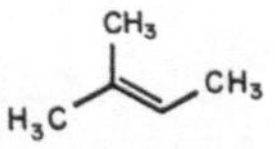


J. Chem. Soc. 1625 (1961)

▶ $C_6H_6$

$HC\equiv CCH_2CH_2C\equiv CH$

Rec. Trav. Chim. 81:635 (1962)

▶ $C_5H_{10}$

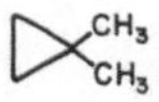


J. Org. Chem. 27:2722 (1962)

▶ $C_6H_6$

J. Am. Chem. Soc. 2649 (1962)
Can. J. Chem. 40:1759 (1962)
J. Chem. Phys. 36:2443, 2644 (1962); 37:2198 (1962)

▶ $C_5H_{10}$

Mol. Phys. 4:361 (1962)
J. Am. Chem. Soc. 83:1226 (1961)
Helv. Chim. Acta 44:1755 (1961)
J. Chem. Phys. 35:1174 (1961)

▶ $C_6H_6 \cdot Cr(CO)_3$

J. Chem. Soc. 3660 (1962)

▶ $C_5H_{12}$

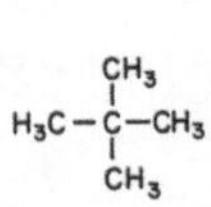


J. Chem. Phys. 35:1174 (1961); 37:2198 (1962)
J. Chem. Soc. 476 (1962)
Mol. Phys. 4:361 (1962)

▶ $C_6H_6 \cdot Fe_2(CO)_6$

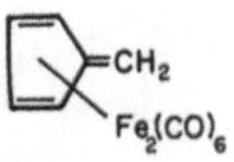


Chem. Ber. 95:1155 (1962)

▶ $C_6H_7 \cdot Fe(CO)_3BF_4$

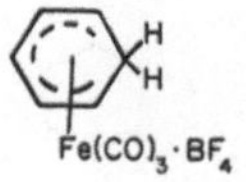


J. Chem. Soc. 4463 (1962)

▶ $C_6H_8$

J. Chem. Soc. 594 (1961)

▶ $C_6H_8$

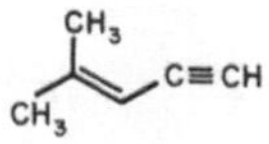


J. Chem. Soc. 1625 (1961)

▶ $C_6H_8 \cdot Fe(CO)_3$

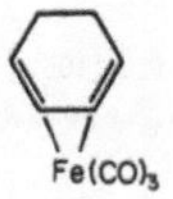


J. Chem. Soc. 594 (1961)

▶ $C_6H_9 \cdot SbF_6$

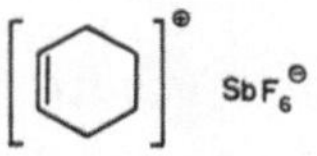


J. Am. Chem. Soc. 83:5031 (1961); 84:1498 (1962)

▶ $C_6H_{10}$

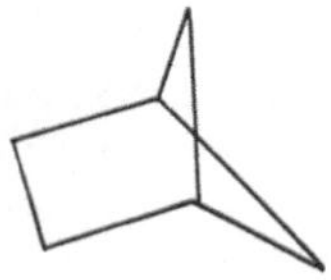

J. Am. Chem. Soc. 83:2590, 4923 (1961)

▶ $C_6H_{10}$

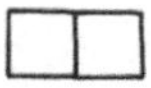

J. Am. Chem. Soc. 83:4923 (1961)

▶ $C_6H_{10}$

J. Chem. Phys. 37:2198 (1962)
J. Am. Chem. Soc. 83:1226, 4095 (1961)

▶ $C_6H_{10}$

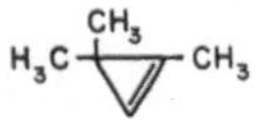


Proc. Chem. Soc. 152 (1962)
J. Am. Chem. Soc. 83:1003 (1961)

▶ $C_6H_{10}$

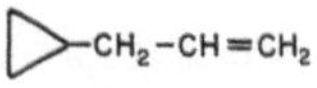


J. Am. Chem. Soc. 83:4923 (1961)

▶ $C_6H_{10}$

$HC{\equiv}C(CH_2)_3CH_3$

J. Am. Chem. Soc. 83:1978 (1961)
Rec. Trav. Chim. 81:635 (1962)

▶ $C_6H_{10}$

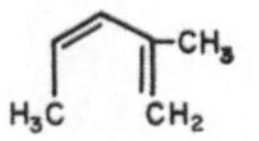

cis and trans

J. Am. Chem. Soc. 84:2775 (1962)

▶ $C_6H_{12}$

Proc. Chem. Soc. 418 (1961)
J. Am. Chem. Soc.83:1226, 1671 (1961); 84:386(1962)
Mol. Phys. 5:361 (1962)
J. Chem. Phys. 36:2644 (1962)
Helv. Chim. Acta 44:1755 (1961)
Bull. Soc. Chim. France 754 (1962)

▶ $C_6H_{10}$

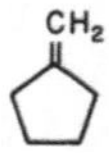

Varian 132

▶ $C_6H_{10} \cdot Pd_2Cl_2$

$[PdCl(CH_2{=}CHCH_2)]_2$

J. Am. Chem. Soc. 83:1601 (1961)
Chem. & Ind. 1190 (1962)

▶ $C_6H_{12}$

$(CH_3)_2CHCH{=}CHCH_3$

Varian 471

▶ $C_6H_{12}$

$CH_2{=}CHCH_2CH(CH_3)CH_3$

J. Am. Chem. Soc. 84:2748 (1962)

▶ $C_6H_{12}Si$

$CH_2{=}CHSi(CH_3)_2CH{=}CH_2$

Varian 476

▶ $C_6H_{12}$

$CH_2{=}C(CH_3){-}CH(CH_3)CH_3$

J. Org. Chem. 26:4194 (1961)

▶ $C_6H_{14}Mg$

$(CH_3CH_2CH_2)_2Mg$

J. Chem. Soc. 5127 (1962)

▶ $C_6H_{12}$

$CH_3CH_2CH_2CH_2CH{=}CH_2$

J. Am. Chem. Soc. 83:231 (1961)

▶ $C_6H_{15}Al$

$(C_2H_5)_3Al$

Rec. Trav. Chim. 81:729 (1962)

▶ $C_6H_{12}$

$CH_3C(H_3C){=}C(CH_3)CH_3$

J. Org. Chem. 26:4194 (1961)

▶ $C_6H_{15}B$

$(C_2H_5)_3B$

J. Am. Chem. Soc. 84:4715 (1962)
Bull. Chem. Soc. Japan 35:1317 (1962)

▶ $C_6H_{15}Tl$

$(CH_3CH_2)_3Tl^{205}$

J. Am. Chem. Soc. 83:4473 (1961)

▶ $C_6H_{18}Al_2$

$Al_2(CH_3)_6$

J. Chem. Phys. 37:2198 (1962)

▶ $C_6H_{18}Sn_2$

$(CH_3)_6Sn_2$

J. Am. Chem. Soc. 83:1514 (1961)

▶ $C_7H_7 \cdot BF_4$

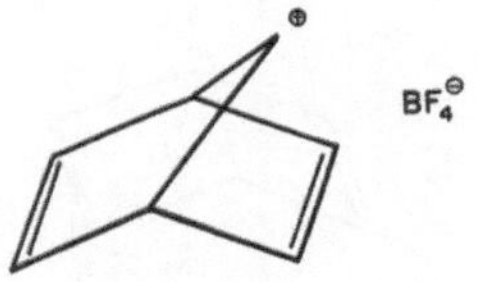


J. Am. Chem. Soc. 84:4876 (1962)

▶ $C_7H_7 \cdot V(CO)_3$

J. Am. Chem. Soc. 83:2024 (1961)

▶ $C_7H_7D$

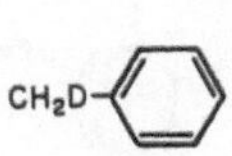


J. Chem. Phys. 37:3012 (1962)

▶ $C_7H_7D$

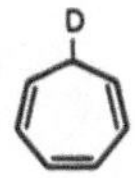


Proc. Chem. Soc. 359 (1962)

▶ $C_7H_8$

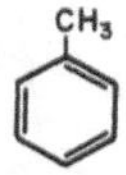


Varian 157
J. Chem. Phys. 34:1099 (1961); 36:2644 (1962); 37:2198 (1962)
J. Org. Chem. 26:4984 (1961)
Bull. Chem. Soc. Japan 34:143 (1961)

▶ $C_7H_8$

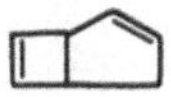

Tetrahedron 12:186 (1961)

▶ $C_7H_8$

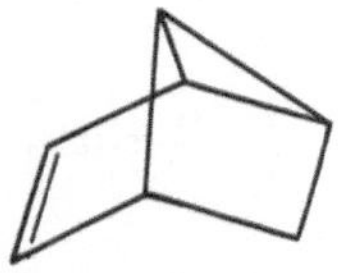

J. Am. Chem. Soc. 83:3347 (1961)

▶ $C_7H_8$

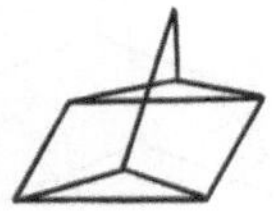

Tetrahedron 15:198 (1961)
J. Am. Chem. Soc. 83:4674 (1961)

▶ $C_7H_8$

Varian 158
J. Chem. Soc. 594, 2037 (1961)

▶ $C_7H_8$

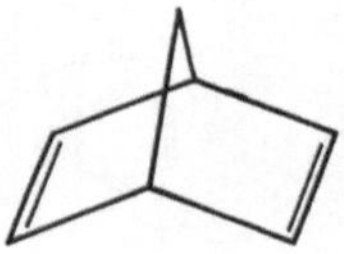

Varian 487
J. Org. Chem. 27:3360 (1962)
J. Am. Chem. Soc. 83:4674 (1961)
J. Chem. Soc. 2037 (1961)

▶ $C_7H_8 \cdot Cr(CO)_3$

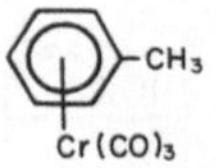


J. Chem. Soc. 3660 (1962)

▶ $C_7H_8 \cdot Cr(CO)_3$

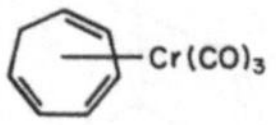


J. Chem. Soc. 2037 (1961)

▶ $C_7H_8 \cdot Cr(CO)_4$

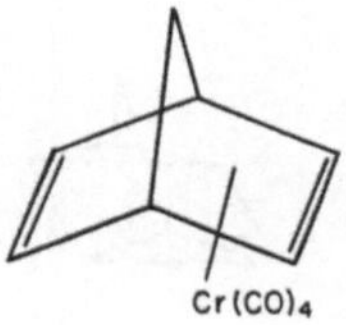


J. Chem. Soc. 2037 (1961)

▶ $C_7H_8 \cdot Fe(CO)_3$

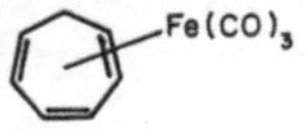


J. Chem. Soc. 594 (1961)
J. Am. Chem. Soc. 83:497 (1961)

▶ $C_7H_8 \cdot Mo(CO)_3$

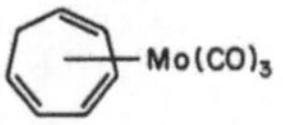


J. Am. Chem. Soc. 83:3590 (1961)
J. Chem. Soc. 2037 (1961)

▶ $C_7H_8 \cdot Mo(CO)_4$

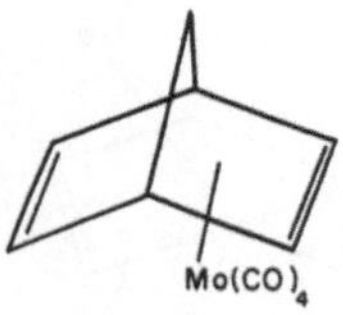


J. Am. Chem. Soc. 83:3590 (1961)
J. Chem. Soc. 2037 (1961)

▶ $C_7H_8 \cdot W(CO)_3$

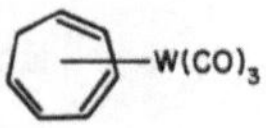


J. Chem. Soc. 2037 (1961)

▶ $C_7H_8D \cdot Fe(CO)_3$

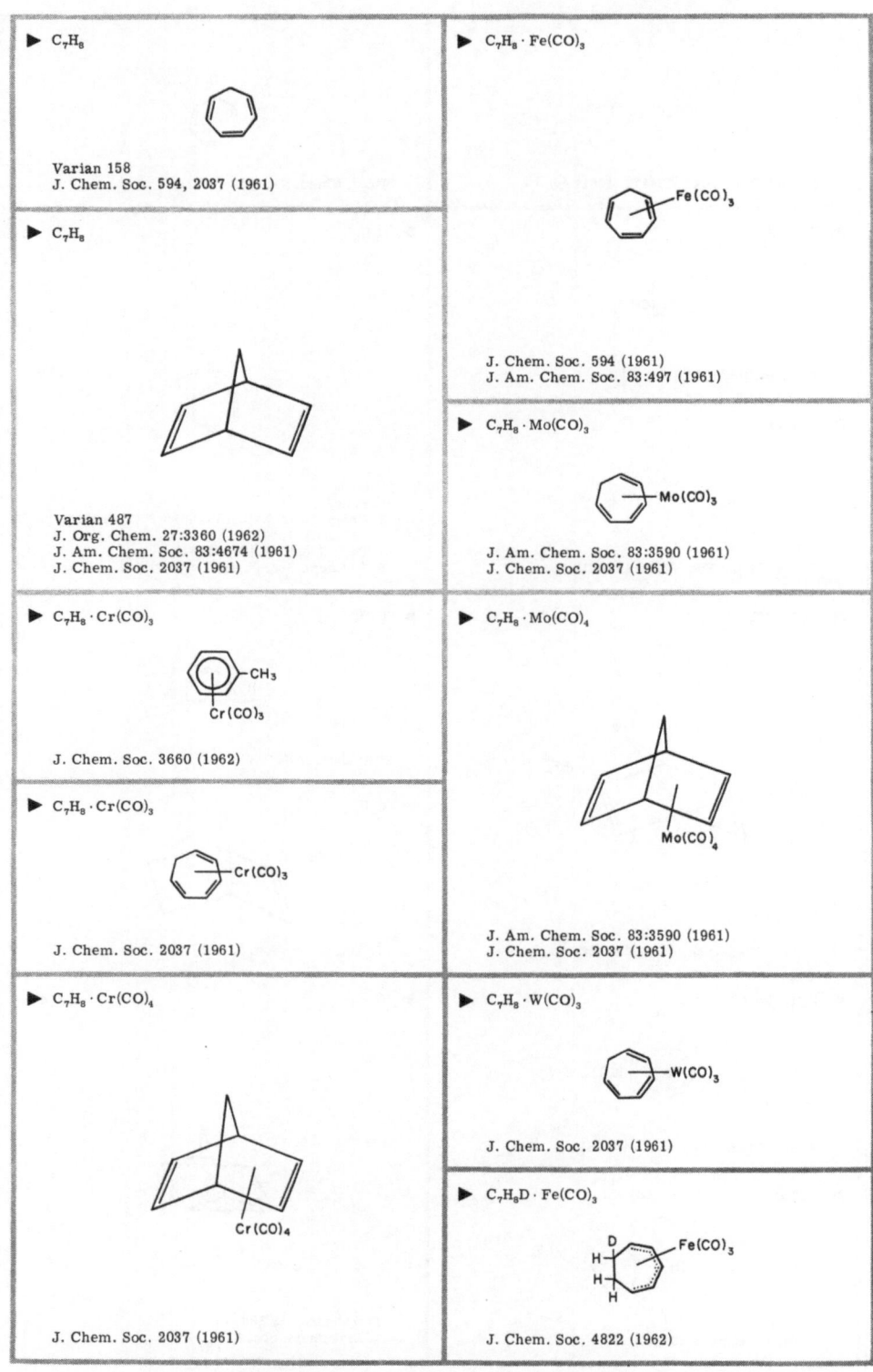

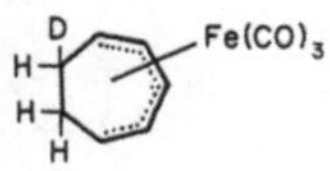


J. Chem. Soc. 4822 (1962)

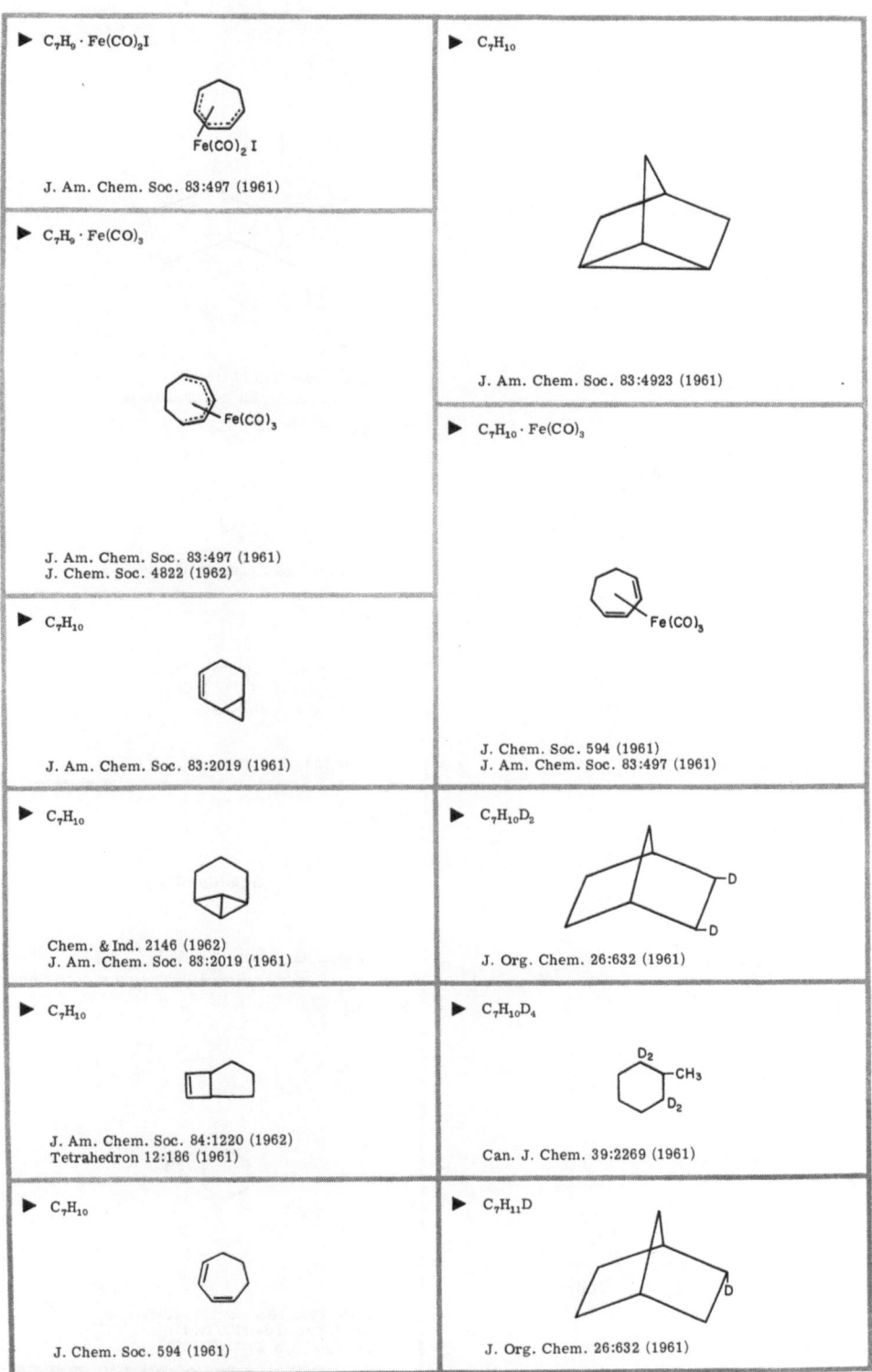

▶ $C_7H_9 \cdot Fe(CO)_2I$

J. Am. Chem. Soc. 83:497 (1961)

▶ $C_7H_9 \cdot Fe(CO)_3$

J. Am. Chem. Soc. 83:497 (1961)
J. Chem. Soc. 4822 (1962)

▶ $C_7H_{10}$

J. Am. Chem. Soc. 83:2019 (1961)

▶ $C_7H_{10}$

Chem. & Ind. 2146 (1962)
J. Am. Chem. Soc. 83:2019 (1961)

▶ $C_7H_{10}$

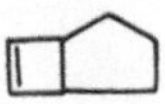

J. Am. Chem. Soc. 84:1220 (1962)
Tetrahedron 12:186 (1961)

▶ $C_7H_{10}$

J. Chem. Soc. 594 (1961)

▶ $C_7H_{10}$

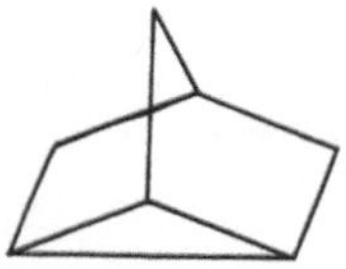

J. Am. Chem. Soc. 83:4923 (1961)

▶ $C_7H_{10} \cdot Fe(CO)_3$

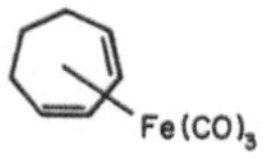


J. Chem. Soc. 594 (1961)
J. Am. Chem. Soc. 83:497 (1961)

▶ $C_7H_{10}D_2$

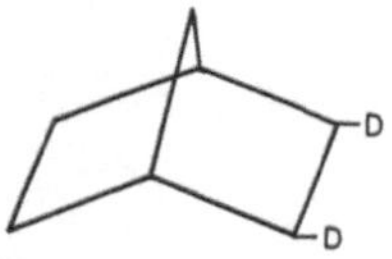


J. Org. Chem. 26:632 (1961)

▶ $C_7H_{10}D_4$

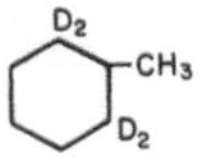


Can. J. Chem. 39:2269 (1961)

▶ $C_7H_{11}D$

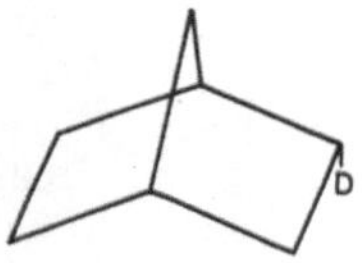


J. Org. Chem. 26:632 (1961)

▶ $C_7H_{12}$

$HC{\equiv}C(CH_2)_4CH_3$

J. Am. Chem. Soc. 83:1978 (1961)

▶ $C_7H_{12}$

$(H_3C)_2C{=}C{=}C(CH_3)_2$

J. Chem. Phys. 36:2644 (1962)

▶ $C_7H_{12}$

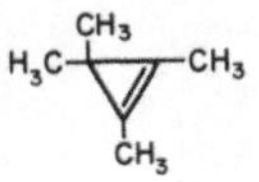


J. Am. Chem. Soc. 83:1003 (1961)

▶ $C_7H_{12}$

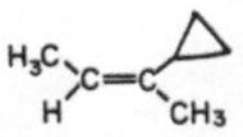


Chem. Ber. 95:2280 (1962)

▶ $C_7H_{12}$

J. Am. Chem. Soc. 83:1226 (1961)

▶ $C_7H_{12}$

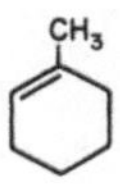


J. Org. Chem. 27:4243 (1962)

▶ $C_7H_{12}$

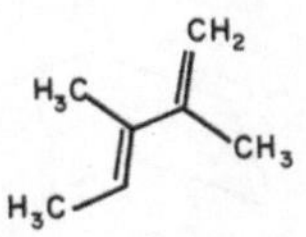


Bull. Chem. Soc. Japan 35:572 (1962)

▶ $C_7H_{12}$

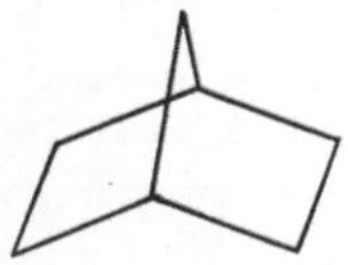

J. Org. Chem. 26:632 (1961)

▶ $C_7H_{12}$

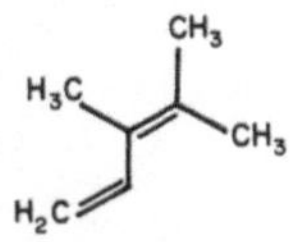


Bull. Chem. Soc. Japan 35:572 (1962)

▶ $C_7H_{12}$

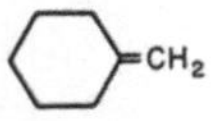


Varian 180

▶ $C_7H_{12}Si$

$CH_3Si(CH{=}CH_2)_3$

J. Chem. Phys. 37:2053 (1962)

▶ $C_7H_{14}$

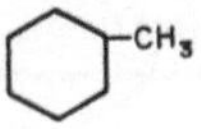


J. Am. Chem. Soc. 83:1146 (1961)
Can. J. Chem. 39:2269 (1961)
J. Chem. Phys. 37:1167 (1962)

▶ $C_7H_{14}$

J. Am. Chem. Soc. 83:1226 (1961)

▶ $C_7H_{14}$

$$CH_2{=}CHCH_2C(CH_3)_3$$

J. Am. Chem. Soc. 84:2748 (1962)

▶ $C_7H_{16}$

$(CH_3)_2CHCH_2CH(CH_3)_2$

J. Am. Chem. Soc. 83:1230 (1961)

▶ $C_7H_{16}$

$(CH_3)_2CHC^{13}H_2CH(CH_3)_2$

J. Am. Chem. Soc. 83:1230 (1961)

▶ $C_7H_{20}Sn_2$

$(CH_3)_3SnCH_2Sn(CH_3)_3$

J. Am. Chem. Soc. 83:514 (1961)

▶ $C_8H_6$

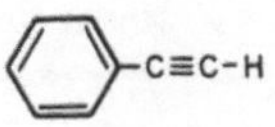

Varian 186
Trans. Faraday Soc. 57:28 (1961)
J. Chem. Phys. 37:2198 (1962)
J. Am. Chem. Soc. 83:1978 (1961); 84:3553 (1962)
Rec. Trav. Chim. 81:635 (1962)

▶ $C_8H_6D_2$

$C_6H_5CD{=}CHD$

Bull. Chem. Soc. Japan 34:560 (1961)

▶ $C_8H_6Li_2$

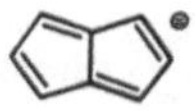 $(Li^{\oplus})_2$

J. Am. Chem. Soc. 84:865 (1962)

▶ $C_8H_7D$

$C_6H_5CH{=}CHD$

Bull. Chem. Soc. Japan 34:560 (1961)

▶ $C_8H_8$

$C_6H_5CH{=}CH_2$

J. Am. Chem. Soc. 83:231 (1961)
Bull. Chem. Soc. Japan 34:560 (1961)

▶ $C_8H_8$

J. Am. Chem. Soc. 84:4307 (1962)
J. Am. Chem. Soc. 83:2944 (1961); 84:671 (1962)
Chem. Ber. 95:158 (1962)

▶ $C_8H_8 \cdot AgNO_3$

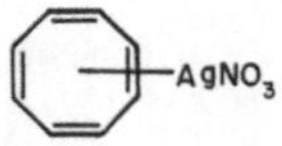


Chem. Ber. 95:158 (1962)

▶ $C_8H_8 \cdot Fe(CO)_3$

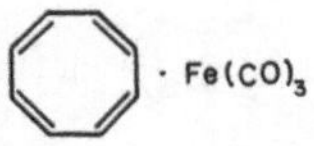


J. Chem. Soc. 851 (1962)
Chem. Ber. 95:158 (1962)

▶ $C_8H_8 \cdot [Fe(CO)_3]_2$

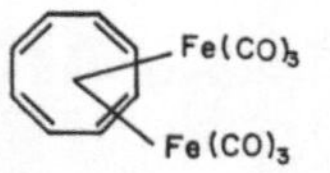


Chem. Ber. 95:158 (1962)

▶ $C_8H_8 \cdot PdCl_2$

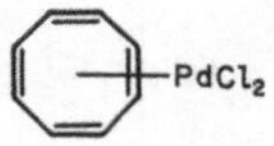


Chem. Ber. 95:158 (1962)

▶ $C_8H_8 \cdot PtI_2$

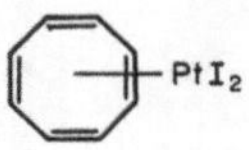


Chem. Ber. 95:158 (1962)

▶ $C_8H_8D \cdot Fe(CO)_3$

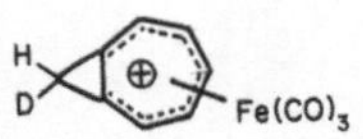


J. Chem. Soc. 4822 (1962)

▶ $C_8H_8K_2$

Chem. Ber. 95:158 (1962)

▶ $C_8H_9 \cdot Fe(CO)_3$

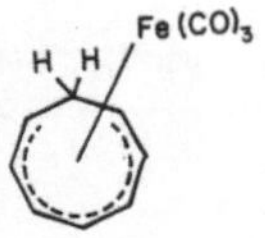


J. Am. Chem. Soc. 83:2966 (1961)

▶ $C_8H_9 \cdot Fe(CO)_3$

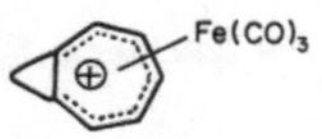


J. Chem. Soc. 4822 (1962)

▶ $C_8H_9 \cdot HSO_4$

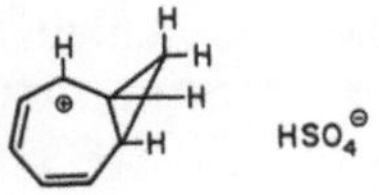


J. Am. Chem. Soc. 84:2842 (1962)

▶ $C_8H_{10}$

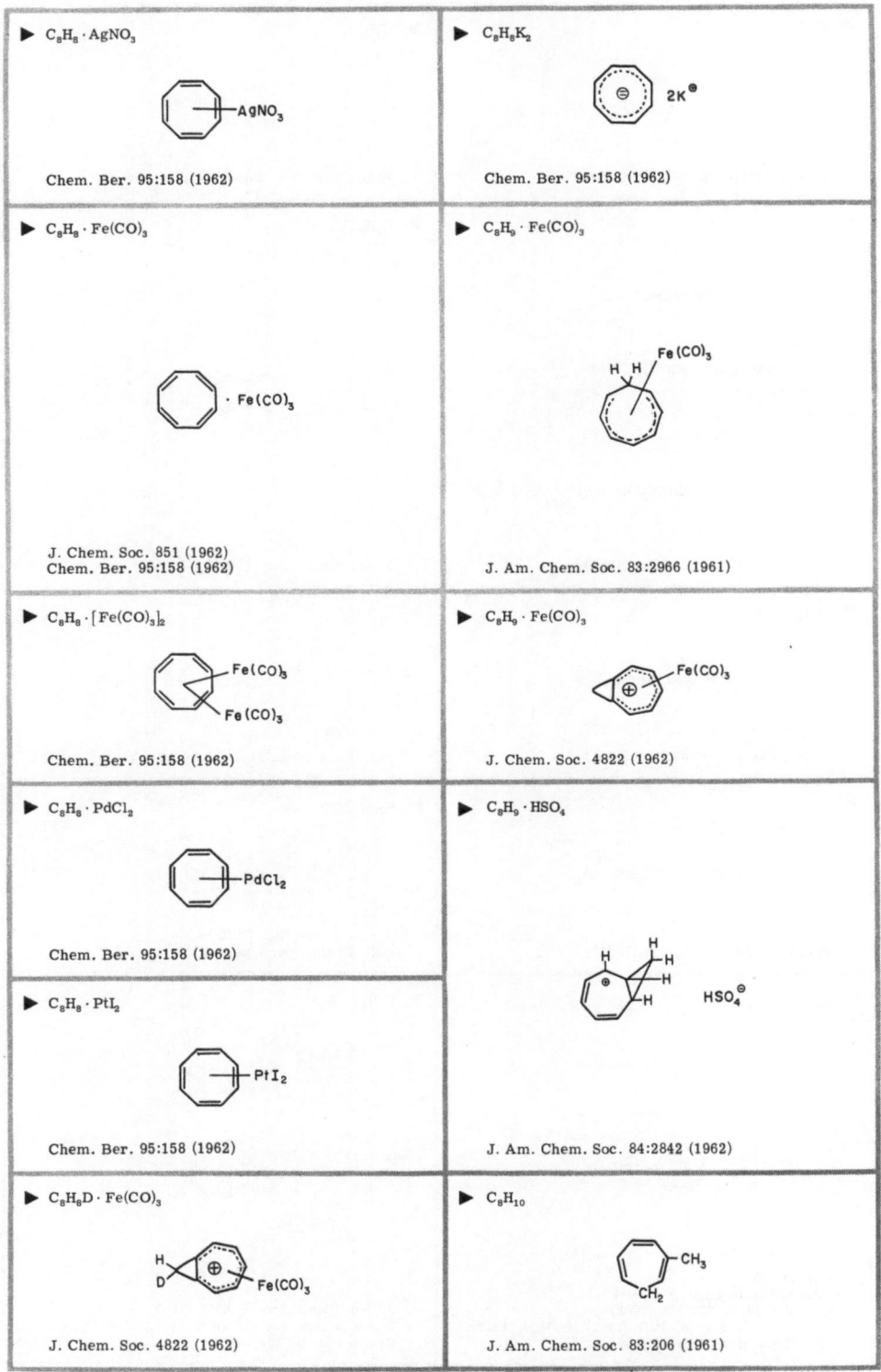

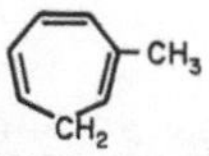


J. Am. Chem. Soc. 83:206 (1961)

▶ $C_8H_{10}$

J. Am. Chem. Soc. 83:206 (1961)

▶ $C_8H_{10}$

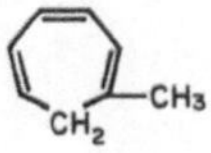


J. Am. Chem. Soc. 83:206 (1961)

▶ $C_8H_{10}$

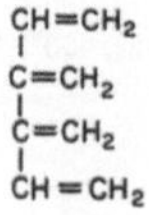


J. Org. Chem. 27:3090 (1962)

▶ $C_8H_{10}$

$C_2H_5-C_6H_5$

Varian 505
J. Chem. Phys. 34:1099 (1961)

▶ $C_8H_{10}$

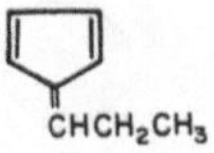


Helv. Chim. Acta 45:1406 (1962)

▶ $C_8H_{10}$

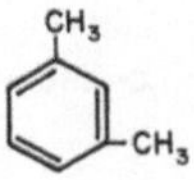


Varian 202
Bull. Chem. Soc. Japan 34:143 (1961)

▶ $C_8H_{10}$

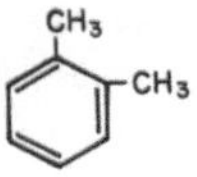


Varian 201
J. Chem. Phys. 37:2594 (1962)
Bull. Chem. Soc. Japan 34:143 (1961)

▶ $C_8H_{10}$

Varian 203
Bull. Chem. Soc. Japan 34:143 (1961)
J. Chem. Phys. 37:2594 (1962)
Helv. Chim. Acta 45:568 (1962)

▶ $C_8H_{10} \cdot Fe(CO)_3$

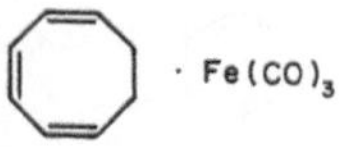


J. Chem. Soc. Japan 82:1389 (1961)

▶ $C_8H_{10} \cdot Fe(CO)_3$

J. Chem. Soc. 4822 (1962)

▶ $C_8H_{10} \cdot Fe(CO)_3$

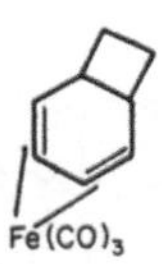


J. Chem. Soc. 594 (1961)

► $C_8H_{10}Ni$

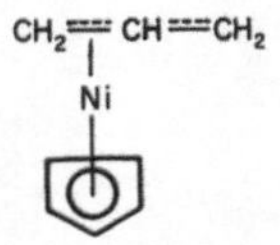

J. Am. Chem. Soc. 83:1601 (1961)

► $C_8H_{10}Pd$

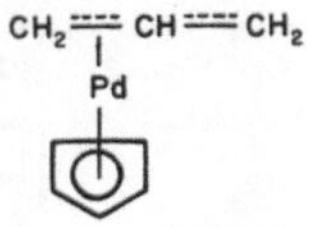

J. Am. Chem. Soc. 83:1601 (1961)

► $C_8H_{10}Pt$

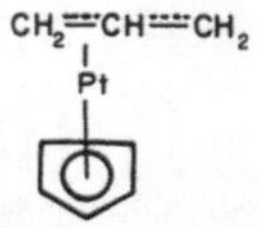

J. Chem. Soc. 4736 (1962)

► $C_8H_{12}$

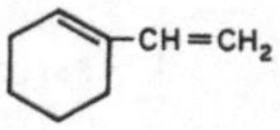

J. Chem. Phys. 35:1900 (1961)

► $C_8H_{12}$

Varian 209

► $C_8H_{12}$

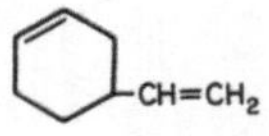

Varian 210

► $C_8H_{12}$

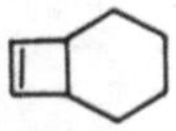

Chem. & Ind. 1198 (1962)
J. Org. Chem. 27:1910 (1962)

► $C_8H_{12}$

Varian 511

► $C_8H_{12}Si$

$(CH_2{=}CH)_4Si$

J. Chem. Phys. 35:1900 (1961); 37:1951 (1962)

► $C_8H_{12}Sn$

$Sn(-CH{=}CH_2)_4$

J. Phys. Chem. 65:224 (1961)

► $C_8H_{14}$

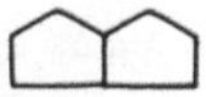

J. Am. Chem. Soc. 83:1671 (1961)

► $C_8H_{14}$

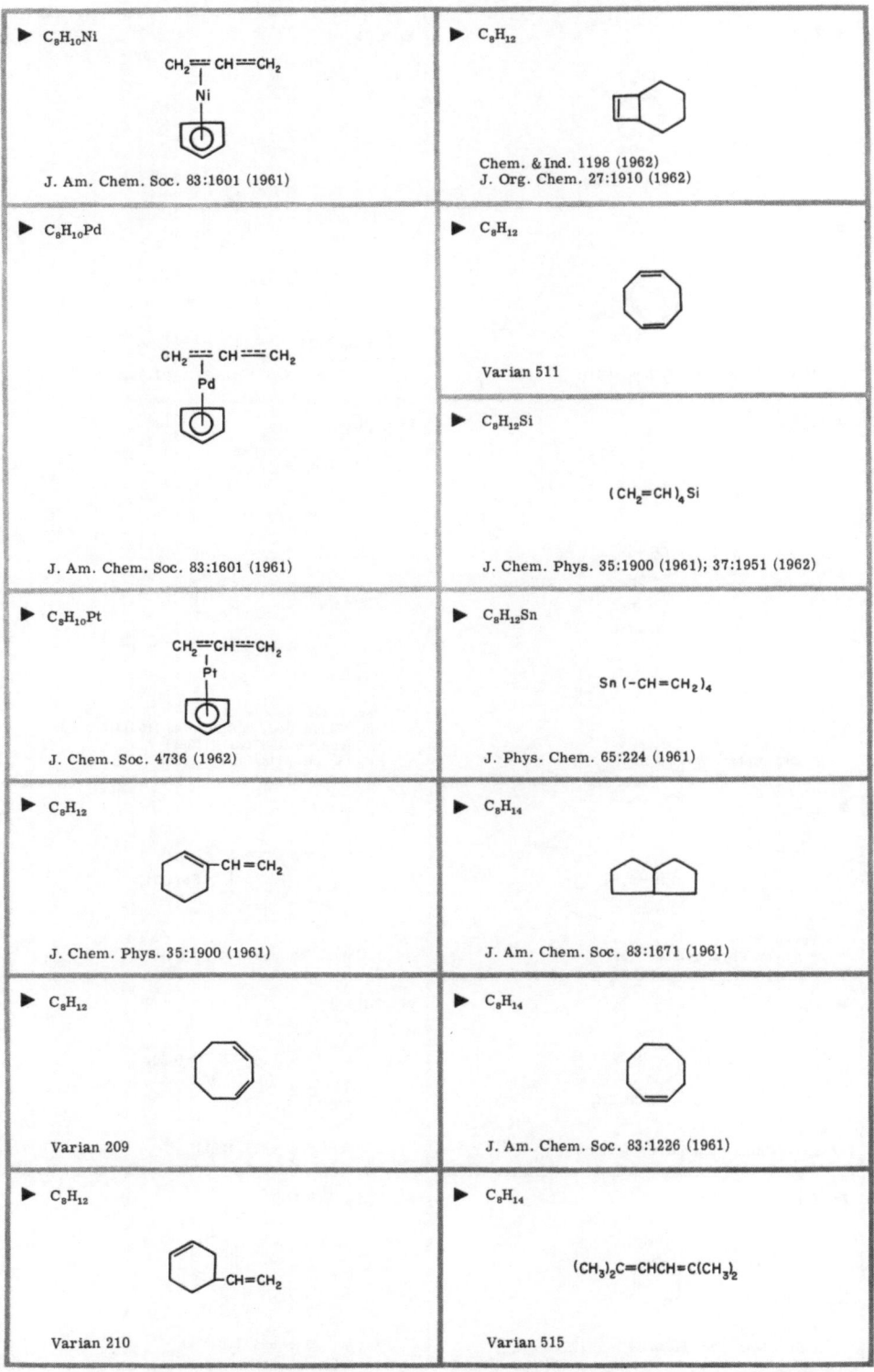

J. Am. Chem. Soc. 83:1226 (1961)

► $C_8H_{14}$

$(CH_3)_2C{=}CHCH{=}C(CH_3)_2$

Varian 515

▶ $C_8H_{16}$

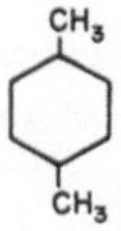


J. Am. Chem. Soc. 83:1146 (1961)
J. Chem. Phys. 37:1167 (1962)

▶ $C_8H_{16}$

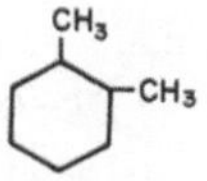


J. Chem. Phys. 37:1167 (1962)

▶ $C_8H_{16}$

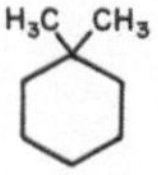


J. Chem. Phys. 37:1167 (1962)

▶ $C_8H_{16}$

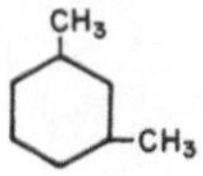


J. Chem. Phys. 37:1167 (1962)

▶ $C_8H_{16}$

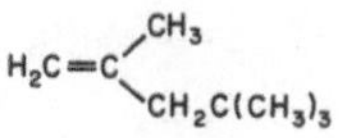


Arkiv Kemi 17:1 (1961)

▶ $C_8H_{16}$

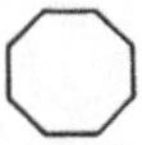

Proc. Chem. Soc. 418 (1961)
J. Am. Chem. Soc. 83:1226 (1961)

▶ $C_8H_{16}$

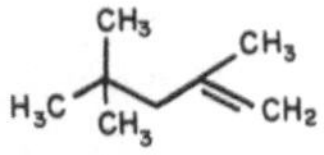


Bull. Chem. Soc. Japan 35:1899 (1962)

▶ $C_8H_{16}$

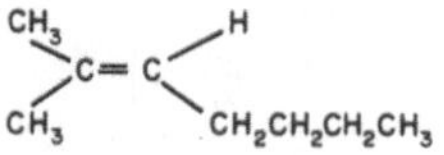


Chem. & Ind. 1020 (1962)

▶ $C_8H_{18}$

$(CH_3)_3C-C(CH_3)_3$

Mol. Phys. 4:361 (1962)

▶ $C_8H_{18}$

$CH_3CH_2(CH_2)_4CH_2CH_3$

Varian 216

▶ $C_8H_{18}AsBr$

$[(CH_3CH_2)_3AsCH=CH_2]$ Br

J. Chem. Soc. 2762 (1961)

▶ $C_8H_{20}Pb$

$Pb(C_2H_5)_4$

J. Chem. Phys. 34:1049 (1961)
J. Am. Chem. Soc. 83:4473 (1961)

▶ $C_8H_{20}Si$

$(C_2H_5)_4Si$

Can. J. Chem. 40:1956 (1962)

▶ $C_8H_{20}Si_2$

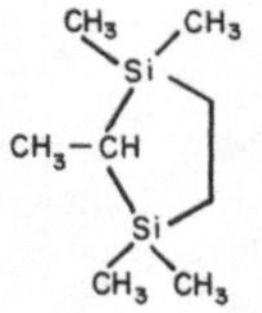

J. Org. Chem. 26:1308 (1961)

▶ $C_8H_{20}Si_2$

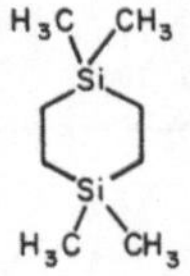

J. Org. Chem. 26:1308 (1961)

▶ $C_8H_{20}Sn$

$Sn(C_2H_5)_4$

J. Chem. Phys. 34:1049 (1961)

▶ $C_8H_{20}Sn$

$(CH_3CH_2)_4Sn^{117}$

J. Am. Chem. Soc. 83:4473 (1961)

▶ $C_8H_{20}Sn$

$(CH_3CH_2)_4Sn^{119}$

J. Am. Chem. Soc. 83:4473 (1961)

▶ $C_9H_8$

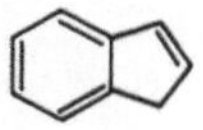

Varian 227
J. Chem. Phys. 36:2346 (1962)

▶ $C_9H_9D$

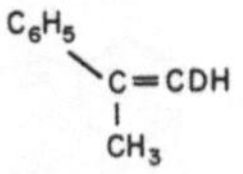

J. Am. Chem. Soc. 84:2252 (1962)

▶ $C_9H_{10}$

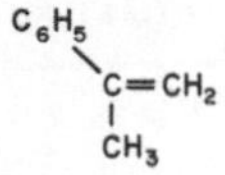

Varian 232
J. Am. Chem. Soc. 84:2252 (1962)

▶ $C_9H_{10}$

$C_6H_5-CH=CHCH_3$

trans

J. Am. Chem. Soc. 83:4420 (1961)

▶ $C_9H_{10}$

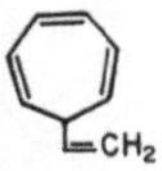

J. Am. Chem. Soc. 83:3590 (1961)

▶ $C_9H_{10}$

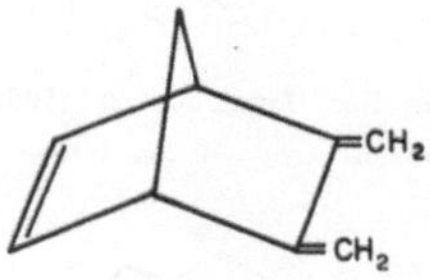


J. Am. Chem. Soc. 83:3590 (1961)

▶ $C_9H_{10}$

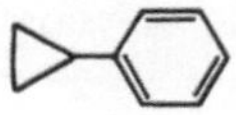

Varian 528
J. Am. Chem. Soc. 83:4420 (1961)
Chem. Ber. 95:2280 (1962)

▶ $C_9H_{10}$

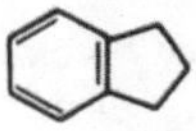

Varian 527

▶ $C_9H_{10} \cdot Cr(CO)_3$

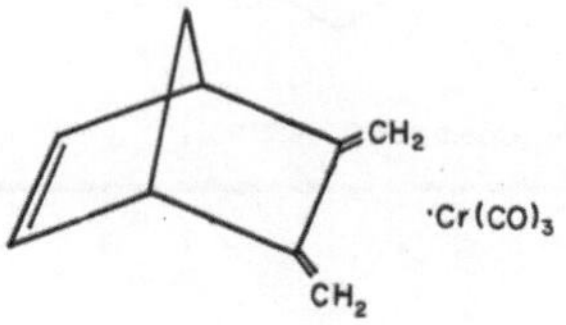


J. Am. Chem. Soc. 83:3590 (1961)

▶ $C_9H_{10} \cdot Mo(CO)_3$

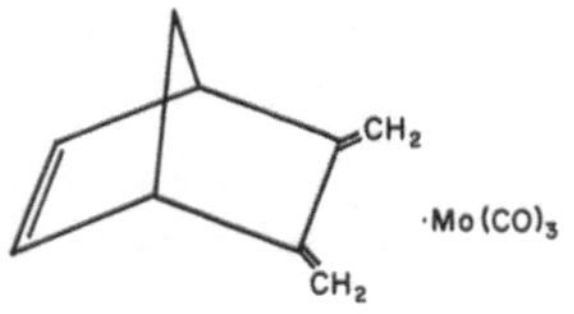


J. Am. Chem. Soc. 83:3590 (1961)

▶ $C_9H_{10} \cdot W(CO)_3$

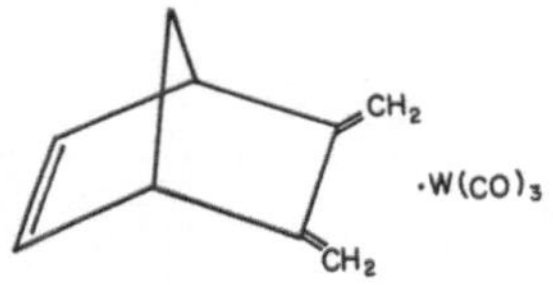


J. Am. Chem. Soc. 83:3590 (1961)

▶ $C_9H_{12}$

$CH_3$
|
$CH-C_6H_5$
|
$CH_3$

Varian 240
J. Chem. Phys. 34:1099 (1961)

▶ $C_9H_{12}$

$CH_3CH_2CH_2-C_6H_5$

J. Chem. Phys. 34:1094, 1099 (1961)

▶ $C_9H_{12}$

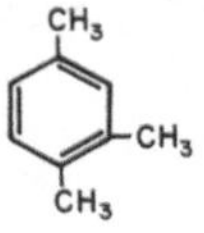


J. Chem. Phys. 34:2208 (1961)

▶ $C_9H_{12}$

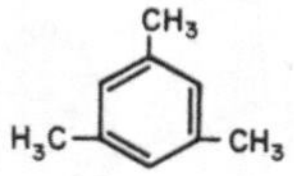


Varian 241
Mol. Phys. 5:1 (1962)
J. Chem. Phys. 34:2208 (1961); 37:2198 (1962)

▶ $C_9H_{12} \cdot Cr(CO)_3$

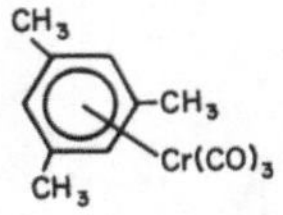


J. Chem. Soc. 3660 (1962)

▶ $C_9H_{15} \cdot ClO_4$

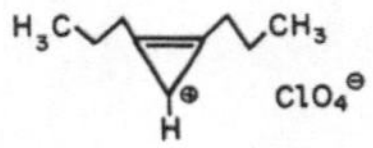


J. Am. Chem. Soc. 84:3168 (1962)

▶ $C_9H_{16}$

$CH_2{=}CHCH_2CH_2\overset{CH_3}{\overset{|}{C}}{=}C(CH_3)_2$

J. Am. Chem. Soc. 83:4923 (1961)

▶ $C_9H_{16}$

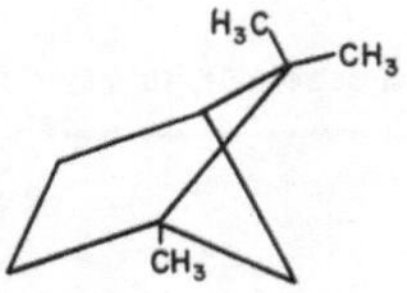


J. Am. Chem. Soc. 83:4923 (1961)

▶ $C_9H_{16}$

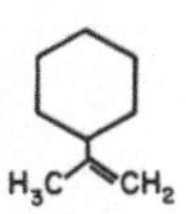


Bull. Chem. Soc. Japan 35:1899 (1962)

▶ $C_9H_{16}$

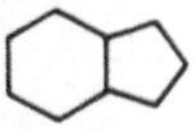

cis and trans

J. Am. Chem. Soc. 83:1671 (1961)

▶ $C_9H_{20}$

$C(C_2H_5)_4$

J. Am. Chem. Soc. 84:673 (1962)

▶ $C_9H_{20}Si_2$

$(CH_3)_3SiCH_2C{\equiv}CSi(CH_3)_3$

Varian 549

▶ $C_{10}H_6$

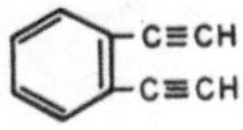


J. Org. Chem. 26:3583 (1961)

▶ $C_{10}H_8$

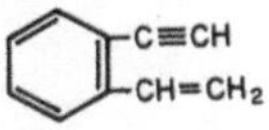


J. Org. Chem. 26:3584 (1961)

▶ $C_{10}H_8$

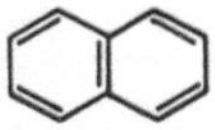

J. Chem. Phys. 36:2443 (1962); 37:1180,2198 (1962)
J. Phys. Chem. 66:2652 (1962)

▶ $C_{10}H_8$

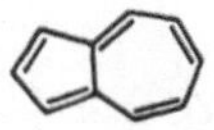

AND CATION

Varian 551
Can. J. Chem. 40:1778 (1962)
Helv. Chim. Acta 45:1965 (1962)
J. Chem. Phys. 37:1180 (1962)

▶ $C_{10}H_{10} \cdot Fe(CO)_4HPF_6$

$[\pi\text{-}C_5H_5Fe(CO)_2]_2H^{\oplus}PF_6^{\ominus}$

J. Chem. Soc. 3654 (1962)

▶ $C_{10}H_{10} \cdot Fe_2(CO)_4H$

$[\pi\text{-}C_5H_5Fe(CO)_2]_2H^{\oplus}$

J. Chem. Soc. 3654 (1962)

▶ $C_{10}H_{10} \cdot MoH_2$

$(C_5H_5)_2MoH_2$

AND SALT

J. Chem. Soc. 4854 (1961)

▶ $C_{10}H_{10} \cdot MoW(CO)_6H$

$[\pi\text{-}(C_5H_5)_2MoW(CO)_6]H^{\oplus}$

J. Chem. Soc. 3654 (1962)

▶ $C_{10}H_{10} \cdot Mo_2(CO)_6H$

$[\pi\text{-}C_5H_5Mo(CO)_3]_2H^{\oplus}$

J. Chem. Soc. 3654 (1962)

▶ $C_{10}H_{10} \cdot Ru_2(CO)_4H$

$[\pi\text{-}C_5H_5Ru(CO)_2]_2H^{\oplus}$

J. Chem. Soc. 3654 (1962)

▶ $C_{10}H_{10} \cdot TaH_3$

$(C_5H_5)_2TaH_3$

J. Chem. Soc. 4854 (1961)

▶ $C_{10}H_{10} \cdot WH_2$

$(C_5H_5)_2WH_2$

AND SALT

J. Chem. Soc. 4854 (1961)

▶ $C_{10}H_{10} \cdot W_2(CO)_6H$

$[\pi\text{-}C_5H_5W(CO)_3]_2H^{\oplus}$

J. Chem. Soc. 3654 (1962)

▶ $C_{10}H_{10} \cdot W_2(CO)_6HPF_6$

$[\pi\text{-}C_5H_5W(CO)_3]_2H^{\oplus}PF_6^{\ominus}$

J. Chem. Soc. 3654 (1962)

▶ $C_{10}H_{10} \cdot W_2(CO)_6Hg$

$[\pi\text{-}C_5H_5W(CO)_3]_2Hg$

J. Chem. Soc. 3660 (1962)

▶ $C_{10}H_{12}$

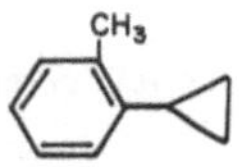


J. Am. Chem. Soc. 83:4423 (1961)

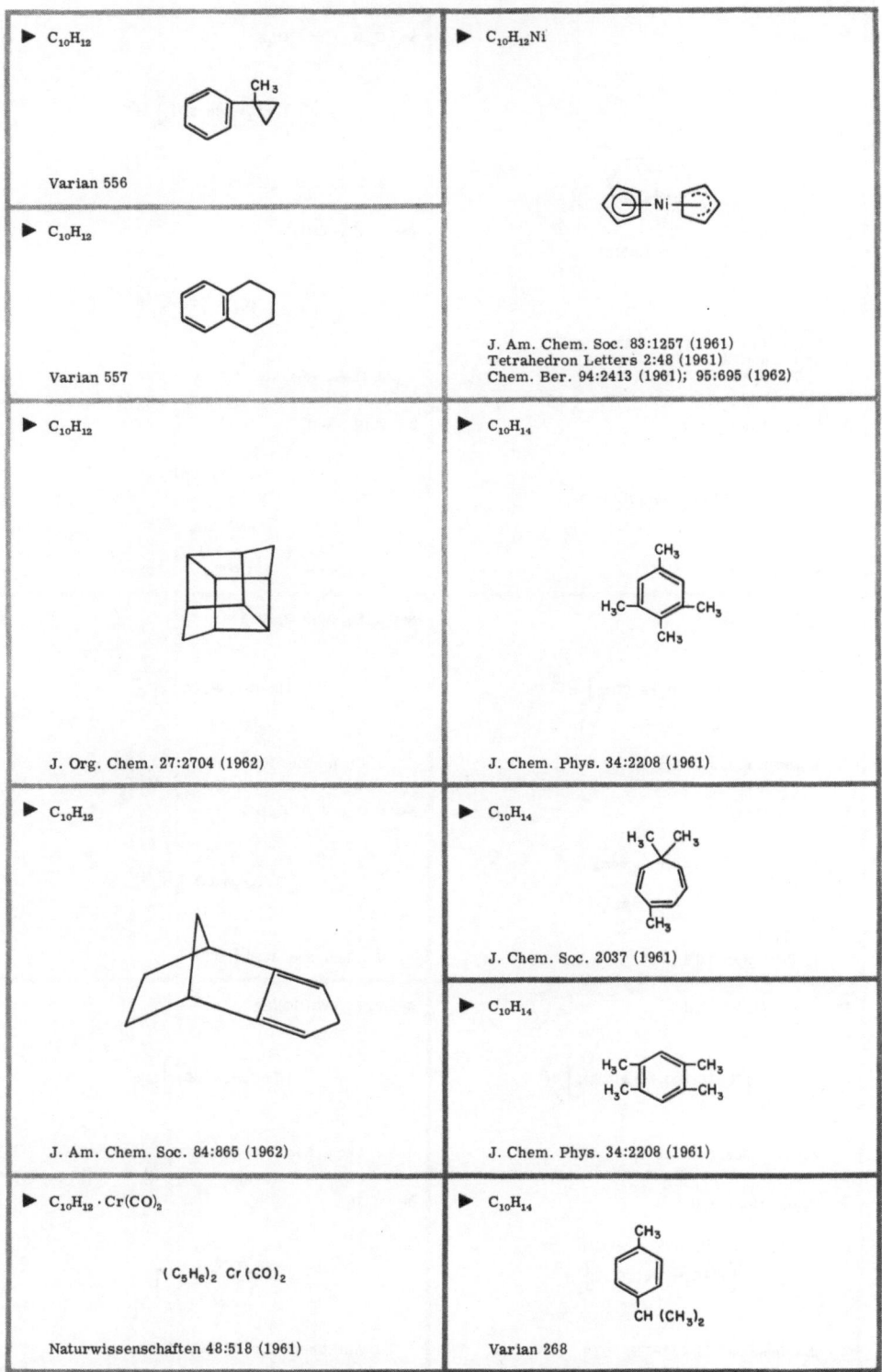

▶ $C_{10}H_{12}$

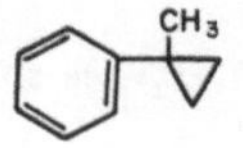


Varian 556

▶ $C_{10}H_{12}$

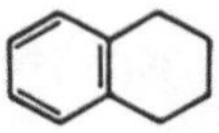

Varian 557

▶ $C_{10}H_{12}$

J. Org. Chem. 27:2704 (1962)

▶ $C_{10}H_{12}$

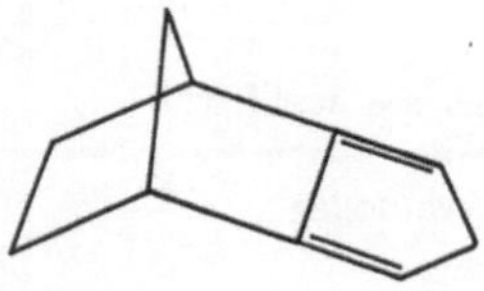

J. Am. Chem. Soc. 84:865 (1962)

▶ $C_{10}H_{12} \cdot Cr(CO)_2$

$(C_5H_6)_2\ Cr(CO)_2$

Naturwissenschaften 48:518 (1961)

▶ $C_{10}H_{12}Ni$

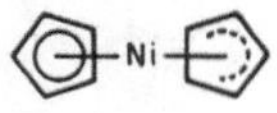


J. Am. Chem. Soc. 83:1257 (1961)
Tetrahedron Letters 2:48 (1961)
Chem. Ber. 94:2413 (1961); 95:695 (1962)

▶ $C_{10}H_{14}$

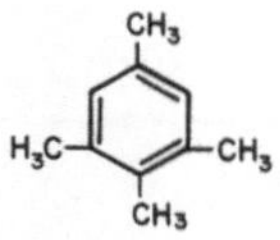


J. Chem. Phys. 34:2208 (1961)

▶ $C_{10}H_{14}$

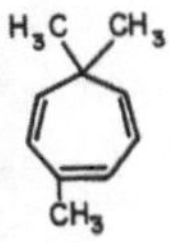


J. Chem. Soc. 2037 (1961)

▶ $C_{10}H_{14}$

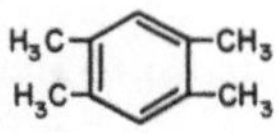


J. Chem. Phys. 34:2208 (1961)

▶ $C_{10}H_{14}$

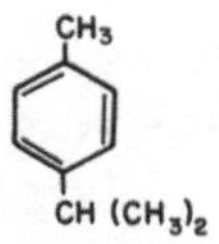


Varian 268

▶ $C_{10}H_{14} \cdot Mo(CO)_3$

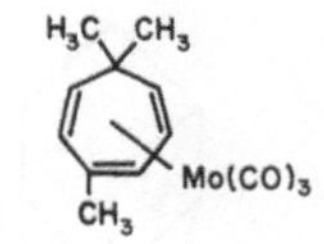

J. Chem. Soc. 2037 (1961)

▶ $C_{10}H_{16}$

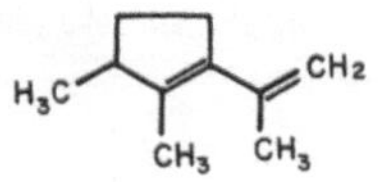

J. Am. Chem. Soc. 84:2775 (1962)
Bull. Chem. Soc. Japan 35:1899 (1962)

▶ $C_{10}H_{15} \cdot HSO_4$

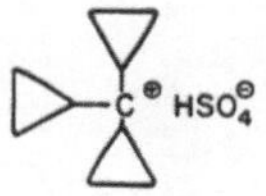

J. Am. Chem. Soc. 84:2462 (1962)

▶ $C_{10}H_{16}$

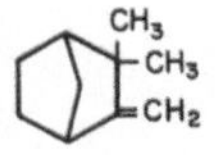

J. Org. Chem. 26:707 (1961)

▶ $C_{10}H_{16}$

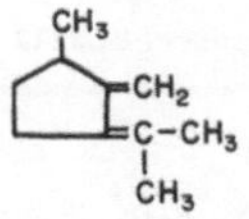

J. Am. Chem. Soc. 84:2775 (1962)

▶ $C_{10}H_{16}$

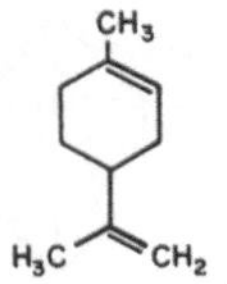

Bull. Chem. Soc. Japan 35:1849 (1962)

▶ $C_{10}H_{16}$

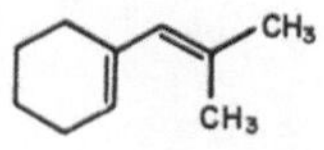

J. Am. Chem. Soc. 84:2775 (1962)

▶ $C_{10}H_{16}$

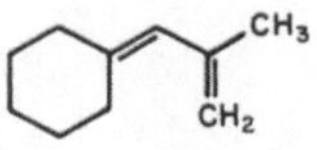

J. Am. Chem. Soc. 84:2775 (1962)

▶ $C_{10}H_{16}$

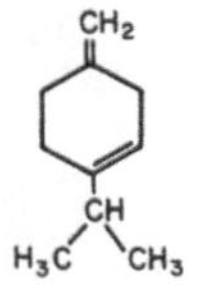

J. Org. Chem. 27:3359 (1962)

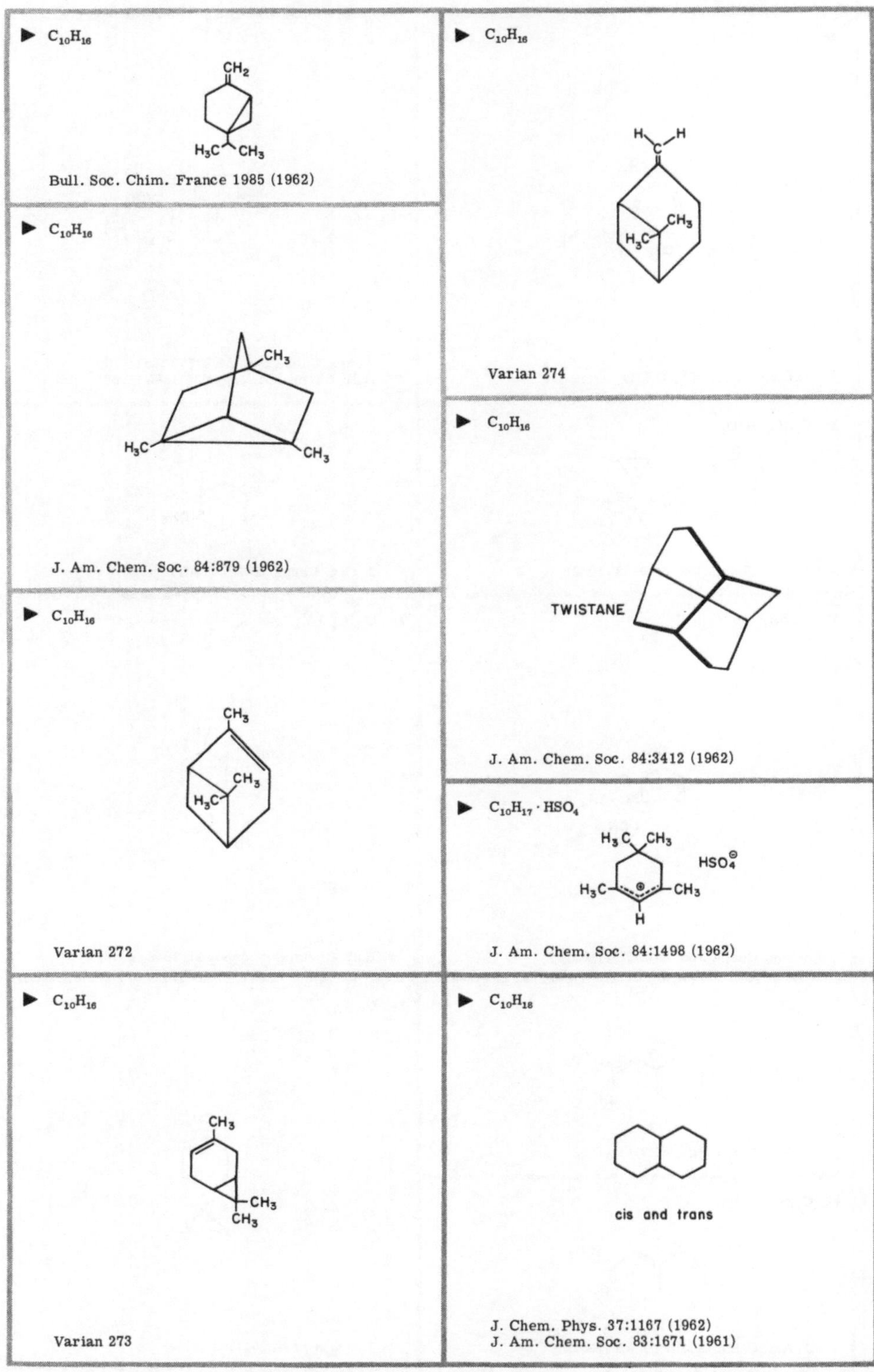

► $C_{10}H_{16}$

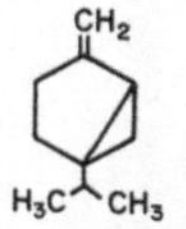


Bull. Soc. Chim. France 1985 (1962)

► $C_{10}H_{16}$

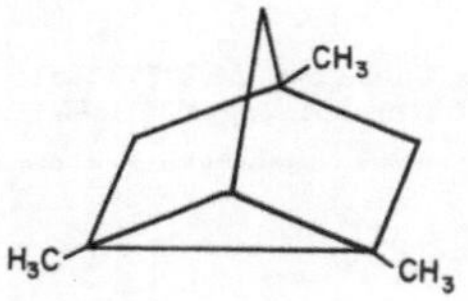


J. Am. Chem. Soc. 84:879 (1962)

► $C_{10}H_{16}$

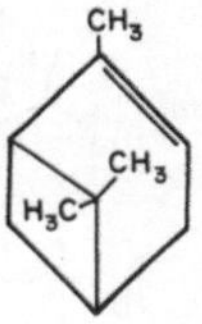


Varian 272

► $C_{10}H_{16}$

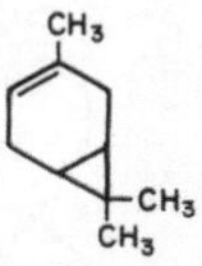


Varian 273

► $C_{10}H_{16}$

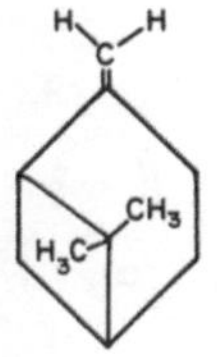


Varian 274

► $C_{10}H_{16}$

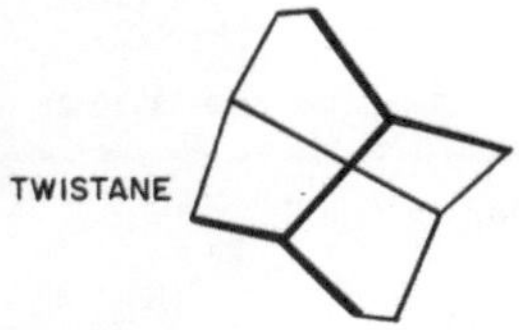


J. Am. Chem. Soc. 84:3412 (1962)

► $C_{10}H_{17} \cdot HSO_4$

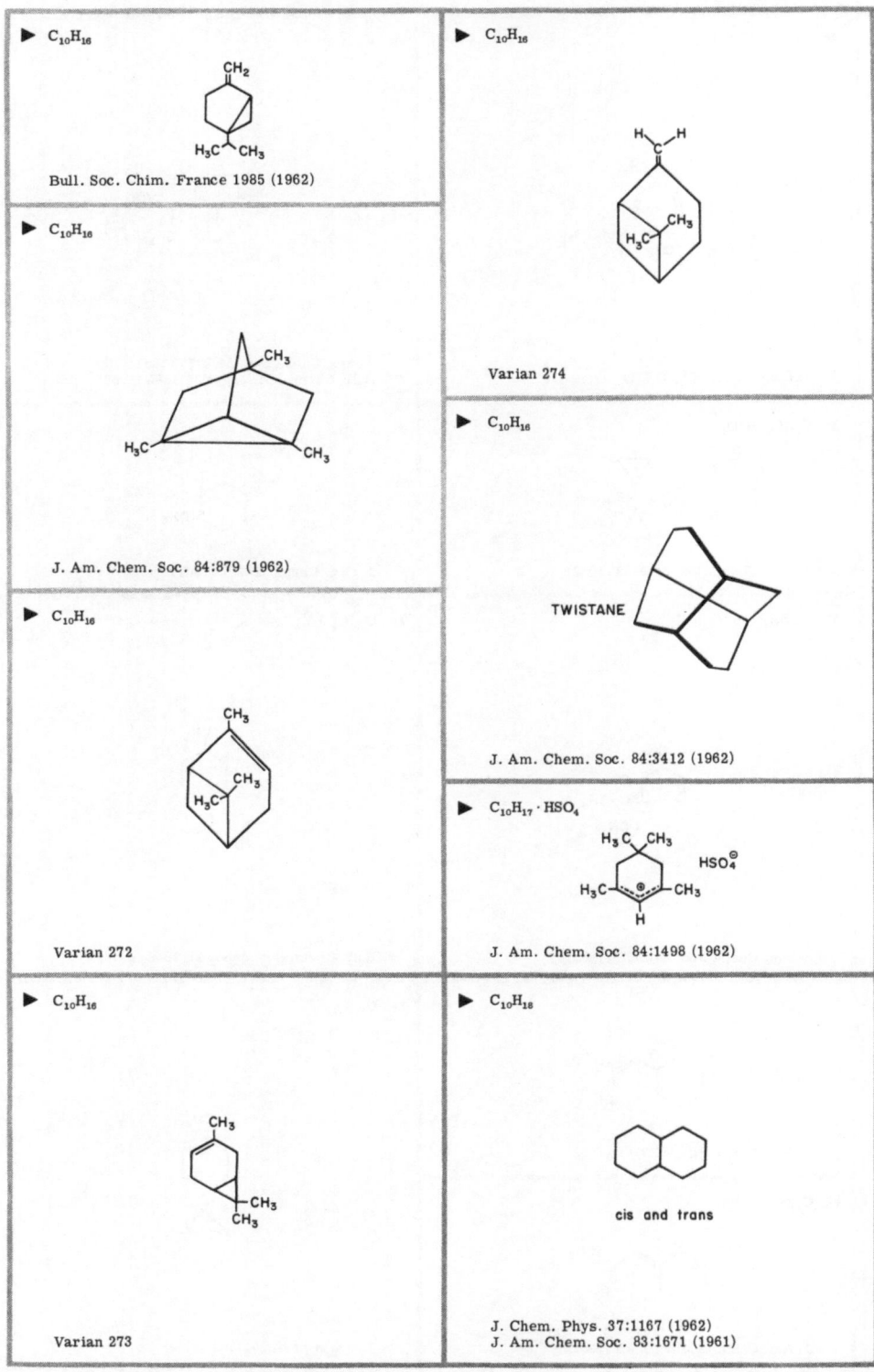

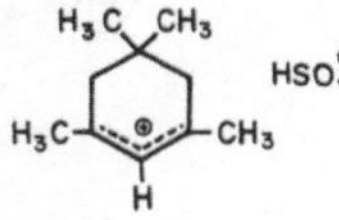


J. Am. Chem. Soc. 84:1498 (1962)

► $C_{10}H_{18}$



cis and trans

J. Chem. Phys. 37:1167 (1962)
J. Am. Chem. Soc. 83:1671 (1961)

▶ $C_{10}H_{18}$

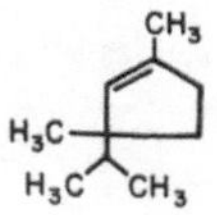


Acta Chem. Scand. 16:640 (1962)

▶ $C_{10}H_{18}$

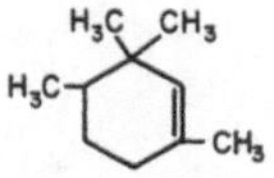


Bull. Chem. Soc. Japan 35:818 (1962)

▶ $C_{10}H_{20}$

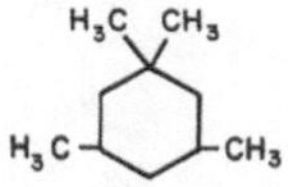


J. Chem. Phys. 37:1167 (1962)

▶ $C_{10}H_{23}B$

$t-C_4H_9B(CH_2CH_2CH_3)_2$

J. Am. Chem. Soc. 83:2541 (1961)

▶ $C_{10}H_{23}BSi$

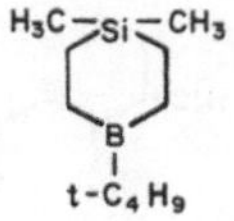


J. Am. Chem. Soc. 83:2541 (1961)

▶ $C_{11}H_{10}$

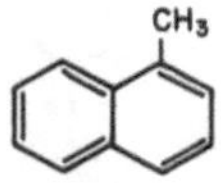


Tetrahedron 18:841 (1962)
Anal. Chem. 33:1085 (1961)

▶ $C_{11}H_{10}$

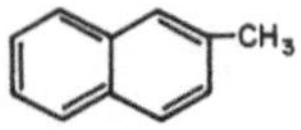


Tetrahedron 18:841 (1962)
Anal. Chem. 33:1085 (1961)

▶ $C_{11}H_{10}$

J. Am. Chem. Soc. 83:4990 (1961)

▶ $C_{11}H_{10}$

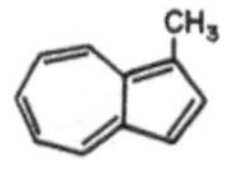


J. Org. Chem. 27:4535 (1962)

▶ $C_{11}H_{11}DFe$

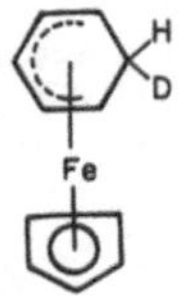


J. Chem. Soc. 4463 (1962)

► $C_{11}H_{12}Fe$

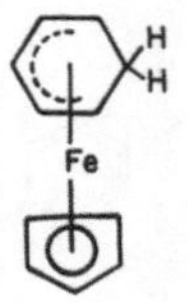


Chem. & Ind. 1408 (1961)
J. Chem. Soc. 4462 (1962)

► $C_{11}H_{13}Co$

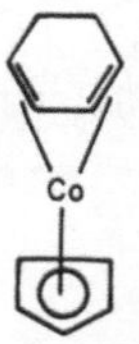


J. Am. Chem. Soc. 83:3593 (1961)

► $C_{11}H_{14}Pd$

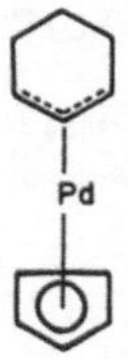


Tetrahedron Letters 2:49 (1961)
Chem. Ber. 95:695 (1962)

► $C_{11}H_{16}$

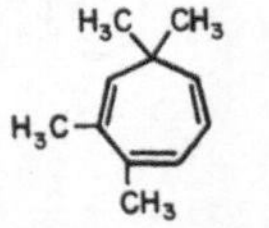


J. Am. Chem. Soc. 83:2958 (1961)
J. Chem. Soc. 2037 (1961)

► $C_{11}H_{16}$

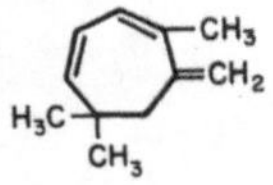


J. Am. Chem. Soc. 83:2958 (1961)

► $C_{11}H_{16}$

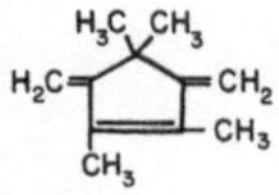


J. Am. Chem. Soc. 83:2392 (1961)

► $C_{11}H_{16}$

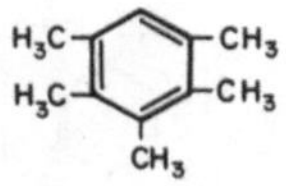


Varian 287

► $C_{11}H_{16}$

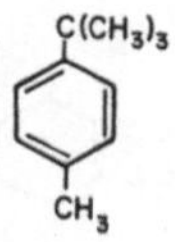


Varian 586

► $C_{11}H_{16} \cdot Mo(CO)_3$

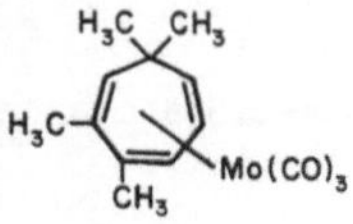


J. Chem. Soc. 2037 (1961)

► $C_{11}H_{18}$

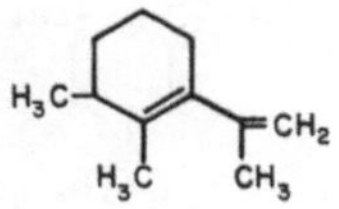


Bull. Chem. Soc. Japan 35:1899 (1962)

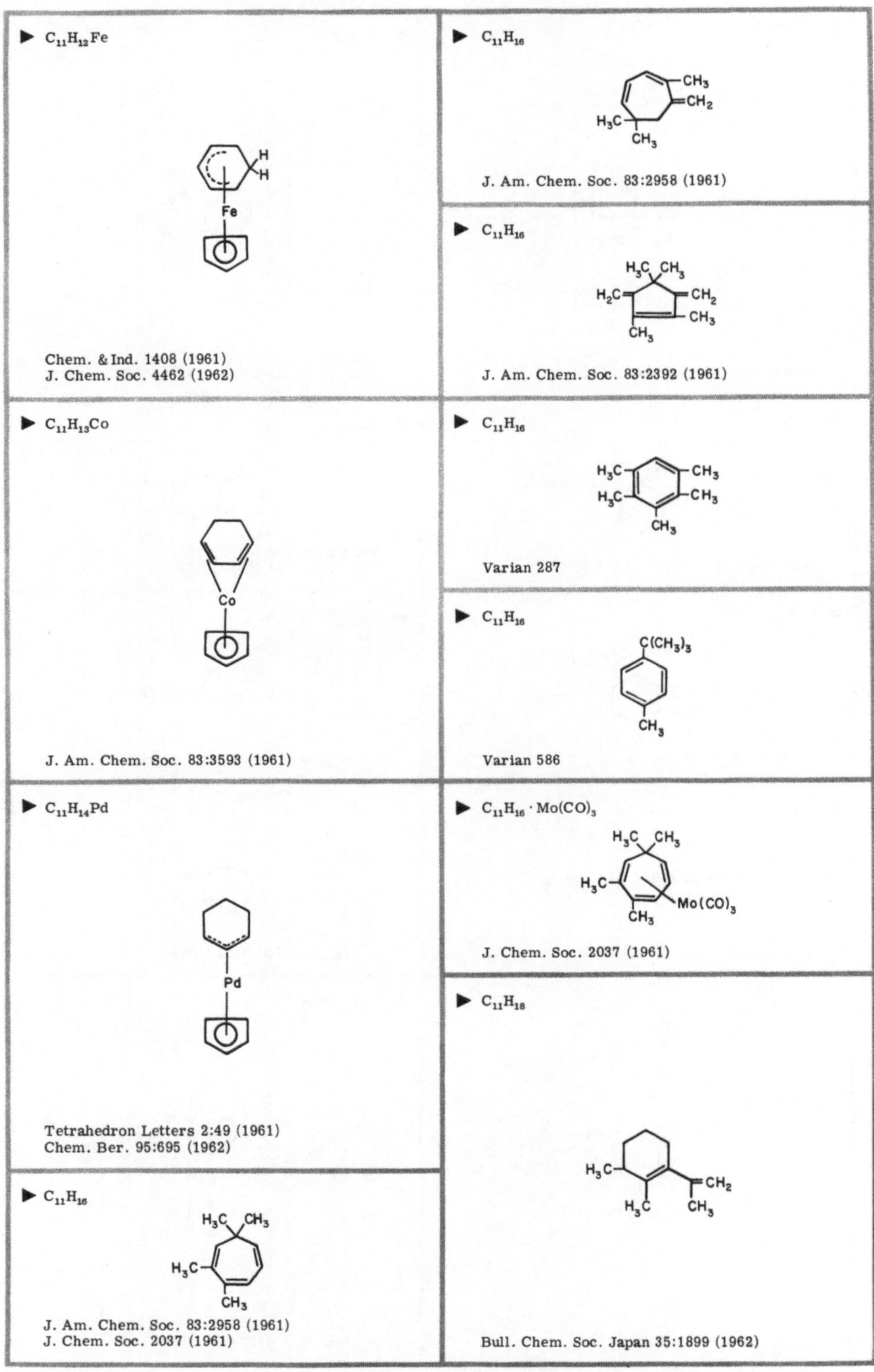

▶ $C_{11}H_{18}$

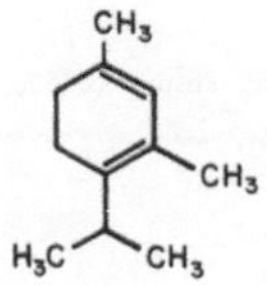


J. Org. Chem. 27:1982 (1962)

▶ $C_{11}H_{18}$

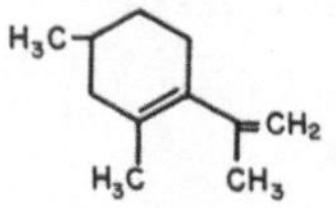


Bull. Chem. Soc. Japan 35:1899 (1962)

▶ $C_{11}H_{22}$

$$CH_2{=}CH\underset{}{C}H\overset{C(CH_3)_3}{C}HC(CH_3)_3$$

J. Am. Chem. Soc. 84:2748 (1962)

▶ $C_{11}H_{22}$

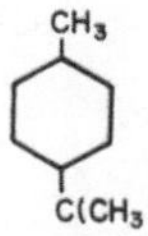


J. Chem. Phys. 37:1167 (1962)

▶ $C_{12}H_{10}$

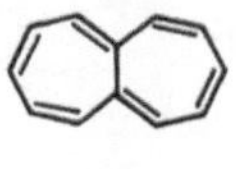

J. Am. Chem. Soc. 83:4659 (1961)

▶ $C_{12}H_{10}$

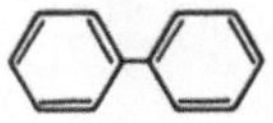

Varian 289

▶ $C_{12}H_{11} \cdot BF_4$

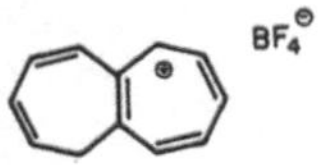


J. Am. Chem. Soc. 83:4657 (1961)

▶ $C_{12}H_{11} \cdot Mn(CO)_3$

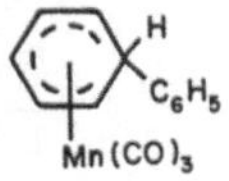


J. Chem. Soc. 4463 (1962)

▶ $C_{12}H_{12}$

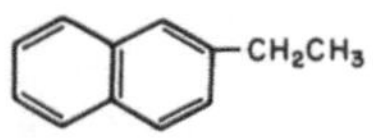


Varian 292

▶ $C_{12}H_{12}Tc \cdot PF_6$

$[Tc(C_6H_6)_2]\,PF_6$

Tetrahedron Letters No. 6:254 (1962)

▶ $C_{12}H_{13}Re$

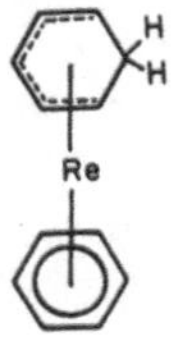


Chem. & Ind. 1408 (1961)
J. Chem. Soc. 4463 (1962)

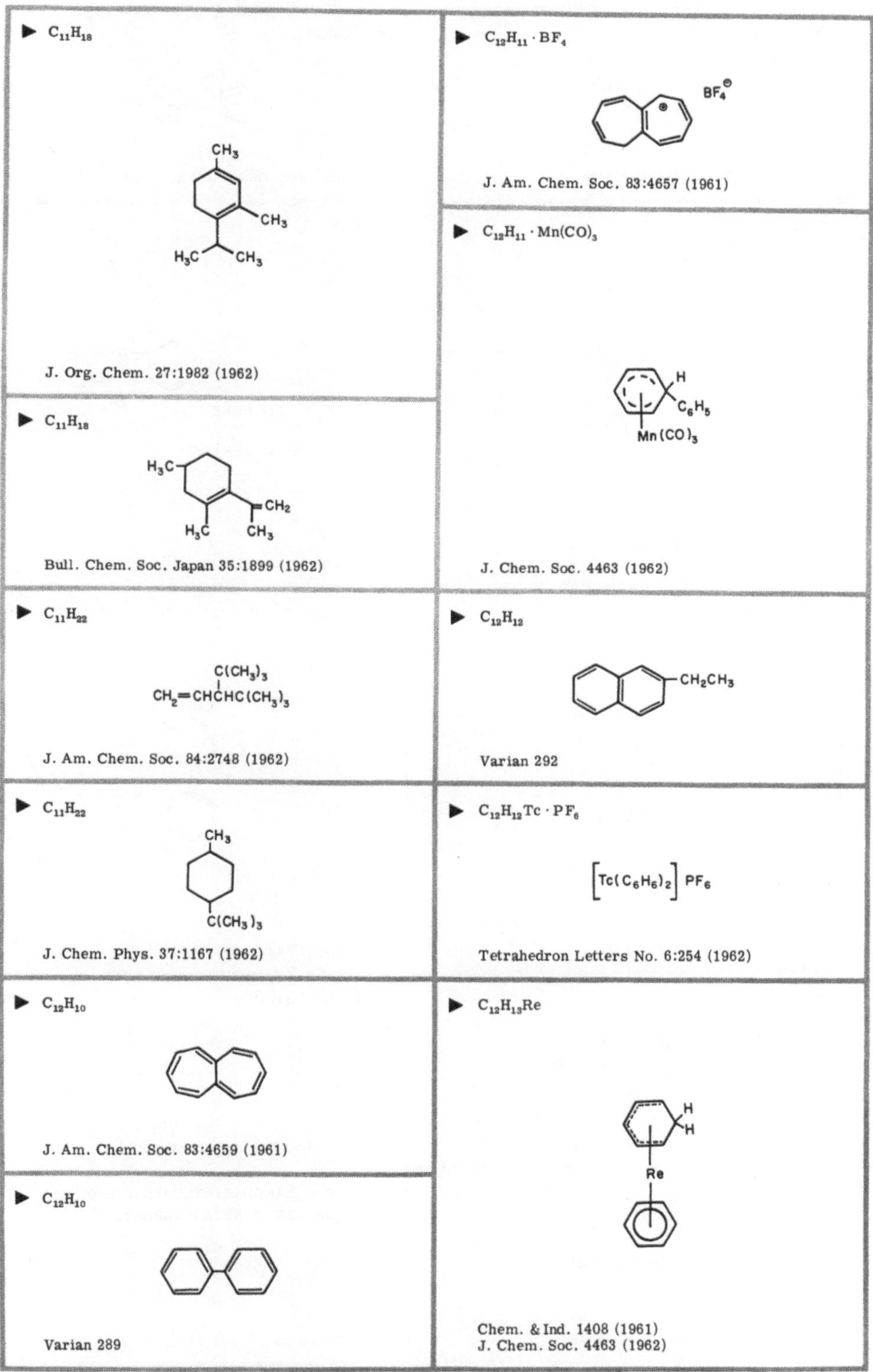

▶ $C_{12}H_{14}$

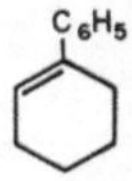


J. Org. Chem. 27:4243 (1962)

▶ $C_{12}H_{16}Ni$

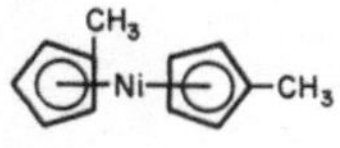


Chem. Ber. 95:695 (1962)

▶ $C_{12}H_{14}Ru$

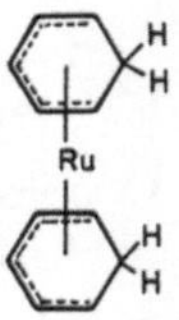


Chem. & Ind. 1408 (1961)
J. Chem. Soc. 4463 (1962)

▶ $C_{12}H_{18}$

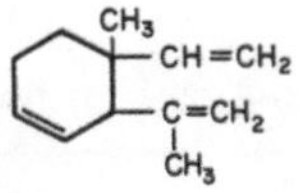


GEIJERENE
ALSO
ISOGEIJERENE

J. Chem. Soc. 2286 (1961)

▶ $C_{12}H_{14}Ru$

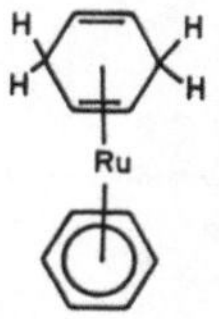


J. Chem. Soc. 4463 (1962)

▶ $C_{12}H_{18}$

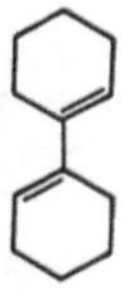

J. Org. Chem. 27:4243 (1962)

▶ $C_{12}H_{16}$

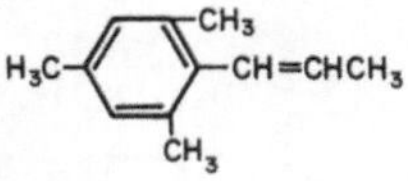


J. Am. Chem. Soc. 83:4423 (1961)

▶ $C_{12}H_{16} \cdot Mo(CO)_2$

$(C_6H_8)_2Mo(CO)_2$

Naturwissenschaften 48:518 (1961)

▶ $C_{12}H_{20} \cdot Co_2(CO)_4$

$Co_2(CO)_4(CH_2{=}C(CH_3){-}C(CH_3){=}CH_2)_2$

DI-(2,3-DIMETHYLBUTA-1,3-DIENE)
DICOBALT TETRACARBONYL

J. Chem. Soc. 602 (1961)

▶ $C_{12}H_{21} \cdot ClO_4$

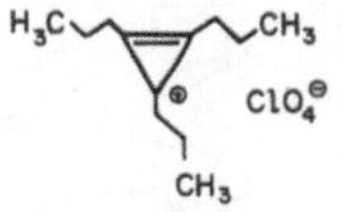


J. Am. Chem. Soc. 84:3168 (1962)

▶ $C_{12}H_{22}$

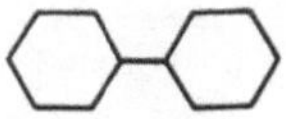

J. Chem. Phys. 37:1167 (1962)

▶ $C_{12}H_{23}B$

$t-C_4H_9B(CH=CHCH_2CH_3)_2$

J. Am. Chem. Soc. 83:2541 (1961)

▶ $C_{12}H_{24}$

$(CH_3)_2CHCH_2CH(CH_3)(CH_2)_4CH=CH_2$

Varian 298

▶ $C_{12}H_{36} \cdot Pt_4I_4$

$[(CH_3)_3PtI]_4$

J. Chem. Soc. 4736 (1962)

▶ $C_{13}H_{10}$

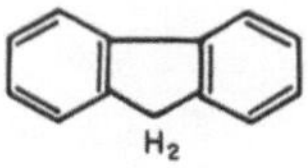


Can. J. Chem. 39:1388 (1961)

▶ $C_{13}H_{12}$

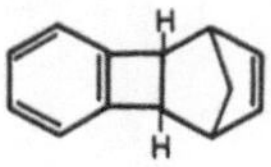


J. Am. Chem. Soc. 83:1657 (1961)

▶ $C_{13}H_{12}$

$C_6H_5C^{13}H_2C_6H_5$

J. Am. Chem. Soc. 83:4479 (1961)

▶ $C_{13}H_{12}$

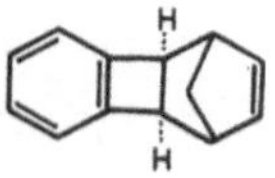


J. Am. Chem. Soc. 83:1657 (1961)

▶ $C_{13}H_{12}$

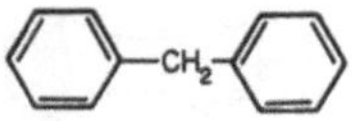


J. Chem. Phys. 37:2198 (1962)
J. Org. Chem. 26:4984 (1961)

▶ $C_{13}H_{13}Co$

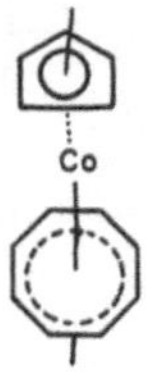


J. Chem. Soc. 4823 (1962)
J. Chem. Soc. Japan 82:1389 (1961)
Chem. Ber. 95:158 (1962)
J. Am. Chem. Soc. 83:3593 (1961)

▶ $C_{13}H_{13}Rh$

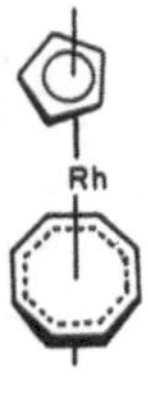


J. Chem. Soc. 4823 (1962)

▶ $C_{13}H_{14}$

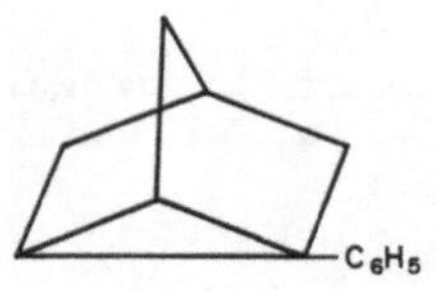


J. Am. Chem. Soc. 83:2329 (1961)

▶ $C_{13}H_{14}Fe$

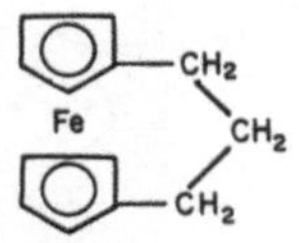


Tetrahedron Letters No. 10:425 (1962)

▶ $C_{13}H_{14}Si$

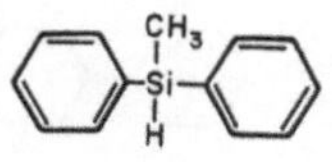


Varian 609

▶ $C_{13}H_{14}$

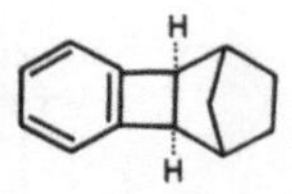


J. Am. Chem. Soc. 83:1657 (1961)

▶ $C_{13}H_{15}Co$

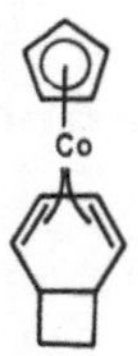


J. Am. Chem. Soc. 83:3593 (1961)

▶ $C_{13}H_{14}$

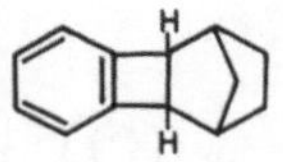


J. Am. Chem. Soc. 83:1657 (1961)

▶ $C_{13}H_{14}$

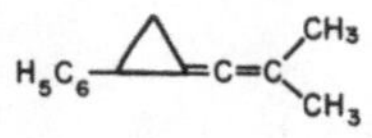


J. Am. Chem. Soc. 83:4990 (1961)

▶ $C_{13}H_{16}$

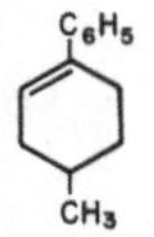


J. Org. Chem. 27:4243 (1962)

▶ $C_{13}H_{14}$

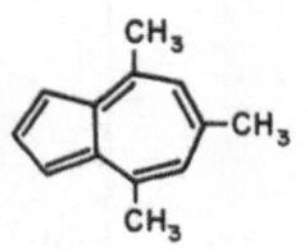


AND CATION

Can. J. Chem. 40:1780 (1962)

▶ $C_{13}H_{16}$

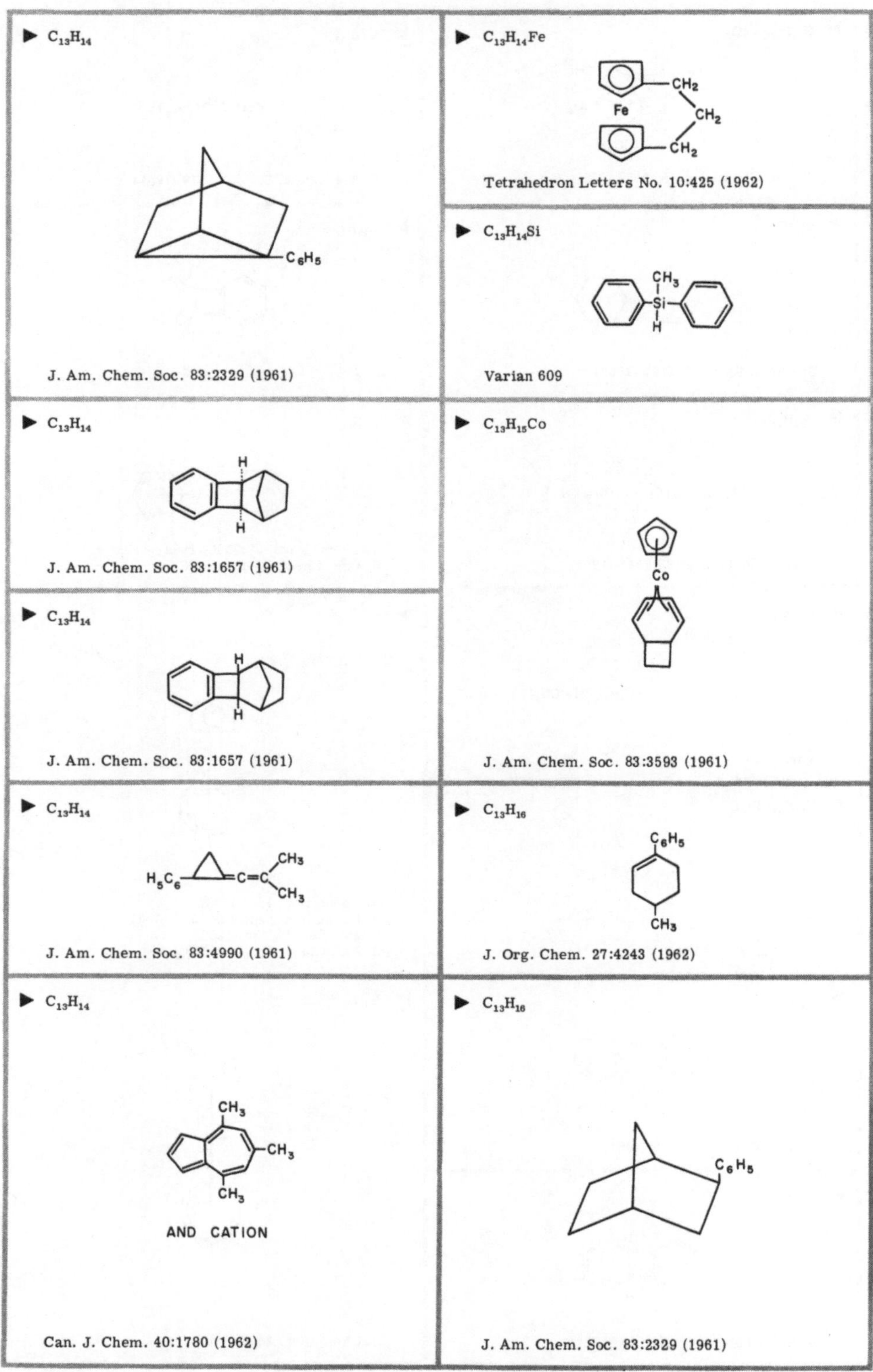

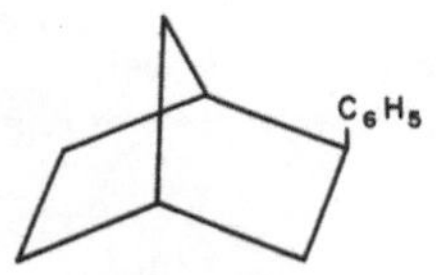


J. Am. Chem. Soc. 83:2329 (1961)

$C_{13}H_{16}$
$C_6H_5$
$CH_3$
J. Org. Chem. 27:4243 (1962)
$C_{13}H_{17}Co$
Co
J. Am. Chem. Soc. 83:3593 (1961)
$C_{13}H_{16}$
$CH_3$
J. Org. Chem. 27:4243 (1962)
$C_{13}H_{21} \cdot Cl$
$H_3C$ $CH_3$
$H_3C$
$CH_3$
$H_3C$
$CH_3$ $Cl^{\ominus}$
$CH_3$
J. Am. Chem. Soc. 84:1498 (1962)
$C_{13}H_{16}$
$C_6H_5$
J. Org. Chem. 26:3747 (1961)
$C_{13}H_{22}$
J. Org. Chem. 26:3124 (1961)
$C_{13}H_{26}$
$(CH_3)_2C{=}CHCHC(CH_3)_3$
$C(CH_3)_3$
J. Am. Chem. Soc. 84:2748 (1962)
$C_{13}H_{16}$
$C_6H_5$
$H_3C$
J. Org. Chem. 27:4243 (1962)
$C_{14}H_{10}$
H H
C C
H H
H H
H H
C C
H H
1,8-BISDEHYDRO [14]-ANNULENE
J. Am. Chem. Soc. 84:4594 (1962)

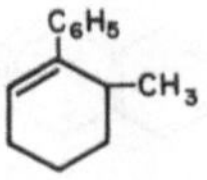

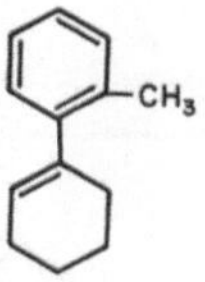

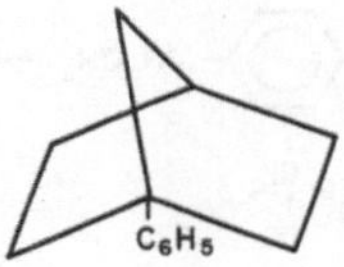

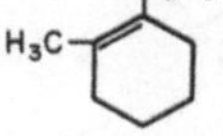

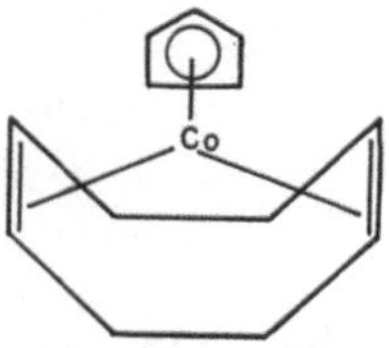

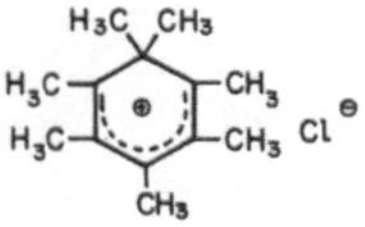

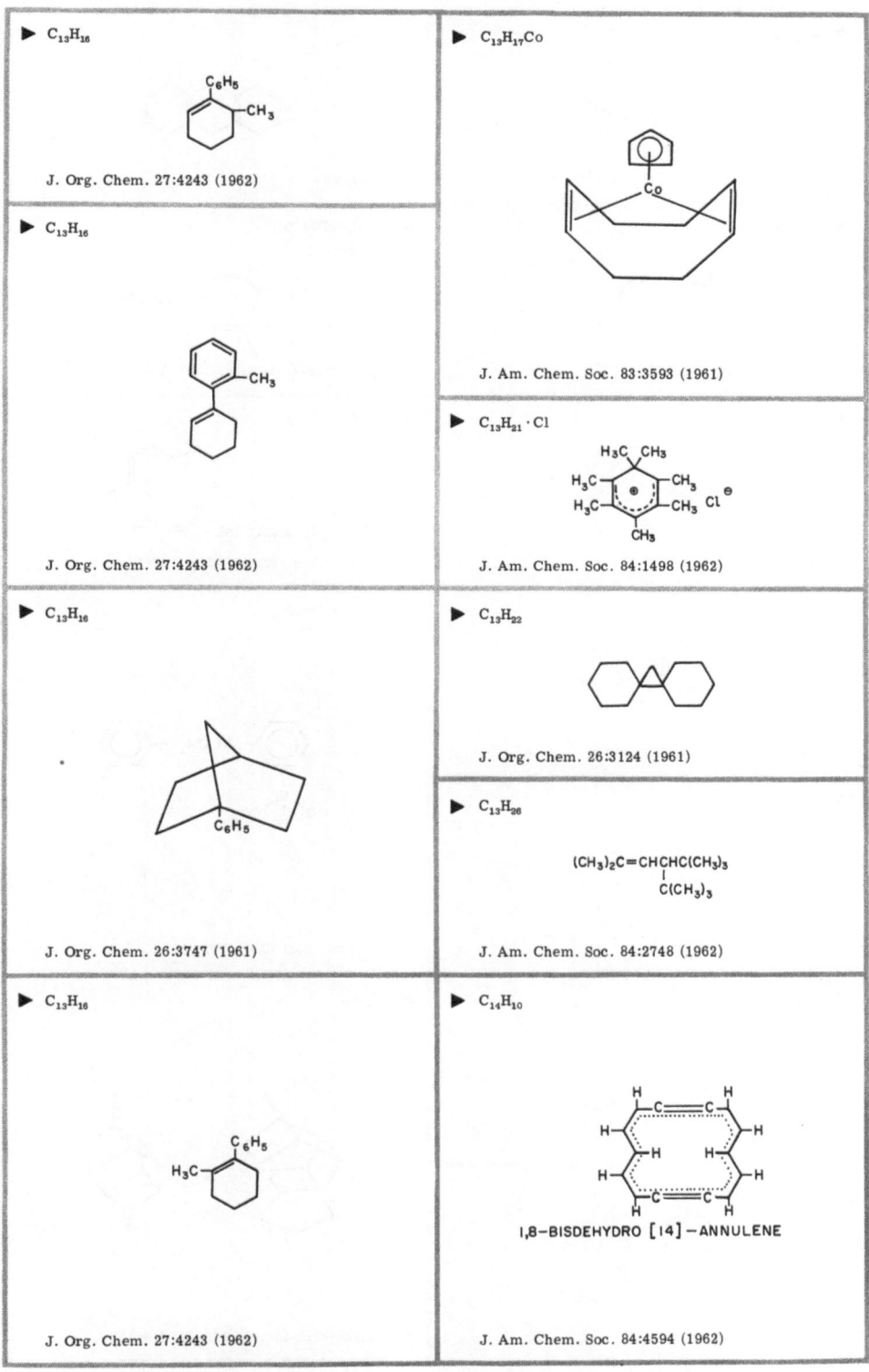

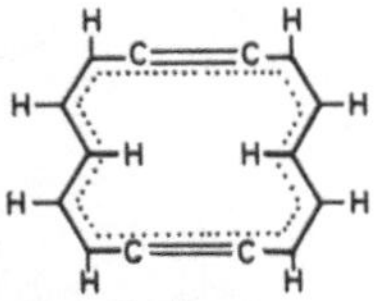

► $C_{14}H_{10}$

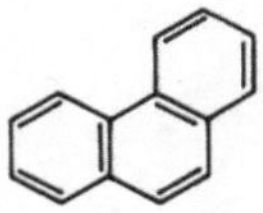

J. Chem. Phys. 36:2443 (1962); 37:1180 (1962)

► $C_{14}H_{10}$

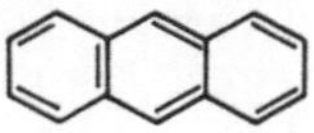

J. Chem. Phys. 36:2443 (1962)

► $C_{14}H_{12}$

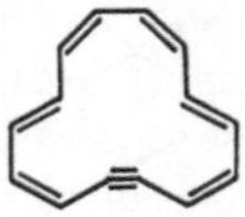

J. Am. Chem. Soc. 84:4307 (1962)

► $C_{14}H_{12}$

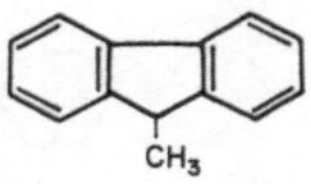


Can. J. Chem. 39:1388 (1961)

► $C_{14}H_{12}$

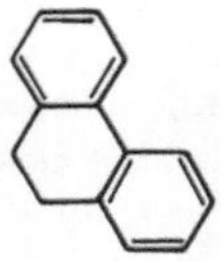

Varian 622

► $C_{14}H_{12}$

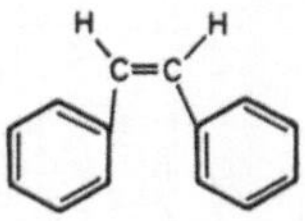


Varian 305

► $C_{14}H_{12}$

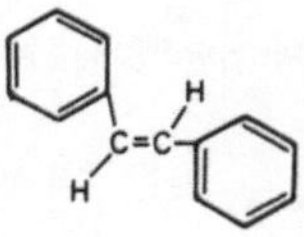


Varian 306

► $C_{14}H_{12}$

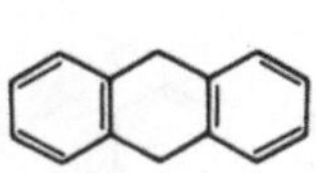

Varian 307

► $C_{14}H_{12} \cdot Fe(CO)_3$

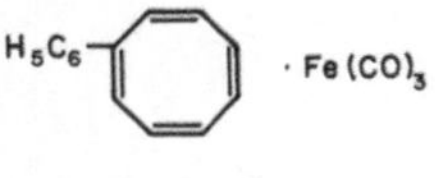


J. Chem. Soc. Japan 82:1387 (1961)

► $C_{14}H_{14}$

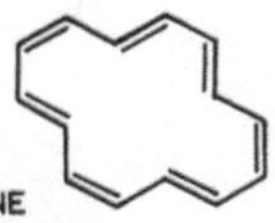

(14)-ANNULENE

J. Am. Chem. Soc. 84:4307 (1962)

► $C_{14}H_{14} \cdot Cr(CO)_3$

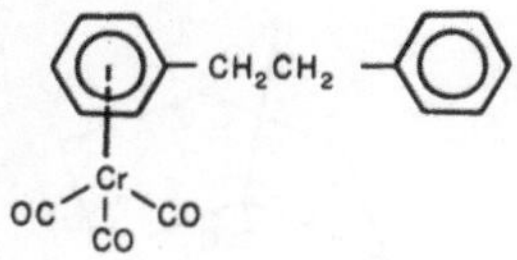


J. Chem. Soc. 3660 (1962)

► $C_{14}H_{16}$

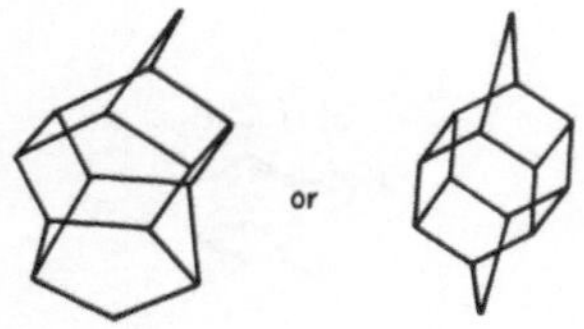


Tetrahedron Letters 368, 373 (1961)

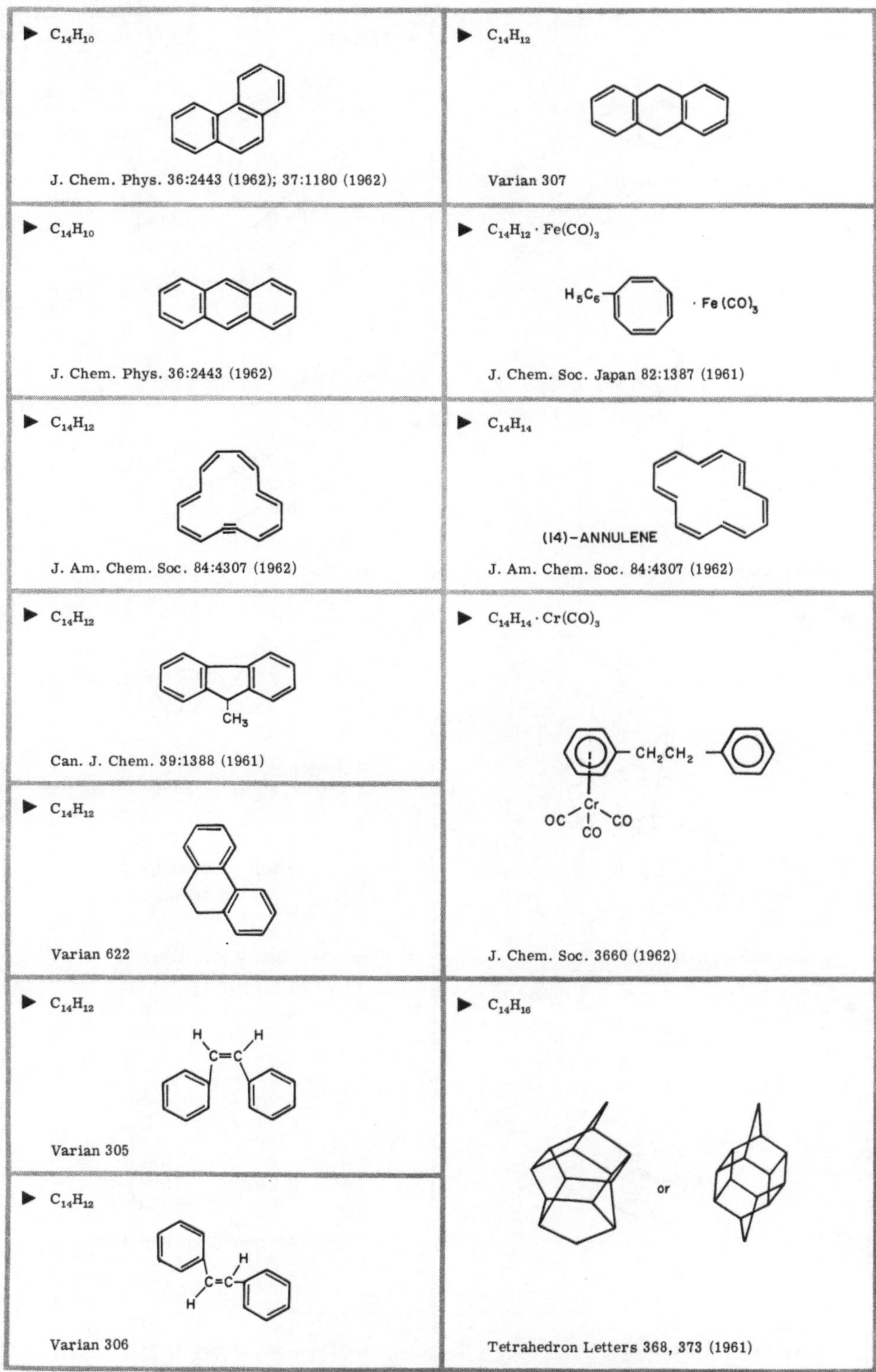

▶ $C_{14}H_{16}$

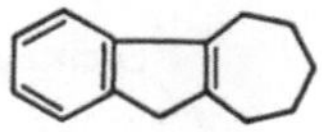

J. Am. Chem. Soc. 83:1752 (1961)

▶ $C_{14}H_{16}Fe$

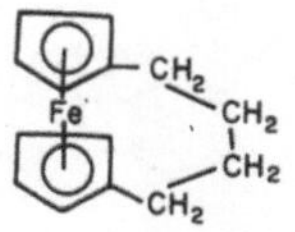


Tetrahedron Letters No. 10:426 (1962)

▶ $C_{14}H_{18}$

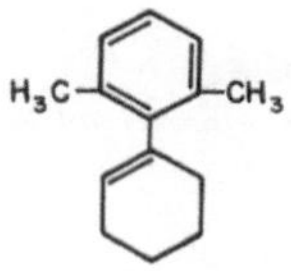


J. Org. Chem. 27:4243 (1962)

▶ $C_{14}H_{18}$

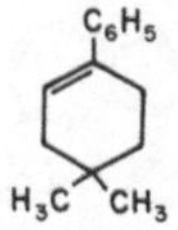


J. Org. Chem. 27:4243 (1962)

▶ $C_{14}H_{18}$

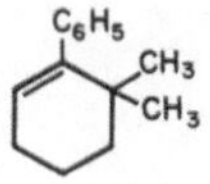


J. Org. Chem. 27:4243 (1962)

▶ $C_{14}H_{20}$

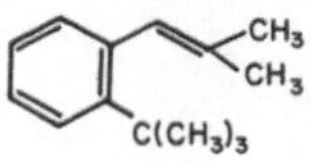


Can. J. Chem. 40:1666 (1962)

▶ $C_{14}H_{22}$

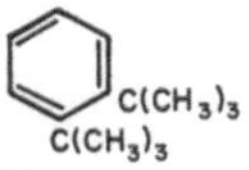


Chem. & Ind. 2008 (1961)
Can. J. Chem. 40:1665 (1962)

▶ $C_{14}H_{24}$

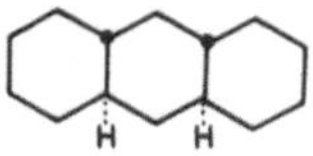


J. Am. Chem. Soc. 83:965 (1961)

▶ $C_{14}H_{24}$

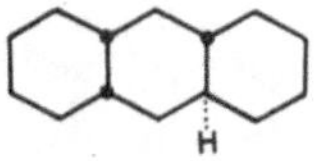


J. Am. Chem. Soc. 83:965 (1961)

▶ $C_{14}H_{24}$

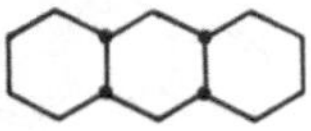

J. Am. Chem. Soc. 83:965 (1961)

▶ $C_{14}H_{24}$

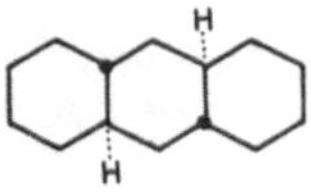


J. Am. Chem. Soc. 83:965 (1961)

▶ $C_{14}H_{24}$

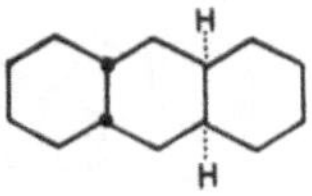


J. Am. Chem. Soc. 83:965 (1961)

▶ $C_{14}H_{26}$

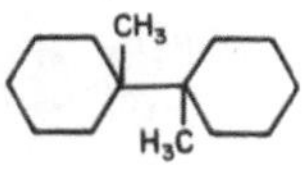


J. Org. Chem. 27:4636 (1962)

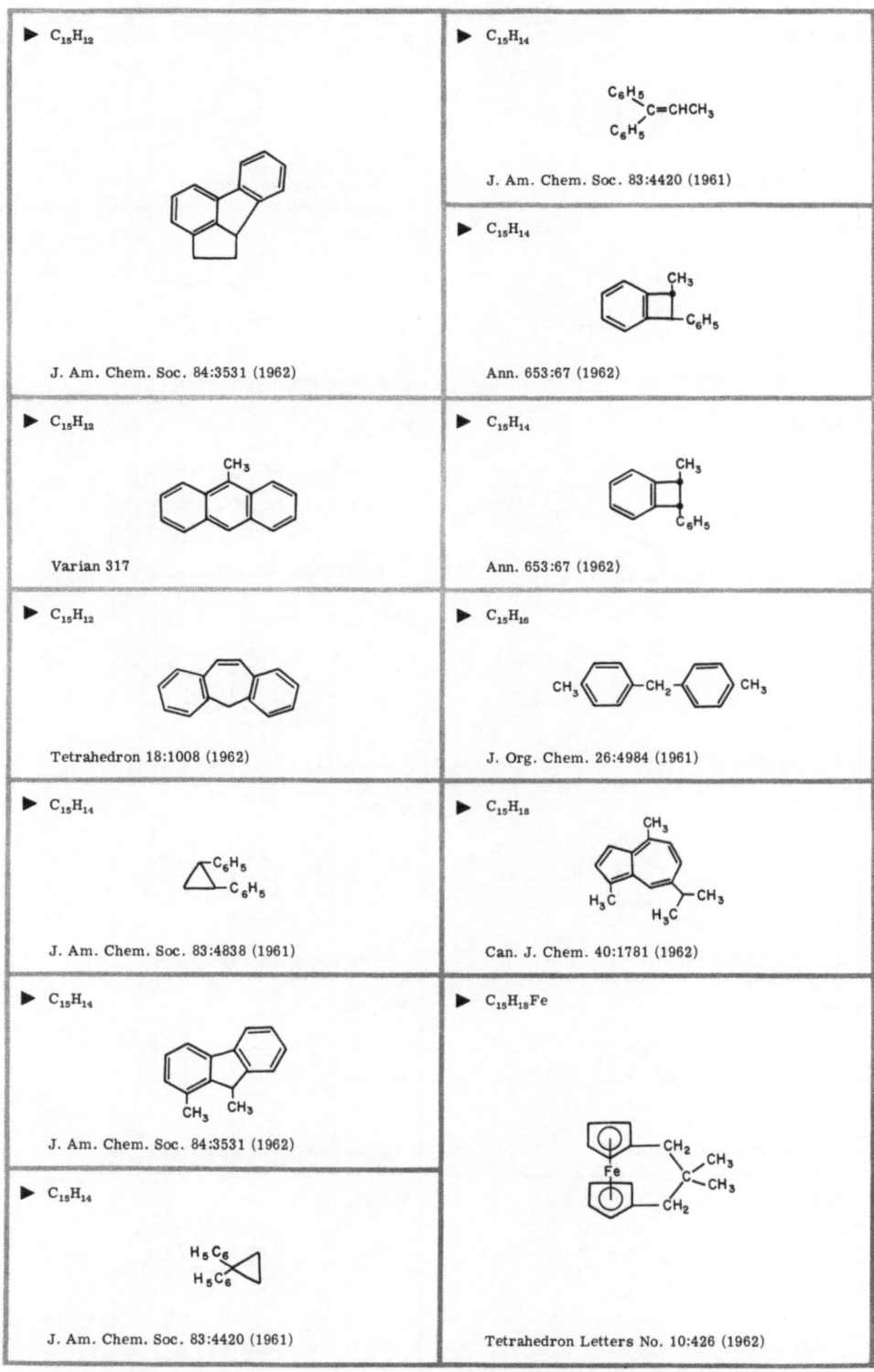

▶ $C_{15}H_{12}$

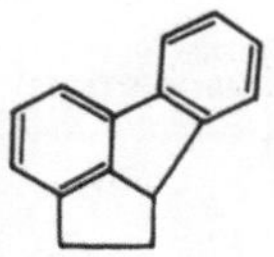

J. Am. Chem. Soc. 84:3531 (1962)

▶ $C_{15}H_{12}$

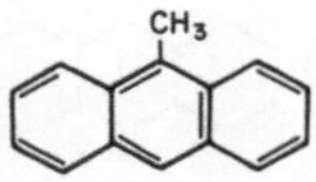


Varian 317

▶ $C_{15}H_{12}$

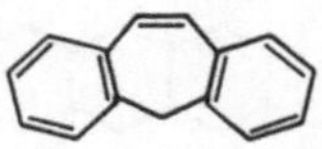

Tetrahedron 18:1008 (1962)

▶ $C_{15}H_{14}$

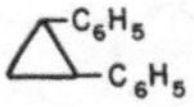


J. Am. Chem. Soc. 83:4838 (1961)

▶ $C_{15}H_{14}$

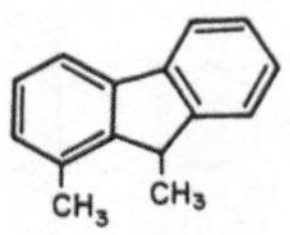


J. Am. Chem. Soc. 84:3531 (1962)

▶ $C_{15}H_{14}$

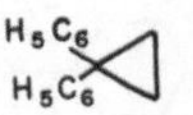


J. Am. Chem. Soc. 83:4420 (1961)

▶ $C_{15}H_{14}$

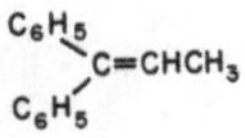


J. Am. Chem. Soc. 83:4420 (1961)

▶ $C_{15}H_{14}$

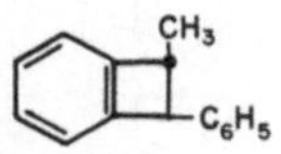


Ann. 653:67 (1962)

▶ $C_{15}H_{14}$

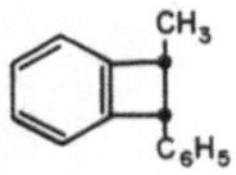


Ann. 653:67 (1962)

▶ $C_{15}H_{16}$

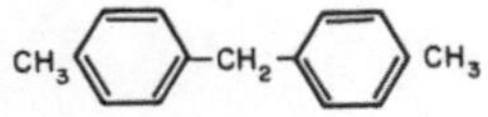


J. Org. Chem. 26:4984 (1961)

▶ $C_{15}H_{18}$

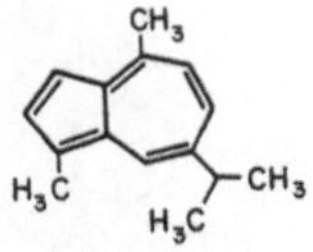


Can. J. Chem. 40:1781 (1962)

▶ $C_{15}H_{18}Fe$

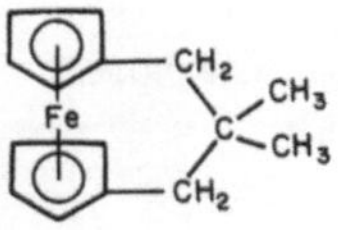


Tetrahedron Letters No. 10:426 (1962)

▶ $C_{15}H_{20}$

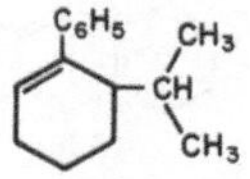


J. Org. Chem. 27:4243 (1962)

▶ $C_{15}H_{20}$

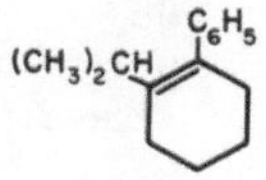


J. Org. Chem. 27:4243 (1962)

▶ $C_{15}H_{24}$

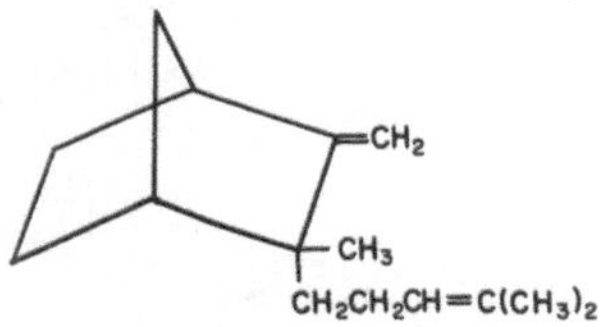


J. Am. Chem. Soc. 84:2611 (1962)

▶ $C_{15}H_{24}$

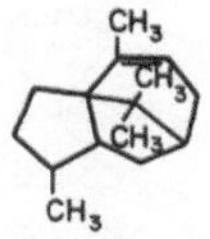


J. Am. Chem. Soc. 84:3205 (1962)

▶ $C_{15}H_{24}$

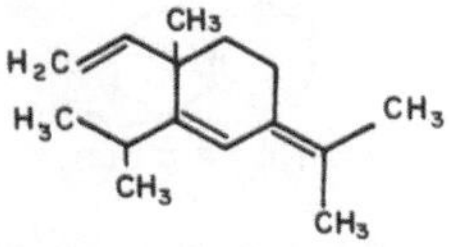


Tetrahedron 18:1509 (1962)

▶ $C_{15}H_{24}$

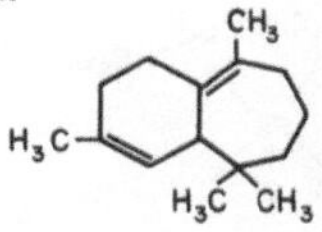


β-HIMACHALENE

Tetrahedron Letters 217 (1961)

▶ $C_{15}H_{24}$

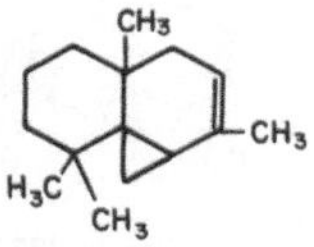


Acta Chem. Scand. 15:592 (1961)

▶ $C_{15}H_{24}$

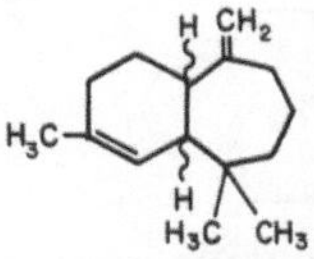


α-HIMACHALENE

Tetrahedron Letters 217 (1961)

▶ $C_{15}H_{24}$

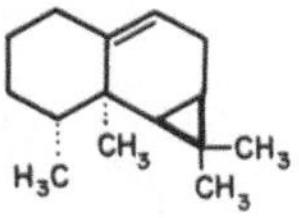


Tetrahedron Letters No. 18:828 (1962)

▶ $C_{15}H_{24}$

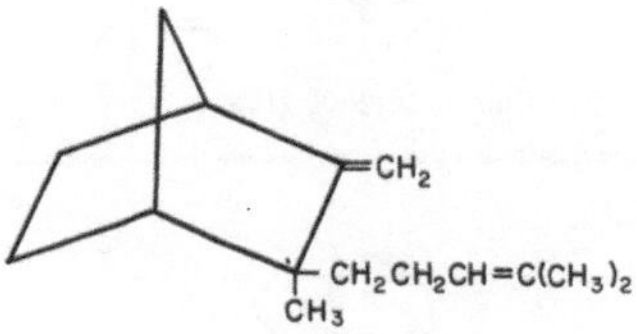


J. Am. Chem. Soc. 84:2611 (1962)

▶ $C_{15}H_{24}$

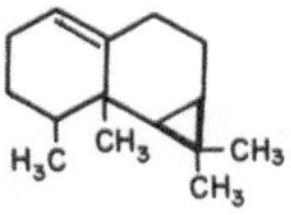


Tetrahedron Letters No. 18:828 (1962)

▶ $C_{15}H_{24}$

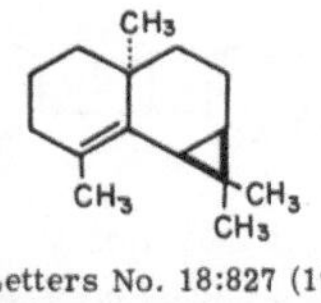


Tetrahedron Letters No. 18:827 (1962)

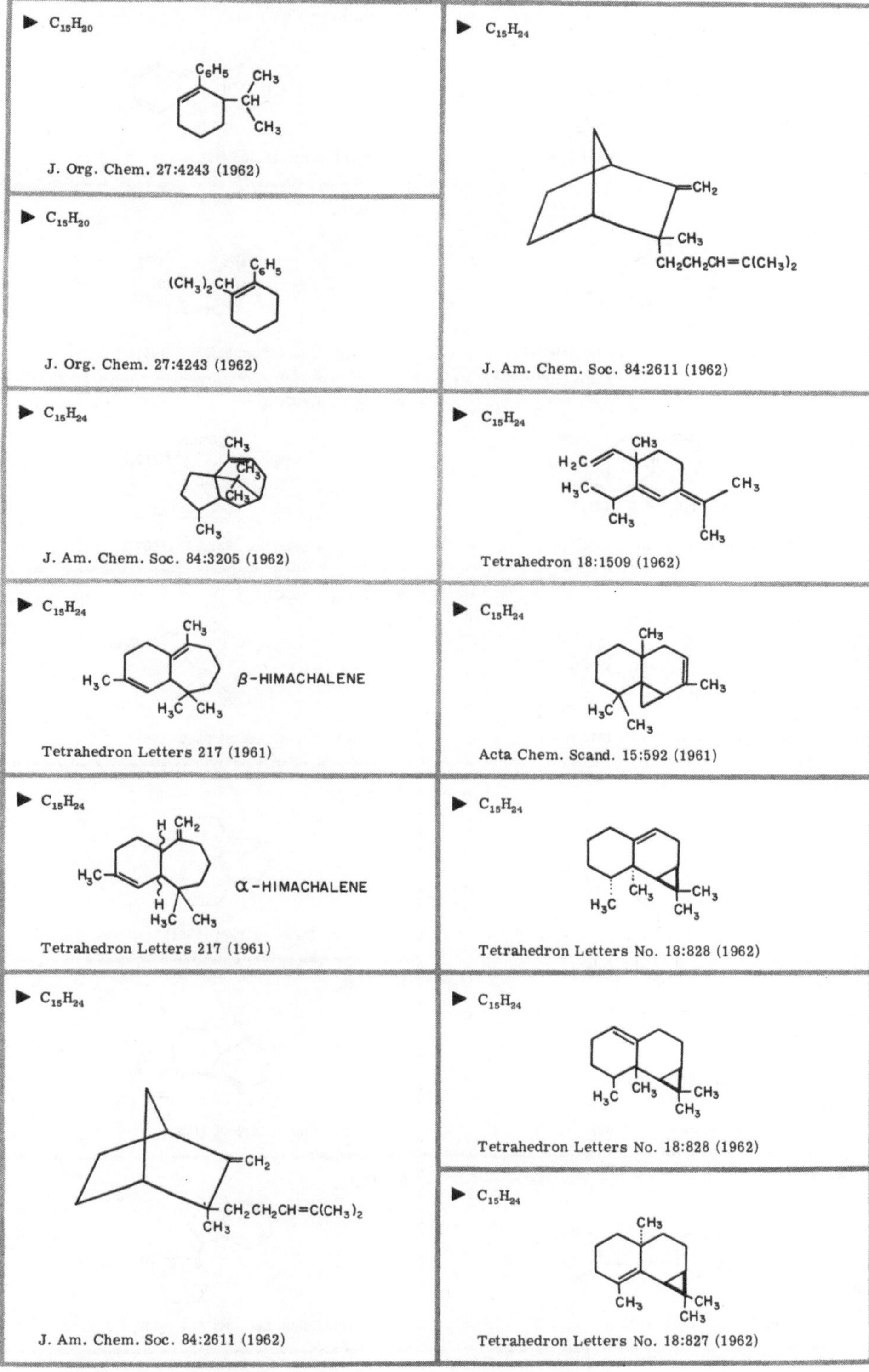

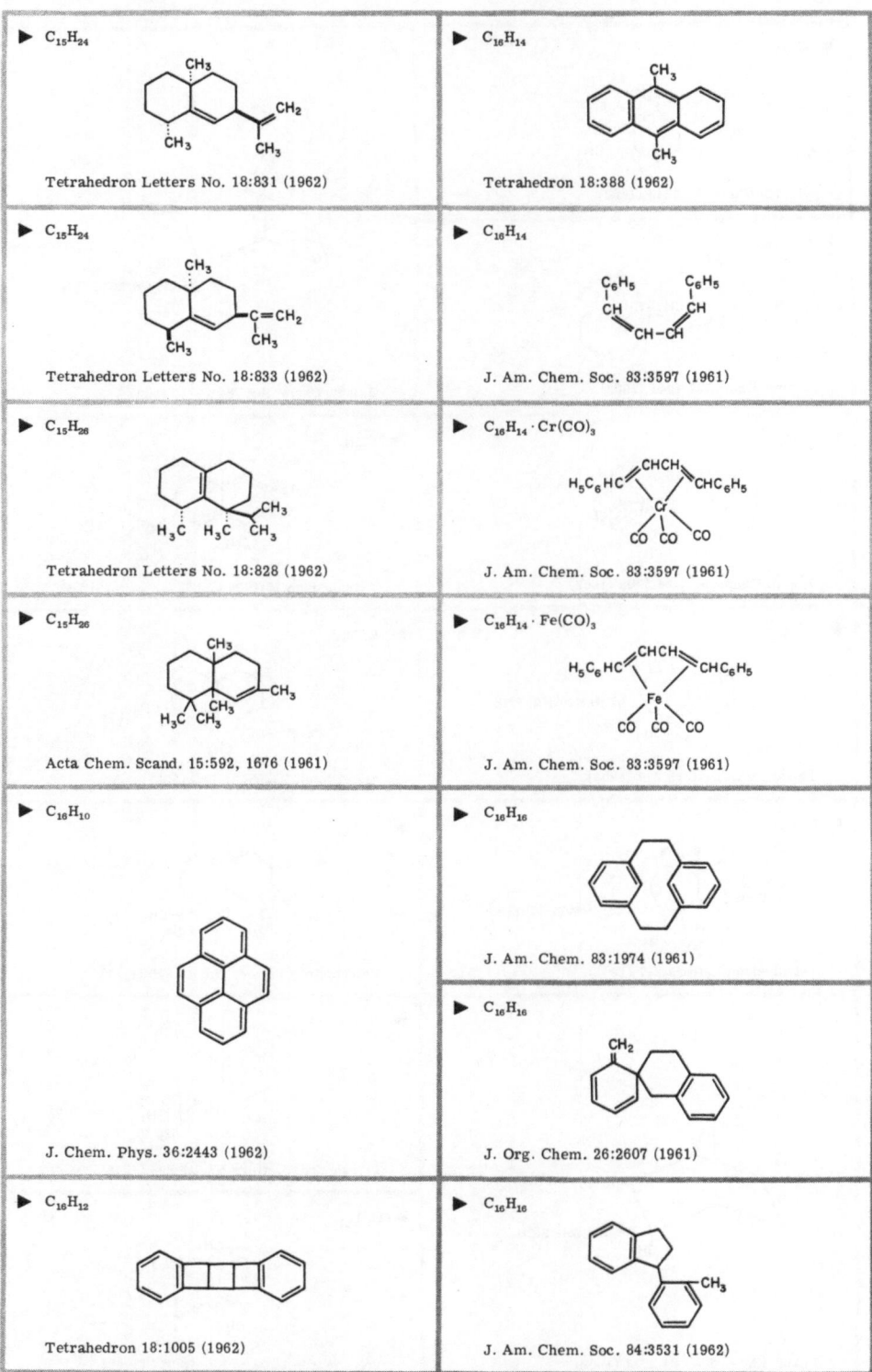

▶ $C_{15}H_{24}$

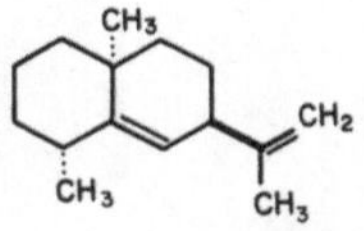

Tetrahedron Letters No. 18:831 (1962)

▶ $C_{16}H_{14}$

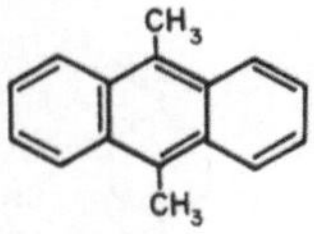

Tetrahedron 18:388 (1962)

▶ $C_{15}H_{24}$

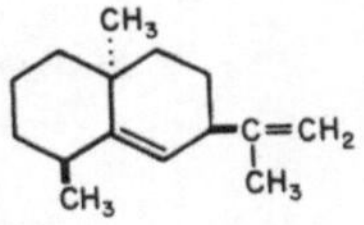

Tetrahedron Letters No. 18:833 (1962)

▶ $C_{16}H_{14}$

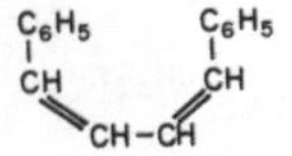

J. Am. Chem. Soc. 83:3597 (1961)

▶ $C_{15}H_{26}$

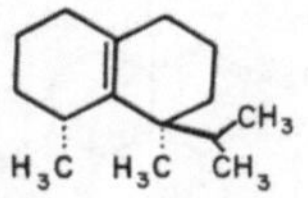

Tetrahedron Letters No. 18:828 (1962)

▶ $C_{16}H_{14} \cdot Cr(CO)_3$

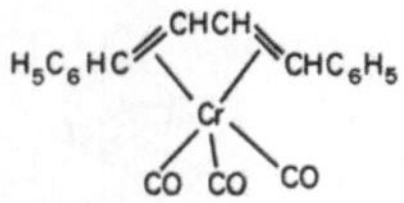

J. Am. Chem. Soc. 83:3597 (1961)

▶ $C_{15}H_{26}$

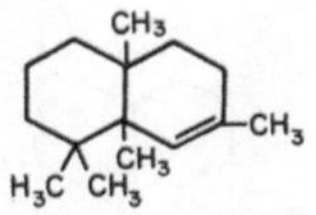

Acta Chem. Scand. 15:592, 1676 (1961)

▶ $C_{16}H_{14} \cdot Fe(CO)_3$

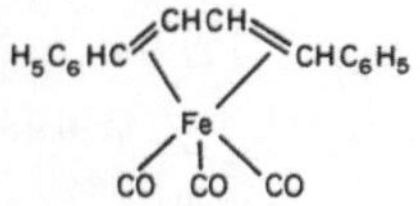

J. Am. Chem. Soc. 83:3597 (1961)

▶ $C_{16}H_{10}$

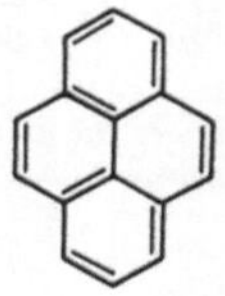

J. Chem. Phys. 36:2443 (1962)

▶ $C_{16}H_{16}$

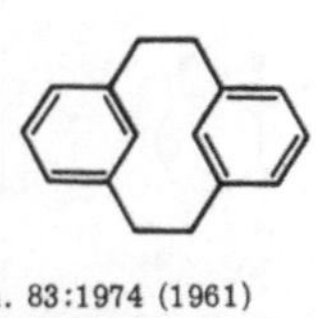

J. Am. Chem. 83:1974 (1961)

▶ $C_{16}H_{16}$

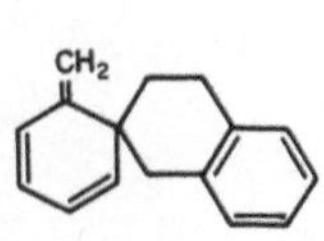

J. Org. Chem. 26:2607 (1961)

▶ $C_{16}H_{12}$

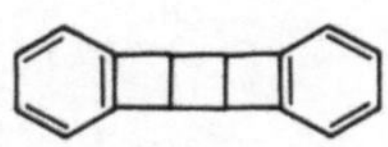

Tetrahedron 18:1005 (1962)

▶ $C_{16}H_{16}$

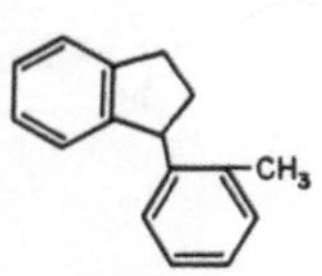

J. Am. Chem. Soc. 84:3531 (1962)

▶ $C_{16}H_{16}$

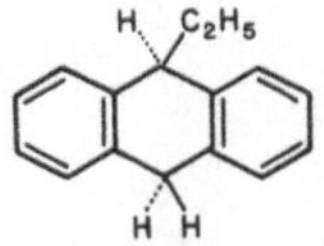


J. Chem. Soc. 3783 (1962)

▶ $C_{16}H_{18}$

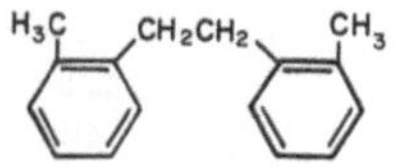


J. Am. Chem. Soc. 83:960 (1961)

▶ $C_{16}H_{16}$

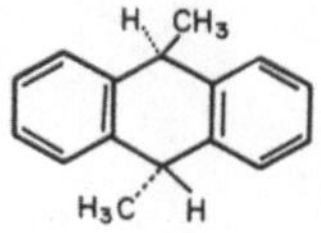


J. Chem. Soc. 3783 (1962)

▶ $C_{16}H_{18}$

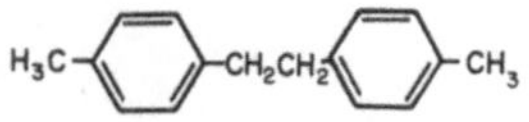


J. Org. Chem. 26:4984 (1961)

▶ $C_{16}H_{16}$

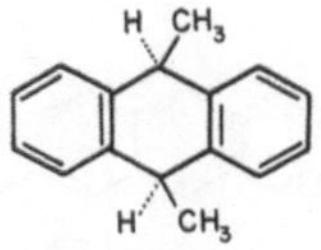


J. Chem. Soc. 3783 (1962)

▶ $C_{16}H_{18}$

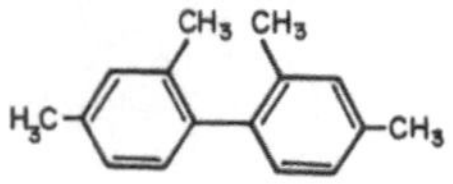


Varian 659

▶ $C_{16}H_{16}$

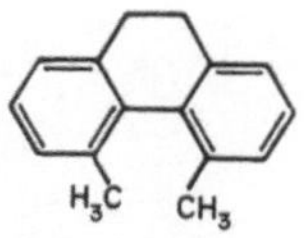


Varian 654

▶ $C_{16}H_{16} \cdot Rh_2Cl_2$

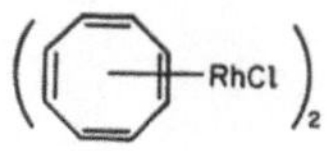


Chem. Ber. 95:158 (1962)

▶ $C_{16}H_{18}Fe$

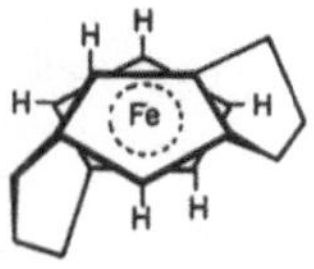


Tetrahedron Letters No. 10:426 (1962)

▶ $C_{16}H_{16}D_2$

$$\begin{array}{c} CH_3CD - CD - CH_3 \\ | \quad\quad | \\ H_5C_6 \quad C_6H_5 \end{array}$$

J. Am. Chem. Soc. 83:2625 (1961)

▶ $C_{16}H_{22}$

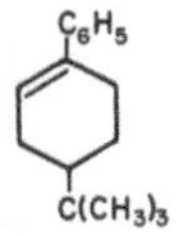


J. Org. Chem. 27:4243 (1962)

▶ $C_{16}H_{18}$

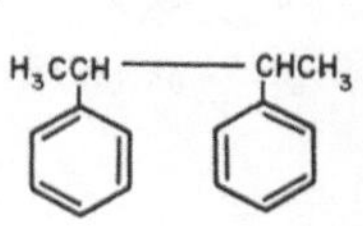


meso and dl

J. Am. Chem. Soc. 83:2625 (1961); 84:743 (1962)

▶ $C_{16}H_{22}$

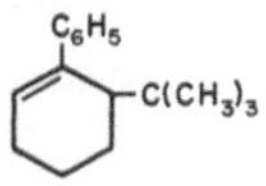


J. Org. Chem. 27:4243 (1962)

▶ $C_{16}H_{22}$

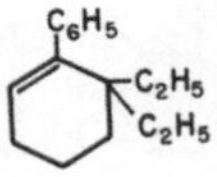


J. Org. Chem. 27:4243 (1962)

▶ $C_{16}H_{27}B$

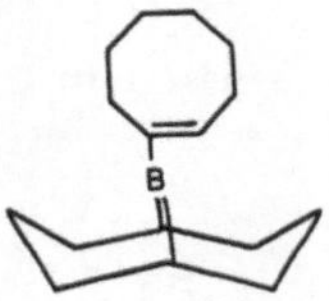


Chem. & Ind. 2087 (1962)

▶ $C_{17}H_{16}Fe$

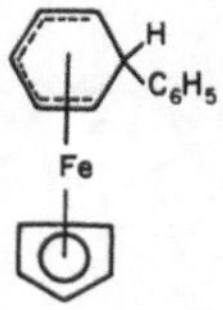


Chem. & Ind. 1408 (1961)
J. Chem. Soc. 4463 (1962)

▶ $C_{17}H_{18}$

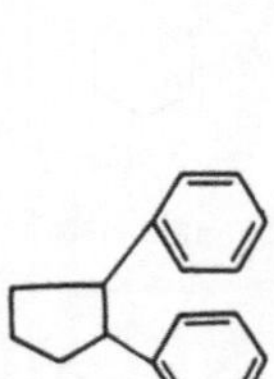

J. Am. Chem. Soc. 83:4838 (1961); 84:859 (1962)

▶ $C_{17}H_{24}$

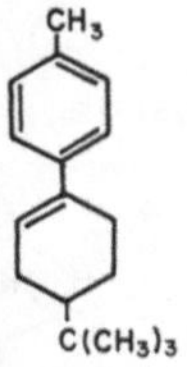


J. Org. Chem. 27:4243 (1962)

▶ $C_{18}H_{12}$

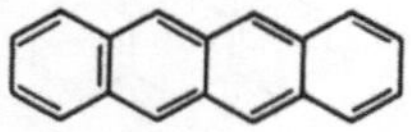

J. Chem. Phys. 37:1180 (1962)

▶ $C_{18}H_{12}$

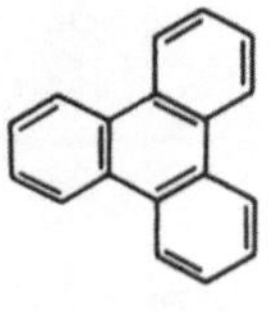

J. Chem. Phys. 36:2443 (1962)

▶ $C_{18}H_{12}$

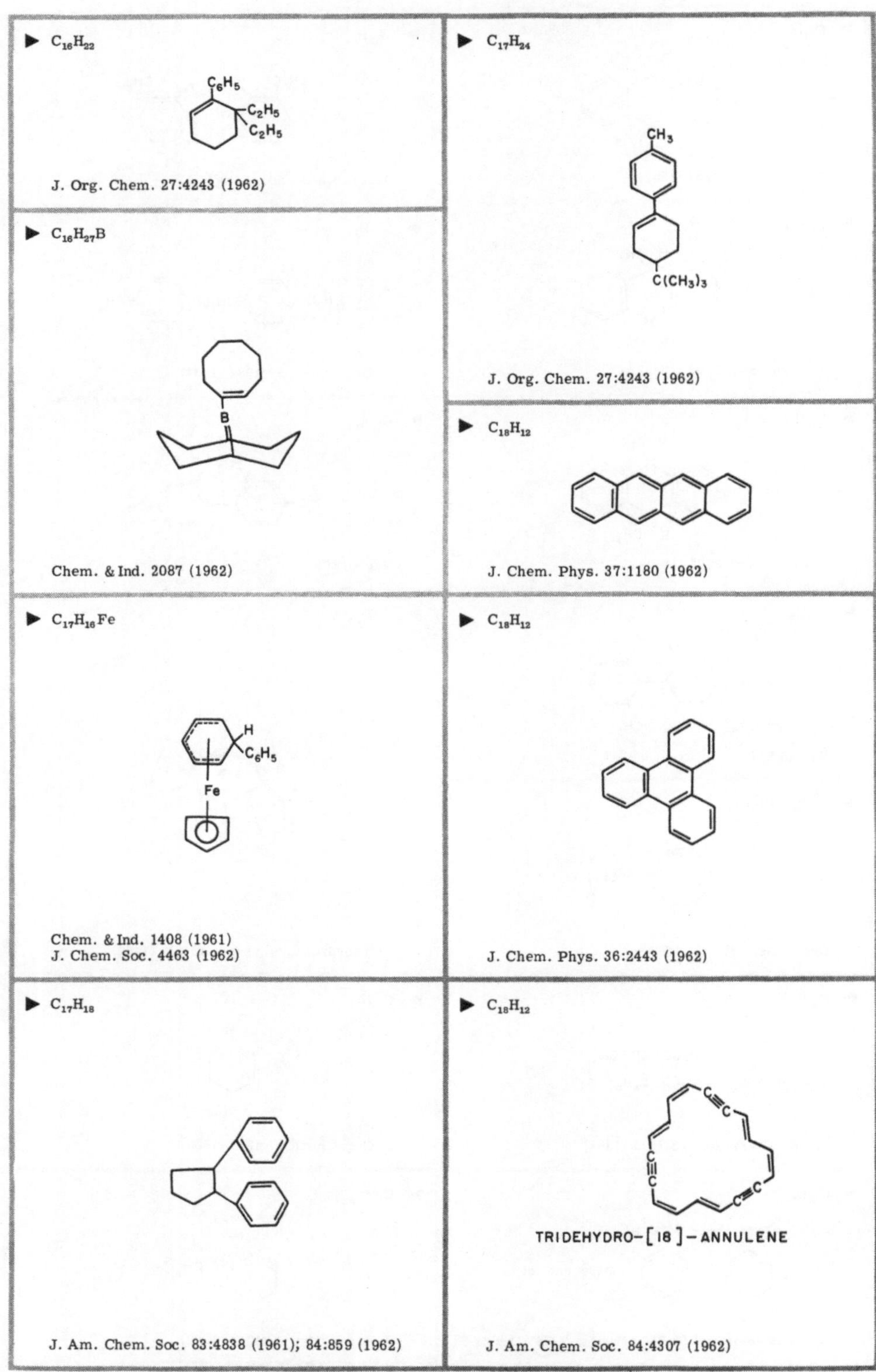

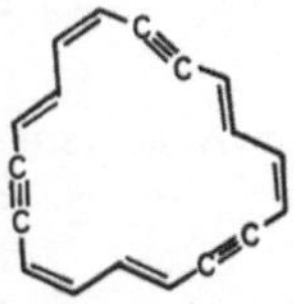

TRIDEHYDRO-[18]-ANNULENE

J. Am. Chem. Soc. 84:4307 (1962)

▶ $C_{18}H_{14}$

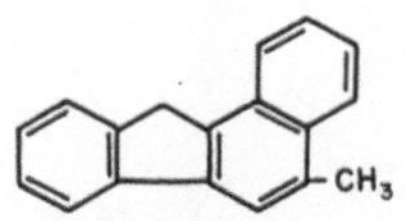


J. Am. Chem. Soc. 83:193 (1961)

▶ $C_{18}H_{16}$

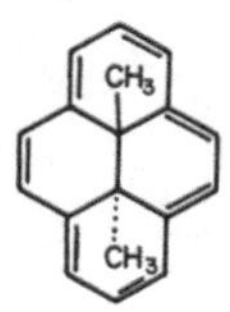


Varian 672

▶ $C_{18}H_{14}$

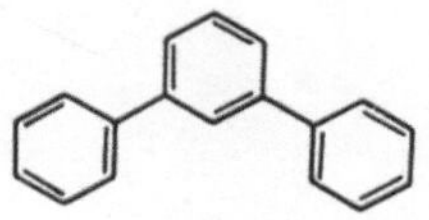

Varian 669

▶ $C_{18}H_{14}$

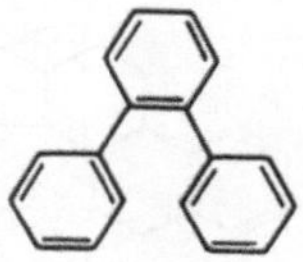

Varian 670

▶ $C_{18}H_{15}As \cdot Fe(CO)_4$

$HFe(CO)_4\,As(C_6H_5)_3^{\oplus}$

J. Chem. Soc. 3661 (1962)

▶ $C_{18}H_{14}$

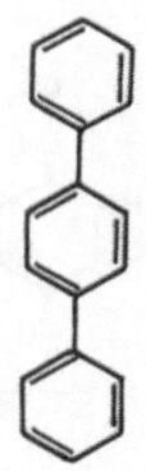

Varian 671

▶ $C_{18}H_{16}Si$

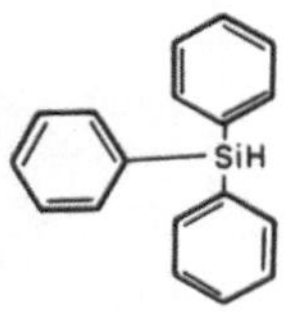


Varian 331

▶ $C_{18}H_{16}$

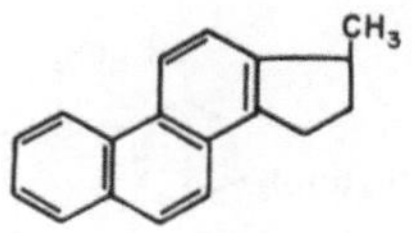


Helv. Chim. Acta 45:103 (1962)

▶ $C_{18}H_{17} \cdot BF_4$

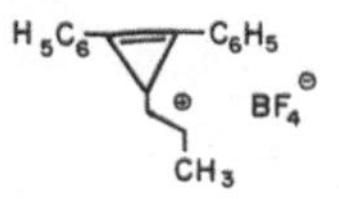


J. Am. Chem. Soc. 84:3168 (1962)

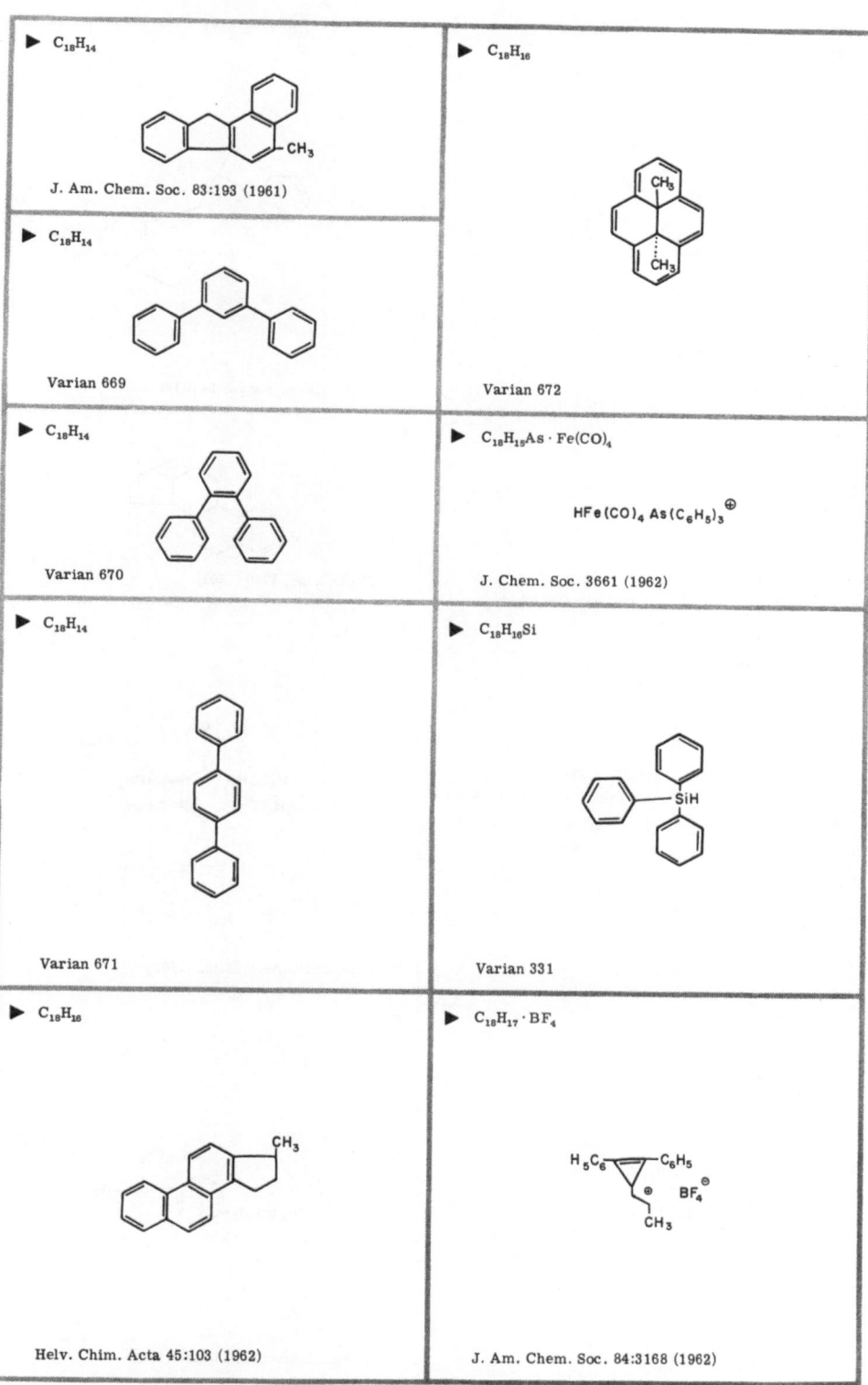

► $C_{18}H_{17}D$

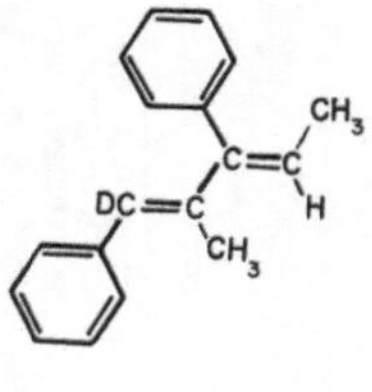


Varian 674

► $C_{18}H_{20}$

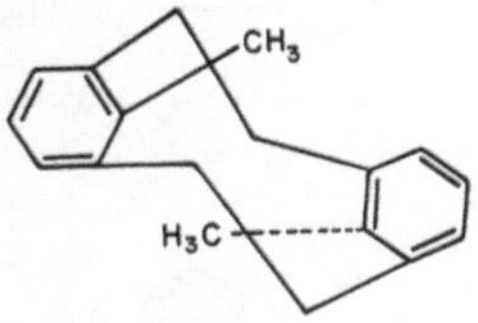


J. Am. Chem. Soc. 83:944 (1961)

► $C_{18}H_{18}$

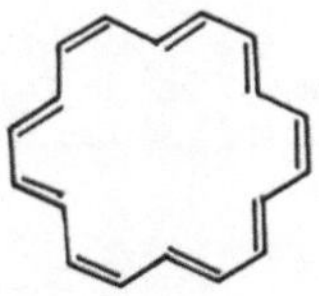

[18]-ANNULENE

J. Am. Chem. Soc. 84:4307 (1962)

► $C_{18}H_{24}$

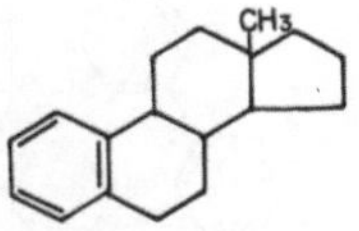


Chem. & Ind. 1716 (1962)

► $C_{18}H_{18}$

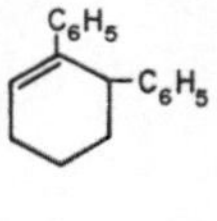


J. Org. Chem. 27:4243 (1962)

► $C_{18}H_{24}$

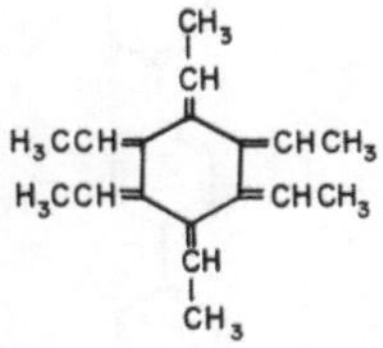


Helv. Chim. Acta 44:383 (1961)

► $C_{18}H_{18}Co_2$

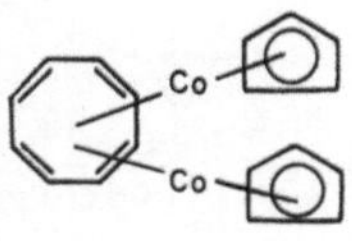


Chem. Ber. 95:158 (1962)

► $C_{18}H_{30}$

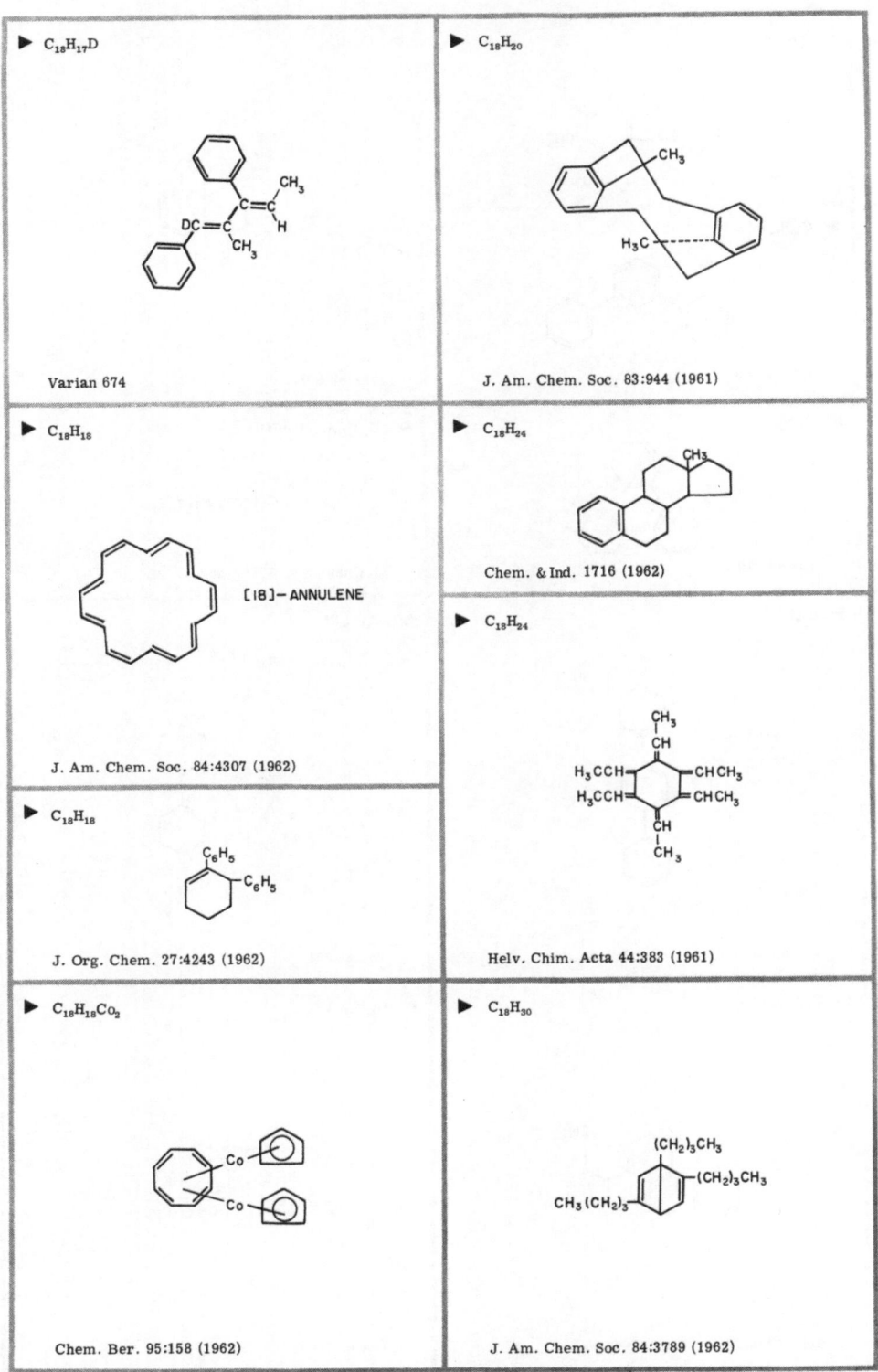

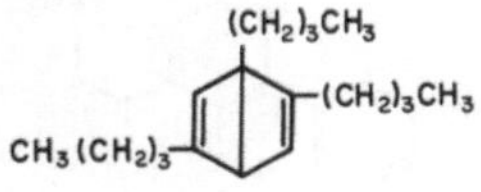


J. Am. Chem. Soc. 84:3789 (1962)

▶ $C_{18}H_{30}$

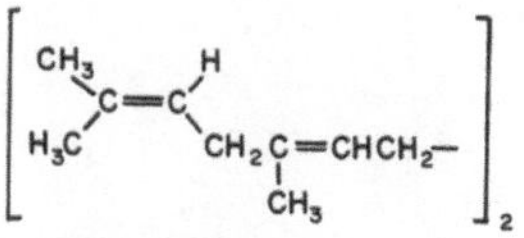


Chem. & Ind. 1020 (1962)

▶ $C_{19}H_{18}$

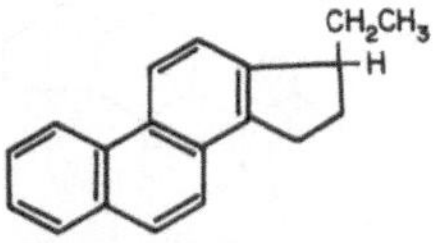


Helv. Chim. Acta 45:103 (1962)

▶ $C_{19}H_{14}$

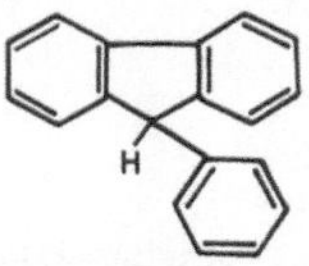


Can. J. Chem. 39:1388 (1961)

▶ $C_{19}H_{18}$

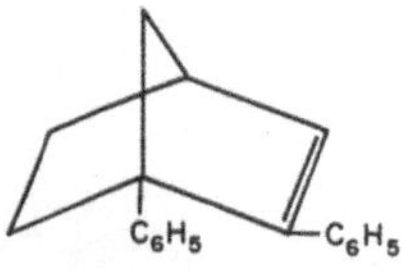


J. Org. Chem. 26:3750 (1961)

▶ $C_{19}H_{15}$

$(C_6H_5)_3C^{\oplus}$

J. Chem. Phys. 34:1460 (1961)

▶ $C_{19}H_{18}$

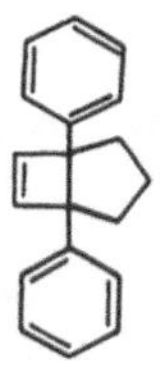

Tetrahedron Letters 98 (1961)

▶ $C_{19}H_{15}D$

$(C_6H_5)_3CD$

J. Org. Chem. 27:4146 (1962)

▶ $C_{19}H_{16}$

$(C_6H_5)_3CH$

J. Org. Chem. 26:4984 (1961)

▶ $C_{19}H_{20}$

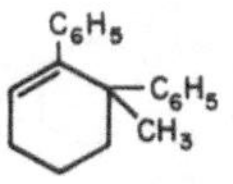


J. Org. Chem. 27:4243 (1962)

▶ $C_{19}H_{17}Co$

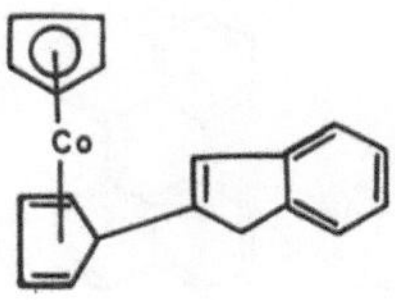


Chem. Ber. 95:2259 (1962)

▶ $C_{19}H_{22}$

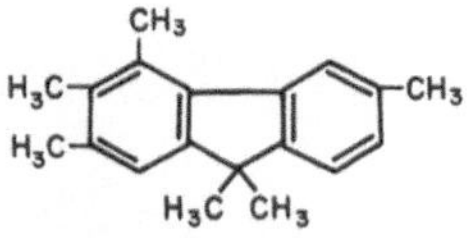


Experientia 18:549 (1962)

▶ $C_{19}H_{22}$

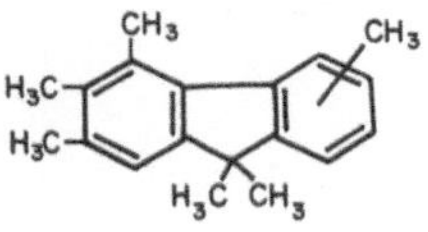


Experientia 18:111 (1962)

▶ $C_{19}H_{26}$

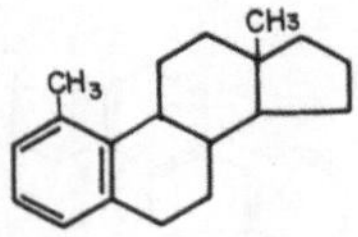


Chem. & Ind. 1716 (1962)

▶ $C_{19}H_{26}$

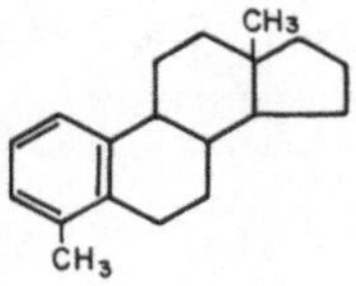


Chem. & Ind. 1716 (1962)

▶ $C_{19}H_{28}$

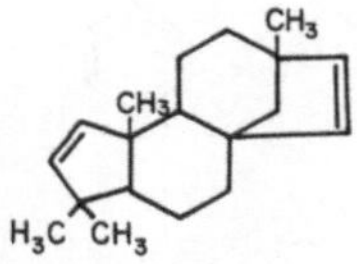


J. Chem. Soc. 4049 (1962)

▶ $C_{19}H_{30}$

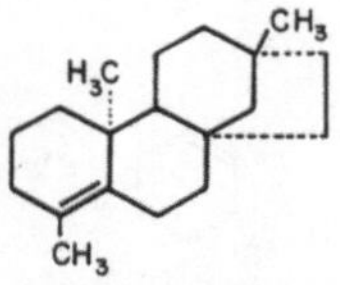


J. Org. Chem. 27:4689 (1962)

▶ $C_{19}H_{32}$

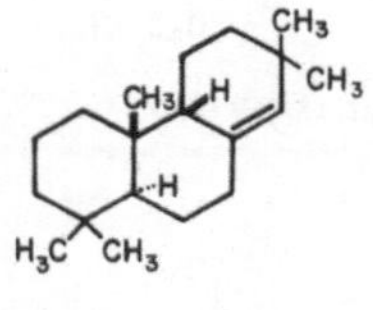


Tetrahedron Letters 496 (1961)

▶ $C_{19}H_{32}$

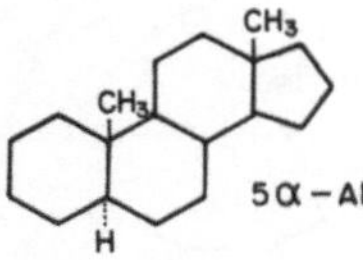


5α – ANDROSTANE

Helv. Chim. Acta 44:1380, 1755 (1961)
J. Org. Chem. 26:2437 (1961)

▶ $C_{19}H_{32}$

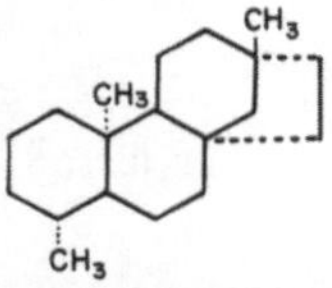


J. Org. Chem. 27:4689 (1962)

▶ $C_{19}H_{32}$

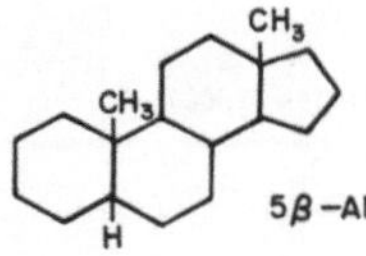


5β –ANDROSTANE

Helv. Chim. Acta 44:1380, 1755 (1961)
J. Org. Chem. 26:2437 (1961)

▶ $C_{20}H_{12}$

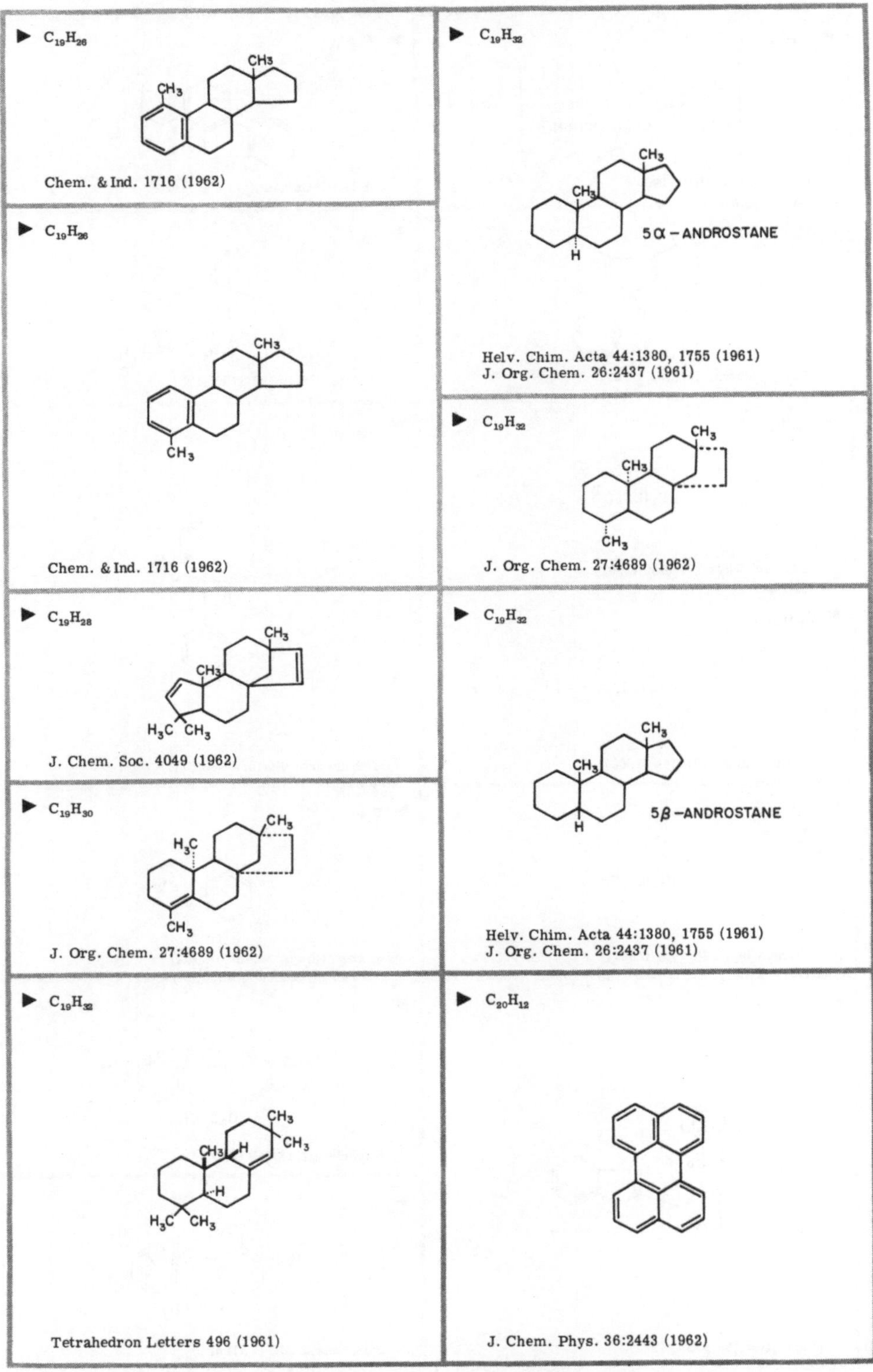

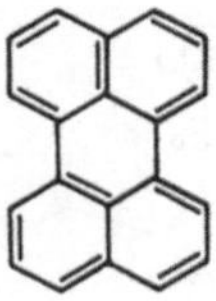

J. Chem. Phys. 36:2443 (1962)

▶ $C_{20}H_{16}$

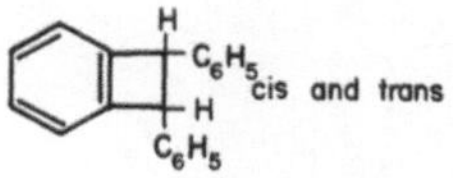


J. Am. Chem. Soc. 84:2196 (1962)

▶ $C_{20}H_{16}$

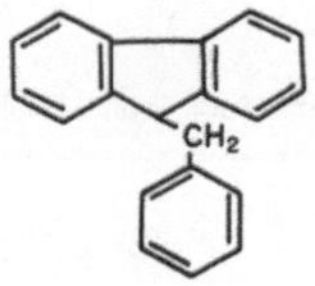


Can. J. Chem. 39:1388 (1961)

▶ $C_{20}H_{20}$

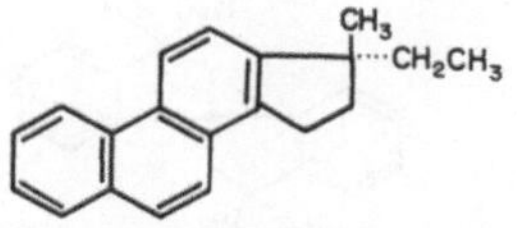


Helv. Chim. Acta 45:103 (1962)

▶ $C_{20}H_{20}$

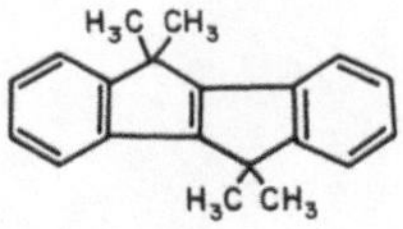


J. Org. Chem. 26:4351 (1961)

▶ $C_{20}H_{20}Mo$

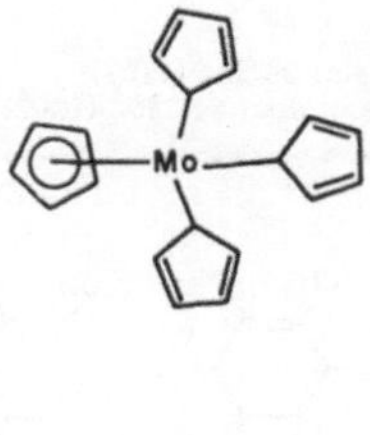


Chem. Ber. 95:253 (1962)

▶ $C_{20}H_{22}$

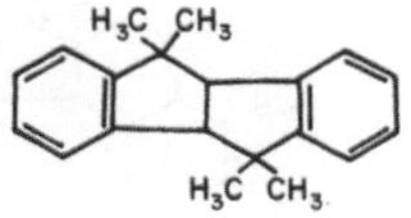


J. Org. Chem. 26:4351 (1961)

▶ $C_{20}H_{22}$

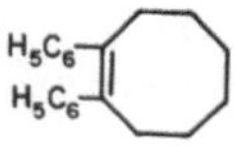


J. Am. Chem. Soc. 84:4855 (1962)

▶ $C_{20}H_{22}$

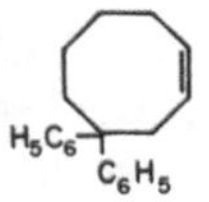


J. Am. Chem. Soc. 82:4655 (1962)

▶ $C_{20}H_{22}$

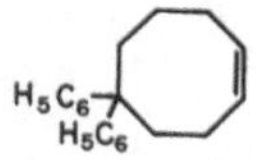


J. Am. Chem. Soc. 84:4855 (1962)

▶ $C_{20}H_{24}$

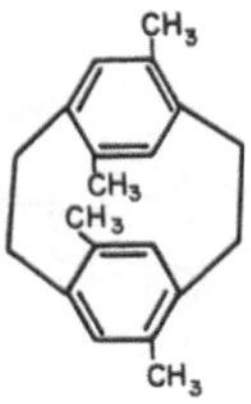


J. Am. Chem. Soc. 84:1507 (1962)

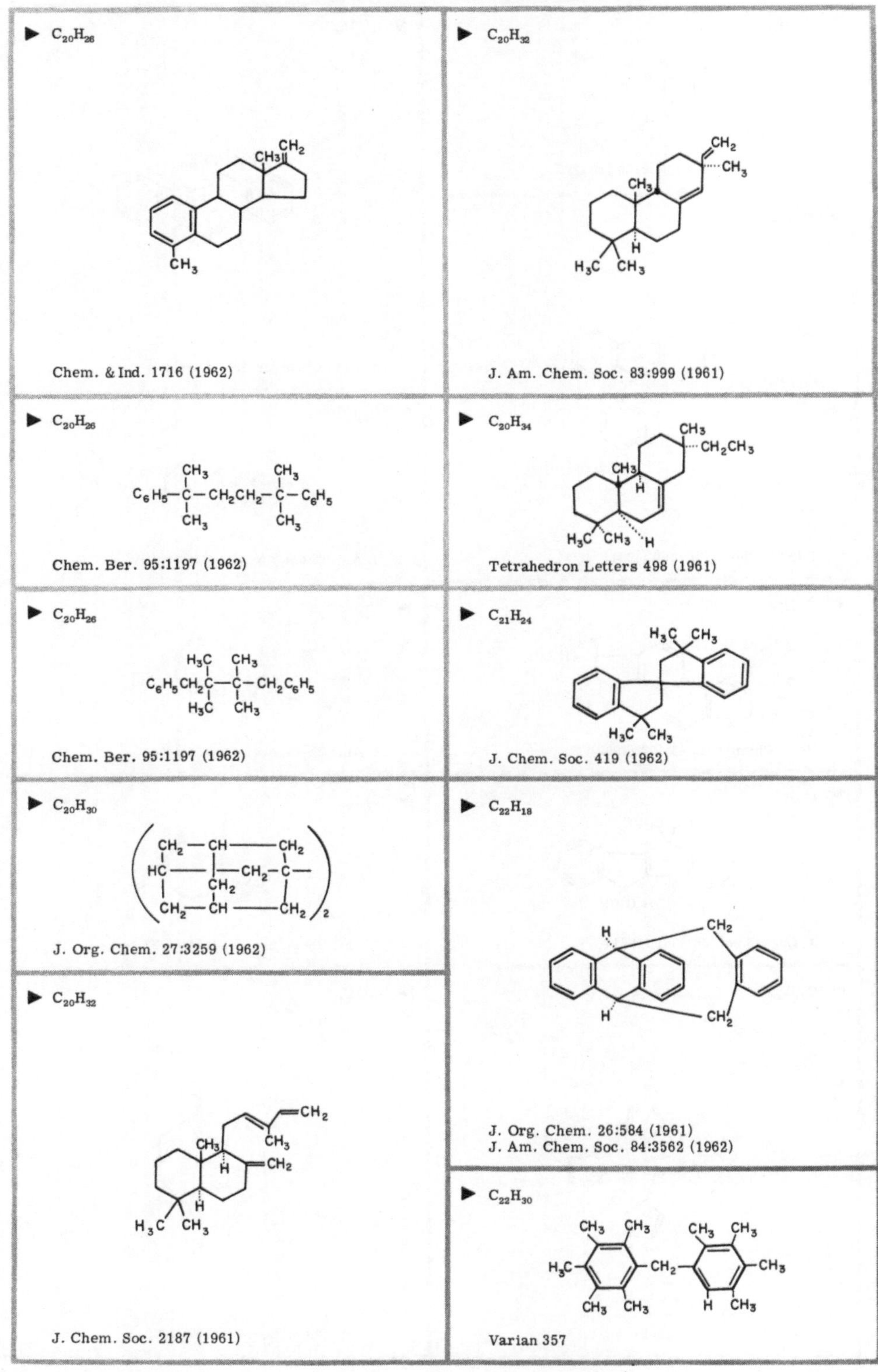

$C_{20}H_{28}$

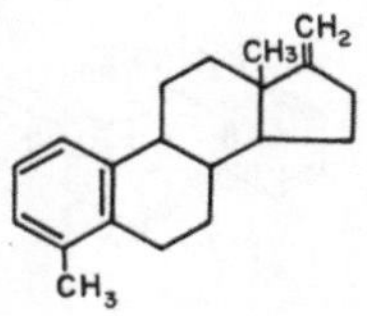

Chem. & Ind. 1716 (1962)

$C_{20}H_{32}$

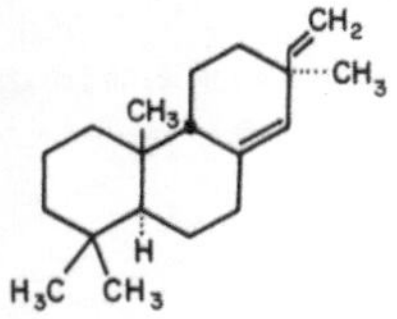

J. Am. Chem. Soc. 83:999 (1961)

$C_{20}H_{26}$

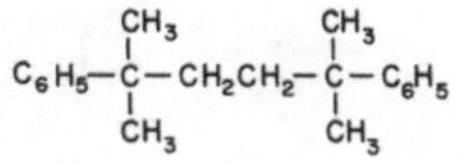

Chem. Ber. 95:1197 (1962)

$C_{20}H_{34}$

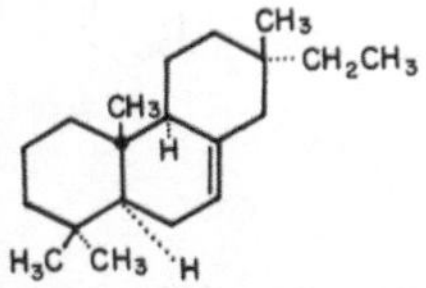

Tetrahedron Letters 498 (1961)

$C_{20}H_{26}$

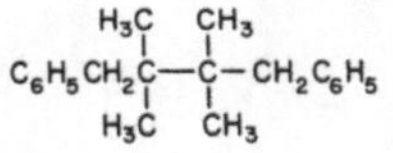

Chem. Ber. 95:1197 (1962)

$C_{21}H_{24}$

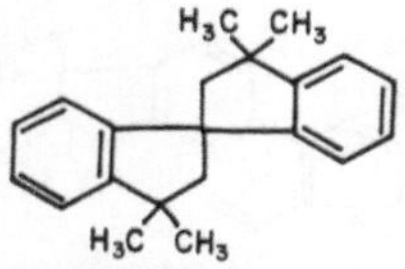

J. Chem. Soc. 419 (1962)

$C_{20}H_{30}$

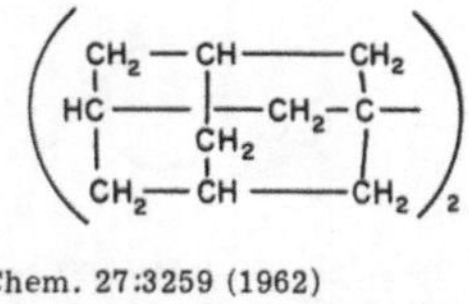

J. Org. Chem. 27:3259 (1962)

$C_{22}H_{18}$

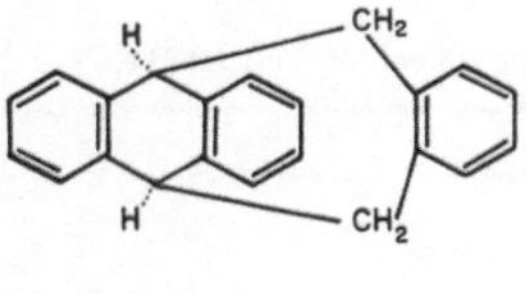

J. Org. Chem. 26:584 (1961)
J. Am. Chem. Soc. 84:3562 (1962)

$C_{20}H_{32}$

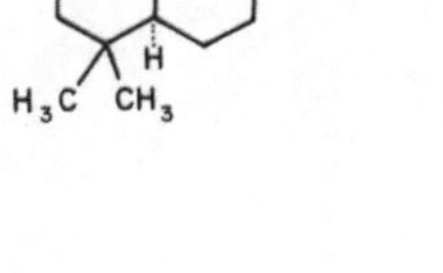

J. Chem. Soc. 2187 (1961)

$C_{22}H_{30}$

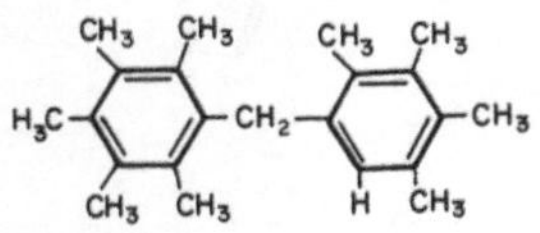

Varian 357

► $C_{22}H_{32}$

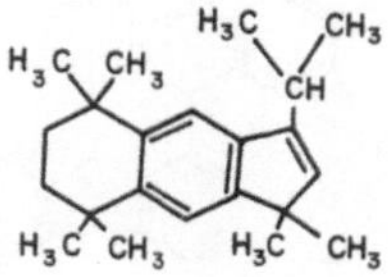


Can. J. Chem. 39:873 (1961)

► $C_{24}H_{16}$

TETRADEHYDRO—[24] ANNULENE

J. Am. Chem. Soc. 84:4307 (1962)

► $C_{22}H_{38}$

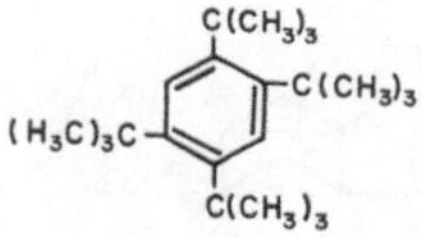


Tetrahedron Letters 640 (1961)

► $C_{24}H_{18}$

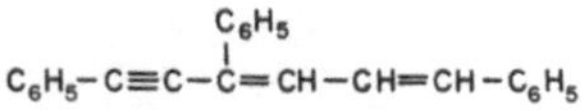


J. Org. Chem. 26:5161 (1961)

► $C_{24}H_{22}Ru$

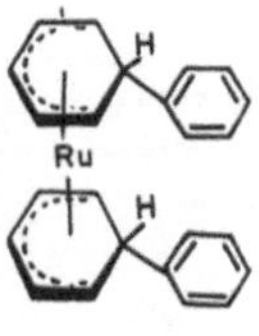


J. Chem. Soc. 4463 (1962)

► $C_{23}H_{18}$

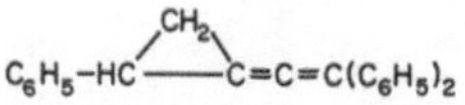


J. Am. Chem. Soc. 83:4990 (1961)

► $C_{24}H_{12}$

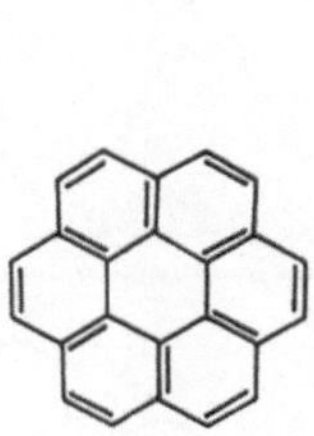

J. Chem. Phys. 36:2443 (1962)

► $C_{24}H_{24}$

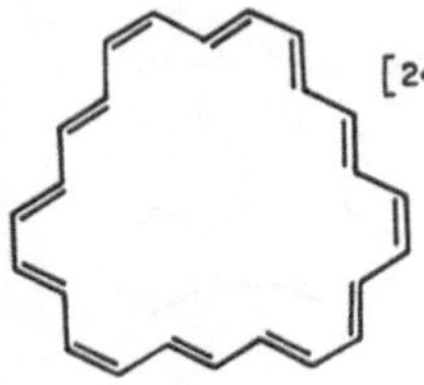

[24] ANNULENE

J. Am. Chem. Soc. 84:4307 (1962)

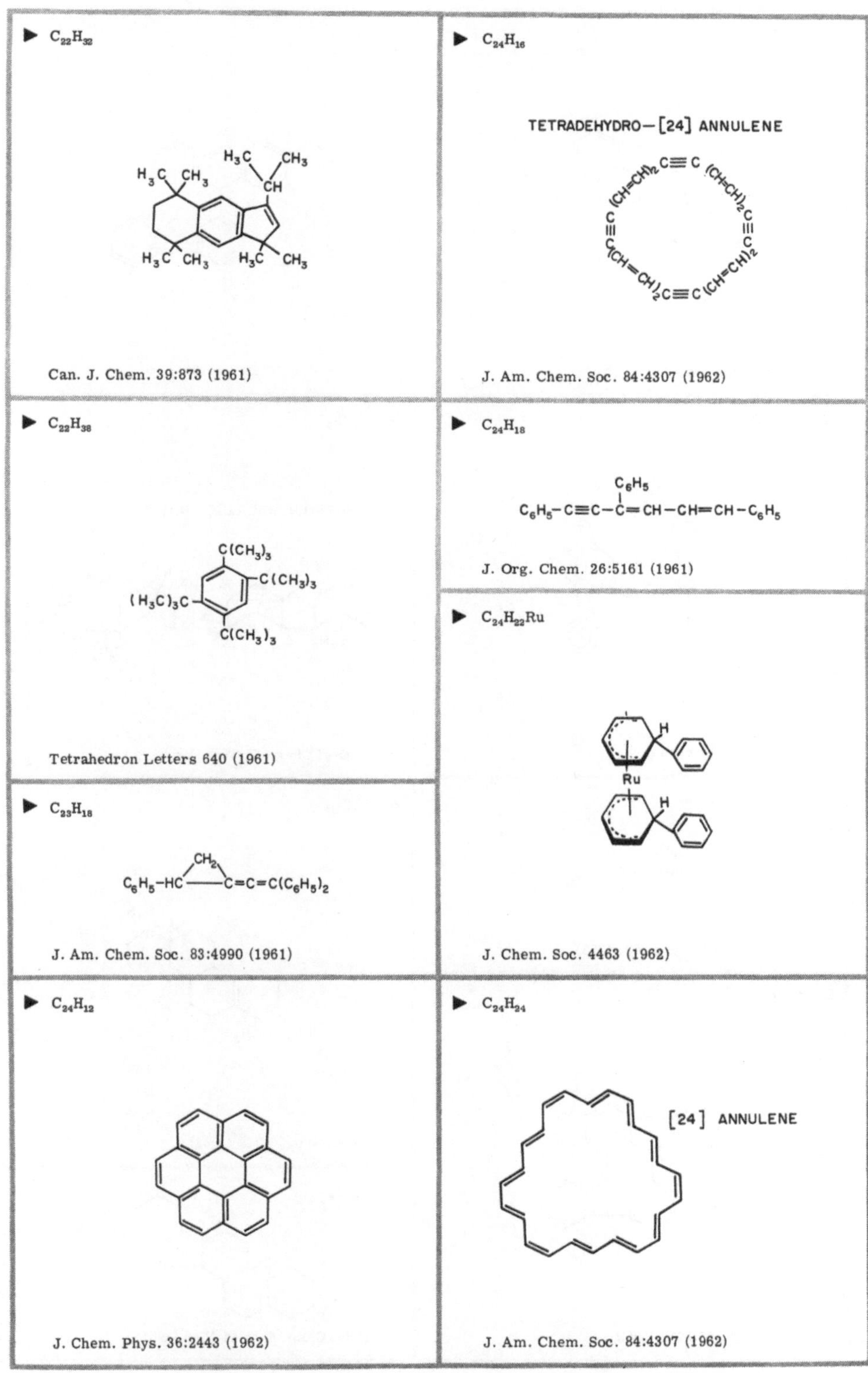

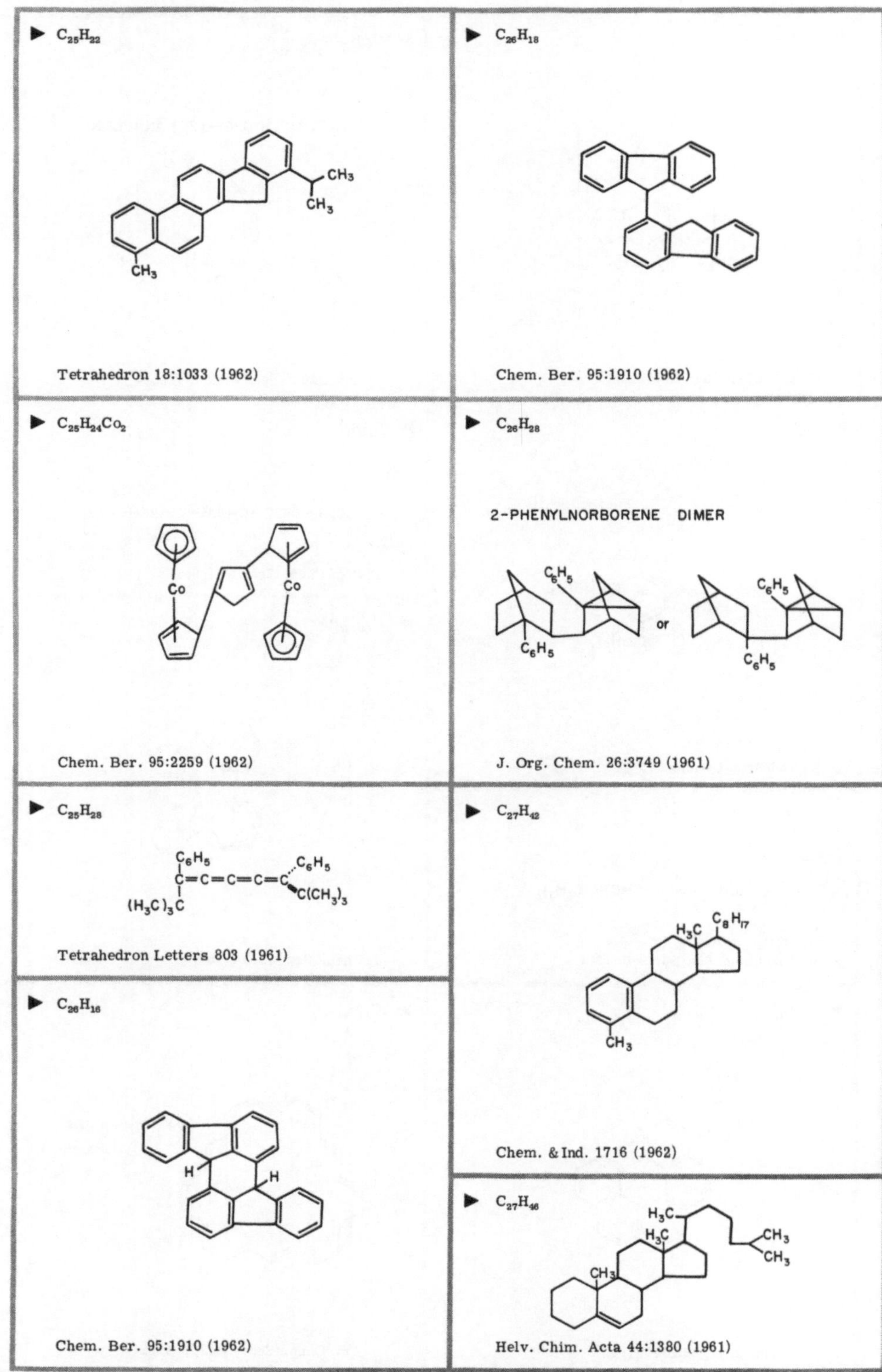

▶ $C_{25}H_{22}$

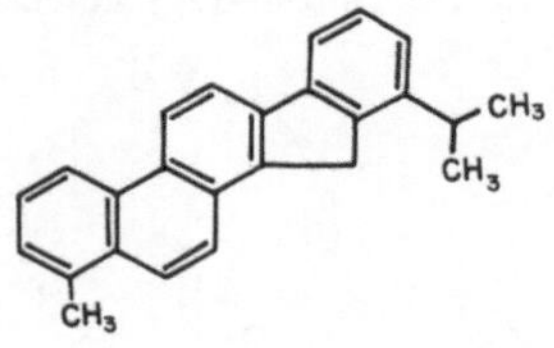

Tetrahedron 18:1033 (1962)

▶ $C_{25}H_{24}Co_2$

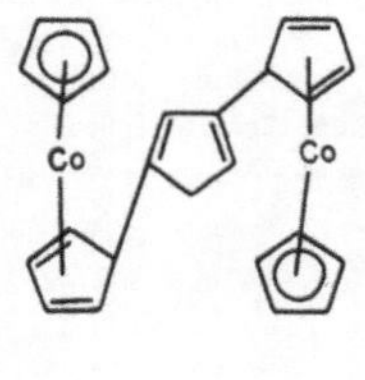

Chem. Ber. 95:2259 (1962)

▶ $C_{25}H_{28}$

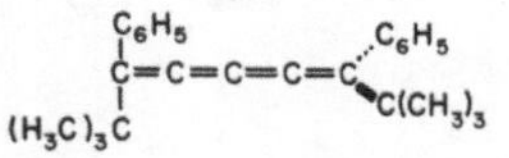

Tetrahedron Letters 803 (1961)

▶ $C_{26}H_{16}$

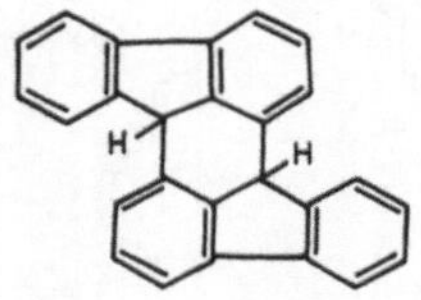

Chem. Ber. 95:1910 (1962)

▶ $C_{26}H_{18}$

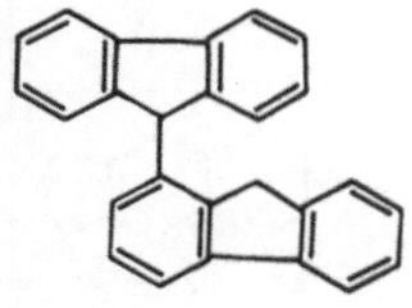

Chem. Ber. 95:1910 (1962)

▶ $C_{26}H_{28}$

2-PHENYLNORBORENE DIMER

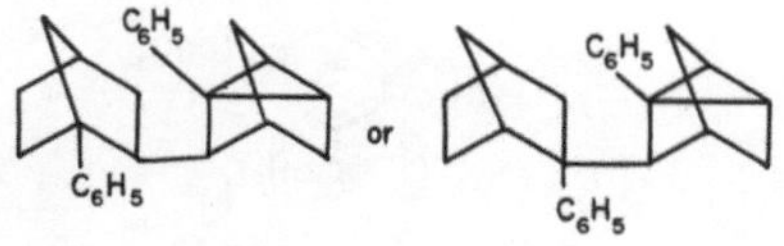

J. Org. Chem. 26:3749 (1961)

▶ $C_{27}H_{42}$

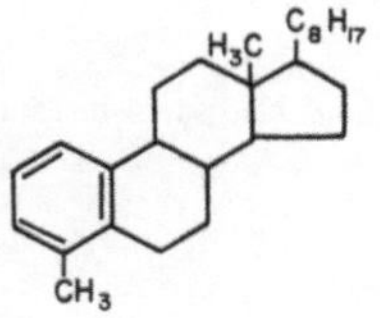

Chem. & Ind. 1716 (1962)

▶ $C_{27}H_{46}$

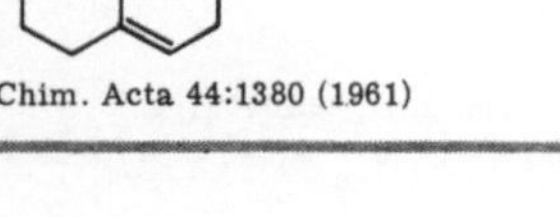

Helv. Chim. Acta 44:1380 (1961)

▶ $C_{27}H_{46}$

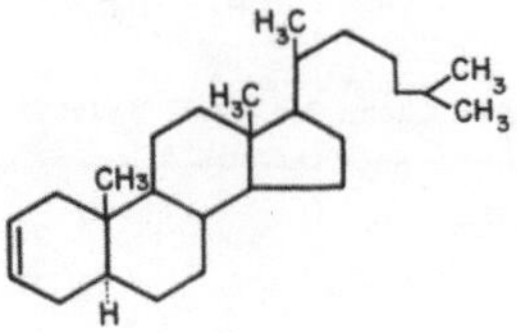


Helv. Chim. Acta 44:1380 (1961)

▶ $C_{28}H_{20}$

$(C_6H_5)_2C{=}C{=}C{=}C(C_6H_5)_2$

J. Am. Chem. Soc. 84:3278 (1962)

▶ $C_{28}H_{20} \cdot SnCl_6$

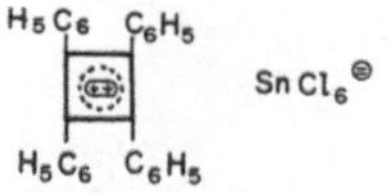


J. Am. Chem. Soc. 84:4166 (1962)

▶ $C_{28}H_{22}$

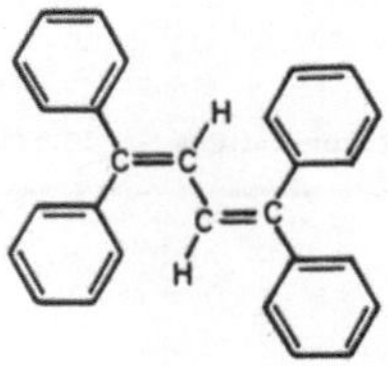


Varian 699

▶ $C_{28}H_{22}$

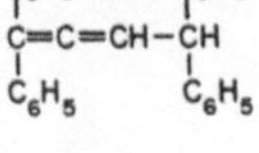


Chem. Ber. 94:3060 (1961)

▶ $C_{28}H_{22}$

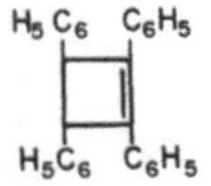


cis-trans MIXTURE

J. Am. Chem. Soc. 83:2194 (1961)

▶ $C_{28}H_{22}$

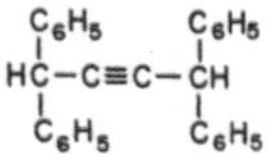


J. Am. Chem. Soc. 84:3278 (1962)

▶ $C_{28}H_{22}$

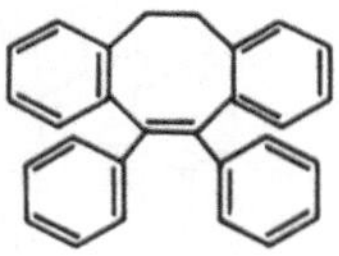

J. Org. Chem. 4715 (1962)

▶ $C_{28}H_{24}$

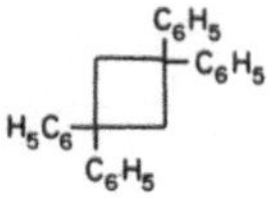


J. Org. Chem. 26:5184 (1961)

▶ $C_{28}H_{26}$

meso and dl forms

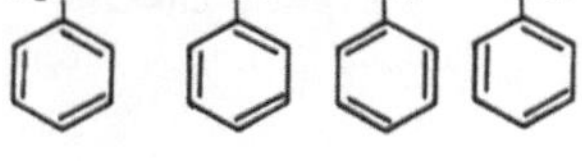


J. Org. Chem. 27:3712 (1962)

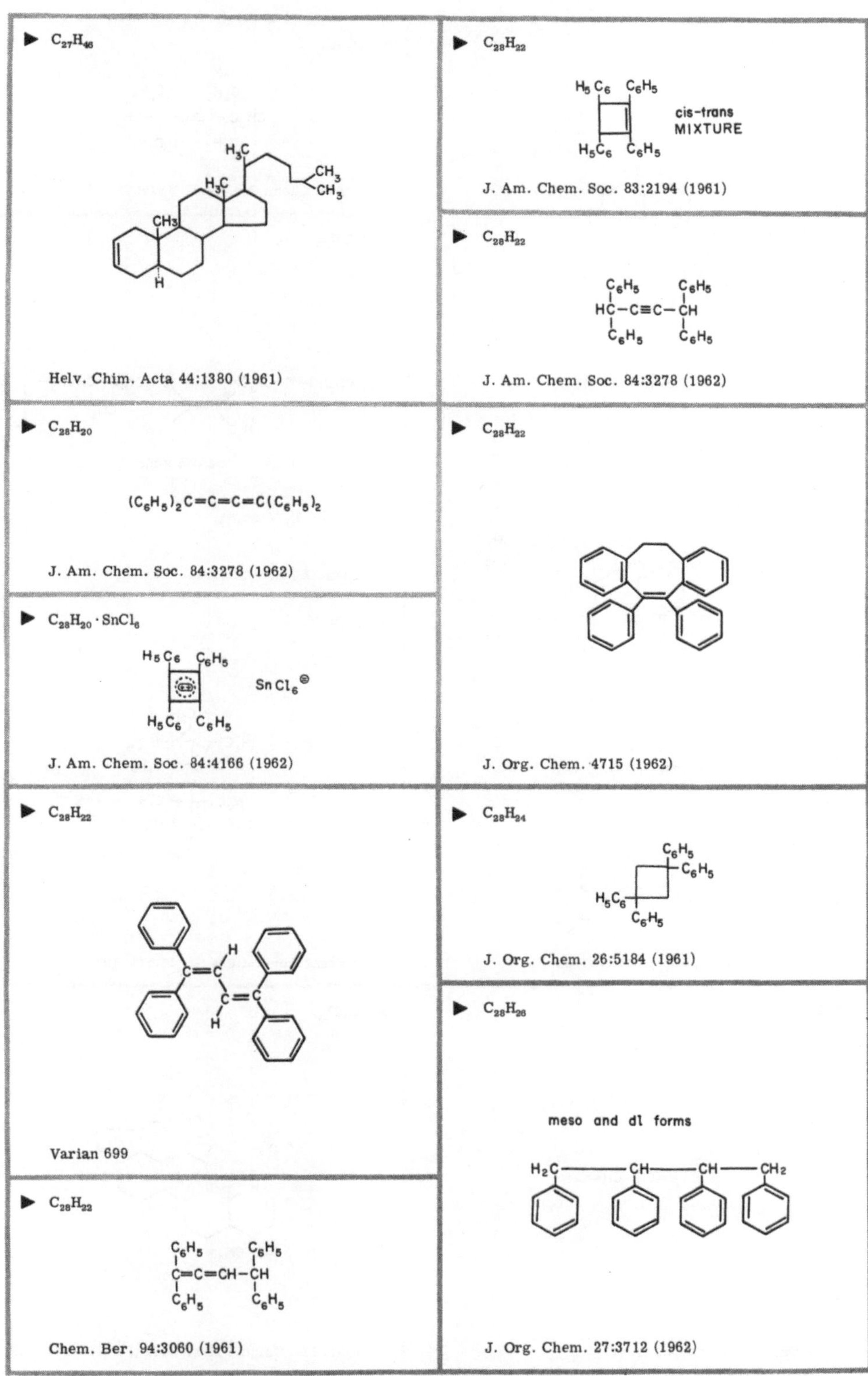

▶ $C_{28}H_{40}$

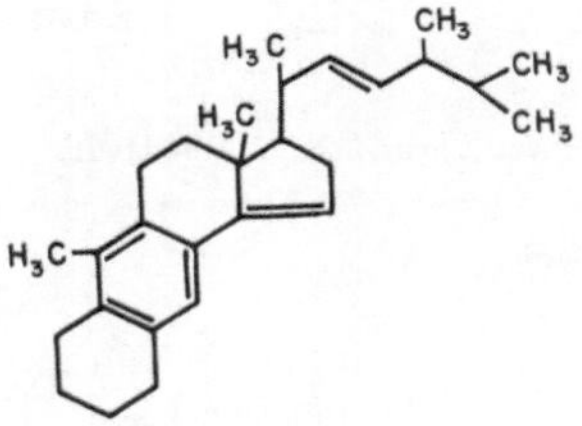

J. Am. Chem. Soc. 83:4811 (1961)

▶ $C_{28}H_{44}$

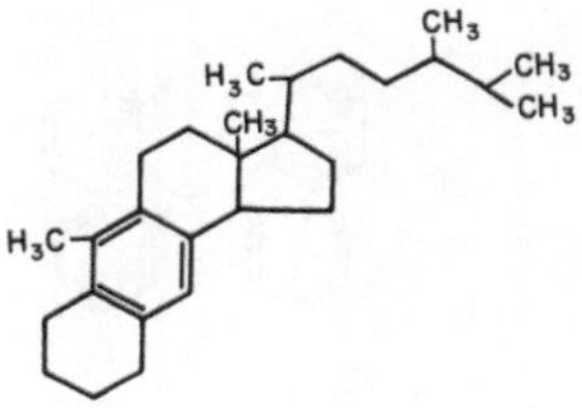

J. Am. Chem. Soc. 83:4811 (1961)

▶ $C_{30}H_{20}$

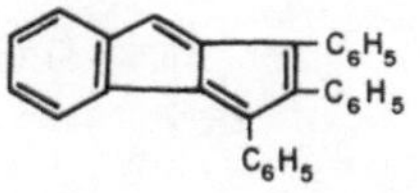

J. Am. Chem. Soc. 84:1505 (1962)

▶ $C_{30}H_{22}$

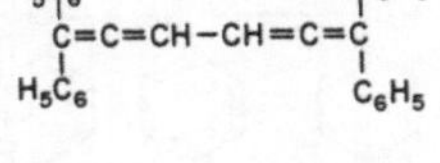

Chem. Ber. 94:3060 (1961)

▶ $C_{30}H_{26}$

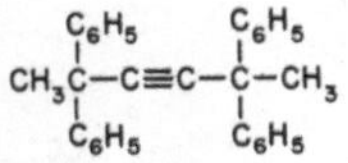

J. Am. Chem. Soc. 84:3278 (1962)

▶ $C_{30}H_{50}$

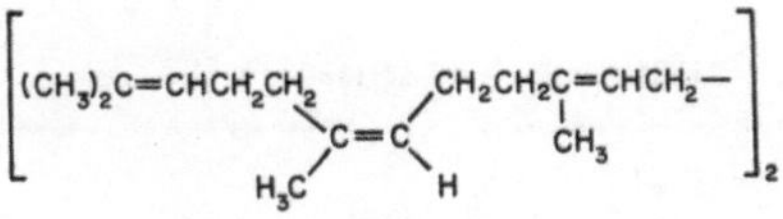

cis and trans

Chem. & Ind. 1020 (1962)

▶ $C_{30}H_{50}Fe$

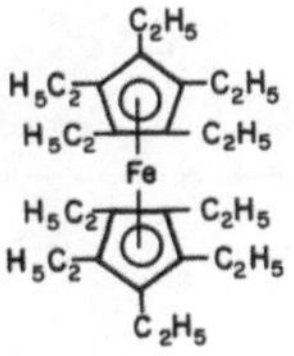

Tetrahedron Letters No. 13:575 (1962)

▶ $C_{30}H_{52}$

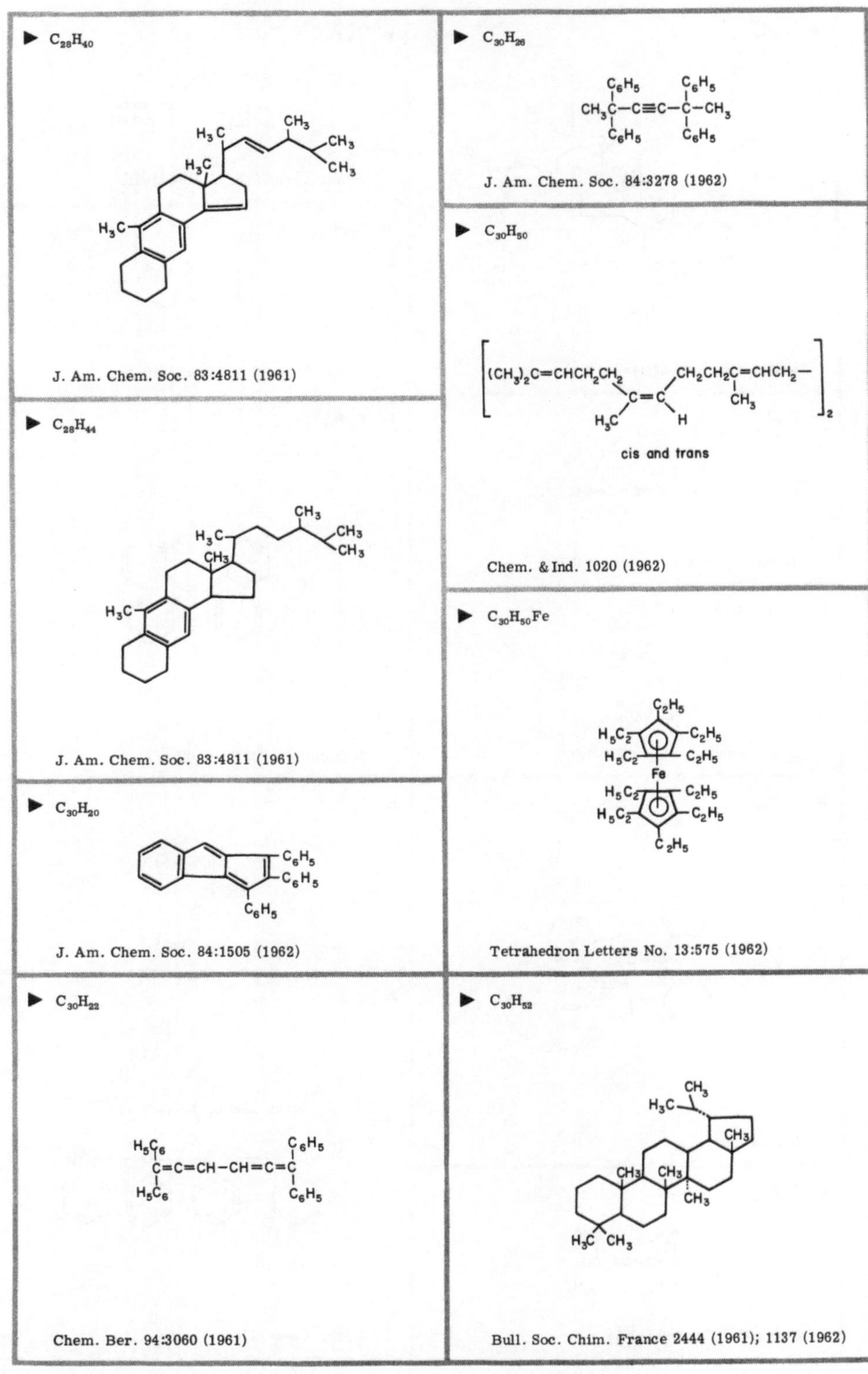

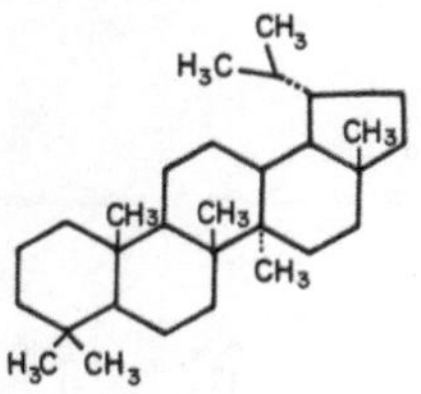

Bull. Soc. Chim. France 2444 (1961); 1137 (1962)

▶ $C_{32}H_{26}Si$

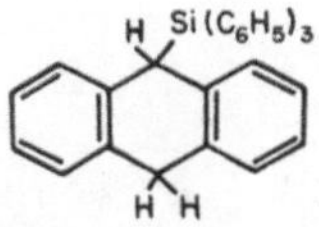


J. Org. Chem. 27:1880 (1962)

▶ $C_{32}H_{30}$

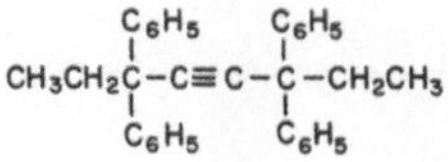

J. Am. Chem. Soc. 84:3278 (1962)

▶ $C_{32}H_{32}$

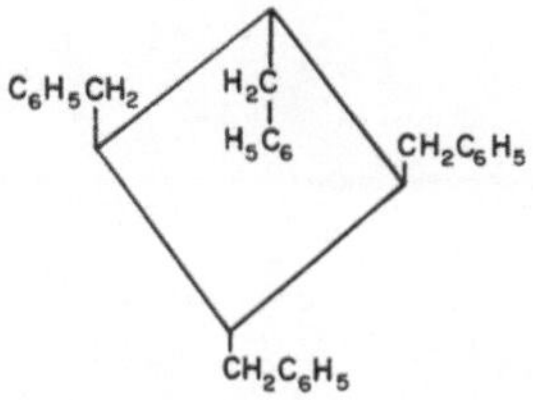

J. Am. Chem. Soc. 84:1008 (1962)

▶ $C_{33}H_{22}$

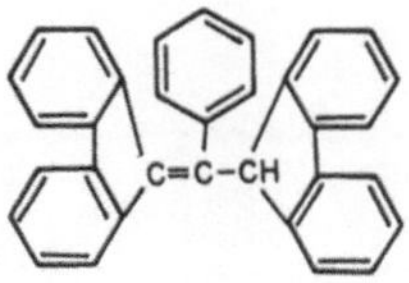


J. Chem. Phys. 35:1527 (1961)

▶ $C_{33}H_{25}Co$

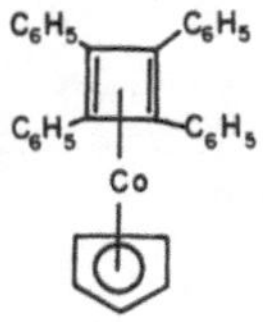


J. Chem. Soc. 3491 (1962)

▶ $C_{33}H_{52}$

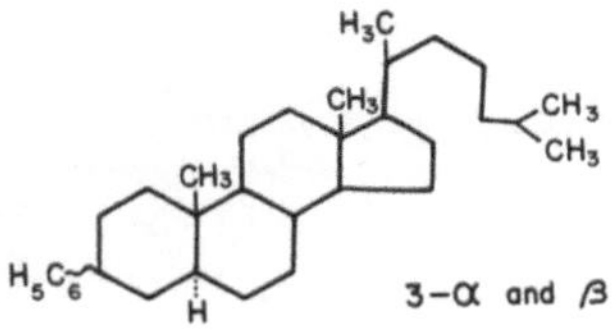


J. Org. Chem. 27:3363 (1962)

▶ $C_{34}H_{30}$

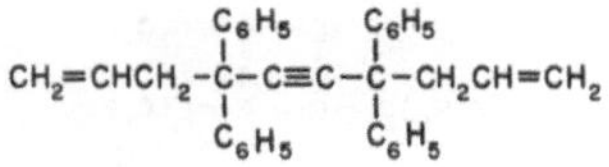

J. Am. Chem. Soc. 84:3278 (1962)

▶ $C_{35}H_{24}$

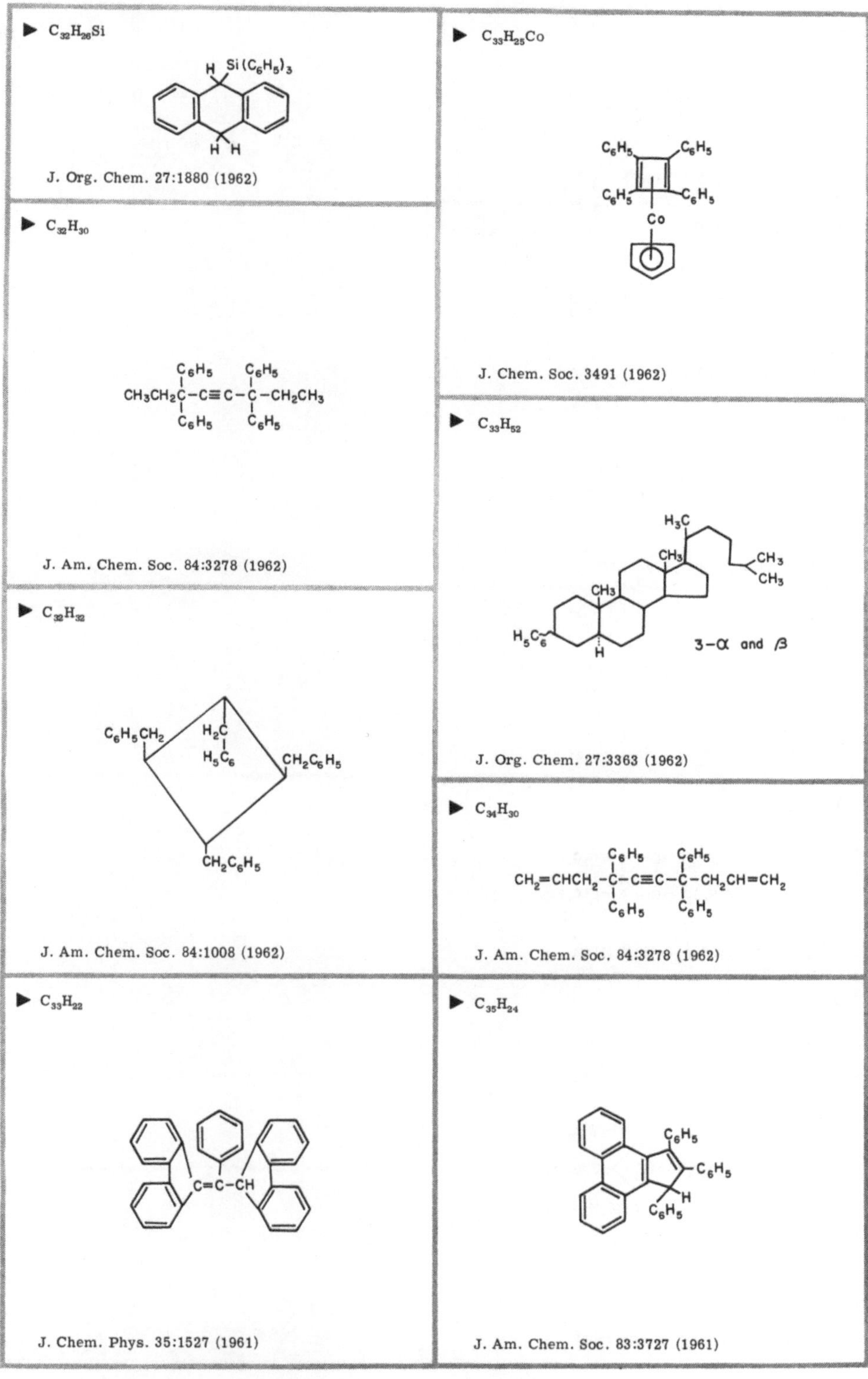

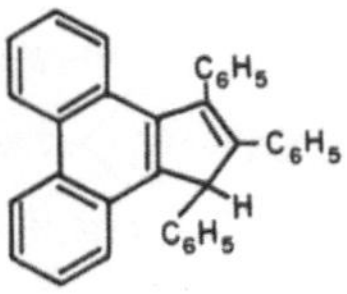

J. Am. Chem. Soc. 83:3727 (1961)

| | |
|---|---|
| ▶ $C_{36}H_{28}$ <br> J. Org. Chem. 27:18 (1962) | |
| ▶ $C_{36}H_{30}As_2 \cdot FeH(CO)_3$ <br> $HFe(CO)_3[As(C_6H_5)_3]_2^{\oplus}$ <br> J. Chem. Soc. 3661 (1962) | |
| ▶ $C_{42}H_{34}$ <br> $C_6H_5CH_2C(C_6H_5)_2-C\equiv C-C(C_6H_5)_2-CH_2C_6H_5$ <br> J. Am. Chem. Soc. 84:3278 (1962) | |
| ▶ $C_{49}H_{35}D$ <br> J. Am. Chem. Soc. 84:1484 (1962) | |
| ▶ $C_{56}H_{40}$ <br> J. Am. Chem. Soc. 84:3397 (1962) | |
| ▶ $C_{56}H_{48}$ <br> J. Am. Chem. Soc. 83:2725 (1961) | |

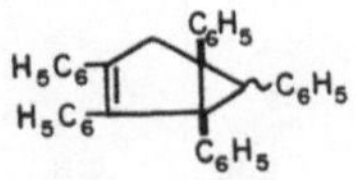

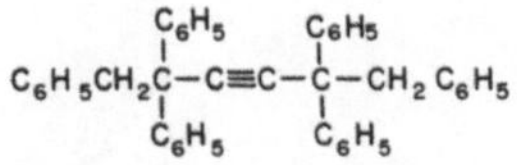

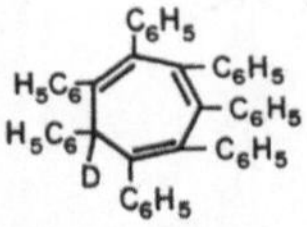

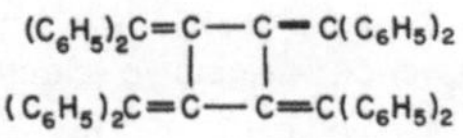

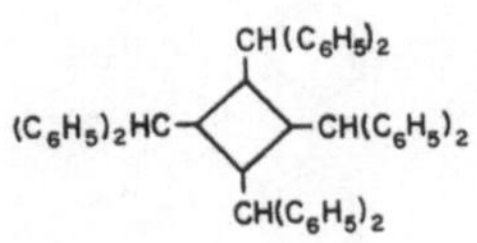